History OF THE Theory OF Numbers

Volume II
Diophantine Analysis

Leonard Eugene Dickson

Dover Publications, Inc.
Mineola, New York

Bibliographical Note

This Dover edition, first published in 2005, is an unabridged republication of the work originally published as Publication Number 256 in Washington, D.C. by the Carnegie Institute of Washington in 1919.

Library of Congress Cataloging-in-Publication Data

Dickson, Leonard E. (Leonard Eugene), 1874–1954.
    History of the theory of numbers / Leonard Eugene Dickson.
        p. cm.
    Originally published: Washington : Carnegie Institute of Washington, 1919–1923, in series: Carnegie Institute of Washington publication ; no. 256, v. 1–3.
    Includes bibliographical references and index.
    Contents:–v. 2. Diophantine analysis
    ISBN-13: 978-0-486-44233-4
    ISBN-10: 0-486-44233-0 (pbk.)
    1. Number theory. 2. Mathematics—History. I. Title.

QA241.D5 2005
512.7—dc22

2005041272

Manufactured in the United States by RR Donnelley
44233003   2015
www.doverpublications.com

PREFACE.

Diophantine analysis was named after the Greek Diophantus, of the third century, who proposed many indeterminate problems in his arithmetic. For example, he desired three rational numbers, the product of any two of which increased by the third shall be a square. Again, he required that certain combinations of the sides, area, and perimeter of a right triangle shall be squares or cubes. He was content with a single numerical rational solution, although his problems usually have an infinitude of such solutions. Many later writers required solutions in integers (whole numbers), so that the term Diophantine analysis is used also in this altered sense. For the case of homogeneous equations, the two subjects coincide. But in the contrary case, the search for all integral solutions is more difficult than that for all rational solutions. In his first course in the theory of numbers, a student is surprised at the elaborate theory relating to the equation which in analytic geometry represents a conic; but it is a real difficulty to pick out those points of the conic whose coordinates are rational and a greater difficulty to pick out those points whose coordinates are integral.

Our subject has appeared not only in works on arithmetic and geometry, but also in algebras; to it was devoted the larger part of Euler's famous Algebra. Some of its topics, as the theory of partitions, belong equally well to analysis. Although most of the problems in this domain may be stated in simple language free of technical mathematics, their investigation has quite often required the aid of many branches of advanced mathematics. A mere reference to the extensive subject index will show how frequently use has been made of elliptic functions and integrals, infinite series and products, algebraic and complex numbers, covariants, invariants, and seminvariants, Cremona and birational transformations, geometrical methods, matrices, gamma and theta functions, cyclotomy, linear differential and difference equations, integration, approximation, limits, minima, asymptotic and mean values.

Following the plan used in Volume I, we proceed to give an account in untechnical language of the main landmarks in the successive chapters. If a reader will not pause to read this entire introduction, let him sample it by selecting the account of the final chapter. This introduction is followed by an explanation of the author's point of view in producing a work quite different from conventional histories.

The notion of triangular numbers 1, 3, 6, . . . goes back to Pythagoras, who represented them by points arranged as are the shot in the base of a triangular pile of shot. The number of shot in such a pile is called a tetrahedral number. In an analogous manner we may define a polygonal number of m sides (m-gonal number) and a pyramidal number. Simple theorems concerning these numbers occur in the Greek arithmetics of Theon of Smyrna, Nicomachus (each about 100 A.D.), and Diophantus (250 A.D.), who wrote also a special tract about them. They were treated two centuries later by Roman and Hindu writers. The most important theorem on the subject is that first stated by Fermat: Every positive integer is either triangular or a sum of 2 or 3 triangular numbers; every positive integer is either a square or a sum of 2, 3, or 4 squares; either pentagonal or a sum of 2, 3, 4, or 5 pentagonal numbers; and similarly for any polygonal numbers. Throughout his half century of mathematical activity, the great Euler was engaged on the subject of polygonal numbers and solved many questions concerning them, but was able to prove Fermat's above theorem only for the case of squares, and noted that the theorem for the case of triangular numbers is equivalent to the fact that every positive integer of the form $8n + 3$ is a sum of three squares. This fact is a case of the theorem that every positive integer, not of one of the forms $8n + 7$ and $4n$, is a sum of three squares, which was proved in a complicated manner by Legendre in 1798 and more clearly by Gauss in 1801, by means of the theory of ternary quadratic forms. Gauss showed how to find the number of ways in which a number N is a sum of three triangular numbers, by means of the number of classes of binary quadratic forms of determinant $- 8N - 3$.

Cauchy gave in 1813–15 the first proof of Fermat's theorem that every number is a sum of m m-gonal numbers (all but four of which may be taken to be 0 or 1). Legendre immediately simplified this proof and showed that every sufficiently large number is a sum of four or five m-gonal numbers according as m is odd or even. In 1892 Pepin gave another proof of Cauchy's result. In 1873 Réalis proved that every positive integer is a sum of four pentagonal or hexagonal numbers extended to negative arguments. In 1895–96 Maillet proved that every integer exceeding a certain function of the relatively prime odd integers α and β is a sum of four numbers of the form $\frac{1}{2}(\alpha x^2 + \beta x)$; also, if $\phi(x) = a_0 x^5 + \ldots + a_5$, where the a's are given rational numbers, is integral and positive for every integer x sufficiently large, then every integer exceeding a fixed function of the a's is a sum of at most ν positive numbers $\phi(x)$ and a limited number of units, where $\nu = 6, 12, 96$, or 192, according as the degree of ϕ is 2, 3, 4, or 5.

From formulas in his treatise on elliptic functions of 1828, Legendre concluded that the number of ways in which N is a sum of four triangular numbers equals the sum of the divisors of $2N + 1$, and found the number of ways in which N is a sum of eight triangular numbers. In 1918 Ramanujan obtained expressions for the number of representations of any number as a sum of $2s$ triangular numbers.

In 1772 J. A. Euler, the son of L. Euler, remarked that, to express every number as a sum of squares of triangular numbers, at least twelve terms are required, and stated that, to express every number as a sum of figurate numbers

$$1, \qquad n+a, \qquad \frac{(n+1)(n+2a)}{1 \cdot 2}, \qquad \frac{(n+1)(n+2)(n+3a)}{1 \cdot 2 \cdot 3}, \qquad \ldots,$$

at least $a + 2n - 2$ terms are necessary. About the same time, N. Beguelin stated erroneously that at most $a + 2n - 2$ terms are sufficient. In 1851

Pollock stated that 5, 7, 9, 13, 21, 11 terms are needed to express every number as a sum of tetrahedral, octahedral, cubic, icosahedral, dodecahedral, and squares of triangular, numbers, and related facts. In 1862–63 Liouville proved that the only linear combinations of three triangular numbers Δ which represent all numbers are $\Delta+\Delta'+c\Delta''$ ($c=1, 2, 4, 5$) and $\Delta+2\Delta'+d\Delta''$ ($d=2, 3, 4$).

Chapter II opens with an account of the method of solving $ax+by=c$ given by the Hindu Brahmegupta in the seventh century. It was based on the mutual division of a and b, as in Euclid's process of finding their greatest common divisor. Essentially the same method was rediscovered in Europe by Bachet de Méziriac in 1612, and expressed in the convenient notation of the development of a/b into a continued fraction by Saunderson in England in 1740 and by Lagrange in France in 1767. The simplest proof that the equation is solvable when a and b are relatively prime is that given by Euler in 1760, who noted that, on dividing $c-ax$ ($x=0, 1, \ldots, b-1$) by b, we obtain b distinct remainders which are therefore $0, 1, \ldots, b-1$ in some order, the remainder zero leading to a solution. Since the same principle underlies the most elegant proof of Euler's generalization $a^\beta\equiv1$ (mod b) for $\beta=\phi(b)$ of Fermat's theorem, it was a simple step to solve our equation, or—what is the same thing—the congruence $ax\equiv c$ (mod b), by multiplying its members by $a^{\beta-1}$. This step was made about 1829 by Binet, Libri, and Cauchy. Or we may evidently employ Wilson's generalized theorem, which states that the product of the positive integers less than and prime to b is $\equiv\pm1$ (mod b). In 1905 Lerch expressed the solution of $ax\equiv1$ (mod b) as a sum involving the greatest integer function.

In the Chinese arithmetic of Sun-Tsŭ, about the first century, occurs the problem of finding a number having the remainders 2, 3, 2 when divided by 3, 5, 7, respectively, with a rule leading to the answers $23+3\cdot5\cdot7n$. The same problem and answer 23 occur in the Greek arithmetic of Nicomachus, about 100 A.D. The rule is essentially the following, given centuries later by Beveridge, Euler, and Gauss: To obtain a number x having the remainders $r_1, r_2, \ldots$ when divided by $m_1, m_2, \ldots$, respectively, where $m_1, m_2, \ldots$ are relatively prime in pairs, find numbers $\alpha_1, \alpha_2, \ldots$ such that $\alpha_i\equiv1$ (mod m_i), $\alpha_i\equiv0$ (mod m/m_i), where m is the product $m_1m_2\ldots$; then $x=\alpha_1r_1+\alpha_2r_2+\ldots$ is an answer. In the seventh century, the Chinese priest Yih-hing extended this rule to the case in which $m_1, m_2, \ldots$ are any integers: express the least common multiple of $m_1, m_2, \ldots$ as a product $m=\mu_1\mu_2\ldots$ of factors relatively prime in pairs (some of which may be unity), such that μ_i divides m_i, and find $\alpha_1, \alpha_2, \ldots$ such that $\alpha_i\equiv1$ (mod μ_i), $\alpha_i\equiv0$ (mod m/μ_i); then $x=\alpha_1r_1+\alpha_2r_2+\ldots$.

The Hindus Brahmegupta and Bháscara found the correct answer 59 to the "popular problem" of finding a number having the remainders 5, 4, 3, 2 when divided by 6, 5, 4, 3, respectively; Leonardo Pisano in 1202 added the condition that the number be a multiple of 7. He treated the problem of Ibn al-Haitam (about 1000 A.D.) of finding a multiple of 7 which has the remainder unity when divided by 2, 3, 4, 5 or 6, a problem occurring in many later books. This subject of the Chinese remainder problem found application in questions on the calendar; for example, to find the year x of the Julian period when the solar cycle, lunar cycle, and Roman indiction are given numbers r_1, r_2, r_3, we seek a number which has the remainders r_1, r_2, r_3 when divided by 28, 19, 15, respectively, these being the periods of the solar, lunar, and indiction, cycles.

The problem of finding the number of positive integral solutions of $ax + by = c$, where a, b, c are positive integers, was treated by Paoli in 1780, Hermite in 1855–58, and many others. There is the corresponding question for a system of such equations.

Systems of equations of the type $x + y + z = m$, $ax + by + cz = n$, where m, n, a, b, c are given positive integers and the unknowns are to have positive integral values, occurred in Chinese and Arabic manuscripts of the sixth and tenth centuries respectively, in Leonardo Pisano's writings, and in many of the early printed books on algebra and arithmetic. The usual method of solution, which began with the elimination of one unknown, was called *regula Coeci*, or the rule of the virgins, a term later applied to a system of any number of linear equations in any number of unknowns with positive integral coefficients. The most important papers on general systems of linear equations or congruences are those by Heger (1858), H. J. S. Smith (1859, 1861, 1871), Weber (1872, 1896), Frobenius (1878–79), Kronecker (1886), and Steinitz (1896).

Chapter II closes with a series of modern theorems, such as the fact that, if ω is irrational, there exist infinitely many pairs of integers x, y, for which $y - \omega x$ is numerically less than the reciprocal of $\sqrt{5}\,x$; and Minkowski's theorem (of prime importance for the theory of algebraic numbers) that, if $f_1, \ldots, f_n$ are linear homogeneous functions of $x_1, \ldots, x_n$ with any real coefficients whose determinant is unity, we can assign integral values not all zero to $x_1, \ldots, x_n$, such that each f_i taken positively does not exceed unity.

Chapter III treats of partitions, which have important applications to symmetric functions and algebraic invariants. The first investigation was that by Euler in 1741, who discussed the two problems of finding the number of ways in which a number n (as 6) is a sum of a given number m (as 2) of distinct parts ($6 = 5 + 1 = 4 + 2$), and the number of ways n is a sum of m equal or distinct parts (so that also $6 = 3 + 3$ is counted). The numbers in question are the coefficients of x^n in the expansions of $x^{m(m+1)/2}/D$ and x^m/D, respectively, into series of powers of x, where

$$D = (1-x)(1-x^2) \ldots (1-x^m).$$

Functions like these which serve to enumerate all the partitions of a specified kind are now called generating functions. In his more attractive exposition in his Introductio in Analysin Infinitorum of 1748, Euler noted that $1/D$ is the generating function giving the number of partitions of n into parts $\leqq m$ which need not be distinct. For $n = 5$, $m = 3$, these partitions are $3 + 2$, $3 + 1 + 1$, $2 + 2 + 1$, $2 + 1 + 1 + 1$, $1 + 1 + 1 + 1 + 1$. Similarly, the reciprocal of $\prod_{j=1}^{j=\infty}(1-x^j)$ is the generating function for the number of unrestricted partitions of n, where now also 5 and 4+1 are counted. Again, the number of partitions of n into m or fewer parts $\leqq t$ is the coefficient of x^n in the expansion of

$$(1-x^{t+1})(1-x^{t+2}) \ldots (1-x^{t+m})/D,$$

where D is the above product. Euler stated empirically the important fact that

$$\prod_{k=1}^{\infty}(1-x^k) = \sum_{n=-\infty}^{+\infty} (-1)^n x^{(3n^2\pm n)/2},$$

which has since been proved by many writers, in particular by Jacobi in his Fundamenta Nova of 1829, where he made important applications of elliptic functions to the theory of partitions. As noted by Legendre in 1830, the last formula implies that every number, not a pentagonal number $(3n^2\pm n)/2$, can be partitioned into an even number of distinct integers as often as into an odd number, while $(3n^2\pm n)/2$ can be partitioned into an even number of parts once oftener or once fewer times than into an odd number of parts, according as n is even or odd. Jacobi in 1846 extended this result to partitions into any given distinct elements.

In 1853 Ferrers gave a diagram which establishes a reciprocity between the partitions of the same number. The partition 3+3+2+1 is represented by four rows of dots containing 3, 3, 2, 1 dots, respectively, such that the left-hand dots are in the same vertical column. Reading the diagram by columns, we get the partition 4+3+2.

Sylvester stated in 1857 that the number of partitions of n into given positive integral elements $a_1, \ldots, a_r$ with repetitions allowed is ΣW_q, where the "wave" W_q is the coefficient of $1/t$ in the development in ascending powers of t of

$$\Sigma \rho^{-n} e^{nt} \prod_{j=1}^{r}(1 - \rho^{a_j} e^{-a_j t})^{-1},$$

the summation extending over the various primitive qth roots ρ of unity. Proofs were soon given by Battaglini, Brioschi, Roberts, and Trudi; Sylvester published his own method in 1882. Cayley wrote several papers on the theory and its applications.

During the years 1882–84, Sylvester and his pupils at Johns Hopkins University published many papers on partitions, in particular on their graphical representation, with the aim to derive the chief theorems constructively without the aid of analysis.

Beginning with his paper of 1886 on perfect partitions, Major MacMahon has made numerous contributions to the thoery of partitions and the more general subject of combinatory analysis, culminating in his treatise in two volumes published in 1915–16 (see the report, pp. 161–2).

Vahlen proved in 1893 that, among the partitions of s into distinct parts the sum of whose absolutely least residues modulo 3 equals a given integer h, there occur as many partitions into an even number of parts as into an odd number of parts, except only when s is the pentagonal number $(3h^2-h)/2$, for which there exists an additional partition into an even or odd number of parts according as h is even or odd. This implies the corollary of Legendre mentioned above. Analogous theorems were obtained by von Sterneck in 1897 and 1900.

Mention should be made of the various papers by Glaisher of 1875–76 and 1909–10, that of Csorba of 1914, and the asymptotic formulas obtained by Hardy and Ramanujan jointly in 1917–18.

Chapter IV reports on the extensive, mostly old, literature on rational right triangles, a subject which was the source of various problems treated in later chapters. Diophantus knew that if the sides of a right triangle are expressed by rational numbers they are proportional to $2mn$, m^2-n^2, m^2+n^2, and referred to the right triangle having the latter sides as that "formed from the two numbers m and n." Pythagoras and Plato had given special cases. Among the many problems on rational right triangles treated by Diophantus, Vieta, Bachet, Girard, Fermat, Frenicle, De Billy, Ozanam, Euler, and others, are the following: Find $n(n \geqq 3)$ rational right triangles of equal areas; two whose areas have a given ratio; one whose area is given or becomes a square on adding a given number or a certain function of the sides; one whose legs exceed the area by squares; one whose legs differ by unity or by a given number; right triangles the sum of whose legs is given; or with a rational angle-bisector.

Chapter V deals with rational triangles, whose sides and area are rational, and rational quadrilaterals, having also rational diagonals. By the juxtaposition of two rational right triangles with a common leg, we obtain a rational triangle. During 1773–82, Euler wrote a series of four papers on triangles whose sides and medians are all rational, while Bachet in 1621 had been content when a single median or single angle-bisector is rational. The Hindus Brahmegupta and Bháscara showed how to form a rational quadrilateral by juxtaposing four right triangles with pairs of equal legs such that the right angles have a common vertex and do not overlap. In 1848 Kummer showed how to obtain all rational quadrilaterals. Euler gave (p. 221) a construction for a polygon of n sides inscribed in a circle of radius unity such that the sides, diagonals, and the area are all rational. No mention will be made of the 160 further papers reported on in this chapter, which closes with the papers on rational pyramids, trihedral angles, and spherical triangles.

Chapters VI–IX deal with the specially interesting literature on the representation of numbers as sums of 2, 3, 4, n squares. Diophantus knew how to express the product of two sums of two squares as a sum of two squares in two ways:

$$(a^2+b^2)(c^2+d^2) = (ac \pm bd)^2 + (ad \mp bc)^2.$$

He knew that no number of the form $4n-1$ is a sum of two squares. But Girard in 1625 and Fermat a few years later were the first to recognize that a number is a sum of two squares if, and only if, its quotient by the largest square dividing it is a product of primes of the form $4n+1$ or the double of such a product.   Fermat also knew how to determine the number of ways in which a given number of the proper form is a sum of two squares. He stated that he could prove that every prime $4n+1$ is a sum of two squares by the method of indefinite descent, i.e., if a prime $4n+1$ is not a sum of two squares there exists a smaller prime of the same nature, etc., until 5 is reached.   Euler wrestled with this theorem for seven years before he succeeded in finding a complete proof in 1749.   He published more elegant proofs in 1773 and 1783.   In the meantime, Lagrange gave several proofs in 1771–75.   An expression for the number of representations of an integer as a sum of two squares was given by Legendre in 1798 and by Gauss in 1801, while a more elegant expression was deduced by Jacobi in 1829 from infinite series for elliptic functions and proved arithmetically by him in 1834 and by Dirichlet in 1840.   In a posthumous paper, Gauss left a formula for the number of sets of integers x, y for which $x^2 + y^2 \leq A$, i.e., the number of lattice points inside or on the circumference of a given circle; the same subject was studied by Eisenstein in 1844, Suhle in 1853, Cayley in 1857, Ahlborn in 1881, and Hermite in 1884 and 1887, while asymptotic formulas were proved by Sierpinski in 1906, Landau in 1912–13, Hardy in 1915–19, and Szilysen in 1917.

Diophantus stated in effect that no number of the form $8m+7$ is a sum of three squares, a fact easily verified by Descartes.   Fermat gave in effect the complete criterion that a number is a sum of three squares if, and only if, it is not of the form $4^n(8m+7)$.   For many years Euler tried in vain to prove this theorem, nor did Lagrange find a proof for all cases.   In 1798 Legendre gave a complicated proof by means of theorems on the quadratic divisors of t^2+cu^2.   In 1801 Gauss published a proof which also expresses the number of ways a number n is a sum of three squares in terms of the number of classes in the principal genus of the properly primitive binary quadratic forms of determinant $-n$.   Other such expressions were obtained by Dirichlet in 1840 by means of his formulas for the number of classes of binary quadratic forms; also by Kronecker in 1860 by use of series for elliptic functions and in 1883 by means of the number of classes of bilinear forms in two pairs of cogredient variables.   In 1850 Dirichlet gave an elegant proof of Fermat's criterion by means of reduced ternary quadratic forms.   Many writers have discussed the solution of $x^2+y^2+z^2=n^2$; a simple expression for the number of solutions was given by A. Hurwitz in 1907.   The problem of the number of integers $\leq x$ which are sums of three squares was investigated by Landau in 1908, while he (in 1912) and Sierpinski in 1909 found asymptotic formulas for the number of sets of integers u, v, w for which $u^2+v^2+w^2 \leq x$.

In the three problems in which Diophantus employed sums of four squares, he expressed 5, 13, and 30 as sums of four rational squares in two ways without mention of any condition on a number in order that it be a sum of four squares, although he gave necessary conditions for representation as a sum of two or three squares in the problems where the latter occur.   Hence Bachet and Fermat ascribed to Diophantus a knowledge of the beautiful theorem that every positive integer is a sum of four integral squares.   In 1621 Bachet verified this theorem for integers up to 325.   The theorem was stated to be true by Girard in 1625 and as an unproved fact by Descartes in 1638.   Fermat stated that he possessed a proof by indefinite descent.

This theorem engaged the serious attention of Euler for more than forty years, as appears from his life-long correspondence with Goldbach; in vain did he convert the problem into an equivalent, but equally baffling, question.   Not until twenty years after he began the study of the theorem did he publish in 1751 some important facts bearing on it, including his formula which expresses the product of two sums of four squares as such a sum.   The first proof published was that by Lagrange in 1772, who acknowledged his indebtedness to ideas in Euler's paper.   The next year Euler published an elegant proof, which is much simpler than Lagrange's and which has not been improved upon to date.   Gauss noted in 1801 that the theorem follows readily from the fact that any number having the remainder 1, 2, 5, or 6, when divided by 8, is a sum of three squares; but the latter fact has not yet been proved in so simple and elementary a manner as the former.   In 1853–54 Hermite gave two proofs by means of the theory of quadratic forms in four variables and a proof by means of a Hermitian form with complex integral coefficients and two pairs of two conjugate complex variables.

In 1828–29 Jacobi compared two infinite series for the same elliptic function to show that, if p is odd and $\sigma(p)$ is the sum of the divisors of p, the number of representations of $2^{\alpha}p$ as a sum of four squares is $8\sigma(p)$ or $24\sigma(p)$, according as $\alpha=0$ or $\alpha>0$, where in a representation the signs of the roots and their arrangement are taken into account.   In a similar manner, he and Legendre proved simultaneously that there are exactly $\sigma(p)$ sets of four positive odd numbers the sum of whose squares is $4p$. For the latter theorem Jacobi gave an arithmetical proof in 1834, which was simplified by Dirichlet in 1856 and by Pepin in 1883 and 1890.   For the former theorem on the representations of $2^{\alpha}p$, elementary proofs have been given by Stern in 1889, Vahlen in 1893, Gegenbauer in 1894, and L. Aubry in 1914, while Mordell gave in 1915 a proof by means of theta functions.

Cauchy proved in 1813 that any odd number k is a sum of four squares the algebraic sum of whose roots equals any assigned odd number between $\sqrt{3k-2}-1$ and $\sqrt{4k}$.   In 1873 Réalis proved also that every number $N=4n+2$ is a sum of four squares the algebraic sum of whose roots is any assigned one of the numbers 0, 2, 4, . . ., 2μ, where μ^2 is the largest square $<N$.   Mention should be made of papers by Torelli (p. 294), Glaisher (p. 296, p. 301), and Petr (p. 300).

Many of the papers in this long Chapter VIII prove the existence of solutions of the congruence $ax^2+by^2+cz^2\equiv0$ (mod p), in which a, b, c are not divisible by the prime p, while some determine the number of sets of solutions.   The corresponding question for n unknowns is discussed in the brief Chapter X.

In Chapter IX the material on representation as sums of n squares is separated from the reports on the more elementary papers giving relations between squares and mainly concerning n squares whose sum is a square. Following a hint by Jacobi, Eisenstein stated in 1847 that the number of representations of an odd number as a sum of eight squares equals 16 times the sum of the cubes of its divisors, and theorems almost as simple for six and ten squares. He also gave, without proof, formulas which express the number of representations of m by 5 and 7 squares as sums of Legendre-Jacobi symbols of quadratic residue character modulo m. In 1860–65 Liouville stated various theorems on representation by 10 and 12 squares, which he apparently deduced from series for elliptic functions, and which have been so proved and generalized by Bell in 1919, and were proved by means of theta functions by Humbert and Petr in 1907. In 1867 H. J. S. Smith stated general results on representation by 5 and 7 squares. This paper was unknown to the members of the commission whose recommendation led the Paris Academy of Sciences to propose for its grand prix des sciences mathématiques for 1882 the subject of representation by 5 squares. Prizes of the full amount were awarded both to Smith and to Minkowski (the latter being then 18 years of age), each of whom developed the theory of quadratic forms in n variables and evaluated the number of representations by 5 squares. There are further papers on the last topic by Stieltjes, Hermite, Pepin, and Hurwitz (pp. 310–1). Mention should be made of the papers by Gegenbauer (p. 313), Boulyguine (p. 317), Mordell, Hardy, and Ramanujan (p. 318) on representation by n squares.

Chapter XI, which is closely related to the last topic, gives a summary of Liouville's series of eighteen articles published in 1858–65, in which he stated results (apparently found from expansions of elliptic functions) which express many equalities between sums of the values of quite general arithmetical functions when the arguments of the functions involve the divisors of two (or more) numbers whose sum is given. The chapter closes with a citation of papers which together give proofs of all the formulas, except only (Q) of the sixth article, besides proving a few related theorems.

The sixty pages of Chapter XII give reports on more than 300 papers on $ax^2+bx+c=y^2$. Diophantus was led to such an equation in at least forty of his problems. He was content with rational solutions, which he showed how to find if a or c is a square, or if $b=0$ and one set of solutions is known. It is a remarkable fact that the Hindu Brahmegupta in the seventh century gave a tentative method of solving $ax^2+c=y^2$ in integers, which is a far more difficult problem than its solution in rational numbers. His method was explained more clearly by the Hindu Bháscara in the twelfth century. Much earlier, the Greeks had given approximations to square roots which may be interpreted as yielding solutions of $ax^2+1=y^2$ for $a=2$ and $a=3$. Moreover, the famous cattle problem of Archimedes, which imposed nine conditions upon eight unknowns, leads in its final analysis to the difficult equation $ax^2+1=y^2$, where $a=4729494$, and has been solved in modern times.

Such an equation $x^2 - Ay^2 = 1$ has long borne the name Pellian equation, after John Pell, due to a confusion on the part of Euler; it would have been more appropriately named after Fermat, who stated in 1657 that it has an infinitude of integral solutions if A is any positive integer not a square, and who stated in 1659 that he possessed a proof by indefinite descent. He proposed it as a challenge problem to the English mathematicians Lord Brouncker and John Wallis, who finally succeeded in discovering a tentative method of solution, without giving a proof of the existence of an infinitude of solutions. This theorem is really only the simplest and first known case of Dirichlet's elegant and very general theorem on the existence of units in any algebraic field or domain. The former theorem is also of great importance in the theory of binary quadratic forms. Moreover, the problem to find all the rational solutions of the most general equation of the second degree in two unknowns reduces readily to that for $x^2 - Ay^2 = B$, all of whose solutions follow from one solution and the solutions of $x^2 - Ay^2 = 1$.

In 1765 Euler exhibited the method of solving a Pellian equation due to Brouncker and Wallis in a more convenient form by use of the continued fraction for $\sqrt{A}$ and found various important facts, but gave no proof that the process leads always to a solution in positive integers. This fundamental fact of the existence of solutions was first proved by Lagrange a year or two later; while in 1769 and 1770 he brought out his classic memoirs which give a direct method to find all integral solutions of $x^2 - Ay^2 = B$, as well as of an equation of degree n, by developing its real roots into continued fractions.

Of the further extensive literature on the Pellian equation, the most notable papers are those by Legendre, Gauss, Dirichlet, Jacobi, and Perott; limits for the least positive solution were obtained by Tchebychef in 1851 and by Remak, Perron, Schmitz, and Schur in 1913–18. Useful tables have been given by Euler, Legendre, Degen, Tenner, Koenig, Arndt, Cayley, Stern, Seeling, Roberts, Bickmore, Cunningham, and Whitford.

Chapter XIII treats of further single equations of the second degree, including $axy + bx + cy + d = 0$, $x^2 - y^2 = g$, $ax^2 + bxy + cy^2 = dz^2$ or d, the most general equation of the second degree in x, y, and its homogeneous form $aX^2 + bY^2 + cZ^2 + dXY + eXZ + fYZ = 0$. Criteria for integral solutions of the latter were stated by H. J. S. Smith (p. 431) and proved by Meyer for the case of an odd determinant, while its complete solution was given by Desboves (p. 432) when one solution is known. Lagrange's method for $x^2 - Ay^2 = B$, cited above, was employed by Legendre in 1785 to prove the important theorem that, if no two of the integers a, b, c have a common factor and if each is neither zero nor divisible by a square, then $ax^2 + by^2 + cz^2 = 0$ has integral solutions not all zero if, and only if, $-bc$, $-ac$, $-ab$ are quadratic residues of a, b, c, respectively, and a, b, c are not all of the same sign. Gauss gave a proof by means of ternary quadratic forms, while a generalization was made by Dirichlet (p. 423) and Goldscheider (p. 426). Meyer gave criteria (pp. 432–3) for integral solutions of $f = 0$, where f is any quadratic form in four variables, with simple criteria in the case of $ax^2+by^2+cz^2+du^2=0$; and noted that, when there is a fifth term ev^2, the equation is solvable in integers not all zero if the coefficients are odd and not all of the same sign. Minkowski (p. 433) proved the generalization that zero can be represented rationally by every indefinite quadratic form in five or more variables, and gave invariantive criteria for four or fewer variables.

Chapter XIV reports on many elementary papers on squares in arithmetical or geometrical progression. While there is a simple, general, formula for three squares in arithmetical progression, known by Vieta, Fermat, and Frenicle, there do not exist four distinct squares in arithmetical progression.

Chapter XV opens with a collection of the problems from Diophantus, in which it is a question of finding values of the unknowns for which several linear functions of them become equal to squares. Such problems were treated by Brahmegupta in the seventh century, by Vieta in 1591, and by Bachet, Fermat, Prestet, Ozanam, and others, in the seventeenth century. One of the problems studied most frequently is that of finding three numbers such that the sum and difference of any two of them are squares; it was treated by Petrus in 1674, Leibniz in 1676, Rolle in 1682, Landen in 1775, by Euler in his Algebra and elsewhere, as well as by various later writers (all cited in note 28, p. 448).

The story of congruent numbers, given in Chapter XVI, is a long one, beginning with Diophantus. If x and k are rational numbers such that x^2+k and x^2-k are both rational squares, k is called a congruent number. Diophantus knew that $x^2+y^2=z^2$ implies $z^2\pm2xy=(x\pm y)^2$, so that $2xy$ is a congruent number. This topic was the chief subject of two Arabic manuscripts of the tenth century. Leonardo Pisano, in his Liber Quadratorum of 1225, treated the subject at length and with skill, making repeated use of the fact that any integral square is a sum of consecutive odd numbers beginning with unity. In particular, he stated, but did not completely prove, that no congruent number is a square, which implies that the area of a rational right triangle is never a square and that the difference of two biquadrates is not a square, results of special importance historically. Although part of Leonardo's work was incorporated in the arithmetics of Luca Paciuolo, Ghaligai, Feliciano, and Tartaglia, the original seemed to be lost and Cossali made a laborious, but unsuccessful, attempt to reconstruct it. The original was found and published by Prince Boncompagni in 1854 and in the Scritti di Leonardo Pisano, II, 1862. The most important later papers on congruent numbers are those by Euler, Genocchi, Woepcke, Collins, and Lucas.

The related problem of concordant forms is to make x^2+my^2 and x^2+ny^2 both squares and was studied by the same writers, especially by Euler in several of his memoirs. The remaining problems of this chapter and those of Chapter XVII relate to special systems of two quadratic functions or equations and do not possess sufficient general interest to warrant mention here. The last remark applies also to Chapter XVIII, which treats of three or more quadratic functions.

Chapter XIX begins with the history of the problem of finding three integers x, y, z such that x^2+y^2, x^2+z^2, y^2+z^2 are all perfect squares. Solutions involving arbitrary parameters, but obtained under special assumptions, were found by Saunderson (who was blind from infancy) and Euler in their Algebras of 1740 and 1770. The problem is equivalent to that of finding a rectangular parallelopiped having rational values for the edges and the diagonals of the faces. If we impose the further restriction that also a diagonal of the solid shall be rational, we have a difficult problem which has been recently attacked but not solved.

The problem of finding n squares the sum of any $n-1$ of which is a square was treated at length by Euler for $n=4$, and for any n by Gill by use of trigonometric functions. The problem of finding three squares the sum of any two of which exceeds the third by a square was treated by four special methods by Euler in a posthumous paper, as well as by Legendre and others. The problem of making a quadratic form in x and y, one in x and z, and one in y and z simultaneously equal to squares has received much attention during the past hundred years. Beginning with Diophantus, there is an extensive early literature on the problem of finding n numbers such that the product of any two of them increased by a given number shall be a square.

Euler developed an interesting method (p. 522) to make several functions simultaneously equal to squares. He selected a suitable auxiliary function f such that solutions of $f=0$ can be readily found. For any set of solutions, P^2-f is evidently a square, whatever be the function P. Many further problems occur in this long chapter, which closes with an account of rational orthogonal substitutions.

The nature of Chapter XX will be illustrated by means of an example of considerable interest for the history of algebraic numbers. Fermat stated that he had a proof that 25 is the only integral square which if increased by 2 becomes a cube. Euler, in attempting a proof in his Algebra of 1770, assumed that $x^2+2=t^3$ implies that each factor $x\pm\sqrt{-2}$ is the cube of a number $p+q\sqrt{-2}$, where p and q are integers, although he knew that a like assumption is not valid when 2 is replaced by other numbers. The justification of his assumption in the first example is due to the fact that for these numbers $p+q\sqrt{-2}$ factorization into primes is unique and to the further fact that ±1 are the only ones of these numbers which divide unity. Instead of this explanation by means of algebraic numbers, we may employ the theory of classes of binary quadratic forms, as was done by Pepin (p. 541).

In the 69 pages of Chapter XXI report is made on about 500 papers on Diophantine equations of degree 3. The method by which Diophantus expressed the difference of two given rational cubes as a sum of two positive rational cubes was given in his Porisms, a work which has not been preserved. The formula (p. 550) which Vieta used in 1591 for this purpose is valid only when the greater of the given cubes exceeds the double of the smaller. While also Bachet could solve only this case, Girard and Fermat showed how, by employing Vieta's three formulas in turn, to solve the remaining case as well as the problem to express a sum of two given rational cubes as another such sum. The last problem had been proposed by Fermat to the English mathematicians Brouncker and Wallis, who gave merely solutions derived from known solutions by multiplication by a constant. The general solution in integers of this problem was first given by Euler in 1756–57. His solution was expressed in a simpler form by Binet in 1841 and deduced elegantly by Hermite in 1872 by means of the ruled lines on the corresponding cubic surface (a method extended to a certain equation of degree n by Brunel, p. 556). Report is made on pp. 560–1 on Japanese writings during 1826–45 on this subject. The related problem of finding three equal sums of two cubes arose in the question of finding four integers the sum of any two of which is a cube.

There are many minor papers of recent decades which give relations between five or more cubes, or express a sum of three cubes as a square. The problem of making a binary cubic form equal to a cube was treated by obvious elementary methods by Fermat and Euler, and recently by birational transformation by von Sz. Nagy, and by covariants by Haentzschel. To make a binary cubic form equal to a square, Fermat and Euler equated it to the square of a linear or quadratic function, and Lagrange used the norm of an algebraic number (p. 570), while Mordell in 1913 employed the theory of invariants.

Since every rational number is a sum of three rational cubes (p. 726), it is an interesting question to determine the rational numbers which are sums of two rational cubes, or, if we prefer, the integers A for which $x^3 + y^3 = Az^3$ is solvable in integers. Reports on fifty papers on this subject are given on pp. 572–8. Euler proved that the problem is impossible if $A = 1$ and $A = 4$, and that $x = \pm y$ if $A = 2$. Legendre erred in his statement that it is impossible if $A = 6$. In 1856 Sylvester stated that it is impossible if $A = p$, $2p$, $4p^2$, $4q$, q^2, $2q^2$, where p and q are primes of the respective forms $18l + 5$ and $18l + 11$. In 1870 Pepin proved these and similar results. Using also analogous facts proved by Sylvester in 1879, we can state whether or not any proposed number, not exceeding 100, is a sum of two rational cubes.

There are 42 papers (pp. 582–8) on the problems of finding numbers in arithmetical progression the sum of whose cubes is a cube or a square.

If $F(x, y, z) = 0$ is a homogeneous cubic equation with rational coefficients and if P is a rational point (i.e., having rational coordinates) on the curve $F = 0$, the tangent at P cuts the curve in a new rational point, called the tangential to P. Similarly, the secant through two rational points on the curve cuts it in a third rational point. Curiously enough, the analytic equivalents of these facts were obtained by Cauchy in 1826 without their geometrical setting. Levi in 1906–9 defined a configuration of rational points on a cubic curve without double points to be the set of all rational points which can be derived from one or more rational points by the operations of finding the tangential to a point of the set and of finding the third intersection of the curve and the secant joining two points of the set. In 1917 A. Hurwitz called such a set of points a complete set and obtained theorems on the number of rational points on the cubic curve. Mordell made use of the invariants of F.

The problem of finding n rational numbers the cube of whose sum increased (or decreased) by any one of the numbers gives a cube was treated for $n = 3$ by Diophantus and his commentators, by Ludolph van Ceulen in his Dutch work on the circle, by van Schooten, J. Pell and others—the simplest answer being that by Hart (p. 611).

Chapter XXII devotes 57 pages to reports on 400 papers on Diophantine equations of degree 4. Fermat's proof of his challenge theorem that no rational right triangle has an area which is a rational square is of special interest, as it illustrates in detail his method of indefinite descent; his proof also shows that the difference of two biquadrates is never a square. Leibniz left a manuscript giving a proof.

Fermat affirmed that the smallest rational right triangle whose hypotenuse and the sum of whose legs are squares has its sides expressed by numbers of thirteen digits. The problem is equivalent to that of finding two numbers (for n numbers, pp. 665–7) whose sum is a square and whose sum of squares is a biquadrate, and was proposed in this form by Leibniz and treated several times by Euler, and at great length by Lagrange in 1777, who found it necessary to solve several equations of the form $ax^4 + by^4 = cz^2$. The extensive literature on the latter equation is reviewed on pp. 627–634; some of the methods employed apply also when there occurs a term dx^2y^2 in the equation (pp. 634–9).

Just as in algebra no general equation of degree exceeding 4 can be solved by radicals, so in Diophantine analysis nearly all the problems for which solutions have been found are those which reduce finally to the question of making a given binary form f of degree $\leqq 4$ equal to a square or higher power. Among the methods (pp. 639–644) of making a quartic function $f(x)$ of special type equal to a square are the rather obvious methods of Fermat; the method of Euler of reducing f to the form $P^2 + QR$, where P, Q, R are quadratic functions of x, so that $f = (P + Qy)^2$ becomes an equation quadratic in x and in y; and the invariantive methods of Mordell and Haentzschel. Euler's method is similar to that employed by him in the problem of the multiplication of an elliptic integral; Jacobi noted a generalization by use of Abel's theorem (p. 641).

Euler, after solving $A^4 + B^4 = C^4 + D^4$ by several methods, stated (p. 648) that it is impossible to find three biquadrates whose sum is a biquadrate, and that he believed it possible to assign four biquadrates whose sum is a biquadrate. But his investigation was incomplete and led to no example.

The first example, $30^4+120^4+272^4+315^4=353^4$, was found by Norrie in 1911. In the meantime various writers gave examples of five or more biquadrates whose sum is a biquadrate and cases of equal sums of biquadrates.

Chapter XXIII, on equations of degree >4, will doubtless be more useful than any other chapter in the volume since it reports on the papers which offer general methods of attacking Diophantine equations. Lagrange showed how to use continued fractions to solve $f=c$, where f is a binary form of any degree. Runge and Maillet obtained conditions for the existence of infinitely many pairs of integral solutions of $f(x, y) = 0$, where f is an irreducible polynomial with integral coefficients. Thue proved the useful theorem that, if $U(x, y)$ is an irreducible homogeneous polynomial of degree >2 with integral coefficients and c is a given constant, $U=c$ has only a finite number of pairs of integral solutions. Maillet gave a generalization (p. 675) to non-homogeneous polynomials U.

Hilbert and Hurwitz, in their joint paper of 1890–1, proved that any homogeneous equation with integral coefficients which represents a curve of genus zero can be transformed birationally into a linear or quadratic equation. Poincaré in 1901 proved the same theorem and found when a curve of genus unity can be transformed birationally into a curve of order p. The related later papers are cited on p. 677.

It is convenient to define at this point the product

$$F(x, y, \ldots, z) \equiv \Pi(x+\alpha y+ \ldots +\alpha^{n-1}z),$$

extended over all the roots $\alpha, \ldots$ of any irreducible equation of degree n with integral coefficients, to be the norm of the general number $x+\alpha y+ \ldots$ of the algebraic field determined by α. Dirichlet noted that $F=1$ has infinitude of integral solutions except when the field is an imaginary quadratic field. If the field is real and if F can take a given value, it takes that value for an infinitude of sets of integers $x, \ldots, z$. Also Poincaré (p. 678) discussed this problem $F=g$. Lagrange (pp. 570, 691) proved in effect that the norm of a product equals the product of the norms of the factors and hence solved $F(X, Y, \ldots, Z) = V^m$, where $V=F(x, y, \ldots, z)$. This method is of considerable power in seeking special solutions of various types of equations. The particular case $x^3+ny^3+n^2z^3-3nxyz$ occurs in the papers on pp. 593–5. This case is also a special case of another type of equations of general degree obtained by Maillet from the theory of recurring series (p. 695). A. Hurwitz's complete discussion (p. 697) of the positive integral solutions of $x_1^2+ \ldots +x_n^2=xx_1 \ldots x_n$ furnishes a model for thoroughness which may well be imitated by writers on Diophantine equations, too many of whom seem to be content with a special solution of their problems.

Chapter XXIV deals with sets of integers with equal sums of like powers. For example, a, b, c and $a+b+c$ have the same sum and same sum of squares as $a+b$, $a+c$, $b+c$. Of the seventy papers on this topic, only five are prior to 1878. On pp. 714–6 is noted the connection of this problem with the older one of rapidly converging series convenient for the computation of logarithms, in which we desire two polynomials in x which differ only in their constant terms and have exclusively integers as their roots.

Chapter XXV furnishes a typical example in the theory of numbers of the contrast between the ease with which empirical theorems are discovered and the difficulty attending a complete mathematical proof. On the basis of numerical experiments, Waring announced in 1770 the empirical theorem that every positive integer is a sum of at most 9 positive cubes, a sum of at most 19 biquadrates, and in general a sum of a limited number of positive mth powers. The last fact was first proved in 1909 by Hilbert, although his investigation does not determine the precise value of the number N_m such that every positive integer is a sum of at most N_m positive mth powers. About the year 1772, J. A. Euler stated that $N_m \geqq \nu + 2^m - 2$, where ν is the largest integer $< (3/2)^m$. Just before 1859, Liouville proved that $N_4 \leqq 53$ by means of an identity equivalent to

$$6(x_1^2 + x_2^2 + x_3^2 + x_4^2)^2 = \sum_6 (x_1 + x_2)^4 + \sum_6 (x_1 - x_2)^4$$

and the fact that any positive integer n is expressible in the form $x_1^2 + x_2^2 + x_3^2 + x_4^2$, so that $6n^2$ is a sum of 12 biquadrates. But any positive integer is of one of the six forms $6p$, $6p+1$, . . ., $6p+5$, while $p = n_1^2 + n_2^2 + n_3^2 + n_4^2$. Thus $6p$ is a sum of 4×12 biquadrates. Since 1, . . ., 5 are sums of as many units, each a biquadrate, we have $N_4 \leqq 4 \times 12 + 5$. Maillet was the first to prove, in 1895, that N_3 is finite, in fact $\leqq 21$. Later writers succeeded in proving that $N_3 = 9$. In his proof that N_m is finite, Hilbert employed a five-fold integral, while later writers have given an algebraic proof. Quite recently, Hardy and Littlewood gave a proof by use of the theory of analytic functions and showed that $N_m \leqq (m-2)2^{m-1} + 5$, which gives 9 cubes, 21 biquadrates, 53 fifth powers, etc. Earlier papers (pp. 726–9) gave elementary proofs that every positive rational number is a sum of three rational cubes and a sum of four positive rational cubes.

The final chapter devotes 46 pages to reports on more than 300 papers on Fermat's last theorem, which states that it is impossible to separate any power higher than the second into two powers of like degree, and the more general trinomial equation $ax^r + by^s = cz^t$, and congruence of the same form. In letters and in annotations to his copy of Diophantus, Fermat announced many interesting discoveries in the theory of numbers, usually with the statement that he possessed a proof. All of these facts have since been proved with the exception of his "last theorem" above, for which he stated that he had found a truly remarkable proof. If there was an oversight in his proof it was certainly not one of the foolish errors committed in the past decade in the thousands of efforts to secure a large cash prize. Fermat proposed the cases of exponents 3 and 4 (p. 545, pp. 616–7) as challenge problems to the mathematicians of his time. The general case has remained a challenge problem to the mathematicians of the subsequent three centuries. At intervals during the past century, leading scientific academies offered one of their prizes for a proof. The dignity of this famous theorem was injured by the offer of a very large prize in 1908. Since only printed proofs may compete, the gain thus far has gone to the printers; in this history no mention will be made of the very numerous false proofs called forth by this last prize.

Fermat's last theorem is not of special importance in itself, and the publication of a complete proof would deprive it of its chief claim to attention for its own sake. But the theorem has acquired an important position in the history of mathematics on account of its having afforded the inspiration which led Kummer to his invention of his ideal numbers, out of which grew the general theory of algebraic numbers, which is one of the most important branches of modern mathematics.

Although Gauss had proved in 1832 that the laws of elementary arithmetic hold also for complex integers (numbers like $5+7\sqrt{-1}$) and made a brilliant application of them in his investigation of biquadratic residues, the theory of algebraic numbers was really born in the year 1847. For it was then (pp. 739, 740) that the mathematical world became definitely conscious of the fact that complex integers $a_0+a_1r+\ldots+a_{n-1}r^{n-1}$, where the a's are ordinary integers and r is an imaginary nth root of unity, do not in general decompose into complex primes in a single manner, do not possess a greatest common divisor, and hence do not obey the laws of elementary arithmetic. This historical fact came to light through discussions of lacunæ in the attempted proof by Lamé that, if n is an odd prime, $x^n+y^n=z^n$ is not satisfied by such complex integers. Other errors of the same nature were made in the same year by Wantzel and by so great a mathematician as Cauchy. Curiously enough, Kummer himself made the error, in a letter of about 1843 to Dirichlet, of assuming that factorization is unique, so that his initial proof of Fermat's last theorem was incomplete. But Kummer did not stop with the mere recognition of the fact that algebraic numbers do not obey the laws of arithmetic; he succeeded in restoring those laws by the introduction of ideal elements, this restoration of law in the midst of chaos being one of the chief scientific triumphs of the past century.

Although the theory of algebraic numbers appears to be a powerful tool especially adapted to attack Fermat's last theorem, it has not yet led to a complete proof of it. Numerous facts have been obtained by a variety of more elementary methods. Until the theorem is actually proved, it will obviously be unwise to attempt to weigh the importance of any particular fact or method. Hence no further analysis will be given here of the contents of the long Chapter XXVI which is itself a condensed history of Fermat's last theorem. Moreover this subject is one of those for which the subject index gives a rather minute classification of the subject matter.

In the preceding summary mention was made of only the most important of the upwards of 5,000 writings upon which report has been made in the text. While many of these papers are of minor importance, the aim has been to give an exhaustive account of the literature on the subject rather than a selective account reflecting the author's imperfect views as to relative importance. This work is intended as a source book not merely for the fastidious professional mathematician, but also for the larger number of amateurs who find endless fascination for the "queen of the sciences," whose rule began centuries ago and has continued without interruption to the present.

Unfortunately, following the practice of Diophantus, many writers on this subject have been content with a special solution of their problem, obtained by making various assumptions which simplify the analysis. A report which would give merely the final formulas in such a paper, without indicating also the restrictive assumptions, would be useless. Instead, there is given here a summary of the essential steps in the proof, and this plan is followed especially in the case of papers not to be found in the average large library. These papers which give only special solutions of the problem attacked have at least the value of showing that the problem is not impossible. Moreover, an examination of many such papers reveals the fact that there are a few constantly recurring types of auxiliary Diophantine problems (such as that of making a quartic function equal to a square), whose complete solution would permit the complete treatment of a very large number of problems, and hence suggest specially useful subjects for thorough investigation. Since there already exist too many papers on Diophantine analysis which give only special solutions, it is hoped that all devotees of this subject will in future refrain from publication until they obtain general theorems on the problem attacked if not a complete solution of it. Only in this way will the subject be able to retain its proper position by the side of other virile branches of mathematics.

It was initially planned to give this work the title "topical history of the theory of numbers"; but the word topical was omitted at the advice of a prominent historian. It is inconceivable that any one would desire this vast amount of material arranged other than by topics. Again, conventional histories take for granted that each fact has been discovered by a natural series of deductions from earlier facts and devote considerable space in the attempt to trace the sequence. But men experienced in research know that at least the germs of many important results are discovered by a sudden and mysterious intuition, perhaps the result of subconscious mental effort, even though such intuitions have to be subjected later to the sorting processes of the critical faculties. What is generally wanted is a full and correct statement of the facts, not an historian's personal explanation of those facts. The more completely the historian remains in the background or the less conscious the reader is of the historian's personality, the better the history. Before writing such a history, he must have made a more thorough search for all the facts than is necessary for the conventional history. With such a view of the ideal self-effacement of the historian, what induced the author to interrupt his own investigations

for the greater part of the past nine years to write this history? Because
it fitted in with his conviction that every person should aim to perform at
some time in his life some serious, useful work for which it is highly im-
probable that there will be any reward whatever other than his satisfaction
therefrom. Certainly, the eight mathematicians mentioned below, who co-
operated with the author, are justly entitled to enjoy the same satisfaction
from their work.

Concerning the various sources of references consulted and the various
libraries in America and Europe in which the material was collected, the
remarks made on page XI of the Preface to Volume I apply also to the
present volume. In particular, those references in the Subject Index of the
Royal Society Catalogue of Scientific Papers, Volume I, 1908, which relate
to Diophantine analysis were used not only in the preparation of the manuscript, but were checked on the proof-sheets. The references to Diophantus
follow the usual numbering and hence not that in the second edition by
Heath.

The reports in Chapters XI–XXVI have been checked by the original
papers in case they are to be found in Chicago. The computations occurring
in the reports in Chapters XXI–XXIV were checked by the author and
various errors in the original papers were detected. Moreover the reports
in four chapters were read carefully and critically by an authority on the
subject of the chapter as follows: Chapter III on partitions by Major
P. A. Mac Mahon, Chapter XXIV on sets of integers with equal sums of
like powers by E. B. Escott, Chapter XXV on Waring's problem by A. J.
Kempner, and Chapter XXVI on Fermat's last theorem by H. S. Vandiver.
A high degree of accuracy and clearness for these Chapters III, XXI–XXVI
was especially desired since they are the ones which will be most frequently
consulted. Also Chapters I–XII were read minutely by Kempner, thanks
to whom various imperfections and errors have been removed. Further-
more, the proof-sheets of the entire volume were read by R. D. Carmichael,
A. Cunningham, E. B. Escott, A. Gérardin, and E. Maillet, each of whom
has written extensively on Diophantine equations and made very valuable
suggestions on the present work. To these eight experts, who gave so
generously of their time to perfect this volume, is due the gratitude not
merely of the author but also of every devotee of Diophantine analysis who
may derive benefit or pleasure from this history.

Miss Minna J. Schick read the proof-sheets of the first eleven chapters
and compared them with the original manuscript, for which purpose the
authorities at the University of Chicago considerately relieved her of the
duties connected with her fellowship in mathematics. Mrs. Louise M.
Swain, who had just completed a year of postgraduate studies in mathematics at the University of Chicago, read the proof-sheets of the last
fifteen chapters, checked the many cross-references throughout the volume,
constructed and checked the author indexes, helped to check the references
with the Royal Society Catalogue, and checked the page-proofs with the
galleys and separately for various types of faults. The author is under

great obligations to these gifted young women for the many improvements in the book due to their accuracy and alertness.   In addition to all this help, the author has devoted a large part of his time for fifteen months to the proof-sheets, comparing them with his original notes, checking computations, comparing reports and readers' suggestions with the original papers, adding reports on current papers, repeating the work done on the manuscript of examining minutely all the reports for results needing citation elsewhere by cross-reference, and inspecting every change made in the proof.

Readers are requested to supply, for insertion in a concluding Volume III on quadratic and higher forms, residues, and reciprocity laws, notices of errata or omissions, as well as abstracts of the few papers marked by the symbol * before authors' names to signify that the papers were not available for report.

L. E. DICKSON
Professor of Mathematics
University of Chicago

APRIL, 1920.

TABLE OF CONTENTS.

TABLE OF CONTENTS.

CHAPTER I.

POLYGONAL, PYRAMIDAL AND FIGURATE NUMBERS.

The formation of *triangular* numbers 1, 1 + 2, 1 + 2 + 3, $\cdots$, and of square numbers 1, 1 + 3, 1 + 3 + 5, $\cdots$, by the successive addition of numbers in arithmetical progression, called *gnomons*, is of geometric origin and goes back to Pythagoras[1] (570–501 B.C.):

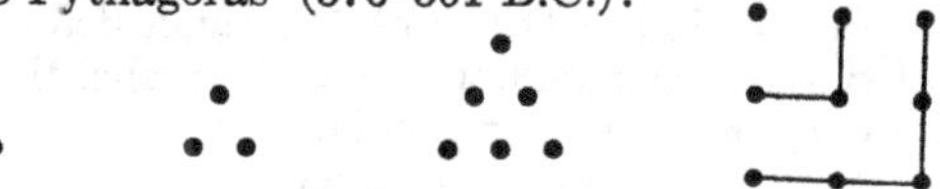

If the gnomons added are 4, 7, 10, $\cdots$ (of common difference 3), the resulting numbers 1, 5, 12, 22, $\cdots$ are *pentagonal*. If the common difference of the gnomons is $m - 2$, we obtain *m-gonal* numbers or *polygonal* numbers with m sides.

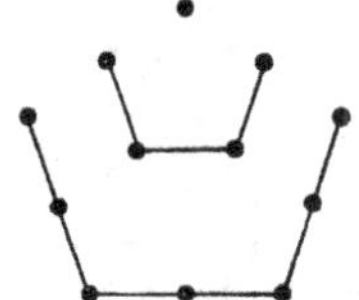

In the cattle problem of Archimedes (third century B.C.), the sum of two of the eight unknowns is to be a triangular number (see Ch. XII).

Speusippus,[2] nephew of Plato, mentioned polygonal and pyramidal numbers: 1 is point, 2 is line, 3 triangle, 4 pyramid, and each of these numbers is the first of its kind; also, 1 + 2 + 3 + 4 = 10.

About 175 B.C., Hypsicles gave a definition of polygonal numbers which was quoted by Diophantus[3] in his Polygonal Numbers, "If there are as many numbers as we please beginning with one and increasing by the same common difference, then when the common difference is 1, the sum of all the terms is a triangular number; when 2, a square; when 3, a pentagonal number. And the number of the angles is called after the number exceeding the common difference by 2, and the side after the number of terms including 1." Given therefore an arithmetical progression with the first term 1 and common difference $m - 2$, the sum of r terms is the r-th m-gonal number[3] p_m^r.

The arithmetic of Theon of Smyrna[4] (about 100 or 130 A.D.) contains 32 chapters. In Ch. 15, p. 41, the squares are obtained from 1 + 3 = 4,

---

[1] F. Hoefer, Histoire des mathématiques, Paris, ed. 2, 1879, ed. 5, 1902, 96–121; W. W. R. Ball, Math. Gazette, 8, 1915, 5–12; M. Cantor, Geschichte Math., 1, ed. 3, 1907, 160–3, 252.

[2] Theologumena arithmeticae, ed. by F. Ast, Leipzig, 1817, 61, 62. For a French transl. and notes, see P. Tannery, Pour l'histoire de la science Hellène, Paris, 1887, 386–390 (374).

[3] Denoted by $P_r^{(m)}$ in Encyc. Sc. Math., I, 1_1, p. 30.

[4] Theonis Smyrnaei Platonici, Latin transl. by Ismael Bullialdi, 1644. Cf. Expositio rerum mathematicarum ad legendum Platonem utilium, ed., E. Hiller, pp. 31–40.

$1 + 3 + 5 = 9$, etc. In Ch. 19, p. 47, the triangular numbers are defined to be $1, 1 + 2, 1 + 2 + 3, \cdots$. In Ch. 20, p. 52, the squares are obtained as before and the pentagonal numbers are obtained by addition of 1, 4, 7, 10, $\cdots$. In Chapters 26 and 27, pp. 62–64, pentagonal and hexagonal numbers are shown by dots forming regular pentagons (as in the figure on the preceding page) or hexagons. Ch. 28, p. 65, gives the theorem that the sum of two consecutive triangular numbers is a square. In Ch. 30, p. 66, is defined the pyramidal number $P_m^r = p_m^1 + p_m^2 + \cdots + p_m^r$.

Nicomachus[5] (about 100 A.D.) gave the same definitions and results as did Theon of Smyrna and perhaps gave them slightly earlier. Ch. 12 gives the theorem on consecutive triangular numbers:

$$\frac{(r - 1)r}{2} + \frac{r(r + 1)}{2} = r^2;$$

also the corresponding theorem that the sum of the rth square and $(r - 1)$th triangular number is the rth pentagonal number, just as a pentagon is obtained by annexing a triangle to a square. He gave the generalization (apart from the notation):

$$p_m^r + p_3^{r-1} = p_{m+1}^r.$$

These theorems are illustrated by means of the following table:

Triangles	1	3	6	10	15	21	28	36	45	55
Squares	1	4	9	16	25	36	49	64	81	100
Pentagons	1	5	12	22	35	51	70	92	117	145
Hexagons	1	6	15	28	45	66	91	120	153	190
Heptagons	1	7	18	34	55	81	112	148	189	235.

Each polygon equals the sum of the polygon immediately above it in the table and the triangle with 1 less in its side [triangle in the preceding column]; for example, heptagon 148 is the sum of hexagon 120 and triangle 28.

Each vertical column is an arithmetical progression whose common difference is the triangle in the preceding column.

In Ch. 13 he remarked that just as polygonal numbers arise by summing the simple arithmetical progressions, so by summing the polygonal numbers one obtains the like named pyramidal numbers,—triangular pyramid from the triangular numbers, pyramid with square base from the squares, etc., the base being the largest polygon.

---

[5] Introductio arithmetica (ed., Hoche), 2, 1866, Book 2, Chs. 8–20. Cf. G. H. F. Nesselmann, Algebra der Griechen, 1842, 202.

Plutarch,[6] a contemporary of Nicomachus, gave the theorem that if we multiply a triangular number by 8 and add 1, we obtain a square:

$$8\frac{r(r+1)}{2} + 1 = (2r+1)^2.$$

This theorem was given by Iamblichus[7] (about 283–330 A.D.), who treated at length (pp. 82–176) polygonal and pyramidal numbers.

Diophantus[8] (about 250 A.D.) generalized this theorem and proved by a cumbersome geometric method that

$$(1) \qquad 8(m-2)p_m^r + (m-4)^2 = \{(m-2)(2r-1)+2\}^2,$$

and spoke of this result as a new definition of p equivalent to that of Hypsicles. Diophantus gave a rule for finding r, equivalent to the solution of (1) for r, and a rule for finding p equivalent to

$$(2) \qquad p_m^r = \frac{[(m-2)(2r-1)+2]^2 - (m-4)^2}{8(m-2)},$$

but did not give the equal simpler expression

$$(3) \qquad p_m^r = \tfrac{1}{2}r\{2 + (m-2)(r-1)\}.$$

In fact, starting with (2), he gave a long geometric discussion to find the number of ways a given number can be polygonal, but made little headway before the abrupt termination of the fragment. G. Wertheim[9] gave a lengthy continuation in the same geometric style which eventually leads to the geometric equivalent to (3) and remarked that we can readily find from (3) the ways in which a given number p can be polygonal: Express $2p$ as a product of two factors > 1 in all possible ways; call the smaller factor r; subtract 2 from the larger factor and find whether or not the difference is divisible by $r-1$; if it is, the quotient is $m-2$, and p is a p_m^r. Since $m-2$ equals $2(p-r)/[r(r-1)]$, the latter must be an integer $\geqq 1$, so that

$$r \leqq \tfrac{1}{2}(\sqrt{8p+1} - 1).$$

For example, if $p = 36$, then $r \leqq 8$. Since r divides $2p = 72$, we have $r = 2, 4, 8, 3, 6$, of which $r = 4$ is excluded. We get

$$36 = p_{36}^2 = p_{13}^3 = p_4^6 = p_3^8.$$

In the Roman Codex Arcerianus[10] (450 A.D.?) occur a number of special cases of the remarkable formula for pyramidal numbers

$$P_m^r = \frac{r+1}{6}(2p_m^r + r).$$

---

[6] Platonicae quaestion., II, 1003.

[7] In Nicomachi Geraseni arith. introd., ed., S. Tennulius, 1668, 127.

[8] Polygonal Numbers. Greek text by P. Tannery, 1893, 1895. Engl. transl. by T. L. Heath, Cambridge, 1885, 1910; German transl. by F. T. Poselger, 1810, J. O. L. Schulz, 1822, and G. Wertheim, 1890; French transl. by G. Massoutié, Paris, 1911. Cf. Nesselmann, Algebra der Griechen, 1842, 462–476; M. Cantor, Geschichte Math., ed. 3, I, 485–7.

[9] Zeitschrift für Math. Physik, Hist. Lit. Abt. 1897, 121–6. Reproduced by T. L. Heath, Diophantus, ed. 2, 1910, 256, where doubt is expressed as to the validity of the restoration in view of the ease with which the geometric equivalent of (3) can be derived geometrically from that of (2).

[10] Cf. M. Cantor, Die Römischen Agrimensoren, Leipzig, 1875, 95–127.

It gave $p_5^r = \frac{1}{2}(3r^2 + r)$, $p_6^r = \frac{1}{2}(4r^2 + 2r)$, where the plus signs should be minus. M. Cantor[11] suggested the following probable derivation. By factoring the numerator of (2), we obtain

$$p_m^r = \frac{(m-2)}{2}r^2 - \frac{(m-4)}{2}r,$$

$$P_m^r = \frac{(m-2)}{2}(1^2 + 2^2 + \cdots + r^2) - \frac{(m-4)}{2}(1 + 2 + \cdots + r).$$

As known by Archimedes (b. Syracuse about 287 B.C.),

$$1 + 2 + \cdots + r = \frac{r(r+1)}{2}, \qquad 1^2 + 2^2 + \cdots + r^2 = \frac{r(r+1)(2r+1)}{6}.$$

Hence

$$P_m^r = \frac{r+1}{6}\left[\frac{2(m-2)}{2}r^2 - \frac{2(m-4)}{2}r + r\right] = \frac{r+1}{6}(2p_m^r + r).$$

The Hindu Aryabhatta[12] (b. 476 A.D.) gave the formula

$$1 + 3 + 6 + \cdots + \frac{r(r+1)}{2} = \frac{r(r+1)(r+2)}{6} = \frac{(r+1)^3 - (r+1)}{6}$$

for the number of spheres in a triangular pile, and hence for the rth pyramidal number P_3^r of order 3, called also a tetrahedral number. The Hindus of his time knew[13] also that $P_4^r = P_3^r + P_3^{r-1}$, whence

$$6P_4^r = r(r+1)(2r+1).$$

The above general formulas relating to polygonal and pyramidal numbers were collected about 983 A.D. by Gerbert[14] (Pope Sylvester II).

Yang Hui[15] gave in his Suan-fa, 1261, the formulas

$$1 + (1+2) + (1+2+3) + \cdots + (1+2+\cdots+n) = n(n+1)(n+2)/6,$$
$$1^2 + 2^2 + \cdots + n^2 = \tfrac{1}{3}n(n+\tfrac{1}{2})(n+1)$$

for the sums of triangular numbers and squares.

Chu Shih-chieh,[16] in 1303, tabulated in the form of a triangle the binomial coefficients as far as those for eighth powers. This arithmetical triangle was known[17] to the Arabs at the end of the eleventh century. Such a triangle was published by Petrus Apianus.[18]

Many of the early arithmetics mentioned (some with fuller titles) in Vol. I, Ch. I, of this History, gave definitions and simple properties of

---

[11] Die Römischen Agrimensoren, 1875, 122; Geschichte der Math., 1, ed. 2, 519; ed. 3, 558. Cf. H. G. Zeuthen, Bibliotheca Mathematica, (3), 5, 1904, 103.

[12] French transl. by L. Rodet, Jour. Asiatique, 13, 1879; Leçons de calcul d'Aryabhatta, p. 13, p. 35.

[13] E. Lucas, La Nature (Revue des Sciences), 14, 1886, II, 282–6: L'Arithmétique en Batons dans l'Inde au temps de Clovis.

[14] Geometrie, Chs. 55–65.

[15] Y. Mikami, Abh. Geschichte Math. Wiss., 30, 1912, 85.

[16] Ibid., 90. Cf. K. L. Biernatzki, Jour. für Math., 52, 1856, 87; Stifel.[24]

[17] M. Cantor, Geschichte der Math., 1, ed. 3, 1907, 687.

[18] Ein newe . . . Kauffmans Rechnung . . ., Ingolstadt, 1527, title page. The latter was reproduced by D. E. Smith, Rara Arith., 1908, 156, who remarked that he knew of no earlier publication of this Pascal triangle.

polygonal numbers; for example, Boethius,[19] G. Valla,[20] Martinus,[21] Cardan,[22] J. de Muris,[22a] Willichius,[23] Michael Stifel,[24] who gave a table of figurate numbers (binomial coefficients), Faber Stapulensis,[25] and F. Maurolycus,[26] who gave

$$p_5^r = 3p_3^{r-1} + r, \qquad p_6^r = 2p_3^{r-1} + r^2,$$
$$P_3^n + P_3^{n-1} = P_4^n, \qquad P_5^n = P_3^n + 2P_3^{n-1}, \qquad P_6^n = P_5^n + P_3^{n-1},$$

and treated (pp. 32–74) polygonal numbers of the second order or central polygonal numbers (the pentagonal being 1, 6, 16, 31, 51, 76, $\cdots$, when in the second are counted the vertices and center of a pentagon), as well as central pyramidal numbers (the pentagonal being 1, 7, 23, 54, 105, $\cdots$). Also I. Unicornus,[27] and G. Henischiib.[28]

Johann Faulhaber[29] treated polygonal and pyramidal numbers.

Johann Benzius[30] devoted twenty chapters to these and figurate numbers.

J. Rudolff von Graffenried[31] noted that

$$(p_3^r)^2 - (p_3^{r-1})^2 = r^3, \qquad (p_3^r)^2 + (p_3^{r-1})^2 = p_3^{r^2},$$

the final number being 666 for $r = 6$.

C. G. Bachet[32] wrote a supplement of two books to the Polygonal Numbers of Diophantus. The most important ones of his theorems (when expressed as formulas) are as follows:

I, 10. $\qquad p_m^{k+r} = p_m^k + p_m^r + kr(m - 2), \qquad p_m^r = p_3^r + (m - 3)p_3^{r-1}.$

II, 18. $\qquad p_m^r + p_m^{2r} + \cdots + p_m^{nr} = p_m^r p_3^n + r^2(m - 2)(p_3^1 + p_3^2 + \cdots + p_3^{n-1}).$

II, 21. $\qquad 3(p_m^r + p_m^{2r} + \cdots + p_m^{nr}) = p_m^r p_3^n + (n + 1)p_m^{rn}.$

II, 25. $\qquad 1^3 + 2^3 + \cdots + n^3 = \left[\dfrac{n(n + 1)}{2} \right]^2 = (p_3^n)^2.$

II, 28. $\qquad n^3 + 6p_3^n + 1 = (n + 1)^3.$

II, 31, 32. $k^3 + (2k)^3 + \cdots + (nk)^3 = k^3(p_3^n)^2 = k(k + 2k + \cdots + nk)^2.$

[19] Arithmetica boetij, 1488, etc., Lib. 2, Caps. 7–17.

[20] De expetendis et fvgiendis rebvs opvs, Aldus, 1501, Lib. III.

[21] Ars Arithmetica, 1513, 1514; Arithmetica, 1519, 15–18.

[22] Practica Arith., 1537, etc.

[22a] Arith. Speculativae, 1538, 53–62.

[23] I. Vvillichii Reselliani, Arith. libri tres, 1540, 95–111.

[24] Arith. Integra, 1544. See references 16–18, 50–52.

[25] Stapulensis, Jacobi Fabri, Arith. Boëthi epitome, 1553, 54–65.

[26] Arith. libri dvo, 1575, 6–8, 14–21. Historical remarks on same by M. Fontana, Memorie dell' Istituto Nazionale Ital., Mat., 2, Pt. 1, 1808, 275–296.

[27] De l'Arithmetica Vniversale, 1598, 67–70.

[28] Arith. Perfecta et Demonstrata [1605], 1609, 133.

[29] Cubicoss Lustgarten, 1604 (also in part 2 of Petrum Rothen, Arithmetica Philosophica, Nürnberg, 1608); Neuer Math. Kunstspiegel, Ulm, 1612, which notes that 1335 (mentioned in the Bible, Daniel, XII, 12) is a pentagonal number whose root 30 is a pronic[35] number with the pentagonal root 5 whose root 2 is pronic, while 2300 (Daniel, VIII, 14) is tetradecagonal whose root 20 is pronic, etc.; Numerus Figuratus, 1614, 24 pp.; Miracula Arithmetica, Augspurg, 1622, a book chiefly on arithmetical combinations giving the "Wunder Zahl" 666, the Apocalyptic number mentioned in the Bible, Revelations, XIII, 18; cf. Remmelin,[35] A. G. Kästner, Geschichte Math., III, 111–52.

[30] Manuductio ad Nvmervm Geometricvm, Kempten, 1621.

[31] Arith. Logistica Popularis, 1618, 238, 627.

[32] Diophanti Alex. Arith., 1621.

His II, 27, relates to the formula of Nicomachus[5] (Ch. 20)

$$1 = 1^3, \qquad 3+5 = 2^3, \qquad 7+9+11 = 3^3, \qquad 13+15+17+19 = 4^3, \quad \cdots,$$

from which follows the above formula II, 25, by addition (as in the Codex Arcerianus[10]). Fermat[33] generalized this proposition by introducing "colonne": In the arithmetical progression $1, 1 + (m-2), 1 + 2(m-2),$ $\cdots$ leading to m-gonal numbers, the first term 1 gives the first colonne; the sum of the next two terms diminished by $m - 4$ times the first triangular number 1 gives the second colonne $2m$; the sum of the fourth, fifth and sixth terms diminished by $m - 4$ times the second triangular number 3 gives the third colonne $9m - 9$; similarly, the fourth colonne is $8(3m - 4)$ and the rth is $r^2 + r^2(r - 1)(m - 2)/2$. It follows (as noted by Editor Tannery) that the rth colonne is the product of the rth m-gonal number by r, and for $m = 4$ is r^3. The term colonne was not coined by Fermat, as Tannery thought, but[34] was used by Maurolycus.[26]

J. Remmelin[35] noted that 666 (cf. Faulhaber[29]) is a triangular number with the root 36, which is a square with the root 6, while 6 is a pronic number [of the form $n^2 + n$] whose base 2 is also a pronic number.

Later we shall quote Bachet's empirical theorem that any integer is the sum of four squares, made à propos of Diophantus IV, 31. In this connection Fermat[36] made the famous comment: "I was the first to discover the very beautiful and entirely general theorem that every number is either triangular or the sum of 2 or 3 triangular numbers; every number is either a square or the sum of 2, 3 or 4 squares; either pentagonal or the sum of 2, 3, 4 or 5 pentagonal numbers; and so on ad infinitum, whether it is a question of hexagonal, heptagonal or any polygonal numbers. I can not give the proof here, which depends upon numerous and abstruse mysteries of numbers; for I intend to devote an entire book to this subject and to effect in this part of arithmetic astonishing advances over the previously known limits." But such a book was not published. Fermat[37] stated the theorem in a letter to Mersenne, Sept., 1636 (to be proposed to St. Croix); to[38] Pascal, Sept. 25, 1654, and Digby, June 19, 1658. The theorem was attributed to St. Croix by Descartes[39] in a letter to Mersenne, July 27, 1638. Descartes[40] gave an algebraic proof of Plutarch's[6] theorem that $8\triangle_r + 1 = (2r + 1)^2$. We shall often write $\triangle_r$ or $\triangle(r)$ for the rth triangular number $r(r + 1)/2$, $\triangle$ or $\triangle'$ for any triangular number, $\square$ for any square, $\boxed{2}$, $\boxed{3}$ or $\boxed{4}$ for a sum of two, three or four squares.

---

[33] Oeuvres, I, 341.

[34] Wertheim, Zeitschr. Math. Phys., 43, 1898, Hist.-Lit. Abt., 41–42.

[35] Johanne Lvdovico Remmelino, Structura Tabularvm qvadratarvm, 1627, Preface. The book treats magic squares at length.

[36] Oeuvres, I, 305; French transl., III, 252. E. Brassinne, Précis des Oeuvres Math. de P. Fermat, Mém. Acad. Imp. Sc. Toulouse, (4), 3, 1853, 82.

[37] Oeuvres, II, 1894, 65; III, 287.

[38] Oeuvres de Fermat, II, 313, 404; III, 315.

[39] Oeuvres de Descartes, II, 1898, 256, 277–8 (editors' comments); X, 297 (statement of the theorem in a posthumous MS.).

[40] Oeuvres, X, 298 (posth. MS.).

The rth *figurate* number of order n is the binomial coefficient

$$f_n^r = \binom{r + n - 1}{n} = \frac{(r + n - 1)(r + n - 2)\cdots r}{1\cdot 2 \cdots n}.$$

Thus f_2^r is the rth triangular number p_3^r, while f_3^r is the rth pyramidal or tetrahedral number P_3^r. In a comment on the Polygonal Numbers of Diophantus, Fermat[41] stated a theorem which, in the present notation, is

$$rf_n^{r+1} = (n + 1)f_{n+1}^r$$

and called f_4^r the rth triangulo-triangular number.

In April, 1638, St. Croix proposed to Descartes the problem: "Trouver un trigone [triangular number] qui, plus un trigone tétragone, fasse un tétragone [square], et de rechef, et que de la somme des côtés des tétragones résulte le premier des trigones et de la multiplication d'elle par son milieu le second. J'ai donné 15 et 120. J'attends que quelqu'un y satisfasse par d'autres nombres ou qu'il montre que la chose est impossible." The problem, without the example, was proposed to Fermat (Oeuvres, II, 63) in 1636, who did not solve it.

Descartes[42] understood a trigone tétragone to be the square $\triangle^2$ of a triangular number, and proved that 15, 120 is the only solution if the problem is understood to require two triangular numbers such that, if either be added to the same $\triangle^2$, the sum is a square; while if one is permitted to add both $\triangle^2$ and a new $\triangle'^2$ to the second required triangular number, the two triangular numbers may be taken to be 45 and 1035, since

$$45 + 6^2 = 9^2, \qquad 1035 + 6^2 + 15^2 = 36^2, \qquad 36 + 9 = 45, \qquad 45\cdot 46/2 = 1035.$$

St. Croix did not admit the validity of Descartes' solution, and probably meant a trigone tétragone to be a number both triangular and square (like 1, 36). The question would then be to find two numbers of the form $n(n + 1)/2$ such that, if a number both triangular and square be added to each, there result two squares; further, the sum of the square roots of these squares must equal the first required triangular number and must also be the first factor n used in forming the second triangular number. If, as seems intended, the numbers to be added to the triangular numbers are to be identical, the only solution is 15, 120. Cf. Gérardin.[220]

Fermat[43] proposed to Frenicle the problem to find a number which shall be polygonal in a given number of ways. Neither gave a solution. [Cf. Euler,[59] end.]

John Wallis[44] derived by summation the expression for the general triangular number (p. 139), pyramidal number with triangular base P_3^r (p. 143), the sum (called trianguli-pyramidal number) of the latter for $r = 1, 2, \cdots, l$ (p. 145), and the sum (called pyramidi-pyramidal number) of these last for $l = 1, 2, \cdots$. The values found are the expanded forms of

---

[41] Oeuvres, I, 341; French transl., III, 273. Also, II, 70, 84–5; French transl., III, 291–2; letters to Mersenne, Sept., 1636, and to Roberval, Nov. 4, 1636.

[42] Oeuvres, II, 1898, 158–165, letter from Descartes to Mersenne, June 3, 1638.

[43] Oeuvres, II, 225, 230, 435, June and Aug., 1641, Aug., 1659.

[44] Arithmetica Infinitorvm, Oxford, 1656.

the figurate numbers $f_2^r, f_3^r, f_4^r, f_5^r$, so that his work amounts to a verification of cases of

$$f_{n+1}^r = f^1 + f^2 + \cdots + f_n^r.$$

Frans van Schooten[45] quoted three of Bachet's rules, proving one. On certain hexagons whose sum is a cube, see Frenicle[6] of Ch. XXI.

Fermat[46] proposed that Brouncker and Wallis find a proof of the proposition (which he himself could prove): There is no triangular number, other than unity, which is a biquadrate.

Diophantus, IV, 44, desired three numbers which if multiplied in turn by their sum give a triangular number, a square, and a cube. Let the sum be x^2. Then the numbers are $\alpha(\alpha + 1)/(2x^2)$, β^2/x^2, γ^3/x^2. Thus

$$\triangle_a + \beta^2 + \gamma^3 = x^4.$$

Take $\beta = x^2 - 1$. Then $\triangle_a = 2x^2 - \gamma^3 - 1$. But

$$8\triangle_a + 1 = (2\alpha + 1)^2 = 16x^2 - 8\gamma^3 - 7 = (4x - \delta)^2,$$

if $x = (8\gamma^3 + \delta^2 + 7)/(8\delta)$. Take $\gamma = 2$, $\delta = 1$; then $x = 9$ and the desired numbers are 153/81, 6400/81, 8/81.

Bachet convinced himself by trial that δ must be unity in order that $\alpha = (8\gamma^3 + 7 - \delta^2 - 2\delta)/(4\delta)$ be integral.

Fermat remarked that "Bachet's conclusion is not rigorous. Indeed, let γ be any number of the form $3n + 1$, say $\gamma = 7$. To make

$$2x^2 - 7^3 - 1 = \triangle$$

and hence $16x^2 - 8 \cdot 7^3 - 7 = \square$, we may take the latter to be the square of $4x - 3$ [whence $x = 115$, $\delta = 3$]. Nothing prevents us from generalizing the method, taking instead of 3 any odd number and making a suitable choice of γ."

G. Loria[47] remarked that the solution becomes evident if we replace x^2 by x; the problem did not require that the sum of the numbers be a square.

Bachet[32] (p. 274) proposed the problem to find five numbers which if multiplied in turn by their sum give a triangular number $\triangle$, a square, a cube, a pentagonal number, and a biquadrate. The sum of the latter shall be x^4. Let the square be $(x^2 - 1)^2$, the cube 8, the pentagonal number 5, and the biquadrate 1. Then $\triangle = 2x^2 - 15$. Thus

$$8\triangle + 1 = 16x^2 - 119 = \square,$$

say $(4x - 1)^2$. Hence $x = 15$.

René F. de Sluse[48] (1622–1685) employed the triangular number q, the square b^2 and cube z^3. Then $q + b^2 + z^3 = \square = (b + n)^2$, whence

$$b = (q + z^3 - n^2)/(2n).$$

---

[45] Exercitationum Math., 1657, Lib. V, 442–5.

[46] Oeuvres, III, 317, letter to Digby, June, 1658.

[47] Le scienze esatte nell' antica Grecia, Libro V, 138.

[48] Renati Francisci Slusii, Mesolabum, $\cdots$, accessit pars altera de analysi et miscellanea, Leodii Eburonum, 1668, 175.

Hence we may assign any values to q, z^3, n and find b. Likewise for Bachet's generalization, we may assign any values to all five products other than the square b^2, and find b.

A. Gérardin[49] noted that the simplest solution of Diophantus' problem is furnished by the three numbers $(x^2 + 1)/2$, θ^2, x, with $\alpha = x^2$, $\beta = x\theta$, $\gamma = x$,

$$\tfrac{1}{2}(x^2 + 1) + \theta^2 + x = x^2.$$

Set $x = 2H + 1$. Then $\theta^2 - 2H^2 = -1$, with the solutions $(H, \theta) = (1, 1)$, $(5, 7)$, $(29, 41)$, etc., giving the numbers 5, 1, 3; 61, 49, 11; 1741, 1681, 59.

René F. de Sluse[50] gave the table [cf. Stifel[24]]

0	1	1	1	1	1
1	2	3	4	5	
1	3	6	10		
1	4	10			
1	5				
1					

in which the numbers (like 1, 3, 3, 1) in a diagonal are binomial coefficients, those in the third column are triangular numbers, those in the fourth column are pyramidal numbers with triangular base, those in the fifth are triangular pyramids of the second order.

B. Pascal[51] gave the same table and noted that any number in it is the sum of the numbers in the preceding column and hence (p. 504) is the sum of the number above it and that immediately to its left. He noted (p. 533) that $n(n + 1) \cdots (n + k - 1)$ is divisible by $k!$, the quotient being a figurate number.

G. W. Leibniz[52] gave a table formed by the diagonals (as 1, 2, 1) of the above table.

J. Ozanam[53] found pairs* of triangular numbers 15 and 21, 780 and 990, 1747515 and 2185095, whose sum and difference are triangular. Their sides are 5 and 6, 39 and 44, 1869 and 2090. Polygonal numbers are treated in the English translation by C. Hutton, London, 1803, pp. 40–47, p. 60.

Pierre Rémond de Montmort[54] cited special cases of (1), due to Diophantus.

F. C. Mayer[55] defined "generalized figurate" numbers

$$a\frac{x(x + 1)\cdots(x + n - 1)}{1 \cdot 2 \cdots n} + (1 - a)\frac{x(x - 1)\cdots(x + n - 2)}{1 \cdot 2 \cdots (n - 1)},$$

[49] Sphinx-Oedipe, 6, 1911, 42.

[50] MS. 10248 du fonds latin, Bibliothèque Nationale de Paris, f. 187.

[51] Traité du triangle arith., Paris, 1665 (written 1354); Oeuvres, III, 1908, 466–7.

[52] Leibniz Math. Schriften, ed., C. I. Gerhardt, VII, 101.

* Others are 171 and 105, 3741 and 2145. Gérardin gave a general discussion in Sphinx-Oedipe, 1914, 113.

[53] Recreations math. et phys., 1, 1696, 20; new eds., 1723, etc.

[54] Mém. Acad. Roy. Sc., 1701. Essai d'Analyse sur les Jeux de Hazards, 1708; ed. 2, 1713, 17.

[55] Maiero, Comm. Acad. Petrop., 3, ad annum 1728 [1726], 52.

which for $n = 2, 3, 4$ include the polygonal numbers and the pyramidal numbers of the first and second kind, the number of sides being $a + 2$.

L. Euler[56] investigated polygonal numbers which are also squares. The problem is a special case of that to make a quadratic function a square. The triangular numbers equal to squares are those with sides 0, 1, 8, 49, 288, 1681, 9800, $\cdots$ and equal the squares of 0, 1, 6, 35, 204, 1189, 6930, $\cdots$. The xth polygonal number with l sides is $\{(l - 2)x^2 - (l - 4)x\}/2$. To make it a square, set $2(l - 2)p^2 + 1 = q^2$. Then the product of the polygonal number by 4 is the square of 0, $(l - 4)p$, $2(l - 4)pq$, $\cdots$ if

$$(4) \qquad x = 0, \qquad \frac{-(l - 4)}{2(l - 2)}(q - 1), \qquad \frac{-(l - 4)}{l - 2}(q^2 - 1), \quad \cdots.$$

Euler gave a law for the derivation of any solution x in terms of two solutions. It remains to make the expressions (4) integers. For $l = 5$, q is to be chosen from 1, 5, 49, $\cdots$ and hence p from 0, 2, 20, $\cdots$. The first fraction (4) is here* $(1 - q)/6$ and is an integer for $q = 49$, whence $x = -8$. But Euler had previously stated that, for $l > 4$, q was to be taken negative. The value $q = -5$ gives $x = 1$ and the pentagonal number 1.

Euler[57] proved Fermat's theorems that no triangular number except unity is a cube (since $x^6 \pm y^6$ is not a square), and no triangular number $x(x + 1)/2 > 1$ is a fourth power. According as x is even or odd, $x/2$ or $(x + 1)/2$ must equal a fourth power m^4, if the $\triangle$ is to be a fourth power. Thus $2m^4 \pm 1 = n^4$. But he had just proved that $2n^4 \pm 2 = \square$ only when $n = 1$, whence $m = 0$ or 1, $x = 0$ or 1.

Abbé Deidier[58] gave the simplest properties of polygonal numbers and derived central polygonal numbers as follows: adding unity to the products of the triangular numbers 0, 1, 3, 6, 10, $\cdots$ by 3, 4 or 5, we get central triangular, square or pentagonal numbers, respectively.

We shall now quote from the correspondence[59] between Euler and Goldbach remarks on polygonal numbers, reserving for later use the comments in which the interest is chiefly on sums of squares. June 25, 1730 (p. 31), Euler noted that $(x^2 + x)/2$ equals $(6/7)^4$ for $x = 32/49$, but said this does not disprove Fermat's assertion that no (integral) triangular number is a biquadrate. Aug. 10, 1730 (p. 36), Euler noted that if

$$a = (3 + 2\sqrt{2})^n, \qquad b = (3 - 2\sqrt{2})^n,$$

the square of $(a - b)/(4\sqrt{2})$ is a triangular number with the side $(a + b - 2)/4$ [evident since $ab = 1$]. Chr. Goldbach stated April 12, 1742 (p. 122) that $4mn - m - n^a \neq \triangle$. Euler remarked May 8, 1742 (p. 123) that $4mn - m - n$ is not a heptagonal number. June 7, 1742 (p. 126), Gold-

---

[56] Comm. Acad. Petrop., 6, 1732–3, 175; Comm. Arith. Coll., I, 9. Cf. Euler.[79]

* Thus $q = 1 - 6x$ so that $6p^2 + 1 = q^2$ becomes $p^2 = -2x + 6x^2$. Hence $p = 2P$, $P^2 = (3x^2 - x)/2$, and we have returned to the problem from which we started.

[57] Comm. Acad. Petrop., 10, 1738, 125; Comm. Arith. Coll., I, 30, 34. Proof republished by E. Waring, Medit. Algebr., ed. 3, 1782, 373.

[58] Suite de l'arithmétique des géomètres, Paris, 1739, 352–365.

[59] Correspondance Mathématique et Physique (ed., P. H. Fuss), St. Pétersbourg, 1, 1843.

bach inferred that every number is of the form $2\triangle \pm \square$, and incorrectly (Euler, p. 134) that every number is a sum of three triangular numbers. Euler, June 30, 1742 (p. 133) noted that every number is of the form $y^2 + y - x^2 = 2\triangle_y - x^2$. April 6, 1748 (pp. 447–9, 468), Goldbach stated that every number can be expressed in each of the eight forms

$$\square + 2\square' + \triangle, \qquad \square + 2\square' + 2\triangle, \qquad \square + \square' + 2\triangle,$$
$$2\square + \triangle + 2\triangle', \quad \text{etc.}$$

June 25, 1748 (pp. 458–460), Euler gave the identity

$$\frac{a^2 + a}{2} + \frac{b^2 + b}{2} = e^2 + 2\left(\frac{d^2 + d}{2}\right), \qquad \text{for } a = d + e, \qquad b = d - e.$$

Hence [Fermat's[36] theorem] every n is a sum of three $\triangle$'s implies

$$n = \square + 2\triangle + \triangle'.$$

Euler expressed his belief that every number of the form $4n + 1$ is a sum [3] of three squares, whence $n = \square + \square' + 2\triangle$. Replacing n by $2n$, we see that every $n = \square + \square' + \triangle$. Euler gave fourteen such formulas. June 9, 1750 (p. 521), Euler remarked that an algebraic discussion of the theorem that any number n is a sum of three triangular numbers is of no help, since the theorem is not true if n is fractional (unlike the theorem on [4]). Dec. 16, 1752 (p. 597), Euler noted as facts, of which he had no proof, that every prime $8n + 1$ or $8n + 3$ is of the form $x^2 + 2y^2$, whence if $n \neq \square + \triangle$, $8n + 1 \neq$ prime, and if $n \neq 2\triangle + \triangle'$, $8n + 3 \neq$ prime. Also (p. 630), if $n \neq \square + 2\triangle$, $4n + 1 \neq$ prime.

April 3, 1753 (pp. 608–9), Euler treated the problem [of Fermat[43]] to find a number z which is polygonal in a given number of ways. Let n be the number of sides of the polygonal number, x its root. Then

$$2z = (n - 2)x^2 - (n - 4)x, \qquad n = 2 - \frac{2z}{x} + \frac{2(z - 1)}{x - 1}.$$

Thus $2z$ must be divisible by x, and $2z - 2$ by $x - 1$. Hence we desire two numbers differing by 2 which have divisors differing by 1. For example, 450 and 448 have such divisors 3 and 2, 5 and 4, 9 and 8, 15 and 14. Thus 225 is a square, 8-gon, 24-gon, and 76-gon.

Euler[60] noted that, if $4n + 1$ is a sum of two squares, $8n + 2$ is a sum of two odd squares $(2x + 1)^2$, $(2y + 1)^2$, whence $n = \triangle_x + \triangle_y$. S. Réalis[60a] noted that conversely this expression for n implies

$$4n + 1 = (x + y + 1)^2 + (x - y)^2.$$

In Ch. III are cited Euler's[3] theorem $\prod(1 - x^k) = \sum(- 1)^j x^p$, where $p = (3j^2 \pm j)/2$ is a pentagonal number, and theorems by Legendre,[23] Vahlen,[150] and von Sterneck,[169] on the partitions of N, in which an exceptional rôle is played by the N's which are pentagonal or triangular.

---

[60] Novi Comm. Acad. Petrop., 4, 1752–3 (1749), 3–40, § 34; Comm. Arith. Coll., I, 164.
[60a] Nouv. Ann. Math., (3), 4, 1885, 367–8; Oeuvres de Fermat, IV, 218–20.

G. W. Kraft[61] and A. G. Kästner[62] proved that

$$\frac{2^{4m+1} - 2^{2m} - 1}{9} = \frac{(2N)(2N + 1)}{2} = \triangle,$$

since $(2^{2m} - 1)/3$ is an integer N.

M. Gallimard[63] obtained "central polygons" by multiplying each term of 0, 1, 3, 6, 10, 15, $\cdots$ by the number n of angles of any polygon whatever and adding unity to each product. Given a central polygon, he treated the problem to find the number of angles if the side be given, or vice versa.

L. Euler[64] proved that a number not the sum of a square and a triangular number $\triangle$ is composite; one not $\triangle + 2\triangle'$ is composite.

Nicolas Engelhard[65] treated Plutarch's[6] questions on triangular numbers.

Elie de Joncourt[66] gave a table of triangular numbers $N(N + 1)/2$, N up to 20000, and showed how the table may be used to test if a number less than a hundred million is a square or not, and to extract square roots approximately.

L. Euler[67] noted that, if $N - ab = \triangle_p + \triangle_q + \triangle_r$, and $p - q = a - b$, then $N = \triangle_{p+b} + \triangle_{p-a} + \triangle_r$. N. Fuss, I, (pp. 191–6) also gave an incomplete argument to show that N is a sum of three triangular numbers if every integer $< N$ is. Let $N - p = \triangle_a + \triangle_b + \triangle_c$ and

$$p = (b - a)n + n^2$$

[a restriction]; then $N = \triangle_{a-n} + \triangle_{b+n} + \triangle_c$. He gave a similar incomplete discussion of the problem to express N as a sum of m m-gonal numbers, given that every integer $< N$ is such a sum. He noted (p. 201) that $9n + 5, 8$; $49n + 5, 19, 26, 33, 40, 47$; $81n + 47, 74$; etc., are not sums of two triangular numbers; thus, $49n + 19 = \triangle_a + \triangle_b$ would imply $(2a + 1)^2 + (2b + 1)^2 = 8(49n + 19) + 2$, whereas the factor 7 of the latter is not a divisor of a sum of two squares. L. Euler (p. 214) noted that $\triangle_x \triangle_y = \triangle_z$ is satisfied* if $px(y + 1) = 2qz$, $qy(x + 1) = p(z + 1)$; the resulting values of z are equal if $\{(2q^2 - p^2)x + 2q^2\}y = p^2x + 2pq$. L. Euler (pp. 264–5, about 1775) noted that

$$9\triangle_a + 1 = \triangle_{3a+1}, \qquad 49\triangle_a + 6 = \triangle_{7a+3},$$
$$25\triangle_a + 3 = \triangle_{5a+2}, \qquad 81\triangle_a + 10 = \triangle_{9a+4}.$$

J. A. Euler[68] (the son of L. Euler) stated that to express every number as a sum of terms of $1^2, 3^2, 6^2, 10^2, 15^2, \cdots$, at least 12 terms are required.

[61] Novi Comm. Acad. Petrop., 3, ad annum 1750 et 1751, 112.
[62] Comm. Soc. Sc. Gottingensis, 1, 1751, 198. Cf. T. Pepin, Atti Accad. Nuovi Lincei, 32, 1878–9, 298.
[63] L'Algèbre ou la Science du Calcul litteral, Paris, 2, 1751, 131–143.
[64] Novi Comm. Acad. Petrop., 6, 1756–7 [1754], 185; Comm. Arith. Coll., I, 192.
[65] Verhandel. Hollandse Maatschappy Weetenschappe te Harlem, 3 Deel, 1757, 223–230; 4 Deel, 1758, 21 (correction to p. 224).
[66] De Natura et Praeclaro Usu Simplicissimae Speciei Numerorum Trigonalium, Hagae Comitum, 1762, 267 pp.
[67] Opera postuma, 1, 1862, 190 (about 1767).
* The least solution is $x = 2$, $y = 5$, $z = 9$, Sphinx-Oedipe, 1913, 90; 1914, 145.
[68] Ibid., pp. 203–4, about 1772.

To express every number as a sum of figurate numbers

$$1, \quad \frac{n+a}{1}, \quad \frac{(n+1)(n+2a)}{1\cdot 2}, \quad \frac{(n+1)(n+2)(n+3a)}{1\cdot 2\cdot 3},$$

$$\frac{(n+1)(n+2)(n+3)(n+4a)}{1\cdot 2\cdot 3\cdot 4}, \quad \cdots,$$

at least $a + 2n - 2$ terms are necessary. Cf. Beguelin,[72] Pollock,[117] Maillet.[181–2]

L. Euler[69] remarked that Fermat's[36] theorem that every integer is a sum of m m-gonal numbers would follow if we could prove that every integer occurs among the exponents in the expansion of the mth power of $1 + x + x^m + x^{3m-3} + \cdots$, whose exponents are the m-gonal numbers. Fermat's theorem that every integer is a sum of three triangular numbers would follow if it were shown that in

$$1/\{(1 - z)(1 - xz)(1 - x^3z)(1 - x^6z)\cdots\} = 1 + Pz + Qz^2 + Rz^3 + \cdots,$$

all integers occur as exponents of x in the series for R.

Euler[70] found squares which are triangular or pentagonal. If $\triangle_z = x^2$, then $y^2 = 8x^2 + 1$ for $y = 2z + 1$. If $(3z^2 - z)/2 = x^2$, $y^2 = 24x^2 + 1$ for $y = 6z - 1$. If $(3q^2 - q)/2 = \triangle_p$, $(6q - 1)^2 = 3x^2 - 2$ for $x = 2p + 1$. Special solutions of the three equations $y^2 = ax^2 + b$ are found by his general method of treating the latter (Ch. XII, below).

Euler[71] admitted that he had no proof of Fermat's assertion that every number is a sum of three or fewer triangular numbers and noted that this is true only of whole numbers, since no one of $\frac{1}{2}, \frac{3}{2}, \frac{5}{2}, \frac{7}{2}$, etc., can be resolved into three triangular numbers. There are no rational solutions x, y, z of

$$\tfrac{1}{2} = \frac{x^2 + x}{2} + \frac{y^2 + y}{2} + \frac{z^2 + z}{2}.$$

Nicolas Beguelin[72] attempted to prove Fermat's theorem that every integer is a sum of s polygonal numbers of s sides. For $s = d + 2$, the latter are 0 and 1, $A = d + 2$, $B = 3d + 3$, $C = 6d + 4$, $D = 10d + 5$, $\cdots$, a series whose second order of differences are d. Let t be the number of terms > 0 needed to produce a given sum e. For $1 < e < A$, evidently $t \leqq A - 1$. For $e = A + \epsilon$, where $1 \leqq \epsilon \leqq A - 1$, $t \leqq A$; for $e = 2A + \epsilon$, $0 \leqq \epsilon \leqq d - 2$, $t \leqq d$. Next, let $B < e < C$. For $e = B + \epsilon$, $1 \leqq \epsilon \leqq A - 1$, $t \leqq A$; for $e = B + A + \epsilon$, $0 \leqq \epsilon \leqq A - 2$, $t \leqq A$; for the "doubtful case" $e = B + A + A - 1$, we replace B by its equal $2A + d - 1$ and have $e = 4A + d - 2$, $t = d + 2$; finally, for $e = B + 2A + \epsilon$,

[69] Novi Comm. Acad. Petrop., 14, I, 1769, 168; Comm. Arith. Coll., I, 399–400.

[70] Algebra, 2, 1770, §§ 88–91; French transl., 2, 1774, pp. 105–9 (Vol. I, Ch. V, pp. 341–354, for definitions of polygonal numbers). Opera omnia, (1), I, 373–5, 159–64.

[71] Acta Eruditorum, Lips., 1773, 193; Acta Acad. Petrop., I, 2, 1775 (1772), 48; Comm. Arith. Coll., I, 548.

[72] Nouv. Mém. Acad. Sc. Berlin, année 1772, 1774, 387–413.

$0 \leqq \epsilon \leqq d - 4$, $t \gtreqqless d - 1$. After this expansion of the argument by Beguelin, we are ready to admit that if e is in one of the intervals 1 to A, A to B, B to C, it is a sum of $d + 2$ or fewer terms 1, A, B. He treated four more intervals with a rapidly increasing number of "doubtful cases" for which linear relations between the polygonal numbers were employed, and found in every case that $t \leqq d + 2$. But he finally admitted (p. 405) that this method does not lead to a proof of the general theorem of Fermat.

On p. 411, Beguelin stated without proof the erroneous generalization [cf. J. A. Euler,[68] L. Euler[73]] that any number is the sum of at most $t = d + 2n - 2$ terms of the series

$$1, \qquad n + d, \qquad \frac{(n + 1)(n + 2d)}{2}, \qquad \frac{(n + 1)(n + 2)(n + 3d)}{1 \cdot 2 \cdot 3}, \qquad \cdots,$$

a series whose nth order of differences are constant and equal to d. For $n = 2$, we have the case of polygonal numbers just considered. For $n = 3$, we have the pyramidal numbers P^r_{d+2} for $r = 1, 2, 3, \cdots$; for $n = 4$, their sums, etc. For $n = 4$, $d = 1$, the series is 1, 5, 15, 35, 70, $\cdots$ and the theorem gives $t = 7$, whereas 8 terms are evidently required to produce the sum 64 (since 4 terms must be unity), as expressly mentioned on p. 412. Thus Beguelin contradicts himself in his generalization of Fermat's theorem to pyramidal and figurate numbers.

L. Euler[73] probably overlooked the last remark, since he stated that the unproved generalization merits great attention. He extended Beguelin's tentative process to any series 1, A, B, $\cdots$. We must employ $A + n - 2$ summands 1, A to produce $nA - 1$. Thus if

$$nA - 1 \leqq B < (n + 1)A - 1,$$

we need $A + n - 2$ summands 1, A to produce all numbers 1, 2, $\cdots$, B. Then

$$A - 1 + \frac{B - 2A + 1}{A} < A + n - 2 \leqq A - 1 + \frac{B - A + 1}{A}.$$

Denote by $\{x\}$ the least integer $> x$, and by t_1 the number of terms 1, A needed to produce 1, $\cdots$, B. Hence

$$t_1 = A - 1 + \left\{ \frac{B - 2A + 1}{A} \right\}.$$

Bringing in also the summand B, let b be the least positive integer such that $B + b$ requires $t_1 + 1$ summands 1, A, B. If $C < B + b$, we need only t_1 summands to produce the numbers $\leqq C$. But if $C \geqq B + b$, let

$$(m + 1)B + b \geqq C > mB + b.$$

To produce the numbers $\leqq C$ from 1, A, B, we need

$$t_1 + m = t_1 + \left\{ \frac{C - B - b}{B} \right\} \equiv t_2$$

---

[73] Opusc. Anal., 1, 1783 (1773), 296; Comm. Arith. Coll., II, 27.

summands. Next, bring in the summand C and let c be the least positive integer such that $C + c$ requires $t_2 + 1$ summands from 1, A, B, C. To produce the numbers $\leqq D$, we need

$$t_2 + \left\{ \frac{D - C - c}{D} \right\} = t_3$$

summands, etc. In the case of an infinite series 1, A, B, $\cdots$, the process furnishes a lower limit to the number t of summands. Euler showed that, for series whose nth order of differences are constant, Beguelin's rule is often quite erroneous, but did not treat the series 1, $n + d$, $\cdots$ of polygonal and pyramidal numbers.

N. Beguelin[74] made a puerile illogical attempt to prove that every number is the sum of three triangular numbers. Admitting the last theorem, Beguelin[75] deduced Bachet's theorem that every integer is a ▣. For,

(5) $n = \Sigma(a^2 + a)/2$ implies $8n + 3 = \Sigma(2a + 1)^2$.

Adding 1, we conclude that $8n + 4$ is a ▣. But it is known that the half or double of a ▣ is a ▣. Hence $2n + 1$ and its product by any power of 2 are ▣. Since Lagrange had given in 1770 an independent proof of this theorem of Bachet, Beguelin next attempted, but failed completely, to deduce from it the result that every integer is a sum of three triangular numbers $\triangle$. On p. 338, he gave the equivalent formulas

$$q = \frac{a^2 + a}{2} + \frac{b^2 + b}{2}, \qquad 4q + 1 = (a - b)^2 + (a + b + 1)^2.$$

He concluded without adequate proof that every number is a sum of a $\triangle$ and two squares, and also is $\triangle + 2\triangle' + 2\triangle''$ (p. 345); further, that every integer $\equiv 1, 2, 3, 5$ or $6 \pmod 8$ is a ▢, later proved by Legendre[19] of Ch. VII. A fitting sample of the lack of insight of Beguelin is furnished by his final theorem* (p. 368): If any number $4m + 1$ is a sum of three squares [each $\neq 0$], it is composite [but $17 = 9 + 4 + 4$ is prime]. Curiously enough, he supposed he had verified the theorem for all numbers < 200; but his tables (pp. 363–4) imply that he assumed that a number can be expressed in a single way as a sum of squares. On this he based a new "proof" that every prime $4m + 1$ is a ▢.

L. Euler[76] noted the result (5).

Euler[77] noted that $\frac{1}{2}$ is not the sum of three fractional triangular numbers $(x^2 + x)/2$, since 7 is not the sum of three odd squares $(2x + 1)^2$. But every

<hr>

[74] Nouv. Mém. Acad. Berlin, année 1773, 1775, 203–215.

[75] Nouv. Mém. Acad. Berlin, année 1774, 1776, 313–369.

* One error is that if the sum of three $\triangle$'s, each $\neq 0$, is of the form $3v + 2$, then v is not divisible by 3, assumed to follow from the *converse* in § 50. But $45 + 10 + 1 \equiv 2 \pmod 9$.

[76] Acta Acad. Petrop., 4, II, 1780 [1775], 38; Comm. Arith. Coll., II, 137. Euler[11] of Ch. VII.

[77] Opusc. Anal., 2, 1785 (1774), 3; Comm. Arith. Coll., II, p. 92.

number N is the sum of four fractional pentagonal numbers $(3x^2 - x)/2$, since

$$24N + 4 = \Sigma a^2 = \Sigma(6x - 1)^2, \qquad x = \frac{a+1}{6}, \quad \cdots.$$

To prove the theorem that any number is the sum of three integral triangular numbers $\triangle$, it would be sufficient to show that the coefficient of every term x^k in the expansion of

$$(1 + x + x^3 + x^6 + \cdots + x^\triangle + \cdots)^3$$

is not zero; similarly for squares, pentagons, etc. [Euler[69]]. Let the polygonal numbers with π sides be 0, $\alpha = 1$, $\beta = \pi$, $\gamma = 3\pi - 3$, $\cdots$, and denote by $[n]$ the coefficient of x^n in $(1 + x^\alpha + x^\beta + \cdots)^\pi$. Euler proved by logarithmic differentiation the recursion formula

$$n[n] = \sum_{j=\alpha,\,\beta,\,\ldots} \{\pi j - (n - j)\}[n - j].$$

F. W. Marpurg[78] treated (pp. 185–257) polygonal numbers, giving special cases of formula (1) of Diophantus, pyramidal numbers and central polygonal numbers, viz., unity more than the number of the angles and division points on m-gons drawn about a common mid point. Also (p. 307) polyhedral numbers, the rth hexahedral, octahedral, dodecahedral and icosahedral being

$$r^3, \qquad \frac{r}{3}(2r^2 + 1), \qquad \frac{r}{2}(9r^2 - 9r + 2), \qquad \frac{r}{2}(5r^2 - 5r + 2).$$

Euler[79] proved that $(x^2 + x)/2$ is a square y^2 only when

$$x = \frac{\alpha + \beta - 2}{4}, \qquad y = \frac{\alpha - \beta}{4\sqrt{2}}, \qquad \alpha = (3 + 2\sqrt{2})^n, \qquad \beta = (3 - 2\sqrt{2})^n.$$

For $n = 0, 1, 2$, we get $x = 0, 1, 8$; $y = 0, 1, 6$. We have the recursion formulas

$$x_n = 6x_{n-1} - x_{n-2} + 2, \qquad y_n = 6y_{n-1} - y_{n-2}.$$

Certain squares x^2 which exceed $(y^2 + y)/2$ by unity are given by

$$x = \frac{(2\sqrt{2} + 1)\alpha + (2\sqrt{2} - 1)\beta}{4\sqrt{2}}, \qquad y = \frac{(2\sqrt{2} + 1)\alpha - (2\sqrt{2} - 1)\beta}{4} - \frac{1}{2}.$$

For $n = 0$, $x = 1$, $y = 0$; for $n = 1$, $x = 4$, $y = 5$. The recursion formula is

$$x_n = 6x_{n-1} - x_{n-2}, \qquad y_n = 6y_{n-1} - y_{n-2} + 2.$$

A second series of solutions is obtained by use of these formulas for negative n's. Thus $x_{-1} = 2$, $y_{-1} = -3$; $x_{-2} = 11$, $y_{-2} = -16$. Since the triangular number $\triangle_{-m}$ equals $\triangle_{m-1}$, we replace $y = -m$ by $m - 1$ and get the sets of positive solutions 2, 2; 11, 15; etc.

---

[78] Anfangsgründe des Progressional Calculs, Berlin, 1774, Book 2.
[79] Mém. Acad. St. Pétersbourg, 4, 1811 [1778], 3; Comm. Arith. Coll., II, 267–9.

To find triangular numbers whose triples are triangular, Euler proved that $3(x^2 + x) = y^2 + y$ has only the solutions

$$x = \frac{r + s}{4\sqrt{3}} - \frac{1}{2}, \qquad r = (3\sqrt{3} + 5)(2 + \sqrt{3})^n,$$

$$y = \frac{r - s}{4} - \frac{1}{2}, \qquad s = (3\sqrt{3} - 5)(2 - \sqrt{3})^n,$$

for $n = 0, \pm 1, \pm 2, \cdots$. Examples are $x = 1, y = 2$; $x = 5, y = 9$.

It was proposed as a prize problem in the Ladies' Diary for 1792 to find n ($n > 1$) such that $1^2 + 2^2 + \cdots + n^2 = \Box$. The sum is $n(n + 1)(2n + 1)/6$. T. Leybourn[80] took $2n + 1 = z^2$, whence $(z^4 - 1)/24$ is to be a square y^2. Thus $z^4 = 24y^2 + 1 = \Box = (xy - 1)^2$, say. Thus $y = 2x/(x^2 - 24) > 0$. It is stated that $x = 5$ or 6. Since $x = 6$ is excluded, $n = 24$. C. Brady took $n = 6r^2$. Then the condition is $(6r^2 + 1)(12r^2 + 1) = \Box$. Thus $(9r^2 + 1)^2 - (3r^2)^2 = \Box$, so that $9r^2 + 1$ and $3r^2$ equal the hypotenuse and one leg of a right triangle. Thus the other leg is $9r^2 - 1$, whence $r = 2, n = 24$.

A. M. Legendre[81] proved Fermat's theorems that no triangular number $x(x + 1)/2$, except unity, is a fourth power or cube. For, in the first problem, x or $x - 1$ is of the form $2m^4$, whence either $1 = n^4 - 2m^4$, contrary to $1 + 2m^4 \neq \Box$, or $1 = 2m^4 - n^4$, $m^8 - n^4 = (m^4 - 1)^2$, contrary to $p^4 - n^4 \neq \Box$ unless $m = 1 = x$. In the second problem, one of $1 + x$, x is a cube and the other the double of a cube, whence $n^3 \pm 1 = 2m^3$, which is impossible if $n \neq 1$.

C. F. Gauss[82] proved by means of the theory of ternary quadratic forms that every number $n = 8M + 3$ is a sum of three odd squares, so that, by (5), M is a sum of three triangular numbers. The number of ways M can be so decomposed depends in a definite manner on the prime factors of n and the number of classes of binary quadratic forms of determinant $- n$.

G. S. Klügel[83] gave an account of figurate, polygonal, polyhedral, and pyramidal numbers P_m^r of the first order, those of the second order being $P_m^1 + \cdots + P_m^r$, etc.

John Gough[84] attempted to prove Fermat's theorem that every number is a sum of m m-gonal numbers. P. Barlow[85] noted that the first three propositions by Gough are correct, but are not used in his defective proof of Fermat's theorem, while various points are not proved, as the Cor. 2 to Prop. 4: every number is a sum of a limited number of polygonal numbers. As to Gough's reply (pp. 241–5), Barlow[86] stated that the defense

[80] Ladies' Diary, 1793, p. 45, Quest. 953. T. Leybourn's Math. Quest. proposed in the Ladies' Diary, 3, 1817, 256–7. Cf. Lucas, papers 130–8.

[81] Théorie des nombres, 1798, 406, 409; ed. 2, 1808, 345, 348; ed. 3, II, 1830, arts. 329, 335; pp. 7, 11. German transl. by Maser, 1893, II, 8, 13.

[82] Disquis. Arith., 1801, art. 293; Werke, I, 1863, 348; German transl. by Maser, 1889, p. 334.

[83] Math. Wörterbuch, 2, 1805, 245–253; 3, 1808, 825–8, 931.

[84] Jour. Nat. Phil. Chem., Arts (ed., Nicholson), 20, 1808, 161.

[85] Ibid., 21, 1808, 118–121.

[86] Ibid., 22, 1809, 33–35.

is on grounds not proved.   As to the revised version by Gough[87], Barlow noted (p. 44) that the argument is correct and trivial to within 12 lines of the end; the proof is valid for numbers $\leqq 3m$, but not for those $> 3m$.

E. Barruel[88] noted that sums of 1, 2, 3, $\cdots$ give the triangular numbers 1, 3, 6, $\cdots$, whose sums give the pyramidal numbers 1, 4, 10, 20, $\cdots$, etc. Forming these sums, we get the general triangular and pyramidal numbers $n(n + 1)/2$, $n(n + 1)(n + 2)/6$, etc.   Application is made to prove the ordinary rule for deriving a binomial coefficient from the preceding coefficients.

F. T. Poselger[89] gave (pp. 19–31) various properties of numbers from the writings of Theon of Smyrna, and (pp. 32–60) gave algebraic expressions for polygonal and figurate numbers, with a discussion of arithmetical series of general order.

P. Barlow[90] noted that, if N is a sum of five pentagons $(3u^2 - u)/2$, and M a sum of six hexagons $2x^2 - x$, then

$$24N + 5 = \sum_{i=1}^{5} (6u_i - 1)^2, \qquad 8M + 6 = \sum_{i=1}^{6} (4x_i - 1)^2.$$

In general, if P is a polygonal number of $\alpha + 2$ sides, Fermat's[36] theorem is equivalent to

$$8\alpha P + (\alpha + 2)(\alpha - 2)^2 = \sum_{i=1}^{\alpha+2} (2\alpha x_i - \alpha + 2)^2.$$

He erred[100, 107] (p. 258) in saying that no triangular number > 1 is pentagonal.

J. Struve[91] discussed figurate numbers (binomial coefficients).

J. D. Gergonne[92] noted that the number of terms of a polynomial of degree m in n unknowns is $(m + n)! \div (m!\, n!)$.   If the latter be designated (m, n), then $(m, n) = (m - 1, n) + (m, n - 1)$.

A. Cauchy[93] gave the first proof of Fermat's theorem that every number is a sum of m m-gonal numbers.   The proof shows that all but four of the m-gons may be taken to be 0 or 1.   The auxiliary theorems on sums of four squares will be quoted in Ch. VIII.   In the simplified proof by Legendre,[94] the case $m = 3$ is not presupposed, as was done by Cauchy. Moreover, Legendre proved (p. 22) in effect that every integer $> 28(m - 2)^3$ is a sum of four m-gonal numbers if m is odd; while, for m even, every integer $> 7(m - 2)^3$ is a sum of five m-gonal numbers one of which is 0 or 1.

---

[87] New Series of the Math. Repository (ed., T. Leybourn), 3, 1814, II, 1–7.

[88] Correspondance sur l'Ecole Imp. Polytechnique, Paris, 2, 1809–13, 220–7.

[89] Diophantus über die Polygonzahlen uebersetzt, mit Zusätzen, Leipzig, 1810.

[90] Theory of Numbers, 1811, 219.   Minor applications in papers 17-19 of Ch. IX.

[91] Über die gewöhnlichen fig. Zahlen, Progr. Altona, 1812.

[92] Annales de Math. (ed., Gergonne), 4, 1813–4, 115–122.

[93] Mém. Sc. Math. et Phys. de l'Institut de France, (1), 14, 1813–15, 177–220; same in Exercices de Math., Paris, 1, 1826, 265–296.   Reprinted in Oeuvres de Cauchy, (2), VI, 320–353.   J. des Mines, 38, 1815, 395.   Report by Cauchy, Bull. Sc. par Soc. Philomatique de Paris, (3), 2, 1815, 196–7.

[94] Théorie des nombres, 1st supplement, 1816, to the 2d edition, 1808, 13–27; 3d ed., 1830, I, 218; II, 340; German transl. by Maser, II, 332.

Cauchy[95] denoted the xth polygonal number of order $m + 2$ by

$$\overline{x}^m = \frac{x(x - 1)}{2} m + x,$$

and proved that if $A, B, \cdots, F$ are integers, not divisible by the odd prime p, there exist integers $x_1, \cdots, x_n$, such that

$$A\overline{x}_1^m + B\overline{x}_2^m + \cdots + E\overline{x}_n^m + F \equiv 0 \pmod{p},$$

where $n = m$ if m is even, and $n = 2m$ if m is odd.   The case $m = 2$ shows that there exist integral solutions of [Lagrange,[9] etc., of Ch. VIII]

$$Ax_1^2 + Bx_2^2 + C \equiv 0 \pmod{p}.$$

L. M. P. Coste[96] showed that the problem to make two integral functions of one variable equal to polygonal numbers of a given order can be reduced to the problem to make two functions equal to squares.   Let

$$P(Z) = (pZ^2 + qZ)/2, \qquad f_1 = Az^2 + A'z + P(a), \qquad f_2 = Bz^2 + B'z + P(b).$$

Then to make f_1 and f_2 equal to numbers $P(Z)$, take $Z = \alpha z + a$ and $Z = \beta z + b$ in the respective cases.   We obtain a quadratic equation for α and one for β, each linear in z.   Solving for α and β, we require that the quantities under the radical signs be squares, viz., $8pf_1 + q^2 = \square$, $8pf_2 + q^2 = \square$.   Next, if f_1 and f_2 are of the form $2a^2pz^2 + Az + A'$, use $Z = 2az + \alpha$.   We can make two quadratic functions equal to $P(Z)$ if a particular solution is known.

Several solvers[97] readily found two pentagonal numbers $P_x = (3x^2 - x)/2$ and P_y whose sum and difference are triangular by solving

$$8(P_x \pm P_y) + 1 = \square.$$

A. M. Legendre[98] concluded from the formula

$$(6) \qquad (1 + q + q^3 + q^6 + q^{10} + \cdots)^4 = \frac{1}{1 - q} + \frac{3q}{1 - q^3} + \frac{5q^2}{1 - q^5} + \cdots$$

that every integer N is a sum of four triangular numbers in $\sigma(2N + 1)$ ways, where $\sigma(k)$ denotes the sum of the divisors of k.   He gave an identity which shows the number of ways N is a sum of eight triangular numbers. Cauchy[99] gave (6); it was attributed to Jacobi by Bouniakowsky (see Vol. I, Ch. X[12, 19] of this History).   Cf. Plana.[123]

Several[100] found numbers > 1 which are simultaneously triangular, pentagonal and hexagonal.   Let $\frac{1}{2}m(m + 1) = \frac{1}{2}(3n^2 - n) = 2p^2 - p$. Then $m = 2p - 1$, $n = (1 + R)/6$, where $R^2 = 48p^2 - 24p + 1$.   Thus $1 + R = 6kp$, whence $p = (2 - k)/(4 - 3k^2)$.   Take $k = b/a$.   Then p is integral if $4a^2 - 3b^2 = 1$.   By the continued fraction for $\sqrt{3}$, we get

[95] Jour. de l'Ecole Polyt., Cah. 16, Vol. 9, 1813, 116–123; Oeuvres, (2), I, 59–63.
[96] Annales de Math. (ed., Gergonne), 10, 1819–20, 101–122.
[97] The Gentleman's Math. Companion, London, 5, No. 30, 1827, 558–9.
[98] Traité des fonctions elliptiques, 3, 1828, 133–4.
[99] Comptes Rendus Paris, 17, 1843, 572; Oeuvres, (1), VIII, 64.
[100] Ladies' Diary, 1828, 36–7, Quest. 1468.

$(2a, b) = (2, 1), (26, 15), (362, 209), (5042, 2911), \cdots,$ whence $p = 1, 143, 27693, \cdots,$ so that answers are $1, 40755, 1533776801, \cdots.$

J. Whitley[101] found pairs of pentagonal numbers p, q whose sum and difference are pentagonal. The conditions are

$$24p + 1 = x^2, \qquad 24q + 1 = y^2, \qquad 24(p + q) + 1 = z^2, \qquad 24(p-q)+1=v^2.$$

Hence $z^2 = x^2 + y^2 - 1, v^2 = x^2 - y^2 + 1.$ Let $x^2 = n^2 + m^2, y^2 = 2nm + 1.$ Then $z = n + m$, $v = n - m$. Take $n = r^2 - s^2$, $m = 2rs$, whence $x = r^2 + s^2$. There remains the condition

$$4rs(r^2 - s^2) + 1 = \square = y^2.$$

This is said to hold if $r = \frac{1}{2}(\phi^5 - \phi)$, $s = \frac{1}{2}(\phi^5 - 3\phi)$, which lead to larger numbers than those found by trial, using $(r, s) = (3, 2), (6, 1), (8, 5), (13, 2), (13, 8), (19, 14).$ [But the resulting numbers $p = 7, 37, 330, \cdots$ are not pentagonal.] See Gill.[108]

C. G. J. Jacobi[22] of Ch. VII gave in 1829 the result

$$\left\{ \sum_{m=-\infty}^{+\infty} (-1)^m x^{(3m^2+m)/2} \right\}^3 = \sum_{n=0}^{\infty} (-1)^n (2n + 1) x^{(n^2+n)/2},$$

the exponents on the left being pentagonal numbers for m negative, and those on the right triangular. Polygonal numbers appear incidentally in Jacobi's paper of 1848 [see Ch. III].

J. Huntington,[102] given a pentagonal number $P = r(3r - 1)/2$ of n digits, found another number p also of n digits such that if p is prefixed to P there results a pentagonal number. Let x be the root of the latter. Then shall $10^n p + P = x(3x - 1)/2$. Taking $p = x - r$, we get

$$x = \tfrac{1}{3}(2 \cdot 10^n + 1) - r.$$

For example, let $r = 1$; then $n = 1$, $x = 6$, $p = 5$ and 51 is pentagonal.

A. Cauchy[103] defined triangular and pyramidal numbers as binomial coefficients.

J. Baines[104] found two squares x^2, y^2 whose sum and difference are hexagonal. Take $8(x^2 - y^2) + 1 = \{2(x + y) \pm 1\}^2$, whence $x = 3y \pm 1$. Then $8(x^2 + y^2) + 1 = 80y^2 \pm 48y + 9 = (ny \pm 3)^2$ determines y.

A. Bernerie[105] gave a table of triangular numbers.

A. Casinelli[106] noted that every triangular number is of one of the three forms

$$(9m^2 + 3m)/2 = \triangle_{m-1} + 2\triangle_{2m}, \qquad (9m^2 + 9m + 2)/2 = \triangle_m + \triangle_{2m} + \triangle_{2m+1},$$
$$(9m^2 + 15m + 6)/2 = \triangle_{m+1} + 2\triangle_{2m+1},$$

also a sum of four $\triangle$'s, and hence a sum of any number of $\triangle$'s. By adding

[101] Ladies' Diary, 1829, 39–40, Quest. 1489.
[102] Ladies' Diary, 1832, 36–7, Quest. 1530.
[103] Résumés Analyt., Turin, 1, 1833, 5.
[104] The Gentleman's Diary, or Math. Repository, London, 1835, 33, Quest. 1320.
[105] Nouv. table des triangulaires, Bordeaux, 1835.
[106] Novi Comm. Acad. Sc. Inst. Bononiensis, 2, 1836, 415–34.

the first two or the second and third equations, we get

$$(3m + 1)^2 = m^2 + (2m + 1)^2 + 2\triangle_{2m},$$
$$(3m + 2)^2 = (m + 1)^2 + 2\triangle_{2m+1} + (2m + 1)^2.$$

Also, $(3m + 3)^2 = (m + 1)^2 + 2(2m + 2)^2$. Hence every square is a sum of three squares or a sum of two squares and two $\triangle$'s. Further, every $\triangle$ is a sum of a square and two equal $\triangle$'s. Next,

$$\triangle_m + \triangle_n + mn = \triangle_{m+n}, \qquad \triangle_m + \triangle_n + (m + 1)(n + 1) = \triangle_{m+n+1},$$

and similarly for three or more $\triangle$'s. Also, $\triangle_m + \triangle_n - m(n + 1) = \triangle_{n-m}$.

C. Gill[107] found numbers both m-gonal and n-gonal, and the generalization

$$T = ax^2 - a'x = by^2 - b'y,$$

where a, a', b, b' are given integers with no common factor. Take

$$ax - a' = yp/q, \qquad x = (by - b')q/p,$$

so that

$$x = q(b'p + c'bq)/N, \qquad y = q(a'p + b'aq)/N, \qquad - N = p^2 - abq^2.$$

Let p', q' give a particular solution of the last equation such that

$$A = (a'p' + ab'q')/N, \qquad B = (b'p' + ba'q')/N$$

are integers. Take $p = p't + abq'u$, $q = q't + p'u$. Then

$$p^2 - abq^2 = - NF,$$

where $F = t^2 - abu^2$, and $x = q(Bt + Abu)/F$, $y = q(At + Bau)/F$. From the initial solution $t_0 = 1$, $u_0 = 0$, of $F = 1$, we get as usual the solution

$$t_i = 2t_1t_{i-1} - t_{i-2}, \qquad u_i = 2t_1u_{i-1} - u_{i-2}.$$

It remains only to find a solution p', q' of $p^2 - abq^2 = - N$. While one may employ the continued fraction for $\sqrt{ab}$, it suffices for our initial problem to note the solution $p' = a - a'$, $q' = 1$, for the case $a - a' = b - b'$; then $N = ab' + ba' - a'b'$, $A = B = 1$. First, if m and n are both odd, we may take

$$a = m - 2, \qquad a' = m - 4, \qquad b = n - 2, \qquad b' = n - 4,$$

which have no common factor. Then $a - a' = b - b' = 2$. For $P_i = \tfrac{1}{2}T_i$,

$$P_0 = 1, \qquad P_i = 2t_4P_{i-1} - P_{i-2} + (2d - 1)(t_4 - 1),$$

$$d = \frac{(m + n - 4)(mn - 2m - 2n + 8)}{16(m - 2)(n - 2)}.$$

But if m and n are both even, take $a = \tfrac{1}{2}m - 1$, $a' = \tfrac{1}{2}m - 2$, $b = \tfrac{1}{2}n - 1$, $b' = \tfrac{1}{2}n - 2$, whence $a - a' = b - b' = 1$, and $P_i = T_i$ satisfies the same recursion formula. Also,

$$P_1 = \tfrac{1}{2}(t_4 + 1) + d(t_4 - 1) + \frac{1}{e}mnu_4,$$

---

[107] Math. Miscellany, Flushing, N. Y., 1, 1836, 220-5.

where $e = 8$ in the former case and $e = 16$ in the present case. For example, 1, 210, 40755 are both triangular and pentagonal, whereas Barlow[90] stated that this is true only for unity.

Gill[108] found n-gonal numbers whose sum and difference are n-gonal, i. e., $P_x + P_y = P_z$, $P_x - P_y = P_v$, where $P_x = (n - 2)x^2 - (n - 4)x$. As a generalization, take $P_x = mx^2 - m'x$, where m and m' are relatively prime. The first condition is satisfied if

$$z - y = \frac{b}{a}(mx - m'), \qquad m(z + y) - m' = \frac{a}{b}x.$$

Each of these linear equations is solved separately and the resulting x's equated. The second of our conditions is treated similarly and the two sets of values of x and y are compared. But the resulting solution does not lead to "convenient numbers." Another method is to assume that $x = aw - h$, $y = bw$, $z = cw - h$, where

$$a^2 + b^2 = c^2, \qquad 2mh(c - a) = m'(a + b - c),$$

whence our first condition is satisfied. Thus take

$$a = 2kl, \qquad b = k^2 - l^2, \qquad c = k^2 + l^2, \qquad l = mh, \qquad k = mh + m'.$$

The second of our initial conditions now becomes

$$4m^2(d^2 - 2m'^2)w^2 - 4m(2mh + m')dw + (2mh + m')^2 = (2mv - m')^2,$$

where $d = a - m'^2$. Take $2mv - m' = 2wt/u + 2mh + m'$. We get w and then v rationally. By choice of the denominator $t^2 - (d^2 - 2m'^2)m^2u^2$ we get integral answers unless $m' = 0$.

Many[109] found two squares x^2, y^2 such that $x^2 \pm y^2$ are pentagonal. Let $24(x^2 - y^2) + 1 = \{4(x + y) \pm 1\}^2$, whence $x = 5y \mp 1$. Then

$$24(x^2 + y^2) + 1 = 624y^2 \mp 240y + 25 = (5 - yr/s)^2$$

determines y. Again, to find pentagonal numbers p, q whose sum and difference are squares x^2, y^2, take $12(x^2 - y^2) + 1 = \{3(x + y) \pm 1\}^2$ and $12(x^2 + y^2) + 1 = (7 - yr/s)^2$, whence $x = 7y \pm 2$.

O. Terquem[110] proved that no triangular number > 1 is a biquadrate.

The ordinary definitions of polygonal and figurate numbers as sums of series were repeated by F. Stegmann,[111] George Peacock,[112] A. Transon,[113] H. F. Th. Ludwig,[114] Albert Dilling,[115] and V. A. Lebesgue.[116]

F. Pollock[117] stated that every integer is a sum of at most 10 odd squares, and a sum of at most 11 triangular numbers 1, 10, 28, 55, $\cdots$ of rank $3n+1$,

[108] Math. Miscellany, Flushing, N. Y., 1, 1836, 225–230.
[109] The Lady's and Gentleman's Diary, London, 1842, 41–3, Quest. 1677.
[110] Nouv. Ann. Math., 5, 1846, 70–78.
[111] Archiv Math. Phys., 5, 1844, 82–89.
[112] Encyclopaedia Metropolitana, London, 1, 1845, 422.
[113] Nouv. Ann. Math., 9, 1850, 257–9.
[114] Ueber fig. Zahlen u. arith. Reihen, Progr. Chemnitz, Leipzig, 1853.
[115] Die Progressionen, fig. u. polyg. Z., Progr. Muehlhausen, 1855.
[116] Exercices d'analyse numérique, 1859, 17–20.
[117] Proc. Roy. Soc. London, 5, 1851, 922–4. Cf. Euler,[66, 73] Beguelin.[72]

while 5, 7, 9, 13, 21, 11 terms are needed to express every number as a sum of tetrahedral, octahedral, cubic, icosahedral, dodecahedral, and squares of triangular, numbers. Legendre had proved that $8n + 3$ is a sum of three odd squares, each being $8\triangle + 1$. Pollock gave the generalization that, if F_x is any figurate number of order x, $8F_x + 3$ is a sum of 3 or $3 + 8$, $\cdots$, or $3 + 8n$ terms of a series whose general term is $8F_y + 1$.

V. Bouniakowsky[118] employed (1) and (9) of Vol. 1, Ch. X, of this History, to prove that every odd pentagonal number can be expressed as a sum of another pentagonal number and either a square or the double of a square; every odd square not a triangular number is a sum of double a triangular number and either a square or the double of a square. Similarly,

$$(1, 2)a^2 = \triangle_\lambda + (1, 2)u^2, \qquad \triangle_\lambda = \triangle_\mu + (1, 2)u^2,$$

the factor $(1, 2)$ denoting 1 or 2.

F. Pollock[119] stated without proof that any integer between two consecutive triangular numbers is the sum of four triangular numbers the sum of whose bases is constant.

J. B. Sturm[120] gave the relations

$$(2n + 1)^2 + (4\triangle_n)^2 = (4\triangle_n + 1)^2,$$
$$(2n + 1)^2(2m + 1)^2 + (4\triangle_n - 4\triangle_m)^2 = (4\triangle_n + 4\triangle_m + 1)^2.$$

V. A. Lebesgue[121] gave two proofs of the final theorem under Wallis.[44]

J. Liouville[122] proved readily that the only forms $a\triangle + b\triangle' + c\triangle''$ which represent all numbers, where the $\triangle$'s are triangular numbers and a, b, c are positive integers, are $\triangle + \triangle' + c\triangle''$ ($c = 1, 2, 4, 5$) and $\triangle + 2\triangle' + d\triangle''$ ($d = 2, 3, 4$). That conversely each of these seven forms represents all numbers is proved by use of Legendre's theorem that a number $\equiv 1, 2, 3, 5, 6$ (mod 8) is a $\boxed{3}$. The case $c = 1$ was treated by Gauss.[82] Next,

$$2(2n + 1) = 4u^2 + (2t + 1)^2 + (2z + 1)^2$$
$$8n + 4 = (2u + 2t + 1)^2 + (2t - 2u + 1)^2 + 2(2z + 1)^2,$$
$$n = \triangle_{u+t} + \triangle_{t-u} + 2\triangle_z,$$

proves the case $c = 2$. Next,

$$8n + 6 = (2x + 1)^2 + (2y + 1)^2 + 4(2z + 1)^2, \qquad n = \triangle_x + \triangle_y + 4\triangle_z,$$
$$8n + 5 = (2x + 1)^2 + 4(2s + 1)^2 + 16t^2$$
$$= (2x + 1)^2 + 2(2s + 1 + 2t)^2 + 2(2s + 1 - 2t)^2,$$
$$n = \triangle_x + 2\triangle_{s+t} + 2\triangle_{s-t},$$

or case $d = 2$. Next, as shown by Gauss,

$$8n + 7 = \square + \square + 2\square = (2x + 1)^2 + 4(2z + 1)^2 + 2(2y + 1)^2,$$
$$n = \triangle_x + 2\triangle_y + 4\triangle_z.$$

The proofs for the remaining cases $c = 5$ and $d = 3$ are longer.

[118] Mém. Acad. Sc. St. Pétersbourg, (6), 5, 1853, 303–322.
[119] Phil. Trans. Roy. Soc. London, 144, 1854, 311.
[120] Archiv Math. Phys., 33, 1859, 92–3.
[121] Introduction à la théorie des nombres, Paris, 1862, 17–20 (26–8).
[122] Jour. de Math., (2), 7, 1862, 407; 8, 1863, 73.

J. Plana[123] wrote ξ^4 for the left member of (6). By expanding the second member as a power series in q and examining the earlier terms, he verified that

$$\xi^4 = 1 + \sum_{n=1}^{\infty} q^n \sigma(2n+1),$$

where $\sigma(k)$ is the sum of the divisors of k. Hence any integer n is a sum of 4 triangular numbers in $\sigma(2n+1)$ ways. Give to ξ^3 the notation of a power series in q, multiply it by ξ and compare with the above series for ξ^4; we get a recursion formula for the coefficients of ξ^3. He states without proof that the coefficient of every power of q is not zero, and so concludes that every integer is a sum of three triangular numbers.

F. Pollock[124] verified for small values that any number may be expressed in the form $s - s'$, where s and s' are sums of two triangular numbers. Now s is always the sum of a square and the double of a triangular number. Thus the theorem is that

$$(7) \qquad a^2 + a + b^2 - (m^2 + m + n^2)$$

represents any number. Take $p^2 - c^2 - c + q$ as the number. Then

$$a^2 + a + b^2 + c^2 + c = m^2 + m + n^2 + p^2 + q.$$

Double and add unity. Thus $A = M + 2q$, where

$$A = 2a^2 + 2a + 1 + 2b^2 + 2c^2 + 2c, \qquad M = 2m^2 + 2m + 1 + 2n^2 + 2p^2.$$

Since q is arbitrary, it is concluded that any odd number can be represented by either of the forms A or M. But M is the sum of four squares.

Again, represent $p^2 - \frac{1}{2}(c^2 + c) + q$ by (7). As before,

$$2a^2 + 2a + 1 + 2b^2 + c^2 + c$$

represents any odd number $2n + 1$. But $a^2 + a + b^2$ is the sum of two triangular numbers. Hence n is the sum of three triangular numbers.

Pollock[53] of Ch. VIII noted that the theorem that every number $4n + 2$ is a sum of four squares implies that every integer n is a sum of four $\triangle$'s.

J. Liouville[125] considered the partition of any number into a sum of ten triangular numbers.

S. Bills[126] solved $\triangle_x + \triangle_y = \triangle_a$ by setting $y = a - xr/s$ and finding x rationally.

E. Lionnet stated and V. A. Lebesgue and S. Réalis[127] proved that every integer is a sum of a square and two $\triangle$'s, also a sum of two squares and a $\triangle$.

A. Hochheim[128] gave linear relations between polygonal and polyhedral numbers.

[123] Mém. Acad. Turin, (2), 20, 1863, 147.
[124] Proc. Roy. Soc. London, 13, 1864, 542–5.
[125] Comptes Rendus Paris, 62, 1866, 771.
[126] Math. Quest. Educ. Times, 6, 1866, 18.
[127] Nouv. Ann. Math., (2), 11, 1872, 95–6, 516–9; (2), 12, 1873, 217.
[128] Archiv Math. Phys., 55, 1873, 189–192.

S. Réalis[129] proved that every integer is a sum of four numbers of the form $(3z^2 \pm z)/2$ and also of four of the form $2z^2 \pm z$, i. e., pentagons and hexagons extended to negative arguments.   Use was made of the theorems that any odd number ω is a sum of four squares the algebraic sum of whose roots is 1 or 3, and its double 2ω is a sum of four squares the algebraic sum of whose roots is zero.   Further, every odd number divisible by h, or the double of every odd number divisible by the even number h, is a sum of four polygonal numbers of order $h + 2$, extended to negative arguments.

E. Lucas[130] stated that [cf. Leybourn[80]] $1^2 + \cdots + n^2$ is a square only when $n = 24$ [and $n = 1$], and is never a cube or fifth power.   A triangular number [> 1] is never a cube, biquadrate or fifth power [Euler[57]].   No pyramidal number is a cube or fifth power, or a square with the exception of

$$\frac{2 \cdot 3 \cdot 4}{1 \cdot 2 \cdot 3} = 2^2, \qquad \frac{48 \cdot 49 \cdot 50}{1 \cdot 2 \cdot 3} = 140^2.$$

Hence except for these and for the pile $24 \cdot 25 \cdot 49/6 = 70^2$ with a square base, no pile of bullets with a triangular or square base contains a number of bullets equal to a square, cube or fifth power.

Lucas[131] stated and proved incompletely that the [pyramidal] number $x(x+1)(2x+1)/6$ of bullets in a pile, whose base is a square with x to a side, is a square only when $x = 1$ or 24 (see papers 130, 132, 137–8).

T. Pepin[132] noted that one case of Lucas' proof of the last result leads to an equation $9r^4 - 12f^2r^2 - 4f^4 = R^2$, not treated by Lucas when f and R are divisible by 3.   Pepin found an infinitude of solutions in this case. G. N. Watson[132a] noted the solution $r = 5$, $f = 3$, $R = 51$, and[132b] proved Lucas'[131] theorem by use of elliptic functions.

Lucas[133] stated that the number of bullets in a pile with a square or triangular base is never a cube or fifth power.   Moret-Blanc[134] gave a proof.

Moret-Blanc[135] noted that the tetrahedral number $n(n + 1)(n + 2)/6$ is a square for $n = 1, 2, 48$.   Lucas stated that it is a square only then, a fact proved by A. Meyl.[136]

E. Fauquembergue[137] and N. Alliston[138] proved that $1^2 + \cdots + n^2 \neq \square$ if $n > 24$.   Cf. Lucas[131] and the papers cited on p. 26.

[129] Nouv. Ann. Math., (2), 12, 1873, 212; Nouv. Corresp. Math., 4, 1878, 27–30.

[130] Recherches sur l'analyse indéterminée, Moulins, 1873, 90; extracted from Bulletin de la société d'émulation Dept. de l'Allier, Sc. Bell. Let., 12, 1873, 530.

[131] Nouv. Ann. Math., (2), 14, 1875, 240; (2), 16, 1877, 429–432. The proof by Moret-Blanc, (2), 15, 1876, 46–8, is incomplete (as noted p. 528).

[132] Atti Accad. Pont. Nuovi Lincei, 32, 1878–9, 292–8.

[132a] Proc. London Math. Soc., Record of Meeting, March 14, 1918.

[132b] Messenger of Math., 48, 1918, 1–22.

[133] Nouv. Ann. Math., (2), 15, 1876, 144 (Nouv. Corresp. Math., 2, 1876, 64; 3, 1877, 247–8, 433, and p. 166 for incomplete proof by H. Brocard).

[134] Ibid., (2), 20, 1881, 330–2.

[135] Ibid., (2), 15, 1876, 46.

[136] Ibid., (2), 17, 1878, 464–7.

[137] L'intermédiaire des math., 4, 1897, 71.

[138] Math. Quest. Educ. Times, 29, 1916, 82–3 (for $n < 10^{21}$ by J. M. Child, 26, 1914, 72–3; for $n < 10^{12}$ by G. Heppel, 34, 1881, 106–7).

For analogous theorems on sums of consecutive squares or the sum of the squares of the first n odd numbers see papers 70, 76, 81, 86, 87, 100, and 103 of Ch. IX, and Brocard[92] of Ch. XXIII.

W. Göring[139] proved by use of infinite series that $2\triangle + 6\triangle' + 1$ can always be represented by the form $a^2 + 3b^2$.

J. W. L. Glaisher[140] noted that every representation of an odd number as a sum of an even square and two triangular numbers corresponds to a representation in which the square is odd, since

$$m^2 + \frac{p(p+1)}{2} + \frac{q(q+1)}{2}$$

$$\equiv \left(\frac{p-q}{2}\right)^2 + \sum \frac{1}{2}\left(\pm m + \frac{p+q}{2}\right)\left(\pm m + \frac{p+q}{2} + 1\right),$$

for p, q both even or both odd, with a similar identity if one is even and the other odd.

Glaisher[141] stated that every triangular number is a sum of three pentagonal numbers.

D. Marchand[142] noted the relations

$$p_5^r = p_3^{r-1} + r^2, \qquad p_6^r = 2p_3^{r-1} + r^2, \qquad p_5^1 + p_5^2 + \cdots + p_5^r = rp_3^r.$$

Marchand[143] gave identities like

$$\triangle(3y+1) = \triangle(y) + (2y+1)^2,$$
$$(x+1)^5 - x^5 = \triangle(y) + \triangle(3y+1) = 2\triangle(y) + (2y+1)^2,$$

where $y = x^2 + x$, and (p. 105) discussed triangular numbers which are squares.

E. Lucas[144] asked when $(\triangle_1^2 + \cdots + \triangle_n^2)/(\triangle_1 + \cdots + \triangle_n)$ is a square.

S. Réalis, E. Catalan and others[145] investigated numbers simultaneously squares and triangular. S. Réalis stated and E. Cesàro[146] proved that the square of every odd multiple of 3 is a difference of two $\triangle$'s prime to 3, $9(2n+1)^2 = \triangle(9n+4) - \triangle(3n+1)$. D. Marchand[147] gave the generalization that the square of any odd number is the difference of two relatively prime triangular numbers (with sides $3x+1$ and x). C. Henry[148] proved a like result for the product of any odd square by any number.

S. Réalis[149] stated that the theorem that every integer n is a sum of three $\triangle$'s implies that n is a sum of four $\triangle$'s of which two are consecutive and that n is a sum of four $\triangle$'s two of which are equal.

[139] Math. Annalen, 7, 1874, 386.
[140] Phil. Mag., London, (5), 1, 1876, 48.
[141] Messenger Math., 5, 1876, 164-5.
[142] Les Mondes, 42, 1877, 164-170.
[143] La Science des nombres, 1877.
[144] Nouv. Corresp. Math., 3, 1877, 433.
[145] Ibid., 4, 1878, 167; 5, 1879, 285-7; Math. Quest. Educ. Times, 30, 1879, 37.
[146] Nouv. Corresp. Math., 4, 1878, 156.
[147] Nouv. Ann. Math., (2), 17, 1878, 463.
[148] Ibid., (2), 19, 1880, 517.
[149] Ibid., (2), 17, 1878, 381.

E. Lucas[150] listed values of A for which $xy(x + y) = Az^3$ has no distinct rational solutions $\neq 0$. Taking $y = 1$ and $y = x + 1$, we obtain theorems on triangular numbers and numbers $x(x + 1)(2x + 1)$.

Lucas stated and Moret-Blanc[151] proved that $1^2 + \cdots + x^2 = ky^2$ and $\triangle_1^2 + \cdots + \triangle_n^2 = ky^2$ are impossible if $k = 2, 3, 6$.

S. Roberts[152] proved by use of $y^2 - 2z^2 = 1$ [Euler[70]] that the $\triangle$'s which are squares are

$$\left\{ \frac{(1 + \sqrt{2})^{2m} - (1 - \sqrt{2})^{2m}}{4\sqrt{2}} \right\}^2 .$$

J. Neuberg stated and E. Cesàro[153] proved that the sum of the squares of $n + 1$ consecutive integers, beginning with the $2n$th triangular number, equals the sum of the squares of the n succeeding integers, each being divisible by $1^2 + \cdots + n^2$. Cf. Dostor[75] of Ch. IX.

E. Lionnet[154] stated that unity is the only triangular number $\triangle$ which equals the sum of the squares of two consecutive integers; 10 is the only $\triangle$ equal to the sum of the squares of two consecutive odd integers; when $\triangle$ is a product of two consecutive integers of which the least is double a triangular number, then $4\triangle + 1$ (and its square root) is a sum of squares of two consecutive integers.

Moret-Blanc[55] proved the preceding theorems stated by Lionnet.

E. Cesàro[156] noted that no triangular number ends with 2, 4, 7, 9.

S. Réalis[157] noted that

$$\triangle(5p + 1) = \triangle(4p + 1) + \triangle(3p), \qquad \triangle(5p + 3) = \triangle(4p+2) + \triangle(3p+2),$$
$$\triangle(k + \alpha) = \triangle(k) + \triangle(2\alpha p + \alpha), \qquad k = 2\alpha p^2 + (2\alpha + 1)p.$$

E. Lionnet[158] noted that 0, 1, 6 are the only $\triangle$'s whose squares are $\triangle$'s. He stated and E. Cesàro[159] proved that there is at least one and at most two $\triangle$'s between any two consecutive squares $\neq 0$; at most one square between two consecutive $\triangle$'s; if there are exactly two $\triangle$'s between $(a + 1)^2$ and $(a + 2)^2$, where $a > 0$, there is just one $\triangle$ between a^2 and $(a + 1)^2$, and just one $\triangle$ between $(a + 2)^2$ and $(a + 3)^2$.

E. Cesàro[160] denoted by $\nabla(n)$ the number of the first $2n$ triangular numbers which are relatively prime to n. Let $\Psi(n)$ be the number of products $1\cdot2$, $2\cdot3$, $3\cdot4$, $\cdots$, $n(n + 1)$ which are prime to n. Then if ν is the largest odd divisor of n,

$$\frac{\nabla(n)}{n} = \frac{\Psi(n)}{n} + \frac{\Psi(\nu)}{\nu}, \qquad \frac{\Psi(n)}{n} = \frac{\nabla(n)}{n} - \tfrac{1}{2}\frac{\nabla(\nu)}{\nu} .$$

[150] Nouv. Ann. Math., (2), 17, 1878, 513.
[151] *Ibid.*, 527; (2), 18, 1879, 470–4.
[152] Math. Quest. Educ. Times, 30, 1879, 37.
[153] Nouv. Corresp. Math., 6, 1880, 232.
[154] Nouv. Ann. Math., (2), 20, 1881, 514.
[155] *Ibid.*, (3), 1, 1882, 357.
[156] Mathesis, 4, 1884, 70.
[157] Jour. de math. spéc., 1884, 6.
[158] Nouv. Ann. Math., (3), 1, 1882, 336. Proof by H. Brocard, (3), 15, 1896, 93–6.
[159] *Ibid.*, (3), 2, 1883, 432 (misprints); 5, 1886, 209–213.
[160] Annali di Mat., (2), 14, 1886–7, 150–3.

The mean value of $\nabla(n)$ is three times that of $\Psi(n)$. He found that the probability that two triangular numbers taken at random shall be relatively prime is

$$\frac{3}{4}\left(1-\frac{4}{3^2}\right)\left(1-\frac{4}{5^2}\right)\left(1-\frac{4}{7^2}\right)\left(1-\frac{4}{11^2}\right)\cdots.$$

Cesàro[161] stated and E. Fauquembergue[161] proved that 5 and 17 are the only integers whose cubes diminished by 13 are quadruples of triangular numbers.

G. de Rocquigny[162] noted that, if $k = (a^2 + b^2 + a + b)/2$,

$$\triangle_k = \triangle_{k-1} + \triangle_a + \triangle_b, \qquad (a^2 + 1)^2 = 1 + \triangle(a^2 + a) + \triangle(a^2 - a).$$

S. Réalis[163] used the known fact that, if p is a product of primes $8q + 1$, $2x^2 + y^2 = p$ has integral solutions. Thus

$$3(8q + 1) = 2(2a + 1)^2 + (2b + 1)^2,$$

so that $3q = 2\triangle + \triangle'$.

Réalis[164] gave various sums like

$$\triangle_1 + \triangle_3 + \triangle_5 + \cdots + \triangle_{2n-1} = \tfrac{1}{6}n(n + 1)(4n - 1),$$
$$\triangle_2 + \triangle_4 + \triangle_6 + \cdots + \triangle_{2n} = \tfrac{1}{6}n(n + 1)(4n + 5),$$
$$\triangle_3 + \triangle_6 + \triangle_9 + \cdots + \triangle_{3n} = \tfrac{3}{2}n(n + 1)^2.$$

E. Cesàro[165] noted that $(n^5 - 1)/4 = \triangle_p + \triangle_q$, $n \neq 5$, implies that $2p + 1$ or $2q + 1$ is composite.

S. Tebay and others[166] found that the least heptagonal number $\tfrac{1}{2}(5x^2 - 3x)$ which when increased by a^2 is equal to a square is given by $x = 24(19a - 9)$.

C. A. Laisant[167] wrote a_α for the ath $(\alpha + 2)$-gonal number $p_{\alpha+2}^a$ and gave

$$(a + b)_\alpha = a_\alpha + b_\alpha + \alpha ab, \qquad (a + \cdots + l)_\alpha = \Sigma a_\alpha + \alpha\Sigma ab.$$

E. Cesàro[168] noted that the number of $\triangle$'s prime to n and $< 2n(n + 1)$ is $k = n\Pi(1 - 2/p)$ or $2k$ according as n is even or odd, where p ranges over the odd prime factors of n.

E. Catalan[169] proved that every $\triangle > 1$ is a sum of six pentagonal numbers. For,[170] $6(2n + 1)^2 = (6x \mp 1)^2 + (6y \mp 1)^2 + 4(6z \mp 1)^2$, whence

$$\frac{n(n + 1)}{2} = \frac{3x^2 \mp x}{2} + \frac{3y^2 \mp y}{2} + 4\left(\frac{3z^2 \mp z}{2}\right).$$

[161] Mathesis, 6, 1886, 23; 7, 1887, 257-9.
[162] Ibid., 6, 1886, 224.
[163] Nouv. Ann. Math., (3), 5, 1886, 113.
[164] Jour. de math. spéc., 1888, 94.
[165] Mathesis, 8, 1888, 75.
[166] Math. Quest. Educ. Times, 50, 1889, 84-5.
[167] Bull. Soc. Philomathique de Paris, (8), 3, 1890-1, 29-30.
[168] Mathesis, (2), 1, 1891, 95-96.
[169] Assoc. franç. av. sc., 1891, II, 201-2.
[170] Recherches sur quelques prod. indéf., Mém. Acad. Roy. Belgique, 40, 1873, 61-191, formula

[But the denominator 2 on the left should be suppressed. Legendre[94] had already proved a more general theorem.]

E. Lucas[171] collected results, mostly algebraic, on triangular and figurate numbers.

E. Catalan[172] stated that every $\triangle$, not pentagonal, is a sum of fewer than 7 pentagonal numbers. [Catalan.[169]]

T. Pepin[173] gave a proof of Cauchy's formulation of Fermat's theorem that every integer A is a sum of $m + 2$ polygonal numbers $\frac{1}{2}m(x^2 - x) + x$ of order $m + 2$ of which $m - 2$ are 0 or 1. We are to prove that

$$A = \tfrac{1}{2}m(a - b) + b + r,$$

where $a = \alpha^2 + \cdots + \delta^2$, $b = \alpha + \cdots + \delta$, $0 \leqq r \leqq m - 2$, whence $a \equiv b \pmod 2$. Since r can take the values 0 and 1, we may take b odd, whence

$$4a - b^2 = 8l + 3 = x^2 + y^2 + z^2,$$

$x > y > z > 0$. Determine integers $\alpha, \cdots, \delta$ so that

$$\alpha + \beta - \gamma - \delta = x, \qquad \alpha + \delta - \beta - \gamma = \pm z,$$
$$\alpha + \gamma - \beta - \delta = y, \qquad \alpha + \beta + \gamma + \delta = b.$$

Then $a = \Sigma\alpha^2$ is satisfied. The condition $b^2 < 4a$ is satisfied if $B > 110$, where $A = mB + c, 0 < c \leqq m$. Hence the theorem is true for all numbers $A > 110m$. It was verified separately for all numbers $\leqq 120m + 16$.

G. Musso[174] proved, by use of geometrical representations, Bachet's[32] second formula I, 10, and the generalizations

$$p_q^n = \frac{s - 1}{2} p_3^n + (q - s)p_3^{n-1} + \frac{s - 3}{2} p_3^{n-2} - \frac{s - 3}{2} \quad (s \text{ odd}),$$

$$p_q^n = \frac{s - 2}{2} p_3^n + (q - s)p_3^{n-1} + \frac{s - 2}{2} p_3^{n-2} + n - \frac{s - 2}{2} \quad (s \text{ even});$$

also

$$p_3^n = n^2 - (n - 1)^2 + (n - 2)^2 - \cdots \pm 1,$$
$$p_q^n = 2p_s^n + \{q - (2s - 1)\}p_3^{n-1} + p_3^{n-2} - 1.$$

E. Catalan[175] gave a shorter proof of Bachet's same formula.

G. de Rocquigny[176] noted that, if $a + b + c = \alpha + \beta + \gamma = 0$,

$$P = (\triangle_a + \triangle_b + \triangle_c)(\triangle_\alpha + \triangle_\beta + \triangle_\gamma), \qquad (\triangle_n + \triangle_{n+1})(\triangle_p + \triangle_{p+2}),$$
$$6n^4, \qquad 6n^4 + 1, \qquad 6n^4 + 2n^2 + 1, \qquad \cdots$$

are sums of three $\triangle$'s, while $n^2 + (2n - 1)^2 + (2n + 1)^2$ is a sum of two.

[171] Théorie des nombres, 1891, 52–62, 83.
[172] Jour. de math. spéc., 1892, No. 353.
[173] Atti Accad. Pont. Nuovi Lincei, 46, 1892–3, 119–131.
[174] Giornale di Mat., 31, 1893, 173–8. His P_n^q is our p_q^n.
[175] Ibid., p. 227.
[176] Mathesis, (2) 4, 1894, 123 171, 211; (2), 5, 1895, 23, 150, 211–2. C : Curjel.[180]

The sum of $2n + 1$ consecutive $\triangle$'s equals the product of the middle one by $2n + 1$, increased by $1^2 + \cdots + n^2$. He asked when

$$\triangle_1 + \cdots + \triangle_n = \triangle.$$

E. Barbette[177] noted that the sum T_k of the kth powers of the first n triangular numbers equals $S^k(S + 1)^k/2^k$ symbolically, where, after expansion, S^t is to be replaced by S_t, the sum of the tth powers of $1, \cdots, n$. The values of T_1, T_2, T_3 are given as functions of n and as functions of the S's. It is shown that $T_k^x = T_r^y$ implies $x = y, k = r$.

A. Boutin[178] noted that $\triangle_x = p\triangle_y$ has an infinitude of solutions if p is not a square. Let $2x = k - 1, 2y = z - 1$. Then $k^2 - pz^2 = 1 - p$. Let $k = \alpha + p\beta, z = \beta + \alpha$. Then $\alpha^2 - p\beta^2 = 1$, having an infinitude of solutions if p is not a square. If $p = m^2$, the problem has only a finite number of solutions if any. It is impossible if $m = 3, 4, 5, 7, 8, 9, 11, 12, 13, 15, 16, 17$. If $m = 4\lambda + 2, x = 4\lambda^2 + 4\lambda, y = \lambda$ is a solution.

Several[179] solved $\triangle_x + \triangle_y = 2\triangle_z$, i. e.,

$$(2x + 1)^2 + (2y + 1)^2 = 2(2z + 1)^2.$$

See Ch. XIV.

H. W. Curjel,[180] to prove de Rocquigny's[176] first statement, took

$$a = y - z, \quad b = z - x, \quad c = x - y, \quad \alpha = \eta - \zeta, \quad \beta = \zeta - \xi, \quad \gamma = \xi - \eta,$$

$$X = x\xi + z\eta + y\zeta, \quad Y = z\xi + y\eta + x\zeta, \quad Z = y\xi + x\eta + z\zeta, \text{ and got}$$

$$P = \triangle(Y - Z) + \triangle(Z - X) + \triangle(X - Y).$$

E. Maillet[181] proved the following generalization of Fermat's[36] theorem on polygonal numbers: If α and β are relatively prime odd numbers, $\alpha > 0$, every integer A exceeding a certain limit (function of α, β) is a sum of four numbers of the form $(\alpha x^2 + \beta x)/2$. We can assign an inferior limit to A such that this decomposition can be made in any assigned number of ways. A like theorem holds if $\alpha/2$ is an odd integer and one of A, $\beta/2$ is odd and the other even, provided $\alpha/2$ and $\beta/2$ are relatively prime; also if $\alpha/2$ is even and $\beta/2$ and A both odd. He proved three complicated theorems stating that every number with certain residues modulo 6 is a sum of at most $\delta < 59$ (or $\delta < 53$) numbers of the form $(\alpha x^4 + \beta x^2)/2$. Later Maillet[182] proved that if $\phi(x) = a_0 x^5 + \cdots + a_5$, in which the a's are given rational numbers, is integral and positive for every integer $x \geq \mu$, every integer exceeding a fixed limit, depending on the a's, is the sum of at most ν positive numbers $\phi(x)$ and a limited number of units, where $\nu = 6, 12, 96$, or 192, according as the degree of ϕ is 2, 3, 4 or 5. Every integer ≥ 19272 is a sum (p. 372) of at most 12 pyramidal numbers $(x^3 - x)/6$.

[177] Mathesis, (2), 5, 1895, 111–2.
[178] Jour. de math. élém., (4), 4, 1895, 179–180.
[179] Math. Quest. Educ. Times, 63, 1895, 40.
[180] Ibid., 33–4. Other proofs, (2), 20, 1911, 78–9.
[181] Bull. Soc. Math. de France, 23, 1895, 40–49.
[182] Jour. de Math., (5), 2, 1896, 363–380.

Several writers[133] found the first six integers n making $n(n + 1)/2$ a square.   Several[184] proved that the difference of the roots of two successive triangular numbers, each a square, equals the sum of two successive integers the sum of whose squares is a square.

A. Boutin[185] reduced $x^2 = \triangle_y + 1$ to $p^2 - 2q^2 = -1$ by setting $2x = 3q \mp p,\ y = (k - 1)/2,\ k = 3p \mp 2q$ [Euler[79]].

G. de Rocquigny[186] noted the identities

$$(5n \pm 1)^2 = \frac{n(n \mp 1)}{2} + \frac{(7n \pm 1)(7n \pm 2)}{2},$$

$$\{\triangle(a^2 + a - 1) + \triangle(a^2 - a - 1)\}\{\triangle(b^2 + b - 1) + \triangle(b^2 - b - 1)\}$$
$$= \triangle(a^2b^2 + ab - 1) + \triangle(a^2b^2 - ab - 1),$$

$$\{\triangle(7a + 1) + \triangle(a - 1)\}\{\triangle(7b + 1) + \triangle(b - 1)\}$$
$$= \triangle(7c + 1) + \triangle(c - 1), \qquad c = 5ab + a + b,$$

and expressed n^6, $n + n^2 + n^3 + n^4$, $n^3 + n^4 + n^5 + n^6$ and $n + \cdots + n^6$ as sums of three triangular numbers, etc.

A. Boutin[187] solved $\triangle_{x-1} + \triangle_n = y^2$ by setting $x = an - b,\ y = \alpha n - \beta$. Then

$$a^2 + 1 = 2\alpha^2, \qquad b^2 + b = 2\beta^2, \qquad 4\alpha\beta + 1 = a(2b + 1),$$

which are solved by means of recursion formulas.

A. Berger[188] proved many relations and inequalities involving the rth m-gonal number (3) designated by $P(m, r)$.   If $|x| < 1$,

$$\sum_{r=1}^{\infty} P(a, r)x^r = \frac{x + (a - 3)x^2}{(1 - x)^3}, \qquad \sum_{r=1}^{\infty} P(a, -r)x^r = \frac{x^2 + (a - 3)x}{(1 - x)^3}.$$

He evaluated $\Sigma 1/P(a, r)$, where r ranges over all integers for which $P(a, r)$ takes positive values and each but once.   If $a \geqq 3$, $|x| < 1$, $\epsilon = \pm 1$,

$$\prod_{r=1}^{\infty} (1 - x^{(a-2)r})(1 + \epsilon x^{(a-2)r-a+3})(1 + \epsilon x^{(a-2)r-1}) = \sum_{r=-\infty}^{+\infty} \epsilon^r x^{P(a, r)},$$

combinations of special cases of which give

$$\prod_{r=1}^{\infty} (1 - x^r)^{(-1)^r} = \sum_{r=-\infty}^{\infty} x^{P(6, r)}, \qquad \prod_{r=1}^{\infty} (1 - x^r) = \sum_{r=-\infty}^{\infty} (-1)^r x^{P(5, r)}.$$

Let $\sigma(k)$ be the sum of the divisors of k, and $\psi(k)$ the excess of the sum of the odd divisors of k over the sum of the even divisors.   Then

$$\log \sum_{r=-\infty}^{\infty} (-1)^r x^{P(5, r)} = -\sum_{k=1}^{\infty} \frac{\sigma(k)x^k}{k}, \qquad \log \sum_{r=0}^{\infty} x^{P(3, r)} = \sum_{k=1}^{\infty} \frac{\psi(k)x^k}{k}.$$

[133] Amer. Math. Monthly, 3, 1896, 81–2; Math. Quest. Educ. Times, 65, 1896, 53; 69, 1898, 51.

[184] Amer. Math. Monthly, 4, 1897, 187–9.

[185] Mathesis, (2), 6, 1896, 28–29.

[186] Mathesis, (2), 7, 1897, 217–221.

[187] Ibid., 269–270.

[188] Nova Acta Soc. Sc. Upsaliensis, (3), 17, 1898, No. 3.

He studied (pp. 20–25) the number $\phi(a, k)$ of polygonal divisors of order a of a positive integer k; every integer has in mean two triangular divisors, $\pi^2/6$ square divisors, etc.

A. Goulard and A. Emmerich[189] found two consecutive integers of which one is a square and the other triangular. In $x^2 - \frac{1}{2}y(y + 1) = \pm 1$, set $2x = z$, $2y + 1 = t$, whence $2z^2 - t^2 = 7$ or -9, which are reduced to the Pell equations $u^2 - 2v^2 = -1$, or $+1$, and solved.

P. Bachmann[190] gave an excellent exposition of Cauchy's[93] proof of his modification of Fermat's theorem: every integer is a sum of m m-gonal numbers of which all but four are 0 or 1.

R. W. D. Christie[191] noted two formulas of the type

$$\sum_{i=1}^{4} \triangle(a_i + n) = \sum_{i=1}^{4} \triangle(\sigma - a_i + n), \qquad \sigma = \frac{1}{2}\sum_{i=1}^{4} a_i.$$

J. W. West[192] noted that if $\triangle_a = 6\triangle_b + 1$, $\triangle_a$ is not a square.

R. W. D. Christie[193] proved that, if p_r^m is the mth r-gonal number,

$$(2n)^3(r - 2)p_r^m + \frac{1}{3}(4n^3 - n)y^2 = (x - y)^2 + (x - 3y)^2 + (x - 5y)^2 + \cdots,$$
$$x = 2mn(r - 2), \qquad y = r - 4.$$

W. A. Whitworth and A. Cunningham[194] noted that if $N = \triangle_m + \triangle_n$, $4N + 1 = (m + n + 1)^2 + (m - n)^2$; conversely, if $4N + 1$ has no prime factor $4k - 1$, it is a sum of two squares and hence N is a sum of two $\triangle$'s.

Crofton[195] noted that

$$9\triangle_k + 1 = \triangle(3k + 1), \qquad 4\triangle_k + 4\triangle_l + 1 = (k - l)^2 + (k + l + 1)^2.$$

Christie employed $\triangle_m + \triangle_{m+1} = (m + 1)^2$ to get

$$\begin{aligned}
\triangle_n + A^2 + B^2 + \cdots &= \triangle_n + (\triangle_{n+1} + \triangle_{n+2}) + \cdots + (\triangle_{2n-2} + \triangle_{2n-1}) \\
&= (\triangle_n + \triangle_{n+1}) + (\triangle_{n+2} + \triangle_{n+3}) + \cdots + \triangle_{2n-1} \\
&= \triangle_{2n-1} + \alpha^2 + \beta^2 + \cdots.
\end{aligned}$$

W. A. Whitworth[196] gave rules, depending on the convergents to the continued fraction for $\sqrt{2}$, to solve $\triangle = \square$ or $\triangle = 2\triangle'$, equivalent to known rules to solve $u^2 - 2v^2 = \pm 1$.

E. Lemoine[197] called a number N decomposed into its maximum triangular numbers $\triangle$, and m the index of N, if $N = A_1 + \cdots + A_m$, where A_1 is the largest $\triangle \leqq N$, A_2 the largest $\triangle \leqq N - A_1$, A_3 the largest $\triangle \leqq N - A_1 - A_2$, etc. If Y_m is the least number of index m,

$$Y_m = \frac{1}{2}Y_{m-1}(Y_{m-1} + 3), \qquad 2^{m-1}Y_m = (Y_1 + 3)(Y_2 + 3)\cdots(Y_{m-1} + 3).$$

[189] Mathesis, (2), 8, 1898, 52–4. Cf. Tits.[223]
[190] Die Arith. der Quadratischen Formen, 1, 1898, 154–162.
[191] Math. Quest. Educ. Times, 68, 1898, 84.
[192] Ibid., 69, 1898, 114.
[193] Ibid., 70, 1899, 119.
[194] Ibid., 71, 1899, 33.
[195] Ibid., 69.
[196] Ibid., 73, 1900, 32–3.
[197] Assoc. franç. avanc. sc., 1900, II, 72.

E. Grigorief[198] discussed Fermat's theorem that every number is a sum of three $\triangle$'s.

L. Kronecker[199] gave a brief history of polygonal numbers.

J. J. Barniville[200] evaluated series involving figurate numbers, such as
$$1^3 + (1^3 + 3^3)2^{-1} + (1^3 + 3^3 + 6^3)2^{-2} + (1^3 + 3^3 + 6^3 + 10^3)2^{-3} + \cdots = 6416.$$

A. Cunningham[201] noted that $\triangle_x^2 + \triangle_z^2 = 2\triangle_y^2$ if $\triangle_x = \xi\triangle_z$, $\triangle_y = \eta\triangle_z$, where $(\xi, \eta) = (1, 1)$, $(7, 5)$, $(41, 29)$, $(239, 169)$, etc.

Cunningham and Christie[202] solved $\mu\triangle_x = \nu\triangle_y$, which is equivalent to $\mu(2x + 1)^2 - \nu(2y + 1)^2 = \mu - \nu$, by use of a solution of $\xi^2 - \mu\nu\eta^2 = 1$.

R. W. D. Christie[203] argued that no $\triangle$ is a cube > 1.

A. Cunningham[204] noted that $\triangle_a\triangle_x = \triangle_a\triangle_y$ is equivalent to
$$\triangle_a(X^2 - 1) = \triangle_a(Y^2 - 1),$$
whose solutions follow from the least solution of $\xi^2 - \triangle_a\triangle_a\eta^2 = 1$.

Christie[205] noted that $N = \triangle_{2a} + \triangle_{2b} + \triangle_{2c}$ implies
$$2N + 1 = (a + b + c + 1)^2 + (a - b - c)^2 + (a + b - c)^2 + (a - b + c)^2,$$
and similar formulas in which some of $2a$, $2b$, $2c$ are replaced by odd numbers.

Cunningham[206] noted that, if $x = \tfrac{2}{3}(10^n - 1)$, $\triangle_x = 2\cdots21\cdots1$ (n two's and n one's).

E. B. Escott[207] proved that 55, 66 and 666 are the only triangular numbers, with fewer than 30 digits, consisting of a single repeated digit.

F. Hromádko[208] noted that if $\triangle_1, \cdots, \triangle_n$ are any consecutive $\triangle$'s,
$$\triangle_n^2 - \triangle_1^2 = (\triangle_2 - \triangle_1)^3 + (\triangle_3 - \triangle_2)^3 + \cdots + (\triangle_n - \triangle_{n-1})^3.$$

L. von Schrutka[209] proved that, if $l \equiv p_m^a \pmod{k}$, then
$$\left(\frac{m}{2} - 1\right)l + \left(\frac{m - 4}{4}\right)^2 \equiv \left\{\left(\frac{m}{2} - 1\right)a - \left(\frac{m}{4} - 1\right)\right\}^2 \pmod{k},$$
and conversely if $m/2 - 1$ is prime to k, so that k is called regular. The question of polygonal residues thus reduces to that of quadratic residues. Irregular moduli k are treated on pp. 190–3.

J. Blaikie[210] noted that $\tfrac{1}{2}n(n + 1)$ is also a pentagonal number $\tfrac{1}{2}m(3m - 1)$ if $3y^2 - x^2 = 2$, where $x = 6m - 1$, $y = 2n + 1$. From solutions of the Pell equation $p^2 - 3q^2 = 1$, we get solutions $x = 123q \pm 71p$, $y = 41p \pm 71q$ of the former.

---

[198] Kazan Izv. fiz. mat. obsc. (= Bull. Math. Phys. Soc. Kasan), 11, 1901, No. 2, 64–69 (in Russian).

[199] Vorlesungen über Zahlentheorie, 1901, 17–22.

[200] Math. Quest. Educ. Times, 74, 1901, 80.

[201] *Ibid.*, 65–6.

[202] *Ibid.*, 87–8.

[203] *Ibid.*, 75, 1901, 36.

[204] *Ibid.*, 120–1.

[205] *Ibid.*, (2), 1, 1902, 94–5; 6, 1904, 85–6.

[206] *Ibid.*, 8, 1905, 25.

[207] *Ibid.*, 33–4.

[208] Zeitschr. Math. Naturw. Unterricht, 35, 1904, 306.

[209] Monatshefte Math. Phys., 16, 1905, 167–193.

[210] Math. Quest. Educ. Times, (2), 9, 1906, 40–41.

It is stated[211] that every nth power is a sum of n $\triangle$'s $\neq 0$; for example,

$$3^4 = 55 + 15 + 10 + 1,$$
$$5^5 = 2850 + 210 + 45 + 2\cdot10 = 3003 + 105 + 10 + 6 + 1.$$

G. Nicolosi[212] gave an elementary proof of Cantor's result that

$$\tfrac{1}{2}(x + y)(x + y + 1) + y = a$$

has one and but one set of integral solutions. Solving for y we see that $8x + 8a + 9$ must be a square u^2. Thus x is integral only if $u^2 - 1 = 8\theta$, whence $\theta = t(t + 1)/2$.

C. Burali-Forti[213] noted relations like

$$p_m^n - p_r^n = (m - r)p_3^{n-1}, \qquad p_m^n + p_m^r = p_m^{n+r} - nr(m - 2),$$
$$mp_m^n - np_n^m = \tfrac{1}{2}mn(m - n).$$

A. Cunningham[214] gave a method of expressing an integer as a sum of three triangular numbers.

P. Bachmann[215] gave an introduction to polygonal and figurate numbers.

T. Hayashi[216] proved that the quadruple of a number $\alpha(\alpha + \beta)(\alpha + 2\beta)/6$ and hence of a pyramidal number is not a cube, by making use of the known impossibility of $x^3 + y^3 = 3z^3$.

E. Barbette[217] summed the pth powers of consecutive n-gonal numbers, found sums of pth powers of n-gonal numbers equal to a pth power of an n-gonal number, found cases with $n \leqq 6$ in which a sum of two n-gonal numbers is n-gonal, and gave a table of the first 5000 triangular numbers.

H. Brocard[218] solved $10\triangle_x + \triangle_y = z^2$ for x and made the radical rational.

L. Aubry[219] noted that $\triangle_{x-1}\triangle_x\triangle_{x+1} = \square$ if $(x - 1)(x + 2) = 2y^2$, whence $u^2 - 8v^2 = 1$, where $2x + 1 = 3u$, $y = 3v$. The solutions are known to be $u = 1, 3, 17, \cdots, u_n = 6u_{n-1} - u_{n-2}$.

A. Gérardin[220] collected recent problems on triangular and pentagonal numbers. He noted (p. 70) that

$$3^{2n}x = \triangle_a - \triangle_b, \qquad a = 3^nx + (3^n - 1)/2, \qquad b = 3^nx - (3^n + 1)/2.$$

Let a, b become c, d when $x = y^2$; then

$$\triangle(c) + \triangle(d) = \triangle(d - 3^ny) + \cdot\triangle(d + 3^ny).$$

---

[211] Sphinx-Oedipe, 1906–7, 31, 46.

[212] Il Pitagora, Palermo, 15, 1908–9, 15–17. In Suppl. al Periodico di Mat., 1908, fasc. 5–6; there is a proof by triangular numbers.

[213] *Ibid.*, 16, 1909–10, 135–6.

[214] Math. Quest. Educ. Times, (2), 15, 1909, 44–5.

[215] Niedere Zahlentheorie, 2, 1910, 1–14.

[216] Nouv. Ann. Math., (4), 10, 1910, 83.

[217] Les sommes de p-ièmes puissances distinctes égales à une p-ième puissance, Liège, 1910, 154 pp. Extract by Barbette.[224]

[218] Sphinx-Oedipe, 6, 1911, 29–30.

[219] *Ibid.*, 187–8. Problem of Lionnet, Nouv. Ann. Math., (3), 2, 1883, 310.

[220] Sphinx-Oedipe, 1911, 40–3, 57–8, 81–6, 113–21, 129–32.

He treated (pp. 97–101) the decomposition of various types of numbers into a sum of three triangular numbers.

The ordinary definitions of polygonal and figurate numbers were repeated by L. Tenca[221] and E. A. Engler.[222]

L. Tits[223] solved Emmerich's[189] equation for y, made the radical rational and was led in both cases to $t^2 - 8v^2 = 1$.

E. Barbette[224] gave many numerical examples in which a sum of n-gonal numbers equals an n-gonal number.

L. von Schrutka[225] found that $\frac{1}{2}\{(\frac{1}{2}T)^2 + U^2\}$ is not expressible in a single way as a sum of two numbers of the form $T(x^2 + x)/2 + Ux$ unless $T/2 = 3$ or 5. In the first case it is shown that, if p is a prime $\equiv 5$ (mod 12), $(p - 2)/3$ can be expressed in one and but one way as a sum of two 8-gonal numbers $3x^2 - 2x$. He gave an analogous theorem for 12-gonal numbers $5x^2 - 4x$, and one for numbers $5x^2 - 2x$.

A. Gérardin[226] solved $n^2 + 2^\theta n = \triangle_x$ for x by setting $n = xp/q$. He (p. 128) reduced $\triangle_x \triangle_y = \triangle_{x^2+x}$ to $2\triangle_x + 1 = \triangle_y$ and noted the solutions $\triangle_x = 10, 45$, $\triangle_y = 21, 91$.

L. Bastien[227] noted that $x^4 - y^4 = \triangle_z$ if $z = x^2 + y^2$ and $x^2 - 3y^2 = 1$, or if $z = (x^2 + y^2)/\lambda$, $z + 1 = 2\lambda(x^2 - y^2)$ or vice versa, whence

$$(2\lambda^2 - 1)x^2 - (2\lambda^2 + 1)y^2 = \pm \lambda.$$

G. Métrod[228] noted that $\triangle_u - \triangle_v = x^3$ if $(u - v)(u + v + 1) = 2x^3$, whence $2x^3$ is to be expressed as a product of two distinct factors, one even and one odd.

F. Mariares[229] noted that the sum of 1, 2, $\cdots$, n is $n(n + 1)/2$ since the sum duplicated makes a rectangle of n by $n + 1$. Again,

$$1 + 3 + 6 + \cdots + \frac{n(n + 1)}{2} = 2^2 + 4^2 + \cdots + \left(2 \cdot \frac{n}{2}\right)^2$$

or

$$1 + 3 + 6 + \cdots = 1^2 + 3^2 + \cdots + \left(2 \cdot \frac{n + 1}{2} - 1\right)^2,$$

according as n is even or odd. Hence

$$\triangle_1 + \triangle_2 + \cdots + \triangle_{n-1} + \triangle_1 + \cdots + \triangle_n = \sum_{k=1}^{n} k^2.$$

Numbers simultaneously triangular and pentagonal have been treated.[230]

[221] Il Boll. di Mat. Sc. Fis. Nat., 12, 1910–11, No. 1, p. 16, No. 3, p. 24.

[222] Trans. St. Louis Acad. Sc., 20, 1911, 37–57.

[223] Mathesis, (4), 1, 1911, 74–5.

[224] L'enseignement math., 14, 1912, 19–30. Cf. Barbette.[217]

[225] Monatshefte Math. Phys., 23, 1912, 267–273.

[226] Sphinx-Oedipe, 8, 1913, 110, 121–2 (1907–8, 173; 1911, 75).

[227] Ibid., 156, 172–3.

[228] Ibid., 174.

[229] Revista Soc. Mat. Española, 2, 1913, 333–5.

[230] Mathesis, (4), 3, 1913, 20–22, 80–81. Cf. Euler.[70]

*S. Minetola[231] gave a combinatory definition of the numbers in Tartaglia's triangle.

N. Alliston and J. M. Child[232] proved that no triangular number > 1 is a biquadrate.

An anonymous writer[233] proved that if a number $4n + 1$ is a $\boxed{2}$, it is a sum of two triangular numbers $c(c + 1)/2$ and $d(d + 1)/2$, where d may be negative; then c and d are of the same parity.

G. Métrod[234] stated that, if p_q^n is the nth polygonal number of q sides, the g.c.d. of p_q^n and p_q^{n+1} is the g.c.d. of $n + 1$ and $q - 3$ unless the latter are even and then is the g.c.d. of $(n + 1)/2$ and $q - 3$. The g.c.d. of p_q^n and p_{q+1}^n is n or $n/2$ according as n is even or odd.

A. Gérardin[234a] noted that $2\triangle_x - 1$ is a prime for $x \leqq 9$. He[234b] gave a series for $\triangle_x \cdot \triangle_y = \triangle_z^2$ with the law of recurrence $z_{n+1} = 6z_n - z_{n-1} + 2$, $z_0 = 3$, $z_1 = 20$. He[234c] gave a general solution of $\triangle_a + \triangle_b = e^2 + 2\triangle_d$, a special case having been noted by Euler[59], and noted the examples $a = 2s+1$, $b = 4s$, $d = 3s$, $e = s + 1$; $a = 6s + 2$, $b = 4s - 1$, $d = 5s + 1$, $e = s - 1$.

E. Bahier[235] found sets of three m-gonal numbers in arithmetical progression: $p_m^\lambda + p_m^\nu = 2p_m^\mu$. Multiply each p by $8(m - 2)$ and add $(m - 4)^2$ to each product. By Diophantus' relation (1), we get

$$P_\lambda^2 + P_\nu^2 = 2P_\mu^2, \qquad P_\lambda \equiv (m - 2)(2\lambda - 1) + 2.$$

Hence, by Ch. XIV,

$$P_\lambda = \pm (x^2 - 2xy - y^2), \qquad P_\mu = x^2 + y^2, \qquad P_\nu = x^2 + 2xy - y^2.$$

Then λ, μ, ν are found in terms of x, y, m by use of the above equation defining P_λ. The conditions that λ, μ, ν be positive integers are discussed at length.

S. Ramanujan[235a] obtained expressions for the number of representations of n as a sum of $2s$ triangular numbers.

Notes* from l'Intermédiaire des Mathematiciens.

A. Boutin,[236] 1, 1894, 91; 2, 1895, 31, noted that the square of each term of the series 0, 1, 6, 35, $\cdots$, $u_n = 6u_{n-1} - u_{n-2}$, $\cdots$ is a triangular number $\triangle$, and stated that the $\triangle$'s in this series (viz., 0, 1, 6 up to u_{24}) are the only $\triangle$'s whose square is a $\triangle$. He gave all solutions $x = 8, 800, \cdots$ of $x^2 + \triangle_x = \square$ and stated that $y^3 \pm 1 = \triangle_x$ only for $y = 1, 3, 16, 20$; $x = 0, 1, 7, 90, 126$. An incorrect solution of the latter by E. Fauquem-

---

[231] Boll. di Matematica, Roma, 12, 1913, 214–22.
[232] Math. Quest. Educ. Times, 25, 1914, 83–5.
[233] Nouv. Ann. Math., (4), 14, 1914, 16–18.
[234] Sphinx-Oedipe, 9, 1914, 5.
[234a] Ibid., p. 41.
[234b] Ibid., p. 75, p. 146.
[234c] Ibid., p. 129.
[235] Recherche . . . Triangles Rectangles en Nombres Entiers, 1916, 217–233.
[235a] Trans. Cambridge Phil. Soc., 22, 1918, 269–272.
*For a more extended account see Gérardin.[220] The present notes were obtained independently.
[236] Jour. de math. élém., (4), 4, 1895, 222. Cf. Lionnet.[158]

bergue, 4, 1897, 159–162, was corrected later, 5, 1898, 257.   P. F. Teilhet, 11, 1904, 11–12, verified that aside from 0, 1, 6 there is no $\triangle$ with fewer than 660 digits whose square is a $\triangle$.

G. de Rocquigny, 2, 1895, 394, noted that every triangular number except 1 and 6 is a sum of three, each $\neq$ 0, since[237]

$$\triangle(3p - 1) = 2\triangle(2p - 1) + \triangle(p), \qquad \triangle(3p) = 2\triangle(2p) + \triangle(p - 1),$$
$$\triangle(3p + 1) = \triangle(2p) + \triangle(2p + 1) + \triangle(p).$$

On $\triangle$'s expressed as a sum of two or three $\triangle$'s, see 4, 1897, 158.   For solutions of $(x + 1)^3 - x^3 = \triangle_y$, see 4, 1897, 262–4; 5, 1898, 18, 110–1 (and Mathesis, (2), 8, 1898, 126).

E. Fauquembergue, 4, 1897, 209, noted that $\triangle_x + \triangle_y = z^2$ is equivalent to $(2x + 1)^2 + (2y + 1)^2 = (2z + 1)^2 + (2z - 1)^2$, which by Euler's formula for the product of two sums of two squares has the solution $2x + 1 = ac + bd$, $2y + 1 = bc - ad$, if $bc + ad = ac - bd + 2$.   Cf. Gérardin[12] of Ch. XXIV.   A. Palmström, 210, noted that the problem is equivalent to

$$(x + y)(x - y + 1) = 2(z + y)(z - y).$$

On $\triangle_x + \triangle_y = z^3$ see 7, 1900, 250.   E. B. Escott, 11, 1904, 82, noted that $\triangle_x + \triangle_y = z^5$ is equivalent to $(2x + 1)^2 + (2y + 1)^2 = 2(4z^5 + 1)$, a necessary and sufficient condition for which is that every prime factor of $4z^5 + 1$ be of the form $4n + 1$; and gave solutions for $z = 1, 4, 6, 9, 12, 16$.

On $\triangle_x = y^2 + z^2$ see 3, 1896, 248; 4, 1897, 129–132, 255.

Any number N is a sum of three pentagons $(3x^2 \pm x)/2$ since

$$24N + 3 = \sum_3 (6x \pm 1)^2$$

is solvable, 4, 1897, 157.   On $\triangle' + \triangle'' = \triangle$, 4, 1897, 158; 5, 1898, 70. The sum $n(n + 1)(n + 2)/6$ of the first n $\triangle$'s is a $\triangle$ for $n = 1, 3, 8, 20, 34$, but for no further $n < 316$, 4, 1897, 159; 6, 1899, 176; 7, 1900, 192; 16, 1909, 236; 17, 1910, 110; and is a square for $n = 1, 2, 48$, but for no others $< 10^{12}$, 9, 1902, 279; 10, 1903, 235.   The sum $n(n + 1)(2n + 1)/6$ of the first n squares is a $\triangle$ for $n = 1, 5, 6, 85$ by 6, 1899, 175; 7, 1900, 211; 9, 1902, 278.

P. Tannery, 5, 1898, 280, and C. Berdellé, 7, 1900, 279, gave algebraic and geometric proofs that, aside from 6, every p-gonal number is a sum of $p - 2$ triangular numbers > 0.

A prime $6n + 1 = 3p^2 + q^2$ is a sum of 3 $\triangle$'s > 0, 4, 1897, 119.   Since a prime $8n + 1$ equals $8m^2 + (2p + 1)^2$, it equals $\triangle_{2m} + \square + P$, where $P = m(6m - 1)$ is pentagonal, 8, 1901, 183.

G. de Rocquigny, 7, 1900, 65, 195; 8, 1901, 52; 9, 1902, 116, 176, 230; 10, 1903, 5–6, 40, 122, 205–6, 285, 300–2; 11, 1904, 99, 150, 158, 163–4, 189, 214, 237; 15, 1908, 181, stated many theorems of the following type: every sixth power is a sum of a square, cube and triangular (or hexagonal) number; every number > 7 is a sum of three $\triangle$'s and three squares each $\neq$ 0.   A. Gérardin, 18, 1911, 177–184, 199, 275, discussed these theorems.

---

[237] Same by R. W. D. Christie, Math. Quest. Educ. Times, 69, 1898, 48.

G. Picou, 9, 1902, 115, noted that $a^2 - b(b+1)/2 = 2^{2n}$ for

$$a = 2^{n+1} + 2^n \pm 1, \qquad b = 2^{n+2} + 1 \qquad \text{or} \qquad 2^{n+2} - 2.$$

H. Brocard, 10, 1903, 24–6, noted the solution

$$a = (9\cdot 2^n - 2)/7, \qquad b = 8(2^n - 1)/7.$$

P. F. Teilhet, 10, 1903, 240–1, gave a somewhat general discussion.

That $1 + 6\triangle \neq$ cube $\neq 0$, see 10, 1903, 97, 197.

P. Jolivald, 12, 1905, 16, 152, gave an erroneous proof that unity is the only number simultaneously a $\triangle$, square and hexagon. As noted by M. Rignaux, 24, 1917, 80–1, a hexagonal number $r(2r-1)$ is triangular, so that we have only to solve $r(2r-1) = y^2$, whence $8y^2 + 1 = \square$, whose solution is known.

A product of 3 consecutive $\triangle$'s may be a square, 11, 1904, 158.

H. B. Mathieu, 16, 1909, 34, gave identities showing that the square of any number $\neq 1$ [4, 16], and not a multiple of 3, is a sum of a $\triangle$ and a square, each not zero [three $\triangle$'s].

A. Arnaudeau, 18, 1911, 132, deposited with the library of the Institute of France the manuscript of his unpublished table of triangular numbers.

A. Gérardin, 1911, 205–7, gave solutions of $x^4 + y^4 + z^4 = 2T^2/t^2$, where T and t are triangular numbers $\triangle$. He cited, 273, Fuss'[67] note giving $9x+5$, $9x + 8$ as linear forms of numbers not a sum of two $\triangle$'s.

To decompose $(n + 1)^5 - n^5$ into three $\triangle$'s see 19, 1912, 37, 104–5. For

$$(x + 1)^3 + (x + 2)^3 + \cdots + (x + m)^3 = \triangle_{x+m}^2 - \triangle_x^2,$$

see 19, 1912, 114. L. Aubry, 19, 1912, 231; 20, 1913, 108, noted that $\triangle_x^2 - \triangle_y^2 = z^3$ for x, $y = (8m^4 \pm 12m^3 - 4m^2 - 1)/3$. A. S. Monteiro, 20, 1913, 18–20, obtained solutions from the fact that the sum of the cubes of any number of consecutive integers equals the difference of the squares of two $\triangle$'s.

R. Niewiadomski,[238] 20, 1913, 5–6, gave many algebraic identities between polygonal numbers, also expressions for n^k, $n^3 + 1$, $n^3 + (n + 1)^3$, etc., as polygonal numbers.

U. Alemtejano (a pseudonym), 21, 1914, 169, stated that if $4m + 1$ is a sum of two squares, m is a sum of two $\triangle$'s, and conversely, since

$$4(\triangle_n + \triangle_a) + 1 = (n + a + 1)^2 + (n - a)^2.$$

Also, 9 is the only number $4\triangle_n + 5$ which is a square of a prime and not a sum of two squares. Again, $\triangle_{2n+a} + \triangle_{a-1} - n = n^2 + (n + a)^2$. Proofs by L. Aubry, 22, 1915, 69. Alemtejano, 22, 1915, 8, gave

$$\{4(\triangle_n + \triangle_a) + 1\}^2 = (2a + 1)^2(2n + 1)^2 + \{4(\triangle_n - \triangle_a)\}^2.$$

Several, 22, 1915, 167–8, proved that every square is expressible in the form $\triangle_u - 2\triangle_v$ in an infinitude of ways. On the last digits of $\triangle$, see 22, 1915, 235–6.

---

[238] Also in Wiadomsci Mat., Warsaw, 17, 1913, 91–98.

A. Gérardin, 21, 1914, 133–5, considered numbers expressible simultaneously in the form $(x + 1)(x + 2) \cdots (x + p)/p!$ for $p = 2, 3, \cdots$. He, 22, 1915, 203–5, considered the representation of numbers by $x^2 + y^2 + z^2 + w$, where w is polygonal.

The question [Meyl[136]] of tetrahedral numbers which are squares reduces to $N^3 - N = 6n^2$, which was treated incompletely by L. Aubry, 26, 1919, 85–87.

A. Boutin, 26, 1919, 35, 123, proved that no number is simultaneously triangular, hexagonal, and a square.

For minor remarks on triangular numbers, see Glaisher[88] of Ch. III; Euler[12] of Ch. VII; Réalis[53] and paper 8 of Ch. XIII; Pepin,[193] and Cunningham[282] (on $\triangle_x = c\triangle_y$) of Ch. XXI; Mathieu[282] of Ch. XXII.

In Vol. I of this History were quoted theorems on triangular numbers by G. W. Leibniz, p. 59; V. Bouniakowsky, pp. 283–4; R. Lipschitz, pp. 291–2; E. Barbette, p. 373; H. Brocard, p. 425; and P. Jolivald, p. 427.

Papers on polygonal or figurate numbers not available for report.

G. U. A. Vieth, Ueber fig. Zahlen, Progr., Dessau, 1817.
J. P. L. A. Roche, Dém. nouv. des formules des piles de boulets, Toulon, 1827.
H. Anton, Arith. Reihen höh. Ord. u. die fig. Z., Progr. Öls, 1850.
A. Wiegand, Trigonaltriaden in arith. Progres., Halle, 1850.
J. Van Cleeff, Verhandeling over de polygonaal of veelhoekige getallen, Groningen, 1855.
N. Nicolaïdès, Les Mondes, 7, 1865, 693; 8, 1865, 615, 708.
J. L. A. Le Cointe, Les Mondes, 8, 1865, 707.
Soufflet, Les Mondes, 13, 1867, 336 [last 3 papers on fig. numbers].
J. Talir, Arith. Reihe höh. Ord. u. fig. Z., Progr., Waidhofen, 1872.
G. de Rocquigny-Adanson, Les nombres triang., Moulins, 1896.

CHAPTER II.

LINEAR DIOPHANTINE EQUATIONS AND CONGRUENCES.

SOLUTION OF $ax + by = c$.

The Hindu Aryabhatta[1] (fifth century or earlier) knew a general method of solving indeterminate equations of the first degree. The original of his treatise (on astronomy mainly) has been lost. Such a method of solution is given in outline by Brahmegupta without the clear details of the later presentation by Bháscara.

Brahmegupta[2] (born 598 A.D.) gave the following rule to find a constant " pulverizer." From the given multiplier and divisor, remove their greatest common divisor (found by mutual division). The thus reduced multiplier and divisor are mutually divided until the residue unity is obtained, and the quotients are written in order. Multiply the residue unity by a number chosen so that the product less one (or plus one, if there be an odd number of quotients) shall be exactly divisible by the divisor which produced the residue unity. After the above listed quotients place this chosen number and after it the quotient just obtained. To the ultimate add the product of the penultimate by the next preceding term [etc.]. The number found, or its residue after division by the reduced divisor, is the constant pulverizer.

Thus if 3 and 1096 are the reduced multiplier and divisor, the single quotient is 365. Multiply the residue unity by the chosen number 2 and add 1. Dividing the sum by 3, we get the quotient 1. Hence the series is 365, 2, 1, so that the pulverizer is $1 + 2 \cdot 365 = 731$. [We have $3 \cdot 731 - 1 = 2 \cdot 1096$.]

Again (§ 27, p. 336), let the reduced dividend [multiplier] and divisor be 137 and 60, while the augment or additive quantity is 10. By reciprocal division of 137 and 60, we get the quotients 0, 2, 3, 1, 1 and last two remainders 8 and 1. Since the augment is now positive and the number of quotients is odd and since $1 \cdot 9 - 1$ is divisible by 8, we select 9 as the chosen number. The constant pulverizer is said to be found as before. Its product by 10 is divided by 60 to give the desired multiplier 10; $10 \cdot 137 + 10 = 60 \cdot 23$.

There occur various problems (§§ 52–60, pp. 348–360) on astronomical time leading to a linear equation in two or more variables, special values being arbitrarily assigned to all but two of the variables. One equation is $6y - 136c = 266$; without detail, the constant pulverizer is said to be 2 and the multiplier $4 = c$, whence the quotient gives $y = 135$.

Mahāvirācārya[3] (about 850 A.D.) gave a process essentially that due to Brahmegupta, though not requiring that the initial division be continued until the remainder unity is reached. To find x such that $31x - 3$ is

---

[1] Algebra, with arithmetic and mensuration, from the Sanscrit of Brahmegupta and Bháscara, translated by H. T. Colebrooke, London, 1817, p. x.

[2] Brahme-sphut'a-sidd'hánta, Ch. 18 (Cuttaca = algebra), §§ 11–14. Colebrooke,[1] pp. 330–1.

[3] Ganita-Sara-Sangraha; described by P. V. S. Aiyar, Jour. Indian Math. Club, 2, 1910, 216–8.

divisible by 73, employ

$$31 = 0 \cdot 73 + 31, \qquad 73 = 2 \cdot 31 + 11, \qquad 31 = 2 \cdot 11 + 9,$$
$$11 = 1 \cdot 9 + 2, \qquad 9 = 4 \cdot 2 + 1.$$

The least remainder of odd rank is 1. Choose a number $a = 5$ such that $a \cdot 1 - 3$ is divisible by the last divisor 2, the quotient being 1. By use of 5, 1 and the quotients 2, 2, 1, 4 after the first, we derive

$$\begin{array}{ccccc} 2 & 2 & 1 & 4 & 5 \quad 1 \end{array}$$
$$172 = 2 \cdot 73 + 26, \qquad 73 = 2 \cdot 26 + 21, \qquad 26 = 1 \cdot 21 + 5, \qquad 21 = 4 \cdot 5 + 1.$$

A smaller answer than 172 is given by $172 - 2 \cdot 73 = 26$.

In the second example, $63x + 7$ is to be made a multiple of 23. Here

$$63 = 2 \cdot 23 + 17, \qquad 23 = 1 \cdot 17 + 6, \qquad 17 = 2 \cdot 6 + 5,$$
$$6 = 1 \cdot 5 + 1, \qquad 5 = 4 \cdot 1 + 1,$$

the division being carried an extra step so as to yield the last remainder of odd rank. Here $a = 1$ makes $a \cdot 1 + 7$ divisible by the last divisor 1. Discarding the first quotient, we have 1, 2, 1, 4, 1, 8 and then get 51, 38 13, 12. Since $51 = 2 \cdot 23 + 5$, an answer is 5.

Bháscara Áchárya[4] (born, 1114) gave detailed methods of finding a pulverizing multiplier (Cuttaca) such that if a given dividend be multiplied by it and the product added to a given additive quantity, the sum will be exactly divisible by a given divisor.

First (§§ 248–252), we reduce the dividend, divisor and additive by their g.c.d. If a common divisor of the dividend and divisor does not divide also the additive, the problem is impossible.

Next (§§ 249–251), divide mutually the reduced dividend and divisor until the remainder unity is obtained. Write the quotients in order, after them write the additive, and after it zero. To the last term add the product of the penult by the next preceding number. Reject the last term and repeat the operation until only two numbers are left. The first of these is abraded by the reduced dividend, and the remainder is the desired quotient. The second of the two, abraded by the reduced divisor, is the desired multiplier.

Example (§ 253): Dividend 17, Divisor 15, Additive 5. The quotients are 1, 7, so that the series is 1, 7, 5, 0. Since $0 + 7 \cdot 5 = 35$, the new series is 1, 35, 5. The final series is 40, 35. Abrading them by multiples of 17 and 15 respectively, we get 6 and 5 as the desired quotient and multiplier $[17 \cdot 5 + 5 = 15 \cdot 6]$.

---

[4] Lílávatí (Arithmetic), Ch. 12, §§ 248–266, Colebrooke[1], pp. 112–122. [It is nearly word for word the same as Ch. II of Bháscara's Víja-gañita (Algebra), §§ 53–74, Colebrooke,[1] pp. 156–169; Bija Ganita or the Algebra of the Hindus, transl. into English by E. Strachey of the Persian transl. of 1634 by Ata Alla Rasheedee of Bhascara Acharya, London, 1813, Ch. √ of Introduction, pp. 29–36. Lilawati or a Treatise on Arith. & Geom. by Bhascara Acharya, transl. from the original Sanskrit by John Taylor, Bombay, 1816, Part III, Sect. I, p. 111; the Persian transl. in 1587 by Fyzi omitted the chapters on indeterminate problems. Lilawati was the name of Bhascara's daughter.]

In case (§ 252) the number of quotients is odd, the numbers found by the above rule must be subtracted from their respective abraders to give the true quotient and multiplier. Thus (§ 255) for Dividend 10, Divisor 63, Additive 9, the successive series are 0, 6, 3, 9, 0 [and 0, 6, 27, 9 and 0, 171, 27, and $27 = 2 \cdot 10 + 7$, $171 = 2 \cdot 63 + 45$], so that $10 - 7 = 3$ is the quotient and $63 - 45 = 18$ is the multiplier [check: $10 \cdot 18 + 9 = 3 \cdot 63$].

Concerning a "constant pulverizer" (§ 263, pp. 119–120), we may solve the first example above by first treating the related problem: Dividend 17, Divisor 15, Additive 1, then multiply the deduced multiplier 7 and quotient 8 by the former additive 5, abrade and get 6 and 5 as the quotient and multiplier when the additive is 5.

As to a "conjunct pulverizer" (§§ 265–6, p. 122), if there be a fixed divisor and several multipliers, make the sum of the latter the dividend, the sum of the remainders the subtractive quantity, and proceed as before. Thus, to find a number whose products by 5 and 10 give the respective remainders 7 and 14 when divided by 63, take Dividend $5 + 10$, Divisor 63, Subtractive $7 + 14$; reduced Dividend, Divisor and Subtractive are 5, 21, 7; the desired multiplier is 14.

Bháscara[5] gave a rule for solving linear equations in two or more unknowns. In case there are k equations, eliminate $k - 1$ of the unknowns and proceed with the single resulting equation as follows. Assign arbitrarily special values to all but two of the unknowns. In the resulting equation in two unknowns, solve for one in terms of the other and render it integral by use of the pulverizer.

For example, of two equally rich men, one has 5 rubies, 8 sapphires, 7 pearls and 90 species; the other has 7, 9, 6 and 62 species; find the prices (y, c, n) of the respective gems in species. Thus

$$5y + 8c + 7n + 90 = 7y + 9c + 6n + 62, \qquad y = \frac{-c + n + 28}{2}.$$

Take $n = 1$, and use the method of a pulverizer to find c so that $y = (-c + 29)/2$ shall be integral. We get

$$c = 1 + 2p, \qquad y = 14 - p,$$

where p is arbitrary. For $p = 0, 1$, we get $(y, c, n) = (14, 1, 1)$, $(13, 3, 1)$.

Again (§ 161, pp. 237–8), what three numbers being multiplied by 5, 7, 9 respectively, and the products divided by 20, have remainders in arithmetical progression with the common difference 1, and quotients equal to remainders? Call the numbers c, n, p; the remainders $y, y + 1, y + 2$. Thus

$$5c - 20y = y, \qquad y = 5c/21;$$
$$7n - 20(y + 1) = y + 1, \qquad y = (7n - 21)/21;$$
$$9p - 20(y + 2) = y + 2, \qquad y = (9p - 42)/21.$$

By the first two values of y, $c = (7n - 21)/5$. By the last two,

$$n = (9p - 21)/7,$$

---

[5] Víja-gañita (Algebra), §§ 153–6; Colebrooke,[1] pp. 227–232.

which by use of the pulverizer gives $n = 9l + 6$, $p = 7l + 7$. Then $c = (63l + 21)/5$, which by the pulverizer gives $c = 63h + 42$, $l = 5h + 3$. Hence $n = 45h + 33$, $p = 35h + 28$, $y = 15h + 10$. Since the quotient equals the remainder, which cannot exceed the divisor, we must take $h = 0$.

What two numbers, except 6 and 8, being divided by 5 and 6 have the respective remainders 1 and 2; while their difference divided by 3 has the remainder 2; their sum divided by 9 has the remainder 5; and their product divided by 7 leaves 6 (§ 163, p. 239)? The conditions other than the last give $45p + 6$ and $54p + 8$ as the numbers. As the product is quadratic, take $p = 1$ [provisionally]. Abrading the product by multiples of 7, we get $3p + 2$, which must equal $7l + 6$. By the pulverizer, $p = 7h + 6$, and the second number is $378h + 332$. The additive $(45p)$ of the first number multiplied by $7h$ is its present additive, so that the first number is $315h + 51$.

What number multiplied by 9 and 7 and the products divided by 30 yields remainders whose sum increased by the sum of the quotients is 26 (§ 164, p. 240)? Answer, 27.

What number multiplied by 23 and the product divided by 60 and 80 has 100 as the sum of the remainders (§§ 166–7, p. 241)? Taking 40 and 60 as the remainders, we get the number $240l + 20$. Taking 30 and 70, we get $240l + 90$; etc.

Bachet de Méziriac[6] stated that if A and B are any relatively prime integers, we can find a least integral multiple of A which exceeds an integral multiple of B by a given integer J [i. e., solve $Ax = By + J$]. Proof was given in the 1624 edition, pp. 18–24. It suffices to solve $AX = BY + 1$. Bachet employed notations for 18 quantities, making it difficult to hold in mind the relations between them and so obtain a true insight into his correct process. Hence we shall here carry out in clearer form his process for his example $A = 67$, $B = 60$. Subtract the smaller number B as many times as possible from the larger number A, to give a positive remainder C. If $C = 1$, A itself is the desired multiple of A which exceeds a multiple of B by unity. Next, let $C > 1$ and subtract C from B as many times as possible, continuing until the remainder 1 is reached:

$$(1) \quad 67 = 1 \cdot 60 + 7, \qquad 60 = 8 \cdot 7 + 4, \qquad 7 = 1 \cdot 4 + 3, \qquad 4 = 1 \cdot 3 + 1.$$

From the last equation we deduce

$$(2) \qquad\qquad 3 \cdot 3 = 2 \cdot 4 + 1,$$

by the rule that if $a = mb + 1$ then $ab + 1 - a$ is the least multiple of b which exceeds by unity a multiple of a. Multiply the third equation (1) by first coefficient 3 in (2) and eliminate the term $3 \cdot 3$ by use of (2); we get

$$(3) \qquad\qquad 3 \cdot 7 = 5 \cdot 4 + 1.$$

---

[6] Clavde Gaspar Bachet, Problemes Plaisans et Delectables, Qui se font par les Nombres, ed. 1, Lyon, 1612, Prob. 5; ed. 2, Lyon, 1624; ed. 3, Paris, 1874, 227–233; ed. 4, 1879; ed. 5, 1884; abridged ed., 1905. See Lagrange.[19]

Multiply the second equation (1) by the coefficient 5 in (3), and eliminate
the term $5 \cdot 4$ by use of (3); we get

(4) $43 \cdot 7 = 5 \cdot 60 + 1.$

Finally, multiply the first equation (1) by the coefficient 43 in (4) and
eliminate the term $43 \cdot 7$ by use of (4); we get

(5) $43 \cdot 67 = 48 \cdot 60 + 1,$

so that the least X is 43 and the corresponding Y is 48.

Bachet's first step, leading to (1), is Euclid's algorithm for finding the
greatest common divisor of A and B. His next steps are the elimination
from equations (1) of the terms in 3, 4, 7, respectively, in a special way
so that negative quantities are not introduced.

John Kersey[7] treated Problems 18 and 21 of Bachet,[6] but "without
following Bachet's very tedious and obscure method of solution." To
solve $9a + 6 = 7b$, start with 6 and by successive additions of 9 form the
series 15, 24, 33, 42, $\cdots$; next, form similarly the multiples 7, 14, 21, 28,
35, 42, $\cdots$ of 7; the common number 42 yields $a = 4$, $b = 6$. Another
method is used for $49a + 6 = 13b$; find the multiple (65) of 13 which just
exceeds $49 + 6$; divide 49 by 13; since in $55 = 65 - 10$, $49 = 39 + 10$,
we have remainders differing only in sign, we add and get 104; then
$b = 104/13$, $a = 2$. If one remainder had been merely a divisor of the
other remainder, we first multiply one of the equations. Neither of these
cases arises for $121a + 5 = 93b$. Then $126 = 186 - 60$, $121 = 93 + 28$,
and we seek c and d such that $93c + 60 = 28d$. After the latter is solved
by the former process, we deduce a and b as in the preceding case. In a
new type of problem, the constant term occurs in the member with the
smaller coefficient, as in $71a + 3 = 173b$. Take $2 \cdot 71$, which increased by 3,
gives a sum < 173. Since $145 = 173 - 28$, solve $173A + 1 = 71B$ as
above to obtain $A = 16$, $B = 39$. Multiply the latter equation by 28
and subtract the former. Thus $173(16 \cdot 28 + 1) = 71 \cdot 39 \cdot 28 + 145$, whence
$b = 16 \cdot 28 + 1 = 449$, $a = 1094$.

Michel Rolle[8] (1652–1719) gave a rule to find integral solutions; he
applied it as follows. For $12z = 221h + 512$, divide the larger coefficient
221 by the smaller 12; the largest integer in the quotient is 18. Set
$z = 18h + p$; we get $12p = 5h + 512$. By the same method [dividing
12 by 5], $h = 2p + s$, $2p = 5s + 512$. By the same method, $p = 2s + m$.
Then $2m = s + 512$, and we have now reached a coefficient which is unity.
Eliminating s and p from

$$s = 2m - 512, \qquad p = 2s + m, \qquad h = 2p + s, \qquad z = 18h + p,$$

we get the desired solution

$$z = 221m - 47104, \qquad h = 12m - 2560.$$

---

[7] The Elements of Algebra, London, I, 1673, 301.
[8] Traité d'Algebre; ou Principes generaux pour resoudre les questions de mathématique,
Paris, 1690, Bk 1. Ch. 7 ("eviter les fractions") pp. 69, 78

But for $111x - 301y = 222$, it is simpler to begin with $301 = 3 \cdot 111 - 32$, rather than with $301 = 2 \cdot 111 + 79$.

Thomas Fantet de Lagny[9] gave examples of a "new method" of solving indeterminate equations. To make $m = (19n - 3)/28$ an integer, double $28n - (19n - 3)$ and subtract the result $18n + 6$ from $19n - 3$; thus $n - 9$ is to be divisible by 28. Hence $n = 9 + 28f$, where f is any integer. Later, he[10] gave (pp. 587–595) the following rule for solving $y = (ax + q)/p$, where a and p are relatively prime, and (as may be assumed) $a < p$, $q < p$: Take a from p as many times as possible and call the remainder r; take r from a as many times as possible and call the remainder t; etc., until the remainder 1 is reached. Then make the same divisions for q and p as were made for a and p, having regard to the signs. According as the last remainder is $- s$ or $+ s$, we have $x = s$ or $p - s$.

L. Euler[11] gave a process to find an integer m such that $(ma + v)/b$ is integral, where $v > 0$. Set $a = \alpha b + c$. Then $A = (mc + v)/b$ must be an integer. Thus $m = (Ab - v)/c$. First, if v is divisible by c, we get a solution by taking $A = 0$. Second, if v is not divisible by c, set $b = \beta c + d$. Then m will be integral if $(Ad - v)/c$ is integral. Thus we set $c = \gamma d + e$, etc. Euler remarked that the process is therefore that of finding the greatest common divisor of a, b, continued until we reach a remainder which divides v. His formula for a solution of $ma + v = nb$ is equivalent to

$$ n = av\left(\frac{1}{ab} - \frac{1}{bc} + \frac{1}{cd} - \frac{1}{de} + \cdots\right), \qquad m = - bv\left(\frac{1}{bc} - \frac{1}{cd} + \frac{1}{de} - \cdots\right), $$

in which the series are continued until we reach a remainder dividing v. For the case a, b relatively prime, these results have been given by C. Moriconi.[12]

N. Saunderson[13] (blind from infancy) gave a method to solve $ax - by = c$, where c is the g.c.d. of a, b. Let $a = 270$, $b = 112$, whence $c = 2$. He employed the equations and successive quotients

$$
\begin{aligned}
1a - 0b &= 270, & 5a - 12b &= 6, \quad 3;\\
0a - 1b &= -112, \ 2; & 17a - 41b &= -2, \quad 2;\\
a - 2b &= 46, \ 2; & 39a - 94b &= 2.\\
2a - 5b &= -20, \ 2;
\end{aligned}
$$

Divide the term 270 of the first equation by the absolute value 112 of the term of the second, to obtain the quotient 2. Multiply the second equation by 2 and add to the first; we get the third equation. The division of 112

[9] Nouveaux Elemens d'Arithmetique et d'Algebre, ou Introduction aux Mathematiques, Paris, 1697, 426–435.

[10] Analyse générale; ou méthodes nouvelles pour résoudre les problèmes de tous les Genres & de tous les Degrez à l'infini, Paris, 1733, 612 pp. Same in Mém. Acad. Roy. des Sciences, 11, 1666–1699 [1733], année 1720, p. 178.

[11] Comm. Acad. Petrop., 7, 1734–5, 46–66; Comm. Arith. Coll., I, 11–20.

[12] Periodico di Mat., 2, 1887, 33–40. Cf. C. Spelta, Giornali di Mat., 33, 1895, 125.

[13] The Elements of Algebra, Cambridge, 1, 1740, 275–288. The solution of the first problem was reproduced by de la Bottiere, Mém. de Math. et Phys., présentés . . . divers savans, 4, 1763, 33–41. Cf. Lagrange.[22]

by 46 gives the quotient 2; the product of the third equation by 2 when added to the second gives the fourth equation; etc. But on dividing 6 by 2 we use 2 and not the exact quotient 3, since the latter would lead to an equation $56a - 135b = 0$ with constant term zero.

Our sixth and seventh equations each solve the problem. Other solutions follow by adding to either equation $56a - 135b$ one or more times.

The process must succeed since the formation of the constant terms is identical with Euclid's process to find the g.c.d. of a, b.

The determinant of the coefficients in any two successive equations of the above set is ± 1. From the pairs of coefficients form the fractions*

$$\frac{0}{1}, \ \frac{1}{0}, \ \frac{2}{1}, \ \frac{5}{2}, \ \frac{12}{5}, \ \frac{41}{17}, \ \frac{94}{39}.$$

They are alternately less than and greater than a/b and converge to it; if f and F are two successive fractions of the set, a/b lies between them and differs less from F than from f. Also a/b is nearer to F than to any fraction whose denominator is less than that of F. This method of approximating to fractions is attributed to Cotes and is said to be simpler than the methods of Wallis and Huygens.[17]

L. Euler[14] proved that if n and d are relatively prime, $a + kd\,(k = 0, 1, \cdots,$ $n - 1)$ give n distinct remainders when divided by n, so that the remainders are $0, 1, \cdots, n - 1$ in some order. Since one remainder is zero, $a + xd = yn$ is solvable in integers.

W. Emerson[15] used the first method of de Lagny[9] to solve $ax = by + c$. Let d and f be the remainders obtained by dividing b and c by a. Subtract some multiple of $(dy + f)/a$ from the nearest multiple of y. The resulting "abridged" fraction or some multiple of it is to be subtracted from the nearest multiple of y, etc., until the coefficient of y is unity. Thus $x = (14y - 11)/19$ is subtracted from y; the product of the difference by 4 is subtracted from y; we get $(y + 6)/19$, an integer p, whence $y = 19p - 6$. The same rule and same example was given by John Bonnycastle.[16]

J. L. Lagrange,[17] to find integers p_1 and q_1 satisfying $pq_1 - qp_1 = \pm 1$, where p, q are relatively prime, reduced p/q to a continued fraction (§ 29, p. 423). As noted by Chr. Huygens, De scriptio automati planetarii, 1703, we get a series of fractions converging towards p/q, alternately less than and greater than p/q. Hence take p_1 equal to the numerator and q_1 equal to the denominator of the convergent immediately preceding p/q. Then $pq_1 - qp_1 = + 1$ or $- 1$ according as $p_1/q_1 <$ or $> p/q$. To apply (§ 8) to $py - qx = r$, where p, q may be assumed relatively prime, multiply the former equation by $\pm r$ and subtract. Thus

$$x = mp \pm rp_1, \qquad y = mq \pm rq_1.$$

---

* The last is replaced by a/b if the final quotient had been taken as 3.

[14] Novi Comm. Acad. Petrop., 8, 1760–1, 74; Comm. Arith. Coll., I, 275.

[15] A Treatise of Algebra, London, 1764, p. 215; same paging in 1808 ed.

[16] Introduction to Algebra, ed. 6, 1803, London, 133.

[17] Mém. Acad. Berlin, 23, année 1767, 1769, § 7; Oeuvres, 2, 1868, 386–8.

Lagrange[18] proved as had Euler[14] that, if b and c are relatively prime, there exist integers y and z such that $by - cz = a$. Next, if $y = p$, $z = q$ is one set of solutions, every set of solutions is given by $y = p + mc$, $z = q + mb$. If $a = a'd$, $c = c'd$, where a' and c' are relatively prime, then p is divisible by d, say $p = p'd$. As in the proof of the initial theorem, we can find m such that $p' + mc'$ is divisible by a'. Hence we can always find a value of y which is a multiple ar of a; then z is a multiple $a's$ of a', and $br - c's = 1$. From a set of solutions r, s of the latter, we get $y = ra + mc$, $z = sa' + mb$.

Lagrange[19] noted that his[17] method is "essentially the same as Bachet's,[6] as are also all methods proposed by other mathematicians." To solve $39x - 56y = 11$, employ

$$56 = 1 \cdot 39 + 17, \qquad 39 = 2 \cdot 17 + 5, \qquad 17 = 3 \cdot 5 + 2, \qquad 5 = 2 \cdot 2 + 1, \qquad 2 = 2 \cdot 1.$$

By means of the quotients 1, 2, 3, 2, 2, we get the convergents

$$\frac{1}{1}, \quad \frac{3}{2}, \quad \frac{10}{7}, \quad \frac{23}{16}, \quad \frac{56}{39}.$$

Thus $x = 23 \cdot 11 + 56m$, $y = 16 \cdot 11 + 39m$.

L. Euler[20] employed the method of always dividing by the smaller coefficient, thus following Rolle[8] in essence. For $5x = 7y + 3$,

$$x = y + \frac{2y + 3}{5}.$$

The numerator must be a multiple of 5. Thus $2y + 3 = 5z$,

$$y = 2z + \frac{z - 3}{2}, \qquad z - 3 = 2u,$$

whence $y = 5u + 6$, $x = 7u + 9$. He showed that the process is equivalent to that for finding the greatest common divisor of 5 and 7:

$$
\begin{aligned}
7 &= 1 \cdot 5 + 2, & x &= 1 \cdot y + z, \\
5 &= 2 \cdot 2 + 1, & y &= 2 \cdot z + u, \\
2 &= 2 \cdot 1 + 0, & z &= 2 \cdot u + 3.
\end{aligned}
$$

Jean Bernoulli[21] applied Lagrange's[19] method to find the least integer u giving an integral solution of $A = Bt - Cu$, when B, C are relatively prime, in the special cases $A = \frac{1}{2}C$, $\frac{1}{2}C + 1$, $\frac{1}{2}(C \pm 1)$. For example, if C is even and $A = C/2$, then $u = (B - 1)/2$, $t = C/2$. If C is odd and $A = \frac{1}{2}(C + 1)$, then $u = \frac{1}{2}(B + s - 1)$, $t = \frac{1}{2}(C + r)$, where $Br - Cs = 1$, r/s being the convergent just preceding C/B in the continued fraction for the latter.

[18] Mém. Acad. Berlin, 24, année 1768, 1770, 184–7; Oeuvres, II, 659.

[19] *Ibid.*, 220–3; Oeuvres, II, 696–9. Additions by Lagrange to Vol. 2 of the transl. by Jean III Bernoulli of Euler's Algebra, Lyon, 1774, 517–523 (Euler's Opera Omnia, (1), 1, 1911, 574–7; Oeuvres de Lagrange, VII, 89–95).

[20] Algebra, 2, 1770, §§ 4–23; French transl., Lyon, 2, 1774, pp. 5–29; Opera Omnia, (1), I, 326–339.

[21] Nouv. Mém. Acad. Roy. Berlin, année 1772, 1774, 283–5.

J. L. Lagrange[22] used the method of Saunderson[13] and noted that the process is equivalent to the usual one of converting b/a into a continued fraction. He[23] gave a more popular account [results as in Lagrange[17]].

C. F. Gauss[24] employed the notations

$$B=[\alpha, \beta]=\beta\alpha+1, \qquad C=[\alpha, \beta, \gamma]=\gamma B+\alpha, \qquad [\alpha, \beta, \gamma, \delta]=\delta C+B, \quad \cdots.$$

Apply the g.c.d. process to a and b which are relatively prime and positive, with $a \geqq b$; let $a = \alpha b + c$, $b = \beta c + d$, $c = \gamma d + e$, $\cdots$, $m = \mu n + 1$, so that

$$a = [n, \mu, \cdots, \gamma, \beta, \alpha], \qquad b = [n, \mu, \cdots, \gamma, \beta].$$

Take $x = [\mu, \cdots, \gamma, \beta]$, $y = [\mu, \cdots, \gamma, \beta, \alpha]$. Then $ax = by + (-1)^k$ if k is the number of the terms $\alpha, \beta, \cdots, \mu, n$. Cf. Euler.[38]

Pilatte[25] solved $a_1 x + a x_1 = b$, where a_1 and a are relatively prime, $a > a_1$, by applying the greatest common divisor process:

$$a = a_1 q_1 + a_2, \qquad a_1 = a_2 q_2 + a_3, \qquad \cdots, \qquad a_{n-1} = q_n.$$

Replacing a by its value, we get $x = x_2 - q_1 x_1$, where $x_2 = (b - a_2 x_1)/a_1$ must be integral. Thus $a_2 x_1 + a_1 x_2 = b$. Proceeding similarly with the latter equation, we get $a_3 x_2 + a_2 x_3 = b$, $\cdots$, $x_{n-1} + a_{n-1} x_n = b$. Eliminating $x_{n-1}, x_{n-2}, \cdots$, we get $x = \pm \alpha b \mp a x_n$, where α is an integer determined by the process.

P. Nicholson[26] gave a method best explained by his example

$$y = \frac{500 - 11x}{35} = 14 - \frac{11x - r}{35}, \qquad r = 10.$$

Divide $35x$ by $11x - r$ to get the remainder $2x + 3r$. Then divide $11x - r$ by $2x + 3r$ to get the remainder $x - 16r$, in which the coefficient of x is unity. The remainder 20 from the division of $16r = 160$ by 35 is the least positive x. But in the example

$$y = \frac{200 - 5x}{11} = 18 - \frac{5x - r}{11}, \qquad r = 2,$$

we reach the remainder $x + 2r$ in which the sign is plus; thus $11 - 2r = 7$ is the least x.

G. Libri[27] gave as the least positive integral solution x of $ax + b = cy$, where a and c are relatively prime,

$$\frac{c-1}{2} + \frac{1}{2} \sum_{u=1}^{c-1} \frac{\sin\left\{ 2u\left(b - \frac{a}{2}\right)\frac{\pi}{c} \right\}}{\sin \dfrac{ua\pi}{c}}.$$

[22] Jour. de l'école polyt., cah. 5, 1798, 93–114; Oeuvres, VII, 291–313.

[23] *Ibid.*, cahs. 7, 8, 1812, 174–9, 208–9; Reprint of Leçons élém. sur math., Séances des écoles normales, 1794–5; Oeuvres, VII, 184–9, 216–9.

[24] Disq. Arith., 1801, § 27; Werke, I, 1863, 20; German transl., Maser, 1889, 12–13.

[25] Annales de Math. (ed., Gergonne), 2, 1811–12, 230–7. Cf. E. Catalan, Nouv. Ann. Math., 3, 1844, 97–101.

[26] The Gentleman's Math. Companion, London, 4, No. 22, 1819, 849–60.

[27] Mém. présentés pars divers savants à l'acad. roy. sc. de l'Institut de France, 5, 1833, 32–7 (read 1825); extr. in Annales de Math., ed., Gergonne, 16, 1825–6, 297–307; Jour. für Math., 9, 1832, 172. Cf. A. Genocchi, Nouv. Corresp. Math., 4, 1878, 319–323.

The number of integral solutions x, $0 \leqq x < c$, is

$$\frac{1}{c} \sum_{x=0}^{c-1} \sum_{u=0}^{c-1} \cos \frac{2u(ax + b)\pi}{c}.$$

A. L. Crelle,[28] after proving the existence of solutions of $a_2 x_1 = a_1 x_2 + k$, where a_1 and a_2 are relatively prime [Euler[14]], solved it by setting

$$a_1 = p_1 a_2 + a_3, \qquad a_2 = p_2 a_3 + a_4, \quad \cdots; \qquad x_1 = p_1 x_2 + x_3, \qquad x_2 = p_2 x_3 + x_4, \quad \cdots;$$

also by the modified equations in which the left members are all a_1, or x_1. There are given three more such methods. The sixth method uses a prime factor α_1 of $a_1 = \alpha_1 \beta_1$, and a primitive root π_1 of α_1. There exists an integer ϵ_1 such that $a_2 \pi_1^{\epsilon_1} = z_1 \alpha_1 \pm 1$. Multiply the proposed equation by $\pi_1^{\epsilon_1}$. Thus $z_1 x_1 = \beta_1 \pi_1^{\epsilon_1} x_2 + x_3$, where $x_3 = (k\pi_1^{\epsilon_1} \mp x_1)/\alpha_1$ is an integer. The latter gives $x_1 = \mp (\alpha_1 x_3 - k\pi_1^{\epsilon_1})$. Here x_3 must satisfy $a_2 x_3 = \mp \beta_1 x_2 + z_1 k$, which is treated as was the initial equation.

Crelle[29] considered $ay = bx + 1$, where a, b are relatively prime and > 1. If x_0, y_0 give the least positive solution, the general solution is $x_\mu = \mu a + x_0$, $y_\mu = \mu b + y_0$ ($\mu = 0, \pm 1, \pm 2, \cdots$). If $y_0 < b/2$, the numerators of

$$\frac{y_0}{x_0}, \quad \frac{y_{-1}}{x_{-1}}, \quad \frac{y_1}{x_1}, \quad \frac{y_{-2}}{x_{-2}}, \quad \frac{y_2}{x_2}, \quad \cdots$$

increase alternately by $b - 2y_0$ and $2y_0$, and the denominators alternately by $a - 2x_0$ and $2x_0$, and no one of these fractions differs more from a/b than the next fraction. There are similar theorems on series of fractions involving only positive or only negative subscripts. If $y_\mu/x_\mu - b/a = k > 0$, $v/u - b/a = \lambda > 0$, where $|v| < |y_{\mu+1}|$, $|u| < |x_{\mu+1}|$, then $\lambda > k$. If $\lambda < 0$, he found the number of fractions v/u for which $k > \lambda$, μ being given.

J. P. M. Binet[30] treated $ax - Ay = 1$, $A > a$, by a process for finding the g.c.d. of a and A in which A is always the dividend. On dividing A by a, a_1, a_2, $\cdots$, let p, p_1, p_2, $\cdots$ be the quotients and $- a_1$, $- a_2$, $- a_3$, $\cdots$ the remainders. Then

$$(6) \qquad app_1 \cdots p_{i-1} = a_i + A\{1 + p_{i-1} + p_{i-1}p_{i-2} + \cdots + p_{i-1} \cdots p_2 p_1\}.$$

Let a_n be the divisor when the remainder is zero. Since a_n divides A, it is the g.c.d. of A and a if it divides a. But if a_n is not a divisor of a, proceed as above with a and a_n and call the remainders $- b_1$, $- b_2$, $\cdots$, $- b_{n'}$, the last corresponding to the remainder zero. Then a_n, $b_{n'}$, $c_{n''}$, $\cdots$ form a rapidly decreasing series and one of them will be ± 1. If $a_n = \pm 1$, (6) for $i = n$ gives a relation of the form $aP = \pm 1 + AP_1$.

E. Midy[31] used Euler's[14] result to solve $by - cz = a$ by trial.

---

[28] Abh. Akad. Wiss. Berlin (Math.), 1836, 1–53.

[29] *Ibid.*, 1840, 1–57.

[30] Comptes Rendus Paris, 13, 1841, 349–353; Jour. de Math., 6, 1841, 449–494.

[31] Nouv. Ann. Math., 4, 1845, 146; C. A. W. Berkhan, Lehrbuch der Unbest. Analytik, Halle, 1, 1855, 144; A. D. Wheeler, Math. Monthly (ed., Runkle), 2, 1860, 23, 55, 402–6; L. H. Bie, Nyt Tidsskrift for Mat., Kjobenhavn, (4), 2, 1878, 164; J. P. Gram, *ibid.*, 3, B, 1892, 57, 73; E. W. Grebe, Archiv Math. Phys., 14, 1850, 333–5.

J. A. Grunert[32] solved $bx - ay = 1$ by a process for finding the greatest common divisor of b, a in which the divisor is always a, while the dividend is the sum of b and the preceding remainder, a process due to Poinsot, Jour. de Math., 10, 1845, 48.

V. Bouniakowsky[33] would solve $ax + by = k$ by adjoining $a'x + b'y = h'$, whose coefficients are arbitrary. Set $D = ab' - ba'$, $p = b'/D$, $q = h'/D$, $r = a'/D$. Then $x = kp - bq$, $y = aq - kr$, subject to $ap - br = 1$.

A. L. Crelle[34] gave over 4000 pairs of positive integral solutions $x_1 < a_1$, $x_2 < a_2$, of $a_1 x_2 = a_2 x_1 + 1$, for $a_1 \leqq 120$, $0 < a_2 < a_1$, with a_2 prime to a_1, and indicated methods used to simplify the calculation of the table.

V. Bouniakowsky[35] integrated by parts

$$\int (ax + b)^{m-1}(a'x + b')^{n-1}dx$$

to obtain an identity giving a solution of $b^m X - b'^n Y = 1$, where $x = a'$, $y = a$ is a particular solution of $bx - b'y = 1$. For $m = n = 2$, the identity is

$$(3a^2 a'b - a^3 b')b'^2 - (3aa'^2 b' - a'^3 b)b^2 = (a'b - ab')^3.$$

H. J. S. Smith[36] reported on a recent method to solve $ax = 1 + Py$ [no reference]. Join the origin to the point (a, P). No lattice point (i. e., with integral coordinates) lies on this segment; but on each side of it there is a point lying nearer to it than any other. Let (ξ_1, η_1) and (ξ_2, η_2) be these two points and let $\xi_1/\eta_1 < \xi_2/\eta_2$. Then the ξ's and η's are the least positive solutions of $a\eta_1 - P\xi_1 = 1$, $a\eta_2 - P\xi_2 = -1$.

G. L. Dirichlet[37] solved $ax - by = 1$ by continued fractions, using the algorithm due to Euler.[38]

C. G. Reuschle[39] found the general solution of $ax + by = c$ by combining it with $\alpha x + \beta y = m$, where m is an arbitrary integer, while α and β are integers determined so that $a\beta - b\alpha = \pm 1$ [cf. Bouniakowsky[33]].

J. J. Sylvester[40] noted that the number of positive integers $< pq$ which are neither multiples of p or q nor can be made up by adding together multiples of p and q is $\frac{1}{2}(p - 1)(q - 1)$ if p and q are relatively prime.

H. Brocard[41] solved $ax + by = 1$ by a process of reduction. It suffices to find the residue of a modulo $a - b$ to obtain an equation $x + y = f$ consistent with the given one. Thus, if $b = 563036$, $a = b + 7$, then $a \equiv b \equiv 5 \pmod{7}$, $3 \cdot 5 \equiv 1$, and the given equation may be combined with $x + y = 3$ to get integral solutions x, y. A table gives the successive

[32] Archiv Math. Phys., 7, 1846, 162.
[33] Bull. Acad. Sc. St. Pétersbourg, 6, 1848, 199.
[34] Bericht Akad. Wiss. Berlin, 1850, 141–5; Jour. für Math., 42, 1851, 299–313.
[35] Bull. Cl. Phys.-Math. Acad. Sc. St. Pétersbourg, 11, 1853, 65.
[36] Report British Assoc. for 1859, 228–267, § 8; Coll. Math. Papers, I, 43.
[37] Zahlentheorie, §§ 23–24, 1863; ed. 2, 1871; ed. 3, 1879; ed. 4, 1894.
[38] Comm. Acad. Petrop., 7, 1734–5, 46 (Euler[96]). Novi Comm. Acad. Petrop., 11, 1765, 28; see Euler,[72] Ch. XII. Cf. Gauss.[24]
[39] Zeitschrift Math. Phys., 19, 1874, 272. Same by J. Slavik, Casopis, Prag, 14, 1885, 137; V. Schäwen, Zeitschrift Math. Naturw. Unterricht, 9, 1878, 107 [194, 367].
[40] Math. Quest. Educ. Times, 41, 1884, 21.
[41] Mém. Acad. Sc. Lettres Montpellier, Sec. Sc., 11, 1885–6, 139–234. See p. 153.

values of $x + y$ when $a - b = 1, 2, \cdots, 100$. The paper ends with a twenty page bibliography and history of linear diophantine equations.

C. A. Laisant[42] constructed the points having as abscissas $1, 2, \cdots, p$ and as ordinates the corresponding residues $< p$ modulo p of $r, 2r, \cdots, pr$ (r prime to p). The lattice defined by these points leads to an immediate solution of $rx - pz = a$ since every point of the lattice has the coordinates $x, y = rx - pz$.

W. F. Schüler[43] gave a collection of 374 problems on linear Diophantine equations and an extract from Bachet[6] with a German translation.

E. W. Davis[44] used points with integral coordinates to solve $ay - bx = k$.

P. Bachmann[45] gave an extended account of Euclid's g.c.d. algorithm, continued fractions, and related questions.

A. Pleskot[46] treated $13x + 23y = c$ somewhat as had Rolle[8]:

$$c = 13(x + 2y) - 3y = 3(4x + 7y) + x + 2y,$$
$$4x + 7y = t, \qquad x + 2y = c - 3t, \qquad x = -7c + 23t, \qquad y = 4c - 13t.$$

J. Kraus[47] solved $\alpha x - \alpha' y = c$, where $\alpha' - \alpha = k$ exceeds α and c, by use of $\alpha r_\lambda - r_{\lambda+1} = ka_\lambda$, $0 < r_\lambda < k$, $0 \leqq a_\lambda < \alpha$, $\lambda = 1, 2, \cdots$, thus representing r_λ/k as a number with the digits $a_\lambda, a_{\lambda+1}, \cdots$ to the base α.

P. A. MacMahon[48] proved that, if the continued fraction for λ/μ has a reciprocal series $a_1, a_2, \cdots, a_2, a_1$ of partial quotients, $2i - 1$ in number, then the fundamental (ground) solutions of $\lambda x = \mu y + z$ are $(x_j, y_j, y_{\sigma+1-j})$, $j = 1, \cdots, \sigma$, if $\lambda > \mu$, where $\sigma = 1 + a_1 + a_3 + a_5 + \cdots + a_5 + a_3 + a_1$; but are $(x_j, y_j, x_{\sigma-j})$ and $(x_\sigma, y_\sigma, 0)$, $j = 1, \cdots, \sigma - 1$, if $\lambda < \mu$, where $\sigma = 1 + a_2 + a_4 + \cdots + a_4 + a_2 + 1$, not including a_i twice. When the partial quotients are even in number, the fundamental solutions depend upon both the ascending and descending sets of intermediate convergents to λ/μ. He[48a] had proved that the fundamental solutions are always (x_j, y_j, z_j), $j = 1, \cdots, \sigma$, where the y_j/x_j are the ascending intermediate convergents to λ/μ.

A. Aubry[49] plotted the points with integral coordinates $0 \leqq x < n$, $0 \leqq y < n$, as well as the lines $y = x$, $y = ax$, $y = bx$, $\cdots$, where $1, a, b, \cdots$ are the integers $< n$ and prime to n. Thus we can read off the integer $\equiv y/x \pmod{n}$ and hence solve $ax - nz = g$.

N. P. Bertelsen[49a] solved $bx - cy = \pm z$, $1 \leqq y < b$, $0 \leqq x \leqq c$, $1 \leqq z < b$, by use of the convergents b_r/c_r to the continued fraction $(a_0, a_1, \cdots, a_n)$ for b/c. Then y is a linear function with positive integral coefficients of $b_r + kb_{r+1}$ ($k = 1, 2, \cdots, a_{r+2} - 1$), and x is the same function of the $c_r + kc_{r+1}$.

[42] Assoc. franç. av. sc., 16, II, 1887, 218–235.

[43] Lehrbuch der unbestimmten Gl. 1 Grades, Stuttgart, 1, 1891, 176 pp. (Kleyers Encykl.).

[44] Amer. Jour. Math., 15, 1893, 84.

[45] Niedere Zahlentheorie, 1, 1902, 99–153.

[46] Zeit. Math. Naturw. Unterricht, 36, 1905, 403 [33, 1902, 47].

[47] Archiv Math. Phys., 9, 1905, 204.

[48] Quar. Jour. Math., 36, 1905, 80–93.

[48a] Trans. Cambridge Phil. Soc., 19, 1901, I.

[49] L'enseignement math., 13, 1911, 187–203. Cf. G. Arnoux, Arith. Graphique, 1894, 1906.

[49a] Nyt Tidsskrift for Mat., B, 24, 1913, 33–53.

Papers without novelty.

Abbé Bossut, Cours de Math., II, 1773; ed. 3, I, 1781, 418 (g.c.d.).

P. Paoli, Elementi d'algebra, Pisa, 1794, I, 159 (Rolle's[8] method).

J. C. L. Hellwig, Anfangsgründe unbest. Analytik, Braunschweig, 1803, 1–80.

S. F. Lacroix, Complément des Elémens d'Algèbre, ed. 3, 1804, 273–9; ed. 4, 1817, 287–292 (g.c.d. and continued fractions).

F. Pezzi, Memorie di Mat. e Fis. Soc. Ital. Sc., Modena, 11, 1804, 410–25 (cont. fr.).

J. G. Garnier, Analyse Algébrique, Paris, ed. 2, 1814, 58–65 (cont. fr.).

L. Casterman, Annales Acad. Leodiensis, Liège, 1819–20 (cont. fr.).

P. N. C. Egen, Handbuch der Allgemeinen Arith., Berlin, 1819–20; ed. 2, 1833–4; ed. 3, II, 1849, 431.

M. W. Grebel, Ueber die unbest. Gl. 1 Gr., Progr. Glogau, 1827 (cont. fr.).

A. J. Chevillard, Nouv. Ann. Math., 2, 1843, 471–3 (cont. fr.).

F. Heime, Arith. Untersuchungen, Progr. Berlin, 1850 (Euler[14]).

T. Dieu, Nouv. Ann. Math., 9, 1850, 67 (g.c.d.).

H. Scheffler, Unbestim. Analytik, Hanover, 1854 (cont. fr.).

F. Thaarup, Nyt Tidsskrift for Mat., A, I, 1 (g.c.d.).

V. A. Lebesgue, Exercices d'analyse numérique, Paris, 1859, 48–54.

Lebesgue, Introduction à la théorie des nombres, 1862, 39–47 (g.c.d.).

J. J. Nejedli, Euler's Auflösungs-Methode unbest. Gl. 1 Gr., Progr. Laibach, 1863.

J. A. Temme, Bemerkungen . . . unbest. Gl., Progr. Münster, 1865 (Euler[20]).

B. I. Clasen, Ann. de l'école norm. sup., 4, 1867, 347 (g.c.d.).

O. Porcelli, Giornale di Mat., 10, 1872, 37–46 (cont. fr.).

J. Knirr, Auflösung der unbest. Gl., Progr. Wien, 1873 (Euler[20]).

C. de Comberousse, Algèbre supérieure, I, 1887, 161–173 (g.c.d., cont. fr.).

L. Matthiessen, Kommentar zur Sammlung . . . Aufgaben . . . E. Heis, ed. 4, 1902, 98–9; 1897, 221.

H. Schubert, Niedere Analysis, I, 1902, 116–126 (cont. fr.); ed. 2, 1908.

G. Calvitti, Suppl. al. Periodico di Mat., 1905 (Euler[20]).

M. Morale, ibid., 1909; Periodico di Mat., 25, 1910, 182–3 (cont. fr.).

A. Bindoni, Il Boll. di Matematica, 11, 1912, 151–3 (Euler[20]).

E. Cahen, Théorie des nombres, I, 1914, 90–108 (also graphic).

A. Sartori, Il Boll. di matematiche e sc. fis., 18, 1916, 2–10.

Papers not available for report.

Bertrand, Analyse indéterminée du premier degré.

L. Casterman, Petitur ut aequationes indet. 1 Gr., Grand, 1823.

C. L. A. Kunze, Einfache u. leichte Methode die unbest. Gl. des 1 Grads mit 2 unbekannten Z. aufzulösen, Progr. Weimar, Eisenach, 1851.

W. Korn (Dorn?), Die Lehre von den unbest. Gl. des 1 u. 2 Grades mit 2 veränderlichen Grössen, Progr. Innsbruck, 1856.

J. Hermes, Gl. 1 u. 2 Grades schematisch aufgelöst, Leipzig, 1862.

G. Elowson, Om indet. Eqvationer af 1 Graden, Progr. Lulea, 1865.

K. Weihrauch, Beitrag zur Lehre unbest. Gl. 1 Gr., Arensberg, 1866.

H. Dembschick, Unbest. Gl. 1 u. 2 Grades, Progr. Straubing, 1876.

Ferrent, Jour. de math. élém., (2), 3, 1884, 121, 155, 169, 193, 217, 241.

E. Sanczer, New method for indeter. first degree (Polish), Cracow, 1887.

G. M. Testi, Sulla ricerca di una soluzione di una equazione di primo grado a due incognite, Livorno, 1902, 4 pp.

E. Ducci, Le mie lezioni di analisi indeterminata di primo grado . . ., Bologna, 1903.

S. Soschino, Suppl. al Periodico di Mat., 12, 1908–9, 20–22.

R. Guatteri, ibid., 52–3; 13, 1909–10, 76–9 (both on Euler[20]).

G. Bernardi, Nuovo metodo di risoluzione dell'equazione $ax + by = c$ in numeri interi e positivi . . ., Bologna, 1913, 27 pp.

L. Carlini, Il Boll. di Matematica, 12, 1913, 128–136.

F. Palatini, ibid., 284.

L. Struiste, Die linearen diophant. Gl., Progr., Innsbruck, 1913.

SOLUTION OF $ax \equiv b$ (MOD m) WITHOUT FERMAT'S THEOREM.

C. F. Gauss[50] noted that $ax \equiv b$ (mod m) is solvable if and only if b is divisible by the g.c.d. d of $a = de$ and $m = df$. Let $b = dk$. Then x is a root of the proposed congruence if and only if $ex \equiv k$ (mod f), while the latter has a unique root modulo f. For a compoiste modulus mn, a second method is often preferable. First, employ the modulus m as above and let $x \equiv v$ (mod m/d), where d is the g.c.d. of m and a. Then $x = v + x'm/d$ is a root of $ax \equiv b$ (mod mn) if and only if $x'a/d \equiv (b - av)/m$ (mod n).

P. L. Tchebychef[51] proved that, if the g.c.d. d of a and p divides b, $ax \equiv b$ (mod p) has the d roots α, $\alpha + p/d$, $\cdots$, $\alpha + p(d - 1)/d$, where $\alpha a/d \equiv b/d$ (mod p/d).

C. Sardi[52] considered the congruence $a_1 x \equiv b$ (mod p) in which p is a prime not dividing a_1, and $b < p$. Dividing p by a_1, let a_2 be the remainder and $[p/a_1]$ the quotient, where $[n]$ is the greatest integer $\leqq n$. Multiply our congruence by $[p/a_1]$; we get

$$a_2 x \equiv - b [p/a_1] \quad (\text{mod } p).$$

Let a_3 be the remainder when p is divided by a_2. Let the decreasing series a_1, a_2, a_3, $\cdots$ end with $a_s = 1$. Then

$$x \equiv (- 1)^{s+1} b \left[\frac{p}{a_1} \right]\left[\frac{p}{a_2} \right] \cdots \left[\frac{p}{a_{s-1}} \right] \quad (\text{mod } p).$$

C. Ladd[53] showed that if a is prime to $M = M_1 \cdots M_k$ and if z_i is determined by $az_i + 1 \equiv 0$ (mod M_i), the root of $ax + b \equiv 0$ (mod M) is

$$x = b\{\Sigma z_i + a\Sigma z_i z_j + a^2 \Sigma z_i z_j z_k + \cdots\}.$$

L. Kronecker[54] reduced the solution of $ax \equiv b$ (mod m), where a is prime to $m = \Pi p_i^{r_i}$, to the case in which m is a power p^r of a prime. Then a root can be expressed to the base p in the form

$$\xi = \xi_0 + \xi_1 p + \cdots + \xi_{r-1} p^{r-1},$$

where each ξ_i is an integer chosen from $0, 1, \cdots, p - 1$. First find the root of $a\xi_0 \equiv b$ (mod p). Then seek ξ_1 from $a(\xi_0 + \xi_1 p) \equiv b$ (mod p^2), whence $a\xi_1 \equiv (b - a\xi_0)/p$ (mod p), etc. Again, if N is the denominator of the next to the last convergent in the continued fraction for a/m, then $x \equiv \pm bN$ (mod m).

M. Lerch[55] showed that, if p is a prime,

$$\frac{1}{a} \equiv a - 12 \sum_{v=1}^{p-1} v \left[\frac{av}{p} \right] \quad (\text{mod } p),$$

where $[t]$ is the greatest integer $\leqq t$, and hence solved $ax - py = 1$. If m

[50] Disq. arith., 1801, Arts. 29, 30; Werke, I, 1863, 20–3; Maser's German transl., 13–15.
[51] Theorie der Congruenzen, in Russian, 1849; in German, 1889, § 16, pp. 58–63.
[52] Giornale di Mat., 7, 1869, 115–6.
[53] Math. Quest. Educ. Times, 30, 1879, 41–2.
[54] Vorlesungen über Zahlentheorie, 1, 1901, 108–120.
[55] Math. Annalen, 60, 1905, 483.

is any odd number relatively prime to* $\phi(m)$, then

$$\frac{1}{a} \equiv a - \frac{12}{P(m)} \Sigma b \left[\frac{ab}{m}\right] \pmod{m},$$

where the summation extends over all positive integers b which are $< m$ and prime to m, while $P(m) = (1 - p)(1 - p') \cdots$, if $p, p', \cdots$ are the distinct prime factors of m.

E. Busche[56] obtained graphically the number of solutions of $az \equiv b \pmod{m}$, including solutions called improper or transfinite,[57] introduced when a and m have a common factor > 1. As the ordinary (proper) solutions may be restricted to the integers $0, 1, \cdots, m - 1$, we are at liberty to designate the improper solutions by numbers $\geqq m$. The simplest case is one like $3z \equiv b \pmod{15}$, in which 3 and 15/3 are relatively prime; then there is defined an improper solution designated by 15 if $b = 0$, $15 + j$ if $b = j$ $(j = 1, \cdots, 4)$, 15 if $b = 5$, $15 + j$ if $b = 5 + j$ $(j = 1, \cdots, 4)$, etc.

Solution of $ax \equiv b$ (mod m) by Fermat's or Wilson's Theorem.

J. P. M. Binet[58] noted that, if a is a prime not dividing b, $bx - ay = 1$ has the solution $x = b^{a-2}$, the corresponding y being integral; while, if $p, p', \cdots$ are the equal or distinct prime factors of a,

$$bx = 1 - (1 - b^{p-1})(1 - b^{p'-1}) \cdots$$

gives an integer x, leading to an integer y, such that x, y satisfy the same equation. The same method was found independently by G. Libri.[59]

A. Cauchy[60] expressed Binet's method in the following form: let

$$n = a^\alpha b^\beta \cdots, \qquad (1 - k^{a-1})^\alpha (1 - k^{b-1})^\beta \cdots = 1 - kK.$$

Then for k prime to n, $1 - kK$ is divisible by n, so that $kx \equiv h \pmod{n}$ has the solution $x \equiv hK \pmod{n}$.

V. Bouniakowsky[61] proved that, if a, b are relatively prime positive integers, $ax \mp by = c$ has the integral solutions*

$$x = ca^{\phi(b)-1}, \qquad y = \frac{\pm c}{b}(a^{\phi(b)} - 1).$$

G. de Paoli[62] gave the last solution, with $\phi(b)$ replaced by $\phi(b)/2$ when b is divisible by 4. To solve $ax - by - cz = e$, where a, b, c, e have no common divisor, let $a = dA$, $b = dB$, where d is the g.c.d. of a, b; then $e + cz$

---

*By $\phi(m)$ is meant the number of integers $< m$ which are prime to m.

[56] Mitt. Math. Gesell. Hamburg, 4, 1908, 355–380.

[57] Imaginary by Gauss, Disq. Arith., Art. 31; G. Arnoux, Arithmétique graphique, 2, 1906, 20. Both excluded such solutions.

[58] Jour. de l'école polyt., cah. 20, 1831, 292 [read 1827]; communicated to the Société Philomatique before 1827.

[59] Mémoires de Math. et de Phys., Florence, 1829, 65–7. Cf. Libri[148] of Ch. XXIII.

[60] Exercices de Math., 1829, 231–   ; Oeuvres, (2), IX, 296.

[61] Mém. Acad. Sc. St. Pétersbourg (Math. Phys.), (6), 1, 1831, 143–4 [read Apr. 1, 1829].

[62] Opuscoli Mat. e Fis. di Diversi Autori, Milano, 1, 1832, 269. He stated that the paper was written in 1830 without knowledge of that by Binet.

must be a multiple du of d; the equations $Ax - By = u$, $du - cz = e$ are each solved by Fermat's theorem; similarly for n variables (pp. 327–338).

A. L. Crelle[63] noted that $ax \equiv 1 \pmod{m}$ has the solution $a^{\phi(m)-1}$.

A. Cauchy[64] obtained independently the result of Bouniakowsky.[61]

J. P. M. Binet[65] employed Wilson's theorem to solve $ax = 1 + py$, when p is a prime. We may take $0 < a < p$. Then $x = -(p-1)!/a$. Whether p is prime or composite, we may also proceed as follows. Divide p by a and call the quotient q and remainder a_1; divide p by a_1 and call the quotient q_1 and the remainder a_2; etc., until the remainder $a_n = 1$ is reached. Then

$$aqq_1\cdots q_{n-1} + (-1)^{n+1} = pM, \qquad x = (-1)^n qq_1\cdots q_{n-1}.$$

V. Bouniakowsky[66] employed $(p, n) = p(p-1)\cdots(p-n+1)$. Then, if $b < p$,

$$(p+b, p) = (p, p) + \binom{p}{1}(p, p-1)(b, 1) + \binom{p}{2}(p, p-2)(b, 2)$$
$$+ \cdots + \binom{p}{b}(p, p-b)(b, b).$$

Divide by (p, p) and write $a = p + b$. We get $aE = 1 + pK$, where E and K are integers* if p is a prime. Hence we have solved $ax = 1 + py$ in integers if $a > p$ and p is a prime. To solve

$$Mx - Ny = 1, \qquad N = p^\lambda q^\mu r^\nu \cdots,$$

where p, q, r, $\cdots$ are distinct primes, determine α_1, β_1, $\cdots$ so that

$$M\alpha_1 - p\beta_1 = 1, \qquad M\alpha_2 - q\beta_2 = 1, \qquad M\alpha_3 - r\beta_3 = 1, \qquad \cdots,$$

as above. Raise $M\alpha_1 - 1$, $M\alpha_2 - 1$, $\cdots$ to the powers λ, μ, $\cdots$. Then

$$Me_1 + (-1)^\lambda = p^\lambda\beta_1^\lambda, \qquad Me_2 + (-1)^\mu = q^\mu\beta_2^\mu, \qquad \cdots,$$

where e_1, e_2, $\cdots$, and A below are integers. By multiplication,

$$MA + (-1)^{\lambda+\mu+\nu+\cdots} = NB, \qquad B = \beta_1^\lambda\beta_2^\mu\cdots.$$

According as $\lambda + \mu + \nu + \cdots$ is odd or even, $y = B$ or $-B$.

L. Poinsot[67] noted that $Lx - My = 1$ has the solution $x = L^{m-1}$ if $L^m \equiv 1 \pmod{M}$, e. g., if $m = \phi(M)$. He also expressed the method in terms of regular polygons. Thus, for $12x - 7y = 1$, take 7 points P_1, $\cdots$, P_7. Take the first, the fifth after the first, etc. (5 being $12 - 7$); we get $P_1P_6P_4P_2P_7P_5P_3$. Since P_2 is now the third point after P_1, we have $x = 3$. We get y from the equation or by use of 12 points.

---

[63] Abh. Akad. Wiss. Berlin (Math.), 1836, 52.

[64] Comptes Rendus Paris 12, 1841, 813; Oeuvres, (1), VI, 113. Exercices d'Analyse et de Physique Math., 2, 1841, 1; Oeuvres, (2), XII. See Vol. I, p. 187, of this History. Cf. report by J. A. Grunert, Archiv Math. Phys., 3, 1843, 203.

[65] Comptes Rendus Paris, 13, 1841, 210–3.

[66] Mém. Acad. Sc. St. Pétersbourg (Math. Phys.), (6), 3, 1844, 287.

* $E = (p + b - 1)! \div \{p!\, b!\}$ is an integer by Catalan,[21] p. 265 of Vol. I of this History.

[67] Jour. de Math., (1), 10, 1845, 55–59.

J. G. Zehfuss[68] gave the formula of Cauchy[60] and noted that, if $\mu = \alpha^m \beta^n \cdots$, and if A is not divisible by the prime α, B not by β, $\cdots$, then

$$\left(\frac{A\mu}{\alpha^m}\right)^{(\alpha-1)\alpha^{m-1}} + \left(\frac{B\mu}{\beta^n}\right)^{(\beta-1)\beta^{n-1}} + \cdots \equiv 1 \pmod{\mu}.$$

For $A = B = \cdots = a$, let the left member become k. Then $ax \equiv b$ $\pmod{\mu}$ has the root kb/a. It also has the root $(1 - AB\cdots)b/a$, where

$$A = \left(1 + a\frac{(\alpha-1)!}{a_\alpha}\right)^m \equiv 0 \pmod{\alpha^m},$$

$$B = \left(1 + a\frac{(\beta-1)!}{a_\beta}\right)^n \equiv 0 \pmod{\beta^n}, \quad \cdots,$$

where a_α is the least positive residue of a modulo α, since, by Wilson's theorem, $a_\alpha + (\alpha - 1)!\, a$ is divisible by the prime α.

M. F. Daniëls[69] noted that, if $\rho_1 \cdots \rho_n \equiv \pm 1 \pmod{k}$ by Wilson's generalized theorem, then $\rho_i x \equiv 1 \pmod{k}$ has the root $\pm \rho_1 \cdots \rho_{i-1} \rho_{i+1} \cdots \rho_n$. Further, if $k = p'q'' \cdots$ and if $ac_1 \equiv 1 \pmod{p}$, $ac_2 \equiv 1 \pmod{q}$, $\cdots$, then $ax \equiv 1 \pmod{k}$ has the root

$$x = \frac{1}{a}\left\{1 - (1 - ac_1)^{\nu}(1 - ac_2)^{\mu}\cdots\right\}.$$

J. Perott[70] noted that if a and u are relatively prime and if a belongs to the exponent t modulo u, $ax \equiv 1 \pmod{u}$ has the unique solution $x \equiv a^{t-1}$ $\pmod{u}$. He admitted he was anticipated by Cauchy.

Chinese Problem of Remainders.

Sun-Tsŭ,[71] in a Chinese work Suan-ching (arithmetic), about the first century A.D., gave in the form of an obscure verse a rule called t'ai-yen (great generalisation) to determine a number having the remainders 2, 3, 2, when divided by 3, 5, 7, respectively. He determined the auxiliary numbers 70, 21, 15, multiples of $5\cdot7$, $3\cdot7$, $3\cdot5$ and having the remainder 1 when divided by 3, 5, 7, respectively. The sum $2\cdot70 + 3\cdot21 + 2\cdot15 = 233$ is one answer. Casting out a multiple of $3\cdot5\cdot7$ we obtain the least answer 23. The rule became known in Europe through an article, "Jottings on the science of Chinese arithmetic," by Alexander Wylie,[72] a part of which was translated into German by K. L. Biernatzki.[73] A faulty rendition by

[68] Diss. (Heidelberg), Darmstadt, 1857; Archiv Math. Phys., 32, 1859, 422.
[69] Lineaire Congruenties, Diss., Amsterdam, 1890, 114, 90.
[70] Bull. des Sc. Math., (2), 17, I, 1893, 73–4.
[71] Y. Mikami, Abh. Geschichte Math. Wiss., 30, 1912, 32.
[72] North China Herald, 1852; Shanghai Almanac for 1853. Cf. remark by G. Vacca, Bibliotheca Math., (3), 2, 1901, 143; H. Cordier, Jour. Asiatic Soc., (2), 19, 1887, 358.
[73] Jour. für Math., 52, 1856, 59–94. French transl. by O. Terquem, Nouv. Ann. Math., (2), 1, 1862 (Bull. Bibl. Hist.), 35–44; 2, 1863, 529–540; and by J. Bertrand, Journal des Savants, 1869. Cf. Matthiessen.[79]

.the latter caused M. Cantor[74] to criticize the validity of the rule. The rule was defended by L. Matthiessen,[75] who pointed out its identity with the following statement by C. F. Gauss.[76] If $m = m_1 m_2 m_3 \cdots$, where m_1, m_2, m_3, $\cdots$ are relatively prime in pairs, and if

$$\alpha_i \equiv 0 \ (\text{mod } m/m_i), \qquad \alpha_i \equiv 1 \ (\text{mod } m_i) \quad (i = 1, 2, 3, \cdots),$$

then $x = \alpha_1 r_1 + \alpha_2 r_2 + \cdots$ is a solution of

$$x \equiv r_1 \ (\text{mod } m_1), \qquad x \equiv r_2 \ (\text{mod } m_2), \qquad \cdots.$$

This method is very convenient when one has to treat several problems with fixed m_1, m_2, $\cdots$, but varying r_1, r_2, $\cdots$.

Nicomachus[77] (about 100 A.D.) gave the same[71] problem and solution 23.

Brahmegupta[78] (born, 598 A.D.) gave a rule which becomes clearer when applied to an example: find a number having the remainder 29 when divided by 30 and the remainder 3 when divided by 4. Dividing 30 by 4, we get the residue 2. Dividing 4 by 2, we get the residue zero and quotient 2. Dividing the difference $3 - 29$ by the residue 2, we get $- 13$. Multiply the quotient 2 by any assumed multiplier 7 and add the product to $- 13$; we get 1. Then $1 \cdot 30 + 29 = 59$ is the desired number.

This problem forms the second stage of the solution of the "popular" problem (§ 7, p. 326): find a number having the remainders 5, 4, 3, 2 when divided by 6, 5, 4, 3, respectively. The answer is stated correctly to be 59.

The priest Yih-hing[79] († 717 A.D.) in his book t'ai-yen-lei-schu gave a generalization to the case in which the moduli m_i are not relatively prime. Express the l.c.m. of m_1, m_2, $\cdots$, m_k as a product $m = \mu_1 \mu_2 \cdots \mu_k$ of relatively prime factors, including unity, such that μ_i divides m_i. Then, if

$$\alpha_i \equiv 0 \ (\text{mod } m/\mu_i), \qquad \alpha_i \equiv 1 \ (\text{mod } \mu_i) \qquad (i = 1, \cdots, k),$$

$x = \alpha_1 r_1 + \alpha_2 r_2 + \cdots$ is a solution. Other solutions are obtained by subtracting multiples of m. Yih-hing proposed to find the number of completed units of work, the same number x of units to be performed by each of four sets of 2, 3, 6, 12 workmen, such that after certain whole days' work, there remain 1, 2, 5, 5 units not completed by the respective sets. The l.c.m. of $m_1 = 2$, $\cdots$, $m_4 = 12$ is $m = 12$. Taking $\mu_1 = \mu_2 = 1$, $\mu_3 = 3$, $\mu_4 = 4$, we get $\alpha_1 = \alpha_2 = 12$, $\alpha_3 = 4$, $\alpha_4 = 9$,

$$x \equiv 1 \cdot 12 + 2 \cdot 12 + 5 \cdot 4 + 5 \cdot 9 = 101 \equiv 17 \ (\text{mod } 12).$$

---

[74] Zeitschrift Math. Phys., 3, 1858, 336; not repeated in his Geschichte der Math., ed. 2, I, 643. H. Hankel, Geschichte d. Math. in Alterthum u. Mittelalter, 1874, erred in identifying the Chinese rule with the Indian cuttaca.[4]

[75] Zeitschrift Math. Phys., 19, 1874, 270–1; Zeitschrift Math. Naturw. Unterricht, 7, 1876, 80.

[76] Disq. Arith., art. 36; Werke, I, 26. Cf. Euler.[96]

[77] Pythagorei introd. arith. libri duo, rec. R. Hoche, Leipzig, 1866, Supplement, prob. V.

[78] Brahme-sphut'a-sidd'hánta, Ch. 18 (Cuttaca = algebra), §§ 3–6, Colebrooke,[1] pp. 325–6.

[79] L. Matthiessen, Comptes Rendus Paris, 92, 1881, 291; Jour. für Math., 91, 1881, 254–261; Zeitschr. Math. Phys., 26, 1881, Hist.-Lit. Abt., 33–37 (correction of Biernatzki[73]).

For $x = 17$, the completed part is $8 \cdot 2 + 5 \cdot 3 + 2 \cdot 6 + 1 \cdot 12 = 55$. We may equally well take $\mu_1 = 1$, $\mu_2 = 3$, $\mu_3 = 1$, $\mu_4 = 4$ and get $\alpha_1 = 12$, $\alpha_2 = 4$, $\alpha_3 = 12$, $\alpha_4 = 9$, $\Sigma \alpha_i r_i = 125 \equiv 17 \pmod{12}$.

A condition on the solvability of the problem is that $r_i - r_j$ be divisible by the g.c.d. of m_i, m_j.

Ibn al-Haitam[80] (about 1000) gave two methods to find a number, divisible by 7, which has the remainder 1 when divided by 2, 3, 4, 5 or 6. The first method gives the one solution $1 + 2 \cdot 3 \cdot 4 \cdot 5 \cdot 6 = 721$. The second method gives a series of solutions 301, etc.; in effect $\frac{3}{4}(6 + 2n \cdot 7)20 + 1$, where n is an integer such that $6 + 2n \cdot 7$ is a multiple of 4.

Bháscara[81] (born, 1114 A.D.) treated the problem to find the number having the remainders 5, 4, 3, 2 when divided by 6, 5, 4, 3 respectively. By the first two conditions, the number is $6c + 5 = 5n + 4$. By use of the "pulverizer," the integral value of $c = (5n - 1)/6$ is $c = 5p + 4$. The number $6c + 5 = 30p + 29$ must equal $4l + 3$. Hence $p = (4l - 26)/30$, which is converted by the pulverizer into $2h + 1$. Thus

$$30p + 29 = 60h + 59$$

is the answer.

Again (§ 162, p. 238), what number being divided by 2, 3, 5 has the respective remainders 1, 2, 3, while the quotients divided by 2, 3, 5 respectively have the remainders 1, 2, 3? Call the quotients $2c + 1$, $3n + 2$, $5l + 3$. Then the number is $4c + 3 = 9n + 8 = 25l + 18$. Applying the pulverizer to the first equality, we get $c = 9p + 8$. The resulting number $36p + 35$ must equal $25l + 18$, whence $p = 25h + 3$ and the answer is $900h + 143$.

Leonardo Pisano[82] treated (p. 281) the problem to find a number N, divisible by 7, which gives the remainder 1 when divided by 2, 3, 4, 5 or 6. By the latter condition, N exceeds 1 by a multiple of 60; but 60 has the remainder 4 when divided by 7, while we need the remainder 6; thus we multiply 60 by 2, 3, $\cdots$ until we reach 60×5 with the remainder 6. Thus $N = 301$, to which we may add a multiple of $420 = 60 \cdot 7$. Similarly, 25201 is the multiple of 11 having the remainder 1 when divided by 2, $\cdots$, 10.

To find (p. 282) a multiple of 7 having the remainders 1, 2, 3, 4, 5 when divided by 2, 3, 4, 5, 6, we take 1 from a multiple of 60 such that the difference is divisible by 7; the result is $2 \cdot 60 - 1 = 119$. Similarly, to find a multiple of 11 having the remainders 1, 2, $\cdots$, 9 when divided by 2, 3, $\cdots$, 10, we subtract 1 from the least common multiple 2520 of 2, $\cdots$, 10 and get 2519, which being a multiple of 11 is the answer.

He employed[83] (p. 304) in effect the rule t'ai-yen[71] to tell what number not exceeding 105 a person has in mind if the latter gives the remainders

---

[80] Arabic MS. in Indian Office, London. Cf. E. Wiedemann, Sitzungsber. Phys. Medic. Soc. Erlangen, 24, 1892, 83.

[81] Víja-gańita (algebra), § 160, Colebrooke,[1] pp. 235–7.

[82] Liber Abbaci (1202, revised 1228), pub. by B. Boncompagni, Rome, 1, 1857.

[83] M. Curtze, Zeitschrift Math. Phys., 41, 1896, Hist. Lit. Abt., 81–2, remarked that if Leonardo had found the rule independently, he would have so stated and would have given a proof.

(say 2, 3, 4) obtained by dividing it by 3, 5, 7:

$$2 \cdot 70 + 3 \cdot 21 + 4 \cdot 15 = 263, \qquad 263 - 2 \cdot 105 = 53 = \text{ans.}$$

Similarly for the number not exceeding 315, given the remainders upon division by 5, 7, 9: the remainders are to be multiplied by 126, 225, 280, and from the sum of the products is to be subtracted a multiple of 315.

Ch'in Chiu-shao[84] gave a method applicable to the problem to find a number x having the remainders $r_1, \cdots, r_n$ when divided by $m_1, \cdots, m_n$, which are relatively prime in pairs. Let M be any one of the quotients $M_k = m_1 \cdots m_n / m_k$, and seek ρ so that $M\rho \equiv 1 \pmod{m = m_k}$. We may replace M by its residue R modulo m. On dividing m by R, let the quotient be Q_1 and the positive remainder be $r_1 \leqq R$. Divide R by r_1 to get the quotient Q_2 and positive remainder $r_2 \leqq r_1$; divide r_1 by r_2 to get the quotient Q_3 and remainder $\leqq r_2$; proceed until we reach an $r_i = 1$. Let $A_1 = Q_1$, $A_2 = A_1 Q_2 + 1$, $A_3 = A_2 Q_3 + A_1$, $A_4 = A_3 Q_4 + A_2$, $\cdots$. Then $\rho = A_i$, and $x = r_1 M_1 \rho_1 + r_2 M_2 \rho_2 + \cdots + r_n M_n \rho_n$.

A German MS.[85] of the fifteenth century proved a general rule corresponding to the Chinese t'ai-yen rule.

Regiomontanus[86] (1436–1476) proposed in a letter the problem to find a number with the remainders 3, 11, 15 when divided by 10, 13, 17. It is possible[87] that he got acquainted in Italy with the work of L. Pisano.

Elia Misrachi[88] (1455–1526) reproduced L. Pisano[82] (pp. 281–2) and gave answers to similar problems.

Michael Stifel[89] gave the correct result that if x has the remainders r and s when divided by a and $a + 1$, respectively, then x has a remainder $(a + 1)r + a^2 s$ when divided by $a(a + 1)$.

Pin Kue[90] treated in 1593 the problem given by Sun-Tsŭ.[71]

The problem to find a multiple of 7 having the remainder 1 when divided by 2, 3, 4, 5 or 6 was treated also by Casper Ens[91] and Daniel Schwenter.[92]

Frans van Schooten[93] treated the problem to find a multiple of 7 having the remainder 1 when divided by 2, 3 or 5. He used $30k + 1$, where $k = 3$ is chosen so that the number is divisible by 7. He gave what is really the t'ai-yen rule, but attributed it to Nicolaus Huberti; it leads here to the multipliers $3 \cdot 5 \cdot 7 = 105$, $2 \cdot 5 \cdot 7 = 70$, $3(2 \cdot 3 \cdot 7) = 126$, each with the remainder 1 when divided by 2, 3, 5, respectively.

---

[84] Nine Sections of Math. (about 1247). Cf. Mikami,[71] pp. 65–9.

[85] M. Curtze, Abh. Geschichte der Math., 7, 1895, 65–7.

[86] C. T. de Murr, Memorabilia Bibl. publ. Norimbergensium et Universitatis Altdorfinae, Pars I, 1786, p. 99.

[87] Cantor, Geschichte der Math., ed. 1, II, 263.

[88] G. Wertheim, Die Arithmetik des E. Misrachi, 1893, ed. 2, 1896, 60–61.

[89] Arithmetica integra, 1544, Book I, fol. 38v. Die Coss Christoffs Rudolffs, Die Schönen Exempeln der Coss Durch Michael Stifel Gebessert, Königsperg, 1553, 1571.

[90] Swan fa tong tsong, Ch. 5, p. 29, MS. in Bibl. Nat. Paris; abstract by E. Biot, Jour. Asiatique, (3), 7, 1839, 193–218.

[91] Thaumaturgus Math., Munich, 1636, 70–71.

[92] Deliciae Physico-Math. oder Math.-u. Phil. Erquickstunden, Nürnberg, 1, 1636, 41.

[93] Exercitationum math. libri quinque, Lugd. Batav., 1657, 407–410.

W. Beveridge[94] treated the problem to find the least number P which has given remainders K and L when divided by A and B, when the latter are relatively prime. Let D be the least multiple of B which has the remainder 1 when divided by A; let C be the least multiple of A which divided by B leaves 1. Then $P = DK + CL$, as shown by a two page proof.

To find the least number P which has the given remainders K, L, Z when divided by the relatively prime numbers M, B, A, first find the least multiple F of AB, least multiple N of AM, least multiple Q of BM, which have the remainder 1 when divided by M, B, A, respectively. Then $P = KF + LN + ZQ$.

This is precisely the rule as given later by Euler[96] and Gauss.[76]

*J. Wallis[95] gave an empirical solution of the problem of the Julian period.

T. F. de Lagny[9] treated the problem to find the year x of the Julian period when the solar cycle is 13, the lunar cycle is 10 and the "indiction" is 7; thus if x is divided by 28, 19, 15, the remainder is 13, 10, 7, respectively. From $x = 28m + 13 = 19n + 10$, he found[9] that $n = 9 + 28f$, where f is an integer. Thus $x = 19n + 10 = 181 + 532f$. Since $x - 7$ is to be divisible by 15, the least f is 3.

L. Euler[96] treated the problem to find an integer z which has the remainders p and q when divided by a and b, respectively, where $a > b$. Thus $z = ma + p = nb + q$. He solved the second equation by use of the process for the greatest common divisor for a, b, continued until one of the remainders c, d, e, $\cdots$ is reached which divides $v = p - q$. He thus deduced the result

$$z = q + abv \left(\frac{1}{ab} - \frac{1}{bc} + \frac{1}{cd} - \frac{1}{de} + \cdots \right),$$

in which the series is continued until we reach a remainder dividing v. At the end of the paper, Euler gave a rule generally attributed to Gauss.[76] To find a number which has the respective remainders p, q, r, s, t when divided by a, b, c, d, e, which are relatively prime in pairs. An answer is $Ap + Bq + Cr + Ds + Et + Mabcde$, where

$$A \equiv 0 \ (\text{mod } bcde), \qquad A \equiv 1 \ (\text{mod } a); \qquad B \equiv 0 \ (\text{mod } acde),$$
$$B \equiv 1 \ (\text{mod } b); \quad \cdots \qquad E \equiv 0 \ (\text{mod } abcd), \qquad E \equiv 1 \ (\text{mod } e).$$

C. von Clausberg[97] found a multiple of 7 having the remainder 10 when divided by 15.

N. Saunderson[93] treated the problem to find a number which has the remainders d and e when divided by a and b, $a > b$. Let l be the g.c.d.

[94] Institutionum Chronologicarum libri II. Unà cum totidem Arithmeticcs Chronologicae Libellis. Per Guilielm Beveregium, Londini, 1669, lib. II, pp. 253–6.
[95] Opera, 2, 1693, 451–5. Cf. Hutton.[101]
[96] Comm. Acad. Petrop., 7, 1734–5, 46–66; Comm. Arith. Coll., I, 11–20.
[97] Demonstrative Rechnenkunst, 1732, § 1366, § 1493.
[98] The Elements of Algebra, Cambridge, 1, 1740, 316–329. Reproduced by de la Bottiere, Mém. de math. phys., présentés . . . divers savans, 4, 1763, 41–65.

of a and b. Evidently l must divide $d - e$. Let this condition be satisfied and determine A and B so that $Aa - Bb = -l$. Multiply the last equation by $(d - e)/l$. Then

$$Aa \cdot \frac{d - e}{l} + d = Bb \cdot \frac{d - e}{l} + e$$

is an answer. Other answers follow by adding any multiple of the l.c.m. M of a, b. Next, let there be three divisors a, b, c and corresponding remainders d, e, f. By the first problem find a number g having the remainders d and e when divided by a and b, and then a number h having the remainders g and f when divided by M and c. From the answer h we obtain others by adding any multiple of the l.c.m. of a, b, c.

*A. G. Kästner,[99] *Lüdicke[100] and C. Hutton[101] treated problems on the Julian period. To find the year x of Christ in which the solar and lunar cycles are 18 and 8, and Roman indiction is 10, Hutton noted that the year before the Christian era was the ninth of the solar cycle, first of the lunar and third of the indiction. Hence the remainders on dividing $x + 9$, $x + 1$, $x + 3$ by 28, 19, 15 respectively (the periods of the solar, lunar and indiction cycles) must be 18, 8, 10. Thus $x = 7980p + 1717$.

To apply[102] the rule in J. Keill's Astronomy Lectures, p. 380, divide $18 \cdot 4845 + 8 \cdot 4200 + 10 \cdot 6916$ by 7980; the remainder 6430 is the year of the Julian period; subtract 4713, the Julian year at the birth of Christ.

A. Thacker[103] proved the last rule, starting as had Hutton.[101]

The least number[104] with the remainders 1, 2, 3, 4, 5 when divided by 2, 3, 4, 5, 6 is $60 - 1$ [L. Pisano[82]].

R. Robinson[105] found a number x which has the remainders 19, 18, $\cdots$, 1 when divided by 20, 19, $\cdots$, 2. Since $x = 2a + 1 = 3b + 2 = \cdots = 20A + 19$, $b = 2m - 1$, $a = 3m - 1$; then use $x = 4c + 3$, etc. Hence $x = 232792560B - 1$, the least being given by $B = 1$.

J. L. Lagrange[106] determined n so that it shall have given remainders N, N_1, N_2, $\cdots$, when divided by M, M_1, M_2, $\cdots$ respectively. Let P be the l.c.m. of M, M_1, M_2, $\cdots$; Q that of M, M_2, M_3, $\cdots$ (omitting M_1); Q_1 that of M, M_1, M_3, $\cdots$ (omitting M_2); etc. Then seek (Lagrange[17]) integers μ, ν, μ_1, ν_1, $\cdots$ such that

$$\mu Q - \nu M_1 = N_1 - N, \qquad \mu_1 Q_1 - \nu_1 M_2 = N_2 - N, \qquad \mu_2 Q_2 - \nu_2 M_3 = N_3 - N, \quad \cdots$$

[99] Angewandte Math. in der Chronologie.

[100] Archiv der Math. (ed., Hindenburg), 2, 1745, 206.

[101] The Diarian Repository, or Math. Register, by a Society of Mathematicians, London, 1774, 306; The Diarian Miscellany, extracted from Ladies' Diary, London, 2, 1775, 33–4; Leybourn's Math. Quest. proposed in Ladies' Diary, 1, 1817, 232–3.

[102] Ladies' Diary, 1735, 33–4, Quest. 175.

[103] A Miscellany of Math. Problems, Birmingham, 1, 1743, 167–8.

[104] Ladies' Diary, 1749, 21, Quest. 296; Diarian Repository . . . by a Society of Mathematicians, London, 1774, 501–2; C. Hutton's Diarian Miscellany, 2, 1775, 264–5; Leybourn's Math. Quest. L. D., 2, 1817, 2.

[105] The Gentleman's Diary, or Math. Repository, 1748; A. Davis' ed., London, 1, 1814, 154–5.

[106] Mém. Acad. Roy. Sc. Berlin, 23, année 1767 (1769); Oeuvres, II, 519–20.

Then $n = \lambda P + N + \mu Q + \mu_1 Q_1 + \mu_2 Q_2 + \cdots$, where λ is any integer. The first of the above set of equations has an infinitude of solutions if Q and M_1 are relatively prime, but no solution in the contrary case unless $N_1 - N$ be divisible by the g.c.d. of Q, M_1.

Lagrange[107] noted that the problem is to make $Mt + N$, $M_1 u + N_1$, $M_2 x + N_2$, $\cdots$ equal. The general value of t making the first two equal is $t = Ar + M_1 m$, where $A = N_1 - N$, r is fixed and m is arbitrary. The next step is to solve

$$M(Ar + M_1 m) + N = M_2 x + N_2$$

for m, x; etc.

K. F. Hindenburg[108] gave a method of "cyclic periods" to find, for example, a number x having the remainders 1 and 2 when divided by $\alpha = 2$ and $\beta = 3$. The numbers $1, 2, \cdots, \alpha$ are written in a column and repeated β times; similarly $1, 2, \cdots, \beta$ are written in a second column and repeated α times. The given remainders appear in the 5th row; hence $x = 5$.

$$
\begin{array}{cc}
1 & 1 \\
2 & 2 \\
1 & 3 \\
2 & 1 \\
1 & 2 \\
2 & 3 \\
\end{array}
$$

C. F. Gauss,[109] to find z with the remainders a and b when divided by A and B, solved $z = Ax + a \equiv b \pmod{B}$, obtaining $x \equiv v \pmod{B/\delta}$, if δ is the g.c.d. of A, B. Hence $z \equiv Av + a \pmod{M}$ is the complete solution of the problem, where $M = AB/\delta$ is the l.c.m. of A, B. If we add the condition that $z \equiv c \pmod{C}$, we get the complete solution

$$z \equiv Mw + Av + a \pmod{M'},$$

where $M' = ABC/\delta\epsilon$ is the l.c.m. of A, B, C, while ϵ is the g.c.d. of M, C.

We may replace $z \equiv a \pmod{A}$ by $z \equiv a \pmod{A'}$, $z \equiv a \pmod{A''}$, $\cdots$, where $A'A'' \cdots = A$ and $A', A'', \cdots$ are powers of distinct primes. Similarly, let $B = B'B'' \cdots$. In case $B' = p^r$, $A' = p^s$, $r \geq s$, the problem is impossible unless $b \equiv a \pmod{A'}$, while if this is satisfied the condition $z \equiv a \pmod{A'}$ may be dropped. In this way we can derive an equivalent set of congruences in which the moduli are relatively prime in pairs and proceed as above or as in Gauss[76] [due to Euler[96]].

A. D. Wheeler[110] noted that the least integer k which has the given remainders $r, r', \cdots$ when divided by the given numbers $d, d', \cdots$ is found by reducing $(x - r)/d$, $(x - r')/d'$, $\cdots$ to equivalent fractions with a

[107] Mém. Acad. Roy. Sc. Berlin, 24, année 1768 (1770), 222; Oeuvres, II, 698.

[108] Leipziger Magazin reine u. angewandte Math., 1786, 281–324; extr. by Lorentz, Lehrbegriff der Math., ed. 2, I, 406–442, and by C. A. W. Berkhan, Lehrbuch der Unbestimmten Analytik, Halle, 1, 1855, 124–144.

[109] Disq. Arith., 1801, arts. 32–5; Werke, I, 1863, 23–6; Maser's German transl., 15–18.

[110] The Math. Monthly (ed., Runkle), New York, 2, 1860, 410.

common denominator and taking a linear combination $x - k$ of the new numerators such that the coefficient of x is unity.

L. Matthiessen[111] discussed the Chinese rules in modern form.

M. F. Daniëls[69] noted that if a, b, $\cdots$ are relatively prime integers, $x \equiv A \pmod{a}$, $x \equiv B \pmod{b}$, $\cdots$ have the solution

$$x \equiv \left(\frac{k}{a}\right)^{\phi(a)} A + \left(\frac{k}{b}\right)^{\phi(b)} B + \cdots \pmod{k = ab\cdots}.$$

T. J. Stieltjes[112] noted that the congruences $x \equiv \alpha \pmod{A}$, $\cdots$, $x \equiv \lambda \pmod{L}$ have a common solution if and only if $\alpha - \beta$, $\alpha - \gamma$, $\beta - \gamma$, $\cdots$ are divisible by (A, B), (A, C), (B, C), $\cdots$, respectively, where (A, B) denotes the g.c.d. of A, B. The case in which A, $\cdots$, L are not relatively prime in pairs can be reduced [Yih-hing[79]] to the contrary case by writing the l.c.m. of the moduli in the form $M = A'B'\cdots L'$, where A', $\cdots$, L' are relatively prime in pairs and divide A, $\cdots$, L respectively. Then any solution of the initial congruences satisfies also $x \equiv \alpha \pmod{A'}$, $\cdots$, $x \equiv \lambda \pmod{L'}$, whence $x \equiv a \pmod{M}$. Conversely, the last x satisfies the initial congruences if they are solvable.

H. J. Woodall[113] found numbers with given remainders when divided by 3, 5, 7, 11, 13.

J. Cullen[114] gave a graphical method to solve $x \equiv \alpha \pmod{P}$, $\cdots$, $x \equiv \lambda \pmod{L}$, useful when P, $\cdots$, L are very large.

G. Arnoux[115] gave implicitly the theorem that, if m_1, $\cdots$, m_n are relatively prime in pairs, $M = m_1\cdots m_n$, $\mu_i = M/m_i$, and if a_1, $\cdots$, a_m are integers such that $a_i\mu_i \equiv r \pmod{m_i}$ for $i = 1$, $\cdots$, n, then $a_1\mu_1 + \cdots + a_n\mu_n \equiv r \pmod{M}$. Proofs were given by C. A. Laisant[116] and T. Hayashi.[116]

ARTICLES ON THE PROBLEM OF REMAINDERS WITHOUT NOVELTY.

G. S. Klügel, Math. Wörterbuch, 3, 1808, 792–800.
J. C. Schäfer, Die Wunder der Rechenkunst, Weimar, 1831, 1842, Prob. 60.
H. Kaiser, Archiv Math. Phys., 25, 1855, 76.
G. Dostor, *ibid.*, 63, 1879, 224.
V. A. Lebesgue, Exercices d'analyse numérique, Paris, 1859, 54–8.
Szenic, Von der Kongruenz der Zahlen, Progr. Schrimm, 1873.
A. Domingues, Les Mondes (Revue Hebdom. des Sciences et Arts), Paris, 55, 1881, 62.
G. de Rocquigny, *ibid.*, 54, 1881, 304.
D. Marchand, *ibid.*, 54, 1881, 437.

NUMBER ω OF POSITIVE INTEGRAL SOLUTIONS OF $ax + by = n$, WHERE a AND b ARE POSITIVE AND RELATIVELY PRIME.

P. Paoli[117] noted that if $ax + by = n$ has integral solutions, any common factor of a and b must divide n and hence can be removed from every term.

[111] Zeitschr. Math. Naturw. Unterricht, 10, 1879, 106–110; 13, 1882, 187–190.
[112] Annales Fac. Sc. Toulouse, 4, 1890, final paper, pp. 31–32.
[113] Math. Quest. Educ. Times, 73, 1900, 67.
[114] Proc. London Math. Soc., 34, 1901–2, 323–34; (2), 2, 1905, 138–141.
[115] Arith. Graphique, Paris, 1906, 29–31.
[116] L'enseignement math., 10, 1908, 220–5; 12, 1910, 141–2.
[117] Opuscula analytica, Liburni, 1780, 114. In one place in the text and in his example, he erroneously took β between $- b/2$ and $b/2$, instead of positive.

Let henceforth a and b be relatively prime and positive. Let β denote the least positive integer such that $n - a\beta$ is divisible by b. Then every solution is given by

$$x = \beta + bm, \qquad y = \frac{n - a\beta}{b} - am.$$

The values of m making x and y positive are $0, 1, \cdots, E$, where E is the largest integer less than $(n - a\beta)/(ab)$. Thus there are $\omega = E + 1$ sets of positive integral solutions x, y.

P. Barlow[118] employed positive integers p, q such that $aq - bp = + 1$. Then all solutions of $ax + by = n$ are given by

$$x = nq - mb, \qquad y = ma - np.$$

Let $[t]$ denote the greatest integer $\leqq t$. Then

$$\omega = \left[\frac{nq}{b}\right] - \left[\frac{np}{a}\right]$$

or one less according as nq/b is not or is an integer. In fact, m must be less than nq/b and $> np/a$ to make x and y positive.

Libri[27] expressed ω as a sum of trigonometric functions.

C. Hermite[119] employed the integers

$$n' = a\left[\frac{n}{a}\right] + b\left[\frac{n}{b}\right] - n, \qquad n'' = a\left[\frac{n'}{a}\right] + b\left[\frac{n'}{b}\right] - n', \qquad \cdots.$$

Then every positive integral solution of $ax + by = n$ is given by

$$x = \left[\frac{n}{a}\right] - \left[\frac{n'}{a}\right] + \left[\frac{n''}{a}\right] - \cdots + (-1)^{i-1}\left[\frac{n^{(i-1)}}{a}\right] + (-1)^i b\xi,$$

$$y = \left[\frac{n}{b}\right] - \left[\frac{n'}{b}\right] + \left[\frac{n''}{b}\right] - \cdots + (-1)^{i-1}\left[\frac{n^{(i-1)}}{b}\right] + (-1)^i a\eta,$$

where ξ, η take the $\omega + 1$ sets of integral values $\geqq 0$ which satisfy $\xi + \eta = \omega$. Here ω is such that $n^{(i)} = \omega ab$. Thus if τ is the greatest integer $\leqq n/(ab)$, and $n = \tau ab + v$, then $\omega = \tau$ or $\tau + 1$, according as $ax + by = v$ has positive integral solutions or not.

M. A. Stern[120] gave Barlow's[118] result.

A. D. Wheeler[121] noted that if $ax+by=c$ has the least positive solution $x = v$, it has the solutions $x = v + b$, etc., and hence n positive solutions if $c > nab$. The least and greatest values of c for n positive solutions are $(n - 1)ab + a + b$ and $(n + 1)ab$. If $c = nab$ there are exactly $n - 1$ solutions. If $c = nab + ax' + by'$, there are $n + 1$ solutions.

[118] Theory of Numbers, London, 1811, 324.

[119] Quar. Jour. Math., 1, 1855–7, 370–3; Nouv. Ann. Math., 17, 1858, 127–130. Oeuvres, I, 440. Cf. Crocchi.[136]

[120] Jour. für Math., 55, 1858, 210.

[121] The Math. Monthly (ed., Runkle), New York, 2, 1860, 56, 193–4.

J. J. Sylvester[122] stated two theorems on the number $(n; a, b)$ of positive integral solutions of $ax + by = r$ for the values $r = 0, 1, \cdots, n$:

$$(n; a, b) = \tfrac{1}{2}k(kab + a + b + 2n' - 1) + (n'; a, b),$$

if k and n' are positive integers for which $n + 1 = kab + n'$;

$$(\nu; a, b) = (\nu'; a', b') - \left(\nu' - \left[\frac{a'\nu}{a}\right]\right) \cdot \left[\frac{a\nu' - \nu a' + 1}{a'}\right], \quad \nu < ab, \quad \nu' = \left[\frac{b'\nu}{b}\right],$$

where a', b' are positive integers such that $ab' - ba' = 1$, $a' < a$, $b' < b$.

E. Catalan[123] made use of the known fact that the solutions of

$$ax + by = n$$

are $x = \alpha - b\theta$, $y = \beta + a\theta$, if α, β is one set of positive integral solutions. Let a, b, n be positive. Then the positive solutions have $\theta < \alpha/b, \theta > -\beta/a$, which are equivalent to $\theta < a\alpha/(ab), \theta > (a\alpha - n)/(ab)$. Hence $\omega = [n/ab]$ or $[n/ab]+1$. Writing $n = abq + n'$, $0 \leqq n' < ab$, he proved that $ax + by = n$ has $q + 1$ or q positive solutions according as $ax' + by' = n'$ has a positive solution or none.

C. de Polignac[124] remarked that $ax + by = n$ may be solved graphically by means of a lattice whose initial rectangle has the base a and altitude b. He concluded that, if $\tau = [n/ab]$, $\omega = \tau$ if the remainder obtained by dividing n by ab is $< b\beta$, where β is the least positive y, while $\omega = \tau + 1$ in the contrary case.

E. Catalan[125] stated and E. Cesàro proved that, if we count the integral solutions $\geqq 0$ of each of the equations $x + 2y = n - 1, 2x + 3y = n - 3, 3x + 4y = n - 5, \cdots$, the total number of solutions equals the excess of $n+2$ over the number of divisors of $n+2$. For, $px + (p+1)y = n - (2p-1)$ has

$$\left[\frac{n+1}{p}\right] - \left[\frac{n+1}{p+1}\right] - \epsilon$$

solutions $\geqq 0$, where $\epsilon = 1$ or 0 according as $p + 1$ is or is not a divisor of $n + 2$.

E. Cesàro stated and J. Gillet[126] proved that if we count the integral solutions $\geqq 0$ of each of the equations $x + 4y = 3n - 1, 4x + 9y = 5n - 4, 9x + 16y = 7n - 9, \cdots$, the total number of solutions is n.

E. Catalan stated and E. Cesàro and H. Schoentjes[127] proved that if we count the integral solutions $\geqq 0$ of each of the $n + 1$ equations

[122] Comptes Rendus Paris, 50, 1860, 367; Coll. Math. Papers, II, 176.
[123] Mélanges Math., 1868, 21–23; Mém. Soc. Sc. Liège, (2), 12, 1885, 23 (Mélanges Math. I). Mathesis, 10, 1890, 220–2.
[124] Bull. Math. Soc. France, 6, 1877–8, 158. E. M. Laquière, *ibid.*, 7, 1878–9, 89, simplified Polignac's work. A resumé of both is given by S. Günther, Zeitschr. Math. Naturw. Unterricht, 13, 1882, 98–101.
[125] Nouv. Ann. Math., (3), 1, 1882, 528; (3), 2, 1883, 380–2.
[126] Mathesis, 2, 1882, 208; 5, 1885, 59–60.
[127] *Ibid.*, 2, 1882, 158; 3, 1883, 87–91.

$x + 2y = n$, $2x + 3y = n - 1$, $\cdots$, $(n + 1)x + (n + 2)y = 0$, the total number of solutions is $n + 1$.

Cesàro[128] proved the last theorem with n replaced by $n - 1$, by showing that $p(x + y + 1) + y = n$ has exactly $N_p = [n/p] - [n/(p + 1)]$ integral solutions ≥ 0. Also,

$$N_p + N_{p+1} + \cdots + N_n = \left[\frac{n}{p}\right],$$

while $N_1 + N_3 + N_5 + \cdots$ equals the difference between the number of odd divisors and the number of even divisors of $1, 2, \cdots, n$. The number of integral solutions ≥ 0 of $x + 2y = 2(n - 1)$, $2x + 3y = 2(n - 2)$, $\cdots$, $nx + (n + 1)y = 0$ is the number of non-divisors of $2n + 1$. As a generalization, $px + (p - 1)y = k(n - p)$, for $p = 1, \cdots, n$, have

$$M = M_1 + \cdots + M_n$$

integral solutions ≥ 0, where $M_p = [kn/p] - [(kn + k - 1)/(p + 1)]$ is the number of solutions of the equation written; for $k = 3$, M equals the sum of the numbers of divisors of $3n + 1$ and $3n + 2$. [The preceding results are special cases of a formula given by Lerch in 1888; cf. Gegenbauer,[29] p. 227 of Vol. I of this History.] As a generalization of Catalan's[127] theorem, the total number of integral solutions ≥ 0 of

$$(1 + jk)x + (1 + \overline{j + 1}\, k)y = k(n - j - 1) \quad (j = 0, 1, \cdots)$$

is n. Given a set $x = -\alpha$, $y = \beta$ of integral solutions of $ax + by = n$, the number of integral solutions ≥ 0 is $[\beta/a] - [(\alpha - 1)/b]$.

Consider a set $u_1, u_2, \cdots$ of positive integers each prime to the term following it. Let $v_1, v_2, \cdots$ be integers and determine a series of w's by

$$w_p = v_p u_{p+1} - (1 + v_{p+1})u_p.$$

If w_r is the first negative term, the total number of integral solutions ≥ 0 of

$$u_p x + u_{p+1} y = w_p \qquad (p = 1, \cdots, r - 1)$$

is $[v_1/u_1] - [v_r/u_r]$, since the equation written has $[v_p/u_p] - [v_{p+1}/u_{p+1}]$ solutions ≥ 0. The case $v_p = n$, $u_p = p^2$, gives the result of Cesàro.[126]

He quoted (p. 273) from a letter from Hermite the result that

$$\left[\frac{n - b}{a}\right] + \left[\frac{n - 2b}{a}\right] + \left[\frac{n - 3b}{a}\right] + \cdots = \left[\frac{n - a}{b}\right] + \left[\frac{n - 2a}{b}\right] + \cdots,$$

each member being the number μ of sets of positive integers for which $ax + by \leq n$. Henceforth, let a and b be relatively prime. Then the number of integral solutions ≥ 0 of $ax + by = n$ is known to be $N_n = [n/ab] + r$, where $r = 0$ or 1. Cesàro noted (p. 278) that $r = 1$ if the remainder R obtained by dividing n by ab is of the form $\rho a + \sigma b$, where ρ, σ are integers ≥ 0, and $r = 0$ in the contrary case [Catalan[123]]. This theorem, which may be expressed in the form $N_n - N_R = [n/ab]$, is

<hr>

[128] Mém. Soc. Roy. Sc. de Liège, (2), 10, 1883, 263–283.

proved in two ways, one by use of a geometric process communicated to him by Lucas: Given one point on the line $ax + by = n$ with integral coordinates $\geqq 0$, it is easy to find all such points. If M is the point with the maximum abscissa, we get a second point M' by subtracting b from the abscissa of M and adding a to the ordinate of M. From M' we obtain similarly a new point, etc.

Cesàro stated and N. Goffart[129] proved that the total number of integral solutions $\geqq 0$ of

$$x + 4y = 3(n - 1), \qquad 4x + 9y = 5(n - 2), \qquad 9x + 16y = 7(n - 3), \qquad \cdots$$

is n.

J. Gillet[130] stated that the sum of the numbers of solutions of

$$p^m x + (p + 1)^m y = \{(p + 1)^m - p^m\}n - p^m \qquad (p = 1, \cdots, n)$$

is n, a generalization of the theorems by Cesàro[126] and Catalan.[127]

E. Lucas[131] proved Catalan's[123] result and added the remark that there are $\frac{1}{2}(a - 1)(b - 1)$ values of his n' for which $ax + by = n'$ has no solutions $\geqq 0$. In the continued fraction for a/b, let α/β be the convergent of rank $n - 1$ immediately preceding a/b. Then, writing r for n', we have the solution $x_0 = (-1)^n r\beta$, $y_0 = -(-1)^n r\alpha$ of $ax + by = r$. The sum of the squares of the values $x = x_0 + bt$, $y = y_0 - at$, giving the general solution, is a minimum for $t = s/(a^2 + b^2)$, where $s = (-1)^{n-1}(a\alpha + b\beta)r$. Let ρ_1 be the least positive remainder and $-\rho_2$ the greatest negative remainder when s is divided by $k = a^2 + b^2$. Then the sets of minimum solutions are given by

$$kx_1 = ar - b\rho_1, \qquad ky_1 = br + a\rho_1, \qquad kx_2 = ar + b\rho_2, \qquad ky_2 = br - a\rho_2.$$

In only one of the sets are the unknowns $\geqq 0$. Hence $ax + by = r$ is solvable in integers $\geqq 0$ if and only if one of $ar - b\rho_1$ and $br - a\rho_2$ is not negative.

E. Catalan[132] showed by an example that Lucas' last method requires long computations. He noted (*ibid.*, 241-3) that, if $\omega(n)$ denotes the number of integral solutions $\geqq 0$ of $px + qy = n$,

$$1 + 2\omega(1) + 2^2\omega(2) + \cdots + 2^{pq-p-q}\omega(pq - p - q) = \frac{2^{pq} - 1}{(2^p - 1)(2^q - 1)}.$$

A. S. Werebrusow[133] noted that $\omega = (n - b\beta - a\alpha)/ab$, if β is the least positive y, and α the greatest negative x.

L. Salkin[134] employed the argument of Catalan[123] to show that $\omega = q$ or $q + 1$, according as $d \leqq d'$ or $d > d'$, where, if $-l, m$ is one set of solutions, $d = l/b - [l/b]$, $d' = m/a - [m/a]$.

[129] Nouv. Ann. Math., (3), 3, 1884, 399, 539-40.
[130] Mathesis, 6, 1886, 32.
[131] Mathesis, 10, 1890, 129-132; Théorie des nombres, 1891, 479-484; Jour. de math. spéc., 1886, 20-22.
[132] Mathesis, 10, 1890, 197-9.
[133] Spaczinski's Bote Math., Odessa, 1901, Nos. 298, 299.
[134] Mathesis, (3), 2, 1902, 107-9.

V. Bernardi[135] would find the positive integral solutions of $ax + by = k$ by employing the remainders r_1', r_1'' and quotients q_1', q_1'' obtained on dividing $k - b$ by a and $k - a$ by b. Thus

$$ax_1 + by_1 = k_1, \qquad k_1 = k - a - b - r_1' - r_1''.$$

Similarly, $ax_2 + by_2 = k_2, \cdots, ax_m + by_m = k_m = k_{m-1} - a - b - r_m' - r_m''$, where r_m', r_m'' are the remainders and q_m', q_m'' the quotients obtained on dividing $k_{m-1} - b$ by a and $k_{m-1} - a$ by b. In this way we find a value u of m such that a zero remainder results from that one of the two divisions in which the divisor is the smaller of a, b, or such that the remainder from the other division is zero or is divisible by the smaller coefficient. Then k_u is divisible by the larger or the smaller of a, b in the respective cases. The positive integral solutions with k_u divisible by a are

$$x_u = k_u/a - nb, \qquad y_u = na \quad (n = 0, 1, \cdots, [k_u/ab]).$$

Then all positive integral solutions of the given equation are

$$x = (-1)^u x_u + q_1' - q_2' + \cdots + (-1)^{u-1} q_u',$$
$$y = (-1)^u y_u + q_1'' - q_2'' + \cdots + (-1)^{u-1} q_u''.$$

Cf. Hermite.[119]

L. Crocchi[136] noted that Hermite's[119] formulas do not give merely the integral solutions. Thus, if $n < a$, $n < b$, they give $x = \pm b\xi$, $y = \pm a\eta$, $\xi + \eta = \pm n/(ab)$, which lead to fractional solutions of $ax + by = n$. Crocchi therefore transformed Hermite's formulas so that the resulting formulas give merely positive integral solutions. Set

$$n = \left[\frac{n}{a}\right] a + r = \left[\frac{n}{b}\right] b + s, \qquad n' = n - r - s, \qquad n' = \left[\frac{n'}{a}\right] a + r', \qquad \cdots.$$

Then

$$\frac{s}{a} = \left[\frac{n}{a}\right] - \frac{n'}{a} = \left[\frac{n}{a}\right] - \left[\frac{n'}{a}\right] - \frac{r'}{a}, \qquad \left[\frac{n'}{a}\right] = \left[\frac{n}{a}\right] - \left[\frac{s}{a}\right]_+,$$

where $[s/a]_+$ is the quotient by excess of s by a. Similarly,

$$\left[\frac{n''}{a}\right] = \left[\frac{n}{a}\right] - \left[\frac{s}{a}\right]_+ - \left[\frac{s'}{a}\right]_+.$$

Taking alternate signs and adding, we get, for m even,

$$x' = \left[\frac{n}{a}\right] - \left\{\left[\frac{s'}{a}\right]_+ + \left[\frac{s'''}{a}\right]_+ + \cdots + \left[\frac{s^{(m)}}{a}\right]_+\right\},$$
$$y' = \left[\frac{n}{b}\right] - \left\{\left[\frac{r'}{b}\right]_+ + \left[\frac{r'''}{b}\right]_+ + \cdots + \left[\frac{r^{(m)}}{b}\right]_+\right\},$$

[135] Atti società italiana per il progresso delle scienze, 2, 1908, 317-8.
[136] Il Boll. di Matematica Gior. Sc.-Didat., 7, 1908, 229-236.

and, for m odd,

$$x' = \left[\frac{n}{a}\right]_+ + \left[\frac{s''}{a}\right]_+ + \cdots + \left[\frac{s^{(m-1)}}{a}\right]_+,$$

$$y' = \left[\frac{r}{b}\right]_+ + \left[\frac{r''}{b}\right]_+ + \cdots + \left[\frac{r^{(m-1)}}{b}\right]_+.$$

Then $x = x' + (-1)^{m+1}b\xi$, $y = y' + (-1)^{m+1}a\eta$, $\xi + \eta = n^{(m)}/(ab)$.

L. Crocchi[137] noted that, if in Hermite's[119] process we have reached the dividend $n^{(p)} = aQ + r_p = bQ' + r'_p$, then $n^{(p+1)} = n^{(p)} - r_p - r'_p$. For example, consider $5x + 11y = 488$.

	Residues		Quotients	
Dividends	by 5	by 11	by 5	by 11
488	3	4	97	44
481	1	8	96	43
472	2	10	94	42
460	0	9	92	41
451	1	0	90	41
450	0	10	90	40
440	0	0	88	40

Here $481 = 488 - 3 - 4$, etc. Thus

$$x' = 97 - 96 + 94 - 92 + 90 - 90 + 88 = 91,$$
$$y' = 1 + 1 + 1 + 40 = 43, \qquad x = 91 - 11m,$$
$$y = 43 - 5n, \qquad\qquad m + n = 440/(5 \cdot 11) = 8.$$

To find x' more readily, use the second, fourth and sixth entries 8, 9, 10 in the third column and set

$$I_2 = 1 + \left[\frac{8}{5}\right] = 2, \qquad I_6 = \left[\frac{10}{5}\right] = 2,$$

$$I_4 = 1 + \left[\frac{9}{5}\right] = 2, \qquad x' = 97 - I_2 - I_4 - I_6 = 91.$$

Similarly, from the second column, $y' = 44 - 1 - 0 - 0 = 43$. But if the number of operations had been even, we would have used I_1, I_3, I_5.

L. Rassicod,[138] V. A. Lebesgue,[139] G. Chrystal,[140] L. Aubry[141] and E. Cesàro[142] evaluated ω by known methods. Cf. Laguerre[91] of Ch. III.

---

[137] Il Pitagora, Palermo, 15, 1908–9, 29–33.
[138] Nouv. Ann. Math., 17, 1858, 126–7.
[139] Exercices d'analyse numérique, 1859, 52–3.
[140] Algebra, 2, 1889, 445–9; ed. 2, vol. 2, 1900, 473–6.
[141] L'enseignement math., 9, 1907, 302.
[142] Mém. Soc. Roy. Sc. de Liège, (3), 9, 1912, No. 13.

G. B. Mathews[142a] proved that, if $\psi(n)$ is the number of positive integral solutions of $x + y = n$ in which $3x \geqq 4y$, $2x \leqq 7y$, then

$$\Sigma \psi(n)x^n = (1 + x^3 + \cdots + x^{13})/\{(1 - x^7)(1 - x^9)\}.$$

For the problem in n instead of two unknowns, see Ch. III.

One Linear Equation in three Unknowns.

T. F. de Lagny[10] (p. 595) treated $py = ax + z$ by giving values to z which are the successive multiples of the g.c.d. of p and a. The methods of de Paoli[62] and Mac Mahon[48] were given above.

Several[143] found the 12 sets of positive integral solutions of

$$10x + 11y + 12z = 200.$$

L. Euler[144] treated $Aa + Bb + Cc = 0$. For example,

$$49a + 59b + 75c = 0.$$

Divide by 49 and set $a + b + c = d$. Thus $10b + 26c + 49d = 0$. Divide by 10 and proceed as before. We ultimately get all integral solutions:

$$a = -8e - 7f, \qquad b = 13e + 2f, \qquad c = 3f - 5e.$$

P. Paoli[145] solved $5x + 8y + 7z = 50$ by successive substitutions:

$$\begin{aligned}
x + y &= t, & 5t + 3y + 7z &= 50, \\
y + t &= t', & 3t' + 2t + 7z &= 50, \\
t + t' &= t'', & 2t'' + t' + 7z &= 50.
\end{aligned}$$

Since a coefficient is now unity, the solution is evident.

A. Cauchy[146] proved that every solution of $ax + by + cz = 0$ is given by

$$x = bw - cv, \qquad y = cu - aw, \qquad z = av - bu,$$

if the g.c.d. of a, b, c is unity.

V. Bouniakowsky[147] proved Cauchy's[146] result by solving

$$ax + by + cz = 0, \qquad a'x + b'y + c'z = h', \qquad a''x + b''y + c''z = h'',$$

the two adjoined equations having arbitrary coefficients. Then

$$x = bw - cv,$$

etc., where $u = (a'h'' - h'a'')/\triangle$, etc., $\triangle$ being a determinant of order three.

[142a] Math. Quest. and Solutions, 6, 1918, 62–64.

[143] The Gentleman's Diary, or Math. Repository, 1743; Davis' ed., London, 1, 1814, 45–7.

[144] Opus. anal., 2, 1785 [1775], 91; Comm. Arith. Coll., II, 99.

[145] Elementi d'Algebra di Pietro Paoli, Pisa, 1, 1794, 162.

[146] Exercices de math., 1, 1826, 234. Oeuvres de Cauchy, (2), 6, 1887, 287. Extr. by J. A. Grunert in Archiv Math. Phys., 7, 1846, 305–8.

[147] Bull. Acad. Sc. St. Pétersbourg, 6, 1848, 196–9.

V. A. Lebesgue[148] noted that, if the g.c.d. of a, b, c is unity, all solutions of

$$(1) \qquad ax + by + cz = d$$

are given by

$$x = da\delta + c\alpha u + vb/D, \qquad y = d\beta\delta + c\beta u - va/D, \qquad z = d\gamma - Du,$$

where u and v are arbitrary, $a\alpha + b\beta = D$, D being the g.c.d. of a, b, and $D\delta + c\gamma = 1$.

H. J. S. Smith[149] stated that if a, b, c and a', b', c' are two sets of solutions of $Ax + By + Cz = 0$, where A, B, C have no common divisor, the complete solution is

$$x = at + a'u, \qquad y = bt + b'u, \qquad z = ct + c'u,$$

if and only if there is no common divisor of

$$bc' - b'c, \qquad ca' - ac', \qquad ab' - a'b.$$

A. D. Wheeler[150] treated (1) by taking 1, 2, $\cdots$ for z until we reach a value for which $ax_1 + by_1 = d - cz < a + b$ and hence is not solvable. By simplifying this method, he found the number of solutions.

L. H. Bie[151] expressed the general solution of (1) in terms of the residues of $d - pc$ modulo b.

C. de Comberousse[152] employed the g.c.d. δ of a, b. Let the g.c.d. of δ and c divide d. Then $d - cz = \delta\theta$ has an infinitude of solutions z, θ. For each θ, $xa/\delta + yb/\delta = \theta$ has an infinitude of solutions. If α, β, γ is one set of solutions of (1), every solution is given by

$$x = \alpha - b\theta + c\theta', \ y = \beta + a\theta + c\theta'', \ z = \gamma - a\theta' - b\theta'' \ (\theta, \theta', \theta'' \text{ arbitrary}).$$

A. Pleskot[153] treated (1) by continued fractions.

While in various books[154] on algebra the solution of (1) involves three parameters, that by G. M. Testi[155] involves only two. Let the greatest common divisor δ of a and b be prime to c. Then

$$\frac{a}{\delta}x + \frac{b}{\delta}y = t, \qquad \delta t + cz = d.$$

The second has the general solution $t_0 - c\phi$, $z_0 + \delta\phi$, if t_0, z_0 is one solution. All solutions of the first are given by $x = x_0 t - \theta b/\delta$, $y = y_0 t + \theta a/\delta$, if x_0, y_0 is a solution of

$$\frac{a}{\delta}x + \frac{b}{\delta}y = 1.$$

---

[148] Exercices d'analyse numérique, Paris, 1859, 60.

[149] British Assoc. Report, 1860, II, 6; Coll. Math. Papers, I, 365–6.

[150] The Math. Monthly (ed., Runkle), New York, 2, 1860, 407–410.

[151] Tidsskrift for Mat., 2, 1878, 168–78.

[152] Algèbre supérieure, 1, 1887, 179–183.

[153] Casopis, Prag, 22, 1893, 71.

[154] Cf. J. Bertrand, Traité élém. d'algèbre, 1850; transl. by E. Betti, Florence, 1862, 285.

[155] Periodico di Mat., 13, 1898, 177.

Thus (1) has the solution $\alpha = x_0 t_0$, $\beta = y_0 t_0$, $\gamma = z_0$, and also

$$x = \alpha - cx_0\phi - \theta b/\delta,$$
$$y = \beta - cy_0\phi + \theta a/\delta,$$
$$z = \gamma + \delta\phi.$$

The latter give all integral solutions of (1) when ϕ and θ take all positive and negative integral values and zero. A like result was given by F. Giudice.[156]

*H. Ruoss[157] showed graphically how to find those values of x, y, z in (1) which satisfy certain restrictions, e.g., are all positive.

One Linear Equation in $n > 3$ Unknowns.

Brahmegupta[2] and Bháscara[5] assigned values to all but two of the unknowns.

T. Moss[158] tabulated the 412 sets of positive integral solutions of

$$17v + 21x + 27y + 36z = 1000.$$

C. F. Gauss[159] noted that, if the constant term is a multiple of the g.c.d. g of the coefficients of the unknowns, then g is a linear function of those coefficients, and the equation is solvable in integers.

V. Bouniakowsky[147] would solve $ax + by + cz + du = 0$ by adjoining three equations $c'x + \cdots = h'$, etc., and solving the system. The general solution of the given equation is thus

$$x = dp - cq + br, \qquad z = dr' - bq' + aq,$$
$$y = - dp' + cq' - ar, \qquad u = - cr' + bp' - ap,$$

where p, q, r, p', q', r' are arbitrary. He gave a like result for five unknowns and outlined the law for n unknowns. V. Schäwen[160] gave the same method.

B. Jaufroid[161] assumed that there is no common divisor of $a, \cdots, m$ in

$$(1) \qquad\qquad ax + by + cz + \cdots + mu + n = 0.$$

First, let a and b be relatively prime. Then

$$aA + bA_1 + c = 0, \qquad \cdots, \qquad aL + bL_1 + m = 0, \qquad aM + bM_1 + n = 0$$

are solvable for $A, A_1, \cdots$, so that (1) is satisfied by

$$x = Az + Bv + \cdots + Lu + M - bt,$$
$$y = A_1z + B_1v + \cdots + L_1u + M_1 + at.$$

Second, let δ be the g.c.d. of a, b and let δ and c be relatively prime. Set $a = a_1\delta$, $b = b_1\delta$, and

$$(2) \qquad\qquad a_1x + b_1y = p.$$

[156] Giornale di Mat., 36, 1898, 227.
[157] Korresp. Bl. f. d. höheren Schulen Württembergs, Stuttgart, 9, 1912, 481–4.
[158] Ladies' Diary, 1774, 35–6, Quest. 658; T. Leybourn's Math. Quest. L. D., 2, 1817, 374–6.
[159] Disquisitiones Arith., 1801, art. 40; Werke, I, 32.
[160] Zeitschr. Math. Naturw. Unterricht, 9, 1878, 111–8.
[161] Nouv. Ann. Math., 11, 1852, 158.

Then (1) becomes $\delta p + cz + \cdots + mu + n = 0$ and is satisfied by

$$z = B_2 v + \cdots + L_2 u + M_2 + \delta t, \qquad p = B_3 v + \cdots + L_3 u + M_3 - ct.$$

For these values, (2) becomes $a_1 x + b_1 y - B_3 v - \cdots = 0$. Thus by the first case we obtain solutions x, y, z in terms of v, $\cdots$, u, t, t'. A similar method applies when the g.c.d. of a, b, c is prime to d, etc.

V. A. Lebesgue[162] noted that if a and b are relatively prime, we may set $a\alpha + b\beta = 1$; then the general solution of (1) is

$$x = Q\alpha + bw, \qquad y = Q\beta - aw, \qquad Q = -n - cz - \cdots - mu.$$

But if no two of the coefficients are relatively prime, proceed as for

$$a_1 x_1 + \cdots + a_5 x_5 = a_6.$$

Set $\Delta_1 = a_1$, and let Δ_i be the g.c.d. of a_1, $\cdots$, a_i for $i = 2$, $\cdots$, 5. Remove from a_6 the necessary factor Δ_5, so that now $\Delta_5 = 1$. Determine integers α_i, β_i such that $\Delta_i \beta_i + a_{i+1}\alpha_{i+1} = \Delta_{i+1}$ $(i = 1, \cdots, 4)$. Solve each of $\Delta_{i-1} y_{i-1} + a_i x_i = \Delta_i y_i$ $(i = 2, 3, 4)$, where $y_1 = x_1$, and $\Delta_4 y_4 + a_5 x_5 = a_6$. Thus

$$y_{i-1} = \beta_{i-1} y_i + z_i a_i / \Delta_i, \qquad y_4 = \beta_4 a_6 + a_5 z_5,$$
$$x_i = \alpha_i y_i - z_i \Delta_{i-1} / \Delta_i, \qquad x_5 = \alpha_5 a_6 - \Delta_4 z_5,$$

for $i = 2$, 3, 4. Eliminating the y's, we get x_1, $\cdots$, x_5 in terms of the parameters z_2, $\cdots$, z_5.

E. Betti[154] gave without proof a formula for the integral solutions of $a_1 x_1 + \cdots + a_n x_n = a$, where a_1, $\cdots$, a_n have no common divisor, and α_1, $\cdots$, α_n is a particular set of solutions:

$$x_1 = \alpha_1 + a_2\theta_2 + a_3\theta_3 + \cdots + a_{n-1}\theta_{n-1} + a_n\theta_n,$$
$$x_2 = \alpha_2 - a_1\theta_2 + a_3\theta_3' + \cdots + a_{n-1}\theta_{n-1}' + a_n\theta_n',$$
$$x_3 = \alpha_3 - a_1\theta_3 - a_2\theta_3' + \cdots + a_{n-1}\theta_{n-1}'' + a_n\theta_n'',$$
$$\cdots\cdots\cdots\cdots\cdots\cdots\cdots\cdots\cdots\cdots\cdots\cdots\cdots$$
$$x_{n-1} = \alpha_{n-1} - a_1\theta_{n-1} - a_2\theta_{n-1}' - \cdots - a_{n-2}\theta_{n-1}^{(n-3)} + a_n\theta_n^{(n-2)},$$
$$x_n = \alpha_n - a_1\theta_n - a_2\theta_n' - \cdots - a_{n-2}\theta_n^{(n-3)} - a_{n-1}\theta_n^{(n-2)},$$

where the $n(n-1)/2$ numbers $\theta_j^{(i)}$ are arbitrary. F. Giudice[156] proved that every solution is of this form and gave a method (based upon equations in two variables) of obtaining the general solution in terms of $n - 1$ parameters.

C. G. J. Jacobi[163] treated in several ways the solution of

$$\alpha_1 x_1 + \alpha_2 x_2 + \cdots + \alpha_n x_n = fu,$$

where f is the g.c.d. of α_1, $\cdots$, α_n. Let $[a, b]$ be the g.c.d. of a, b. One obtains easily all solutions of

$$\alpha_1 x_1 + \alpha_2 x_2 = f_2 y_2, \qquad f_2 = [\alpha_1, \alpha_2],$$
$$f_2 y_2 + \alpha_3 x_3 = f_3 y_3, \qquad f_3 = [f_2, \alpha_3],$$
$$f_3 y_3 + \alpha_4 x_4 = f_4 y_4, \qquad f_4 = [f_3, \alpha_4], \qquad \cdots.$$

[162] Exercices d'analyse numérique, 1859, 58.
[163] Jour. für Math., 69, 1868, 1–28; Werke, VI, 355–384 (431).

Adding, we get the given equation since $f_n = f$. His second method consists in treating these equations taken in reverse order, after each is divided by the f_i in the second member. He noted that the method of Euler[144] is applicable also to $Aa + Bb + Cc = u$.

K. Weihrauch[164] denoted by $E(M : N)$ the integral part obtained on dividing M by N, and by $R(M : N)$ the remainder. Thus [if $A_1 \neq 0$]

$$(3) \qquad A_1x_1 + A_2x_2 + \cdots + A_nx_n = A$$

gives

$$x_1 = E(A : A_1) - x_2E(A_2 : A_1) - \cdots - x_nE(A_n : A_1) + t_1,$$

$$t_1 = \frac{1}{A_1}\{R(A : A_1) - x_2R(A_2 : A_1) - \cdots - x_nR(A_n : A_1)\}.$$

Treating the latter similarly, we get x_2. Finally, we get a relation between x_{n-1}, x_n, whose solution involves a new parameter t_{n-1}. Thus

$$(4) \qquad x_i = M_i + a_{i1}t_1 + \cdots + a_{in-1}t_{n-1} \qquad (i = 1, \cdots, n),$$

in which $M_1, \cdots, M_n$ is a set of solutions of (3), and

$$A_1a_{1j} + A_2a_{2j} + \cdots + A_na_{nj} = 0 \quad (j = 1, \cdots, n - 1).$$

The condition that (4) shall give all solutions of (3) is

$$\frac{1}{A_1}|\, a_{ij}\,| = \pm 1 \quad (i = 2, \cdots, n;\ j = 1, \cdots, n - 1),$$

where the symbol denotes an $(n - 1)$-rowed determinant.

T. J. Stieltjes[165] reduced $a_1x_1 + \cdots + a_{n+1}x_{n+1} = u$ to an equation in one variable. If $\lambda = (a_1, a_2)$ is the g.c.d. of a_1, a_2, we can find relatively prime integers α, γ such that $a_1\alpha + a_2\gamma = \lambda$. Taking $\beta = - a_2/\lambda$, $\delta = a_1/\lambda$, we have $\alpha\delta - \beta\gamma = 1$. Set

$$x_1 = \alpha x_1' + \beta x_2', \qquad x_2 = \gamma x_1' + \delta x_2'.$$

Then the initial equation is equivalent to

$$(a_1, a_2)x_1' + a_3x_3 + \cdots + a_{n+1}x_{n+1} = u.$$

Similarly, we can replace the first two new terms by $(a_1, a_2, a_3)x_1''$, etc., and finally get $dx_1^{(n)} = u$, where $d = (a_1, \cdots, a_{n+1})$ is the g.c.d. of $a_1, \cdots, a_{n+1}$. Giving to $x_2', \cdots, x_{n+1}'$ all sets of integral values, we get all solutions of the initial equation if it be solvable, viz., if u be divisible by d. A system of n independent sets of solutions is fundamental (Smith[207]) if the g.c.d. of the $n + 1$ n-rowed determinants is unity.

W. F. Meyer[166] solved (3) by use of recurring series obtained by simplifying and extending C. G. J. Jacobi's[167] generalized continued fraction algorithm.

<hr>

[164] Untersuchungen über eine Gl. 1 Gr., Diss. Dorpat, 1869. Zeitschrift Math. Phys., 19, 1874, 53.

[165] Annales Fac. Sc. Toulouse, 4, 1890, final paper, pp. 38–47.

[166] Verhand. des ersten Intern. Math.-Kongresses, 1897, Leipzig, 1898, 168–181.

[167] Jour. für Math., 69, 1868, 29–64; Werke, VI, 385–426.

R. Ayza[168] treated $ax + by + cz + du + \cdots = k$ by means of

$$ax + by = k_1, \qquad cz + du = k_2, \qquad \cdots, \qquad k_1 = k - k_2 - k_3 - \cdots,$$

where $k_2, k_3, \cdots$ are arbitrary. For m linear equations in $m + n$ variables, successive elimination gives one equation in $m + n$ variables, one in $m + n - 1$ variables, $\cdots$, one in $n + 1$ variables, which is solved as above.

A. P. Ochitowitsch[169] treated $\Sigma a_i y_i = 0$. If a_p, a_q are relatively prime,

$$y_p = za_q + ry_p', \qquad y_q = - za_p + ry_q', \qquad a_p y_p' + a_q y_q' + 1 = 0,$$

where $r = \Sigma a_i y_i$ for $i \neq p, q$. For $a_1 = p_1^{m_1} \cdots p_n^{m_n}$, where $p_1, \cdots, p_n$ are distinct primes, a set of solutions of $1 + a_1 x_1 + a_2 x_2 = 0$ is given by

$$x_1 = - \left(\frac{1 + a_2 z_1}{p_1} \right)^{m_1} \left(\frac{1 + a_2 p_1 z_2}{p_2} \right)^{m_2} \cdots \left(\frac{1 + a_2 p_1 p_2 \cdots p_{n-1} z_n}{p_n} \right)^{m_n},$$

where $z_1, \cdots, z_n$ are to be chosen to make the indicated fractions integral.

E. B. Elliott[170] recalled the fact that all sets of positive integral solutions of a linear diophantine equation in n variables are linear combinations of a finite number of " simple " [fundamental] sets of solutions $(\alpha_1, \cdots, \alpha_n)$, $\cdots$, $(\omega_1, \cdots, \omega_n)$ and that a linear combination of these simple sets is always a solution. He noted that two such combinations may give the same solution since the simple sets are usually connected by syzygies. For example, the three simple sets (103), (230), (111) of solutions of $3x = 2y + z$ are connected by the syzygy $(103) + (230) = 3(111)$, so that

$$x = t_1 + 2t_2 + t_3, \qquad y = 0t_1 + 3t_2 + t_3, \qquad z = 3t_1 + 0t_2 + t_3$$

give duplicate solutions unless we restrict t_3 to the values 0, 1, 2. In this sense, he obtained formulas giving each solution of an equation in n variables once and but once, making use of generating functions.

G. Bonfantini[171] noted that, if $a, \cdots, l, k$ have no common factor, $ax + by + \cdots + lu = k$ has integral solutions if and only if $a, \cdots, l$ have no common factor.

Several[172] found all positive integral solutions of

$$13k + 21l + 29m + 37n = 300.$$

* P. B. Villagrasa[173] treated (3).

D. N. Lehmer[173a] proved that (3) with $A = 1$ is satisfied by the co-factors of the elements of the last row of a certain determinant of value unity, those elements being any integers whose g.c.d. is 1. The general solution of (3) is deduced.

[168] Archivo de Matematicas, Madrid, 2, 1897, 21–25.
[169] Text on linear equations, Kasan, 1900.
[170] Quar. Jour. Math., 34, 1903, 348–377.
[171] Il Boll. di Matematica, Gior. Sc. Didattico, 3, 1904, 45–47.
[172] Math. Quest. Educ. Times, 7, 1905, 21–22.
[173] Revista de la Sociedad Mat. Española, 3, 1914, 149–156.
[173a] Proc. Nat. Acad. Sc., 5, 1919, 111–4; Amer. Math. Monthly, 26, 1919, 365–6.

Since the problem to solve $n = ax + by + \cdots$ in positive integers is the same as to partition n into parts a, b, $\cdots$, reference should be made to Ch. III, in particular for theorems on the number of solutions.

System of linear equations.

Chang Ch'iu-chien[174] (sixth century A.D.) treated a problem equivalent to

$$x + y + z = 100, \qquad 5x + 3y + \tfrac{1}{3}z = 100,$$

and gave the answers (4, 18, 78), (8, 11, 81), (12, 4, 84),

Mahāvīrācārya[175] (about 850 A.D.) treated special cases of

$$x + y + z + w = n, \qquad ax + by + cz + dw = p.$$

Shodja B. Aslam[176] (about 900 A.D.), an Arab known as Abū Kamil, found positive integral solutions of $x + y + z = 100, 5x + y/20 + z = 100$, whence $y = 4x + 4x/19$, $x = 19$; $x + y + z = 100 = \tfrac{1}{3}x + \tfrac{1}{2}y + 2z$, whence $x = 60 - 9y/10$, $y = 10m$, $m = 1, \cdots, 6$;

$$x + y + z + u = 100, \qquad 4x + \tfrac{1}{10}y + \tfrac{1}{2}z + u = 100,$$

whence $x = \tfrac{3}{10}y + \tfrac{1}{6}z$, with 98 sets of solutions (two of which are omitted). When the last equation is changed to $2x + \tfrac{1}{2}y + \tfrac{1}{3}z + u = 100$, there are 304 sets of solutions. There is no solution of

$$x + y + z = 100 = 3x + y/20 + \tfrac{1}{3}z.$$

There are 2676 sets of positive integral solutions of

$$x + y + z + u + v = 100, \qquad 2x + \tfrac{1}{2}y + \tfrac{1}{3}z + \tfrac{1}{4}u + v = 100.$$

Alhacan Alkarkhi[177] (eleventh or twelfth century) treated the system

$$\tfrac{1}{2}x + w = \tfrac{1}{2}s, \qquad \tfrac{2}{3}y + w = \tfrac{1}{3}s, \qquad \tfrac{5}{6}z + w = \tfrac{1}{6}s,$$

$$s \equiv x + y + z, \qquad w \equiv \tfrac{1}{3}\left(\frac{x}{2} + \frac{y}{3} + \frac{z}{6}\right),$$

by taking $z = 1$, whence $x = 33$, $y = 13$. He treated the problems of Diophantus I, 24–28, as had Diophantus, by making the indeterminate problems determinate by assigning a value to one unknown.

Leonardo Pisano,[178] in 1228, treated various linear systems, the first being that of Alkarkhi[177] without the final condition:

$$x + y + z = t, \qquad \frac{t}{2} = \frac{x}{2} + u, \qquad \frac{t}{3} = \frac{2y}{3} + u, \qquad \frac{t}{6} = \frac{5z}{6} + u.$$

---

[174] Suan-ching (Arith.). Cf. Mikami,[n] 43–44.

[175] Ganita-Sara-Sangraha.[3] Cf. D. E. Smith, Bibliotheca Math., (3), 9, 1909, 106–10.

[176] H. Suter, Bibliotheca Math., (3), 11, 1911, 110–20, gave a German transl. of a MS. copy of about 1211–8 A.D.

[177] Extrait du Fakhrî, French transl. by F. Woepcke, Paris, 1853, 90, 95–100.

[178] Scritti di L. Pisano, 2, 1862, 234–6. Cf. A. Genocchi, Annali di Sc. Mat. e Fis., 6, 1855, 169; O. Terquem, ibid., 7, 1856, 119–36; Nouv. Ann. Math., Bull. Bibl. Hist., 14, 1855, 173–9; 15, 1856, 1–11, 42–71.

These determine x, y, z, t in terms of u. Since $7t = 47u$, he took $u = 7$, whence $t = 47$, $x = 33$, $y = 13$, $z = 1$. His next indeterminate problem[179] is

$$t + x_1 = 2(x_2 + x_3), \qquad t + x_3 = 4(x_4 + x_1),$$
$$t + x_2 = 3(x_3 + x_4), \qquad t + x_4 = 5(x_1 + x_2).$$

Since the problem is impossible if x_1 and x_2 are positive, change x_1 to $- x_1$. Now $x_2 = 4x_1$. Take $x_2 = 4$, whence $x_1 = x_3 = 1$, $x_4 = 4$, $t = 11$.

For[180] $x + y + z = 30$, $\frac{1}{3}x + \frac{1}{2}y + 2z = 30$, we have $y + 10z = 120$, $y + z < 30$, $z \geqq 9$. The case $z = 10$ is impossible. For $z = 11$, we get $y = 10$, $x = 9$. The same problem with the constant term 30 replaced by 29 or 15 is treated similarly.

Finally,[181] consider the system

$$x + y + z + t = 24, \qquad \frac{x}{5} + \frac{y}{3} + 2z + 3t = 24.$$

Hence $2y + 27z + 42t = 288$, $y + z + t < 20$. Thus z is even and < 10. The cases $z = 6$, $z = 8$ are impossible. Thus there are only two solutions:

$$z = 2, \quad t = 5, \quad y = 12, \quad x = 5; \qquad z = 4, \quad t = 4, \quad y = 6, \quad x = 10.$$

Regiomontanus (1436–1476) proposed in a letter (cf. de Murr,[86] p. 144) the problem to solve in integers

$$x + y + z = 240, \qquad 97x + 56y + 3z = 16047.$$

J. von Speyer gave the solution 114, 87, 39 (de Murr, p. 167).

Estienne de la Roche[182] treated the solution in integers of

$$x + y + z = a, \qquad mx + ny + pz = b.$$

His rule [applied to the case $a = b = 60$, $m = 3$, $n = 2$, $p = \frac{1}{2}$] is as follows. Let p be the least of m, n, p. From the second equation subtract the product of the first by p; we get

$$(m - p)x + (n - p)y = b - ap \qquad [\tfrac{5}{2}x + \tfrac{3}{2}y = 30].$$

To avoid fractions, multiply by 2. Thus $5x + 3y = 60$. Although $x = 60/5$ gives an integral solution, the corresponding y is zero and is excluded. The next smaller values 11 and 10 for x lead to fractions for y, while $x = 9$ gives $y = 5$ [whence $z = 46$]. For $x = 1, 2, \cdots$, the least x yielding an integer for y is $x = 3$, whence $y = 15$, $z = 42$. The problem may be impossible, as shown by the case $a = b = 20$, $m = 5$, $n = 2$, $p = \frac{1}{2}$, whence $9x + 3y = 20$.

---

[179] Scritti, II, 238–9 (De quatuor hominibus et bursa). Genocchi,[178] 172–4. Three misprints in the account by Terquem.

[180] Scritti, II, 247–8 (De auibus emendis). Genocchi, 218–22. For analogous problems, see Liber Abbaci, Scritti, 1, 1857, 165–6.

[181] Scritti, II, 249 (Item passeres). Genocchi, 222–4.

[182] Larismetique & Geometrie, Lyon, 1520, fol. 28; 1538. Cf. L. Rodet, Bull. Math. Soc. France, 7, 1879, 171 [162].

Luca Paciuolo[183] treated the solution of

$$p + c + \pi + a = 100, \qquad \tfrac{1}{2}p + \tfrac{1}{3}c + \pi + 3a = 100,$$

giving the single solution $p = 8$, $c = 51$, $\pi = 22$, $a = 19$. Many solutions were found by P. A. Cataldi.[184]

Christoff Rudolff[185] stated the following problem. To find the number of men, women and maidens in a company of 20 persons if together they pay 20 pfennige, each man paying 3, each woman 2 and each maiden $\tfrac{1}{2}$. The answer is given to be 1 man, 5 women, 14 maidens. [The only solution of $x + y + z = 20$, $3x + 2y + \tfrac{1}{2}z = 20$ in positive integers is $x = 1$, $y = 5$, $z = 14$.] The solution is said to be found by the rule called Cecis or Virginum.

C. G. Bachet de Méziriac[186] solved in integers the system of equations

$$x + y + z = 41, \qquad 4x + 3y + \tfrac{1}{3}z = 40.$$

Multiplying the second by 3 and subtracting the first, he obtained $11x + 8y = 79$. Since $y = 9\tfrac{7}{8} - 1\tfrac{3}{8}x$, x must have one of the values 1, $\cdots$, 7. By the value of $8z$ in terms of x, $1 + 3x$ must be divisible by 8. Hence $x = 5$, so that $y = 3$, $z = 33$. He treated Rudolff's[185] and a similar system and found 81 sets of positive integral solutions of

$$x + y + z + w = 100, \qquad 3x + y + \tfrac{1}{2}z + \tfrac{1}{7}w = 100.$$

J. W. Lauremberg[187] described and illustrated by examples the rule called Cecis [Coeci] or Virginum[188] for solving indeterminate linear equations, referring to the Arabs [although known to the Indians].

René François de Sluse[189] (1622–1685) treated the problem to divide a given number b into three parts the sum of whose products by given numbers z, g, n shall be p. Call the first and second parts a and e. Then

$$za + ge + n(b - a - e) = p, \qquad a = \frac{p - nb + ne - ge}{z - n}.$$

Take $p = b = 20$, $z = 4$, $g = \tfrac{1}{2}$, $n = \tfrac{1}{4}$. Then $a = (60 - e)/15$.

Johann Prätorius[190] solved the following problem: Anna took to market 10 eggs, Barbara 30, Christina 50. Each sold a part of her eggs at the same price per egg and later sold the remainder at another price. Each

---

[183] Summa de Arithmetica, 1523, fol. 105; [Suma . . ., Venice, 1494]; same solution by N. Tartaglia, General Trattato di Nvmeri . . ., I, 1556.

[184] Regola della Quantita o Cosa di Casa, Bologna, 1618, 16–28.

[185] Künstliche Rechnung, 1526; Nürnberg, 1534, f. nvij a and b; Nürnberg, 1553 and Vienna, 1557, f. Rvii a and b.

[186] Diophantus Alex. Arith., 1621, 261–6; comment on Dioph., IV, 41.

[187] Arithmetica, Sorae, Denmark, 1643, 132–3. Cf. H. G. Zeuthen, l'intermédiaire des math., 3, 1896, 152–3.

[188] According to O. Terquem, Nouv. Ann. Math., 18, 1859, Bull. Bibl., 1–2, the term problem of the virgins arose from the 45 arithmetical Greek epigrams, Bachet,[186] pp. 349–370, and J. C. Heilbronner, Historia Math. Universae, 1742, 845. Cf. Sylvester[54] of Ch. III.

[189] MS. No. 10248 du fonds latin, Bibliothèque Nationale de Paris, f. 194, "De problematibus arith. indefinites," Prob. 2.

[190] Abentheuerlicher Glückstopf, 1669, 440. Cf. Kästner.[197]

received the same total amount of money. How many did each sell at first and what were the two prices? The answer given is that at first A sold 7, B 28, C 49 at 7 eggs per kreuzer; the remainder were sold for 3 kreuzer per egg. Thus they received 1 + 9, 4 + 6, 7 + 3 kreuzer each.

There[191] are eleven sets of positive integral solutions of

$$x + y + z = 56, \qquad 32x + 20y + 16z = 22 \cdot 56.$$

T. F. de Lagny[10] (p. 583) treated the problem of Diophantus[192] II, 18, to find three numbers such that if the first gives to the second $\frac{1}{5}$ of itself + 6, the second gives to the third $\frac{1}{6}$ of itself + 7, the third gives to the first $\frac{1}{7}$ of itself + 8, the results after each has given and taken shall be equal. To avoid fractions call the numbers $5x, 6y, 7z$. Then the first gives $x + 6$ and receives $z + 8$ and becomes $4x + z + 2$. Thus

$$4x + z + 2 = 5y + x - 1 = 6z + y - 1.$$

Eliminating z and y in turn, we get

$$y = \frac{19x + 18}{26}, \qquad z = \frac{17x + 12}{26}.$$

Their difference $(2x + 6)/26$ must be an integer. Multiply it by 8 and subtract from z; thus $x - 36$ and hence $x - 10$ is divisible by 26. Since $2x + 6$ and $2(x - 10)$ are divisible by 26, while their difference is 26, the problem is possible. We may take $x = 10 + 26k$ and get an infinity of integral solutions. He employed the same method to treat any such " double equalities " of the first degree, which may be reduced to

$$y = \frac{\pm ax \pm q}{p}, \qquad z = \frac{\pm bx \pm d}{p}.$$

The principle is to get $x \pm c$ by elimination.

N. Saunderson[13] (pp. 337–354) and A. Thacker[193] treated two equations in x, y, z in the usual way.

L. Euler[194] discussed the regula Coeci. Given

$$p + q + r = 30, \qquad 3p + 2q + r = 50,$$

eliminate r. Thus $2p + q = 20$, whence p may have any value $\leqq 10$. In general, for

$$(1) \qquad x + y + z = a, \qquad fx + gy + hz = b, \qquad f \geqq g \geqq h,$$
$$b \leqq f(x + y + z) = fa, \qquad b \geqq h(x + y + z) = ha,$$

while b must not be too near these limits fa, ha. By eliminating z, we get $\alpha x + \beta y = c$, where α and β are positive. A similar pair of equations in

[191] Ladies' Diary, 1709–10, Quest. 8; C. Hutton's Diarian Miscellany, 1, 1775, 52–3; T. Leybourn's Math. Quest. L. D., 1, 1817, 5.

[192] Diophantus used $5x, 6x, 7x$ and got $x = 18/7$. G. Wertheim, in his edition of Diophantus, 1890, proceeded as had de Lagny.

[193] A Miscellany of Math. Problems, Birmingham, 1, 1743, 161–9.

[194] Algebra, II, 1770, Cap. 2, §§ 24–30; 1774, pp. 30–41; Opera omnia, ser. 1, 1, 1911, 339–344.

four variables is treated; also

$$3x - 5y + 7z = 560, \qquad 9x + 25y + 49z = 2920.$$

E. Bézout[195] solved $x + y + z = 41$, $24x + 19y + 10z = 741$ by eliminating x and showing that the integral solutions of $5y + 14z = 243$ are $z = 5u - 3$, $y = 57 - 14u$.

Abbé Bossut[196] solved by eliminating z

$$x + y + z = 22, \qquad 24x + 12y + 6z = 36.$$

A. G. Kästner[197] treated the problem of Prätorius[190] and its generalization: Three peasants have a, b, c eggs, respectively, where a, b, c are distinct numbers. They sold x, y, z eggs respectively at the price m per egg and the remainder at n. Each received the same total amount of money. Find x, y, z, m/n. We have

$$mx + n(a - x) = my + n(b - y) = mz + n(c - z),$$

where x, $a - x$, etc., are to be positive integers. We get

$$\frac{m}{n} = \frac{b - a}{x - y} + 1 = \frac{c - b}{y - z} + 1, \qquad z = \frac{(b - c)x + (c - a)y}{(b - a)}.$$

Give successive values to x and solve the equation in y, z.

A. G. Kästner[198] discussed the " Regel Cöci." From (1),

$$y = \frac{b - ah - (f - h)x}{g - h},$$

whence

$$x \leqq \frac{b - ah}{f - h}.$$

Also, $ag + (f - g)x \geqq b$, so that we have limits for x.

J. D. Gergonne[199] considered n equations in $m > n$ variables,

$$a_{i1}x_1 + \cdots + a_{im}x_m = k_i \qquad (i = 1, \cdots, n),$$

with integral coefficients, and stated a priori that

$$x_j = T_j + A_j\alpha + B_j\beta + \cdots$$

where α, β, $\cdots$ are parameters in number $m - n$ at least. Substitute these expressions for the x's into the given equations and equate the coefficients of α, of β, etc. Some of the resulting conditions show that T_1, T_2, $\cdots$ is a set of solutions of the given equations. The remaining conditions show that the A's, the B's, $\cdots$ are sets of solutions of

$$a_{i1}x_1 + \cdots + a_{im}x_m = 0 \qquad (i = 1, \cdots, n),$$

[195] Cours de Math., 2, 1770, 94–6.
[196] Cours de Math., II, 1773; ed. 3, I, Paris, 1781, 414.
[197] Leipziger Magazin für reine u. angew. Math., 1788, 215–227.
[198] Math. Anfangsgründe, I, 2 (Fortsetzung der Rechenkunst, ed. 2, 1801, 530).
[199] Annales de Math. (ed., Gergonne), 3, 1812–13, 147–158.

and hence are determined by the matrix (a_{ij}). The same discussion was given by J. G. Garnier,[200] who remarked that the determination of the A's, B's, $\cdots$ is facilitated by the use of determinants.

J. Struve[201] reduced the solution of (1) to an equation in 2 variables.

V. Bouniakowsky[202] discussed the solution of one or more indeterminate equations, chiefly of linear type.

G. Bianchi[203] treated three linear equations in x, y, z, u, solving by determinants for x, y, z as linear functions of u and determining by inspection what positive integral values, if any, may be given to u such that the expressions for x, y, z become integers.

C. A. W. Berkhan[204] noted that if (1) have positive integral solutions, the x's are in arithmetical progression, with the common difference $g - h$.

I. Heger[205] considered a system of homogeneous equations

$$(2) \qquad k_{i1}x_1 + \cdots + k_{i\,m+n}x_{m+n} = 0 \qquad (i = 1, \cdots, n),$$

with integral coefficients. Let x_{11} be the numerically least value $\neq 0$ of x_1 in all possible sets of integral solutions, and let $x_{11}, \cdots, x_{1\,m+n}$ be one such set. Their products by ξ_1 give a set of solutions. The only possible x_1's are multiples of x_{11}. In (2) set

$$x_1 = x_{11}\xi_1, \qquad x_i = x_{1i}\xi_1 + x_i' \quad (i = 2, \cdots, m + n).$$

Then

$$k_{i2}x_2' + \cdots + k_{i\,m+n}x_{m+n}' = 0 \qquad (i = 1, \cdots, n).$$

As before, $x_2' = x_{22}\xi_2$, where x_{22} is the numerically least value $\neq 0$ of x_2' in all sets of integral solutions. Let $x_{22}, \cdots, x_{2\,m+n}$ be such a set. Proceeding in this manner, we get

$$x_1 = x_{11}\xi_1,$$
$$x_2 = x_{12}\xi_1 + x_{22}\xi_2,$$
$$x_3 = x_{13}\xi_1 + x_{23}\xi_2 + x_{33}\xi_3,$$
$$\cdots \cdots \cdots \cdots$$
$$x_m = x_{1m}\xi_1 + x_{2m}\xi_2 + x_{3m}\xi_3 + \cdots + x_{mm}\xi_m.$$

If the determinant of the coefficients of $x_{m+1}, \cdots, x_{m+n}$ in (2) is not zero, those variables are definite linear functions of $x_1, \cdots, x_m$, whence

$$x_{m+j} = x_{1\,m+j}\xi_1 + \cdots + x_{m\,m+j}\xi_m \qquad (j = 1, \cdots, n),$$

where the $x_{i\,m+j}$ may be taken integral. Giving arbitrary integral values to $\xi_1, \cdots, \xi_m$, we obtain all integral solutions of (2).

For n non-homogeneous equations in m variables, $n < m$, let all the determinants D of order n of the matrix of coefficients have the g.c.d. f;

[200] Cours d'Analyse Algébrique, ed. 2, Paris, 1814, 67–79.

[201] Erläuterung einer Regel für unbest. Aufgaben . . ., Altona, 1819.

[202] Bull. phys. math. acad. sc. St. Pétersbourg, 6, 1848, 196.

[203] Memorie di Mat. e Fis. Soc. Italiana Sc., Modena, 24, II, 1850, 280–9.

[204] Lehrbuch der Unbest. Analytik, Halle, I, 1855, 46–53.

[205] Denkschriften Akad. Wiss. Wien (Math. Nat.), 14, II, 1858, 1–122. Extract in Sitzungsber. Akad. Wiss. Wien (Math.), 21, 1856, 550–60.

consider the determinants K in which the constant terms appear in one column, and let F be the g.c.d. of the D's and K's. There exist integral solutions if and only if $f = F$; while f/F is the least common denominator of all sets of fractional solutions. Cf. Smith[207] and Frobenius.[210]

V. A. Lebesgue[206] would select, if possible, two equations $ax_1 = F(x_2, \cdots, x_n)$ and $a'x_1 = F_1(x_2, \cdots, x_n)$ from the system of linear equations such that a, a' are relatively prime. Determine r, s, p, q so that $ar - a's = 1$, $ap - a'q = 0$. Then $x_1 = rF - sF_1$, $pF - qF_1 = 0$, whence the system is reduced to the former and equations in $x_2, \cdots, x_n$ only. To solve $ax + by = cz$, $a'x + b'y = c't$, where the g.c.d. of a, b, c is unity, we may set $z = Du$, where $D = a\alpha + b\beta$ is the g.c.d. of a, b. Thus $x = c\alpha u + bv/D$, $y = c\beta u - av/D$. Then the second equation becomes $Au + Bv = c't$, which may be treated as was the first. Given a system of m linear equations in $m + n$ unknowns in which an m-rowed minor D is not zero, we get $Dx_i = f_i(y_1, \cdots, y_n)$, $i = 1, \cdots, m$. It remains to solve the congruences $f_i \equiv 0 \pmod{D}$, which can be treated by the method for linear equations.

H. J. S. Smith[207] proved that if the excess of the number of unknowns above the number of linearly independent equations is m, we can assign m sets of integral solutions (called a fundamental system of sets of solutions) such that the determinants of the matrix formed by them admit no common divisor > 1. Every set of integral solutions of the equations can be expressed linearly and with integral multipliers in terms of the fundamental system. By use of this concept he proved the theorem of Heger:[205] A system of linear equations is or is not solvable in integers according as the g.c.d. of the determinants of the matrix of the coefficients is or is not equal to the g.c.d. for the augmented matrix obtained by annexing a column composed of the constant terms (cf. Frobenius[210]). Use is made of the important elementary divisors.

H. Weber[208] considered the system of equations

$$h_i = m_1\sigma_{1i} + \cdots + m_p\sigma_{pi} + \lambda_i \qquad (i = 1, \cdots, p)$$

with integral coefficients σ_{ji} of determinant δ. If $\delta \neq 0$ we obtain every set of integers $h_1, \cdots, h_p$ and each δ^{p-1} times if we take all possible combinations of integers for $m_1, \cdots, m_p$ and let $\lambda_1, \cdots, \lambda_p$ run independently through a complete set of residues modulo δ. If $\delta = 0$, we can apply to the m's such a substitution of determinant ± 1 that the matrix (σ_{ji}) is transformed into one with columns of zeros at the right. Then by a linear substitution on $h_1, \cdots, h_p$ of determinant ± 1, we get a matrix having zeros except in the q-rowed minor in the upper left-hand corner.

E. d'Ovidio[209] treated algebraically a system of $n - r$ independent linear homogeneous equations in n unknowns and the conditions that it have the same ∞^r solutions as a second such system.

[206] Exercices d'analyse numérique, Paris, 1859, 66–75.

[207] Phil. Trans. London, 151, 1861, 293–326; abstr. in Proc. Roy. Soc., 11, 1861, 87–9. Coll. Math. Papers, I, 367–409.

[208] Jour. für Math., 74, 1872, 81.

[209] Atti R. Accad. Sc. Torino, 12, 1876–7, 334–350.

G. Frobenius[210] proved the following generalization of the theorem of Heger:[205] Several non-homogeneous linear equations have integral solutions if and only if the rank l and the g.c.d. of the l-rowed determinants of the matrix of the coefficients of the unknowns are the same as for the augmented matrix obtained by annexing a column formed by the constant terms. Again, sets of integral solutions of m independent linear homogeneous equations in n unknowns $(n > m)$ form a fundamental system if and only if the $(n - m)$-rowed determinants formed from them have no common divisor. He discussed (pp. 194–202) the equivalence under linear transformation of determinant ± 1 of two systems of m linear forms in n variables; on this subject, see Smith,[207] G. Eisenstein,[211] and G. Frobenius.[212]

Ch. Méray[213] considered a system of m linear forms

$$(3) \qquad \phi_i = a_i x + b_i y + \cdots + j_i v \qquad (i = 1, \cdots, m)$$

in $n > m$ unknowns. Multiplication of this system by the matrix

$$(4) \qquad \begin{pmatrix} \lambda_1 & \mu_1 & \cdots & \omega_1 \\ \cdot & \cdot & \cdot & \cdot \\ \lambda_m & \mu_m & \cdots & \omega_m \end{pmatrix}$$

is defined to be the operation of forming the system of m forms

$$\psi_1 = \lambda_1 \phi_1 + \lambda_2 \phi_2 + \cdots + \lambda_m \phi_m, \cdots, \psi_m = \omega_1 \phi_1 + \omega_2 \phi_2 + \cdots + \omega_m \phi_m.$$

If we multiply the latter system by a second matrix, we get a system which can be derived from (3) by multiplication by the product of the two matrices. Given a system of m forms (3) with integral coefficients, the m-rowed determinants of whose matrix of coefficients are not all zero and have the g.c.d. d, we can assign a matrix (4) of rational elements of determinant $1/d$, and a linear substitution on n variables with integral coefficients of determinant unity, such that after multiplication by the matrix and transformation by the substitution, we obtain a system of forms $\pm x_1$, $\pm x_2$, $\cdots$, $\pm x_m$. Then the system $\phi_i + k_i = 0$ $(i = 1, \cdots, m)$ have integral solutions if and only if the m-rowed determinants of the coefficients of the ϕ's have for their g.c.d. a number d dividing all the m-rowed determinants obtained from the preceding determinants by replacing the elements of an arbitrary column by the k's [Heger[205]]. When the equations have a set of integral solutions $\xi, \cdots, \psi$, all sets of integral solutions are given without duplication by

$$x = \xi + x_1 \theta_1 + \cdots + x_{n-m} \theta_{n-m}, \qquad \cdots, \qquad v = \psi + v_1 \theta_1 + \cdots + v_{n-m} \theta_{n-m},$$

where the θ's are arbitrary integers and the coefficients of any θ_j satisfy the system $\phi_i = 0$ $(i = 1, \cdots, m)$.

A. Cayley[214] suggested that, to solve a system of linear homogeneous

---

[210] Jour. für Math., 86, 1878, 171–3. Cf. Kronecker.[218]

[211] Berichte Akad. Wiss. Berlin, 1852, 350.

[212] Jour. für Math., 88, 1879, 96–116.

[213] Annales sc. de l'école normale sup., (2), 12, 1883, 89–104; Comptes Rendus Paris, 94, 1882, 1167.

[214] Quar. Jour. Math., 19, 1883, 38–40; Coll. Math. Papers, XII, 19–21.

equations in the unknowns A, B, $\cdots$, we first equate to zero as many unknowns (say A, $\cdots$, E) as possible such that there exists a solution with $F \neq 0$; we may take $F = 1$ and have a solution " beginning with $F = 1$." Next, set $F = 0$ in the initial equations and equate to zero as many of the earlier unknowns (say A, B, C) as possible such that there exists a solution with $D \neq 0$; we may take $D = 1$ and have a solution beginning with $D = 1$ and having $F = 0$. The third step might lead to a solution with $A = 1$, $D = F = 0$. Then we have a system of three standard solutions.

E. de Jonquières[215] discussed the equations, arising in Cremona transformations,

$$\sum_{i=1}^{n-1} i\alpha_i = 3(n - 1), \qquad \Sigma\, i^2\alpha_i = n^2 - 1.$$

G. Chrystal[216] proved that if x', y', z' form a particular set of solutions of

$$ax + by + cz = d, \qquad a'x + b'y + c'z = d',$$

and if ϵ is the g.c.d. of the determinants (bc'), (ca'), (ab'), while u is an arbitrary integer, all solutions are given by

$$x = x' + (bc')u/\epsilon, \qquad y = y' + (ca')u/\epsilon, \qquad z = z' + (ab')u/\epsilon.$$

T. J. Stieltjes[217] gave an exposition of the results by H. J. S. Smith[207].

L. Kronecker[218] gave a simple proof by induction of the theorem due to Frobenius[210] that every n-rowed square matrix with integral elements can be reduced by elementary transformations (interchange of rows or columns and simultaneous change of sign of one row or column, and the addition of one row or column to another) to a matrix in which every element outside the diagonal is zero while every element $\neq 0$ in the diagonal is positive and a divisor of the following element. A matrix has a single such reduced form.

P. Bachmann[219] gave a detailed account of the theory of systems of linear forms, equations and congruences. For a summary account, see Encyclopédie des Sc. Math., tome I, vol. 3, 76–89.

J. H. Grace and A. Young[220] gave a simple proof that any system of linear homogeneous equations with integral coefficients has only a finite number of irreducible solutions in integers $\geqq 0$, a solution being called irreducible if not the sum of two solutions in smaller integers $\geqq 0$.

J. König[221] treated, from the standpoint of modular systems, systems of linear equations and congruences whose coefficients are polynomials in assigned variables.

[215] Giornale di Mat., 24, 1886, 1; Comptes Rendus Paris, 101, 1885, 720, 857, 921. *Pamphlet, Mode de solution d'une question d'analyse indéterminée . . . théorie des transformations de Cremona, Paris, 1885.
[216] Algebra, 2, 1889, 449; ed. 2, vol. 2, 1900, 477–8.
[217] Annales Fac. Sc. Toulouse, 4, 1890, final paper, pp. 49–103.
[218] Jour. für Math., 107, 1891, 135–6.
[219] Arith. der Quadratischen Formen, 1898, 288–370.
[220] Algebra of Invariants, 1903, 102–7.
[221] Einleitung . . . Algebraischen Gröszen, Leipzig, 1903, 347–460.

A. Châtelet[222] gave a brief summary of results, especially Heger's.[205]

E. Cahen[223] gave an extended treatment of systems of linear equations, congruences, and linear forms.

M. d'Ocagne[224] solved $x + y + z + t = n$, $5x + 2y + z + \frac{1}{2}t = n$ to find the number of ways to pay a sum of n francs with 5, 2, 1, $\frac{1}{2}$ franc coins, n in all. For similar problems, see Schubert[143] and d'Ocagne[178] of Ch. III.

One linear congruence in two or more unknowns.

Th. Schönemann[225] considered the number Q of sets of solutions of

$$a_1\xi_1 + \cdots + a_m\xi_m \equiv 0 \quad (\bmod\ p),$$

with $\xi_1, \cdots, \xi_m$ distinct and with the understanding that the solutions obtained by permuting equal elements a count as a single solution, and p is prime. Let μ of the a's be equal, ν further a's be equal, etc. If $a_1 + \cdots + a_m \not\equiv 0 \ (\bmod\ p)$ and $m \leqq p$,

$$Q = \frac{(p-1)(p-2)\cdots(p-m+1)}{\mu!\,\nu!\,\cdots}.$$

But if $a_1 + \cdots + a_m \equiv 0 \ (\bmod\ p)$, while the sum of fewer a's is not divisible by p,

$$Q = \frac{(m-1)!(p-1)(-1)^{m-1}}{\mu!\,\nu!\,\cdots} + \frac{(p-1)\cdots(p-m+1)}{\mu!\,\nu!\,\cdots}.$$

V. A. Lebesgue,[226] by specialization of his[17] result in Ch. VIII of Vol. I of this History, found that, if ρ is a primitive root of the prime p,

$$\rho^b x_1 + \rho^c x_2 + \cdots + \rho^i x_k \equiv 0, \qquad \rho^a + \rho^b x_1 + \cdots + \rho^i x_k \equiv 0 \quad (\bmod\ p)$$

each have p^{k-1} sets $x_1, \cdots, x_k$ of solutions $\geqq 0$, but have

$$\frac{1}{p}(p-1)\{(p-1)^{k-1} - (-1)^{k-1}\}, \qquad \frac{1}{p}\{(p-1)^k - (-1)^k\}$$

sets of solutions > 0, respectively.

M. A. Stern[227] proved that, if p is an odd prime, any integer can be expressed modulo p in exactly $P = (2^{p-1} - 1)/p$ ways as one or the sum of several distinct numbers chosen from the set $1, 2, \cdots, p - 1$. For example, $3 \equiv 1 + 2 \equiv 1 + 3 + 4 \ (\bmod\ 5)$. Restricting ourselves to an even number of summands, we find that zero can be expressed in $\frac{1}{2}(P + p - 2)$ ways, while $1, 2, \cdots,$ or $p - 1$ can be expressed in $\frac{1}{2}(P - 1)$ ways. We shall report in the chapter on quadratic residues on his results when the set is $1, 2, \cdots, (p - 1)/2$.

[222] Leçons sur la théorie des nombres, 1913, 55–8.
[223] Théorie des nombres, 1, 1914, 110–85, 204–62, 278, 299–315, 383–7, 405–6.
[224] L'enseignement math., 18, 1916, 45–7. Cf. Amer. Math. Monthly, 26, 1919, 215–8.
[225] Jour. für Math., 19, 1839, 292.
[226] Jour. de Math., (2), 4, 1859, 366.
[227] Jour. für Math., 61, 1863, 66.

E. Lucas[228] noted that, if a is prime to n, the points (x, y), where $x = 0$, $1, \cdots, n$ and y is the residue modulo n of ax, lie on a lattice (composed of equal parallelograms), and are said to form a *satin* n_a. These satins lead graphically to all solutions of $mx + ny \equiv 0 \pmod{p}$.

L. Gegenbauer[229] gave a direct proof of Lebesgue's[226] results. Let the number of sets of solutions each $\not\equiv 0$ of $a_1x_1 + \cdots + a_kx_k + b \equiv 0 \pmod{p}$, where each a is not divisible by the prime p, be S'_k or S_k according as b is or is not divisible by p. Let N be the number of all sets of solutions. Since $a_kx_k + b$ ranges with x_k over a complete set of residues modulo p, N is the sum of the numbers of sets of solutions of the p congruences $a_1x_1 + \cdots + a_{k-1}x_{k-1} + c \equiv 0 \pmod{p}$, $c = 0, 1, \cdots, p - 1$; while the number of those of these sets of solutions whose elements are prime to p equals the sum of the numbers of the sets of solutions of like property of the $p - 1$ congruences $a_1x_1 + \cdots + a_{k-1}x_{k-1} + c' \equiv 0$, $c' = 0, \cdots, b - 1$, $b + 1, \cdots, p - 1$. Hence

$$N = p^{k-1}, \qquad S'_k = (p - 1)S_{k-1}, \qquad S_k = S'_{k-1} + (p - 2)S_{k-1}.$$

K. Zsigmondy[230] proved that, according as α is not or is divisible by the prime p, $k_0 + k_1 + \cdots + k_{p-1} \equiv \alpha \pmod{p}$ has $\psi(p - 1)$ or $\psi(p - 1) - 1$ sets of solutions in which each k_i is prime to p, where $\psi(n)$ is the number of congruences of degree n with no integral root modulo p. The system of congruences

$$k_0 + \cdots + k_{p-1} \equiv 0, \qquad k_1 + 2k_2 + \cdots + (p - 1)k_{p-1} \equiv \alpha \pmod{p}$$

has $\psi(p - 2)$ or $\psi(p - 2) + p - 1$ sets of solutions prime to p according as $\alpha \not\equiv 0$ or $\alpha \equiv 0$.

R. D. von Sterneck[231] found the number $(n)_i$ of additive compositions of n modulo M formed of i summands which are incongruent modulo M, i. e., the number of solutions of

$$n \equiv x_1 + x_2 + \cdots + x_i \pmod{M}, \qquad 0 \leqq x_1 < x_2 < \cdots < x_i < M.$$

Let $(n)_i^0$ denote the corresponding number when each summand is not divisible by M, so that $0 < x_1 < \cdots < x_i < M$. Define $f(n, d)$ to be zero if any prime occurs in d with an exponent which exceeds by at least 2 its exponent in n; but when the primes $p_1, \cdots, p_j$ occur in d with exponents which exceed by unity their exponents in n, and the remaining prime factors of d occur in n at least to the same power as in d, let

$$f(n, d) = \frac{(-1)^j \phi(d)}{(p_1 - 1)\cdots(p_j - 1)},$$

[228] Application de l'arith. à la construction de l'armure des satins réguliers, Paris, 1868. Principii fondamentali della geometria dei tessuti, l'Ingegnere Civile, Turin, 1880; French transl. in Assoc. franç. av. sc., 40, 1911, 72–87. See S. Günther, Zeitschr. Math. Naturw. Unterricht, 13, 1882, 93–110; A. Aubry, l'enseignement math., 13, 1911, 187–203; Lucas[106] of Ch. VI.

[229] Sitzungsber. Akad. Wiss. Wien (Math.), 99, IIa, 1890, 793–4.

[230] Monatshefte Math. Phys., 8, 1897, 40–1.

[231] Sitzungsber. Akad. Wiss. Wien (Math.), 111, IIa, 1902, 1567–1601. By simpler methods, and removal of the restriction on the modulus M, ibid., 113, IIa, 1904, 326–340.

where ϕ is Euler's function; finally, let $f(n, d) = \phi(d)$ if no prime occurs in d to a higher power than in n, so that n is divisible by d. Then

$$(n)_i = \frac{(-1)^i}{M} \Sigma f(n, d)(-1)^{i/d}\binom{M/d}{i/d},$$

$$(n)_i^0 = \frac{(-1)^i}{M} \Sigma f(n, d)(-1)^{[i/d]}\binom{M/d - 1}{[i/d]},$$

summed for all the divisors d of M, where $\binom{k}{j}$ is a binomial coefficient and is zero if j is not an integer. By the second formula, $f(n, M)$ equals the difference between the numbers of representations of n by an odd and by an even number of summands not divisible by the modulus M.

Von Sterneck[232] proved that the number $[n]_i$ of representations of n as the residue modulo M of a sum of i elements chosen from $0, 1, \cdots, M - 1$, repetitions allowed, is

$$[n]_i = \frac{1}{M} \Sigma f(n, d)\binom{(M + i)/d - 1}{i/d},$$

summed for all the divisors d of M. If the elements are chosen from the numbers $e_1, \cdots, e_\nu$ incongruent modulo M, then

$$i[n]_i = \sum_{\lambda=1}^{i} \sum_{e=e_1}^{e_\nu} [n - \lambda e]_{i-\lambda}, \qquad i(n)_i = \sum_{\lambda=1}^{i} (-1)^{\lambda-1} \sum_{e} (n - \lambda e)_{i-\lambda}.$$

Von Sterneck[233] determined $(n)_i$ and $[n]_i$ for a prime power modulus.

O. E. Glenn[234] found the number of sets of solutions of $\lambda + \mu + \nu \equiv 0$ (mod $p - 1$) and of $\lambda + \mu + \nu + \xi \equiv 0$ (mod $p - 1$), the order of $\lambda, \mu, \cdots$ being disregarded, and p being prime.

D. N. Lehmer[235] proved that $a_1x_1 + \cdots + a_nx_n + a_{n+1} \equiv 0$ (mod m) has $m^{n-1}\delta$ solutions or no solution according as the g.c.d. δ of $a_1, \cdots, a_n, m$ does or does not divide a_{n+1}.

L. Aubry[236] noted that if A is prime to N and if $B/\sqrt{N}$ is not integral, $Ax \equiv By$ (mod N) is solvable in integers $\neq 0$ numerically $< \sqrt{N}$.

SYSTEM OF LINEAR CONGRUENCES.

A. M. Legendre[237] treated the problem to find integers x such that, if a and b, a' and b', $\cdots$ are relatively prime,

$$\frac{ax - c}{b}, \qquad \frac{a'x - c'}{b'}, \qquad \cdots$$

are all integers. The first condition gives $x = m + bz$. Then the second condition requires that $a'bz + a'm - c'$ be divisible by b', which is im-

[232] Sitzungsber. Akad. Wiss. Wien (Math.), 114, IIa, 1905, 711–730.
[233] *Ibid.*, 118, IIa, 1909, 119–132.
[234] Amer. Math. Monthly, 13, 1906, 59–60, 112–4.
[235] *Ibid.*, 20, 1913, 155–6.
[236] Mathesis, (4), 3, 1913, 33–5.
[237] Théorie des nombres, 1798, 33; ed. 2, 1805, 25; ed. 3, 1830, I, 29; Maser, I, p. 29.

possible if the g.c.d. θ of b and b' is not a divisor of $a'm - c'$; but, if θ be such a divisor, the general solution is of the form $z = n + z'b'/\theta$. Thus $x = m' + B'z$, where B' is the l.c.m. of b, b'. Similarly, also the third fraction will be integral if $x = M + Bz$, where B is the l.c.m. of b, b', b''.

* M. Fekete[238] treated the general system of linear congruences in one unknown.

C. F. Gauss[239] discussed at length the solution of n linear congruences in n unknowns. His second (more typical) example is

$$3x + 5y + z \equiv 4, \qquad 2x + 3y + 2z \equiv 7, \qquad 5x + y + 3z \equiv 6 \quad (\text{mod } 12).$$

We first seek integers[240] ξ, ξ', ξ'' without a common factor such that the sum of their products by the coefficients of y (and by those of z) is congruent to zero:

$$5\xi + 3\xi' + \xi'' \equiv 0, \qquad \xi + 2\xi' + 3\xi'' \equiv 0 \quad (\text{mod } 12).$$

Thus $\xi = 1$, $\xi' = -2$, $\xi'' = 1$. Multiplying the congruences by these, and adding, we get $4x \equiv -4$ (mod 12). Similarly, the multipliers 1, 1, -1 give $7y \equiv 5$, while the multipliers -13, 22, -1 give $28z \equiv 96$. Thus $x = 2 + 3t$, $y = 11$ (or $11 + 12r$), $z = 3u$. The proposed congruences now give three equivalent to

$$19 + 3t + u \equiv 0, \qquad 10 + 2t + 2u \equiv 0, \qquad 5 + 5t + 3u \equiv 0 \quad (\text{mod } 4),$$

which are all satisfied if and only if $u \equiv t + 1$ (mod 4). Thus

$$(x, y, z) \equiv (2, 11, 3), (5, 11, 6), (8, 11, 9), (11, 11, 0) \quad (\text{mod } 12).$$

H. J. S. Smith[241] noted that the theory was left imperfect by Gauss. In

$$(1) \qquad A_{i1}x_1 + \cdots + A_{in}x_n \equiv A_{in+1} \quad (\text{mod } M) \quad (i = 1, \cdots, n),$$

denote the determinant $|A_{ij}|$ by D. If D is prime to M, there is one and but one set of solutions. Next, let D be not prime to $M = p_1^{m_1}p_2^{m_2}\cdots$, where the p's are distinct primes. A necessary condition for solvability is that there be solutions for each modulus $p_i^{m_i}$. Conversely, if there be P_i sets of solutions for modulus $p_i^{m_i}$, there are $P_1P_2\cdots$ sets of solutions modulo M. Hence consider (1) for the modulus p^m, and let I_r be the exponent of the highest power of p dividing all the r-rowed minors of D. Then, if $I_n - I_{n-1} \leq m$, the congruences, if solvable, have p^{I_n} sets of solutions. But if $I_n - I_{n-1} > m$, we can assign a value of r such that

$$I_{r+1} - I_r > m \geq I_r - I_{r-1}$$

and then the number (if any) of sets of solutions is p^k, where

$$k = I_r + (n - r)m.$$

[238] Math. és Phys. Lapok, Budapest, 17, 1908, 328–49.

[239] Disq. Arith., Art. 37; Werke, I, 27–30.

[240] F. J. Studnička, Sitzungsberichte, Akad. Wiss., Prag, 1875, 114, noted that they are proportional to the signed minors of the coefficients of the first column in the determinant of the coefficients.

[241] Report British Assoc. for 1859, 228–67; Coll. Math. Papers, I, 43–5.

Smith[207] wrote ∇_n for the determinant $|A_{ij}|$ of (1), ∇_{n-1} for the g.c.d. of its first minors, $\cdots$, ∇_1 for the g.c.d. of the elements A_{ij}, and set $\nabla_0 = 1$. Let D_n, D_{n-1}, $\cdots$, D_0 be the corresponding g.c.d.'s for the augmented matrix. Let δ_i and d_i denote respectively the g.c.d. of M, ∇_i/∇_{i-1}, and M, D_i/D_{i-1}. Set $d = d_1 \cdots d_n$, $\delta = \delta_1 \cdots \delta_n$. Then the system of congruences (1) is solvable if and only if $d = \delta$; when this condition is satisfied the number of incongruent sets of solutions is d. There are similar theorems (pp. 402–4) when the number of unknowns is either less or greater than the number of congruences.

Smith[242] employed a prime factor p of M, and the exponents μ, a_s, α_s of the highest powers of p dividing M, D_s, ∇_s, respectively. He proved that his preceding theorems can be replaced by the following: For the modulus p^μ, the congruences (1) are solvable if and only if $a_\sigma = \alpha_\sigma$, where $a_\sigma - a_{\sigma-1}$ is the first term of $a_n - a_{n-1}$, $a_{n-1} - a_{n-2}$, $\cdots$ which is $< \mu$; when this condition is satisfied the number of incongruent sets of solutions is p^k, where $k = a_\sigma + (n - \sigma)\mu$.

G. Frobenius[210] (pp. 185–194) proved that the congruences (1) have M^{n-1} incongruent sets of solutions if the l-rowed determinants of A have with M no common divisor and if the $(l + 1)$-rowed determinants of the augmented matrix of all coefficients are divisible by M, where l is the rank of the matrix A of the coefficients of the unknowns. If the rank of the augmented matrix is $l + 1$ and the g.c.d. of the $(l + 1)$-rowed determinants is d', while the rank of A is l and the g.c.d. of the l-rowed determinants is d, the congruences (1) have no solutions if the modulus M is not a divisor of d'/d. The number of incongruent sets of solutions of the homogeneous congruences $A_{i1}x_1 + \cdots + A_{in}x_n \equiv 0 \pmod{M}$ equals $s_1 s_2 \cdots s_n$, where s_λ is the g.c.d. of M and the λth elementary divisor of the matrix (A_{ij}).

Frobenius[212] proved that a system of linear homogeneous congruences modulo M in n unknowns has a fundamental system of $n - s$ sets of solutions, but none of fewer than $n - s$, if the determinants of order $s + 1$ have with M a common divisor but the determinants of order s do not. He investigated the rank and equivalence of linear forms modulo M.

F. Jorcke[243] treated systems of linear congruences without novelty.

D. de Gyergyószentmiklos[244] considered the congruences

$$\sum_{j=1}^{n} a_{\rho j}x_j \equiv u_\rho \pmod{m} \qquad (\rho = 1, \cdots, n).$$

Let $D = |a_{\rho j}|$, and V_k be the determinant derived from D by putting the u's in the kth column. Let δ be the g.c.d. of m and D. If any V_k is not divisible by δ, there is no solution. Next, let each V_k be divisible by δ. Then $Dx_k \equiv V_k \pmod{m}$ uniquely determines $x_k \equiv \alpha_k$ modulo m/δ. Set $x_k = \alpha_k + t_k m/\delta$ in the initial congruences. Thus

$$(a_{\rho 1}t_1 + \cdots + a_{\rho n}t_n)m/\delta \equiv u_\rho - a_{\rho 1}\alpha_1 - \cdots - \alpha_{\rho n}\alpha_n \pmod{m},$$
$$a_{\rho 1}t_1 + \cdots + a_{\rho n}t_n \equiv w_\rho \pmod{\delta}.$$

---

[242] Proc. London Math. Soc., 4, 1871–3, 241–9; Coll. Math. Papers, II, 71–80.

[243] Ueber Zahlenkongruenzen und einige Anwendungen derselben, Progr. Fraustadt, 1878.

[244] Comptes Rendus Paris, 88, 1879, 1311.

For the latter system, the modulus δ divides the determinant D. Hence if some minor of order $n - \nu$ is not divisible by δ, while all minors of higher order are divisible by δ, the solution involves exactly ν arbitrary parameters and there are δ^ν sets of solutions.

L. Kronecker[245] deduced from his theory of modular systems the theorem that, for p a prime, the general solution of

$$\sum_{k=1}^{\tau} V_{ik}X_k \equiv 0 \pmod{p} \qquad (i = 1, \cdots, t)$$

involves $\tau - r$ independent parameters if the matrix of the $t\tau$ numbers V_{ik} is of rank r modulo p.

K. Hensel[246] considered a system of m linear homogeneous congruences in n unknowns in which the coefficients and the modulus P are either integers or rational integral functions of one variable. We may replace the system by an equivalent system whose modulus divides P and hence finally obtain modulus unity.

E. Busche[247] proved that the number of solutions of a system of n linear homogeneous congruences in n unknowns equals the modulus if the latter divides the determinant of the system. This theorem is equivalent to the following. Write $a \sim b$ if $a - b$ is an integer. If the a_{ij} are integers of determinant $\neq 0$, the number of non-equivalent solutions $x_1, \cdots, x_n$ of $a_{i1}x_1 + \cdots + a_{in}x_n \sim 0$ $(i = 1, \cdots, n)$ is the absolute value of the determinant $|a_{ij}|$.

G. B. Mathews[248] noted that a system of n linear congruences in which the moduli are $m_1, \cdots, m_n$ respectively may be reduced to a system with the same modulus m (the l.c.m. of $m_1, \cdots, m_n$), by multiplication by m/m_1, $\cdots, m/m_n$ respectively. For the case of a common modulus m the method is to derive an equivalent system of congruences involving respectively $n, n - 1, \cdots, 1$ unknowns. Details are given only for the example

$$ax + by + cz \equiv d, \ a'x + b'y + c'z \equiv d', \ a''x + b''y + c''z \equiv d'' \pmod{m}.$$

Let θ be the g.c.d. of a, a', a'' and let $\theta = pa + qa' + ra''$. Multiplying the congruences by p, q, r respectively and adding, we get a congruence $\theta x + \beta y + \gamma z \equiv \delta$. If, for example, p is prime to m, we get an equivalent system by taking the latter in place of the first congruence of the system. Then eliminate x from the second and third by means of $\theta x + \cdots$.

L. Gegenbauer[249] showed that the system of linear congruences

$$\sum_{k=0}^{p-2} b_{k+\rho}y_k \equiv 0 \pmod{p} \qquad (\rho = 0, \cdots, p - 2)$$

has as many linearly independent sets of solutions as

$$\sum_{k=0}^{p-2} b_k x^k \equiv 0 \pmod{p}$$

[245] Jour. für Math., 99, 1886, 344; Werke, 3, I, 167. Cf. papers 24–26, p. 226, and 43, p. 232 of Vol. I of this History.

[246] Jour. für Math., 107, 1891, 241.

[247] Mitt. Math. Gesell. Hamburg, 3, 1891, 3–7.

[248] Theory of Numbers, 1892, 13–14.

[249] Monatshefte Math. Phys., 5, 1894, 230. Further report on p. 229 of Vol. I of this History.

has distinct roots not divisible by p. Such a system of linear congruences has been discussed by W. Burnside.[250]

E. Steinitz[251] stated that all theorems on linear congruences follow easily from one: Given k linear congruences in n variables modulo m, the k sets of coefficients form the basis of a Dedekind Modul. If $e_1, \cdots, e_n$ are the invariants of this Modul (the last $n - r$ e's are zero if the rank r is $< n$) and if $[e_i, m]$ is the g.c.d. of e_i and m, then the totality of sets of solutions of the k congruences represent a Modul with the invariants

$$\frac{m}{[e_n, m]}, \cdots, \frac{m}{[e_1, m]}.$$

Expositions in the texts by Bachmann,[219] J. König,[221] and Cahen[223] have been cited. Zsigmondy[230] found the number of solutions of a system of two special congruences.

H. Weber[252] made a direct examination of the conditions under which

$$(2) \qquad a_{1j}y_1 + a_{2j}y_2 + \cdots + a_{\rho j}y_\rho \equiv 0 \ (\mathrm{mod}\ p^\tau) \qquad (j = 1, \cdots, \mu)$$

shall require that each y_i be divisible by p^τ, where p is a prime. It is assumed that not every a_{ij} is divisible by p (otherwise a solution is obtained by taking each y_i to be any multiple of $p^{\tau-1}$). We may assume that $\Delta = |a_{ij}|_{i, j=1, \ldots, \tau}$ is not divisible by p, while every $(\tau + 1)$-rowed determinant of the matrix (a_{ij}) is divisible by p. Denote the signed minors of Δ by Δ_{kh} and set

$$D_{ks} = \Delta_{k1}a_{s1} + \Delta_{k2}a_{s2} + \cdots + \Delta_{k\tau}a_{s\tau}.$$

Thus $D_{ks} = \Delta$ if $k = s$; $D_{ks} = 0$ if $s \leqq \tau$, $s \neq k$; while, if $s > \tau$, D_{ks} is a τ-rowed determinant of (a_{ij}). Applying Cramer's rule to the first τ congruences (2), we get

$$(3) \qquad \Delta y_j + D_{j, \tau+1}y_{\tau+1} + \cdots + D_{j\rho}y_\rho \equiv 0 \ (\mathrm{mod}\ p^\tau) \qquad (j = 1, \cdots, \tau).$$
Hence

$$\Delta(a_{1r}y_1 + \cdots + a_{\rho r}y_\rho) \equiv A_{\tau+1, r}y_{\tau+1} + \cdots + A_{\rho r}y_\rho \quad (\mathrm{mod}\ p^\tau),$$
where

$$A_{sr} = \Delta a_{sr} - \sum_{k=1}^{\tau} a_{kr}D_{ks}$$

equals a $(\tau + 1)$-rowed determinant of (a_{ij}) and hence is divisible by p. Thus, if $\tau < \rho$, (2) are satisfied when $y_{\tau+1}, \cdots, y_\rho$ are divisible by $p^{\tau-1}$. In order that (2) shall require that each y_i be divisible by p^τ it is therefore necessary that $\tau = \rho$. This condition is also sufficient, since (3) then reduce to $\Delta y_1 \equiv 0, \cdots, \Delta y_\rho \equiv 0$, whence $y_1, \cdots, y_\rho$ are divisible by p^τ.

F. Riesz[253] stated that, if the α_{ik} and β_i are real, the congruences

$$\sum_{k=1}^{n} \alpha_{ik}x_k \equiv \beta_i \quad (\mathrm{mod}\ 1) \qquad\qquad (i = 1, \cdots, m)$$

[250] Messenger Math., 24, 1894, 51.
[251] Jahresbericht d. Deutschen Math.-Verein., 5, 1896 [1901], 87.
[252] Lehrbuch der Algebra, 2, 1896, 87–8; ed. 2, 1899, 94. Cf. Smith.[242]
[253] Comptes Rendus Paris, 139, 1904, 459–462.

are solvable in integers when the β's are arbitrary, with a desired approximation, if and only if $\Sigma\alpha_{ik}x_k \equiv 0$ (mod 1) are not solvable exactly in integers not all zero.

U. Scarpis[254] proved that a system of n linear homogeneous congruences in n unknowns has solutions not all divisible by the modulus M if and only if the determinant Δ of the coefficients is not prime to M. The problem is reduced as usual to the case $M = p^m$, where p is a prime. Then let some ρ-rowed minor of Δ be prime to p, but all k-rowed minors ($k \geqq \rho + 1$) be divisible by p. Let p^e be the highest power of p dividing Δ and all its k-rowed minors ($k \geqq \rho + 1$). Then ρ of the congruences are linearly independent. We may assume that $|a_{ij}|$, where $i, j = 1, \cdots, \rho$, is prime to p. Then the last $n - \rho$ of the congruences can be replaced by congruences in $x_{\rho+1}, \cdots, x_n$ in which each coefficient is divisible by p^e. If $m = 1$, no more than ρ of the initial congruences are linearly independent; the values of $x_1, \cdots, x_\rho$ are uniquely determined in terms of $x_{\rho+1}, \cdots, x_n$ which are arbitrary, so that there are $p^{n-\rho}$ sets of solutions.

Linear forms with real coefficents; approximation.

J. L. Lagrange[255] noted that, if a is a given positive real number, we can find relatively prime positive integers p, q such that $p - aq$ shall be numerically smaller than $r - as$ for $r < p$, $s < q$, by taking p/q as any principal convergent to the continued fraction for a with all terms positive.

Lagrange[22] determined a fraction m/a, with given numerator or denominator, which shall approximate as closely as possible to the given fraction $B/A < 1$, where A, B are relatively prime. For example, let m be given. Take as a the quotient found on dividing Am by B. If $\mp C$ is the remainder numerically $< \frac{1}{2}B$. then $Ba - Am = \pm C$, $B/A = m/a \pm C/(Aa)$. Starting with C/A, determine similarly n/b, with n given, by using the quotient b and remainder $\mp D$ when An is divided by C, whence $C/A = n/b \pm D/(Ab)$. Similarly, $D/A = p/c \pm E/(Ac)$. It follows that $m \gtreqqless a$, $n \lesseqqgtr b$, $p \lesseqqgtr c$, $\cdots$ and that $A, B, C, D, \cdots$ form a decreasing series terminating with zero:

$$\frac{B}{A} = \frac{m}{a} \pm \frac{n}{ab} \pm \frac{p}{abc} \pm \cdots.$$

In case the denominators $a, b, c, \cdots$ were given and all equal, we have expressed B/A to the base a. Finally, suppose that neither m nor a is given, but are to be found such that $m < B$, $n < A$, and such that m/a is as close an approximation to B/A as possible. Hence must $C = \pm 1$. Then m and a are found by Euclid's g.c.d. process. Saunderson[13] had already treated the approximation to a fraction and cited earlier writers.

C. G. J. Jacobi[256] proved that integral values not all zero can be assigned to x, y, z such that $ax + a'y + a''z$ and $bx + b'y + b''z$ are less than any assigned quantity. Cf. Sylvester[108] of Ch. III.

[254] Periodico di Mat., 23, 1908, 49–61.
[255] Additions to Euler's Algebra, 2, 1774, 445; Oeuvres, VII, 45–57.
[256] Jour. für Math., 13, 1835, 55; Werke, II, 29–31.

G. L. Dirichlet[257] stated that it has been long known from the theory of continued fractions that, if α is irrational, there exists an infinitude of pairs of integers x, y for which $x - \alpha y$ is numerically $< 1/y$. He proved the following generalization: If $\alpha_1, \cdots, \alpha_m$ are such that

$$f = x_0 + \alpha_1 x_1 + \cdots + \alpha_m x_m$$

vanishes for no set of integral values of $x_0, \cdots, x_m$, not all zero, there exists an infinitude of sets of integers $x_0, \cdots, x_m$, with $x_1, \cdots, x_m$ not all zero, such that f is numerically $< 1/s^m$, where s is the greatest of $x_1, \cdots, x_m$. Similarly for several forms f. For example, if $\alpha = \alpha_1 x_1 + \cdots + \alpha_m x_m$ and $\beta = \beta_1 x_1 + \cdots + \beta_m x_m$ vanish simultaneously only when $x_1, \cdots, x_m$ are all zero, there exists an infinitude of sets of integers $x_1, \cdots, x_m$ not all zero for which $|\alpha| < A/s^u$, $|\beta| < B/s^{m-2-a}$, in which A and B are constants depending on the α_i, β_i, while a is any constant between 0 and $m - 2$.

Ch. Hermite[258] remarked that, if A and B are given irrational numbers, we can readily find the linear relations $Aa + Bb + c = 0$ (if existent), where a, b, c are integers. In fact, $\alpha = mA - m'$ and $\beta = mB - m''$ can be made as small as one pleases [by choice of integers m, m', m'']. Since $a\alpha + b\beta = -am' - bm'' - cm$ is an integer, it cannot be made < 1 without reducing to zero. Thus to find m, m', m'', we have only to convert β/α into a continued fraction to obtain the desired relation.

Hermite[259] proved by means of the minimum of a binary quadratic form that, if a, Δ are real, there exist integers m, n such that

$$(m - an)^2 + n^2/\Delta^2 < \frac{1}{\Delta}\sqrt{\frac{4}{3}},$$

whence $|m - an| < 1/(n\sqrt{3})$. Let m', n' be the integers corresponding to $\Delta' = \Delta + \delta$, where δ is an infinitesimal. Then $mn' - nm' = \pm 1$.

P. L. Tchebychef[260] proved that, if a is irrational and b is given, there exists an infinitude of sets of integers x, y such that $y - ax - b$ is numerically $< 2/|x|$.

Hermite[261] proved that, in Tchebychef's result, we may replace $2/|x|$ by $1/\{2|x|\}$ and in fact by $\sqrt{2/27}/|x|$.

L. Kronecker[262] treated the problem to find integers w, w' such that $aw + a'w'$ takes a value as near as possible to ξ, where a, a', ξ are given real numbers. In general, consider a system of p equations

$$a_{i1}w_1 + \cdots + a_{iq}w_q = \xi_i \qquad (i = 1, \cdots, p),$$

with real coefficients. Let r be the number of these equations whose left

[257] Sitzungsber. Akad. Wiss. Berlin, 1842, 93; Werke, I, 635–8.
[258] Jour. für Math., 40, 1850, 261; Oeuvres, I, 101.
[259] Ibid., 41, 1851, 195–7; Oeuvres, I, 168–171.
[260] Zapiski Acad. nauk St. Pétersbourg, 10, 1866, Suppl. No. 4, p. 50; Oeuvres, 1, 1899, 679.
[261] Jour. für Math., 88, 1879, 10–15; Ouevres, III, 513–9.
[262] Monatsber. Akad. Wiss. Berlin, 1884, 1179–93, 1271–99; Werke, III₁, 47–109. Cf. ibid., 1071–80; Comptes Rendus Paris, 96, 1883, 93–8, 148–52, 216–21; 99, 1884, 765–71, Werke, III₁, 1–44, for application to algebraic units.

members are linearly independent, so that r is the ordinary (absolute) rank of the rectangular matrix

$$(a_{ik}) \qquad (i = 1, \cdots, p; \; k = 1, \cdots, q).$$

This matrix is said to be of relative rank (or rank of rationality) R if R is the least number such that, by means of a linear substitution on the rows with arbitrary coefficients, the matrix can be transformed into a matrix all but r of whose rows contain only zero elements and all but R rows contain only integral elements. Necessary and sufficient conditions for the approximate solution in integers of our equations are expressed in different forms: R of the ξ's can be given arbitrary values, while the choice of $r - R$ of the ξ's is limited only by certain conditions of rationality, while the remaining $p - r$ ξ's are uniquely determined in terms of the earlier r ξ's.

A. Hurwitz[263] proved that if ω is irrational there exists an infinitude of pairs of integers x, y for which $|\, y/x - \omega \,| < 1/(\sqrt{5}x^2)$. Likewise, $|\, y/x - \omega \,| < 1/\{\,\sqrt{8}x^2\}$ if ω is not equivalent to $(1 + \sqrt{5})/2$.

H. Minkowski[264] found by use of lattice points and other geometrical concepts the fundamental theorem that, if $f_1, \cdots, f_n$ are linear homogeneous functions of $x_1, \cdots, x_n$ with any real coefficients whose determinant Δ is not zero, we can assign integral values not all zero to $x_1, \cdots, x_n$ such that $|f_i| \leqq \sqrt[n]{|\Delta|}$ for $i = 1, \cdots, n$. If $a_1, \cdots, a_{n-1}$ are real, we can find integers $x_1, \cdots, x_n$ without a common factor and with $x_n > 0$ such that

$$\left| \frac{x_j}{x_n} - a_j \right| < \frac{1}{kx_n^k}, \qquad k = \frac{n}{n-1}, \qquad (j = 1, \cdots, n-1).$$

For $n > 1$, consider n linear forms $f_1, \cdots, f_n$ in $x_1, \cdots, x_n$ with a determinant $\Delta \neq 0$, such that r of the forms have real coefficients and $s = (n - r)/2$ pairs have conjugate imaginary coefficients, and let p be any real number $\geqq 1$. Then integral values not all zero can be assigned to $x_1, \cdots, x_n$ such that[265]

$$\frac{1}{n}\sum_{j=1}^{n} |f_j|^p < \left\{ \left(\frac{2}{\pi}\right)^s \frac{n^{-n/p}\Gamma(1 + n/p)\,|\Delta|}{\{\Gamma(1 + 1/p)\}^r \, 2^{-2s/p}\{\Gamma(1 + 2/p)\}^s} \right\}^{p/n},$$

except for $p = 1$, $s = 0$, $n = 2$, when the members may be equal; here Γ denotes the ordinary gamma function. He obtained (p. 161) Lagrange's[255] result on the minimum of $x - ay$.

A. Hurwitz[266] gave an elegant analytic proof of Minkowski's[264] theorem, and the fact that the inequality sign may be taken in $n - 1$ of the n relations.

Ch. Hermite[267] remarked that Euclid's g.c.d. process leads to approxi-

[263] Math. Annalen, 39, 1891, 279. This and papers cited on p. 158 of Vol. I of this History give approximations by use of Farey series.

[264] Geometrie der Zahlen, 1896, 104–123. Extracts in Math. Papers Chicago Congress, 1896, 201–7; French transl., Nouv. Ann. Math., (3), 15, 1896, 393–403.

[265] Also in Comptes Rendus Paris, 112, 1891, 209; Werke, I, 261–3.

[266] Göttingen Nachrichten, 1897, 139. French transl., Nouv. Ann. Math., (3), 17, 1898, 64–74. Cf. P. Bachmann, Allgemeine Arith. d. Zahlenkörper, 1905, 335–41; G. Humbert, Annales de la Fac. Sc. Toulouse, (3), 3, 1911, 8–12.

[267] Le Matematiche pure ed applicate, Città di Castello, 1, 1901, 1–2; Werke, IV, 552–3.

mations to a fraction by means of a series of fractions m/n, the error being $< h/n^2$. He gave a slight modification of Dirichlet's[257] method.

H. Minkowski[268] proved that if $\xi = \alpha x + \beta y$ and $\eta = \gamma x + \delta y$ have any real coefficients of determinant $\alpha\delta - \beta\gamma = 1$ and if ξ_0, η_0 are any given real numbers, there exist integers x, y for which $|(\xi - \xi_0)(\eta - \eta_0)| \leqq \frac{1}{4}$. In particular, if a is irrational and b is not an integer, there are integers x, y for which $|(y - ax - b)(x - c)| < \frac{1}{4}$; the case $c = 0$ gives a better approximation than Hermite's,[261] since $1/4 < \sqrt{2/27}$.

E. Cahen[269] discussed the approximate solution in integers of a system of linear equations.

E. Borel[270] proved that if a, b, M are any given real numbers, integers x, y, z, numerically $< M$, can be assigned such that

$$|ax + by + z| < \frac{\theta}{M}\sqrt{a^2 + b^2 + 1},$$

θ being independent of a, b, M (but not found). Again, intervals (A_n, B_n) can be found such that A_n and B_n increase indefinitely with n and such that, if α is any irrational number between 0 and 1, integers p_n, q_n exist for which

$$\left|\frac{p_n}{q_n} - \alpha\right| < \frac{1}{q_n^2 \sqrt{5}}, \qquad A_n < q_n < B_n.$$

At least one of three successive convergents to α satisfies the first inequality [cf. Hurwitz[263]].

Minkowski[271] proved that if a is real we can choose integers x, y such that $|x/y - a| < 1/y^2$ and deduced the existence of solutions of $sx - ry = 1$ if s, r are relatively prime integers. He gave a new proof, suggested by D. Hilbert,[271a] of his[264] theorem on n real linear forms. He discussed (pp. 47–58) the maximum value of the minimum of $|\xi|^p + |\eta|^p$, where ξ and η are real linear forms. He treated (pp. 68-82) the equivalence and minimum of three linear forms ξ, η, ζ, and gave theorems on the values of their sum or product.

B. Levi[272] proved Minkowski's[264] thoerem, and for the limit case in which no integers, not all zero, make each $|f_i| < 1$, proved his result that then at least one of the f_i has integral coefficients.

[268] Math. Annalen, 54, 1901, 91–124, see pp. 108, 116 (Ges. Abhandl., I, 320); French transl., Ann. de l'école normale sup., (3), 13, 1896, 45. For an account of Minkowski's investigations, see Verhandl. des dritten intern. Math. Congresses Heidelberg, 1905, 164. Proof by J. Uspenskij, Applications of continuous parameters in the theory of numbers, St. Petersburg, 1910; cf. Jahrb. Fortschritte Math., 1910, 252.

[269] Bull. Soc. Math. France, 30, 1902, 234–242. He also made additions to the subject in his article in the Encyclopédie des Sc. Math., 1906, tome I, vol. III, 89–97.

[270] Jour. de Math., (5), 9, 1903, 329–375; Comptes Rendus Paris, 163, 1916, 596–8. Leçons sur la théorie de la croissance, 1910, 143–154. Cf. A. Denjoy, Bull. Soc. Math. de France, 39, 1911, 175–222.

[271] Diophantische Approximationen, Leipzig, 1907, 1–19, 28.

[271a] Cf. J. Sommer, Vorlesungen über Zahlentheorie, 1907, 65–72; French transl. by A. Lévy, 1911.

[272] Rendiconti Circolo Mat. Palermo, 31, 1911, 318–340.

S. Kakeya[273] proved the theorem (Minkowski,[264] p. 108) that if $a_1, \cdots, a_n$ are real there exist integers $x_1, \cdots, x_n, z$ such that $x_1 - a_1 z, \cdots, x_n - a_n z$ are as small numerically as we please.   He proved that these forms approach indefinitely any real numbers.   He[274] gave a generalization to any linear functions.

R. Remak[274a] proved arithmetically Minkowski's[268] first theorem.

H. Weber and J. Wellstein[274b] gave a new arithmetical proof of Minkowski's[264] initial theorem for both real and imaginary linear forms.

H. F. Blichfeldt[275] proved a result which in Minkowski's notations becomes

$$| f_1 | + \cdots + | f_n | \leqq \sqrt{\frac{2n}{\pi}} \left\{ \Gamma \left(1 + \frac{n+2}{2} \right) \right\}^{1/n} | \Delta |^{1/n}.$$

For small values of n this limit is higher than Minkowski's[264] (p. 122), but for large n's it is smaller.   Given the positive numbers $\alpha_1, \cdots, \alpha_{n-1}$ and any positive number $b < \frac{1}{2}$, we can find integers $X_1, \cdots, X_{n-1}, Z$ such that the $n - 1$ differences $| X_i/Z - \alpha_i |$ are $\leqq 2b$ and

$$\leqq \frac{\gamma}{Z^{n/(n-1)}} = \frac{(n-1)Z^{-n/(n-1)}}{n \left\{ 1 + \left(\dfrac{n-2}{n} \right)^{n+2} \right\}^{1/(n-1)}}.$$

Except for $n = 2$, this approximation is closer than that obtained by Hermite,[259] Kronecker, and Minkowski.[264]

G. H. Hardy and J. E. Littlewood[276] proved that if $\theta_1, \cdots, \theta_m$ are irrational and connected by no linear relation with integral coefficients not all zero, and if $\alpha_{11}, \cdots, \alpha_{km}$ are any numbers such that $0 \leqq \alpha_{ij} < 1$, there exists a sequence of positive integers $n_1, n_2, \cdots$ such that the fractional part of $n_r^l \theta_p$ approaches α_{lp} as r increases, for $l = 1, \cdots, k; \; p = 1, \cdots, m$ (the case $k = 1$ being due to Kronecker[262]).   Given λ, there is therefore a function Φ of k, m, λ and the θ's and α's such that the difference between the fractional part $(n^l \theta_p)$ of $n^l \theta_p$ and α_{lp} is numerically $< 1/\lambda$ for some $n < \Phi$. When the θ's are given, a Φ can be found independent of the α's.   When all the α's are zero, a Φ can be found independent of the θ's.   An upper bound for Φ, in this last case, was later given by H. T. J. Norton.[277]   H. Weyl[277a] went further by showing that the numbers $(n^l \theta_p)$ are " uniformly distributed " throughout the unit cube $0 \leqq x_{lp} \leqq 1$ in space of km dimensions [i.e., if we associate with n the point whose km coordinates are $x_{lp} = (n^l \theta_p)$ and denote by n_v the number of the first n points which lie inside an assigned part of the cube, of volume V, then $n_v \sim nV$ when $n \to \infty$].

[273] Science Reports Tôhoku University, 2, 1913, 33–54.
[274] Tôhoku Math. Jour., 4, 1913–4, 120–131.
[274a] Jour. für Math., 142, 1913, 278–82.
[274b] Math. Annalen, 73, 1913, 275–85.
[275] Trans. Amer. Math. Soc., 15, 1914, 227–235.

Hardy and Littlewood[276] also considered the same problem when n^t is replaced by an arbitrary increasing sequence λ_n with infinite limit and obtained the same result, but with the exception of a set of values of θ of measure zero.   R. H. Fowler[277b] established uniform distribution, with an upper bound for the error, when λ_n increases as rapidly as an exponential $e^{n\delta}$ ($\delta > 0$).   Weyl[277a] extended the theorem of uniform distribution to all cases in which λ_n increases with tolerable regularity and as fast as $(\log n)^{2+\delta}$ ($\delta > 0$).   These questions are intimately connected with the problem of the behavior of the series $\Sigma_0^N \exp. (2\pi i \lambda_n)$ when $N \to \infty$, which has been considered in detail by Hardy and Littlewood,[277c] and Weyl.[277a]

G. Giraud[278] proved that there exist integral values of the x's and y's for which

$$| x_i - a_{i1}y_1 - \cdots - a_{ip}y_p - A_i | < \epsilon \quad (i = 1, \cdots, n),$$

whatever ϵ be, if and only if all the forms $m_1X_1 + \cdots + m_nX_n$, which take integral values when $X_1, \cdots, X_n$ are replaced in turn by the p sets of values $a_{1j}, \cdots, a_{nj}$ ($j = 1, \cdots, p$), take also integral values when we replace $X_1, \cdots, X_n$ by $A_1, \cdots, A_n$.

S. L. van Oss[279] proved that n real linear functions of $x_1, \cdots, x_n$ of determinant unity have the minimum value unity for integral x's if at least one of the forms has integral coefficients without a common divisor.   This had been proved by Minkowski for $n = 3$.

W. E. H. Berwick[280] gave a method to find which pair of integers x, y ($0 \leqq y < N$) gives the least value for $f = ax + by + c$, where a, b, c are real and not zero.   Thus he finds the point with integral coordinates nearest to the line $f = 0$ and within the strip between $y = 0$ and $y = N$.

A. Brown[281] noted that, to find the fraction whose denominator does not exceed a given integer and which approximates most closely to a given number, Lagrange's theory gives the fraction nearest in defect and the fraction nearest in excess, but does not decide which of them is nearest in absolute value to the given number.   A simple method is here given for deciding between the two fractions.

A. J. Kempner[282] noted that any straight line with an irrational slope has on either side of it an infinitude of points with integral coordinates lying closer to it than any assigned distance.

G. Humbert[283] developed Hermite's[259] method of approximating to an irrational number ω, showed that it differs very little from the method of

[276] Acta Math., 37, 1914, 155–191; Proc. Fifth Internat. Congress Math., 1, 1912, 223–9.
[277] Proc. London Math. Soc., (2), 16, 1917, 294–300.
[277a] Göttingen Nachrichten, 1914, 234–244; Math. Annalen, 77, 1916, 313–352.
[277b] Proc. London Math. Soc., (2), 14, 1915, 189–206.
[277c] Acta Math., 37, 1914, 155–238; Proc. Nat. Acad. Sc., 2, 1916, 583–6; 3, 1917, 84–8.
[278] Soc. Math. France, Comptes Rendus des Séances, 1914, 29–32.
[279] Handelingen XVde Nederlandsch Natuur- en Geneeskundig Congres, 1915, 192–3.
[280] Messenger Math., 45, 1916, 154–160.
[281] Trans. Phil. Soc. South Africa, 5, 1916, 653–7.
[282] Annals of Math., 19, 1917, 127.
[283] Comptes Rendus Paris, 161, 1915, 717–21; 162, 1916, 67; Jour. de Math., (7), 2, 1916, 79–103.

continued fractions, and found necessary and sufficient conditions that a given fraction be in Hermite's series of fractions tending towards ω.   The main condition was generalized by E. Cahen.[284]

Humbert[285] gave simple proofs of the theorems by Hurwitz.[263]

M. Fujiwara[286] supplemented Hurwitz's[263] second theorem as Borel[270] had the first.

J. H. Grace[287] proved that if $\frac{3}{2} \leqq k \leqq 2$ and if x/y and x'/y' are two consecutive rational approximations to an irrational number θ such that

$$\left| \frac{x}{y} - \theta \right| < \frac{1}{ky^2},$$

then $xy' - x'y = \pm 1$ [Hermite[259] for $k = \sqrt{3}$, Minkowski[268] for $k = 2$]. He[288] proved that Minkowski's[268] last result is final, i.e., if $k < \frac{1}{4}$, it is possible to choose a and b such that there is not an infinitude of integers x for which $| y - ax - b | < k/| x |$.

[284] Comptes Rendus Paris, 162, 1916, 779–782.
[285] Jour. de Math., (7), 2, 1916, 155–167.
[286] Tôhoku Math. Jour., 11, 1917, 239–242.   Cf. 14, 1918, 109–115.
[287] Proc. London Math. Soc., (2), 17, 1919, 247–258.
[288] *Ibid.*, 316–9.

CHAPTER III.

PARTITIONS.

G. W. Leibniz[1] asked Bernoulli if he had investigated the number of ways a given number can be separated into two, three or many parts, and remarked that the problem seemed difficult but important. Leibniz[2] used the term number of divulsions for the number of ways a given integer can be expressed as a sum of smaller integers, as 3, $2+1$, $1+1+1$, and noted the connection with the number of symmetric functions of a given degree, as Σa^3, $\Sigma a^2 b$, Σabc.

L. Euler[3] found relations between $A = \Sigma a$, $B = \Sigma a^2$, $C = \Sigma a^3$, $\cdots$, and

$$\alpha = \Sigma a, \qquad \beta = \Sigma ab, \qquad \gamma = \Sigma abc, \qquad \cdots,$$
$$\mathfrak{A} = \Sigma a,$$
$$\mathfrak{B} = a^2 + ab + b^2 + ac + \cdots,$$
$$\mathfrak{C} = a^3 + a^2 b + ab^2 + b^3 + a^2 c + abc + \cdots,$$
$$\mathfrak{D} = a^4 + a^3 b + \cdots + abcd + \cdots, \qquad \cdots.$$

We have

$$P \equiv \Sigma \frac{az}{1 - az} = Az + Bz^2 + Cz^3 + \cdots,$$

$$Q \equiv \frac{zdR}{Rdz} = z\Sigma \frac{a}{1 + az} = Az - Bz^2 + Cz^3 - \cdots,$$

$$R \equiv \Pi(1 + az) = 1 + \alpha z + \beta z^2 + \cdots,$$

$$\frac{zdR}{dz} = \alpha z + 2\beta z^2 + \cdots = RQ.$$

Hence

$$A = \alpha, \qquad \alpha A - B = 2\beta, \qquad \beta A - \alpha B + C = 3\gamma, \cdots.$$

Next, expanding $(1 + az)^{-1}$, we get

$$T \equiv \frac{1}{R} = 1 - \mathfrak{A}z - \mathfrak{B}z^2 - \mathfrak{C}z^3 + \cdots, \qquad \mathfrak{A} - \alpha = 0, \qquad \mathfrak{B} - \alpha\mathfrak{A} + \beta = 0, \qquad \cdots.$$

Now take $a = n$, $b = n^2$, $c = n^3$, $\cdots$. Then

$$A = n/(1 - n), \qquad B = n^2/(1 - n^2), \qquad \cdots.$$

Hence

$$P \equiv \frac{nz}{1 - nz} + \frac{n^2 z}{1 - n^2 z} + \cdots = \frac{nz}{1 - n} + \frac{n^2 z^2}{1 - n^2} + \cdots,$$

$$R \equiv (1 + nz)(1 + n^2 z) \cdots = 1 + \alpha z + \beta z^2 + \cdots,$$

$$\alpha = n + n^2 + n^3 + \cdots,$$

[1] Math. Schriften (ed., Gerhardt), 3, II, 1856, 601; letter to Joh. Bernoulli, 1669.
[2] MS. dated Sept. 2, 1674. Cf. D. Mahnke, Bibliotheca Math., (3), 13, 1912–3, 37.
[3] "Observ. anal. de combinationibus," Comm. Acad. Petrop., 13, ad annum 1741–3, 1751, 64–93.

β, γ, $\cdots$ being the sum of the products of n, n^2, $\cdots$ two, three, $\cdots$ at a time, whence

$$\beta = n^3 + n^4 + 2n^5 + 2n^6 + 3n^7 + \cdots, \qquad \gamma = n^6 + n^7 + 2n^8 + 3n^9 + \cdots.$$

The coefficient of n^s in β, γ, $\cdots$ is the number of ways s is a sum of two, three, $\cdots$ distinct parts. This solves the problem (proposed to Euler by Ph. Naudé) to find the number of ways a number is a sum of a given number of distinct parts.

By the above relations between α, β, $\cdots$, A, B, $\cdots$, we get

$$\alpha = \frac{n}{1 - n} = A, \qquad \beta = \frac{n^3}{(1 - n)(1 - n^2)} = AB,$$

$$\gamma = \frac{n^6}{(1 - n)(1 - n^2)(1 - n^3)} = ABC, \cdots.$$

To give a proof of these results found by induction, write nz for z in R. We get $(1 + n^2 z)(1 + n^3 z) \cdots = 1 + \alpha n z + \beta n^2 z^2 + \cdots$. Its product by $1 + nz$ gives $R = 1 + \alpha z + \cdots$. Hence we get the preceding values of α, β, γ, $\cdots$. Let $m_i^{(\mu)}$ be the number of ways m is a sum of μ distinct integral parts, where the affix i (signifying inaequalus) is omitted if the parts need not be distinct. This $m_i^{(\mu)}$ is the coefficient of n^m in

$$\frac{n^{\mu(\mu+1)/2}}{(1 - n)(1 - n^2) \cdots (1 - n^\mu)},$$

the sum of the μth series α, β, γ, $\cdots$. Replacing the numerator by $n^{\mu(\mu-1)/2}$, we get the series whose general term is $m_i^{(\mu)} n^{m-\mu}$, or, if we prefer, $(m + \mu)_i^{(\mu)} n^m$. Subtract the former fraction from the latter; we get

$$\frac{n^{\mu(\mu-1)/2}}{(1 - n) \cdots (1 - n^{\mu-1})},$$

the general term of the series for which is $m_i^{(\mu-1)} n^m$. Hence, transposing,

$$(1) \qquad (m + \mu)_i^{(\mu)} = m_i^{(\mu)} + m_i^{(\mu-1)},$$

which serves as a recursion formula. Since in the series for $1/\{(1 - n) \cdots (1 - n^\mu)\}$ the coefficient of n^s is the number of ways s is a sum of parts $1, \cdots, \mu$ when the number of parts is not prescribed and the parts may be equal, $m_i^{(\mu)}$ also gives the number of ways $m - \mu(\mu + 1)/2$ can be obtained by addition from $1, \cdots, \mu$.

The second problem proposed by Naudé was to find the number $m^{(\mu)}$ of ways m is a sum of μ equal or distinct parts. To treat it, set

$$\frac{1}{(1 - nz)(1 - n^2 z) \cdots} = 1 + \mathfrak{A}z + \mathfrak{B}z^2 + \cdots.$$

Writing nz in place of z, we get

$$(1 - nz)(1 + \mathfrak{A}z + \mathfrak{B}z^2 + \cdots) = 1 + \mathfrak{A}nz + \mathfrak{B}n^2 z^2 + \cdots,$$

$$\mathfrak{A} = \frac{n}{1 - n}, \qquad \mathfrak{B} = \frac{\mathfrak{A}n}{1 - n^2} = \frac{n^2}{(1 - n)(1 - n^2)}, \qquad \mathfrak{C} = \frac{\mathfrak{B}n}{1 - n^3}, \qquad \cdots.$$

Hence $\alpha = \mathfrak{A}$, $\beta = n\mathfrak{B}$, $\gamma = n^3\mathfrak{C}$, $\delta = n^6\mathfrak{D}$, $\cdots$, where 1, 3, 6, $\cdots$ are the successive triangular numbers. From the above series for α, β, $\cdots$, we see that

$$m^{(\mu)} = \left\{ m + \frac{\mu(\mu-1)}{2} \right\}_i^{(\mu)}, \qquad m_i{}^{(\mu)} = \left\{ m - \frac{\mu(\mu-1)}{2} \right\}^{(\mu)}.$$

Hence $m^{(\mu)}$ is also the number of ways $m - \mu$ can be obtained by addition from 1, $\cdots$, μ. The former recursion formula for $m_i{}^{(\mu)}$ gives

(2) $$m^{(\mu)} = (m - \mu)^{(\mu)} + (m - 1)^{(\mu-1)}.$$

He stated, as a fact he could not prove,

(3) $$p(x) \equiv \prod_{k=1}^{\infty} (1 - x^k) = 1 - x - x^2 + x^5 + x^7 - \cdots + (-1)^n x^{(3n^2 \pm n)/2} + \cdots,$$

and that the reciprocal of the product is $1 + x + 2x^2 + 3x^3 + 5x^4 + \cdots$, the coefficient of x^s being the number of ways s can be partitioned into equal or distinct parts. As to (3), see Euler[1-6], Ch. X, Vol. 1.

Euler,[4] in a letter to N. Bernoulli, Nov. 10, 1742, stated the preceding facts on partitions. The answer to the second problem he stated in the following equivalent form: $m^{(\mu)}$ is the coefficient of n^m in the expansion of $n^\mu / \{(1 - n)(1 - n^2) \cdots (1 - n^\mu)\}$.

Euler[5] gave (3) and $p(x) = 1 - P_1 + P_2 - P_3 + \cdots$ [see Euler[9]].

P. R. Boscovich[6] gave a method of finding all the partitions of a given number n into integral parts > 0. Write down n units in a line. Replace the last two units by 2, then replace two units by 2, etc. Next, write $n - 3$ units and 3; replace two units by 2, etc. Then write $n - 6$ units and two 3's; replace two units by 2, etc. Thus the partitions of 5 are

$$11111, \qquad 1112, \qquad 122, \qquad 113, \qquad 23, \qquad 14, \qquad 5.$$

He applied partitions to find any power of a series in x, also in a paper, *ibid.*, 1748. In his third paper, *ibid.*, 1748, he showed how to list the partitions of n into parts $\leqq m$, by stopping his above process just before a part $m + 1$ would be introduced. He applied the rule also to the case when the parts are any assigned numbers. He treated the problem to find all the ways in which a given integer n can be decomposed in an assigned number m of parts, equal or distinct; but the solution by Hindenburg[16] is much more simple and direct. Boscovich attempted in vain to find a formula for the number of partitions. He gave elsewhere[7] his rule.

K. F. Hindenburg[8] would obtain the partitions of 8 by annexing unity to those of 7, and supplement them with

$$2222, \qquad 224, \qquad 233, \qquad 26, \qquad 35, \qquad 44, \qquad 8.$$

[4] *Opera postuma*, 1, 1862; Corresp. Math. Phys. (ed., Fuss), 2, 1843, 691–700.

[5] Letter to d'Alembert, Dec. 30, 1747; Bull. Bibl. Storia Sc. Mat., 19, 1886, 143.

[6] Giornale de' Letterati, Rome, 1747. Extract by Trudi,[98] pp. 8–10.

[7] Archiv der Math. (ed., Hindenburg), 4, 1747, 402.

[8] *Ibid.*, 392, and Erste Samml. Combinatorisch-Analyt. Abhand., 1796, 183. Quoted from G. S. Klügel's Math. Wörterbuch, 1, 1803, 456–60 (508–11, for references).

L. Euler[9] noted that the coefficient of $x^n z^m$ in the expansion of

$$(1 + x^\alpha z)(1 + x^\beta z)(1 + x^\gamma z) \cdots$$

is the number of different ways n is a sum of m different terms of the set $\alpha, \beta, \gamma, \cdots$. The coefficient of $x^n z^m$ in the series giving the expansion of

$$(1 - x^\alpha z)^{-1}(1 - x^\beta z)^{-1}(1 - x^\gamma z)^{-1} \cdots$$

is the number of different ways n is a sum of m terms of $\alpha, \beta, \cdots$, repetitions allowed. In particular, the coefficient of x^n in

$$\prod_{j=1}^{\infty} (1 - x^j)^{-1} = 1 + x + 2x^2 + 3x^3 + 5x^4 + 7x^5 + \cdots$$

is the number of partitions of n. If the product extends only to $j = m$, the coefficient of x^n is the number of partitions of n into parts $\leqq m$. In

$$Z = \prod_{j=1}^{\infty} (1 + x^j z) = 1 + P_1 z + P_2 z^2 + \cdots,$$

replace z by xz, so that Z becomes $Z/(1 + xz)$. Hence

$$(1 + xz)(1 + P_1 xz + P_2 x^2 z^2 + \cdots) = Z.$$

By comparison of coefficients, we get

$$P_m = \frac{x^{m(m+1)/2}}{(1 - x)(1 - x^2) \cdots (1 - x^m)}.$$

Hence the number of partitions of n into parts $\leqq m$ equals the number of ways of expressing $n + m(m + 1)/2$ as a sum of m distinct parts. Applying the same process to $\Pi(1 - x^j z)^{-1}$, we obtain the series

$$1 + \frac{xz}{1 - x} + \frac{x^2 z^2}{(1 - x)(1 - x^2)} + \frac{x^3 z^3}{(1 - x)(1 - x^2)(1 - x^3)} + \cdots.$$

Hence the number of partitions of n into parts $\leqq m$ equals the number of ways of expressing $n + m$ as a sum of m parts, not necessarily distinct.

If (n, m) is the number of partitions of n into parts $\leqq m$, then

$$(n, m) = (n, m - 1) + (n - m, m).$$

By use of this recursion formula, Euler computed a table of the values of (n, m) for $n \leqq 69$, $m \leqq 11$. The product of

$$P = \prod_{j=1}^{\infty} (1 - x^j), \qquad Q = \prod_{j=1}^{\infty} (1 + x^j)$$

is $\Pi(1 - x^{2j})$, all of whose factors occur in P. Hence [proof by L. Kronecker[10] for $|x| < 1$, to insure absolute convergency],

$$(4) \qquad Q = \frac{PQ}{P} = \frac{1}{(1 - x)(1 - x^3)(1 - x^5) \cdots},$$

---

[9] Introductio in analysin infinitorum, Lausanne, 1, 1748, Cap. 16, 253–275. German transl. by J. A. C. Michelsen, Berlin, 1788–90. French transl. by J. B. Labey, Paris, 1, 1835, 234–256.

[10] Vorlesungen über Zahlentheorie, 1, 1901, 50–56.

so that the number of partitions of n into distinct integers equals the number of partitions of n into odd parts not necessarily distinct.

Replace x by x^2 in (3) .   Since $\Pi(1 - x^{2k}) = PQ$,

$$Q = (1 - x^2 - x^4 + x^{10} + x^{14} - \cdots) \frac{1}{P},$$

$$\frac{1}{P} = 1 + x + 2x^2 + 3x^3 + 5x^4 + \cdots.$$

Hence, by multiplication,

$$Q = 1 + x + x^2 + 2x^3 + 2x^4 + 3x^5 + 4x^6 + \cdots.$$

Thus the coefficient of x^s in this series gives the number of partitions of s into distinct parts.   Since

$$(1+ x)(1 + x^2)(1 + x^4) \cdots = 1 + x + x^2 + x^3 + \cdots,$$
$$(x^{-1} + 1 + x)(x^{-3} + 1 + x^3)(x^{-9} + 1 + x^9) \cdots = 1 + x + x^2 + x^3 + \cdots$$
$$+ x^{-1} + x^{-2} + x^{-3} + \cdots,$$

every integer can be obtained by adding different terms of the progression $1, 2, 4, 8, 16, \cdots$ or of $\pm 1, \pm 3, \pm 3^2, \cdots$.   The latter facts were known by Leonardo Pisano,[11] Michael Stifel,[11a] and Frans van Schooten,[12] who gave a table expressing each number $\leqq 127$ in terms of $1, 2, 4, \cdots$, and every number $\leqq 121$ in terms of $\pm 1, \pm 3, \pm 9, \cdots$.

Euler[13] reproduced essentially his preceding treatment.   He concluded (§ 41, p. 91) that, if $P(n)$ or $n^{(\infty)}$ denotes the number of all partitions of n,

$$P(n) = P(n - 1) + P(n - 2) - P(n - 5) - P(n - 7) + P(n - 12) + \cdots,$$

the numbers subtracted from n being the exponents in (3).   His table of the number $n^{(m)}$ of partitions of n into parts $\leqq m$ here extends to $n \leqq 59$, $m \leqq 20$ and includes $m = \infty$.   He proved again that every integer equals a sum of different terms of $1, 2, 4, 8, \cdots$.

Euler[14] noted that the number (N, n, m) of partitions of N into n parts each $\leqq m$ is the coefficient of x^N in the expansion of $(x + x^2 + \cdots + x^m)^n$. Set

$$(5) \qquad (1 + x + \cdots + x^{m-1})^n = 1 + A_n x + B_n x^2 + \cdots,$$

bring to a common denominator the derivatives of the logarithms of each member and equate the coefficients of like powers of x in the expansions of the numerators.   The resulting linear relations determine $A_n, B_n, \cdots$ in turn, whence

$$\lambda(n + \lambda, n, m) = (n + \lambda - 1)(n + \lambda - 1, n, m)$$
$$- (mn + m - \lambda)(n + \lambda - m, n, m)$$
$$+ (mn - n + m + 1 - \lambda)(n + \lambda - m - 1, n, m).$$

[11] Scritti L. Pisano, I, Liber abbaci, 1202 (revised about 1228), Rome, 1857, 297.
[11a] Die Coss Christoffs Rudolffs . . . durch Michael Stifel gebessert . . ., 1553.
[12] Exercitationum Math., 1657, 410–9.
[13] Novi Comm. Acad. Petrop., 3, ad annum 1750 et 1751, 1753, 125 (summary, pp. 15–18); Comm. Arith. Coll., I, 73–101.
[14] Novi Comm. Acad. Petrop., 14, I, 1769, 168; Comm. Arith. Coll., I, 391–400.

Again, by comparing (5) with the corresponding relation with n replaced by $n + 1$, it is found that

$$(N + 1, n + 1, m) = (N, n + 1, m) + (N, n, m) - (N - m, n, m).$$

Finally, by expanding $(1 - x^m)^n$ and $(1 - x)^{-n}$ by the binomial theorem,

$$(n + \lambda, n, m) = \binom{n + \lambda - 1}{\lambda} - \binom{n}{1}\binom{n + \lambda - m - 1}{\lambda - m}$$
$$+ \binom{n}{2}\binom{n + \lambda - 2m - 1}{\lambda - 2m} - \binom{n}{3}\binom{n + \lambda - 3m - 1}{\lambda - 3m} + \cdots.$$

Euler's proofs were made for $m = 6$ and, except for the third formula, involve incomplete inductions. By evaluating the coefficient of x^N in the expansion of

$$(x + \cdots + x^6)(x + \cdots + x^8)(x + \cdots + x^{12})$$
$$= (x^3 - x^9 - \cdots - x^{29})/(1 - x)^3,$$

Euler found the number of partitions of $N \leqq 26$ into three parts, the first part $\leqq 6$, the second $\leqq 8$, the third $\leqq 12$.

As to the problem known as the rule of the Virgins [cf. Sylvester,[54] and note 188 of Ch. II], the number of sets of integral solutions $p, q, \cdots$, each $\geqq 0$, of the pair of equations

$$ap + bq + \cdots = n, \qquad \alpha p + \beta q + \cdots = \nu,$$

is the coefficient [not determined] of $x^n y^\nu$ in the expansion of

$$(1 - x^a y^\alpha)^{-1}(1 - x^b y^\beta)^{-1}\cdots.$$

K. F. Hindenburg[15] gave a method, different from Boscovich's, for listing all partitions of n. For $n = 5$, the method lists them in the order

$$5, \quad 14, \quad 23, \quad 113, \quad 122, \quad 1112, \quad 11111.$$

Hindenburg[16] gave a method of listing all partitions of n into m parts. The initial partition contains $m - 1$ units and the element $n - m + 1$. To obtain a new partition from a given one, pass over the elements of the latter from right to left, stopping at the first element f which is less, by at least two units, than the final element [$f = 2$ in 1234]. Without altering any element at the left of f, write $f + 1$ in place of f and every element to the right of f with the exception of the final element, in whose place is written the number which when added to all the other new elements gives the sum n. The process to obtain partitions stops when we reach one in which no part is less than the final part by at least two units.

Case $n = 10$, $m = 4$:

1 1 1 7	1 2 3 4
1 1 2 6	1 3 3 3
1 1 3 5	2 2 2 4
1 1 4 4	2 2 3 3
1 2 2 5	

[15] Methodus nova et facilis serierum infinitarum exhibendi dignitates, Leipsae, 1778. Infinitinomii dignitatum historia, leges, ac formulae, Gottingae, 1779, 73–91 (166, tables of partitions). A less interesting method is given in a Progr., 1795.

[16] Exposition by C. Kramp, Élémens d'Arith. Universelle, 1808, § 339. Quoted by Trudi.[98]

P. Paoli[17] noted (p. 38) that n can be separated into m positive integral parts in $\binom{n-1}{m-1}$ ways, if different permutations are counted separately. The number (p. 42) of partitions (different permutations not counted) of n into m parts > 0 is $\phi(1) + \phi(m+1) + \phi(2m+1) + \cdots$, where $\phi(j)$ is the number of partitions of $n - j$ into $m - 1$ parts. The number (p. 53) of ways n can be divided into m distinct parts is $\lambda(m) + \lambda(2m) + \lambda(3m) + \cdots$, if $\lambda(j)$ is the number of ways $n - j$ can be divided into $m - 1$ distinct parts. There are (p. 63) as many divisions of n into m distinct parts as of $n - m(m-1)/2$ into m parts equal or distinct. Let ψ, ϕ, ω be the number of ways $2n + 1$, $2n$, $2n + 1$ can be divided into $2m - 1$, $2m$, $2m + 1$ odd parts, respectively; let $\psi[r]$, $\phi[r]$, $\omega[r]$ be the corresponding numbers when n is replaced by $n - r$. Then

$$\phi = \psi[1] + \psi[2m+1] + \psi[4m+1] + \psi[6m+1] + \cdots,$$
$$\omega = \phi + \phi[2m+1] + \phi[4m+2] + \phi[6m+3] + \cdots.$$

If we impose also the condition that the odd parts be distinct, we have

$$\phi = \psi(2m) + \psi(4m) + \cdots, \quad \omega = \phi(2m) + \phi(4m+1) + \phi(6m+2) + \cdots.$$

The number (p. 76) of ways $2n$ is a sum of m even parts is $\phi(1) + \phi(m+1) + \phi(2m+1) + \cdots$, if $\phi(r)$ is the number of ways $2(n - r)$ is a sum of $m - 1$ even parts. The number (p. 79) of ways n is a sum of m parts is the number of ways $2n$ is a sum of m even parts. The number (p. 80) of ways n is a sum of m distinct parts is the number of ways pn is a sum of m distinct parts multiples of p. The number $P(n, m)$ of partitions of n into parts $\leqq m$ is $\Sigma P(n - j, m - 1)$, summed for $j = 0, m, 2m, \cdots$. The number (p. 85) of partitions of n into m parts equals $P(n - m, m)$. If ϕ, ω denote the number of ways $(m - 1)a + rb$ and $ma + rb$ can be formed additively from m and $m - 1$ terms of the progression $a, a + b, a + 2b, \cdots$, then $\omega = \phi + \phi(m) + \phi(2m) + \cdots$, where $\phi(j)$ is derived from ϕ by replacing r by $r - j$. Similarly (p. 92) when only distinct terms of the progression are used. If (p. 98) ϕ is the number of ways n is a sum of numbers chosen from $a, a + b, \cdots, a + (m - 1)b$, and ω that for $a, \cdots, a + mb$, then

$$\omega = \phi + \phi[a + mb] + \phi[2(a + mb)] + \cdots.$$

Finally (p. 103) the number of ways n is a sum of terms of any given series is discussed. He gave a more extended treatment in his next paper.

Paoli[18] treated linear difference equations with variable coefficients:

$$Z(y, x) = A_x Z(y - 1, x) + B_x Z(y - 2, x) + \cdots + X_x Z(y - x, x) + \cdots$$
$$+ A'_x Z(y, x - 1) + B'_x Z(y - 1, x - 1) + \cdots + X'_x Z(y - x, x - 1) + \cdots,$$

where $A_x, \cdots$ are given functions of x, and y is a function of x. Let the integral be

$$Z(y, x) = ma^y \nabla \alpha_x + nb^y \nabla \beta_x + \cdots, \qquad \nabla \alpha_x \equiv \alpha_1 \alpha_2 \cdots \alpha_x,$$

where $m, a, n, b, \cdots$ are constants. The condition that $a^y \nabla \alpha_x$ shall be an

<hr>

[17] Opuscula analytica, Liburni, 1780, Opusc. II (Meditationes Arith.), § 1.
[18] Memorie di mat. e fis. società Italiana, 2, 1784, 787–845.

integral is

$$\alpha_x = \frac{A_x' + B_x'a^{-1} + \cdots + X_x'a^{-x} + \cdots}{1 - A_x a^{-1} - B_x a^{-2} - \cdots - X_x a^{-x} - \cdots}.$$

Hence we get $\nabla\alpha_x$; let its expansion be

$$\nabla\alpha_x = A + A'a^{-1} + A''a^{-2} + \cdots.$$

Differentiate its logarithm, regarding x as constant and a variable. Thus

$$-\frac{a^2 d\nabla\alpha_x}{\nabla\alpha_x \cdot da} = \frac{A' + 2A''a^{-1} + \cdots}{A + A'a^{-1} + \cdots} = r + r'a^{-1} + r''a^{-2} + \cdots,$$

where $r^{(m)}$ is the excess of the sum of the $(m + 1)$th powers of the roots of the denominator over the sum of the $(m + 1)$th powers of the roots of the numerator in the fractional function of a^{-1} giving $\nabla\alpha_x$. Hence

$$A' = Ar, \qquad 2A'' = A'r + Ar', \qquad 3A''' = A''r + A'r' + Ar'', \qquad \cdots,$$

which give A'/A, A''/A, $\cdots$ as functions of r, r', $\cdots$. Hence, evidently,

$$Z(y, x) = A\phi(y) + A'\phi(y - 1) + A''\phi(y - 2) + \cdots,$$
$$\phi(y) \equiv ma^y + nb^y + \cdots.$$

Consider (pp. 817–21) the number (y, x) of ways y is a sum of x equal or distinct positive integers. Those in which 1 is a part furnish the $(y - 1, x - 1)$ ways $y - 1$ is a sum of $x - 1$ parts; while those in which each part exceeds 1 give, upon subtracting 1 from every part, the $(y - x, x)$ ways $y - x$ is a sum of x parts. Hence

$$(y, x) = (y - x, x) + (y - 1, x - 1).$$

It has the integral $(y, x) = a^y\nabla\alpha_x$ if $\alpha_x = a^{-x}\alpha_x + a^{-1}$, whence

$$\nabla\alpha_x = \frac{a^{-x}}{(1 - a^{-1})(1 - a^{-2})\cdots(1 - a^{-x})}.$$

The sum of the mth powers of the roots of $a = 1$, $a^2 = 1$, $\cdots$, $a^x = 1$ is the sum $\delta(m)$ of those of the numbers m, $m/2$, $m/3$, $\cdots$, m/m which are integral and $\leqq x$. Hence

$$r^{(m)} = \delta(m + 1), \qquad A' = \delta(1), \qquad A'' = \tfrac{1}{2}\{\delta(2) + \delta^2(1)\}, \qquad \cdots,$$

$$A^{(m)} = \frac{\delta(m)}{m} + \frac{\delta(1)\delta(m - 1)}{m} + \tfrac{1}{2}\{\delta(2) + \delta^2(1)\}\frac{\delta(m - 2)}{m}$$
$$+ \left\{\frac{\delta(3)}{3} + \frac{\delta(1)\delta(2)}{3} + \left(\frac{\delta(2)}{2} + \frac{\delta^2(1)}{2}\right)\frac{\delta(1)}{3}\right\}\frac{\delta(m - 3)}{m} + \cdots,$$

$$\nabla\alpha_x = a^{-x} + A'a^{-x-1} + A''a^{-x-2} + \cdots,$$
$$(y, x) = \phi(y - x) + A'\phi(y - x - 1) + A''\phi(y - x - 2) + \cdots.$$

Take $x = 1$. Then $A' = A'' = \cdots = 1$, $(y, 1) = \phi(y - 1) + \phi(y - 2) + \cdots$. Replace y by $y - 1$. Thus $(y - 1, 1) = \phi(y - 2) + \phi(y - 3) + \cdots$. Hence $(y, 1) - (y - 1, 1) = \phi(y - 1)$. By the nature of our problem, $(y, 1) = 1$ or 0 according as $y > 0$, $y \leqq 0$. Hence $\phi(z) = 1$ or 0 according as $z = 0$ or $z \neq 0$. Hence (y, x) reduces to the single term $A^{(y-x)}$, so that

$$(y, x) = \frac{\delta(y - x)}{y - x} + \frac{\delta(1)\delta(y - x - 1)}{y - x} + \tfrac{1}{2}\{\delta(2) + \delta^2(1)\}\frac{\delta(y - x - 2)}{y - x} + \cdots.$$

Next (pp. 821–4), to find the number (y, x) of ways y is a sum of x distinct positive integers, we have $(y, x) = (y - x, x) + (y - x, x - 1)$. Now $\alpha_x = a^{-x}/(1 - a^{-x})$. The values of $r^{(m)}$, $\delta(m)$, $A^{(m)}$ are the same as in the preceding problem. But

$$\nabla\alpha_x = a^{-z} + A'a^{-z-1} + \cdots, \qquad (y, x) = \phi(y - z) + A'\phi(y - z - 1) + \cdots,$$
$$z \equiv \frac{x(x + 1)}{2}.$$

Again, $\phi(y) = 1$ if $y = 0$, $\phi(y) = 0$ if $y \neq 0$. Hence (y, x) is derived from $A^{(m)}$ by replacing m by $y - x(x + 1)/2$. Hence y is a sum of x distinct parts as often as $y - x(x - 1)/2$ is a sum of x equal or distinct parts.

For (pp. 824–7) the number (y, x) of ways y is a sum of x equal or distinct positive odd numbers, it is stated that $(y, x) = (y - 2x, x) + (y - 1, x - 1)$. Here $\alpha_x = a^{-1}/(1 - a^{-2x})$, $(y, x) = \phi(y - x) + A'\phi(y - x - 2) + \cdots$, and $\phi(y) = 0$ unless $y = 0$, $\phi(1) = 1$. Thus (y, x) is obtained from $A^{(m)}$ by taking $m = (y - x)/2$, where y and z are necessarily both even or both odd. If y is partitioned into distinct odd numbers, $(y, x) = (y - 2x, x) + (y - 2x + 1, x - 1)$, and (y, x) is obtained from $A^{(m)}$ by taking $m = (y - x^2)/2$. Hence y is a sum of x distinct odd numbers as often as $y - x(x - 1)$ is a sum of x equal or distinct odd numbers.

The number (y, x) of ways y is a sum of terms chosen from $z_1, \cdots, z_x$ is the number of sets of solutions $p, \cdots, t$ of $y = pz_1 + \cdots + tz_x$. Taking $t = 0, 1, \cdots$ in turn, we get

$$(y, x) = (y, x - 1) + (y - z_x, x - 1) + (y - 2z_x, x - 1)$$
$$+ (y - 3z_x, x - 1) + \cdots.$$

Replace y by $y - z_x$ and subtract. Thus $(y, x) = (y - z_x, x) + (y, x - 1)$. Here $\alpha_x = 1/(1 - a^{-z_x})$. Write $\delta(m)$ for the sum of those terms $m, m/2, \cdots, m/m$ which are integers $\leqq z_x$ and are z's. The formula for $A^{(m)}$ is the same as in the first problem. Since $A^{(z_1)}$ is the first A which is not zero, we have

$$(y, 1) = \phi(y) + \phi(y - z_1) + \phi(y - 2z_1) + \cdots, \quad (y, 1) - (y - z_1, 1) = \phi(y).$$

Thus $\phi(0) = 1$, $\phi(y) = 0$, $y \neq 0$. Hence (y, x) is given by $A^{(y)}$. In particular, if $z_x = x$, we have the number of ways y is a sum of numbers $\leqq x$. Hence by the first problem, y is a sum of x integers as often as $y - x$ is a sum of integers $\leqq x$. For $z_x = n(2x - 1)$, (y, x) is given by $A^{(y/n)}$ if to form $\delta(m)$ we retain only the terms which are integral, odd and $\leqq 2x - 1$.

For the number (y, x) of ways y is a sum of distinct terms chosen from $z_1, \cdots, z_x$, $(y, x) = (y, x - 1) + (y - z_x, x - 1)$. Let $\gamma(m)$ be the sum of those numbers m, $-m/2$, $m/3$, $-m/4$, $\cdots$, $\pm m/m$ which are integers $\leqq z_x$ and are z's. Then $A^{(m)}$ is derived from the $A^{(m)}$ of the first problem by replacing δ's by γ's, while (y, x) is given by $A^{(y)}$. Let $z_x = 2^{x-1}$ and let x increase indefinitely, i. e., use the infinite series 1, 2, 4, 8, $\cdots$. Then $\gamma(m) = 2^m - 2^{m-1} - \cdots - 1 = 1$, $(y, x) = 1$, so that every integer is a sum of terms 1, 2, 4, 8, $\cdots$ in a single way [L. Pisano[11]].

For the number (y, x) of ways y is a sum of x terms of m, $m + n$, $m + 2n$, $\cdots$ or x distinct terms, $(y, x) = (y - nx, x) + (z, x - 1)$, where $z = y - m$ or $y - nx + n - m$, respectively. Then (y, x) is given by $A^{(\mu)}$ for $n\mu = y - mx$ or $y - mx - nx(x - 1)/2$, respectively. Hence y is a sum of x distinct terms of the progression as often as $y - nx(x - 1)/2$ is a sum of x equal or distinct terms.

Finally (pp. 842–5), to reduce the integration of
$$[y, x] = A_x[y - m\phi(x), x] + B_x[y - \psi(x), x - 1]$$
to that of $(y, x) = A_x(y - \phi(x), x) + B_x(y - f(x), x - 1)$, substitute $[y, x] = (\{y - F(x)\}/m, x)$ into the former and compare the result with the latter. The condition for agreement is $F(x) - F(x - 1) = \psi(x) - mf(x)$, whence, for a constant c independent of x,
$$F(x) = \Sigma\{\psi(x + 1) - mf(x + 1)\} + c.$$
Thus, in our second problem, $[y, x] = [y - x, x] + [y - x, x - 1]$, while in the first problem concerning $z_1, \cdots, z_x = x$,
$$(y, x) = (y - x, x) + (y, x - 1).$$
Hence $F(x) = \Sigma(x + 1 - 0) = x(x + 1)/2 + c$ and $c = 0$. Hence
$$[y, x] = (y - x(x + 1)/2, x),$$
so that y is a sum of x distinct parts as often as $y - x(x + 1)/2$ is a sum of parts $\leqq x$. Again, for the equation in our first problem, and
$$[y, x] = [y - 2x, x] + [y - 1, x - 1]$$
of our third problem, we have $F(x) = -x$, $[y, x] = (\{y + x\}/2, x\}$, so that y is a sum of x odd parts as often as $(y + x)/2$ is a sum of x even or odd parts. Finally, for our first and last problems, $F(x) = (m - n)x$, so that y is a sum of x terms of m, $m + n$, $m + 2n$, $\cdots$ as often as $\{y - (m - n)x\}/n$ is a sum of x positive integers.

G. F. Malfatti[19] obtained the general term of a recurring series whose scale of relation has a multiple root. In the Appendix, he treated the number of partitions into x distinct terms of the series 1, 2, $\cdots$, extended either to infinity (as by Paoli) or to a given number p. Taking first the former case, he showed how to pass from any of the series

$x = 1$:	1	1	1	1	1	1	1	1	1	$\cdots$
$x = 2$:	1	1	2	2	3	3	4	4	5	$\cdots$
$x = 3$:	1	1	2	3	4	5	7	8	10	$\cdots$

---

[19] Memorie di mat. e fis. società Italiana, 3, 1786, 571–663.

to the next.   Here the entries for $x = 2$ give the number of partitions of $3 = 1 + 2, 4, 5, \cdots$ into 2 distinct parts; and are the sums of the units in the respective columns in the accompanying scheme of units arranged by

$$
\begin{array}{cccccccc}
1\ 1 & 1\ 1 & 1\ 1 & 1\ 1 & \cdots \\
 & 1\ 1 & 1\ 1 & 1\ 1 & \cdots \\
 & & 1\ 1 & 1\ 1 & \cdots \\
 & & & 1\ 1 & \cdots
\end{array}
$$

twos.   Apply the same process to these numbers for $x = 2$, taking them by threes:

$$
\begin{array}{cccc}
1\ 1\ 2 & 2\ 3\ 3 & 4\ 4\ 5 & \cdots \\
 & 1\ 1\ 2 & 2\ 3\ 3 & \cdots \\
 & & 1\ 1\ 2 & \cdots
\end{array}
$$

Summing the columns, we obtain the number of partitions of $6 = 1 + 2 + 3$, $7, 8, \cdots$ into $x = 3$ distinct parts.   Taking these by fours, we get similarly the series for $x = 4$.   This property shows that

$$(t, x) = (t - x, x) + (t, x - 1),$$

if (t, x) is the tth term of the series for x.

To pass to the number of partitions into x distinct terms of $1, \cdots, p$, we must delete the partition $(p + 1) + 1$ of $p + 2$, and $(p + 1) + 2$, $(p + 2) + 1$ of $p + 3$, etc.   Thus the number of terms in the " first series of subtraction " is

$$
\begin{array}{llcccccc}
x = 1: & 1 & 1 & 1 & 1 & 1 & 1 & \cdots \\
x = 2: & 1 & 2 & 3 & 4 & 5 & 6 & \cdots \\
x = 3: & 1 & 2 & 4 & 6 & 9 & 12 & \cdots
\end{array}
$$

any line of which is formed from the preceding line as in the former problem. Thus $(t, x + 1) = (t - x, x + 1) + (t, x)$.   But for $x = 2$ we counted the partition of $2p + 2$ into parts each $p + 1$.   Hence we must correct our subtractive series by employing the "first additive series":

$$x = 2:\ \ 1\ 1\ 2\ 2\ 3\ 3 \cdots;\qquad x = 3:\ \ 1\ 2\ 4\ 6\ 9\ 12 \cdots;\qquad \cdots$$

leading to $(t, x + 2) = (t - x, x + 2) + (t, x + 1)$.   Then we have a second subtractive series, etc.   The general one of these difference equations is

$$(t, x + \lambda) = (t - x, x + \lambda) + (t, x + \lambda - 1).$$

It has the integral $a'\Pi$, where $\Pi = \Pi_{j=1}^{j=x}\alpha_{j+\lambda}$, if $\alpha_{j+\lambda} = 1/(1 - a^{-j})$.   Thus $\Pi = 1/D$, $D = (1 - a^{-1})(1 - a^{-2})\cdots(1 - a^{-x})$.   If

$$1/D = 1 + A'a^{-1} + A''a^{-2} + \cdots,$$

we find (as by Paoli) that $A' = r$, $2A'' = A'r + r'$, $\cdots$, where $r^{(m)}$ is the sum of the $(m + 1)$th powers of the roots of $D = 0$.   The general integral is $(t, x + \lambda) = \phi(t) + A'\phi(t - 1) + A''\phi(t - 2) + \cdots$.   For $\lambda = 0$, we find by using $x = 1$ (cf. Paoli) that $(t, x) = A^{(t-1)}$.   For general λ, write

$n_\lambda^{(i)}$ in place of $A^{(i)}$. Taking $x = 1$ in the general integral, we see that $(t, 1 + \lambda) - (t - 1, 1 + \lambda) = \phi(t)$, which is shown to be $A^{(t-1)}$ of Paoli's case. Hence $\phi(t) = n_\lambda^{(t-1)}$, and

$$(t, x + \lambda) = n_\lambda^{(t-1)} + A'n_\lambda^{(t-2)} + A''n_\lambda^{(t-3)} + \cdots + A^{(t-x)}.$$

He gave the following results for $x \leqq 4$:

$$(t, 2) = \tfrac{1}{4}\{2t + 1 - (-1)^t\},$$

$$(t, 3) = \frac{6t^2 + 24t + 17}{72} - \frac{(-1)^t}{8} + \frac{\alpha^{t-1} + \alpha_1^{t-1}}{9},$$

$$(t, 4) = \frac{2t^3 + 24t^2 + 81t + 68}{288} - \frac{(t + 4)(-1)^t}{32} - \frac{\alpha^{t+1} + \alpha_1^{t+1}}{27}$$
$$- \frac{2(\alpha^t + \alpha_1^t)}{27} + \frac{\beta^{t-1} + \beta_1^{t-1}}{16},$$

where α and α_1 are the imaginary cube roots of unity, and $\beta = i$, $\beta_1 = -i$. He also gave the general term of the first subtraction series:

$$(t', 2) = t' = t - p + 1;$$

$$(t', 3) = \tfrac{1}{8}\{2t'^2 + 4t' + 1 - (-1)^{t'}\}, \qquad t' = t - p + 2;$$

$$(t', 4) = \frac{4t'^3 + 30t'^2 + 60t' + 25}{144} - \frac{(-1)^{t'}}{16}$$
$$- \frac{(\alpha^{t'+1} + \alpha_1^{t'+1})}{27} - \frac{2(\alpha^{t'} + \alpha_1^{t'})}{27}, \qquad t' = t - p + 3.$$

Earlier in the paper (pp. 618–26), he gave $(t, 5)$ and the general terms of the addition and subtraction series; these and various other results given above occur in his earlier two articles in Prodromo dell' Enciclopedia Italiana and (in more detail) in Antologia Romana, 11, 1784.

V. Brunacci[20] reproduced Paoli's[18] treatment of his first problem.

S. Vince[21] proved by induction that every positive integer is a sum of distinct terms of $1, 2, 4, 8, \cdots$. For, if true for numbers up to $s = 1 + 2 + \cdots + 2^{n-1} = 2^n - 1$, it will be true for the remaining numbers up to $s + 2^n$. The proof for $\pm 1, \pm 3, \pm 3^2, \cdots$ is longer.

S. F. Lacroix[22] reproduced part of the discussion by Euler.[9]

Frégier[22a] proved that a^m equals a sum of a terms of the arithmetical progession whose first term is unity and common difference is

$$2 + 2a + \cdots + 2a^{m-2}.$$

Cf. Volpicelli,[37] Lemoine,[76] Mansion,[87] and Candido.[213]

[20] Corso di Matematica Sublime, Firenze, 1, 1804; §§ 108–9, pp. 237–248. Cf. Compendium del Calc. Subl., 1811, § 114.

[21] Trans. Roy. Irish Acad., 12, 1815, 34–38. Euler.[13]

[22] Traité du Calcul Diff. Int., 3, 1819, 461–6.

[22a] Annales de math. (ed., Gergonne), 9, 1818–9, 211–2.

C. G. J. Jacobi[22b] gave fundamental applications of elliptic function formulas to the theory of partitions. He proved the identical relation

$$1 + q(v + v^{-1}) + q^4(v^2 + v^{-2}) + q^9(v^3 + v^{-3}) + \cdots$$
$$= (1 - q^2)(1 - q^4)(1 - q^6) \cdots \times (1 + qv)(1 + q^3v)(1 + q^5v) \cdots$$
$$\times (1 + qv^{-1})(1 + q^3v^{-1})(1 + q^5v^{-1}) \cdots,$$

if $|q| < 1$, and another deducible from it by writing qv^2 for v and multiplying by $q^{1/4}v$, viz.,

$$q^{1/4}(v + v^{-1}) + q^{9/4}(v^3 + v^{-3}) + q^{25/4}(v^5 + v^{-5}) + \cdots$$
$$= (1 - q^2)(1 - q^4)(1 - q^6) \cdots \times q^{1/4}(v + v^{-1})$$
$$\times (1 + q^2v^2)(1 + q^4v^2)(1 + q^6v^2) \cdots \times (1 + q^2v^{-2})(1 + q^4v^{-2})(1 + q^6v^{-2}) \cdots.$$

From this he inferred, through the intermediary of the four theta functions, the following relations of great importance in the theory of partitions:

$$\sqrt{\frac{2K}{\pi}} = \sum_{m=-\infty}^{+\infty} q^{m^2} = \prod_{m=1}^{\infty} (1 - q^{2m})(1 + q^{2m-1})^2,$$

$$\sqrt{\frac{2\kappa'K}{\pi}} = \sum_{m=-\infty}^{+\infty} (-1)^m q^{m^2} = \prod_{m=1}^{\infty} (1 - q^{2m})(1 - q^{2m-1})^2,$$

$$\sqrt{\frac{2\kappa K}{\pi}} = \sum_{m=-\infty}^{+\infty} q^{(2m+1)^2/4} = 2q^{1/4} \prod_{m=1}^{\infty} (1 - q^{2m})(1 + q^{2m})^2,$$

$$\sqrt{\frac{2\kappa\kappa'K}{\pi}} = \frac{\pi}{2K} \sum_{m=-\infty}^{+\infty} (-1)^m(2m + 1)q^{(2m+1)^2/4} = \frac{\pi}{K} q^{1/4} \prod_{m=1}^{\infty} (1 - q^{2m})^3.$$

For his expansions, as series in q, of powers of these functions see Chs. VI–IX on sums of squares. If in the first identical relation above we write $+z$ and $-z$ in turn for v and multiply the results together, we obtain

$$\Sigma(-1)^m q^{m^2+n^2} z^{m+n} = \Sigma(-1)^{m+n} q^{2(m^2+n^2)} z^{2m}.$$

A. M. Legendre[23] noted that Euler's formula (3) implies that every number, not a pentagonal number $(3n^2 \pm n)/2$, can be partitioned into an even number of distinct integers as often as into an odd number; while $(3n^2 \pm n)/2$ can be partitioned into an even number of parts once oftener or once fewer times than into an odd number, according as n is even or odd. This result was implied by Euler[13] (§ 46).

C. J. Brianchon[24] noted that the literal part of the general term in the expansion of $(a_1 + a_2 + \cdots + a_n)^m$ is of the type $a_1^{\alpha_1} \cdots a_x^{\alpha_x}$, where $\alpha_1 + \cdots + \alpha_x = m$, $x \leqq m$, $x \leqq n$. Thus the terms form as many classes as there are values of x, and the terms of a class form as many groups as there are partitions of m into x numbers α_i. In view of Euler's[9] table we know the number of groups of each class.

[22b] Fundamenta Nova Theoriae Func. Ellip., 1829, 182–4. Werke, I, 234–6. Cf. Jacobi.[30] See the excellent report by H. J. S. Smith, Report British Assoc. for 1865, 322–75; Coll. Math. Papers, I, 289–94, 316–7.

[23] Théorie des nombres, ed. 3, 1830, II, 128–133.

[24] Jour. de l'école polyt., tome 15, cah. 25, 1837, 166.

E. Catalan[25] proved that $x_1 + \cdots + x_n = m$ has $\binom{n+m-1}{m}$ sets of solutions $\geqq 0$.

O. Rodrigues[26] noted that the number $Z_{n,i}$ of ways of permuting n letters, such that there are i inversions in each permutation, is the number of sets of solutions of $x_0 + x_1 + \cdots + x_{n-1} = i$, where x_k takes only the values $0, 1, \cdots, k$ and where the value of x_k for each permutation is the number of inversions produced by x_{k+1}. Thus $Z_{n,i}$ is the coefficient of t^i in the expansion of

$$(1 + t)(1 + t + t^2) \cdots (1 + t + \cdots + t^{n-1}) = (1 - t)^{-n}P,$$

where $P = (1 - t)(1 - t^2) \cdots (1 - t^n)$. Let $E_{n,i}$ be the coefficient of t^i in the expansion of P. Thus $E_{n,i} = E_{n-1,i} - E_{n-1,i-n}$, $E_{n,i} = E_{i,i}$, and

$$Z_{n,i} = E_{n,i} + \binom{n}{1}E_{n,i-1} + \cdots + \binom{n+i-1}{i}E_{n,0}.$$
$$= \binom{n+i-1}{i} + \binom{n+i-2}{i-1}E_{1,1} + \cdots + E_{i,i}.$$

Here $E_{n,i}$ equals the excess of the number of partitions of i in an even number of distinct integers $< n + 1$ over the number in an odd, the number of parts being also $< n + 1$.

M. A. Stern[27] wrote $_nC_q$ (or $_nC_q'$) for the number of combinations without (or with) repetitions with the sum n and class q (i. e., q at a time), meaning the number of partitions of n into q distinct parts (or equal or distinct parts). Evidently $_nC_2' = [n/2]$. Hence, by (2), we get

$$_nC_3' = \left[\frac{n-1}{2}\right] + \left[\frac{n-4}{2}\right] + \left[\frac{n-7}{2}\right] + \cdots,$$

$$_nC_q' = \sum_{k_{q-3}=0}^{\frac{n-1}{q}} \sum_{k_{q-4}=0}^{\frac{n-1}{q-1}} \cdots \sum_{k_1=0}^{\frac{n-1}{4}} \sum_{k=0}^{\frac{n-1}{3}} \left[\tfrac{1}{2}\{n - (3k + 1) - (4k_1 + 1) - \cdots - (qk_{q-3}+1)\}\right].$$

Since $_nC_q = {}_mC_q'$ if $m = n - q(q - 1)/2$, we get by (1),

$$_nC_2 = \left[\frac{n-1}{2}\right], \qquad _nC_3 = \left[\frac{n-4}{2}\right] + \left[\frac{n-7}{2}\right] + \left[\frac{n-10}{2}\right] + \cdots.$$

Again,

$$_nC_3' = \frac{1}{2}\left\{n\left[\frac{n}{3}\right] - \frac{3}{2}\left[\frac{n}{3}\right]^2 + \frac{1}{2}\left[\frac{n}{3}\right] - \left[\frac{n+2}{6}\right] + \left[\frac{n+1}{6}\right] - \left[\frac{n}{6}\right]\right\}$$

If $C(n)$ is the number of all partitions of n into distinct parts,

$$\sum_{y=0}^{2n} (-1)^y C(n - y/2) = (-1)^r \text{ or } 0 \qquad (y \equiv 3z^2 \mp z),$$

according as n is or is not of the form $3r^2 \mp r$. This follows by expanding

$$\frac{1 - x^2}{1 - x} \cdot \frac{1 - x^4}{1 - x^2} \cdot \frac{1 - x^6}{1 - x^3} \cdots$$

---

[25] Jour. de Math., 3, 1838, 111–2.
[26] Jour. de Math., 4, 1839, 236–240.
[27] Jour. für Math., 21, 1840, 91–97, 177–9. Further results were quoted under Stern[15a] of Ch. X in Vol. I of this History.

Also,

$$\sum_{y=0}^{n} (-1)^z C(n - y) = 1 \text{ or } 0 \qquad (y \equiv 3z^2 \mp z),$$

according as n is or is not of the form $z(z + 1)/2$.

A. De Morgan[28] considered the number $u_{x,y}$ of ways x can be formed additively from y and numbers $\leqq y$. Adding y to each such composition of $x - y$, we see that

$$u_{x,y} = u_{x-y,1} + u_{x-y,2} + \cdots + u_{x-y,y}.$$

Subtracting from this the equation obtained by decreasing x and y by unity, we get

(6) $$u_{x,y} - u_{x-1,y-1} = u_{x-y,y}.$$

Regard y as fixed and the second u as a given function, we have a difference equation of order y whose general integral is of the form

$$u_{x,y} = A_{y-1} + A_{a_2}P_2 + \cdots + A_{a_y}P_y,$$

where A_{a_n} is a rational integral function whose degree a_n is the greatest integer in $(n - y)/y$, while P_n is a circulating function with a cycle of n values.   In particular,

$$u_{x,2} = \frac{x}{2} - \tfrac{1}{4} + \tfrac{1}{4}(-1)^x,$$

$$u_{x,3} = \tfrac{1}{72}\{6x^2 - 7 - 9(-1)^x + 8(\beta^x + \gamma^x)\},$$

$$u_{x,4} = \tfrac{1}{864}\{6x^3 + 18x^2 - 27x - 39 + 27(x + 1)(-1)^x$$
$$+ 32(\beta^{x-1} + \gamma^{x-1} - \beta^x - \gamma^x) + 54i^x + 54(-i)^x\},$$

where β, γ are the imaginary cube roots of unity and $i = \sqrt{-1}$. Thus

$$12u_{x,3} = x^2, \; x^2 - 1, \; x^2 - 4, \; x^2 + 3, \; x^2 - 4, \; x^2 - 1,$$

according as $x \equiv 0$, 1, 2, 3, 4, 5 (mod 6). Similarly, $u_{x,4}$ has 12 forms depending on the residues of x modulo 12. Again, $u_{x,3}$ is the integer nearest $x^2/12$, and $u_{x,4}$ that nearest to $(x^3 + 3x^2)/144$ or $(x^3 + 3x^2 - 9x)/144$, according as x is even or odd.

A. Cauchy[29] proved (3) and the other formulas of Euler[3] and the related ones involving a finite number of factors:

$$P(x) = \prod_{j=0}^{n-1}(1 + t^j x) = 1 + \frac{1 - t^n}{1 - t}x + \frac{(1 - t^n)(t - t^n)}{(1 - t)(1 - t^2)}x^2$$
$$+ \frac{(1 - t^n)(t - t^n)(t^2 - t^n)}{(1 - t)(1 - t^2)(1 - t^3)}x^3 + \cdots,$$

$$\frac{1}{P(-x)} = 1 + \frac{1 - t^n}{1 - t}x + \frac{(1 - t^n)(1 - t^{n+1})}{(1 - t)(1 - t^2)}x^2$$
$$+ \frac{(1 - t^n)(1 - t^{n+1})(1 - t^{n+2})}{(1 - t)(1 - t^2)(1 - t^3)}x^3 + \cdots.$$

[28] Cambridge Math. Jour., 4, 1843, 87–90.
[29] Comptes Rendus Paris, 17, 1843, 523; Oeuvres, (1), VIII, 42–50.

C. G. J. Jacobi[30] stated that if we replace q by q^n and set $v = \mp q^m$ in his first formula,[22b] we get

$$(1 \pm q^{n-m})(1 \pm q^{n+m})(1 - q^{2n})(1 \pm q^{3n-m})(1 \pm q^{3n+m})(1 - q^{4n}) \cdots$$

$$\equiv \prod_{t=1}^{\infty} (1 \pm q^{2tn-n-m})(1 \mp q^{2tn-n+m})(1 - q^{2tn}) = \sum_{-\infty}^{\infty} (\pm 1)^i q^{ni^2+mi}.$$

For $m = 1/2$, $n = 3/2$, that with the lower signs becomes Euler's (3). Although he[22b] (pp. 185-6) gave two simple proofs of it, Jacobi here reproduced Euler's proof in essential points, but with a generalization. He gave a proof of Legendre's[23] corollary and proved the following generalization. Let $(P, \alpha, \beta, \cdots)$ be the excess of the number of partitions of P into an even number of the given distinct elements α, β, $\cdots$, each $\neq 0$, over the number of partitions into an odd number of them. Then

$$(P, \alpha, \beta, \gamma, \cdots) = (P, \beta, \gamma, \cdots) - (P - \alpha, \beta, \gamma, \cdots).$$

Let a, a_1, $\cdots$, a_{m-1} form any arithmetical progression, and b_0, b_1, $\cdots$, b_m an arithmetical progression with the common difference $- a$. Set

$$c_i = b_{i+1} - a_{i+1}, \qquad d_i = c_{i+1} - a_{i+1}, \qquad \cdots.$$

Then

$$L \equiv (b_0, a) + (b_1, a, a_1) + (b_2, a, a_1, a_2) + \cdots + (b_{m-1}, a, a_1, \cdots, a_{m-1})$$
$$= \Delta - (b_m, a_1, \cdots, a_{m-1}) + (c_{m-1}, a_2, \cdots, a_{m-2})$$
$$- (d_{m-2}, a_3, \cdots, a_{m-3}) + \cdots,$$
$$\Delta \equiv [b_0] - [c_0] - [c_1] + [d_1] + [d_2] - [e_2] - \cdots.$$

If b_0 and a are positive and $ma > b_0$, L vanishes except when b_1 equals $s_{i-1} + 2s_i$ or $2s_{i-1} + s_i$, and then equals $(- 1)^i$, where

$$s_i = a_1 + a_2 + \cdots + a_i.$$

Jacobi[31] noted that Euler[9] expressed $P \equiv (1 + q)(1 + q^2)(1 + q^3) \cdots$ in the form $f(q^2)/f(q)$, where $f(x)$ is given by (3). Jacobi expressed P in six ways as quotients of two infinite products and expanded each into infinite series; the next to the last case is

$$\frac{(1 + q)(1 + q^2)(1 - q^3)(1 + q^4)(1 + q^5)(1 - q^6) \cdots}{(1 - q^3)^2(1 - q^6)(1 - q^9)^2(1 - q^{12})(1 - q^{15})^2(1 - q^{18}) \cdots} = \frac{\Sigma q^{(3i^2+i)/2}}{\Sigma(- 1)^i q^{3i^2}}.$$

Expressing this in the form $\Sigma_{j=1}^{\infty} C_j q^j$, we conclude that, if C_i is the number of partitions of i into arbitrary distinct integers or into equal or distinct odd integers,

$$C_i = 2\{C_{i-3} - C_{i-12} + C_{i-27} - C_{i-48} + C_{i-75} - \cdots\} + \delta,$$

where $\delta = 1$ or 0 according as i is or is not of the form $(3n^2 \pm n)/2$. He gave

---

[30] Jour. für Math., 32, 1846, 164-175; Werke, 6, 1891, 303-317; Opuscula Math., 1, 1846, 345-356. Cf. Sylvester[117], Goldschmidt.[148]

[31] Jour. für Math., 37, 1848, 67-73, 233; Werke, 2, 1882, 226-233, 267; Opuscula Math., 2, 1851, 73-80, 113.

expansions of P^2 and P^3. Only those m-gonal numbers give the remainder 1, when divided by $m = a^2b$, whose side has the remainder 1 when divided by ab, where a^2 is the greatest square dividing m.

H. Warburton[32] considered the number $[N, p, \eta]$ of partitions of N into p parts each $\geqq \eta$, and proved that

$$[N, p, \eta] - [N, p, \eta + 1] = [N - \eta, p - 1, \eta],$$

$$[N + p, p, 1] = \sum_{z=0}^{p} [N, z, 1], \qquad [N, p, \eta] = \sum_{z=0}^{p} [N - p\eta, z, 1],$$

$$[N, p, 1] = [N - 1, p - 1, 1] + [N - p - 1, p - 1, 1]$$
$$+ [N - 2p - 1, p - 1, 1] + \cdots,$$

to $[N/p]$ terms. He applied these formulas to the construction of a table of partitions and proved that the number of partitions of x into three parts is $3t^2$, $3t^2 \pm t$, $3t^2 \pm 2t$, $3t^2 + 3t + 1$ according as $x = 6t$, $6t \pm 1$, $6t \pm 2$, $6t + 3$ [in accord with De Morgan].

J. F. W. Herschel[33] recalled his[34] earlier notation $s_x = s^{-1}\Sigma\alpha^x$, where α ranges over the sth roots of unity, so that $s_x = 1$ or 0 according as x is or is not divisible by s. Then $A_x s_x + B_x s_{x-1} + \cdots + N_x s_{x-s+1}$ will circulate in its successive values as x increases by units from zero, being A_x when x is divisible by s, but B_x when $x - 1$ is divisible by s, etc. If A_x, etc., are constants, the function is called periodic. He wrote $^s\Pi(x)$ for the number (x, s) of partitions of x into s parts > 0. Starting with

$$(x, s - 1) = \phi(x) + Q_x,$$

where $\phi(x)$ is the non-periodic part and Q_x the periodic or circulating function, and applying the final formula quoted from Warburton, he obtained

$$(x, s) = A + Z, \qquad A = \phi(x - 1) + \phi(x - s - 1) + \cdots,$$
$$Z = Q_{x-1} + Q_{x-s-1} + \cdots,$$

each extending to $[x/s]$ terms. Then A is expressed explicitly in terms of the numbers $\Delta^m 0^r$, giving the mth order of difference of z^n for $z = 0$, while Z is expressed in terms of these numbers and the above circulating functions s_x. He deduced explicit expressions for (x, s), $s = 2, 3, 4, 5$, as $(x, 2) = \frac{1}{2}(x - 2_{x-1})$,

$$(x, 3) = \tfrac{1}{12}\{x^2 - 6_{x-1} - 4\cdot6_{x-2} + 3\cdot6_{x-3} - 4\cdot6_{x-4} - 6_{x-5}\},$$

which, with the expression for $(x, 4)$, are in accord with the results by De Morgan,[28] although the latter was not treating partitions into s parts. While the method of Herschel is laborious, it anticipated to some extent the simpler method of Cayley.[44]

J. J. Sylvester[35] quoted Euler's theorem that the number of partitions of n is the same whether the number of parts is $\leqq m$ or every part is $\leqq m$, and noted that, if we apply the theorem also when the limiting number is

[32] Trans. Cambridge Phil. Soc., 8, 1849, 471–492.
[33] Phil. Trans. Roy. Soc. London, 140, II, 1850, 399–422.
[34] From his paper on circulating functions, *ibid.*, 108, 1818, 144–168.
[35] Phil. Mag., (4), 5, 1853, 199–202; Coll. Math. Papers, I, 595–8.

$m - 1$, we obtain by subtraction the following corollary. The number of partitions of n into m parts equals the number of partitions of n into parts one of which is m and the others are $\leqq m$. Sylvester credited the corollary to N. M. Ferrers who communicated to him the following proof. Take any set A composed of 3, 3, 2, 1, written as

$$
\begin{array}{ccc}
1, & 1, & 1 \\
1, & 1, & 1 \\
1, & 1 & \\
1. & &
\end{array}
$$

Reading it by columns, we get the set B composed of 4, 3, 2. Similarly, every A in which the number of parts is 4 gives rise to a B in which 4 is a part and every part is $\leqq 4$; conversely, every B produces an A. Euler's theorem can be proved by the same diagram. Similarly, the number of partitions of n into m or more parts equals the number of partitions of n into parts the greatest of which is $\geqq m$. If we partition each of i numbers into parts so that the sum of the greatest parts shall not exceed (or be less than) m, the number of ways this can be done is the same as the number of ways these i numbers can be simultaneously partitioned so that the total number of parts shall never exceed (or never be less than) m.

P. Volpicelli[36] arranged the natural numbers n, $n + 1$, $\cdots$ in a rectangle with $k + 1$ rows, each with $h + 1$ numbers, but in reverse order in alternate rows.   For example,

$$
\begin{array}{ccc}
18 & 19 & 20 \\
23 & 22 & 21 \\
24 & 25 & 26.
\end{array}
$$

The successive sums by columns are 65, 66, 67 (of common difference unity) and so always when the number of columns is odd; but, if $k + 1$ is even, the sum of the numbers in each column is constant, being

$$a = \{2n + h(k + 1) + k\}(k + 1)/2,$$

and we have special partitions of a. Given a, to find integral solutions n, h, k, we note that $h = \gamma/\delta$, where $\delta = (k + 1)^2$, while γ and $(2a)^2/\delta$ are integers.   Hence seek those divisors of $(2a)^2$ which are squares δ; for each such δ, we have k and seek integers n for which γ/δ is an integer h.

Volpicelli[37] expressed n^k as a sum of numbers in arithmetical progression.

* P. Bonialli[38] treated partitions.

T. P. Kirkman[39] proved that the number of partitions of N into p parts $\geqq a$ equals the sum of the number of partitions of $N - a$, $N - p - a$, $N - 2p - a$, $\cdots$ into $p - 1$ parts $\geqq 0$.   The case $a = 1$ is the last formula of Warburton.[32]   He gave an analytic expression for the number (x, k) of

[36] Atti Accad. Pont. Nuovi Lincei, 6, 1852–3 (1855), 631; 10, 1856–7, 43–51, 122–131; Annali di sc. mat. e fis., 8, 1857, 22–27

[37] Atti Accad. Pont. Nuovi Lincei, 6, 1852–3, 104–119. Frégier.[22a]

[38] Formole algebriche esprimenti il numero delle partizioni di qualunque intero. Progr., Clusone, 1855.

[39] Mem. Lit. Phil. Soc. Manchester, (2), 12, 1855, 129–145.

partitions of x into k parts > 0, for $k = 2, \cdots, 6$, in terms of the circulator s_e, which is unity if e/s is a positive integer, zero if e/s is fractional or negative. For $k \leqq 5$, his results are identical with those of Herschel,[33] but were obtained by more elementary methods. Kirkman[40] corrected his expression for $(x, 6)$ and found $(x, 7)$. He[41] found $(r^2 - r + 1, r)$.

J. J. Sylvester[42] called the number of ways of composing n with given positive integral summands $a_1, \cdots, a_r$ the *quotity* Q of n with respect to $a_1, \cdots, a_r$. Thus Q is the number of sets of integral solutions $\geqq 0$ of

$$a_1 x_1 + \cdots + a_r x_r = n.$$

He stated that $Q = A + U$, the periodic part U (depending on roots of unity) not being discussed, while the non-periodic part A is the coefficient of $1/t$ in the expansion of

$$e^{nt}(1 - e^{-a_1 t})^{-1} \cdots (1 - e^{-a_r t})^{-1}.$$

Other formulas for A are given. But all these formulas were provisional and were replaced in his next paper by others more expeditious for computation.

Sylvester[43] stated that $Q = \Sigma W_q$, where W_q (called a *wave*) is the coefficient of $1/t$ in the development in ascending powers of t of*

$$\Sigma \rho^{-n} e^{nt} \prod_{j=1}^{r} (1 - \rho^{a_j} e^{-a_j t})^{-1},$$

summed for the various primitive qth roots ρ of unity. Thus $W_q = 0$ except for a q which divides one or more of the a_i. Thus W_1 is his former A. Taking the a's to be $1, \cdots, 6$, Sylvester computed $W_1, \cdots, W_6$ initially in terms of certain $\Sigma \rho^k$ and finally in terms of Herschel's[34] circulating functions, obtaining results agreeing with Cayley's.[44]

But Sylvester did not give a full account[107] of his theorem until 1882.

A. Cayley[44] employed $P(a, b, \cdots)q$, in the sense of Sylvester's Q, to denote the number of partitions of q into the elements $a, b, \cdots$, with repetitions allowed. As known, it is the coefficient of x^q in $\Pi(1 - x^a)^{-1}$. By decomposing the latter into partial fractions, it is shown that

$$P(a, b, \cdots)q = Aq^{k-1} + Bq^{k-2} + \cdots + Lq + M + \Sigma q^r (A_0, A_1, \cdots, A_{l-1}) pcrl_q,$$

where k is the number of the elements $a, b, \cdots$, and l is any divisor > 1 of one or more of these elements, and the summation extends, for each such divisor, from $r = 0$ to $r = x - 1$, if x is the number of elements $a, b, \cdots$ having l as a divisor. Also

$$(A_0, \cdots, A_{l-1}) pcrl_q = A_0 a_q + A_1 a_{q-1} + \cdots + A_{a-1} a_{q-a+1}$$

---

[40] Mem. Lit. Phil. Soc. Manchester, (2), 14, 1857, 137–149.

[41] Proc. and Papers Lancashire and Cheshire Hist. Soc. Liverpool, 9, 1857, 127.

[42] Quar. Jour. Math., 1, 1855 (1857), 81–4; Coll. Math. Papers, II, 86–9.

[43] *Ibid.*, 141–152; Coll. Math. Papers, II, 90–99. An Italian transl. of an extract appeared in Annali di sc. mat. e fis., 8, 1857, 12–21.

* Sylvester's first factor ρ^n has been changed to ρ^{-n} to accord with Battaglini,[48] Brioschi,[49] Roberts,[51] and Trudi.[56]

[44] Phil. Trans. Roy. Soc. London, 146, 1856, 127–140; Coll. Math. Papers, II, 235–249.

is the "prime circulator to the period a," if $a_q = 1$ or 0 according as q is divisible by a or not, and

$$A_i + A_{l+i} + \cdots + A_{(\lambda-1)l+i} = 0 \quad (i = 0, 1, \cdots, l - 1; \ \lambda = a/l).$$

He showed how to evaluate the A's and then the coefficients $A, \cdots, L, M$ of the non-circulating part. Next, he evaluated the number $P(0, 1, \cdots, k)^m q$ of partitions of q into m terms $0, 1, \cdots, k$, with repetitions allowed, known to be the coefficient of $x^q z^m$ in $(1 - z)^{-1}(1 - xz)^{-1} \cdots (1 - x^k z)^{-1}$.

Finally, Cayley proved that the non-circulating part of the fraction $\phi(x)/f(x)$ is the coefficient of $1/t$ in

$$\frac{1}{1 - xe^t} \cdot \frac{\phi(e^{-t})}{f(e^{-t})}.$$

Cayley[45] later considered his last formula for $\phi(x) \equiv 1$, obtaining a formula equivalent to Sylvester's theorem, and applied it to find $P(1, 2, \cdots, 6)q$.

Cayley[46] noted that $P(0, 1, \cdots, m)^\theta q - P(0, \cdots, m)^\theta(q - 1)$ is the number of asyzygetic covariants of degree θ and order q of a binary quantic of order m. Thus it is the coefficient of x^θ in the expansion of a given function. He calculated the literal parts of covariants by Arbogast's method of derivatives.[85a]

F. Brioschi[47] started with Euler's remark that the number C_s of partitions of s into r parts $\leqq n$ is the coefficient of $x^s z^r$ in the expansion of

$$Z = (1 - z)^{-1}(1 - xz)^{-1} \cdots (1 - x^n z)^{-1}.$$

Now $Z = \Sigma \psi(x) z^r$, where

$$\psi(x) = \frac{(1 - x^{n+1})(1 - x^{n+2}) \cdots (1 - x^{n+r})}{(1 - x)(1 - x^2) \cdots (1 - x^r)} \equiv \frac{f(x)}{\phi(x)}.$$

Since $\psi(x)$ is unaltered by the interchange of n and r, C_s equals the number of partitions of s into n parts $\leqq r$. Let $\alpha_1, \alpha_2, \cdots$ be the roots of $f(x) = 0$; $\beta_1, \beta_2, \cdots$ the roots of $\phi(x) = 0$ and set

$$s_m = \Sigma \frac{1}{\beta_1{}^m} - \Sigma \frac{1}{\alpha_1{}^m}, \qquad \psi(x) = 1 + C_1 x + C_2 x^2 + \cdots.$$

Then

$$\frac{\psi'(x)}{\psi(x)} = s_1 + s_2 x + \cdots,$$

$$(7) \quad C_1 = s_1, \quad 2C_2 = C_1 s_1 + s_2, \quad \cdots, \quad pC_p = C_{p-1}s_1 + \cdots + C_1 s_{p-1} + s_p.$$

[45] Phil. Trans. R. Soc. London, 148, I, 1858, 47–52; Coll. Math. Papers, II, 506–512.

[46] Phil. Trans. R. Soc. London, 146, 1856, 101–126; Coll. Math. Papers, II, 250–281. Cf. F. Brioschi, Annali di Mat., 2, 1859, 265–277.

[47] Annali di sc. mat. e fis., 7, 1856, 303–312. Reproduced by Faà di Bruno.[92]

Hence

$$p!\,C_p = \begin{vmatrix} s_1 & -1 & 0 & \cdots & 0 \\ s_2 & s_1 & -2 & \cdots & 0 \\ s_3 & s_2 & s_1 & \cdots & 0 \\ \cdot & \cdot & \cdot & \cdots & \cdot \\ s_p & s_{p-1} & s_{p-2} & \cdots & s_1 \end{vmatrix}.$$

Set $\epsilon(h/k) = 0$ or k, according as h is not or is divisible by k.　Thus

$$s_m = 1 + \epsilon\left(\frac{m}{2}\right) + \epsilon\left(\frac{m}{3}\right) + \cdots + \epsilon\left(\frac{m}{r}\right) - \epsilon\left(\frac{m}{n+1}\right)$$
$$- \epsilon\left(\frac{m}{n+2}\right) - \cdots - \epsilon\left(\frac{m}{n+r}\right).$$

G. Battaglini[48] proved Sylvester's formula for the wave W_q by means of the special case $(a_1 = 1, \cdots, a_r = 1)$ where the coefficient of x^n in $(1 - x)^{-r}$ equals the coefficient of $1/t$ in $e^{nt}(1 - e^{-t})^{-r}$.　To evaluate the waves, we need the value of S:

$$S = \Sigma \frac{F_a}{F_b} x_i^{-n}, \qquad F_a = \Sigma_\mu A_\mu x_i^\mu, \qquad F_b = \Sigma_\nu B_\nu x_i^\nu,$$

where, in S, the summation extends over all imaginary kth roots x_i of unity. We can find c's such that

$$F_a/F_b = c_0 + c_1 x_i + \cdots + c_{k-1} x_i^{k-1}.$$

Since $\Sigma x_i^j = -1$ for $j \leqq k - 1$, we see that S is $\gamma_0, \gamma_1, \cdots, \gamma_{k-1}$, according as $n \equiv 0, 1, \cdots, k - 1 \pmod{k}$, where $\gamma_j = kc_j - c_0 - c_1 - \cdots - c_{k-1}$, and $\Sigma \gamma_j = 0$.　Hence we obtain Cayley's[44] prime circulator [with k_q for a_q]

$$S = \gamma_0 k_n + \gamma_1 k_{n-1} + \cdots + \gamma_{k-1} k_{n-k+1}.$$

F. Brioschi[49] proved Sylvester's[43] theorem by use of Cauchy's theory of residues.　He noted that, if $a_1, \cdots, a_r$ are all primes,

$$W_m = \frac{1}{m} \sum_{s=1}^{l} y_s^{-n} \prod_{i=1}^{r-1} (1 - y_s^{\beta_i})^{-1} \text{ or } 0,$$

according as m is or is not one of the a's.　Here $y_1, \cdots, y_l$ are the primitive mth roots of unity, and $\beta_1, \beta_2, \cdots$ are the a's not divisible by m.　Application is made to $2x_1 + 3x_2 + 5x_3 = n$.

A. Cayley[50] wrote $(p_1^{n_1} \cdots p_r^{n_r})$ for the partition of n into n_1 parts p_1, n_2 parts p_2, etc., where $p_1 > p_2 > \cdots$.　It is conjugate to the partition

$$((n_1 + \cdots + n_r)^{p_r}(n_1 + \cdots + n_{r-1})^{p_{r-1}-p_r} \cdots (n_1 + n_2)^{p_2-p_3} n_1^{p_1-p_2})$$

of n.　For example, $(6\ 3^2\ 2^2)$ and $(5^2\ 3\ 1^3)$ are conjugate partitions.　Given

[48] Memorie della R. Accad. Sc. Napoli, 2, 1855–7 (1857), 353–363.
[49] Annali di sc. mat. e fis., 8, 1857, 5–12.
[50] Phil. Trans. Roy. Soc. London, 147, 1857, 489–499; Coll. Math. Papers, II, 417–439. Reviewed by E. Betti, Annali di mat., 1, 1858, 323–6.

$x^m - a_1 x^{m-1} + \cdots \pm a_m = 0$ with the roots x_i, the symmetric function belonging to the partition $(p_1 \cdots p_m)$ is $\Sigma x_1^{p_1} \cdots x_m^{p_m}$. Part of $a_1^i a_2^s \cdots a_m^p$ is the symmetric function

$$\Sigma x_1^{p+q+\cdots+t} x_2^{p+\cdots+s} \cdots x_m^p$$

to which belongs the partition $(p + \cdots + t, \cdots, p)$ conjugate to $(m^p \cdots 2^s 1^t)$. Thus $a_1^3 a_3$, belonging to (31^3), contains with the coefficient unity the symmetric function belonging to the conjugate partition (41^2), and with other coefficients, the symmetric functions belonging to (321), (2^3), (31^3), $(2^2\,1^2)$, (21^4), (1^6), but not (3^2).

J. J. Sylvester[51] stated that the number of ways n can be composed additively of the positive integers $a_1, \cdots, a_i$, relatively prime in pairs, differs by a periodic quantity depending on the remainder of n modulo $a_1 a_2 \cdots a_i$ from

$$Q_n = \frac{1}{a_1 \cdots a_i} \left\{ \binom{n+i-1}{i-1} + \frac{1}{2}\binom{n+i-1}{i-2} S_1 \right.$$
$$\left. + \frac{1}{4}\binom{n+i-1}{i-3} S_2 + \cdots + \frac{1}{2^{i-1}} S_{i-1} \right\},$$

where $S_1, \cdots, S_{i-1}$ are the coefficients of $x, \cdots, x^{i-1}$ in

$$(x + a_1 - 1)(x + a_2 - 1) \cdots (x + a_i - 1).$$

For systems like $(a_1, \cdots) = (1, 2, 3)$ or $(1, 3, 4)$, the residual periodic quantity lies between $\frac{1}{2}$ and $-\frac{1}{2}$, whence the number of partitions is the integer nearest to Q_n.

Cayley[52] proved that the number of partitions into x parts, such that the first part is unity and no part is greater than the double of the preceding part, equals the number of partitions of $2^{x-1} - 1$ into the parts 1, 1', 2, 4, $\cdots$, 2^{x-2}.

Sylvester[53] gave an explicit expression for $\Sigma x^\alpha y^\beta \cdots w^\lambda$, summed for all N sets of integral solutions of $ax + by + \cdots + lw = n$, where $a, \cdots, l$ are positive integers. The case $\alpha = \beta = \cdots = \lambda = 0$ gives the number N of sets. Let $\Theta(Ft)$ denote the coefficient of $1/t$ in the expansion of Ft in ascending powers of t. Let m be the l.c.m. of $a, \cdots, l$. Then his[43] former theorem may be expressed in the form

$$N = \Sigma\, \Theta \left\{ \frac{\Lambda(-n)}{(1 - \Lambda a) \cdots (1 - \Lambda l)} \right\},$$

summed for the primitive mth roots ρ of unity, where $\Lambda p = \rho e^{-pt}$. Then, for example,

$$\Sigma x^i = \Sigma\, \Theta \left\{ \frac{\Lambda(a)(1 + \Lambda a) \cdots (i - 1 + \Lambda a)\Lambda(-n)}{(1 - \Lambda a)^{i+1}(1 - \Lambda b) \cdots (1 - \Lambda l)} \right\}.$$

[51] Quar. Jour. Math., 1, 1857, 198–9.
[52] Phil. Mag., (4), 13, 1857, 245–8; Coll. Math. Papers, III, 247–9.
[53] Ibid., (4), 16, 1858, 369–371; Coll. Math. Papers, II, 110–2.

Sylvester[54] cited Euler's[14] transformation of the problem of the Virgins and noted that the general form of the problem is to find the number* of ways in which a given set of numbers $l_1, \cdots, l_r$ [an r-partite number] can be made up simultaneously of the compound elements $a_1, \cdots, a_r; b_1, \cdots, b_r;$ etc.   This problem of compound partition can be made to depend on simple partition.   Omitting details, he stated the following theorem: Given r linear equations in n variables with integral coefficients such that the r coefficients of each variable have no common factor, and such that not more than $r - 1$ variables can be simultaneously eliminated from the r equations, then the determination of the number of sets of positive integral solutions may be made to depend on like determinations for each of n derived independent systems each in $n - 1$ variables.   The conditions are satisfied by Euler's equations

$$ax + \cdots + lw = m, \qquad x + \cdots + w = \mu,$$

if $a, \cdots, l$ are distinct.   Sylvester never published an explicit statement of the theorem just sketched, nor of his obscure generalization.   See the following paper.

Cayley[55] called $(a, \alpha) + (b, \beta) + \cdots$ a double partition of (m, μ) if

$$a + b + \cdots = m, \qquad \alpha + \beta + \cdots = \mu.$$

If $a/\alpha, b/\beta, \cdots$ are distinct irreducible fractions and if $\alpha, \beta, \cdots$ are each $< \mu + 2$, the number of such partitions is

$$D(\alpha m - a\mu; \alpha b - a\beta, \alpha c - a\gamma, \cdots)$$
$$+ D(\beta m - b\mu; \beta a - b\alpha, \beta c - b\gamma, \cdots) + \cdots,$$

where the denumerant[107] $D(m; a, b, \cdots)$ is the coefficient of x^m in

$$(1 - x^a)^{-1}(1 - x^b)^{-1} \cdots.$$

He noted that Sylvester apparently eliminated each of the r variables in turn from $ax + by + \cdots = m$, $\alpha x + \beta y + \cdots = \mu$, obtaining r equations of the form

$$(\alpha b - a\beta)y + (\alpha c - a\gamma)z + \cdots = \alpha m - a\mu,$$

from which the above formula follows.

* E. Mortara[56] treated partitions into three distinct elements.

Sylvester[57] delivered seven lectures on partitions in 1859.

G. Bellavitis[58] proved that the number $[\mu, n, p]$ of sets of integral solutions $\geqq 0$ of $\alpha_0 + \alpha_1 + \cdots + \alpha_n = p$, $\alpha_1 + 2\alpha_2 + \cdots + n\alpha_n = \mu$, equals the number $[\mu, p, n]$ of sets of solutions $\geqq 0$ of $\beta_0 + \beta_1 + \cdots + \beta_p = n$, $\beta_1 + 2\beta_2 + \cdots + p\beta_p = \mu$.   For, every set of solutions of the first pair

* Number of sets of integral solutions $\geqq 0$ of $a_i x + b_i y + \cdots = l_i$ $(i = 1, \cdots, r)$.

[54] Phil. Mag., (4), 16, 1858, 371–6; Coll. Math. Papers, II, 113–7.

[55] Phil. Mag., (4), 20, 1860, 337–341; Coll. Math. Papers, IV, 166–170.

[56] Le partizioni di un numero in 3 parti differenti, Parma, 1858.

[57] Outlines of the lectures were printed privately in 1859 and republished in Proc. London Math. Soc., 28, 1897, 33–96; Coll. Math. Papers, II, 119–175.

[58] Annali di mat., 2, 1859, 137–147.

of equations consists of a partition of μ into α_n numbers n, $\cdots$, α_1 numbers 1, where p is the total number of parts.   To such a partition corresponds as conjugate

$$(\alpha_n + \alpha_{n-1} + \cdots + \alpha_1) + (\alpha_n + \cdots + \alpha_2) + \cdots + (\alpha_n + \alpha_{n-1}) + \alpha_n = \mu,$$

which gives a partition of μ into n parts $\leqq p$.   These parts occur in the second pair of equations as β_p numbers p, $\cdots$, β_1 numbers 1.   Again,

$$[\mu, n, p] = [\mu, n - 1, p] + [\mu - n, n, p - 1],$$
$$[\mu, n, p] = [np - \mu, n, p].$$

There are $[\mu, n, p]$ partitions of N into p parts from $c, c + d, \cdots, c + nd$, if $\mu = (N - cp)/d$, since if each part be diminished by c and the remainder be divided by d, we get the parts $0, 1, \cdots, n$ whose sum is μ.   Application is made to seminvariants.

L. Oettinger[59] stated and J. Derbès[59] proved that $(k - 1)^\nu k^{r-\nu}$ is the maximum of the products of the r equal or distinct integers into which the positive integer $N = rk - \nu$ can be partitioned, where ν is the least positive integer such that k is integral.

Sylvester[60] noted that Bellavitis'[58] first theorem reduces for p infinite to Euler's theorem that the number of partitions of μ into parts $\leqq n$ equals the number of partitions of μ into n or fewer parts.   Bellavitis' theorem, which is capable of intuitive proof by Ferrer's[35] method, may be stated as follows: The number of distinct combinations of $a_0, \cdots, a_n$ figuring in the coefficient of x^μ in $(a_0 + a_1x + \cdots + a_nx^n)^p$ is the same as the number of distinct combinations of $b_0, \cdots, b_p$ in the coefficient of x^μ in $(b_0 + b_1x + \cdots + b_px^p)^n$.

S. Roberts[61] proved Sylvester's[43] formula for waves.

Sylvester[62] noted that, if $\Pi n = n!$,

$$\sum \frac{1}{\Pi\alpha \cdot a^\alpha \cdot \Pi\beta \cdot b^\beta \cdots} = 1,$$

where the summation extends over all ways of expressing n as a sum of α parts each a, β parts each b, etc.

E. Fergola[63] proved the analogous result:

$$\sum \frac{\Pi n}{\Pi 1^{\alpha_1} \cdot \Pi 2^{\alpha_2} \cdots \Pi n^{\alpha_n} \cdot \Pi\alpha_1 \cdots \Pi\alpha_n} = \frac{\Delta^\alpha 0^n}{\Pi\alpha},$$

summed for all positive integers satisfying

$$\alpha_1 + \cdots + \alpha_n = \alpha, \qquad \alpha_1 + 2\alpha_2 + \cdots + n\alpha_n = n,$$

where $\Delta^\alpha 0^n$ denotes the αth order of difference of x^n for $x = 0$.   He evaluated

[59] Nouv. Ann. Math., 18, 1859, 442; 19, 1860, 117–8.
[60] Phil. Mag., (4), 18, 1859, 283–4, under pseud. Lanavicensis.
[61] Quar. Jour. Math., 4, 1861, 155–8.
[62] Comptes Rendus Paris, 53, 1861, 644; Phil. Mag., 22, 1861, 378; Coll. Math. Papers, II, 245, 290.
[63] Rendiconto dell'Accad. Sc. Fis. e Mat., Napoli, 2, 1863, 262–8.

sums in which the preceding summand is multiplied by $\Pi(\alpha - 1)y^a$ or $\Pi(\alpha)y^a$.

Fergola[64] stated that the number of sets of positive integral solutions of

$$a_1 x_1 + \cdots + a_n x_n = n$$

is $\Delta/(n!)$, where

$$\Delta = \begin{vmatrix} \sigma_1\sigma_{n-1} + \sigma_n & -\sigma_1 & -\sigma_2 & -\sigma_3 & \cdots & -\sigma_{n-3} & -\sigma_{n-2} \\ \sigma_1\sigma_{n-2} + \sigma_{n-1} & n-1 & -\sigma_1 & -\sigma_2 & \cdots & -\sigma_{n-4} & -\sigma_{n-3} \\ \sigma_1\sigma_{n-3} + \sigma_{n-2} & 0 & n-2 & -\sigma_1 & \cdots & -\sigma_{n-5} & -\sigma_{n-4} \\ \sigma_1\sigma_{n-4} + \sigma_{n-3} & 0 & 0 & n-3 & \cdots & -\sigma_{n-6} & -\sigma_{n-5} \\ \cdot & \cdot & \cdot & \cdot & \cdot & \cdot & \cdot \\ \sigma_1\sigma_2 + \sigma_3 & 0 & 0 & 0 & \cdots & 3 & -\sigma_1 \\ \sigma_1\sigma_1 + \sigma_2 & 0 & 0 & 0 & \cdots & 0 & 2 \end{vmatrix},$$

while σ_r is the sum of those divisors of r which occur among the positive integers a_1, a_2, $\cdots$. When $a_i = i$ ($i = 1, \cdots, n$), σ_r becomes the sum $\sigma(r)$ of all the divisors of r. If in Δ we change the sign of the second components in the first column and change the sign before each σ above the main diagonal, we obtain a determinant equal to $(-1)^k n!$ when n is of the form $k(3k \pm 1)/2$, but equal to zero when n is not of that form.

C. Sardi[65] proved the preceding theorems.

N. Trudi[66] proved Sylvester's formula ΣW_q for the number $P_n(\alpha, \cdots, \lambda)$ of partitions of n into elements $\alpha, \cdots, \lambda$. He also showed that $W_q = \Sigma F(\rho)$, summed for the primitive qth roots ρ of unity, where $F(\rho)$ is the coefficient of $1/t$ in

$$-(\rho + t)^{-n-1}\{1 - (\rho + t)^a\}^{-1} \cdots \{1 - (\rho + t)^\lambda\}^{-1}.$$

Let $a_1, \cdots, a_r$ be those of the numbers $\alpha, \cdots, \lambda$ which are divisible by q, and $b_1, \cdots, b_s$ the remaining numbers. Let

$$\frac{e^{nt}}{\Pi(1 - e^{-at})\Pi(1 - \rho^b e^{-bt})} = \frac{1 + A_1 t + A_2 t^2 + \cdots}{t^r a_1 \cdots a_r \Pi(1 - \rho^b)},$$

upon writing the denominator on the left as the exponential of its logarithm and expanding the exponentials. Laws are given to determine the A's. From the coefficient of t^{-1} we see that $P_n = \Sigma V_{r, q}$, summed for the various divisors q of the various elements $\alpha, \cdots, \lambda$, where

$$V_{r, q} = \frac{1}{a_1 \cdots a_r} \Sigma \frac{A_{r-1}\rho^{-n}}{(1 - \rho^{b_1}) \cdots (1 - \rho^{b_s})},$$

summed for all the primitive qth roots ρ of unity. Simplifications are given in three cases: $m = 1$, $m = 2$, $r = 1$. He tabulated results for

$$P_n(1, 3, 6, 8), \; P_n(1, 2, 3, 6, 8, 10), \; P_n(1, 2, \cdots, q), \; P_n(2, 3, \cdots, q), \; q \leqq 8.$$

---

[64] Giornale di Mat., 1, 1863, 63–64.

[65] *Ibid.*, 3, 1865, 94–99, 377–380.

[66] Atti Accad. Sc. Fis. e Mat. Napoli, 2, 1865, No. 23, 50 pp.

A. Cayley,[67] denoting by P_i the number of partitions of n into i parts, proved that

$$1 - P_2 + 1 \cdot 2 P_3 - \cdots \pm (n-1)! P_n = 0.$$

For, the number of partitions $n = a\alpha + b\beta + \cdots$ is

$$\frac{n!}{a! \, b! \cdots (\alpha!)^a (\beta!)^b \cdots}.$$

Multiply this by $(-1)^{p-1}(p-1)!$ and sum for the sets of solutions of $p = \alpha + \beta + \cdots$; we get the initial theorem.

A. Vachette[68] stated that one of n^2, $n^2 - 1$, $n^2 - 4$, $n^2 + 3$ is divisible by 12 and the quotient is the number of sets of integral solutions > 0 of $x + y + z = n$ [De Morgan[28]].

L. Bignon[69] noted that the respective cases occur for $n = 6n'$, $6n' + 1$ or 5, $6n' + 2$ or 4, $6n' + 3$. For $n = 6n'$, for example, he separated the sets of solutions into $n/3$ sets each with $y - x$ a constant 0, 1, $\cdots$, $\frac{1}{2}n - 2$, and exhibited the solutions of each set.

E. Catalan[70] noted that $x_1 + \cdots + x_n = s$ has $\binom{s-1}{n-1}$ sets of positive integral solutions. Subtract unity from each x and apply his[25] former result.

Let[71] (n, q) be the number of partitions of n into q distinct parts, $[n, q]$ into q equal or distinct parts. Proof is given of theorems of Euler:

$$(n, q) = (n - q, q - 1) + (n - q, q), \qquad (n, q) = \left[n - \frac{q(q-1)}{2}, q \right],$$

$$[n, q] = \sum_{i=1}^{q} [n - q, i], \qquad (n, q) = \sum_{i=1}^{p-1} (n - iq, q - 1), \qquad p = \left[\frac{n+1}{q} \right],$$

and the first written for $[\ ]$. Here $n \geqq 2q$.

In $x_1 + \cdots + x_q = n$, $x_1 \leqq x_2 \leqq \cdots \leqq x_q$, take $x_1 = a \leqq [n/q]$, and set $x_i = y_i + a - 1$ $(i = 2, \cdots, q)$. Then

$$y_2 + \cdots + y_q = n - 1 - (a-1)q$$

for y's > 0. Hence[72]

$$[n, q] = \sum_{a=1}^{\alpha} [n - 1 - (a-1)q, q - 1], \qquad \alpha = [n/q].$$

Taking $q = 3$, he deduced the result of De Morgan[28] and Vachette.[68]

C. Hermite[73] stated that the number of sets of positive integral solutions of

$$x + y + z = N, \qquad x \leqq y + z, \qquad y \leqq z + x, \qquad z \leqq x + y$$

[67] Math. Quest. Educ. Times, 7, 1867, 87–8; Coll. Math. Papers, VII, 576–8.
[68] Nouv. Ann. Math., (2), 6, 1867, 478.
[69] Ibid., (2), 8, 1869, 415–7.
[70] Mélanges Math., 1868, 16; Mém. Soc. Roy. Sc. Liège, (2), 12, 1885, No. 2, 19.
[71] Ibid., 62–65; Mém. Liège, 56–58.
[72] Ibid., 305–12; Mém. Liège, 264–71. Nouv. Ann. Math., (2), 8, 1869, 407.
[73] Nouv. Ann. Math., (2), 7, 1868, 335. Solution by V. Schlegel, (2), 8, 1869, 91–3.

is $(N^2 - 1)/8$ or $(N + 2)(N + 4)/8$ according as N is odd or even. An anonymous writer (pp. 93–4) stated that the number of sets of positive integral solutions of $x_1 + \cdots + x_m = N$ is $\{N\} - m\{(N - j)/2\}$, where $\{i\} = \binom{m+i-1}{i}$ and $j = 1$ or 2 according as N is odd or even.

K. Weihrauch[74] discussed the number $f_n(A)$ of sets of solutions of

$$a_1 x_1 + \cdots + a_n x_n = A,$$

where the a's are positive integers. Set

$$P = a_1 a_2 \cdots a_n, \qquad S_i = a_1^i + \cdots + a_n^i, \qquad A = pP + m,$$

where m is one of the integers $1, \cdots, P$. Then

$$f_2(A) = p + f_2(m), \qquad f_3(A) = \frac{p^2 P}{2} + p\left(m - \frac{S_1}{2}\right) + f_3(m),$$

$$f_4(A) = \frac{p^3 P^2}{6} + \frac{p^2 P}{2}\left(m - \frac{S_1}{2}\right) + \frac{p}{2}\left\{\left(m - \frac{S_1}{2}\right)^2 - \frac{S_2}{12}\right\} + f_4(m),$$

$$f_n(A) = f_n(m) + \sum_{r=0}^{n-2} P^{n-r-2} \frac{p^{n-r-1}}{(n - r - 1)!} \sum_{q=0}^{\epsilon} (-1)^q R^{r-2q} \frac{D_{2q}}{(r - 2q)!},$$

the last being stated without proof, where ϵ is the largest integer $\leqq r/2$, $R = m - S_1/2$, and

$$D_{2s} = \sum_{\alpha,\,\beta,\,\cdots} \frac{c_2^\alpha}{\alpha!} \frac{c_4^\beta}{\beta!} \frac{c_6^\gamma}{\gamma!} \cdots \qquad (2\alpha + 4\beta + 6\gamma + \cdots = 2s),$$

$$c_{2r} = \frac{S_{2r} B_{2r-1}}{2r(2r)!} \qquad (B_1 = \tfrac{1}{6}, \quad B_3 = \tfrac{1}{30}, \quad B_5 = \tfrac{1}{42}, \quad \cdots),$$

the B's being Bernouilli numbers. Cf. Meissel[135] and Daniëls.[146]

* E. Meissel[75] treated the partition of very large numbers.

E. Lemoine[76] noted that every power n^μ of an integer n equals a sum of n^k consecutive terms from $1, 3, 5, 7, \cdots$, if $\mu \geqq 2k$. Cf. Frégier.[22a]

G. B. Marsano's[77] Table 1 is an extension of Euler's table of partitions of n into m parts for $n \leqq 103$, $m \leqq 102$. Table 2 gives the coefficients as far as x^{53} of the expansions of

$$S, \quad \frac{S}{1 - x}, \quad \frac{S}{(1 - x)(1 - x^2)}, \quad \cdots, \quad \frac{S}{(1 - x)\cdots(1 - x^{35})}, \quad S \equiv \prod_{j=0}^{\infty}(1 - x^j)^{-1},$$

and the coefficients as far as x^{107} of the expansion of the first ten functions. The results for $S/(1 - x)$ give the number of ways of partitioning a number into parts $1, 1', 2, 3, \cdots$. Those for $S/(1 - x)(1 - x^2)$, into parts $1, 1', 2, 2', 3, 4, \cdots$.

[74] Untersuchungen Gl. 1 Gr., Diss. Dorpat, 1869, 25–43. Zeitschrift Math. Phys., 20, 1875, 97, 112, 314; *ibid.*, 22, 1877, 234 ($n = 4$); 32, 1887, 1–21.

[75] Notiz über die Anzahl aller Zerlegungen sehr grosser ganzer positiver Zahlen in Summen ganzer positiver Zahlen, Progr., Iserlohn, 1870.

[76] Nouv. Ann. Math., (2), 9, 1870, 368–9; de Montferrier, Jour. de math. élém., 1877, 253.

[77] Sulla legge delle derivate generali delle funzioni di funzioni e sulla teoria delle forme di partizione de'numeri interi, Genova, 1870, 281 pp. Described by A. Cayley, Report British Assoc. for 1875 (1876), 322–4; Coll. Math. Papers, IX, 481–3.

G. Silldorf[78] considered the number $f(s, k)$ of decompositions of s into k integral summands $\geqq 0$, and the number $f_r(s, k)$ in which r is the least summand. In the former, 0 occurs in the first place $f(s, k - 1)$ times, 1 occurs $f_1(s - 1, k - 1)$ times, etc. But $f(s, k) = f_r(s + rk, k)$. Hence

$$f(s, k) = f(s, k - 1) + f(s - k, k - 1) + \cdots + f(s - rk, k - 1) + \cdots.$$

Thus $f(s, 2) = \frac{1}{2}(s + 2)$ or $\frac{1}{2}(s + 1)$ according as s is even or odd,

$$f(s, 3) = (s^2 + 6s + 12)/12, \qquad s \equiv 0 \pmod 6,$$

with similar results for $s \equiv 1, \cdots, 5 \pmod 6$. Let $F(s, k)$ be the number of combinations without repetitions of k elements with the sum s. Then

$$F(s, k) = F(s - k, k - 1) + \cdots + F(s - rk, k - 1) + \cdots,$$
$$F(s, k) = f\left(s - \frac{(k + 1)k}{2}, k\right)$$

[Euler,[9] § 315]. There are as many partitions in parts $\leqq m$ as into m or fewer parts. The number of ways s can be expressed as a sum of numbers $\leqq m$, with repetitions allowed, is

$$2^{s-1} - (s - m - 1)_0(s - m + 1)2^{s-m-2} + (s - m - 1)_1 \frac{s - 2m + 2}{2} 2^{s-2m-3}$$

$$- (s - m - 1)_2 \frac{s - 3m + 3}{3} 2^{s-3m-4} + \cdots.$$

F. Gambardella[79] noted that $ax + by + cz = m$ has

$$\tfrac{1}{2}q(2m + a + b + c - abcq) + s + k$$

sets of integral solutions if a, b, c are positive and relatively prime in pairs, and $m > 0$, $m = c\gamma + \lambda$, $\gamma + 1 = qab + r$. Here s is the sum of the quotients and $\rho_1, \cdots, \rho_r$ the remainders upon dividing $\lambda, \lambda + c, \cdots,$ $\lambda + (r - 1)c$ by ab; while k is the number of solvable equations $ax + by = \rho_a$.

T. P. Kirkman,[79a] counting $5 \cdot 1 = 1 \cdot 5 = 1 \cdot 3 + 1 \cdot 2 = 1 \cdot 3 + 2 \cdot 1 = \cdots$ as partitions of 5, evaluated the sum of the reciprocals of $(2e_1)^{m_1}(2e_2)^{m_2}$ $\cdots m_1! \, m_2! \cdots$, for all such partitions $m_1e_1 + m_2e_2 + \cdots$ of R.

J. J. Sylvester[80] noted that a list of all partitions of n may be checked by

$$\Sigma(1 - x + xy - xyz + \cdots) = 0,$$

summed for all the partitions, where in any partition, x is the number of 1's, y the number of 2's, etc.

Von Wasserschleben[81] expressed $60k$ as a sum of four numbers each a prime or product of two equal or distinct primes, for $k = 1, \cdots, 16$.

* L. Jelinek[82] treated a kind of partitions.

[78] Ueber die Zerlegung ganzer Zahlen in Summanden, Progr. Salzwedel, 1870, 17 pp.
[79] Giornale di Mat., 9, 1871, 262–5. Extensions by C. Sardi, 11, 1873, 123.
[79a] Math. Quest. Educ. Times, 15, 1871, 60–3; 16, 1872, 74–5.
[80] Report British Assoc., 41, 1871 (1872), 23–5; Coll. Math. Papers, II, 701–3.
[81] Archiv Math. Phys., 54, 1872, 411–8
[82] Die Würfelzahlen u. die Zerlegung einer Zahl in ganzen Z., deren Summe gegeben ist, Progr. Wiener Neustadt, 1874.

* V. Bouniakowsky[83] treated partitions.

J. W. L. Glaisher[84] considered the number $P(a, \cdots, q)x$ of ways of forming x by addition of the elements $a, \cdots, q$, repetitions allowed, and proved that

$$P(1, 3, 5, \cdots)(2x) = 1 + P(1, 2)(x - 1) + P(1, 2, 3, 4)(x - 2) + \cdots$$
$$+ P(1, 2, \cdots, 2x - 2)1,$$
$$P(1, 3, 5, \cdots)(2x + 1) = 2 + P(1, 2, 3)(x - 1)$$
$$+ P(1, 2, \cdots, 5)(x - 2) + \cdots + P(1, 2, \cdots, 2x - 1)1,$$
$$P(1, 3, 5, \cdots)x = P(1, 2)(x - 1) + P(1, 2, 3, 4)(x - 1 - 2 - 3) + \cdots.$$

Glaisher[85] formed the derivations of a^4 by the rule of L. F. A. Arbogast:[85a]

$$a^4;\ a^3b;\ a^3c,\ a^2b^2;\ a^3d,\ a^2bc,\ ab^3;\ \cdots,$$

omitting coefficients. Each term corresponds to a partition of 4. Thus, if $a = 1$, $b = 2$, $\cdots$, a^3b corresponds to the only partition 1 1 1 2 of 5 into 4 parts > 0. In general, from the derivations of a^n we see that the number of terms of the xth derivations of a^n equals the number of partitions of x into n parts including zero, also equals the number of partitions of $x + n$ into n parts > 0, and finally equals $P(1, \cdots, n)x$.

Glaisher[86] gave formulas for checking the tabulation of partitions. The summations extend over all the N partitions of a given number n, while in any partition, x is the number of 1's, y the number of 2's, etc.

$$\Sigma(1 + x + xy + xyz + \cdots) = \Sigma 2^r, \qquad \Sigma(x - 2xy + 3xyz - \cdots) = \tau(n),$$
$$\Sigma(1 - 2y + 3yz - 4yzw + \cdots) = \tau(n + 1) - \tau(n),$$
$$\Sigma\{x - 1 - (x - 2)y + (x - 3)yz - \cdots\} = N - 1,$$

where r is the number of different elements in a partition, and $\tau(n)$ is the number of divisors of n. If $Q(a, b, \cdots)n$ is the number of partitions without repetitions of n into the elements $a, b, \cdots$, and $S(1, \cdots, r)n$ the number of partitions of n into $1, \cdots, r$ in which all but the highest r appears at least once,

$$2Q(1, 2, \cdots)n = 1 + S(1, 2)n + S(1, 2, 3)n + \cdots,$$
$$Q(1, 3, 5, \cdots)n - Q(1, 3, 5, \cdots)(n - 4)$$
$$- Q(1, 3, 5, \cdots)(n - 8) + Q(1, 3, 5, \cdots)(n - 20) + \cdots = 1 \text{ or } 0,$$

according as n is a triangular number or not. The excess of the number of partitions of n into an even number of parts over an odd number of parts is $(-1)^n Q(1, 3, 5, \cdots)n$. A partition into α 1's, β 3's, γ 5's, etc., is transformable into $\pi = \alpha + 3\beta + 5\gamma + \cdots$. Express α, β, $\cdots$ in the binary scale: $\alpha = 2^a + 2^{a'} + \cdots$, $\beta = 2^b + \cdots$. In the new form of π no two parts are equal. Hence a partition into odd parts is converted into a partition into distinct parts, and conversely.

---

[83] Memoirs Imp. Acad. Sc., St. Petersburg, 18, 1871, 20; 25, 1875 (Suppl.), No. 1 (In Russian).
[84] Phil. Mag., (4), 49, 1875, 307–311.
[85] Report British Assoc. for 1874 (1875), Sect., 11–15; Comptes Rendus Paris, 80, 1875, 255–8.
[85a] Calcul des dérivations, Strasburg, 1800. See papers 46, 102, 198.
[86] Proc. Roy. Soc. London, 24, 1875–6, 250–9.

P. Mansion[87] noted that the kth power of an integer n is the sum of n consecutive odd numbers (those nearest n^{k-1}), as $3^4 = 25 + 27 + 29$.

J. W. L. Glaisher[88] stated that, if C_m is the number of compositions of N into m triangular numbers, and A is the sum of the reciprocals of those divisors of N whose conjugates are odd, B if even, then

$$C_1 - \tfrac{1}{2}C_2 + \tfrac{1}{3}C_3 - \cdots \pm \frac{1}{N}C_N = A - C.$$

Glaisher[89] noted that, if $P(x)$ is the number of partitions of x into 1, 2, 3, $\cdots$, repetitions allowed, and $Q(x)$ is the number of partitions of x into 1, 3, 5, 7, $\cdots$, repetitions excluded, then $Q(x) = \Sigma P\{(x - t)/4\}$, summed for the triangular numbers $t < x$ such that $t \equiv x \pmod 4$.

Glaisher[89a] used an identity due to Jacobi,[22b] p. 185, to show that

$$P(x) + 2P(x - 1) + 2P(x - 4) + 2P(x - 9) + \cdots$$
$$= Q(x) + Q(x - 1) + Q(x - 3) + \cdots + Q(x - \tfrac{1}{2}n(n + 1)) + \cdots,$$

if $P(x)$ is the number of partitions of x into even elements without repetitions, and $Q(x)$ the number into odd elements without repetitions.

A. Cayley[90] denoted by u_n the number of partitions of n with no part < 2 and order attended to. Then $u_2 = u_3 = 1$, $u_n = u_{n-1} + u_{n-2}$.

E. Laguerre[91] started with Euler's result that the number $T(N)$ of sets of positive integral solutions of $ax + by + \cdots = N$ is the coefficient of ξ^n in

$$F(\xi) = \frac{1}{(1 - \xi^a)(1 - \xi^b) \cdots},$$

decomposed the latter into partial fractions, and called the result $\Phi(\xi) + \phi(\xi)$, where $\Phi(\xi)$ is the sum of the simple fractions whose denominator is a power higher than the first of one of the factors in the denominator of $F(\xi)$. Let $\Theta(N)$ denote the coefficient of ξ^N in the expansion of $\Phi(\xi)$. Then

$$T(N) = \Theta(N),$$

with an error which is independent of N. For example, if $ax + by = N$ and a, b are relatively prime, $\Theta(N) = (N + 1)/(ab)$, so that $T(N) = N/(ab)$ approximately [Paoli[117] of Ch. II], the error being < 1. For

$$ax + by + cz = N,$$

the approximation is $N(N + a + b + c)/(2abc)$.

F. Faà di Bruno[92] gave an exposition of Brioschi's[47] work and noted that his linear equations (7) are of the same form as Newton's identities if the sign of s_i be changed. Hence, by Waring's formula,

$$C_p = \Sigma \frac{1}{\lambda_1! \cdots \lambda_p!} \left(\frac{s_1}{1}\right)^{\lambda_1} \cdots \left(\frac{s_p}{p}\right)^{\lambda_p},$$

[87] Messenger Math., 5, 1876, 90. Cf. Frégier.[22a]

[88] *Ibid.*, 91.

[89] *Ibid.*, 164–5.

[89a] Math. Quest. Educ. Times, 24, 1876, 91.

[90] Messenger of Math., 5, 1876, 188; Coll. Math. Papers, X, 16.

[91] Bull. Math. Soc. France, 5, 1876–7, 76–8; Oeuvres, 1, 1898, 218–20.

[92] Théorie des formes binaires, 1876, 157; German transl. by T. Walter, 1881, 127.

summed for all solutions of $\lambda_1 + 2\lambda_2 + \cdots + p\lambda_p = p$. At the end of this § 12, he gave other expressions for C_p. He[93] later transformed the above formula into

$$p!C_p = [x^p]\left(\delta + \frac{s_1}{1}x + \frac{s_2}{2}x^2 + \cdots + \frac{s_p}{p}x^p\right)^p$$

$$= [x^p]\left\{\delta + \log\frac{(1 - x^{n+1})\cdots(1 - x^{n+r})}{(1 - x)\cdots(1 - x^r)}\right\}^p,$$

where $[x^p]\tau$ denotes the coefficient of x^p in τ, while, after the expansion, δ^i is to be replaced by $i!$. Similarly, for the number W_p of sets of positive integral solutions of $a_1x_1 + \cdots + a_nx_n = p$,

$$p!W_p = [x^p]\{\delta - \log(1 - x^{a_1})\cdots(1 - x^{a_n})\}^p,$$

which is much simpler to apply than Sylvester's[43] formula. He stated (p. 1259) the generalization to two variables:

$$[x^py^q]\psi(x, y) = \frac{1}{p!q!}[x^py^q]\{\delta + (\delta + \log\psi)^p\}^q.$$

F. Franklin[94] proved that if, in all the partitions of n which do not contain more than one element 1, each partition containing 1 be counted as unity and each partition not containing 1 be counted as the number of different elements occurring in it, the sum of the numbers so obtained is the number of partitions of $n - 1$. Application is made to the distribution of bonds between atoms.

A. Cayley[95] noted that the partition $abc \cdot def$ of 6 letters into 3's contains 6 duads ab, ac, bc, $\cdots$, while the partition $ab \cdot cd \cdot ef$ into 2's contains 3 duads. Hence if α partitions into 3's and β partitions into 2's contain all 15 duads once and but once, $6\alpha + 3\beta = 15$. The solution $\alpha = 1$, $\beta = 3$, furnishes an answer of the partition problem: $abc \cdot def$, $ad \cdot be \cdot cf$, $ae \cdot bf \cdot cd$, $af \cdot bd \cdot ce$. Likewise for $\alpha = 0$, $\beta = 5$; but not $\alpha = 2$, $\beta = 1$. Similarly for 15 or 30 letters.

J. J. Sylvester[96] considered the $e = (w; i, j)$ partitions of w into j parts $0, 1, \cdots, i$, the elements of a partition being arranged in non-increasing order, as 3, 2, 2. Without computing e and $f = (w - 1; i, j)$ separately, we obtain $e - f = E - F$, by counting the E partitions of w in which the initial two parts are equal, and the F partitions of $w - 1$ in which one element is i. Also,

$$e - f = -(w - i - 1; i, j - 1) + \sum_{q=0}^{i}(w - 2q; q, j - 2).$$

F. Franklin[97] proved this rule of Sylvester's by converting each partition into one consisting of i of the numbers $0, 1, \cdots, j$. Then $e - f = \epsilon - \phi$,

[93] Comptes Rendus Paris, 86, 1878, 1189, 1259; Jour. für Math., 85, 1878, 317–26; Math. Annalen, 14, 1879, 241–7; Quar. Jour. Math., 15, 1878, 272–4.
[94] Amer. Jour. Math., 1, 1878, 365–8.
[95] Messenger Math., 7, 1878, 187–8; Coll. Math. Papers, XI, 61–2.
[96] Ibid., 8, 1879, 1–3; Coll. Math. Papers, III, 241–8.
[97] Amer. Jour. Math., 2, 1879, 187–8.

where ϵ is the number of partitions of w not containing the element 1, and ϕ is the number of partitions of $w - 1$ not containing 0.

N. Trudi[98] gave an account of the early history of partitions, made extensive applications to isobaric functions, and finally enumerated the combinations of n letters into α sets each of p letters, β sets of q letters, etc., first when the n letters are distinct and second for repeated letters.

C. M. Piuma[99] treated the following problem: From an urn containing B balls marked $1, \cdots, B$, three are drawn and the three numbers written on them are added; find the number of times the sum is $\leq C$. To find the number S_H of sets of solutions of $\phi + \psi + \chi = H$ with $0 < \phi < \psi < \chi \leq B$. First, let $C < B + 4$. Then every solution satisfies the inequalities. Of the six cases $H = 6h + j$ $(j = 0, \cdots, 5)$, let $H = 6h + 4$ and set $\psi - \phi = x$, $\chi - \phi = y$. Then $x + y = 6h - 3\phi + 4$, $0 < x < y$. If ϕ is even, $\phi = 2\alpha$, there are evidently $3h - 3\alpha + 1$ sets of solutions x, y, and h is shown to be the largest α giving a solution. Thus there are

$$\Sigma_{a=1}^{h}(3h - 3\phi + 1) = h(3h - 1)/2$$

sets ϕ, ψ, χ. For ϕ odd, we get $h(3h + 3)/2$ sets. Adding, we get $S_{6h+4} = h(3h + 1)$. Then $T_C = \Sigma_{H=6}^{C} S_H$ is found by treating six cases; for example, $T_{6c} = c(12c^2 - 15c + 5)/2$. Finally, there is treated the case $C \geq B + 4$.

P. Boschi[100] treated partitions into s parts from $1, \cdots, n$. Let

$$S_{1,r} = x^r + x^{r+1} + \cdots + x^n,$$
$$S_{2,r} = x^r S_{1, r+1} + x^{r+1} S_{1, r+2} + \cdots + x^{n-1} S_{1, n},$$
$$S_{3,r} = x^r S_{2, r+1} + x^{r+1} S_{2, r+2} + \cdots + x^{n-2} S_{2, n-1}, \quad \cdots$$

Expand and collect the terms of $S_{u, r}$; the coefficient of x^P is the number of ways P is a sum of distinct numbers chosen from $r, r + 1, \cdots, n$. It is proved by induction that

$$S_{u, r} = x^{(2r+u-1)u/2} T_{u, r}, \qquad T_{u, r} \equiv \frac{(1 - x^{n-r+1})(1 - x^{n-r}) \cdots (1 - x^{n-r-u+2})}{(1 - x)(1 - x^2) \cdots (1 - x^u)}.$$

Thus the coefficient of x^P in $S_{u, 1} = x^{(u+1)u/2} T_{u, 1}$ is the number of ways P is a sum of s different terms of $1, \cdots, n$. For $u = 2$,

$$T_{2, 1} = A_0 + A_1 x + \cdots + A_{2n-4} x^{2n-4},$$

where A_s is the number of ways $s + 3$ is a sum of two numbers of $1, \cdots, n$. Then $A_r = A_{2n-r-4}$,

$$A_r = \tfrac{1}{4}\{2r + 3 + (- 1)^r\} \text{ if } 2 \leq r \leq n - 2;$$
$$A_r = n - 2 + r + \tfrac{1}{4}\{2r + 3 + (- 1)^r\} \text{ if } n - 2 < r \leq 2n - 4.$$

<hr>

[98] Atti R. Accad. Sc. Fis. Mat. Napoli, 8, 1879, No. 1, 88 pp.
[99] Giornale di Mat., 17, 1879, 360–372.
[100] Memorie Accad. Sc. Ist. Bologna, 1, 1880, 555–571.

Let U_r be the number of pairs from $1, \cdots, n$ whose sums are $\leq r$.   Then

$$U_r = \sum_{s=0}^{r-3} A_s;$$
$$U_r = \tfrac{1}{4}\{r(r-2) + \tfrac{1}{2}[1 - (-1)^r]\}, \quad 3 \leq r \leq n+1;$$
$$U_r = \tfrac{1}{2}n(n-1) - U_{2n-r+1}, \quad n+1 < r \leq 2n-1.$$

Similar applications are made to the cases $u = 3$, $u = 4$.

J. W. L. Glaisher[101] noted that, if $P(u)$ is the number of partitions of u into the elements $1, \cdots, n$, each partition containing exactly r parts, order attended to and repetitions not excluded, then

$$P(r+k) + P(r+n+k) + P(r+2n+k) + \cdots = n^{r-1}$$
$$(k = 0, 1, \cdots, n-1).$$

E. A. A. David[102] noted that Arbogast's[85a] law of derivatives gives

$$\frac{a_1^n}{n!} + \frac{a_1^{n-2}a_2}{(n-2)!} + \frac{a_1^{n-3}a_3}{(n-3)!} + \frac{a_1^{n-4}(a_4 + a_2^2/2)}{(n-4)!}$$
$$+ \frac{a_1^{n-5}(a_2a_3 + a_5)}{(n-5)!} + \cdots = \Sigma \frac{a_1^{p_1}}{p_1!} \frac{a_2^{p_2}}{p_2!} \cdots,$$

summed for all sets of positive integral solutions of

$$p_1 + 2p_2 + 3p_3 + \cdots = n.$$

The latter sets are all given by the exponents of the terms in the left member.

A. Cayley[103] tabulated all partitions of $1, \cdots, 18$, where in each partition $1, 2, \cdots$ are designated by $a, b, \cdots$, so as to give the literal terms in the coefficients of any covariant of a binary quantic.

G. B. Marsano[104] treated the number of combinations 2 or 3 at a time of $1, 2, \cdots, m$ to give a sum $\leq C$.   Simpler and more general results were given by Gigli.[181]

F. Franklin[105] proved Euler's formula (3).   The coefficient of x^w in the left member is evidently the excess E of the number of partitions of w into an even number of distinct parts over that into an odd number of parts. To find E, write $\{a\}$ for a number $\geq a$, and let the parts of each partition be in ascending order.   Consider a partition with r parts, the first being 1; deleting 1 and adding 1 to the final part, we get a partition into $r-1$ parts, the first being $\{2\}$, and without two consecutive numbers at the end, and conversely.   These two types of partitions do not affect the required E, one being of even order and one of odd order.   Hence we need consider only partitions commencing with $\{2\}$ and ending with two consecutive numbers.   Consider any one of these with r parts, the first being 2; deleting 2 and adding 1 to each of the last two parts, we get a partition into $r-1$

<hr>

[101] Messenger Math., 9, 1880, 47–8.

[102] Comptes Rendus Paris, 90, 1880, 1344–6; 91, 1880, 621–2; Jour. de Math., (3), 8, 1882, 61–72.

[103] Amer. Jour. Math., 4, 1881, 248–255; Coll. Math. Papers, XI, 357–364.

[104] Giornale di Mat., 19, 1881, 156–170; 20, 1882, 249–270.

[105] Comptes Rendus Paris, 92, 1881, 448–450.   Cf. Sylvester,[117] 11–13.

parts, the first one being {3} and without three consecutive numbers at the
end. We may suppress these partitions. In general, consider a partition
commencing with {n} and ending with n consecutive numbers. If the
first term is n, efface it and add 1 to each of the last n numbers, which can
be done unless the number of parts is $\leqq n$, whence $w = n(3n - 1)/2$. If
the first term is $n + 1$ and if the last $n + 1$ terms are not consecutive,
reduce by 1 each of the last n and place n before the first part, which can
be done unless the number of parts is n, whence $w = n(3n + 1)/2$. Hence
$E = 0$ unless $w = n(3n \pm 1)/2$, and in that case $E = 1$, there remaining
a single partition into n parts. For an exposition of this proof, with illus-
trative graphs, see E. Netto, Lehrbuch der Combinatorik, 1901, 165–7.

A. Capelli[106] considered a matrix (α_{ij}) of n^2 integers $\geqq 0$ such that the
sum of the numbers in each row or column is always m:

$$\alpha_{i1} + \alpha_{i2} + \cdots + \alpha_{in} = \alpha_{1j} + \alpha_{2j} + \cdots + \alpha_{nj} = m.$$

The number of these matrices equals the number of linearly independent
forms derived from the general form in n sets of variables, homogeneous
and of degree m in each set of variables, by means of the operation $\Sigma \eta_i \partial/\partial \xi_i$,
where the ξ and η are two of the n sets.

Several[106a] found the number of ways 34 is a sum of four distinct positive
integers.

J. J. Sylvester[107] gave an exposition of the theory previously only
sketched by him.[43] Employing Cauchy's term residue to denote the coefficient of $1/x$ in the expansion of a function of x in ascending powers of x, he
considered any proper rational function $F(x)$, so that the degree of the
numerator is less than that of the denominator. Then we may write

$$F(x) = \sum_{\lambda \leqq 1} \sum_{\nu=1}^{j} \frac{c_{\lambda,\mu}}{(a_\mu - x)^\lambda} + \sum_\lambda \frac{\gamma_\lambda}{x^\lambda}.$$

The residue of $\Sigma_{\nu=1}^{\nu=j} F(a_\nu e^x)$ is easily seen to be the constant term of $- F(x)$.
Hence if $x^{-n} f(x)$ is a proper rational function, the coefficient of x^n in the
rational function $f(x)$ is the residue of $\Sigma r^{-n} e^{nx} f(re^{-x})$, summed for each
value $r \neq 0$ of x making $f(x)$ infinite [as the a's for $F(x)$]. The " denumerant to the equation $ax + \cdots + lt = n$," denoted by

$$\frac{n}{a,\, b,\, \cdots,\, l,}\, ,$$

is the number of sets of integral solutions $\geqq 0$ of the equation, and equals
the coefficient of x^n in the expansion of

$$F(x) = (1 - x^a)^{-1} \cdots (1 - x^l)^{-1}.$$

Let $\delta_1 = 1, \delta_2, \cdots, \delta_\mu$ be the integers dividing one or more of the numbers

<hr>

[106] Giornale di Mat., 19, 1881, 87–115.
[106a] Math. Quest. Educ. Times, 34, 1881, 51.
[107] Amer. Jour. Math., 5, 1882, 119–136 (Excursus on rational fractions and partitions).
    Johns Hopkins Univ. Circ., 2, 1883, 22 (for the first theorem). Coll. Math. Papers,
    III, 605–622; 658–660.

$a, \cdots, l$. The denumerant thus equals $\Sigma_{i=1}^{i=\mu} W_i$, where the wave W_i is the residue of

$$\Sigma r_q^{-n} e^{nx} F(r_q e^{-x}) = \Sigma r_q^{n} e^{nx} F(r_q^{-1} e^{-x}),$$

summed for the primitive δ_i-th roots r_q of unity (or for their reciprocals). Now make the important substitution $\nu = n + (a + \cdots + l)/2$. Then

$$W_i = \text{residue of } \Sigma r_q^{\nu} e^{\nu x} / \Pi(r_q^{a/2} e^{ax/2} - r_q^{-(a/2)} e^{-(ax/2)}),$$

the product extended over the similar terms in $a, b, \cdots, l$. Expanding the summands into power series, we see that each wave and hence the denumerant is a sum of products of polynomials in ν each multiplied by a quantity $c\Sigma(r^{\nu+\delta} \pm r^{\nu-\delta})$, where δ is one-half of the number $\phi(i)$ of integers $< i$ and prime to i (since W_i becomes $\pm W_i$ when ν is changed in sign). Give to each such term of the denumerant an undetermined coefficient, as

$$\frac{n}{1, 2, 3,} = A\nu^2 + B + (-1)^{\nu}C + D\Sigma(r^{\nu+1} + r^{\nu-1}), \qquad r^2 + r + 1 = 0.$$

Write $s = a + \cdots + l$ ($s = 6$ in this case). It is shown that the denumerant is zero for all values of ν from 0 to $\tfrac{1}{2}s - 1$ inclusive if s be even, and for all values from $\tfrac{1}{2}$ to $\tfrac{1}{2}s - 1$ inclusive if s be odd. This fact serves to determine uniquely the ratios of undetermined coefficients. For example, in the above case, $\nu = 0, 1, 2$, and

$$B + C - 2D = 0, \qquad A + B - C + D = 0, \qquad 4A + B + C + D = 0,$$

whence $A = 6\sigma$, $B = -7\sigma$, $C = -9\sigma$, $D = -8\sigma$. The value $9A + B - C - 2D$ for $\nu = 3$ must be unity. Hence $\sigma = 1/72$. Since $\nu = n + 3$, the result agrees with that given by De Morgan.[28] The case of the elements 1, 2, 3, 4 is treated similarly. The wave W_1 is discussed in detail. Application is made to the number of sets of solutions of

$$a_1 x_1 + \cdots + a_i x_i < \mu a_1 \cdots a_i,$$

where $a_1, \cdots, a_i$ are relatively prime in pairs. For $i = 2$, the number is $(a_1 a_2 - a_1 - a_2 - 1)/2$.

Sylvester[108] noted that there is a one to one correspondence between the indefinite partitions of n with parts in ascending order and the series $0, \cdots, n$ such that each term is not greater than the mean between its antecedent and consequent.

If a and b are incommensurable, integers x, y can be found such that $ax + by + c$ is indefinitely small. If it be impossible to find integers λ, μ, ν such that

$$\lambda(b\gamma - c\alpha) + \mu(c\alpha - a\gamma) + \nu(a\gamma - b\alpha) = 0,$$

$ax + by + cz + d$ and $\alpha x + \beta y + \gamma z + \delta$ may simultaneously be made arbitrarily small by choice of integers x, y, z. Cf. Jacobi[256] of Ch. II.

---

[108] Johns Hopkins Univ. Circ., 1, 1882, 179–180; Coll. Math. Papers, III, 634–9. First theorem also in Math. Quest. Educ. Times, 37, 1882, 101–2.

O. H. Mitchell[109] wrote $(w; i, j)$ for the number of partitions of w into j or fewer parts each $\leqq i$. Let $\phi_j(w)$ be the largest integer $\leqq (j - 1)w/j$. Then

$$(w; i, j) = \sum_{x=w-i}^{\phi_j(w)} (x; w - x, j - 1).$$

By successive applications of this formula, j can be reduced to unity. Hence

$$(w; i, j) = \sum_{x_1=w-i}^{\phi_j(w)} \sum_{x_2=2x_1-w_1}^{\phi_{j-1}(x_1)} \sum_{x_3=2x_2-x_1}^{\phi_{j-2}(x_2)} \cdots \sum_{x_{j-1}=2x_{j-2}-x_{j-3}}^{\phi_2(x_{j-2})} (1),$$

where the final $\Sigma(1)$ denotes $1 + \phi_2(x_{j-2}) - (2x_{j-2} - x_{j-3})$, i. e., as many units as values of the summation index. There is given the long expression equivalent to the last two signs of summation. This is said to furnish a proof of the final result by Sylvester.[96]

G. S. Ely[110] noted that Euler's[13] table of partitions

	0	1	2	3	4	5	6
1	1	1	1	1	1	1	1
2	1	1	2	2	3	3	4
3	1	1	2	3	4	5	7

may be constructed by use of columns instead of rows: To get the ith element in the jth column, add to the $(i - 1)$th element in the jth column the ith element in the $(j - i)$th column. Euler had noted that the number $(w; w, j)$ of partitions of w into j or fewer parts is given by the number in line j and column w. The number $(w; i, j)$ of partitions of w into j parts $\leqq i$ can be found from this table when the greater of i and j is $\geqq (w - 4)/2$ by the following rule: Since $(w; i, j) = (w; j, i)$, let $i \geqq j$. Then to get $(w; i, j)$ subtract from the tabulated value of $(w; w, j)$ the sum of the first $w - i$ elements in the $(j - 1)$th row and add to the result 0, 1 or 2, according as $i \geqq (w - 2)/2$, $= (w - 3)/2$ or $= (w - 4)/2$. Next, the number of expressions $(w; i, j)$ is

$$N = \frac{w^2 - 2w + t}{2} - \sum_{n=2}^{\infty} \left[\frac{w - n^2 - n - 1}{n + 1} \right],$$

where $t = 6$ if w is even, $t = 5$ if w is odd. Let $s = 24$ or 27 in the respective cases. Then

$$\sum_{w=1}^{n} N = \frac{2n^3 - 3n^2 + 28n - s}{12} - \frac{1}{2} \sum_{i=3}^{\infty} \left\{ ia_i(a_i - 1) + 2a_i \left(n - i \left[\frac{n - 1}{i} \right] \right) \right\},$$

$$a_i \equiv \left[\frac{n - i^2 + i - 1}{i} \right].$$

W. P. Durfee[111] defined a self-opposite or self-conjugate partition to be one such that, if exhibited as an array of units (an element n being repre-

---

[109] Johns Hopkins Univ. Circ., 1, 1882, 210.
[110] *Ibid.*, 211 (in full).
[111] *Ibid.*, 2, Dec., 1882, 23 (in full).

sented by n units in a row), the sums of the columns reproduce the original partition. Thus 4 3 2 1 is a self-conjugate partition of 10. Evidently

$$
\begin{array}{cccc|c}
1 & 1 & 1 & 1 & 4 \\
1 & 1 & 1 & & 3 \\
1 & 1 & & & 2 \\
1 & & & & 1 \\
\hline
4 & 3 & 2 & 1
\end{array}
$$

every such array contains a central square of q^2 units (4 in the diagram), where q is odd or even, according as the partitioned number n is odd or even, since of the $n - q^2$ units outside the square half are at the right and half below the square. The partition remains self-conjugate under any rearrangement of the $(n - q^2)/2$ units to the right, provided those below be arranged symmetrically. The number $\{\frac{1}{2}(n - q^2); q\}$ of such rearrangements is the number of ways of dividing $\frac{1}{2}(n - q^2)$ into q or fewer parts. In the above diagram we may replace the double row of three dots to the right of the square by a single row of three dots and derive the only other self-conjugate partition of 10. In general, the number of self-conjugate partitions of n is $\Sigma\{\frac{1}{2}(n - q^2); q\}$, summed for all odd or all even integers $q < \sqrt{n}$, according as n is odd or even.

J. J. Sylvester[112] noted that Durfee's[111] theorem may be expressed in the following form: The number of self-conjugate partitions of n (or of symmetrical partition graphs for n) is the coefficient of x^n in

$$
1 + \cdots + \frac{x^{i^2}}{(1 - x^2)(1 - x^4)\cdots(1 - x^{2i})} + \cdots = (1 + x)(1 + x^3)(1 + x^5)\cdots
$$

and hence is the number of partitions of n into unrepeated odd integers. He gave a modification of Franklin's[105] proof of (3).

Sylvester[113] proved Brioschi's[47] formula $Z = \Sigma\psi(r)z^r$.

Sylvester[114] proved by use of the binary scale Euler's theorem that the number of partitions of n into odd parts equals the number of its partitions into distinct parts [Glaisher[86]]. Of graphical methods in partitions, he called Ferrers'[35] method transversion and Durfee's[111] method apocopation. He gave a graphical proof of Euler's (3).

F. Franklin[115] noted that, since the number $(w; i, j)$ of ways w can be partitioned into i or fewer parts $\leqq j$ is the coefficient of $a^j x^w$ in the development of the reciprocal of $(1 - a)(1 - ax)\cdots(1 - ax^i)$, the coefficient of a^j in its development in ascending powers of a is the generating function F in which the coefficient of x^w is $(w; i, j)$. To obtain F directly, note that the number of ways of forming w with i or fewer parts of which at least one is a number $> j$, say $j + k$, equals the number of ways of forming $w - (j + k)$ with $i - 1$ or fewer parts; the number of partitions in which

[112] Johns Hopkins Univ. Circ., 2, 1882–3, 23–24, 42–4; Coll. Math. Papers, III, 661–671.

[113] *Ibid.*, 2, 1883, 46; Coll. Math. Papers, III, 677–9; Amer. Jour. Math., 5, 1882, 271–2; Coll. Math. Papers, IV, 21–23.

[114] *Ibid.*, 70–71; Coll. Math. Papers, III, 680–6. Cf. Coll. Math. Papers, IV, 13–18.

[115] *Ibid.*, 72 (in full).

at least two of the parts are $> j$, say $j + k$, $j + k'$, equals the number of partitions of $w - (j + k) - (j + k')$ into $i - 2$ parts; etc.   Hence

$$F = \frac{1}{(1 - x)(1 - x^2)\cdots(1 - x^i)} - \frac{x^{j+1} + x^{j+2} + \cdots}{(1 - x)\cdots(1 - x^{i-1})}$$

$$+ \frac{x^{j+1}(x^{j+2} + x^{j+3} + \cdots) + x^{j+2}(x^{j+3} + \cdots) + \cdots}{(1 - x)\cdots(1 - x^{i-2})}$$

$$- \frac{x^{j+1}x^{j+2}(x^{j+3} + \cdots) + x^{j+2}x^{j+3}(x^{j+4} + \cdots) + \cdots}{(1 - x)\cdots(1 - x^{i-3})} + \cdots$$

$$= \frac{1}{(1 - x)\cdots(1 - x^i)} \{1 - (1 - x^i)\Sigma_1(x^{j+1},\, x^{j+2},\, \cdots)$$

$$+ (1 - x^{i-1})(1 - x^i)\Sigma_2(x^{j+1},\, \cdots) - \cdots\},$$

where $\Sigma_m(x^{j+1},\, \cdots)$ is the sum of the m-ary combinations of x^{j+1}, x^{j+2}, $\cdots$. By induction,

$$(1 - x^i)\cdots(1 - x^{i-m+1})\Sigma_m(x^{j+1},\, \cdots) = \Sigma_m(x^{j+1},\, \cdots,\, x^{j+i}).$$

Hence

$$F = \frac{(1 - x^{j+1})(1 - x^{j+2})\cdots(1 - x^{j+i})}{(1 - x)(1 - x^2)\cdots(1 - x^i)}.$$

Euler's theorem that a number can be partitioned into odd parts as often as into any distinct parts is proved constructively and extended. The number of ways of forming w additively with an indefinite number of parts not divisible by k and with m distinct parts (each repeated indefinitely) divisible by k is equal to the number of ways of forming w with an indefinite number of parts each occurring fewer than k times and with m distinct parts each occurring k or more times.   The proof is made for $k = 10$, though the argument is general.   First, let $m = 0$.   Consider any partition consisting only of parts not divisible by 10 and let the number of times any such part λ occurs be written in the decimal notation, say $\cdots cba$; then if in place of $\cdots cba$ times λ we write a times λ, b times 10λ, c times 100λ, $\cdots$, we get a partition in which no part occurs as many as 10 times, and the correspondence is 1 to 1, so that the theorem is proved if $m = 0$.   Next, if along with the non-tenfold parts we introduce m distinct parts each divisible by 10 and at the same time introduce in the corresponding partition of the other set 10 times these same parts, each divided by 10, the partitions of the second set will contain m parts occurring 10 or more times, while the 1 to 1 correspondence will not be disturbed.

A. Cayley[116] remarked that Franklin's[105] theory does more than group the partitions into pairs.   In addition to the existing division $E + O$ of the partitions into even and odd, it establishes a new division $I + D$ of the same partitions into increasible and decreasible.   There is thus a fourfold division $EI,\ OI,\ ED,\ OD$.   For instance, if $N = 10$, the arrangement is

$EI : 8 + 2,\ 7 + 3,\ 6 + 4$	$OI : 10,\ 5 + 3 + 2$
$ED : 9 + 1,\ 4 + 3 + 2 + 1$	$OD : 7 + 2 + 1,\ 6 + 3 + 1,\ 5 + 4 + 1$

[116] Johns Hopkins Univ. Circ., 86 (in full).

where the EI and OD, each taken in order, pair with each other, and similarly for the OI and ED.   Of course for the exceptional numbers 1, 2, 5, 7, 12, $\cdots$, there is just one partition which is neither I nor D, and, according as it is O or E, we have in the product a coefficient -1 or $+1$.

J. J. Sylvester[117] called a partition regularized if its parts be written in their order of magnitude, represented each part p by p points (nodes) in a horizontal line, and noted that the conjugate partition is obtained by counting the nodes by columns [Ferrers[35]].   There is given (pp. 4–7) a method due to Franklin to construct the partitions which are to be eliminated from the indefinite partitions of n into j parts, including zero, so as to obtain the partitions of n into j parts $\leqq i$, and hence to obtain the generating function enumerating the latter partitions; also (pp. 18–21) his constructive proof for the generating functions for partitions into repeated or unrepeated parts limited in number and magnitude.   Sylvester (p. 7) gave his own construction of partitions of n into j parts chosen from 0, 1, $\cdots$, i by employing a square matrix M_1 of order j in which the diagonal elements are all $i + 1$, the elements below the diagonal are all unity and those above the diagonal all zero.   For $1 \leqq q \leqq j$, let M_q be the matrix whose $\binom{j}{q}$ rows are obtained by adding the rows of M_1 in sets of q.   Denote the rth row of M_q by (r, q) and the sum of its elements by $[r, q]$.   To each regularized partition of $n - [r, q]$ into j parts $\geqq 0$, add (r, q) term to term.   The partitions of n into j parts so obtained from M_q for all values of r are said to form the system P_q.   If P is the system of all partitions of n into j parts, the complete system of partitions of n into j parts $\leqq i$ is

$$S = P - P_1 + P_2 - \cdots + (-1)^j P_j,$$

where the minus sign denotes cancellation, and the system may involve duplicates as well as non-regularized partitions.   It remained to prove that a partition of n, in which the number μ of different parts is $> i$, occurs $\binom{\mu}{q}$ times in P_q and hence $(1 - 1)^\mu$ times in S; this was proved later by M. Jenkins.[118]   Hence the number of partitions of n into j parts $\leqq i$ is the coefficient of x^n in

$$(1 - x^{i+1})(1 - x^{i+2}) \cdots (1 - x^{i+j})/\{(1 - x) \cdots (1 - x^j)\}.$$

Any integer N can be expressed (p. 15) as a sum of consecutive integers in as many ways as N has odd factors; Sylvester[119] also stated this elsewhere.   Cf. Barbette,[201] Agronomov,[204] and Mason.[207]

The subsequent topics treated are: generating functions, correspondence (p. 24, p. 38) between partitions into odd parts and partitions into distinct parts,[119a] and graphical conversion of continued products into series.   Then he noted (p. 60) that if in Jacobi's[30] formula we use the lower signs and take

<hr>

[117] A Constructive Theory of Partitions . . ., Amer. Jour. Math., 5, 1882, 251–330; 6, 1884, 334–6 (for list of errata noted by M. Jenkins).   Coll. Math. Papers, IV, 1–83 (with the errata noted by Jenkins corrected in the text), to which the page citations refer.

[118] Ibid., 6, 1884, 331–3.

[119] Comptes Rendus Paris, 96, 1883, 674–5; Coll. Math. Papers, IV, 92.   Math. Quest. Educ. Times, 39, 1883, 122; 48, 1888, 48–49.

[119a] Comptes Rendus Paris, 96, 1883, 1110–2; Coll. Math. Papers, IV, 95–96.

$n = \frac{1}{2}$, $m = \frac{1}{2} + \epsilon$, where ϵ is infinitesimal, we get

$$\{(1-q)(1-q^2)(1-q^3)\cdots\}^3 = 1 - 3q + 5q^3 - \cdots + (-1)^n(2n+1)q^{n(n+1)/2} + \cdots,$$

a result due to Jacobi[10] of Ch. X in Vol. I of this History. Sylvester wrote Jacobi's initial formula in an equivalent form by setting $n - m = a$, $n + m = b$, and discussed at length (here and elsewhere[120]) the new formula from the standpoint of arrangements of three kinds of elements. He noted (p. 53, p. 70) that Euler's formula (3) is the special case $a = -1$ of

$$(1 + ax)(1 + ax^2)(1 + ax^3) \cdots = 1 + \frac{1 + ax^2}{1 - x}xa + \cdots$$

$$+ \frac{(1 + ax)\cdots(1 + ax^{j-1})}{(1 - x)\cdots(1 - x^{j-1})} \cdot \frac{(1 + ax^{2j})}{1 - x^j}x^{\frac{3j^2-j}{2}}a^j + \cdots,$$

which was given elsewhere by Sylvester[121] and proved also by Cayley.[122]

Chr. Zeller[123] stated Euler's[13] recursion formula for $P(n)$ and expressed the number $\sigma(n)$ of divisors of n in terms of the $P(j)$, $j < n$. [See Vol. I of this History, p. 290, Catalan,[42] p. 292, p. 312, Glaisher,[55, 114] p. 303, Stern.[85]]

E. Cesàro[124] noted that $a_1x_1 + \cdots + a_kx_k = n$ has $n^{k-1}/\{a_1\cdots a_k(k-1)!\}$ sets of positive integral solutions, in mean.

J. W. L. Glaisher[125] noted that Euler's theorem that there are as many partitions without repetitions as into odd parts follows from the case $r = 2$ of the fact that the number of partitions of n, in each of which a part occurs at least r times, equals the number of partitions of n in each of which either r or a multiple of r occurs. In the proof, a repeated term is replaced by its expression to base r (Glaisher[86]). If $P(n)$ is the total number of partitions of n, and $Q_r(n)$ is the number of partitions of n in which no part occurs more than r times,

$$P(n) - P(n-r) - P(n-2r) + P(n-5r) + P(n-7r) - \cdots = Q_{r-1}(n),$$
$$Q_r(n) - Q_r(n-1) - Q_r(n-2) + Q_r(n-5) + Q_r(n-7) - \cdots = 0 \text{ or } (-1)^m,$$

according as n is or is not of the form $(3m^2 \pm m)(r + 1)/2$, and

$$P(0) = Q_r(0) = 1.$$

Write $Q = Q_1$. There are given recursion formulas for Q, and

$$Q(2m) = P(n) + P(n - 3) + P(n - 5) + P(n - 14) + \cdots,$$

involving halves of triangular numbers; similarly for $Q(2m + 1)$.

M. A. Stern[126] proved that the number of variations [with attention to the arrangement of the parts] with the sum n formed from two elements 1 and m equals the number of variations with the sum $n + m$ formed from all elements $\geq m$. This is the analogue of Euler's[9] second theorem.

[120] Comptes Rendus Paris, 96, 1883, 1276–80; Coll. Math. Papers, IV, 97–100.
[121] Comptes Rendus Paris, 96, 1883, 674, 743–5; Coll. Math. Papers, IV, 91, 93–4.
[122] Amer. Jour. Math., 6, 1884, 63–4; Coll. Math. Papers, XII, 217–9.
[123] Acta Math., 4, 1884, 415–6.
[124] Mém. Soc. R. Sc. de Liège, (2), 10, 1883, No. 6, 229.
[125] Messenger Math., 12, 1883, 158–170.
[126] Jour. für Math., 95, 1883, 102–4.

G. S. Ely[127] noted that the partitions of $n + 1$ can be derived from those of n by adding unity to each of the parts in turn or adding a new part unity. Hence every partition of n into parts of which ν are distinct gives $\nu + 1$ partitions of $n + 1$. If the total number of partitions of n be of parity opposite to that of the number of partitions of $n + 1$, there has been a gain in the self-conjugate partitions of $n + 1$ over those of n, if $n > 1$.

A. Cayley[127a] wrote the article on partitions in the Encyclopaedia Britannica. The article on combinatory analysis was by P. A. MacMahon.[127b]

G. S. Ely[128] called a compound[54] partition of N,

$$a_1 a_2 \cdots a_\alpha \mid b_1 \cdots b_\beta \mid \cdots \mid e_1 \cdots e_\epsilon,$$

regular if $a_i \geqq b_i \geqq \cdots \geqq e_i$ for every i. A graph is obtained by representing each portion by an array of points in a plane and superimposing the planes in order. Thus any compound partition may be read in six ways. If $(w; n; i, j)$ is the number of regular compound partitions of w, the number of portions being $\leqq n$, and each portion being partitioned into i or fewer parts $\leqq j$, the symbol is unaltered by any of the six rearrangements of n, i, j.

G. Chrystal[129] gave a recursion formula which may be used to form mechanically a double entry table for the number $_nP_r$ of partitions of r obtained from $2, 3, \cdots, n$. Since

$$\frac{1}{(1 - x^2)(1 - x^3) \cdots (1 - x^n)} = \prod_{i=1}^{n} (1 + x^i + x^{2i} + \cdots)$$
$$= 1 + {_nP_1}x + \cdots + {_nP_r}x^r + \cdots,$$

we see by changing n to $n + 1$ that

$$(1 - x^{n+1})(1 + {_{n+1}P_1}x + {_{n+1}P_2}x^2 + \cdots) = 1 + {_nP_1}x + \cdots,$$

whence

$$_{n+1}P_s = {_nP_s}\ (s = 1, \cdots, n), \qquad _{n+1}P_{n+1} = {_nP_{n+1}} + 1,$$
$$_{n+1}P_{n+r} = {_nP_{n+r}} + {_nP_{n-r+1}} \qquad (r \geqq 2).$$

He noted that Tait[138] had recently communicated similar results.

J. J. Sylvester stated and W. J. C. Sharp[129a] proved the double theorem that, if ν [and ν_j] is the number of ways n is a sum of i distinct positive integers [and $\leqq j$], then

$$\Sigma x^n = (1 - \dot{x}^j)(1 - x^{j-1}) \cdots (1 - x^{j-i+1}) \Sigma \nu x^n.$$

M. Jenkins[129b] evaluated the number of partitions of n into three parts.

A. Cayley[130] employed non-unitary partitions (into parts > 1) and gave the developments up to x^{100} of the reciprocals of (2), (2)(3), $\cdots$, (2)$\cdots$(6),

[127] Johns Hopkins Univ. Circ., 3, 1884, 76–7.
[127a] Ed. 9, 17, 1884, 614; ed. 11, 19, 1911, 865.   Coll. Math. Papers, XI, 589–91.
[127b] Ed. 11, 6, 1911, 752–8; ed. 9, Supplement, 3 (= ed. 10, vol. 27), 1902, 152–9.
[128] Amer. Jour. Math., 6, 1884, 382–4.
[129] Proc. Edinburgh Math. Soc., 2, 1884, 49–50.
[129a] Math. Quest. Educ. Times, 41, 1884, 66–7.
[129b] Ibid., 107.
[130] Amer. Jour. Math., 7, 1885, 57–8; Coll. Math. Papers, XII, 273–4.

where $(k) = 1 - x^k$, for application to seminvariants.

M. Jenkins[131] gave a method to examine bends of a graph of a partition without actually constructing the graphs (cf. Sylvester[117]), and discussed the addition of two regularized graphs, row to row, in order.

J. B. Pomey[132] wrote A_n^m for the number of sets of values $\lambda_i = 0$ or 1 satisfying $\lambda_1 + 2\lambda_2 + \cdots + m\lambda_m = n$. Then

$$f(x) \equiv (1 + x)(1 + x^2) \cdots (1 + x^m) = \sum_{i=0}^{\mu} A_i^m x^i, \qquad \mu = m(m + 1)/2.$$

It follows readily that

$$A_n^m = A_{\mu-n}^m, \qquad A_n^m = A_n^{m-1} + A_{n-m}^{m-1}, \qquad \sum_{i=0}^{\mu} i A_i^m = 2^{m-1}\mu, \qquad \sum_{i=0}^{\mu} A_i^m = 2^m,$$

$$\frac{1}{f(x)} = \sum_{i=0}^{\infty} C_i^m x^i, \qquad C_i^m = \Sigma(-1)^{\lambda_1 + \cdots + \lambda_m},$$

summed for all positive solutions of $\lambda_1 + 2\lambda_2 + \cdots + m\lambda_m = i$. Thus C_i^m is the excess of the number of partitions into an even number of parts over that into an odd number. Also,

$$\sum_{j=0}^{i} C_j^m A_{i-j}^m = 0, \qquad C_i^n + \sum_{j=1}^{n} C_{i-j}^j = 0.$$

D. Bancroft[133] considered the $(w; i, j)$ partitions of w into j parts $\leqq i$. Then

$$(w; i, j) = (w - j; i - 1, j) + (w; i, j - 1).$$

Taking $j = w - k$ and summing for $k = 0, \cdots, k$, we get

$$(w; i, w) = (w; i, w - k - 1) + \sum_{x=0}^{k} (x; i - 1, w - x).$$

Hence, if $k \leqq w/2$, $(w; i, w - k - 1)$ is expressed in terms of $r_j = (r; r, j)$. If $k = \frac{1}{2}w + a$, where w is even and $0 < a \leqq (w + 4)/6$,

$$(w; i, \tfrac{1}{2}w - a - 1) = w_i - \sum_{x=0}^{k} x x_{i-1} + a(0_{i-2} + 1_{i-2})$$

$$+ (a - 1)(2_{i-2} + 3_{i-2}) + \cdots + (2a - 2)_{i-2} + (2a - 1)_{i-2}.$$

This and a like formula include the rule by Ely.[110]

E. Catalan[134] noted that, if (N, p) is the number of partitions of N into p distinct parts, and $\tau(k)$ is the number of divisors of k,

$$(N, 1) - 2(N, 2) + 3(N, 3) - \cdots$$

$$= \tau(N) - \tau(N - 1) - \tau(N - 2) + \tau(N - 5) + \tau(N - 7) - \cdots.$$

[131] Amer. Jour. Math., 7, 1885, 74–81.
[132] Nouv. Ann. Math., (3), 4, 1885, 408–417.
[133] Johns Hopkins Univ. Circ., 5, 1886, 64.
[134] Assoc. franç. av. sc., 15, 1886, I, 86.

E. Meissel[135] gave the formulas of Weihrauch[74] for $n = 3, 4, 5$ and noted that a synthesis of these cases gives

$$f_n(pP + m) - f_n(m) = \frac{1}{P} \frac{\partial f_{n+1}(pP + m)}{\partial p},$$

provided the final term of the derivative be omitted.

P. A. MacMahon[136] called a partition perfect if it contains one and only one partition of every lower integer; sub-perfect if, when each part is taken positive or negative (but not both), it is possible to compose every lower number in only one way. Thus, $3 + 1$ is sub-perfect since $2 = 3 - 1$, $3 = 3, 4 = 3 + 1$. Any factorization

$$\phi_{u, 1} = \phi_{l, \lambda} \phi_{m, \mu} \cdots, \qquad \phi_{p, q} \equiv 1 + x^q + x^{2q} + \cdots + x^{pq},$$

leads to the perfect partition $(\lambda^l \mu^m \cdots)$ of u; then

$$u + 1 = (l + 1)(m + 1) \cdots, \qquad u + 1 = (l + 1)\lambda, \qquad \lambda = (m + 1)\mu, \qquad \cdots.$$

Formulas involving the number of partitions of u are given. For sub-perfect partitions, use

$$\psi_{p, q} = x^{-pq} - x^{-(p-1)q} + \cdots + x^{-q} + 1 + x^q + \cdots + x^{pq}$$

instead of ϕ, and divisors of $2u + 1$ instead of those of $u + 1$.

E. Catalan[137] noted that

$$\log (1 + x + x^2 + \cdots) = - \log (1 - x) = x + \frac{x^2}{2} + \frac{x^3}{3} + \cdots,$$

$$1 + x + x^2 + \cdots = e^x e^{x^2/2} e^{x^3/3} \cdots.$$

Developing each exponential, we get Jacobi's result (Jour. für Math., 22, 1841, 372–4)

$$\sum \frac{1}{2^b 3^c 4^d \cdots \Gamma(a + 1)\Gamma(b + 1)\Gamma(c + 1) \cdots} = 1,$$

where the summation extends over all solutions $\geqq 0$ of

$$a + 2b + 3c + \cdots = n.$$

Since the denominator equals $1 \cdot 2 \cdots a \cdot 2 \cdot 4 \cdot 6 \cdots 2b \cdot 3 \cdot 6 \cdots 3c \cdots$, we see that if n is partitioned in all ways into parts $\alpha, \beta, \gamma, \cdots$ belonging to progressions with the differences $1, 2, 3, \cdots$, the sum of the fractions $1/(\alpha\beta\gamma\cdots)$ is unity.

W. J. C. Sharp stated and H. W. Lloyd Tanner[137a] proved that, if P_n or Q_n be the number of partitions of n without or with repetitions, then

$$Q_n = P_n + P_{n-2}Q_1 + P_{n-4}Q_2 + \cdots,$$

[135] Über die Anzahl der Darstellungen einer gegebenen Zahl A durch die Form $A = \Sigma p_n x_n$, in welcher die p gegebene, unter sich verschiedene Primzahlen sind, Progr. Kiel, 1886. His f_{n-1} has been changed to f_n to conform to Weihrauch's notation.

[136] Quar. Jour. Math., 21, 1886, 367–373.

[137] Mém. Soc. Roy. Sc. de Liège, (2), 13, 1886, 314–8 (= Mélanges Math. II).

[137a] Math. Quest. Educ. Times, 45, 1886, 123.

and two similar relations.　There is a list of unsolved questions on partitions by Sylvester.[137b]

P. G. Tait[138] considered in connection with knots of order n those partitions of $2n$ with no part $> n$ and no part < 2.　After the largest part is removed, the numbers left form the partitions $p_n^n, p_{n+1}^{n-1}, \cdots, p_{2n-2}^2$, where p_s^r is the number of partitions of s with no part $> r$ and none < 2.　If $r > s$, $p_s^r = p_s'$.　If $r < s$, the above argument shows that

$$p_s^r = p_{s-r}^r + p_{s-r-1}^{r-1} + \cdots + p_{s-2}^2.$$

There is a table of values of p_s^r for $r \leqq 17$, $s \leqq 32$.

E. Pascal[139] used n numerical functions $f_i(x)$ which increase when x increases.　Let the difference of two values of f_1 for two successive integral values of x_1 be unity.　If $x_{k-1} < x_k$ and

$$f_k(x+1) - f_k(x) > f_{k-1}(x), \quad f_{k-1}(x_{k-1}) < f_k(x_k + 1) - f_k(x_k),$$

every number is expressible in the form $f_1(x_1) + \cdots + f_n(x_n)$ in one and but one way.　As corollaries, every number N can be expressed in one and but one way as a sum of n decreasing binomial coefficients:

$$N = (x_1)_1 + (x_2)_2 + \cdots + (x_n)_n, \quad x_k < x_{k+1};$$

also as a sum of n increasing binomial coefficients:

$$N = [2]_{x_1} + [3]_{x_2} + \cdots + [n+1]_{x_n}, \quad x_k < x_{k+1}.$$

E. Sadun[140] considered the number $s(n, r)$ of sets of integral solutions $\geqq 0$ of the pair of equations, in which $r \leqq n$,

$$\lambda_1 + \lambda_2 + \cdots + \lambda_n = r, \quad \lambda_1 + 2\lambda_2 + \cdots + n\lambda_n = n.$$

Set $S(n) = s(n, 1) + \cdots + s(n, n)$.　If $r \geqq [n/2]$, $S(n - r) = s(n, r)$. For $r \leqq n$, the pair of equations have as many solutions as the equation

$$\alpha_1 + 2\alpha_2 + \cdots + r\alpha_r = n$$

has integral solutions $\geqq 0$ with $\alpha_r > 0$, or as the system

$$\alpha_1 + 2\alpha_2 + \cdots + (r-1)\alpha_{r-1} = n - tr \quad (t = 1, 2, \cdots, [n/r])$$

has solutions.　Hence we can compute $s(n, r)$.　For $r = 1$, the equation is $\alpha_1 = n$, $\alpha_1 > 0$, whence $s(n, 1) = 1$.　For $r = 2$, the system is

$$\alpha_1 = n - 2, \quad \alpha_1 = n - 4, \quad \cdots, \quad \alpha_1 = n - \left[\frac{n}{2}\right] \cdot 2,$$

whence $s(n, 2) = [n/2]$.　Finally, he identified $s(n, r)$ with a function connected with a linear differential equation of order n.

P. A. MacMahon[141] employed symmetric functions as an instrument for the study of partitions and other problems of combinations.　He considered n objects specified by $(pqr\cdots)$, $p + q + \cdots = n$, meaning that p objects

[137b] Math. Quest. Educ. Times, 45, 1886, 133–7.　One is proved by Sharp, 47, 1887, 139–140.
[138] Trans. Roy. Soc. Edinburgh, 32, 1887, 340–2.
[139] Giornale di Mat., 25, 1887, 45–9.
[140] Annali di Mat., (2), 15, 1887–8, 209–221.
[141] Proc. London Math. Soc., 19, 1887–8, 220–256.　Cf. 28, 1896–7, 9–10.

are of one kind, q of another kind, etc.   The general problem of combinatory analysis is to enumerate, under various imposed conditions, the distributions of the n objects amongst the m parcels specified by

$$(p_1 q_1 \cdots), \qquad p_1 + q_1 + \cdots = m,$$

when the arrangement of the objects in a parcel is immaterial, and when the arrangement is material.   The solution is effected by identities between symmetric functions.   To pass to the special case of partitions of n into m parts, consider the distributions of n similar objects (n) into m similar parcels (m), it being allowed to place more than one object in a parcel.   In the partitions of multipartite numbers, we distribute objects $(pqr \cdots)$ into parcels (m).

G. Platner[142] found for $r \leqq 6$ the number $\phi(r, n)$ or $\psi(r, n)$ of ways of forming a sum n or a sum $\leqq n$ from r terms of $1, 2, 3, \cdots$.   For $r = 2$, the result is $q + x - 1$ or $q^2 + (x - 1)q$, respectively, if $n = 2q + x$, $x < 2$.   In the second paper, he expressed the results as functions of n. For example, the number of pairs with the sum n is $(n - k)/2$, $k = 2$ or 1 according as n is even or odd; the number of pairs with a sum $\leqq n$ is $(n^2 - 2n + l)/4$, $l = 0$ or 1 according as n is even or odd.   For $r = 3, 4, 5, 6$ the formulas involve a parameter with listed values for the least positive residues of n modulo $6, 12, 60, 60$, respectively.   It is proved that

$$f(r, n + r) = f(r, n) + f(r - 1, n), \qquad f = \phi \text{ or } \psi.$$

[All the results for ϕ are due to De Morgan,[28] Herschel,[33] Kirkman,[39] etc.; while the results for ψ follow readily from those for ϕ.]

Schubert[143] noted that $10m$ Pfennige can be made up of $1, 2, 5$ and 10 Pfennige coins in $1 + 10m_1 + 19m_2 + 10m_3$ ways, if $m_i = \binom{m}{i}$, and treated two similar problems.

G. Chrystal[144] collected theorems on partitions and introduced various notations.

Bellens and Verniory[145] found the number of sets of solutions of $x + y + z = n + 2$, x, y, z chosen from $1, \cdots, n$, by grouping the solutions corresponding to a fixed x, and separating the cases $n \equiv 0, \cdots, 5 \pmod{6}$.

M. F. Daniëls[146] obtained the results of Weihrauch[74] another way.

P. A. MacMahon[147] enumerated the perfect and sub-perfect partitions. For example, if a is a prime, there are $2^{\alpha-1}$ perfect partitions of $a^\alpha - 1$.   If $a, b, \cdots$ are primes, $a^\alpha b^\beta \cdots - 1$ has as many perfect partitions as the multipartite number $(\alpha, \beta, \cdots)$ possesses compositions (partitions with attention to order).

S. Tebay[147a] found the number of ways s is a sum of i distinct integers, also when each part is $\leqq q$.

[142] Rendiconti R. Ist. Lombardo di Sc. Let., (2), 21, 1888, 690–5, 702–8.
[143] Mitt. Math. Gesell. Hamburg, 1, 1889, 269   Cf. d'Ocagne[224] of Ch. II.
[144] Algebra, 2, 1889, 527–537; ed. 2, vol. 2, 1900, 555–565.
[145] Mathesis, 9, 1889, 125–7.
[146] Lineaire Congruenties, Diss., Amsterdam, 1890, 120–135.
[147] Messenger Math., 20, 1891, 103–119.   Cf. MacMahon.[136]
[147a] Math. Quest. Educ. Times, 56, 1892, 34–37.

L. Goldschmidt[148] gave an elementary proof of Jacobi's[30] theorem on the excess $(P, \alpha, \beta, \cdots)$ of the number of partitions of P into an even number of the $\alpha, \beta, \cdots$ over those into an odd number of them, and showed that

$$(P, 1, 2, \cdots, m - 1) = (P, 1, 2, 3, \cdots) + (P - m, 2, 3, \cdots)$$
$$+ (P - 2m, 3, 4, \cdots) + \cdots.$$

His proof of Euler's formula (3) is essentially the same as Franklin's,[105] as admitted, *ibid.*, 39, 1894, 212.

J. Zuchristian[149] proved, by means of Euler's recursion formula for the number n_k of partitions of n into k parts, that n_3 is the integer nearest to $(n + 3)^2/12$, while

$$n_4 = \left[\frac{(n + 1)^3}{144}\right] - \left[\frac{n + 1}{12}\right] \quad \text{or} \quad \left[\frac{(n + 2)^3}{144}\right] - \left[\frac{(n + 2)^2}{48}\right] + \eta,$$

according as n is congruent to an odd or even number k modulo 12, while $\eta = 0$ if $k \neq 8$, $\eta = 1$ if $k = 8$.

K. Th. Vahlen[150] wrote $N(s = \Sigma a_i)$ for the number of partitions $s = \Sigma a_i$. Consider a partition $s = \Sigma e_i a_i$ where the ν elements a_i are distinct. If we select λ of these a's, say $\bar{a}_1, \cdots, \bar{a}_\lambda$, the partition may be written

$$(8) \qquad s = \sum_1^\lambda \bar{a}_i + \Sigma k_i a_i,$$

Consider all possible partitions (8). The excess of the number of those for which λ is even over the number for which λ is odd is denoted by

$$N(s = \sum_1^\lambda \bar{a}_i + \Sigma k_i a_i; \ (-1)^\lambda),$$

and is proved to be zero. It suffices to prove this for the partitions (8) which arise for any one $s = \Sigma e_i a_i$. From the latter we get $\binom{\nu}{\lambda}$ partitions (8) for each λ; since λ has the values $0, 1, \cdots, \nu$,

$$N = 1 - \binom{\nu}{1} + \binom{\nu}{2} - \cdots + (-1)^\nu \binom{\nu}{\nu} = (1 - 1)^\nu = 0.$$

He proved analogous formulas. Next (p. 10), from the theory of elliptic functions, we have

$$\prod_{n=1}^\infty (1 - x^{3n-2}z)(1 - x^{3n-1}z^{-1})(1 - x^{3n}) = \sum_{h=-\infty}^{+\infty} (- z)^h x^{(3h^2-h)/2}.$$

which, if $R(n)$ denotes the absolutely least residue of n modulo 3, may be written

$$\prod_{n=1}^\infty (1 - x^n z^{R(n)}) = \sum_{h=-\infty}^{+\infty} (- z)^h x^{(3h^2-h)/2}.$$

Hence $N(s = \sum_{i=1}^{i=k} n_i; \ (-1)^k)$, for $\Sigma R(n_i) = h$, equals 0 unless $s = (3h^2 - h)/2$, and then equals $(-1)^h$. Or, in words, among those partitions of s into

[148] Zeitschrift Math. Phys., 38, 1893, 121–8; Progr. d. höheren Handelsschule, Gotha, 1892.
[149] Monatshefte Math. Phys., 4, 1893, 185–9. Cf. Glösel.[166]
[150] Jour. für Math., 112, 1893, 1–36. Cf. von Schrutka.[218]

distinct positive summands in which the sum of the absolutely least residues modulo 3 of the summands equals a given positive or negative number h, there occur as many partitions into an even number of summands as into an odd number, except only when s is the pentagonal number $(3h^2 - h)/2$, for which there exists an additional partition into an even or odd number of parts according as h is even or odd. Also a purely arithmetical proof is given. If we employ this theorem for each of the permissible values of h and add the results, we get Legendre's[23] result:

$$N\left(s = \sum_1^\lambda a_i;\ (-1)^\lambda\right) = N\left(s = \frac{3k^2 - k}{2};\ (-1)^k\right).$$

These theorems are extended (pp. 16–17) to m-gonal numbers.

T. P. Kirkman[151] took all partitions of x into k parts $\geqq 0$, as 0 0 5, 1 1 3, 2 2 1, 0 1 4, 0 2 3 for $x = 5$, $k = 3$, formed their permutation symbols, $3a^2b + 2abc$, counted their permutations $3 \cdot 3 + 2 \cdot 6 = 21 = \binom{7}{2}$, and stated that the result is always $\binom{x+k-1}{k-1}$. There is a question on the partition of a polygon of r sides into k parts, treated later (*ibid.*, 8, 1894, 109–129); cf. Cayley.[152]

P. Bachmann[153] gave an exposition of the work by Euler.

P. A. MacMahon[154] considered compositions, i. e., partitions in which the arrangement of the parts is essential. The number of compositions of n into p parts > 0 is the binomial coefficient $\binom{n-1}{p-1}$. The total number of compositions of n is 2^{n-1}. If the parts are $\leqq s$, the number is the coefficient of x^n in $(x + x^2 + \cdots + x^s)^p$. A multipartite number $\overline{p_1p_2\cdots}$ specifies $p_1 + p_2 + \cdots$ numbers (or things), p_1 of one sort, p_2 of a second sort, etc. The number of its compositions into r parts is the number of distributions of the $p_1 + p_2 + \cdots$ numbers into r parcels and is the coefficient of $\alpha_1^{p_1}\alpha_2^{p_2}\cdots$ in the expansion of $(h_1 + h_2 + \cdots)^r$, where h_s is the sum of the homogeneous products of degree s of $\alpha_1, \alpha_2, \cdots$. The graph of a composition (2, 1, 4) of 7 is given by placing nodes at points P, Q on the line AB divided into 7 equal segments, so that in moving from A to B by steps proceeding from node to node, 2, 1 and 4 segments of the line are passed over in succession. The graph of a composition of a bipartite number $\overline{pq}$ is derived by placing nodes at suitable points on $q + 1$ similar graphs of p placed parallel and equidistant and with corresponding points joined by a second set of parallels. Let A and B be opposite vertices of the resulting total parallelogram [see figure, MacMahon[168]]. Pass from A to B by successive steps, each consisting in moving a certain number of segments parallel to AK and then moving a certain number of segments parallel to KB. The successive steps are marked by nodes, which define the graph of a composition. An essential node is where the course changes from the

[151] Mem. and Proc. Manchester Lit. Phil. Soc., (4), 7, 1893, 211–3. Math. Quest. Educ. Times, 60, 1894, 98–102.

[152] Proc. London Math. Soc., 22, 1891, 237–262; Coll. Math. Papers, XIII, 93–113.

[153] Zahlentheorie, 2, 1894, Ch. 2, 13–45.

[154] Phil. Trans. Roy. Soc. London for 1893, 184, A, 1894, 835–901.

KB direction to the AK direction. The number of different lines of route with exactly s essential nodes is $\binom{p}{s}\binom{q}{s}$. Each of these lines of route represents $2^{p+q-s-1}$ compositions. For tripartite numbers, we need three dimensions. Generating functions were found for the number of all compositions of multipartite numbers; he[155, 194] treated this topic also later.

K. Zsigmondy[156] partitioned m into distinct parts each unity or a product of distinct ones of the first s primes; for example, the parts may be 1, 2, 3, 5, 2·3, 7, 2·5, 11. If the partition has an even number of parts, consider the excess of E of the number of parts with an odd number of prime factors over the number of terms with an even number of prime factors, unity being a possible term. Thus for $11 = 2·3 + 5$, $E = 0$; for $2·5 + 1$, $E = -2$; for $5 + 3 + 2 + 1$, $E = 2$. But if the partition has an odd number of parts, let E be the excess of the number of parts with even over that with odd number of prime factors. Thus for $2·3 + 3 + 2$, or $7 + 3 + 1$ or 11, $E = -1$. The sum Σ_{11} of the E's for these 6 partitions of 11 is $0 - 2 + 2 - 1 - 1 - 1 = -3$. Next, $\sigma_{11} = 3 - 3 = 0$ is the excess of the number of the partitions of 11 into an odd number of parts over those into an even number of parts. He proved that, if $m > 1$, $\Sigma_m + \sigma_{m-1} = 1$ or 0, according as m is the $(s+1)$th prime p or is $< p$. For example, if $p = 13$, $m = 11$, we had $\Sigma_m = -3$, while $\sigma_{m-1} = 3$ since the partitions of 10 into an odd number of parts are $2·5$, $7 + 2 + 1$, $2·3 + 3 + 1$ and $5 + 3 + 2$, while $7 + 3$ is the only partition into an even number of parts.

W. J. C. Sharp stated and H. J. Woodall[156a] proved that, if P_n is the number of partitions of n without repetitions and Q_n is the number of partitions into odd parts, then $P_n = Q_n + Q_{n-2}P_1 + Q_{n-4}P_2 + \cdots$, and that the same formula holds when P_n and Q_n denote the number of such partitions with repetitions.

L. Eamonson[156b] expressed the number of partitions of $2n$ into two primes in terms of the number of odd primes $\leqq k$ for various values of k.

L. J. Rogers[156c] established the important identities

$$1 + \frac{q}{1-q} + \frac{q^4}{(1-q)(1-q^2)} + \frac{q^9}{(1-q)(1-q^2)(1-q^3)} + \cdots$$
$$= \frac{1}{(1-q)\prod_{n=1}^{\infty}(1-q^{5n-1})(1-q^{5n+1})},$$

---

[155] Phil. Trans. Roy. Soc. London for 1894, 185, A, 111–160.
[156] Monatshefte Math. Phys., 5, 1894, 123–8.
[156a] Math. Quest. Educ. Times, 60, 1894, 41.
[156b] Ibid., 63, 1895, 116–7.
[156c] Proc. London Math. Soc., (1), 25, 1894, 328–9, formulas (1), (2). Cf. papers 226–8.

$$1 + \frac{q^2}{1-q} + \frac{q^6}{(1-q)(1-q^2)} + \frac{q^{12}}{(1-q)(1-q^2)(1-q^3)} + \cdots$$

$$= \frac{1}{(1-q^2)\prod_{n=1}^{\infty}(1-q^{5n-2})(1-q^{5n+2})},$$

where, on the left, the exponents in the numerators are n^2 and $n(n+1)$.

G. Brunel[157] considered two sets of n points such that from each point of each set issue two bonds connecting it with two points or a single point of the other set. Each such configuration can be considered as the result of the juxtaposition of polygons of $2k_1, \cdots, 2k_r$ sides, where

$$k_1 + \cdots + k_r = n.$$

Regard two configurations as identical if, after a permutation of the points of each set, the bonds are in the same order in the two. For the number $h_{n,\,r}$ of configurations relative to n and r,

$$h_{n,\,r} = h_{n-1,\,r-1} + h_{n-r,\,r}.$$

J. Hermes[158] noted that the number of compositions [as by Mac-Mahon[154]] of m into k parts $\geqq \rho$ is $\binom{m+k-1-k\rho}{k-1}$. There are 2^{m-1} compositions of m; each defines the elements of a Gauss Klammer $[\alpha, \cdots, \rho]$, occurring in continued fractions (Gauss[24] of Ch. II); they give the 2^{m-2} Farey numbers of the $(m-1)$th set, each taken twice [see Vol. 1, p. 158 of this History].

Hermes[159] generalized Euler's[9] formulas on the number of partitions. If s, t, n are integers $\geqq 0$, let $E_{s,\,t}(n) = E_{t,\,s}(n)$ be an integer such that $E(0) = 1$, $E_{00}(n) = 0$ if $n > 0$, and

$$E_{s,\,t}(n) = E_{s,\,t}(n-t) + E_{s,\,t-1}(n).$$

For $t = 0$, $E_{s,\,0}(n)$ is the number of partitions of $n + s$ into s positive parts. Several recursion formulas are proved, including

$$E_{s,\,t}(n) = \sum_{h=0}^{s} E_{s-h,\,t}(n - s + h),$$

$$\sum_{k=0}^{d-1} E_{s-1,\,t}(x - ks) = E_{s,\,t}(x) - E_{s,\,t}(x - ds).$$

The number of partitions of $n + x - 1$ into $x - 1$ terms chosen from $1, \cdots, s + 1$ is

$$A_{s,\,x}(n) = \sum_{k=0}^{s} (-1)^h E_{s-h,\,h}\left(n - h\left(x + \frac{h-1}{2}\right)\right) = A_{x-1,\,s+1}(n),$$

unless $n > sx - \varepsilon$, when the sum is zero. Properties of the A's are given.

A. Thorin[160] asked for the integer k for which the number of sets of positive integral solutions of $a_1x_1 + \cdots + a_nx_n = b$, $x_1 + \cdots + x_n = k$ is a maximum.

---

[157] Procès-verbaux des séances soc. des sc. phys. nat. de Bordeaux, 1894–5, 24–7.
[158] Math. Annalen, 45, 1894, 370–80.
[159] *Ibid.*, 47, 1896, 281–297.
[160] L'intermédiaire des math., 1, 1894, 181–2.

"Rotciv"[161] treated the last question for $n = 3$.   Take the greatest integer $X_2 \leqq (b - a_1 - a_3)/a_2$.   In the first of the pair of equations, replace x_2 by X_2.   Then if $a_1x_1 + a_3x_3 = b - a_2X_2$ has integral solutions X_1, X_3, the required k is $X_1 + X_2 + X_3$.

M. Kuschniriuk[162] proved that, if $\Gamma_h(m)$ is the number of partitions of m into h parts > 0, then

$$\sum_{\lambda=0}^{h-1} (-1)^\lambda \binom{h-1}{\lambda} \Gamma_h(m - \lambda H) = \frac{H^{h-1}}{h!}.$$

R. D. von Sterneck[163] considered the number $\{n\}$ of ways of obtaining n additively from $a_1, a_2, \cdots$, using a_1 at most k_1 times, a_2 at most k_2 times, etc.   The number of these representations of n in which the element a_i occurs at least once is

$$\sum_{\lambda \geqq 0} \{n - (\lambda k_i' + 1)a_i\} - \sum_{\lambda \geqq 1} \{n - \lambda k_i'a_i\},$$

where $k_i' = k_i + 1$.   This is used to prove that the number of representations of n as a sum of an odd number of distinct summands is odd if and only if in the decomposition of $24n + 1$ into primes either a single exponent is odd and of the form $4t + 1$ or no exponent is odd and there is an odd value to the half sum of the exponents of those primes which are $\equiv 1, 5, 7, 11$ (mod 24).   He also found the condition that there be an odd number of those representations of n by distinct summands whose number is an odd multiple of 3 (or of 5 or of 7).   Finally, he drew similar conclusions from a general theorem due to Vahlen.[150]

A. R. Forsyth[164] expanded the reciprocal of the product

$$(1 - ax)\left(1 - \frac{x}{a}\right) \cdot (1 - abx^2)\left(1 - \frac{x^2}{ab}\right) \cdot (1 - abcx^3)\left(1 - \frac{x^3}{abc}\right) \cdots$$

of n pairs of factors, suppressed every term with a negative exponent for any of the symbols $a, b, \cdots$, and in the surviving terms replaced each $a, b, \cdots$ by unity, and proved (in accord with a conjecture communicated privately by MacMahon) that the sum of the resulting series is the reciprocal of

$$(1 - x)(1 - x^2)^2(1 - x^3)^2 \cdots (1 - x^n)^2(1 - x^{n+1}).$$

He gave a similar theorem when each pair of factors is replaced by $r + 1$ factors.

G. B. Mathews[165] showed that the problem of multipartite partition is reducible in an infinitude of ways to a problem in simple partition.   For example, every set of integral solutions $\geqq 0$ of

$$ax + by + cz + dw = m, \qquad a'x + b'y + c'z + d'w = m'$$

[161] L'intermédiaire des math., 3, 1896, 249–250.
[162] Progr., Mähr.-Trübau, 1895.   Quoted from Netto,[180] 128–130.
[163] Sitzungsber. Akad. Wiss. Wien (Math.), 105, IIa, 1896, 875–899.
[164] Proc. London Math. Soc., 27, 1895–6, 18–35.
[165] *Ibid.*, 28, 1896–7, 486–490.

is a set of solutions of

$$(\lambda a + \mu a')x + \cdots + (\lambda d + \mu d')w = \lambda m + \mu m'.$$

Conversely, if λ, μ are suitably chosen positive integers, every set of solutions ≥ 0 of the latter is a set of solutions of the pair of equations.

K. Glösel[166] considered the number $C_r(\sigma)$ of ways of expressing σ as a sum of r distinct positive integers, gave a new proof of De Morgan's[28] formulas for $r = 2, 3$, and, for $r = 4$, simpler expressions than Zuchristian's.[149]   If $\{\alpha\}$ is the integer nearest to α,

$$C_4(2k + 1) = \left\{\frac{2k(k - 3)^2}{36}\right\}, \qquad C_4(2k) = \left\{\frac{(2k - 3)(k - 3)^2}{36}\right\},$$

which may be combined into

$$C_4(\sigma) = \left\{\left[\frac{\sigma - 6}{2}\right]^2\left(3\left[\frac{\sigma - 1}{2}\right] - \left[\frac{\sigma}{2}\right]\right)\Big/ 36\right\}.$$

The complicated expression for $C_5(\sigma)$ was simplified on page 290.

P. A. MacMahon[167] gave a report on combinatory analysis and partitions. He suggested (pp. 30–1) a method of enumerating multipartite partitions.

MacMahon[168] noted that a partition $(p_1 \cdots p_5)$ has the " separations " $(p_1p_2)(p_3p_4)(p_5)$, $(p_1p_2p_3)(p_4p_5)$, etc., the numbers in any parenthesis being considered as a partition with those parts. It is easily proved that the number of separations of the partition $(p_1^{\pi_1}p_2^{\pi_2}\cdots)$, where π_1 indicates the number of repetitions of the part p_1, is identical with the number of partitions of the multipartite number $\overline{\pi_1\pi_1\cdots}$. Sylvester's method of graphical representation of partitions can not be simply extended to multipartite

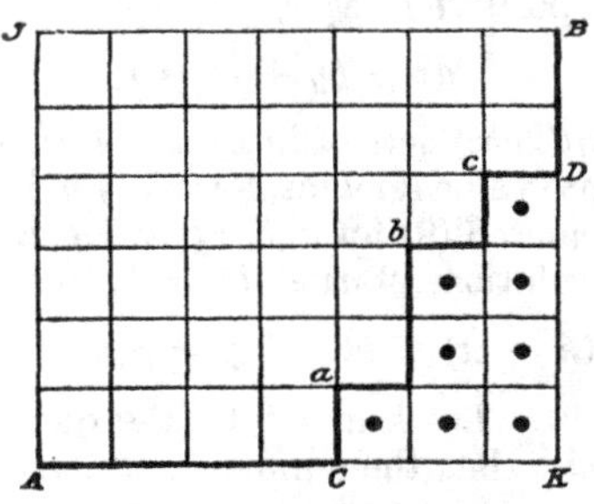

partitions. But there is a correspondence between m-partite partitions and $(m + 1)$-partite compositions. For example, let $m = 1$ and consider the graph of the bipartite number $\overline{76}$. Each composition has a line of route through the lattice [as MacMahon[154]], a, b, c being the essential nodes of the line of route shown in the figure. The principal composition

[166] Monatshefte Math. Phys., 7, 1896, 133–141.

[167] Proc. London Math. Soc., 28, 1896–7, 5–32.

[168] Phil. Trans. Roy. Soc. London, for 1896, 187, A, 1897, 619–673.   Memoir I on Partitions.

is $(\overline{41}\ \overline{12}\ \overline{11}\ \overline{12})$, since 4, 1 are the coordinates of a referred to the origin A, 1, 2, the coordinates of b referred to the origin a, and of B referred to the origin c. The nodes in the lower portion $Ca\cdots cDK$ form a Sylvester regularized graph of the partition $(3\ 2^2\ 1)$; similarly for the nodes in the upper portion.

Again, we may think of Sylvester's graph $\because\ \cdot\,\cdot\,$, not as representing the partition $(3\ 2)$, but as representing the multipartite number $\overline{4, 2}$. Then consider the partition $(\overline{42},\overline{31})$ of the multipartite number $\overline{4+3, 2+1}$. By placing the graph of $\overline{3, 1}$ upon the former graph, we obtain a three-dimensional graph of the partition. Such a graph can in general be read in six ways. At the end of the memoir are conjectures as to the generating functions of partitions whose three-dimensional graphs are limited in height, breadth and length.

R. D. von Sterneck[169] proved Legendre's[23] theorem and deduced from it in a simple way Vahlen's[150] extension. He proved also that, if k is not a triangular number and if we represent k as a sum of integers so that the same part is not used oftener than 3 times in the same representation, then among the representations which contain ρ distinct parts less often than 3 times there are as many sums of even as of odd parts. If $\frac{1}{3}(n - h)$ is not triangular, among the representations of n by distinct summands for which the sum of the absolutely least residues of the summands is $\equiv h \pmod 3$ and in which occur ρ pairs, each pair being two of three numbers of the form $3m - 1, 3m, 3m + 1$, there are as many sums of even as of odd parts. Corresponding to the last two theorems there are more complicated ones for triangular numbers.

J. Franel[170] stated that, if a, b, c are positive integers, relatively prime by twos, and if n is a positive integer,

$$(9) \qquad\qquad ax + by + cz = n$$

has $n(n + a + b + c)/(2abc)$ sets of integral solutions $\geqq 0$, if we neglect a quantity whose absolute value remains, for every n, less than a fixed number.

E. Barbette[171] considered (9) for a, b, c positive, a and b relatively prime. If α, β are particular solutions of $ax + by = 1$, then

$$x = \alpha(n - cz) + b\theta, \qquad y = \beta(n - cz) - a\theta$$

are the solutions of (9). Let k and h be the quotients obtained when n and c are divided by ab; then the number ω of positive integral solutions is

$$\tfrac{1}{2}[2k - (q + 1)h - 2]q,$$

where q is the largest integer $\leqq n/c$. If n is divisible by b, and c is divisible by ab, set $H = c/ab$ and call K the largest integer $\leqq n/ab$; then

$$\omega = \tfrac{1}{2}[2K - (q + 1)H]q.$$

---

[169] Sitzungsber. Akad. Wiss. Wien (Math.), 106, IIa, 1897, 115–122.
[170] L'intermédiaire des math., 5, 1898, 54.
[171] Mathesis, (3), 5, 1905, 125–7.

P. A. MacMahon[172] found the number of ways n is a sum of 8 numbers

$$n_1 \quad n_2 \quad n_3 \quad n_4$$
$$m_1 \quad m_2 \quad m_3 \quad m_4,$$

two solutions being identified if one can be derived from the other by a permutation of the two rows or of the four columns. This question of bipartition is solved also when the number of columns is arbitrary.

H. Wolff[173] evaluated the number $F_\mu(n)$ of partitions of n into μ positive integers x_i arranged in order of magnitude, $x_0 \leqq x_1 \leqq x_2 \leqq \cdots \leqq x_{\mu-1}$, and proved that

$$F_\mu(n) = \Sigma \frac{\phi(n)}{f!\,\xi'^g!\,\eta^g \cdots} \qquad (\mu = f\xi + g\eta + \cdots),$$

where the summation extends over all decompositions $\mu = f\xi + g\eta + \cdots$, while, for each, $\phi(n)$ is the number of partitions of n into f sets of ξ successive equal parts, followed by g sets of η successive equal parts, etc., the various groups not being arranged according to the magnitude of the parts. Thus, for example, $n = 4 = 0 + 0 + 2 + 2$ and $2 + 2 + 0 + 0$ are counted as distinct in computing $\phi(n)$.

The number of decompositions of n into λ equal parts is evidently 1 or 0 according as n is or is not divisible by λ, and hence is

$$\rho(n, \lambda) = \frac{1}{\lambda}\left\{ - R\left(\frac{n}{\lambda}\right) + R\left(\frac{n-1}{\lambda}\right) + 1 \right\}$$

if $R(n/\lambda)$ denotes the least positive remainder on the division of n by λ. If λ is the g.c.d. of $\xi, \eta, \cdots$, the above $\phi(n)$ equals the product of $\rho(n, \lambda)$ by the number $\phi(n/\lambda)$ of partitions $n = f\xi/\lambda + g\eta/\lambda + \cdots$. Again, the number of decompositions $n = f\xi + g\eta$ is $\rho(n, \eta) + [n\xi'/\xi] - [n\eta'/\eta]$, if ξ, η are relatively prime and $\xi\eta' - \eta\xi' = \mp 1$. Recursion formulas for the ϕ's are found and the $F_\mu(n)$ evaluated for $\mu \leqq 6$ as explicit functions of n. By means of Bernoullian functions, $F_\mu(n)$ is expressed as a polynomial in n whose coefficients are linear functions of the coefficients of $F_{\mu-1}(n)$.

*G. Csorba[173a] made an addition to the theory of partitions.

P. A. MacMahon[174] generalized the concept of a partition into parts $\alpha_1, \alpha_2, \cdots, \alpha_s$ by replacing the conditions $\alpha_1 \geqq \alpha_2 \geqq \cdots$ by the conditions

$$A_1^{(i)}\alpha_1 + A_2^{(i)}\alpha_2 + \cdots + A_s^{(i)}\alpha_s \geqq 0 \qquad (i = 1, \cdots, r),$$

where at least one of the integers A is positive. There is a finite number of fundamental solutions $(\alpha_1^{(j)}, \cdots, \alpha_s^{(j)})$ for $j = 1, \cdots, m$ of these conditions, such that every solution is of the form $\alpha_i = \lambda_1\alpha_i^{(1)} + \cdots + \lambda_m\alpha_i^{(m)}$ for $i = 1, \cdots, s$, where the λ's are positive integers.

MacMahon[175] treated the generating functions for the enumeration of three-dimensional graphs possessing either xy-symmetry (when each layer·

[172] Bull. Soc. Math. France, 26, 1898, 57–64; M. d'Ocagne, p. 16, for $n = 3, 4$.
[173] Über die Anzahl der Zerlegungen einer ganzen Zahl in Summanden, Diss., Halle, 1899.
[173a] Math. és termés értesitö (Hungarian Acad. Sc.), 17, 1899, 189.
[174] Phil. Trans. Roy. Soc. London, 192, A, 1899, 351–401. Memoir II on Partitions.
[175] Trans. Cambridge Phil. Soc., 17, 1899, 149–170.

of nodes is symmetrical in two dimensions) or xyz-symmetry (when the six forms obtained by rotations about the various axes are identical).

MacMahon,[176] to enumerate the combinations defined by certain laws, would find an operation and a function such that the result of performing the operation on the function gives the number of combinations. Thus, operating with $(d/dx)^n$ on x^n we get the number $n!$ of permutations of n distinct letters. Again, let $d_1 = d/da_1 + a_1 d/da_2 + a_2 d/da_3 + \cdots$, where the a's are the elementary symmetric functions of $\alpha_1, \cdots, \alpha_n$. Using symbolic multiplication as in Taylor's theorem, write $D_s = d_1^s/s!$. Then operating with $D_{\pi_1} \cdots D_{\pi_n}$ on $(\alpha_1 + \cdots + \alpha_n)^n$ we get the number of permutations of $\alpha_1^{\pi_1} \cdots \alpha_n^{\pi_n}$ where $\Sigma \pi_i = n$. Finally, if we apply $D_3 D_2^2 D_1$ to the symmetric function $(1^4)(1^3)(1)$, where (1^s) denotes $a_s = \Sigma \alpha_1 \cdots \alpha_s$ in partition notation, we get the Sylvester-Ferrers' graph of the partition $(3 \; 2^2 \; 1)$ or its conjugate $(4 \; 3 \; 1)$, according as it is read by rows or columns. The method is successful in solving the problem of the Latin Square[189] in its most general aspect. Cf. Hammond.[217a]

R. D. von Sterneck,[177] to extend Vahlen's[150] work from modulus 3 to modulus 5, considered the excess $\{n\}^h$ of the number of representations of n by an even number of summands over the number by an odd number of summands, where the summands are distinct and the sum of their absolutely least residues $(-2, -1, 0, 1, 2)$ modulo 5 has the value h. He proved the recursion formulas

$$\{k\}^h = \{k - 2h + 3\}^{3-h}, \qquad \{k\}^h = - \{k - 5h + 15\}^{h-5}.$$

By successive applications of the second, we get

$$\{k\}^h = (-1)^\tau \left\{ k - 5\tau \left(h - \frac{5\tau + 1}{2} \right) \right\}^{h-5\tau}.$$

Hence its value depends on certain $\{l\}^j$ for $j = 0, \pm 1, \pm 2$. By Lagrange's theorem, $\Sigma \{k\}^h = 0$ or $(-1)^t$ for $k \neq$ or $k = (3t^2 \pm t)/2$,
where h ranges over the integers $\equiv k \pmod 5$. This gives a recursion formula for $\{k\}^j$, $j = 0, \pm 1, \pm 2$. Hence we can compute any $\{k\}^h$.

M. d'Ocagne[178] found the number of ways s francs can be formed with s French silver coins $(5, 2, 1, \frac{1}{2}, \frac{1}{5}$ francs), also when the number of smallest coins is fixed.

R. D. von Sterneck[179] gave an elementary derivation of the number of decompositions of n into six or fewer equal or distinct positive integral summands, distinguishing 29 types like $n = \alpha + \alpha + \beta + \beta$, often with various sub-cases. Thus the results are expressed by many formulas.

E. Netto[180] employed eight symbols for the various types of combinations and variations, with a prescribed sum, of given numbers taken k at a time,

[176] Trans. Cambridge Phil. Soc., 16, 1898, 262; Phil. Trans. Roy. Soc. London, 194, A, 1900, 361.
[177] Sitzungsber. Akad. Wiss. Wien (Math.), 109, IIa, 1900, 28–43.
[178] Bull. Soc. Math. France, 28, 1900, 157–168.
[179] Archiv Math. Phys., (3), 3, 1901, 195–216.
[180] Lehrbuch der Combinatorik, 1901.

with or without repetitions.   In Ch. 6, he gave an exposition of Euler's work on partitions and Sylvester's theory of waves, illustrated by examples. In Ch. 7 it is noted that any relation between two partitions of n leads to an identity between two infinite series.

A. S. Werebrusow[180a] noted that if $a, b, \cdots, k, l$ are positive integers and if $\{n\}$ denotes the number of sets of positive integral solutions of

$$f \equiv ax + by + \cdots + kt = n,$$

the number of sets for $f + lu = m$ is $\{m - l\} + \{m - 2l\} + \{m - 3l\} + \cdots$.

D. Gigli[181] considered the number N_s of combinations of $1, \cdots, m$ taken n at a time with the sum s.   The least s is $L = n(n + 1)/2$ and the greatest is $G = mn - n(n - 1)/2$.   It is shown by induction that $N_L, N_{L+1}, \cdots, N_G$ are the coefficients of the powers of x in the expansion of

$$(m, n) = \frac{(1 - x^m)(1 - x^{m-1}) \cdots (1 - x^{m-n+1})}{(1 - x)(1 - x^2) \cdots (1 - x^n)}.$$

C. F. Gauss[182] had treated this function without reference to partitions and noted that

$$(m, n) = (m, m - n), \qquad (m, \mu + 1) = \sum_{i=\mu}^{m-1} x^{i-\mu}(i, \mu).$$

Gigli tabulated the N's for $m = 10$, $n = 2, 3, \cdots$, and proved that

$$(m, n) = \sum_{p=1}^{m-n+1} x^{n(p-1)}(m - p, n - 1).$$

T. Muir[183] noted that there are $C_{n-kr+k,\,r}$ combinations of n elements taken r at a time such that no element is taken along with any one of the k elements immediately following it in the initial set.   The number of sets of r things obtained from n by omitting $n - r$ of them so chosen that they form $(n - r)/k$ sets of k consecutive things is $C_{s,\,r}$, where $s = (n + kr - r)/k$.

E. Landau[184] discussed the maximum order of literal substitutions on a given number n of letters.   It is a question of the maximum of the l.c.m. of $a_1, \cdots, a_\nu$ in all decompositions $n = a_1 + \cdots + a_\nu$ of n into positive integral summands.   Cf. Landau.[196]

E. Netto[185] found the number of cyclic decompositions obtained by arranging in a circle each of the $\binom{n-1}{\rho-1}$ decompositions of n into ρ summands with attention to order.

L. Brusotti[186] proved the result of Catalan's.[25]

F. H. Jackson[187] wrote P^x for $p_1^{x_1} \cdots p_m^{x_m}$ and $[p^x z]^n$ for

$$\lim_{k \to \infty} \frac{(1 + P^{x+(n-1)l}z)(1 + P^{x+(n-2)l}z) \cdots (1 + P^{x+(n-k)l}z)}{(1 + P^{x-l}z)(1 + P^{x-2l}z) \cdots (1 + P^{x-kl}z)},$$

[180a] Spaczinski's Bote, Odessa, 1901, Nos. 298–9, pp. 224–9, 250–4.
[181] Rendiconti Circ. Mat. Palermo, 16, 1902, 280–5.
[182] Comm. Soc. Gotting., 1, 1811; Werke, II, 16–17.
[183] Proc. Roy. Soc. Edinburgh, 24, 1901–3, 102–4.
[184] Archiv Math. Phys., (3), 5, 1903, 92–103.
[185] Ibid., 185–196.
[186] Periodico di Mat., 17, 1903, 191–2.
[187] Proc. London Math. Soc., (2), 1, 1903–4, 63–88.

which reduces by cancellation to $(1 + P^x z)(1 + P^{x+l} z) \cdots (1 + P^{x+(n-1)l} z)$
if n is a positive integer.   The simplest of the general formulas proved is

$$[P^x z]^n = 1 + \Sigma P^{rx+r(r-1)l/2} \frac{(P^{nl} - 1)(P^{(n-1)l} - 1) \cdots (P^{(n-r+1)l} - 1)}{(p^l - 1)(P^{2l} - 1) \cdots (P^{rl} - 1)} z^r,$$

which includes as special cases formulas of Euler[3, 9] and Cauchy.[29]

A. S. Werebrusow[188] gave a recursion formula for the number of sets of
positive integral solutions of $a_1 x_1 + \cdots + a_n x_n = A$, where the positive
integers a have no common factor.   Then he considered the number of
sets when at least one x is $\leqq 0$.

P. A. MacMahon[189] treated a " general magic square," consisting of n^2
integers (zeros and repetitions permitted) arranged in a square such that
the rows, columns and diagonals contain partitions of the same number
(whereas in an ordinary magic square the n^2 integers are 1, 2, $\cdots$, n^2).
The treatment applies to all arrangements of integers which are defined by
linear homogeneous Diophantine equations or inequalities such that the
sums of corresponding elements of two solutions give a solution [cf. Mac-
Mahon[174]].

O. Meissner[190] noted that to decompose n into positive integral summands whose product is a maximum, the summands must be equal or differ
at most by unity, and must include as many threes as possible.

G. Mignosi[191] wrote c_n for the number of sets of integral solutions $\geqq 0$
of $a_1 x_1 + \cdots + a_m x_m = n$, and $\sigma(j)$ for the sum of those of $a_1, \cdots, a_m$
which are divisors of j, and proved the recursion formula

$$\sigma(1)c_{i-1} + \sigma(2)c_{i-2} + \cdots + \sigma(i)c_0 = ic_i, \qquad c_0 = 1.$$

Taking $i = 1, \cdots, n$, we obtain $n! c_n$ as a determinant of order n.   If each
$a_i = 1$, then $\sigma(i) = m$ and c_n is the number of combinations of $m + n - 1$
things taken n at a time.

S. Minetola[192] wrote $R_{m,\,n}$ for the number of different ways m distinct
objects can be separated into n groups, where $n \leqq m$.   For example,
$R_{4,\,2} = 7$, the separations being $a_1 - a_2 a_3 a_4, \cdots, a_4 - a_1 a_2 a_3, a_1 a_2 - a_3 a_4,$
$a_1 a_3 - a_2 a_4, a_1 a_4 - a_2 a_3.$ We have

$$R_{m,\,n} = nR_{m-1,\,n} + R_{m-1,\,n-1}, \qquad R_{m,\,2} = 1 + 2 + 2^2 + \cdots + 2^{m-2}.$$

$$\binom{n}{k} R_{m,\,n} = \binom{m}{m-k} R_{m-k,\,n-k} R_{k,\,k}$$

$$+ \binom{m}{m-k-1} R_{m-k-1,\,n-k} R_{k+1,\,k} + \cdots + \binom{m}{n-k} R_{n-k,\,n-k} R_{m-n+k,\,k},$$

<hr>

[188] Matem. Sbornik (Math. Soc. Moscow), 24, 1904, 662–688.

[189] Phil. Trans. Roy. Soc. London, 205, A, 1906, 37–59.   Memoir III on Partitions.   Abstract
in Proc. Roy. Soc., 74, 1905, 318.

[190] Math. Naturw. Blätter, 4, 1907, 85.

[191] Periodico di Mat., 23, 1908, 173–6.

[192] Giornale di Mat., 45, 1907, 333–366; 47, 1909, 173–200.   Corrections, generalizations and
simplifications in Il Boll. di Matematica Gior. Sc.-Didat., Rome, 11, 1912, 34–50, with
errata corrected pp. 121–2.

which for $k = 1$ becomes

$$R_{m,\,n} = \frac{1}{n}\left\{ \binom{m}{m-1} R_{m-1,\,n-1} + \binom{m}{m-2} R_{m-2,\,n-1} + \cdots \right.$$
$$\left. + \binom{m}{n-1} R_{n-1,\,n-1} \right\}.$$

The number $\bar{R}_{m,\,n}$ of ways of separating m like objects into n groups is the number of partitions of m into n parts > 0. Let $k = m - n$. Then

$$\bar{R}_{m,\,n} = \sum_{j=1}^{n} \bar{R}_{k,\,j} \quad (k \geqq n), \qquad \bar{R}_{m,\,n} = \sum_{j=1}^{k} \bar{R}_{k,\,j} \quad (k < n).$$

There are as many partitions of m as partitions of $2m$ into m parts. Recursion formulas are found for the number N of ways of separating into n groups $m = l + \alpha_1 + \cdots + \alpha_h$ objects[192a] of which l are distinct, but one is repeated α_1 times, and the last α_h times. Thus if the objects are a, a, a, b, b, c, d, then $l = 4$, $\alpha_1 = 2$, $\alpha_2 = 1$. There are N factorizations into n positive integral factors of a number which is a product of m primes not necessarily distinct.

Minetola[193] proved by use of $(2n + 1)(2n' + 1) = 2k + 1$, etc., that if $2k + 1$ is decomposed into a product of h primes, the $h - 1$ equations

$$2nn' + \Sigma n = k, \qquad 2^2 n_1 n_1' n_1'' + 2\Sigma n_1 n_1' + \Sigma n_1 = k, \quad \cdots$$

admit $R_{h,\,2}$, $R_{h,\,3}$, $\cdots$ sets of positive integral solutions, respectively.

P. A. MacMahon[194] used the example of a permutation 3, 1 | 4 | 5, 2 of the first five integers separated into compartments with the numbers in each arranged in descending order; the succession of numbers 2, 1, 2 giving the size of the compartments is a composition of 5. He found the number $N(a, b, \cdots)$ of permutations of $1, \cdots, n$ having as the descending specification (corresponding to 2, 1, 2 in the example) a given composition $(a, b, \cdots)$ of n. He proved that

$$\binom{n}{a_1 + \cdots + a_s} N(a_1 \cdots a_s) N(a_{s+1} \cdots a_{s+t})$$
$$= N(a_1 \cdots a_{s+t}) + N(a_1 \cdots a_{s-1}, a_s + a_{s+1}, a_{s+2}, \cdots, a_{s+t})$$

and similar formulas. He found the number of permutations of $1, \cdots, n$ whose descending specifications contain a given number of integers. He treated the analogous problems for permutations of numbers not all different, and problems on packs of cards. The number of permutations of $\alpha_1^{p_1} \cdots \alpha_k^{p_k}$ with descending specifications of m parts is the coefficient of $\lambda^{m-1} \alpha_1^{p_1} \cdots \alpha_k^{p_k}$ in the reciprocal of

$$1 - \Sigma\alpha_1 + (1 - \lambda)\Sigma\alpha_1\alpha_2 - (1 - \lambda^2)\Sigma\alpha_1\alpha_2\alpha_3 + \cdots.$$

His[154] study of this generating function is continued here.

[192a] Giornale di Mat., 47, 1909, 43–54, for the number of combinations of these m objects n at a time.

[193] *Ibid.*, 47, 1909, 305–320.

[194] Phil. Trans. Roy. Soc. London, 207, A, 1908, •65–134. Abstract, Proc. Roy. Soc., 78, 1907, 459–60.

MacMahon[195] applied his[174] second memoir to find the probability that in the election of P by m votes to Q's n votes ($m > n$) the order of the ballots is such that P has at all times more votes than Q, and similarly for n candidates.

Start with any Ferrers' graph of an ordinary partition and place the

parts of the partition at the nodes so that the numbers in a row, read from west to east, and in columns, read from north to south, are in descending order.   We obtain a two-dimensional partition of 19:

$$\begin{array}{cccc} 3 & 2 & 2 & 2 \\ 2 & 1 & 1 & 1 \\ 2 & 1 & & \\ 2 & & & \end{array}$$

E. Landau[196] considered the maximum value $f(n)$ of the l.c.m. of a_1, $\cdots$, a_ρ in all the partitions of n into positive parts, $n = a_1 + \cdots + a_\rho$ ($\rho \leqq n$).   Thus, for $n = 5 = 4 + 1 = 2 + 3$, $f(5) = 6$.   He proved that

$$\lim_{x=\infty} \frac{\log f(x)}{\sqrt{x} \log x} = 1.$$

R. W. D. Christie[197] noted that, if $1 \leqq M \leqq 5$, $6N + M$ has

$$\nu = (3N + M)(N + 1)$$

partitions into parts $\leqq 3$, and $\nu + 1$ partitions if $M = 0$.

J. W. L. Glaisher[198] treated various questions of partitions by solving equations in finite differences which were constructed by means of L. F. A. Arbogast's[85a] rule of derivations.   The capital letters A, B, C, $\cdots$ signify any distinct numbers in ascending order of magnitude, while Greek letters denote any distinct numbers.   The only partition of 8 of the form A^2BC is 1, 1, 2, 4; the only one of the form AB^2C is 1, 2, 2, 3; while either partition is of the form $\alpha^2\beta\gamma$.   Denote by $P_n(i, j, k, \cdots; A^pB^q\cdots)x$ the number of partitions of x into the elements i, j, k, $\cdots$, each partition consisting of n parts and being of the form $A^pB^q\cdots$.   When the elements are 0, 1, 2, $\cdots$, this number P is the number $G_n(x, A^pB^q\cdots)$ of terms of that form in the xth derivation of a^n; its values for $n = 2, 3, 4$ and all possible forms are tabulated (p. 67), and by simple additions, we deduce

$$P_n(0, 1, \cdots; \alpha^p\beta^q\cdots)x = G_n(x, \alpha^p\beta^q\cdots).$$

The latter are computed for $n \leqq 7$; likewise $G_n(x) = P(1, 2, \cdots, n)x$ and $P_n(1, 2, \cdots)$ for $n \leqq 9$, and $P_n(1, 2, \cdots; \alpha^p\beta^q\cdots)x$ for $n \leqq 7$.   It is

---

[195] Phil. Trans. Roy. Soc. London, 209, A, 1909, 153–175.   Memoir IV on Partitions.
[196] Handbuch . . . Verteilung der Primzahlen, 1, 1909, 222–9.   Cf. Landau.[184]
[197] Math. Quest. Educ. Times, (2), 16, 1909, 104.
[198] Quar. Jour. Math., 40, 1909, 57–143.

proved that the last circulator of $G_n(x, \alpha^p\beta^q \cdots)$ is the same for all forms $\alpha^p\beta^q \cdots$, and hence need be computed for the form α^n only, which case is treated at length.

Glaisher[199] proved Sylvester's theorem on waves, developed the formulas for waves of periods 3, 4, 5, 6, and treated the non-periodic terms.

Glaisher[200] noted that his[198] formulas for the number $P(1, \cdots, n)x$ of partitions of x into 1, $\cdots$, n, repetitions allowed, are greatly simplified if expressed in terms of $\xi = x + \frac{1}{4}n(n + 1)$ instead of x and gave the simplified formulas for $n \leqq 9$, and also those in terms of $X = 2\xi$ for $n = 2, 5, 6, 9$. He proved (p. 104) that

$$(- 1)^{n-1}P(1, \cdots, n)(- x) = P(1, \cdots, n)\{x - \tfrac{1}{2}n(n + 1)\} = Q_n(1, 2, \cdots)x,$$

where Q is the number of partitions of x into elements 1, 2, $\cdots$, unlimited in number, each partition containing exactly n parts without repetition. He proved (p. 106) that, if in the circulators occurring in the ξ-formulas, the order of the elements be reversed, the original circulator is reproduced except as to sign. Finally, he gave the leading circulator in each wave $W_m(1, 2, \cdots, mh + r)$.

E. Barbette[201] noted that there are exactly $2(2^{x-2} - 1)$ ways of partitioning $x + \alpha$ into distinct parts the greatest of which is x, where

$$\alpha = S_x - R, \qquad 1 \leqq R \leqq \tfrac{1}{2}x(x - 1) - 1, \qquad S_x \equiv 1 + 2 + \cdots + x.$$

In fact, such a partition of $x + \alpha$ corresponds to a partition of α into distinct parts each $< x$. Next, to find all the partitions of N into distinct parts, let x be the least integer for which $S_x \geqq N$, and convert S_x, S_{x+1}, $\cdots$, S_{N-1} into sums of distinct numbers of which the greatest is N and such that all the other parts are less than x, $x + 1$, $\cdots$, $N - 1$, respectively. Suppress the parts in common to two members of the resulting equalities. Finally, to find all sets of consecutive integers whose sum is N (as $8 + 9 = N$), write 1, 2, 3, $\cdots$ along the diagonal of a square; above x in the diagonal write the sum $2x - 1$ of x and the preceding term $x - 1$; above that sum write the sum $3x - 3$ of it and the number $x - 2$ preceding it in the former list; etc., until 1 is added. Cf. Sylvester.[119]

P. Bachmann[202] gave an extended clear account of the literature on partitions. He inserted (pp. 109–110) a theorem communicated to him by J. Schur: If S is any set of positive integers not divisible by r, and R is the set of numbers obtained by multiplying the numbers in S by 1, r, r^2, $\cdots$, then any positive integer can be partitioned into equal or distinct parts chosen from S as often as into parts chosen from R, each occurring at most $r - 1$ times. The case $r = 2$ gives Euler's theorem that any integer can be partitioned into equal or distinct odd integers as often as into any distinct parts.

[199] Quar. Jour. Math., 40, 1909, 275–348.

[200] *Ibid.*, 41, 1910, 94–112.

[201] Les sommes de p-ièmes puissances distinctes égales à une p-ième puissance, Liège, 1910, 12–19.

[202] Niedere Zahlentheorie, 2, 1910, 102–283.

R. D. von Sterneck[203] proved De Morgan's[28] result that the number of partitions of n into 3 parts is the integer nearest to $n^2/12$ by use of three coordinate axes every pair of which make an angle $< 60°$ and counting the lattice points inside or on the triangle cut out of the plane $x + y + z = n$ by the coordinate planes. Similar use is made of 4-dimensional space to show that the number of partitions of n into 4 parts is the integer nearest $(n^3 + 3n^2 - 4)/144$.

N. Agronomov[204] noted that $N = 2^a p_1^{a_1} \cdots p_k^{a_k}$ is representable as a sum of consecutive integers in $(\alpha_1 + 1) \cdots (\alpha_k + 1)$ ways [Sylvester[119]].

P. A. MacMahon[205] noted that his three-dimensional graphs of plane partitions admit not only of 1, 3 or 6 readings, but may admit just two readings if the weight be $\geqq 13$. Let each part be $\leqq l$ and be placed at a node of a two-dimensional lattice with m rows and n columns. The generating function giving as the coefficient of x^w the number of partitions of weight w is expressible in six ways, one of them being

$$\prod_{s=1}^{n} \frac{(l + s)(l + s + 1) \cdots (l + m + s - 1)}{(s)(s + 1) \cdots (m + s - 1)}, \qquad (t) \equiv 1 - x^t,$$

and the other five being derived from this by permuting l, m, n. A general proof is here first given. The theory of generating functions, especially for $l = \infty$, is developed further here and in his next paper.[206]

T. E. Mason[207] proved that $2^a p_1^{a_1} \cdots p_r^{a_r}$, where the p's are distinct odd primes, can be represented as a sum of consecutive integers not necessarily positive in $2(\alpha_1 + 1) \cdots (\alpha_r + 1)$ ways. In just one half the representations there is an even number of terms, and in just one half are the terms all positive [Sylvester[119]].

W. J. Greenstreet[208] proved that $x + 2y + 3z = 6n$ has $3n^2 + 3n + 1$ integral solutions $\geqq 0$.

MacMahon[209] showed that the enumeration of partitions of multipartite numbers may be made to depend upon his[141] theory of distributions and symmetric functions of a single system of quantities.

A. J. Kempner[210] proved that, if 1, c_1, c_2, $\cdots$ form a set of increasing positive integers such that every positive integer is a sum of k or fewer of them, the radius of the circle of convergence of $1 + c_1 x + c_2 x^2 + \cdots$ is unity. Let every positive integer be a sum of at most k terms of a given set $a_1, a_2, \cdots$; let α_i, β_i be integers such that $0 < \alpha_i \leqq R$, $|\beta_i| \leqq S$, where R and S are any given positive integers; then every positive integer is a sum of fewer than $R!(2kS + k + 1)$ terms of the set 1, $\alpha_1 a_1 + \beta_1$, $\alpha_2 a_2 + \beta_2$, $\cdots$. Finally, the known theorems that any positive integer n is a sum of four squares and that $x^2 = 1 + 3 + 5 + \cdots + (2x - 1)$ imply

[203] Rendiconti Circ. Mat. Palermo, 32, 1911, 88–94.
[204] Math. Unterr. 2, 1912, 70–2 (Russian).
[205] Phil. Trans. Roy. Soc. London, 211, A, 1912, 75–110.　Memoir V on Partitions.
[206] *Ibid.*, 345–373.　Memoir VI on Partitons.
[207] Amer. Math. Monthly, 19, 1912, 46–50.　Cf. Sylvester.[119]
[208] *Ibid.*, 50–1.
[209] Trans. Cambridge Phil. Soc., 22, 1912, 1–13.
[210] Über das Waringsche Problem . . ., Diss. Göttingen, 1912.

that $n = u_1 \cdot 1 + u_2 \cdot 3 + u_3 \cdot 5 + \cdots$ is solvable in integers such that $4 \geqq u_1 \geqq u_2 \geqq u_3 \cdots \geqq 0$. A generalization is noted.

S. Minetola[211] investigated the number $R(t; \alpha_1, \cdots, \alpha_p; n)$ of ways of separating into n groups $m = t + \alpha_1 + \cdots + \alpha_p$ objects of which t are not repeated, while p further objects, distinct from each other and from the preceding t, are repeated $\alpha_1, \cdots, \alpha_p$ times, respectively. After finding recursion formulas for R, he proved theorems on the maximum value of R when m and n vary, but so that $m - n$ remains constant. Finally, he studied $R(1; m; n)$, so that one object is taken single and another is repeated m times. It is the coefficient of x^{m+1} in

$$x^n / \{(1 - x)^2(1 - x^2)(1 - x^3)\cdots(1 - x^{n-1})\}.$$

Its recursion formula is

$$R(1; m; n) = R(1; m - 1; n - 1) + R(1; m - n + 1; n).$$

G. Scorza[212] evaluated sums of reciprocals of products, each summation extended over all the partitions of a given arbitrary integer.

G. Candido[213] noted that a^m is a sum of a consecutive odd integers. For $m = 3$, this was also proved by J. W. N. le Heux.[214] Cf. Frégier.[22a]

G. Csorba[215] stated that all questions concerning partitions can be reduced to a single one, viz., the question of the number of ways A can be obtained from $a_1, \cdots, a_n$ by addition, repetitions allowed. Cayley[44] had expressed this number of partitions of A in the form

$$c_0(A) + Ac_1(A) + A^2c_2(A) + \cdots + A^{n-1}c_{n-1}(A),$$

where $c_i(A)$ is a periodic function of A; but essentially proved only the existence of such a representation. Csorba gave for $c_i(A)$ an explicit formula involving Bernoullian numbers and the g.c.d. d of all the a's except $a_{t_1}, \cdots, a_{t_m}$, and involving summations extended over all solutions ξ_e of the congruence $\Sigma_{e=1}^{e=n} a_{i_e}\xi_e \equiv A \pmod{d}$.

*Csorba[216] treated multiple partition.

P. A. MacMahon[217] has given an extended account of the theory of partitions as a branch of combinatory analysis. A small part of Vol. I and nearly the whole of Vol. II are taken up with theories more or less connected with the partitions of numbers. The theory is investigated from the standpoint of a new definition of a partition. A partition is defined as a set of positive integers $\alpha_1, \alpha_2, \cdots, \alpha_n$, whose sum is n, such that $\alpha_1 \geqq \alpha_2 \geqq \cdots \geqq \alpha_n$. The importation of linear Diophantine inequalities leads to a syzygetic theory and thence to the determination of ground forms connected by various orders of syzygies as in the theory of algebraic invariants. A generalization is made by considering one or more general

[211] Periodico di Mat., 29, 1913, 67–82.
[212] Rendiconti Circolo Mat. Palermo, 36, 1913, 163–170.
[213] Suppl. al Periodico di Mat., 17, 1914, 116–7.
[214] Wiskundig Tijdschrift, 12, 1915–6, 97–8.
[215] Math. Annalen, 75, 1914, 545–568.
[216] Math. és termés értesitö (Hungarian Acad. Sc.), 32, 1914, 565–601.
[217] Combinatory Analysis, Cambridge, I, 1915; II, 1916.

linear inequalities connected with a number of linear relations. Such theories are grouped under the title " partition analysis." As regards the simple partition of numbers the idea results in laying foundations deeper than the intuitive generating functions which served Euler and his successors as points of departure.   There is an extension in the direction of two dimensions in such wise that the parts are laid out in the compartments of a chess board of any dimensions, a partition being defined as a distribution of numbers such that in every row and in every column of the board a descending order of part magnitude is in evidence.   The complete enumerative solution of this question for a complete or incomplete lattice or chess board is reached.   The solution depends upon the idea of a lattice permutation and of an associated lattice function.   An assemblage of letters $a_1^{a_1} a_2^{a_2} \cdots a_s^{a_s}$ is said to be a lattice assemblage when the repetitional exponents satisfy the condition $\alpha_1 \geqq \alpha_2 \geqq \cdots \geqq \alpha_s$; and of this assemblage a permutation is a lattice permutation iⁱ the first k letters (k being any number $< s$) of the permutation constitute a permutation of a lattice assemblage $a_1^{\beta_1} a_{2}^{\beta_2} \cdots a_s^{\beta_s}$.   These permutations have been enumerated, but the theory of the derived lattice functions is not yet complete.   The theory of partitions in three dimensions is completed in this book only as far as the simplest case when the parts are placed at the angular points of a single cube.   The enumeration of the partitions of multipartite numbers is investigated principally by means of J. Hammond's[217a] differential operators [MacMahon[176]].   The problem of enumerating partitions which do not involve sequences of parts is considered in Vol. I.

   * L. von Schrutka[218] gave an extended account of methods employed to further develop Vahlen's[150] results.

   R. Goormaghtigh[219] noted that, if N is the sum of the consecutive integers comprised between $\nu + 1$ and n, then $2N = (n - \nu)(n + \nu + 1)$ and the number of couples n, ν is the number of odd divisors > 1 of N.

   G. H. Hardy and S. Ramanujan[220] proved that the logarithm of the number $p(n)$ of partitions of n is asymptotic to $\pi \sqrt{2n/3}$, and the logarithm of the number of partitions of n into distinct positive integers is asymptotic to $\pi \sqrt{n/3}$.   They[221] developed a general method for the discussion of these, and analogous problems of combinatory analysis, by means of the methods of the theory of functions of a complex variable.   This method is, within limits, applicable to the study of all numerical functions which occur as coefficients in power series possessing the unit circle as a natural boundary. In this particular problem it leads to the result that

$$p(n) = \frac{1}{2\pi \sqrt{2}} \frac{d}{dn} \frac{e^{\pi \sqrt{2(n-1/24)/3}}}{\sqrt{n - 1/24}} + O(e^{k\sqrt{n}}), \qquad k < \pi/\sqrt{6},$$

[217a] Proc. London Math. Soc., 13, 1882, 79; 14, 1883, 119.
[218] Jour. für Math., 146, 1915–6, 245–254.   Sitzungsber. Akad. Wiss. Wien (Math.), 126, IIa, 1917, 1081–1163.
[219] L'intermédiaire des math., 24, 1917, 95.
[220] Proc. London Math. Soc., (2), 16, 1917, 131.
[221] Comptes Rendus Paris, 164, 1917, 35–38.   Proc. London Math. Soc., (2), 17, 1918, 75–115.

and to still more exact results in which the sum of a number of approximating functions appears on the right hand side. Six terms of the series thus obtained give $p(200) = 3972999029388$, with an error of .004, a result confirmed by MacMahon by direct calculation. Here $O(g(t))$ denotes a function whose quotient by $g(t)$ remains numerically under a fixed finite value for all sufficiently large values of t. At the end of the paper occurs a table, calculated by MacMahon, of the number of partitions of n for $n \leqq 200$.

P. A. MacMahon[222] proved that, if $p_1, \cdots, p_t$ are integers in descending order of magnitude and $(m_1 \cdots m_s)$ is the partition conjugate to $(p_1 \cdots p_t)$, the number of ways of distributing n objects of specification (n) into boxes of specification $(m_1 \cdots m_s)$ is the coefficient of x^n in the expansion of

$$1 \div \{(1 - x)^{p_1}(1 - x^2)^{p_2} \cdots (1 - x^t)^{p_t}\}.$$

MacMahon[223] established a $(1, 1)$ correspondence between combinations derived from m identical sets of n distinct letters and general magic squares of order n in which the numbers in any row or column have the sum m [MacMahon[189]].

S. Ramanujan[224] proved that, if $p(n)$ is the number of partitions of n,

$$p(5m + 4) \equiv 0 \ (\mathrm{mod}\ 5), \qquad p(7m + 5) \equiv 0 \ (\mathrm{mod}\ 7),$$
$$p(35m + 19) \equiv 0 \ (\mathrm{mod}\ 35), \qquad p(25m + 24) \equiv 0 \ (\mathrm{mod}\ 25),$$
$$p(49n + 47) \equiv 0 \ (\mathrm{mod}\ 49);$$

$$p(4) + p(9)x + p(14)x^2 + \cdots = 5\frac{\{(1 - x^5)(1 - x^{10})(1 - x^{15}) \cdots\}^5}{\{(1 - x)(1 - x^2)(1 - x^3) \cdots\}^6},$$

$$p(5) + p(12)x + p(19)x^2 + \cdots = 7\frac{\{(1 - x^7)(1 - x^{14})(1 - x^{21}) \cdots\}^3}{\{(1 - x)(1 - x^2)(1 - x^3) \cdots\}^4}$$
$$+ 49x\frac{\{(1 - x^7)(1 - x^{14})(1 - x^{21}) \cdots\}^7}{\{(1 - x)(1 - x^2)(1 - x^3) \cdots\}^8},$$

which imply the first two congruence theorems.

H. B. C. Darling[225] gave elementary proofs of the first two of Ramanujan's[224] congruence theorems.

L. J. Rogers[226] gave a new proof of his[156c] two identities. J. Schur[227] gave two proofs. Finally, Rogers[228] and S. Ramanujan[228] each gave a proof which is much simpler than all earlier proofs.

P. A. MacMahon[229] solved the problem of multipartite partition.

[222] Proc. London Math. Soc., (2), 16, 1918, 352–4.
[223] *Ibid.*, (2), 17, 1918, 25–41
[224] Proc. Cambridge Phil. Soc., 19, 1919, 207–210.
[225] *Ibid.*, pp. 217–8.
[226] Proc. London Math. Soc., (2), 16, 1917, 315–7.
[227] Situngsber. Akad. Wiss. Berlin (Math.), 1917, 302–321.
[228] Proc. Cambridge Phil. Soc., 19, 1919, 211–6.
[229] Phil. Trans. Roy. Soc. London, 217, A, 1916–7, 81–113.   Memoir VII on Partitions.

A. Tanturri[230] gave expressions for the number of partitions of n into 2, 3, 4 or 5 distinct parts, and recursion formulas. He[231] investigated the number D_n of partitions of n into powers of 2 and the number $D(2^p, n)$ of partitions of n into powers of 2 of which 2^p is the maximum. The first function can be computed from the second. In the second paper occur recursion formulas for the second function, and expressions for $D(2^p, 2^p k)$ and $D(2^p, 2^p k + 2^{p-1})$ in terms of binomial coefficients.

On the number of positive integral solutions of $ax + by = n$, see papers 117–142a of Ch. II. Cesàro, Vol. I, p. 306, gave relations involving the number of positive integral solutions of $\xi_1 + 2\xi_2 + \cdots + \nu\xi_\nu = n$.

Von Sterneck, Vol. I, p. 427, used partitions into elements formed from the first s primes.

---

[230] Atti R. Accad. Sc. Torino, 52, 1916–7, 902–918. In Peano's symbolism, with a translation of most of the results.

[231] *Ibid.*, Dec. 1, 1918. Continued in Atti R. Accad. Lincei, Rendiconti, 27, II, 1918, 399–403. In Peano's symbolism with partial translation.

CHAPTER IV.

RATIONAL RIGHT TRIANGLES.

METHODS OF SOLVING $x^2 + y^2 = z^2$ IN INTEGERS.

According to Proclus,[1] Pythagoras represented the smaller leg by $x = 2\alpha + 1$, the larger leg by $y = 2\alpha^2 + 2\alpha$, and the hypotenuse by $z = y + 1$. Plato[1] took the difference $z - y$ to be 2 (instead of 1) and obtained[2] $x = 2\alpha$, $y = \alpha^2 - 1$, $z = \alpha^2 + 1$.

The Hindus Baudhâyana and Apastamba,[3] about the fifth century B.C., obtained independently[4] (?) of the Greeks the solutions (3, 4, 5), (5, 12, 13), (7, 24, 25), which are cases of the rule of Pythagoras, and (8, 15, 17), (12, 35, 37), cases of the rule of Plato.

Euclid[5] gave the set of solutions

$$\alpha\beta\gamma, \qquad \tfrac{1}{2}\alpha(\beta^2 - \gamma^2), \qquad \tfrac{1}{2}\alpha(\beta^2 + \gamma^2),$$

as well as (II, 5; X, 30) the related set

$$\sqrt{mn}, \qquad \tfrac{1}{2}(m - n), \qquad \tfrac{1}{2}(m + n).$$

Marcus Junius Nipsus,[6] at least a century before Diophantus, gave two rules to find right triangles with integral sides, one leg being given. Expressed algebraically, his rules give, as solutions of $z^2 - y^2 = x^2$,

$$z = \tfrac{1}{2}(x^2 + 1), \qquad y = \tfrac{1}{2}(x^2 - 1), \qquad \text{for } x \text{ odd;}$$
$$z = \tfrac{1}{4}x^2 + 1, \qquad y = \tfrac{1}{4}x^2 - 1, \qquad \text{for } x \text{ even,}$$

formulas equivalent to those of Pythagoras and Plato, respectively.

Diophantus[7] took a given value (in fact, 4) for z and required that $z^2 - x^2$ shall be a square of the form $(mx - z)^2$. Thus

$$x = \frac{2mz}{m^2 + 1}, \qquad y = mx - z = \left(\frac{m^2 - 1}{m^2 + 1}\right)z.$$

Here m is any rational number; replacing it by m/n, and taking $z = m^2 + n^2$, we get

$$(1) \qquad x = 2mn, \qquad y = m^2 - n^2, \qquad z = m^2 + n^2.$$

---

[1] Proclus Diadochus, primum Euclidis elem. libr. comm. (5th cent.), ed. by G. Friedlein, Leipzig, 1873, 428. Eléments d' Euclide avec les Comm. de Proclus, 1533, 111; Latin trans. by F. Barocius, 1560, 269. M. Cantor, Geschichte Math., ed. 3, I, 1907, 185–7, 224. G. J. Allman, Greek Geometry from Thales to Euclid, 1889, 34.

[2] Cited by Heron of Alexandria, Geometrie, p. 57; Boethius (6th cent.), Geometrie, lib. 2.

[3] Sulbasûtra, publ. by A. Bürk with German transl., Zeitschrift der deutschen morgenländischen Gesell., 55, 1901, 327–91, 543–91.

[4] Bürk.[3] H. G. Zeuthen, Bibliotheca Math., (3), 5, 1904, 105–7. M. Cantor, Geschichte Math., ed. 3, I, 1907, 636–45; 96 for $3^2 + 4^2 = 5^2$ in Egypt.

[5] Elementa, X, 28, 29, lemma 1; Opera, ed. by J. L. Heiberg, 3, 1886, 80. M. Cantor, Geschichte Math., ed. 3, I, 1907, 270–1, 482.

[6] Cf. J. B. Biot, Jour. des Savants, 1849, 250–1; Comptes Rendus Paris, 28, 1849, 576–81 (Sphinx-Oedipe, 4, 1909, 47–8). M. Cantor, Die römischen Agrim . . . Feldmess., 1875, 103, 112, 165. C. Henry, Bull. Bibl. Storia Sc. Mat. Fis., 20, 1887, 401–2.

[7] Arith., II, 8; Opera, ed. by P. Tannery, 1, 1893, 90; T. L. Heath, 1910, 145.

Diophantus (III, 22, etc.) referred to the right triangle with these sides as that formed from the two numbers m, n.

Brahmegupta[8] (born 598 A.D.) gave explicitly the solution (1).

An anonymous Arabic manuscript[9] of 972 stated that in every primitive right triangle (i. e., with relatively prime integral sides), the sides are given by (1). Necessary conditions that (1) give a primitive triangle are that m, n be relatively prime and $m + n$ be odd. The hypotenuse of a primitive right triangle is a sum of two squares and is of the form $12k + 1$ or $12k + 5$, though not all such numbers are sums of two squares. But 65^2 is a sum of two squares in two ways: $63^2 + 16^2 = 33^2 + 56^2$. To find a triangle with a given hypotenuse h, we need an expeditious method to find two numbers the sum of whose squares equals h. If the last digit d of h is 1, the two squares end in 5 and 6 or in 00 and 1. If $d = 3$, they end in 4 and 9; if $d = 7$, in 1 and 6; if $d = 5$, in 00 and 5, 1 and 4, or 6 and 9; if $d = 9$, in 00 and 9, or 4 and 5; with similar rules if d is even.

The Arab Ben Alhocain[10] (tenth cent.) gave a geometrical proof that (1) give the sides of a right triangle, and noted that if the hypotenuse is even, also both legs are even. Rules equivalent to that by Pythagoras are given; also false theorems on triangles formed from several consecutive numbers.

Alkarkhi[11] (end of tenth cent.) derived the solution 3, 4, 5 of $x^2 + y^2 = z^2$ by setting $y = x + 1$, $z = 2x - 1$.

Bháscara[12] (born 1114) gave (1) and employed it, as had Brahmegupta, to find the second leg $(m^2/n - n)/2$ and hypotenuse, $(m^2/n + n)/2$, given one leg m. Given the hypotenuse h, the legs are[12a] $l = 2hb/(b^2 + 1)$ and $lb - h$ or $h - q$ and bq, where $q = 2h/(b^2 + 1)$. To find (p. 201) a right triangle whose area equals the hypotenuse take $3x$, $4x$, $5x$ as the sides.

Leonardo Pisano[13] employed the fact that the sum $1 + 3 + \cdots$ of n consecutive odd numbers is n^2 to find two squares whose sum is a square. First, if one square a^2 is odd, take the other to be $1 + 3 + \cdots + (a^2 - 2)$; their sum $1 + 3 + \cdots + a^2$ is a square. If one square is even, as 36, add and subtract unity from its half, obtaining the consecutive odd numbers 17 and 19; then $1 + 3 + \cdots + 15 = 64$ and

$$64 + 36 = 1 + \cdots + 15 + 17 + 19 = 10^2.$$

[8] Brahme-sphut'a-sidd'hánta; Algebra with Arithmetic and Mensuration, from the Sanskrit of Brahmegupta and Bháscara, transl. by H. T. Colebrooke, London, 1817, 306–7, 363–72.

[9] French transl. by F. Woepcke, Atti Accad. Pont. Nuovi Lincei, 14, 1860–1, 213–227, 241–269 (M. Cantor, Geschichte Math., ed. 3, I, 1907, 751–2).

[10] *Ibid.*, 301–24, 343–56.

[11] Extrait du Fakhrî, French transl. by F. Woepcke, Paris, 1853, 89.

[12] Colebrooke,[8] pp. 61–63. John Taylor's transl. of Brahme . . . ,[8] Bombay, 1816, p. 71.

[12a] Same in Ladies' Diary, 1745, 14, Quest. 254; T. Leybourn's Math. Quest. Ladies' Diary, 1, 1817, 366–7; C. Hutton's Diarian Miscellany, 2, 1775, 200.

[13] Liber quadratorum L. Pisano, 1225, in Tre Scritti inediti, 1854, 56–66, 70–5; Scritti L. Pisano, 2, 1862, 253–4. Cf. A. Genocchi, Annali Sc. Mat. Fis., 6, 1855, 234–5; P. Volpicelli, Atti Accad. Pont. Nuovi Lincei, 6, 1852–3, 82–3; P. Cossali, Origine, Trasporto in Italia . . . Algebra, 1, 1797, 97–102, 118–9.

[These correspond to the rules of Pythagoras and Plato.] Leonardo[14] obtained rational solutions of $x^2 + y^2 = a^2$ by a method quite different from that of Diophantus; starting with any known rational triangle for which $\alpha^2 + \beta^2 = \gamma^2$, he took $x = a\alpha/\gamma$, $y = a\beta/\gamma$.

F. Vieta[15] (1540–1613) used the method of Leonardo, last cited, and that of Diophantus.

M. Stifel[16] called $a \cdot b$ a diametral number if $a^2 + b^2 = c^2$ and stated incorrectly that $a \cdot b$ is a diametral number if and only if a/b belongs to one of the series $1\frac{1}{3}$, $2\frac{2}{5}$, $3\frac{3}{7}$, $\cdots$ and $1\frac{7}{8}$, $2\frac{11}{12}$, $3\frac{15}{16}$, $\cdots$, and hence in effect that $a : b = 2n^2 + 2n : 2n + 1$ or $a : b = 4n^2 + 8n + 3 : 4n + 4$ [cf. Meyer[46]], which correspond to the solutions of $a^2 + b^2 = c^2$ by Pythagoras and Plato. These diametral numbers are not those defined by Theon of Smyrna[2] of Ch. XII.

The Japanese manuscript of Matsunago[17] of the first half of the eighteenth century contains three proofs of (1).

T. Fantet de Lagny[18] replaced m by $d + n$ in (1) and obtained

$$x = 2n(d + n), \qquad y = d(d + 2n), \qquad z = x + d^2 = y + 2n^2.$$

Taking $d = 1$ or $n = 1$, we obtain the rule of Pythagoras or that of Plato.

C. A. Koerbero[19] proved that the sides of any rational right triangle are proportional to the numbers (1).

L. Euler[20] expressed the hypotenuse c as $b + an/m$. By $a^2 + b^2 = c^2$, $b : a = m^2 - n^2 : 2mn$. Hence a, b, c are proportional to the numbers (1) with $m > n > 0$.

Euler[21] noted that the sum of the squares of $x + 1/x$ and $y + 1/y$ is a square if

$$y = \frac{px - 1}{x + p}, \qquad (x + p)^2(px - 1)^2 + x^2(p^2 + 1)^2 = \square,$$

the latter being true if $(p^2 - 1)x = 4p$.

J. P. Grüson[22] noted that $n + 1$ and n generate a triangle [of Pythagoras' type] whose larger leg $y = 2n^2 + 2n$ and hypotenuse $y + 1$ generate a new triangle whose least side is a square $[2y + 1 = (2n + 1)^2]$.

L. Poinsot[23] noted that every set of integral solutions of $z^2 - y^2 = x^2$ is given by $z = (p + q)/2$, $y = (p - q)/2$, where x^2 has been expressed in every way as a product of two integers p and q, both odd and relatively prime or both even, but with no common factor > 2.

[14] Liber Abbaci, Ch. 15 (Scritti L. Pisano, Rome, 1, 1857).

[15] Franciscus Vieta, Zetetica, 1591, IV, 1; Opera Math., 1646, 62.

[16] Arith. Integra, Nürnberg, 1544, f. 14v–f. 15v. Copied by Ioseppo Vnicorno, De l'Arith. Universale, Venetia, 1598, 62.

[17] Y. Mikami, Abh. Gesch. Math. Wiss., 30, 1912, 229. Report by K. Yanagihara, Tôhoku Math. Jour., 6, 1914–5, 120–3; continued, 9, 1916, 80–7 (by use of progressions).

[18] Hist. Acad. Sc. Paris, 1729, 318.

[19] Nova trianguli rectanguli analysis, Halae Magd., 1738, 8.

[20] Comm. Acad. Petrop., 10, 1738, 125; Comm. Arith., 1, 1849, 24.

[21] Opusc. anal., 1, 1783, 329; Comm. Arith., II, 46.

[22] Enthüllte Zaubereyen u. Geheimnisse der Arith., Berlin, 1796, 104–6.

[23] Comptes Rendus Paris, 28, 1849, 581–3; also p. 579 by J. B. Biot.

P. Volpicelli[24] noted that $z = a^2 + b^2 = \alpha^2 + \beta^2$ imply that

$$x = \pm (a\alpha \mp b\beta), \qquad y = \pm (a\beta \pm b\alpha)$$

are solutions of $x^2 + y^2 = z^2$ and stated incorrectly that they give all the solutions, whereas formulas (1) do not.   As to J. Liouville's[25] remark that, for z given, $x^2 + y^2 = z^2$ has relatively prime solutions if and only if z is a product of primes $4n + 1$, the solutions $x = 1020$, $y = 425$, $z = 5 \cdot 13 \cdot 17$ are not relatively prime.

Volpicelli[26] distinguished k types of solutions of $x^2 + y^2 = z^2$, where $z = h_1 \cdots h_k$, $h_j = a_j^2 + b_j^2$.   The k solutions of the first type are $q(a_j^2 - b_j^2)$, $2qa_jb_j$, where $q = z/h_j$.   The $k(k-1)$ solutions of the second type are

$$q\{(a_i^2 - b_i^2)(a_j^2 - b_j^2) \pm 4a_ia_jb_ib_j\}, \qquad q\{2(a_ia_j \pm b_ib_j)(a_jb_i \mp a_ib_j)\},$$

where $q = z/(h_ih_j)$, the quantities x_2, y_2 in brackets being such that $x_2^2 + y_2^2 = h_i^2 h_j^2$.   From

$$x_3^2 + y_3^2 = (x_2^2 + y_2^2)\{(a_t^2 - b_t^2)^2 + (2a_tb_t)^2\} = h_i^2 h_j^2 h_t^2,$$

we obtain the $4\binom{k}{3}$ solutions qx_3, qy_3 of the third type, where $q = z/(h_ih_jh_t)$. Thus the total number of solutions is

$$\sum_{s=1}^{k} \frac{2^{s-1}k(k-1)\cdots(k-s+1)}{1 \cdot 2 \cdots s} = \tfrac{1}{2}(3^k - 1).$$

Volpicelli[27] noted that all solutions of $x^2 + y^2 = z^2$ depend on the solutions of $x^2 + y^2 = z_j^2$ $(j = 1, \cdots, k)$, where $z_1, \cdots, z_k = z$ are the products of the factors of z taken $1, 2, \cdots, k$ at a time.   For $z^2 = (a^2 + b^2)^k$, a solution is

$$x = a^k - \binom{k}{2}a^{k-2}b^2 + \binom{k}{4}a^{k-4}b^4 - \cdots, \qquad y = \binom{k}{1}a^{k-1}b - \binom{k}{3}a^{k-3}b^3 + \cdots.$$

For, if $(a + ib)^k = A + iB$, $(a^2 + b^2)^k = A^2 + B^2$, which was verified without using $i = \sqrt{-1}$.   Also $a^2 - b^2$ is a factor of B if $k = 4h$, but is a factor of A if $k = 4h + 2$.

C. A. W. Berkhan[28] gave nineteen methods of finding two numbers the sum of whose squares is a square, with references on several proofs.

E. de Jonquières[29] discussed Volpicelli's[26] topic.

A. J. F. Meyl[30] noted that, according to an argument by de Jonquières,[29]

$$(x + 3)^2 + (x + 4)^2 = [(y + 1)^2 + (y + 2)^2]^2$$

has only the solutions $x + 3 = 3$ or -4, whereas $x + 3 = 0$ or -1 also.

C. de Polignac[31] used a rectangular lattice to prove that (1) gives all integral solutions of $x^2 + y^2 = z^2$.

[24] Giornale Arcadico di Sc. Let. ed Arti, Rome, 119, 1849–50, 27.   Annali di Sc. Mat. Fis., 1, 1850, 159–166, 369, 443.

[25] Comptes Rendus Paris, 28, 1849, 687.

[26] Atti Accad. Pont. Nuovi Lincei, 4, 1850–1, 124–140, 346–377, 508–510.

[27] *Ibid.*, 5, 1851–2, 315–352; Comptes Rendus Paris, 36, 1853, 443–5.   Extract in Annali di Sc. Mat. Fis., 3, 1852, 130–3; 4, 1853, 286–297.

[28] Die merkwürdigen Eigenschaften der Pythag. Zahlen, Eisleben, 1853.

[29] Nouv. Ann. Math., (2), 17, 1878, 241–7, 289.   Cf. papers 26–31 of Ch. XVII.

[30] *Ibid.*, (2), 18, 1879, 332–3.

[31] Bull. Math. Soc. France, 6, 1877–8, 162.

C. M. Piuma[32] quoted the known result that all relatively prime integral solutions of $x^2 + y^2 = z^2$ are given by

$$x = mn, \qquad y = \frac{m^2 - n^2}{2}, \qquad z = \frac{m^2 + n^2}{2},$$

where m and n are relatively prime odd integers, and proved conversely that then these three expressions are relatively prime in pairs, by showing by use of congruences that no two are divisible by the same power of a prime.

D. S. Hart[33] proved for $n \leqq 4$ that, if z is a product of n primes each a sum of two squares ☐, z^2 is a ☐ in $(3^n - 1)/2$ ways [Volpicelli[26]].

L. E. Dickson[34] obtained, as a solution equivalent to (1), $r + s$, $r + t$, $r + s + t$, where $r^2 = 2st$ is a square. The same rule was given later by P. G. Egidi,[35] D. Gambioli,[36] A. Bottari,[39] and H. Schotten.[36a]

Graeber[37] noted that if the point of tangency of a circle inscribed in a right triangle divides the hypotenuse z into the segments k and m, while n and m are the corresponding segments of leg y, then

$$(k + m)^2 = (m + n)^2 + (n + k)^2, \qquad k = (n^2 + mn)/(m - n).$$

Thus x, y, z are proportional to (1). The sides if integral are shown by a long proof to be (1).

L. Kronecker[38] proved that all positive integral solutions of $x^2 + y^2 = z^2$ are given without duplication by

$$x = 2pqt, \qquad y = t(p^2 - q^2), \qquad z = t(p^2 + q^2), \qquad p > q > 0, \qquad t > 0,$$

p and q being relatively prime and not both odd. The reason why every solution is obtained once and but once is due to the fact that the circle $\xi^2 + \eta^2 = 1$ is of genus zero, all its points being expressible rationally in $\tau = \tan \omega/2$:

$$\xi = \cos \omega = \frac{1 - \tau^2}{1 + \tau^2}, \qquad \eta = \sin \omega = \frac{2\tau}{1 + \tau^2}.$$

A. Bottari[39] proved that all integral solutions of $x^2 + y^2 = z^2$ are given by $x = u + w$, $y = v + w$, $z = u + v + w$, where $u = p^2k$, $v = 2^{2s-1}q^2k$, $w = 2^s pqk$, p and q being relatively prime odd integers. Thus xy is not a square.

P. Cattaneo[40] gave a simple proof of Bottari's theorem.

P. Reutzel[41] noted that, if $a > 2$, we can solve $c^2 - b^2 = a^2$. Set $c = b + v$. Then $b = (a^2 - v^2)/(2v)$ is an integer if $v = 1$ and a is odd, or if $v = 2$ and c is even. We may take v to be any divisor a/n of a; then $b = (n^2 - 1)v/2$, $c = (n^2 + 1)v/2$.

[32] Giornale di Mat., 19, 1881, 311–5.
[33] Math. Quest. Educ. Times, 39, 1883, 47–8.
[34] Amer. Math. Monthly, 1, 1894, 8.
[35] Atti Accad. Pont. Nuovi Lincei, 50, 1897, 103.
[36] Periodico di Mat., 16, 1901, 151–5.
[36a] Zeitschrift Math. Naturw. Unterricht, 47, 1916, 181–2.
[37] Archiv Math. Phys., (2), 17, 1900, 36.
[38] Vorlesungen über Zahlentheorie, 1, 1901, 31–35.
[39] Periodico di Mat., 23, 1908, 104–110. Cf. Dickson.[34]
[40] *Ibid.*, 218.
[41] Zeitschrift Vermessungswesen d. Deutschen Geometervereins, Stuttgart, 38, 1909, 208–11.

J. Gediking[42] noted that, for relatively prime solutions of $x^2 - y^2 = z^2$, we may take as $x - y$ any number of the form $(2n + 1)^2$ or $2n^2$, but no other.   Then $x + y = (2m + 1)^2$ or $2m^2$, with $2m + 1$ and $2n + 1$ or m and n relatively prime. [It was overlooked that we may restrict to one of the two cases.] All solutions < 1000 are given. J. C. Milborn (pp. 167-9) erred in saying that this method does not give all solutions. T. Boelen (pp. 238-40) noted that we may take as z any integer > 2, if solutions with a common factor are allowed.

C. J. van der Burg[43] gave an incomplete proof of (1).

Fitting[44] discussed the relatively prime solutions of $x^2 + y^2 = z^2$ by setting $z = x + a$, whence $y^2 = a(2x + a)$. Without loss of generality we may take a to be an odd square 1, 9, 25, $\cdots$, and equate $2x + a$ to the successive odd squares.

W. Kluge[45] noted that $x^2 + y^2 = z^2$ is satisfied by

$$x^2 = d\zeta, \qquad d < x, \qquad y = \frac{\zeta - d}{2}, \qquad z = \frac{\zeta + d}{2},$$

and gave recursion formulas for computing successive solutions.

E. Meyer[46] noted that Stifel's[16] formulas for diametral numbers do not give all, for example not $33 \cdot 56$, and that he should have used

$$a : b = m^2 - n^2 : 2mn.$$

He compared many known ways of solving $x^2 + y^2 = z^2$.

P. Lambert[47] solved $x^2 + y^2 = z^2$ by use of numbers $a + bi$.

N. Gennimatás[48] would solve $x^2 + y^2 = a^2$ by setting $2a = c + d$, where cd is a square x^2, whence $y = a - d$.

*E. Haentzschel[48a] noted that from one rational right triangle we can derive an infinity by use of the formulas for $\sin n\alpha$ and $\cos n\alpha$ [cf. Vieta,[4] Ch. VI]. From two right triangles whose hypotenuses are primes of the form $4k+1$, we can derive an infinity by use of the addition theorem for sine and cosine. By means of these theorems we can arrange in order the proper solutions of $x^2 + y^2 = z^2$.

P. Quintili[49] attributed to F. Klein (!) the solution (1) of $x^2 + y^2 = z^2$.

A. E. Jones[50] discussed right triangles whose three sides are of the form $x^2 - 1$.

C. A. Laisant[51] noted that MQ, $2PN$, $P^2 + N^2$ are sides of a right triangle if M, N, P, Q are four consecutive terms of Fibonacci's series (Vol. I, Ch. XVII of this History), so that $P = M + N$, $Q = N + P$.

---

[42] Vriend der Wiskunde, 25, 1910, 86-96.
[43] *Ibid.*, 26, 1911, 188-191.
[44] L'intermédiaire des math., 18, 1911, 87-90 (233-4).
[45] Verhandlungen der Versamm. deutscher Philologen u. Schulmänner, Leipzig, 51, 1911, 137. Unterrichtsblätter Math. Naturwiss., Berlin, 19, 1913, 11.
[46] Zeitschrift Math. Naturw. Unterricht, 43, 1912, 281-7.
[47] Nouv. Ann. Math., (4), 12, 1912, 408-421.
[48] Zeitschr. Math. Naturw. Unterricht, 44, 1913, 14-15.
[48a] Blätter für d. Fortbildung d. Lehrers u. Leherin, Berlin, 6, 1913, 395-6.
[49] Il Boll. Mat. Sc. Fis. Nat., 16, 1915, 69-71.
[50] Math. Quest. and Solutions (contin. of Math. Quest. Educ. Times), 2, 1916, 18.
[51] Comptes Rendus des Sc. Soc. Math. France, 1917, 18-19.

Papers without Novelty.

G. Oughtred, Opuscula Math., Oxonii, 1677, 130–8.
A. Thacker, A Miscellany of Math. Problems, Birmingham, 1, 1743, 171–8 [Proof of (1)].
Anonymous, Ladies' Diary, 1752, 39, Quest. 344 [Proof of (1)].
A. D. Wheeler, Amer. Jour. Arts. Sc. (ed., Silliman), 20, 1831, 295 [Plato's rule].
J. A. Grunert, Klügel's Math. Wörterbuch, 5, 1831, 1141–3 [Euler[20]].
C. M. Ingleby and S. Bills, Math. Quest. Educ. Times, 6, 1866, 39–40 [Proof of (1)].
M. A. Gruber, Amer. Math. Monthly, 4, 1897, 106–8.
H. Schubert, Niedere Analysis, 1, 1902, 159–162 [Proof of (1)].
F. Thaarup, Nyt Tidsskrift for Mat., 15, A, 1904, 33 [Proof of (1)].
A. Aubry, Mathesis, 5, 1905, 6–13 [historical].
A. Holm, Math. Quest. Educ. Times, (2), 9, 1906, 92; 10, 1906, 56 [Proof of (1)].
V. Varali-Thevenet, Rivista Fis. Mat. Sc. Nat., 8, I, 1906, 422–3.
C. Botto, Giornale di Mat., 46, 1908, 297–8 [Poinsot[23]].
P. Richert, Unterrichtsblätter Math., 14, 1908, 55–7, 87.
C. Botto, Suppl. al Periodico di Mat., 12, 1908–9, 68–74.
T. S. Rao, Jour. Indian Math. Club, Madras, 1, 1909, 130–4.
School Sc. and Math., 10, 1910, 683; 11, 1911, 293–4; 13, 1913, 320–2.

Sides of a right triangle divisible by 3, 4, or 5.

Frenicle de Bessy[52] († 1675) noted that if the g.c.d. of the integral sides of a right triangle is a square or the double of a square, the sides are of the form (1), and that one of the sides is divisible by 5, one of the legs by 3 and one by 4. If the sides are relatively prime, the sum and difference of the legs are of the forms $8k \pm 1$.

P. Lenthéric[53] noted that the product xyz of the numbers (1) is divisible by 60, since $mn(m^2 - n^2)$ is divisible by 6 and if no one of m, n, $m \pm n$ is divisible by 5, $m^2 + n^2$ is. F. Paulet added (p. 382) the remark that $m^4 - n^4$ is divisible by 5 if neither m nor n is, since $m^4 = 10k + 1$ or $10k + 6$.

L. Poinsot[23] stated, as if new, that if x, y, z are relatively prime solutions of $x^2 + y^2 = z^2$, 3 is a factor of x or y, 4 a factor of x or y, and 5 a factor of x, y or z. This was proved by E. R. Grenoble[54] by considering the residues modulo 3, 4 or 5, and by J. Binet (pp. 686–7, 755) by use of Fermat's theorem. J. Liouville remarked (p. 687) that x, y, $x + y$ or $x - y$ is divisible by 7. Bourdat[55] stated that he had found these facts in 1839 and added that, if $x^2 + y^2 = z^4$, 5 is a factor of x, y or z, likewise 7 and 24. If $x^2 + y^2 = z^8$, one of the numbers has the factor $2^4 \cdot 3 \cdot 7$.

A. Vermehren[56] proved that xyz is divisible by 60.

A. Lévy[57] noted that in $a^2 + b^2 = c^2$, 7 divides $a + b$ or $a - b$ if 7 is prime to a, b, c; 11 divides one of $5a \pm b$, $5b \pm a$ if 11 is prime to a, b, c.

[52] Traité des triangles rectangles en nombres, I, Paris, 1676, §§ 24–25, pp. 59–61. Reprinted with part II in 1677 at end of Problèmes d'Architecture de Blondel. Both parts in Mém. Acad. R. Sc. Paris, 5, 1666–99; éd. Paris, 1729, pp. 146–7. C. Henry, Bull. Bibl. Storia Sc. Mat. Fis., 12, 1879, 691–2, gave a list of Frenicle's writings; cf. Nouv. Ann. Math., 8, 1849, 364–5.
[53] Annales de Math. (ed., Gergonne), 20, 1829–30, 376–382; 21, 1830–1, 96–98. Cf. Jour. für Math., 5, 1830, 386; Jour. de math. élém. spéc., 1880, 261.
[54] Comptes Rendus Paris, 28, 1849, 665–6.
[55] Bull. de l'Acad. Delphinale, Grenoble, 3, 1850, 37–43.
[56] Die Pythagoräischen Zahlen, Progr. Domschule, Güstrow, 1863.
[57] Bull. de math. élém., 15, 1909–10, 277.

Number of right triangles with a given side.

Report has been given above of the papers by Volpicelli,[26] Hart[33] and de Jonquières.[29] See Fermat[10] and Frenicle[17] of Ch. VI and papers 19–32 of Ch. XIII.

F. Gauss[58] noted that to every hypotenuse composed of k distinct primes belong

$$\left[\frac{k}{1}\right] + 2\left[\frac{k}{2}\right] + 2^2\left[\frac{k}{3}\right] + \cdots + 2^{k-1}\left[\frac{k}{k}\right]$$

different pairs of legs, where $[x]$ is the largest integer $\leqq x$. The legs are relatively prime for 2^{k-1} pairs.[59]

D. N. Lehmer[60] proved that the number N of right triangles whose sides are integers with no common divisor, and whose hypotenuse is $\leqq n$, is asymptotically $n/(2\pi)$. But, if the sum of the three sides is $\leqq n$, $N = n (\log 2)/\pi^2$, asymptotically.

O. Meissner[61] stated that the number P of integral right triangles with one leg $x = 2^m p_1^{m_1} \cdots p_n^{m_n}$ (p's distinct primes) is:

$$P = P_2 + \frac{m-1}{\left[1 + \dfrac{1}{m}\right]}(2P_2 + 1), \qquad P_2 \equiv \frac{1}{2}\left\{\prod_{\nu=1}^{n}(2m_\nu + 1) - 1\right\},$$

where $[a]$ is the largest integer $\leqq a$. Also $P + 1$ is the number of sets of positive integral solutions z, y of $z^2 - y^2 = x^2$ (x given).

E. Bahier[62] noted that if $A, B, \cdots, P$ are distinct odd primes the number of right triangles one of whose legs is $A^\alpha B^\beta \cdots P^\pi$ is

$$\Sigma\alpha + 2\Sigma\alpha\beta + 2^2\Sigma\alpha\beta\gamma + \cdots + 2^{k-1}\alpha\beta\gamma \cdots \pi.$$

If $A = 2$, we have only to replace α by $\alpha - 1$ in the last result.

Right triangles of equal area.

Diophantus, V, 8, required three rational right triangles of equal areas. If, as in V, 7, $ab + a^2 + b^2 = c^2$, the right triangles formed[7] from c, a; c, b; $c, a + b$ have the same area $abc(a + b)$. The chosen example has $a = 3$, $b = 5$, $c = 7$. This solution was given in general form by F. Vieta, Zetetica, IV, 11. Fermat[62a] observed that if z is the hypotenuse and b, d the legs of a rational right triangle, we obtain a new right triangle of the same area by forming the triangle from $z^2, 2bd$ and dividing its sides by $2z(b^2 - d^2)$. From this new triangle we may derive similarly a third, etc. Apart from notation, this method is the same as the " construction " in

[58] Über die Pythag. Zahlen, Progr. Bunzlau, 1894, p. 15.
[59] If the hypotenuse is 65, the legs are 25, 60; 16, 63; 33, 56; or 39, 52.
[60] Amer. Jour. Math., 22, 1900, 327–8.
[61] Archiv Math. Phys., (3), 8, 1904, 181.
[62] Recherche Méthodique et Propriétés des Triangles Rectangles en Nombres Entiers, Paris, 1916, 21–27.
[62a] Oeuvres, III, 254–5; S. Fermat's Diophanti Alex. Arith., 1670, 220.

the second part of Frenicle's[52] Traité; this process has been summarized by A. Cunningham.[63]

Fermat[64] stated that he could give five right triangles of equal area and had a method to find as many as one pleases, whereas Diophantus, V, 8, and Vieta, Zetetica, IV, 11, gave only three.

J. de Billy[65] noted that the right triangle with the legs $3r$, $(x + 4)r$ will have the same area 6 as $(3, 4, 5)$ if $\frac{3}{2}(x + 4)r^2 = 6$. Thus $x + 4$ and $9 + (x + 4)^2$ must be squares, which is the case if $x = -\, 6725600/2405601$.

John Kersey[66] discussed the problem to deduce a rational right triangle with the same area as a given one, and stated many problems on areas.

L. Euler[67] discussed the solution of

$$pr(p^2 - r^2) = qs(q^2 - s^2),$$

noting the case $p = 11$, $r = -\, 35$, $q = -\, 23$, $s = 33$. Hence the right triangles formed from 11, 35 and 23, 33 have equal areas.

Euler[68] noted that if we take $q = p$, $p^2 = r^2 + rs + s^2$, we get

$$2r + s = \sqrt{4p^2 - 3s^2} = 2p - sf/g \quad \text{if } \frac{p}{s} = \frac{f^2 + 3g^2}{4fg}.$$

Take $p = f^2 + 3g^2$, $s = 4fg$. Hence the values $x = f^2 + 3g^2$, $y = 4fg$ or $3g^2 - f^2 \pm 2fg$ give the sides $2xy$, $x^2 - y^2$, $x^2 + y^2$ of three right triangles with the same area $xy(x^2 - y^2)$.

Grüson[22] (pp. 109–114) and Young[134] of Ch. XIX discussed the determination of three right triangles of equal area.

J. Collins[69] employed the three right triangles with the legs

$$v^2 - x^2,\ 2vx; \qquad v^2 - y^2,\ 2vy; \qquad z^2 - v^2,\ 2zv.$$

The first two are of equal area if $v^2 = x^2 + xy + y^2$. Set $v = x - t$. Then $x = (t^2 - y^2)/(y + 2t)$. The first and third have equal areas if $v^2 = x^2 - xz + z^2$. Set $v = x - s$. Then $x = (s^2 - z^2)/(2s - z)$. To make the values of x equal, take $t = m + n$, $y = m - n$, $s = p + q$, $z = p - q$. Then $mn/(3m + n) = pq/(3p + q)$ determines m in terms of p, q, n.

J. Cunliffe[70] treated the problem to find k rational right triangles of equal areas. For $k = 3$, let $m^2 \pm n^2$, $2mn$ be the sides of one triangle. In

$$mn(m^2 - n^2) = pq(p^2 - q^2),$$

set $p = m + r$, $q = n - r$, and solve the resulting quadratic for r. Thus $4r = \pm\, \sqrt{R} - 3(m - n)$, where $R = m^2 + 14mn + n^2$. Set

$$R = (m + n + s)^2,$$

---

[63] Math. Quest. Educ. Times, 72, 1900, 31–2.

[64] Oeuvres, II, 263, letter to Mersenne, Sept. 1, 1643. He had asked (p. 259) for four.

[65] Inventum Novum, I, § 38, Oeuvres de Fermat, III, 348. In his Diophantus Geometra, Paris, 1660, 108, 121, de Billy treated the problems of Diophantus V, 8, VI, 3.

[66] The Elements of Algebra, London, Books 3 and 4, 1674, 94, 124–142.

[67] Nova Acta Acad. Petrop., 13, 1795 (1778), 45; Comm. Arith., II, 285.

[68] Opera postuma, 1, 1862, 250–2 (about 1781).

[69] The Gentleman's Math. Companion, 2, No. 11, 1808, 123.

[70] New Series of the Math. Repository (ed., Th. Leybourn), 3, II, 1814, 60.

thus determining m rationally. Hence we get two new rational right triangles. For any k, let a, b, h be the legs and hypotenuse of one right triangle; another of equal area has the sides

$$\frac{2abh}{2b^2 - h^2}, \qquad \frac{2b^2 - h^2}{2h}, \qquad \frac{h^4 + 4b^2h^2 - 4b^4}{2h(2b^2 - h^2)}.$$

From this, we obtain a third, etc. To find any number of rational squares h^2, h'^2, $\cdots$ and a number N which if added to or subtracted from each of the squares yields sums and differences which are rational squares, use right triangles of equal area and take $h^2 = a^2 + b^2$, $h'^2 = a'^2 + b'^2$, $\cdots$, $N = 2ab = 2a'b' = \cdots$. Cf. Ch. XVI.

D. S. Hart[71] repeated the method of Diophantus V, 8.

A. Martin,[72] using the 3 triangles of Collins,[69] concluded that the conditions reduce to $x = z - y$, $v^2 = z^2 - zy + y^2$, which is satisfied if $y = m^2 - n^2$, $z = 2mn + m^2$, $v = m^2 + mn + n^2$.

C. E. Hillyer[73] noted that equal right triangles are formed[7] from

$$k^2 + kl + l^2, \ k^2 - l^2; \qquad k^2 + kl + l^2, \ 2kl + l^2; \qquad k^2 + 2kl, \ k^2 + kl + l^2.$$

C. Tweedie,[74] to find all rational right triangles of area A, discussed $\alpha^2 + \beta^2 = \gamma^2$, $\alpha\beta = 2A$, whence $x_1^2 + y_1^2 = 1$, $\gamma^2 x_1 y_1 = 2A$. Thus

$$x_1 = \frac{2m}{1 + m^2}, \qquad y_1 = \frac{1 - m^2}{1 + m^2}, \qquad 2\gamma^2 m(1 - m^2) = 2A(1 + m^2)^2.$$

Write $x = m$, $y = (1 + m^2)/\gamma$. Hence we seek the rational points on

$$(2) \qquad x(1 - x^2) = Ay^2.$$

To apply Cauchy's tangential method (papers 287, 296, etc. of Ch. XXI), start with any right triangle with sides α, β, γ and derive the corresponding rational point (x, y). The tangent there cuts the cubic at a new rational point, which corresponds to a new right triangle with the legs $2\alpha\beta\gamma/(\alpha^2 - \beta^2)$, $(\alpha^2 - \beta^2)/(2\gamma)$. From it we get a third right triangle. The problem is also treated by Cauchy's second method (the line joining two rational points of a cubic determines a third).

E. Bahier,[62] pp. 149–168, treated the subject.

Two right triangles whose areas have a given ratio.

Diophantus, V, 24, asked for three squares x_i^2 such that $x_1^2 x_2^2 x_3^2 + x_i^2$ are squares for $i = 1, 2, 3$. A solution will be $x_i = sb_i/p_i$ if three right triangles (p_i, b_i, h_i) are found such that $p_1 p_2 p_3 = s^2 b_1 b_2 b_3$, since

$$x_1 x_2 x_3 = s, \qquad s^2 + x_i^2 = s^2\left(1 + \frac{b_i^2}{p_i^2}\right) = \left(\frac{sh_i}{p_i}\right)^2.$$

[71] Math. Visitor, 2, 1882, 17–18.
[72] Math. Quest. Educ. Times, 48, 1888, 118–9.
[73] Math. Quest. Educ. Times, 72, 1900, 30.
[74] Proc. Edinb. Math. Soc., 24, 1905–6, 7–19. He quoted from "Life and Letters of Lewis Carroll," p. 343, that the triangles (20, 21, 29) and (12, 35, 37) are equal, but failed to find three.

Diophantus took (3, 4, 5) as one triangle and stated that it is easy to find two triangles such that the product of the legs of one is 12 (or 3) times that of the other, as (9, 40, 41), (8, 15, 17).   C. G. Bachet[74a] chose an arbitrary triangle (p_1, b_1, h_1) and the two triangles formed[7] from b_1, h_1 and p_1, h_1, obtaining $s = p_1/(2h_1)$.   Fermat[75] gave general rules for finding two right triangles whose areas are in a given ratio r/s, where $r > s$, viz., form the triangles from $2r \pm s$, $r \mp s$ and $2s \pm r$, $r \mp s$; or from $6r$, $2r - s$ and $4r + s$, $4r - 2s$; or from $r + 4s$, $2r - 4s$ and $6s$, $r - 2s$.   Thus to find three right triangles whose areas are proportional to given numbers r, s, t, such that $r + t = 4s$, $r \geq t$, form the triangles from $r + 4s$, $2r - 4s$; $6s$, $r - 2s$; $4s + t$, $4s - 2t$.   The areas of the triangles formed from 49, 2; 47, 2; 48, 1 are themselves the sides of a right triangle.[76]

L. Euler[77] found ten types of pairs of right triangles whose areas $A = pq(p^2 - q^2)$ and $B = rs(r^2 - s^2)$ have a given ratio $a : b$.   He equated r and s to two of the numbers $p, 2p, q, 2q, p \pm q$.   For example, $r = p$, $s = p - q$ give $p + q : 2p - q = a : b$, whence $p : q = a + b : 2a - b$; taking $r = p = a + b$, we get $q = 2a - b$, $s = 2b - a$.   He gave (pp. 222–3) several methods to make A/B a square (cf. Euler[33] of Ch. XV, Euler[81] of Ch. XVI, Euler[18, 19] of Ch. XVIII and Euler[253] of Ch. XXII).

A. Holm[78] noted that the problem leads to a cubic curve with two given rational points, whence the chord determines a third.

Other problems involving only area.

An anonymous[79] Greek manuscript, probably dating between Euclid and Diophantus, found the sides of a right triangle with the area 5 by seeking a product of 5 and a square 36, divisible by 6, such that the product $5 \cdot 36$ is the area of a right triangle with the sides 9, 40, 41, and reduced them in the ratio 1 : 6,—which shows a knowledge of the fact that the area of a right triangle with integral sides is a multiple of 6 (L. Pisano,[14] Ch. XVI).

Diophantus, VI, 3, required a right triangle whose area increased by a given number g yields a square.   Take $g = 5$ and denote the triangle by (hx, px, bx); we are to choose x so that $\frac{1}{2}pbx^2 + 5 = n^2x^2$.   Let (h, p, b) be formed from $m, 1/m$ and take $n = m + 2 \cdot 5/m$.   Then $\frac{1}{2}pb = m^2 - 1/m^2$.   When this is subtracted from n^2, the difference shall be 5 times a square.   Hence $100m^2 + 505 = \square$, say $(10m + 5)^2$.   Thus $m = 24/5$, $n = 413/60$, $x = 24/53$.   F. Vieta (Zetetica, V, 9) took $g = r^2 + s^2$, formed the triangle from $(r + s)^2$, $(r - s)^2$, and divided its sides by $2(r + s)(r - s)^2$; the area is now $2rs(r^2 + s^2)/(r - s)^2$, which added to g yields the square of $(r^2 + s^2)/(r - s)$.   C. G. Bachet[74a] remarked that g need not be the sum of two squares

[74a] Diophanti Alex. Arith. . . . Commentariis . . . Avctore C. G. Bacheto, 1621, 333.

[75] Oeuvres, I, 319; French transl., III, 259.   Cf., II, 224–6.

[76] Other solutions, Oeuvres de Fermat, II, 93, 250, 277; Oeuvres de Descartes, II, 165. De Billy gave the triangles formed from 6, 1; 7, 6; 8, 1; Oeuvres de Fermat, IV, 1912, 139; Bull. Bibl. Storia Sc. Mat. Fis., 12, 1879, 517.

[77] Opera postuma, 1, 1862, 224–7 (about 1773).

[78] Proc. Edinburgh Math. Soc., 22, 1903–4, 48.

[79] With German transl. by J. L. Heiberg and comments by H. G. Zeuthen, Bibliotheca Math., (3), 8, 1907–8, 121–131.

and solved the problem when $g = 6$. Fermat (Oeuvres, III, 265) pointed out the probable origin of Vieta's unnecessary assumption on g. Let the triangle be formed from ax^2, a; its area $x^2a^4(x^4 - 1)$ increased by $5z^2$ shall give a square. Since 5 is a sum of two squares, we can determine y so that $5y^2 - 1 = \square$. Take $y = x + 1$; then $x^4 - 1 + 5y^2$ can readily be made a square. But Vieta did not observe that the problem can be solved when $x^4 - 1$ is replaced by $1 - x^4$ since we can solve $gy^2 + 1 = \square$. Fermat found the triangle (9/3, 40/3, 41/3) whose area 20 increased by 5 gives 5^2.

The history of the theorem that the area of a rational right triangle is never a square or double a square is given in Ch. XXII, where are given Bachet's and Vieta's comments on the problem to find a right triangle with a given area.

Fermat[80] stated that the area of the right triangle with the sides 2896804, 7216803, 7776485 is of the form $6u^2$; likewise for the triangle with the sides 3, 4, 5. E. Lucas[81] obtained these triangles and that with the sides 49, 1200, 1201 and area $6(70)^2$. He noted that the area of a right triangle is never a square, nor the double, triple or quintuple of a square.

Fermat's problem to find three right triangles the sum of whose areas by twos are sides of a right triangle was solved by Gillot at the request of Descartes.[82] The triangles

$$\left(\frac{24}{5}, \frac{35}{12}, \frac{337}{60}\right), \qquad \left(\frac{8}{3}, \frac{21}{2}, \frac{65}{6}\right), \qquad \left(12, \frac{7}{2}, \frac{25}{2}\right)$$

have the areas 7, 14, 21, whose sums by twos are the sides 35, 28, 21 of a right triangle. Gillot gave also the areas 15, 30, 45 and 7 more sets.

MISCELLANEOUS PROBLEMS INVOLVING THE AREA AND OTHER ELEMENTS.

In an early Greek manuscript[79] there occurs the problem to find the integral legs a, b and hypotenuse c of a right triangle such that the sum of the area T and perimeter $2s$ is a given number A. The solution given for $A = 8 \cdot 35$, $6 \cdot 45$, $5 \cdot 20$, $5 \cdot 18$ is made clear if we introduce the radius r of the inscribed circle, whence $T = rs = ab/2$, $r + s = a + b$, $c = s - r$. Separate A into two factors s, $r + 2$ such that $(r + s)^2 - 8rs$ is a perfect square n^2. Then $2a$, $2b = r + s \pm n$. Cf. E. Bahier,[62] pp. 190–9.

Diophantus VI, 6–9 relate to right triangles whose areas increased or diminished by one leg or by the sum of both legs shall be a given number g. To solve the first two problems, Fermat formed the triangle from g, 1 and divided the sides by $g + 1$ or $g - 1$; he enunciated the problems to find a right triangle such that one leg or the sum of the legs diminished by the area is a given number. Cf. E. Bahier,[62] pp. 170–190.

Diophantus, VI, 10 [11], found a right triangle $(28x, 45x, 53x)$ whose area increased [diminished] by the sum of the hypotenuse and one leg is 4.

---

[80] Oeuvres, III, 256, 348; comment on Diophantus V, 8 and Inventum Novum, I, § 38, Cf. A. Genocchi, Annali Sc. Mat. Fis., 6, 1855, 319–20.
[81] Bull. Bibl. Storia Sc. Mat., 10, 1877, 290.
[82] Oeuvres, II, 179; letter from Descartes to Mersenne, June 29, 1638. Cf. Oeuvres de Fermat, IV, 1912, 56.

Fermat asked that the sum of the hypotenuse and one leg, diminished by the area, shall be 4; the answer (17/3, 15/3, 8/3) is given in the Inventum Novum, III, 33 (Oeuvres de Fermat, III, 389). Bachet found a right triangle whose area increased (or decreased) by the hypotenuse is 4.

Diophantus VI, 13 relates to a right triangle (px, bx, hx) whose area increased by either leg is a square. Let $A = pb/2$. From $Ax^2 + bx = m^2x^2$, $x = b/(m^2 - A)$. Then $Ax^2 + px = \square$ requires that

$$pbm^2 + Ab(b - p) = \square.$$

As in VI, 12, we may choose (p, b, h) similar to (3, 4, 5) so that the greater leg b, $b - p$ and $p + A$ are all squares, say $b - p = m^2$. The preceding condition is thus satisfied. Fermat's method (Oeuvres, III, 267) yields an infinitude of triangles not similar to (3, 4, 5).

Diophantus, VI, 15 [17], gave a right triangle $(8x, 15x, 17x)$ whose area diminished [increased] either by the hypotenuse or one leg is a square. Fermat[83] required that on subtracting the area from the hypotenuse or one leg each difference be a square.

Diophantus, VI, 19 [20], required a right triangle the sum of whose area and hypotenuse is a square [cube], and perimeter a cube [square]. His solution and various related papers are considered in Ch. XX.

Diophantus, VI, 21 [22], required a right triangle the sum of whose area and one leg is a square [cube], and perimeter a cube [square]. Use a triangle given by the rule of Pythagoras,[1] after dividing its sides by $\alpha + 1$. The perimeter $4\alpha + 2$ is to be a cube. By the other condition, $2\alpha + 1 = \square$. But 8 is the only cube which is double a square. Hence $\alpha = 3/2$.

Diophantus VI, 23 [24][84] relates to a right triangle the sum of whose area and perimeter is a cube [square], and perimeter a square [cube]. Use a triangle given by the rule of Plato.[1] The perimeter $p = 2\alpha^2 + 2\alpha$ is a square for $\alpha = 2/(m^2 - 2)$. Then $\alpha(\alpha^2 - 1) + p$ and hence $2m$ is to be a cube for $2 < m^2 < 4$, which is the case when $m = 27/16$.

Bháscara[12] found a right triangle whose area equals the hypotenuse.

C. G. Bachet, at the end of book VI of his edition of Diophantus, added 22 problems. In the first 13, we are given the perimeter, or hypotenuse or area of a rational right triangle and seek the maximum or minimum of some specified function of the sides. In 14–18, we seek the sides, given the sum of the legs or perimeter p, or p and the area A, or p and the product of the sides. In 19, p and $p \pm A$ are to be squares. In 21 and 22, we are given p or A and the perpendicular from the right angle to the hypotenuse.

J. de Billy[85] found a right triangle in which one leg, the sum of the legs, and the excess of each leg over double the area are all squares. If x and $y = 1 - x$ are the legs, the conditions are that y and $x^2 + y^2$ be squares, as is true if $x = 40/49$. If we formulate the problem algebraically and then

---

[83] His solution is in Inventum Novum, I, 26, 40; Oeuvres, III, 341, 349.

[84] For VI, 24, see T. L. Heath, Diophantus, 1885, 236–7; 1910, 244–5; P. Mansion, Mathesis, (4), 4, 1914, 145–9.

[85] Inventum Novum, I, § 52; Oeuvres de Fermat, III, 359.

interpret as the hypotenuse the letter which stood for one leg, we have a new problem solved by A. Cunningham.[86]

Fermat[87] proposed that St. Martin find two right triangles whose areas are in a given ratio and such that the two legs of the larger triangle differ by unity.

Fermat[88] noted that if in (205769, 190281, 78320) we add the area to the square of the sum of the legs, we get a square.

Frenicle[89] stated the last result without comment; also that the sum of the area and hypotenuse of (17, 144, 145) is a square; while the first three right triangles in which the sum of the area and smaller leg is a square are (3, 4, 5), (16, 30, 34), (105, 208, 233).

J. de Billy[90] treated a large number of problems on rational right triangles. In the first 44, a prescribed multiple of the area when added to or subtracted from certain sides gives squares. The next five involve the perimeter. In Prob. 58, the cube of the sum of the hypotenuse and one leg when increased by a given multiple of the area shall be a cube, while 55–67 are analogous. In Prob. 68, the areas of $(30 \cdot 2^{3n}, 18 \cdot 2^{3n}, 24 \cdot 2^{3n})$ are seen to form a geometrical progression of ratio 2^6, while 69–73 are similar. The 120 problems of Ch. 2 do not involve areas, but make certain functions of the sides squares and cubes.

J. Ozanam[91] found that in the right triangle whose sides are the ratios of 2264592, 18325825 and 18465217 to 20590417 each side exceeds double the area by a square. This problem was proposed in obscure verse in the Ladies' Diary for 1728 as Question 133; a modified uninteresting problem was solved in 1729.

C. Wildbore[92] took x and $1 - x$ as the legs; they exceed the double area by x^2 and $(1 - x)^2$. Equating the hypotenuse h to $v(1 - x) + x$, we get $x = (1 - v^2)/(1 + 2v - v^2)$. The condition $h - (x - x^2) = \square$ becomes $1 + 4v^3 - v^4 = \square$. First, take $v = b/a$, $b = d - 3$, $a = d + 5$; then $4d^4 + \cdots = \square = (2d^2 - 260d - 2)^2$ for $d = 4223/66$, which yields Ozanam's answer. The next value of v is said to be 491050/555466, which gives $x = 8426546832/76616941657$. Elsewhere[93] he took

$$1 + 4v^3 - v^4 = (1 + nv^2)^2.$$

By the radical in the solution for v,

$$2(1 - n)(2 + n + n^2) = \square = 4r^2(1 - n)^2,$$

say. Solving for n, we see that $4r^4 + 12r^2 - 7 = \square$. Take $r = a/b$,

---

[86] Math. Quest. and Solutions, 3, 1917, 79–80.

[87] Oeuvres, II, 252, letter to Mersenne, Feb. 16, 1643.

[88] Oeuvres, II, 263 (260, 3°), letter to Mersenne, Sept. 1, 1643.

[89] Methode pour trouver la solution des problèmes par les exclusions, Ouvrages de Math., Paris, 1693; Mém. Acad. R. Sc. Paris, 5, 1666–99 (1676), éd. 1729, 56.

[90] Diophanti Redivivi, Lvgdvni, 1670, Pars Prior, pp. 1–302.

[91] Nouveaux élémens d'algèbre, 1702, 604.

[92] Ladies' Diary, 1772, 40–1, Quest. 638; T. Leybourn's Math. Quest. from Ladies' Diary, 2, 1817, 342–5; C. Hutton's Diarian Miscellany, 3, 1775, 356–7.

[93] C. Hutton's Miscellanea Math., London, 1775, 163–4; Leybourn's Math. Quest. L. D., 2, 1817, 342–5.

$a = d + 1$, $b = d - 1$ and equate the quartic in d to the square of $3 + 22d/3 - 43d^2/27$; thus $d = 202752/179200$, which gives the last answer. A longer analogous discussion led to the new value $r = 50929/46200$, which yields an answer involving numbers of ten digits.

T. Leybourn[94] took $x/(x + y)$ and $y/(x + y)$ as the legs, since each exceeds double the area by a square. Take $x = m^2 - n^2$, $y = 2mn$. Then the hypotenuse exceeds double the area by a square if $m^4 + 4mn^3 - n^4 = \square$. Take $m = 1 + v$, $n = 4$, and equate the quartic in v to the square of $v^2 - 130v + 1$, whence $v = 4223/66$. Or take $m = v + 5$, $n = v - 3$, and equate the quartic in v to $(2v^2 - 236v - 2)^2$, whence $v = 7619/176$.

Malézieux[95] proposed to find two right triangles the sum or difference of whose perimeters is a square; the difference of the areas a square; the difference of the least side of the first and the least side of the second equals the difference of the two largest sides of the first or of the two largest sides of the second, the difference being a cube; the difference of the largest leg of the first and the least leg of the second is a square; and the sum of the least side of the first and the medium side of the second is a square.

L. Euler[96] discussed the problem proposed by Fermat (on the margin of his Diophantus VI, 14): Find a right triangle such that each leg exceeds the area by a square. Euler denoted the legs by $2x/z$, y/z, where $x = ab$, $y = a^2 - b^2$. Subtract the area xy/z^2. Hence $2xz - xy$ and $yz - xy$ are to be squares. Let their product be the square of $xy - yzp/q$. Hence

$$z - x = x^2y(p - q)^2/k, \qquad 2z - y = x(2qx - py)^2/k, \qquad k = 2q^2x^2 - p^2yx.$$

It remains only to make k a square, say r^2x^2. Thus $x : y = p^2 : 2q^2 - r^2$. Taking the proportionality factor with z, we may set $x = p^2$, $y = 2q^2 - r^2$. Then $z = p^2 + (p - q)^2(2q^2 - r^2)/r^2$. The condition $4x^2 + y^2 = \square$ becomes $E \equiv 4p^4 + (2q^2 - r^2)^2 = \square$. Special solutions are obtained by setting $\sqrt{E} = 2p^2 \mp r^2$, $2p^2 \pm 2q^2$ or $r^2 + 2q^2 \pm 2p^2$. Returning in § 20 to the general case, Euler expressed $k = r^2x^2$ in the form

$$ab(a^2 - b^2) = \frac{a^2b^2}{p^2}(2q^2 - r^2) = 2t^2 - u^2.$$

Every product of primes 2, $8m \pm 1$ and a square is of the form $2t^2 - u^2$ and only such products. Moreover, if a product of two numbers whose g.c.d. is 1 or 2 is of the form $2t^2 - u^2$, each factor is. Hence a, b, $a + b$, $a - b$ must each be of the form $2t^2 - u^2$. Conversely, when this is the case, solutions of the initial problem can be readily found. Euler tabulated the permissible values $a < 200$ for each permissible $b < 100$, and gave formulas for p, q, r, z.

To find a right triangle whose area increased by the square of the hypotenuse is a square, J. Whitley[97] wrote $rs(r^2 - s^2) + (r^2 + s^2)^2 = a^2$

[94] Math. Quest. from Ladies' Diary, 1, 1817, 173–5.

[95] Éléments de Geométrie de M. le Duc de Bourgogne, par de Malézieux, 1722. Solved by E. Fauquembergue, Sphinx-Oedipe, 2, 1907–8, 15–16.

[96] Novi Comm. Acad. Petrop., 2, 1749, 49; Comm. Arith., I, 62.

[97] The Gentleman's Math. Companion, 2, No. 10, 1807, 69.

and took $r = t - 8s$, $a = t^2 - mts + 61s^2$, and found $t = 3839s/488$, $r = -65s/488$. J. Wright took $a = r^2 + s^2 + \frac{1}{2}rs$ and found $r = -8s$, which does not give positive answers. Hence set $r = t - 8s$.

"Calculator"[98] found three right triangles of equal perimeters and areas in arithmetical progression. The areas are proportional to the radii r of the inscribed circles; for the sides $2amn$, $a(m^2 \pm n^2)$, $r = an(m - n)$. A long computation yielded triangles all of whose sides involve eight digits:

$$(18601944,\ 13951458,\ 23252430), \qquad (18559223,\ 13999464,\ 23247145),$$
$$(18515584,\ 14048388,\ 23241860).$$

W. Wright[99] found a right triangle whose perimeter is a square and area a cube by taking $m^2 \pm n^2$, $2mn$ as the sides. Let the perimeter equal q^2m^2, whence $m = 2n/(q^2 - 2)$. Then the area is a cube if

$$8n - 2n(q^2 - 2)^2 = s^3,$$

which gives n. "Epsilon" took $p(m^2 \pm n^2)$, $2pmn$ as the sides. The perimeter is a square if $p = 2m(m + n)$. The area is a cube if $4n(m - n)$ is, whence either n is a cube and the double of $m - n$ is a cube or vice versa.

To find a right triangle the sum of whose sides equals the area, many solvers[100] noted that $2s^2 + 2rs = rs(s^2 - r^2)$ implies $-2 = r^2 - sr$. The root r involves the radical $\sqrt{s^2 - 8}$, which is equated to $s - x$, giving $s = (8 + x^2)/(2x)$. For integral solutions we have $x < s$, whence $x = 4$, $s = 3$, $r = 2$ or 1 and the only triangles are $(13, 5, 12)$, $(10, 8, 6)$.

J. Baines,[101] to find two right triangles the differences between whose bases, perpendiculars, hypotenuses, perimeters and diameters of inscribed circles are all squares, and difference of areas a cube, took $25m^2 - n^2$, $10mn$ and $25m^2 + n^2$ as base, perpendicular and hypotenuse of one, and $9m^2 - n^2$, $6mn$, $9m^2 + n^2$ for the other, so that we have only to make $4mn$ and $A = 32m^2 + 4mn$ squares and $B = 98m^3n - 2mn^3$ a cube. Take $mn = a^2$. Then $A = \square$ if $8a^2 + n^2 = \square = (2ar/s + n)^2$. Take $r = s = 1$, whence $n = a = m$. Then $B = 96a^4$ is a cube if $a = b^3/96$. G. Heald took the triangles $(10x^2, 24x^2, 26x^2)$ and $(6x^2, 8x^2, 10x^2)$. All but the last condition is satisfied identically. The difference $96x^4$ of the areas is a cube if $x = p^3/12$.

J. Davey[102] found a right triangle whose perimeter is a square p^2 such that p^3 equals the area. Take pr, ps, pt as the sides. Then $r = p - s - t$, $s = 2p/t$, and $r^2 = s^2 + t^2$, which gives $p = 2t(t - 2)/(t - 4)$.

Many[103] found the sides a, b and hypotenuse c of a right triangle such that a, $c + b$, $c - b$ are integral cubes, say p^3, m^3, n^3. Then $c^2 - b^2 = a^2$ gives $mn = p^2$.

---

[98] The Gentleman's Math. Companion, 4, No. 22, 1819, 861–4. Cf. Perkins.[104]

[99] Ibid., 5, No. 28, 1825, 371–3.

[100] Ladies' Diary, 1828, 34, Quest. 1465.

[101] Ladies' Diary, 1830, 37, Quest. 1500.

[102] The Lady's and Gentleman's Diary, London, 1841, 58 (Quest. 1416 of Gentleman's Diary, 1840).

[103] Ibid., 1845, 51–2, Quest. 1722.

G. R. Perkins[104] noted that the triangles (40, 30, 50), (45, 24, 51), (48, 20, 52) have equal perimeters and areas 600, 540, 480 in arithmetical progression.

V. J. Knisely[105] found the same result as had Perkins, by taking as the sides

$$(p^2 + 2pq)a, \qquad (2pq + 2q^2)a, \qquad (p^2 + 2pq + 2q^2)a,$$
$$(p^2 - q^2)b, \qquad 2pqb, \qquad (p^2 + q^2)b,$$
$$(p^2 - 4q^2)c, \qquad 4pqc, \qquad (p^2 + 4q^2)c.$$

The conditions reduce to

$$(p + 2q)a = pb, \qquad (p + q)a = pc, \qquad 2(p - q)b = pa + 2pc - 4qc.$$

Substitute into the third the values of b, c given by the first two conditions; we get $p = 4q$, whence $b = 6a/4$, $c = 5a/4$. For $q = 1$, $p = a = 4$, $c = 5$, $b = 6$, we get the answer cited. A. B. Evans gave a long discussion said to give the complete solution; but his numerical example involves very large numbers.

E. Lucas proposed and Moret-Blanc[106] solved the problems to find a right triangle such that the square of the hypotenuse increased or diminished by the area (or by double the area) is a square.

Lucas[107] showed that the method of descent leads to a complete solution of the second (double area) of the last two problems.

C. de Comberousse[108] discussed rational right triangles whose area and perimeter are equal. Eliminating z between $x^2 + y^2 = z^2$ and

$$x + y + z = xy/2,$$

we get $y = 4 + 8/(x - 4)$. Thus $x - 4$ is a divisor of 8, and the only solutions are $(x, y, z) = (5, 12, 13)$, $(6, 8, 10)$.

A. Holm[109] discussed a problem including the cases of Diophantus VI, 6–11, and the additions by Bachet and Fermat: Find a rational right triangle such that the sum of given multiples of the area and three sides shall be a given number. Taking $(x^2 \pm 1)/y$, $2x/y$ as the sides, the condition is

$$a\frac{x(x^2 - 1)}{y^2} + b\left(\frac{x^2 + 1}{y}\right) + c\left(\frac{x^2 - 1}{y}\right) + d\left(\frac{2x}{y}\right) = e.$$

The discriminant of this quadratic for y is a quartic function $Q(x)$ in which the coefficient of x^4 and the constant term are squares. There are many known methods of making $Q(x)$ a square.

Right triangles whose legs differ by unity.

A. Girard[109a] gave fourteen such triangles in which the least leg is 3, 20, 119, 696, 4059, 23660, 137903, 803760, $\cdots$, 31509019100.

---

[104] The Analyst, Des Moines, 1, 1874, 151–4. Cf. Calculator.[98]

[105] Math. Quest. Educ. Times, 20, 1874, 81–3.

[106] Nouv. Ann. Math., (2), 14, 1875, 510; (2), 20, 1881, 155–160.

[107] Bull. Bibl. Storia Sc. Mat., 10, 1877, 291–3.

[108] Algèbre supérieure, 1, 1887, 190–1.

[109] Proc. Edinburgh Math. Soc., 22, 1903–4, 45–8; Math. Quest. Educ. Times, (2), 10, 1906, 47–8.

[109a] L'arith. de Simon Stevin $\cdots$ par A. Girard, 1625, 629; Oeuvres, 1634, 158, col. 1.

From one right triangle $(x, x + 1, z)$ whose legs are consecutive integers, Fermat[110] deduced the second triangle $(X, X + 1, Z)$, where

$$X = 2z + 3x + 1, \qquad Z = 3z + 4x + 2.$$

For example, we have the series $(3, 4, 5)$, $(20, 21, 29)$, $(119, 120, 169)$, $\cdots$. The alternate triangles give solutions of the problem to find right triangles whose least side differs from the other two sides by squares. He noted later (pp. 232–3) that such a triangle is formed from $r^2 + s^2$, $2s(r - s)$.

Fermat[111] noted that the sixth such triangle is $(23660, 23661, 33461)$. From the first such triangle $(3, 4, 5)$, we get the second by taking the double (viz., 24) of the sum of the three sides and subtracting separately the legs and adding the hypotenuse.

J. Ozanam[112] gave the first six such triangles. If one is formed (Diophantus[7]) from m, n, where $m > n$, the next is formed from m, $2m + n$. In the edition by J. E. Montucla, 1, 1790, 48, the triangle is formed from any two consecutive terms of $1, 2, 5, 12, 29, 70, \cdots, k$, where k is such that one of the two numbers $2k^2 \pm 1$ is a square. The same rule was given by Grüson.[22]

C. Hutton[113] noted that, if p_r/q_r is the rth convergent to $\sqrt{2}$, then $p_r p_{r+1}$ and $2q_r q_{r+1}$ are consecutive integers the sum of whose squares is a square q_{2r+2}^2.

Du Hays[114] gave triangles the difference of whose legs is 1 (or 7).

L. Brown[115] gave the first six and the eleventh such triangles.

G. H. Hopkins and M. Jenkins[116] reduced the problem to $x^2 - 2y^2 = \pm 1$, and gave recursion formulas for the solutions. A. B. Evans used the continued fraction for $\sqrt{2}$. Cf. Moret-Blanc[154] of Ch. XII.

Judge Scott[117] gave the first eight and the eleventh.

A. Martin[118] employed the legs $\frac{1}{2}(x \pm 1)$, whence $x^2 - 2y^2 = -1$, and the odd convergents x_n/y_n to the continued fraction for $\sqrt{2}$. Thus $x_n = 6x_{n-1} - x_{n-2}$ and likewise for the y's. Also

$$2x_n, \quad 2\sqrt{2}\,y_n = (1 + \sqrt{2})^{2n+1} \pm (1 - \sqrt{2})^{2n+1}.$$

The eightieth such triangle is found.

T. T. Wilkinson stated and J. Wolstenholme[119] proved a rule equivalent to a recursion formula for the solutions of $x^2 - 2y^2 = 1$.

D. S. Hart[120] took x and $x + 1$ as the legs. Then

$$2x^2 + 2x + 1 = \square = (xp/q - 1)^2$$

---

[110] Oeuvres, II, 224–5. Reproduced in Sphinx-Oedipe, 7, 1912, 103–4.

[111] Oeuvres, II, 258; letter to St. Martin, May 31, 1643; reproduced, Sphinx-Oedipe, 7, 1912, 104.

[112] Recreations Math., 1, 1723; 1724; 1735, 51; etc. (first ed., 1696).

[113] English transl. of Ozanam's Recreations, 1, 1814, 46.

[114] Jour. de Math., 7, 1842, 331.

[115] Math. Monthly (ed., Runkle), Cambridge, Mass., 2, 1860, 394.

[116] Math. Quest. Educ. Times, 12, 1869, 104–6.

[117] Of commensurable right-angled triangles . . . , Bucyrus, Ohio, 1871, 23 pp.

[118] Math. Quest. Educ. Times, 14, 1871, 89–91; 16, 1872, 107; 19, 1873, 89; 20, 1874, 21, 42–4.

[119] Ibid., 20, 1874, 97–99.

[120] Ibid., 63–4.

gives $x = (2pq + 2q^2)/d$, where $d = p^2 - 2q^2$. He made $d = \pm 1$ by use of the theory of Pell's equation.

A. Martin[121] gave the nth triangle for $n = 80$ and 100.

P. Bachmann[122] proved that the only integral solutions of $x^2 + y^2 = z^2$ in which $z > 0$, while x and y are consecutive integers, are those given by

$$x + y + z\sqrt{2} = (1 + \sqrt{2}) \cdot (3 + 2\sqrt{2})^k \quad (k = 0, 1, 2, \cdots).$$

Several writers[123] obtained the first six triangles.

R. W. D. Christie[124] noted that the solution of $x^2 + (x + 1)^2 = y^2$ in integers is

$$x = 2_0 + 2_1 + \cdots + 2_{2r-1}, \qquad y = 2_{2r},$$

where 2_r is the simple continuant of order r all of whose diagonal elements are 2. This was proved by T. Muir,[125] who cited Fermat's[110] rule.

A. Martin[126] noted various methods. The first three methods are based on the solution of $2k^2 \pm 1 = \square$ [Ozanam,[112] Hutton,[113] Bachmann[122]]. Fermat's method was used to compute a table (p. 322) of the first forty such triangles.

A. Lévy[127] found when two of the numbers (1) are consecutive. Evidently $z - y = 2n^2 \mp 1$. Next, $z - x = (m - n)^2 = 1$ for $m = n + 1$. Finally, $y - x = \pm 1$ gives $(m - n)^2 - 2n^2 = \pm 1$. Write $(1 - \sqrt{2})^p$ in the form $a - b\sqrt{2}$; then a, b are integral solutions of $a^2 - 2b^2 = (-1)^p$, and all solutions of $u^2 - 2v^2 = \pm 1$ are said to be obtained in this way by using all integral values of p. Or we may compute the solutions of the latter by the recursion formulas of G. Fontené[284] of Ch. XII. We get

(3, 4, 5), (21, 20, 29), (119, 120, 169), (697, 696, 985), (4060, 4059, 5741).

G. A. Osborne[128] discussed the problem. Cf. Barisien[100] of Ch. IX. Several[129] made use of $x^2 - 2y^2 = -1$. F. Nicita[130] employed recurring series.

Right triangles the difference d or sum of whose legs is given.

Frenicle[131] stated that every number is the difference of the legs in an infinitude of ways, every prime $8n + 1$ or product of such primes is the difference of the legs of an infinitude of primitive triangles. To find all triangles with $d = 7$, start with (5, 12, 13) formed from 3, 2, and take that formed from 3, $2 \cdot 3 + 2$, etc. A second series is found similarly from (8, 15, 17), formed from 4, 1. He discussed right triangles the sum of whose legs is given.

---

[121] The Analyst, Des Moines, 3, 1876, 47–50; Math. Visitor, 1, 1879, 56, 122 (erroneous values for $n = 5$, 6 occur on pp. 55–6).
[122] Zahlentheorie, 1, 1892, 194–6; Niedere Z., 2, 1910, 436.
[123] Amer. Math. Monthly, 4, 1897, 24–28.
[124] Math. Gazette, 1, 1896–1900, 394.
[125] Proc. Roy. Soc. Edinburgh, 23, 1899–1901, 264–7.
[126] Math. Magazine, 2, 1910, 301–24.
[127] Bull. de math. élémentaires, 15, 1909–10, 165–6.
[128] Amer. Math. Monthly, 21, 1914, 148–150.
[129] L'intermédiaire des math., 22, 1915, 139–144, 185–8.
[130] Periodico di Math., 32, 1917, 200–210.
[131] Oeuvres de Fermat, II, 235–6, 238–41, letter to Fermat, Sept. 6, 1641.

Fermat[132] noted that $d = 7$ for $(5, 12, 13)$ and $(8, 15, 17)$; from these we get all by his[111] rule.

Frenicle[89] examined the 16 triangles with hypotenuses < 100 and found that $d = 1, 7, 7^2, 17, 23, 31, 41$, each of the form $8n \pm 1$. The triangles formed from n, m and m, $2m + n$ have the same difference of legs.[132a]

T. T. Wilkinson[133] would start with a solution of $a^2 + b^2 = c^2$ and form $\alpha = a + c$, $\beta = b + c$, $\gamma = a + b + c$; repeat the process; we obtain

$$a' = \alpha + \gamma = 2a + b + 2c, \qquad b' = \beta + \gamma = a + 2b + 2c,$$
$$c' = \alpha + \beta + \gamma = 2a + 2b + 3c,$$

which are sides of a new triangle with $a' - b' = a - b$. H. S. Monck (pp. 20–21, 76) failed in his attempt to prove that if we start with $(3n, 4n, 5n)$ and apply the process repeatedly we obtain all triangles with the same difference of legs. J. W. L. Glaisher (p. 54) noted that the proof is inadequate. Proof was given by S. Tebay (p. 99) and P. Mansion.[134]

T. Pepin[135] considered the problem of Fermat (Oeuvres, II, 231) to find the number of right triangles the sum of whose legs is a given number A. To the resulting condition $x^2 - 2y^2 = A$ we may apply the theory of quadratic forms and show that, if $A = a^\alpha \cdots c^\gamma$, where $a, \cdots, c$ are primes $8l \pm 1$, the total number of primitive triangles whose sum of legs is A is $\frac{1}{2}\{(2\alpha + 1)\cdots(2\gamma + 1) - 1\}$.

J. H. Drummond and M. A. Gruber[136] found solutions when d is given. Several[137] treated the case $d = 7$.

E. Bahier,[62] pp. 72–120, treated the problem at length by recurring series.

TWO RIGHT TRIANGLES WITH EQUAL DIFFERENCES OF LEGS, AND LARGER LEG OF ONE EQUAL TO THE HYPOTENUSE OF THE OTHER.

Frenicle[138] proposed the problem to J. Wallis. Wallis (Aug., 1661) took two overlapping triangles BAC and BCE with the respective hypotenuses $BC = 5 + x$ and BE. Take $BA = 5 - x$. Then $BC^2 - BA^2 = 20x$ is a square if $5x$ is; take $5 = ba^2$, $x = be^2$. Then

$$BC = ba^2 + be^2, \qquad BA = ba^2 - be^2, \qquad AC = 2bae.$$

On AB lay off $AD = AC$; on BC lay off $B\delta = BD$. Since

$$BC - CE = AB - AC = BD$$

---

[132] Oeuvres, II, 258–9; letter to St. Martin, May 31, 1643.
[132a] Oeuvres de Fermat, II, 235–7.
[133] Math. Quest. Educ. Times, 20, 1874, 20, 100. G. H. Hopkins, p. 22. On the proof sheets, E. B. Escott noted that " this process can be applied to other triangles than right-angled triangles. Under this transformation, $c^2 - 2ab$ as well as $a - b$ is invariant. Cf. Dickson.[34]"
[134] Mathesis, (3), 6, 1906, 113.
[135] Mem. Pont. Accad. Nuovi Lincei, 8, 1892, 84–108; extract, Oeuvres de Fermat, 4, 1912, 205–7; cf. 253.
[136] Amer. Math. Monthly, 9, 1902, 230, 292–3.
[137] Math. Quest. Educ. Times, (2), 7, 1905, 88–9.
[138] Cf. C. Henry, Bull. Bibl. Storia Sc. Mat. Fis., 12, 1879, 695; 13, 1880, 446; 17, 1884, 351–2.

by hypothesis, $CE = \delta C = 2be^2 + 2bae$.   Hence

$$BE^2 = BC^2 + CE^2 = b^2 f, \qquad f = a^4 + 5e^4 + 6a^2e^2 + 8ae^3.$$

It remains to make f a square, which Wallis suspected to be impossible. Frenicle (Dec. 20, 1661) took $a = 2$, $e = 4$, whence $f = 52^2$ [whereas we desire $a > e$]. Fermat[139] formed the first triangle from $N + 1$ and 2. Then the legs of the second triangle are $N^2 + 2N + 5$ and $4N + 12$; by their sum of squares,

$$N^4 + 4N^3 + 30N^2 + 116N + 169 = \square = \left(13 + \frac{58}{13} N - N^2\right)^2,$$

say. Thus $N = -1525/546$. Hence we use as the first triangle that formed from $+\,979$ and $2 \cdot 546$. The resulting triangles are

$$(2150905, 2138136, 234023), \qquad (2165017, 2150905, 246792).$$

If we had used the sum of the legs instead of their difference, we would obtain the simpler solution $(1517, 1508, 165)$ and $(1525, 1517, 156)$.

T. Pepin[140] noted that the initial problem is equivalent to

$$(3) \qquad x^2 + y^2 = z^2, \qquad u^2 + v^2 = x^2, \qquad u - v = x - y > 0.$$

We have $u, v = a^2 - e^2, 2ae$; $x = a^2 + e^2$. According as u is odd or even, $y = 2e(a + e)$ or $2a(a - e)$. Then the first condition becomes

$$z^2 = a^4 + 5e^4 + 6a^2e^2 + 8ae^3 \qquad \text{or} \qquad z^2 = 5a^4 + e^4 + 6a^2e^2 - 8a^3e,$$

according as the larger leg of the smaller triangle is odd or even. Contrary to Frenicle's solution $a = 2$, $e = 4$, the geometry requires $a > e$. But we can satisfy the first condition by taking $x = d(m^2 - n^2)$, $y = 2dmn$, $z = d(m^2 + n^2)$, where $d = 1$ if x, y are relatively prime, and $d = 2$ if x, y are even, while m, n are relatively prime and one is even. Then $2dmn = 2e(a + e)$, which is completely solved by

$$(4) \quad m = \alpha\beta, \quad n = hk, \quad e = \beta k, \quad a + e = \alpha h, \quad \text{or} \quad e = \alpha h, \quad a + e = 2\beta k,$$

according as $d = 1$ or $d = 2$, where α, β, h, k are relatively prime in pairs, the first three being odd and k even. Whether $d = 1$ or $d = 2$,

$$d(m^2 - n^2) = a^2 + e^2$$

gives

$$(5) \qquad k^2(h^2 + 2\beta^2) - 2\alpha\beta hk + \alpha^2(h^2 - \beta^2) = 0.$$

Solving this for k/α or h/β, and making the radicals rational, we get $2\beta^4 - h^4 = \square$, $\alpha^4 - 2k^4 = \square$, which have been completely solved by Lagrange[54] of Ch. XXII, so that we know all solutions under a given limit. Then (4) give solutions of the proposed problem. We may also solve (5) by a method equivalent to that of Euler[143-145] of Ch. XXII. Set $h/\beta = \xi$, $k/\alpha = \eta$; then

$$(\xi^2 + 2)\eta^2 - 2\xi\eta + \xi^2 - 1 = 0.$$

---

[139] Inventum Novum of de Billy, in S. Fermat's Diophanti Alex. Arith., 1670, 34–35. Oeuvres de Fermat 3, 1896, 393–4; 4, 1912, 132.

[140] Atti Accad. Pont. Nuovi Lincei, 33, 1879–80, 284–9; extract in Oeuvres de Fermat, 4, 1912, 219–220.

Call η, η' the values corresponding to the same ξ; and ξ, ξ' the values corresponding to the same η.   Hence

$$\eta + \eta' = \frac{2\xi}{\xi^2 + 2}, \qquad \xi + \xi' = \frac{2\eta}{\eta^2 + 1}.$$

Hence all solutions follow from the primitive solution $\eta = 0$, $\xi = 1$:

$$\xi = 1, \ \eta = \frac{2}{3}; \qquad \xi_1 = -\frac{1}{13}, \ \eta_1 = -\frac{84}{113}; \qquad \xi_2 = -\frac{1343}{1525}, \ \cdots$$

The second set is said to furnish the least positive solution of (3):

$$x = 2150905, \quad y = 246792, \quad z = 2165017, \quad u = 2138136, \quad v = 234023.$$

M. Martone[141] satisfied the first equation (3) by taking $x = 2ab$, $y = a^2 - b^2$, $z^2 = a^2 + b^2$.   From the square of the third given equation, we get $x^2 - 2uv = z^2 - 2xy$.   Thus we have uv and $u - v$ expressed in terms of a, b.   Thus

$$2v = a^2 - 2ab - b^2 \pm r, \qquad r^2 = 8a^2b^2 - (2ab - a^2 + b^2)^2.$$

Taking $a = 5b$, we get $r^2 = 4b^4$, $\qquad (v, u) = (6b^2, - 8b^2)$ or $(8b^2, - 6b^2)$.

MISCELLANEOUS PROBLEMS INVOLVING THE SIDES, BUT NOT THE AREA.

Diophantus V, 25 relates to $x_1^2 x_2^2 x_3^2 - x_i^2 = \square$ for $i = 1, 2, 3$.   A solution will be $x_i = tb_i/h_i$ if three right triangles (p_i, b_i, h_i) are found such that $h_1 h_2 h_3 = t^2 b_1 b_2 b_3$.   He took $(3, 4, 5)$ as the first triangle and $b_3 = 4$.   From the triangles $(13, 5, 12)$ and $(5, 3, 4)$, the ratio of whose areas is $5 : 1$, we can find two triangles such that the product of the hypotenuse and base of one is 5 times that of the other.   Indeed,[142] he knew how to deduce from a right triangle (α, β, γ) a triangle (a, b, c) with $ac = \beta\gamma/2$, where α and a are the hypotenuses.   He took $a = \alpha/2$, $b = (\beta^2 - \gamma^2)/(2\alpha)$, $c = \beta\gamma/\alpha$. From $(13, 5, 12)$ and $(5, 3, 4)$ he thus deduced $(6\frac{1}{2}, 119/26, 60/13)$ and $(2\frac{1}{2}, 7/10, 12/5)$, the product of the hypotenuse and final leg being 30 and 6, respectively.   Fermat[143] gave two such triangles for which the ratio in question is $5 : 1$, the sides being numbers of 10 and 11 figures (Oeuvres, I, 325; III, 263).

Fermat,[144] to find two right triangles (p, b, h), (p', b', h') for which $p - b = b' - h'$ and $b - h = p' - b'$, took three squares r^2, s^2, t^2 in arithmetical progression and formed the triangles from $r + s$, s and $s + t$, s. From $r = 1$, $s = 5$, $t = 7$, we get $(11, 60, 61)$, $(119, 120, 169)$.   We may also take $r = 7$, $s = 13$, $t = 17$.

---

[141] Sopra un problema di analisi indeterminata, Catanzaro, 1887.

[142] Restoration of the obscure text by J. O. L. Schulz, "Diophantus," 1822, 546–61.

[143] P. Tannery, Bull. Math. Soc. France, 14, 1885–6, 41–5 (reproduced in Sphinx-Oedipe, 4, 1909, 185–7), concludes that Fermat was aided by chance in obtaining his solution, which is not general and contains an error of sign. S. Roberts, Assoc. franç. av. sc., 15, II, 1886, 43–9, discussed the problem. Both papers are reprinted in Oeuvres de Fermat, 4, 1912, 168–180. This problem of Fermat's has been treated by A. Holm and A. Cunningham, Math. Quest. Educ. Times, (2), 11, 1907, 27–29; special cases by K. J. Sanjána and Cunningham, ibid., (2), 13, 1908, 24–26; E. Fauquembergue, l'intermédiaire des math., 24, 1917, 30–1; cf. 25, 1918, 130–1.

[144] Oeuvres, II, 225, letter to Frenicle, June 15, 1641.   Cf. II, 229, 232.

Saint-Martin asked how many ways 1803601800 is the difference of the [larger] sides of a right triangle whose least side differs from the other sides by squares. Fermat[145] replied that there are exactly 243 such triangles.

Fermat[146] asked for two right triangles such that the product of the hypotenuse and least leg of one shall have a given ratio to the corresponding product for the other triangle.

Under $2x^4 - y^4 = \square$ in Ch. XXII are discussed right triangles whose hypotenuse is a square and either the sum of the legs is a square or the least side differs by a square from each of the remaining sides.

Fermat[147] gave (156, 1517, 1525) in reply to Frenicle's question to find a right triangle in which the square of the difference of the legs exceeds the double of the square of the least leg by a square. A. Aubry[148] obtained an infinity of solutions by descent.

Frenicle[89] noted (pp. 71-8) that if the hypotenuse and perimeter of a right triangle are squares, the perimeter has at least 13 digits.

J. Ozanam[149] gave a rule to find a right triangle whose hypotenuse exceeds the larger leg by unity [Pythagoras[1]]. From the lengths of its legs form a new triangle; its hypotenuse is a square. He found right triangles whose base and hypotenuse are triangular numbers and altitude is a cube.

Wm. Wright[150] found a right triangle the sum of whose perimeter and square of any side is a square. Let the sides be ax, bx, cx, where $a^2 + b^2 = c^2$. Then $x^2 + px$, $x^2 + qx$, $x^2 + rx$ are made squares in the usual way (Ch. XVIII), where $p = s/a^2, q = s/b^2, r = s/c^2, s = a + b + c$. He and others[151] gave a similar treatment to find a right triangle such that the square of any side exceeds that side by a square.

Several[152] found a right triangle whose perimeter is a square, also the sum of the square of any side and the remaining two sides, also the sum of any side and the square of the sum of the remaining two sides. These seven conditions are satisfied if the sum of the sides is 1/4. Take $f(p^2 \mp q^2)$, $2fpq$ as the sides. Equating the sum to 1/4, we get f.

R. Tucker and S. Bills[153] found a right triangle with perimeter a square and diameter of the inscribed circle a cube [or vice versa]. Let the sides be $(p^2 \pm q^2)x$, $2pqx$. Then $2p(p + q)x = \square = r^2$, and the diameter $2q(p - q)x$ is to be a cube, say r^3/s^3. From the two values of x we get r in terms of q, s.

A. B. Evans[153a] found a right triangle with integral values for the sides a, b, c, diameter d of the inscribed circle and side s of the inscribed square having one angle coincident with the right angle of the triangle and having

[145] Oeuvres, II, 250, letter to Mersenne, Jan. 27, 1643.
[146] Oeuvres, II, 252, letter to Mersenne, Feb. 16, 1643.
[147] Oeuvres, II, 265, letter to Carcavi, 1644.
[148] L'intermédiaire des math., 20, 1913, 141-4.
[149] Recreations Math., 1, 1723, 1735, 52-5.
[150] The Gentleman's Math. Companion, London, 5, No. 24, 1821, 59-60.
[151] Ibid., 5, No. 27, 1824, 312-6.
[152] Ibid., 5, No. 25, 1822, 157-9.
[153] Math. Quest. Educ. Times, 19, 1873, 82.
[153a] Ibid., 21, 1874, 103-4.

the opposite vertex on the hypotenuse c, and such that $d + s$ is a square. Take the sides to be the products of the numbers (1) by uv. Then $s = abc/(ab + c^2)$ equals uv times a fractional function of m, n, whose denominator is taken as u. Since $d = a + b - c = 2n(m - n)uv$, the condition $d + s = \square$ is of the form $Av = \square$ and holds if $v = A$.

S. Tebay[154] noted the existence of an infinitude of pairs of right triangles with the same hypotenuse such that the differences between the hypotenuse and the legs are a square and double a square.

G. de Longchamps[154a] stated and Svechnikoff proved that $x^2 = y^2 + z^2$ has an infinitude of solutions for which $x + y$ is a biquadrate.

Several[155] found right triangles with the base 105, and two right triangles with the same base which is a mean proportion between the two perpendiculars.

To find any number of dissimilar rational triangles of equal perimeter, R. W. D. Christie[156] multiplied the sides of special triangles by suitable common factors, while A. Cunningham employed (1) and solved

$$m(m + n) = \text{const.}$$

A. Gérardin[157] noted that, to find two right triangles having the same sum of squares of the hypotenuse and one leg, we have to solve

$$(x^2 + y^2)^2 + (2xy)^2 = (\alpha^2 + \beta^2)^2 + (2\alpha\beta)^2,$$

and gave a solution in which x, y, α, β are functions of the seventh degree of two parameters.

R. Janculescu[158] noted that the problem to find a right triangle with integral values for the sides and perpendicular from the right-angle leads to $1/x^2 + 1/y^2 = 1/z^2$. Thus $x^2 + y^2 = t^2$. Let d be the g.c.d. of $x = d\alpha$, $y = d\beta$, $t = d\gamma$. Then $z = \pm d\alpha\beta/\gamma$, so that d must be a multiple of γ.

E. Turrière[159] discussed right triangles each of whose sides is a sum of two squares, as $9 = 3^2$, $40 = 2^2 + 6^2$, $41 = 4^2 + 5^2$.

E. Bahier,[62] pp. 122–148, investigated right triangles with a given perimeter.

RIGHT TRIANGLE WITH A RATIONAL ANGLE-BISECTOR.

Diophantus, VI, 18, found a rational right triangle with the bisector of one acute angle rational. Let the bisector be $5N$, altitude $4N$, so that one segment of the base is $3N$. The other segment is taken to be $3 - 3N$. Then (by proportion) the hypotenuse is $4 - 4N$. Equating its square to $(4N)^2 + 3^2$, we get $N = 7/32$. Multiply all our numbers by 32. Then the sides are 28, 96, 100, and the bisector is 35.

C. G. Bachet[74a] in his commentary on the preceding noted that no rational right triangle has a rational bisector of the right angle.

---

[154] Math. Quest. Educ. Times, 55, 1891, 99–101.
[154a] Jour. de math. élém., 1892, 282.
[155] Amer. Math. Monthly, 5, 1898, 51–4, 277–9.
[156] Math. Quest. Educ. Times, (2), 14, 1908, 19–21.
[157] Sphinx-Oedipe, 5, 1910, 187.
[158] Mathesis, (4), 3, 1913, 119–20.
[159] L'enseignement math., 19, 1917, 247–252.

J. Kersey[66] (p. 143) took the right triangle with the rational sides

$$AC = p(p^2 + b^2), \qquad AB = p(p^2 - b^2), \qquad BC = p(2bp).$$

The bisector AD of angle A divides the base into two segments

$$CD = b(p^2 + b^2), \qquad BD = b(p^2 - b^2)$$

proportional to AC and AB. Since $AB : BD = p : b$, we have

$$AD = h(p^2 - b^2),$$

if h, p, b are sides of any rational right triangle.

Several[160] writers found a right triangle with a rational bisector of one acute angle.

E. Turrière[161] found a rational right triangle with rational interior and exterior bisectors of an acute angle.

Tables of right triangles with integral sides.

The tables are usually arranged according to the magnitude of the hypotenuse h or the area A.

An Arab manuscript[9] of 972 gave a brief table (see Ch. XVI).

J. Kersey, Elements of Algebra, Books 3, 4, 1674, 8, $h \leqq 265$.

J. C. Schulze, Sammlung Log., Trig. $\cdots$ Tafeln, Berlin, II, 1778, 308, gave the decimal values of $\tan \omega/2 = m/n$ for 200 pairs of relatively prime integers m, n each $\leqq 25$, $m < n$; also right triangles with an angle ω.

A. Aida[17] (1747–1817) listed the 292 primitive triangles with $h < 2000$.

Le père Saorgio, Mém. Acad. Sc. Turin, 6, années 1792–1800, 1801, 239–252, quoted a table of primitive right triangles from Schulze.

C. A. Bretschneider, Archiv Math. Phys., 1, 1841, 96, $h \leqq 1201$.

Du Hays, Jour. de Math., 7, 1842, 331–4, gave four tables each with 32 entries to illustrate the systematic tabulation of primitive right triangles, using (1) with m, n relatively prime, $m > n$. First, give to m the values 2, 3, $\cdots$ and to n the values $< m$ and prime to m, such that one of m, n is even. Second, take 1, 3, 5, $\cdots$ as the odd side and factor each into two factors $m \pm n$. Third, begin with the even side $2mn$. Fourth, take a sum of two squares as the hypotenuse.

A. Wiegand, Sammlung Trig. Aufgaben, Leipzig, 1852, 131 triangles and their angles.

D. W. Hoyt, Math. Monthly (ed., Runkle), Cambridge, Mass., 2, 1860, 264–5, $h < 100$.

E. Sang, Trans. Roy. Soc. Edinburgh, 23, III, 1864, 757, $h \leqq 1105$.

S. Tebay, Elements of Mensuration, London and Cambridge, 1868, 111–2, gave an incomplete table arranged according to area A, the largest A, 863550, being an error for 934800. Reprinted by G. B. Halsted, Metrical Geometry, 1881, 147–9.

H. Rath, Archiv Math. Phys., 56, 1874, 188–224, used formulas [due to de Lagny[18]] to form a double-entry table, and noted an error by Berkhan.[23]

---

[160] Amer. Math. Monthly, 7, 1900, 83–5.
[161] L'enseignement math., 18, 1916, 407–8.

W. A. Whitworth, Proc. Lit. Phil. Soc. Liverpool, 29, 1875, 237, $h < 2500$.

Whitworth and G. H. Hopkins, Math. Quest. Educ. Times, 31, 1879, 67–70; D. S. Hart, Math. Visitor, 1, 1880, 99, forty triangles with

$$h = 5 \cdot 13 \cdot 17 \cdot 29.$$

N. Fitz, Math. Magazine, 1, 1884, 163, primitive with $h < 500$.

G. B. Airy, Nature, 33, 1886, 532, $h < 100$.

A. Tiebe, Zeitschr. Math. Naturw. Unterricht, 18, 1887, 178, 420, solved $a^2 + x^2 = h^2$ by setting $h = x + y$, whence $2x = a^2/y - y$, so that y is to be chosen as a divisor of a^2 ($a > 2$) such that the difference is even. Whence he constructed a table with $h < 100$. Cf. T. Meyer, *ibid.*, 36, 1905, 339.

H. Lieber and F. von Lühmann, Trig. Aufgaben, ed. 3, Berlin, 1889, 287–9, gave the 131 primitive triangles with $h < 999$.

P. G. Egidi, Atti Accad. Pont. Nuovi Lincei, 50, 1897, 126–7, $h \leqq 320$.

J. Sachs, Tafeln zum Math. Unterricht, Wiss. Beilage zum Jahresbericht Gym. Baden-Baden, 1905, $h < 2000$; $2000 < h < 5000$, h a product of primes $4n + 1$; one side < 500.

J. Gediking,[42] $h < 1000$.

A. Martin, Math. Mag., 2, 1910, 301–324 (preface, 2, 1904, 297–300), tabulated the values of $p^2 \pm q^2$, $2pq$ and area $A = pq(p^2 - q^2)$ for $p \leqq 65$, $q < p$, q prime to p, q even if p is odd. Omitting the entries $p = 33$, $q = 22$, and $p = 35$, $q = 14$, we have 862 triangles of which 443 have $A \leqq 934800$ (the largest A of the 178 triangles in Tebay's table). There is a table of the sides $p^2 \pm q^2$, $2pq$ of triangles for which $p = q + 1 \leqq 157$ and those with $p \leqq 312$, $q = 1$, whence h exceeds a leg by 1 or 2 respectively.

P. Barbarin, l'intermédiaire des math., 18, 1911, 117–120, gave the 35 pairs of primitive triangles with the same $h < 1000$. A. Martin, *ibid.*, 19, 1912, 41, 134, noted the omission of one pair and stated that there are 41 pairs with $1000 < h < 2000$.

A. Martin, Proc. Fifth Internat. Congress Math., 2, 1912, 40–58, gave the primitive triangles with $h < 3000$, noting two omissions by Sang. He listed many sets of k ($k \leqq 15$) triangles whose h's are consecutive integers; also sets of three triangles whose h's are sides of a right or scalene triangle. A product of n distinct primes $4m + 1$ is the hypotenuse of $(3^n - 1)/2$ different right triangles, only 2^{n-1} of which are primitive.

W. Könnemann, Rationale Lösungen Aufgaben, Berlin, 1915, $h < 1000$ (adverse review, Zeitschrift Math. Naturw. Unterricht, 46, 1915, 390).

E. Bahier,[62] pp. 255–9, tabulated the primitive triangles with a leg $\leqq 300$.

On systems of equations including $x^2 + y^2 = z^2$ see papers 76, 77, 80, 46, 84, 89, 139, 140 of Ch. XVI; 5 of Ch. XVII; 51, 146 of Ch. XIX; 354, 357, 360, 362, 366, 369–71, 436 of Ch. XXI; 109, 113, 313 of Ch. XXII; 207 of Ch. XXIII.

PAPERS NOT AVAILABLE FOR REPORT.

G. M. Pagnini, Collezione d' Opuscoli Sc., Firenze, 3, 1807, 3–24; Giornale di Fisica, Chimica e Storia Nat., Pavia, 3, 1810, 193–207. [Series of rational right triangles.]

Gruhl, Die Aufstellung Pythagoreischer Zahlen, Blätter Fortbildung d. Lehrers u. d. Lehrerin, Berlin, 4, 1911, 998–1000.

CHAPTER V.

TRIANGLES, QUADRILATERALS AND TETRAHEDRA WITH RATIONAL SIDES.

RATIONAL OR HERON TRIANGLES.

Heron of Alexandria gave the well known formula for the area of a triangle in terms of the sides and noted that when the sides are 13, 14, 15, the area is 84. A triangle with rational sides and rational area is called a rational triangle or Heron triangle.

Brahmegupta[1] (born 598 A.D.) noted that, if a, b, c are any rational numbers,

$$\frac{1}{2}\left(\frac{a^2}{b} + b\right), \qquad \frac{1}{2}\left(\frac{a^2}{c} + c\right), \qquad \frac{1}{2}\left(\frac{a^2}{b} - b\right) + \frac{1}{2}\left(\frac{a^2}{c} - c\right)$$

are sides of an oblique triangle [whose[2] altitudes and area are rational and which is formed by the juxtaposition of two right triangles with the common leg a].

S. Curtius[2a] proposed the following question: Three archers A, B, and C stand at the same distance from a parrot, B being 66 feet from C, B 50 feet from A, and A 104 feet from C; if the parrot rises 156 feet from the ground, how far must the archers shoot to reach the parrot? He noted that they stand at the vertices of a triangle the radius of whose circumscribed circle is 65 feet, while the parrot is 156 feet above its center. Since $65^2 + 156^2 = 169^2$, each archer is 169 feet from the parrot. It is stated to be difficult to explain why the radius turns out to be an integer. Cf. Gauss.[14a] [The triangle is rational since its area is $2^3 \cdot 3 \cdot 5 \cdot 11 = 1320$.]

C. G. Bachet,[3] in his comments on Diophantus VI, 18, treated several problems, the second of which is to find a triangle with rational sides and a rational altitude (and hence a Heron triangle). Taking a right triangle ADC with the sides 10, 8, 6, he found $BD = N$ such that $N^2 + 8^2$ shall be the square of a rational number (AB). Assuming first that angle BAC is acute, so that $DC : AD < AD : BD$, we must have $6N < 64$, whence $N < 32/3$. Let $N^2 + 8^2$ be the square of $8 - xN$; then

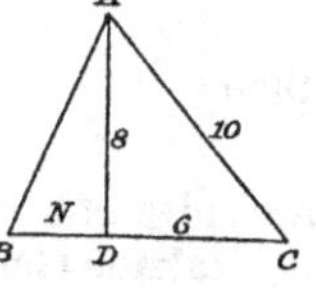

$$\frac{16x}{x^2 - 1} = N < \frac{32}{3}, \qquad x = \frac{x^2 - 1}{16} N < \frac{2}{3}(x^2 - 1), \qquad 3x + 2 < 2x^2,$$

---

[1] Brahme-Sphut'a-Sidd'hánta, Ch. 12, Sec. 4, §34 Algebra with Arith. and Mensuration, from the Sanscrit of Brahmegupta and Bháscara, transl. by H. T. Colebrooke, London, 1807, 306.

[2] E. E. Kummer. Jour. für Math., 37, 1848, 1.

[2a] Tractatus geometricus . . . , Amsterdam, 1617. Quoted by A. G. Kästner, Geschichte der Math., III, 294.

[3] Diophanti Alex. Arithmeticorum . . . Commentariis . . . Avctore C. G. Bacheto, 1621, 416. Diophanti Alex. Arithmeticorum, cum Commentariis C. G. Bacheti & Observationibus D. P. de Fermat (ed., S. Fermat), Tolosae, 1670, 315.

whence $x > 2$. Taking $x = 5$, we have
$$N^2 + 8^2 = (8 - 5N)^2, \qquad N = \tfrac{10}{3},$$
and the sides are 10, $9\tfrac{1}{3}$, $8\tfrac{2}{3}$, while the altitude is 8.

If BAC is oblique, $N > 32/3$. He took
$$N^2 + 8^2 = \left(8 - \frac{3}{2}N\right)^2, \qquad N = \frac{96}{5}.$$

Bachet's second method of solution is of greater importance, since it consists in juxtaposing two rational right triangles having a common side AD. Take as the latter any number, as 12. Seek two squares such that the sum of each and 12^2 is a square: $35^2 + 12^2 = 37^2$, $16^2 + 12^2 = 20^2$. Hence by juxtaposition, we get a rational triangle with the sides 37, 20, $35 + 16 = 51$, and altitude 12. Using the first relation with $9^2 + 12^2 = 15^2$ or $5^2 + 12^2 = 13^2$, we get the rational triangle $(37, 15, 35 + 9)$ or $(37, 13, 35 + 5)$.

F. Vieta[4] started with a given right triangle with legs B, D and hypotenuse Z, and formed (Diophantus[7] of Ch. IV) a second right triangle from $F + D$ and B, having therefore the altitude $A = 2B(F + D)$, and multiplied its sides by D, and the sides of the given triangle by A. Juxtaposing the resulting two triangles with the common altitude AD, we obtain a rational triangle with the sides AZ, $D(F + D)^2 + B^2D$, $D(F + D)^2 - B^2D + BA$, whose angle at the vertex is acute or obtuse according as $F < Z$ or $F > Z$.

Frans van Schooten[5] used the juxtaposition of right triangles.

The Japanese manuscript of Matsunago,[6] first half of the eighteenth century, started with any two right triangles with integral sides and multiplied the sides of each by the hypotenuse of the other and then juxtaposed the triangles. The sides below 1000 of the resulting oblique triangles were tabulated. Removing common factors, he obtained a table of primitive triangles. From Kurushima († 1757) he quoted the result that, if
$$n_3 : d_3 = d_1 d_2 - n_1 n_2 : n_1 d_2 + n_2 d_1,$$
then
$$n_1(n_2 d_3 + n_3 d_2), \qquad n_2(n_3 d_1 + n_1 d_3), \qquad n_3(n_1 d_2 + n_2 d_1)$$
are sides of a triangle with rational area.

Nakane Genkei[6a] in 1722 considered triangles whose sides are consecutive integers such that the perpendicular upon the longest side from the opposite vertex shall be rational. Denote the solutions $(3, 4, 5)$, $(13, 14, 15)$, $(51, 52, 53)$ and $(193, 194, 195)$ by (a_j, b_j, c_j), $j = 1, 2, 3, 4$. Then
$$a_{z+1} = 4a_z + 2 - a_{z-1},$$
and similarly for the b's and c's. Whether or not he made the induction complete does not, however, appear.

[4] Ad Logisticem Speciosam Notae Priores, Prop. 55, Opera Math., 1646. French transl. by F. Ritter, Bull. Bibl. Storia Sc. Mat., 1, 1868, 274–5.

[5] Exercitationum Math., Lugd. Batav., 1657, 426–432.

[6] Y. Mikami, Abh. Gesch. Math. Wiss., 30, 1912, 230–1.

[6a] D. E. Smith and Y. Mikami, A History of Japanese Mathematics, Chicago, 1914, 168.

L. Euler[7] noted that in any triangle with rational sides a, b, c, and rational area,

$$(1) \qquad a : b : c = \frac{(ps \pm qr)(pr \mp qs)}{pqrs} : \frac{p^2 + q^2}{pq} : \frac{r^2 + s^2}{rs},$$

and that every pair of sides are in the ratio of two numbers of the form $(\alpha^2 + \beta^2)/\alpha\beta$, since

$$a : b = \frac{r^2 + s^2}{rs} : \frac{x^2 + y^2}{xy}, \qquad \text{if} \qquad x = ps \pm qr, \qquad y = pr \mp qs,$$

whence

$$x^2 + y^2 = (p^2 + q^2)(r^2 + s^2).$$

The portion of Euler's paper containing his derivation of (1) is missing. It is probable that he employed Bachet's method of juxtaposing two right triangles, using those with the sides

$$2, \quad \frac{p^2 + q^2}{pq}, \quad \frac{p^2 - q^2}{pq}; \qquad 2, \quad \frac{r^2 + s^2}{rs}, \quad \frac{r^2 - s^2}{rs},$$

and obtaining (1) with the upper or lower signs according as the component triangles do not or do overlap.

J. Cunliffe[8] juxtaposed two right triangles with a common side $2rs = 2mn$ and hypotenuses $r^2 + s^2$, $m^2 + n^2$.

J. Davey[9] found three triangles with integral sides and areas having equal perimeters and areas in the ratio of $a = 2$, $b = 7$, $c = 15$. Let the triangles be AFB, BFC, CFD with collinear bases and the common altitude FE. Take

$$AF = \frac{r^2 + 1}{2r} \cdot v, \ BF = \frac{s^2 + 1}{2s} \cdot v, \ CF = \frac{t^2 + 1}{2t} \cdot v, \ DF = \frac{u^2 + 1}{2u} \cdot v, \ EF = v.$$

Then $AE = (r^2 - 1)v/(2r)$, etc.   By the equality of the perimeters,

$$\frac{s^2 - 1}{s} = r - \frac{1}{t}, \qquad \frac{t^2 - 1}{t} = s - \frac{1}{u}.$$

Then the conditions that the bases be proportional to a, b, c reduce to $(ar^2 + b)/r = (at^2 + b)/t$, whence $r = b/(at)$ (since $r \neq t$), and to $u = c/(bs)$. Eliminating r, u, s between our four relations in r, u, s, t, we get

$$t^4 - (d^2 + de + 2)t^2 + de + 1 = 0, \qquad d = \frac{c - b}{c}, \qquad e = \frac{b - a}{a}.$$

For $a = 2$, $b = 7$, $c = 15$, we get the rational root $t = 5/3$. Taking $v = 420$, we have $AF = 541$, $BF = 525$, $CF = 476$, $DF = 421$, $AB = 26$, $BC = 91$, $CD = 195$, perim. $= 1092$.

To find a triangle ABC with integral sides and area such that the distances from A, B, C to the center O of the inscribed circle shall be integers,

[7] Comm. Arith. Coll., II, 1849, 648, posthumous fragment. Same in Opera postuma, 1, 1862, 101.

[8] The Gentleman's Math. Companion, London, 3, No. 15, 1812, 398.

[9] Ladies' Diary, 1821, 36–7, Quest. 1364.

C. Gill[10] made a computation which (although not so stated) in effect
consists in finding three right triangles AOF, BOF, COE (see the figure

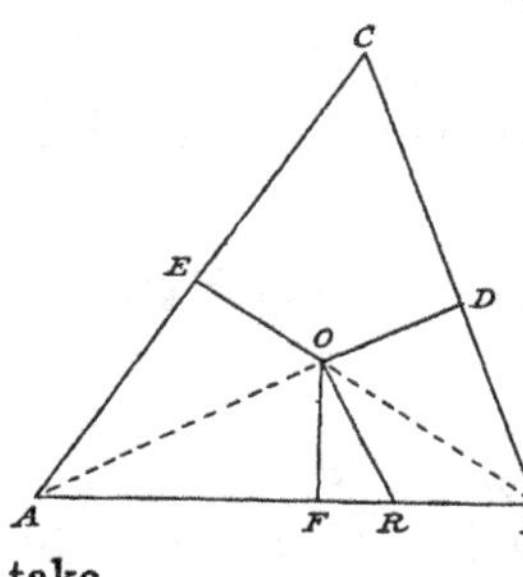

below) with integral sides such that $OF = OE$
$= OD = r$ is the radius of the inscribed circle, but omitted the condition that the sum of
their angles at O shall be two right angles.
If $AF = m$, $BF = n$, $CE = s$, this condition
is $mns = r^2(m + n + s)$. Thus his solution
fails.

A. Cook[11] gave the following solution. Draw
OR perpendicular to AO to meet AB at R.
[The sides of any rational right triangle are
proportional to $r^2 \pm a^2$, $2ra$.] Hence we may
take

$$AF = (r^2 - a^2)/(2a), \qquad AO = (r^2 + a^2)/(2a), \qquad BF = (r^2 - b^2)/(2b),$$
$$BO = (r^2 + b^2)/(2b), \qquad OF = r.$$

By similar triangles,

$$AF : FO :: FO : FR = \frac{2ar^2}{r^2 - a^2} :: AO : RO = \frac{r(r^2 + a^2)}{r^2 - a^2}.$$

Hence we have $RB = BF - FR$. Since angles BOR and OCB are equal,
the same lettered triangles are similar. Hence

$$BR : BO :: RO : OC = \frac{r(r^2 + a^2)(r^2 + b^2)}{d} :: BO : BC = \frac{(r^2 + b^2)^2(r^2 - a^2)}{2bd},$$

where $d = (r^2 - a^2)(r^2 - b^2) - 4abr^2$. Hence

$$DC = BC - BD = 2r^2\{(r^2 - a^2)b + (r^2 - b^2)a\}/d.$$

We may assign any values to a, b and any value, exceeding a and b, to r.
For $a = 16$, $b = 18$, $r = 72$, we get $AF = 154$, $AO = 170$, $BF = 135$,
$BO = 153$, $OC = 120$, $CD = 96$, $AB = 289$, $AC = 250$, $BC = 231$.

Several[12] employed Heron's formula

$$\Delta^2 = (B + S + s)(B + S - s)(B - S + s)(- B + S + s)$$

for the square of the area Δ of a triangle with sides $2B$, $2S$, $2s$. T. Baker
wrote x, y, z for the last three factors of Δ^2. Then $\Delta^2 = xyz(x + y + z)$.
Let $\Delta = (axz)^2$. We get x rationally. " A. B. L." took $B = x - y$, $S = x$,
$s = x + y$; then $3x^2 - 12y^2 = \square$, whence $u^2 - 3v^2 = 1$. C. Holt equated
the last three factors of Δ^2 to $4p^2q^2$, $(q^2 + r^2 - p^2)^2$, $4p^2r^2$; by addition,
$B + S + s = (q^2 + r^2 + p^2)^2$. J. Anderson equated the product of the
four factors to s^2x^2. Hence

$$\{B^2 - (S^2 + s^2)\}^2 = s^2(4S^2 - x^2) = s^2(2S - y)^2,$$

say; hence we get S and then B^2.

---

[10] Ladies' Diary, 1824, 43, Quest. 1416.

[11] Ladies' Diary, 1825, 34-5.

[12] The Gentleman's Math. Companion, London, 5, No. 27, 1824, 289-292. Report in changed
notations in Math. Mag., 2, 1898, 224-5.

C. Gill[13] found integral sides x, y, z of a triangle the four diameters of whose inscribed and escribed circles are integral squares r^2 and R^2, R_1^2, R_2^2. Take $x + y + z = a^2$, $y + z - x = b^2$, $x + z - y = c^2$, $x + y - z = d^2$. Let the condition $a^2 = b^2 + c^2 + d^2$ be satisfied. We get $x = (a^2 - b^2)/2$, y, z. It is known that $4\Delta = r^2a^2 = R^2b^2 = R_1^2c^2 = R_2^2d^2$.

C. L. A. Kunze[14] derived eight rational triangles from the two rational right triangles (3, 4, 5) and (5, 12, 13) by reducing their sides in proper ratios so that any chosen leg of one shall equal any chosen leg of the other and then juxtaposing the resulting triangles either with or without overlapping. Schlömilch[36] noted that we may start with any two rational right triangles.

C. F. Gauss,[14a] whose attention had been called to Curtius'[2a] problem by Schumacher, stated that the sides of every triangle such that each side and the radius r of the circumscribed circle are integers are of the form

$$4abfg(a^2 + b^2), \qquad \pm 4ab(f + g)(a^2f - b^2g), \qquad 4ab(a^2f^2 + b^2g^2),$$

where a, b, f, g are positive integers, while $r = (a^2 + b^2)(a^2f^2 + b^2g^2)$. We obtain Curtius' numbers by taking $a = g = 1$, $b = 2$, $f = 10$, and deleting the common factor 8. Many writers[14b] derived Gauss' formula.

E. W. Grebe[15] tabulated for 46 rational triangles the 12 rational values of the segments of the altitudes and the segments of the sides cut off by the altitudes.

Grebe[16] gave a table of 496 rational triangles, showing also the area, perimeter, altitudes, and diameter of the circumscribed circle. He began with 32 rational right triangles (4, 3, 5), $\cdots$, (195, 28, 197) with small ratios of sides, took each pair of these triangles and multiplied their sides by such factors as produce two triangles whose larger legs are equal. By juxtaposition he formed a rational acute triangle.

To find a triangle with integral sides whose area and perimeter are equal, B. Yates[17] took, in accord with (1), the sides to be $pq(r^2 + s^2)/n$, $rs(p^2 + q^2)/n$, $(ps + qr)(pr - qs)/n$. The latter multiplied by $pqrs/n$ is the area. Equating the area to the perimeter $2pr(ps + qr)/n$, we get

$$qs(pr - qs) = 2n.$$

Integral solutions are found when $n = 1, 2, 8$. Many solvers used the segments l, m, n into which the sides a, b, c are divided at the points of contact of the inscribed circle of radius r. Thus $l + m = a$, $l + n = b$, $m + n = c$. If s is the semi-perimeter, $rs = 2s$, whence $r = 2$. But $r^2s^2 = slmn$. Hence $4(l + m + n) = lmn$. The least side exceeds $2r = 4$. Hence we may take $l + m = 5, 6, \cdots$ and find integral solutions.

[13] The Gentleman's Math. Companion, London, 5, No. 29, 1826, 509–512.
[14] Lehrbuch der Geometrie, Jena, 1842, 205.
[14a] Briefwechsel zwischen C. F. Gauss and H. C. Schumacher (ed., C. A. F. Peters), Altona, 5, 1863, 375; letter of Oct. 21, 1847. Quoted in Archiv Math. Phys., 44, 1865, 504–6.
[14b] Archiv Math. Phys., 45, 1866, 220–231.
[15] Eine Gruppe von Aufgaben über das geradlinige Dreieck, Progr., Marburg, 1856.
[16] Zusammenstellung von Stücken rationaler ebener Dreiecke, Halle, 1864, 248 pp.
[17] The Lady's and Gentleman's Diary, London, 1865, 49–50, Quest. 2019.

Many[18] proved that if the sides and area be integers, the area is divisible by 6. Take the sides to be the products of (1) by $pqrs$. Then the area is $pqrs(ps + qr)(pr - qs)$.

S. Tebay[19] tabulated 237 rational triangles arranged according to the magnitude of the area, the greatest area being 46410 (cf. Martin[46]).

J. Wolstenholme[20] found a triangle whose sides and area are in arithmetical progression. Take $a - b$, a, $a + b$ as the sides, $a + 2b$ as the area. Then

$$2b = \frac{a(3a^2 - 16)}{16 + 3a^2}.$$

W. Ligowski[21] found a triangle whose sides a, b, c, area F, and radii r and ρ of circumscribed and inscribed circles, are all rational. He assumed that $s - a = \rho x$, $s - b = \rho y$, $s - c = \rho z$, where s is the semi-perimeter, and readily proved that the sides are proportional to

$$a = x(y^2 + 1), \qquad b = y(x^2 + 1), \qquad c = (x + y)(xy - 1),$$

whence

$$\rho = xy - 1, \qquad r = \tfrac{1}{4}(x^2 + 1)(y^2 + 1), \qquad F = xy(x + y)(xy - 1).$$

W. Šimerka[21a] gave several methods of finding rational triangles and a table of the 173 having sides $\leqq 100$, showing also the area, tangents of the half angles, and the coordinates of the vertices (cf. Scherrer[62a]). He proved that the perimeter is always even.

H. Rath[22] employed the segments α, β, γ of the sides determined by the points of tangency of the inscribed circle. Then the sides are $\alpha + \beta$, $\alpha + \gamma$, $\beta + \gamma$ and the square of the area is $\alpha\beta\gamma(\alpha + \beta + \gamma)$. The latter is a rational square only for $\alpha = dj^2$, $\beta = \delta B$, $\gamma = \delta C$, where B and C are any two positive relatively prime integers, and likewise for k and j, while d/δ is the value of the fraction

$$\frac{BC(B + C)}{k^2 - BCj^2}$$

when reduced to its lowest terms. Each resulting set of rational numbers α, β, γ defines a rational triangle, the condition that the sum of any two sides shall exceed the third being evidently satisfied. His final tables show relatively prime integral sides, the triangles whose area is a multiple of some side being listed separate from the others. He gave (p. 218) nine rational triangles whose sides form an arithmetical progression, the common difference being here given as a subscript:

$$(3, 4, 5)_1, \quad (13, 14, 15)_1, \quad (15, 26, 37)_{11}, \quad (75, 86, 97)_{11},$$
$$(25, 38, 51)_{13}, \quad (61, 74, 87)_{13}, \quad (15, 28, 41)_{13}.$$

---

[18] The Lady's and Gentleman's Diary, London, 1866, 61, Quest. 2044.

[19] Elements of Mensuration, London and Cambridge, 1868, 113–5. Table reprinted by G. B. Halsted, Metrical Geometry, 1881, 167–170.

[20] Math. Quest. Educ. Times, 13, 1870, 89–90. Same by D. S. Hart, 20, 1874, 56.

[21] Archiv Math. Phys., 46, 1866, 503–4.

[21a] Ibid., 51, 1870, 196–240.

[22] Archiv Math. Phys., 56, 1874, 188–224. See the compact exposition by P. Bachmann, Niedere Zahlentheorie, 2, 1910, 440–1. Cf. Kommerell[270] of Ch. XXII.

D. S. Hart[23] juxtaposed two right triangles with the common leg $2pr$ and further legs $r(p^2 - 1)$, $p(r^2 - 1)$, and obtained

$$(p + r)(pr - 1), \qquad r(p^2 + 1), \qquad p(r^2 + 1),$$

viz., (1) for the case of the upper signs and $q = s = 1$. The last assumption does not restrict the generality of the result.

Hart[24] noted that the triangle with the sides $w - 1$, w, $w + 1$ has a rational area if $3w^2 - 12 = \square$. He obtained $w = n/d$, $d = x^2 - 3y^2$ and took $d = 1$, whose general set of solutions is known.

A. B. Evans[25] found a triangle whose sides a, b, c, radii

$$x = \tfrac{1}{2}r(1 + \tan \tfrac{1}{4}A)(1 + \tan \tfrac{1}{4}B)/(1 + \tan \tfrac{1}{4}C),$$

y, z of Malfatti's circles, and radius r of the inscribed circle are all rational. Take $\cot \tfrac{1}{2}A = m/n$, $\cot \tfrac{1}{2}B = p/q$, $m^2 + n^2 = \square$, $p^2 + q^2 = \square$. Then $\tan \tfrac{1}{4}A$, etc., are rational. A. Martin took $\cot \tfrac{1}{4}C = 3$, $\cot \tfrac{1}{4}B = 4$; then the ratios of x, y, z, a, b, c to r are known.

H. S. Monck[26] showed how to deduce a second from one triangle with integral sides, two differing by unity.

J. L. McKenzie[27] found a triangle whose area and sides are integers, semi-perimeter is a square, two sides having a given common difference.

D. S. Hart[28] discussed rational triangles two of whose sides differ by unity.

R. Hoppe[29] discussed triangles with the sides $n - r$, n, $n + r$ and rational area Δ. Thus $\Delta = \tfrac{3}{4}mn$, where $3m^2 = n^2 - 4r^2$. Hence n is even, $n = 2p$, and $m = 2q$, whence $p^2 - 3q^2 = r^2$. First, let $r = 1$. If p_k, q_k is a solution in integers, then is also

$$p_{k+1} = 2p_k + 3q_k, \qquad q_{k+1} = p_k + 2q_k.$$

Further, $p_{k+1} - 4p_k + p_{k-1} = 0$ and similarly for the q's. Hence

$$s_k \equiv p_{k+1} - (2 + \sqrt{3})p_k = \frac{p_k - (2 + \sqrt{3})p_{k-1}}{2 + \sqrt{3}},$$

$$s_k(2 + \sqrt{3})^k = s_c.$$

The resulting values of n, Δ are

$$n = (2 + \sqrt{3})^k + (2 - \sqrt{3})^k, \qquad \Delta = \frac{\sqrt{3}}{4}\{(2 + \sqrt{3})^{2k} - (2 - \sqrt{3})^{2k}\},$$

for $k = 0, 1, \cdots$. It is proved that there are no further solutions.

Next, let r be undetermined. Then $p : r = 3\lambda^2 + \mu^2 : 3\lambda^2 - \mu^2$, where λ and μ are relatively prime integers. Thus the sides are

$$3(\lambda^2 + \mu^2), \qquad 2(3\lambda^2 + \mu^2), \qquad 9\lambda^2 + \mu^2.$$

[23] Math. Quest. Educ. Times, 23, 1875, 108.
[24] Ibid., 23, 1875, 83–4.
[25] Ibid., 22, 1875, 70–1.
[26] Ibid., 24, 1876, 36–8.
[27] Ibid., 25, 1876, 105–6.
[28] Ibid., 28, 1878, 66–7.
[29] Archiv Math. Phys., 64, 1879, 441.

W. A. Whitworth[30] noted that the triangle with the altitude 12 and sides 13, 14, 15 is the only one in which the altitude and sides are consecutive integers.

G. Heppel[31] noted that there are 220 triangles with integral sides $\leqq 100$ and integral areas, but repeated (39, 41, 50). He listed only 55 rational scalene triangles with relatively prime sides.

Worpitzky[32] gave without proof a formula equivalent to (1).

R. Müller[33] considered rational triangles whose sides are consecutive integers $x - 1$, x, $x + 1$. Since the area is to be rational, $x^2 - 4 = 3y^2$, whence $x = 2u$, $y = 2v$, $u^2 - 3v^2 = 1$. Hence the triangles are (3, 4, 5), (13, 14, 15), etc.

A. Martin[34] noted that the triangle with the sides $2m^2 + 1$, $2m^2 + 2$, $4m^2 + 1$ has a rational area.

T. Pepin[35] gave a historical note on rational triangles.

O. Schlömilch[36] gave the same method and results as Hart.[23]

C. A. Roberts[37] noted that, if u is a square and w the double of a square, $u + w$, $u + 2w$, $2u + w$ are the sides of a triangle with rational area $(u + w)\sqrt{2uw}$ and listed many triangles with sides < 500. The triangle is special since one side equals one-third of the sum of the remaining two.

S. Robins[38] tabulated rational triangles with a given base and a given difference between the remaining two sides; also (pp. 262–3) rational triangles with sides x, $x + n$, $2x - n$ for given n's.

H. F. Blichfeldt[39] derived (1) by use of Heron's formula for area.

S. Robins[40] found rational triangles whose sides are consecutive integers by taking $x - 2$ and $x + 2$ as the segments of the base made by the perpendicular to the base. The altitude is $(3x^2 - 3)^{\frac{1}{2}}$, which is made rational by choice of x by means of convergents to the continued fraction for $\sqrt{3}$.

A. Martin[41] juxtaposed two right triangles in various ways to obtain rational triangles. From Heron's formula for the area Δ of a triangle with the sides x, y, z,

$$\frac{1}{16}(z^2 - x^2 - y^2)^2 = \tfrac{1}{4}x^2y^2 - \Delta^2 = (\tfrac{1}{2}xy - \Delta q/p)^2, \quad \text{if} \quad \Delta = \frac{pqxy}{p^2 + q^2}.$$

Then

$$z^2 = x^2 + y^2 \pm \frac{2(p^2 - q^2)xy}{p^2 + q^2} = \left(\frac{r}{s}y - x\right)^2$$

<hr>

[30] Math. Quest. Educ. Times, 36, 1881, 42.
[31] *Ibid.*, 39, 1883, 37–8. Cf. Martin.[59]
[32] Zeitschr. Math. Naturw. Unterricht, 17, 1886, 256.
[33] Archiv Math. Phys., (2), 5, 1887, 111–2.
[34] Math. Magazine, 2, 1890, 6.
[35] Mem. Accad. Pont. Nuovi Lincei, 8, 1892, 85.
[36] Zeitschr. Math. Naturw. Unterricht, 24, 1893, 401–9.
[37] Math. Magazine, 2, 1893, 136.
[38] Amer. Math. Monthly, 1, 1894, 13–14, 402–3 (for base 9).
[39] Annals of Math., 11, 1896–7, 57–60.
[40] Amer. Math. Monthly, 5, 1898, 150–2.
[41] Math. Magazine, 2, 1898, 221–236.

determines x/y. Taking x to be the numerator of the resulting fraction, we have
$$x = (p^2 + q^2)(r^2 - s^2), \qquad y = 2rs(p^2 + q^2) \pm 2s^2(p^2 - q^2),$$
$$z = (p^2 + q^2)(r^2 + s^2) \pm 2rs(p^2 - q^2).$$

He discussed at length rational triangles two of whose sides differ by a given integer, making use of a Pell equation $gq^2 - p^2 = \pm 1$.

T. H. Safford[42] juxtaposed the right triangles (5, 12, 13), (9, 12, 15) of areas 30 and 54 to obtain Heron's triangle (13, 14, 15) of area 84, also to obtain (4, 13, 15) of area $54-30$. He listed 37 rational right triangles.

D. N. Lehmer[43] derived (1) by use of the rationality of the sines and cosines of the three angles, a necessary and sufficient condition for the rationality of the triangle.

Rational triangles with consecutive integral sides have been found.[44]

W. A. Whitworth and D. Biddle[45] proved that there are only five triangles with integral sides whose area equals the perimeter: (5, 12, 13), (6, 8, 10), (6, 25, 29), (7, 15, 20), (9, 10, 17).

A. Martin[46] formed rational triangles by the juxtaposition of two rational right triangles. He tabulated 168 rational triangles of area $\leqq 46410$ not found in Tebay's[19] table.

H. Schubert[47] considered a Heron triangle with integral sides a, b, c and area J. If α, β, γ are the angles, $f = \tan \alpha/2$ and hence also $\sin \alpha$ and $\cos \alpha$ must be rational (such an angle α being called a Heron angle). Set $f = n/m$, where n and m are relatively prime integers. Then
$$\sin \alpha = \frac{2mn}{m^2 + n^2}, \qquad \sin \beta = \frac{2pq}{p^2 + q^2}, \qquad \sin \gamma = \frac{2(mq + np)(mp - nq)}{(m^2 + n^2)(p^2 + q^2)},$$
since $\tan \gamma/2 = \cot(\alpha + \beta)/2$. By $a = 2r \sin \alpha$, etc.,
$$4r = (m^2 + n^2)(p^2 + q^2).$$
Hence
$$a = mn(p^2 + q^2), \quad b = pq(m^2 + n^2), \quad c = (mq + np)(mp - nq), \quad J = mnpqc.$$

J. Sachs[48] gave tables of rational triangles with altitudes < 100; acute rational triangles with altitudes 100, $\cdots$, 500; rational triangles arranged according to the least side and according to the greatest side. The last tables are convenient for the formation by juxtaposition of rational quadrilaterals, pentagons, etc.

T. Harmuth[49] considered rational triangles with sides a, $a + d$, $a + 2d$.

---

[42] Trans. Wisconsin Acad. Sc., 12, 1898–9, 505–8.
[43] Annals of Math., (2), 1, 1899–1900, 97–102.
[44] Amer. Math. Monthly, 10, 1903, 172–3.
[45] Math. Quest. Educ. Times, 5, 1904, 54–6, 62–3.
[46] Math. Magazine, 2, 1904, 275–284.
[47] Die Ganzzahligkeit in der algebraischen Geometrie, Leipzig, 1905, 1–16. Festgabe 48 Versammlung d. Philologen u. Schulmänner zu Hamburg, 1905. Reprinted in Auslese aus meiner Unterrichts- u. Vorlesungspraxis, Leipzig, 2, 1905, 1–23.
[48] Tafeln zum Math. Unterricht, Progr. 794, Baden-Baden, Leipzig, 1908.
[49] Unterrichtsblätter für Math. u. Naturwiss., 15, 1909, 105–6.

Its area is rational if $(a + 3d)(a - d) = 3y^2$. Hence decompose $3y^2$ in every way into two factors congruent modulo 4.

E. N. Barisien[50] noted that, if $2p$ is the perimeter, the area is an integer if

$$p = (\alpha n + 1)(\beta n + 1), \qquad p - b = \lambda n(\gamma n + 1),$$
$$p - a = (\alpha n + 1)(\gamma n + 1), \qquad p - c = \mu n(\beta n + 1),$$

and $\lambda\mu = k^2$. The condition $p = \Sigma(p - a)$ is satisfied if $4\gamma + \beta = 5\alpha$, $\beta - \gamma = 5$. If in $p - b$ and $p - c$ we replace λn and μn by $\delta n + 1$, the area is integral and the condition $p = \Sigma(p - a)$ gives for $\delta n + 1$ a value which is integral if $\beta + \gamma = 2\alpha$; then B is a right angle. A. Gérardin noted that we may set $p = (\alpha n + t)(\beta n + t)$, etc., and take $\beta + \gamma = 2\alpha$, $\beta - \gamma = 2\rho$, $t = (\rho - \delta)n$.

L. Aubry[51] noted that the triangle with the sides $x - 1$, x, $x + 1$ has an integral area if $(x/2)^2 - 3y^2 = 1$, i. e., if

$$x = 2, 4, 14, \cdots, \qquad x_n = 4x_{n-1} - x_{n-2}.$$

The area[52] of any triangle with integral sides and area is a multiple of 6.

B. Hecht[53] discussed triangles whose sides are integers, also the area or the four radii of the escribed and inscribed circles.

A. Martin[54] proved that in any primitive rational triangle two sides are odd, the least side is > 2, the difference between the sum of the two smaller sides and the largest side is not unity, and the area is a multiple of 6. Every integer > 2 is the least side of an infinitude of primitive rational triangles.

E. N. Barisien[55] noted that the triangle with the sides 7, 15, 20 has its area and perimeter each 42. Multiplying the sides by 10, we get a triangle with integral altitudes.

* H. Böttcher[56] gave rational triangles with an angle 60° or 120°.

Barisien[57] gave complicated formulas for the integral sides of a triangle, with integral values for the altitudes, area, radius of circumscribed circle, radii of tritangent circles, segments of the sides made by the altitudes, and segments of the altitudes made by the orthocenter.

Of several triangles[58] with integral sides, area and one altitude, the least appears to have the sides 4, 13, 14, area 24 and altitude (to side 4) 12.

A. Martin[59] added 61 rational scalene triangles to Heppel's[31] list.

N. Gennimatas[60] proved that any rational triangle is similar to one with

---

[50] Sphinx-Oedipe, 5, 1910, 57–9.
[51] *Ibid.*, 6, 1911, 188.
[52] Math. Quest. Educ. Times, 21, 1912, 17–8. See paper 18 above.
[53] Ueber rationale Dreiecke, Wiss. Beil. z. Jahresber. Städt Realschule in Königsberg, 1912, 7 pp.
[54] School Science and Math., 13, 1913, 323–6.
[55] Mathesis, (4), 3, 1913, 14, 67.
[56] Unterrichtsblätter für Math. u. Naturwiss., 19, 1913, 132–3.
[57] Sphinx-Oedipe, 8, 1913, 182–3; 9, 1914, 74–5, 91, 94. Assoc. franç. av. sc., 43, 1914, 48–57. Mathesis, (4), 4, 1914, 114–6 for 7 examples.
[58] L'intermédiaire des math., 21, 1914, 76, 143, 186–8; 22, 1915, 119–120.
[59] Math. Quest. Educ. Times, 25, 1914, 76–8.
[60] L'enseignement math., 16, 1914, 48–53.

the sides $x^2 + y^2$, $(1 + y^2)x$, $c = (1 + x)(y^2 - x)$. Conversely, if x, y, $y^2 - x$ are positive, these numbers are the sides of a triangle, of area cxy.

E. Turrière[61] noted several methods to find Heron triangles. There is an infinitude of Heron triangles with sides in arithmetical progression such that no two are similar. He investigated Heron triangles in which the semi-perimeter p and $p - a$, $p - b$, $p - c$ are all rational squares, and the analogous problem for inscriptible quadrilaterals. He[62] found Heron triangles in which the sum of the squares of two sides is a square.

F. R. Scherrer[62a] made use of the theory of complex integers $a + bi$ to obtain the coordinates of the vertices, of the centers of the circumscribed, inscribed, escribed and Feuerbach circles, of the intersection of the altitudes, etc., of primitive Heron triangles. Cf. Šimerka.[21a]

M. Rignaux[63] stated the final formulas of Schubert.[47]

E. T. Bell stated and W. Hoover[64] proved incompletely that if $u_0 = 2$, $u_1 = 4$, $\cdots$, $u_{n+2} = 4u_{n+1} - u_n$, then $u_n - 1$, u_n, $u_n + 1$ are the consecutive sides of a triangle with integral area, and all such triangles are given by this method.

PAIRS OF RATIONAL TRIANGLES.

Frans van Schooten[5] found two isosceles rational triangles with equal perimeters and equal areas. Divide each into halves and let the right triangles be formed from a, b and k, d respectively. By the perimeters,

$$2(a^2 + b^2) + 2(2ab) = 2(k^2 + d^2) + 2(2kd), \qquad a + b = k + d.$$

Set $k = a + x$, $d = b - x$. The equality of the areas requires

$$2x^2 + 3(a - b)x + a^2 - 4ab + b^2 = 0, \qquad x = \tfrac{1}{4}(r + 3b - 3a),$$

where $r^2 = a^2 + b^2 + 14ab$. Set $r = a + b + c$. Thus

$$a = \frac{c^2 + 2bc}{12b - 2c}.$$

The general solution thus involves the parameters b, c. For $b = 1$, $c = 3$, we get $a = 5/2$, $x = 1/2$. Multiply the sides by 4. We get the right triangles (20, 21, 29) and (12, 35, 37). Their doubles have the perimeter 98 and area 420.

J. H. Rahn[65] devoted 8 pages to this problem, and J. Pell 62 pages. There is first given the above solution by van Schooten, attributed to Descartes.

Several[66] gave straightforward solutions to van Schooten's problem.

J. Cunliffe[67] treated the problem to find two triangles with rational altitudes and segments of sides and with equal perimeters and equal areas.

[61] L'enseignement math., 18, 1916, 95–110.

[62] *Ibid.*, 19, 1917, 259–261. Cf. Euler[21] of Ch. IV.

[62a] Zeitschrift Math. Naturw. Unterricht, 47, 1916, 513–30.

[63] L'intermédiaire des math., 24, 1917, 86.

[64] Amer. Math. Monthly, 24, 1917, 295, 471. Cf. Hoppe.[29]

[65] Algebra, Zürich, 1659. Engl. transl. by T. Brancker, augmented by D. P., London, 1668, 131–192.

[66] The Gentleman's Math. Companion, London, 5, No. 26, 1823, 183–5.

[67] New Series of the Math. Repository (ed., Th. Leybourn), 2, 1809, II, 54–7.

He found a pentagon inscribed in a circle with rational sides and areas for all the triangles into which the pentagon can be divided by diagonals.

TRIANGLES ALL OF WHOSE SIDES AND MEDIANS ARE RATIONAL.

L. Euler[68] denoted the sides by $2a$, $2b$, $2c$, and the medians by f, g, h. Then

$$2b^2 + 2c^2 - a^2 = f^2, \text{ etc.}, \qquad 2g^2 + 2h^2 - f^2 = 9a^2, \text{ etc.}$$

Hence, if $2f$, $2g$, $2h$ be taken as sides of a triangle, its medians are $3a$, $3b$, $3c$. Write $\sigma = a + b + c$. Then

$$(b - c)^2 + \sigma(b + c - a) = f^2, \qquad (a - c)^2 + \sigma(a + c - b) = g^2.$$

Set $f = b - c + \sigma p$, $g = a - c + \sigma q$. Then

$$b + c - a = 2(b - c)p + \sigma p^2, \qquad a + c - b = 2(a - c)q + \sigma q^2.$$

Solving each for c and adding $a + b$, we have two expressions for σ. Equating these, we get the ratio $a' : b'$ of $a : b$. Euler took $a' = a$ and got

$$a = 1 + q - p^2 - 2pq - p^2q + 2pq^2, \qquad b = 1 + p - q^2 - 2pq - pq^2 + 2p^2q.$$

Then $\sigma/2 = 1 + p + q - 3pq$, so that c is known and hence also f, g. Next,

$$h^2 = (a - b)^2 + \sigma(a + b - c) = A^2q^4 + 2Bq^3 + Cq^2 + 2Dq + E^2,$$

where

$$A = 1 + 3p, \qquad\qquad\qquad B = -1 + 11p - 9p^2 - 9p^3,$$
$$C = -3(1 + 2p - 2p^2 + 6p^3 - 3p^4), \quad D = 2 - 9p - 3p^2 + 11p^3 + 3p^4,$$
$$E = 2 + p - p^2.$$

We can obtain rational solutions by setting

$$h = Aq^2 + \frac{B}{A}q \pm E \qquad \text{or} \qquad Aq^2 \pm \frac{D}{E}q \pm E.$$

Euler examined the simplest cases $p = \pm 2$ ($p = 0$ or ± 1 being excluded). For $p = -2$, we have $A = -5$, $B = 13$, $C = 321$, $D = -32$, $E = -4$. Taking the second expression for h, we have

$$h = -5q^2 - 8q + 4, \qquad q = \tfrac{11}{2}, \qquad h = \tfrac{765}{4}.$$

Multiplying the resulting values of a, b, $\cdots$ by $4/3$, we get

$$a = 158, \qquad b = 127, \qquad c = 131, \qquad f = 204, \qquad g = 261, \qquad h = 255.$$

Since $\tfrac{2}{3}f$, $\tfrac{2}{3}g$, $\tfrac{2}{3}h$ are sides of a triangle with the medians a, b, c as remarked at the outset, we get the new solution

$$a = 68, \qquad b = 87, \qquad c = 85, \qquad f = 158, \qquad g = 127, \qquad h = 131.$$

Euler's[69] paper of 1778 deals with triangles in which the distances of the vertices from the center of gravity are rational and the sides are rational.

---

[68] Novi Comm. Acad. Petrop., 18, 1773, 171; Comm. Arith. Coll., 1, 1849, 507–15.
[69] Nova Acta Acad. Petrop., 12, 1794, 101; Comm. Arith., II, 294–301.

We have $g^2 - h^2 = 3(c^2 - b^2)$. Euler took

$$g + h = 3pq, \qquad g - h = rs, \qquad c + b = pr, \qquad c - b = qs.$$

From

$$g^2 + h^2 = 4a^2 + b^2 + c^2, \qquad f^2 = 2c^2 + 2b^2 - a^2,$$

we get, on setting $p = x + y$, $s = x - y$,

$$\frac{a^2}{q^2} = x^2 + y^2 + 2Mxy, \qquad \frac{f^2}{r^2} = x^2 + y^2 + 2Nxy,$$

$$M = \frac{5q^2 - r^2}{4q^2}, \qquad\qquad N = \frac{5r^2 - 9q^2}{4r^2}.$$

Take $a/q = x + ty$, $f/r = x + uy$. Then

$$\frac{x}{y} = \frac{1 - t^2}{2(t - M)} = \frac{1 - u^2}{2(u - N)}.$$

All conditions are satisfied if we take

$$u = -t = \frac{N - M}{2}, \qquad \frac{x}{y} = \frac{(M - N)^2 - 4}{4(M + N)}.$$

The cases $r = q$ and $r = 3q$ are excluded since $M + N \neq 0$. For $q = 1, r = 2$, we obtain the solution given above. For $q = 2, r = 1$, we get

$$a = 404, \qquad b = 377, \qquad c = 619, \qquad f = 3 \cdot 314, \qquad g = 3 \cdot 325, \qquad h = 3 \cdot 159.$$

Euler's[70] paper of 1779 does not differ materially from the preceding.

Euler's[71] paper of 1782 avoided the earlier restrictions on the generality of the solution. Changing the notations to conform with his earlier ones, we may set

$$h + g = \frac{3\alpha}{\beta}(b - c), \qquad h - g = \frac{\beta}{\alpha}(b + c).$$

From

$$(h + g)^2 + (h - g)^2 = 2h^2 + 2g^2 = 8a^2 + (b + c)^2 + (b - c)^2,$$

we get

$$8a^2 = \left(\frac{9\alpha^2 - \beta^2}{\beta^2}\right)(b - c)^2 + \left(\frac{\beta^2 - \alpha^2}{\alpha^2}\right)(b + c)^2.$$

Then $f^2 = (b + c)^2 + (b - c)^2 - a^2$ gives a similar formula for f^2. Write

$$b + c = \alpha(\gamma + \delta), \quad b - c = \beta(\gamma - \delta), \quad P = \frac{\beta^2 - 5\alpha^2}{4\alpha^2}, \quad Q = \frac{9\alpha^2 - 5\beta^2}{4\beta^2}.$$

Then

$$(2) \qquad\qquad \frac{a^2}{\alpha^2} = \gamma^2 + \delta^2 + 2P\gamma\delta, \qquad \frac{f^2}{\beta^2} = \gamma^2 + \delta^2 + 2Q\gamma\delta.$$

Take $\gamma = 4(P + Q)$, $\delta = (P - Q)^2 - 4$. Then (2) are the squares of

$$(P - Q)(3P + Q) - 4, \qquad (Q - P)(3Q + P) - 4.$$

[70] Mém. Acad. Petrop., 2, 1807–8, 10; Comm. Arith., II, 362–5.
[71] Mém. Acad. Petrop., 7, 1820, 3; Comm. Arith., II, 488–91.

Set $PQ + 1 = n(P + Q)$. We may discard the common factor $P + Q$ of γ and δ, thus altering α and β in the same ratio, and set $\gamma = 4$, $\delta = P + Q - 4n$. The first expression (2) is the square of

$$(P - Q)(P + Q) + 2P(P - Q) - 4 = (P + Q)(3P - Q - 4n),$$

which is to be divided by $P + Q$. Hence

$$\frac{a}{\alpha} = 3P - Q - 4n, \qquad \frac{f}{\beta} = 3Q - P - 4n.$$

From the above expressions for P, Q, we readily get $n = -5/4$. Set

$$C = 16\alpha^2\beta^2, \qquad D = (9\alpha^2 + \beta^2)(\alpha^2 + \beta^2), \qquad F = 2(9\alpha^4 - \beta^4).$$

Then $\gamma = 4$, $\delta = D/(4\alpha^2\beta^2)$. Suppressing the common denominator $4\alpha^2\beta^2$ in a, $b \pm c$, f, $h \pm g$, we get

$$a = \alpha(D - F), \qquad b + c = \alpha(C + D), \qquad b - c = \beta(C - D),$$
$$f = \beta(D + F), \qquad h + g = 3\alpha(C - D), \qquad h - g = \beta(C + D).$$

Euler[72] noted that $2a^2 + 2b^2 - c^2$ is a square if

$$a = (m + n)p - (m - n)q, \quad b = (m - n)p + (m + n)q, \quad c = 2mp - 2nq.$$

It suffices to make the product of the remaining two medians a square. We obtain a homogeneous quartic in p, q. A special set of values making it a square is found to be

$$p = (m^2 + n^2)(9m^2 - n^2), \qquad q = 2mn(9m^2 + n^2).$$

Euler deduced his[68] two solutions and three others:

$$207, 328, 145; \qquad 881, 640, 569; \qquad 463, 142, 529.$$

To make $\alpha = 2x^2 + 2y^2 - z^2$, $\beta = 2x^2 + 2z^2 - y^2$, $\gamma = 2y^2 + 2z^2 - x^2$ squares, "Atticus"[73] took $x = 5n - 4m$, $y = 2m$, $z = 2m + n$. Then $\alpha = (7n - 6m)^2$, and $\gamma = 48mn - 23n^2 = p^2$ determines m. Also,

$$64n^2\beta = p^4 - 50p^2n^2 + 1649n^4 = \square$$

if $p = n$, whence $m = n/2$.

J. Cunliffe[74] treated the problem subject to very special assumptions and obtained for the halves of the sides 807, 466, 491. Later, he[75] gave another very special treatment and obtained the sides 884, 510, 466 and medians 208, 659, 683.

N. Fuss[76] reproduced the solution in Euler's paper of 1782 with α replaced by $r - s$, β by $r + s$, γ by p, etc.

J. Cunliffe[77] wrote $x = AC$, $y = BC$, $z = AB$ for the sides, and BE, AF, CD for the medians. Take $z = x + y - d$. Then

$$4AF^2 = 2(AB^2 + AC^2) - BC^2 = 4x^2 + 4xy + y^2 - 4d(x + y) + 2d^2.$$

[72] Posthumous paper. Comm. Arith. Coll., 2, 1849, p. 649; Opera postuma, 1, 1862, 102–3.
[73] The Gentleman's Math. Companion, London, 2, No. 9, 1806, 17.
[74] New Series of the Math. Repository (ed., Leybourn), London, 1, 1806, II, 44.
[75] Ibid., 2, 1809, II, 31–4.
[76] Mém. Acad. Sc. St. Petersburg, 4, 1813, 247–252.
[77] The Gentleman's Math. Companion, London, 5, No. 27, 1824, 349–53. Extract in l'intermédiaire des math., 5, 1898, 10–11.

Equate it to $(2x + y - m)^2$ and the similar expression for $4BE^2$ to $(x + 2y - n)^2$. Solve the two resulting linear equations for x, y in terms of d, m, n. Reject the common denominator. Thus

$$x = d^2(4n - 2m) + 2d(m^2 - n^2) - mn(2m - n),$$
$$y = d^2(4m - 2n) - 2d(m^2 - n^2) + mn(m - 2n),$$
$$z = 2d^2(m + n) - 6mnd + mn(m + n).$$

Then $4CD^2 = 2(x^2 + y^2) - z^2$ becomes a quartic in d which is a square for

$$d = \frac{3(m + n)(m - n)^2(2m^2 - 5mn + 2n^2)}{10(m - n)^4 - mn(m^2 + n^2)}$$

C. Gill[78] gave a solution in which the sides are proportional to expressions in the sines and cosines of two of the angles A, B, subject to the condition that $\tan A/2$ equals one of four complicated functions of $\sin B$ and $\cos B$. The numerical example is the same as the first one of Euler's[68] paper of 1773.

E. W. Grebe[79] thought the problem was a new one. Changing his notations to conform with Euler's, we see that $2b^2 + 2c^2 - a^2 = f^2$ implies

$$(b + c + f)(b + c - f) = (a + b - c)(a - b + c).$$

From this and a similar formula involving g, we get

$$b + c + f = m(a + b - c), \qquad b + c - f = \frac{1}{m}(a - b + c),$$

$$c + a + g = p(b + c - a), \qquad c + a - g = \frac{1}{p}(b - c + a),$$

where m and p are unknowns. These four relations determine the ratios of a, b, c, f, g as rational functions of m and p. Then $2a^2 + 2b^2 - c^2$ (which is to equal h^2) is made a rational square by choice of p rationally in terms of m. Then the sides and medians are quintic functions of m.

C. L. A. Kunze[80] gave essentially the solution in Euler's[71] paper of 1782. J. W. Tesch[81] gave Cunliffe's[75] solution.

* E. Haentzschel[82] and Schubert[83] treated the problem. Cf. papers 101, 106.

The medians of a triangle with rational sides a, b, c are proportional to the sides if and only if $a^2 + c^2 = 2b^2$; such a triangle is called automédian. Reports of many papers on this equation are given in Ch. XIV.

Triangles with a Rational Median and Rational Sides; Parallelograms with Rational Sides and Diagonals.

C. G. Bachet's[3] fourth problem, added to his comment on Diophantus, VI, 18, was to find a rational triangle with one rational median. First

[78] Application of the angular analysis to the solution of indeterminate problems of the second degree, New York, 1848, 50–2. Results quoted in l'intermédiaire des math., 5, 1898, 10. Cf. A. Martin, Math. Quest. Educ. Times, 25, 1876, 96–7; E. Turrière, l'enseignement math., 19, 1917, 267–272.
[79] Archiv Math. Phys., 17, 1851, 463–74.
[80] Ueber einige Aufgaben aus der Dioph. Analysis, Progr. Weimar, 1862, 9.
[81] L'intermédiaire des math., 3, 1896, 237. Repeated, 20, 1913, 219.
[82] Jahresber. d. Deutschen Math.-Vereinigung, 25, 1916, 333–351.

let the angle A from which the median AD is drawn be acute. Let BC denote the side whose mid point is D. Take any number, as 13, which is a sum of two squares, $2^2 + 3^2$, and take $DC = 2$, $AD = 3$. Then $AB^2 + AC^2 = 2AD^2 + 2DC^2 = 2\cdot 13 = 5^2 + 1^2$, since the double of a sum of two squares is a sum of two squares. But 5 and 1 are not values of AB, AC. Hence we divide $5^2 + 1^2$ into a sum of two other squares by Diophantus II, 10, viz., $(5 - N)^2 + (1 + 2N)^2$, whence $N = 6/5$, $AB = 3\frac{4}{5}$, $AC = 3\frac{2}{5}$. Multiplying all by 5, we get $AB = 19$, $AC = 17$, $BC = 20$, $AD = 15$.

If A is obtuse, take $DC = 3$, $AD = 2$. We get the same values of AB and AC as before, while $BC = 30$, $AD = 10$ (in place of the misprint 12).

T. F. de Lagny[83] proved that in any parallelogram the sum of the squares of the two diagonals equals the sum of the squares of the four sides and noted the examples $9^2 + 13^2 = 2(5^2 + 10^2)$, $17^2 + 31^2 = 2(15^2 + 20^2)$. To solve $x^2 + y^2 = 2(a^2 + b^2)$ in integers, we may, for $a = b$, take $y = 2a - xb/c$, whence $x = 4abc/(b^2 + c^2)$. Next, a special solution of

$$x^2 + y^2 = 2\{a^2 + (a + b)^2\}$$

is given by $x = b$, $y = b + 2a$; to find the general solution, set $c = 2a + b$, $x = c \pm z$, $y = b \mp zd/e$; then $z = (\pm\, 2bde \mp 2ce^2)/(d^2 + e^2)$.

B. A. Gould[84] found a parallelogram with rational sides a, b and diagonals x, y. The condition is $x^2 + y^2 = 2(a^2 + b^2)$. Set $a + b = s$, $a - b = t$, whence $x^2 + y^2 = t^2 + s^2$. A solution is $fx = sd + te$, $fy = se - td$, if $f^2 = d^2 + e^2$. Wm. Lenhart called the sides $a \pm a'$ and diagonals $2b$, $2b'$, whence $a^2 - b^2 = b'^2 - a'^2$, which is satisfied if

$$a,\, b = nn' \pm mm'; \qquad b',\, a' = nm' \pm mn'.$$

J. Maurin[85] gave Gould's solution.

E. Hénet[86] noted that in the triangle with the sides $x = \rho v + u$, $y = \rho u - v$, $z = u + v + \rho(u - v)$, where $u > v$, $\rho > 1$, the median m_z is rational: $2m_z = \rho(u + v) - u + v$. Also m_y is rational if

$$u : v = (\mu - 2)(4\rho - \mu) : (\mu - 4)(2\rho + \mu), \qquad 4 < \mu < 3 + \rho.$$

M. A. Gruber[87] solved $2(a^2 + b^2) = c^2 + d^2$ by setting $b = a + p$, $c = 2a + q$, whence a follows rationally. W. F. King (pp. 320-2) proceeded as had Gould.[84]

H. Schubert[88] discussed triangles with rational sides a, b, c, and one or more rational medians, that to side a being designated by t_a. Since

$$(2t_a)^2 - (b - c)^2 = (b + c)^2 - a^2 = 4s(s - a), \qquad s = \tfrac{1}{2}(a + b + c),$$

the rationality of t_a implies that of x, where

$$\pm\, t_a - \tfrac{1}{2}(b - c) = sx, \qquad \pm\, t_a + \tfrac{1}{2}(b - c) = (s - a)/x.$$

---

[83] Hist. Acad. Roy. Sc. avec les Mém., année 1706, Paris, 1731, 319-333 (Hist., 83-99).
[84] Cambridge Miscellany, 1, 1843, 14.
[85] L'intermédiaire des math., 3, 1896, 210.
[86] Ibid., 240.
[87] Amer. Math. Monthly, 3, 1896, 219-221.
[88] Auslese Unterrichts- u. Vorlesungspraxis, Leipzig, 2, 1905, 68-92; same in Schubert,[47] 33-50.

Subtract, replace sx by $(s-a)x + (s-b)x + (s-c)x$, and $c-b$ by $s-b-(s-c)$. Thus

$$\frac{s-a}{x} + \frac{s-b}{x+1} + \frac{s-c}{x-1} = 0.$$

Since $s-a$, etc. shall be positive $-1 < x < 1$. Similarly, the rationality of t_b implies the existence of a rational value y, $-1 < y < 1$, for which

$$\frac{s-b}{y} + \frac{s-c}{y+1} + \frac{s-a}{y-1} = 0.$$

The two equations determine the ratios of $s-a$, $\cdots$. We may set

$$s-a = (x+2y+1)x(1-y) = A, \qquad s-b = (2x+y-1)(1+x)y = B,$$
$$s-c = (x-y+1)(1-x)(1+y) = C.$$

By addition, $s = 3xy + x - y + 1$. Hence for any proper fractions x, y, we have rational values of a, b, c, and of

$$\pm 2t_a = sx + (s-a)/x, \qquad \pm 2t_b = sy + (s-b)/y.$$

For $x = \frac{1}{2}$, $y = \frac{1}{3}$, we find $a = 17$, $b = 27$, $c = 16$, $2t_a = 41$, $2t_b = 19$.

If also t_c is to be rational, we must have a rational solution z, $-1 < z < 1$, of

$$\frac{s-c}{z} + \frac{s-a}{z+1} + \frac{s-b}{z-1} = 0.$$

Replacing $s-a$, $s-b$, $s-c$ by their values A, B, C, we obtain a relation R between x, y, z, quadratic in each. Now the pair of equations

$$\frac{3yx(1-y)}{1+z} - \frac{B}{1-z} = 0, \qquad \frac{(x-y+1)x(1-y)}{1+z} + \frac{C}{z} = 0$$

have the sum R and are such that the elimination of z gives

$$y = (7 - 4x - 2x^2)/(10x - 5).$$

He gave eight further pairs of equations with the sum R such that the elimination of z yields an equation linear in x or y. For the problem of three rational medians, this method lacks the generality and simplicity of Euler's.[71]

Heron triangles with a rational median; Heron parallelograms.

H. Schubert[89] defined a Heron parallelogram to be one whose sides, diagonals and area are rational. Call α, β the angles made by a diagonal with the concurring sides a, b; and θ the angle between the diagonals and opposite b. Then

$$a : b = \sin(\theta + \beta) : \sin(\theta - \alpha), \qquad a \sin \alpha = b \sin \beta,$$

the second following from the equal areas on each side of our diagonal. Hence

$$2 \cot \theta = \cot \alpha - \cot \beta.$$

<hr>

[89] Auslese Unterrichts- u. Vorlesungspraxis, Leipzig, 2, 1905, 36–45. Unterrichtsblätter Math. u. Naturw., 6, 1900, 70–1. Schubert,[47] 21–26.

The area being rational, we may set (Schubert[47])

$$\tan \tfrac{1}{2}\alpha = \frac{n}{m}, \qquad \tan \tfrac{1}{2}\beta = \frac{q}{p}, \qquad \tan \tfrac{1}{2}\theta = \frac{y}{x},$$

where m, n are relatively prime integers, etc.   Hence

$$2 \cdot \frac{x^2 - y^2}{2xy} = \frac{m^2 - n^2}{2mn} - \frac{p^2 - q^2}{2pq},$$

$$2(x^2 - y^2)mnpq = xy(mp + nq)(mq - np).$$

It is concluded erroneously[90] that the only integral solutions are

$$(x, y) = (mq, np) \quad \text{or} \quad (mp, nq).$$

Hence there remains in doubt his conclusion that no Heron triangle has more than one rational median.

R. Güntsche[91] considered a triangle ABC whose sides a, b, c, area I and median CF are rational.   If s is the semi-perimeter and ρ the radius of the inscribed circle,

$$\cot \tfrac{1}{2}A = s(s - a)/I, \qquad s\rho = I,$$

so that the cotangents α, β, γ of $\tfrac{1}{2}A$, $\tfrac{1}{2}B$, $\tfrac{1}{2}C$ must be rational.   Also, $\alpha + \beta + \gamma = \alpha\beta\gamma$.   Taking $\rho = (\alpha\beta - 1)/(\alpha\beta)$, we have

$$s = \alpha + \beta, \qquad a = \beta + \frac{1}{\beta}, \qquad b = \alpha + \frac{1}{\alpha}, \qquad c = s - \frac{1}{\alpha} - \frac{1}{\beta} = I.$$

Let F be the center of AB and $\nu = \cot \tfrac{1}{2}(CFB)$.   From triangles CAF and CFB we obtain the two values of $c/2$:

$$(3) \qquad \alpha - \frac{1}{\alpha} + \frac{1}{\nu} - \nu = \nu - \frac{1}{\nu} + \beta - \frac{1}{\beta}.$$

To secure symmetry, set $\beta' = 1/\beta$.   We obtain

$$(4) \qquad 2\nu^2\beta'\alpha - \nu(\beta'^2\alpha + \beta'\alpha^2 - \beta' - \alpha) - 2\beta'\alpha = 0,$$

which is quadratic in each of ν, α, β'.   Taking α as a parameter, we may treat the equation in ν, β' by Euler's[144] method of Ch. XXII.   But the second value of ν belonging to β' is $- 1/\nu$, so that the corresponding angle has been increased by π.   To obtain an essentially new solution, introduce the variable $\xi = \nu\beta'$ in place of ν before applying Euler's process.   A similar remark holds for the more general equation

$$(5) \qquad px^2y + qxy^2 + rxy + hqx + hpy = 0.$$

---

90 Other sets of solutions are $m = 2, n = 1, p = -2, q = 1, x = 2, y = 1$ or $x = 1, y = -2$; $m = 2, n = 1, p = 3, q = 1, x = 3, y = 4$ or $x = 4, y = -3$.   For $x = mq, y = np$, the factor $mq - np$ may be cancelled from the equation in the text, giving $p(m - 2n) = q(2m - n)$.   Hence $2m - n = lp, m - 2n = lq$, where $l = 1$ or 3 (m and n being relatively prime).   Schubert erroneously excluded $l = 3$, an example for which is $p = 3$, $q = 1, m = 5, n = 1$; this however does not affect the relation between $\tan (\alpha/2)$ and $\tan (\beta/2)$.

91 Sitzungsber. Berlin Math. Gesell., 4, 1905, 27–38.

which includes the case treated by Kummer.[133]   To simplify Euler's process, set

$$\theta(\xi) = \frac{q\xi + ph}{p\xi + qh}, \qquad X_i = \frac{h}{x_i}, \qquad Y_i = \frac{h}{y_i}.$$

From the initial pair $x = x_0$, $y = y_0$, we form

$$x_1 = y_0\theta_1, \quad y_1 = X_0\theta_1, \quad \theta_1 \equiv \theta(x_0y_0); \quad x_2 = y_1\theta_2, \quad y_2 = X_1\theta_2, \quad \theta_2 \equiv \theta(x_1y_1); \quad \cdots.$$

Then x_i, y_i is a new pair of solutions of (5). Similarly, we may start with x_0, Y_0. For (4),

$$h = -1, \quad p = 2\alpha, \quad q = -\alpha, \quad r = 1 - \alpha^2, \quad \theta(\xi) = -(\xi + 2)/(2\xi + 1).$$

Hence from the initial pair ν_0, β_0', we get $\nu_1 = \beta_0'\theta(\nu_0\beta_0')$, $\beta_1' = -\nu_0^{-1}\theta(\nu_0\beta_0')$. From the trivial solution $\alpha = p$, $\nu_0 = 1$, $\beta_0' = 1/p$, we get

$$\alpha = p, \qquad \nu_1 = \frac{-(2p+1)}{p(p+2)}, \qquad \beta_1' = \frac{2p+1}{p+2}.$$

From these we obtain a new set; etc. We may replace ν by $-1/\nu$, since (3) remains unaltered; we obtain the solution

$$\alpha = p, \qquad \nu = \frac{p(p+2)}{2p+1}, \qquad \beta = \frac{p+2}{2p+1}, \qquad a = b\{(p+2)^2 + (2p+1)^2\},$$

$$b = (p+2)(2p+1)(p^2+1), \qquad c = 2(p^2-1)(p^2+p+1),$$

$$CF = p^2(p+2)^2 + (2p+1)^2, \quad I = \tfrac{1}{2}p(p^2-1)(p+2)(2p+1)(p^2+p+1).$$

E. Haentzschel[92] repeated Güntsche's deduction of (3), with α, β interchanged. For symmetry replace the new α by its reciprocal. Hence

$$\frac{\alpha^2 - 1}{2\alpha} + \frac{\beta^2 - 1}{2\beta} = \frac{\nu^2 - 1}{\nu}.$$

The value obtained by solving for ν will be rational if

$$\{\beta^2\alpha + \beta(\alpha^2 - 1) - \alpha\}^2 + (4\beta\alpha)^2 = \square.$$

This quartic in β is treated by use of Weierstrass's elliptic $\wp$-function [cf. Haentzschel[82] of Ch. XV]. There result various particular types of Heron parallelograms.

TRIANGLES WITH RATIONAL SIDES AND ONE OR MORE RATIONAL ANGLE-BISECTORS.

C. G. Bachet[93] gave a long construction and discussion leading to the special acute angled triangle with the sides (reduced 1 : 4) 20, 20, 5 and having 6 as the bisector of either equal angle; also the oblique angled triangle with the sides 80, 125, 164 and having 60 as the angle-bisector drawn to the side 164.   [The area of each triangle is irrational.]

J. Kersey[94] discussed oblique triangles with rational sides and area and one rational angle-bisector or median.

[92] Sitzungsber. Berlin Math. Gesell., 13, 1913-4, 80-9.
[93] Diophanti Alex. Arith.[2] . . . ., 1621, 419-21.   Ed. by S. Fermat, 1670, 317-9.
[94] The Elements of Algebra, London, Books 3 and 4, 1674, 144-8.

N. Fuss[95] investigated triangles with rational sides a, b, c, rational angle-bisectors α, β, γ and rational area σ. The altitudes are then rational. Set

$$b + c - a = 2f, \qquad a + c - b = 2g, \qquad a + b - c = 2h.$$

Then

$$a + b + c = 2(f + g + h), \qquad \sigma^2 = (f + g + h)fgh.$$

He took $f = pq$, $g = qr$, $h = pr$. Then σ is rational if

$$pq + pr + qr = s^2,$$

where s is rational. Since $a = g + h$, $b = f + h$, $c = f + g$, we get

$$\alpha = \frac{\sqrt{bc(b + c + a)(b + c - a)}}{b + c} = \frac{2pqs}{pq + s^2} \cdot \sqrt{(p + r)(q + r)}.$$

The quantity under the last radical equals $r^2 + s^2$, which is therefore to be a square. Similarly, $p^2 + s^2$ and $q^2 + s^2$ are to be squares. Set $p = ls$, $q = ms$, $r = ns$. Then $1 + l^2$, etc., are to be squares, while

$$lm + ln + mn = 1.$$

These conditions are satisfied if

$$l = \frac{P^2 - Q^2}{2PQ}, \qquad m = \frac{R^2 - S^2}{2RS}, \qquad n = \frac{1 - lm}{l + m}.$$

For example, let $P = R = 2$, $Q = S = 1$. Then $l = m = 3/4$, $n = 7/24$. Take $s = 1$ and multiply a, α, etc. by 32. We get

$$a = 14, \qquad b = c = 25, \qquad \alpha = 24, \qquad \beta = \gamma = \tfrac{560}{39}.$$

J. Cunliffe[96] noted that the triangle with the sides

$$mn(m^2 - n^2)(r^2 + s^2)^2, \qquad rs(r^2 - s^2)(m^2 + n^2)^2,$$
$$\{mn(r^2 - s^2) - rs(m^2 - n^2)\}\{(r^2 - s^2)(m^2 - n^2) + 4rsmn\}$$

has rational area and angle-bisectors. He[97] obtained such a triangle with the sides 39, 150, 175 by taking three right triangles ($m^2 + n^2$, $m^2 - n^2$,

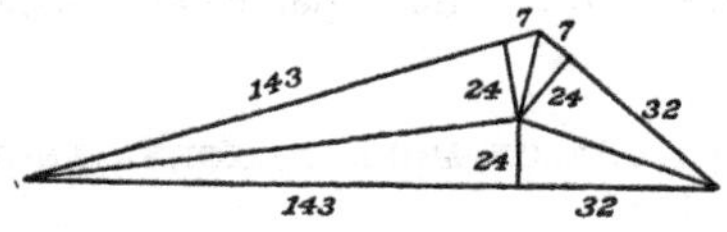

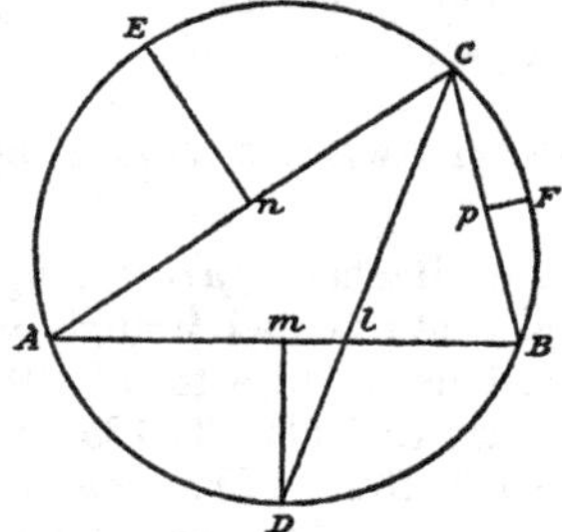

$2mn$) with a common leg $2mn$, where m, $n = 12$, 1; 6, 2; 4, 3 and using each right triangle twice.

<hr>

[95] Mém. Acad. Sc. St. Pétersbourg, 4, 1813, 240-7.
[96] New Series of the Math. Repository, London, 3, Pt. 2, 1814, 13–15.
[97] Ibid., 4, Pt. 2, 1819, 64.

Cunliffe[98] found a rational triangle ABC with rational values for the sides, altitudes and angle-bisectors. Circumscribe the circle with the rational diameter d. Let the perpendicular bisectors mD, nE, pF of the sides meet the circle at D, E, F (see right-hand figure on p. 210). Set $a = AD = DB$, $b = AE$, $c = BF$. Then $Dm = a^2/d$, $En = b^2/d$, $Fp = c^2/d$,

$$Am = \frac{a}{d}\sqrt{d^2 - a^2}, \qquad Cn = \frac{b}{d}\sqrt{d^2 - b^2}, \qquad Cp = \frac{c}{d}\sqrt{d^2 - c^2}.$$

Since the chords a, b, c subtend arcs whose sum is the semi-circle, they serve to form with the diameter an inscribed quadrilateral with sides $a = MP$, $c = PQ$, $b = QN$, $d = MN$. Hence

$$NP^2 = d^2 - a^2, \qquad MQ^2 = d^2 - b^2,$$

and

$$MP \cdot NQ + MN \cdot PQ = NP \cdot MQ, \qquad cd = \sqrt{d^2 - a^2} \cdot \sqrt{d^2 - b^2} - ab.$$

Hence if $d^2 - a^2$ and $d^2 - b^2$ are rational squares, c as well as a and b are rational. By the inscribed quadrilateral $ACBD$,

$$AB \cdot DC = DB \cdot AC + AD \cdot BC;$$

hence DC is rational. Thus the angle-bisector $Dl = (DB)^2/DC$ is rational. A second solution employs the inscribed circle with radius r and center S, lengths a, b, c of the tangents from A, B, C, and foot T of the perpendicular from S to AB. Then $AS^2 = AT^2 + ST^2 = a^2 + r^2$. To satisfy it in integers, take $a = 2mnr/(m^2 - n^2)$. Similarly, satisfy $BS^2 = b^2 + r^2$. It is proved that $CS^2 = c^2 + r^2$ is a rational square by use of

$$abc = r^2(a + b + c).$$

W. Wright and C. Gill[99] employed an isosceles triangle with the equal sides CA and CB, altitude CD, and intersection O of the angle-bisectors AP and BQ. Set $x = AD$, $a = AC + x =$ semi-perimeter. Then

$$CD = \sqrt{a^2 - 2ax}$$

will have a rational value ap if $x = \frac{1}{2}a(1 - p^2)$. Then

$$OD = \frac{AD \cdot CD}{AD + AC} = \frac{1}{2}ap(1 - p^2), \qquad AO = \frac{1}{2}a(1 - p^2)\sqrt{1 + p^2}.$$

It follows from certain proportions that CP is rational, while AP involves $\sqrt{1 + p^2}$. Hence the problem is solved if $1 + p^2 = \square = (1 - qp)^2$, say, which gives $p = 2q/(q^2 - 1)$. Taking $q = 3, 4, 5, 7$, we get four isosceles triangles with the same perimeter and having rational sides, areas and angle-bisectors.

S. Jones[100] found a triangle whose sides x, y, z and angle-bisectors are rational. Let nx and ny be the segments of z made by the bisector of the opposite angle; mx and mz those of y. Hence $y = (1 + n)mx/(1 - mn)$, $z = (1 + m)nx/(1 - mn)$. The square of the bisector of angle (x, y) is

[98] The Gentleman's Math. Companion, London, 5, No. 27, 1824, 344–9.
[99] *Ibid.*, 5, No. 30, 1827, 588–9.
[100] The Gentleman's Diary, or Math. Repository, London, 1840, 33–5, Quest. 1400.

$xy(1 - n^2)$, which is a square if $(1 - n)m(1 - mn) = \square = m^2n^2$, say, whence $m = (1 - n)/n$. Then $y = (1 - n^2)x/n^2$, $z = x/n$. Then the bisectors of angles (x, z) and (y, z) are rational if $2n^2 - n$ and $2n^2 + n$ are squares. Equating the first to p^2n^2, we get n. Then $2n^2 + n = \square$ if $4 - p^2 = \square = (2 - pq)^2$, which determines p. W. Rutherford called the sides a, b, c; the square of the bisector AD of angle A equals $4bcs(s - a)/(b + c)^2$, where $s = (a + b + c)/2$. Thus $bcs(s - a) = \square$. Similarly, $abs(s - c) = \square$, $acs(s - b) = \square$. Hence $s(s - a)(s - b)(s - c) = \square$ and the area is rational. Thus the problem is that treated by Cunliffe.[98]

J. Davey[101] found a triangle ABC in which the sides, the angle-bisector CD, the median CE, and the segments $AE = EB$ and ED of the base are all integers. Take

$$AC = (m + 1)p, \qquad BC = (m - 1)p, \qquad AD = (m+1)q, \qquad BD = (m-1)q.$$

Then $AE = mq$, $ED = q$, $CD^2 = (m^2 - 1)(p^2 - q^2)$. Take

$$CD = (m^2 - 1)(p - q),$$

whence $p = m^2q/(m^2 - 2)$. Then

$$CE^2 = (m^2 + 1)p^2 - m^2q^2 = \left(\frac{mq}{m^2 - 2}\right)^2 (5m^2 - 4).$$

Hence take $5m^2 - 4$ to be the square of $5(m - 1)r/s - 1$, thus obtaining m rationally.

Feldhoff[102] treated 31 problems on triangles in which certain elements (area, perimeter, side) are rational, are equal, or are squares. In the triangle formed by the juxtaposition of two rational right triangles, the angle-bisectors are rational if two expressions of the form $x^2 + 1$ are squares.[103]

Worpitzky[32] stated that, if the rational triangle with sides (1) has its angle-bisectors rational, then $p = \mu^2 - \nu^2$, $q = 2\mu\nu$, $r = \rho^2 - \sigma^2$, $s = 2\rho\sigma$.

D. Biddle[104] found special oblique triangles having integral values for the sides, area, altitude from one vertex and bisector of the angle at that vertex. Use is made of 3 right triangles with a common side.

R. Chartres[105] and others found integral values for the sides and the bisector g of the largest angle such that the perimeter equals mg.

* P. Dolgušin[106] gave examples, but no general solution, of the problem to find all triangles whose area, bisectors, medians, etc. are all rational.

---

[101] The Lady's and Gentleman's Diary, London, 1842, 69. He noted that J. Holroyd's solution, 1841, 57–8, leads only to degenerate triangles whose base equals the difference of the other sides.

[102] Einige Sätze über das Rationale Dreieck, Progr., Osnabrück, 1860.

[103] For if $2rs = 2mn$ is the common side, so that the composite triangle has the sides $b = m^2 + n^2$, $c = r^2 + s^2$, $a = m^2 - n^2 + r^2 - s^2$, then

$$a + b + c = 2(r^2 + m^2), \qquad b + c - a = 2(s^2 + n^2).$$

Hence the quantity under the radical in the expression for the bisector α (Fuss[95]) is a product of four sums of two squares and hence equals such a sum. In the expression for the bisector β occurs the square root of $E = ac(a + b + c)(a + c - b) = 4(r^2 + s^2) \times (r^2 + m^2)a(r^2 - n^2)$. Replacing s by mn/r in a, we get $a = (r^2 - n^2)(1 + m^2/r^2)$. Hence E is a sum of two squares. The product $\alpha\beta\gamma$ is rational since the area is rational.

[104] Math. Quest. Educ. Times, 57, 1892, 32.

[105] Ibid., 66, 1897, 102–3.

[106] Věst. opytn. Fiziki (Spaczinski's Bote), Odessa, 1903, No. 355, 145–157 (Russian).

H. Schubert[47] (pp. 17–21, or Schubert,[88] 27–36) considered a Heron triangle ABC in which the bisector w_a of angle A is rational. Since it divides the triangle into two Heron triangles we need only take $A/2$ and B to be Heron angles, i. e.,

$$\sin \frac{A}{2} = \frac{2uv}{u^2 + v^2}, \quad \cos \frac{A}{2} = \frac{u^2 - v^2}{u^2 + v^2}, \quad \sin B = \frac{2pq}{p^2 + q^2}, \quad \cos B = \frac{p^2 - q^2}{p^2 + q^2}.$$

Thus $\sin A$ and $\cos A$ are rational, so that in his[47] formulas for the sides of a Heron triangle we need only take $m = u^2 - v^2$, $n = 2uv$. To make w_a and w_b (and hence w_c) rational, take both $A/2$ and $B/2$ as Heron angles. He considered (§ 6) Heron triangles with both a rational bisector and a rational median.

An anonymous writer[107] gave three large integers which are the sides of a triangle having integral values for the area, three interior and three exterior angle-bisectors and the 12 segments cut off by them on the opposite sides. Also a triangle having integral values for the sides, area, altitude and two bisectors from the vertex, and the four segments of the base cut off by the two bisectors. M. Rignaux[108] gave a solution in smaller integers of the last problem.

E. Turrière[109] considered a triangle with rational values for the sides a, b, c and bisector d of the interior angle A. Thus

$$y^2 = nx^2 + 1, \qquad y = \frac{b + c}{a}, \qquad x = \frac{b + c}{a} \cdot d, \qquad n = \frac{1}{bc}.$$

The rational solutions of this Pell equation are

$$y = \frac{t^2 + n}{t^2 - n}, \qquad x = \frac{2t}{t^2 - n}.$$

Hence the desired triangle is obtained by assigning any rational values to b, c and taking $a = (bc - t^2)(b + c)/q$, $d = 2bct/q$, $q = bc + t^2$. In a Heron triangle, the bisector of angle A is rational if and only if $\tan \frac{1}{4}A$ is rational. Every Heron triangle whose bisectors are rational is the pedal triangle to a Heron triangle.

*O. Schulz[157] (pp. 72–3) treated rational triangles with three rational angle-bisectors.

TRIANGLES WITH RATIONAL SIDES AND A LINEAR RELATION BETWEEN
THE ANGLES.

K. Schwering[110] discussed triangles with integral sides one of whose angles is double another.

J. Heinrichs[111] generalized the problem, taking the relation $\alpha = n\beta + \gamma$ between the angles. Set $B = \beta/2$. Then

$$a : c : b = \cos (n - 1)B : \cos (n + 1) B : 2 \cos B \sqrt{1 - \cos^2 B}.$$

[107] L'intermédiaire des math., 23, 1916, 51–2, 73.
[108] Ibid., 234–7.
[109] L'enseignement math., 18, 1916, 397–407.
[110] Gymn. Progr., Coesfeld, 1886.
[111] Zeitschr. Math. Naturwiss. Unterricht, 42, 1911, 148–153.

Use may be made of the expansion of $\cos kB$ in terms of $\cos B$ or of

$$2 \cos kB = (x + \sqrt{x^2 - 1})^k + (x - \sqrt{x^2 - 1})^k, \qquad x = \cos B.$$

K. Schwering[112] took any linear relation between the angles.

Miscellaneous Results on Triangles whose Area need not be Rational.

A. Girard[112a] noted that $z = B^2 + BD + D^2$, $x = 2BD + D^2$, $y = 2BD + B^2$ are sides of a triangle in which an angle is 60° [i.e., satisfy $z^2 = x^2 - xy + y^2$], and the same is true of z, $x_1 = B^2 - D^2$, y. Also z, x, x_1 are sides of a triangle in which an angle is 120° [i.e., $z^2 = x^2 + xx_1 + x_1^2$].

To find integral sides a, b, c of a triangle ABC such that, if P is the point within it from which the sides subtend equal angles, the distances $x = AP$, $y = BP$, $z = CP$ are expressed by integers, we have

$$c^2 = x^2 + xy + y^2, \qquad b^2 = x^2 + xz + z^2, \qquad a^2 = y^2 + yz + z^2.$$

Many solvers[113] took $c = x + y - m$, $b = x + z - n$ and obtained two values for x, from which we get $z = (hy - mn)/(y - k)$, where h and k are known. Then

$$(y - k)^2 a^2 = y^4 + \cdots = \left(y^2 + \frac{h - 2k}{2} y + mn \right)^2$$

determines y rationally. Cf. papers 116 and 123; also 65, 67, 68, 70–73 of Ch. XIX.

Berton stated and J. de Virieu[114] proved that the area of a triangle is not rational if the sum of the sides, without a common factor 2, is odd.

W. S. B. Woolhouse[115] proved that, if three numbers $\leqq n$ are taken at random from a list of such triples and if p_n is the probability they will be sides of a possible triangle, then p_n, p_{n+1}, p_{n+2} are in arithmetical progression if n is even. He found the probability that three integers $\leqq n$ named by three different persons or by the same person will be proportional to the sides of a real triangle.

S. Bills[116] found the least integral sides BC, CA, AB of a triangle for which $x = OA$, $y = OB$ and $z = OC$ make equal angles and are measured by integers.[113] First, $AB^2 = x^2 + xy + y^2 = \square$, $AC^2 = x^2 + xz + z^2 = \square$ if

$$y = \frac{p^2 - 1}{2p + 1} x, \qquad z = \frac{q^2 - 1}{2q + 1} x.$$

Take $q = 2$. Then $BC^2 = y^2 + yz + z^2 = \square$ if $25p^4 + \cdots = \square$, which holds if $p = 9/4$, whence $x = 440$, $y = 325$, $z = 264$.

H. S. Monck[117] gave a very special discussion of the problem to find the least triangle with sides in arithmetical progression and altitudes in

[112] Archiv Math. Phys., (3), 21, 1913, 129–136.
[112a] L'Arith. de S. Stevin . . . par A. Girard, Leide, 1625, 676; Les Oeuvres Math. de S. Stevin, par A. Girard, 1634, 169.
[113] The Lady's and Gentleman's Diary, London, 1844, 50–1, Quest. 1705.
[114] Nouv. Ann. Math., (2), 3, 1864, 168–170.
[115] Math. Quest. Educ. Times, 9, 1868, 63–5, 91–2.
[116] Ibid., 20, 1874, 60–1.
[117] Ibid., 21, 1874, 108–9.

harmonical progression.   Its sides are the halves of the sides of a triangle whose area is divisible by each side.   A. B. Evans[118] noted that the altitudes p_i vary inversely as the sides a, b, c, whence the condition is $a + c = 2b$. Let $x = \cot \frac{1}{2}A$, $y = \cot \frac{1}{2}B$. Thus $2y = x + (x + y)/(xy - 1)$, which gives y rationally.   Then, if r is the radius of the inscribed circle,

$$a = r \left(\cot \tfrac{1}{2}B + \cot \tfrac{1}{2}C\right), \; \cdots, \qquad p_1 = r(a + b + c)/a, \; \cdots.$$

Evans and A. Martin[119] found rational triangles with integral sides and lines from the vertices to the center O of the inscribed circle, by use of $OA = r \csc \frac{1}{2}A$.

M. Weill stated and E. Cesàro[120] proved that $(4, 5, 6)$ is the only triangle whose sides are consecutive integers and the ratio of two of whose angles is an integer.

K. Schwering[121] noted that the ratios of the sines of the three angles α, β, γ are rational if the sides are rational.   Assigning values to $\tan \alpha/2$ and $\tan \beta/2$, whose ratio is rational, we have $\tan \gamma/2$ and hence the ratios of $a \pm b \pm c$ and therefore the ratios of a, b, c.   He discussed the problem to find a point O inside an equilateral triangle with the given rational side a such that the distances from O to the vertices shall be rational.

Züge[122] gave the general solution of $z^2 = x^2 + y^2 - 2xy \cos \alpha$, where $\cos \alpha$ is rational.   [But the topic is of little interest since we obtain a triangle with rational sides x, y, z by assigning to them any rational values such that $x + y > z$, etc.]

A. B. Evans[123] noted that, if $BC = 399, AC = 455, AB = 511, CO = 195, BO = 264, AO = 325$, the lines joining O to the vertices of triangle ABC make equal angles.[113]

Several[124] gave triangles with integral sides and an angle 60°.

A. Martin[125] discussed the last problem.

R. A. Johnson[126] gave expressions for the integral sides of any triangle with a given rational value for the cosine of one angle.

Several[127] gave pairs of triangles with integral sides having a common base and equal altitudes.

E. Turrière[128] found points whose distances from the three vertices of a given triangle with rational sides are all rational.

N. Alliston[129] gave special triangles with integral sides and points whose distances from the vertices are integers.

[118] Math. Quest. Educ. Times, 22, 1875, 54.
[119] Ibid., 102–3.
[120] Mathesis, 9, 1889, 142–3.   Also proof by Weill, Nouv. Ann. Math., (4), 14, 1914, 526–7.
[121] Geom. Aufgaben mit rationalen Lösungen, Progr. Düren, 1898.
[122] Archiv Math. Phys., (2), 17, 1900, 354.
[123] Math. Quest. Educ. Times, 72, 1900, 77.
[124] Zeitschrift Math. Naturw. Unterricht, 45, 1914, 184–5.
[125] Amer. Math. Monthly, 21, 1914, 98–9.   Cf. Neuberg[36, 40] of Ch. XIII.
[126] Ibid., 22, 1915, 27–30.
[127] Math. Quest. Educ. Times, 27, 1915, 91–2.
[128] L'enseignement math., 19, 1917, 262–7.
[129] Math. Quest. and Solutions, 5, 1918, 37.

On the ratios of the sides to the radius of the inscribed circle see
Gerono,[150] Ch. XXIII.

The following papers were not available for report:

C. Klobassa, Über Pythagoreische u. Heronische Zahlen, Progr., Troppau, 1908.

E. Haentzschel, Das Rationale in der algebraischen Geometrie [an address], Unterrichtsblätter Math. Naturw., 21, 1915, 1–5.

RATIONAL QUADRILATERALS.

A rational quadrilateral is one whose sides, diagonals and area are
expressed by rational numbers.

Brahmegupta[1] (§ 38) stated that "the legs of two right triangles multiplied reciprocally by the hypotenuses give the four sides of a trapezium."

Bháscara[130] (born 1114) illustrated this construction of a rational quadrilateral by starting with the right triangles $(3, 4, 5)$, $(5, 12, 13)$. Multiplying

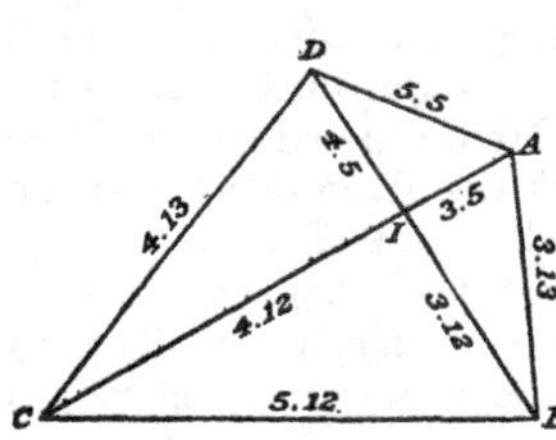

the legs of the first by the hypotenuse of the
second, we get two opposite sides of the quadrilateral; multiplying the legs of the second by
the hypotenuse of the first, we get the two remaining sides of the quadrilateral. One diagonal is the sum $4 \cdot 12 + 3 \cdot 5 = 63$ of the products
of the legs of one triangle by the corresponding
legs of the other. The other diagonal is

$$4 \cdot 5 + 3 \cdot 12 = 56.$$

As the Commentator Gaṇésa (1545 A.D.) indicated (p. 81), the quadrilateral is formed by the juxtaposition of four right triangles obtained by
multiplying the sides of each given triangle by the perpendicular and base
of the other. Bháscara noted that if we take the sides of the quadrilateral
in the new sequence 25, 39, 52, 60, one diagonal is still 56, but the other
is now the product 65 of the two hypotenuses. He noted (§§ 179–184, pp.
76–8) that the quadrilateral with the sides 40, 51, 68, 75 and diagonals 77,
85 has the area 3234.

M. Chasles[131] made clear the true sense of Brahmegupta's theorem.
Let a, b, c, d, e be integers, such as 3, 4, 5, 12, 13, for which $a^2 + b^2 = c^2$,
$c^2 + d^2 = e^2$. Construct the quadrilateral $ABCD$ with perpendicular diagonals AC, BD, crossing at I (see figure above), with

$$AI = ac, \qquad CI = bd, \qquad BI = ad, \qquad DI = bc.$$

Then

$$AB = ae, \qquad BC = cd, \qquad CD = be, \qquad AD = c^2.$$

Hence the sides are rational and the quadrilateral is inscriptible in a circle,
since $AI \cdot CI = BI \cdot DI$; its diameter is $ce/2$. The area is

$$\tfrac{1}{2}(ac + bd)(bc + ad).$$

---

[130] Lílávatí, § 191–2; Colebrooke,[1] pp. 80–83.

[131] Aperçu historique, Bruxelles, 1837, Note 12, p. 440; ed. 2, Paris, 1875; ed. 3, Paris, 1889, p. 421. Cf. O. Terquem, Nouv. Ann. Math., 5, 1846, 636; H. G. Zeuthen, Bibliotheca Math., (3), 5, 1904, 108.

From one inscriptible quadrilateral we get two others (but not with perpendicular diagonals) by permuting the sides. The area of each of the three quadrilaterals is the product of the three distinct diagonals divided by double the area of the circumscribed circle (A. Girard; proof by Grebe, Manuel de Géom., 1831, 435).

L. N. M. Carnot[132] noted that the segments of the diagonals of a quadrilateral are expressible rationally in terms of the sides and diagonals.

E. E. Kummer[133] noted that Chasles unriddled the obscurity of Brahmegupta without perceiving the method used by the latter, and expressed Brahmegupta's theorem in the following form. If the four sides of a quadrilateral, inscriptible in a circle, have the values

$$(a^2 + b^2)(c^2 - d^2), \qquad (a_2 - b^2)(c^2 + d^2), \qquad 2cd(a^2 + b^2), \qquad 2ab(c^2 + d^2),$$

where a, b, c, d are rational, then both diagonals (perpendicular to each other), the segments of them, the area of the quadrilateral and the diameter of the circumscribed circle are all rational.

Kummer showed how to obtain all rational quadrilaterals.   Let $ABCD$ have rational sides and diagonals.   Then the segments α, β, γ, δ of the diagonals are rational.   For, by

$$b^2 = a^2 + AC^2 - 2a \cdot AC \cos u,$$

$\cos u$ is rational; likewise $\cos v$ and $\cos(u + v)$. Hence $\sin u \sin v$ is rational; also $\sin^2 u$ and therefore $\sin u/\sin v$. But

$$\frac{a}{\beta} = \frac{\sin w}{\sin u}, \qquad \frac{d}{\delta} = \frac{\sin w}{\sin v}, \qquad \frac{\beta}{\delta} = \frac{a}{d} \cdot \frac{\sin u}{\sin v}.$$

Hence β/δ, $1 + \beta/\delta = BD/\delta$, δ and β are rational.   Similarly, α and γ are rational.   Next, $c = \cos w$ is rational, in view of

$$(1) \qquad a^2 = \alpha^2 + \beta^2 - 2\alpha\beta c.$$

Set $c = m/n$, where m, n are relatively prime. Without loss of generality, we may assume that a, α, β are integers with no common factor. To treat one of two analogous cases leading to like results, let n be odd. Then n must divide $\alpha\beta$. Thus $\alpha = r\alpha_1$, $\beta = s\beta_1$, $n = rs$,

$$(2) \qquad a^2 = r^2\alpha_1^2 + s^2\beta_1^2 - 2m\alpha_1\beta_1.$$

Now α_1, β_1 are relatively prime, since a common factor would divide a. We may take β_1 odd. The product of (2) by r^2 may be given the form

$$F_1 F_2 = (n^2 - m^2)\beta_1^2, \qquad F_1 = ar + r^2\alpha_1 - m\beta_1, \qquad F_2 = ar - r^2\alpha_1 + m\beta_1.$$

If F_1 and F_2 were both divisible by a prime factor p of β_1, then $2r^2\alpha_1$ and hence $r\alpha_1$ would be divisible by p, likewise a by (2), whereas a, α, β do not

<hr>

[132] Géométrie de position, Paris, 1803, 391-3.
[133] Jour. für Math., 37, 1848, 1-20.

have the common factor p.   Hence

$$F_1 = fy^2, \qquad F_2 = gz^2, \qquad yz = \beta_1, \qquad fg = n^2 - m^2,$$

$$\frac{F_1 - F_2}{\beta_1} = \frac{2r^2\alpha_1}{\beta_1} - 2m = \frac{fy}{z} - \frac{gz}{y}.$$

Divide the latter equation by n and set $\xi = fy/(nz)$.   Thus

$$\frac{2\alpha}{\beta} = 2c + \xi - \frac{1}{\xi} + \frac{c^2}{\xi}, \qquad \frac{\alpha}{\beta} = \frac{(\xi + c)^2 - 1}{2\xi}.$$

The rationality of ξ is thus a necessary condition for the rationality of the ratios of the sides of triangle AEB.   It is a sufficient condition, since

$$\frac{a}{\beta} = \frac{\xi^2 - c^2 + 1}{2\xi},$$

by (1).   There are similar formulas for the remaining three triangles whose angles at E are w and $\pi - w$.   Taking β as the unit of length, we have

$$(3) \qquad \alpha = \frac{(\xi + c)^2 - 1}{2\xi}, \qquad \gamma = \frac{(\eta - c)^2 - 1}{2\eta},$$

$$(4) \qquad \frac{\delta}{\alpha} = \frac{(x - c)^2 - 1}{2x}, \qquad \frac{\delta}{\gamma} = \frac{(y + c)^2 - 1}{2y},$$

where ξ, η, x, y are rational.   By multiplication, we obtain two values for δ. Hence we have the condition

$$(5) \qquad \frac{(\xi + c)^2 - 1}{2\xi} \cdot \frac{(x - c)^2 - 1}{2x} = \frac{(\eta - c)^2 - 1}{2\eta} \cdot \frac{(y + c)^2 - 1}{2y}.$$

Hence for any set of rational solutions of (5), such that $|c| < 1$, we obtain a quadrilateral with rational diagonals and rational sides

$$(6) \quad AB = \frac{\xi^2 + t}{2\xi}, \quad BC = \frac{\eta^2 + t}{2\eta}, \quad CD = \gamma\left(\frac{y^2 + t}{2y}\right), \quad DA = \alpha\left(\frac{x^2 + t}{2x}\right),$$

where $t = 1 - c^2$, while α, γ are given by (3).

Let also the area $\frac{1}{2}(\alpha\beta + \beta\gamma + \gamma\delta + \delta\alpha)\sin w$ of the quadrilateral be rational, and hence also $\sin w$.   The rational solutions of $\sin^2 w + c^2 = 1$ are

$$\sin w = \frac{2\lambda}{\lambda^2 + 1}, \qquad c = \frac{\lambda^2 - 1}{\lambda^2 + 1}.$$

Hence to obtain all rational quadrilaterals we have only to seek the rational solutions c, ξ, η, x, y of (5) for which c is of the form $(\lambda^2 - 1)/(\lambda^2 + 1)$. Now (5) is a quadratic equation in y whose discriminant must be a square:

$$(7) \qquad \{ax^2 - 2c(\alpha + \gamma)x - \alpha t\}^2 + 4t\gamma^2 x^2 = z^2.$$

Hence we may obtain all rational quadrilaterals as follows: Give arbitrary rational values to ξ, η, λ and set

$$c = \frac{\lambda^2 - 1}{\lambda^2 + 1}, \qquad t = 1 - c^2.$$

Determine all rational solutions[134] x, z of (7). Then (5) determines two rational values of y, and (3), (4), (6) give the segments of the diagonals and the sides as rational numbers.

W. Ligowski[135] and J. Cunliffe[135a] gave special rational inscribed quadrilaterals.

D. S. Hart[136] desired an inscriptible quadrilateral with integral sides a, b, c, d and diagonals x, y. Thus $xy = ac + bd$, $x : y = bc + ad : ab + cd$, so that the product of the three sums is to be a square, say the square of $abc + d(a^2 + b^2 + c^2)/2$, which determines d. A. B. Evans took the sides $AB = x$, $BC = mx$, $CD = nx$, $AD = px$. As known,

$$AC^2 = (mp + n)\alpha x^2, \qquad BD^2 = (mp + n)x^2/\alpha, \qquad \alpha = \frac{p + mn}{m + pn}.$$

The last gives p rationally. Let

$$\alpha = a^2, \qquad n = q^2, \qquad mp + n = \{q + my/(a^2q^2 - 1)\}^2,$$

which gives m. Hart[137] found a trapezoid with integral values for the sides, diagonals, area and perpendicular between the parallel sides.

G. Darboux[138] based a geometrical theory of quadrilaterals upon two equations

$$at_1 + bt_2 + ct_3 + dt_4 = 0, \qquad \frac{a}{t_1} + \frac{b}{t_2} + \frac{c}{t_3} + \frac{d}{t_4} = 0,$$

where a, b, c, d are the sides, and $t_j = e^{i\omega_j}$, ω_j being the angle between the jth side and any line in the plane. Regarding the t's as homogeneous coordinates, we have a plane cubic curve.

O. Schlömilch,[139] started with two right triangles $T_a = (1 - \alpha^2, 2\alpha, 1 + \alpha^2)$ and T_β, reduced their sides proportionally to obtain a common leg, and juxtaposed them to obtain a triangle with the sides $(1 + \alpha^2)\beta$, $(\alpha + \beta)(1 - \alpha\beta)$, $(1 + \beta^2)\alpha$. Treating two such oblique triangles similarly, we obtain a quadrilateral with the sides $(1 + \alpha^2)\beta$, $(1 + \beta^2)\alpha$, $(1 + \gamma^2)\delta\epsilon$, $(1 + \delta^2)\gamma\epsilon$, where

$$\epsilon = \frac{(\alpha + \beta)(1 - \alpha\beta)}{(\gamma + \delta)(1 - \gamma\delta)}.$$

The sides, diagonals and area are rational if $\alpha, \cdots, \delta$ are.

S. Robins[140] listed rational trapeziums whose area equals the square root of the product of the four sides, found by use of convergents to $\sqrt{a^2 + 1}$.

H. Schubert[88] (pp. 49–54) considered quadrilaterals inscribed in a circle of radius r. Let $2\alpha_1, \cdots, 2\alpha_4$ be the arcs subtended by the sides. Then

[134] From simple solutions of (7), Kummer obtained new solutions by the method of Euler[142–145] of Ch. XXII and thus deduced various rules for forming rational quadrilaterals.

[135] Archiv Math. Phys., 47, 1867, 113–6.

[135a] New Series of Math. Repository (ed., T. Leybourn), 2, 1809, I, 74–5, 225–6.

[136] Math. Quest. Educ. Times, 20, 1874, 64–5.

[137] Ibid., 80–81. For history of inscriptible quadrilaterals with given sides, 21, 1874, 29–35.

[138] Bull. Sc. Math. Astr., (2), 3, I, 1879, 109–128; Comptes Rendus Paris, 88, 1879, 1183, 1252.

[139] Zeitschr. Math. Naturw. Unterricht, 24, 1893, 401–9.

[140] Amer. Math. Monthly, 5, 1898, 181–2.

the sides are $2r \sin \alpha_i$, the diagonals are

$$e = 2r \sin (\alpha_1 + \alpha_2), \qquad f = 2r \sin (\alpha_2 + \alpha_3).$$

The area is $\tfrac{1}{2}ef \sin (\alpha_1 + \alpha_3)$. In the very special case in which the tangents of $\tfrac{1}{2}\alpha_1$, $\tfrac{1}{2}\alpha_2$, $\tfrac{1}{2}\alpha_3$ are rational, as well as one side or r, the four sides, diagonals and area will be rational.

A. Gérardin[141] juxtaposed two right triangles with a common hypotenuse to obtain a quadrilateral whose sides have the values quoted from Brahmegupta by Kummer; also a second quadrilateral.

E. N. Barisien[142] noted the inscriptible quadrilateral with the sides $AB = 75$, $BC = 68$, $CD = 40$, $DA = 51$, segments of diagonals (at right angles) $AI = 45$, $BI = 60$, $CI = 32$, $DI = 24$, and diameter 85 of circumscribed circle.

F. Neiss[142a] treated rational triangles and rational quadrilaterals.

I. Newton[142b] treated the problem to find the diameter $x = DA$ of a circle having an inscribed quadrilateral $ABCD$, three of whose consecutive sides $a = AB$, $b = BC$, $c = CD$ are given, while the fourth side is the diameter. We have $x^3 - (a^2 + b^2 + c^2)x - 2abc = 0$. E. Haentzschel and E. Lampe[142c] found rational quadrilaterals of this type by the method of Kummer.[133]

E. Haentzschel[143] treated rational quadrilaterals with perpendicular diagonals by setting $c = 0$, $t = 1$, in Kummer's work. Condition (7) is now $\alpha^2(x^2 - 1)^2 + 4\gamma^2x^2 = z^2$; methods of finding rational solutions are developed. An evident special solution is obtained by taking $\xi = \eta$; then $y = x$, $AB = BC$, $CD = AD$, and the quadrilateral is given by the juxtaposition of two congruent rational triangles. Next, taking $x = \eta = c/d$, $y = \xi = a/b$, we get Brahmegupta's quadrilateral as quoted by Kummer. More general solutions are found by use of Weierstrass' $\wp$-function.

Haentzschel[144] noted that the determination of a quadrilateral with rational sides, diagonals, area, and radii of the inscribed and circumscribed circles, depends on the rational solution of

$$(\mu^2 + 1)(\nu^2 + 1)\{(\mu^2 + 1)(\nu^2 + 1) + 4\mu\nu\} = \square.$$

By use of Weierstrass' $\wp$-function, he found two infinite sets of rational solutions, including the special solutions by O. Schulz[157] (pp. 98–103). Ankum's method to deduce a rational tetrahedron from a rational quadrilateral is applied to the quadrilaterals found here.

E. N. Barisien[145] noted that in the quadrilateral with the sides

$$AB = 1625, \qquad BC = 2535, \qquad CD = 3900, \qquad DA = 3380,$$

<hr>

[141] Sphinx-Oedipe, 6, 1911, 187.

[142] Mathesis, (4), 3, 1913, 263. He noted (p. 14) the quadrilateral with successive sides 15, 20, 24, 7, diagonals 20, 25, and area 234.

[142a] Rationale Dreiecke, Vierecke . . . , Diss., Leipzig, 1914.

[142b] Arithmetica universalis, Amsterdam, 1, 1761, IV, Ch. 1, 140–150.

[142c] Zeitschrift Math. Naturw. Unterricht, 46, 1915, 190–4; 49, 1918, 139–144, 144–5.

[143] Sitzungsber. Berlin Math. Gesell., 14, 1915, 23–31.

[144] Ibid., 14, 1915, 85–94.

[145] L'intermédiaire des math., 23, 1916, 195–6.

with diagonals crossing at right angles at I, and having E, F, G, H as the projections of I on the sides and K, L, M, N as its projections on EF, FG, GH, HE, there are integral values for the distances from I to these 12 points, for the 8 segments on the sides, and for the 8 segments on EF, FG, GH, HE.

E. Turrière[146] gave known results on inscriptible quadrilaterals with rational sides and diagonals.

W. F. Beard stated and G. N. Watson[147] proved that if two circles with centers O and O', and radii R and R', are such that quadrilaterals can be inscribed in the first and circumscribed about the second circle, then the least integral values for R, R', $c = OO'$ are 35, 24, 5, while general solutions follow from

$$(R^2 - c^2)^2 = \{R'(R + c)\}^2 + \{R'(R - c)\}^2.$$

For rational quadrilaterals, see Turrière,[61–62] Euler[148] and Schwering[150]; also, Euler[32] of Ch. XV.   Cf. Berton[49] of Ch. XXIII.

Rational Inscribed Polygons.

L. Euler[148] gave a construction to find a polygon with any number n of sides, inscribed in a circle with center O and radius unity, such that the sides, all diagonals, and the area are rational.   Employ $n - 1$ arbitrary angles $2A$, $2B$, $\cdots$, and take as the nth angle one whose sine and cosine equal the sine and negative of the cosine of the sum of those $n - 1$ angles. Take arc $AB = 2A$, arc $BC = 2B$, arc $CD = 2C$, etc.   Hence side AB is $2 \sin A$, side BC is $2 \sin B$, $\cdots$, diagonal AC is $2 \sin (A + B)$, $\cdots$.   To make all the sines and cosines rational, take $\sin A = 2ab/(a^2 + b^2)$, etc. Since triangle AOB equals $\sin A \cos A$, the area is rational.   He gave complicated expressions which serve as rational sides and diagonals of an inscribed quadrilateral, but do not make the area rational.

H. Schubert[47] (pp. 28–38, or Schubert,[88] pp. 55–67) considered an inscribed polygon with the sides a_1, $\cdots$, a_n.   Let $2\alpha_i$ be the arc subtended by a_i.   Let $n - 1$ of the α's (and hence all) be Heron angles.[47]   Let

$$\tan \tfrac{1}{2}\alpha_i = q_i/p_i, \qquad 4r = \prod_{i=1}^{n-1} (p_i^2 + q_i^2),$$

so that r is the radius of the circumscribed circle.   Then the sides $a_i = 2r \sin \alpha_i$ are rational, also all diagonals since the ratio of any one to $2r$ is the sine of a sum of certain α's.   The area $(\sin 2\alpha_1 + \cdots + \sin 2\alpha_n)r^2/2$ is rational.

J. Cunliffe[67] found rational inscribed pentagons.

Rational Pyramids; Rational Trihedral Angles.

A rational pyramid is one whose edges and volume V are rational.

R. Hoppe[149] considered a rational trihedral angle (one having rational sines and cosines of the face and dihedral angles).   Let a, b, c be the tangents

[146] L'enseignement math., 18, 1916, 408–410.
[147] Math. Quest. and Solutions, 4, 1917, 31–2.
[148] Opera postuma, 1, 1862, 229 (about 1781).
[149] Archiv Math. u. Phys., 61, 1877, 86–98.

of the half face angles. Then the cosine of the dihedral angle (b, c) is $[b^2 + c^2 - a^2(1 + b^2c^2)]/\{2bc(1 + a^2)\}$. Adding and subtracting 1, we obtain as factors of the numerators

$$D = a + b + c - abc, \qquad A = -a + b + c + abc,$$
$$B = a - b + c + abc, \qquad C = a + b - c + abc.$$

Hence $s = \sin(b, c) = \sqrt{ABCD}/\{2bc(1 + a^2)\}$. If f, g, h are the tangents of the half dihedral angles, then

$$s = 2f/(1 + f^2), \qquad (1 + b^2)(1 + h^2)/bh = (1 + c^2)(1 + g^2)/cg,$$

etc. If the latter equation has rational solutions, we obtain 32 distinct rational trihedrals, since we may replace b by its reciprocal, etc.

To obtain a rational tetrahedron, we may take two rational trihedrals having a common dihedral angle and subject to the condition that the edges converge (in the earlier notation, $bb' < 1$, $cc' < 1$, $f = f'$). While the tetrahedron now has a rational volume, it remains to make the sixth edge rational. The condition is that $b_1^2 + b_2^2 + c_1^2 + c_2^2 - 2 - 2b_1b_2c_1c_2 - 2m$ be a square, where

$$b_1 = \frac{1 + bb'}{1 - bb'}, \qquad b_2 = \frac{b - b'}{b + b'}, \qquad c_1 = \frac{1 + cc'}{1 - cc'}, \qquad c_2 = \frac{c - c'}{c + c'},$$

$$m = \frac{16bcb'c'(1 - f^2)}{(1 - bb')(1 - cc')(b + b')(c + c')(1 + f^2)}.$$

K. Schwering[150] discussed rational tetrahedra by use of the formula

$$36V^2 = f^2g^2h^2F, \qquad F = (1 - \cos^2 \alpha)(1 - \cos^2 \beta) - (\cos \gamma - \cos \alpha \cos \beta)^2,$$

where f, g, h are the edges from the vertex D, and α, β, γ are the face angles at D, while a, b, c are the sides of the base of the tetrahedron. The first problem is to choose rational values of the cosines such that F shall be the square of a rational number. The first term of F must be the sum of two squares. Give $1 - \cos^2 \alpha$ the form of a fraction whose denominator is a perfect square. Then its numerator is a divisor of a sum of the squares of two integers and hence is itself the sum of two squares. Thus $1 - \cos^2 \alpha$ equals the sum of two rational squares. Hence $\cos^2 \alpha$ is one of three rational squares whose sum is unity; likewise for $\cos^2 \beta$. Consider the integral squares equal to the sums of the squares of three integers; for instance

$$(m^2 + n^2 + p^2 + q^2)^2 = (m^2 + n^2 - p^2 - q^2)^2 + (2mp + 2nq)^2 + (2mq - 2np)^2.$$

If

$$Q^2 = M^2 + N^2 + P^2, \qquad Q_1^2 = M_1^2 + N_1^2 + P_1^2,$$

we take

$$\cos \alpha = \frac{M}{Q}, \qquad \cos \beta = \frac{M_1}{Q_1}, \qquad \cos \gamma = \frac{MM_1 - NP_1 + PN_1}{QQ_1}$$

and find that F is the square of $(NN_1 + PP_1)/QQ_1$.

---

[150] Jour. für Math., 115, 1895, 301–7.

The next problem is to find rational solutions of

$$a^2 = g^2 + h^2 - 2gh \cos \alpha, \quad b^2 = h^2 + f^2 - 2hf \cos \beta, \quad c^2 = f^2 + g^2 - 2fg \cos \gamma,$$

where the cosines are given rational numbers. Set

$$a = \lambda g + h, \qquad b = \mu f + h, \qquad c = \nu g + f.$$

Then

$$g(1 - \lambda^2) = 2h(\lambda + \cos \alpha), \qquad f(1 - \mu^2) = 2h(\mu + \cos \beta),$$
$$g(1 - \nu^2) = 2f(\nu + \cos \gamma).$$

Hence g/f has the value

$$q = \frac{1 - \mu^2}{1 - \lambda^2} \cdot \frac{\lambda + \cos \alpha}{\mu + \cos \beta} = \frac{2(\nu + \cos \gamma)}{1 - \nu^2}.$$

If $\cos \alpha = \cos \beta = \cos \gamma = 0$, we have a rectangular tetrahedron and the problem reduces to that treated by Euler[3] of Ch. XIX to find three squares such that their sums by pairs are squares. This process of Euler leads in the general problem to

$$- \lambda = \frac{p^2(1 + \cos \gamma) + 2p(\cos \alpha + \cos \beta) + 1 - \cos \gamma + 2 \cos \alpha \cos \beta}{4(p + \cos \beta)}.$$

For example, let $M = N = 0$, $P = Q = 3$, $M_1 = P_1 = 2$, $N_1 = - 1$, $Q_1 = 3$. Then

$$\cos \alpha = 0, \qquad \cos \beta = \frac{2}{3}, \qquad \cos \gamma = \frac{-1}{3}, \qquad - \lambda = \frac{p^2 + 2p + 2}{6p + 4}.$$

Thus f, g, h are proportional to $(6p + 4)(p^2 - 2p - 4)(5p^2 + 2p - 2)$, $6(6p + 4)(p^2 - 1)(p^2 + 2p + 2)$, $3(p^2 - 1)(p^2 - 4p - 2)(p^2 + 8p + 6)$. For $p = 0$, we remove the factor 4, and get $f = 8$, $g = - 12$, $h = 9$, $a = 15, b = - 7, c = 12, V = 96$, in which the signs may be taken positive. For $p = - 2$ we get $f = 112, g = 72, h = 135, a = 153, b = 103, c = 152$, $V = 120960$.

To obtain a rational quadrilateral, set $\beta + \gamma = \alpha$ or $2\pi - \alpha$. For example, for $\cos \alpha = \cos \beta = \cos \gamma = - \frac{1}{2}$, we have

$$f = (7p^2 - 4)(p^2 - 4)(2p - 1), \qquad g = 8(p^2 - 1)(p^2 + 2)(2p - 1),$$
$$h = p(p^2 - 1)(p + 4)(p^2 - 12p + 8).$$

Thus, for $p = - \frac{1}{2}$, we have the rational quadrilateral $ABCD$, in which

$$AB = 138, BC = 192, CD = 168, DA = 127, AC = 283, DB = 120.$$

We obtain a simpler solution by taking $\lambda = \mu$. Thus, for $\cos \alpha = - 3/7$, $\cos \beta = 0$, $\cos \gamma = 2/7$, $\nu = - 2$, we have $\lambda = - 3$ and $f = 6, g = 7, h = 8$, $a = 9, b = 10, c = 11, V = 48$. For $\cos \alpha = \cos \gamma = \frac{1}{2}$, $\cos \beta = - \frac{1}{2}$, $\nu = 2$, we get the rational quadrilateral $AB = 48, BC = 57, CD = 73$, $DA = 80, AC = 63, BD = 112$.

H. Schubert[47] (pp. 50–7, or Schubert,[88] 92–104) employed a rational polygon inscribed in a circle of radius r and center C. Draw a perpendicular to its plane at C and of length h such that in the right triangle of legs h and r the angle opposite h is a Heron angle[47] μ. Thus we have a rational pyramid. For example, if the sides of the triangular base are 13, 14, 15, take $\cos \mu = 65/97$, $\sin \mu = 72/97$; then the altitude is $h = 9$, lateral edge $97/8$, and volume 252.

Schubert[151] discussed rational spherical triangles, i. e., having rational values for the tangents of half of each side and angle.

R. Güntsche[152] made use of F. Bessell's[153] relations between the face and trihedral angles and reduced the problem of the rational tetrahedron to a diophantine equation quadratic in q and quadratic in r with coefficients involving an arbitrary parameter p. Euler's[144] process of Ch. XXII is used to find solutions q, r rational in p, so that the six edges, the surface areas and volume are expressed rationally in p.

Güntsche[154] considered tetrahedra whose edges, surface areas and volume are all rational and having all faces congruent. He reduced the problem to the solution of

$$\psi\theta(\psi\theta + \psi + \theta - 1)(\psi\theta - \psi - \theta - 1) = h^2,$$

but did not solve it in general. But seven particular sets of solutions involving an arbitrary parameter are found.[155] The tetrahedra of Hoppe[149] are all of the type here considered.

E. Haentzschel[156] wrote Güntsche's cubic function in the form

$$\psi^3(\theta^3 - \theta) - 4\psi^2\theta^2 - \psi(\theta^3 - \theta)$$

and reduced it to Weierstrass' normal form $4\Pi(s - e_i)$ by the substitution

$$\psi = \frac{4(s + \theta^2/3)}{\theta^3 - \theta},$$

obtaining $e_1 = -\theta^2/3$; $e_2, e_3 = \mp \theta^3/4 + \theta^2/6 \mp \theta/4$. By use of Weierstrass' $\wp$-function, he solved $4\Pi(s - e_i) = v^2$. The case $\theta = 7/3$ is treated in detail.

* O. Schulz[157] treated rational tetrahedra.

For special tetrahedra, see papers 30-31 of Ch. XIX.

<hr>

[151] Auslese . . . Unterrichts- und Vorlesungspraxis, 3, 1906, 202-250.

[152] Sitzungsber. Berlin Math. Gesell., 6, 1907, 2-16.

[153] Archiv Math. Phys., 65, 1880, 363-372, on spherical triangles with rational values for the sines and cosines of the angles and sides. Cf. M. Bambas, (3), 26, 1918, 195-6.

[154] Sitzungsber. Berlin Math. Gesell., 6, 1907, 38-53.

[155] He gave two such sets in Archiv Math. Phys., (3), 11, 1907, 371.

[156] Sitzungsber. Berlin Math. Gesell., 12, 1913, 101-8. Continued, 17, 1918, 37-9.

[157] Ueber Tetraeder mit rationalen Masszahlen der Kantenlängen und des Volumens, Halle, 1914, 292 pp. Cf. Haentzschel.[144]

CHAPTER VI.

SUM OF TWO SQUARES.

Diophantus, II, 10, divided a given number $13 = 2^2 + 3^2$, which is a sum of two squares, into two other squares, $(z + 2)^2 + (mz - 3)^2$, by taking $m = 2$, whence $z = 8/5$. In III, 22, Diophantus required four numbers x_i such that each of the eight expressions $E = (\Sigma x_i)^2 \pm x_i$ shall be a square. In any right triangle (p, b, h), $h^2 \pm 2pb = \square$. [If $h^2 = p_i^2 + b_i^2$ $(i = 1, \cdots, 4)$, take $x_i = 2p_i b_i x^2$, $\Sigma x_i = hx$; then $E = x^2(h^2 \pm 2p_i b_i) = \square$.] Hence we seek four right triangles with equal hypotenuses. We must therefore find a square which can be expressed as a sum of two squares in four ways. Take the right triangles (3, 4, 5) and (5, 12, 13); multiply the sides of each by the hypotenuse of the other. We obtain the triangles (39, 52, 65) and (25, 60, 65) with equal hypotenuses. The number 65 can be expressed as a sum of two squares in two ways: $65 = 4^2 + 7^2 = 1^2 + 8^2$, since 65 is the product of 13 and 5, each a sum of two squares. Now form[*] the right triangle (33, 56, 65) from 7, 4 and (16, 63, 65) from 8, 1. We now have four right triangles with equal hypotenuses. [If we carry out the corresponding process on the right triangles $(a^2 - b^2, 2ab, a^2 + b^2)$, $(c^2 - d^2, 2cd, c^2 + d^2)$, we obtain by multiplication two triangles with the hypotenuse[†]

$$(1) \qquad (a^2 + b^2)(c^2 + d^2) = (ac \pm bd)^2 + (ad \mp bc)^2.$$

The right triangles formed from $ac \pm bd$ and $ad \mp bc$ give two new triangles with the same hypotenuse, provided c/d is distinct from a/b, b/a, $(a \pm b)/(a \mp b)$.]

Diophantus, V, 12, treated the division of unity into two parts such that, if a given number a is added to each part, the sums are (rational) squares. The problem is equivalent to the representation of $2a + 1$ as a sum of two squares. It is stated that a must not be odd [so that no number $4n - 1$ is a sum of two squares]. Unfortunately the text of the second part of the necessary condition is very obscure. C. G. J. Jacobi[1] emended it to read that $2a + 1$ must have no factor of the form $4n - 1$; P. Tannery and T. L. Heath, in their editions of Diophantus, read prime factor; but neither correction makes the criterion exact.

Diophantus, VI, 15, stated that 15 is not a sum of two (rational) squares.

Mohammed Ben Alhocain,[2] an Arab of the tenth century, gave a table of numbers equal to a sum of two squares, formed by adding each square to itself and to the larger squares. It is stated falsely that if an even number is a sum of two squares, one of them is unity.

---

[*] See Ch. IV, Diophantus.[7]

[†] For a like composition of factors $a^2 - eb^2$, see Euler[66] of Ch. XII.

[1] Berichte Akad. Wiss. Berlin, 1847, 265–278; Werke, 7, 1891, 332–344 (report below). Same by H. Hankel, Zur Geschichte der Math., 1874, 169.

[2] Cf. F. Woepcke, Atti Accad. Pont. Nuovi Lincei, 14, 1860–1, 306–9.

Leonardo Pisano,[3] in his Liber Quadratorum of 1225, proved (1) and used it to solve $x^2 + y^2 = a^2 + b^2$, given a solution of $c^2 + d^2 = e^2$:

$$x = (ac + bd)/e, \qquad y = (ad - bc)/e.$$

This solution was reproduced without proof by Lucas Paciuolo and Cardan in their arithmetics (full titles on p. 6 and p. 8 of Vol. I).

F. Vieta[4] noted that $X^2 = F^2 + G^2$, $Z^2 = B^2 + D^2$ imply

$$(1') \qquad\qquad (XZ)^2 = (BG \pm DF)^2 + (BF \mp DG)^2.$$

If B and D are the hypotenuses, M, N and MD/B, ND/B the pairs of legs of two similar right triangles, a third right triangle with the legs $(BM \pm DN)/B$ and $(BN \mp DM)/B$ has the hypotenuse $\sqrt{B^2 + D^2}$. In the special case $F = B$, $G = D$, (1') becomes $(X^2)^2 = (2BD)^2 + (B^2 - D^2)^2$; the right triangle with the sides $2BD$, $D^2 - B^2$, X^2 is called the triangle of double angle. Using the latter and the given triangle (B, D, X), and applying the same rule, we obtain the triangle $(3BD^2 - B^3, D^3 - 3B^2D, X^3)$ of triple angle, etc. [equivalent to De Moivre's formulas for $\cos na$, $\sin na$ in terms of $\cos a$, $\sin a$].

Vieta,[5] to express $Z^2 = B^2 + D^2$ as the sum of two new squares, employed a second right triangle (F, G, X) to obtain (1'), whence [cf. L. Pisano[3]]

$$Z^2 = \left(\frac{BG \pm DF}{X}\right)^2 + \left(\frac{BF \mp DG}{X}\right)^2.$$

He noted that the method of Diophantus II, 10 consists in denoting the sides of the required squares by $A + B$, $SA/R - D$. Thus

$$A = \frac{2SRD - 2R^2B}{S^2 + R^2}, \qquad A + B = \frac{2SRD + B(S^2 - R^2)}{S^2 + R^2}.$$

Hence from (B, D, Z) and the triangle $(2SR, S^2 - R^2, S^2 + R^2)$, formed from S, R, construct a third triangle by (1') and reduce the sides in the ratio $R^2 + S^2$.

G. Xylander,[6] in his comment on Diophantus V, 12, stated incorrectly that a must be the double of a prime.

C. G. Bachet[7] remarked that 10 is the double of a prime, while $2 \cdot 10 + 1 = 21$ is neither a square nor the sum of two integral squares, and expressed his belief that 21 is not the sum of two rational squares. While Diophantus seemed to infer that the double of the even number a, increased by unity, should be a prime, this would exclude 22, 58, 62, for which $2 \cdot 22 + 1 = 45 = 36 + 9$, etc., whereas 45, 117 are not primes.

---

[3] Tre Scritti inediti, 1854, 66–70, 74–5; Scritti L. Pisano, 2, 1862, 256. Review by O. Terquem, Annali Sc. Mat. Fis., 7, 1856, 138; Nouv. Ann. Math., 15, 1856, Bull. Bibl. Hist., 61. Cf. Woepcke, Jour. de Math., 20, 1855, 57; A. Genocchi, Annali Sc. Mat. Fis., 6, 1855, 241–4; M. Chasles, Jour. de Math., 2, 1837, 42–9, who gave a geometrical proof.

[4] Ad Logisticem Speciosam Notae Priores, Props. 46–48; Opera Math., 1646, 34. French transl. by F. Ritter, Bull. Bibl. Storia Sc. Mat., 1, 1868, 267–9.

[5] Zetetica, 1591, IV, 2, 3; Opera Math., 1646, 62–3.

[6] Diophanti Alex. Rerum Arith. Libri sex, Basel, 1575, 129, l. 9.

[7] Diophanti Alex. Arith., 1621, 301–4.

He treated the generalization to divide any number (as 2) into two parts such that, if a given number (as 4) is added to each part, the sums are squares,—whence 10 is to be expressed as a sum of two squares each > 4.

Fermat's[8] comment was: " The true condition (namely, that which is general and which excludes all the numbers which are inadmissible) is that the given number a must not be odd and that $2a + 1$, when divided by the largest square entering it as a factor, must not be divisible by a prime $4n - 1$."

A. Girard[9] († Dec. 9, 1632) had already made a determination of the numbers expressible as a sum of two integral squares: every square, every prime $4n + 1$, a product formed of such numbers, and the double of one of the foregoing.

Bachet[7] (p. 173) in his comment on Diophantus III, 22 found that 5525 is the sum of the squares of 55 and 50, 62 and 41, 70 and 25, 71 and 22, 73 and 14, 74 and 7.   Also $1073 = 32^2 + 7^2 = 28^2 + 17^2$ is a sum of two squares in four ways.   Thus $5525 \cdot 1073$ is a sum of two squares in 24 ways, all being given.   He stated and proved (1) in his Porisms, III, 7.

Fermat[10] made, apropos of Bachet's preceding comments, the remarks:

(A) Every prime of the form $4n + 1$ is the hypotenuse of a right triangle in a single way, its square in two ways, its cube in three, its biquadrate in four, and so on indefinitely.

(B) The same prime $[4n + 1]$ and its square are the sums of two squares in a single way, its cube and biquadrate in two ways, its fifth and sixth powers in three ways, and so on indefinitely.

(C) If a prime which is the sum of two squares be multiplied by another prime also the sum of two squares, the product will be the sum of two squares in two distinct ways; if the first prime be multiplied by the square of the second prime, the product will be the sum of two squares in three distinct ways; if the first prime be multiplied by the cube of the second, the product will be the sum of two squares in four distinct ways, and so on indefinitely.

(D) It is now easy to determine in how many ways w a given number can be the hypotenuse of a right triangle.   For the number $p^a q^b r^c s$, where p, q, r are primes of the form $4n + 1$, while s is a square having no such prime factor,

$$w = 2c(2ab + a + b) + 2ab + a + b + c.$$

Here, and in (E), Fermat used numerical values.

(E) To find a number which is an hypotenuse in an assigned number w of ways, take the prime factors of $2w + 1$, subtract 1 from each and

[8] Oeuvres, III, 256.

[9] L'arith. de Simon Stevin $\cdots$ annotations par A. Girard, Leide, 1625, 622; Oeuvres Math. de Simon Stevin par Albert Girard, 1634, p. 156, col. 1, note on Diophantus V, 12.   Cf. G. Vacca, Bibliotheca Math., (3), 2, 1901, 358–9.   Cf. G. Maupin, Opinions et Curiosités touchant la Mathématique, Paris, 2, 1902, 158–325.

[10] Oeuvres, I, 293; III, 243–6.   Diophanti Alex. Arith., ed., S. Fermat, 1670, 127.

take half of the remainder as the exponent of any prime $4n + 1$.   [Since

$$2w + 1 = (2a + 1)(2b + 1)(2c + 1)\cdots,$$

by D.]   For $w = 7$, $15 = (2 + 1)(2 \cdot 2 + 1)$, and pq^2 answers the question.

(F) To find a number which shall be the sum of two squares in any assigned number w of ways.   For $w = 10$, set $2w = 2 \cdot 2 \cdot 5$.   Subtracting 1 from each prime factor, we get 1, 1, 4.   Take three primes of the form $4n + 1$; for example, 3, 13, 17.   The number sought is the product of two of these by the fourth power of the third.

(G) Conversely, to find in how many ways a given number, say 325, is the sum of two squares, consider its prime factors of the form $4n + 1$. Since $325 = 5^2 \cdot 13$, we take $\frac{1}{2}\{2 \cdot 1 + 2 + 1 + 1\} = 3$.   Then 325 is the sum of two squares in three ways.   For three exponents a, b, c, the number of ways is $k/2$ if $k = (a + 1)(b + 1)(c + 1)$ is even, but is $(k - 1)/2$ if k is odd.

(H) To find an integer which is the hypotenuse of any assigned number w of right triangles, and which if increased by a given number a becomes a square.   The question is difficult.   If $w = a = 2$, 2023 and 3362 satisfy the conditions, as do also an infinitude of numbers.

That no number $4n - 1$ is a square or a sum of two rational squares was communicated to Descartes March 22, 1638, as having been proved by Fermat.   Descartes[11] proved this for integral squares by observing that a square is of the form $4k$ or $8k + 1$.

Fermat[12] stated that he had proved that a number is neither a square nor the sum of two squares, integral or fractional, if its quotient by the largest square dividing it contains a prime factor $4n - 1$; and that $x^2 + y^2$ is divisible by no prime $4n - 1$ if x and y are relatively prime.

Fermat (Oeuvres, II, 213) stated the contents of A, B, D, E in a letter to Mersenne, Dec. 25, 1640.   Frenicle, in a letter to Fermat (*ibid.*, 241), Sept. 6, 1641, proposed the problem to find the least number in F.   T. Pepin[13] noted that this problem and D are answered by the theory of quadratic forms.

Fermat[14] called the theorem that every prime $4n + 1$ is a sum of two squares [cited henceforth as Girard's[9] theorem] the fundamental theorem on right triangles.   He[15] stated that he possessed an irrefutable proof. Elsewhere he[16] stated that his proof was by the method of indefinite descent: "If a prime $4n + 1$ is not a sum of two squares, there exists a smaller prime of the same nature, then a third still smaller, etc., until the number 5 is reached," thus leading to a contradiction.   He found it much more difficult to apply the method to such an affirmative question than to negative theorems (cf. Fermat,[2] etc., Ch. XXII); for the former, "the method had to be supplemented by some new principles."

[11] Oeuvres de Descartes, II, 92; letter to Mersenne, March 31, 1638. Cf. p. 195.

[12] Oeuvres, II, 203–4; letter to Roberval, Aug., 1640.

[13] Memorie Accad. Pont. Nuovi Lincei, 8, 1892, 84–108; Oeuvres de Fermat, 4, 1912, 205–7.

[14] Oeuvres, II, 221; letter to Frenicle, June 15, 1641.

[15] Oeuvres, II, 313, 403; III, 315; letters to Pascal, Sept. 25, 1654, and to Digby, June 19, 1658.

[16] Oeuvres, II, 432; letter to Carcavi, communicated to Huygens, Aug. 14, 1659.

Frenicle[17] concluded from numerical tables that, if p_1, p_2, $\cdots$ are distinct primes, each the hypotenuse of a right triangle (a necessary and sufficient condition being that the prime is of the form $4k + 1$), a number $N = p_1^{e_1} \cdots p_n^{e_n}$ is the hypotenuse of exactly 2^{n-1} primitive right triangles (i. e., with relatively prime legs). He recognized that the problem reduces to the question of the number of ways in which the proposed number N can be expressed as the product of two relatively prime factors. The nonprimitive triangles are obtained from the primitive triangles whose hypotenuses are the factors of N. Fermat's rule D is given. Problem G is discussed (pp. 34–46).

John Kersey,[18] to treat $x^2 + y^2 = d^2 + b^2$ of Diophantus II, 10, set $x = ra + b$, $y = sa - d$. Thus $a = 2(sd - rb)/(s^2 + r^2)$, so that the values of x, y follow. He also treated the problem [Bachet,[7] 304] with the restriction that x or y shall fall within given limits.

Claude Jaquemet,[19] in a letter Jan. 26, 1690, proved that an integer not a square, which divides no sum of two squares without dividing each square, is not a sum of two squares, integral or fractional. A manuscript by Jaquemet or N. Malebranche proved also that a number which divides a sum of two relatively prime squares is itself a sum of two squares; but the later proof by Euler[24] is far simpler. Cf. Bháscara,[30] § 88, of Ch. XII.

The Japanese Matsunago,[20] the first half of the 18th century, would solve $x^2 + y^2 = k$ by setting $k/2 = r^2 + R$, where r^2 is the greatest square contained in k, and forming the equations

$$a_1 = 2r - 1, \qquad a_2 = a_1 - 2, \qquad a_3 = a_2 - 2, \qquad \cdots,$$
$$b_1 = 2r + 1, \qquad b_2 = b_1 + 2, \qquad b_3 = b_2 + 2, \qquad \cdots.$$

From $2R$ subtract successively b_1, b_2, $\cdots$. When a difference is negative, add the corresponding a_i. If the remainder zero is reached, and a', b' are the values last employed, a solution is

$$x = \tfrac{1}{2}(a' + 1), \qquad y = \tfrac{1}{2}(b' - 1).$$

It was stated that a set of solutions of $x^2 + y^2 = z^3$ is given by

$$x = (m^2 - 3n^2)m, \qquad y = (3m^2 - n^2)n, \qquad z = m^2 + n^2.$$

L. Euler[21] proved that, if neither a nor b is divisible by the prime $p = 4n - 1$, then $a^2 + b^2$ is not divisible by p. For, $a^{4n-2} - b^{4n-2}$ is divisible by p and hence $a^{4n-2} + b^{4n-2}$ is not; thus the factor $a^2 + b^2$ of the latter is not divisible by p.

Euler[22] stated that if $4m + 1$ is composite it is either not a sum $\boxed{2}$ of two squares or is so in more than one way; if ab and a are $\boxed{2}$, b is a $\boxed{2}$.

---

[17] Mém. Acad. Roy. Sc., 5, 1666–99, éd. Paris, 1729, 22–34, 156–163.

[18] The Elements of Algebra, London, Book 3, 1674, 9–17, 20–23.

[19] Bull. Bibl. Storia Sc. Mat. e Fis., 12, 1879, 890–4, 644; 13, 1880, 444.

[20] Y. Mikami, Abh. Geschichte Math. Wiss., 30, 1912, 233.

[21] Correspondence Math. Phys. (ed., Fuss), 1, 1843, 117; letter to Goldbach, March 6, 1742. Novi Comm. Acad. Petrop, 1, 1747–8, 20; Comm. Arith., I, 53, § 16. French transl. in Nouv. Ann. Math., 12, 1853, 46.

[22] Corresp. Math. Phys. (ed., Fuss), 1, 1843, 134, letter to Goldbach, June 30, 1742.

He stated he had a rigorous proof.   He stated Feb. 16, 1745 (p. 312) that it has not yet been proved that the sum of the squares of two relatively prime integers has no divisor other than a $\square$, nor that every prime $4n + 1$ is a $\square$, uniquely.

Chr. Goldbach[23] proved Fermat's statement that a prime $4k - 1$ cannot divide the sum of two relatively prime squares.   Let a^2 be the minimum square of the form $(4n - 1)m - 1$.   Set $\nu = 4n - 1$.   Then

$$\nu(m - 2a + \nu) - 1 = (a - \nu)^2,$$

so that $a^2 \leqq (a - \nu)^2$, whence $\nu \geqq 2a$.   Similarly,

$$\{4(n - a + m) - 1\}m - 1 = (a - 2m)^2, \qquad a^2 \leqq (a - 2m)^2, \qquad m \geqq a.$$

Thus $a^2 + 1 = \nu m \geqq 2am \geqq 2a^2$, $a = 0$ or 1, values leading to contradictions.

Euler[24] proved the Lemma: Every divisor of the sum of two relatively prime squares is itself the sum of two squares.

It is first shown that, if $p = c^2 + d^2$ is a prime and $pq = a^2 + b^2$, then q is a $\square$.   Since $c^2(a^2 + b^2) - a^2(c^2 + d^2)$ is divisible by p, one of the factors $bc \pm ad$ is of the form mp.   Set $b = mc + x$, $a = \pm md + y$.   Then $cx \pm dy = 0$.   But c is prime to d.   Thus* $x = nd$, $y = \mp nc$.   Hence

$$pq = (m^2 + n^2)(c^2 + d^2), \qquad q = m^2 + n^2.$$

It now follows from (1) that, if the primes $p_1, \cdots, p_k$ and the product $p_1 \cdots p_k q$ are all $\square$, then q is a $\square$.   Hence if pq, but not q, is a $\square$, p has a prime factor not a $\square$.

Let p divide $a^2 + b^2$, where a, b are relatively prime, while p is not a $\square$.   Set $a = mp \pm c$, $b = np \pm d$, $0 \leqq c \leqq \tfrac{1}{2}p$, $0 \leqq d \leqq \tfrac{1}{2}p$.   Then $c^2 + d^2 = pq \leqq \tfrac{1}{2}p^2$.   Hence q has a prime factor $r \leqq \tfrac{1}{2}p$, not a $\square$.   As before, the divisor r of $c^2 + d^2$ divides a sum $e^2 + f^2 \leqq \tfrac{1}{2}r^2$, and $e^2 + f^2$ has a prime factor $\leqq \tfrac{1}{2}r$ not a $\square$.   Proceeding in this manner we ultimately reach a contradiction with the fact that the sum of two sufficiently small squares has all its prime factors sums of two squares.

Euler gave a " tentative proof " of Girard's theorem that every prime $p = 4n + 1$ is a $\square$.   If neither a nor b is divisible by p, $a^{4n} - b^{4n}$ is divisible by p.   If p divides the factor $a^{2n} + b^{2n}$, a $\square$, then p is a $\square$.   It remains to show† that $a^{2n} - b^{2n}$ is not divisible by p for some pair of values of a, b [proved later by Euler[25]].

Since $p = a^2 + b^2$ implies $2p = (a + b)^2 + (a - b)^2$, and conversely $2p = a^2 + b^2$ implies $p = \alpha^2 + \beta^2$, where $\alpha = (a + b)/2$, $\beta = (a - b)/2$ are integers, there are as many representations of p as of $2p$ as a sum of two squares (including the case in which one square is zero).

---

[23] Corresp. Math. Phys. (ed., Fuss), 1, 1843, 255, letter to Euler, Sept. 28, 1743.   Euler, p. 258, expressed surprise at the simplicity of the proof.

[24] Ibid., 416–9; letter to Goldbach, May 6, 1747.   Novi Comm. Acad. Petrop., 4, 1752–3 (1749), 3–40; Comm. Arith., I, 155–173.

* In the letter, it is concluded from $bc \pm ad = m\,(c^2 + d^2)$ that $md \mp a$ is divisible by c; Thus $\mp a = cn - dm$, $b = cm + dn$.

† In the letter, it is stated that there are innumerable cases in which $a^{2n} - b^{2n}$ is not divisible by $4n + 1$.

From Girard's theorem and (1) it was concluded that any number is a $\boxed{2}$ if it has the form $2^j a^2 b$, where each prime factor of b is of the form $4k + 1$.

Euler[25] later succeeded in establishing the point which he could not prove in his preceding paper.[24] If the differences $(a + 1)^{2n} - a^{2n}$ of the first order of $1, 2^{2n}, 3^{2n}, \cdots, (4n)^{2n}$ were all divisible by p, the differences of order $2n$ would be divisible by p, whereas they equal $(2n)!$. This point can also be proved by means of Euler's[26] criterion for quadratic residues; however, Euler proved this criterion by the method of differences. In the former[25] paper (§ 70), Euler noted that the negative of a residue of a square when divided by a prime $4n - 1$ is not the residue of a square, whence $a^2 + b^2$ is not divisible by $4n - 1$ if a and b are not. Since a product of primes of the form $4k + 1$ is of that form, it follows (§ 73) that $4n - 1$, whether prime or composite, is not a divisor of a sum of two relatively prime squares.

Lagrange[9] of Ch. VIII proved that if a $\boxed{2}$ divides a $\boxed{2}$ the quotient is a $\boxed{2}$.

Euler[27] proved (1) by multiplying $(a + bi)(c + di)$ by its conjugate.

Euler[28] gave a more elegant proof of the Lemma.[24] Let N divide $P^2 + Q^2$, where P and Q are relatively prime. Set

$$P = fN \pm p, \qquad Q = gN \pm q, \qquad 0 \le p \le \tfrac{1}{2}N, \qquad 0 \le q \le \tfrac{1}{2}N.$$

Then $p^2 + q^2 = Nn$, where $n \le \tfrac{1}{2}N$. Set $p = \alpha n + a$, $q = \beta n + b$, where a and b are numerically $\le \tfrac{1}{2}n$. Set $A = a\alpha + b\beta$. Then

$$Nn = n^2(\alpha^2 + \beta^2) + 2nA + a^2 + b^2.$$

Hence $a^2 + b^2 = nn'$, $n' \le \tfrac{1}{2}n$. Thus $N = n(\alpha^2 + \beta^2) + 2A + n'$. By (1),

$$nn'(\alpha^2 + \beta^2) = (a^2 + b^2)(\alpha^2 + \beta^2) = A^2 + B^2, \qquad B = a\beta - b\alpha.$$

Hence $Nn' = (n' + A)^2 + B^2$. Just as this was derived from $Nn = p^2 + q^2$, so from it we get $Nn'' = \boxed{2}$, $n'' \le \tfrac{1}{2}n'$, etc., finally $N \cdot 1 = \boxed{2}$.

C. G. J. Jacobi[1] (p. 341) repeated this proof and stated that, while it contained nothing not known to Diophantus, there is no ground for the assumption that the latter actually possessed the proof.

Euler[29] gave a second proof of Girard's theorem. Since -1 is a quadratic residue of every prime $p = 4n + 1$, there exists a square b^2 with the residue -1, so that p divides $1 + b^2$. Hence, by the Lemma, p is a $\boxed{2}$.

In a posthumous manuscript, Euler[30] proved the first step in the above Lemma. Let $P = p^2 + q^2$ be divisible by $A = a^2 + b^2$, where a is prime

[25] Corresp. Math. Phys. (ed., Fuss), 1, 1843, 493; letter to Goldbach, April 12, 1749. Novi Comm. Acad. Petrop., 5, 1754–5 (1751), 3; Comm. Arith., I, 210.

[26] Novi Comm. Acad. Petrop., 7, 1758–9 (1755), 49, seq., § 78; Comm. Arith., I, 273.

[27] Algebra, St. Petersburg, 2, 1770, §§ 168–172. French transl., Lyon, 2, 1774, pp. 201–8. Opera Omnia, (1), I, 417–420.

[28] Acta Eruditorum Lips., 1773, 193; Acta Acad. Petrop., I, 2, 1780 (1772), 48; Comm. Arith., I, 540. Proof reproduced by Weber-Wellstein, Encyklopädie der Elem. Math., I (Alg. und Analysis), 1903, 244–250.

[29] Opusc. anal., 1, 1783 (1772), p. 64 seq., § 36; Comm. Arith., I, 483.

[30] Tractatus de numerorum, §§ 564–7; Comm. Arith., II, 572. Same in Opera Postuma, 1, 1862, 72.

to b.   Since A is prime to a and b, we may set $p = mA \pm fa$, $q = nA \pm gb$. Thus $f^2a^2 + g^2b^2$ is divisible by A.   The error in the conclusion that $g = f$ was pointed out in a marginal note by means of the case $p = 17$, $q = 6$, $a = 7$, $b = 4$.   However, $(g^2 - f^2)b^2$ and hence $g^2 - f^2$ is divisible by A. If we assume that A is a prime, we see that $g \pm f$ is divisible by A, so that $q = \nu A \pm fb$.   Hence

$$P/A = (f \pm ma \pm \nu b)^2 + (\pm \nu a \mp mb)^2.$$

Thus Euler's proof of the first step in the Lemma is valid if A is a prime. He gave (p. 570) another proof by setting $p = ma - nb$, $q = na + mb + s$. Then $P = A(m^2 + n^2) + sk$, $k = 2(na + mb) + s$.   Since A is a prime, either $s = tA$ or $k = -tA$.   In either case,

$$P/A = (m + bt)^2 + (n + at)^2.$$

J. L. Lagrange[31] deduced from Wilson's theorem the fact that the prime $4n + 1$ divides $(1 \cdot 2 \cdots 2n)^2 + 1$.   He[32] proved the Lemma in connection with the general problem to find the form of the divisors of numbers represented by $Bt^2 + Ctu + Du^2$.   He[33] deduced Girard's theorem from the fact that a prime p of the form $4n + 1$ divides $x^{2n} + 1$ for $2n$ integral values of x numerically $< \frac{1}{2}p$ (it being a factor of $x^{p-1} - 1$).

Beguelin[75] of Ch. I failed in his attempt to prove Girard's theorem.

P. S. Laplace[34] remarked that every prime $4n + 1$ will be a ▣ if proved to divide a ▣, in view of Lagrange.[32]   But $4n + 1$ divides $(a^{2n} + 1)(a^{2n} - 1)$ and not the last factor for every a, since

$$(2n)! = \{(2n + 1)^{2n} - 1\} - 2n\{(2n)^{2n} - 1\} + \cdots,$$

by the formula for the $2n$th order of differences of $x^{2n} - 1$ for $x = 1$ [Euler[25]].

J. Leslie[34a] solved $x^2 + y^2 = a^2 + b^2$ by setting

$$x + a = (b - y)m, \qquad x - a = (b + y)/m.$$

C. F. Kausler[35] gave tentative numerical methods of expressing a given number A as a sum of 2, 3 or 4 squares.

Let $A = 4C + 1 = (2P)^2 + (2Q + 1)^2$.   Then $C = P^2 + Q(Q + 1)$. If $C = 2D + 1$, then $P = 2T + 1$ and $D - \frac{1}{2}Q(Q + 1) = 2T(T + 1)$. Hence we subtract from D in turn the halves of the *pronic* numbers $Q(Q + 1)$, given by a table (extending to $Q = 225$), and note if any remainder is double a pronic number.   If $C = 2D$, then $P = 2T$ and we use $D - \frac{1}{2}Q(Q + 1) = 2T^2$.

A number $A = 4B + 2$ can only be the sum of two odd squares, whence

$$B = P(P + 1) + Q(Q + 1).$$

Thus $B = 2C$.   Set $P = Q + R$.   Solving the quadratic for Q, we see

---

[31] Nouv. Mém. Acad. Berlin, année 1771 (1773), 125; Oeuvres, III, 431.

[32] *Ibid.*, année 1773, 275; Oeuvres, III, 707.

[33] *Ibid.*, année 1775, 351; Oeuvres, III, 789–790.

[34] Théorie abrégée des nombres premiers, 1776, p. 24.

[34a] Trans. Roy. Soc. Edinburgh, 2, 1790, 193.

[35] Nova Acta Acad. Petrop., 11, ad annum 1793 (1798), Histoire, 125–156.

that $4C^2 + 1 - R^2$ must be a square. The problem thus reduces to finding two squares with the sum $4C^2 + 1$, treated in the first case.

The methods employed to express A as a ③ or ④ are no better than the similar one of subtracting from A in turn squares, or sums of two squares, and ascertaining if the remainder is a ②.

Kausler[36] extended his table of pronic numbers to $Q = 1000$, and gave their halves and quarters, and applied them as in the former paper. Given $A = a^2 + b^2$, to solve $x^2 + y^2 = A$, set $x = a + 2m\alpha$, $y = 2n\alpha - b$. Then $\alpha = (nb - ma)/(m^2 + n^2)$ is to be integral. Let m, n be relatively prime. Then $b = \alpha n + \beta m$, where $\beta = (\alpha m + a)/n$ is an integer. The latter gives $n = pa + \mu\alpha$, $m = qa + \mu\beta$, where p/q is a convergent to α/β. Then the former gives a relation between α, β, p, q, μ which is not solved.

C. F. Gauss[37] applied the theory of binary quadratic forms to prove that every prime $4n + 1$ is a ② in a single way. In a foot-note he considered $M = 2^\mu S a^\alpha b^\beta \cdots$, where a, b, $\cdots$ are distinct primes of the form $4n + 1$, and S is the product of all the prime factors $4n + 3$ of M. If S is not a square, M is not a ②. It is stated that, if S is a square, there are

$$k = \tfrac{1}{2}(\alpha + 1)(\beta + 1) \cdots$$

decompositions of M into a sum of two squares, when one of the exponents α, β, $\cdots$ is odd; but $k + \tfrac{1}{2}$ if α, β, $\cdots$ are all even. Here the squares and not their roots are counted.

A. M. Legendre[38] had already given the last result.

Legendre[39] developed $\sqrt{p}$ into a continued fraction with the

$$\begin{array}{llllllllll}
\text{quotients} & a & \alpha & \beta & \cdots & \mu & \mu & \cdots & \beta & \alpha & 2a \cdots, \\
\text{convergents} & \dfrac{1}{0} & \dfrac{a}{1} & & \cdots & \dfrac{m_0}{n_0} & \dfrac{m}{n} & \cdots & & \dfrac{f_0}{g_0} & \dfrac{f}{g} \cdots,
\end{array}$$

where $f^2 - pg^2 = -1$. Then by use of the convergents corresponding to μ, μ,

$$\frac{f}{g} = \frac{m(n/n_0) + m_0}{n(n/n_0) + n_0}, \qquad f = mn + m_0 n_0, \qquad g = n^2 + n_0^2.$$

Substituting these values into $f^2 - pg^2 = -(mn_0 - m_0 n)^2$, we get

$$m^2 - pn^2 = -(m_0^2 - pn_0^2).$$

But if $(\sqrt{p} + I_0)/D_0$ and $(\sqrt{p} + I)/D$ are the complete quotients corresponding to m_0/n_0, m/n, then

$$m^2 - pn^2 = (mn_0 - m_0 n)D, \qquad m_0^2 - pn_0^2 = -(mn_0 - m_0 n)D_0.$$

Hence $D_0 = D$, so that $DD_0 + I^2 = p$ gives $p = D^2 + I^2$.

---

[36] Nova Acta Acad. Petrop., 14, ad annos 1797–8 (1805), 232–267.
[37] Disquisitiones Arith., 1801, Art. 182; Werke, I, 1863, 159–163.
[38] Théorie des nombres, 1798, p. 293; ed. 3, 1830, I, 314 (transl. by Maser, I, 309).
[39] Théorie des nombres, ed. 2, 1808, 59–60; ed. 3, 1830, I, 70–1. (Maser, I, 71–73). Cf. Dirichlet,[38] § 83, long footnote. Cf. Euler[72] (end), of Ch. XII.

Legendre[40] stated that every divisor of a sum of two relatively prime squares is a sum of two relatively prime squares. P. Volpicelli[41] noted that the latter need not be relatively prime since $d = 2197 = 39^2 + 26^2$ is a divisor of $13d = 119^2 + 120^2$ [but d also equals $9^2 + 46^2$].

P. Barlow[42] stated that a number $4n + 1$ is a prime if a ☐ in one way only. [He should have said relatively prime squares; $45 = 36 + 9$ is a ☐ in a single way. For Euler's proofs of the correct theorem see Ch. XIV of Vol. I].

A. Cauchy[43] obtained (1) by taking the norm of the product of two complex numbers.

C. F. Gauss[44] stated that, if a prime $p = 4k + 1$ is expressed in the form $e^2 + f^2$, e odd, f even, then $\pm e$ and $\pm f$ equal the minimum residues (i. e., between $- p/2$ and $+ p/2$) modulo p of $\frac{1}{2}r/(k!)$ and $\frac{1}{2}r^2$, respectively, where

$$r = (k + 1)(k + 2) \cdots (2k).$$

The residue of $\pm e$ is positive or negative according as the positive value of e is of the form $4m + 1$ or $4m + 3$. But there is given no general rule as to the sign of $\pm f$ (cf. Goldscheider[130]).

Gauss[45] noted that the number of sets of integers x, y for which $x^2 + y^2 \leqq A$ is

$$4q^2 + 1 + 4[\sqrt{A}] + 8 \sum_{j=q+1}^{r} [\sqrt{A - j^2}]$$
$$= 1 + 4\{[A] - [A/3] + [A/5] - [A/7] + \cdots\},$$

where $q = [\sqrt{A/2}]$, $r = q + [\sqrt{A}]$, and $[t]$ denotes the greatest integer $\leqq t$. Denote by $f(A)$ the number of representations of A by $x^2 + y^2$, which is 8 if A is a prime $4n + 1$, while for $A = 2^\mu S a^\alpha b^\beta \cdots$ (as in Gauss[37])

$$f(A) = 4(\alpha + 1)(\beta + 1) \cdots$$

or 0, according as S is a square or not. The mean of $f(A)$ is π. Set $f'(m) = f(m) + f(3m)$; the mean of $f'(m)$ is $4\pi/3$. Set

$$f''(m) = f'(5m) - f'(m);$$

the mean of $f''(m)$ is $16\pi/15$. Proceeding, we approach the mean 4 and find that

$$4 = \pi \cdot \tfrac{4}{3} \cdot \tfrac{4}{5} \cdot \tfrac{8}{7} \cdot \tfrac{12}{11} \cdots \text{ (to infinity),}$$

the denominators being the successive odd primes p, and the numerators being $p \pm 1$.

---

[40] Théorie des nombres, 1798, 190; ed. 2, 1808, 175; ed. 3, 1830, I, 203 (Maser, I, 204).

[41] Annali di Sc. Mat. Fis., 4, 1853, 296.

[42] Theory of Numbers, London, 1811, p. 205.

[43] Cours d'analyse de l'école polyt., 1, 1821, 181.

[44] Gött. gelehrte Anz., 1, 1825; Comm. soc. sc. Gott. recent., 6, 1828; Werke, II, 1863, 168, 90–1. Cf. Bachmann,[95] Kreisteilung, Ch. X.

[45] Posth. MS., Werke, II, 1863, 269–275, 292; Gauss-Maser, Höhere Arith., 1889, 656–661. Cf. Eisenstein,[56] Hermite.[117, 127]

C. G. J. Jacobi[46] stated in a letter to Legendre, Sept. 9, 1828, that the theorems relative to numbers represented as a ▣ follow from

$$(1 + 2q + 2q^4 + 2q^9 + \cdots)^2 = 1 + \frac{4q}{1-q} - \frac{4q^3}{1-q^3} + \frac{4q^5}{1-q^5} - \cdots$$

$$= 1 + \frac{4q}{1-q} - \frac{4q^3}{1+q^2} - \frac{4q^6}{1-q^3} + \frac{4q^{10}}{1+q^4} + \cdots.$$

A. Genocchi[75] noted the conclusion that, if $x^2 + y^2 = n$ has N_1 (0 or 2) sets of solutions with x or y zero, and N_2 other sets, $N_1 + 2N_2$ is double the excess of the number of divisors $4m + 1$ of n over the number of divisors $4m + 3$ of n.

Jacobi[47] gave the formulae

$$\frac{2kK}{\pi} = \frac{4q^{1/2}}{1-q} - \frac{4q^{3/2}}{1-q^3} + \frac{4q^{5/2}}{1-q^5} + \cdots = 4\Sigma\psi(n)q^{m^2 n/2},$$

where m, n range over all odd integers such that all prime factors of m are $\equiv 3 \pmod 4$, all of n are $\equiv 1 \pmod 4$, while $\psi(n)$ is the number of factors of n and hence is the excess of the number of divisors $4k + 1$ of $m^2 n$ over the number of divisors $4k + 3$ of $m^2 n$;

$$\left(\frac{2kK}{\pi}\right)^{1/2} = 2q^{1/4} + 2q^{9/4} + 2q^{25/4} + \cdots.$$

A comparison of the square of the latter series with the former shows that the number of representations of $2m^2 n$ as a sum of two odd squares is the excess of the number of divisors $4k + 1$ of $m^2 n$ over the divisors $4k + 3$.

Jacobi[48] proved that

$$\frac{2K}{\pi} = 1 + 4\sum_{x=1}^{\infty} A^{(x)}q^x, \qquad \left(\frac{2K}{\pi}\right)^{1/2} = \sum_{n=-\infty}^{+\infty} q^{n^2},$$

where $A^{(x)}$ is the excess of the number of divisors $4m + 1$ of x over the number of divisors $4m + 3$. Although not explicitly stated by Jacobi, it follows that the number of representations of x as a ▣ is $4A^{(x)}$ [cf. Dirichlet[52]]. Evident corollaries relating to the evenness or oddness of the two squares were noted by J. W. L. Glaisher.[49]

Jacobi[50] gave an arithmetical proof of his[47] first theorem: If p is odd, the number of sets of positive integral solutions of $y^2 + z^2 = 2p$ is the excess E of the number of factors $4m + 1$ of p over the number of factors $4m + 3$.   Let

$$p = \alpha^A \cdots \rho^R \alpha'^{A'} \cdots \sigma'^{S'},$$

where $\alpha, \cdots, \rho$ are primes $4m + 1$, and $\alpha', \cdots, \sigma'$ are primes $4m + 3$. The factors of p are the terms of the product

$$(1 + \alpha + \cdots + \alpha^A) \cdots (1 + \rho + \cdots + \rho^R)(1 + \alpha' + \cdots + \alpha'^{A'}) \cdots.$$

[46] Jour. für Math., 80, 1875, 241; Werke, I, 424.

[47] Fundamenta Nova Func. Ellip., 1829, 106 (31), 107, 103(5), 184(7); Werke, I, 162(31), 163, 159(5), 235(7).   Cf. Jacobi[22b] of Ch. III.

[48] Fund. Nova Func. Ellip., 107, 184 (6); Werke, I, 162-3, 235(6).

[49] Quar. Jour. Math., 38, 1907, 7.

[50] Jour. für Math., 12, 1834, 167-9; Werke, VI, 245-7.

Set $\alpha = \cdots = \rho = 1$, $\alpha' = \cdots = \sigma' = -1$. Then a factor $4m + 1$ is replaced by $+1$, a factor $4m + 3$ by -1. Hence the product is replaced by E. Thus

$$E = (1 + A) \cdots (1 + R) \left\{ \frac{1 + (-1)^{A'}}{2} \right\} \cdots \left\{ \frac{1 + (-1)^{S'}}{2} \right\}.$$

Hence $E = 0$ unless $A', \cdots, S'$ are all even. If the latter are all even, E is the number of factors of $n = \alpha^A \cdots \rho^R$, while $p = nQ^2$, where every prime factor of Q is of the form $4m + 3$. Now $2p$ is not a ▢ unless p is of this form nQ^2. Also $2nQ^2 = y^2 + z^2$ requires that y and z be divisible by Q, while $2n = w^2 + x^2$ has as many sets of positive solutions as n has factors (all the factors of n being of the form $4m + 1$).

A. D. Wheeler[51] gave trivial or known results on ▢.

G. L. Dirichlet[52] obtained, as a special case of a general thoerem on quadratic forms, Jacobi's[48] result that, if n is odd and positive, the number of sets of solutions of $x^2 + y^2 = n$ is the quadruple of the excess of the number of divisors $4k + 1$ of n over the number of divisors $4k + 3$.

A. Cauchy[53] proved Gauss'[44] result that, if $p = x^2 + y^2$,

$$x \equiv -\frac{1}{2} \frac{(2\omega)!}{(\omega!)^2} \pmod{p}, \qquad \omega = \frac{p - 1}{4}.$$

Cauchy[54] proved identities of the type

$$(1 + 2t + 2t^4 + 2t^9 + \cdots)^2 = (1 + 2t^2 + 2t^8 + \cdots)^2$$
$$+ 4t(1 + t^4 + t^{12} + t^{24} + \cdots)^2.$$

G. Eisenstein[55] gave the values of A, B in $p = 4n + 1 = A^2 + B^2$ and $p = 3n + 1 = A^2 - AB + B^2$, where p is a prime. He[56] stated that the number of lattice points inside and on the circumference of a circle of radius $\sqrt{m}$ and center at the origin is

$$1 + 4\left\{ [m] - \left[\frac{m}{3}\right] + \left[\frac{m}{5}\right] - \left[\frac{m}{7}\right] + \cdots \right\}.$$

C. G. J. Jacobi[57] gave the representation as a ▢ of each prime $4n + 1 \lessgtr 11981$.

Jacobi[1] noted in 1847 that an insignificant change in the text of Diophantus V, 12 gives the result that, if a number without a square factor is a ▢, neither itself nor a factor of it has the form $4n - 1$, and expressed his belief that Diophantus had a proof, though he gave none, since all that is essential to a proof was in the Greek mathematics and is

[51] Amer. Jour. Sc. and Arts (ed., B. Silliman), 25, 1834, 87.
[52] Jour. für Math., 21, 1840, 3; Werke, I, 463. Zahlentheorie, § 91.
[53] Mém. Ac. Sc. Paris, 17, 1840, 726; Oeuvres, (1), 3, 1911, 414.
[54] Comptes Rendus Paris, 17, 1843, 523, 567; Oeuvres, (1), VIII, 50, 54.
[55] Jour. für Math., 27, 1844, 274.
[56] Ibid., 28, 1844, 248. Cf. Gauss,[45] Suhle[73a] and Cayley.[81] Proved also by H. Ahlborn, Ueber Berechnung von Summen von grössten Ganzen auf geometrischem Wege, Progr. Hamburg, 1881, 18.
[57] Jour. für Math., 30, 1846, 174-6; Werke, VI, 265-7. Errata, Mess. Math., 34, 1904, 132.

in the spirit of their method. From this point of view, Jacobi proved that, if a given odd number N is the sum of the squares of two integers b and c having no common factor $4n - 1$, every prime factor p of N is of the form $4k + 1$. When b, c are relatively prime, the proof shows that p and hence every divisor of N is a sum of two rational squares. The fact that every divisor is a sum of two integral squares is established by an argument perhaps not known to Diophantus and not necessary for his assertion.

F. Arndt,[58] using continued fractions as had Legendre[39] for the case of a prime, proved that the hth power of a prime $4n + 1$ is a $\boxed{2}$ in 2^{h-1} ways.

J. B. Kulik[59] gave the representation as a $\boxed{2}$ of each prime $\leqq 10529$.

V. A. Lebesgue[60] noted that $x^2 + y^2 = z^2 + t^2$ becomes $pq = rs$ if we set

$$2x = p + q + r - s, \qquad 2y = p + q - r + s,$$
$$2z = p - q + r + s, \qquad 2t = p - q - r - s.$$

C. Hermite[61] developed a/p into a continued fraction, where $a^2 \equiv -1$ (mod p), and employed two consecutive convergents m/n, m'/n', such that $n < \sqrt{p}$, $n' > \sqrt{p}$. Then

$$\frac{a}{p} = \frac{m}{n} + \frac{\epsilon}{nn'}, \qquad \epsilon < 1; \qquad (na - mp)^2 = \epsilon^2 p^2/n'^2 < p.$$

Since $(na - mp)^2 + n^2$ is a multiple of p and is $< 2p$, it equals p.

J. A. Serret[62] employed $q^2 \equiv -1$ (mod p), $q < p$, and developed p/q into a continued fraction so that the number of quotients is even (replacing if necessary the last quotient Q by $Q - 1 + 1$). In the series of quotients the terms equidistant from the extremes are shown to be equal. Let m/n be the convergent which includes the quotients of the first half of the series, and m_0/n_0 the preceding convergent. Then the continued fraction whose quotients are those of the second half of the series has the value m/m_0. If ω is the common middle quotient, the convergent following m/n equals

$$\frac{m\omega + m_0}{n\omega + n_0}.$$

Replacing ω by m/m_0, we get the entire continued fraction. Thus

$$\frac{p}{q} = \frac{m^2 + m_0^2}{mn + m_0 n_0}, \qquad p = m^2 + m_0^2.$$

L. Wantzel[63] stated that the use of complex integers affords the simplest proof that every prime divisor of a $\boxed{2}$ is a $\boxed{2}$. He proved that no complex prime $a + bi$ divides a product without dividing one factor [due to Gauss].

[58] Jour. für Math., 31, 1846, 343-358; extract of Diss., Sundiae, 1845. Arndt,[124] Ch. XII.

[59] Tafeln der Quadrat- und Kubik-Zahlen aller Zahlen bis Hundert Tausend . . . , Leipzig, 1848, Table 2.

[60] Nouv. Ann. Math., 7, 1848, 37.

[61] Jour. de Math., 13, 1848, 15; Oeuvres, I, 264; Nouv. Ann. Math., 12, 1853, 45; Société philomatique de Paris, 1848, 13-14.

[62] Algèbre Supér., ed. 1, 1849, 331; Jour. de math., (1), 13, 1848, 12-14; Nouv. Ann. Math., 12, 1853, 12; Société philomatique de Paris, 1848, 12-13,

[63] Société philomatique de Paris, 1848, 19-22.

P. Volpicelli[64] noted that, if $z = a_j^2 + b_j^2$ $(j = 1, \cdots, m)$, (1) shows that z^2 is a sum of two squares in $m(m - 1)$ ways, not necessarily distinct. If $z = m^2 + n^2 = p^2 + q^2$, then

$$z = (a_1^2 + b_1^2)(a_2^2 + b_2^2), \qquad a_1a_2 = \frac{m + p}{2}, \qquad b_1b_2 = \frac{p - m}{2},$$

$$a_2b_1 = \frac{n + q}{2}, \qquad a_1b_2 = \frac{n - q}{2}.$$

To show that a number having a prime factor $p = 4n + 3$ is not a sum of two relatively prime squares, raise $a^2 = pq - b^2$ to the power $2n + 1$, whence $s = a^{p-1} + b^{p-1}$ is a multiple of p, whereas $s \equiv 2 \pmod{p}$ by Fermat's theorem. In attempting to prove that every prime $p = 4n + 1$ is a ▢, he employed relatively prime integers x, y, not divisible by p and one even. By Fermat's theorem, $x^{4n} - y^{4n} = pQ$. Since every odd number can be expressed as a difference of two squares, he claimed that we can satisfy $x^{2n} - y^{2n} = Q$, whence $p = (x^n)^2 + (y^n)^2$. By use of (1), a product of k distinct primes of the form $4n + 1$ is a sum of two squares in 2^{k-1} ways, and only in that many ways. Several examples illustrate the method to express A as a ▢ by use of the continued fraction for $\sqrt{A}$. The nth power of a ▢ is a ▢ in $n/2$ or $(n + 1)/2$ ways, according as n is even or odd.

Volpicelli[65] considered the number ν of ways of expressing z as a ▢, when each prime factor of z is a ▢. When z is a product of k distinct primes, $\nu = 2^{k-1}$. When just two of these k primes have exponents m and m', his three formulas can be combined into the single one $\nu = 2^{k-3+\mu+\mu'}$, where $\mu = m/2$ or $(m + 1)/2$ according as m is even or odd, and similarly for μ'. When the roots of the two squares are given double signs, the number is 4ν.

Volpicelli[66] considered Gauss'[37] theorem on the number ν of the ways of expressing $P = a^\alpha b^\beta \cdots$ as a ▢, when $a, b, \cdots$ are distinct primes of the form $4n + 1$. Let $N = (\alpha + 1)(\beta + 1) \cdots$ be the number of divisors of P. Let N' be the number of ways of expressing P as a product of two factors A, B. Then $N' = (N + 1)/2$ or $N/2$ according as $\alpha, \beta, \cdots$ are all even or not all even. If P is a product of two distinct factors > 1 each expressible as a ▢, the product theorem (1) yields two expressions for P as ▢, and conversely. Thus if P is not a square, $\nu = N' = N/2$. If P is a square, $\nu - 1 = N' - 2$, $\nu = (N - 1)/2$, whereas Gauss gave $\nu = (N + 1)/2$. [It is merely a question as to the inclusion or exclusion of $P = P + 0$, cf. Genocchi.[75]] The special cases in which P is a power of a prime or a product of distinct primes are treated (pp. 71-81). He[67] insisted until[76] 1854 that there is a misprint in Gauss' formula.

[64] Raccolta di Lettere . . . Fis. ed Mat. (Palomba), Roma, 5, 1849, 263, 313, 392, 402.

[65] Giornale Arcadico di Sc., Let. ed Arti, Roma, 119, 1849-50, 20-26; Annali di Sc. Mat. e Fis., 1, 1850, 156.

[66] Atti Accad. Pont. Nuovi Lincei, 4, 1850-1, 22-31. Same by Volpicelli.[67]

[67] Nouv. Ann. Math., 9, 1850, 305-8; Annali di Sc. Mat. e Fis., 1, 1850, 527-531; 2, 1851, 61-4.

V. A. Lebesgue[68] proved that $y^2 + 1 \neq x^m$ if $y \neq 0$, $m > 1$, by use of complex numbers.

G. Bellavitis[69] stated that every solution of $x^2 + y^2 = 5\cdot 13\cdot 17$ is given by

$$x + yi = (2 \pm i)(3 \pm 2i)(4 \pm i).$$

If each c_i is a prime $4k + 1$, $x^2 + y^2 = c_1^{m_1}c_2^{m_2}\cdots$ has $k = \frac{1}{2}(m_1+1)(m_2+1)\cdots$ or $k - \frac{1}{2}$ essentially different sets of solutions, according as $y = 0$ gives no solution or a solution.

E. Prouhet[70] proved Gauss'[37] formula.

D. Chelini[71] gave an "elegant proof" of Gauss' formula by noting that every solution of $x^2 + y^2 = (a^2 + b^2)^m(a_1^2 + b_1^2)^{m_1}\cdots$ is given by the development of

$$x + yi = (a + bi)^n(a - bi)^{m-n}(a_1 + b_1i)^{n_1}(a_1 - b_1i)^{m_1-n_1}\cdots,$$

where $n = 0, 1, \cdots, m$; $n_1 = 0, 1, \cdots, m_1$; etc.

A. Genocchi[72] noted that Chelini[71] did not prove that the solutions obtained are all different, nor that no other solutions exist.

V. Bouniakowsky[73] proved that every prime $8k + 5$ is a $\boxed{2}$ by use of his formula (10), Ch. X, Vol. I, involving sums of divisors.

H. Suhle[73a] noted that Jacobi's[48] theorem implies the generalization that the number of positive solutions x, y of $x^2 + y^2 = p$ is the excess of the number of divisors $4m + 1$ of p over the number of divisors $4m + 3$. He proved Eisenstein's[56] result.

C. Hermite[74] noted that, to express as a $\boxed{2}$ a number A for which $\alpha^2 \equiv -1 \pmod{A}$ is solvable, it suffices to consider the form

$$Ax^2 + 2\alpha xy + A^{-1}(\alpha^2 + 1)y^2,$$

which is reducible to $X^2 + Y^2$.

A. Genocchi[75] considered the number of representations of n by $u^2 + v^2$. By the remark of Euler[24] (end), it suffices to take n odd. Let t be the g.c.d. of u, v. If n has a prime factor $p = 4m + 3$, set $n = p^\pi n'$, $t = p^\rho t'$, where n' and t' are prime to p. Since p cannot divide a $\boxed{2}$, $\pi = 2\rho$, so that the product of all the prime divisors $4m + 3$ of n is a square which divides u^2 and v^2. It thus suffices to treat the case in which every prime factor of n is of the form $4m + 1$. For such an n, set $n = p^\pi n'$, p being a prime not dividing n'. Then

$$(u + iv)(u - iv) = (q + ir)^\pi(q - ir)^\pi n', \qquad q^2 + r^2 = p.$$

Now $q \pm ir$ are complex primes, and decomposition into such primes is unique. Thus

$$u + iv = i^t(q + ir)^h(q - ir)^k(u' + iv'),$$

[68] Nouv. Ann. Math., 9, 1850, 178–181.
[69] Annali di Sc. Mat. e Fis., 1, 1850, 422–5.
[70] Comptes Rendus Paris, 33, 1851, 225–6.
[71] Annali di Sc. Mat. e Fis., 3, 1852, 126–9.
[72] Nouv. Ann. Math., 12, 1853, 235–6.
[73] Mém. Ac. Sc. St. Pétersbourg, (6), 5, 1853, 303.
[73a] De quorundam theoriae numerorum theorematum applicatione, Berlin, 1853, 18, 26.
[74] Jour. für Math., 47, 1854, 345; Oeuvres, I, 237.
[75] Nouv. Ann. Math., 13, 1854, 158–170.

where the final factor divides n'. Multiplying by the conjugate, we get $n = p^{h+k} (u'^2 + v'^2)$. Hence $h + k = \pi$, $n' = u'^2 + v'^2$. The multiplication of $u + iv$ by i^{-t} at most interchanges u^2 and v^2. Hence the effective solutions u, v are given by

$$u + iv = (q + ir)^h (q - ir)^{\pi-h} (u' + iv') \qquad (h = 0, 1, \cdots, \pi),$$

where u', v' range over the N' solutions of $u'^2 + v'^2 = n'$. If we change the sign of v' and replace h by $\pi - h$, we get $u - iv$. If π is even, and n' is a square u'^2, the representation $n = (p^{\pi/2} u')^2$ is excluded. Hence the number of representations of n as a $\boxed{2}$ is $\frac{1}{2}(\pi + 1)N'$, unless π is even and n is a square, and then is $\frac{1}{2}\{(\pi + 1)N' - 1\}$. The number of representations of n' is $\frac{1}{2}N'$ or $\frac{1}{2}(N' - 1)$ according as N' is even or odd. Hence by induction we obtain Gauss' [37] result that if a, b, $\cdots$ are distinct primes $4m + 1$, the number of representations of $n = a^{\alpha}b^{\beta} \cdots$ as a $\boxed{2}$ is $\frac{1}{2}N$ or $\frac{1}{2}(N - 1)$, according as n is or is not a square, where $N = (\alpha + 1)(\beta + 1) \cdots$. The second would be $\frac{1}{2}(N + 1)$ if we count also the case of $n + 0$. Hence the " correction " by Volpicelli[67] is unnecessary.

P. Volpicelli[76] retracted his[67] claim of an error on the part of Gauss[37] and Legendre,[38] but gave $k - \frac{1}{2}$ as the number of representations of M as a $\boxed{2}$ when μ and α, β, $\cdots$ are all even, i. e., when M itself is a square. Concerning Euler's remark, quoted by Genocchi,[75] that an integer and its double have the same number of representations as a $\boxed{2}$, Volpicelli (p. 185) stated that $p = 4225$ has only four [omitting $p = 65^2 + 0$], while $2p$ has five, representations.

A. Genocchi[77] answered the latter objection by noting that zero is to be counted as an integer. He remarked (p. 495) that the " new " case noted by Volpicelli (that of M a square) had been treated by Fermat, who discussed the number of ways a number is the hypotenuse of a rational right triangle.

A. Cayley[78] noted that a formula of Dirichlet's[52] becomes, for $D = -1$,

$$(1 + 2q^4 + 2q^{16} + 2q^{36} + \cdots)(q + q^9 + q^{25} + \cdots)$$

$$= \frac{q}{1 - q^2} - \frac{q^3}{1 - q^6} + \frac{q^5}{1 - q^{10}} - \frac{q^7}{1 - q^{14}} + \cdots.$$

H. J. S. Smith,[79] in accord with Gauss[24] of Ch. II, denoted by $[q_1 \cdots q_n]$ the numerator of the common fraction equal to the continued fraction

$$q_1 + \frac{1}{q_2 +} \frac{1}{q_3 +} \cdots + \frac{1}{q_n},$$

and employed Euler's[72] relations (Ch. XII)

$$(2) \qquad\qquad [q_1 q_2 \cdots q_{i-1} q_i] = [q_i q_{i-1} \cdots q_2 q_1],$$

$$(3) \qquad [q_1 \cdots q_n] = [q_1 \cdots q_i][q_{i+1} \cdots q_n] + [q_1 \cdots q_{i-1}][q_{i+2} \cdots q_n].$$

[76] Annali di Sc. Mat. e Fis., 5, 1854, 176–186; Jour. für Math., 49, 1855, 119–122.

[77] Annali di Sc. Mat. e Fis., 5, 1854, 491–8.

[78] Cambridge and Dublin Math. Jour., 9, 1854, 163–5.

[79] Jour. für Math., 50, 1855, 91–2; Coll. Papers, I, 33–4. Reproduced by Borel and Drach, Introduction à la théorie des nombres, 1895, 109–12; Chrystal, Algebra, ed. 1, II, 1889, 471; ed. 2, II, 499.

For p a given integer, let $\mu_1, \cdots, \mu_s$ denote the integers prime to p and $< \frac{1}{2}p$. In the continued fraction for p/μ_k, $[q_1 \cdots q_n]$ is now p. In view of (2), $[q_n \cdots q_1]$ arises from some $p/\mu_{k'}$. Let p be a prime $4\lambda + 1$, so that $s = 2\lambda$. Hence there is some $\mu_k \neq 1$ which coincides with $\mu_{k'}$ and thus there is a set of quotients $q_1, \cdots, q_n$ symmetrical from the ends. If n were odd, $n = 2i - 1 \geqq 3$, $p = [q_1 \cdots q_{i-1}q_iq_{i-1} \cdots q_1]$ has the factor $[q_1 \cdots q_{i-1}]$ by (3). Hence $n = 2i$ and

$$p = [q_1 \cdots q_iq_i \cdots q_1] = [q_1 \cdots q_i]^2 + [q_1 \cdots q_{i-1}]^2.$$

C. G. Reuschle[80] expressed as a sum of two squares each prime $4n + 1$ up to 12377, and to 24917 for those primes for which 10 is a quadratic residue.

A. Cayley[81] wrote $E'(n/k) = 1$ or 0 according as n/k is an integer or not and proved that the number of ways the integer n is a ▢ is

$$\nu = E'(n) - E'(n/3) + E'(n/5) - E'(n/7) + \cdots,$$

if $n = \alpha^2 + \beta^2$ is counted twice when $\alpha \neq \beta$. Hence ν is the number of lattice points on the quadrant of the circle with radius $\sqrt{n}$ and center at the origin. Eisenstein's[56] formula follows readily.

J. Liouville[82] stated the formula

$$\Sigma(- 1)^{(s-1)/2} \left[\frac{n}{s} \right] = \Sigma[\sqrt{n - \theta^2}],$$

summed for $s = 1, 3, 5, \cdots$ and for $\theta = 0, 1, 2, \cdots, [\sqrt{n}]$, and implied that it is connected with sums of two squares. It was proved geometrically by L. Goldschmidt,[83] who showed that the right member is the number of lattice points in a quadrant of the circle $r^2 + \theta^2 = n$.

F. Unferdinger[84] proved, by use of norms of complex numbers, that a product of n sums of two squares can be expressed as a ▢ in 2^{n-1} ways, distinct in general.

S. Kaminsky[85] proved that $x^2 + y^2 = pz^2$ is impossible in integers if p is a prime $4n + 3$.

F. Woepcke[86] proved by induction from p, p^n, p^{n+1} to p^{n+2} that any power of a prime $4m + 1$ can be expressed in one and but one way as a sum of two relatively prime squares. The proof shows that the number of all decompositions (primitive or not) of p^λ as a ▢ is $(\lambda + 1)/2$ if p is odd, $\lambda/2$ if $p = 2$. Hence follows Gauss'[37] formula. Also the number of primitive decompositions of $p_1^{a_1} \cdots p_\nu^{a_\nu}$ is $2^{\nu-1}$, if each p_i is of the form $4m + 1$.

J. Plana[87] used Jacobi's[46] formula to prove Gauss'[37] result on the number of ways of expressing $N = 2^\mu S^2 p^\alpha p'^\beta \cdots$ as $a^2 + b^2$, where p, p', $\cdots$ are

[80] Math. Abh., Neue Zahlenth. Tabellen, Progr. Stuttgart, 1856. Errata by Cunningham, Mess. Math., 34, 1904–5, 133–5.
[81] Quar. Jour. Math., 1, 1857, 186–191.
[82] Jour. de Math., (2), 5, 1860, 287–8.
[83] Beiträge zur Theorie der quad. Formen, Diss. Göttingen, Sondershausen, 1881.
[84] Archiv Math. Phys., 34, 1860, 83–100.
[85] Nouv. Ann. Math., (1), 20, 1861, 97–9.
[86] Atti Accad. Pont. Nuovi Lincei, 14, 1860–1, 311–5.
[87] Mem. Accad. Turin, (2), 20, 1863, 123–6.

primes $4k + 1$. To find a, b without trial, express p, p' as $\boxdot$ by continued fractions and apply (1) and

$$(P^2 + Q^2)^t = G^2 + H^2, \qquad G = P^t - \binom{t}{2} P^{t-2}Q^2 + \binom{t}{4} P^{t-4}Q^4 - \cdots,$$

$$H = tP^{t-1}Q - \binom{t}{3} P^{t-3}Q^3 + \binom{t}{5} P^{t-5}Q^5 - \cdots.$$

G. L. Dirichlet[88] used the theory of binary quadratic forms to prove that, if m is a product of powers of μ primes $4h + 1$, the number of sets of relatively prime solutions x, y of $x^2 + y^2 = m$ is $2^{\mu+2}$. The number (§ 91) of all sets of solutions is the quadruple of the excess of the number of its divisors $4h + 1$ over the number of its divisors $4h + 3$.

A. Vermehren,[89] to express z^3 as a sum of two squares, put $z = u + v$; then $z^3 = u^2(u + 3v) + v^2(3u + v)$. He took $u + 3v = 4n^2$, $3u + v = 4m^2$.

F. Unferdinger[90] noted that the product of the expansions of $(a \pm bi)^m$ gives $(a^2 + b^2)^m = A^2 + B^2$, where A, B are known polynomials. He[84] had shown that a product P of n sums of two squares can be expressed as a $\boxdot$ in 2^{n-1} ways distinct in general. The same result therefore holds for P^m.

G. C. Gerono[90a] proved that every divisor of a sum of two relatively prime squares is a sum of two relatively prime squares.

V. Eugenio[91] proved the Lemma[24] as follows. Let M divide $P^2 + Q^2$, where P is prime to Q, and call P'/Q' the next to the last convergent of the continued fraction for P/Q. Then $PQ' - P'Q = \pm 1$. By (1), M divides $(PP' + QQ')^2 + 1$. Thus M divides $N^2 + 1$, where N is an integer $< M$. Express M/N as a continued fraction with an even number of quotients:

$$a + \cfrac{1}{a_1 + \cdots + \cfrac{1}{a_{n-1}}},$$

where $n = 2s$. Let $M_1/N_1, \cdots, M_n/N_n \equiv M/N$ be the successive convergents. Then

(4)　$M_{i+1} = M_i a_i + M_{i-1}, \ N_{i+1} = N_i a_i + N_{i-1}, \ MN_{n-1} - NM_{n-1} = (-1)^n$,

(5)　$$\frac{M}{M_{n-1}} = a_{n-1} + \cfrac{1}{a_{n-2} + \cdots + \cfrac{1}{a_1 + \cfrac{1}{a}}}.$$

Now $N^2 + 1 = MN'$. Thus by (4_3), $M(N' - N_{n-1}) = N(N - M_{n-1})$. Thus M divides $N - M_{n-1} < M$. Hence $M_{n-1} = N$. Thus (5) equals M/N, and $a = a_{n-1}$, etc. Hence

$$\frac{M}{N} = a + \cfrac{1}{a_1 + \cdots + \cfrac{1}{a_{s-1} + \cfrac{1}{a_{s-1} + \cdots + \cfrac{1}{a_1 + \cfrac{1}{a}}}}} = \frac{M_s}{N_s} + 1 \Big/ \left(\frac{M_s}{M_{s-1}}\right).$$

But $M_{s-1} = N_s$. Thus $M/N = (M_s^2 + N_s^2)/M_s N_s$, $M = M_s^2 + N_s^2$.

[88] Zahlentheorie, § 68, 1863; ed. 2, 1871; ed. 3, 1879; ed. 4, 1894.
[89] Die Pythagoräischen Zahlen, Progr. Domschule, Güstrow, 1863.
[90] Archiv Math. Phys., 49, 1869, 116–7.
[90a] Nouv. Ann. Math., (2), 8, 1869, 454–6, 559.
[91] Giornale di Mat., 8, 1870, 162–5.

P. Seeling[92] proved that if A is a prime $4m + 1$ the period of the continued fraction for $\sqrt{A}$ has an odd number of terms. Hence A is a $\square$.

J. Petersen[93] reproduced Euler's[24] proof that every divisor of a sum of two relatively prime squares is a $\square$. Then by Wilson's theorem, every prime $4n + 1$ is a $\square$. He proved Gauss'[37] result on the number of solutions of $x^2 + y^2 = A$.

L. Lorenz[94] proved that

$$\sum_{m,\,n=-\infty}^{+\infty} q^{m^2+n^2} = 1 + 4 \sum_{m=0}^{\infty} \sum_{n=1}^{\infty} \{q^{(4m+1)n} - q^{(4m+3)n}\},$$

whence $m^2 + n^2 = N$ has $4(a_N - b_N)$ solutions if a_N is the number of divisors of the form $4m + 1$ of N, and b_N the number of divisors of the form $4m + 3$.

P. Bachmann[95] employed the theory of roots of unity to prove that every prime $p = 4n + 1$ is a sum of two squares, to compute the squares, and to prove Gauss'[44] result.

J. W. L. Glaisher[96] would strike out of the list of numbers

$$
\begin{array}{rrrrrrl}
1 & 2 & 3 & 4 & 5 & 6 & \cdots \\
-3 & -6 & -9 & -12 & -15 & -18 & \cdots \\
5 & 10 & 15 & 20 & 25 & 30 & \cdots \\
-7 & -14 & -21 & -28 & -35 & -42 & \cdots \\
9 & 18 & 27 & 36 & 45 & 54 & \cdots \\
\cdot & \cdot & \cdot & \cdot & \cdot & \cdot & \cdots \\
\cdot & \cdot & \cdot & \cdot & \cdot & \cdot & \cdots
\end{array}
$$

every one whose negative occurs in the list. Each remaining positive number $1, 2, 4, 5, 8, 9, 10, \cdots$ is a $\square$ and every $\square$ occurs in the final set. The proof is by Jacobi's[46] formula. He gave a like scheme to obtain the numbers expressible as a sum of two odd squares.

R. Hoppe[97] proved that every prime $p = 4n + 1$ is a $\square$. The values of $r = x^2$ for $x = 1, \cdots, 2n$ are incongruent modulo p. But $r^{2n} \equiv 1$ has only $2n$ roots and $-r$ is a root. Hence to each x corresponds an integer y such that $y^2 \equiv -r$. Thus $x^2 + y^2 = pq$. If p_1 is a factor of q, we get $x_1^2 + y_1^2 = p_1 q_1$. Since the q's decrease, we finally get a $q_k = 1$, whence $x_k^2 + y_k^2 = p_k$. The remaining factors of q_{k-1} are $\square$, whence q_{k-1} is a $\square$. Then $p_{k-1} = \square/q_{k-1} = \square$, etc. Finally, p is a $\square$.

F. L. F. Chavannes[98] considered an integer N whose prime factors are distinct and each of the form $4e + 1$ and hence a $\square$. Thus $N = \Pi(\alpha^2 + \beta^2)$. Set $N_1 = (\alpha^2 + \beta^2)(\gamma^2 + \delta^2)$, $N_2 = N_1(\epsilon^2 + \zeta^2)$, $\cdots$, whence $N_1 = x_1^2 + y_1^2$ for $x_1 = \alpha\gamma \pm \beta\delta$, $y_1 = \beta\gamma \mp \alpha\delta$. Similarly, each pair x_1, y_1 yields two

[92] Archiv Math. Phys., 52, 1871, 40–9.

[93] Tidsskrift for Math., (3), 1, 1871, 80–4.

[94] *Ibid.*, 97.

[95] Die Lehre von der Kreistheilung, 1872, 122–137, 235.

[96] Math. Quest. Educ. Times, 20, 1873, 87; British Assoc. Report, 46, 1873, 10–12 (Trans. Sect.).

[97] Archiv Math. Phys., 56, 1874, 223.

[98] Bull. Soc. Vaudoise des Sc. Naturelles, Lausanne, 13, 1874–5, 477–509.

sets of solutions x_2, y_2 of $N_2 = x_2^2 + y_2^2 = (x_1^2 + y_1^2)(\epsilon^2 + \zeta^2)$.   Then $N_3 = x_3^2 + y_3^2$ has 8 sets, etc. It is proved (pp. 503–6) that if p and p' are primes $4e - 1$, no one of p, p' or pp' is a $\boxed{2}$.

V. Schlegel[99] stated that the numbers $(8\lambda + 7)4^\mu$ are the only ones not a sum of fewer than four squares; the numbers $(4\lambda + 3)2^\mu$ and the products of two relatively prime numbers of that form are the only numbers not a sum of fewer than three squares.   The numbers representable as a $\boxed{2}$ are $s \cdot 2^\mu$, where $s = 4(\lambda^2 + \nu^2 + \nu) + 1$.   The numbers representable in n ways as a $\boxed{2}$ are 2^μ times the product of n factors s.

T. Muir[100] noted that by Lagrange's theorem any integer A is of the form $x^2 + y^2$ if in the continued fraction for $\sqrt{A}$ the period of the partial denominators has an odd number of terms.   Muir[101] gave formulas for x and y.   For, the general expression for such an integer is $A = R^2 + S$,

$$R = \tfrac{1}{2}K(a_1 a_2 \cdots a_2 a_1)M + \tfrac{1}{2}K(a_1 a_2 \cdots a_2)K(a_2 a_3 \cdots a_3 a_2),$$
$$S = K(a_1 a_2 \cdots a_2)M + K(a_2 \cdots a_2)^2,$$

where $a_1 a_2 \cdots a_n a_n \cdots a_2 a_1$ is the period, while K is a continuant.   For example,

$$K(a_1 a_2 a_3 a_4) = \begin{vmatrix} a_1 & 1 & 0 & 0 \\ -1 & a_2 & 1 & 0 \\ 0 & -1 & a_3 & 1 \\ 0 & 0 & -1 & a_4 \end{vmatrix}.$$

Then $A = x^2 + y^2$,

$$2x = \{K(a_1 \cdots a_n)^2 - K(a_1 \cdots a_{n-1})^2\}M + K(a_1 \cdots a_n)K(a_2 \cdots a_n)^3$$
$$- K(a_1 \cdots a_{n-1})K(a_2 \cdots a_{n-1})^3 + (-1)^n 3K(a_2 \cdots a_{n-1})K(a_2 \cdots a_n),$$
$$y = \{K(a_1 \cdots a_n)K(a_1 \cdots a_{n-1})\}M + K(a_1 \cdots a_n)K(a_2 \cdots a_{n-1})^3$$
$$+ K(a_1 \cdots a_{n-1})K(a_2 \cdots a_n)^3.$$

When $M = K(a_1 \cdots a_2)$, $A = x^2 + y^2$ is also the sum of 3 squares.

E. Lucas[102] gave the complete solution of $u^2 + v^2 = y^4$ and stated that the same process applies to $u^2 + v^2 = y^{2^n}$.

S. Roberts[103] derived all the decompositions into the sum of two squares of an odd positive integer D, containing no square factor, and such that $t^2 - Du^2 = -1$ is solvable in integers, by developing into a continued fraction $\sqrt{N/M}$, where M and N are complementary factors of D and $M < \sqrt{D}$.   For D odd, we take $M < \sqrt{D/2}$.

G. H. Halphen[104] considered the sum $s(x)$ of the positive divisors d of a positive integer x such that x/d is odd.   Then

$$\tfrac{1}{2}s(x) = s(x - 1) - s(x - 4) + s(x - 9) - \cdots \pm s(x - n^2) + \cdots,$$

[99] Zeitschrift Math. Phys., 21, 1876, 79–80.
[100] Proc. London Math. Soc., 8, 1876–7, 215–9.   The Expression of a Quadratic Surd as a Continued Fraction, Glasgow, 1874, § 51.   Euler[72] of Ch. XII wrote (a, b) for $K(a, b)$.
[101] Proc. Roy. Soc. Edinb., 1873–4, 234.
[102] Bull. Bibl. Storia Sc. Mat. Fis., 10, 1877, 243.   Cf. J. Bertrand, Traité élém. d'algèbre, Paris, 1850, 244; 1851, 224.   Cf. Lucas[57] of Ch. XXII.
[103] Proc. London Math. Soc., 9, 1877–8, 187–196.
[104] Bull. Soc. Math. France, 6, 1877–8, 119–120, 179–180.

the series being continued as long as $x - n^2$ is positive; if x is a square, $s(0)$ is replaced by $x/2$. The proof is by use of the series for

$$Q \equiv (1 - q)(1 - q^2)(1 - q^3) \cdots = (1 + q)(1 + q^2) \cdots (1 - 2q + 2q^4 - 2q^9 + \cdots).$$

Hence if x is not a square and no $x - n^2$ is a square, $s(x)$ is a multiple of 4. Thus $s(x)$ is a multiple of 4 when x is not a square or a □. If also x is a prime, x is of the form $4m - 1$, since $s(x) = x + 1$. Hence every prime not a □ is of the form $4m - 1$, so that every prime $4m + 1$ is a □.

S. Réalis[105] proved that every prime $4n + 1$ is the quotient of $x^2 + y^2$ by the common factor of x^2 and y^2, where

$$x = \alpha^2 + \beta^2 - \gamma^2, \qquad y = (\gamma - \alpha)^2 + (\gamma - \beta)^2 - \gamma^2.$$

For the latter values and

$$u = \alpha^2 + (\alpha - \gamma)^2 - (\alpha - \beta)^2, \qquad v = \beta^2 + (\beta - \gamma)^2 - (\beta - \alpha)^2,$$

we have $x^2 + y^2 = u^2 + v^2$, identically, and they furnish all the solutions.

E. Lucas[106] proved that every prime $4k + 1$ is a □ by use of " satins " n_a formed of the points (x, y) with $x = 0, 1, \cdots, n$ such that y is the residue of ax modulo n where a is prime to n and $a < n$. Since each parallel to the y-axis contains one and but one point of the satin, $\alpha x \equiv 1 \pmod{n}$ has a unique solution. If $f^2 + 1 \equiv 0$ is solvable, $y \equiv fx$ gives $fy \equiv f^2 x \equiv -x$, and the satin n_f is unaltered by a rotation through a right angle and is a square satin. If n is a prime $p = 4k + 1$, we can separate $2, 3, \cdots, p - 2$ into $(p - 5)/4$ sets of four numbers like $a, \alpha, p - a, p - \alpha$, where $a\alpha \equiv 1 \pmod{p}$, and one set $p, p - f$, such that $f(p - f) \equiv 1$, whence $f^2 + 1 \equiv 0$ is solvable. Thus p divides a sum of two squares. Since the satin is formed of squares having p as a side, p is a sum of two squares.

T. Harmuth[107] proved that every prime $p = 4n + 1$ divides a sum of two relatively prime squares. Let g be an odd primitive root of p and set $g^\lambda \equiv 2 \pmod{p}$. Then $g^{2e} + 2^2 \equiv 0 \pmod{p}$, $e = \lambda + (p - 1)/4$.

S. Günther[108] proved (1) by use of lattice (gitter) points. No three lattice points are vertices of a regular triangle. The geometrical proof by Lucas shows that

$$x^2 + y^2 = u^2 + v^2 = 2(ux + vy)$$

have no rational solutions. If a^2 is a □, a is a □.

For the knight's path problem in chess, we have (pp. 14–16) the system of equations

$$(x_i - x_{i+1})^2 + (y_i - y_{i+1})^2 = 5 \quad (i = 1, 2, \cdots, n^2 - 1),$$

and, if the path is closed, also

$$(x_{n^2} - x_1)^2 + (y_{n^2} - y_1)^2 = 5.$$

[105] Nouv. Ann. Math., (2), 18, 1879, 500–4.

[106] L'Ingegnere Civile, Turin, 1880; French transl., Assoc. franç., 40, 1911, 72–87. Cf. A. Aubry, l'enseignement math., 13, 1911, 200; Sphinx-Oedipe, numéro spécial, Jan., 1912, 10–13.

[107] Archiv Math. Phys., 66, 1881, 327–8.

[108] Zeitschrift Math. Naturw. Unterricht, 13, 1882, 94–98, 102.

If the path is symmetrical, there are further conditions.   He gave a history of the subject.

N. V. Bougaief[109] applied elliptic functions to the decomposition of numbers into squares (with relation to Jacobi's[47] Fundamenta Nova).

E. Fauquembergue[110] noted that a cube $\neq 1$ is never a sum of squares of two consecutive integers.

E. Cesàro[111] considered the function $\psi(n) = \Sigma f(a)$, where a ranges over all the positive integers for which $n - a^2$ is a square.   Then

$$\sum_{j=1}^{n} \psi(j) = \sum_{j=1}^{\mu} r_j f(j), \qquad r_j = [\sqrt{n - j^2}], \qquad \mu = [\sqrt{n}].$$

For $f(x) = 1$, $\psi(n)$ is the number of positive integral solutions of $x^2 + y^2 = n$; then $\Sigma\psi(j)$ equals $n\pi/4$ asymptotically, whence the number of ways of decomposing a number into a sum of two squares is in mean $\pi/4$.

T. J. Stieltjes[112] states that if $f(n)$ is the number of solutions of $x^2 + y^2 = n$, and if μ is the largest odd integer $\leqq \sqrt{n}$, then

$$f(2 \cdot 1) + f(2 \cdot 5) + \cdots + f(2 \cdot n)$$
$$= 8 \sum_{t=0} (-1)^t \left[\frac{n - (2t + 1)^2}{4(2t + 1)} \right] + 4 \cos^2 \frac{(\mu - 1)\pi}{4}, \quad n \equiv 1 \pmod 4,$$

$$f(1) + f(9) + f(17) + \cdots + f(n)$$
$$= 8 \sum_{t=0} (-1)^t \left[\frac{n - (2t + 1)^2}{8(2t + 1} \right] + 4 \cos^2 \frac{(\mu - 1)\pi}{4}, \quad n \equiv 1 \pmod 8,$$

$$f(5) + f(13) + f(21) + \cdots + f(n)$$
$$= 8 \sum (-1)^t \left[\frac{n - (2t + 1)(2t + 5)}{8(2t + 1)} \right] + \sin^2 \frac{k\pi}{2}, \quad n \equiv 5 \pmod 8,$$

where, in the last, $k = [\tfrac{1}{2}(\sqrt{n + 4} - 1)]$.   If $\phi(x)$ is the sum of the odd divisors of x,

$$\phi(1) + \phi(5) + \cdots + \phi(4n + 1), \qquad \phi(1) + \phi(3) + \cdots + \phi(2n - 1),$$
$$\phi(1) + \phi(2) + \cdots + \phi(n)$$

are expressed as sums of greatest integers.

T. Pepin[113] proved that, if m is an odd number not a square,

$$m\sigma(m) = 2 \sum_{n} \{2 + (-1)^{m-n}\}(5n^2 - m)X(m - n^2),$$

where $X(k)$ is the sum of the odd divisors of k and $\sigma(k)$ is the sum of all the divisors of k.   Let m be a prime $4l + 1$.   Hence

$$1 \equiv \Sigma(20\mu^2 - m)\sigma(m - 4\mu^2) \pmod 2.$$

Thus among the differences $m - 4\mu^2$ occur an odd number of squares, so that m is a ▣.

[109] Math. Soc. Moscow, 11, 1883, 200–312, 415–456, 515–602; 12, 1885, 1–21.
[110] Nouv. Ann. Math., (3), 2, 1883, 430.
[111] Mém. Soc. Roy. Sc. de Liège, (2), 10, 1883, No. 6, pp. 192–4, 224.
[112] Comptes Rendus Paris, 97, 1883, 889–891.
[113] Atti Accad. Pont. Nuovi Lincei, 37, 1883–4, 41.

E. Catalan[114] expressed $s = x^{4n+2} + y^{4n+2}$ as the sum of the squares of two polynomials, and s^2 as such a sum in two ways (p. 51). By use (p. 63) of $(x \pm iy)(x^2 \pm iy^2) \cdots (x^{2^{n-1}} \pm iy^{2^{n-1}}) = P + iQ$, we get 2^{n-1} decompositions of $(x^2 + y^2)(x^4 + y^4) \cdots (x^{2^n} + y^{2^n})$ as a $\boxed{2}$.

Catalan[115] noted that, if $a + b = \boxed{2}$, and $n = 2^p$,

$$a^{n-1} + a^{n-2}b + \cdots + b^{n-1} = \boxed{2}.$$

C. Hermite[116] stated that if $f(n)$ is the number of solutions of $x^2 + y^2 = n$,

$$f(2) + f(6) + \cdots + f(4n + 2)$$
$$= 4 \left\{ E_1\left(\frac{2n + 1}{2}\right) + E_1\left(\frac{2n + 2}{6}\right) + \cdots + E_1\left(\frac{4n + 1}{4n + 2}\right) \right\},$$

where $E_1(x) = [x + \frac{1}{2}] - [x] = [2x] - 2[x]$ is the function used by Gauss.

Hermite[117] proved by use of expansions of elliptic functions

$$s \equiv f(1) + f(2) + \cdots + f(C) = 4\Sigma(-1)^{(a-1)/2}[C/a],$$
$$t \equiv f(2) + f(10) + \cdots + f(8C + 2) = 4\Sigma(-1)^{c-1}[(2C + c)/(2c - 1)],$$

summed for $a = 1, 3, 5, \cdots$; $c = 1, 2, 3, \cdots$. He stated that

$$\tfrac{1}{4}s = \left[\frac{C}{1}\right] - \left[\frac{C}{3}\right] + \cdots - (-1)^n \left[\frac{C}{2n - 1}\right] + E_1\left[\frac{C + 1}{4}\right]$$
$$+ E_1\left(\frac{C + 2}{8}\right) + \cdots + E_1\left(\frac{C + n}{4n}\right) - n \sin^2\frac{n\pi}{2},$$

where $n = [(\sqrt{8C + 1} + 1)/4]$. Also, for $n = [(\sqrt{4C + 1} + 1)/2]$,

$$t = 8 \left\{ \left[\frac{C}{1}\right] - \left[\frac{C - 1 \cdot 2}{3}\right] + \left[\frac{C - 2 \cdot 3}{5}\right] - \cdots \right.$$
$$\left. - (-1)^n \left[\frac{C - n^2 + n}{2n - 1}\right] \right\} + 4 \sin^2\frac{n\pi}{2}.$$

He proved Gauss'[45] result for s; also, J. Liouville's[118] result

$$t = 4 \sum \left[\tfrac{1}{2}(\sqrt{4n + 2 - a^2} + 1)\right].$$

L. Gegenbauer[119] concluded from a general theorem on quadratic forms that the number of ways any number r which is odd or the double of an odd number can be represented as a sum of two squares is the quadruple of the number of decompositions into two relatively prime factors of those divisors of r which have only prime factors of the form $4s + 1$ and a square as complementary factor. The number of representations by $x^2 + y^2$ of those divisors of r whose complementary divisor is a product of

[114] Atti Accad. Pont. Nuovi Lincei, 37, 1883–4, 80.
[115] Mathesis, 4, 1884, 70.
[116] Amer. Jour. Math., 6, 1884, 173–4.
[117] Bull. Ac. Sc. St. Pétersbourg, 29, 1884, 343–7 (Oeuvres, IV, 159–163); reprinted, Acta Math., 5, 1884–5, 320.
[118] Jour. de Math., (2), 5, 1860, 287–8.
[119] Sitzungsber. Akad. Wiss. Wien (Math.), 90, II, 1884, 438.

an even number of primes exceeds the number of representations of the remaining divisors by the excess of the number of those divisors, with complementary square divisor, of the form $4s + 1$ over the number of such divisors of the form $4s - 1$.

T. Pepin[120] quoted Dirichlet's[52] theorem that the number of representations of an odd number n by $x^2 + y^2$ is 4ρ, where

$$\rho = \sum_{i|n} \left(\frac{-1}{i} \right)$$

is a sum of Legendre-Jacobi symbols. It follows readily that the number of representations of $2n$ is 4ρ and the number of decompositions is ρ. Since ρ is the excess of the number of divisors $4l + 1$ over the number of divisors $4l + 3$, we have Jacobi's[50] theorem that the number of decompositions of $2n$ is that excess. Likewise, $2^a n = x^2 + y^2$ has 4ρ solutions.

S. Réalis[121] noted that if p is a prime or a product of primes of the form $4q + 1$, all integral solutions of $x^2 + y^2 = p$ are found from the identity

$$(a + b + 1)^2 + (a - b)^2 = 4\left(\frac{a^2 + a}{2} + \frac{b^2 + b}{2} \right) + 1,$$

by giving to a and b such integral values that the second member takes the value p. Thus the problem reduces to that of expressing q as a sum of two triangular numbers. If p is odd or the double of an odd number and if $p = x^2 + y^2$, where x and y are relatively prime, then

$$x, y = p - \left(\frac{m^2 - m}{2} + \frac{n^2 \mp n}{2} \right).$$

J. W. Bock[122] employed the $n(2n - 1)$ pairs formed by two of $1^2, 2^2, \cdots, (2n)^2$. From any pair x_1^2, y_1^2, whose sum is not divisible by the prime $p = 4n + 1$, we obtain $2n$ incongruent sums $\nu^2 x_1^2 + \nu^2 y_1^2$, $\nu = 1, \cdots, 2n$. If $x_2^2 + y_2^2$ is not congruent to one of these sums, nor to zero, it leads to $2n$ new sums $\nu^2 x_2^2 + \nu^2 y_2^2$; etc. But $2n$ does not divide $n(2n - 1)$. Hence there exists a sum $s = A^2 + B^2$ divisible by p, $0 < A < \frac{1}{2}p$, $0 < B < \frac{1}{2}p$. In the attempt to prove that, if s is divisible by a prime $q = a^2 + b^2$, the quotient is a sum of two squares, the quotient is taken to be $c^2 + d^2$, c and d not being assumed integral. By (1), $q(c^2 + d^2)$ is of the form $x^2 + y^2$. From $s = x^2 + y^2$, it is concluded erroneously that $A = x$ or y, $B = y$ or x.

R. Lipschitz[123] noted that all real substitutions of determinant unity for which $x_1^2 + x_2^2 = y_1^2 + y_2^2$ (i. e., automorphs) are given by multiplying

$$(\lambda_0 + i\lambda_{12})(x_1 + ix_2) = (\lambda_0 - i\lambda_{12})(y_1 + iy_2)$$

by $\lambda_0 - i\lambda_{12}$ and equating the real terms and the imaginary terms, and

[120] Atti Accad. Nuovi Lincei, 38, 1884–5, 166.
[121] Nouv. Ann. Math., (3), 4, 1885, 367–9; Oeuvres de Fermat, IV, 218–220.
[122] Mitt. Math. Gesell. Hamburg, 1, 1885, 101–4.
[123] Untersuchungen über die Summen von Quadraten, Bonn, 1886, 147 pp. French transl. by J. Molk, Jour. de Math., (4), 2, 1886, 373–439. Summary in Bull. des Sc. Math. Astr., (2), 10, I, 1886, 163–183.

conversely. In particular, all rational automorphs of $x_1^2 + x_2^2$ are derived by taking λ_0 and λ_{12} to be relatively prime integers. To show (p. 384) that every prime $p = 4r + 1$ is a $\boxed{2}$, use a solution of $\omega^2 + 1 \equiv 0 \pmod{p}$ and set $\xi_1 = \omega\xi_2$, where ξ_2 is any integer not divisible by p. We can choose relatively prime integers ρ_0, ρ_{21} such that $\tau\rho_0$ and $\tau\rho_{21}$ are numerically $< p/2$ and congruent modulo p to ξ_1 and ξ_2 respectively. Take $\rho_{12} = -\rho_{21}$. Then $\tau^2(\rho_0^2 + \rho_{12}^2)$ is $< \frac{1}{2}p^2$ and is divisible by p. Hence $\rho_0^2 + \rho_{12}^2 = pt$, where $t < p/2$. Determine ϕ_0 and ϕ_{12} numerically $< t/2$ and congruent modulo t to ρ_0 and ρ_{12} respectively. Then $\phi_0^2 + \phi_{12}^2 = tt'$, where $t' \leqq t/2$. Then

$$(\phi_0 - i\phi_{12})(\rho_0 + i\rho_{12}) = \tau't(\rho_0' + i\rho_{12}'),$$

where ρ_0', ρ_{12}' are relatively prime. Hence $\rho_0'^2 + \rho_{12}'^2 = pk$, $k = t'/\tau'^2 \leqq t/2$. Repeating this process, we finally get $\lambda_0 = \rho_0^{(s)}$, $\lambda_{12} = \rho_{12}^{(s)}$, such that $\lambda_0^2 + \lambda_{12}^2 = p$, and

$$(6)\qquad \lambda_0\xi_1 - \lambda_{12}\xi_2 \equiv 0, \qquad \lambda_{12}\xi_1 + \lambda_0\xi_2 \equiv 0 \pmod{p}.$$

Similarly we can find a complex integer with relatively prime coordinates λ_0, λ_{12}, whose norm is any power p^γ of p and which satisfies (6) modulo p^γ. If $m = p^\gamma q^\delta \cdots$, where p, q, $\cdots$ are primes $\equiv 1 \pmod 4$, or if m is the double of such a product, apply the preceding discussion for each p^γ and take the product of the resulting complex integers. By using all sets of solutions of $\xi_1^2 + \xi_2^2 \equiv 0 \pmod{p^\gamma}$, we get every proper representation of m as a $\boxed{2}$ and each once and but once.

C. Hermite[124] proved by use of elliptic functions that, if $M = 4n + 1$,

$$S = f(1) + f(5) + f(9) + \cdots + f(M)$$
$$= 4\Sigma(-1)^{(m-1)/2} + 8\Sigma(-1)^{(m-1)/2}\left[\frac{M - m^2}{4m}\right],$$

summed for $m = 1, 3, 5, \cdots$, where $f(n)$ is the number of representations of n as a $\boxed{2}$. The asymptotic value of S is $\frac{1}{2}M\pi$.

A. Berger[125] gave an elementary proof of the theorem that, if n is a positive odd integer, the number of all sets of solutions of $x^2 + y^2 = n$ is $4\Sigma(-1)^{(\delta-1)/2}$, where δ ranges over all positive divisors of n. While Dirichlet's proof was by transcendental analysis, Berger uses only the known number (Dirichlet[88]) of relatively prime sets of solutions.

Berger[126] proved that if n is a positive integer the number of sets of integers x, y for which $x^2 + y^2 = n$ is $4\Sigma \sin \delta\pi/2$ (Berger[125]).

C. Hermite[127] proved Gauss'[45] formula for the number of sets of integers x, y for which $x^2 + y^2 \leqq A$.

E. Catalan[128] noted that, if $x^2 + y^2 + z^2$ is a square,

$$\{(x^2 + z^2)p - (y^2 + z^2)q\}^2 + 4x^2y^2pq = \boxed{2}.$$

If $B^2 - AC = -m^2$, $(Ca - Ac)^2 - 4(Bc - Cb)(Ab - Ba) = \boxed{2}$.

[124] Jour. für Math., 99, 1886, 324–8; Oeuvres, IV, 209–214. Cf. Gegenbauer.[131]
[125] Acta Math., 9, 1886–7, 301–7.
[126] Öfversigt af Kongl. Vetenskaps-Akad. Förhandl., 44, 1887, 153–8.
[127] Amer. Jour. Math., 9, 1887, 381–8; Oeuvres, IV, 241–250.
[128] Mathesis, 7, 1887, 120, 144.

J. W. L. Glaisher[129] wrote $4G(n)$ for the excess of the number of representations of n in the form $(6r)^2 + (6s + 1)^2$ over the number of those in the form $(6r + 2)^2 + (6s + 3)^2$, provided $n \equiv 1 \pmod{12}$, whence the representations of n as a ▨ are of one of those two types. If p, q are relatively prime numbers $12k + 1$, $G(pr) = G(p)G(r)$. He evaluated $G(a^\alpha)$, a being a prime. The number of representations of n as a ▨ is $4E(n)$, where $E(n)$ is the excess of the number of divisors $4k + 1$ of n over the number of divisors $4k + 3$. There are noted simple relations between $E(n)$ and $G(n)$. It is shown (p. 195) by elliptic functions that the number of representations of $4n + 1$ as a sum of an even and an odd square is $4E(4n + 1)$; the number of representations of $8n + 2$ as a sum of two odd squares is $4E(4n + 1)$. Hence if $n \equiv 1 \pmod 4$, n and $2n$ have the same number of representations as ▨. Next, $E(36n + 9) = E(4n + 1)$. The number of compositions of a number as a sum of two squares, both of the form $(12n + 1)^2$ or both of the form $(12n + 5)^2$, or one of each form, is expressed in terms of functions E and G. Similarly for representations by the forms at the beginning of this summary. Let (pp. 211–3) m be odd, a even, b odd and not divisible by 3, $c \equiv 1$, $d \equiv 5 \pmod{12}$; then the number of representations by $3a^2 + b^2$, $3a^2 + c^2$, $3a^2 + d^2$, $3m^2 + c^2$ or $3m^2 + d^2$ is expressed in terms of G and the excess $H(n)$ of the number of divisors $\equiv 1 \pmod 3$ of n over the number of divisors $\equiv 2 \pmod 3$.

F. Goldscheider[130] discussed the sign of f, not determined by Gauss.[44]

L. Gegenbauer[131] noted that Hermite's[124] formula is one of a set which follows from a general formula for the sum of the values taken by an arbitrary function $f(y)$ when y ranges over all those divisors $\leqq \sqrt{k}$ of $k = 4n + 1$ or $4n + 3$.

E. Lucas[132] gave two proofs by use of continued fractions that every divisor of a sum of two relatively prime squares is a ▨.

K. Th. Vahlen[133] deduced from the theory of partitions the fact that every odd integer is a ▨ in E ways, if $g^2 + u^2$ and $(-g)^2 + u^2$ are regarded as different ways, while E is the excess of the number of factors $4m + 1$ over the number of factors $4m + 3$. He noted that this fact is equivalent to the theorem of Jacobi[50] in view of a remark by Euler[24] (end). Since every integer N is the product of an even power of 2 by an odd integer or by the double of an odd integer, the number of sets of solutions $\geqq 0$ of $x^2 + y^2 = N$ is E. He gave a summation formula for the number of primitive representations as a ▨.

From a representation $a^2 + b^2 + c^2 + d^2$ of an odd prime p we obtain a multiple of 32 representations by permuting $a, \cdots, d$ or changing their signs, except when two are zero, the factor being then $12 \cdot 4$. But there are $8\sigma(p)$ representations of p. Thus if $p = a^2 + b^2$ has N sets of solutions

[129] Proc. London Math. Soc., 21, 1889–90, 182–215.
[130] Das Reziprozitätsgesetz der achten Potenzreste, Progr. Berlin, 1889, 26–29.
[131] Sitzungsber. Akad. Wiss. Wien (Math.), 99, IIa, 1890, 387–403.
[132] Théorie des nombres, 1891, 454–6.
[133] Jour. für Math., 112, 1893, 25–32.

$b > a > 0$, then $8\sigma(p) \equiv 48N \pmod{32}$. For $p = 4n + 1$,

$$\sigma(p) = 2(2n + 1)$$

and N is odd.

A. Matrot[134] noted that, if $p = 2h + 1$ is a prime, and a is not divisible by p, $a^h \equiv \pm 1 \pmod{p}$ by Fermat's theorem. If the upper sign held for every a,

$$s_h = 1^h + \cdots + (p - 1)^h \equiv p - 1 \pmod{p},$$

whereas, for $q < p - 1$, $s_q \equiv 0 \pmod{p}$, as shown by induction. Hence there exists an a for which $a^h \equiv -1$. Let $h = 2k$. Thus p divides a $\boxed{2}$. That p is a $\boxed{2}$ follows as in his 1891 paper on $\boxed{4}$.

E. Catalan[135] repeated the proof by Eugenio.[91]

H. Weber[136] proved that every prime $n = 4f + 1$ is a $\boxed{2}$ by use of the four periods each of f terms of nth roots of unity.

C. Störmer[137] proved that $1 + x^2 \neq 2y^n$ if $|x| > 1$ and n has an odd divisor > 1.

Several[138] treated $x^2 + (x + 1)^2 = y^4$, whence $t^2 - 2u^2 = -1$ if

$$t = 2x + 1, \qquad u = y^2.$$

Störmer[139] applied a theorem on Pell's equation (Störmer[230] of Ch. XII) to find the complete solution of $1 + x^2 = kA_1^{z_1} \cdots A_n^{z_n}$ in positive integers, where $k, A_1, \cdots, A_n$ are given positive integers. In particular, there is a new proof that $1 + x^2 = y^n$ or $2y^n$ is impossible if $x > 1$, $y > 1$, n being an odd prime.

M. A. Gruber[140] gave a table and identities for $4n + 1 = \boxed{2}$.

Several writers[141] discussed $x^2 + p^2 = y^3$ for p a prime.

G. de Longchamps[142] noted that N^4 is a $\boxed{2}$ or $\boxed{3}$ if $N/\lambda - 1$ is a square or $\boxed{2}$, since

$$N^4 \equiv 16\lambda(N - \lambda)(N - 2\lambda)^2 + (N^2 - 8\lambda N + 8\lambda^2)^2.$$

R. Hoppe[143] used Girard's theorem to prove that a number is a $\boxed{2}$ or not according as it has no prime factor of the form $4n - 1$ to an odd power or at least one such prime power factor.

J. H. McDonald[144] gave a direct proof of Jacobi's[48] result on the number of representations of an odd positive number as a $\boxed{2}$.

C. A. Laisant[145] noted that $(a^{4n+2} + 1)/(a^2 + 1)$ is always a $\boxed{2}$.

[134] Jour. de math. élém., (4), 2, 1893, 73.
[135] Mém. Acad. Roy. Belgique, 52, 1893–4, 17.
[136] Lehrbuch der Algebra, I, 1895, 583–5; ed. 2, I, 1898, 632–4.
[137] L'intermédiaire des math., 3, 1896, 171; 5, 1898, 94 for $n = 2^m$.
[138] Ibid., 4, 1897, 212–5.
[139] Videnskabs-Selskabets Skrifter, Christiania, 1897, No. 2.
[140] Amer. Math. Monthly, 5, 1898, 240–3.
[141] L'intermédiaire des math., 5, 1898, 157–9; 16, 1909, 177.
[142] Ibid., 7, 1900, 65. Misprint of $2N - \lambda$ for $N - 2\lambda$.
[143] Archiv Math. Phys., (2), 17, 1900, 128, 333.
[144] Proc. and Trans. Roy. Soc. Canada, (2), 6, 1900, Sec. III, 77–8.
[145] Nouv. Ann. Math., (4), 1, 1901, 239–240.

H. Schubert[146] noted that, if in $x^2 + y^2 = u^2 + z^2$ the unknowns have no common factor, either all four are odd or in each member one number is odd and one even. In the first case,

$$\tfrac{1}{2}(x + z) \cdot \tfrac{1}{2}(x - z) = \tfrac{1}{2}(u + y) \cdot \tfrac{1}{2}(u - y),$$

whence we must factor an arbitrary number g in two ways with always one factor even and the other odd. In the second case, g must be a product of two even factors and also a product of an even and an odd factor.

R. E. Moritz[146a] proved that every rational number not a square can be expressed in an infinitude of ways as a quotient of two sums or two differences of two squares, and gave one such expression for each such number < 100.

A. Palmström[147] noted that $x^3 = y^2 + z^2$ implies $x = a^2 + b^2$, whence $y = a^3 + ab^2$ or $a^3 - 3ab^2$ [provided y and z are relatively prime]. P. F. Teilhet[148] obtained all the solutions.

A. Thue[149] proved that a prime divisor of a $\boxed{2}$ is a $\boxed{2}$.

Several[149a] found three consecutive integers each a $\boxed{2}$, including $(2n)^2 + (2n)^2$, $8n^2 + 1$, $(2n - 1)^2 + (2n + 1)^2$, provided the second be a $\boxed{2}$, i. e., n be triangular, $n = (m^2 + m)/2$.

L. E. Dickson[150] proved that all factors of a sum of two relatively prime squares are sums of two squares by use of the theorem that if a and b are relatively prime every prime divisor of $a^2 + b^2$ is of the form $4n + 1$ and the theorem that every prime $4n + 1$ is a sum of squares of two relatively prime integers.

G. Fontené[151] proved Gauss' [37] theorem by showing that, if A, B, $\cdots$ are primes $4h + 1$, there is a $(1, 1)$ correspondence between the decompositions of $A^\alpha B^\beta \cdots$ as a product of two factors and its decomposition into a sum of two squares, provided we fix the order of the two squares whose sum is A, or B, etc.

A. Cunningham[152] expressed each prime $4n + 1 < 100000$ as a $\boxed{2}$.

P. Pasternak[153] proved that all solutions of $x^2 + y^2 = v^2 + w^2$ are

$$x = m\omega + np, \quad v = m\omega - np, \quad y = n\omega - mp, \quad w = n\omega + mp,$$

whence

$$x^2 + y^2 = (m^2 + n^2)(\omega^2 + p^2).$$

Thus every integer which can be expressed as a $\boxed{2}$ in more than one way is itself a product of two sums of two squares. From known theorems it is said to now follow that no prime $4n + 1$ is a $\boxed{2}$ in more than one way.

---

[146] Niedere Anal., 1, 1902, 167–171; ed. 2, 1908.
[146a] Ueber Continuanten . . ., Diss. Strassburg, Göttingen, 1902. Cf. Moritz[40] of Ch. IX.
[147] L'intermédiaire des math., 8, 1901, 302.
[148] *Ibid.*, 10, 1903, 210–1.
[149] Oversigt D. Viden. Selsk. Förh., Kristiania, 1902, No. 7.
[149a] Math. Quest. Educ. Times, (2), 3, 1903, 41–3.
[150] Amer. Math. Monthly, 10, 1903, 23.
[151] Nouv. Ann. Math., (4), 3, 1903, 108–115.
[152] Quadratic Partitions, London, 1904. Errata, Mess. Math., 34, 1904–5, 132.
[153] Zeitschr. Math. Naturw. Unterricht, 37, 1906, 33–35.

A. Gérardin[154] discussed the solution of

$$(10x + m)^2 + (10y + p)^2 = 100a, \qquad a = b^2 + d^2, \qquad m < 10, p < 10.$$

Since $m^2 + p^2 = 20h$, we have $m = 2$, $p = 4$ or 6; $m = 4$ or 6, $p = 8$. These cases are treated in turn. To solve (pp. 89–90) $x^2 + y^2 = a^2 + b^2$, set $x = a + mh$, $b = y + h$, $m(x + a) = b + y$. Then

$$h = 2(y - am)/(m^2 - 1),$$

and the general solution is said to be

$$(am^2 - 2my + a)^2 + y^2(m^2 - 1)^2 = a^2(m^2 - 1)^2 + (ym^2 - 2am + y)^2.$$

W. Sierpinski[155] gave a long proof that, if $A(x)$ is the number of pairs of integers u, v for which $u^2 + v^2 \leqq x$, $A(x) = \pi x + O(x^{1/3})$, for O defined as in Landau,[179] while π is the usual constant.

E. Jacobsthal[156] proved that, if p is a prime $\equiv 1 \pmod 4$, $p = a^2 + b^2$, where, in terms of Legendre's symbols,

$$a = \tfrac{1}{2}\phi(r), \qquad b = \tfrac{1}{2}\phi(n), \qquad \phi(e) = \sum_{m=1}^{p} \left(\frac{m}{p}\right)\left(\frac{m^2 + e}{p}\right),$$

where r is any quadratic residue (as -1) of p, and n any non-residue. Also, $a \equiv (p - 3)/2 \pmod 8$. Proof is given of formulas, equivalent to Gauss',[44] for the residues of a, b modulo p.

Identities[157] solving $a^2 + b^2 = 2c^n$ have been given.

W. Sierpinski[158] evaluated sums like

$$\sum_{n=1}^{x} \tau(n^2), \qquad \sum \tau^2(n), \qquad \sum \tau_8(n),$$

where $\tau(n)$ and $\tau_8(n)$ denote the number of decompositions of n into 2 and 8 squares.

* E. N. Barisien[159] expressed 2^n as a ratio of two ▢.

J. Sommer[160] applied ideals to show that every prime $4n + 1$ is a ▢.

L. Aubry[161] cited known results.

G. Bisconcini[162] proved that n is a ▢ if and only if n contains no odd power of a prime $4k - 1$, and deduced all decompositions of p^r as a ▢, given that of the prime $p = 4k + 1$. He[163] proved that, if p_i is a prime $4k + 1$, $p_1^{a_1} \cdots p_m^{a_m}$ has 2^{m-1} proper decompositions into ▢; also Gauss'[37] theorem. He treated (pp. 68–80) the decomposition of fractions into one of the forms $x^2 \pm y^2$.

[154] Sphinx-Oedipe, 1906–7, 112–9.

[155] Prace mat.-fiz., Warsaw, 17, 1906, 77–118 (Polish). See papers 179, 180, 189, 198–203.

[156] Anwendungen einer Formel aus der Theorie der quadratischen Reste, Diss. Berlin, 1906, 13; Jour. für Math., 132, 1907, 238–245.

[157] L'intermédiaire des math., 13, 1906, 62, 184; 14, 1907, 72.

[158] Prace mat.-fiz., Warsaw, 18, 1907, 1–60 (Polish). Reviewed in Jahrb. Fortschritte Math., 38, 319–21; Bull. des Sc. Math., (2), 37, II, 1913, 30–31.

[159] Bull. Sc. Math. Élém., 12, 1907, 262–6.

[160] Vorlesungen über Zahlentheorie, 1907, 112, 123–4. French transl. (of revised text) by A. Lévy, 1911, 105, 117–9.

[161] L'enseignement math., 9, 1907, 421.

[162] Periodico di Mat., 22, 1907, 270–285.

[163] Ibid., 23, 1908, 9–23.

F. Ferrari[164] found the known solution of $x^2 + y^2 = z^n$ by use of $z = r + si$.

H. Brocard[165] noted that $n^2 + (n + 1)^2 = m^k$ has solutions for $k = 2$, but not for $k = 3$.

E. Landau[166] considered the number $B(x)$ of positive integers $\leq x$ which are ☐ and gave a long proof that

$$\lim_{x = \infty} \frac{B(x) \cdot \sqrt{\log x}}{x} = \frac{1}{\sqrt{2}} \sqrt{\Pi\left(1 - \frac{1}{r^2}\right)^{-1}},$$

where r ranges over all primes of the form $4m + 3$.

E. Landau[167] applied binary quadratic forms to show that a number is a ☐ if and only if it has no prime factor $4m + 3$ to an odd power.

E. N. Barisien[168] used the epicycloid to derive the identity

$$(8t^3 - 6t^2 - 6t + 3)^2 + 4(1 - t^2)(1 + 3t - 4t^2)^2 = 13 - 12t,$$

whence $12 - 13t$ is a ☐ if $t = (1 - \theta^2)/(1 + \theta^2)$.

M. Kaba and L. E. Dickson[169] deduced, by use of special theta functions,

$$\sqrt{\frac{2K}{\pi}} = 1 + 2q + 2q^4 + \cdots, \qquad \frac{2K}{\pi} = 1 + 4\left(\frac{q}{1 - q} - \frac{q^3}{1 - q^3} + \cdots\right).$$

Hence there is no representation as a ☐ of a number having a prime factor $4m + 3$ with an odd exponent, and no proper representation when such a factor has an even exponent. If $P = p_1^{\pi_1} \cdots p_s^{\pi_s}$, where $p_1, \cdots, p_s$ are all the distinct primes of the form $4m + 3$ which divide e, and if $\pi_1, \cdots, \pi_s$ are all even, there are as many improper representations of e as there are representations of e/P; every representation of e is of the type $(P^{1/2}x)^2 + (P^{1/2}y)^2$. Hence the problem reduces to the case in which every prime factor of e is of the form $4m + 1$. Then the number of representations of e as a ☐ is $(\pi_1 + 1) \cdots (\pi_n + 1)$.

P. Bachmann[170] gave an exposition of the work of Lagrange[32] and Vahlen.[133]

Welsch[171] stated that the general solution of $u^2 + x^2 = y^2 + z^2$ is

$$2x = ab + cd, \qquad 2y = ac + bd, \qquad 2z = ab - cd, \qquad 2u = ac - bd,$$

where a, d are even, or b, c are even, or all four are odd.

L. Aubry[172] proved that $x^2 + (x + 1)^2 \neq m^k$ if k is not a power of 2.

A. Deltour[173] applied continuants (Muir[101]) to prove that a prime $4h + 1$ is a ☐ in one and but one way.

---

[164] Periodico di Mat., 25, 1909–10, 59–66; Supplem. al Period. di Mat., 12, 1908–9, 132–4.

[165] L'intermédiaire des math., 15, 1908, 18–19.

[166] Archiv Math. Phys., (3), 13, 1908, 305–12.

[167] Handbuch . . . Verteilung der Primzahlen, 1, 1909, 549–550.

[168] Assoc. franç. av. sc., 38, 1909, 101–7.

[169] Amer. Math. Monthly, 16, 1909, 85–7.

[170] Niedere Zahlentheorie, 2, 1910, 304–319 (477).

[171] L'intermédiaire des math., 17, 1910, 96, 118, 205.

[172] Ibid., 18, 1911, 8–9; errata, 113; Sphinx-Oedipe, numéro spécial, March, 1914, 15–16; errata, 39.

[173] Nouv. Ann. Math., (4), 11, 1911, 116.

Marchand[174] presented the known application of complex integers $a + bi$ to find all decompositions of a product of primes $4n + 1$ as a $\boxed{2}$.

Paulmier[175] gave solutions of $x^2 + y^2 = A^3$ for five special values of A.

Several writers[176] found x such that $x + 1$ and $x^2 + 2$ are sums of two squares.

J. K. Heydon[177] noted that, if $a, b, \cdots$ are distinct primes,

$$a^{2^p-1} b^{2^q-1} \cdots = \boxed{2} \text{ in } 2^{p+q+\cdots-1} \text{ or } 0 \text{ ways.}$$

P. Lambert[178] applied complex integers $a + bi$. He gave two proofs that a divisor of a $\boxed{2}$ is a $\boxed{2}$.

E. Landau[179] proved that, if $A(x)$ is the number of pairs of integers u, v for which $u^2 + v^2 \leqq x$, then $A(x) = \pi x + O(x^{1/3+\epsilon})$, for every $\epsilon > 0$. Here $f(x) = O(g(x))$ means a function such that there exist two numbers ξ and A for which $|f(x)| < A g(x)$ when $x \geqq \xi$. Although the result is not quite as sharp as that by Sierpinski,[155] the proof is much shorter.

Landau[180] gave a new proof of the theorem due to Sierpinski.[155]

R. Bricard[181] gave an elementary proof that every prime $p = 4n + 1$ is a $\boxed{2}$. By Wilson's theorem, $m^2 + 1 \equiv 0 \pmod{p}$ for $m = [(p - 1)/2]!$. Write x_i for the minimum residue of mi modulo p. Consider the $p - 1$ points $M_i = (x_i, i)$. The square of the distance $M_i M_j$ between any two of these points is divisible by p. It is shown that the least of these squares is $< 2p$ if $p > 32$ and hence equals p. A like proof shows that every prime $8q \pm 1$ is of the form $x^2 - 2y^2$.

F. Ferrari[182] noted that the least number decomposable in 2^n distinct ways as a sum of two relatively prime squares $\neq 0$ is the product, found by (1), of the first $n + 1$ primes of the form $4k + 1$. For this least $x = p_i^2 + q_i^2$ $(i = 1, \cdots, 2^n)$, set $y_i = p_i^2 - q_i^2$, $z_i = 2p_i q_i$; then $x^2 = y_i^2 + z_i^2$ is the least square decomposable in 2^n ways as a $\boxed{2}$. To find the least $(p + 1)$th power decomposable in 2^p ways as a $\boxed{2}$, use $P = b_i^2 + c_i^2$ $(i = 1, \cdots, 2^p)$, whence $\Pi(b_i^2 + c_i^2) = P^{p+1}$ has 2^p decompositions.

A. Aubry[183] noted that (1) can be derived from Brahmegupta's (Ch. V) inscribed quadrilateral $ABCD$ whose diagonals meet at right angles at O, by evaluating the perpendiculars BE and OJ to DC.

E. Haentzschel[184] noted that his [152] method in Ch. XXI to deduce a new solution of $ax^3 + \cdots + d = y^3$ from one solution may be applied to $x^2 + y^2 = z^3$ in two ways according as x or y is taken as the variable. He

[174] L'intermédiaire des math., 18, 1911, 228–232.
[175] *Ibid.*, 19, 1912, 151.
[176] *Ibid.*, 55–7, 257.
[177] Math. Quest. Educ. Times, (2), 21, 1912, 98–9.
[178] Nouv. Ann. Math., (4), 12, 1912, 408–421.
[179] Göttingen Nachrichten, 1912, 691–2. Giornale di Mat., 51, 1913, 73–81.
[180] Annali di Mat., (3), 20, 1913, 1–28; Sitzungsber. Akad. Wiss. Wien (Math.), 121, 1912, IIa, 2298–2328.
[181] Nouv. Ann. Math., (4), 13, 1913, 558–562.
[182] Periodico di Mat., 28, 1913, 71–8.
[183] Sphinx-Oedipe, numéro spécial, June, 1913, 23–24.
[184] Sitzungsber. Berlin Math. Gesell., 13, 1914, 92–6.

quoted from A. Fleck[185] the solution

$$(a^2c + 2abd - b^2c)^2 + (b^2d + 2abc - a^2d)^2 = (a^2 + b^2)^3, \quad a^2 + b^2 = c^2 + d^2,$$

which includes the primitive solution $(a^3 - 3ab^2)^2 + (3a^2b - b^3)^2 = (a^2 + b^2)^3$ by Euler[6] of Ch. XX.

* Hesse[186] gave the general solution of $x^2 + y^2 = z^n$.

Several writers[187] found solutions of $x^2 + y^2 = z^4$.

* J. G. van der Corput[188] treated sums of two squares.

G. H. Hardy[189] wrote $r(n)$ and $R(n)$ for the number of integral solutions of $\mu^2 + \nu^2 = n$ and of $\mu^2 + \nu^2 \leqq n$, respectively, and set $R(x) = \pi x + P(x)$. He proved the existence of a positive constant K such that each of

$$P(x) > Kx^{1/4}, \qquad P(x) < - Kx^{1/4}$$

is satisfied by values of x surpassing all limit. Hence in Sierpinski's[155] result $P(x) = O(x^{1/3})$, with O defined as by Landau,[179] the exponent $\frac{1}{3}$ cannot be replaced by a number $< \frac{1}{4}$. He gave an explicit analytic expression for $P(x)$ in terms of Bessel's functions.

Hardy[190] proved that, for every positive ϵ, $P(x)$ is on the average $O(x^{1/4+\epsilon})$, i. e.,

$$\frac{1}{x} \int_1^x |P(\tau)| \, d\tau = O(x^{1/4+\epsilon}).$$

G. Bonfantini[191] proved that, if a number n not a prime is a $\square$, it equals either a product of several factors each a $\square$ or such a product multiplied by a square which is a common factor of the given squares whose sum is n. Conversely, if m is a product of several sums of two squares and if m is not an even power of 2, m is a $\square$.

G. Koenigs and L. Bastien[192] discussed the number of decompositions of $(a^2 + b^2)^5$ as a $\square$.

A. Gérardin[193] noted that $t^2 - 2hu^2 = 1$ implies

$$\{(h - 1)t\}^2 + \{(h - 1)^2u^2 - 1\}^2 = 1 + \{(h - 1)^2u^2 + h - 1\}^2.$$

By means of the fact that every prime of the form $4n + 1$ is a factor of a number $t^2 + 1$, R. D. Carmichael[193a] proved by Fermat's method of infinite descent that such a prime is a $\square$.

* A. L. Bartelds[194] gave an elementary proof of Girard's theorem.

T. Hayashi[195] proved that $y^2 + 1 \neq z^3$ if $y \neq 0$.

<hr>

[185] Vossische Zeitung zu Berlin, June 1, 1913.

[186] Unterrichtsblätter für Math. u. Naturwiss., 20, 1914, 16. Haentzschel, p. 55, discussed Hesse's paper.

[187] Amer. Math. Monthly, 21, 1914, 199–201.

[188] Nieuw Archief voor Wiskunde, 11, 1914–5, 61.

[189] Quar. Jour. Math., 46, 1915, 263–283; Proc. London Math. Soc., (2), 15, 1916, 15–16.

[190] Proc. London Math. Soc., (2), 15, 1916, 192–213.

[191] Suppl. al Periodico di Mat., 18, 1915, 81–6. By use of Bonfantini[142] of Ch. XIII.

[192] L'intermédiaire des math., 22, 1915, 253–4; 23, 1916, 34–5.

[193] *Ibid.*, 22, 1915, 57.

[193a] Diophantine Analysis, 1915, 39–40.

[194] Wiskundig Tijdschrift, 12, 1915–6, 159–166.

[195] Nouv. Ann. Math., (4), 16, 1916, 150.

M. Weill[196] noted that the product of p sums of two squares is a sum of two squares in 2^{p-1} distinct ways.

M. Chalaux[197] proved Girard's theorem by induction using the fact that if a prime is a ▣ and divides a sum of two relatively prime squares, the quotient is a sum of two relatively prime squares.

E. Landau[198] proved his[179] former theorem by means of a new simplification of Pfeiffer's method (cf. pp. 305, 322 of Vol. 1 of this History). He[199] next considered the lower limit α of the constants for which $A(x) = \pi x + O(x^\alpha)$, and proved that $\alpha \geqq \frac{1}{4}$. Later he[200] proved a theorem on the number of lattice points in certain regions which is a generalization of the main theorem applied in his[179] above papers.

*K. Szilysen[201] stated empirically an asymptotic formula for the number of pairs of integers for which $x^2 + y^2 \leqq N$, a formula already proved by Lipschitz.

M. Rignaux[202] announced a table in manuscript of the decompositions as a ▣ of the 3908 decomposable numbers < 10000.

G. H. Hardy[203] deduced Landau's[179] theorem very simply by two different methods from the theorems in Hardy's[190] former paper. If[204] $a_1, \cdots\cdot a_m$ are primes of the form $4k + 1$, there are $4(n + 1)^m$ sets of solutions of $x^2 + y^2 = (a_1\, a_2 \cdots a_m)^n$, in 2^{m+2} of which x and y are relatively prime.

On the number of solutions of $x^2 + (4y)^2 = n$, see Nasimoff[68] of Ch. XIII. On $x^2 + y^2 = (m^2 + n^2)\, z^2$, see papers 142–5 of Ch. XIII and the cross-references given there. On $1 + x^2 = 2y^4$, see Euler[7] of Ch. XIV and Cunningham[79] of Ch. XX. In Ch. XVII are given reports on papers on a number and its square both sums of two consecutive squares; cf. Meyl[30] of Ch. IV. On $x^2 + n^2 \neq y^3$, see Pepin[10] and Hayashi[61] of Ch. XX. On $x + y = \square$, $x^2 + y^2 = z^4$, see papers 40, 48, 50, 52, 54–56, 63 of Ch. XXII. On systems of equations including $x^2 + y^2 = z^3$, see papers 353, 363, 368 of Ch. XXI. Equal sums of two squares occur on p. 37, p. 206; in paper 107a of Ch. VII; 18 of Ch. XIII; papers 21, 35, 45, 62, 80 of Ch. XV; 7, 9, 18, 20 of Ch. XVIII; 4, 13, 15, 33, 37, 42, 46-50, 75, 102, 133, 149 of Ch. XIX; 177 of Ch. XXII; 45 of Ch. XXIV. In Vol. I were cited the papers by Euler[3, 7] and Gauss,[13] pp. 381–2, containing tables of primes and factors of numbers $x^2 + y^2$; by Lucas[53] and Catalan,[61] pp. 402–3, on special numbers which are ▣; by Liouville[28], p. 286; and various papers, pp. 360–1, on factoring numbers which are ▣ in two ways.

[196] Nouv. Ann. Math., (4), 16, 1916, 311–4.
[197] Ibid., (4), 17, 1917, 305–8.
[198] Göttingen Nachrichten, 1915, 148–160.
[199] Ibid., 161–171.
[200] Ibid., 209–244; 1917, 96–101. Cf. Revue semestrielle, 27, I, 1918, 16, 18.
[201] Math. és termés. értesitö (Hungarian Acad. Sc.), 35, 1917, 54–6.
[202] L'intermédiare des math., 25, 1918, 143; 26, 1919, 54–55.
[203] Proc. London Math. Soc., (2), 18, 1919, 201–4.
[204] Amer. Math. Monthly, 26, 1919, 367–8.

CHAPTER VII.

SUM OF THREE SQUARES.

Diophantus V, 14 relates to the division of unity into three parts such that if the same given number a be added to each part the sums will be squares. This problem is equivalent to the determination of three squares, each $> a$, whose sum is $3a + 1$. Diophantus stated that a must not be of the form $8l + 2$.

C. G. Bachet[1] stated that this condition is not sufficient and gave as a sufficient condition that a must not be of the form $8k + 2$ or $32k + 9$, stating that he had tested the numbers $a < 325$. He also divided 5 into three parts such that each increased by 3 is a square; since

$$3 \cdot 3 + 5 = 1 + 2^2 + 3^2,$$

he took the sides of the squares to be $1 + 7N$, $2 + N$, $3 - 5N$, whence $N = 4/25$.

Fermat[2] remarked that Bachet's condition fails to exclude $a = 37, 149$, etc., and himself gave the correct sufficient condition that a must not be of one of the forms

$$8k + 2, \qquad 4 \cdot 8k + 2 \cdot 4 + 1, \qquad 4^2 \cdot 8k + 2 \cdot 4^2 + 4 + 1,$$
$$4^3 \cdot 8k + 2 \cdot 4^3 + 4^2 + 4 + 1, \qquad \cdots.$$

[Thus a must not equal

$$4^n \cdot 8k + 2 \cdot 4^n + (4^n - 1)/3 = [(24k + 7)4^n - 1]/3,$$

so that $3a + 1$ must not be of the form $(24k + 7)4^n$ and hence not $(8m + 7)4^n$, since m is a multiple of 3 if $3a + 1$ is of the latter form.]

Regiomontanus[3] (Johannes Müller, 1436–1476) proposed in a letter the problem of solving the pair of equations

$$x + y + z = 116, \qquad x^2 + y^2 + z^2 = 4624 = 68^2.$$

Fermat[4] stated that no integer $8k + 7$ is the sum of three rational squares. Descartes[5] proved this for integral squares by noting that a square is of one of the forms $4k$ or $8k + 1$.

Fermat[6] treated the problem to find two numbers each of which, as well as their sum, is composed of three squares only [not composed of one or two squares]. He took any such number, as 11, and multiplied it by two squares whose sum is a square, for example, 9 and 16. The problem was proposed by Sainte-Croix to Descartes in April, 1638, with the illustra-

---

[1] Diophanti Alex. Arith., 1621, 310–3.

[2] Oeuvres, I, 314–5; French transl., III, 257–8.

[3] C. T. de Murr, Memorabilia Bibl., 1, 1786, 145.

[4] Oeuvres, II, 66; III, 287; letter to Mersenne, Sept. or Oct., 1636. The latter communicated it to Descartes.

[5] Oeuvres, II, 92; letter from Descartes to Mersenne, March 31, 1638. See also p. 195.

[6] Oeuvres, II, 29, 57; letters to Mersenne, July 15 and Sept. 2, 1636.

tion 3, 11. In his reply, Descartes[7] gave $a^2 + 2$, $b^2 + 2$ (a and b odd); he[8] took the interpretation that each required number and their sum shall be the sum of three squares in one and but one way, and gave nine examples including

$$22 = 9 + 9 + 4, \qquad 35 = 25 + 9 + 1, \qquad 57 = 49 + 4 + 4.$$

But Sainte-Croix desired that each be the sum of 3, but not of 4, squares.

Fermat[9] asserted that the double of any prime $8n - 1$ is the sum of three squares; he desired that Brouncker and Wallis seek a proof. Reference will be made under the subject of binary quadratic forms to the assertion of Fermat and proof by Lagrange that any prime $8h + 1$ or $8h + 3$ is expressible in one and but one way as the sum of a square and double of a square.

The Japanese Matsunago[10] in the first half of the eighteenth century solved $x^2 + y^2 + z^2 = u^2$ by taking x and y at pleasure, expressing $x^2 + y^2$ as a product of two factors and equating the latter to $u - z$ and $u + z$. He noted that $x^2 + y^2 + z^2 = u^4$ has the solutions

$$x = m^4 - n^4, \qquad y = 4m^2n^2, \qquad z = 2(m^2 - n^2)mn, \qquad u = m^2 + n^2.$$

L. Euler[11] noted that if Fermat's theorem that every number x is a sum of three triangular numbers $(a^2 + a)/2$ is true, then every number $8x + 3$ is a sum of three squares $(2a + 1)^2$.

Euler[12] noted that, to prove that a prime $8m + 3$ is of the form $2a^2 + b^2$, one needs the theorems (of which he had no proofs): If the integer n is not a sum of two integral squares, then no integer np^2 is a sum of two integral squares; if n is not a sum of three integral squares, it is not a sum of three fractional squares.

May 6, 1747 (p. 414), Euler wrote that he had verified for small integers m that there always exists a triangular number $\Delta = (x^2 + x)/2$ such that $4(m - \Delta) + 1$ is a prime. If this be true, set $n = m - \Delta$; then $4n + 1$ is a $\boxed{2}$ and $2(4n + 1)$ is a $\boxed{2}$. Set $a = 2x + 1$. Then $n = m - \Delta$ gives $8m + 1 = 8n + a^2$. Hence $8m + 3 = 2(4n + 1) + a^2$ is a $\boxed{3}$. On pp. 442–5, Euler and Chr. Goldbach discussed without result the problem to express $8m + 3$ as a $\boxed{3}$. June 25, 1748 (pp. 458–460), Euler expressed his belief that any number $4n + 1$ or $4n + 2$ is a $\boxed{3}$. The latter would give

$$4n + 2 = (2a)^2 + (2b + 1)^2 + (2c + 1)^2, \qquad 2n + 1 = 2a^2 + (2e)^2 + (2d+1)^2,$$

for $b = d + e$, $c = d - e$, whence any odd number is of the form $2x^2 + y^2 + z^2$.

March 24, 1750 (p. 512), Goldbach gave the identity

$$\beta^2 + \gamma^2 + (3\delta - \beta - \gamma)^2 \equiv (2\delta - \beta)^2 + (2\delta - \gamma)^2 + (\delta - \beta - \gamma)^2.$$

---

[7] Oeuvres, II, 167; letter to Mersenne, June 3, 1638; Oeuvres de Fermat, 4, 1912, 57.

[8] Oeuvres de Descartes, II, 180–2.

[9] Oeuvres, II, 405; III, 318; letter to K. Digby, June, 1658.

[10] Y. Mikami, Abh. Geschichte Math. Wiss., 30, 1912, 233.

[11] Corresp. Math. Phys. (ed., Fuss), 1, 1843, 45; letter to Goldbach, Oct. 17, 1730.

[12] Ibid., 263; Oct. 15, 1743.

June 9, 1750 (p. 515), Euler expressed this as the first of the following:

$$a^2 + b^2 + c^2 = (2m - a)^2 + (2m - b)^2 + (2m - c)^2, \text{ if } a + b + c = 3m;$$
$$a^2 + b^2 + c^2 = (m - a)^2 + (m - b)^2 + (2m - c)^2, \quad \text{ if } a + b + 2c = 3m;$$
$$a^2 + b^2 + c^2 = (2m - a)^2 + (4m - b)^2 + (4m - c)^2, \text{ if } a + 2b + 2c = 9m;$$

and gave five more such formulae and similar ones for $\boxed{4}$.

Euler[13] verified that if $m \leqq 187$ and m is of the form $8N + 3$, then m is the sum of an odd square and the double of a prime $4n + 1$. Since $4n + 1 = a^2 + b^2$, $2(4n + 1) = (a + b)^2 + (a - b)^2$, and the m's in question are $\boxed{3}$.

J. L. Lagrange[14] remarked that a prime $8n - 1$ is of the form $24n - 1$ or $24n + 7$. Since he had proved that any prime $24n + 7$ is of the form $y^2 + 6z^2$, its double equals $(y + 2z)^2 + (y - 2z)^2 + (2z)^2$. He added that he did not see a proof of Fermat's[9] assertion for the remaining case of primes $24n - 1$.

J. A. Euler[15] used $(a^2 - 1)^2 + 4a^2 = (a^2 + 1)^2$ for $a = p, q$, to prove the identity

$$(p^2 + 1)^2(q^2 + 1)^2 = (q^2 - 1)^2(p^2 + 1)^2 + 4q^2(p^2 - 1)^2 + (4pq)^2.$$

A. M. Legendre[16] remarked that Fermat's[9] assertion is true not only of primes but of all odd numbers, and stated that either every number or its double is a $\boxed{3}$. His proof[17] (pp. 545–8) was based on empirical theorems on the quadratic divisors of $t^2 + cu^2$. He was led (pp. 530–542) to the empirical theorem that, if c is a prime $8m - 3$ or $8m + 1$, the number of decompositions of c into a sum of three squares (ignoring the order and signs of the roots) is the number of reduced quadratic divisors of the form $4n + 1$ (or of the form $4n - 1$); while for a prime $c = 8m + 3$, it is the number of reduced quadratic divisors.

P. Cossali[18] noted that the sum of the squares of n, $n + 1$, $n(n + 1)$ equals the square of $n^2 + n + 1$. This result has been attributed[100] to Diophantus, who in III, 5 noted that $2^2 + 3^2 + 6^2 = \square$.

Legendre[19] proved [the statement of Beguelin[75] of Ch. I] that every positive integer, not of the form $8n + 7$ or $4n$, is a sum of three squares having no common factor; the proof is by means of theorems on reciprocal (p. 367) quadratic divisors of $t^2 + cu^2$. In $2(2a + 1) = x^2 + y^2 + z^2$, two of the squares must be odd and the third even. Hence we may set $x = p + q$, $y = p - q$, $z = 2r$ and get $2a + 1 = p^2 + q^2 + 2r^2$. Again, any integer is of the form $2^{2n}(2a + 1)$ or $2^{2n} \cdot 2(2a + 1)$, and the latter is a $\boxed{3}$; hence either any integer or its double is a $\boxed{3}$. The product (p. 198) of two $\boxed{3}$ is not in general a $\boxed{3}$, since $(1 + 1 + 1)(16 + 4 + 1)$ is not a $\boxed{3}$.

[13] Acta Acad. Petrop., 4, II, 1780 (1775), 38; Comm. Arith., II, 138.

[14] Nouv. Mém. Acad. Roy. Berlin, année 1775, 356–7; Oeuvres, III, 795. In the quotation from Fermat, sum of a square and a double square should read sum of three squares.

[15] Acta Acad. Petrop., 3, 1779, 40–8. L. Euler's Comm. Arith., II, 463.

[16] Hist. et Mém. Acad. Roy. Sc. Paris, 1785, 514–5.

[17] Incomplete. Cf. A. Genocchi, Atti Accad. Sc. Torino, 15, 1879–80, 803; Gauss.[20]

[18] Origine, Trasporto in Italia. . . Algebra, 1, 1797, 97.

[19] Théorie des nombres, 1798, 398–9 (stated p. 202); ed. 2, 1808, 336–9 (p. 186); ed. 3, I, 1830, 393–5 (German transl. by Maser, I, 1893, 386–8).

C. F. Gauss[20] determined the number $\phi(m)$ of proper representations x, y, z, without common factor (and counted as different from y, x, z and from $- x, y, z$; etc.) of an integer m as a ☒. Let h be the number of classes, in the principal genus, of the properly primitive binary quadratic forms of determinant $- m$. Let μ be the number of distinct prime factors of m. Then

$$\phi(m) = 3 \cdot 2^{\mu+2}h \text{ if } m \equiv 1, 2, 5, 6 \pmod{8},$$
$$\phi(m) = 2^{\mu+2}h \quad \text{if } m \equiv 3 \pmod{8}.$$

In particular, we have Legendre's[19] theorem. But the squares of x, y, z; $- x, y, z$; y, x, z; etc. give the same decomposition of m into a ☒. The resulting number of decompositions (art. 292) of m agrees with that derived by (incomplete) induction by Legendre[16] for the case m a prime.

A. Cauchy[21] noted, as a corollary to Legendre's theorem,[19] that if a is any integer and if 4^α is the highest power of 4 dividing a, then a is a ☒ if and only if $a/4^\alpha$ is not of the form $8n + 7$.

J. R. Young[21a] solved $x^2 + y^2 + z^2 = w^2$ by taking $w = x + p$ and finding x rationally, or by setting $y^2 = 2xz$. Then if w is given, take $y = pz$, whence z is found in terms of p. To find (p. 346) three numbers in harmonical progression whose sum of squares is a square, take $1/(x \pm y)$, $1/x$ as the three numbers; the condition $3x^4 + y^4 = \square$ is satisfied if $x = 2$, $y = 1$.

C. Gill[21b] noted that the sum of the squares of $2mn(k^2 + l^2)$, $2kl(m^2 - n^2)$ and $(k^2 - l^2)(m^2 - n^2)$ equals the square of $(k^2 + l^2)(m^2 + n^2)$.

C. G. J. Jacobi[22] proved by use of elliptic functions that

$$(1) \qquad \left\{ \sum_{m=-\infty}^{+\infty} (-1)^m x^{(3m^2+m)/2} \right\}^3 = \sum_{n=0}^{\infty} (-1)^n(2n+1)x^{(n^2+n)/2},$$

a result occurring also in Gauss' posthumous papers.

Jacobi[23] gave an elementary proof of (1). Replace x by x^{24} and multiply the resulting equation by x^3; we get

$$(2) \qquad \left\{ \sum_{m=-\infty}^{+\infty} (-1)^m x^{(6m+1)^2} \right\}^3 = \sum_b (-1)^{(b-1)/2} b x^{3b^2} \quad (b \text{ odd}, b > 0).$$

For m positive, set $a = 6m + 1$; for m negative, set $a = - 6m - 1$; thus

$$\left(\sum_a \pm x^{a^2} \right)^3 = \sum_b (-1)^{(b-1)/2} b x^{3b^2},$$

where a and b range over all positive odd integers such that a is not divisible by 3. The sign in the left member is $+$ if $a = 12k \pm 1$, $-$ if $a = 12k \pm 5$. The expansion gives the following theorem: If a number $24k + 3$, not of the form $3b^2$, be expressed as a sum of three squares $(6m \pm 1)^2$ in all possible

---

[20] Disq. Arith., 1801, Art. 291; Werke, I, 1863, 343; German transl. by H. Maser, pp. 329–33.
 Cf. H. J. S. Smith, British Assoc. Report, 1865; Coll. Math. Papers, I, 324.

[21] Mém. Sc. Math. Phys. de l'Institut de France, (1), 14, 1813–5, 177; Oeuvres, (2), VI, 323.

[21a] Algebra, 1816; S. Ward's Amer. ed., 1832, 326–7.

[21b] The Gentleman's Math. Companion, London, 5, No. 29, 1826, 364.

[22] Fund. nova func. ellip., 1829, § 66(7); Werke, I, p. 237 (7).

[23] Jour. für Math., 21, 1840, 13–32; Werke, VI, 281–302. French transl., Jour. de Math., 7, 1842, 85–109.

ways, counting two for each case of three distinct squares, then the number of decompositions in which one or three of the m's are even equals that in which one or three of the m's are odd. But for $3b^2$ the first number exceeds the second if and only if $b \equiv 1 \pmod{4}$, the excess being always $[b/3]$.

If N is any odd integer, (2) shows that

$$3N^2 = (6m + 1)^2 + (6m_1 + 1)^2 + (6m_2 + 1)^2$$

in more than one way if $N > 1$, so that the squares need not all be equal. Thus

$$N^2 = n^2 + 2n_1^2 + 6n_2^2, \qquad n = 2(m + m_1 + m_2) + 1,$$
$$n_1 = 2m - m_1 - m_2, \qquad n_2 = m_1 - m_2,$$

where n_1 and n_2 are not both zero. By changing the sign of n if necessary, we may assume that $N - n$ is divisible by 4. Let N be a prime. Then $(N - n)/4$ and $(N + n)/2$ are relatively prime and each divides $n_1^2 + 3n_2^2$, whence each are of the latter form:

$$\tfrac{1}{2}(N + n) = \alpha^2 + 3\gamma^2, \qquad \tfrac{1}{4}(N - n) = \beta^2 + 3\delta^2.$$

Hence every prime can be expressed in the form $\alpha^2 + 2\beta^2 + 3\gamma^2 + 6\delta^2$. Since the product of two such expressions is of the same form, every number can be expressed in that form.

G. L. Dirichlet[24] remarked that, by use of his formulas for the number h of classes of binary quadratic forms, one can give a new expression for the number $\phi(m)$ of proper representations of m as a $\boxed{3}$ (Gauss[20]). According to G. Eisenstein,[25] the result is

$$\phi(m) = 24 \sum_{s=1}^{[m/4]} \left(\frac{s}{m} \right), \text{ if } m \equiv 1 \pmod{4};$$

$$\phi(m) = 8 \sum_{s=1}^{[m/2]} \left(\frac{s}{m} \right), \text{ if } m \equiv 3 \pmod{4},$$

where (s/m) is Jacobi's symbol and is 0 if s, m have a common factor.

T. Weddle[26] noted that, if (a, p, z), (b, q, z') and (c, r, z'') are the extremities of a system of conjugate semi-axes of an ellipsoid,

$$(a^2 + b^2 + c^2)(p^2 + q^2 + r^2) = (aq - bp)^2 + (ar - cp)^2 + (br - cq)^2.$$

J. R. Young[27] noted that the last formula follows by taking $d = s = 0$, $ap + bq + cr = 0$ in Euler's formula (1) of Ch. VIII. But if we take $d = s = 0$, $a/p = b/q$, we get

$$(a^2 + b^2 + c^2)(p^2 + q^2 + r^2) = (ap + bq + cr)^2 + (ar - cp)^2 + (br - cq)^2.$$

G. L. Dirichlet[28] gave an elegant proof of Legendre's[19] theorem. Let a be a positive integer not of one of the forms $4n$, $8n + 7$. It suffices to

[24] Jour. für Math., 21, 1840, 155; Werke, 1, 1889, 496.

[25] Jour. für Math., 35, 1847, 368. Cf. T. Pepin, Atti Accad. Pont. Nuovi Lincei, 37, 1883–4, 44.

[26] Cambridge and Dublin Math. Jour., 2, 1847, 13–19.

[27] Trans. Irish Acad., 21, II, 1848, 330.

[28] Jour. für Math., 40, 1850, 228–232; Werke, 2, 1897, 91. French transl. by J. Hoüel, Jour. de Math., (2), 4, 1859, 233.

show that there exists a positive ternary quadratic form F of determinant $+1$ whose first coefficient is a. Indeed, such a form is equivalent to $x^2 + y^2 + z^2$, so that the latter can be transformed into F by a substitution of determinant unity; thus a is the sum of the squares of three of the coefficients (having no common factor) of the substitution. Now the ternary form

$$ax^2 + by^2 + cz^2 + 2a'yz + 2xz \qquad (\Delta = bc - a'^2)$$

has the determinant $+1$ if $b = a\Delta - 1$. The form is positive if $\Delta > 0$. It suffices to show that a positive value of Δ can be found for which $-\Delta$ is a quadratic residue of b, so that c and a' may be determined to satisfy $a'^2 - bc = -\Delta$. For $a = 4k + 2$, we take Δ odd. Then $b \equiv 1 \pmod 4$. We seek a suitable prime b. Since, for Jacobi symbols,

$$\left(\frac{-1}{\Delta}\right) = \left(\frac{b}{\Delta}\right) = \left(\frac{\Delta}{b}\right) = \left(\frac{-\Delta}{b}\right) = +1,$$

Δ must be of the form $4t + 1$, whence $b = 4at + a - 1$. The latter is the general term of an arithmetical progression, containing primes. For $a = 8k + 1$, we take $\Delta = 8t + 3$, and seek a prime p for which $2p = b$. Since $2p = a\Delta - 1$, $p \equiv 1 \pmod 4$,

$$1 = \left(\frac{-2}{\Delta}\right) = \left(\frac{p}{\Delta}\right) = \left(\frac{\Delta}{p}\right) = \left(\frac{-\Delta}{p}\right) = \left(\frac{-\Delta}{b}\right).$$

There exists a prime in the progression $p = 4at + \frac{1}{2}(3a - 1)$. A like result follows for $a = 8k + 3$, $\Delta = 8t + 1$, and for $a = 8k + 5$, $\Delta = 8t + 3$.

H. Burhenne[29] noted that $x^2 + y^2 + z^2 = (a^2 + b^2 + c^2)s^2$ if

$$s = m^2 + n^2 + p^2$$

and

$$x = 2ml - as, \qquad y = 2nl - bs, \qquad z = 2pl - cs, \qquad l = am + bn + cp.$$

H. Faure[30] noted that no number $m^2(8x + 7)$ is a $\boxed{3}$.

V. A. Lebesgue[31] proved that every odd number p is of the form $x^2 + y^2 + 2z^2$, where x, y, z are integers with no common factor. The method is that of Dirichlet.[28] It follows that

$$2p = (x + y)^2 + (x - y)^2 + (2z)^2.$$

J. Liouville[32] denoted the number of sets of integral solutions of $x^2 + y^2 + z^2 = \mu$ by $\psi(\mu)$. Set $n = 2^a m$, m odd, $\alpha > 0$. Let ω be the greatest integer $\leqq \sqrt{n}$. Then

$$\sum (As^4 + Bs^2 + C)\psi(n - s^2) = (3An^2 + 6Bn + 24C)\sigma(m)$$

$$(s = 0, \pm 1, \cdots, \pm \omega),$$

where $\sigma(m)$ is the sum of the divisors of m.

[29] Archiv Math. Phys., 20, 1853, 466–8.
[30] Nouv. Ann. Math., 12, 1853, 336.
[31] Jour. de Math., (2), 2, 1857, 149–152.
[32] Jour. de Math., (2), 5, 1860, 141–2.

L. Kronecker[33] proved by use of series for elliptic functions that the number of representations of n as a ③ is $24F(n) - 12G(n)$, where $G(n)$ is the number of classes of binary quadratic forms of determinant $- n$, and $F(n)$ is the number of classes of such forms of determinant $- n$ in which at least one of the two outer coefficients is odd. This result gives the theorem of Gauss[20] since $G(n) = F(n)$ if $n \equiv 1$ or $2 \pmod 4$; $G(n) = 2F(n)$ if $n \equiv 7 \pmod 8$, $3G(n) = 4F(n) + t$ if $n \equiv 3 \pmod 8$, where $t = 2$ if n is the triple of an odd square, $t = 0$ in the remaining case.

J. Liouville[33a] noted that, if $m \equiv 3 \pmod 8$, the number of solutions of $m = i^2 + i_1^2 + i_2^2$, where i, i_1, i_2 are odd and positive, is

$$\rho\left(\frac{m - 1^2}{2}\right) + \rho\left(\frac{m - 3^2}{2}\right) + \cdots$$
$$= \rho'(m) + 2\rho'(m - 4\cdot 1^2) + 2\rho'(m - 4\cdot 2^2) + \cdots,$$

where $\rho'(n)$ is the excess of the number of divisors $< \sqrt{n}$ of n which are of the form $4\mu + 1$ over the number of such divisors of the form $4\mu + 3$, while $\rho(n)$ is the corresponding excess for all the divisors of n.

V. A. Lebesgue[34] stated that every solution of $t^2 = x^2 + y^2 + z^2$ is given by

$$t = G(e^2 A + f^2 C), \qquad x = G(e^2 A - f^2 C), \qquad y^2 + z^2 = 4e^2 f^2 G^2 A C,$$

where $G = g^2 + h^2$, $A = a^2 + b^2$, $C = c^2 + d^2$. In the identity

$$t^2 - x^2 = y^2 + z^2,$$

set $g = 1$, $h = 0$, and replace ae, be, cf, df by α, β, γ, δ; we get

$$(3) \quad (\alpha^2 + \beta^2 + \gamma^2 + \delta^2)^2 = (\alpha^2 + \beta^2 - \gamma^2 - \delta^2)^2$$
$$+ 4(\alpha\gamma + \beta\delta)^2 + 4(\alpha\delta - \beta\gamma)^2,$$

a special case[35] of Euler's formula (1) of Ch. VIII. Since every integer n is a ④, n^2 is a sum of three squares [each $\neq 0$, in general].

A. Genocchi[36] proved Fermat's statement that the double of any prime $8k - 1$ is a ③.

J. Liouville[37] stated that, if $m \equiv 1 \pmod 4$ and F is any function,

$$\Sigma(-1)^{s + (i^2 - 1)/8} F(\omega) = \Sigma(-1)^{s_1} F(\omega_1),$$

summed for all the decompositions $i^2 + \omega^2 + 16s^2 = m = i_1^2 + \omega_1^2 + 8s_1^2$ in which i and i_1 are odd and positive, while ω and ω_1 are even. G. Zolotaref[38] gave a proof by use of elliptic functions.

[33] Jour. für Math., 57, 1860, 253. French transl., Jour. de Math., (2), 5, 1860, 297. Cf. Mordell.[112] For $n \equiv 3 \pmod 8$, C. Hermite, Jour. de Math., (2), 7, 1862, 38; Comptes Rendus Paris, 53, 1861, 214; Oeuvres, II, 109.
[33a] Jour. de Math., (2), 7, 1862, 43–44. Cf. Liouville[7] of Ch. XI.
[34] Comptes Rendus Paris, 66, 1868, 396–8.
[35] Also given in Bellacchi's Algebra, 1, 1869, 105.
[36] Annali di Mat., (2), 2, 1868–9, 256.
[37] Jour. de Math., (2), 15, 1870, 133–6.
[38] Bull. Acad. Sc. St. Pétersbourg, 16, 1870–1, 85–7.

E. Catalan[39] noted that the excess of the number of even values of $x + y + z$ in

$$(6x \pm 1)^2 + (6y \pm 1)^2 + (6z \pm 1)^2 = 3(2n + 1)^2$$

over the number of odd values of $x + y + z$ is $(2n + 1)(- 1)^n$. There are at least $[(2n + 1)/6]$ decompositions of $3(2n + 1)^2$ into a $\boxed{3}$. The sextuple[40] of an odd square is a sum of three squares, two of which are of the form $(6\mu \pm 1)^2$ and the third is $4(6k \pm 1)^2$. The excess of the number of even values of x in

$$4x^2 + 4y^2 + (2z + 1)^2 = (2n + 1)^2$$

over the number of odd values is $\{(2n + 1)(- 1)^n - 1\}/4$. If a prime p is not a $\boxed{2}$, then p^2 is a $\boxed{3}$.

Catalan stated and V. A. Lebesgue[41] proved that the square of a $\boxed{3}$ is a $\boxed{3}$, since (3) for $\delta = 0$ becomes

$$(4) \qquad (\alpha^2 + \beta^2 + \gamma^2)^2 = (\alpha^2 + \beta^2 - \gamma^2)^2 + (2\alpha\gamma)^2 + (2\beta\gamma)^2.$$

This formula was employed by Euler[308] of Ch. XXII.

J. Neuberg[42] also gave (4).

Catalan[43] gave the identity

$$(a^2 + b^2 + c^2 + ab + bc + ac)^2$$
$$= (a + c)^2(a + b)^2 + (b + c)^2(a + b)^2 + (c^2 + ac + bc - ab)^2$$

and by a change of notation deduced

$$(f^2 + 2g^2 + h^2)^2 = (f^2 - h^2)^2 + \{2g(f + h)\}^2 + (2fh - 2g^2)^2$$
$$= \{2g(f + h)\}^2 + \{2g(f - h)\}^2 + (f^2 - 2g^2 + h^2)^2.$$

Catalan[44] stated empirically that the triple of any odd square not divisible by 5 is a sum of squares of three primes other than 2 and 3.

G. H. Halphen[45] proved that every prime $8m + 3$ is a $\boxed{3}$ by means of his[104] recursion formula (Ch. VI) for the sum $s(x)$ of the divisors of x whose complementary divisors are odd. Let x be not a square, $\boxed{2}$ or $\boxed{3}$; then no one of the arguments $x - n^2$ is a $\boxed{2}$, so that $s(x)$ is divisible by 8. Let also x be a prime, so that $s(x) = x + 1$. Hence a prime not a $\boxed{2}$ or $\boxed{3}$ is of the form $8m - 1$.

U. Dainelli[46] derived by integration the case $c = 0$ of Catalan's[43] formula

$$(a^2 + ab + b^2)^2 = (ab)^2 + \{a(a + b)\}^2 + \{b(a + b)\}^2.$$

S. Réalis[47] noted that $kz^2 = z_1^2 + z_2^2 + z_3^2$ if

$$k = A^2 + B^2 + C^2, \ z = \alpha^2 + \beta^2 + \gamma^2, \ z_1 = A(\beta^2 + \gamma^2 - \alpha^2) - 2\alpha(B\beta + C\gamma),$$
$$z_2 = B(\alpha^2 - \beta^2 + \gamma^2) - 2\beta(C\gamma + A\alpha), \ z_3 = C(\alpha^2 + \beta^2 - \gamma^2) - 2\gamma(A\alpha + B\beta).$$

[39] Recherches sur quelques produits indéfinis, Mém. Acad. Roy. Belgique, 40, 1873, 61–191; extract in Nouv. Ann. Math., (2), 13, 1874, 518–523.
[40] Repeated by Catalan, Nouv. Ann. Math., (2), 14, 1875, 428.
[41] Nouv. Ann. Math., (2), 13, 1874, 64, 111.
[42] Nouv. Corresp. Math., 1, 1874–5, 195–6.
[43] Ibid., 153; 2, 1876, 117.
[44] Nouv. Corresp. Math., 3, 1877, 29.
[45] Bull. Soc. Math. France, 6, 1877–8, 180.
[46] Giornale di Mat., 15, 1877, 378.
[47] Nouv. Corresp. Math., 4, 1878, 325. Cf. Malfatti[19] of Ch. VIII.

The case $A = 1$, $B = C = 0$ expresses the square of a $\boxed{3}$ as a $\boxed{3}$. The case $A = \gamma$, $B = \beta$, $C = \alpha$ expresses the cube of a $\boxed{3}$ as a $\boxed{3}$.

H. S. Monck[43] noted that if a, b, c are integral edges of a rectangular parallelopiped and the diagonal d is an integer, then $a^2 + b^2 + c^2 = d^2$, and another has the edges $a + b + d$, $a + c + d$, $b + c + d$ and diagonal $a + b + c + 2d$. From $a = 1$, $b = -2$, $c = 2$, $d = 3$, we get the new edges 2, 3, 6 and diagonal 7. Cf. papers 25–29 of Ch. XIX.

S. Réalis[49] gave a complicated identity

$$x^2 + y^2 + z^2 = t^2 + u^2 + v^2, \qquad x = \alpha^2 + \beta^2 + \gamma^2 - \delta^2 - \epsilon^2, \qquad \cdots,$$

said to give all solutions of the equation. He gave a similar identity which is said to give all solutions of $\boxed{4} = \boxed{4}$. Supplementing the theorem that N is a $\boxed{3}$ if N has no square factor and is of one of the forms $4p + 1$, $4p + 2$, $8p + 3$, he stated that N is the quotient of $x^2 + y^2 + z^2$ by the factor common to x^2, y^2, z^2, where x, y, z are given above.

F. Pisani[50] discussed $u^2 + (u + 1)^2 = (x - 1)^2 + x^2 + (x + 1)^2$, whence $(2u + 1)^2 = 6x^2 + 3$. Thus $2u + 1 = 3y$, $2x^2 - 3y^2 = -1$. An infinitude of solutions is found from the continued fraction for $\sqrt{3/2}$.

S. Réalis[51] expressed as a $\boxed{2}$ the sum of the three squares of

$$2(\alpha^2 - \beta^2 - \gamma^2 + \delta^2) + 2\alpha(2\beta + 3\gamma + 4\delta)$$

and two similar expressions. He gave (p. 501) expressions for $9P^n$ and $18P^n$ as $\boxed{3}$ if $P = a^2 + b^2$.

E. Catalan stated and Réalis[52] proved that every power of 3 is a sum of three squares prime to 3. Réalis (p. 75) expressed $n^2(x^2 + y^2 + z^2)$ as a $\boxed{3}$ when $n = a^2 + ab + b^2$.

Catalan[53] proved that, if $a \equiv b \pmod 3$, $a^2 + b^2$ is a sum of three squares $\neq 0$; also if $a \equiv b \pmod{x + y}$ and $2xy = \square$. Also that every power of 3 is a sum of three squares prime to 3. He[54] proved that every even power of $a^2 + ab + b^2$ is a $\boxed{3}$ and gave special identities $\boxed{3} \cdot \boxed{3} = \boxed{3}$.

O. Schier[55] solved $x^2 + y^2 + z^2 = u^2$ by setting $y = x + \beta$, $z = x + \gamma$, $u = x + \delta$, and taking $\beta + \gamma = \delta$. Then

$$2x^2 = \delta^2 - \beta^2 - \gamma^2, \qquad x^2 = \beta\gamma = (y - x)(z - x),$$

whence $x = yz/(y + z)$. Multiplying the values by $y + z$, we get the identity of Dainelli.[46]

J. Neuberg[56] noted that $x^2 + y^2 + z^2 = X^2 + Y^2 + Z^2$ for

$$x/a = y/b = z/c = k^2 + 3, \qquad X = a(k^2 - 1) + 2b(k + 1) - 2c(k - 1),$$

Y and Z being derived from X by permuting a, b, c cyclically.

[48] Math. Quest. Educ. Times, 29, 1878, 74.
[49] Nouv. Ann. Math., (2), 18, 1879, 505–6.
[50] Nouv. Ann. Math., (2), 19, 1880, 524–6. Same in Zeitschr. Math. Naturw. Unterricht, 12, 1881, 268. Cf. Lionnet[183] of Ch. XII.
[51] Ibid., (2), 20, 1881, 335–6.
[52] Mathesis, 1, 1881, 73, 87.
[53] Atti Accad. Pont. Nuovi Lincei, 34, 1880–1, 63–4, 135–6.
[54] Ibid., 35, 1881–2, 103–114. Extract in Sphinx-Oedipe, 5, 1910, 54–55.
[55] Sitzungsber. Akad. Wiss. Wien (Math.), 82, II, 1881, 890–1.
[56] Mathesis, 2, 1882, 116; (4), 4, 1914, 116–7.

S. Réalis[57] gave expressions involving five parameters satisfying

$$X^2 + Y^2 + Z^2 = k(x^2 + y^2 + z^2)$$

for $k = 7, 19, 67$, and formulas to deduce solutions from a given one.

L. Kronecker[58] employed the number of classes of bilinear forms in two pairs of cogredient variables to find the number of ways any integer is a ▣, in accord with Gauss.[20]

E. Catalan[59] stated that all solutions of $x^2 + y^2 = u^2 + v^2 + w^2$ are given without repetition by $u = x + \alpha$, $v = y - \beta$, $x = sp + \beta\theta$, $y = sq + \alpha\theta$, where $2s = \alpha^2 + \beta^2 + w^2$ and α, β are relatively prime integers, while $\beta q - \alpha p = 1$. If $r, s = \pm a + \sqrt{a^2 + b^2}$, and $n > 1$, then[60]

$$(r^{2n-1} + s^{2n-1})/(r + s)$$

is a ▣ and ▣. Hence the same is true of $x^{4n} - x^{4n-2}y^2 + \cdots + y^{4n}$ for x, y relatively prime integers > 1.

G. C. Gerono[61] noted that if N^2 is a sum of squares of two consecutive integers, N is a sum of squares of three integers of which two are consecutive, as $29^2 = 20^2 + 21^2$, $29 = 2^2 + 3^2 + 4^2$.

Catalan[62] noted that every power of a ▣ is a ▣ since

$$(x^2 + y^2 + z^2)^3 = y^2(3z^2 - x^2 - y^2)^2 + x^2(3z^2 - x^2 - y^2)^2 + z^2(z^2 - 3x^2 - 3y^2)^2.$$

To solve (p. 103) $x^2 + y^2 = u^2 + v^2 + w^2$, set $u = x + \alpha$, $v = y - \beta$. Then $\beta y - \alpha x = s$, where $s = \frac{1}{2}(\alpha^2 + \beta^2 + w^2)$. Take α, β relatively prime and w such that s is an integer. For $\beta q - \alpha p = 1$, all solutions are given without repetition by $x = sp + \beta\theta$, $y = sq + \alpha\theta$. [Catalan[59]].

Catalan stated and E. Fauquembergue[63] proved that, unless $x = 1$ or $4a^2 + 1$, $(a^2 + 1)x^2 = y^2 + 1$ implies that x is a ▣, since all solutions (if any) of $y^2 - Ax^2 = -1$ are given by the terms of convergents of even rank in the continued fraction for $\sqrt{A}$. The latter proved (p. 346) that $x^2 + y^2 = u^2 + v^2 + 1$ is satisfied by $2\alpha + 1$, $\alpha - 1$, $\alpha + 1$, 2α and by $2\alpha^2 + 1$, $\beta^2 - 1$, $2\alpha^2 - \beta^2 + 1$, $2\alpha\beta$.

J. W. L. Glaisher[64] proved that, if the number of representations of $8n + 1$ by

$$(2p + 1)^2 + (4r)^2 + (4s)^2, \qquad (2p + 1)^2 + (4r + 2)^2 + (4s + 2)^2$$

is R_1, R_2, respectively, then $R_1 = R_2$ unless $8n + 1$ is a square, while if $8n + 1 = t^2$,

$$R_1 - R_2 = 6t(-1)^{(t-1)/2}.$$

[57] Mathesis, 2, 1882, 64–7.
[58] Abh. Akad. Berlin (Math.), 2, 1883, 52; Werke, 2, 1897, 483.
[59] Assoc. franç. av. sc., 12, 1883, 98–101.
[60] Also stated Nouv. Ann. Math., (3), 3, 1884, 342; Mathesis, 6, 1886, 65, 113.
[61] Nouv. Ann. Math., (3), 2, 1883, 329.
[62] Atti Accad. Pont. Nuovi Lincei, 37, 1883–4, 54–6.
[63] Nouv. Ann. Math., (3), 3, 1884, 538. Cf. Catalan[191] of Ch. XII.
[64] Quar. Jour. Math., 20, 1885, 94.

Catalan[65] noted that (3) with $\delta = 0$ does not give all solutions of $u^2 = x^2 + y^2 + z^2$, for example not that with $u = 27$. But all primitive solutions (u, x, y, z with no common factor) are said to be given by (3). There are several identities giving an infinitude of (but not all) solutions of $(x^2 + y^2 + z^2)^2 = \boxed{3}$.

A. Desboves[66] noted that the complete solution in integers of

$$X^2 + Y^2 + Z^2 = U^2$$

is given by the identity

$$[2(p^2 + q^2 - s^2)]^2 + \{2[(p - s)^2 - q^2 + p(q - s)]\}^2$$
$$+ [(q - s)^2 - p^2 + 4q(p - s)]^2 = \{3[(p - s)^2 + q^2] + 2s(p - q)\}^2.$$

Catalan[67] noted that, if $x^2 + y^2 + z^2 = 1$, $xx' + yy' + zz' = 0$,

$$(x'^2 + y'^2 + z'^2)\{(yz'' - zy'')^2 + (zx'' - xz'')^2 + (xy'' - yx'')^2\}$$
$$= (x'x'' + y'y'' + z'z'')^2 + \{(yz'' - zy'')x' + (zx'' - xz'')y' + (xy'' - yx'')z'\}^2.$$

Catalan[68] treated $u^2 = x^2 + y^2 + z^2$. Since a prime $4\mu + 1$ is of the form $y^2 + z^2$, one solution is given by $u = 2\mu + 1$, $x = 2\mu$. We may set $u + x = \alpha^2 + \beta^2$, $u - x = \gamma^2 + \delta^2$ and obtain a solution leading to the identity (3).

C. Hermite[69] expressed the number of decompositions of an integer into 3 and 5 squares in terms of the number of classes of binary quadratic forms.

J. W. L. Glaisher[70] considered the compositions $a^2 + b^2 + c^2$, $a^2 + b^2 + b^2$, $a^2 + a^2 + a^2$ of n as a sum of three squares when $n \equiv 3 \pmod{4}$, a, b, c being distinct odd numbers, and formed from them the respective quantities $8a\alpha + 8b\beta + 8c\gamma$, $4a\alpha + 8b\beta$, $4a\alpha$, where $\alpha = (-1)^{(a-1)/2}$, $\cdots$, $\gamma = (-1)^{(c-1)/2}$. The sum of the quantities so derived from all the compositions of n equals the expression

$$\sigma(n) - 2\sigma(n - 4) + 2\sigma(n - 16) - 2\sigma(n - 36) + \cdots,$$

where $\sigma(k)$ is the sum of the divisors of k. This result holds also when $n \equiv 1 \pmod{4}$ if we use the quantities $8a\alpha$, $4a\alpha$, $4a\alpha$, $a\alpha$ for the respective compositions $a^2 + b^2 + c^2$, $a^2 + b^2 + 0$, $a^2 + b^2 + b^2$, $a^2 + 0 + 0$, where a is odd, b and c are even, distinct and $\neq 0$. The number of representations of n as a sum of three squares is expressed in several ways as a series involving the number of representations of k as a sum of two squares.

E. Catalan[71] noted that

$$3\{(a + 2b - 1)^2 + (b + 2a - 1)^2 + (a - b)^2\}$$
$$= (3a - 1)^2 + (3b - 1)^2 + (3a + 3b - 2)^2,$$
$$(x^2 + y^2 + z^2)(x'^2 + y'^2 + z'^2) = \sum_{(3)} (yz'' - zy'')^2,$$

if $x'x'' + \cdots = 1$, $x = x' - x''\Sigma x'^2$, $\cdots$.

[65] Bull. Acad. Roy. Belgique, (3), 9, 1885, 531.
[66] Nouv. Ann. Math., (3), 5, 1886, 232.
[67] Mém. Soc. Roy. Sc. Liège, (2), 13, 1886, 34–9 (Mélanges Math. III).
[68] Ibid., (2), 15, 1888, 73–5, 211, 259 (Mélanges Math. III, 1885, 120).
[69] Jour. für Math., 100, 1887, 60, 63; Oeuvres, IV, 233, 237.
[70] Messenger Math., 21, 1891–2, 122–130.
[71] Assoc. franç. av. sc., 1891, II, 195–7.

De Rocquigny[72] obtained a solution of $\boxed{3} \cdot \boxed{3} = \boxed{3}$ by use of

$$(a^2 + \lambda b^2)(a_1^2 + \lambda b_1^2) = (aa_1 + \lambda bb_1)^2 + \lambda(ab_1 - a_1b)^2, \qquad \lambda = c^2 + d^2.$$

Catalan[73] took the fourth variables zero in Euler's (1) of Ch. VIII and got

$$\begin{aligned} P &\equiv (x^2 + y^2 + z^2)(x_1^2 + y_1^2 + z_1^2) \\ &= (xx_1 + yy_1 + zz_1)^2 + (xy_1 - yx_1)^2 + (yz_1 - zy_1)^2 + (zx_1 - xz_1)^2. \end{aligned}$$

Taking $x : x_1 = y : y_1$ [Young[27]], we get $P = \boxed{3}$; but the condition is not necessary in view of $(9 + 4 + 1)(1 + 1 + 1) = 25 + 16 + 1$.

K. Th. Vahlen[74] deduced (1) from the theory of partitions. The identity

$$\alpha^2 + 2\beta^2 + 3\gamma^2 + 6\delta^2 = \alpha^2 + (\beta + \gamma + \delta)^2 + (-\beta + \gamma + \delta)^2 + (\gamma - 2\delta)^2$$

and Jacobi's[23] final result shows that every number is a $\boxed{4}$.

Catalan[75] proved that if p is not a $\boxed{2}$, then p^2 is a $\boxed{3}$. For, if

$$p = a^2 + b^2 + c^2, \qquad p^2 = (a^2 + b^2 - c^2)^2 + (2ac)^2 + (2bc)^2.$$

If $p = a^2 + b^2 + c^2 + d^2$, then

$$p^2 = (a^2 + b^2 - c^2 - d^2)^2 + 4(a^2 + b^2)(c^2 + d^2).$$

Catalan[76] noted that every odd square > 1 is a sum of 2 or 3 squares.

P. Bachmann[77] considered the number A of decompositions of s into three distinct squares $\alpha^2 + \alpha_1^2 + \alpha_2^2$ where one (or three) of α, α_1, α_2 is of the form $12k \pm 1$ and the others are of the form $12k \pm 7$; the number A' of decompositions into three distinct squares for which the reverse is true; the number B of decompositions $s = \alpha^2 + 2\alpha_1^2$ in which α, α_1 are distinct and α is of the form $12k \pm 1$; and the number B' of such decompositions in which $\alpha = 12k \pm 7$. He proved that $2A + B = 2A' + B' + D$, where $D = 0$ or $\{(-1)^i(2i + 1) - j\}/3$, according as s is not or is of the form $3(2i + 1)^2$, and j is the absolutely least residue modulo 3 of $(-1)^i(2i + 1)$.

Bachmann[78] gave an exposition of the theory of $\boxed{3}$.

J. F. d'Avillez[79] applied Catalan's[43] formula to express the squares of 1, 3, 6, 11, 17, 25, 34, 45, $\cdots$ as $\boxed{3}$.

We may express 1521 as a $\boxed{3}$ in 7 ways.[80] Many identities giving equal sums of three squares have been noted.[81]

M. A. Gruber[82] tabulated solutions of $3^{2n} = \boxed{3}$ for $n \leqq 6$.

R. D. von Sterneck[83] gave an elementary proof of (1).

[72] Mathesis, (2), 2, 1892, 136.
[73] Ibid., (2), 3, 1893, 105–6.
[74] Jour. für Math., 112, 1893, 23.
[75] Mém. Acad. Roy. Belgique, 52, 1893–4, 21.
[76] Mathesis, (2), 4, 1894, 27, 52–53.
[77] Die Analytische Zahlentheorie, 1894, 37–9.
[78] Arith. der Quadrat. Formen, 1898, 139–162, 600; Niedere Zahlentheorie, 2, 1910, 320–323.
[79] Jornal de Sc. Math. Phys. e Nat., (2), 5, 1897–8, 90–2.
[80] Amer. Math. Monthly, 5, 1898, 214.
[81] Ibid., 6, 1899, 17–20.
[82] Ibid., 8, 1901, 49–50.
[83] Sitzungsber. Akad. Wiss. Wien (Math.), 109, IIa, 1900, 28–43.

H. Schubert[84] treated $x^2 + y^2 + z^2 = u^2$, where x, y, z have no common factor. They are not all odd, as seen by their residues modulo 4. Hence we may assume that x and y are even, and z and u odd. Thus $(x/2)^2 + (y/2)^2$ is to be factored into $\frac{1}{2}(u + z)$, $\frac{1}{2}(u - z)$, which is done by trial.

P. Whitworth[34a] tabulated the number of ways each integer ≤ 64 is a sum of three squares each > 0. R. W. D. Christie noted cases of equal sums of three squares.

E. Grigorief[85] noted that [by (3)] $x^2 + y^2 + 1 = z^2$ is satisfied if

$$2x = p^2 - q^2 + r^2 - s^2, \quad y = pq + rs, \quad 2z = p^2 + q^2 + r^2 + s^2, \quad ps - rq = 1,$$

when $p + q + r + s$ is even. Escott (p. 285) listed 34 values < 500 of z.

F. Hromádko[86] noted that $n^2 + (n + 1)^2 + x^2 = (x + 1)^2$ for

$$x = n(n + 1),$$

while $a^2 + b^2 + x^2 = z^2$ for $z = x + a - b$, $(a - b)x = ab$.

Haag[87] stated that every number not of the form $(8n - 1)p^2$ is a $\boxed{3}$.

H. B. Mathieu[88] noted the identity

$$(\alpha^2 + \beta^2 + \gamma^2)[a^2\gamma^2 + b^2\gamma^2 + (a\alpha + b\beta)^2]$$
$$= [a\alpha\beta + b(\beta^2 + \gamma^2)]^2 + [a(\alpha^2 + \gamma^2) + b\alpha\beta]^2 + (\alpha\beta\gamma - b\alpha\gamma)^2.$$

G. Humbert[89] gave theorems on the decomposition of $M + P\rho$ into a sum of three squares of such complex integers, where $\rho = (1 + \sqrt{5})/2$.

A. Hurwitz[90] noted that, if $n = 2^\kappa m q_1^{a_1} q_2^{a_2} \cdots$, where q_1, q_2, $\cdots$ are primes $4k + 3$, and m is a product of powers of primes $4k + 1$,

$$n^2 = x^2 + y^2 + z^2$$

has

$$6m \prod \left(q_i^{a_i} + 2 \frac{q_i^{a_i} - 1}{q_i - 1} \right)$$

solutions. It has solutions each $\neq 0$ except for $n^2 = 2^{2\mu}$, $5^2 \cdot 2^{2\mu}$, since $n^2 = x^2 + y^2$ has $4\sigma(n^2)$ solutions.

A. S. Werebrusow[91] expressed a $\boxed{3}$ as the cube of a $\boxed{3}$, but made errors.

G. Bisconcini[92] gave a table of solutions of (4).

E. Landau[93] considered the number $C(x)$ of integers $\leq x$ which are $\boxed{3}$. Since a positive integer is a $\boxed{3}$ if and only if it is not of the form

$$f = 4^a(8b + 7), \quad a \geq 0, \quad b \geq 0,$$

[84] Niedere Analysis, 1, 1902, 165–6.

[34a] Math. Quest. Educ. Times, (2), 1, 1902, 94–5.

[85] L'intermédiaire des math., 10, 1903, 245.

[86] Zeitschr. Math. Naturw. Unterricht., 34, 1903, 258; 35, 1904, 305.

[87] Ibid., 35, 1904, 57.

[88] L'intermédiaire des math., 11, 1904, 273. Taking $\alpha = \beta = \gamma = 1$ and replacing b by $b + a$, we get the identity on p. 163.

[89] Comptes Rendus Paris, 142, 1906, 537.

[90] L'intermédiaire des math., 14, 1907, 107.

[91] Ibid., 15, 1908, 275–6; cf. 16, 1909, 135, 256.

[92] Periodico di Mat., 22, 1907, 28–32.

[93] Archiv Math. Phys., (3), 13, 1908, 305.

the number of integers $\leq x$ of one of the forms f is $[x] - C(x)$. Since there are $[(x+1)/8]$ integers $8b + 7 \leq x$,

$$[x] - C(x) = \sum_{j=0}^{\infty} \left[\frac{1 + x/4^j}{8} \right], \qquad \lim_{x=\infty} \frac{C(x)}{x} = \frac{5}{6}.$$

A. Gérardin[94] noted that

$$(mx - ny)^2 + (nx + 2my)^2 = (mx + ny)^2 + (nx)^2 + (2my)^2,$$
$$(x - 1)^2 + x^2 + (x + 1)^2 = 1 + t^2, \text{ if } t^2 = 3x^2 + 1,$$

as for $(x, t) = (0, 1)$, $(1, 2)$, $(4, 7)$, $(15, 26)$, $(56, 97)$, $\cdots$. To Lucas is attributed

$$(12m \pm 2)^2 + 1 = (8m \pm 2)^2 + (8m \pm 1)^2 + (4m)^2.$$

W. Sierpinski[95] noted that if k is a $\boxed{3}$ in $\tau_3(k)$ ways,

$$S(x) \equiv \sum \frac{1}{l^2 + m^2 + n^2} = \sum_{k=1}^{[x]} \frac{\tau_3(k)}{k}, \qquad \lim_{n=\infty} \{S(x) - 4\pi \sqrt{x}\} = \text{const.},$$

where $0 < l^2 + m^2 + n^2 \leq x$. The number of sets of integers l, m, n satisfying that inequality is $\frac{4}{3}\pi x^{3/2} + O(x^{5/6})$, for O as in Landau[179] of Ch. VI.

E. Landau[96] proved that every positive integer not of the form $4^a(8m + 7)$ is a $\boxed{3}$, using the equivalence of every positive ternary quadratic form of discriminant unity to $x^2 + y^2 + z^2$.

K. J. Sanjana[97] found solutions of the system of equations

$$x^2 = y^2 + z^2 + u^2, \qquad x + y + z + u = 100.$$

Let $x = a + b, y = a - b$. Then $z^2 + u^2 = 4ab, 2a = 100 - z - u$. Hence

$$(z + b)^2 + (u + b)^2 = 2b^2 + 200b.$$

He took $u + b = z - b$, whence $z^2 = 100b$. Taking $b = 1, 4, 9, \cdots$, he found the solutions 42, 40, 10, 8 and 38, 30, 20, 12. The solution 39, 34, 14, 13 was noted by N. B. Pendse.

H. B. Mathieu[98] stated that the general solution of $\boxed{3} = \boxed{3}$ is

$$lA \pm rB \pm pD, \qquad pA + qB \mp lD, \qquad rA \mp lB - qD.$$

Welsch[99] gave $l \pm mv, n \mp pv, lm - np \mp v$ as the general solution.

A. Gérardin[100] gave the identity

$$(7a^2 + 7b^2 - 12ab)^2 = (6a^2 + 6b^2 - 14ab)^2 + (3a^2 - 3b^2)^2 + (2a^2 - 2b^2)^2.$$

L. Aubry[101] noted the existence of an infinitude of primes each a sum of three distinct squares. Every prime $p = 12n + 5 > 17$ gives a solution.

[94] Assoc. franç., 38, 1909, 143–5.
[95] Spraw. Towarz. Nauk (Proc. Sc. Soc. Warsaw), 2, 1909, 117–9.
[96] Handbuch . . . Verteilung der Primzahlen, 1, 1909, 545–505.
[97] Jour. Indian Math. Club, 2, 1910, 202.
[98] L'intermédiaire des math., 17, 1910, 288. On pp. 72, 166 it is shown that his earlier solution, 16, 1909, 220, is not general.
[99] *Ibid.*, 18, 1911, 62. Gleizes, 21, 1914, 156–7, stated that we may need to give fractional values to l, m, n, p, v.
[100] *Ibid.*, 17, 1910, 278; Sphinx-Oedipe, 1907–8, 27.
[101] Sphinx-Oedipe, 6, 1911, 25–26. Proposed bv F. Proth, Nouv. Corresp. Math., 4, 1878, 95.

We have $p = a^2 + b^2$, where a and b are prime to 3, so that we can set $a + b \equiv 0 \pmod 3$,

$$a^2 + b^2 = \left(\frac{2a - b}{3}\right)^2 + \left(\frac{2a + 2b}{3}\right)^2 + \left(\frac{2b - a}{3}\right)^2,$$

where the three squares are distinct if $p > 17$.

L. Aubry[102] proved that not all decompositions of the square of a $\boxed{3}$ into a $\boxed{3}$ are given by (4). Expressions for $x^2 + y^2$ or $x^2 + 2y^2$ as a $\boxed{3}$ are given on p. 124 and 19, 1912, 11, 188–190.

H. C. Pocklington[103] noted that, if $N = 4m + 1$ or $4m + 2$, there are properly primitive forms of determinant $- N$ that have the quadratic character $- 1$; while if $N = 8m + 3$ there are improperly primitive forms of determinant $- N$ which have the character $- 2$. Transform such a form into (b, f, c), where b is prime to N. Solve $bg^2 \equiv - 1 \pmod N$ for g and let $bg^2 + 1 = aN$. Then

$$N = (a, b, c, f, g, 0)(bc - f^2, fg, - bg)$$

is a representation of N by a definite ternary quadratic form of determinant unity. Reducing it in the ordinary way, we get $N = \boxed{3}$.

R. F. Davis[104] noted that, if $p + q + r = 1$, $1/p + 1/q + 1/r = 0$, then

$$a^2 + b^2 + c^2 = (pa + qb + rc)^2 + (qa + rb + pc)^2 + (ra + pb + qc)^2.$$

E. Landau[105] proved that the number of sets of integers u, v, w for which $u^2 + v^2 + w^2 \leqq x$ is $\frac{4}{3}\pi x^{3/2} + O(x^{3/4+\epsilon})$, for $\epsilon > 0$. Application is made to the number of classes of positive forms of given discriminant.

L. Aubry[106] proved that $pA^2 = B^2 + C^2 + D^2$ implies that p is a sum of three squares; similarly for four squares.

E. N. Barisien[107] noted various special cases of (3).

*G. Mühle[107a] solved $x^2+y^2+z^2=g^2$, where g is given; also, $x^2+y^2=g^2$ and $x^2+y^2=z^2+w^2$.

G. Humbert,[138] by use of an identity involving theta-functions, proved that if $f(x)$ is any even function of x,

$$\Sigma f(t) = \Sigma(- 1)^{(d-1)/2} f(d + 2h),$$

where t ranges over the integers occurring in the decomposition of a given number $8M + 3$ into $t^2 + t_1^2 + t_2^2$, each t an odd integer > 0, while in the second member the summation extends over the decompositions

$$8M + 3 = 4h^2 + dd_1 \ (d_1 > d > 0).$$

The case $f = 1$ is due to Hermite.[69] He gave a similar result and

$$\Sigma f(t) = 2\Sigma(d_1 + d - 4h)f(d + 2h),$$
$$4N + 3 = t^2 + t_1^2 + t_2^2 + 4l^2 + 4l_1^2 = 4h^2 + dd_1 \ (t, t_1, t_2 \text{ odd}).$$

<hr>

[102] L'intermédiaire des math., 18, 1911, 236. Cf. M. Rignaux, 24, 1917, 35–6.
[103] Proc. Cambr. Phil. Soc., 16, 1911, 19.
[104] Math. Quest. Educ. Times, (2), 21, 1912, 23.
[105] Göttingen Nachr., 1912, 693, 764–9. Cf. Sierpinski.[95]
[106] Sphinx-Oedipe, 7, 1912, 81.
[107] *Ibid.*, 8, 1913, 142, 175.
[107a] Ein Beitrag zur Lehre von den pythagoreischen Zahlen, Progr., Wollstein, 1913.
[108] Comptes Rendus Paris, 158, 1914, 220–6; errata, 380. Cf. 157, 1913, 1361–2.

W. C. Eells,[109] to solve $x^2 + y^2 + z^2 = a^2$, took $x = 2MN$, $y = M^2 - N^2$, $a = m^2 + n^2$, and gave to $M^2 + N^2$, z the values $m^2 - n^2$, $2mn$ in either order. He tabulated 125 sets of solutions arranged according to the size of a.

A. Gérardin and E. Miot[110] gave many identities $x^2 + y^2 = u^2 + v^2 + w^2$.

L. Aubry[111] gave a very long, but elementary proof, by use of theorems on divisors of numbers $x^2 + my^2$, that every number not of the form $4^r(8n + 7)$ is a $\boxdot$.

L. J. Mordell[112] proved Kronecker's[33] theorem by use of theta functions.

A. S. Werebrusow[113] noted that the problem to find two equal sums of three squares is evidently equivalent to $mm' + nn' + pp' = 0$, the general solution of which is stated to be

$$m = a\beta - b\alpha, \qquad n = a\gamma - c\alpha, \qquad p = a\delta - d\alpha,$$
$$m' = c\delta - d\gamma, \qquad n' = d\beta - b\delta, \qquad p' = b\gamma - c\beta.$$

He gave long formulas said to solve $x^2 + y^2 = u^2 + v^2 + w^2$ completely.

E. Bahier[114] found solutions of $a^2 + b^2 + c^2 = d^2$ in which $a + b = d$, $d = c + 1$, $d^2 = c^2 + \gamma^2$, or a and b are given. He discussed the nature of numbers d such that d^2 is a sum of three squares $\neq 0$.

E. Turrière[115] derived (4) geometrically and showed how to deduce new solutions of $x_1^2 + \cdots + x_n^2 = R^2$ from a given solution.

W. de Tannenberg[116] found real polynomials of degree $2n$ in a variable θ satisfying $x^2 + y^2 + z^2 = P^2$, where P is a given polynomial of degree $2n$ in θ, not zero for any real θ. Hence set $P = (a_1^2 - t_1^2) \cdots (a_n^2 - t_n^2)$, $t_p = i(\theta + b_p)$. For arbitrary parameters $\alpha_0, \cdots, \alpha_n$, define two sets of functions by

$$u_p = (a_p u_{p-1} + t_p v_{p-1})e^{ia_p}, \qquad v_p = (a_p v_{p-1} + t_p u_{p-1})e^{-ia_p} \qquad (p = 1, \cdots, n),$$

$u_0 = e^{ia_0}$, $v_0 = e^{-ia_0}$. Let the u, v become u', v' when $t_1, \cdots, t_n$ are changed in sign. Define x, y, z by means of

$$P - z = 2u_n v_n', \qquad P + z = 2v_n u_n', \qquad x + iy = 2u_n u_n', \qquad x - iy = 2v_n v_n',$$

which are consistent since $u_n v_n' + v_n u_n' = P$. Take $t_p = i(\theta + b_p)$.

On two equal sums of three squares, see papers 19 and 86 of Ch. VIII. By Cesàro[26] of Ch. IX there are in mean $\frac{1}{4}\pi n^{1/2}$ representations of n as a $\boxdot$. On a $\boxdot$ equal to $2v^2$, v^2 or v^4, see papers 171 of Ch. XIII, 69 of Ch. XV, 312 of Ch. XXII. On numbers not a $\boxdot$, papers 4, 5 of Ch. VIII. On systems of equations including $\boxdot = \square$, papers 97 of Ch. VII, 94 of Ch. IX, 32–39a, 51, 146, 165, 168 of Ch. XIX, 390–8 of Ch. XXI, 308–9 of Ch. XXII. On systems including $\boxdot = u^3$ or u^5, papers 95, 97 of Ch. XX, 353, 392, 402–3 of Ch. XXI.

<hr>

[109] Amer. Math. Monthly, 21, 1914, 269–273.
[110] L'intermédiaire des math., 21, 1914, 190–2.
[111] Sphinx-Oedipe, numéro spécial, Jan., 1914, 1–24.
[112] Mess. Math., 45, 1915, 78.
[113] L'intermédiaire des math., 23, 1916, 12–13, 17–18.
[114] Recherche . . . Triangles Rectangles en Nombres Entiers, 1916, 234–254.
[115] L'enseignement math., 18, 1918, 90–5.
[116] Comptes Rendus Paris, 165, 1917, 783–4.

CHAPTER VIII.

SUM OF FOUR SQUARES.

Diophantus, IV, 31 [32], desired four numbers x_i such that the sum of their squares increased [diminished] by the sum of the numbers is a given number n. He took $n = 12$ [$n = 4$]. Since $x^2 \pm x + \frac{1}{4}$ is a square, $\Sigma x_i^2 \pm \Sigma x_i + 1$ is the sum of four squares, here 13 [5]. Hence we have to divide 13 [5] into four squares and subtract $\frac{1}{2}$ from [add $\frac{1}{2}$ to] each of their sides to obtain the sides of the required squares. Since

$$13 = 4 + 9 = \frac{64}{25} + \frac{36}{25} + \frac{144}{25} + \frac{81}{25}, \qquad \left[5 = \frac{9}{25} + \frac{16}{25} + \frac{64}{25} + \frac{36}{25} \right],$$

the sides of the required squares are

$$\frac{11}{10}, \frac{7}{10}, \frac{19}{10}, \frac{13}{10}; \qquad \left[\frac{11}{10}, \frac{13}{10}, \frac{21}{10}, \frac{17}{10} \right].$$

G. Xylander[1] noted that if we take 1430 in place of 4 in the second problem, we get the solution 6^2, 11^2, 21^2, 30^2.

C. G. Bachet[1a] remarked that Diophantus apparently assumed here and occasionally in Book V that any number is either a square or the sum of 2, 3 or 4 squares [Bachet's theorem], and added that he himself had verified this proposition for all numbers up to 325 and would welcome a proof; he gave decompositions into 4 or fewer squares of each number up to 120. He mentioned the generalization of Diophantus IV, 31 to the problem to find k numbers such that the sum of their squares increased by the sum of the numbers is a given number n. Thus $n + k/4$ is to be the sum of k squares. Bachet stated that if $k \geqq 4$ there is no condition.

Fermat, in his comment quoted in Ch. I[36], stated that he possessed a proof that every number is a sum of four squares. In stating the theorem elsewhere, Fermat[2] remarked that Diophantus seems to have known the theorem.

The reason for ascribing a knowledge of this theorem to Diophantus lies in the fact that he made no mention of a condition on a number in order that it be a sum of four squares, in the three cases IV, 31, 32 and V, 17, in which he mentioned the subject, but that he gave necessary conditions for representation as a sum of two or three squares (Chs. VI, VII).

Diophantus, V, 17, sought to divide a given number into four parts such that the sum of any three of the parts is a square. Thus three times the sum of the four parts is the sum of four squares. Let the given number

---

[1] Diophanti Alexandrini Rerum Arith., . . . , G. Xylandro, Basileae, 1575, 104.

[1a] Diophanti Alex. Arith., 1621, 241-2.

[2] Oeuvres, II, 65; III, 287; letter to Mersenne, Sept. or Oct., 1636; to be proposed for solution to Sainte-Croix. Mersenne communicated it to Descartes, March 22, 1638. The latter ascribed the theorem to St. Croix (Oeuvres de Descartes, II, 256). Fermat, Oeuvres, II, 403-4; III, 315, letter to Digby, June, 1658, proposed that Brouncker and Wallis seek a proof of the theorem.

be 10. Then 30 is to be divided into four squares each < 10. Since $30 = 16 + 9 + 4 + 1$, we take 9 and 4 as two of the squares and divide 17 into two squares each < 10 [the squares of 1016/349 and 1019/349]. If we subtract each of the resulting four squares from 10, we obtain the required parts 1, 6, etc. In V, 16, the number 10 is divided into three such parts. For a generalization to n parts, see Kausler[47] of Ch. XV.

Regiomontanus[3] (J. Müller) proposed in a letter the problems to find four squares whose sum is a square and twenty squares whose sum is a square > 300000.

Jakob von Speyer[3a] gave

$$1 + 2^2 + 4^2 + 10^2 = 11^2, \qquad 2^2 + 4^2 + 7^2 + 10^2 = 13^2.$$

A. Girard,[4] in commenting on Diophantus V, 15, stated that there are numbers, as 7, 15, 23, 28, 31, 39, not a sum of three squares, but that any integer is a sum of four squares.

R. Descartes[5] announced the theorem ("whose demonstration he judged so difficult that he dared not undertake to find it"): Any number which is the sum of three squares and exceeds 41 can be expressed also as the sum of four squares, excepting only the products of 6 or 14 by 4, 4^2, 4^3, $\cdots$. There are no other numbers which are not composed of four squares, except $2 \cdot 4^n$, which is not a square, nor composed of three or four squares, but only of two.

Fermat[6] stated that he had much trouble in finding the new principles needed to apply his method of infinite descent to show that every number is a square or the sum of 2, 3 or 4 squares; but stated that he had finally proved that if a given number is not of this nature there would exist a smaller which is not.

L. Euler[7] admitted that he could not prove Bachet's theorem that every integer is a ▣, nor give a general rule to express $n^2 + 7$ as a ▣. Oct. 17, 1730 (p. 45), he noted that, if Fermat's theorem that every integer x is a sum of three triangular numbers $(a^2 + a)/2$ is true, then $8x + 3$ is the sum of the three squares $(2a + 1)^2$. Hence $8x + 4$ and $8x + 7$ are ▣. [Cf. Beguelin[75] of Ch. I.] Since $m^2(8x + 4) = k^2(2x + 1)$, it remains only to prove that $4x + 2$ is a ▣. Oct. 15, 1743 (p. 263), Euler noted that, if np^2 is a ▣, n is a sum of four integral squares. Thus if it be true that $8m + 3$ is a ▣, $8m + 4$ is a ▣ and also $2m + 1$, so that every integer is a ▣. May 6, 1747 (p. 419), he stated that Bachet's theorem depends on the unproved fact that every number $4m + 2$ is the sum of two numbers $4x + 1$ and $4y + 1$, neither having a factor $4p - 1$ [and hence each a ▣]. For,

[3] C. T. de Murr, Memorabilia Bibl., 1, 1786, 160, 201.

[3a] *Ibid.*, 168.

[4] L'arith. de Simon Stevin . . . annotations par A. Girard, Leide, 1625, p. 626; Oeuvres math. de S. Stevin par A. Girard, 1634, p. 157.

[5] Oeuvres, 2, 1898, 256, 337–8, letters to Mersenne, July 27 and Aug. 23, 1638. The limit 33 given in the first letter was changed to 41 in the second.

[6] Oeuvres, II, 433, letter to Carcavi, communicated Aug. 14, 1659, to Huygens.

[7] Corresp. Math. et Phys. (ed., P. H. Fuss), St. Petersburg, 1, 1843, 24, 30, 35; letters to Goldbach, June 4, June 25, Aug. 30, 1730.

then $2(4m + 2)$ is a $\boxed{4}$ and hence $2m + 1$ is a $\boxed{4}$.   May 4, 1748 (p. 452), he gave the fundamental formula (Cf. Euler[165] of Ch. XIX)

$$(1) \quad \begin{cases} (a^2 + b^2 + c^2 + d^2)(p^2 + q^2 + r^2 + s^2) = x^2 + y^2 + z^2 + v^2, \\ x = ap + bq + cr + ds, \quad y = aq - bp \pm cs \mp dr, \\ z = ar \mp bs - cp \pm dq, \quad v = as \pm br \mp cq - dp, \end{cases}$$

and stated (p. 454, and Aug. 17, 1750, p. 531) that Bachet's theorem would follow if the fourth power of $1 + x + x^4 + x^9 + x^{16} + \cdots$ contained x^n with a coefficient $\neq 0$.   April 12, 1749 (pp. 495–7), he stated that he had a proof that, if p is any prime, there exist four integers $a, \cdots, d$, each not divisible by p, such that $a^2 + \cdots + d^2$ is divisible by p.   Set $a = \alpha p \pm x$, $\cdots, d = \delta p \pm v$, where $0 \leqq x \leqq \frac{1}{2}p, \cdots, 0 \leqq v \leqq \frac{1}{2}p$.   Hence $x^2 + \cdots + v^2$ is divisible by p.   If p is odd, $x < \frac{1}{2}p, \cdots$, so that $x^2 + \cdots + v^2 < p^2$. To prove that every prime is a $\boxed{4}$, suppose there is a minimum prime p not a $\boxed{4}$.   But $x^2 + \cdots + v^2 = pq, q < p$.   Euler believed, but could not prove, that if $pq = \boxed{4}$, $p \neq \boxed{4}$, then $q \neq \boxed{4}$.   Admitting this, we would have a contradiction with the assumption about the minimum p.   Thus every prime is a $\boxed{4}$ and hence by (1) every integer is a $\boxed{4}$.

On the point here left in doubt that $pq = \boxed{4}$ and $q = \boxed{4}$ imply $p = \boxed{4}$, Euler proved, July 26, 1749, pp. 505–10, that, if* $m \leqq 7$, $mA = \boxed{4}$ and $m = \boxed{4}$ imply $A = \boxed{4}$.   Set

$$m = a^2 + b^2 + c^2 + d^2,$$
$$mA = (j + mp)^2 + (g + mq)^2 + (h + mr)^2 + (k + ms)^2,$$

[where $f, \cdots, k$ are numerically $\leqq m/2$].   Then $f^2 + \cdots + k^2$ is divisible by m.   For $m \leqq 7$, the quotient was verified to be a $\boxed{4}$,

$$f^2 + g^2 + h^2 + k^2 = m[X^2 + Y^2 + Z^2 + V^2],$$

and [in accord with, but not a consequence of, (1)]

$$f = aX + bY + cZ + dV, \quad g = bX - aY - dZ + cV,$$
$$h = cX + dY - aZ - bV, \quad k = dX - cY + bZ - aV,$$

$$A = X^2 + Y^2 + Z^2 + V^2 + 2(fp + gq + hr + ks) + m(p^2 + q^2 + r^2 + s^2)$$
$$= (x + X)^2 + (y - Y)^2 + (z - Z)^2 + (v - V)^2$$

where $x, \cdots, v$ are given by (1) with the upper signs.   Moreover, he gave a proof of Chr. Goldbach's assertion of June 16 (p. 503) that the sum s of four odd squares can be expressed as a sum of four even squares.   Since

$$\tfrac{1}{2}(2p + 1)^2 + \tfrac{1}{2}(2q + 1)^2 = (p + q + 1)^2 + (p - q)^2,$$
$$\frac{s}{2} = (a + b + 1)^2 + (a - b)^2 + (c + d + 1)^2 + (c - d)^2.$$

The last sum involves two even and two odd squares since $s = 8m + 4$.

---

* For the general case Euler[8] admitted in 1751 that he had no proof.

Hence

$$\frac{s}{2} = (2p + 1)^2 + (2q + 1)^2 + 4r^2 + 4s^2,$$

$$\frac{s}{4} = (p + q + 1)^2 + (p - q)^2 + (r + s)^2 + (r - s)^2.$$

As a corollary, $2A = \boxed{4}$ implies $A = \boxed{4}$.

On March 24, 1750 (p. 513), Goldbach had stated that there is a definite connection between the sets of four squares whose sums are $2m - 1$ and $2m + 1$, as derived from $8m + 3 = \boxed{3}$. June 9, 1750 (p. 518), Euler interpreted this as follows: From $8m - 5 = a^2 + b^2 + c^2$, where a, b, c are odd,

$$4m - 2 = \left(\frac{1 + a}{2}\right)^2 + \left(\frac{a - 1}{2}\right)^2 + \left(\frac{b - c}{2}\right)^2 + \left(\frac{b + c}{2}\right)^2,$$

where two of the squares are even. Set $2p = (a + 1)/2$, $2q = (b + c)/2$. Then

$$4m - 2 = (2p)^2 + (2q)^2 + r^2 + s^2,$$

$$2m - 1 = (p + q)^2 + (p - q)^2 + \left(\frac{r + s}{2}\right)^2 + \left(\frac{r - s}{2}\right)^2$$

$$= \Sigma\left(\frac{a \pm b \pm c \pm 1}{2}\right)^2,$$

where two or four signs are $+$. From $8m + 4 = 9 + a^2 + b^2 + c^2$,

$$4m + 2 = \left(\frac{a + 3}{2}\right)^2 + \left(\frac{a - 3}{2}\right)^2 + \left(\frac{b + c}{2}\right)^2 + \left(\frac{b - c}{2}\right)^2,$$

$$2m + 1 = \Sigma\left(\frac{a \pm b \pm c \pm 3}{2}\right)^2,$$

where two or four signs are $+$. Hence, from $8m - 5 = \boxed{3}$,

$$2m - 1 = p^2 + q^2 + r^2 + s^2,$$
$$2m + 1 = (p + 1)^2 + (q + 1)^2 + (r - 1)^2 + (s - 1)^2.$$

Thus $r + s - p - q = 1$ and we can express any odd number as a sum of four squares the algebraic sum of whose roots is unity. [Cf. Cauchy, 1813]. Euler stated (p. 521, p. 527, and again on Dec. 4, 1751, p. 559) that while he had proved that any rational number is the sum of four rational squares, he had not proved the theorem for integral squares.

Goldbach (p. 526) noted that α, β, γ, $\alpha + \beta + \gamma + 2\delta$, and $\alpha + \beta + \delta$, $\alpha + \gamma + \delta$, $\beta + \gamma + \delta$, δ, and $\alpha + \delta$, $\beta + \delta$, $\gamma + \delta$, $\alpha + \beta + \gamma + \delta$ have the same sum of squares.

Euler, July 3, 1751, p. 542, discussed the problem to make

$$s = \alpha^2 + \beta^2 + \gamma^2 + \delta^2 + e$$

a $\boxed{4}$. Call the roots $\alpha - kx$, $\beta - mx$, $\gamma - nx$, $\delta + x$. Then

$$\delta = A - \tfrac{1}{2}Bx + \frac{e}{2x}, \qquad A \equiv k\alpha + m\beta + n\gamma, \qquad B \equiv k^2 + m^2 + n^2 + 1.$$

Resolve $e \cdot B$ into two factors a, b, both even or both odd. Then for $x = e/a$, $\delta = A + (a - b)/2$. Take k, m, n arbitrarily and determine $a - b$, or conversely. The case $e = 8$ was partially treated by Goldbach, pp. 540, 546–8, 555 and by Euler, p. 557. A $\boxed{4}$ with the sum of the roots zero is a $\boxed{3}$ since (pp. 548–9)

$$a^2 + b^2 + c^2 + (a + b + c)^2 = (a + b)^2 + (a + c)^2 + (b + c)^2.$$

Goldbach (p. 548) noted that

$$8n + 4 = a^2 + b^2 + c^2 + d^2, \qquad a + b + c + d = 2.$$

Euler, Sept. 4, 1751, p. 551, deduced this from

$$8n + 3 = \boxed{3} = (a + b - 1)^2 + (a + c - 1)^2 + (b + c - 1)^2.$$

Euler[8] published some results on Bachet's theorem. He proved

Theorem I. There exist integers a, b for which $1 + a^2 + b^2$ is divisible by a given prime p. For, if -1 is a quadratic residue of p, there is an integer a for which $1 + a^2$ is divisible by p. Next, let -1 be a non-residue and suppose the theorem is false. Then $1 + 1 - 2 = 0$ shows that -2 is a non-residue and hence $+2$ a residue; then $1 + 2 - 3 = 0$ shows that -3 is a non-residue and hence $+3$ a residue; and in this way $1, 2, \cdots, p - 1$ would all be residues.

If $A = a^2 + \cdots + d^2$, $P = p^2 + \cdots$, then $A/P = AP/P^2 = (x/P)^2 + \cdots$ by (1), so that A/P is the sum of four rational squares. Euler admitted he was unable to prove that, if A is divisible by P, A/P is the sum of four integral squares. If this were proved, Bachet's theorem would follow. But it is readily proved that every integer is a sum of four rational squares. For, if p be the least prime not such a sum, there exists (Theorem I) an integer $A = a^2 + b^2 + c^2$ divisible by p, where a, b, c are $< p/2$. Then $A/p < \frac{3}{4}p$, and yet A/p was seen to be the sum of four rational squares.

J. L. Lagrange[9] gave the first proof of the theorem of Bachet and acknowledged his indebtedness to ideas in the preceding paper by Euler. The steps are as follows:

(i) If $p^2 + q^2 = t\rho$ and $r^2 + s^2 = u\rho$, where p, q, r, s have no common divisor, then t and u are sums of two squares.

For, call M the g.c.d. of $p = Mp_1$ and $q = Mq_1$; N that of $r = Nr_1$ and $s = Ns_1$. Then M and N are relatively prime. Call μ the g.c.d. of M^2 and $\rho = \mu\rho_1$. Since

$$(2) \qquad M^2(p_1^2 + q_1^2) = t\mu\rho_1,$$

ρ_1 divides the sum $p_1^2 + q_1^2$ of two relatively prime squares. By Euler's[24] theorem of Ch. VI, the quotient is a sum $c^2 + d^2$ of two squares. Set $\mu = \nu^2\mu_1$, where μ_1 has no square factor. Then M is divisible by $\nu\mu_1$, $M = K\nu\mu_1$. Now $N^2(r_1^2 + s_1^2) = u\mu\rho_1$. Since μ divides M^2, it is prime to N^2 and hence divides $r_1^2 + s_1^2$. As before, $\mu_1 = e^2 + f^2$. Then, by (2),

$$t = (c^2 + d^2)M^2/\mu = (c^2 + d^2)K^2\mu_1 = K^2(ec + fd)^2 + K^2(ed - fc)^2.$$

[8] Novi Comm. Acad. Petrop., 5, 1754–5 (1751), 3; Comm. Arith., I, 230–233.

[9] Nouv. Mém. Acad. Roy. Sc. de Berlin, année 1770, Berlin, 1772, 123–133; Oeuvres, 3, 1869, 189–201. Cf. G. Wertheim's Diophantus, pp. 324–330.

(ii) If $\gamma^2 + \delta^2$ is divisible by $m^2 + n^2$, the quotient t is a sum of two squares.

Let l be the g.c.d. of $\gamma = lp$, $\delta = lq$, $m = lr$, $n = ls$. Then $p^2 + q^2$ is divisible by $r^2 + s^2 = \rho$. Hence, by (i), $t = (p^2 + q^2)/\rho$ is a sum of two squares.

(iii) If $P = p^2 + q^2 + r^2 + s^2$ is divisible by a prime $A > \sqrt{P}$, then A is a sum of four squares.

Set $P = Aa$. Then $a < A$. A common divisor d of p, q, r, s is $< A$, so that d^2 divides a and may be deleted from a, p^2, $\cdots$, s^2. Let therefore $d = 1$.

Let ρ be the g.c.d. of $a = b\rho$ and $p^2 + q^2 = t\rho$. Then $(r^2 + s^2)/\rho$ is an integer u. By (i), $t = m^2 + n^2$, $u = h^2 + l^2$. Thus
$$tu = x^2 + y^2, \qquad x = mh + nl, \qquad y = ml - nh.$$
From $P = Aa$ follows
$$Ab = t + u, \qquad Abt = t^2 + x^2 + y^2.$$
Since b is prime to t, there exist integers α, $\cdots$, δ such that
$$x = \alpha t + \gamma b, \qquad y = \beta t + \delta b, \qquad |\alpha| < \tfrac{1}{2}b, \qquad |\beta| < \tfrac{1}{2}b,$$
$$(3) \quad Abt = kt^2 + 2\alpha\gamma tb + 2\beta\delta tb + (\gamma^2 + \delta^2)b^2, \qquad k \equiv 1 + \alpha^2 + \beta^2.$$
Hence kt^2 is divisible by b. Thus $k = a_1 b$, where $a_1 < b/2 + 1/b$. Then
$$At = a_1 t^2 + 2\alpha\gamma t + 2\beta\delta t + (\gamma^2 + \delta^2)b,$$
$$a_1 At = (a_1 t + \alpha\gamma + \beta\delta)^2 + \gamma^2(a_1 b - \alpha^2) + \delta^2(a_1 b - \beta^2) - 2\alpha\beta\gamma\delta.$$
Replacing $a_1 b$ by $1 + \alpha^2 + \beta^2$, we get
$$a_1 At = (a_1 t + \alpha\gamma + \beta\delta)^2 + (\beta\gamma - \alpha\delta)^2 + \gamma^2 + \delta^2.$$
By (3), $\gamma^2 + \delta^2$ is divisible by $t = m^2 + n^2$. By the last equation and (ii),
$$\gamma^2 + \delta^2 = t(p_1^2 + q_1^2), \qquad (a_1 t + \alpha\gamma + \beta\delta)^2 + (\beta\gamma - \alpha\delta)^2 = t(r_1^2 + s_1^2),$$
$$a_1 A = p_1^2 + q_1^2 + r_1^2 + s_1^2.$$
If $a = b\rho$ is > 1, $a_1 < b/2 + 1/b < a$. Similarly, if $a_1 > 1$, $a_2 A$ is the sum of four squares, where $a_2 < a_1$, etc. But each $a_i \geqq 1$. Thus a certain $a_k = 1$, and $a_k A = A$ is the sum of four squares.

(iv) Any prime which divides the sum of four or fewer squares which have no common factor is itself the sum of four or fewer squares.

If the prime A divides $p^2 + q^2 + r^2 + s^2$, it divides the sum obtained by replacing p by $\pm (p - mA)$, where m is such that $0 \leqq |p - mA| < \tfrac{1}{2}A$, etc. The sum of the four new squares is $< A^2$ and is divisible by A. Then (iii) may be applied, even if some of the four squares are zero.

(v) If B and C are integers not divisible by the odd prime A, there exist integers p and q such that $p^2 - Bq^2 - C$ is divisible by A.

Suppose that there is no integer q which makes $b = Bq^2 + C$ divisible by A (since otherwise we may take $p = 0$). For
$$P = p^{A-3} + bp^{A-5} + b^2 p^{A-7} + \cdots + b^{(A-3)/2},$$
$$(p^2 - b)P = p^{A-1} - 1 - (b^{(A-1)/2} - 1).$$

Multiply the last equation by $Q = b^{(A-1)/2} + 1$. If p and q can be chosen so that pPQ is not divisible by A, then $p^2 - b$ will be divisible by A, as shown by using Fermat's theorem. For q constant and $p = 1, \cdots, A - 2$, let P become $P_1, \cdots, P_{A-2}$. Then by the theory of differences,

$$P_1 - (A - 3)P_2 + \tfrac{1}{2}(A - 3)(A - 4)P_3 - \cdots + P_{A-2} = (A - 3)\,!.$$

Hence at least one P_i is not divisible by A. Set $m = \tfrac{1}{2}(A - 1)$. Then

$$Q = q^2R + C^m + 1, \qquad R = B^m q^{A-3} + mB^{m-1}q^{A-5}C + \cdots + mBC^{m-1}.$$

If $C^m + 1$ is not divisible by A, it suffices to take $q = 0$. In the contrary case, we note that if R becomes R_i for $q = i$,

$$R_1 - (A - 3)R_2 + \tfrac{1}{2}(A - 3)(A - 4)R_3 - \cdots + R_{A-2} = (A - 3)\,! \, B^m,$$

so that at least one R_i is not divisible by A. Hence by (iv) every prime is a $\boxed{4}$.

(vi) Every positive integer is the sum of four or fewer squares.

This follows from Euler's relation (1). Lagrange added the generalization

$$(p^2 - Bq^2 - Cr^2 + BCs^2)(p_1^2 - Bq_1^2 - Cr_1^2 + BCs_1^2)$$

$$(4) \qquad = \{pp_1 + Bqq_1 \pm C(rr_1 + Bss_1)\}^2 - B\{pq_1 + qp_1 \pm C(rs_1 + sr_1)\}^2$$
$$- C\{pr_1 - Bqs_1 \pm rp_1 \mp Bsq_1\}^2 + BC\{qr_1 - ps_1 \pm sp_1 \mp rq_1\}^2.$$

L. Euler's[10] proof is much simpler than Lagrange's. It is shown that if N divides $P = p^2 + q^2 + r^2 + s^2$, but not all the numbers $p, \cdots, s$, then N is a sum of four squares. Set $P = Nn$. Determine a, b, c, d, each numerically $\leqq \tfrac{1}{2}n$, so that

$$p = a + n\alpha, \qquad q = b + n\beta, \qquad r = c + n\gamma, \qquad s = d + n\delta.$$

Set $\sigma = a^2 + b^2 + c^2 + d^2$. Then $\sigma \leqq n^2$. We readily dispose of the case* $\sigma = n^2$. [If n is odd, $a, \cdots, d$ may be chosen numerically $< n/2$, whence $\sigma < n^2$. If n is even, we have $\sigma < n^2$ unless $a, \cdots, d$ numerically equal $n/2$, whence $p \pm q$ and $r \pm s$ are divisible by n and are even. But $Nn = P = \Sigma p^2$, whence

$$(5) \qquad \tfrac{1}{2}nN = \left(\frac{p + q}{2}\right)^2 + \left(\frac{p - q}{2}\right)^2 + \left(\frac{r + s}{2}\right)^2 + \left(\frac{r - s}{2}\right)^2$$

may be used in place of the initial multiple P of N.] Hence let $\sigma < n^2$. Then

$$Nn = \sigma + 2nA + n^2t, \qquad A \equiv a\alpha + b\beta + c\gamma + d\delta, \qquad t \equiv \alpha^2 + \beta^2 + \gamma^2 + \delta^2.$$

Thus σ is divisible by n. Set $\sigma = nn'$, so that $n' < n$. By (1),

$$\sigma t = A^2 + B^2 + C^2 + D^2.$$

Multiply $N = n' + 2A + nt$ by n'. Then

$$Nn' = (n' + A)^2 + B^2 + C^2 + D^2.$$

---

[10] Acta Erudit. Lips., 1773, 193; Acta Acad. Petrop., 1, II, 1775 [1772], 48; Comm. Arith., I, 543–4. Euler's Opera postuma, 1, 1862, 198–201. He first repeated Lagrange's proof and his[8] proof of Theorem I.

* Stated to occur only when $a = b = c = d = \tfrac{1}{2}n = 1$, whence $p, \cdots, r$ are odd and $N = \tfrac{1}{4}P$ equals the right member of (5).

In the same way, Nn'' $(n'' < n')$ is the sum of four squares; etc., finally $N \cdot 1$ is a sum of four squares.

He proved that, if N is a prime not dividing the given integers λ, μ, ν, we can find integers x, y, z not divisible by N such that $s = \lambda x^2 + \mu y^2 + \nu z^2$ is divisible by N. Since λ is prime to N, we can determine integers m and n such that $\lambda m \equiv -\mu$, $\lambda n \equiv -\nu$ (mod N). Then $s \equiv 0$ is equivalent to $a \equiv mb + nc$ (mod N) for quadratic residues a, b, c. If the latter is impossible, then $mb + n$ is a non-residue for each of the $(N - 1)/2$ residues b and hence gives all the non-residues. Then if d is any residue, bd is a residue e, so that $me + dn$ must be a non-residue. This exceeds the non-residue $me + n$ by $n(d - 1) = \omega$. For $d \not\equiv 1$, ω is prime to N. Thus, if α is any non-residue, $\alpha + \omega$ is a non-residue. But α, $\alpha + \omega$, $\cdots$, $\alpha + (N - 1)\omega$ are congruent to 0, 1, $\cdots$, $N - 1$ in some order and hence are not all non-residues.

Euler[11] gave a slight modification of his preceding proof. We may assume that p, q, r, s in $Nn = p^2 + q^2 + r^2 + s^2$ are numerically $< \frac{1}{2}N$, where N is a prime. Then $n < N$ and we can find integers a, α, $\cdots$, d, δ, such that

$$p = Na + n\alpha, \qquad q = Nb + n\beta, \qquad r = Nc + n\gamma, \qquad s = Nd + n\delta,$$

where a, b, c, d are numerically $< \frac{1}{2}n$. Then $Nn = N^2\sigma + 2NnA + n^2t$, so that $\sigma = nn'$, $n' < n$. Multiplying by n'/n, we get

$$Nn' = (Nn' + A)^2 + B^2 + C^2 + D^2.$$

Euler[12] noted that $a^2 + b^2 + c^2 = 4(x^2 + 3y^2) = \boxed{4}$ for

$$a = 2m(ps + qr) + 2n(3qs - pr),$$
$$b, c = m\{(3q \pm p)s + (q \mp p)r\} + n\{3(q \mp p)s - (3q \pm p)r\}.$$

Euler[13] remarked that the sum of two primes of the form $4n + 1$ is a $\boxed{4}$ since each is a $\boxed{2}$, and verified that every number $4k + 2 \leqq 110$ is a sum of two primes $4n + 1$.

A. M. Legendre[14] remarked that a proof of Fermat's assertion that every prime $8n - 1$ is of the form $p^2 + q^2 + 2r^2$ would complete the proof that every number is a $\boxed{4}$. For, any prime $8n - 3$ is of form $p^2 + q^2$, any prime $8n + 3$ is of form $p^2 + 2q^2$, any prime $8n + 1$ is simultaneously of the last two forms.

Legendre[15] reproduced Euler's[10] proof, using in place of Theorem I its generalization by Lagrange.

C. F. Gauss[16] subtracted from the given number $4n + 2$ any square less than it, from $4n + 1$ an even square, from $4n + 3$ an odd square. The remainder is $\equiv 1, 2, 5$ or 6 (mod 8) and hence is a sum of 3 squares. Thus

---

[11] Opera postuma, 1, 1862, 197–8 (about 1773).

[12] Novi Comm. Acad. Petrop., 18, 1773, 171; Comm. Arith., I, 515.

[13] Acta Acad. Petrop., 4, II, 1780 (1775), 38; Comm. Arith., II, 134–9.

[14] Mém. Acad. Roy. Sc. Paris, 1785, 514 Cf. Pollock[47]; also Euler,[12] Lebesgue[31] of Ch. VII.

[15] Essai sur la théorie des nombres, Paris, 1798, 198; ed. 2, 1808, 182; ed. 3, 1830, I, 211–6, Nos. 151–4 (Maser, I, pp. 212–6).

[16] Disq. Arith., 1801, art. 293; Werke, I, 1863, 348.

the given number is a sum of 4 squares.   Finally, a multiple of 4 is of the form $4^\mu N$, where N is one of the preceding three types.

Gauss[17] noted that the theorem (1) that a product of two sums of four squares is a $\boxed{4}$ is represented in the simplest way by

$$(Nl + Nm)(N\lambda + N\mu) = N(l\lambda + m\mu) + N(l\mu' - m\lambda'),$$

where N denotes the norm and l, m, λ, μ, λ', μ' are complex numbers, λ, λ' and μ, μ' being conjugate imaginaries.   He noted (p. 447) that [cf. Glaisher,[59] Hermite[69]]

$$(6) \quad (1 + 2y + 2y^4 + 2y^9 + \cdots)^4$$
$$= (1 - 2y + 2y^4 - \cdots)^4 + (2y^{1/4} + 2y^{9/4} + \cdots)^4.$$

He noted (p. 445) that [cf. Legendre,[23] Jacobi,[24] and Genocchi[39]]

$$(7) \quad (1 + 2y + 2y^4 + 2y^9 + \cdots)^4 = 1 + 8\left(\frac{y}{1-y} + \frac{2y^2}{1+y^2} + \frac{3y^3}{1-y^3} + \cdots\right),$$

$$(8) \quad (q + q^9 + q^{25} + q^{49} + \cdots)^4 = \frac{q^4}{1-q^8} + \frac{3q^{12}}{1-q^{24}} + \frac{5q^{20}}{1-q^{40}} + \cdots.$$

Gauss[18] noted that every decomposition of a multiple of a prime p into $a^2 + b^2 + c^2 + d^2$ corresponds to a solution of $x^2 + y^2 + z^2 \equiv 0 \pmod{p}$ proportional to $a^2 + b^2$, $ac + bd$, $ad - bc$ or to the sets derived by interchanging b and c or b and d.   For $p \equiv 3 \pmod 4$, the solutions of $1 + x^2 + y^2 \equiv 0 \pmod{p}$ coincide with those of $1 + (x + iy)^{p+1} \equiv 0$. From one value of $x + iy$ we get all by using

$$(x + iy)(u + i)/(u - i) \qquad (u = 0, 1, \cdots, p - 1).$$

For $p \equiv 1 \pmod 4$, $p = a^2 + b^2$; then $b(u + i)/\{a(u - i)\}$ give all values of $x + iy$ if we exclude the values a/b and b/a of u.

G. F. Malfatti[19] did not prove as he promised to do that every integer is a $\boxed{4}$.   After verifying this for about 50 small numbers, he considered the equation $Kn^2 = p^2 + q^2$, where K is a given integer.   If we admit his assertion that K must be a $\boxed{2}$, the equation has evident solutions with $n = 1$.   Taking $K = a^2 + b^2$, he found an infinitude of solutions, with f and g arbitrary, by setting

$$\frac{an - q}{g} = \frac{p - bn}{f}, \qquad g(an + q) = f(p + bn).$$

The equation obtained by eliminating p is satisfied if we take

$$n = f^2 + g^2, \qquad q = (f^2 - g^2)a + 2fgb.$$

Next, $Kn^2 = p^2 + q^2 + r^2$, in which we may limit K to be odd or the double of an odd number, and n to be odd, is said without adequate proof to be

---

[17] Posth. MS., Werke, 3, 1876, 383–4.
[18] Posth. paper, Werke, 8, 1900, 3.
[19] Memorie di Mat. e Fis. Soc. Italiana Sc., Modena, (1), 12, pt. 1, 1805, 296–317.

impossible unless K is a ▣. For $K = a^2 + b^2 + c^2$, the equation becomes

$$Hh(an + r) = Ff(q + bn) + Gg(p + cn),$$

$$H = \frac{an - r}{h}, \qquad F = \frac{q - bn}{f}, \qquad G = \frac{p - cn}{g}.$$

It is stated that $H = F = G$, and that the linear equation in n, r, derived by eliminating p, q requires $n = f^2 + g^2 + h^2$, whence

$$p = (f^2 - g^2 + h^2)c + 2gha - 2fgb,$$
$$q = (-f^2 + g^2 + h^2)b + 2fha - 2fgc,$$
$$r = (f^2 + g^2 - h^2)a + 2fhb + 2ghc.$$

[For these, $n^2(a^2 + b^2 + c^2) = p^2 + q^2 + r^2$, identically in f, g, h, a, b, c.] There is a similar treatment of the corresponding problem for 4 or 5 squares. If Malfatti had proved his statement that K must be a sum of the like number of squares, he could have deduced Bachet's theorem from Euler's[8] result that every integer is a sum of four rational squares.

P. Barlow[20] gave a " simplification of Legendre's[15] proof." To show that any prime A divides a sum of ▣, he proved at length that $x^2 + w^2 - 1 = mA$ is solvable [evidently by $x = 1$, $w = 0$!] and stated that a like proof shows that $y^2 + z^2 + 1 = nA$ is solvable. The proof probably meant for the latter is as follows. If $p \equiv y^2 \pmod{A}$, either $- (p + 1)$ is a quadratic residue ($\equiv z^2$) and the result follows, or it is a non-residue and hence $p + 1$ a residue, since $- 1$ is a non-residue (otherwise our equation holds for $y = 0$). But p, $p + 1$, $p + 2$, $\cdots$ are not all residues. The proof is thus only a slight modification of that by Euler.[10]

A. Cauchy's proof in 1813 of Fermat's theorem on 3 triangular numbers, 4 squares, 5 pentagons, etc., was considered in Ch. I. It is in place to mention here the theorems on sums of squares upon which his proof rests, especially since special cases were cited above from the correspondence of Euler and Goldbach. If

$$(9) \qquad k = t^2 + u^2 + v^2 + w^2, \qquad s = t + u + v + w,$$

then

$$(10) \quad 4k - s^2 = (t + u - v - w)^2 + (t - u + v - w)^2 + (t - u - v + w)^2.$$

But if 4^a is the highest power of 4 dividing a, then a is a ▣ if and only if $a/4^a$ is not of the form $8n + 7$. If k is even, the three sums in (10) are even, so that $k - s^2/4$ is a ▣. By (9), $k \equiv s \pmod 2$. Cauchy proved that, if k is even, sufficient conditions for (9) are that s be even and between $\sqrt{3k - 1}$ and $\sqrt{4k}$, and $k - s^2/4 \neq 4^a(8n + 7)$. With the exception of $s > \sqrt{3k - 1}$, these were seen above to be necessary conditions. For k odd, sufficient conditions for (9) are that s be odd and between $\sqrt{3k - 2} - 1$ and $\sqrt{4k}$; there exists such an s for any k. As to the former case, he proved that for any k there exists an integer between $\sqrt{3k}$ and $\sqrt{4k}$ and congruent to k modulo 2 except when $k = 1, 5, 9, 11, 17, 19, 29, 41, 2, 6, 8, 14, 22, 24, 34$.

---

[20] New Series of Math. Repository (ed., Leybourn), 2, 1809, II, 70; Theory of Numbers, London, 1811, 212.

Cauchy[21] noted that if p is a prime and α, β are integers for which $\alpha + \beta + 1 \leqq p$, and if A ranges over $\alpha + 1$ distinct values modulo p, and B over $\beta + 1$ values, then $A + B$ takes at least $\alpha + \beta + 1$ distinct values modulo p. For A and B not divisible by p, Ax^2 and $By^2 + C$ each take $(p + 1)/2$ distinct values modulo p, when p is a prime > 2. Hence $Ax^2 + By^2 + C$ takes all p distinct values modulo p and therefore the value zero. Cf. Cauchy[95] of Ch. I.

Cauchy[22] noted [the case $d = s = 0$ of (1)]

$$(a^2 + b^2 + c^2)(p^2 + q^2 + r^2)$$
$$= (ap + bq + cr)^2 + (aq - bp)^2 + (ar - cp)^2 + (br - cq)^2,$$

and a like formula with n squares instead of 3 [see Cauchy[61] of Ch. IX].

A. M. Legendre[23] gave (8) and concluded that every number of the form $8n + 4$ is a sum of four odd squares in $\sigma(2n + 1)$ ways, where $\sigma(k)$ is the sum of the divisors of k. It is said to follow readily that every integer is a ④.

C. G. J. Jacobi[24] proved Bachet's theorem by comparing the formulas

$$\sqrt{2K/\pi} = 1 + 2q + 2q^4 + 2q^9 + \cdots = \sum_{n=-\infty}^{+\infty} q^{n^2},$$

$$(2K/\pi)^2 = 1 + 8\left\{ \frac{q}{1-q} + \frac{2q^2}{1+q^2} + \frac{3q^3}{1-q^3} + \cdots \right\}$$
$$= 1 + 8\Sigma\sigma(p)(q^p + 3q^{2p} + 3q^{4p} + 3q^{8p} + \cdots),$$

including (7), where p ranges over the positive odd numbers, and $\sigma(p)$ denotes the sum of the divisors of p. At the same time we obtain the theorem: The number of representations[76] of $2^\alpha p$ as a sum of 4 squares is $8\sigma(p)$ or $24\sigma(p)$, according as $\alpha = 0$ or $\alpha > 0$. Cf. Jacobi[22b] of Ch. III.

Jacobi[25] compared the formulas[26]

$$(2kK/\pi)^2 = 16\Sigma\sigma(p)q^p, \qquad \sqrt{2kK/\pi} = 2q^{1/4} + 2q^{9/4} + 2q^{25/4} + \cdots,$$

where p ranges over the odd positive numbers, and concluded that there are $\sigma(p)$ sets of four positive odd numbers the sum of whose squares is $4p$ [see papers 23, 30, 42, 52, 69, 72, 82, 91].

V. Bouniakowsky[27] proved that, if A, B, C are integers not divisible by the prime p, we can give to x, y such integral values that $Ax^2 + By^2 - C$ is divisible by p. He first found the conditions that x or y can be a multiple of p; then noted that, if neither can be a multiple of p, the congruence can be written $\rho^M + \rho^N - 1 \equiv 0 \pmod{p}$, where ρ is a primitive root of p,

---

[21] Jour. de l'école polyt., vol. 9 (cah. 16), 1813, 104–116; Oeuvres, (2), I, 39–63.

[22] Cours d'analyse de l'école polyt., 1, 1821, 457.

[23] Traité des fonctions elliptiques, 3, 1828, 133. Stated in Legendre's Théorie des nombres, ed. 3, I, 1830, 216, No. 154 (Maser, I, 217); not in eds. 1, 2. Cf. Bouniakowsky, Vol. I, p. 283. Cf. Jacobi.[25]

[24] Werke, I, 423–4; Jour. für Math., 80, 1875, 241–2; Bull. des sc. math. astr., 9, 1875, 67–9; letter, Sept. 9, 1828, Jacobi to Legendre. Jacobi, Fundamenta Nova Funct. Ellipt., Konigsberg, 1829, p. 188, p. 106 (34), p. 184 (6); Werke, I, 239. Cf. J. Tannery and J. Molk, Élém. théorie fonct. ell., 4, 1902, 260–3; J. W. L. Glaisher, Quar. Jour. Math., 38, 1907, 8; papers 51–2, 81, 88, 110–1.

[25] Jour. für Math., 3, 1828, 191; Werke, I, 247. Cf. Liouville[1] and Deltour[29] of Ch. XI.

[26] Fundamenta Nova Funct. Ellipt., 1829, 106 (35), 184 (7); Werke, I, 162, 235.

[27] Mém. Acad. Sc. St. Pétersbourg (Math.), (6), 1, 1831, 565–581.

and M and N are odd. The latter congruence can be solved. Or the theorem can be derived by multiplication from Lagrange's case $A = 1$.

If N is any odd integer or the double of an odd integer, while A, B, C are integers prime to N, $Ax^2 + By^2 - C \equiv 0 \pmod{N}$ is solvable.

Given two arithmetical progressions whose first terms α, β are arbitrary and whose common differences A, B are not divisible by the prime p, we can choose n and n' so that the total sum of n terms of the first, n' terms of the second, and any given integer E, is divisible by p:

$$\tfrac{1}{2}\{2\alpha + (n - 1)A\}n + \tfrac{1}{2}\{2\beta + (n' - 1)B\}n' + E \equiv 0 \pmod{p}.$$

For, this can be reduced to the above congruence.

F. Minding[28] noted that integers u and v can be chosen so that $u^2 - Bv^2 - C$ is divisible by the prime p, if neither B nor C is divisible by p. In fact, for $v = 0, 1, \cdots, (p - 1)/2$, the function $Bv^2 + C$ takes $(p + 1)/2$ distinct values modulo p, and at least one must be congruent to one of the $(p + 1)/2$ values of u^2, since otherwise there would be $p + 1$ residues modulo p. Hence we can choose u and v less than $p/2$ so that $u^2 + v^2 + 1$ is divisible by p. The proof that p is a ▣ is that by Euler.[10]

G. Libri[29] proved that there are $n \pm 1$ sets of solutions $< n$ of

$$x^2 + ay^2 + b \equiv 0 \pmod{n},$$

if a, b are not divisible by the prime n. He first expressed the number of sets of solutions as a double sum involving roots of unity.

C. G. J. Jacobi[30] gave an arithmetical proof of his[25] theorem on the number μ of sets of positive odd solutions $w, \cdots, z$ of

(11) $$w^2 + x^2 + y^2 + z^2 = 4p,$$

where p is a given positive odd number. Two distinct permutations of the same numbers are counted as different solutions. For such a set,

$$w^2 + x^2 = 2p', \qquad y^2 + z^2 = 2p'', \qquad p' + p'' = 2p,$$

where p' and p'' are odd. Conversely, these equations imply (11). Hence

$$\mu = \sum_{p',\,p''} N[2p' = w^2 + x^2] \cdot N[2p'' = y^2 + z^2], \qquad p' + p'' = 2p;\; p',\, p''\,\text{odd},$$

where $N[2p' = w^2 + x^2]$ denotes the number of positive solutions w, x of $2p' = w^2 + x^2$. The latter number is $N[p' = a\alpha] - N[p' = a\alpha']$, where α ranges over the factors of the form $4m + 1$ of p' and α' over the factors $4m + 3$. Let β and β' range over the factors $4m + 1$ and $4m + 3$, respectively, of p''. Then

$$N[2p'' = y^2 + z^2] = N[p'' = b\beta] - N[p'' = b\beta'].$$

Set $N[u] = N[2p = u]$. Then

$$\Sigma N[p' = a\alpha] \cdot N[p'' = b\beta] = N[a\alpha + b\beta], \text{ etc.},$$
$$\mu = N[a\alpha + b\beta] + N[a\alpha' + b\beta'] - 2N[a\alpha + b\beta'].$$

[28] Anfangsgründe der höheren Arith., Berlin, 1832, 191–3.
[29] Jour. für Math., 9, 1832, 182. See Libri[147–8] of Ch. XXIII.
[30] Jour. für Math., 12, 1834, 167–172; Werke, 6, 1891, 245–251.

Unless $\alpha = \beta$, $\alpha' = \beta'$, we may set $\beta = \alpha + 4A$, $\beta' = \alpha' + 4A$, $A > 0$, if the term be repeated. Thus

$$\mu = N[\alpha(a + b)] + N[\alpha'(a + b)] - 2N[a\alpha + b\beta']$$
$$+ 2N[\alpha(a + b) + 4bA] + 2N[\alpha'(a + b) + 4bA].$$

Let c range over both the α and α' numbers. Then

$$\mu = N[c(a + b)] + 2N[c(a + b) + 4bA] - 2N[a\alpha + b\beta'].$$

In the second term set $c = d + 4AB$, $d < 4A$, $B \geqq 0$. Now $a + b$ may represent any even number $2C$, and $b + B(a + b)$ any odd number e. Thus

$$\mu = N[c(a + b)] + 2N[2Cd + 4Ae] - 2N[a\alpha + b\beta'].$$

Since $\alpha + \beta' \equiv 0 \pmod 4$, $a \neq b$. Thus the second member of

$$2N[a\alpha + b\beta'] = N[a\alpha + b\beta'] + N[a\beta' + b\alpha]$$

is twice the like sum with $b > a$. Set $b = a + 2G$, $\alpha + \beta' = 4A$. Then

$$N[a\alpha + b\beta'] = N[2\beta'G + 4Aa] + N[2\alpha G + 4Aa] = N[2dG + 4Aa],$$

where $d < 4A$. Hence $\mu = N[c(a + b)]$. Here c ranges over all the divisors of p. If $p = cf$, the equation $2p = c(a + b)$ becomes $2f = a + b$, which has f sets of odd solutions. But $\Sigma p/c$ is the sum of the divisors of p. Thus $\mu = \sigma(p)$.

T. Schönemann[31] used the notation $\cos n$, $\sin n$ for a pair of solutions of $x^2 + y^2 \equiv 1 \pmod p$. If $\cos m$, $\sin m$ is the notation for a second pair of solutions, then the expansions of $\cos (n + m)$, $\sin (n + m)$ give a third pair of solutions. Then, for α an integer,

$$(\cos n + i \sin n)^\alpha \equiv \cos \alpha n + i \sin \alpha n \pmod p.$$

If p is a prime, $\cos pn \equiv \cos n$, $\sin pn \equiv (-1)^{(p-1)/2} \sin n \pmod p$. Hence $\cos (p \mp 1)n \equiv 1$ if $p = 4k \pm 1$. An integer a is put into "class A" if $1 - a^2$ is a quadratic residue of p, otherwise into class B. It is proved that if $\cos n$ belongs to class A and if α is the least integer for which $\cos \alpha n \equiv 1 \pmod p$, then α is a divisor of $p \mp 1$ when $p = 4k \pm 1$; then $\cos n$ is said to belong to the number α. There exist $\phi(p \pm 1)$ "primitive" cosines which belong to $p \pm 1$. For $p = 4n + 1$, $\cos n$ is primitive, so that all sets of real solutions of $x^2 + y^2 \equiv 1 \pmod p$ are given by $\cos tn$, $\sin tn$ for $t = 1, 2, \cdots, p - 1$; the cases of coincidence are found. The result is that for any prime $8m \pm 1$, $8m + 3$ or $8m + 5$, there are m essentially different sets of solutions, provided $0^2 + 1^2 \equiv 1$ is excluded. The same ideas are applied to the determination of the quadratic character of 2, 3, 5.

G. Eisenstein[32] stated without proof that the number of all representations of an odd integer m as a ▣ is $8\sigma(m)$ [Jacobi[24]], and that, if

[31] Jour. für Math., 19, 1839, 93–110.

[32] Jour. für Math., 35, 1847, 133; Math. Abhandlungen, 1847, 193. In Jour. de Math., 17, 1852, 477, the first result is said to follow from a property of ternary quadratic forms.

$m = a^\alpha b^\beta \cdots$, where a, b, $\cdots$ are distinct primes, the number of proper representations is

$$8m(1 + 1/a)(1 + 1/b) \cdots.$$

P. L. Tchebychef[33] proved that $x^2 - Ay^2 - B \equiv 0 \pmod{p}$ is solvable if A is not divisible by the prime p. Proof is needed only when $p > 2$ and $Ay^2 + B$ is never divisible by p, whence

$$(Ay^2 + B)^{(p-1)/2} + 1 \equiv 0 \pmod{p}.$$

This congruence of degree $p - 1$ is not satisfied by all the values $0, 1, \cdots$, $p - 1$ of y, so that for one of them $Ay^2 + B$ is a quadratic residue of p.

F. Pollock[34] noted that if any odd square $16n^2 \pm 8n + 1$ is increased by 3 the sum is $3(4n^2 \pm 4n + 1) + (4n^2 \mp 4n + 1)$, and hence is the sum of four odd squares. By adding also 8, the new sum is divisible into four odd squares, with a like result for each addition of 8. He stated that every number $8k + 4$ is reached in this way. Since every number $8k + 4$ is thus a $\boxed{4}$, Bachet's theorem is said to follow.

C. Hermite[35] showed that, if A is odd or the double of an odd number,

$$(12) \qquad\qquad \alpha^2 + \beta^2 + 1 \equiv 0 \pmod{A}$$

has integral solutions. First, let $A \equiv \epsilon \pmod{4}$, $\epsilon = \pm 1$. The arithmetical progression with the general term $4Az + 2\epsilon A - 1$ contains by Dirichlet's theorem an infinitude of primes, each $\equiv 1 \pmod{4}$ and hence the sum of two squares $\alpha^2 + \beta^2$. Next, let $A \equiv 2 \pmod{4}$; we employ similarly the progression $2Az + A - 1$.

For integral solutions α, β of (12), the definite form

$$f = (Ax + \alpha z + \beta u)^2 + (Ay - \beta z + \alpha u)^2 + z^2 + u^2$$

has as the numerical value of the invariant Δ the value A^4 (being the product of the square of the determinant A^2 of the four linear functions by the value 1 of Δ for the sum of 4 squares) and hence its minimum for integral values of the variables x, $\cdots$, u is $< (\frac{4}{3})^{3/2}\Delta^{1/4} < 2A$. Since f represents only multiples of A, the minimum is A itself. Thus A can be represented by f and hence is a sum of four squares.

Hermite[36] repeated the preceding proof and gave the following. The form

$$\frac{1}{A}f = A(x^2 + y^2) + 2\alpha(zx + yu) + 2\beta(xu - zy) + \frac{1}{A}(\alpha^2 + \beta^2 + 1)(z^2 + u^2)$$

has integral coefficients, and $\Delta = 1$. Hence it is equivalent to

$$X^2 + Y^2 + Z^2 + U^2,$$

the single reduced definite quaternary form with $\Delta = 1$. Hence in the four linear functions X, $\cdots$, U of x, $\cdots$, u, the sum of the squares of the coefficients of x or of y equals A.

[33] Theorie der Congruenzen, in Russian, 1849; in German, 1889, 207–9.
[34] Proc. Roy. Soc. London, 6, 1851, 132–3.
[35] Comptes Rendus Paris, 37, 1853, 133–4; Oeuvres, I, 288–9.
[36] Jour. für Math., 47, 1854, 343–5, 364–8; Oeuvres, I, 234–7, 258–263.

For M an odd integer, the Hermitian form

$$MV\overline{V} + (\alpha + \beta i)V\overline{U} + (\alpha - \beta i)\overline{V}U + \frac{1}{M}(\alpha^2 + \beta^2 + 1)U\overline{U},$$

with complex integral coefficients, has for the invariant Δ the value -1, and hence is equivalent to $v\bar{v} + u\bar{u}$, the single reduced form with $\Delta = -1$. Let the latter be transformed into the former by

$$v = aV + bU, \qquad u = cV + dU, \qquad ad - bc = 1,$$

$a, \cdots, d$ being complex integers. Then $M = a\bar{a} + c\bar{c}$, where a and c are relatively prime. Thus any odd integer is the sum of four squares such that the sum of two of the squares is prime to the sum of the two remaining squares.[37]

By considering the proper and improper representations of M by $vv + uu$, he obtained Jacobi's formula $8\Pi(p_i + 1)$ for the number of representations as a sum of 4 squares of $M = \Pi p_i$, when M is not divisible by the square of a prime.

F. Pollock[38] proved Cauchy's theorem (1813) that any odd number $2p + 1$ is a sum of four squares the algebraic sum of whose roots is any assigned odd number from 1 to the maximum. For, p is a sum of three or fewer triangular numbers. If $p = (q^2 + q)/2$, then whether $q = 2n$ or $2n - 1$, we have $2p + 1 = 4n^2 \pm 2n + 1$, which is the sum of the squares of n, $-n$, $\mp n$, $\pm(n \pm 1)$. If $p = (q^2 + q)/2 + (r^2 + r)/2$, then p is of the form $a^2 + a + b^2$, and $2p + 1$ is the sum of the squares of $a + 1$, $-a$, b, $-b$. If p is the sum of three triangular numbers,

$$p = a^2 + a + b^2 + \tfrac{1}{2}(m^2 + m),$$
$$2p + 1 = 2(a^2 + a + b^2) + 4n^2 \pm 2n + 1,$$

the latter being the sum of the squares of $b \mp n$, $-b \mp n$, $-a \pm n$, $a \pm n + 1$. In every case the algebraic sum of the four roots is unity.

A. Genocchi[39] " recalled " (without reference) formulas (7) and (8) and noted that the second implies that the number of representations of $4n$ as a ④ is $\sigma(n)$ when n is odd, and that the first implies

$$N_1 + 2N_2 + 4N_3 + 8N_4 = 4(D_1 + D_2 - D_4),$$

where D_1 is the sum of the odd divisors of n, D_2 (or D_4) the sum of the even divisors d of n with n/d odd (or even), while $N_1, \cdots, N_4$ is the number of solutions of $x_1^2 + \cdots + x_4^2 = n$ with 3, 2, 1, 0 unknowns zero. For another similar formula see Cesàro[30] of Ch. IX.

A. Desboves[40] stated empirically that the double of any odd integer is a sum of two primes $4n + 1$. Such a prime is a ②. Hence every integer is a ④.

[37] E. Picard, the editor of Hermite's Oeuvres, 1, p. 259, noted that when a and c are relatively prime, $a\bar{a}$ and $c\bar{c}$ are not necessarily so; but that the theorem in the text is probably true.
[38] Phil. Trans. Roy. Soc. London, 144, 1854, 311–9.
[39] Nouv. Ann. Math., 13, 1854, 169.
[40] Nouv. Ann. Math., 14, 1855, 293–5.

C. A. W. Berkhan[41] decomposed the integers < 360 into four rational or integral squares, and into two or three squares if possible.

G. L. Dirichlet[42] gave a simplification of Jacobi's[30] proof. According as a factor a' of $p' = a'q'$ has the form $4m + 1$ or $4m + 3$, set $\delta' = + 1$ or $- 1$. Then the number of positive solutions of $2p' = w^2 + x^2$ is $\Sigma\delta'$. Hence each couple p', p'' furnishes $\Sigma\delta' \cdot \Sigma\delta'' = \Sigma\eta$ solutions of (11), where $\eta = + 1$ or $- 1$ according as $a' - a''$ is or is not divisible by 4. Thus $\mu = \Sigma\eta$, obtained by varying also p', p'', so that there is a term η for each set of odd solutions a', a'' of

$$(13) \qquad\qquad a'q' + a''q'' = 2p.$$

Let η' be a term obtained when $a' = a''$, η'' one when $a' > a''$. Then $\mu = \Sigma\eta' + 2\Sigma\eta''$. From one set of odd solutions of (13), we obtain the new odd solutions

$$A' = q''(x + 1) + q'(x + 2),$$
$$A'' = q''x + q'(x + 1),$$
$$Q' = - a'x + a''(x + 1) = a'' - (a' - a'')x,$$
$$Q'' = a'(x + 1) - a''(x + 2) = (a' - a'')(x + 1) - a''.$$

Let $a' > a''$. In order that Q' and Q'' be positive, $(a' - a'')x$ must be the least multiple of $a' - a''$ less than a''. Then x is uniquely determined and $A' > A'' > 0$. If we repeat the process, starting with A', Q', A'', Q'', we obtain merely the initial set a', q', a'', q'', since the preceding equations hold after the interchange of a' with A', q' with Q', etc. Since

$$a' - a'' = Q' + Q'',$$

two such sets of solutions give values of η'' differing in sign. Indeed, one and but one of the even numbers $a' - a''$ and $q' + q''$ is divisible by 4, since $a' \equiv \pm a''$, $q' \equiv \mp q''$ (mod 4) contradicts (13). Hence $\Sigma\eta'' = 0$. Thus $\mu = \Sigma\eta'$, with each $\eta' = + 1$, so that $\mu = N[a'(q' + q'')] = \sigma(p)$, as above. Cf. Pepin.[72]

J. J. Sylvester[43] employed the lemma that, if $3M = p^2 + q^2 + r^2 + s^2$, M is a sum of four squares. We may assume that p is divisible by 3 and, by a proper choice of the signs of q, r, s, take $q \equiv r \equiv s$ (mod 3). Then M is the sum of the squares of the integers

$$\tfrac{1}{3}(q + r + s), \qquad \tfrac{1}{3}(p + r - s), \qquad \tfrac{1}{3}(p - q + s), \qquad \tfrac{1}{3}(p + q - r).$$

For $N \equiv 1$ (mod 4), the function $3^{2x+1}N - 2$ of x is not rationally decomposable and has no constant divisor; it is assumed to represent a prime T for some integer x. Since $T \equiv 1$ (mod 4), T is the sum of two squares. Hence $T + 2 = 3^{2x+1}N$ is the sum of four squares. The same is true of N by the lemma.

For $N \equiv 3$ (mod 4), $3^{2x}N - 2$ is employed similarly. For N even, it suffices to treat $N \equiv 2$ (mod 4), by use of $3^xN - 1$, since the theorem is true for $4N$ if true for N.

[41] Lehrbuch der Unbestimmten Analytik, Halle, 2, 1856, 286.
[42] Jour. de Math., (2), 1, 1856, 210–214; Werke, 2, 1897, 201–8.
[43] Quar. Jour. Math., 1, 1857, 196–7; Coll. Math. Papers, 2, 1908, 101–2.

J. Liouville[44] considered an integer m all of whose prime factors are $\equiv 1 \pmod 4$. Express $4m$ in all possible ways in the form $(u^2 + v^2)(u_1^2 + v_1^2)$, where $u, \cdots, v_1$ are odd and positive, and call two such decompositions identical if and only if $u = u', \cdots, v_1 = v_1'$. Denote the first factor $u^2 + v^2$ by $2a$. It is stated that Σa equals the number of decompositions of $16m$ as a product of two sums of four positive odd squares. The latter number exceeds Σa if m has a prime factor $\equiv 3 \pmod 4$.

Liouville[45] considered the N representations of a given even integer n as a sum $s_i^2 + t_i^2 + u_i^2 + v_i^2$ of four squares, where $s_i, \cdots, v_i$ may be positive, negative or zero, and two representations are distinct unless $s_1 = s_2, \cdots$, $v_1 = v_2$. For the first squares s_i^2, we have

$$\sum_{i=1}^{N} s_i^\mu = 0 \ (\mu \text{ odd}), \qquad \sum_{i=1}^{N} s_i^2 = \frac{n}{4} N, \qquad \sum_{i=1}^{N} s_i^4 = \frac{n^2}{8} N.$$

The second follows from $nN = \Sigma s_i^2 + \cdots + \Sigma v_i^2$ and $\Sigma s_i^2 = \Sigma t_i^2$, etc. The third was verified for small values of n [proved by Stern[81]]. By means of it and $n^2 N = \Sigma(s_i^2 + \cdots + v_i^2)^2$, we get $\Sigma_{i=1}^{i=N} s_i^2 t_i^2 = n^2 N/24$.

J. G. Zehfuss[46] noted the identity

$$(2a)^2 + (2b)^2 + (2c)^2 + (2d)^2 = (a + b + c \pm d)^2 + (a + b - c \mp d)^2$$
$$+ (a - b + c \mp d)^2 + (a - b - c \pm d)^2.$$

F. Pollock[47] stated that any odd number is the sum of four squares the roots of two of which differ by any assigned number d from zero to the maximum. For $d = 0$, we use $a^2 + b^2 + 2c^2$ (Legendre, Théorie des nombres, I, 186; II, 398). Next, let $d = 1$. Since $4n + 1$ is a sum of three squares, only one being odd,

$$4n + 1 = (2a)^2 + (2b)^2 + (2c + 1)^2,$$
$$2n + 1 = (a + b)^2 + (a - b)^2 + c^2 + (c + 1)^2.$$

The case in which d is general is discussed by means of a special arithmetical series with the general term $2n^2 + 1$.

C. Souillart[48] proved Euler's formula (1) by multiplying

$$(a^2 + b^2 + c^2 + d^2)^2 = \begin{vmatrix} a & b & c & d \\ -b & a & -d & c \\ -c & d & a & -b \\ -d & -c & b & a \end{vmatrix}$$

by the similar determinant with p, q, r, s as first row.

F. Pollock[49] stated that every odd number is a sum of the squares of $a + p + 1, a - p, a + q, a - q$, the sum of two of which exceed the sum of the remaining two by unity; also is a sum of four squares the sum of whose roots is unity.

[44] Jour. de Math., (2), 2, 1857, 351–2.
[45] Ibid., (2), 3, 1858, 357–360.
[46] Archiv Math. Phys., 30, 1858, 466.
[47] Phil. Trans. Roy. Soc. London, 149, 1859, 49–59.
[48] Nouv. Ann. Math., 19, 1860, 321.
[49] Phil. Trans. Roy. Soc. London, 151, 1861, 409–421.

J. Liouville[49a] proved that the number of representations by $x^2+y^2+z^2+4t^2$ of an odd number m is $\{4 + 2(-1)^{(m-1)/2}\}\sigma(m)$, of $2m$ is $12\sigma(m)$, of $4m$ is $8\sigma(m)$, of $2^\alpha m\,(\alpha \geqq 3)$ is $24\sigma(m)$. The number of representations by $x^2 + 4y^2 + 4z^2 + 4t^2$ of $m = 4l + 3$ is zero, of $m = 4l + 1$ is $2\sigma(m)$, of $2m$ is zero, of $4m$ is $8\sigma(m)$, of $2^\alpha m\,(\alpha \geqq 3)$ is $24\sigma(m)$. He found also the number of proper representations by these forms. He[49b] expressed the number of representations of $2^\alpha m$ by $x^2 + ay^2 + bz^2 + 16t^2$ for $(a, b) = (4, 4)$, $(16, 16)$, $(4, 16)$, $(1, 16)$, $(1, 4)$, $(1, 1)$, in terms of $\sigma(m)$ and $\Sigma(-1)^{(i-1)/2}i$, summed for the odd integers i for which $m = i^2 + 4s^2$. From Jacobi's[24] result, he[49c] derived also the number of representations by $x^2 + y^2 + 9z^2 + 9t^2$.

J. Liouville[50] considered an odd integer m and the decompositions

$$4m = i^2 + i_1^2 + i_2^2 + i_3^2, \qquad 2m = r^2 + r_1^2 + 4s^2 + 4s_1^2,$$

where i, i_1, i_2, i_3, r, r_1 are positive odd integers, and stated that

$$\Sigma(-1)^{(ii_1-1)/2}ii_1 = (-1)^{(m-1)/2}\Sigma(-1)^{(rr_1-1)/2}rr_1.$$

J. Plana[51] proved Jacobi's[24] formula

$$(1 + 2q + 2q^4 + 2q^9 + \cdots)^4 = 1 + 8\Sigma\sigma(p)(q^p + 3q^{2p} + 3q^{4p} + \cdots).$$

H. J. S. Smith[52] discussed Jacobi's[24, 25] theorems that the number of representations of an odd number m as a $\boxed{4}$ is $8\sigma(m)$; the number of representations of $4m$ as a sum of four odd squares is $16\sigma(m)$.

F. Pollock[53] stated that the algebraic sum of the roots in some representation of a given odd number as a $\boxed{4}$ will equal any assigned odd number not exceeding the maximum; that the difference of some two of the roots will equal any number not exceeding the maximum. But all that is definitely proved in this paper, dealing with numerical statements, is that any number n is a sum of four triangular numbers, since Bachet's theorem gives

$$4n + 2 = (2a + 1)^2 + (2b + 1)^2 + (2c)^2 + (2d)^2,$$
$$n = (a^2 + a + c^2) + (b^2 + b + d^2).$$

V. Bouniakowsky[54] employed the known result that the quadratic residues of a prime $p = 4n + 1$ may be paired so that the sum of a pair is p, and likewise the non-residues, to obtain relations like

$$10^2 + 11^2 = 2^2 + 3^2 + 8^2 + 12^2, \qquad 6^2 + 7^2 = 1^2 + 2^2 + 4^2 + 8^2 \ (p = 17),$$
$$13^3 = 1^3 + 5^3 + 7^3 + 12^3, \qquad 13^3 + 14^3 = 1^3 + 3^3 + 17^3$$

[the first from $2^2 + 3^2 = 13$, $8^2 + 12^2 \equiv -1 + 1 \pmod{13}$].

[49a] Jour. de Math., (2), 6, 1861, 440–8. Cf. Liouville[2] of Ch. XI.
[49b] Ibid., (2), 7, 1862, 73–6, 77–80, 105–8, 117–20, 157–60, 165–8.
[49c] Ibid., (2), 10, 1865, 14–24.
[50] Jour. de Math., (2), 8, 1863, 431–2.
[51] Mem. Accad. Turin, (2), 20, 1863, 130.
[52] British Assoc. Report, 1865, 337; Coll. Math. Papers, I, 307.
[53] Proc. Roy. Soc. London, 15, 1867, 115–127; 16, 1868, 251–4; abstract of Phil. Trans., 158, 1868, 627–642. His "proof" of Bachet's theorem is given in Ch. 1.[124]
[54] Bull. Acad. Sc. St. Pétersbourg, 13, 1869, 25–31.

F. Unferdinger[55] denoted $a^2 + b^2 + c^2 + d^2$ by Σa^2 and expressed $\Sigma a^2 \cdot \Sigma a_1^2 \cdots \Sigma a_{n-1}^2$ algebraically as a ▣ in 48^{n-1} ways, different in general.

E. Lionnet stated and V. A. Lebesgue[56] proved that every odd number is a sum of four squares of which two are consecutive. For, $4n + 1$ is a ▣, necessarily $4q^2 + 4r^2 + (2s + 1)^2$, whence

$$2n + 1 = (q + r)^2 + (q - r)^2 + s^2 + (s + 1)^2.$$

J. W. L. Glaisher[57] noted that, by an identity in Jacobi's Fund. Nova,

$$48\alpha + 24\alpha_2 + 12\alpha_{22} + 8\alpha_3 + 2\alpha_4 + 24\beta + 12\beta_2 + 4\beta_3 + 6\gamma + 3\gamma_2 + \delta$$

equals $\sigma(N)$ if N is odd, and $3\sigma(N)$ if N is even, where α, α_2, α_{22}, α_3, α_4 is the number of ways N is a sum of four squares all distinct, two equal, two pairs equal, three equal, four equal, respectively, while β, β_2 or β_3 is the number of ways N is a sum of three squares, distinct, two or three equal, and γ, γ_2, δ are the analogous numbers for two squares and one square.

S. Réalis[58] employed $8n + 3 = (2a - 1)^2 + (2b - 1)^2 + (2c - 1)^2$ to show that $2n + 1$ is the sum of the squares of

$$\tfrac{1}{2}\{k \pm (a - b + c)\}, \qquad \tfrac{1}{2}\{k \pm (a + b - c)\},$$
$$\tfrac{1}{2}\{k \pm (- a + b + c)\}, \qquad \tfrac{1}{2}\{k \mp (a + b + c - 2)\},$$

whose sum is unity, where, if $s = a + b + c$ is even, the upper signs are chosen and $k = 0$, while if s is odd, the lower signs are taken and $k = 1$. More generally, every odd number N is a sum of 4 squares, the algebraic sum of whose roots equals any odd number $< 2\sqrt{N}$. Any number $N = 4n + 2$ is a sum $a^2 + b^2 + c^2 + k^2$, where k^2 is any chosen square $< N$; for, according as k is even or odd, $N - k^2$ is of the form $4p + 2$ or $4p + 1$ and hence a ▣. Also [Zehfuss[46]],

$$N = \alpha^2 + \beta^2 + \gamma^2 + \delta^2, \quad 2\alpha = a + b + c + k, \quad 2\beta = - a + b - c + k,$$
$$2\gamma = - a - b + c + k, \quad 2\delta = a - b - c + k, \quad \alpha + \beta + \gamma + \delta = 2k.$$

Hence every number $N = 4n + 2$ is a sum of 4 squares the algebraic sum of whose roots is any assigned one of the numbers 0, 2, 4, $\cdots$, 2μ, where μ^2 is the largest square $< N$. Every number $N = 4n + 1$ (or $4n + 3$) is a sum of 4 squares one of which can be chosen arbitrarily among the even (or odd) squares $< N$.

Glaisher[59] expanded Gauss' proof of (6) and gave an arithmetical proof by showing that, if N is odd, the number of representations of $4N$ as a sum of 4 odd squares equals double the number of representations of N as a sum of 4 or fewer squares.

E. Catalan[60] attributed to J. Neuberg the identity

$$(a^2 + b^2 + c^2 + bc + ca + ab)^2 = (a + b + c)^2(a^2 + b^2 + c^2) + (bc + ca + ab)^2.$$

[55] Sitzungsber. Akad. Wiss. Wien (Math.), 59, II, 1869, 455–464.
[56] Nouv. Ann. Math., (2), 11, 1872, 516–9; same by Réalis.[58]
[57] British Assoc. Report, 46, 1873, 11 (Trans. Sect.).
[58] Nouv. Ann. Math., (2), 12, 1873, 212–23.
[59] Phil. Mag. London, (4), 47, 1874, 443; (5), 1, 1876, 44–7.
[60] Nouv. Corresp. Math., 1, 1874–5, 154–5.

Hence, by a change of notation,

$$(f^2 + 2g^2 + h^2)^2 = (f^2 - g^2)^2 + (f + g)^2(g + h)^2$$
$$+ (f + g)^2(g - h)^2 + (h^2 - 2fg + g^2)^2.$$

Since every odd number is of the form $f^2 + 2g^2 + h^2$, every odd square is a sum of four squares.

S. Réalis[61] used (1) to show that, for any integer p,

$$p^2 = P^2 + Q^2 + R^2 + S^2, \qquad 2p + P + Q + R + S = \Box,$$

and that we can find four integers whose algebraic sum is p and the sum of whose squares is p^2.

Catalan[62] gave the identity

$$\Sigma a^2 \Sigma (b\gamma - c\beta)^2 \Sigma f^2$$
$$= (\Sigma af \Sigma a\alpha - \Sigma \alpha f \Sigma a^2)^2 + \{a\Sigma f(b\gamma - c\beta) + (b\gamma - c\beta)\Sigma af\}^2$$
$$+ \{b\Sigma f(b\gamma - c\beta) + (c\alpha - a\gamma)\Sigma af\}^2 + \{c\Sigma f(b\gamma - c\beta) + (a\beta - b\alpha)\Sigma af\}^2,$$

expressing a product of three $\boxed{3}$ as a $\boxed{4}$.

Réalis[63] noted that, for every odd integer p,

$$p = P + Q + R + S, \qquad p^2 = P^2 + Q^2 + R^2 + S^2,$$

the algebraic sum of three of $P, \cdots, S$ being a square. For,

$$p = x^2 + y^2 + 2z^2$$
$$= (x + z)(x - z) + (x + z)(z + y) + (x + z)(z - y) + (y^2 + z^2 - 2xz).$$

Also, if $p = 4n + 1, 4n + 2$ or $8n + 3$, we can make $P + Q + R + 3S = \Box$. For,

$$p = \boxed{3} = (x^2 - yz) + (y^2 - xz) + (z^2 - xy) + (xy + xz + yz).$$

G. Torelli[64] proved by means of Jacobi's[25] theorem the result (I) that if $2n - 1$ is not divisible by 3 and if p, q are respectively the numbers of sets of distinct odd integral solutions, not all divisible by 3, of

$$2x^2 + y^2 + z^2 = 36(2n - 1), \qquad x^2 + y^2 + z^2 + t^2 = 36(2n - 1),$$

then $p + 2q$ is the sum $\sigma(2n - 1)$ of all the divisors of $2n - 1$. (II) When the second members are replaced by $4 \cdot 3^{h+2}(2m - 1)$, then

$$p + 2q = 3^h \sigma(2m - 1).$$

(III) If k is a prime $12\lambda - 1$ and if $2n - 1$ is not divisible by k, while p, q are respectively the numbers of sets of distinct odd integral solutions not all divisible by k of

$$2x^2 + y^2 + z^2 = 4k^\kappa(2n - 1), \qquad x^2 + y^2 + z^2 + t^2 = 4k^\kappa(2n - 1),$$

then $p + q = k^{\kappa-1}\lambda\sigma(2n - 1)$. (IV) If $M = a^\alpha b^\beta \cdots$, where $a, b, \cdots$ are distinct odd primes, $4M$ is a sum of four odd squares without a common

---

[61] Nouv. Ann. Math., (2), 14, 1875, 90–91.
[62] Nouv. Corresp. Math., 4, 1878, 333, foot-note.
[63] Nouv. Ann. Math., (2), 17, 1878, 45.
[64] Giornale di Mat., 16, 1878, 152–167.

factor in $M(1 + 1/a)(1 + 1/b) \cdots$ ways.   (V) If r_1, p_1, p_2 are the numbers of sets of distinct integral solutions not zero of

$$x^2 + y^2 + z^2 + t^2 = 2(2n - 1), \qquad x^2 + y^2 + z^2 + t^2 = 2n - 1,$$
$$2x^2 + y^2 + z^2 = 2n - 1,$$

then $r_1 = 3p_1 + p_2$.   (VII) If $x^2 + y^2 + z^2 + t^2 = 4(2n - 1)$ has s_1 sets of distinct odd integral solutions and $x^2 + y^2 + z^2 = 2n - 1$ has p_4 sets of distinct solutions $\neq 0$, then $s_1 = 2p_1 + p_4$.   (IX) If Σ_4 denotes $x^2 + y^2 + z^2 + t^2$, the number of sets of solutions of $\Sigma_4 = 2^k(2n - 1)$ is expressed in terms of the numbers of sets of solutions of $\Sigma_4 = 2n - 1$ and $\Sigma_3 = 2n - 1$ and the number of sets of solutions when two or three variables are equal.

E. Fergola[65] had stated the preceding theorem (V), and (I) with the restriction that $2n - 1$ is not a square.

E. Catalan[66] noted that $2p = a + b + c$ implies

$$p^2 + (p - a)^2 + (p - b)^2 + (p - c)^2 = a^2 + b^2 + c^2$$

and gave various identities in a, b, c, which express the square of the sum of three squares as a $\boxed{4}$.

J. J. Sylvester[67] proved that any prime p is a divisor of $x^2 + y^2 + 1$. Assume the contrary.   Then $p \neq 4i + 1$ since p does not divide $x^2 + 1$. Let ρ be any primitive pth root of unity and set $R = \Sigma\rho^{x^2}$, summed for the quadratic residues $x^2 < p$.   Let R' be the period conjugate to R. Expand R^2 as a sum of powers of ρ.   Since $p \neq 4i + 1$, $x^2 + y^2 \neq p$ and no pth power of ρ can occur in the expansion of R^2.   Since, by hypothesis, neither $2x^2$ nor $x^2 + y^2$ is $\equiv -1 \pmod{p}$, no such power as ρ^{p-1} can appear in R^2, while it belongs to R'.   Thus no term of R' appears in R^2.   As each power of ρ in R^2 belonging to the same period must appear a like number of times, we have $R^2 = R(p - 1)/2$, whereas $R \neq 0$ or $(p - 1)/2$.

From this theorem follows Bachet's theorem.   A similar proof shows that $Ax^2 + By^2 + Cz^2 \equiv 0 \pmod{p}$ is solvable.

H. J. S. Smith[68] indicated a proof of Bachet's theorem by continued fractions.

C. Hermite[69] proved (6) by elliptic functions and concluded that the number of decompositions into four squares of any odd integer n equals 8 times the number of decompositions of $4n$ as a sum of four squares whose roots are odd and positive.   Cf. Jacobi.[25]

J. W. L. Glaisher[70] considered the $\sigma(N)$ compositions (allowing permutations) of $4N$ as a sum of 4 odd squares, took the square root of the first square (for example) in each such composition, giving it the sign $\pm$ according as it is of the form $4m \pm 1$, and formed the algebraic sum A of these square roots.   Next, consider the compositions of $2N$ as a sum of 2 odd squares, take the product of the square roots of the two squares in each such

[65] Giornale di Mat., 10, 1872, 54.
[66] Nouv. Corresp. Math., 5, 1879, 92–93.
[67] Amer. Jour. Math., 3, 1880, 390–2; Coll. Math. Papers, 3, 1909, 446–8.
[68] Coll. Math. in memoriam D. Chelini, Milan, 1881, 117; Coll. Math. Papers, II, 309.
[69] Cours, Fac. Sc. Paris, 1882; 1883, 175; ed. 4, 1891, 242.
[70] Quar. Jour. Math., 19, 1883, 212–5; 36, 1905, 342–3.

composition, determine the sign as before, and form the algebraic sum B of the products.   Then $A = B$, as shown by use of infinite series and products.

E. Catalan[71] noted that

$$x^{4n} + y^{4n} = \left(\frac{x^{2n+2} \pm y^{2n+2}}{x^2 + y^2}\right)^2 + 2\left(xy \cdot \frac{x^{2n} \mp y^{2n}}{x^2 + y^2}\right)^2 + \left(x^2 y^2 \cdot \frac{x^{2n-2} \pm y^{2n-2}}{x^2 + y^2}\right)^2.$$

T. Pepin[72] gave a purely arithmetical proof that the number of representations of m as a $\boxed{4}$ is $8\{2 + (-1)^m\}X(m)$, where $X(m)$ is the sum of the odd divisors of m.   The proof is like that by Jacobi[30] and Dirichlet.[42] Pepin[73] gave an exposition of this proof by Dirichlet and noted (p. 173) that the theorem is a special case of one by Liouville; he proved (pp. 176–184) the theorems of Jacobi.[24]

M. Weill[74] noted that Jacobi deduced from the formula $k^2 + k'^2 \equiv 1$ in elliptic functions the result that, if N is odd, the number of representations of $4N$ as a sum of 4 odd squares is double the number of representations of N as a $\boxed{4}$, and gave a direct proof by means of the identity of Zehfuss.[46] By a similar identity, Weill proved that if N is any integer not divisible by 3, and if N and $3N$ admit only decompositions into four distinct squares $\neq 0$, the number of decompositions of $3N$ as a $\boxed{4}$ is double the number of those of N.

G. Frattini[75] proved that the number of pairs of squares for which $x^2 - Dy^2 \equiv \lambda \pmod{p}$ is $\frac{1}{2}\{p - (D/p)\}$, where (D/p) is the quadratic character of D with respect to the prime p.   There is given an elegant proof, due to Bianchi, of the existence of solutions if $p > 3$.   If λ is a residue, take $y = 0$.   If λ is a non-residue, it is shown that, when α ranges over the $(p - 1)/2$ residues, $\alpha - \lambda$ is not always a residue and not always a non-residue.   For, if $e = (p - 1)/2$ and every root of $x^e \equiv 1$ satisfies $(x - \lambda)^e \equiv \pm 1$, it satisfies $(x - \lambda)^e - x^e \equiv 0$ or -2, whereas the degree is less than the number e of the roots.

J. W. L. Glaisher[76] used the term partition (resolution) of N as a sum of squares when we disregard the order in which the squares are placed and the signs of the roots; composition when the order of the squares is taken into account, but not the signs of the roots; representation when both the order and the signs are attended to.   For N odd, $\chi(N)$ denotes the sum of the square roots of the distinct squares appearing in the various partitions of $2N$ into two squares, the sign $+$ or $-$ being prefixed to each root according as its numerical value is of the form $4n + 1$ or $4n + 3$.   An equivalent definition (p. 98) is that $\chi(N)$ is the sum of all the primary complex numbers $a + bi$ of norm $N = a^2 + b^2$.   Two odd squares are said to be of the same class if and only if both are of the form $(8n \pm 1)^2$ or both of the form

[71] Nouv. Ann. Math., (3), 3, 1884, 347.
[72] Atti Accad. Pont. Nuovi Lincei, 37, 1883–4, 12–20.
[73] Ibid., 38, 1884–5, 140–5.
[74] Comptes Rendus Paris, 99, 1884, 859–861; Bull. Soc. Math. France, 13, 1884–5, 28–34.
[75] Rendiconti Reale Accad. Lincei, (4), 1, 1885, 136–9.
[76] Quar. Jour. Math., 20, 1885, 80–167.

$(8n \pm 3)^2$. The following theorems were proved by use of infinite series. If $N = 4n + 1$ and if H_1 (or H_2) denotes the number of compositions of $4N$ as a sum of 4 odd squares of the same class (or not of same class), then $H_1 - \frac{1}{3}H_2 = \chi(N)$. As known, $H_1 + H_2 = \sigma(N)$. If $N = 4n + 1$ and if of the partitions of $4N$ into 4 odd squares of which two are equal, P is the number having the remaining two squares of the form $(8n \pm 1)^2$ and Q the number for which they are of the form $(8m \pm 3)^2$, then $P = Q$ if N is not a square, while

$$P - Q = \frac{1}{2}\left\{\left(\frac{-1}{\nu}\right)\nu + \left(\frac{2}{\nu}\right)\right\}, \qquad N = \nu^2.$$

Write S for $(2p + 1)^2 + (2q + 1)^2$; the number of representations of $8n + 2$ as $S + (4r)^2 + (4s)^2$ or $S + (4r + 2)^2 + (4s + 2)^2$ is respectively

$$12\{\sigma(4n + 1) + \chi(4n + 1)\}, \qquad 12\{\sigma(4n + 1) - \chi(4n + 1)\};$$

while there are $12\sigma(4n + 3)$ representations $8n + 6 = S + (4r)^2 + (4s + 2)^2$.

Let $E(N)$ denote the excess of the number of divisors $4n + 1$ of N over the number of divisors $4n + 3$; then $E(N)$ is the number of primary numbers of norm N. If $n \equiv 1 \pmod 4$,

$$\chi(n) = E(1)E(2n - 1) - E(5)E(2n - 5) + E(9)E(2n - 9) - \cdots$$
$$+ E(2n - 1)E(1),$$
$$\sigma(2m + 1) = E(1)E(4m + 1) + E(5)E(4m - 3) + E(9)E(4m - 7) + \cdots$$
$$+ E(4m + 1)E(1).$$

Call $E_2(n)$ the excess of the sum of the squares of the divisors $4m + 1$ of n over the sum of the squares of the divisors $4m + 3$; $\lambda(n)$ the sum of the squares of the primary numbers of norm n. There are given many formulas serving to evaluate χ, σ, E, E_2, λ, whose values are tabulated for arguments $n \leqq 100$, with citation to longer tables.

R. Lipschitz[77] found the number of sets of solutions of $\xi_1^2 + \xi_2^2 + \xi_3^2 \equiv 0$ (mod p^γ), where p is a prime, and applied the result to find all integral quaternions with a given norm and hence the solutions of $m = \boxed{4}$. He discussed the real and rational automorphs of $x_1^2 + x_2^2 + x_3^2$.

S. Réalis[78] concluded from $pq = \alpha^2 + \cdots + \delta^2$ three sets of fractional expressions for p and q in terms of $\alpha, \cdots, \delta$ and new parameters, but admitted that he was unable to utilize them to prove Bachet's theorem.

A. Puchta[79] repeated Gauss'[17] derivation of Euler's formula (1). To interpret (1), use the four-dimensional regular body bounded by 5 tetrahedra and having as vertices 5 equidistant points P_i. There exists a point O such that $OP_1, \cdots, OP_4$ are perpendicular lines, while the " planes " through O and any three of $P_1, \cdots, P_4$ are perpendicular. We may take O to be the point with the coordinates $x_1 = (a_1 + a_2 + a_3 + a_4)/2$, etc.,

[77] Untersuchungen über die Summen von Quadraten, Bonn, 1886. French transl. by J. Molk, Jour. de Math., (4), 2, 1886, 393–439.

[78] Jour. de math. élém., (2), 10, 1886, 89–91.

[79] Sitzungsber. Akad. Wiss. Wien (Math.), 96, II, 1887, 110.

and get the identity $\Sigma a_i^2 \cdot \Sigma x_i^2 = \Sigma \pi_i^2$, where

$$\pi_1 = \tfrac{1}{2}(- a_1 + a_2 + a_3 + a_4)x_1 + \tfrac{1}{2}(- a_1 - a_2 + a_3 - a_4)x_2 + \cdots,$$

etc. By permuting the a's or changing the signs, we get 96 formulas (1).

E. Catalan[80] made an invalid criticism of Legendre's[15] proof that every prime is a $\boxed{4}$, who is said to have assumed that every integer N has a prime divisor $> \sqrt{N}$. Catalan's remark (p. 164) that if N and A are sums of four integral squares, their quotient N/A is a sum of fractional squares, was known to Euler.[8] Catalan proved that every integer is a sum of four fractional squares in an infinitude of ways and stated (p. 212) that every number $8n + 4$ is a sum of four odd squares of which two are equal.

M. A. Stern[81] gave an elementary proof of Jacobi's[24] theorem. Let m be odd. The number of representations of $2m$ as a $\boxed{4}$ is three times that of m, since $m = p^2 + q^2 + r^2 + s^2$ implies

$$2m = \Sigma(p \pm q)^2 + \Sigma(r \pm s)^2 = \Sigma(p \pm r)^2 + \Sigma(q \pm s)^2$$
$$= \Sigma(p \pm s)^2 + \Sigma(q \pm r)^2.$$

Conversely, if $2m = \alpha^2 + \beta^2 + \gamma^2 + \delta^2$, two of the squares are even and two odd, so that

$$m = \left(\frac{\alpha + \beta}{2}\right)^2 + \left(\frac{\alpha - \beta}{2}\right)^2 + \left(\frac{\gamma + \delta}{2}\right)^2 + \left(\frac{\gamma - \delta}{2}\right)^2.$$

Repeating the process, we get [cf. Zehfuss[46]]

$$4m = (2p)^2 + (2q)^2 + (2r)^2 + (2s)^2,$$
$$(14) \quad 4m = (p + q + r \pm s)^2 + (p + q - r \mp s)^2$$
$$+ (p - q + r \mp s)^2 + (p - q - r \pm s)^2.$$

Conversely, $4m = \Sigma\alpha^2$ implies $2m = \{\tfrac{1}{2}(\alpha + \beta)\}^2 + \cdots$. Hence $4m$ and $2m$ have the same number of representations as a $\boxed{4}$. It is shown that if $2^a m$ and $2^{a+1}m$ have the same number ν of representations, then $2^t m(t \geqq \alpha)$ has ν representations. If $m = 4k + 1$, three of the numbers p, q, r, s are even and the fourth is odd, so that the squares in (14) are all odd. If $m = 4k + 3$, three of p, q, r, s are odd and one is even, and the preceding conclusion holds. By Jacobi's[30] theorem, there are $16\sigma(m)$ representations of $4m$ by four odd squares. Hence if $pqrs \neq 0$, there are $8\sigma(m)$ representations of $4m$ by four even squares and hence $24\sigma(m)$ representations in all. This result is proved to hold also if $pqrs = 0$. Cf. Vahlen.[88]

T. Pepin[82] proved Jacobi's[25] theorem that, if m is odd, the number of decompositions of $4m$ as a sum of 4 odd squares with positive roots is $\sigma(m)$, by taking $t = \pi/2$ in a formula involving sums of sines of multiples of t. The number of representations of $2m$ by $x^2 + y^2 + 4z^2 + 4t^2$ is $4\sigma(m)$. The number of representations of $2m$ by $x^2 + y^2 + z^2 + t^2$ or $x^2 + y^2 + z^2 + 4t^2$, with $x + y \equiv 1 \pmod 2$, is $16\sigma(m)$ or $8\sigma(m)$ respectively. He gave

[80] Mém. Soc. Roy. Sc. de Liège, (2), 15, 1888, 160 (Mélanges Math., III).
[81] Jour. für Math., 105, 1889, 251–262.
[82] Jour. de Math., (4), 6, 1890, 19–20.

various theorems on the representations of $2^k m$ by forms

$$x^2 + (2^\alpha y)^2 + (2^\beta z)^2 + (2^\gamma w)^2.$$

E. Catalan[83] noted that, if $k = 2a^2 + 3$, k is a $\boxed{4}$ and k^2 a $\boxed{3}$.

A. Matrot[84] duplicated in essence the proof by Euler[10] except as regards the theorem that every prime p divides a sum of 2 or 3 squares. Let $p = 2h + 1$. Consider the couples j, $2h - j$ $(j = 1, \cdots, h - 1)$. If both terms α, α_1 of some couple are quadratic residues of p, $\alpha \equiv A^2$, $\alpha_1 \equiv A_1^2$, $A^2 + A_1^2 + 1 \equiv 0 \pmod{p}$. But if no couple is composed of two quadratic residues, the number of residues contained in the couples is $\leqq h - 1$. Hence one of the numbers h, $2h$, not lying in a couple, is a quadratic residue (there being h such). If $h \equiv A^2$, $A^2 + A^2 + 1 \equiv 0 \pmod{p}$. If $2h \equiv A^2$, $A^2 + 1 \equiv 0 \pmod{p}$.

E. Humbert[85] proved that if p is odd and $\neq 3, 9$, at least one of the numbers $\frac{1}{2}(p + 1)$, $\frac{1}{2}(p + 3)$, $\cdots$, $p - 1$ is a square. Hence if the absolutely least quadratic residues of a prime $p > 3$ be arranged in increasing order of numerical value, the series contains negative terms. Hence if $p = 4n + 3$, there exsits a positive residue α followed by the residue $-\alpha - 1$. Then $\alpha \equiv x^2$, $-\alpha - 1 \equiv y^2$, $x^2 + y^2 + 1 \equiv 0 \pmod{p}$.

R. F. Davis[86] noted that, if $s = a + b + c + d$ is even, $a^2 + b^2 + c^2 + d^2$ is expressible as a sum of four new squares by means of the identity of Zehfuss[46] (divided by 4). If s is odd, add m^2 to each member and transform into a $\boxed{4}$. R. W. D. Christie made use of various formulas expressing a $\boxed{3}$ as a $\boxed{3}$ after proper selection of three of four squares.

A. Matrot[87] noted that, if $p = 2h + 1$ is a prime, we can find two consecutive integers α and $\alpha + 1$ satisfying $x^h \equiv 1$ and $x^h \equiv -1 \pmod{p}$, respectively. For, otherwise $1, 2, \cdots, p - 1$ would all satisfy the first. Hence

$$\alpha^{h+1} + (\alpha + 1)^{h+1} + 1 \equiv \alpha - (\alpha + 1) + 1 \equiv 0 \pmod{p}.$$

For $p \equiv 3 \pmod{4}$, $h + 1$ is even, and p divides a $\boxed{3}$. His proof that every prime $p \equiv 1 \pmod{4}$ divides a $\boxed{2}$ was quoted under that topic.

K. Th. Vahlen[88] gave essentially the same argument as had Stern.[81] His proof of Bachet's theorem is given in Ch. VII.[74]

E. Catalan[89] gave Legendre's[15] proof of Bachet's theorem. Euler[8] gave the empirical theorem that an integer is not a sum of four fractional squares unless it is a sum of four integral squares. This is said to be false since every integer is a sum of four fractional squares in an infinitude of ways.

[83] Assoc. franç. av. sc., 20, 1891, II, 198.

[84] Assoc. franç. av. sc. (Limognes), 19, 1890, II, 79–81 [20, 1891, II, 185–191 for historical remarks on the proofs by Lagrange and Euler]; Jour. de math. élém., (3), 5, 1891, 169–74; pamphlet, Paris, Nony, 1891. Reproduced by E. Humbert, Arithmétique, Paris, 1893, 284, and by G. Wertheim, Zeit. Math. Naturw. Unterricht, 22, 1891, 422–3.

[85] Bull. des Sc. Math., (2), 15, I, 1891, 51–2.

[86] Math. Quest. Educ. Times, 57, 1892, 120–2.

[87] Jour. de math. élém., (4), 2, 1893, 73–6.

[88] Jour. für Math., 112, 1893, 29.

[89] Mém. Acad. Roy. Sc. Belgique, 52, 1893–4, 22–28.

F. J. Studnička[90] noted that Euler's (1) includes the formula of Cauchy,[22] and deduced the like formula expressing a product of three sums of 3 squares as a ④.

L. Gegenbauer[91] proved new expressions of Jacobi's theorems. The number of representations of an odd number n as a ④ equals 8 times the number of divisors of the various g.c.d.'s of n with the numbers $\leqq n$; also equals 8 times the sum of the products obtained by multiplying the number of divisors of every factor of n by the number of integers not exceeding the complementary factor and relatively prime to it. The number of proper representations of an odd number n as a ④ equals 8 times the number of decompositions, into two relatively prime factors, of the various g.c.d.'s of n with the integers $\leqq n$; also equals 8 times the sum of the products obtained by multiplying the number of decompositions of every divisor of n into two relatively prime factors by the number of integers relatively prime to and not exceeding the complementary divisor.

B. Sollertinski[92] noted [Catalan[66]] that a ③ is a ④:

$$a^2 + b^2 + c^2 = \left(\frac{am}{p}\right)^2 + \left(\frac{an}{p}\right)^2 + \left(\frac{bm \pm cn}{p}\right)^2 + \left(\frac{bn \mp cm}{p}\right)^2,$$
$$p^2 = m^2 + n^2.$$

E. N. Barisien[93] noted that s^5 is a ④ if $s = x^2 + y^2$, since

$$s^2 = (x^2 - y^2)^2 + 4x^2y^2, \qquad s^3 = (3xy^2 - x^3)^2 + (3x^2y - y^3)^2.$$

[We may conclude that s^5 is a ②, not merely a ④.]

G. Wertheim[94] proved that every prime p divides a ③ as had Matrot,[84] and also by finding how often in the series $1, 2, \cdots, p - 1$ a residue follows a residue, or a quadratic non-residue follows a residue.

L. E. Dickson[95] exhibited all solutions of $x^2 + y^2 \equiv 1 \pmod{p}$ and of $x^2 + y^2 \equiv 0 \pmod{5^4}$.

K. Petr[96] proved two formulas by Gauss (Werke, III, 476) on theta functions by the method outlined by Gauss. From them are derived relations giving the number $\varphi(N)$, $\psi(N)$, $\psi'(N)$ of representations of N by

$$x^2 + y^2 + 9z^2 + 9u^2, \qquad x^2 + y^2 + z^2 + 9u^2, \qquad x^2 + 9y^2 + 9z^2 + 9u^2,$$

respectively. Let $\chi(N)$ be the known number for four squares. Then

$$\varphi(N) = \tfrac{1}{6}\{\chi(N) + 16\Sigma(-1)^{\lfloor(3x+y)/6\rfloor}x\}, \qquad N \not\equiv 0 \pmod{3},$$

summed for all positive odd solutions of $3x^2 + y^2 = 4N$. For N divisible by an odd power of 3, $\varphi(N) = 0$; if by an even power of 3, $\varphi(N) = \chi(N/9)$. Also,

$$\psi(N) + 3\psi'(N) - 3\varphi(N) = \begin{cases} 0, & N \not\equiv 0 \pmod{3} \\ \chi(N/3), & N \equiv 0 \pmod{3}. \end{cases}$$

[90] Prag Sitzungsber. (Math. Naturw.), 1894, XV.
[91] Sitzungsber. Akad. Wiss. Wien (Math.), 103, IIa, 1894, 121.
[92] El Progreso Matemático, 4, 1894, 237.
[93] Le matematiche pure ed applicate, 1, 1901, 182–3.
[94] Anfangsgründe der Zahlenlehre, Braunschweig, 1902, 396.
[95] Amer. Math. Monthly, 11, 1904, 175; 18, 1911, 43–4, 118.
[96] Prag Sitzungsber. (Math. Naturw.), 1904, No. 37, 6 pp.

Now the third form represents N only if N is a quadratic residue 0, 1, 4, 7 of 9. But in these cases, the first form represents N only when x or y is divisible by 3. Thus $\psi'(N)$ is zero except in the following cases:

$$\psi'(N) = \tfrac{1}{2}\varphi(N) \text{ if } N \equiv 1, 4, 7 \pmod 9; \qquad \psi'(N) = \chi(N/9) \text{ if } N \equiv 0.$$

Thus ψ' and hence also ψ is fully determined.

R. D. von Sterneck[97] gave an elementary proof that every prime p divides the sum of two or three squares, no one divisible by p. Let R_j denote a quadratic residue and N_j a non-residue of p. If -1 is a residue of p, a sum $1 + s^2$ is divisible by p. If -1 is a non-residue of p, there exist two residues whose sum is a non-residue. For, if not, the sum of j residues is a residue; in particular, $jR \equiv R_j \pmod p$, which is false when j is a non-residue. From

$$R + R_1 \equiv N, \qquad -N \equiv R_2 \pmod p$$

follows $R + R_1 + R_2 \equiv 0 \pmod p$.

B. Bolzano[98] proved the existence of integers t, u such that

$$t^2 - Bu^2 - C \equiv 0 \pmod p,$$

B and C not being divisible by the prime p [Lagrange[9]]. For $t = 0, 1, \cdots$, $\tfrac{1}{2}(p-1)$, its square t^2 takes $\tfrac{1}{2}(p+1)$ incongruent values modulo p. For $u = 0, 1, \cdots, \tfrac{1}{2}(p-1)$, the sum $Bu^2 + C$ takes $\tfrac{1}{2}(p+1)$ incongruent values. Hence at least one of the first values is congruent to one of the latter, since otherwise there would be $p + 1$ incongruent numbers modulo p.

J. W. L. Glaisher[99] noted that all the partitions $\alpha^2 + \beta^2 + \gamma^2 + \delta^2$ of $4m$ into 4 odd squares can be derived from the partitions $a^2 + b^2 + c^2 + d^2$ of the odd number m by the transformations [cf. Stern[81]]:

$$\alpha = a \pm b + c + d, \qquad \beta = a \mp b - c + d,$$
$$\gamma = a \mp b + c - d, \qquad \delta = a \pm b - c - d.$$

A partition of m produces twice as many representations of $4m$ as of m, and every partition of $4m$ can be derived from one of m by such a transformation. Hence the number of representations of m as a $\boxed{4}$ is 8 times the number of compositions of $4m$ as a sum of 4 odd squares. Here and later, he[100] made a further study of the function $\lambda(m)$ [Glaisher[76]] and the related functions $P(m)$, $Q(m)$, $\Omega(m)$, defined as the sums of the products of the roots (taken in the form $4n + 1$) of the first 2, 3, 4 squares in each composition of $4m$ as a sum of 4 odd squares, $\lambda(m)$ itself being the sum of the roots of the first square in the various compositions.

Glaisher[101] applied elliptic function formulas to find the number of representations of a number as a sum of four squares of which r are even, for $r = 0, 1, 2, 3, 4$.

---

[97] Monatshefte Math. Phys., 15, 1904, 235-8.

[98] Ibid., 237-8 (posthumous paper).

[99] Quar. Jour. Math., 36, 1905, 305-358. Extracts by P. Bachmann, Niedere Zahlentheorie, II, 287-292, 319.

[100] Ibid., 37, 1906, 36-48.

[101] Ibid., 38, 1907, 8-9.

A. Martin[102] noted that, if $t = 2p^2 + 2q^2 - n^2$, the sum of the squares of $t + 4np$, $t - 4np$, $t + 4nq$, $t - 4nq$ equals the square of $4p^2 + 4q^2 + 2n^2$. Also [Aida[59] of Ch. IX],

$$(p^2 + q^2 + r^2 - s^2)^2 + (2ps)^2 + (2qs)^2 + (2rs)^2 = (p^2 + q^2 + r^2 + s^2)^2.$$

P. Bachmann[103] gave an exposition of papers by Glaisher,[70] Dirichlet,[42] and Stern.[81]

L. Aubry[104] proved that every integer N is a ▣. It evidently suffices to treat the case N odd or double an odd number. It is first shown that N divides a certain $X^2 + Y^2 + 1$, where we may take $X \leqq N/2$, $Y \leqq N/2$. Consider therefore the numbers N_1, N_2, $\cdots$ defined by

$$X_i^2 + Y_i^2 + 1 = N_i N_{i+1}, \qquad X_i \leqq N_i/2, \qquad Y_i \leqq N_i/2.$$

The N's form a decreasing series of positive integers. Hence a certain N_n is unity. Then $N_{n-1} = X_{n-1}^2 + Y_{n-1}^2 + 1$. But if

$$X^2 + Y^2 + 1 = DE, \qquad E = p^2 + q^2 + r^2 + s^2, \qquad -pX + rY + s = aE,$$
$$sX + qY + p = cE, \qquad qX - sY + r = dE, \qquad rX + pY - q = bE,$$

then $D = a^2 + b^2 + c^2 + d^2$. Applying this theorem for $p = 1$, $r = 0$, $s = X_{n-1}$, $q = Y_{n-1}$, $X = X_{n-2}$, $Y = Y_{n-2}$, whence $D = N_{n-2}$, $E = N_{n-1}$, we see that N_{n-2} is a ▣. By the same theorem we see by induction that every N_i is a ▣. Hence $N = N_1$ is a ▣. [There is no explicit proof that a, $\cdots$, d may be taken to be integers and hence that the decomposition is not merely into four rational squares.]

E. Dubouis[105] proved that Descartes'[5] statements are true. The numbers not a sum of 4 squares > 0 are 1, 3, 5, 9, 11, 17, 29, 41 and $4^n \lambda$ ($\lambda = 2, 6, 14$), $n \geqq 0$.

S. A. Corey[106] gave a vector interpretation of (1) by use of four pentagons with a common vertex and four consecutive sides in one pentagon parallel to corresponding sides of the others.

C. van E. Tengbergen[107] proved that $x^2 + y^2 + z^2 \equiv 0 \pmod{p}$ has $(p - 1)(p - k)/48$ sets of solutions $< p/2$, where $k = -1, 5, 11, 17$ according as the prime $p = 8v - 1$, $8v - 3$, $8v + 3$, $8v + 1$.

E. Landau[108] proved that the number of sets of integral solutions of $u^2 + v^2 + w^2 + y^2 \leqq x$ is $\frac{1}{2}\pi^2 x^2 + O(x^{1+\epsilon})$, for $\epsilon > 0$ and O as by Landau,[179] Ch. VI.

G. Métrod[109] solved $x^2 + (x + y)^2 + (x + 2y)^2 + (x + 3y)^2 = z^2$ for x; the radical is rational if $z^2 - 5y^2 = u^2$ and hence if $z = a^2 + 5b^2$, $y = 2ab$, $u = a^2 - 5b^2$.

L. Aubry[110] showed how to find all solutions of $a^2 + b^2 + c^2 + d^2 = N$, first when $a^2 + b^2$, $ac + bd$ and N have no common factor, and next when

[102] Amer. Math. Monthly, 16, 1909, 19–20.
[103] Niedere Zahlentheorie, 2, 1910, 286, 323, 348–358.
[104] Assoc. franç., 40, 1911, I, 61–6.
[105] L'intermédiaire des math., 18, 1911, 55–6, 224–5.
[106] Amer. Math. Monthly, 18, 1911, 183.
[107] Wiskundige Opgaven, Amsterdam, 11, 1913, 244–7.
[108] Göttingen Nachrichten, 1912, 765–6.
[109] Sphinx-Oedipe, 8, 1913, 129–130.
[110] *Ibid.*, numéro spécial, March, 1914, 1–14; errata, 39.

their g.c.d. is m, but $a, \cdots, d$ have no common factor.   Combining numerous cases, he obtained Jacobi's[24] theorem on the total number of solutions, and the theorem that, if $N = 2^\alpha p_1^\beta \cdots p_i^\lambda$ and $\alpha \leqq 2$, the number of solutions in which $a, \cdots, d$ have no common factor is

$$8h(p_1 + 1) \cdots (p_i + 1)p_1^{\beta-1} \cdots p_i^{\lambda-1},$$

where $h = 1$ if $\alpha = 0$, $h = 3$ if $\alpha = 1$, $h = 2$ if $\alpha = 2$.   He showed how to find the $4n$ sets of solutions of $x^2 + y^2 + 1 \equiv 0 \pmod{p}$, where p is a prime $4n \pm 1$, also the solutions for any composite modulus.

L. J. Mordell[111] proved by use of theta functions that the number of solutions of $x^2 + y^2 + z^2 + t^2 = m$ is $8\{\Sigma b - \Sigma(-1)^c c\}$, where b and c range over those divisors of m whose complementary divisors are odd and even respectively [equivalent to Jacobi's[24] result].

Mordell[112] proved the conjecture by Glaisher[100] (p. 48) on the derivation of all representations of $4m_1m_2$ as a $\boxed{4}$ from those of $4m_1$ and $4m_2$.

A. S. Werebrusow[113] gave the general solution of $\boxed{4} = \boxed{4}$.

L. E. Dickson[114] gave a history of the proofs of Euler's[7] formula (1), its interpretations and generalization to 8 squares.

For Pellet's proof that $Ax^2 + By^2 + C \equiv 0 \pmod{p}$ is solvable see paper 104 of Ch. XXVI.

For minor results, see papers 12 (end), 31, 49, 106 of Ch. VII; 13, 26, 30, 39, 52, 76, 84, 94, 95 of Ch. IX; 159 of Ch. XIX; 434 of Ch. XXI.

[111] Mess. Math., 45, 1915, 78.
[112] *Ibid.*, 47, 1918, 142–4.
[113] L'intermédiaire des math., 25, 1918, 50–51; extr. from Math. Soc. Moscow.
[114] Annals of Math., (2), 20, 1919, 155–171, 297.

CHAPTER IX.

SUM OF n SQUARES.

REPRESENTATION AS A SUM OF FIVE OR MORE SQUARES.

C. G. J. Jacobi[1] remarked that a comparison of the sixth and eighth powers of two series for $(2K/\pi)^{1/2}$ would yield arithmetical theorems (for that from the fourth powers see Jacobi[24, 25] of Ch. VIII).

G. Eisenstein[2] stated that he had obtained purely arithmetical proofs of these theorems of Jacobi on the representation* of numbers as the sum of six or eight squares and stated the generalizations:

The number of representations of $4r + 1$ as a sum of six squares is $12s$ and that of $4r + 3$ is $-20s$, where $s = \Sigma(d_1^2 - d_3^2)$, d_1 ranging over the divisors of the form $4k + 1$ of the given number, d_3 over the divisors $4k + 3$.

The number of representations of an odd number as a sum of eight squares equals 16 times the sum of the cubes of its divisors.

He stated that there is no analogue for $4r + 1$ of the theorem that the number of representations of $4r + 3$ as a sum of ten squares is $12\Sigma(d_3^4 - d_1^4)$.

Eisenstein[3] stated that, if m is an odd number > 1 having no square factor, the number $\psi(m)$ of representations of m as a sum of five squares is $-80s$, -80σ, $-112s$, 80σ, according as $m \equiv 1, 3, 5, 7 \pmod 8$, where

$$s = \Sigma\left(\frac{\mu}{m}\right)\mu, \qquad \sigma = \Sigma(-1)^\mu\left(\frac{\mu}{m}\right)\mu \qquad \left(\mu = 1, 2, \cdots, \frac{m-1}{2}\right),$$

the symbol being Jacobi's. For proofs see Smith[13, 31] and Minkowski.[28]

Eisenstein[4] stated that the number of solutions of $x_1^2 + \cdots + x_7^2 = m$ is

$$-16\cdot37\Sigma\left(\frac{\mu}{m}\right)\mu^2, \qquad \mu < \frac{m}{2}, \qquad \text{if } m \equiv 7 \pmod 8;$$

$$8\cdot35\left\{\tfrac{1}{3}m^2\Sigma\left(\frac{\mu}{m}\right) - 2\Sigma\left(\frac{\mu}{m}\right)\mu^2\right\}, \qquad \mu < \frac{m}{2}, \qquad \text{if } m \equiv 3 \pmod 8;$$

$$28\Sigma(-1)^{(\mu-1)/2}\left(\frac{\mu}{m}\right)\mu(2m - \mu), \qquad \mu \text{ odd and } < m, \qquad \text{if } m \equiv 1 \pmod 4;$$

provided m has no square factor.

V. A. Lebesgue[5] discussed the decomposition of a prime p or its double into m squares, where m is a divisor > 2 of $p - 1$. Using indices relative to a primitive root of p, divide the indices of $s(s + 1)$ for $s = 1, 2, \cdots, p - 2$ by m and let $a_0, a_1, \cdots, a_{m-1}$ be the number of the indices with the

---

[1] Fundamenta Nova Func. Ellip., 1829, p. 188; Werke, 1, 1881, 239. Cf. H. J. S. Smith, Coll. Math. Papers, 1, 1894, 306–11. Cf. Jacobi[22b] of Ch. III.

[2] Jour. für Math., 35, 1847, 135; Math. Abh., Berlin, 1847, 195.

* One representation yields a new one if the roots of the squares are permuted or changed in sign, while a composition is unaltered.

[3] Jour. für Math., 35, 1847, 368.

[4] Jour. für Math., 39, 1850, 180–2.

[5] Comptes Rendus Paris, 39, 1854, 593–5.

residues $0, 1, \cdots, m-1$ respectively. Write $a_{m+t} = a_t$. For m odd,

$$\sum_{i=0}^{m-1} a_i^2 - p = \Sigma a_i a_{i+1} = \Sigma a_i a_{i+2} = \cdots = \Sigma a_i a_{i+m-1}, \qquad 2p = \sum_{i=0}^{m-1} (a_i - a_{i+k})^2,$$

when $k = 1, \cdots, m-1$. For m even, $\Sigma a_i a_{i+j} = \Sigma a_i a_{i+k}$ if $j - k$ is even, and

$$2p = \sum_{i=0}^{m-1} (a_i - a_{i+2k})^2, \qquad \tfrac{1}{2}m > k > 0.$$

Lebesgue[6] proved his preceding results.

Lebesgue[7] noted that tables of indices lead to integers a_j such that

$$p = f(\rho)f(\rho^{-1}), \qquad f(\rho) = a_0 + a_1\rho + \cdots + a_{m-1}\rho^{m-1}, \qquad \rho^m = 1,$$

where p is a prime $m\omega + 1$, $m > 2$. Set

$$\{f(\rho)\}^k = A_0 + A_1\rho + \cdots + A_{m-1}\rho^{m-1} = F(\rho).$$

Then $p^k = F(\rho)F(\rho^{-1})$. Hence if in the decomposition of $2p$ into a sum of m squares we change a_i into A_i, we get a decomposition of $2p^k$.

J. Liouville[8] stated that the number of representations of the double of an odd number m as a sum of 12 squares is $264\Sigma d^5$, where d ranges over the divisors of m. The number of proper representations is $264Z_5(m)$, where

$$Z_n(m) = \{a^{n\alpha} + a^{n(\alpha-1)}\} \cdots \{c^{n\gamma} + c^{n(\gamma-1)}\}, \qquad m = a^\alpha b^\beta \cdots c^\gamma,$$

$a, \cdots, c$ being distinct primes. If D^2 ranges over the square divisors of m,

$$\sum_D Z_n(m/D^2) = \sum_d d^n.$$

Liouville[9] stated that the number of representations of $2^\alpha m$ ($\alpha > 0$) as a sum of 12 squares is

$$\frac{24}{31}(21 + 2^{5\alpha+1} \cdot 5)\Sigma d^5,$$

summed for the divisors d of m. Proof by Humbert.[48]

Liouville[10] denoted by $N(n, p, q)$ the number of decompositions of n into p squares of which the roots of the first q are taken odd and positive, while the last $p - q$ are even and the roots are taken positive or negative or zero; by $N(n, p)$ the number of representations of n as a sum of p squares. It is stated that

$$(1) \quad N(2m, 12) = 264\{N(2m, 12, 2) + 224N(2m, 12, 6)$$
$$+ 256N(2m, 12, 10)\} \quad (m \text{ odd}).$$

Let m be odd, d any divisor of m, $\delta = m/d$, and set

$$\zeta_\mu(m) = \Sigma d^\mu, \qquad \rho_\mu(m) = \Sigma(-1)^{(\delta-1)/2} d^\mu.$$

---

[6] Jour. de Math., 19, 1854, 298; (2), 2, 1857, 152.

[7] *Ibid.*, 19, 1854, 334–6; Comptes Rendus Paris, 39, 1854, 1069–71.

[8] Jour. de Math., (2), 5, 1860, 143–6.

[9] *Ibid.*, (2), 9, 1864, 296–8.

[10] *Ibid.*, (2), 6, 1861, 233–8. Proof by Bell.[58b]

The following formula is stated:

$$\zeta_{2\nu-1}(m) = \sum_{s=0}^{\nu-1} A_s N(2m, 4\nu, 4s + 2),$$

$$A_0 = 1, \qquad A_{\nu-1} = 16^{\nu-1}, \qquad A_{\nu-s-1} = 16^{\nu-2s-1}A_s.$$

The cases $\nu = 1$, $\nu = 2$ correspond to theorems proved by Jacobi.[1] For $\nu = 3$, (1) gives $N(2m, 12) = 264\zeta_5(m)$. It is stated that

$$N(m, 12) = 8\zeta_5'(m) - 16m^2\zeta_1(m) + 16\Sigma s^4 = 24\zeta_5(m) - 2^{12}N(4m, 12, 12),$$

where Σs^4 is the sum of the squares of the first terms in the various representations of m as a sum of 4 squares $s^2 + s_1^2 + s_2^2 + s_3^2$.

It is stated that

$$\rho_{2\nu}(m) = \sum_{s=0}^{\nu} B_s N(2m, 4\nu + 2, 4s + 2), \qquad B_0 = 1, \quad B_\nu = 0 \quad (\nu > 0),$$

B_s being independent of m, but dependent on ν;

$$\rho_0(m) = N(2m, 2, 2), \qquad \rho_4(m) = N(2m, 10, 2) + 64N(2m, 10, 6).$$

From the latter, $N(2m, 10) = 12 \cdot 17\rho_4(m)$, when $m \equiv 3 \pmod 4$. For such an m, Eisenstein[2] had given $N(m, 10) = 12\rho_4(m)$.

Liouville[11] noted the existence of numbers $a_0 = 1, a_1, \cdots, a_{\nu-1} = 16^{\nu-1}$, $b_0 = 1, b_1, \cdots, b_{\nu-1}$, independent of m and α, but depending on ν, such that, for every odd integer m and every integer $\alpha \geqq 0$,

$$2^{(2\nu+1)\alpha}\zeta_{2\nu+1}(m) = \sum_{s=0}^{\nu-1} a_s N(2^{\alpha+2}m, 4\nu + 4, 4s + 4),$$

$$2^{2\alpha\nu}\rho_{2\nu}(m) = \sum_{s=0}^{\nu-1} b_s N(2^{\alpha+2}m, 4\nu + 2, 4s + 4).$$

These results and those in his[10] preceding paper hold also if N be replaced by M, where $M(n, p, q)$ is the number of solutions of

$$n = i_1^2 + \cdots + i_q^2 + \omega_1^2 + \cdots + \omega_{p-q}^2$$

(i's odd and positive, ω's even) for which $i_1, \cdots, \omega_{p-q}$ have no common factor, and if ζ_μ, ρ_μ be replaced by

$$Z_\mu(m) = \Pi\{P^{r\mu} + P^{(r-1)\mu}\}, \qquad R_\mu(m) = \Pi\{P^{r\mu} + (-1)^{(P-1)/2}P^{(r-1)\mu}\},$$

$$Z_\mu(1) = R_\mu(1) = 1, \qquad m = \Pi P^r,$$

where P ranges over the distinct prime factors of m.

Liouville[11a] noted that, if m is odd, the number of representations of $2^{\alpha+2}m$ by $Q = x^2 + 4(y^2 + z^2 + t^2 + u^2 + v^2)$ evidently equals the number $4\{4^{\alpha+1} - (-1)^{(m-1)/2}\}\rho_2(m)$ of representations of $2^\alpha m$ as a sum of six squares (Jacobi[1]). The number of representations of $n \equiv 1 \pmod 4$ by Q is $\rho_2(n) + 2\Sigma i^2 - n\rho_0(n)$, summed for the odd integers i for which $n = i^2 + 4s^2$. Corresponding results are found for forms like Q in which however only 4, 3, 2, or 1 of the coefficients are 4, and for $x^2 + 4(y^2 + z^2 + t^2 + u^2) + 16v^2$.

[11] Jour. de Math., (2), 6, 1861, 369–376.
[11a] Ibid., (2), 10, 1865, 65–70, 71–2, 77–80, 151–4, 161–8, 203–8.

Liouville[12] stated that the number of representations of $n = 2^a m$ (m odd) as a sum of 10 squares is

$$\tfrac{4}{5}\{16^{a+1} + (-1)^{(m-1)/2}\}\lambda + \tfrac{8}{5}n^2\mu - \tfrac{64}{5}\nu,$$

where λ is the positive value of $\Sigma(d_1^4 - d_3^4)$, where d_1 ranges over the divisors $4l + 1$ of n and d_3 over the divisors $4l + 3$ (λ being the same for m as for n), while μ is the number of integral solutions, positive, negative or zero, of $n = s^2 + s'^2$, and ν is the sum of the products $s^2 s'^2$ for all the solutions.

When $m \equiv 3 \pmod 4$, $\mu = \nu = 0$ and the formula becomes that of Eisenstein[2] if $\alpha = 0$, and that of Liouville[10] for $\alpha = 1$. In the notation of that paper, $\lambda = \rho_4(m)$. Thus

$$N(2^a m, 10) = \tfrac{4}{5}\{16^{a+1} + (-1)^{(m-1)/2}\}\rho_4(m) + \tfrac{16}{5}\sum_{s,\,s'}(s^4 - 3s^2 s'^2) \qquad (s^2 + s'^2 = n).$$

The last sum is multiplied by -4 when α is replaced by $\alpha + 1$. Hence

$$N(2^{a+1}m, 10) + 4N(2^a m, 10) = \{16^{a+2} + 4(-1)^{(m-1)/2}\}\rho_4(m).$$

The values of $N_4 = N(2^{a+2}m, 10, 4)$ and $N_8 = N(2^{a+2}m, 10, 8)$ follow from

$$2^{4a}\rho_4(m) = N_4 + 4N_8, \qquad 4(-1)^{(m-1)/2}\rho_4(m) = 5N(2^a m, 10) - 96N_4 + 256N_8,$$
$$N_4 - 16N_8 = \tfrac{1}{2}\Sigma(s^4 - 3s^2 s'^2) \qquad (s^2 + s'^2 = 2^a m).$$

H. J. S. Smith[13] stated that the principles indicated in his paper enable one to deduce by a uniform method the theorems of Jacobi, Eisenstein and the numerous recent ones by Liouville on the representation of numbers by a sum of four squares and other simple quadratic forms; also the theorems of Jacobi[1] on six and eight squares. In view of Eisenstein's remark that there is a single class of quadratic forms of discriminant unity in $n \leqq 8$ variables, but always more than one class if $n > 8$, the series of theorems relating to representation by sums of n squares ceases when $n > 8$. There remain the cases $n = 5, 7$. Smith gave a description of the general theory on which are based the formulas for the numbers N_5 and N_7 of primitive representations of $4^a \omega^2 \delta$ as a sum of 5 and 7 squares, respectively, where ω is odd and δ has no square factor:

$$N_5 = 5 \cdot 2^{3a} \omega^3 \frac{\eta}{\delta} F_5 \prod \left[1 - \left(\frac{\delta}{q}\right)\frac{1}{q^2}\right],$$

where, as in N_7, the product extends over every prime dividing ω but not δ, while F_5 is defined as follows: For $\delta \equiv 1 \pmod 4$,

$$F_5 = \sum_{s=1}^{\delta}\left(\frac{s}{\delta}\right)s(s - \delta),$$

and $\eta = 12$ if $\delta \equiv 1 \pmod 8$; $\eta = 28$ or 20, if $\delta \equiv 5 \pmod 8$, according as $\alpha = 0$ or $\alpha > 0$; while,* if $\delta = 1$, $\eta\Pi$ is to be replaced by 2. But, if $\delta \not\equiv 1 \pmod 4$,

$$F_5 = \sum_{s=1}^{4\delta}\left(\frac{\delta}{s}\right)s(s - 4\delta),$$

[12] Comptes Rendus Paris, 60, 1865, 1257; Jour. de Math., (2), 11, 1866, 1–8.
[13] Proc. Roy. Soc. London, 16, 1867, 207; Coll. Math. Papers, 1, 1894, 521.
* The $\eta\Sigma$ here used was replaced by $\eta\Pi$ in his[31] later paper giving proofs.

where $\eta = 1$ or $\tfrac{1}{2}$, according as $\alpha = 0$ or $\alpha > 0$. Next,

$$N_7 = 7 \cdot 2^{5\alpha} \omega^5 \frac{\eta}{\delta} F_7 \prod \left[1 - \left(\frac{-\delta}{q} \right) \frac{1}{q^3} \right].$$

For $\delta \equiv 3 \pmod 4$,

$$F_7 = \sum_{s=1}^{\delta} \left(\frac{s}{\delta} \right) s(s - \delta)(2s - \delta),$$

where $\eta = 30$ if $\alpha = 0$, $\Delta \equiv 3 \pmod 8$; $\eta = 74/3$ if $\alpha = 0$, $\Delta \equiv 7 \pmod 8$; $\eta = 140/3$ if $\alpha > 0$. For $\delta \not\equiv 3 \pmod 4$,

$$F_7 = \sum_{s=1}^{4\delta} \left(\frac{-\delta}{s} \right) s(s - 2\delta)(s - 4\delta),$$

where $\eta = 1/3$ or $5/12$ according as $\alpha = 0$, $\alpha > 0$.

J. Liouville[14] stated that, if m is of the form $8k + 7$,

$$\sum_i (m - 7i^2)\rho_2 \left(\frac{m - i^2}{2} \right) = 0, \qquad \rho_2(n) = \Sigma(- 1)^{(d-1)/2} \left(\frac{n}{d} \right)^2,$$

where i ranges over the positive odd integers $< \sqrt{m}$, and d ranges over the divisors of the odd number n.

E. Catalan[15] obtained by means of elliptic functions the result that the number of solutions of $i_1^2 + \cdots + i_8^2 = 8n$ in odd integers $i_1, \cdots, i_8$ equals the sum of the cubes of the divisors of n.

J. W. L. Glaisher[16] stated that, if R_m is the number of representations of N as a sum of m squares (attention being paid to the signs of the roots of the squares), and if P is the sum of the reciprocals of the odd divisors of N, then

$$R_1 - \tfrac{1}{2}R_2 + \tfrac{1}{3}R_3 - \cdots \pm \tfrac{1}{N}R_N = (- 1)^{N-1}2P.$$

C. Sardi[17] stated that the numbers of the form $40m + 63$ are decomposable into seven squares which end with the digit 9. Cf. Santomauro.[19]

G. Torelli[18] noted that the preceding result follows from Fermat's theorem that every number is a sum of m m-gonal numbers, in the equivalent formulation by Barlow[90] of Ch. I, which implies also that $200m + 14283$ is a sum of 27 squares ending in 29, of which 23 equal 529 or 729.

E. Santomauro[19] proved that every integer $40m + 9k$ is a sum of k squares which end with the digit 9 [if $k > 1$, as it fails for $m = 2$, $k = 1$]. Cf. Sardi.[17]

E. Lemoine[20] called $N = a_1^2 + \cdots + a_n^2$ a decomposition of N into maximum squares and n the index of N if a_1^2 is the largest square $\leqq N$,

[14] Jour. de Math., (2), 14, 1869, 302–4.
[15] Recherches sur quelques produits indéfinis, Mém. Ac. Roy. Belgique, 40, 1873, 61–191. Résumé in Nouv. Ann. Math., (2), 13, 1874, 518–23. Cf. Berdellé.[23]
[16] Mess. Math., 5, 1876, 91.
[17] Giornale di Mat., 7, 1869, 115.
[18] Ibid., 16, 1878, 167.
[19] Un teorema d'analisi, 1879, 8 pp.
[20] Comptes Rendus Paris, 95, 1882, 719–22.

a_2^2 the largest square $\leqq N - a_1^2$, etc.   Let y_n be the least number of index n. For n even, y_n ends with 67; for n odd, with 23.   Also,

$$y_{p+1} = \left(\frac{y_p + 3}{2}\right)^2 - 2 \quad (p \geqq 3).$$

M. d'Ocagne[21] stated the empirical generalization that, if $m \geqq 3$, the last $[(m - 1)/2]$ digits of y_m are the same and in the same order as those of y_{m+2}.   Lemoine added the remark that the only possible final squares are R^2, $R^2 + 1$, $R^2 + 1 + 1$, $R^2 + 1 + 1 + 1$, $R^2 + 2^2$, $R^2 + 2^2 + 2^2$, $R^2 + 2^2 + 1$, $R^2 + 2^2 + 1 + 1$, $R^2 + 2^2 + 1 + 1 + 1$, where $R > 2$.

T. J. Stieltjes[22] noted that, in view of Jacobi[25] of Ch. VIII, the number of decompositions of $N \equiv 5 \pmod 8$ as a sum of 5 positive odd squares is

$$\sum_j \sigma\left\{\frac{N - (2j - 1)^2}{4}\right\} = f(N) + 2f(N - 8\cdot1^2) + 2f(N - 8\cdot2^2) + \cdots,$$

where $\sigma(n)$ is the sum of the divisors of n, and $4f(m) = -\Sigma(-1)^{(d^2-1)/8}d$, summed for the divisors d of m.

C. Hermite[23] proved by use of elliptic functions that the number of decompositions of $N \equiv 5 \pmod 8$ as a sum of 5 positive odd squares is

$$\tfrac{1}{2}\chi(N) + \chi(N - 2^2) + \chi(N - 4^2) + \chi(N - 6^2) + \cdots,$$
$$\chi(n) \equiv \Sigma(3d + d')/4,$$

summed for all factorizations $n = dd'$, $d' > 3d$.

Stieltjes[24] noted that the total number $F(n)$ of solutions of

$$n = x_1^2 + \cdots + x_5^2$$

is $24A(n) + 16B(n)$ for n even, and $8A(n) + 48B(n)$ for n odd, where

$$A(n) = X(n) + 2X(n - 4) + 2X(n - 16) + 2X(n - 36) + \cdots,$$
$$B(n) = X(n - 1) + X(n - 9) + X(n - 25) + \cdots,$$

$X(n)$ being the sum of the odd divisors of n.   He expressed $A(n)$ in terms of $B(n)$, and $B(4n)$ in terms of $B(n)$, and therefore $F(4n)$ in terms of $F(n)$. He verified for each odd prime $p < 100$ that $F(p^2) = 10(p^3 - p + 1)$, and for $p = 3, 5, 7$ that

$$F(p^4) = 10\{p(p^2 - 1)(p^3 + 1) + 1\}.$$

T. Pepin[25] expressed the number $N(m, 5)$ of representations of m as a sum of 5 squares in terms of $N(m, 4)$ in the evident way of considering $m - x^2$ as a ⊡.   By use of elliptic functions he evaluated $N_1 - N_2$, where N_1 (or N_2) is the number of representations of m as a sum of 5 squares of which the first is even (or odd); also $P - Q$, where P (or Q) is the number of representations of m as a sum of 5 squares of which the first two have an even (or odd) sum; he also proved that

$$N'(m) - N''(m) = 2(-1)^m(\Sigma b - \Sigma a),$$

[21] L'intermédiaire des math., 1, 1894, 232.
[22] Comptes Rendus Paris, 97, 1883, 981.
[23] *Ibid.*, 982.
[24] *Ibid.*, 1545.
[25] Atti Accad. Pont. Nuovi Lincei, 37, 1883–4, 9–48.

where a ranges over the divisors $8l \pm 1$ of m, and b over the divisors $8l \pm 3$, while N' (or N'') is the number of solutions of $m = x^2 + y^2 + z^2 + 2t^2$ with x^2 even (or odd). For $m = 8l \pm 1$, $N' = 2N''$. He noted the recursion formulas

$$mN(m, 5) = 2 \sum_{n=1}^{\sqrt{m}} (6n^2 - m)N(m - n^2, 5) = 10\Sigma n^2 N(m - n^2, 4).$$

He proved (p. 43) for any odd prime p the statements by Stieltjes[24] concerning $F(p^2)$, $F(p^4)$.

E. Cesàro[26] stated that the number of ways of decomposing n into a sum of p squares is in mean $Cn^{p/2-1}$, where

$$C = \frac{1}{2(p - 2)(p - 4)(p - 6) \cdots} \cdot \left(\frac{\pi}{2}\right)^{[p/2]}.$$

For $p = 3$, $C = \pi/4$. For $p = 4$, $C = \pi^2/16$.

A. Hurwitz[27] proved and generalized the conjectured results by Stieltjes[24] concerning $F(p^2)$ and $F(p^4)$. If $m = 2^k p^\alpha q^\beta \cdots$, where $2, p, q, \cdots$ are distinct primes, the number of decompositions of m^2 into 5 squares is

$$F(m^2) = K \left[p, \alpha\right]\left[q, \beta\right] \cdots, \qquad K = 10 \cdot \frac{2^{3k+3} - 1}{2^3 - 1},$$

$$\left[p, \alpha\right] \equiv \frac{p^{3a+3} - p^{3a+1} + p - 1}{p^3 - 1}.$$

For proof, set $m = 2^k n$. Then by Stieltjes' formula, $F(m^2)$ is K times the sum, for all positive odd integral solutions a, b of $a + b = 2n$,

$$\Sigma X(a, b) \equiv X(n^2) + 2X(n^2 - 2^2) + 2X(n^2 - 4^2) + \cdots.$$

But if γ, δ, ϵ, $\cdots$ are the odd primes dividing both a and b,

$$X(a, b) = X(a)X(b) - \Sigma\gamma X\left(\frac{a}{\gamma}\right) X\left(\frac{b}{\gamma}\right) + \Sigma\gamma\delta X\left(\frac{a}{\gamma\delta}\right) X\left(\frac{b}{\gamma\delta}\right) - \cdots,$$

$$\Sigma X(a, b) = \sum_{a_1, b_1} X(a_1)X(b_1) - \sum p \sum_{a_p, b_p} X(a_p)X(b_p)$$
$$+ \sum pq \sum_{a_{pq}, b_{pq}} X(a_{pq})X(b_{pq}) - \cdots,$$

where the summation with respect to a_t, b_t extends over all positive odd integers a_t, b_t whose sum is $2n/t$. By the known formula

$$X(1)X(2n - 1) + X(3)X(2n - 3) + X(5)X(2n - 5) + \cdots$$
$$+ X(2n - 1)X(1) = \zeta_3(n),$$

viz., the sum of the cubes of the divisors of the odd number n, we get

$$\Sigma X(a, b) = \left[\zeta_3(p^\alpha) - p\zeta_3(p^{\alpha-1})\right]\left[\zeta_3(q^\beta) - q\zeta_3(q^{\beta-1})\right] \cdots,$$

and hence equals $\left[p, \alpha\right]\left[q, \beta\right] \cdots$ in the desired formula for $F(m^2)$. Part of Stieltjes' formulas follow from those of Liouville[5] of Ch. XI.

[26] Mém. Soc. Roy. Sc. de Liège, (2), 10, 1883, No. 6, pp. 199–200.
[27] Comptes Rendus Paris, 98, 1884, 504–7.

T. J. Stieltjes[27a] wrote $F_7(n)$ for the number of decompositions of n into 7 squares and stated that $F_7(4^k m)/F_7(m)$ equals

$$f(k) = \frac{40 \cdot 32^k - 9}{31}, \quad m \equiv 1, 2 \pmod 4; \qquad \frac{32^{k+1} - 1}{31}, \quad m \equiv 3 \pmod 8;$$

$$\frac{28f(k) + 9}{37}, \quad m \equiv 7 \pmod 8.$$

H. Minkowski[28] proved that the numbers of the form $8n + 5$ are sums of 5 odd squares. The number of proper representations of d as a sum of 5 squares, not all odd, is

$$\frac{40}{\pi^2} \{3 - (-1)^{[d/2]}\} \sqrt{d^3} \, \Sigma \left(\frac{d}{m}\right) \frac{1}{m^2},$$

summed for the integers m prime to $2d$. A number $d \equiv 5 \pmod 8$ has

$$\frac{32}{\pi^2} \sqrt{d^3} \, \Sigma \left(\frac{d}{m}\right) \frac{1}{m^3}$$

proper representations as a sum of 5 odd squares.

P. S. Nasimoff[29] proved that the number of decompositions of $n = 2^a m$ (m odd) as a sum of 8 squares is $\frac{16}{7}(8^{a+1} - 15) \zeta_3(m)$, where $\zeta_3(m)$ is the sum of the cubes of the divisors of m. He determined the number of decompositions of any integer into 12 squares.

E. Cesàro[30] noted that the number of decompositions of n into ν squares is $N_1 - N_2 - N_3 + N_4 - N_5 + N_6 + \cdots$, where N_p is the number of positive integral solutions of the system of equations

$$x_1 x_2 \cdots x_\nu = p, \qquad x_1 y_1 + \cdots + x_\nu y_\nu = n.$$

The numbers of decompositions of n into two and four squares increased by double the number into three squares is $M_1 - M_3 + M_5 - M_7 + \cdots$, where M_p is the number of positive integral solutions of $xy = p$, $x\xi + y\eta = n$.

H. J. S. Smith[31] proved the formula for the number of representations as a sum of five squares which had been stated by him in 1867, and deduced therefrom the formulas of Eisenstein.[3] The subject proposed by the Paris Academy of Sciences for the Grand Prix des Sciences Math. for 1882 was the theory of the representation of integers as a sum of 5 squares (with citation of results of Eisenstein). Apparently no member of the commission which proposed the subject of the prize knew of the earlier paper by Smith; nor was the latter mentioned in the report[32] of the commission which recommended that prizes of the full amount be awarded to Smith and to

[27a] Comptes Rendus Paris, 98, 1884, 663–4.

[28] Mém. présentés à l'Acad. Sc. Inst. France, (2), 29, 1884, No. 2. Gesamm. Abh., I, 1911, 118–9, 133–4.

[29] Application of Elliptic Functions to Number Theory, Moscow, 1885. French résumé in Annales sc. de l'école norm. supér., (3), 5, 1888, 36–7.

[30] Giornale di Mat., 23, 1885, 175.

[31] Mém. Savans Etr. Paris Ac. Sc., (2), 29, 1887, No. 1; Coll. Math. Papers, 2, 1894, 623–680; cf. p. 677.

[32] Smith's Coll. Math. Papers, 1, 1894, lxvii–lxxii.

Minkowski,[28] then a student of 18 years of age at the University of Königsberg.

Ch. Berdellé[33] proved that any multiple of 8 is a sum of 8 odd squares. From

$$n = a^2 + b^2 + c^2 + d^2, \qquad 8a^2 = 4a^2 - 4a + 4a^2 + 4a,$$

$8 + 8n$ is the sum of the squares of

$$2a + 1, \quad 2a - 1, \quad 2b + 1, \quad 2b - 1, \quad 2c + 1, \quad 2c - 1, \quad 2d + 1, \quad 2d - 1.$$

If k of the integers a, b, c, d are zero, $2k$ of the 8 squares are unity.

J. W. L. Glaisher[34] noted that, if $\sigma(n)$ is the sum of the divisors of n, the number of representations of n as a sum of five squares is

$$10\{\sigma(n) + 2\sigma(n - 4) + 2\sigma(n - 16) + \cdots\} \quad \text{if } n \equiv 1 \pmod 8,$$

but twice that expression if $n \equiv 3 \pmod 4$.

L. Gegenbauer[35] proved that the number of representations of an odd number n as a sum of eight squares equals $16M$, where M is the number of divisors of the various g.c.d.'s of n with all triples chosen from $1, \cdots, n$. Also M is the sum of the products of the number of divisors of every factor of n by the number of those triples whose elements do not exceed the complementary divisor and form a system relatively prime to it. There are three further theorems on sums of 8 squares, five on sums of 12 squares and two on sums of 6 and 10 squares each. The number of all [or proper] representations of an odd number n as a sum of three squares and double a square is $2\{4 - (2/n)\}\mu$, where the symbol is Jacobi's and μ is the number of all [or proper] representations $x^2 - 2y^2$, $y \geqq 0$, $2x > 3y$, of the various g.c.d.'s of n and the numbers $\leqq n$. There is a similar theorem on a sum of five squares and double a square.

G. B. Mathews[36] noted that the number of sets of solutions of

$$x_1^2 + \cdots + x_k^2 = n$$

is the coefficient c_n of q^n in the expansion of

$$\theta^k = (1 + 2q + 2q^4 + 2q^9 + \cdots)^k, \qquad \theta \equiv \frac{1+q}{1-q} \cdot \frac{1-q^2}{1+q^2} \cdot \frac{1+q^3}{1-q^3} \cdots .$$

By logarithmic differentiation,

$$\frac{1}{\theta}\frac{d\theta}{dq} = -\sum_{n=1}^{\infty} \psi(n)q^{n-1}, \qquad \psi(n) \equiv 2\sum (-1)^{n/\mu} n/\mu,$$

summed for all odd divisors μ of n. For[34] $n = 2^a m$, $\psi(n) = 2^{a+1}\sigma(m)$. By the logarithmic differentiation of $\theta^k = 1 + c_1 q + c_2 q^2 + \cdots$ and comparison

[33] Bull. Soc. Math. de France, 17, 1888–9, 102, 205. Cf. Catalan.[15]
[34] Messenger Math., 21, 1891–2, 129–130.
[35] Sitzungsber. Akad. Wiss. Wien (Math.), 103, IIa, 1894, 122–5.
[36] Proc. London Math. Soc., 27, 1895–6, 55–60.

of coefficients, we get linear equations for the c's, from which

$$c_n = \frac{(-1)^{n(n-1)/2}}{n!} \begin{vmatrix} k\psi(n) & k\psi(n-1) & \cdots & k\psi(2) & k\psi(1) \\ k\psi(n-1) & k\psi(n-2) & \cdots & k\psi(1) & n-1 \\ k\psi(n-2) & k\psi(n-3) & \cdots & n-2 & 0 \\ \cdot & \cdot & \cdots & \cdot & \cdot \\ k\psi(1) & 1 & \cdots & 0 & 0 \end{vmatrix}.$$

P. Bachmann[37] gave an exposition of the work of Smith[13, 31] and Minkowski[28] on sums of 5 squares, and Eisenstein[2-4] on sums of 5, 6, 7, 8 squares.

E. Lemoine stated and L. Ripert[38] proved that every integer equals the sum of p and certain distinct squares, where $p = 0, 1, 2$ or 4.

H. Delannoy[39] proved that every even square > 4 and every 4th power > 1 is a sum of five squares > 0, and that $a(a+2)$ is a sum of 4 or 5 squares > 0.

R. E. Moritz[40] considered the representation of numbers as quotients of sums and differences of squares.

O. Meissner[41] considered the representation of numbers of an algebraic field as a sum of n squares. In particular, the numbers of the field defined by $i\sqrt{z}$ are sums of 5 squares, 4 of which are rational.

J. W. L. Glaisher[42] employed the sums $P(m)$ and $Q(m)$ of the products of the roots (taken in the form $4n + 1$) of the first two and three squares, respectively, in each composition of $4m$ as a sum of 4 odd squares, and proved the following theorems when m is odd. The sum of the odd roots in all the representations of m as a sum of 6 squares, 3 of which are odd and 3 even, is $\pm 120P(m)$, the sign being $+$ or $-$ according as $m \equiv 7$ or 3 (mod 8). If $\alpha^2 + \cdots + \zeta^2$ is any partition of $2m$ into 6 odd squares, where $\alpha, \cdots, \zeta$ are taken in the form $4n + 1$, and if s is the sum of the 15 products of $\alpha, \cdots, \zeta$ taken two at a time, then $\Sigma s = -120\, Q(m)$, summed for all the representations of $2m$ by 6 odd squares. For the partitions of $8N$ into 8 odd squares, where N is even, the corresponding sum Σs is zero. The number of compositions of $8m$ as a sum of 8 odd squares is the sum of the cubes of the divisors of m.

K. Petr[43] proved, by use of theta functions, two hitherto unproved theorems stated by Liouville on the representation of even numbers as a sum of 12 or 10 squares.

E. Jacobsthal[44] proved that every prime $p = 4n + 1$ is a sum

$$p = \Sigma \left\{ \frac{1}{\delta}\, \phi_n(g^\rho) \right\}^2, \qquad \phi_n(a) \equiv \sum_{m=1}^{p-1} \left(\frac{m}{p}\right)\left(\frac{m^n + a}{p}\right),$$

of δ squares, where δ is the g.c.d. of n and $p - 1$, and g is a primitive root of p, while ρ ranges over a complete set of residues modulo δ.

[37] Arith. der Quad. Formen, 1898, 608–22, 652–68.
[38] Nouv. Ann. Math., (3), 17, 1898, 195–6; 19, 1900, 335–6.
[39] L'intermédiaire des math., 7, 1900, 392; 9, 1902, 237, 245.
[40] Univ. Nebraska Studies, 3, 1903, 355. Cf. Moritz[146a] of Ch. VI.
[41] Archiv Math. Phys., (3), 5, 1903, 175–6; 7, 1904, 266–8.
[42] Quar. Jour. Math., 36, 1905, 349–354.
[43] Casopis, Prag, 34, 1905, 224–9. Petr.[49]
[44] Anwendungen . . . quadratischen Reste, Diss. Berlin, 1906, 20. Cf. Jacobsthal,[156] Ch. VI.

J. W. L. Glaisher[45] evaluated the number $R^{(t)}(n)$ of representations of n as a sum of t squares for each even integer $t \leqq 18$. The simplest results are

$$R^{(6)}(n) = 4\{4E'_2(n) - E_2(n)\}, \qquad\qquad R^{(8)}(n) = (-1)^{n-1}16\zeta_3(n),$$
$$R^{(10)}(n) = \tfrac{4}{5}\{E_4(n) + 16E'_4(n) + 8\chi_4(n)\}, \qquad R^{(12)}(2n) = -8\xi_5(2n),$$

the first two of which are due to Eisenstein[2] for n odd and to H. J. S. Smith[1] for any n. Here $E_r(n)$ [or $E'_r(n)$] is the excess of the sum of the rth powers of the divisors of n which [or whose conjugates] are of the form $4k + 1$ over the sum of the rth powers of the divisors of n which [or whose conjugates] are of the form $4k + 3$; also,

$$\zeta_r(n) = \Sigma(-1)^{d-1}d^r, \qquad \xi_r(n) = \Sigma(-1)^{d+d'}d^r \qquad (dd' = n);$$

while $4\chi_4(n)$ is the sum of the fourth powers of all the complex numbers having n as norm. In an addition to this paper, Glaisher[46] evaluated by elliptic modular functions the sum of the rth powers of all primary complex numbers of norm n and (p. 274) evaluated $R^{(14)}(n)$.

W. Sierpinski[47] noted that the number of representations of n as a sum of r squares is

$$\frac{(2r)^n}{n!}\left\{ a_0(n) + \frac{1}{r}a_1(n) + \frac{1}{r^2}a_2(n) + \cdots \right\},$$

where $a_i(n)$ is a polynomial of degree $2i$ with rational coefficients.

G. Humbert[48] derived the formula, in which $\eta_1 = H_1(0)$, $\theta_1 = \Theta_1(0)$, in Jacobi's notations for elliptic functions of the variable q,

$$(2) \qquad\qquad 4\eta_1^6\theta_1^4 + \eta_1^2\theta_1^8 = 4\sum_{m=0}^{\infty}(2m + 1)^4 q^{m+1/2}/(1 + q^{2m+1}).$$

Let $G_{p,q}(a)$ be the number of decompositions of a into $p + q$ squares of which the first p are odd and the last q are even. By equating the coefficients of $q^{N+1/2}$ in the two members of (2) and in the formula obtained by changing q to $-q$, we get

$$4G_{6,4}(4N + 2) + G_{2,8}(4N + 2) = 4(-1)^N\Sigma(-1)^m(2m + 1)^4,$$
$$5G_{10,0}(4N + 2) - 6G_{6,4}(4N + 2) + G_{2,8}(4N + 2) = 4\Sigma(-1)^m(2m + 1)^4,$$

the summations extending over the odd divisors $2m + 1$ of $4N + 2$. If N is odd, $N = 2M + 1$, $G_{10,0}(4N + 2)$ is evidently zero. The preceding equations give

$$G_{6,4}(8M + 6) = G_{2,8}(8M + 6) = \tfrac{4}{5}\Sigma(-1)^{m+1}(2m + 1)^4.$$

The total number of decompositions of $8M + 6$ into ten squares is evidently

$$\frac{10 \cdot 9 \cdot 8 \cdot 7}{1 \cdot 2 \cdot 3 \cdot 4}G_{6,4} + \frac{10 \cdot 9}{1 \cdot 2}G_{2,8},$$

[45] Quar. Jour. Math., 38, 1907, 1–62, 178–236, 289–351; summary in Proc. London Math. Soc., (2), 5, 1907, 479–490.
[46] Quar. Jour. Math., 39, 1908, 266–300.
[47] Wiadomosci Mat., Warsaw, 11, 1907, 225–231.
[48] Comptes Rendus Paris, 144, 1907, 874–8.

and this number equals $204\Sigma(d_3^4 - d_1^4)$, where d_3 ranges over the divisors $4h + 3$ of $8M + 6$, and d_1 over the divisors $4h + 1$.

In (2) replace $\eta_1^2(q)$ by $2\eta_1(q^2)\theta_1(q^2)$ and $\theta_1^2(q)$ by $\theta_1^2(q^2) + \eta_1^2(q^2)$. Then change q^2 into q. We get

$$\eta_1^9\theta_1 + 38\eta_1^5\theta_1^5 + \eta_1\theta_1^9 + 20\eta_1^7\theta_1^3 + 20\eta_1^3\theta_1^7 = 2\Sigma(2m + 1)^4 q^{(2m+1)/4}/(1 + q^{(2m+1)/2}).$$

Equating the coefficients of $q^{N+3/4}$ and those of $q^{N+1/4}$, we get

$$10G_{7,\,3}(4N + 3) + 10G_{3,\,7}(4N + 3) = \Sigma(- 1)^{m+1}(2m + 1)^4,$$
$$G_{1,\,9}(4N + 1) + G_{9,\,1}(4N + 1) + 38G_{5,\,5}(4N + 1) = 2\Sigma(- 1)^m(2m + 1)^4,$$

where $2m + 1$ ranges over the odd divisors of $4N + 3$ and $4N + 1$, respectively. The first formula gives for the total number $120(G_{7,\,3} + G_{3,\,7})$ of decompositions of $4N + 3$ into ten squares the value $12\Sigma(d_3^4 - d_1^4)$, due to Eisenstein.[2]

For 12 squares, it is shown that

$$\eta_1^{10}\theta_1^2 + 14\eta_1^6\theta_1^6 + \eta_1^2\theta_1^{10} = 4\sum_{m=0}^{\infty} (2m + 1)^5 q^{m+1/2}/(1 - q^{2m+1}).$$

Thus the total number $66(G_{10,\,2} + 14G_{6,\,6} + G_{2,\,10})$ of decompositions of $4N + 2$ as a sum of 12 squares equals $264\Sigma d^5$, d ranging over the divisors of $4N + 2$. Changing q into q^2, we find that

$$G_{8,\,4}(8M) = G_{4,\,8}(8M), \qquad G_{8,\,4}(8M + 4) + G_{4,\,8}(8M + 4) = 16\Sigma(2m + 1)^5,$$

summed for the divisors $2m + 1$ of $8M + 4$. Next,

$$\eta_1^8\theta_1^4 + \eta_1^4\theta_1^8 = 16\Sigma m^5 q^m/(1 - q^{2m})$$

gives $G_{8,\,4}(8M) + G_{4,\,8}(8M) = 16\Sigma m^5$, m being such that $2M/m$ is odd. By these and a more complex relation, one may obtain the total number

$$G_{12,\,0} + G_{0,\,12} + 495(G_{8,\,4} + G_{4,\,8})$$

of decompositions of $4N$ into 12 squares, and thus prove Liouville's[9] theorem.

K. Petr[49] proved Liouville's[12] formula for the number of representations of $2^a m$ as a sum of ten squares by use of the theta functions with the characteristics $(1, 1)$, $(1, 0)$, $(0, 1)$, $(0, 0)$ and formulas in Jacobi's Fundamenta Nova (p. 101). Also, Liouville's[9] result on 12 squares by use of the fourth derivatives of $\wp(u)$.

E. Dubouis[50] wrote S_n for a sum of n squares each > 0. For $k > 45$, the odd number $k - 1$ or $k - 4$ is a S_4, whence k is a S_5. The only numbers not S_5's are stated to be $A = 0, 1, 2, 3, 4, 6, 7, 9, 10, 12, 15, 18, 33$. Every number $\neq A + 1$ is a S_6. The numbers not S_6's are stated to be $B = 1, 2, 3, 4, 5, 7, 8, 10, 11, 13, 16, 19$. The only numbers not a S_{6+n} are the $B + n$ and the first n integers.

* J. V. Uspenskij[51] discussed the representation of numbers as sums of squares.

[49] Archiv Math. Phys., (3), 11, 1907, 83–5. Petr.[48]
[50] L'intermédiaire des math., 18, 1911, 55–56.
[51] Math. Soc. Kharkov, (2), 14, 1913, 31–64.

B. Boulyguine[52] employed the notations

$$\phi_k(x, y) = \tfrac{1}{2}\{(x + yi)^{4k} + (x - yi)^{4k}\} = x^{4k} - \binom{4k}{2}x^{4k-2}y^2 + \cdots$$

$$\sum_p^k (n) = \Sigma\phi_k(x_1, x_2),$$

summed for all the $N_p(n)$ integral solutions (positive, negative, or zero) of $x_1^2 + \cdots + x_p^2 = n$. Write $\sigma_k(m)$ for the sum of the kth powers of the divisors of m and

$$\rho_k(m) = \Sigma(-1)^{(m/d-1)/2}d^k$$

for the difference between the sum of the kth powers of the divisors $4h + 1$ of m and the sum of the kth powers of the divisors $4h + 3$. By use of elliptic functions, it is shown that, if $n = 2^a m$, where m is odd,

$$N_{8r+2}(n) = a_r\{2^{4r+4ra} + (-1)^{(m-1)/2}\}\rho_{4r}(m)$$
$$+ a_r^{(1)}\sum_{8r-6}^1 (n) + a_r^{(2)}\sum_{8r-14}^2 (n) + \cdots + a_r^{(r)}\sum_2^r (n).$$

There is given a similar expression for $N_{8r+6}(n)$. Also,

$$N_{8r+8}(n) = d_r(-1)^n \frac{2^{4r+3(1+a)} - 2^{4r+4} + 1}{2^{4r+3} - 1}\zeta_{4r+3}(m)$$
$$+ d_r^{(1)}\sum_{8r}^1 (n) + d_r^{(2)}\sum_{8r-8}^2 (n) + \cdots + d_r^{(r)}\sum_8^r (n),$$

with a similar expression for $N_{8r+4}(n)$. Here the a's and d's are rational numbers not depending on n. It is stated that there result the known formulas for the number of decompositions into 2, 4, 6, 8, 10, or 12 squares and apparently new formulas for 14 or 16 squares.

Boulyguine[53] stated a recursion formula for his[52] $\Sigma(n)$:

$$A_r N_r(n) = F_r(n) + A_{r1}\sum_{r-8}^1 (n) + A_{r2}\sum_{r-16}^2 (n) + A_{r3}\sum_{r-24}^3 (n) + \cdots,$$

for $r = 2, 3, \cdots$, where $A_r, A_{r1}, \cdots$ are independent of n, while $F_r(n)$ is a specified function differing in the three cases r odd, $r = 4k + 2$, $r = 4k + 4$.

S. Ramanujan[54] studied the function $\psi(n)$ for which

$$\sum_{n=0}^\infty \psi(n)x^n = \prod_{i=1}^r f^{a_i}(x^{c_i}), \qquad f(x) \equiv x^{1/24}(1 - x)(1 - x^2)(1 - x^3) \cdots.$$

Special cases of ψ are the functions $\chi(n)$, $P(n)$, $\chi_4(n)$, $\Theta(n)$, $\Omega(n)$ of Glaisher[99] of Ch. VIII. He touched (pp. 179, 183–4) on the number of representations of n as a sum of s squares, $s = 10, 16$, etc.

L. J. Mordell[55] proved that various empirical results of Ramanujan[54] follow from expansions of elliptic modular functions.

[52] Comptes Rendus Paris, 158, 1914, 328–330.
[53] *Ibid.*, 161, 1915, 28–30.
[54] Trans. Cambr. Phil. Soc., 22, 1916, 173–9.
[55] Proc. Cambr. Phil. Soc., 19, 1917, 117–124.

R. Goormaghtigh[56] proved that every power of an even [odd] integer with an exponent $\geqq 3$ is a sum of 5 [6] squares > 0. If n is odd and > 1 and if $a > 0$, n^{4a+1} is a sum of 5 squares > 0.

Mordell[57] employed the theory of modular functions to find the number of representations as a sum of $2r$ squares.

G. H. Hardy[58] deduced from the theory of elliptic functions the number of representations as a sum of 5 or 7 squares. This investigation, continued by S. Ramanujan,[58a] led to a complete solution of the problem of the representation of a number as a sum of n squares for $n < 8$, and to asymptotic formulas for any n. The method used is an application of the general theory cited in Ch. III.[221]

E. T. Bell[58b] proved Liouville's[10, 11] formulas by use of series for elliptic functions and stated that they are only the first cases of an infinitude of similar results which may be found by using higher powers than the first and second, or products, of the series.

On 10 odd squares, see Pollock[117] of Ch. I. On 8 squares, see Sierpinski[158] of Ch. VI. For 5 squares, see Hermite[69] and Humbert[108] of Ch. VII. In Ch. XI are noted Liouville's results on sums of n squares for $n = 8$ and 12 and in papers 6 and 7 minor results for $n = 5$ and 7.

RELATIONS BETWEEN SQUARES.

The Japanese Aida Ammei[59] proved between 1807 and 1817 that

$$x_1 = -a_1^2 + a_2^2 + a_3^2 + \cdots + a_n^2, \qquad x_r = 2a_1a_r \ (r = 2, \cdots, n),$$
$$y = a_1^2 + \cdots + a_n^2$$

satisfy $x_1^2 + \cdots + x_n^2 = y^2$. This result was known to Euler[191, 294, 308] of Ch. XXII. Ajima Chokuyen,[59a] in a manuscript dated 1791, had solved $x_1^2 + \cdots + x_5^2 = y^2$ in integers.

It was proved by J. R. Young,[60] who proved also the identity

$$(x_1^2 + \cdots + x_n^2)(y_1^2 + \cdots + y_n^2) = (x_1y_1 + \cdots + x_ny_n)^2$$
$$+ \Sigma(x_iy_j - x_jy_i)^2 \quad (i, j = 1, \cdots, n; \ i < j).$$

The latter was proved otherwise by A. Cauchy.[61]

Aida's result has been published also by D. S. Hart and A. Martin,[62] E. Catalan,[63] A. Martin,[64] and G. Bisconcini[65] (by geometrical considerations

---

[56] L'intermédiaire des math., 23, 1916, 152–3.

[57] Quar. Jour. Math., 48, 1917, 93–104.

[58] Proc. Nat. Acad. Sc., 4, 1918, 189–193. Proc. London Math. Soc., Records of Meeting, March, 14, 1918.

[58a] Trans. Cambr. Phil. Soc., 22, 1918, 259–276.

[58b] Bull. Amer. Math. Soc., 26, 1919, 19–25.

[59] Y. Mikami, Abh. Gesch. Math. Wiss., 30, 1912, 247. Based on C. Hitomi's article in Jour. Phys. School of Tokyo, 15, 1906, 359–62.

[59a] Jour. Phys. School of Tokyo, 22, 1913, 51.

[60] Trans. Roy. Irish. Acad., 21, II, 1848, 333.

[61] Cours d'analyse de l'école polyt., 1, 1821, 455–7.

[62] Math. Quest. Educ. Times, 20, 1874, 83; 63, 1895, 49, 112.

[63] Bull. Acad. Roy. Sc. Belgique, (3), 27, 1894, 10–15.

[64] Proc. Edinburgh Math. Soc., 14, 1896, 113–5; Math. Mag., 2, 1898, 209.

[65] Periodico di Mat., 22, 1907, 28.

in n-space).   By multiplying a_i by $\sqrt{\alpha_i}$ for $i = 2, \cdots, n$, we get

$$(a^2 + \Sigma\alpha_i a_i^2)^2 = (-a^2 + \Sigma\alpha_i a_i^2)^2 + \Sigma\alpha_i(2aa_i)^2,$$

a formula noted by G. Candido.[66]

M. Moureaux[67] noted that successive applications of Aida's formula gives

$$(a_1^2 + \cdots + a_n^2)^{2^p} = b_1^2 + \cdots + b_n^2.$$

J. Cunliffe[68] noted that we can find any number of rational squares whose sum is a rational square since $n + k^2 = \square$, $k = (4r^2 - n)/(4r)$. Thus, if $n = 1 + 4 + 9 + 16$, take $r = 3$, whence $k = 1/2$, and we have five squares whose sum is a square.

L. Calzolari[69] found special solutions of

$$x_1^2 + \cdots + x_n^2 = y^2$$

by setting $x_i = k + a_i$, $y = k + \Sigma a_i$.   The new equation is linear in each a_i.

E. Lucas[70] noted that the sum of x consecutive squares may be a square for $x = 2, 11, 23, 24$, but for no further value $1 < x \leq 24$; the sum of n consecutive odd squares is $\neq \square$ if $1 < n < 16$.   Cf. papers 76, 81, 86, 87, 100, and 103 below; also papers 80, 130–8 of Ch. I; and Brocard[92] of Ch. XXIII.

H. S. Monek[71] noted that $t^2 = (a^2 + b^2)^2 = (2ab)^2 + (2bc + c^2)^2$ if $a = b + c$.   Hence if

$$a^2 = c_1^2 + \cdots + c_n^2, \qquad t^2 = 4b^2 c_1^2 + \cdots + 4b^2 c_n^2 + (2bc + c^2)^2$$

is a sum of $n + 1$ squares.   Also,[72]

$$\Sigma\alpha_i^2 = \{2s + (n+1)a\}^2, \qquad s = \Sigma c_i, \qquad \alpha_i = 2s + 2a - (n-1)c_i.$$

F. P. Ruffini[73] discussed the positive integral solutions $i_r \leqq i_{r-1} \leqq \cdots \leqq i_1$ of

$$i_1^2 + \cdots + i_r^2 = u, \qquad i_1 + \cdots + i_r = v.$$

Let x_1 be the number of i's with the value 1, and x_2 the number with the value 2.   Set $s = r - x_1 - x_2$.   Then

$$x_1 + 4x_2 + \Sigma i^2 = u, \qquad x_1 + 2x_2 + \Sigma i = v \quad (3 \leqq i_s \leqq i_{s-1} \cdots \leqq i_1).$$

Solve for x_1 and x_2, and require that the values be $\geqq 0$.   By x_1,

$$i_s^2 - 2i_s \geqq V \equiv u - 2v - \Sigma i^2 + 2\Sigma i,$$

where the summations extend over $s - 1$ values of i.   Hence

$$i_s \geqq 1 + \sqrt{1 + V}.$$

The condition $1 + V \geqq 0$ is treated similarly, first solving for i_{s-1}.   For

---

[66] Suppl. al Periodico di Mat., 19, 1916, 97–100.   Case $\alpha_r = r$ by Aida[148] of Ch. XIII.

[67] Comptes Rendus Paris, 118, 1894, 700–1.

[68] The Gentleman's Math. Companion, London, 3, No. 14, 1811, 281–2.

[69] Giornale di Mat., 7, 1869, 313.   Cf. Ch. XIII.[123]

[70] Recherches sur l'analyse indéterminée, Moulins, 1873, 91.   Extract from Bull. Soc. d'Emulaticn Dépt. de l'Allier, Sc. Bell. Lettres, 12, 1873, 530.

[71] Math. Quest. Educ. Times, 20, 1874, 83–4.

[72] *Ibid.*, 30, 1879, 37–8.

[73] Mem. Accad. Sc. Istituto Bologna, 9, 1878, 199–215.   Simpler than his paper, *ibid.*, 8, 1877.

$u = n^2 - 1$, $v = 3(n - 1)$, the initial pair of equations are the conditions on a Cremona transformation. For $u = n^2 - 2$, $v = 3n + 2p - 4$, they are the conditions on the transformation of R. De Paolis, Mem. Accad. Lincei, 1877–8.

J. W. L. Glaisher[74] expressed the sum $\Sigma(a_i - a_j)^2$ of $n(n - 1)/2$ squares as

$$\sum_{i=1}^{\nu} \{(a_1 + a_2 c_i + a_3 c_{2i} + \cdots + a_n c_{(n-1)i})^2 + (a_2 s_i + \cdots + a_n s_{(n-1)i})^2\},$$

where $\nu = (n - 1)/2$ or $n/2 - 1$, according as n is odd or even, and

$$c_m = \cos(2m\pi/n), \qquad s_m = \sin(2m\pi/n).$$

G. Dostor[75] desired $2n + 1$ consecutive integers such that the sum of the squares of the first $n + 1$ of them equals that of the last n, and proved that the first of the numbers is $n(2n + 1)$ or $-n$.

A. Martin[76] proved for $n = 3, 4, 5$ that a sum of n consecutive squares is not a square. Call x^2 the middle square when $n = 3$ or 5; the problem reduces to the fact that $3x^2 + 2 = \square$ or $5(x^2 + 2) = \square$ is impossible.

G. Dostor[77] noted that, if $a_1 + \cdots + a_n = np/2$,

$$a_1^2 + \cdots + a_n^2 = \sum_{i=1}^{n} (p - a_i)^2, \qquad a_1^2 + \cdots + a_{n-1}^2 = p^2 + \sum_{i=1}^{n-1} (p - a_i)^2,$$

the last by setting $a_n = 0$, so that[78] a sum of n or $n - 1$ squares is expressed as a sum of n squares. Also

$$(\Sigma a_i^2 + \Sigma a_i a_j)^2 = (\Sigma a_i)^2 \Sigma a_i^2 + (\Sigma a_i a_j)^2.$$

D. S. Hart[79] found squares whose sum is a square by subtracting $(s + m)^2 - s^2$ from $1^2 + 2^2 + \cdots + n^2$ and, by trial, expressing the difference as a sum of squares, which are then deleted from the n squares.

J. A. Gray[80] noted that we may start with a sum S of squares, choose a divisor a of S and set $S + x^2 = (x + a)^2$, whence $2x = S/a - a$.

Hart[81] considered the sum S of the squares of $2n - 1$ consecutive numbers the middle one of which is x and, for special values $\leqq 181$ of n, made S a square. Cf. Lucas[70].

E. Catalan[82] proved there is a number equal to a sum of p squares and having its square equal to a sum of $2p$ squares, by use of the identity

$$(x^{2n} + x^{2n-2}y^2 + \cdots + y^{2n})^2 = (x^{2n})^2 + (x^{2n-1}y)^2 + \cdots + (y^{2n})^2$$
$$+ [xy(x^{2n-2} + x^{2n-4}y^2 + \cdots + y^{2n-2})]^2.$$

[74] Messenger Math., 8, 1878–9, p. 48.
[75] Archiv Math. Phys., 64, 1879, 350–2. Cf. Zeitschr. Math. Naturw. Unterricht, 12, 1881, 269; E. Collignon, Assoc. franç. av. sc., 25, II, 1896, 17; Cesàro[153] of Ch. I.
[76] Math. Visitor, 1, 1880, 156. Cf. Lucas[70].
[77] Archiv Math. Phys., 67, 1882, 265–8.
[78] For $n = 3$, E. Catalan, Nouv. Corresp. Math., 4, 1878, 3.
[79] Math. Magazine, 1, 1882–4, 8–9.
[80] Ibid., 76.
[81] Ibid., 119–122; errata corrected by Martin, 2, 1892, 94.
[82] Mathesis, 3, 1883, 199.

Catalan[83] proved the last result and (p. 106) gave a long identity furnishing particular solutions of $u^2 = x_1^2 + \cdots + x_5^2$. If an odd number N is a $\boxed{2}$ and if n is the number of equal or distinct prime factors of N, then N^2 is a sum of k squares $\neq 0$, $k = 2, 3, \cdots, n + 1$.

R. W. D. Christie[84] noted equal sums of four or more squares.

A. Martin[85] noted that $2^2 + 3^2 + 6^2 = 7^2$, $1^2 + 2^2 + 4^2 + 6^2 + 8^2 = 11^2$,

$$1^2 + 2^2 + \cdots + 50^2 - 206^2 = 1 + 2^2 + 22^2 = 5^2 + 8^2 + 20^2.$$

He[86] stated that one can find several sets of 50 squares whose sum is 231^2, that $1^2 + 2^2 + \cdots + 24^2 = 70^2$, and similar results. Cf. Lucas.[70]

F. Tano's method to find an infinitude of solutions of

$$x_1^2 + \cdots + x_k^2 - y_1^2 - \cdots - y_{k+1}^2 = a,$$

when k is of the form $(3^n - 1)/2$, is given in Ch. XII.[207]

A. Martin[87] found many sets of squares whose sum is a square by means of the methods of Aida[59] and Gray,[80] and by seeking to express $S_n - b^2$ as a sum of distinct squares $\leqq n^2$, where b^2 lies between n^2 and $S_n = 1^2 + \cdots + n^2$. He noted that the sum of n consecutive squares is not a square for $2 < n < 11$, and gave solutions for $n = 11, 23, 24, 26$, etc. [cf. Lucas[70]]. He gave solutions of

$$S_n - x^2 = \square, \qquad S_n + 1 = \square, \qquad S_n - S_m - x^2 = \square,$$

and tabulated the values of S_n for $n < 400$.

E. Catalan[88] noted that, if $N \pm 1$ are primes and $N \neq 2$, $2N^2 + 2$ is a sum of 2, 3, 4, and 5 squares.

E. Fauquembergue[89] and others noted the identities

$$(a_1^2 + \cdots + a_n^2)^2 = (a_1^2 + \cdots + a_i^2 - a_{i+1}^2 - \cdots - a_n^2)^2 + \sum_{r=1}^{i} \sum_{s=i+1}^{n} (2a_r a_s)^2,$$

$$(a_1^2 + \cdots + a_5^2)^2 = (a_1^2 + a_2^2 + a_3^2 - a_4^2 - a_5^2)^2 + 4(a_1 a_4 \pm a_3 a_5)^2$$
$$+ 4(a_1 a_5 \mp a_3 a_4)^2 + 4a_2^2 a_5^2 + 4a_2^2 a_4^2.$$

P. H. Philbrick[90] noted that we may find n squares whose sum is a square by Aida's[59] method or by starting with a sum S of $n - 1$ squares such that S is a product of two factors a and b, both even or both odd, and applying

$$ab + \left(\frac{a - b}{2}\right)^2 = \left(\frac{a + b}{2}\right)^2.$$

R. J. Adcock[91] noted that, if $s = x + y + z + v$,

$$x^2 s^2 + y^2 s^2 + z^2 s^2 + v^2 s^2 + (xy + xz + xv + yz + yv + zv)^2 = (\Sigma x^2 + \Sigma xy)^2.$$

[83] Atti Accad. Pont. Nuovi Lincei, 37, 1883–4, 53.
[84] Math. Quest. Educ. Times, 49, 1888, 159–173; French transl., Sphinx-Oedipe, 7, 1912, 177–87
[85] Bull. Phil. Soc. Wash., 10, 1888, 107 (Smithsonian Miscel. Coll., 33, 1888).
[86] *Ibid.*, 11, 1892, 580–1.
[87] Math. Mag., 2, 1891–3, 69–76, 89–96, 137–140.
[88] Mathesis, (2), 3, 1893, 235.
[89] Mathesis, (2) 4, 1894, 277; 6, 1896, 101.
[90] Amer. Math. Monthly, 1, 1894, 256–8.
[91] *Ibid.*, 2, 1895, 285.

Several writers[92] found nine integers in arithmetical progression whose sum of squares is a square.

A. Martin[93] noted that the sum of the squares of the nine numbers $x - 4y$, $x - 3y$, $\cdots$, $x + 4y$ in arithmetical progression is a square if $9x^2 + 60y^2 = \square$. Take $y = 3z$, $x^2 + 60z^2 = (x + zp/q)^2$; hence x/z is found rationally.

Various writers[94] made $\Sigma_{i=1}^{i=n} x_i$ and Σx_i^2 squares for $n = 2, 3, 4, 5, 9$.

A. Boutin[95] noted values $n = 4, 9, \cdots, 50$ such that the sum of the squares of n integers in arithmetical progression is a square.

A. Martin[96] solved $b_1^2 + \cdots + b_m^2 = c_1^2 + \cdots + c_n^2$ by setting $c_n = a + b_m$ and finding b_m rationally.

T. Meyer[97] gave solutions of $a^2 + b^2 + \cdots + n^2 + x^2 = z^2$.

G. La Marca[98] proved that $\Sigma a_i^2 = \square$ if $a_1, \cdots, a_n$ are integers such that $a_1 : a_2 = 3 : 4$, $a_i : a_{i+1} = 3 : 5$ $(i = 2, \cdots, n - 1)$. For, by $a_1 = 3q_1$, $a_2 = 4q_1$, $a_2 = 3q_2$, $a_3 = 5q_2$, we have $a_1^2 + a_2^2 = (5q_1)^2$, $5q_1 : a_3 = 3 : 4$, $(5q_1)^2 + a_3^2 = (5z)^2$, etc., where $z = 5q_2/4$ is stated erroneously to be q_2.

Ed. Collignon[99] noted that $x = 2ak(k + 1)$ is a solution of

$$x^2 + (x - a)^2 + \cdots + (x - ka)^2 = (x + a)^2 + \cdots + (x + ka)^2.$$

E. N. Barisien[100] noted that a sum of p consecutive squares is not a square for $p < 20$, except for $p = 2, 11$, without treating the case $p = 13$. First, let $p = 2n + 1$ and denote the middle square by x^2 and the least square by $(x - n)^2$. The sum of the squares is $(2n + 1)\{x^2 + n(n + 1)/3\}$, which is not a square for $n \leqq 4$, $n = 7, 8, 9$. For $n = 5$, $11(x^2 + 10)$ is to be a square, whence $x = 11h \pm 1$. Then $x^2 + 10 = 11m^2$, $h = 2l$, $l \equiv 0$ or 1 (mod 3). A table of 8 solutions includes

$$(x, h, m) = (23, 2, 7), \ (43, 4, 13), \ (461, 42, 139), \ (859, 78, 259).$$

For $p = 2n$, let $(x + n)^2$ be the largest square. Their sum is

$$N = 2nx(x + 1) + n(2n^2 + 1)/3.$$

For $n = 1$, $N = 2x^2 + 2x + 1 = 4T + 1$, where T is a triangular number. Thus $T = 6, 210, 7158, \cdots$, giving

$$3^2 + 4^2 = 5^2, \qquad 20^2 + 21^2 = 29^2, \qquad 119^2 + 120^2 = 169^2.$$

The cases $1 < n \leqq 9$ are impossible. Cf. Lucas.[70]

E. N. Barisien[101] gave the identity

$$(a^2 + b^2 + c^2)^3 = [a(b^2 + c^2 - a^2)]^2 + [b(b^2 + a^2 - 3c^2)]^2$$
$$+ [c(a^2 + c^2 - 3b^2)]^2 + (2a^2b)^2 + (2a^2c)^2 + (4abc)^2,$$

[92] Amer. Math. Monthly, 2, 1895, 129–30, 163.
[93] Math. Quest. Educ. Times, 63, 1895, 111–2.
[94] L'intermédiaire des math., 1, 1897, 42–4.
[95] *Ibid.*, 5, 1898, 75.
[96] Math. Magazine, 2, 1898, 212–3.
[97] Zeitschr. Math. Naturw. Unterricht, 36, 1905, 337–340.
[98] Il Boll. Mat. Giornale Sc. Didat. (ed., Conti), 5, 1906, 152–5.
[99] Sphinx-Oedipe, 1906–7, 129. Case $a = 1$, Dostor.[75]
[100] Sphinx-Oedipe, 1907–8, 121–6. Cf. Martin.[87]
[101] Bull. de math. élém., 15, 1909–10, 181.

and obtained[102] ten decompositions of 266^2 into nine squares by multiplying, two by two, five decompositions of 266 as a sum of three squares.

E. Barbette[103] used the method of Martin[87] to find squares whose sum is a square. He gave (pp. 87, 96) many sets of consecutive squareswhose sum is a square. [cf. Lucas[70]]

E. Miot[104] stated that, if $2^k < m \leqq 2^{k+1}$, the square of a sum of m squares is a sum of $2^k + 1$ squares.

E. N. Barisien[105] noted that the sum of the squares of x^6, $4x^2y^4$, $4xy^5$, $2y^6$ and $2xy(2x^4 + 5x^2y^2 + 2y^4)$ equals the square of $x^6 + 8x^4y^2 + 8x^2y^4 + 2y^6$, and gave seven squares whose sum is a square.

L. E. Dickson[106] gave a history of the problem to express the product of two sums of n squares as a sum of n squares.

On $1^2 + \cdots + x^2 = ky^2$, see Lucas[151] of Ch. I. On $x_1^2 + \cdots + x_n^2 = R^2$, see Turrière[115] of Ch. VII, Escott[261] of Ch. XXI and paper 94, p. 322. On $x_1^2 + \cdots + x_n^2 = y^p$, see papers 96a, 98 of Ch. XX; 268 of Ch. XXI ($p=3$); and papers near the end of Ch. XXII ($p=4$). By Landau[21] of Ch. XXV, every definite polynomial in x is a sum of the squares of 8 polynomials.

---

[102] Mathesis, 10, 1910, 185.

[103] Les sommes de p-ièmes puissances distinctes égales à une p-ième puissance, Liège, 1910, 77–104.

[104] L'intermédiaire des math., 19, 1912, 195.

[105] Sphinx-Oedipe, 8, 1913, 142.

[106] Annals of Math., (2), 20, 1919, 155–171, 297.

CHAPTER X.

NUMBER OF SOLUTIONS OF QUADRATIC CONGRUENCES IN n UNKNOWNS.

For $n \leqq 4$, report was made in Ch. VIII on the papers by Libri,[29] Schönemann,[31] Frattini,[25] Lipschitz,[77] Dickson,[95] Tengbergen,[107] and L. Aubry,[110] and to many papers proving merely the existence of solutions. See also Hermite,[21] Lebesgue,[63] and Pepin[80] of Ch. VIII, Vol. I of this History.

V. A. Lebesgue[1] noted that $F = \Sigma a_i x_i^2 \equiv 0 \pmod{p}$, where p is a prime $2h + 1$, may be reduced by multiplication of the variables by constants to a form

$$(1) \qquad y_1^2 + \cdots + y_f^2 \equiv n(z_1^2 + \cdots + z_i^2) \pmod{p},$$

where $n = 1$ if $p = 4q - 1$, and n is a quadratic non-residue of p if $p = 4q + 1$. Let N_k^0, N_k, N_k' denote the number of sets of solutions of

$$y_1^2 + \cdots + y_k^2 \equiv a \pmod{p},$$

according as $a \equiv 0$, a is a quadratic residue or non-residue of p. In view of his[2] general theorem, the number of sets of solutions of (1) is

$$N_f^0 N_i^0 + h(N_f N_i + N_f' N_i'), \qquad N_f^0 N_i^0 + h(N_f N_i' + N_f' N_i),$$

according as $n = 1$ or a quadratic non-residue of p. Also, if P_0 is the number of solutions of $F \equiv 0$ and π the number of $F - ax^2 \equiv 0$, the number for $F \equiv a$ is $(\pi - P_0)/(p - 1)$. It is proved that, if k is odd,

$$N_k^0 = p^{k-1}, \qquad N_k = p^{k-1} + t, \qquad N_k' = p^{k-1} - t, \qquad t = (-1)^{(p-1)(k-1)/4} p^{(k-1)/2};$$

while, if k is even,

$$N_k^0 = p^{k-1} + (p-1)l, \qquad N_k = N_k' = p^{k-1} - l, \qquad l = (-1)^{(p-1)/2 \cdot k/2} p^{(k/2)-1}.$$

Lebesgue[3] gave a simpler proof of the last results and also found the number of sets of solutions prime to p.

C. Jordan[4] proved by induction from $n = l$ and $n = m$ to $n = l + m$ that, if $a_1 \cdots a_{2n} \not\equiv 0$, $a_1 x_1^2 + \cdots + a_{2n} x_{2n}^2 \equiv k \pmod{p}$, where p is an odd prime, has $p^{2n-1} - p^{n-1}\nu$ sets of solutions if $k \not\equiv 0 \pmod{p}$, and $p^{2n-1} + (p^n - p^{n-1})\nu$ sets if $k \equiv 0$, where

$$\nu = \left(\frac{(-1)^n a_1 \cdots a_{2n}}{p} \right), \qquad \nu' = \left(\frac{(-1)^n a_1 \cdots a_{2n+1} k}{p} \right)$$

are Legendre symbols. Also, $a_1 x_1^2 + \cdots + a_{2n+1} x_{2n+1}^2 \equiv k \pmod{p}$ has $p^{2n} + p^n \nu'$ sets of solutions. As a corollary, there are $(p - 1)/2$ variations

---

[1] Jour. de Math., 2, 1837, 266–275.
[2] Vol. I, pp. 224–5 of this History.
[3] Jour. de Math., 12, 1847, 467–471.
[4] Comptes Rendus Paris, 62, 1866, 687–90; Traité des substitutions, 1870, 156–61 (with a misprint of sign in the theorem on p. 610).

of signs in

$$\left(\frac{1}{p}\right), \qquad \left(\frac{2}{p}\right), \qquad \cdots, \qquad \left(\frac{p-1}{p}\right).$$

V. A. Lebesgue[5] gave two proofs of Jordan's formulas, not using induction. The first proof uses his[1] results for reduced congruences. The second proof is based on his[2] amplification of Libri's method.

H. J. S. Smith[6] proved that if p is an odd prime and m any integer, $xz - y^2 \equiv m \pmod{p}$ has $p\{p + (- m/p)\}$ solutions. Each of the congruences $xz - y^2 \equiv 1, 3, 5, 7 \pmod 8$ has 48 solutions in which x and y are not both even. If p is any prime and $i > 0$, $i' \geqq 0$,

$$xz - y^2 \equiv mp^i \pmod{p^{i+i'}}$$

has $p^{2i+2i'}(1 - 1/p^2)$ solutions in which x, z are not both divisible by p.

C. Jordan[7] proved that $x_1y_1 + \cdots + x_ny_n \equiv 0 \pmod 2$ has $2^{2n-1} + 2^{n-1}$ sets of solutions, while $x_1 + y_1 + x_1y_1 + \cdots + x_ny_n \equiv 0$ has $2^{2n-1} - 2^{n-1}$ sets of solutions.

Jordan[8] determined the number of sets of solutions of $f \equiv c \pmod{M}$, where f is any homogeneous quadratic function of $x_1, \cdots, x_m$. The number is the product of the numbers of solutions for moduli which are the powers of primes whose product is M. Consider

$$f = P^\alpha(a_1x_1^2 + \cdots + a_mx_m^2 + b_{12}x_1x_2 + \cdots) \equiv c \pmod{P^\lambda},$$

where at least one coefficient $a_1, \cdots, a_m, b_{12}, \cdots$ is not divisible by the prime P. First, let $P > 2$. By means of a linear transformation, we may remove the terms x_1x_2, etc., not squares. The problem is reduced to

$$A_1x_1^2 + \cdots + A_px_p^2 + P^\beta(B_1y_1^2 + \cdots + B_qy_q^2) + \cdots \equiv d \pmod{P^\mu}.$$

The number of sets of solutions, in which $x_1, \cdots, x_p$ are not all divisible by P, is P^rU, where $r = (\mu - 1)(n - 1) + n - p$, $n = p + q + \cdots$, and U is the number of sets of solutions of $A_1x_1^2 + \cdots + A_px_p^2 \equiv d \pmod P$, given above.[4] For solutions in which $x_1, \cdots, x_p$ are divisible by P, we can remove a power of P and are led to the preceding case.

For $P = 2$, we can transform f linearly into $2^\alpha\Sigma_a + 2^\beta\Sigma_\beta + \cdots$, where each Σ_ρ is of one of the four types $S_p = x_1y_1 + \cdots + x_py_p$,

$$S_p + Az^2, \qquad S_p + Az^2 + A_1z^2, \qquad S_p + u^2 + uv + v^2,$$

where A and A_1 are odd integers, $A \leqq 7$, and p may be zero. The number of solutions is found by treating these four cases in turn.

T. Pepin[9] proved Jordan's[4] results by expressing the number of solutions in terms of the number for the congruence in which the number of unknowns is less by two.

[5] Comptes Rendus Paris, 62, 1866, 868–72.

[6] Trans. Phil. Soc. London, 157, 1867, 286–7, § 18; Coll. Math. Papers, I, 492–4.

[7] Traité des substitutions, 1870, 198.

[8] Jour. de Math., (2), 17, 1872, 368–402. Comptes Rendus Paris, 74, 1872, 1093.

[9] Nouv. Ann. Math., (2), 10, 1871, 227–234.

H. Minkowski[10] found the number $f\{m;\ N\}$ of sets of solutions of

$$f = \sum_{i,\,k=1}^{n} a_{ik}x_i x_k \equiv m \quad (\mathrm{mod}\ N).$$

If

$$f(h;\ N) = \sum_{m=1}^{N} f\{m;\ N\}\rho^{mh}, \qquad \rho = e^{2\pi i/N},$$

then

$$f\{m;\ N\} = \frac{1}{N}\sum_{h=1}^{N} f(h;\ N)\rho^{-hm},$$

so that the problem remains to find $f(m;\ N)$ whose determination depends upon that of $\Sigma\rho^{mf}$, where $x_1,\ \cdots,\ x_n$ range each over a complete set of residues modulo N. The problem is reduced to the case of a power of prime modulus. The paper is too complicated to admit of a brief report.

L. Gegenbauer[11] considered $f = a_1 x_1^2 + \cdots + a_n x_n^2$, with r of the a's quadratic residues of the odd prime p. Let $\sigma_n'(r)$ be the number of sets of solutions of $f \equiv 0 \pmod{p}$, and $\sigma_n(r)$ the number of those in which no $x \equiv 0$. Let s' and s be the corresponding numbers for $f \equiv 1$ (to which we may reduce $f_1 \equiv b \not\equiv 0$ by multiplication). For $r > 0$,

$$\sigma_n'(r) = \sigma_{n-1}'(r-1) + (p-1)s_{n-1}'(r-1), \qquad \sigma_n'(0) = \sigma_n'(n),$$
$$\sigma_n(r) = (p-1)s_{n-1}(r-1), \qquad\qquad\quad \sigma_n(0) = \sigma_n(n),$$

with more complicated recursion formulas for $s_n'(r)$, $s_n(r)$, which with $\sigma_1'(r) = 1$, $\sigma_1(r) = 0$, $s_1(r) = s_1'(r) = 1 + (2r-1)(-1/p)$ determine the s and σ as by Jordan.[4]

K. Zsigmondy[12] proved the final results of Lebesgue.[1]

P. Bachmann[13] gave an exposition of the subject.

L. E. Dickson[14] gave a generalization of Jordan's[4, 7] work to any finite field and a derivation of canonical forms.

R. Le Vavasseur[15] discussed $f \equiv u \pmod{p}$, where p is a prime and

$$f = ax^2 + bxy + a'y^2 + cx + c'y + d, \quad \Delta = 4aa'd + bcc' - ac'^2 - a'c^2 - db^2,$$
$$\delta = 4aa' - b^2.$$

If δ is a quadratic non-residue of p, $f \equiv \Delta/\delta$ has one and but one solution; for $u \not\equiv \Delta/\delta$, $f \equiv u$ has $p+1$ solutions. If δ is a quadratic residue of p, $f \equiv \Delta/\delta$ has $2p-1$ solutions, $f \equiv u \not\equiv \Delta/\delta$ has $p-1$ solutions. If $\delta \equiv 0$, $f \equiv u$ has p solutions.

J. Klotz[16] found the number of sets of solutions of the general quadratic congruence in any algebraic field.

[10] Mém. présentés à l'Acad. Sc. Inst. France, (2), 29, 1884, No. 2, Arts. 7, 8, 9; Acta Math., 7, 1885, 201–258, espec., pp. 210–37. Gesamm. Abh., 1, 1911, 3, 157.

[11] Sitzungsber. Akad. Wiss. Wien (Math.), 99, IIa, 1890, 795–9.

[12] Monatshefte Math. Phys., 8, 1897, 38.

[13] Arith. der Quadrat. Formen, 1, 1898, 478–515.

[14] Linear Groups, 1901, 46–9, 158, 197–9, 205–6; Madison Colloquium Lectures, Amer. Math. Soc., 1914. Cf. J. E. McAtee, Amer. Jour. Math., 41, 1919, 225–42, on Jordan.[8]

[15] Mém. Acad. Sc. Toulouse, (10), 3, 1903, 44–8.

[16] Vierteljahrsschrift d. naturf. Gesell. Zürich, 58, 1913, 239–68.

CHAPTER XI.

LIOUVILLE'S SERIES OF EIGHTEEN ARTICLES.

J. Liouville enunciated without proof numerous results in a series of eighteen articles, "Sur quelques formules générales qui peuvent être utiles dans la théorie des nombres."

Let m be an odd integer, α an integer $\geqq 1$. Set

$$2^\alpha m = m' + m'', \qquad m = d\delta, \qquad m' = d'\delta', \qquad m'' = d''\delta'',$$

where m' and m'' are odd positive integers. Let $f(x) = f(-x)$ be an even single-valued function. He[1] stated that

(a) $\quad \sum \{ \sum_{d',\,d''} [f(d' - d'') - f(d' + d'')] \} = 2^{\alpha-1} \sum_d d\{f(0) - f(2^\alpha d)\},$

where d, d', d'' range over all the divisors of m, m', m'', respectively, and the first summation extends over all the pairs of positive odd integers m', m'' whose sum is $2^\alpha m$. Taking $f(x) = x^{2\mu}$, we get

$$2^{2\alpha\mu + \alpha - 2} \zeta_{2\mu+1}(m) = 2\mu \Sigma \zeta_1(m') \zeta_{2\mu-1}(m'')$$

$$+ \frac{2\mu(2\mu - 1)(2\mu - 2)}{1 \cdot 2 \cdot 3} \Sigma \zeta_3(m') \zeta_{2\mu-3}(m'') + \cdots + 2\mu \Sigma \zeta_{2\mu-1}(m') \zeta_1(m''),$$

where the coefficients are those of even rank in the binomial formula, and $\zeta_\mu(m)$ denotes the sum of the μth powers of the divisors of m. For $\mu = 1$ and $\mu = 2$, we have

$$2^{3\alpha-3} \zeta_3(m) = \Sigma \zeta_1(m') \zeta_1(m''), \qquad 2^{5\alpha-5} \zeta_5(m) = \Sigma \zeta_1(m') \zeta_3(m'').$$

The first gives the number of decompositions of $4 \cdot 2^\alpha m$ as a sum of 8 odd squares; the second gives the number of decompositions of $8 \cdot 2^\alpha m$ into $s + 2\sigma$, where s is a sum of 8 odd squares such that $s/8$ is odd, while σ is a sum of 4 odd squares.

For $f(x) = \cos xt$, (a) gives

$$\Sigma(\Sigma \sin d't \cdot \Sigma \sin d''t) = 2^{\alpha-1} \Sigma d \sin^2 (2^{\alpha-1} dt).$$

Taking $\alpha = 1$, $t = \pi/2$, or by setting $f(x) = (-1)^{x/2}$, we get

$$\Sigma(\Sigma(-1)^{(d'-1)/2} \cdot \Sigma(-1)^{(d''-1)/2}) = \Sigma d = \zeta_1(m),$$

which yields Jacobi's[25, 30] theorem of Ch. VIII that $4m$ has $\zeta_1(m)$ representations as a sum of four odd squares.

For a function $f(x, y)$ which is unaltered by the change of the sign of x or of y, Liouville stated that

(b) $\quad \sum \{ \sum_{d',\,d''} [f(d' - d'', \delta' + \delta'') - f(\delta' + \delta'', d' - d'')] \}$

$$= 2^{\alpha-1} \sum_d d\{f(0, 2^\alpha d) - f(2^\alpha d, 0)\} = \sigma,$$

(c) $\quad \sum \{ \sum_{d',\,d''} [f(d' - d'', \delta' + \delta'') - f(d' + d'', \delta' - \delta'')] \} = \sigma.$

---

[1] Jour. de Math., (2), 3, 1858, 143–152, 193–200. First and second articles.

Set

$$f(x, y) = \cos xt \cdot \cos yz,$$

$$\psi(m) = \sum_d \sin dt \cdot \cos \delta z, \qquad \omega(m) = \sum_d \cos dt \cdot \sin \delta z \quad (d\delta = m).$$

Then (c) yields the result

$$\Sigma\psi(m')\psi(m'') - \Sigma\omega(m')\omega(m'') = 2^{a-1}\Sigma d\{\sin^2 2^{a-1}dt - \sin^2 2^{a-1}dz\}.$$

We now include the case in which $\alpha = 0$ and set

$$2^a m = 2^{a'}m' + 2^{a''}m'' \quad (\alpha' \geqq 0, \alpha'' \geqq 0, m', m'' \text{ odd}).$$

Let $m = d\delta$, etc., as before. Liouville[2] stated the formula [a case of (e)[3]]

$$\text{(G)} \quad \sum \{ \sum_{d', d''} [f(2^{a'}d' - 2^{a''}d'') - f(2^{a'}d' + 2^{a''}d'')] \}$$
$$= \sum_d (\delta - 2^a d)\{f(2^a d - f(0)\},$$

where d, d', d'' range over all the divisors of m, m', m'', respectively, and the first summation extends over all the pairs of even or odd integers $2^{a'}m'$, $2^{a''}m''$ whose sum is $2^a m$. Consider the case $\alpha = 0$; then α' or α'' is zero; but, by introducing the factor 2 before the first member of (G), we may restrict attention to the case $m = m' + 2^{a''}m''$. Since $\Sigma\delta = \Sigma d$, we get

$$\text{(F)} \quad 2 \sum \{ \sum_{d', d''} [f(d' - 2^{a''}d'') - f(d' + 2^{a''}d'')] \} = \sum_d (\delta - d)f(d),$$

a case of (d). For example, if $f(x) = x^2$,

$$\Sigma\{2^{a''}\zeta_1(m')\zeta_1(m'')\} = \tfrac{1}{8}\{\zeta_3(m) - m\zeta_1(m)\}.$$

For $f(x) = x^2$ or x^4 in (G) we get

$$\Sigma\{2^{a'+a''}\zeta_1(m')\zeta_1(m'')\} = 2^{3a-2}\zeta_3(m) - 2^{2a-2}m\zeta_1(m),$$
$$\Sigma\{2^{3a'+a''}\zeta_3(m')\zeta_1(m'')\} = 2^{5a-4}\zeta_5(m) - 2^{4a-4}m\zeta_3(m).$$

Again using the notation $m = m' + 2^{a''}m''$, Liouville stated the following two cases of (d):

$$\text{(D)} \quad f(0)\zeta_1(m) = \sum_d \{f(0) + 2f(2) + 2f(4) + \cdots + 2f(d - 1)\}$$
$$+ 2 \sum \{ \sum_{d', d''} [f(d' - d'') - f(d' + d'')] \},$$

$$\text{(E)} \quad \sum \{ \sum_{d', d''} [F(d' - d'' + 1) - F(d' - d'' - 1) - F(d' + d'' + 1)$$
$$+ F(d' + d'' - 1)] \} = F(1)\zeta_1(m) - \sum_d F(d),$$

where F is an odd function: $F(-x) = -F(x)$. For $f(x) = (-1)^{x/2}$, (D) gives

$$\tfrac{1}{4}\{\zeta_1(m) - \rho(m)\} = \Sigma\rho(m')\rho(m''), \qquad \rho(m) \equiv \Sigma(-1)^{(d-1)/2}.$$

The first expression is therefore the number of decompositions of $2m$ into

$$\text{(1)} \qquad\qquad y^2 + z^2 + 2^a(u^2 + v^2),$$

with y, z, u, v odd positive integers and $\alpha > 0$; it is also the number of decompositions of m into the form (1) with y and z positive odd numbers,

---

[2] Jour. de Math., (2), 3, 1858, 201-8, 241-250. Third and fourth articles.

and u, v any even integers. For $f(x) = x^2$, we deduce from (D) that

$$\tfrac{1}{24}\{\zeta_3(m) - \zeta_1(m)\} = \Sigma\zeta_1(m')\zeta_1(m''),$$

which gives the number of decompositions of $4m$ into $s + 2^a\sigma$, where s and σ are sums of 4 odd squares.

For m any integer > 1, let $m = m' + m''$. Liouville stated the following case of (f):

$$\text{(H)} \quad \Sigma\left\{ \sum_{d',\,d''} [f(d' - d'') - f(d' + d'')] \right\} = f(0)\{\zeta_1(m) - \zeta(m)\}$$
$$- \Sigma f(d)\{2\zeta(\delta) + d - 2\delta - 1\} - 2\Sigma'\{f(2) + f(3) + \cdots + f(d-1)\},$$

where $\zeta(m)$ is the number of factors of m and the accent on the final summation sign signifies that a term $f(k)$ is to be suppressed when k is a divisor of d. For $f(x) = x^2$ and m a prime, (H) gives

$$\text{(H')} \qquad \Sigma\zeta_1(m')\zeta_1(m'') = \tfrac{1}{12}(m^2 - 1)(5m - 6).$$

This result may be used to prove the theorem of Bouniakowsky that any prime m of the form $16k + 7$ can be decomposed into $2x^2 + p^{4l+1}y^2$ in an odd number of ways, where p is a prime $4\lambda + 1$ not dividing y.

Liouville[3] stated that, if $f(x, y)$ is unaltered by the change of the sign of x or y,

$$\text{(d)} \quad 2\Sigma\left\{ \sum_{d',\,d''} [f(d' - 2^{a''}d'',\, \delta' + \delta'') - f(d' + 2^{a''}d'',\, \delta' - \delta'')] \right\}$$
$$= \sum_d \{f(d, 0) + 2f(d, 2) + 2f(d, 4) + \cdots + 2f(d, \delta - 1) - df(d, 0)\},$$

where the first summation extends over all decompositions $m' + 2^{a''}m''$ of m. If $f(x, y)$ reduces to a function $f(x)$ of x only, (d) becomes (F). If it reduces to $f(y)$, (d) becomes (D). To pass from (D) to (E), take

$$f(x) = F(x + 1) - F(x - 1).$$

In (d) take f to be $(-1)^{y/2}f(x)$, where $f(x)$ is an even function. Then

$$\text{(I)} \quad 2\Sigma\{\Sigma(-1)^{(\delta'-1)/2}(-1)^{(\delta''-1)/2}[f(d' - 2^{a''}d'') + f(d' + 2^{a''}d'')]\}$$
$$= \Sigma df(d) - \Sigma(-1)^{(\delta-1)/2}f(d).$$

For $2^a m = 2^{a'}m' + 2^{a''}m''$,

$$\Sigma\{\Sigma[f(2^{a'}d' - 2^{a''}d'',\, \delta' + \delta'') - f(2^{a'}d' + 2^{a''}d'',\, \delta' - \delta'')]\}$$
$$\text{(e)} \quad = \Sigma d\{f(0, 2d) + 2f(0, 4d) + 4f(0, 8d) + \cdots + 2^{a-1}f(0, 2^a d)\}$$
$$+ \Sigma\{f(2^a d, 0) + 2f(2^a d, 2) + 2f(2^a d, 4) + \cdots$$
$$+ 2f(2^a d, \delta - 1)\} - 2^a\Sigma df(2^a d, 0),$$

which reduces to (G) for $f(x, y) = f(x)$. Formula (H) is a special case of

$$\text{(f)} \quad \sum_{m'+m''=m} \left\{ \Sigma[f(d' - d'',\, \delta' + \delta'') - f(d' + d'',\, \delta' - \delta'')] \right\}$$
$$= \Sigma(d - 1)\{f(0, d) - f(d, 0)\} + 2\Sigma'\{f(\delta, 2) + \cdots$$
$$+ f(\delta, d - 1)\} - 2\Sigma'\{f(2, \delta) + \cdots + f(d - 1, \delta)\},$$

where the accent indicates that $f(\delta, y)$ is to be suppressed if y is a divisor

---

[3] Jour. de Math., (2), 3, 1858, 273–288. Fifth article.

of d, and $f(x, \delta)$ if x divides d. Set $\Delta(x, y) = f(x, y) - f(y, x)$. Then

$$(g) \quad \Sigma\{\Sigma\Delta(d' - d'', \delta' + \delta'')\} = \Sigma(d - 1)\Delta(0, d) \\ + 2\Sigma'\{\Delta(\delta, 2) + \cdots + \Delta(\delta, d - 1)\},$$

where $\Delta(\delta, y)$ is to be suppressed from the final sum if y divides d. The last formula is valid for any function Δ for which $\Delta(x, y) = -\Delta(y, x)$.

Liouville[4] employed in his sixth article two simultaneous partitions

$$2m = m' + m'', \qquad m = m_1 + 2^{a_2}m_2 \quad (m\text{'s odd and} > 0).$$

Set $m_i = d_i\delta_i$, etc.   Let $F(x)$ be a function for which

$$F(0) = 0, \qquad F(-x) = -F(x).$$

He stated that

$$(L) \quad \Sigma\{\Sigma\Sigma(-1)^{(d''-1)/2}[F(d'+d'')+F(d'-d'')]\} = \Sigma F(2d) + 4\Sigma\Sigma\rho(m_2)F(2d_1),$$

where d, d_1, d', d'' range over the divisors of m, m_1, m', m'', and the first summation extends over the m' and m'' whose sum is $2m$. For $F(x) = x$,

$$\Sigma\zeta_1(m')\rho(m'') = \zeta_1(m) + 4\Sigma\zeta_1(m_1)\rho(m_2),$$

so that there are $\zeta_1(m) + 4B$ decompositions of $8m$ into $s + 2\sigma$, where s is the sum of the squares of four odd positive numbers and σ is the sum of the squares of two such, while B is the number of decompositions of $4m$ into $s + 2^a\sigma$.

For a like function $F(x)$, another formula was stated:

$$(M) \quad 8\Sigma\{\Sigma\Sigma\Sigma[F(d' + d'' + d''') + F(d' - d'' - d''') - F(d' + d'' - d''') \\ - F(d' - d'' + d''')]\} = \Sigma(d^2 - 1)F(d) - 24\Sigma\Sigma\zeta_1(m_2)F(d_1),$$

where the two members relate to the respective modes of partitions

$$m = m' + m'' + m''', \qquad m = m_1 + 2^{a_2}m_2.$$

For $F(x) = x^3$ there results the formula

$$192\Sigma\zeta_1(m')\zeta_1(m'')\zeta_1(m''') + 24\Sigma\zeta_3(m_1)\zeta_1(m_2) = \zeta_5(m) - \zeta_3(m).$$

Hence if G is the number of decompositions of $4m$ into a sum of 12 odd squares, and H that of $8m$ into $s + 2^a\sigma$, where s is a sum of 8 odd squares with $s/8$ odd, and σ is a sum of 4 odd squares, then

$$8G + H = \tfrac{1}{24}\{\zeta_5(m) - \zeta_3(m)\}.$$

From (M) and (F), with $f(x) = xF(x)$ is derived

$$(N) \quad 4\Sigma\{\Sigma\Sigma 2^{a_2}d_2[F(d_1 + 2^{a_2}d_2) + F(d_1 - 2^{a_2}d_2)]\} \\ = \Sigma(d^2 - 1)F(d) + 8\Sigma\Sigma(2^{a_2} - 3)\zeta_1(m_2)F(d_1),$$

$$(O) \quad 4\Sigma\{\Sigma\Sigma d_1[F(d_1 - 2^{a_2}d_2) - F(d_1 + 2^{a_2}d_2)]\} \\ = \Sigma(2m - 1 - d^2)F(d) + 8\Sigma\Sigma(2^{a_2} - 3)\zeta_1(m_2)F(d_1),$$

each relating to the single mode of partition $m = m_1 + 2^{a_2}m_2$, $m_i = d_i\delta_i$.

---

[4] Jour. de Math., (2), 3, 1858, 325–336. Sixth article.

Liouville[5] remarked that if we multiply the members of (a) by x^p, where $p = 2^a m$, and sum for $p = 2, 4, 6, \cdots$, we get

$$(\alpha) \qquad \sum_{s', s''}^{1, 3, 5, \cdots} \frac{\{f(s' - s'') - f(s' + s'')\}x^{s'+s''}}{(1 - x^{2s'})(1 - x^{2s''})} = \sum_{s=1}^{\infty} \frac{s\{f(0) - f(2s)\}x^{2s}}{1 - x^{4s}},$$

which includes various formulas of the theory of elliptic functions. He stated that it is easy to prove (α) and then deduce (a), and that he had in his lectures at the Collège de France given a direct, elementary proof of (a), based on Dirichlet[42] of Ch. VIII, the method applying to (b) and with slight changes to the other formulas.

For any integer m, let

$$(2) \qquad m = m'^2 + m'', \qquad m'' = 2^{a''}d''\delta'' > 0 \quad (d'', \delta'' \text{ odd and} > 0),$$

while m' may be negative. Then for $F(-x) = -F(x)$, $F(0) = 0$,

$$(\beta) \qquad \Sigma\Sigma(-1)^{m''-1}F(2^{a''}d'' + m') = \begin{cases} \sqrt{m}\, F(\sqrt{m}) & \text{if } m = \text{square}, \\ 0 & \text{if } m \neq \text{square}. \end{cases}$$

A discussion of the case $F(x) = x$ shows that, if we set

$$\sigma = \zeta_1(m) - 2\zeta_1(m-1) + 2\zeta_1(m-4) - 2\zeta_1(m-9) + 2\zeta_1(m-16) - \cdots,$$

continued as long as the argument of ζ_1 is positive, then for m even,

$$\Sigma 2^a d = \zeta_1(m) - \zeta_1\left(\frac{m}{2}\right),$$

$$\sigma - \zeta_1\left(\frac{m}{2}\right) - 2\zeta_1\left(\frac{m-4}{2}\right) = \begin{cases} -m & \text{if } m = \text{square}, \\ 0 & \text{if } m \neq \text{square}, \end{cases}$$

while for m odd,

$$\sigma + 2\zeta_1\left(\frac{m-1}{2}\right) + 2\zeta_1\left(\frac{m-9}{2}\right) = \begin{cases} m & \text{if } m = \text{square}, \\ 0 & \text{if } m \neq \text{square}. \end{cases}$$

Using the same partitions of m and a function such that

$$\mathscr{F}(x, -y) = \mathscr{F}(x, y), \qquad \mathscr{F}(-x, y) = -\mathscr{F}(x, y), \qquad \mathscr{F}(0, y) = 0,$$

Liouville stated in his eighth article that

$$(\gamma) \quad \begin{aligned} &\Sigma\Sigma(-1)^{m''-1}\mathscr{F}(2^{a''}d'' + m', \delta'' - 2m') \\ &\qquad = 0 \text{ or } \mathscr{F}(\sqrt{m}, 1) + \mathscr{F}(\sqrt{m}, 3) + \cdots + \mathscr{F}(\sqrt{m}, 2\sqrt{m} - 1), \end{aligned}$$

according as m is not or is a square. As a special case,

$$\rho(m) - 2\rho(m-4) + 2\rho(m-16) - \cdots = 0 \text{ or } (-1)^{(\nu-1)/2}\nu, \qquad \nu = \sqrt{m}.$$

For $\mathscr{F}(x, y)$ a function of x only, (γ) reduces to (β).

Set $\mathscr{F}(x, y) = (-1)^{\nu/2}F(x, y)$, so that F is an odd function with respect to x and to y. Then (γ) gives

$$(\epsilon) \quad \begin{aligned} &\Sigma\Sigma(-1)^{(\delta''-1)/2}F(2^{a''}d'' + m', \delta'' - 2m') \\ &\qquad = 0 \text{ or } (-1)^{m+1}\{F(\sqrt{m}, 1) - F(\sqrt{m}, 3) \\ &\qquad\qquad + F(\sqrt{m}, 5) - \cdots \pm F(\sqrt{m}, 2\sqrt{m} - 1)\}, \end{aligned}$$

according as m is not or is a square.

---

[5] Jour. de Math., (2), 4, 1859, 1–8, 72–80. Seventh and eighth articles.

Liouville[6] stated that, for a function $f(x) = f(-x)$,

$$(\zeta) \quad \Sigma(-1)^{m''-1}\delta''f(2^{a''}d'' + m') - \Sigma\zeta_1(m_2)f(m_1) = \begin{cases} mf(\sqrt{m}) & \text{if } m = \text{square,} \\ 0 & \text{if } m \neq \text{square,} \end{cases}$$

where the summations relate to the partitions (2) and $m = m_1^2 + 2m_2$ respectively.

For $m = 8\nu + 5$, $f(x) = x \sin(x\pi/2)$, he derived the relation

$$\rho(m-4) - 4\rho(m-16) + 9\rho(m-36) - \cdots$$
$$= \zeta_1\left(\frac{m-1}{4}\right) - 3\zeta_1\left(\frac{m-9}{4}\right) + 5\zeta_1\left(\frac{m-25}{4}\right) - \cdots.$$

It follows that, if we effect in all possible ways the decompositions

$$m = 4s^2 + s_1^2 + s_2^2, \qquad m = n^2 + 4(n_1^2 + \cdots + n_4^2) \qquad (s > 0,\ n \text{ odd and} > 0),$$
$$\Sigma(-1)^{(n-1)/2}n = 2\Sigma(-1)^{s-1}s^2.$$

If, in place of the second type of decomposition, we employ

$$m = r^2 + r_1^2 + \cdots + r_4^2,$$

where $r, r_1, \cdots, r_4$ are positive and odd, then

$$4\Sigma(-1)^{(r-1)/2}r = \Sigma(-1)^{s-1}s^2.$$

For the same two types of partitions and for a function $f(x, y)$, even with respect to x and to y, Liouville stated in his tenth article that

$$\Sigma\Sigma(-1)^{m''-1}\delta''f(2^{a''}d''+m',\ \delta''-2m') - \Sigma\Sigma(2d_2-\delta_2)f(m_1,\ 2d_2+\delta_2)$$
$$(\eta) \qquad = 0 \quad \text{or} \quad f(\sqrt{m},\ 2\sqrt{m}-1) + 3f(\sqrt{m},\ 2\sqrt{m}-3)$$
$$+ \cdots + (2\sqrt{m}-1)f(\sqrt{m},\ 1),$$

according as m is not or is a square. If $f(x, y)$ is a function of x only, this reduces to (ζ).

For the same partitions and for a function $\mathcal{F}(x, y, z, t)$, even with respect to x, y, z and odd with respect to t, it is stated that

$$\Sigma\Sigma(-1)^{m''-1}\mathcal{F}(2^{a''}d'' + m',\ \delta'' - 2m',\ 2^{a''}d'' + m' - \delta'',\ \delta'')$$
$$(\nu) \quad - \Sigma\Sigma\mathcal{F}(m_1,\ 2d_2 + \delta_2,\ 2d_2 - m_1 - \delta_2,\ 2d_2 - 2m_1 - \delta_2)$$
$$= 0 \text{ or } \sum_j \mathcal{F}(\sqrt{m},\ 2\sqrt{m} - j,\ j - \sqrt{m},\ j) \qquad (j = 1, 3, 5, \cdots, 2\sqrt{m} - 1),$$

according as m is not or is a square. If $\mathcal{F} = tf(x, y)$, (ν) becomes (η). Other noteworthy cases are $\mathcal{F} = tf(z)$ and $\mathcal{F} = F(t)$.

Liouville[7] stated in his eleventh article that, if f is an even function,

$$(\xi) \quad \Sigma\Sigma(-1)^{(\delta''-1)/2}f(\delta'' - 2m') = f(1)\rho(2m-1) + f(3)\rho(2m-9) + \cdots,$$

the summation extending over all integers m' and all divisors δ'' of m'', where

$$(3) \qquad m = 2m'^2 + m'', \qquad m'' = d''\delta'' \quad (m'' \text{ odd and} > 0).$$

---

[6] Jour. de Math., (2), 4, 1859, 111–120, 195–204. Ninth and tenth articles.
[7] Jour. de Math., (2), 4, 1859, 281–304. Eleventh article.

The second member of (ξ) equals $\Sigma f(i)$, the summation extended over all the decompositions

(4) $\qquad 2m = i^2 + i_1^2 + p^2 \quad (i, i_1 \text{ odd and} > 0, p \text{ even}).$

For $f(x) = (-1)^{(x-1)/2}x$, the first member of (ξ) is $\Sigma(-1)^{m'}\zeta_1(m - 2m'^2)$ and equals $\frac{1}{8}E$, where E is the excess of the number of cases in which m' is even over the odd cases in

$$m = 2m'^2 + m_1^2 + \cdots + m_4^2 \quad (m', m_j \text{ any integers}),$$

since $8\zeta_1(m)$ is the number of representations of m as a sum of 4 squares for m odd.

Let $\mathcal{N}_1$ be the number of sets of solutions of

$$m = 2m'^2 + m_1^2 + \cdots + m_6^2$$

in which m' is odd, $\mathcal{N}_2$ the number in which m' is even. Then a discussion of (ξ) for the case $f(x) = x^2$ leads to the result, relating to (4),

$$\tfrac{1}{12}\mathcal{N}_2 - \tfrac{1}{20}\mathcal{N}_1 = \Sigma i^2 - \Sigma p^2 \quad \text{if } m \equiv 1 \pmod 4,$$
$$\tfrac{1}{12}\mathcal{N}_1 - \tfrac{1}{20}\mathcal{N}_2 = \Sigma i^2 - \Sigma p^2 \quad \text{if } m \equiv 3 \pmod 4.$$

If M is the number of solutions of (4),

$$2mM = 2\Sigma i^2 + \Sigma p^2.$$

Let $f(x, y)$ be a function even with respect to x and to y. Then

(π) $\Sigma\Sigma(-1)^{(\delta''-1)/2}f(\delta'' - 2m', 2d'' + 4m') = \Sigma\Sigma(-1)^{(\delta_2-1)/2}f(m_1, d_2 + \delta_2),$

where the summation on the left relates to (3) and that on the right to

$$2m = m_1^2 + m_2, \qquad m_2 = d_2\delta_2 \quad (m_1, m_2, d_2 \text{ odd and} > 0).$$

If f reduces to $f(x)$, (π) becomes (ξ). Also,

(ρ) $4\Sigma\Sigma(-1)^{m'+(\delta''-1)/2}f(2^{a''}d'' + m') - \Sigma\Sigma(-1)^{s}f(s') = 2(-1)^{m-1}f(\sqrt{m})$ or 0,

according as m is a square or not, where m is any integer and

$$m = m'^2 + m'', \qquad m'' = 2^{a''}d''\delta'', \qquad m = s^2 + s'^2 + s''^2,$$

m'', d'', δ'' being positive and the last two odd.

A discussion of the case $m = 8\nu + 7, f(x) = x^2$, shows that $N_1/N_2 = 17/20$, where N_1 is the number of representations of m as a sum of 7 squares in which the first square is odd, and N_2 the number in which the first square is even, including zero.

For m odd and $f(x)$ any even function,

(τ) $\Sigma\Sigma(-1)^{m'+(\delta''-1)/2}f\left(m' + \dfrac{d'' - \delta''}{4}\right) = \begin{cases} (-1)^{(\sqrt{m}-1)/2}\sqrt{m}f(0) & \text{if } m = \text{square}, \\ 0 & \text{if } m \neq \text{square}, \end{cases}$

$$m = 4m'^2 + d''\delta'' \quad (d'', \delta'' \text{ odd and} > 0).$$

For $m = 4\nu + 1$, this formula holds for any function $f(x)$.

Liouville[8] stated that for $F(x, y, z)$ odd with respect to x, y, and z,

(υ) $\Sigma\Sigma F(2^{a''}d'' + m', \delta'' - 2m', 2^{a''+1}d'' + 2m' - \delta'') = 0$ or $\Sigma F(\sqrt{m}, j, j),$

---

[8] Jour. de Math., (2), 5, 1860, 1–8. Twelfth article.

according as m is not or is a square, where $j = 1, 3, 5, \cdots, 2\sqrt{m} - 1$,

$$m = m'^2 + 2^{a''}d''\delta'' \quad (d'', \delta'' \text{ odd and } > 0).$$

This becomes (ϵ) for $F = (-1)^{(1-z)/2}F(x, y)$. Next,

(φ)
$$\Sigma\Sigma F(d'' + m', \delta'' - 2m', 2d'' + 2m' - \delta'')$$
$$= 0 \text{ or } \sum_{s=1}^{2\sqrt{m}-1} F(\sqrt{m}, s, s) - \sum_{t=1}^{\sqrt{m}-1} F(t, 2\sqrt{m}, 2t),$$

according as m is not or is a square, the summation on the left relating to

$$m = m'^2 + d''\delta'' \quad (d'' > 0, \delta'' > 0).$$

For m, d'', δ'' odd and positive,

(χ) $\Sigma\Sigma F(d'' + 2m', \delta'' - 2m', 2m' + d'' - \delta'') = 0 \quad (m = 2m'^2 + d''\delta'')$.

Liouville[9] stated that for a function $F(x, y, z)$ odd with respect to x, y, and z,

(A) $\Sigma F(\delta_3 - 2m_2, d_3 + 2m_2 - m_1, d_3 + 2m_2 + m_1) = 0$,

the summation extending over all partitions of a given integer $m \equiv 3$ (mod 4):

$$m = m_1^2 + 4m_2^2 + 2d_3\delta_3 \quad (m_1, d_3, \delta_3 \text{ odd}, d_3 > 0, \delta_3 > 0).$$

Take

$$F(x, y, z) = \mathscr{F}\left(x, \frac{z+y}{2}\right) - \mathscr{F}\left(x, \frac{z-y}{2}\right),$$

$\mathscr{F}(x, u)$ being odd with respect to x, even with respect to u. Then (A) becomes

(A$_2$) $\Sigma\mathscr{F}(\delta_3 - 2m_2, d_3 + 2m_2) = \Sigma\mathscr{F}(\delta_3 - 2m_2, m_1)$.

With the same notations, Liouville stated in the fourteenth article that

(B) $\Sigma F(\delta_3 - 2m_2, d_3 + 2m_2 - m_1, \delta_3 + m_1) = 0$,

and if $\mathscr{F}(x, y, z, t)$ is changed in sign by a change of sign of x only, or of y only, or of both z and t,

(C) $\Sigma\mathscr{F}(\delta_3 - 2m_2, d_3 + 2m_2 - m_1, d_3 + 2m_2 + m_1, \delta_3 + m_1) = 0$.

When $\mathscr{F}$ is independent of t or z, (C) becomes (A) or (B), respectively.

In the fifteenth article is given the following generalization of (C):

$$\Sigma\mathscr{F}(2^{a_3}\delta_3 - 2m_2, d_3 + 2m_2 - m_1, d_3 + 2m_2 + m_1, 2^{a_3}\delta_3 + m_1)$$

$$= \sum_{\alpha, \beta} \sum_{s=0}^{\frac{1}{2}(\alpha-3)} \mathscr{F}\left(\frac{\alpha - \beta}{2}, \alpha - 2s - 1, \beta + 2s + 1, \frac{\alpha + \beta}{2}\right), \quad (\alpha^2 + \beta^2 = 2m),$$

where $\alpha > 1$ and the sign of β is chosen so that $\frac{1}{2}(\alpha + \beta)$ is odd, while the summation in the first member applies to the partition

$$m = m_1^2 + 4m_2^2 + 2^{a_3+1}d_3\delta_3 \quad (m_1, d_3, \delta_3 \text{ odd}, d_3 > 0, \delta_3 > 0),$$

m being a given odd integer > 1.

---

[9] Jour. de Math., (2), 9, 1864, 249–256, 281–8, 321–336 (13th–15th articles).

Liouville[10] stated that, if $\mathcal{F}(x, y, z, t)$ changes sign with x, or y, or both z and t,

$$\Sigma \mathcal{F}(\delta_3 - 2m_2, d_3 + m_2 - m_1, d_3 + m_2 + m_1, \delta_3 + 2m_1)$$

$$= \sum_{a,\,b} \sum_{s=0}^{a-1} \mathcal{F}(2a - 2s - 1, a - b, a + b, 2b + 2s + 1) \quad (a^2 + b^2 = m, a > 0),$$

where the summation on the left relates to the partitions (of any given integer m)

$$m = m_1^2 + m_2^2 + d_3\delta_3 \quad (d_3 > 0, \delta_3 > 0, \delta_3 \text{ odd}).$$

Liouville[11] stated that if $\psi(x, y)$ is symmetric and even with respect to x,

$$\Sigma(- 1)^{(\delta'-1)/2+(d''-1)/2}\psi(d' - d'', \delta' + \delta'') = \Sigma(- 1)^{(\delta-1)/2}\psi(0, 2d)$$
$$+ 4\Sigma(- 1)^{(\delta_1-1)/2+(\delta_2-1)/2}\psi(2d_1, 2^{a_2+1}d_2),$$

where the summations relate to the partitions, in which m is odd:

$$2m = d'\delta' + d''\delta'', \qquad m = d\delta, \qquad m = d_1\delta_1 + d_2\delta_2,$$

all the symbols being positive integers and, with the exception of α_2, odd.

In the eighteenth article, Liouville employed a function $\mathcal{F}(x, y)$, odd with respect to x and even with respect to y, and stated that

$$\Sigma(- 1)^{(d''-1)/2}\{\mathcal{F}(d' + d'', \delta' - \delta'') + \mathcal{F}(d' - d'', \delta' + \delta'')\}$$
$$= \Sigma\mathcal{F}(2d, 0) + 4\Sigma(- 1)^{(\delta_2-1)/2}\mathcal{F}(2d_1, 2^{a_2+1}d_2).$$

For $\mathcal{F}(x, y) = x$, the latter gives

$$\Sigma\zeta_1(m')\rho(m'') = \zeta_1(m) + 4\Sigma\zeta_1(m_1)\rho(m_2),$$

the summations relating to $2m = m' + m''$, $m = m_1 + 2^{a_2}m_2$, where the m's are all odd and positive.

G. L. Dirichlet[12] proved (a) of Liouville[1] for $\alpha = 1$.   G. Humbert[13] gave a proof by use of infinite series.   G. B. Mathews[14] gave a proof.

J. Liouville[15] stated his[5] formula (γ) and that

$$\Sigma\Sigma(- 1)^m(2^a d + m' - \delta)f(2^a d + m', 2m' - \delta) = \Sigma\Sigma(2^a d - \delta)f(m', 2^a d + \delta),$$

where the double accents on m, α, d, δ have been dropped.

Liouville[16] considered two arbitrary functions $f(m)$ and $F(m)$ having definite values for $m = 1, 2, 3, \cdots$, and set

$$X_\mu(m) = \Sigma d^\mu f(d), \qquad Z_\mu(m) = \Sigma d^\mu F(d),$$

where each summation extends over all divisors d of m.   For any real or complex numbers μ, ν,

$$\Sigma d^{\mu-\nu}X_\nu(d)Z_\mu(\delta) = \Sigma d^{\mu-\nu}Z_\nu(d)X_\mu(\delta) \qquad (\delta = m/d).$$

If we take $f(m)$ and $F(m)$ to be powers of m, we obtain a formula concerning

[10] Jour. de Math., (2), 9, 1864, 389–400.   Sixteenth article.
[11] Jour. de Math., (2), 10, 1865, 135–144, 169–176 (17th and 18th articles).
[12] Bull. des Sc. Math., (2), 33, I, 1909, 58–60; letter to Liouville, Aug. 27, 1858.
[13] Ibid., (2), 34, I, 1910, 29–31.
[14] Proc. London Math. Soc., 25, 1893–4, 85–92.
[15] Bull. des Sc. Math., (2), 33, I, 1909, 61–4; letter to Dirichlet, Oct. 21, 1858.
[16] Jour. de Math., (2), 3, 1858, 63–68.

the sum $\sigma_\mu(k)$ of the μth powers of the divisors of k and given in Ch. X of Vol. I of this History.   From the above formula we readily pass to

$$\Sigma x_\nu(d)z_\mu(\delta) = \Sigma z_\nu(d)x_\mu(\delta), \qquad x_\mu(m) = \Sigma\delta^\mu f(d), \qquad z_\mu(m) = \Sigma\delta^\mu F(d).$$

V. A. Lebesgue[17] noted that for any integer m,

$$\Sigma\zeta_1(m')\zeta_1(m'') = \tfrac{1}{12}\{5\zeta_3(m) - (6m - 1)\zeta_1(m)\},$$

which reduces for the case m a prime to the final formula (H′) of Liouville.[2]

Liouville[18] gave formulas of the type of those in his series of articles.

Liouville[19] noted that, for any integer m,

$$m\zeta_1(m) + 2\sum_{m_1=1}^{[\sqrt{m}]} (m - 5m_1^2)\zeta_1(m - m_1^2) = 0 \text{ or } m(4m - 1)/3,$$

according as m is not or is a square.   This follows from (ϕ) of Liouville[8] with $F(x, y, z) = xyz$.

H. J. S. Smith[19a] gave a proof of (a) and

$$\Sigma f(d' + 2m') = \Sigma f\{\tfrac{1}{2}(d_1 + \delta_1)\},$$

the summations extended respectively over all solutions of

$$m = 2m'^2 + d'\delta', \qquad 2m = m_1^2 + d_1\delta_1,$$

where d', δ', d_1, δ_1, m, m_1 are positive and odd, while $f(x)$ is an odd function.

C. M. Piuma[20] proved (e), (L), (N), (γ), and (ν).

E. Fergola[21] stated and G. Torelli[22] proved a theorem related to one in Liouville's seventh article. Let a_n denote the product of the highest power of 2 dividing n by the sum of the odd divisors of n.   Then

$$a_n - 2a_{n-1} + 2a_{n-4} - 2a_{n-9} + 2a_{n-16} - 2a_{n-25} + \cdots = (-1)^{n-1} n \text{ or } 0,$$

according as n is or is not a square.

S. J. Baskakov[23] proved the formulas in Liouville's twelfth article.

T. Pepin[24] proved all the formulas in Liouville's first five articles except (f) and its specializations (H), (g).

N. V. Bougaief[25] proved some of the theorems in Liouville's series of articles by showing that, if $F(x)$ is an even function, an identity

$$\sum_{m=0}^{\infty} A_m \cos mx = \sum_{n=0}^{\infty} B_n \cos nx$$

implies $SA_m F(m) = SB_n F(n)$, and a similar theorem involving sines and an odd function $F_1(n)$.

[17] Jour. de Math., (2), 7, 1862, 256.

[18] *Ibid.*, 41–8.   To be considered under class number in Vol. III.

[19] Jour. de Math., (2), 7, 1862, 375–6.

[19a] Report British Assoc. for 1865, art. 136; Coll. Math. Papers, I, 346.

[20] Giornale di Mat., 4, 1866, 1–14, 65–75, 193–201.

[21] Giornale di Mat., 10, 1872, 54.

[22] *Ibid.*, 16, 1878, 166–7.

[23] Math. Soc. Moscow, 10, I, 1882–3, 313.

[24] Atti Accad. Pont. Nuovi Lincei, 38, 1884–5, 146–162.

[25] Math. Soc. Moscow, 12, 1885, 1–21.

Pepin[26] proved all the theorems in Liouville's first five and last two articles, and (L) of the sixth.

E. Meissner[27] proved all the theorems in Liouville's articles VII–XVI. Thus there remain unproved essentially only (N) and (Q) of the sixth article. [Piuma,[20] pp. 197–201, proved (N).]

P. Bachmann[28] gave an exposition of selected formulas from Liouville's series.

A. Deltour[29] proved (a) and recalled how it implies that, if m is odd, the number of decompositions of $4m$ (or $8m$) into a sum of 4 (or 8) odd squares equals the sum (or sum of cubes) of the divisors of m.

P. S. Nasimoff[30] proved formulas (a) and (c) of Liouville,[1] (F) of Liouville,[2] (P) of Liouville,[4] one of Liouville,[18] and related results.

---

[26] Jour. de Math., (4), 4, 1888, 83–127.
[27] Zürich Vierteljahr Naturf. Ges., 52, 1907, 156–216 (Diss., Zürich).
[28] Niedere Zahlentheorie, 2, 1910, 365–433.
[29] Nouv. Ann. Math., (4), 11, 1911, 123–9.
[30] Application of Elliptic Functions to Number Theory, Moscow, 1885.   French résumé in Annales sc. de l'école norm. supér., (3), 5, 1888, 147–64.

CHAPTER XII.

PELL EQUATION; $ax^2 + bx + c$ MADE A SQUARE.

The very important equation $x^2 - Dy^2 = 1$, which has long borne the name of Pell, due to a confusion originating with Euler, should have been designated as Fermat's equation (cf. papers 41, 62–64).

There appeared in India and Greece as early as 400 B.C. approximations a/b to $\sqrt{2}$ such that $a^2 - 2b^2 = 1$, and similarly for other square roots, the derivation of successive approximations being in effect a method of solving the Pell equation. For example, Baudhâyana, the Hindu author of the oldest of the works, Sulva-sutras, gave the approximations 17/12 and 577/408 to $\sqrt{2}$. Note that

$$\frac{17}{12} + \frac{-1}{2 \cdot 17 \cdot 12} = \frac{577}{408}, \qquad 17^2 - 2 \cdot 12^2 = 1, \qquad 577^2 - 2 \cdot 408^2 = 1.$$

Proclus[1] (410–485 A.D.) noted that the Pythagoreans made the following construction: On the prolongation of the side AB of a square with the diagonal BE lay off $BC = AB$, $CD = BE$. Then

$$AD^2 + CD^2 = 2AB^2 + 2BD^2.$$

But $CD^2 = BE^2 = 2AB^2$. Hence

$$AD^2 = 2BD^2 = FD^2, \qquad FD = AD = 2AB + EB.$$

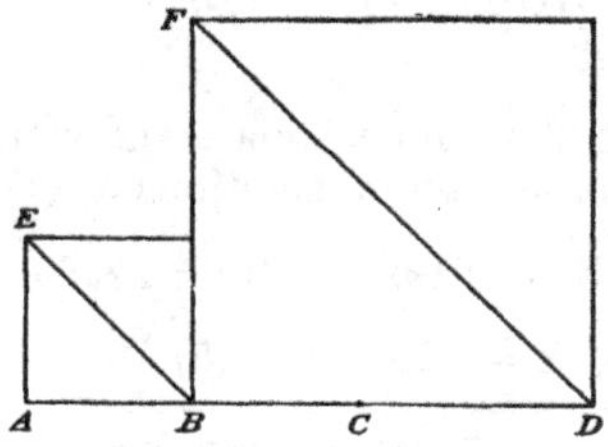

Also $BD = AB + EB$. Write $s_1, s_2, \cdots$ for the sides $AB, BD, \cdots$, and $d_1, d_2, \cdots$ for the diagonals $BE, FD, \cdots$. Then

$$s_{n+1} = s_n + d_n, \qquad d_{n+1} = 2s_n + d_n.$$

Now let $s_1 = 1$ and replace $d_1 = \sqrt{2}$ by the integral approximation $\delta_1 = 1$, and employ our recursion formulæ with d_n replaced by δ_n. We get

$$s_2 = s_1 + \delta_1 = 2, \qquad \delta_2 = 2s_1 + \delta_1 = 3,$$
$$s_3 = s_2 + \delta_2 = 5, \qquad \delta_3 = 2s_2 + \delta_2 = 7, \cdots$$

Then δ_n, s_n give a solution of $\delta^2 - 2s^2 = (-1)^n$.

---

[1] In Platonis rem publicam commentarii, ed., G. Kroll, 2, 1901, 24–9; excurs II (by F. Hultsch), 393–400.

Theon of Smyrna[2] (about 130 A.D.) called the s's and δ's side and diametral (diagonal) numbers and gave the above recursion formulæ without the geometrical interpretation.

Archimedes (third century B.C.) gave the approximations 265/153 and 1351/780 to $\sqrt{3}$, which can be explained in connection with $x^2 - 3y^2 = -2$, $x^2 - 3y^2 = 1$.

Heron of Alexandria used the approximation $a + r/(2a)$ for $\sqrt{a^2 + r}$.

For a more detailed account than what precedes of the connection between the knowledge of the early Greeks and Hindus of approximation to square roots and Pell equations, see H. Konen[3] and E. E. Whitford.[4]

The history of the cattle problem of Archimedes will now be discussed in detail.

In 1773, Gotthold Ephraim Lessing[5] published a Greek epigram in 24 verses, from a manuscript in the Wolfenbüttel library, stating a problem purporting to be one proposed by Archimedes,[6] in a letter to Eratosthenes, to the mathematicians of Alexandria, as well as a scholium giving a false answer, and a long mathematical discussion by Chr. Leiste. The problem is to find the numbers W, X, Y, Z of the white, black (or blue), piebald (or spotted), and yellow (or red) bulls, and the numbers w, x, y, z of the cows of the corresponding colors, when

$$(1)\quad W = (\tfrac{1}{2} + \tfrac{1}{3})X + Z, \qquad (2)\quad X = (\tfrac{1}{4} + \tfrac{1}{5})Y + Z,$$
$$(3)\quad Y = (\tfrac{1}{6} + \tfrac{1}{7})W + Z, \qquad (4)\quad w = (\tfrac{1}{3} + \tfrac{1}{4})(X + x),$$
$$(5)\quad x = (\tfrac{1}{4} + \tfrac{1}{5})(Y + y), \qquad (6)\quad y = (\tfrac{1}{5} + \tfrac{1}{6})(Z + z),$$
$$(7)\quad z = (\tfrac{1}{6} + \tfrac{1}{7})(W + w), \qquad (8)\quad W + X = \square,$$
$$(9)\quad Y + Z = \triangle,$$

the final notations being those for a square and a triangular number.

Leiste found at once the integral solutions of (1), (2), (3):

$$(10)\quad Y = 1580m, \qquad Z = 891m, \qquad W = 2226m, \qquad X = 1602m.$$

Then, by (4), $m = 2p$, $x = 12\alpha$. By (5), $\alpha = 3\beta$, $y = 20(4\beta - 158p)$.

---

[2] Platonici . . . expositio, 1544, 67. Theon Smyrnaeus, ed., E. Hiller, Leipzig, 1878, 43; French transl., by J. Dupuis, Paris, 1893, 71–5.

[3] Geschichte der Gleichung $t^2 - Du^2 = 1$, Leipzig, 1901, 2–17. Reviews by Wertheim, Bibl. Math., (3), 3, 1902, 248–251; and Tannery, Bull. des Sc. Math., 27, II, 1903, 47.

[4] The Pell Equation, Columbia Univ. Diss., New York, 1912, 3–22. The following related papers are not mentioned in the pages just cited: E. S. Unger, Kurzer Abriss der Gesch. Z. von Pythagoras bis Diophant, Progr., Erfurt, 1843; C. Henry, Bull. des Sc. Math. Astr., (2), 3, 1, 1879, 515–20; H. Weissenborn, Die irrationalen Quadratwurzeln bei Archimedes und Heron, Berlin, 1884; Zeitschr. Math. Phys., Hist.-Lit. Abt., 28, 1883, 81; E. Mahler, ibid., 29, 1884, 41–3; W. Schoenborn, 30, 1885, 81–90; C. Demme, 31, 1886, 1–27; K. Hunrath, 33, 1888, 1–11; V. V. Bobynin, 41, 1896, 193–211; M. Curtze, 42, 1897, 113, 145; F. Hultsch, Göttingen Nachr., 1893, 367; G. Wertheim, Abh. Gesch. Math., VIII, 146–160 (in Zeitschr. Math. Phys., 42, 1897); Zeitschr. Math. Naturw. Unterricht, 30, 1899, 253; T. L. Heath, Euclid's Elements, 1, 1908, 398–401.

[5] Zur Geschichte der Literatur, Braunschweig, 2, 1773, No. 13, 421–446. Lessing, Sämmtliche Schriften, Leipzig, 22, 1802, 221; 9, 1855, 285–302; 12, 1897, 100–15; Opera, XIV, 232.

[6] Archimedes opera, ed., J. L. Heiberg, 2, 1881, 450–5; new ed., 2, 1913, 528–34.

By (6), $p = 5q$, $z = 30\gamma$, $y = 11(297q + \gamma)$, whence $11\gamma = 80\beta - 19067q$. Then (7) gives $30\gamma = (1505q + \beta)13/2$, $q = 2r$, $\beta = 2\delta$. Comparing the resulting γ with the earlier γ, we get a linear equation in δ, r, whence

$$r = 4657u, \qquad \delta = 1359235u.$$

By substitutions, we get $m = 93140u$, whence

$$
(10')\quad
\begin{aligned}
W &= 207329640u, & w &= 144127200u \\
X &= 149210280u, & x &= 97864920u \\
Y &= 147161200u, & y &= 70316400u \\
Z &= 82987740u, & z &= 108784260u.
\end{aligned}
$$

For $u = 4$, we get the numbers in the scholium; but they satisfy neither (8) nor (9), since neither $W + X$ nor $8(Y + Z) + 1$ is a square.

Returning to (10'), we note that the greatest common divisor of the numerical factors is 20, whence $u = v/20$, where v is an integer. Then

$$W + X = 4\cdot957\cdot4657v, \qquad v = 957\cdot4657n^2,$$

since $W + X$ is to be a square. Then $Y + Z = (t^2 + t)/2$ gives

$$(2t + 1)^2 = 8(Y + Z) + 1 = an^2 + 1, \qquad a = 410286423278424.$$

Since a is positive and not a square it is possible to choose an integer n so that $an^2 + 1 = \square$ by Euler.[81] If the resulting square is even, we can deduce one making $an^2 + 1$ an odd square (Euler,[83] § 86, § 88).

J. J. I. Hoffmann[7] said the problem was due to a much later computer.

J. Struve[8] gave a 36 page discussion making no advance over Leiste.

Gottfried Hermann[9] made an interpretation which led, not to (8) and (9), but to $W + X = $ a square whose side is of the form $a^2(a - b)$, $Y + Z = \triangle$, $W + X + Y + Z = \triangle_1$. Thus if we take the numbers (10), we must make

$$3828m = \{a^2(a - b)\}^2, \qquad 2471m = \frac{c(c + 1)}{2}, \qquad 6299m = \frac{d(d + 1)}{2}.$$

He stated on the authority of K. B. Mollweide that C. F. Gauss had completely solved the problem under the earlier interpretation, but had not published the solution.

J. Fr. Wurm,[10] in a review of Hermann's paper, replaced (p. 201) condition (8) by the condition that $W + X$ shall be a product of two approximately equal factors. Without returning to this condition, he passed to (9):

$$Y + Z = 2471m = 2471\cdot151t = \triangle.$$

---

[7] Ueber die Arith. der Griechen, Mainz, 1817, Introd., p. xvi (transl. of Delambre).

[8] Altes griechisches Epigramm, mathematischen Inhalts, von Lessing erst einmal zum Drucke befördert jetzt neu abgedruckt und mathematisch und kritisch behandelt von Dr. J. Struve und Dr. K. L. Struve, Vater und Sohn. Altona, 1821, 47 pp.

[9] Ad memoriam Kregelio-Sternbachianam in and. jur. die 17 Julii 1828: De Archimedis Problemate Bovino, Universitäts programm, Leipzig, 1828. Reprinted in Godofredi Hermanni, Opvscvla, Lipsiae, 4, 1831, iii–v, 228–238.

[10] Jahrbücher für Philologie u. Paedagogik (ed., J. C. Jahn), 14, 1830, 194–202.

The least t is 990, the side of $\triangle$ being then 27180. He considered also higher values of t, but gave no final answer to (1)–(9).

G. H. F. Nesselmann[11] argued that the final part of the epigram leading to conditions (8) and (9) was a later addition, partly since he believed that triangular numbers were not employed in Archimedes' time (a view already expressed by G. S. Klügel[12]).

O. Terquem[13] stated that the tenth condition added by Hermann is incompatible with the earlier conditions.

A. J. H. Vincent[14] regarded as spurious the conditions relating to the cows. By the first three conditions, we have (10). Then $Y + Z = 2471m$ is to be a $\triangle$ and this is the case if $m = 99 \cdot 122314$, the side of the $\triangle$ being 244628. Then $4\sqrt{W + X}$ is approximately 861182, which is very nearly the area of Sicily in square *stades*, in accord with Vincent's interpretation of the condition to replace (8).

C. F. Meyer[15] duplicated the paper by Lessing and discussion by Leiste, adding merely that, in attempting to make $an^2 + 1$ a square by the convenient method of Kausler, he had carried the development of $\sqrt{a}$ into a continued fraction to the 240th quotient without finding the period.

A. Amthor[16] showed that Wurm's problem (1)–(7), (9) is satisfied by taking $u = v/20$, $v = 117423$ in Leiste's values of $W, \cdots, z$, since then

$$Y + Z = 1643921 \cdot 1643922/2, \qquad W + X = 1485583 \cdot 1409076.$$

For the main problem (1)–(9), he satisfied (8) by taking $v = f \cdot 4657n^2$, $f = 3 \cdot 11 \cdot 29 = 957$, as in Leiste. Then in (9), viz., $Y + Z = q(q + 1)/2$, set $t = 2q + 1$, $u = 2 \cdot 4657n$. We obtain the Pell equation

$$t^2 - Du^2 = 1, \qquad D = 2 \cdot 7 \cdot f \cdot 353 = 4729494.$$

He found that the continued fraction for $\sqrt{D}$ has a period of 91 terms and obtained as the least solutions

$$T = 109\ 931\ 986\ 732\ 829\ 734\ 979\ 866\ 232\ 821\ 433\ 543\ 901\ 088\ 049,$$
$$U = 50\ 549\ 485\ 234\ 315\ 033\ 074\ 477\ 819\ 735\ 540\ 408\ 986\ 340.$$

It remains to derive the least solutions t, u in which u is divisible by $2 \cdot 4657$, so that n shall be integral. By proving and applying general lemmas concerning $t_k + u_k \sqrt{D} = (T + U \sqrt{D})^k$, he found that, for $k = 2329$, t_k, u_k is the desired pair. He verified that W has 206545 digits.

B. Krumbiegel[17] made a historical and philological discussion of the problem and concluded that, while the epigram itself is probably subsequent to Archimedes, the problem itself is due to him. This accords with the

[11] Die Algebra der Griechen, Berlin, 1842, 488. On p. 485, his $g = 57 \cdots$ should be $54 \cdots$.

[12] Math. Wörterbuch, 1, 1803, 184. Cf. M. Cantor, Geschichte Math., ed. 2, I, 297; ed. 3, I, 312.

[13] Nouv. Ann. Math. 14, 1855, Bull. Bibl., 113–124, 130–1. He at first attributed incorrectly Hermann's paper to F. E. Theime.

[14] Nouv. Ann. Math., 14, 1855, Bull. Bibl., 165–173; 15, 1856, Bull. Bibl., 39–42 (restored Greek text and French transl.).

[15] Ein diophantisches Problem, Progr., Potsdam, 1867, 14 pp.

[16] Zeitschrift Math. Phys., 25, 1880, Hist.-Lit. Abt., 153–171.

[17] Zeitschrift Math. Phys., 25, 1880, Hist.-Lit. Abt., 121–136.

view of J. L. Heiberg,[18] P. Tannery,[19] F. Hultsch,[20] T. L. Heath,[21] and
S. Günther.[22]

A. H. Bell[23] found a "complete solution," based on the $an^2 + 1 = \square$ of
Leiste, involving numbers of 206545 digits, as by Amthor.[16]

G. Loria,[23a] M. Merriman,[23b] and R. C. Archibald[23c] gave accounts of the
cattle problem.

Diophantus (about 250 A.D.) was frequently led to special Pell equations in solving problems in his Arithmetica. In II, 12, 13, 14, 29, he
made $y^2 + 1$, $y^2 + 12$, $y^2 - 1$, $y^2 + 1$, $9y^2 + 9$ equal to a square z^2, by
taking $z = y - 4$, $y - 4$, $y - 2$, $y - 2$, $3y - 4$, respectively, and similarly
in II, 30. In III, 12, 13, he avoided the initial equations $52x^2 + 12 = \square$,
$266x^2 - 10 = \square$, since 52 and 266 are not squares [though $x = 1$ is
a solution of each], and, beginning anew, was led to $y^2 + 12 = \square$,
$77^2z^2 - 160 = \square$, which he solved by equating them to $(y + 3)^2$ and
$(77z - 2)^2$, respectively. In IV, 8, 33, he treated $2x^2 + 4 = \square = (2x - 2)^2$
and $7m^2 + 81 = \square = (8m + 9)^2$. In V, 12, 14, he discussed

$$26x^2 + 1 = \square = (5x + 1)^2$$

and $30x^2 + 1 = \square = (5x + 1)^2$. So far, the problems solved are all of
the form $ax^2 + b = \square$ with either a or b a square. In VI, 12, he stated the
lemma: Given two numbers whose sum is a square, we can find an infinitude
of squares s such that, when the square s is multiplied by one of the given
numbers and the product is added to the other, the result is a square.
Thus, given the numbers 3 and 6, let $s = (x + 1)^2$; then shall

$$3(x + 1)^2 + 6 = 3x^2 + 6x + 9 = \square,$$

say $(3 - 3x)^2$, whence $x = 4$; and an infinitude of other solutions can be
found. This lemma is applied in VI, 13, 14 to $12x^2 + 24 = \square$ to obtain
the solutions $x = 1, 5$. In VI, 15, $15x^2 - 36 = \square$ is said to be impossible
since 15 is not a sum of two squares. In VI, 16, he made the important
statement that, given one solution of $Ax^2 - B = y^2$, we can find a second
solution; thus, given $3 \cdot 5^2 - 11 = 8^2$, set $x = 5 + z$, whence

$$3(5 + z)^2 - 11 = 3z^2 + 30z + 8^2$$

will be the square of $8 - 2z$ for $z = 62$. In VI, 12, he had made the more

---

[18] Questiones Archimedeae, Diss. Hauniae, 1879, 25–27; Philologus, 43, 1884, 486.
[19] Mém. soc. sc. phys. nat. Bordeaux, (2), 3, 1880, 370; Bull. des Sc. Math. et Astr., (2), 5, I,
     1881, 25–30; Bibl. Math., 3, 1902, 174. Reprinted in Tannery's Mémoires scientifiques,
     1, 1912, 103–5, 118–23.
[20] Archimedes, in Pauly-Wissowa's Real-Encyclopädie, II₁, 1896, 534, 1110.
[21] Diophantus, ed. 2, 1900, 11–12, 122, 279; Archimedes, 1897, 319; Archimedes' Werke, 1914,
     471–7.
[22] Die quadr. Irrationalitäten, etc., Zeitschrift Math. Phys., Abh. Gesch. Math., 27, 1882, 92.
     This and K. Hunrath's Ueber das Ausziehen der Quadratwurzel bei Griechen und Indern,
     1883, were reviewed in La Revue Scientifique, 1884, I, 81–3, 499–502.
[23] Math. Magazine, Washington, 2, 1895, 163–4; Amer. Math. Monthly, 2, 1895, 140–1
     (1, 1894, 240).
[23a] Le scienze esatte nell'antica Grecia, ed. 2, 1914, 932–9.
[23b] The Popular Science Monthly, 67, 1905, 660–5.
[23c] Amer. Math. Monthly, 25, 1918, 411–4.

special remark that $6x^2 + 3 = \square$ has an infinitude of solutions, since it has one solution $x = 1$.

Diophantus solved $Ax^2 + Bx + C = y^2$ only in the following cases. (a) If A is a square, a^2, set $y = ax + m$, whence x is found rationally; examples in II, 20, 21, 23, 24, 33, III, 9, 16, 18, IV, 15, 21, V, 3, 4, 18, 20. (b) If $C = c^2$, set $y = mx + c$; examples in II, 17, IV, 9, 10, 12, 14, 45. (c) In IV, 33, $18 + 3x - x^2$ is to be made a square, say m^2x^2, where $(m^2 + 1) 18 + (\frac{3}{2})^2 = \square$. Then, multiplying by 4, $72m^2 + 81 = \square$, say $(8m + 9)^2$, whence $m = 18$, $18 + 3x - 325x^2 = 0$, $x = 6/25$. In general, as remarked by Nesselmann[11] (pp. 333-4), the corresponding condition that the root x of $Ax^2 + Bx + C = m^2x^2$ be rational is $\frac{1}{4}B^2 - AC + Cm^2 = \square$, and, as in (b), can be satisfied if $\frac{1}{4}B^2 - AC$ is a square.

While H. Hankel[24] believed that Diophantus was influenced by Indian sources, M. Cantor[25] took the opposite view except as to integral solutions. P. Tannery[26] went to the extreme of believing that the Greeks influenced the Indians also in the question of integral solutions, while even the cyclic method [next explained] is only a variation of the Greek method of solving $t^2 - Du^2 = 1$, since from the Greek method of deriving from one approximation to $\sqrt{D}$ a closer approximation it is easy to pass to the Indian method.

E. B. Crowell[27] compared the work of Diophantus with that of Brahmegupta,[28] and the first solution by Brouncker[43] with that of Bháscara.[30]

Brahmegupta[28] (born 598 A.D.) gave a rule to find x so that $Cx^2 + 1$ shall be a square. Assume any "least root" L and add to CL^2 such an "additive" number A that the sum is a square G^2; call G the "greatest root" [L and G are values of x, y satisfying $Cx^2 + A = y^2$]. Write L, G, A twice. By cross multiplication, we obtain a least root $LG + GL$, while $CLL + GG$ is a greatest root, for additive AA; dividing these new roots by A, we get roots for additive unity. For details, see Bháscara.[30]

For example (§ 67), let $C = 92$. Take $L = 1$, $A = 8$, whence $G = 10$. Then $2LG = 20$, $92L^2 + G^2 = 192$ are least and greatest roots for additive 64. Dividing them by 8, we get $5/2$ and 24 as roots for additive unity. By composition of the last pair with itself, we get other roots 120 and 1151 for additive unity.

By composition of the roots for additive unity with the roots for additive A (§ 68, p. 364). For example (§ 77, p. 368), from $3 \cdot 30^2 + 900 = 60^2$, $3 \cdot 1^2 + 1 = 2^2$, we get the least root

$$30 \cdot 2 + 1 \cdot 60 = 120$$

and greatest root $3 \cdot 30 \cdot 1 + 60 \cdot 2 = 210$ for $3 \cdot 120^2 + 900 = 210^2$.

---

[24] Zur Geschichte der Math. in Alterthum und Mittelalter, 1874, 204.

[25] Vorles. über Geschichte Math., 1, 1880, 533; ed. 2, 556; ed. 3, 596.

[26] Mém. Soc. Sc. Phys. Nat. Bordeaux, (2), 4, 1882, 325.

[27] M. Elphinstone's History of India, ed. 9, 1905, 142, Note 16 (ed., Crowell).

[28] Brahme-sphut'a-sidd'hánta, Ch. 18 (algebra), §§ 65–66. Algebra, with arith. and mensuration, from the Sanscrit of Brahmegupta and Bháscara, transl. by H. T. Colebrooke, 1817, p. 363. Cf. Simon.[300]

We may deduce roots for additive unity from roots for additive ± 4 (§§ 69–72, pp. 365–6). If $CL^2 + 4 = G^2$, then $L(G^2 - 1)/2$ and $G(G^2 - 3)/2$ are corresponding least and greatest roots for additive unity. If $CL^2 - 4 = G^2$, and we set $p = (G^2 + 1)(G^2 + 3)/2$, then pLG and $(p - 1)(G^2 + 2)$ are corresponding least and greatest roots for additive unity.

If the coefficient C be a square (§ 73, p. 366), divide the additive by any assumed number b. To the quotient add b and from it subtract b and divide by 2. The first result is a greatest root; the second, divided by the square root of C, is the corresponding least root.

If the coefficient be divisible by a square t^2 (§ 75, p. 367), use the quotient as a new coefficient and find roots. If the least root so found is divided by t, we get the desired least root. The greatest root remans the same.

For $C = 3$, $A = -800$ (§ 77, p. 368), remove the factor 20^2. For the new additive -2, we get roots 1 and 1. Their products by 20 are the roots desired.

Alkarkhi[29] (about 1010) solved $x^2 + 5 = y^2$ by setting $y = x + 1$, and $x^2 - 10 = y^2$ by setting $y = x - 1$. To solve $77^2 x^2 - 160 = w^2$, set $w = 77x - 2$. To solve (pp. 72–4) $x^2 + 4x = y^2$, set $y = 2x$; to solve $4x^2 + 16x + 9 = y^2$, set $y = 2x - n$, where $n^2 > 9$, say $n = 5$. As the condition (p. 113) for rational solutions of $\pm (ax - b) - x^2 = \square$, he found that $\frac{1}{4}a^2 \mp b$ must be a sum of two squares. Finally (p. 121), $v^2 - w^2 = \alpha\beta$ for $v = (\alpha + \beta)/2$, $w = (\alpha - \beta)/2$.

Alkarkhi[29a] used the approximation $a + r/(2a + 1)$ for $\sqrt{a^2 + r}$.

Ibn Albanná[29b] (born about 1255) used the same approximation when $r > a$, but for $r \leqq a$ employed $a + r/(2a)$. The latter was used by Heron of Alexandria and by Elia Misrachi (1455–1526) in his Arithmetic (ed., G. Wertheim, 1893, 1896).

Bháscara Achárya[30] (born 1114) gave a method of deducing new sets of solutions of $Cx^2 + 1 = y^2$ from one set found by trial. Take any number $\neq 0$ and call it the "least root" L [for additive A]. By the positive or negative additive quantity A is meant a number which added to or subtracted from CL^2 makes the sum or difference a perfect square, its root being called the "greatest root" G. Thus if $C = 8$, $L = 1$, $A = 1$, then $G = 3$.

Composition (§§ 76–77, p. 171). From these roots L, G and the same or a new set of roots l, g, we obtain by cross multiplication and addition a new least root $\lambda = Lg + lG$, while $\gamma = CLl + Gg$ is the corresponding new greatest root. The product of the two additives gives the new additive. Thus (§ 82) for the former example, take $l = 1$, $g = 3$, $A = 1$; then $\lambda = 6$, $\gamma = 17$. Next, from $L = 1$, $G = 3$ and $\lambda = 6$, $\gamma = 17$, we get the new roots 35, 99 and so on indefinitely by means of composition.

[29] Extrait du Fakhrî, Traité d'algèbre par Ben Alhacan Alkarkhî (Arab MS.), French transl. by F. Woepcke, Paris, 1853, 84, 120.

[29a] Kâfî fîl Hisâb, German transl. by A. Hochheim, II, 14.

[29b] Le Talkhys, p. 23. French transl. by A. Marre, Atti Accad. Pont. Nuovi Lincei, 17, 1864, 311.

[30] Víja-gan'ita (algebra), Ch. 3, §§ 75–99, "Affected square." Colebrooke,[28] 170–184.

Or (§ 78, p. 171) we may take $Lg - lG$ and $CLl - Gg$ as new roots.

A second method (§§ 80–81, p. 172) for additive unity consists in taking the least root to be $2a/(a^2 - C)$, where a is arbitrary, and finding the greatest root. Thus (end of § 82, p. 174), for $C = 8$, take $a = 3$; the least root is 6 and the greatest is the square root 17 of $8 \cdot 6^2 + 1$.

Cyclic method (§§ 83–86, pp. 175–6). Taking the least root, greatest root and additive as dividend, additive and divisor, find the multiplier by use of the pulverizer (see papers 2, 4 of Ch. II). If the excess of the square of that multiplier over the given coefficient C be divided by the original additive, we get a new additive. The quotient corresponding to the multiplier and found from it will be the new least root, from which a greatest root may be deduced. The operation may be repeated. We find integral roots with 4, 2 or 1 for additive, and by composition deduce roots for additive unity from those for additives 4 and 2.

For example (§ 87, pp. 176–8), to make $67x^2 + 1$ a square, take 1 as a least root, $- 3$ as additive, whence 8 is the greatest root. Thus dividend $= 1$, divisor $= - 3$, additive $= 8$. By the pulverizer, a multiplier is 7 and the quotient is $- 5$, a new least root. The new additive is $6 = (7^2 - 67)/(- 3)$. By $67(- 5)^2 + 6 = 41^2$, 41 is the new greatest root. Now start with dividend 5, divisor 6, additive 41, get the multiplier 5, quotient $11 =$ least root, new additive $- 7 = (5^2 - 67)/6$, and greatest root 90. Next, start with dividend 11, divisor $- 7$, additive 90. Reducing the last by multiples of the divisor, we get the abraded additive 6. The multiplier is 2. Adding the negative of the divisor, we get the new multiplier 9 and the quotient 27, giving a least root. The new additive is $(9^2 - 67)/(- 7) = - 2$, and greatest root is 221. By composition of this set of roots with itself, we get $L = 11934$, $G = 97684$, $A = 4$. Divide the roots by the square root of 4. We get $l = 5967$, $g = 48842$ for additive 1.

When unity is subtractive (§§ 88–89, p. 179), the problem is impossible if the coefficient C be not a sum of two squares. In the contrary case, we may take as two least roots the reciprocals of the roots of the two component squares. Thus (§ 90) if $C = 13 = 2^2 + 3^2$, the least root $\frac{1}{2}$ gives the greatest root $\frac{3}{2}$. Doubling and applying the cyclic method, we have dividend 1, divisor $- 2$, additive 3. We deduce the multiplier 3 and quotient $- 3$, the least root. The new additive is 4 and greatest root is 11. Repeating the operation, we get $L = 5$, $G = 18$, $A = - 1$.

When C is a square α^2 (§ 95, p. 182) and the additive is A, least and greatest roots are (for b arbitrary)

$$\frac{1}{2\alpha}\left(\frac{A}{b} - b\right), \qquad \frac{1}{2}\left(\frac{A}{b} + b\right).$$

Bháscara solved various problems by the method of the affected square. For $6y^2 + 2y = c^2$ (§ 177, p. 247), $(6y + 1)^2 = 6c^2 + 1$ for $c = 2$ or 20, $y = \frac{2}{3}$ or 8. To find (§ 178, p. 248) two numbers the square of whose sum added to the cube of their sum equals twice the sum of their cubes, take $y-c$ and $y+c$ as the numbers, whence

$$(2y)^2 + (2y)^3 = 2(2y^3 + 6yc^2), \qquad (2y+1)^2 = 12c^2 + 1,$$
$$c = 2, 28; \qquad y = 3, 48.$$

For $5y^4 - 100y^2 = c^2$ (§ 181, p. 249), divide by y^2. To find (§ 182) two numbers whose difference is a square, and sum of squares a cube, take c and $c - n^2$ as the numbers; the sum $2c^2 - 2cn^2 + n^4$ of their squares is equated to n^6 (a restriction), whence $(2c - n^2)^2 = n^4(2n^2 - 1)$, and $2n^2 - 1$ is made a square. To make (§ 188, p. 253) $y^2 + z^3$ and $y + z$ both squares, treat the first condition by § 95 with z^3 as the additive and z as the arbitrary number b; we get $y = (z^2 - z)/2$; the second condition now becomes $\frac{1}{2}(z^2 + z) = p^2$, or $(2z + 1)^2 = 8p^2 + 1$, which is a square for $p = 6$ or 35. The sum (§ 189, p. 254) of the squares of two numbers increased by their product is to be a square; on adding unity to the product of their sum by the root of that square, the sum shall be a square. The first condition is found to be satisfied by the numbers $\frac{8}{3}c$ and c; then the second condition $(\frac{8}{3}c)(\frac{7}{3}c) + 1 = \square$ holds if $c = 6$ or 180.

E. Strachey[31] translated into English the Persian manuscript of 1634 of Bháscara. To solve $Ax^2 + B = y^2$, take any square f^2 and find a number β such that $Af^2 + \beta$ is a square, say g^2. Then $x' = 2fg$, $y' = Af^2 + g^2$ satisfy $Ax'^2 + \beta' = y'^2$ for $\beta' = \beta^2$; and

$$x'' = x'g \pm y'f, \qquad y'' = y'g \pm Ax'f$$

satisfy

$$Ax''^2 + \beta'' = y''^2, \qquad \beta'' = \beta'\beta.$$

If $\beta'' = Bp^2$, remove the factor p from x'', y''; we get a solution of the proposed equation (if $\beta'' = B/p^2$, multiply by p). Otherwise, we proceed as before. For example, consider $8x^2 + 1 = y^2$. Take $f = 1$; then

$$8f^2 + 1 = 3^2,$$

so that we take $\beta = 1$. Then

$$x' = 2 \cdot 1 \cdot 3 = 6, \qquad y' = 8 \cdot 1^2 + 3^2 = 17, \qquad 8 \cdot 6^2 + 1 = 17^2.$$

A new set of solutions is given by

$$x'' = 6 \cdot 3 + 17 \cdot 1 = 35, \qquad y'' = 17 \cdot 3 + 8 \cdot 6 \cdot 1 = 99.$$

For the cyclic method ("operation of circulation"), choose as before [relatively prime] numbers f and g such that $Af^2 + \beta = g^2$. Then by an earlier rule [for solving a linear Diophantine equation] choose integers X, Y such that $(fX + g)/\beta = Y$. Choose an integer m so that the difference between $(m\beta + X)^2$ and A shall be as small as possible numerically. Now $(m\beta + X)^2 - A$ is divisible by β; call the quotient β'. Set $x' = mf + Y$. Then $Ax'^2 + \beta'$ is a square, say y'^2. Unless $\beta' = Bp^2$ or B/p^2, proceed as before. For example, let $A = 67$, $B = 1$. Take $f = 1$, $\beta = -3$; then $g = 8$, $X = 1$, $Y = -3$, $m = -2$,

$$(m\beta + X)^2 - A = 7^2 - 67 = -18 = \beta\beta', \qquad \beta' = 6, \qquad x' = -5, \qquad y' = 41.$$

---

[31] Bija Ganita, or the algebra of the Hindus, London, 1813, Introduction, pp. 36–53.

The next step gives $\beta'' = -7$; the third, $\beta''' = -2$. His solution of $x^2 - 61y^2 = 1$ is quoted by Whitford[4] (pp. 37–8), who remarked that the wording is clearer than in Colebrooke's translation.

El-Hassar[32] (1432) obtained for $\sqrt{a^2 + r}$, when $a = 2, r = 1$, the approximations $a + \rho = 9/4$, where $\rho = r/(2a)$, and

$$a + \rho - \rho^2/\{2(a + \rho)\} = 161/72.$$

[Note that (9, 4) and (161, 72) are solutions of $x^2 - 5y^2 = 1$.]

Nicolas Chuquet[33] obtained, in 1484, successive approximations to $\sqrt{n}$ for $n \leqq 14$. He began by noting that $\sqrt{6}$ lies between 2 and 3. Their arithmetical mean is $2\frac{1}{2}$; its square $6\frac{1}{4}$ exceeds 6 by $\frac{1}{4}$. Take the next smaller term $\frac{1}{3}$ in the series $\frac{1}{2}, \frac{1}{3}, \frac{1}{4}, \frac{1}{5}, \cdots$. We have $2\frac{1}{3}$, whose square is less than 6. We now have an approximation exceeding the root and one less than it. Adding the numerators and denominators of $\frac{1}{2}$ and $\frac{1}{3}$, we get the new approximation $2\frac{2}{5}$, whose square < 6. Similarly from $2\frac{1}{2}$ and $2\frac{2}{5}$ we get $2\frac{3}{7}$. In this way he obtained the approximations $2 + r$, where

$$r = \tfrac{1}{2}, \tfrac{1}{3}, \tfrac{2}{5}, \tfrac{3}{7}, \tfrac{4}{9}, \tfrac{5}{11}, \tfrac{9}{20}, \tfrac{13}{29}, \tfrac{22}{49}, \tfrac{31}{69}, \tfrac{40}{89}, \tfrac{49}{109}, \tfrac{89}{198}.$$

[For $r = 0, \tfrac{1}{2}, \tfrac{4}{9}, \tfrac{9}{20}, \tfrac{40}{89}, \tfrac{89}{198}$, $2 + r$ gives the successive convergents to the continued fraction for $\sqrt{6}$. To deduce a third convergent p_2/q_2 from two successive ones p_0/q_0, p_1/q_1, the law is $p_2 = p_0 + zp_1$, $q_2 = q_0 + zq_1$. Thus Chuquet's process produced also intermediate fractions, obtained by replacing z by smaller numbers.] Chuquet[34] gave answers to the following problems, but with no details as to solution. Find a square which increased by 7 (or 4) gives a square; answer, 9 (or 9/4). Find three squares whose sum is 13; answer, $11\frac{1}{9}, 1\frac{7}{9}, \frac{1}{9}$. Find three cubes whose sum is 20; answer, $15\frac{5}{8}, 3\frac{3}{8}, 1$.

Jordanus Nemorarius[35] noted that $x(x + 1)$ is neither a square nor a cube [if x is an integer $\neq 0, -1$; for $x = \frac{1}{3}$, it equals $(\frac{2}{3})^2$].

Estienne de la Roche[36] copied the above method of approximation from Chuquet's manuscript.

Juan de Ortega in the later editions (1534, 1537, 1542) of his Arithmetica gave the approximations

$$\sqrt{128} = 11\tfrac{16}{51}, \qquad \sqrt{297} = 17\tfrac{659}{2820}, \qquad \sqrt{300} = 17\tfrac{25}{78},$$
$$\sqrt{375} = 19\tfrac{285}{781}, \qquad \sqrt{135} = 11\tfrac{13}{21}, \qquad \cdots,$$

which correspond[37] to the first solution of $x^2 - Dy^2 = 1$, and

$$\sqrt{80} = 8\tfrac{17}{18}, \quad \sqrt{75} = 8\tfrac{103}{156}, \quad \sqrt{756} = 27\tfrac{109}{220}, \quad \sqrt{231} = 15\tfrac{151}{760},$$

which correspond to the second solution.

[32] H. Suter, Bibliotheca Math., (3), 2, 1901, 37. Also simultaneously by Alkalcâdî, French transl. in Atti Accad. Pont. Nuovi Lincei, 12, 1858–9, 402–4.

[33] Le triparty en la science des nombres, Bull. Bibl. Storia Sc. Mat., 13, 1880, 697–9. Discussed by S. Günther, Zeitschrift für das Realschulwesen, 2, 1877, 430; L. Rodet, Bull. Soc. Math. de France, 7, 1879, 162; P. Tannery, Bibliotheca Math., (2), 1, 1887, 17.

[34] Le triparty . . . , Appendix; Bull. Bibl. Storia Sc. Mat., 14, 1881, 455.

[35] Elementa Arith. decem libris, demonstr. Jacobi Fabri Stapulensis, Paris, 1514, VI, 26.

[36] Larismetique, 1520.

[37] J. Perott, Bull. Bibl. Storia Sc. Mat. Fis., 15, 1882, 169. Cf. P. Tannery, Bibliotheca Math., (2), 1, 1887, 19–20.

J. Buteo[38] gave several approximations for $\sqrt{66}$ all of which give solutions of $x^2 - 66y^2 = 1$, the last one being x/y, $x = 8449$, $y = 1040$. He also made use of Chuquet's method.

P. A. Cataldi[39] gave approximations to $\sqrt{44}$ by the two formulas used by El-Hassar[32] and used implicitly approximations by continued fractions.

Nicolas Rhabdas[40] used the first approximation by El-Hassar. It was used later by Luca Paciuolo, Cardan and Tartaglia (references, Vol. I, Ch. I).

Fermat[41] stated February, 1657, that if D is any number not a square there exists an infinitude of integral solutions of $x^2 - Dy^2 = 1$; for example, $2^2 - 3\cdot1^2 = 1$, $7^2 - 3\cdot4^2 = 1$. He asked for the least solution of $61y^2 + 1 = \square$ and of $109y^2 + 1 = \square$, and a general rule for finding the solutions of $Dy^2 + 1 = \square$.

Although Fermat, in the introductory remarks to his "Second défi," had expressly called for solutions in integers, this introduction was omitted[42] in the copy made for Lord Brouncker by the secretary of K. Digby. This explains why W. Brouncker and John Wallis[43] first gave merely the rational solution

$$x = \frac{4ps}{s^2 - 4p^2n}, \qquad y = \frac{s^2 + 4p^2n}{s^2 - 4p^2n}$$

of $nx^2 + 1 = y^2$, the case $p = 1$, $s = 2r$, giving Brouncker's solution $x = 2r/(r^2 - n)$. The latter had been given by Bháscara[30] (second method), and was obtained by René François de Sluse[44] (1622–1685) by setting $nx^2 + 1 = (1 - rx)^2$.

Fermat[45] was not satisfied with these evident solutions in fractions.

W. Brouncker[46] gave an infinitude of integral solutions x for $n = 2$, 3, 5, 6 and their products by squares; thus, for $n = 2$,

$$x = 2 \times 5\tfrac{1}{4} \times 5\tfrac{5}{6} \times 5\tfrac{29}{35} \times \cdots,$$

each numerator being equal to the corresponding denominator diminished by the preceding denominator, while each denominator equals the numerator of the term immediately preceding when reduced to an improper fraction. [The formula gives $\tfrac{1}{2}x = 1, 6, 35, 204, 1189, \cdots$, with the recursion formula $t_{n+1} = 6t_n - t_{n-1}$.]

Wallis[47] noted that if $x = f$ is one solution, so that $nf^2 + 1 = l^2$, then $x = 2fl$ is a second: $n(2fl)^2 + 1 = (2l^2 - 1)^2$, so that one can get an

<hr>

[38] Ioan. Buteonis Logistica, quae et arith. . . . , Lyons, 1559, 76.

[39] Trattato del Modo Brevissimo di trouare la Radice quadra delli numeri, Bologna, 1613, 12.

[40] P. Tannery, Notice sur les deux arithmétiques de N. Rhabdas, Paris, 1886, 40, 68.

[41] Oeuvres, II, 333–5, letter to Frenicle and "Second défi aux mathématiciens" [Wallis and Brouncker]; French transl. of latter, III, 312–3.

[42] G. Wertheim, Abhandl. Geschichte Math., 9, 1899, 563.

[43] Commercium epistolicum de Wallis, Oxford, 1658, 767; bound with Wallis' Algebra, Oxford, 1685; Wallis' Opera, Oxford, 2, 1693. French transl. in Oeuvres de Fermat, III, 417–8; letter IX, Wallis to Digby, Oct. 7, 1657.

[44] MS. 10247, f. 286 verso, du fonds latin, Bibliothèque Nat. de Paris.

[45] Oeuvres, II, 342, 377; letters to Digby, June 6, 1657, April 7, 1658.

[46] Commercium, 775, letter XIV, Nov. 1, 1657; Oeuvres de Fermat, III, 423.

[47] Letters XVI, XVIII to Digby, Dec. 1, and Dec. 26, 1657; Oeuvres de Fermat, III, 434–5; 480–9.

infinitude of solutions in this way, but not all. He stated that all solutions are obtained from Brouncker's rule by setting $r = \alpha/e$, whence $x = 2\alpha e/(\alpha^2 - ne^2)$, and choosing integers α, e such that $\alpha^2 - ne^2$ divides $2\alpha e$.

Wallis[48] gave a long exposition of results which he implied are essentially due to Brouncker. He gave a tentative method to solve $na^2 + 1 = \square$. For $n = 7$, take the square 3^2 just > 7; then $7 = 3^2 - 2$, $7 \cdot 2^2 = 6^2 - 8$, $7 \cdot 3^2 = 9^2 - 18$, whence we have a number 18 which is double the root 9; hence $7 \cdot 3^2 = (9 - 1)^2 - 1$. In general, use the square c^2 just $> n$ and exceeding n by b. Employ $na^2 = (ca)^2 - ba^2$ for $a = 1, 2, 3, \cdots$, until we reach a value α of a for which $ba^2 \geqq 2ca$, and then replace ca by $(ca - 1) + 1$. For each $a \geqq \alpha$, we thus have two values of na^2. Presently we can make a further reduction of $ca - 1$ to $ca - 2$, etc., etc. It is stated that we finally reach an equation in which the number subtracted is unity and hence a solution. Devices are suggested (pp. 465–74) to abbreviate the long calculations.

Given (pp. 474–8) one solution, $nr^2 + 1 = s^2$, set $t = 2s$; then the values of x in the successive solutions of $nx^2 + 1 = \square$ are $r, rt, r(t^2 - 1)$, $r(t^3 - 2t), \cdots$, while if $r\alpha, r\beta$ are any two consecutive terms, the next term is $r(t\beta - \alpha)$.

Wallis[49] explained in an example Brouncker's method of finding a fundamental solution. The example chosen was $13a^2 + 1 = \square$. Since 13 lies between the squares 9 and 16, set $13a^2 + 1 = (3a + b)^2$, whence

$$4a^2 + 1 = 6ab + b^2, \qquad 2b > a > b.$$

Hence set $a = b + c$, whence $2bc + 4c^2 + 1 = 3b^2$, $2c > b > c$. Set $b = c + d$, $c = d + e$, $d = e + f$. Then $e^2 + 1 = 6ef + 4f^2$, $7f > e > 6f$. Hence set $e = 6f + g$, $f = g + h$, $g = h + i$. Then $4hi + 3i^2 + 1 = 3h^2$. Thus $h > i$. Taking* $h = 2i$, we see that the last equation becomes $11i^2 + 1 = 12i^2$ and holds for $i = 1$, whence $h = 2, \cdots, a = 180$. It is noted (pp. 492–3) that, since $b, c, d, \cdots$ are decreasing integers, we finally reach a term which divides the preceding, as in Euclid's process to find the g.c.d., a process entirely analogous to the present one. If we had proposed the example $13a^2 + 9 = \square$, we would get $11i^2 + 9 = 12i^2$, whence $i = 3$, and similarly for any square in place of 1 or 9. But if k is not a square, $13a^2 + k = \square$ is not always solvable, but when solvable the solution can be found by the above method.

As noted by H. J. S. Smith,[50] Brouncker's method is the same as that given by Euler[65, 72, 81] and really consists in the successive determination

---

[48] Commercium, 789, letter XVII to Brouncker, Dec. 17, 1657; Oeuvres de Fermat, III, 457–480.

[49] Commercium, 804, letter XIX to Brouncker, Jan. 30, 1658; Oeuvres de Fermat, III, 490–503. Cf. Wallis, Algebra, 1693, Ch. 98.

* To proceed as would later writers, set $h = i + j$, whence $- 4i^2 + 2ij + 3j^2 = 1$; then $i = j + k$, whence $j^2 - 6jk - 4k^2 = 1$, with unity as coefficient of a square term, so that $j = 1$, $k = 0$ is an evident solution.

[50] British Assoc. Report, 1861, 313; Coll. Math. Papers, I, 193. Cf. Konen,[3] p. 39; Whitford,[4] pp. 52–6; Wertheim.[42]

of the integral quotients in the development* of T/U into a continued fraction, where $T = 649$, $U = 180$, is the fundamental solution of $T^2 - 13U^2 = 1$. But[51] Brouncker did not prove that his method will always lead to a solution of $T^2 - DU^2 = 1$.

Frenicle[52] cited his table[53] of solutions of $x^2 - Dy^2 = 1$ for all values of D up to 150 which are not squares and suggested that Wallis extend it to 200 or at least solve it for $D = 151$, not to speak of $D = 313$ which is perhaps beyond his ability. In reply, Brouncker[54] stated that within an hour or two he had found by his method that $313a^2 - 1 = b^2$ for $a = 7170685$, $b = 126862368$, whence $x = 2ab$ is the desired solution.

Wallis[55] gave the last solution and $151(140634693)^2 + 1 = (1728148040)^2$.

Fermat[56] was at first satisfied with the solution of $an^2 + 1 = \square$ by Brouncker and Wallis. Later, Fermat[57] stated that he had proved by the method of descent the existence of an infinitude of solutions n of $an^2 + 1 = \square$ when a is any number not a square. He admitted that Frenicle and Wallis had given various special solutions, though not a proof and general construction.

In an anonymous letter to Digby, either by Frenicle[58] or inspired by him, it is stated that Wallis[47] affirmed that he could easily prove the existence of an infinitude of integral solutions of $an^2 + 1 = \square$ and implied that the proof is expressly contained in that passage; "but our analysts recognize no trace of proof there".

N. Malebranche[59] (1638–1715), after stating that he had not seen the work in the Commercium Epist. of Fermat and Wallis on $Ax^2 + 1 = \square$, remarked that we can find a solution if $A = a^2 \pm ka$, $k = 1, 2,$ or $\frac{1}{2}$ (no details given), or if the difference between A and some square t^2 divides $2t$. Thus, if $A = 33$ or 39, $t = 6$, $A - t^2 = \pm 3$, a divisor of $2t$. We have $39x^2 + 1 = (6x + 1)^2$, $x = 4$; $33x^2 + 1 = (6x - 1)^2$, $x = 4$. He treated by a tentative process the new types $A = 13, 19, 21$. For 13, multiply by the squares $1, 4, 9, \cdots$, until we get a product whose difference from the square divides double the root of the same square; since $13 \cdot 25 - 1 = 18^2$, set $325x^2 + 1 = (18x + 1)^2$, whence $x = 36$. Again, $19 \cdot 9 - 13^2 = 2$, whence $171x^2 + 1 = (13x + 1)^2$, $x = 13$. He noted that if $Ax^2 + 1 = y^2$,

---

$$*\frac{T}{a} = \frac{3a + b}{a} = 3 + \frac{b}{b+c} = 3 + 1\Big/\Big(1 + \frac{c}{b}\Big) = 3 + 1\Big/\Big(1 + \frac{c}{c+d}\Big) = \cdots$$

$$= 3 + \frac{1}{1} + \frac{1}{1} + \frac{1}{1} + \frac{1}{1} + \frac{1}{6} + \frac{1}{1} + \frac{1}{1} + \frac{1}{1} + \frac{1}{1} = \frac{649}{180}$$

[51] Also noted Sept. 6, 1658, by Chr. Huygens, Oeuvres complètes, II, 1889, 211.
[52] Commercium, 321, letter XXVI to Digby, sent by the latter to Wallis Feb. 20, 1658; Oeuvres de Fermat, III, 530–3.
[53] Solutio duorum problematum . . ., 1657 (lost work).
[54] Commercium, 823, letter XXVII to Digby, March 23, 1658; Oeuvres de Fermat, III, 536–7.
[55] Letter XXIX to Brouncker, March 29, 1658; Oeuvres de Fermat, III, 542.
[56] Letters from Fermat, June, 1658, and Frenicle to Digby, Oeuvres, III, 314, 577; II, 402 (Latin).
[57] Oeuvres, II, 433, letter to Carcavi, Aug. 1659.
[58] Oeuvres de Fermat, III, 604–5 (French transl., 607–8).
[59] C. Henry, Bull. Bibl. Storia Sc. Mat. Fis., 12, 1879, 696–8.

then $A(2xy)^2 + 1 = \square$, so that we obtain an infinitude of solutions, but not all, from one solution. A. Marre[60] stated that the last result was copied from a letter written by Claude Jaquemet, who gave the second solution $X = 2xy$, $Y = 2Ax^2 + 1$.

Wallis[61] attempted to prove that $t^2 - Du^2 = 1$ always has positive integral solutions, but made use of a lemma which is false [Lagrange[74, 85] and Gauss[93]]: Let m be the integer just $> \sqrt{D}$, whence $m - \sqrt{D} < 1$, and set $p = m - \sqrt{D}$, $r = 1/(2\sqrt{D})$; then it is possible to find two integers z and a such that

$$\frac{z}{a} < p < \frac{\sqrt{z^2 + 4pr} + z}{2a}.$$

But the difference of the fractions in this inequality approaches zero as z and a increase, so that their ratio approaches p.

The name Pell equation for $x^2 - Dy^2 = 1$ originated in the erroneous notion of L. Euler[62] that John Pell was the author of the unique method of solution explained in Wallis' Opera, whereas Wallis gave only Brouncker's method. Nor, as stated by Hankel,[24] had Pell treated the equation in a widely read work, i. e., in his notes to Brancker's[63] English translation of J. H. Rahn's algebra. After examining three copies of this translation, G. Eneström[64] stated that there is nothing relating to this equation. However, $x = 12y^2 - z^2$ is treated in Rahn's[63] Algebra, p. 143.

Euler[62] noted that if $az^2 + bz + c$ is a square l^2 for $z = p$, it is a square for

$$z = \frac{1}{2a}(- b + bR) + pR + \lambda l, \qquad R \equiv \sqrt{1 + a\lambda^2},$$

so that the problem is to make $1 + a\lambda^2$ a square.

Euler[65] again noted that, if $f \equiv ax^2 + bx + c$ is a square m^2 for $x = n$, it is the square of $m' = apn + pb/2 + qm$ for $x = qn + pm + (bq - b)/(2a)$, provided that $q^2 = ap^2 + 1$. In the latter expression for x we replace n by this x and replace m by m' and get

$$x' = 2q^2n + 2pqm + \frac{b}{a}(q^2 - 1) - n,$$

which makes $f = \square$. If A, B are consecutive terms of the series n, x, x', $\cdots$, the next term is $2qB - A + b(q - 1)/a$. In the case $f = ax^2 + 1$, whence $b = 0$, $c = 1$, the series becomes 0, p, $2pq$, $4pq^2 - p$, $\cdots$, A, B,

[60] Bull. Bibl. Storia Sc. Mat. Fis., 12, 1879, 893. Attributed incorrectly to Marquis de l'Hôpital in Comptes Rendus Paris, 88, 1879, 76–7, 223.

[61] Algebra, Oxford, 1685, Ch. 99; Opera, 2, 1693, 427–8. Reproduced by Konen,[3] 43–6.

[62] Letter to Goldbach, Aug. 10, 1730, Correspondance Math. et Physique (ed., P. H. Fuss), St. Petersburg, 1, 1843, 37. Also, Euler.[65, 72] Cf. Euler[56] of Ch. XIII. Cf. P. Tannery, Bull. des Sc. Math., (2), 27, I, 47–9.

[63] An introduction to algebra, translated out of the High Dutch into English by T. Brancker. Much altered and augmented by D. P. London, 1668. On Rahn's algebra of 1659, see Bibliotheca Math., (3), 3, 1902, 125.

[64] Bibliotheca Math., (3), 3, 1902, 204; cf. G. Wertheim, 2, 1901, 360–1.

[65] Comm. Acad. Petrop., 6, 1732–3, 175; Comm. Arith. Coll., 1, 1849, 4; Op. Om., (1), II, 6

$2qB - A, \cdots$.   Hence if one solution $ap^2 + 1 = q^2$ is known, we get an infinitude of solutions $p' = 2pq$, etc.   Euler noted special forms of numbers a for which a solution of $ap^2 + 1 = q^2$ may be given at once, viz., (a, p, q):

$$e^2 - 1,\ 1,\ e;\quad e^2 + 1,\ 2e,\ 2e^2 + 1;\quad \alpha^2 e^{2b} \pm 2\alpha e^{b-1},\ e,\ \alpha e^{b+1} \pm 1;$$
$$(\alpha e^b + \beta e^\mu)^2 + 2\alpha e^{b-1} + 2\beta e^{\mu-1},\ e,\ \alpha e^{b+1} + \beta e^{\mu+1} + 1;$$
$$\tfrac{1}{4}\alpha^2 k^2 e^{2b} \pm \alpha e^{b-1},\ ke,\ \tfrac{1}{2}\alpha k^2 e^{b+1} \pm 1.$$

If a is not of one of these forms, apply the method explained by Wallis, which is here illustrated for $31p^2 + 1 = q^2$.   Euler gave a table showing, for each $a \leqq 68$ not a square, the least positive integer p and the corresponding q satisfying $ap^2 + 1 = q^2$.   From $\sqrt{a} = \sqrt{q^2 - 1}/p$, Euler noted that, if q is sufficiently large, q/p is a close approximation to $\sqrt{a}$; let P be the ith term of the above series $0,\ p,\ 2pq,\ \cdots$ and Q the ith term of the series $1,\ q,\ 2q^2 - 1,\ \cdots$ such that $aP^2 + 1 = Q^2$; then the successive values of Q/P are closer and closer approximations to $\sqrt{a}$.

Euler[66] noted that the least integral solution x of $ax^2 + 1 = \square$ is 226153980 for $a = 61$, and 15140424455100 for $a = 109$, and stated he could shorten very much the work necessary by "Pell's method."   If $x^2 - ey^2 = N$ has the solution a, b, it has also the solution

$$x = a + pz, \qquad y = b + qz, \qquad z = \frac{2ebq - 2ap}{p^2 - eq^2}.$$

Making use of the existence of integral solutions of $p^2 - eq^2 = 1$ for e not a square, we can assign an infinitude of integral solutions of $x^2 - ey^2 = N$, since

$$(11) \qquad N = (a^2 - eb^2)(p^2 - eq^2) = (ap \pm ebq)^2 - e(bp \pm aq)^2.$$

This formula of composition was known[67] by Brahmegupta.[28]

R. Simpson[68] noted that if we are given a and a fraction b/c such that $(b^2 \mp 1)/c^2 = a$, the series of fractions, converging to $\sqrt{a}$,

$$\frac{b}{c},\qquad \frac{d}{e} = \frac{b^2 + acc}{2bc},\qquad \frac{f}{g} = \frac{bd + ace}{cd + be},\qquad \frac{h}{k} = \frac{bf + acg}{cf + bg},\qquad \cdots$$

are such that the numerator of any fraction (as h/k) is the sum of the products of the numerators and the denominators of $b/(ac)$ and the preceding fraction (then f/g), while the denominator (then k) is the sum of the products of the numerators and denominators of c/b and that preceding fraction (f/g).   By (11), every fraction N/D in the series has the property $N^2 - 1 = aD^2$ if $b^2 - 1 = ac^2$; but if $b^2 + 1 = ac^2$ that property holds only for alternate fractions, while $N^2 + 1 = aD^2$ for the others.   He cited the "obscure passage" where A. Girard[68a] gave the approximations 577/408

[66] Corresp. Math. Phys. (ed., Fuss), 1, 1843, 616–7, 629–631; letters to Goldbach, Aug. 4, 1753, Aug. 23, 1755.

[67] Cf. M. Chasles, Jour. de Math., 2, 1837, 37–50.   Reprinted, Sphinx-Oedipe, 5, 1910, 65–75.

[68] Phil. Trans. London, 48, I, 1753, 370–7; abr. ed., 10, 1809, 430–4.

[68a] Les Oeuvres math. de Simon Stevin de Bruges . . . par A. Girard, Leyde, 1634, I, 170.

and 1393/985 to $\sqrt{2}$ and an approximation to $\sqrt{10}$. Jean Plana[69] gave reasons to show that Girard there in effect reduced $\sqrt{A}$ to a continued fraction.

A solution[70] of $44000x^2 + 1 = \square$ is $x = 404482981221781$.

Euler[71] published his formula (11), and treated $ax^2 + bx + c = y^2$, given the solution $x = n$, $y = m$. Set $x = n + \mu z$, $y = m + \nu z$. Then $(\nu^2 - a\mu^2)z = 2a\mu n - 2\nu m + b\mu$. If a is positive and not a square, we can make $\nu^2 - a\mu^2 = 1$ and obtain integral solutions, and then a third set, etc.; if the general set is (x_i, y_i), we have

$$x_{i+2} = 2(\nu^2 + a\mu^2)x_{i+1} - x_i + 2b\mu^2, \qquad y_{i+2} = 2(\nu^2 + a\mu^2)y_{i+1} - y_i.$$

But we may obtain solutions not having $\nu^2 - \alpha\mu^2 = 1$; setting

$$p = \frac{\nu^2 + a\mu^2}{\nu^2 - a\mu^2}, \qquad q = \frac{2\mu\nu}{\nu^2 - a\mu^2},$$

we obtain the first formulas in Euler's[65] earlier paper. Euler proved that if an odd prime, not dividing α, is of the form $b^2 - \alpha a^2$, it is of one of the linear forms $4\alpha n + r^2$, $4\alpha n + r^2 - \alpha$, where r ranges over the odd and even numbers $< \alpha$ and prime to α, respectively. He conjectured, conversely, that if A is a prime or product of primes of these linear forms, then $A = x^2 - \alpha y^2$ is solvable in integers [not always true, Lagrange[76]].

Euler[72] again repeated his initial formulas and added that, if P, Q, R are the values of y in three successive sets of solutions, $R = 2qQ - P$, while the general set of solutions is said to be [after correction of signs]

$$x = \frac{r + s}{4a} - \frac{b}{2a}, \qquad y = \frac{r - s}{4\sqrt{a}}, \qquad r, s \equiv (2an + b \pm 2m\sqrt{a})(q \pm p\sqrt{a})^\mu,$$

where μ is an integer. The method published by Wallis to find integral solutions of $x^2 = ly^2 + 1$, where l is positive and not a square, can be more conveniently exhibited by means of the continued fraction for $\sqrt{l}$. If $x = p$, $y = q$ is a solution, it is stated that $p/q > \sqrt{l}$ and that p/q gives so close an approximation to $\sqrt{l}$ that a closer one cannot be found without using larger numbers. After developing $\sqrt{z}$ into a continued fraction for $z = 13, 61, 67$, he took a general z and set

$$\sqrt{z} = v + \frac{1}{a +} \frac{1}{b +} \frac{1}{c +} \cdots,$$

where v is the largest integer $< \sqrt{z}$, and $a, b, c, \cdots$ are found as follows. In $\sqrt{z} = v + 1/x$, $x = 1/(\sqrt{z} - v) = (\sqrt{z} + v)/\alpha$, where $\alpha = z - v^2$; hence let

[69] *Réflexions nouvelles sur deux mémoires de Lagrange*[74] . . . , Turin, 1859, 24 pp; Memorie R. Accad. Torino, (2), 20, 1863, 87–108.

[70] Ladies' Diary, 1759, pp. 39–41, Quest. 443. The Diarian Repository, or Math. Register . . . by a Society of Mathematicians, London, 1774, 677–9. C. Hutton's Diarian Miscellany, 3, 1775, 81–83. T. Leybourn's Math. Quest. proposed in Ladies' Diary, 2, 1817, 162–4.

[71] Novi Comm. Acad. Petrop., 9, 1762–3 (1759), 3; Comm. Arith. Coll., I, 297–315; Op. Om., (1), II, 576.

[72] Novi Comm. Acad. Petrop., 11, 1765 (1759), 28; Comm. Arith. Coll., I, 316–336; Op. Om., (1), III, 73.

a be the largest integer $\leqq (\sqrt{z}+v)/\alpha$. In $x = a + 1/y$,

$$y = \frac{1}{x-a} = \frac{\alpha}{\sqrt{z}+v-a\alpha} = \frac{\alpha(\sqrt{z}-v+a\alpha)}{z-(v-a\alpha)^2} = \frac{\sqrt{z}+B}{\beta},$$

where $B = a\alpha - v$, $\beta = 1 + 2av - a^2\alpha$. Hence let b be the largest integer $\leqq (\sqrt{z}+B)/\beta$. Taking $y = b + 1/t$, and proceeding similarly, we obtain Euler's table:

$$\text{I}\quad A = v, \qquad\qquad \alpha = z - A^2 = z - v^2, \qquad\qquad a \leqq \frac{v+A}{\alpha},$$

$$\text{II}\quad B = \alpha a - A, \qquad \beta = \frac{z - B^2}{\alpha} = 1 + a(A - B), \qquad b \leqq \frac{v+B}{\beta},$$

$$\text{III}\quad C = \beta b - B, \qquad \gamma = \frac{z - C^2}{\beta} = \alpha + b(B - C), \qquad c \leqq \frac{v+C}{\gamma},$$

$$\text{IV}\quad D = \gamma c - C, \qquad \delta = \frac{z - D^2}{\gamma} = \beta + c(C - D), \qquad d \leqq \frac{v+D}{\delta},$$

etc., where in the last column the equality sign is taken only when the fraction is an integer. It follows that $A, B, C, \cdots$ are $\leqq v$, and the indices $a, b, c, \cdots$ are $\leqq 2v$. Euler observed in many examples that when the value $2v$ is reached, the values $a, b, c, \cdots$ repeat; but no proof[73] is given that the index $2v$ exists [proof by Lagrange[74]]. For each $z \leqq 120$ and not a square, he gave the values of $v, a, b, c, \cdots$ (at least as far as a period), and underneath them the values of $1, \alpha, \beta, \gamma, \cdots$. Such values are given also for certain types of numbers, viz., $z = n^2 + k$, $k = 1, 2, n, 2n - 1, 2n$, and $z = 4n^2 + 4, 9n^2 + 3, 9n^2 + 6$.

The successive convergents $v, (va + 1)/a, \cdots$ to $\sqrt{z}$ are found by the law:

$$v, \quad a, \quad\quad b, \quad\quad\quad\quad c, \quad\quad\quad \cdots, \quad m, \quad n, \quad \cdots,$$
$$\frac{1}{0}, \quad \frac{v}{1}, \quad \frac{av+1}{a}, \quad \frac{(ab+1)v+b}{ab+1}, \quad \cdots, \quad \frac{M}{P}, \quad \frac{N}{Q}, \quad \frac{nN+M}{nQ+P}, \quad \cdots.$$

These convergents are given the symbolic notation

$$\frac{1}{0}, \quad \frac{(v)}{1}, \quad \frac{(v, a)}{(a)}, \quad \frac{(v, a, b)}{(a, b)}, \quad \frac{(v, a, b, c)}{(a, b, c)}, \quad \cdots,$$

where

$$(v) = v, \qquad (v, a) = v(a) + 1, \qquad (v, a, b) = v(a, b) + b,$$
$$(v, a, b, c) = v(a, b, c) + (b, c), \quad \cdots.$$

He stated that

$$(v, a, b, c, d, e) = v(a, b, c, d, e) + (b, c, d, e) = (v, a)(b, c, d, e) + v(c, d, e)$$
$$= (v, a, b)(c, d, e) + (v, a)(d, e) = (v, a, b, c)(d, e) + (v, a, b)(e),$$

<hr>

[73] As remarked by H. J. S. Smith, British Assoc. Report, 1861, § 96, pp. 313–5; Coll. Math. Papers, I, 1894, 194, Euler's paper contains all the elements necessary to give a rigorous proof of this fact and hence that the process always leads to a solution, other than $x = 1$, $y = 0$, of $x^2 - zy^2 = 1$. Plana[69] noted that Euler's proof becomes rigorous if slightly modified as by Legendre.[87] For $(\alpha, \beta, \cdots)$, Gauss[24] of Ch. II wrote $[\alpha, \beta, \cdots]$.
[74] Miscellanea Taurinensia, 4, 1766–9, 41; Oeuvres, 1, 1867, 671–731.

and proved that

$$(v)^2 - z \cdot 1^2 = -\alpha, \qquad (v, a)^2 - z(a)^2 = \beta, \qquad (v, a, b)^2 - z(a, b)^2 = -\gamma,$$
$$(v, a, b, c)^2 - z(a, b, c)^2 = \delta, \qquad (v, a, b, c, d)^2 - z(a, b, c, d)^2 = -\epsilon,$$

so that, for example, $x^2 - zy^2 = -\gamma$ has the solution $x = (v, a, b)$, $y = (a, b)$. No one of β, γ, δ, $\cdots$ equals ± 1 unless the corresponding index is $2v$. Hence if any period contains the index $2v$ and if x/y is the convergent defined by this period, we have $x^2 - zy^2 = -1$ or $+1$, according as the number of indices in the period is odd or even. In the first case, $\xi = 2x^2 + 1$, $\eta = 2xy$ give a solution of $\xi^2 - z\eta^2 = +1$; or we may take two successive periods and apply the second case. He applied this theory to eight special types of periods, such as $v, a, b, b, a, 2v, a, \cdots$. He recognized that we need only use a half period. Thus, for the period just cited, we employ the half period v, a, b of indices and convergents $1/0$, $v/1$, B/β, C/γ. Then $x^2 - zy^2 = -1$ for

$$x = (v, a, b, b, a) = (a, b)(v, a, b) + (a)(v, a) = \gamma C + \beta B,$$
$$y = (a, b, b, a) = (a, b)(a, b) + (a)(a) = \gamma^2 + \beta^2.$$

But if z has the indices $v, a, b, c, b, a, 2v$, with an even number of terms in the period, we use the half period v, a, b, c and the additional convergent D/δ and find that $x^2 - zy^2 = +1$ for

$$x = (a, b)(v, a, b, c) + (a)(v, a, b) = \gamma D + \beta C,$$
$$y = (a, b)(a, b, c) + (a)(a, b) = \gamma\delta + \beta\gamma.$$

As equivalent formulas were restated by Tenner,[118] they are often attributed to him rather than to Euler. The formulas are stated in general form by Muir[160a] and Konen,[3] pp. 55–6.

Finally, he tabulated the least solutions of $p^2 - lq^2 = 1$ for each $l < 100$ which is not a square, and for $l = 103, 109, 113, 157, 367$ [errata for $l = 33$, 83, 85, Cunningham[309]].

J. L. Lagrange[74] gave the first proof that $x^2 - ay^2 = 1$ has integral solutions with $y \neq 0$, if a is any integer not a square. He noted that Wallis[61] committed a petitio principii in attempting a proof, while the method of solution explained by Wallis[49] is tentative and not shown to succeed. Lagrange started with the continued fraction

$$\sqrt{a} = q + \frac{1}{q'} + \frac{1}{q''} + \cdots$$

and its successive convergents m/n, M/N, m'/n', M'/N', $\cdots$. Taking $(x, y) = (M, N), (M', N'), \cdots$, we always obtain positive values $< 2M/N$ for $x^2 - ay^2$. Hence an infinitude of these values are identical. Let $(x, y), (x', y'), (x'', y''), \cdots$ be an infinitude of pairs of integers for which $x^2 - ay^2$ has the same value R. First, let R, a be relatively prime. By multiplication and by elimination of a,

$$\text{(A)} \qquad\qquad R^2 = (xx' \pm ayy')^2 - a(xy' \pm yx')^2,$$
$$\text{(B)} \qquad\qquad R(y'^2 - y^2) = x^2y'^2 - y^2x'^2.$$

If R is a prime, (B) gives $xy' \pm yx' = qR$, whence, by (A), $xx' \pm ayy' = pR$, where q and p are integers. Thus, by (A), $p^2 - aq^2 = 1$. Next, let $R = AB$, where A and B are primes. By (B), one of $xy' + yx'$, $xy' - yx'$ is divisible by AB, or one by A and the other by B. In the first case we have the same result as when R was a prime. In the second case, $xy' \pm yx' = qB$, where q is an integer not divisible by A. Then (A) gives $xx' \pm ayy' = pB$, whence

$$\text{(C)} \qquad\qquad p^2 - aq^2 = A^2.$$

Arguing similarly with a third equation $x''^2 - ay''^2 = R$ of our set, in conjunction with $x^2 - ay^2 = R$, we get $p_1^2 - aq_1^2 = A^2$. Treating this and (C) as we did our first pair, we get a solution of $r^2 - as^2 = 1$. A similar treatment is made for the case in which R is a product of several primes or is an arbitrary number.

Second, let $R = \theta T$, $a = \theta b$ be not relatively prime. To treat the first of two analogous cases, let θ be not divisible by a square. Then $x = \theta u$ and $T = \theta u^2 - by^2$. Hence $T^2 = (\theta u^2 + by^2)^2 - a(2uy)^2$. Since T^2 and a are relatively prime, we may employ this equation in place of the former $x^2 - ay^2 = R$. Hence there exist solutions of $x^2 - ay^2 = 1$ and we have a process to find them.

If $p^2 - aq^2 = 1$, then $x^2 - ay^2 = 1$ for

$$x + y\sqrt{a} = E = (p + q\sqrt{a})^m, \qquad x - y\sqrt{a} = F = (p - q\sqrt{a})^m,$$

and

$$\text{(12)} \qquad\qquad x = \tfrac{1}{2}(E + F), \qquad y = \frac{1}{2\sqrt{a}}(E - F)$$

are expressed as polynomials in p, q, a. If p, q is the least positive solution, then (12) gives all the solutions, m being an integer. All solutions occur among the sets (M, N), (M', N'), $\cdots$ given by the convergents M/N, M'/N', $\cdots$ to $\sqrt{a}$, and each is $> \sqrt{a}$. If m is a prime, and if x, y are given by (12), $x - p$ and $y - qa^{(m-1)/2}$ are divisible by m; hence, if r is the residue (0 or ± 1) of $a^{(m-1)/2}$ modulo m, and if p', q' are given by (12) with m replaced by $m - r$, then $p'^2 - aq'^2 = 1$, and q' is divisible by m, and either $p' - p$ or $p' - 1$ is divisible by m according as $r = 0$ or $r \neq 0$. Likewise when in (12) m is replaced by $M = n(m - r)(m' - r') \cdots$, where $m, m', \cdots$ are odd primes and r' is the residue of $a^{(m'-1)/2}$ modulo m', etc., n being any positive integer, $x^2 - ay^2 = 1$ and y is divisible by $N = mm' \cdots$, and either $x - p$ or $x - 1$ is divisible by N according as M is odd or even. After giving numerical examples illustrating what precedes, Lagrange stated that, if a is not a sum of two squares, no number is simultaneously of the forms $x^2 - ay^2$, $ay_1^2 - x_1^2$; but was not certain of the converse [cf. Legendre[88]]. If $x^2 - ay^2 = R$ and $x_1^2 - ay_1^2 = -R$, and if R is a prime, we can solve $p^2 - aq^2 = -1$, and conclude that every number of the form $x^2 - ay^2$ is also of the form $ay_1^2 - x_1^2$. By squaring $t^2 - au^2 = -1$, we get solutions (12) of $x^2 - ay^2 = 1$; hence $p \pm q\sqrt{a}$ must be the square of a quantity $r \pm s\sqrt{a}$, whence $p = r^2 + as^2$, $q = 2rs$. Hence $t^2 - au^2 = -1$ is impossible unless p, q are of this form; and if they are, the resulting t, u give the least solutions.

Lagrange[75] gave a direct method to solve $a + bt^2 = u^2$ in integers. Removing the factors common to t and u, it suffices to treat

$$(13) \qquad\qquad A = p^2 - Bq^2,$$

where p, q are relatively prime. If B is negative, we may assume that $|A| > -B$, since otherwise $pq = 0$. If B is positive, we here assume that $A^2 > B$, treating later the contrary case. Choose integers p_1, q_1 such that $pq_1 - qp_1 = \pm 1$, and multiply (13) by $A_1 \equiv p_1^2 - Bq_1^2$. Thus $AA_1 = \alpha^2 - B$, where $\alpha = pp_1 - Bqq_1$. Since $\alpha^2 - B$ is divisible by A, $(\mu A \pm \alpha)^2 - B$ is divisible by A, and $\mu A \pm \alpha$ can be made numerically $< |A|/2$ by choice of μ. Hence if $\alpha^2 - B$ is divisible by A for no value of $\alpha < |A|/2$, (13) is not solvable. If such an α exists, the problem reduces to the solution of

$$(14) \qquad\qquad A_1 = p_1^2 - Bq_1^2, \qquad |A_1| < |A|.$$

If solutions of the latter are found, we deduce solutions

$$p = \frac{\alpha p_1 \mp Bq_1}{A_1}, \qquad q = \frac{\alpha q_1 \mp p_1}{A_1}$$

of (13) from $pp_1 - Bqq_1 = \alpha$, $pq_1 - qp_1 = \pm 1$. If, in (14), $B < 0$ or if $B > 0$, $A_1^2 > B$, we proceed as before and see that (14) reduces to the solution of

$$A_2 = p_2^2 - Bq_2^2, \qquad \alpha_1 < \tfrac{1}{2}|A_1|, \qquad |A_2| < |A_1|.$$

The case $B > 0$, $A_1^2 < B$, falls under that treated later. Thus, unless such a postponed case arises at some stage, we shall finally reach, if B is negative $(B = -b)$, a term A_n such that $|A_n| = b$ or $< b$. If $|A_n| = b$, we have $b = p_n^2 + bq_n^2$, whence $q_n = 0$ or 1 and (13) is solved. If $|A_n| < b$, then $q_n = 0$. But, if B is positive, we reach a term $\alpha_n = e$, where $e < \sqrt{B}$, and $A_n A_{n+1} = e^2 - B$. Thus $A_n = \pm E$, $A_{n+1} = \mp D$, where D and E are positive and $DE = B - e^2$. Moreover,

$$\mp D = \rho^2 - B\sigma^2, \qquad \pm E = r^2 - Bs^2,$$

the solution of one of which implies that of the other. Since $DE < B$, one of the equations is of the next type.

The postponed type is $\pm E = r^2 - Bs^2$, where $E < \sqrt{B}$, $B > 0$. We first seek (§ 34, p. 435) an integer ϵ, $\sqrt{B} > \epsilon > \sqrt{B} - E$, such that $B - \epsilon^2$ is divisible by E. If no such ϵ exists, the equation is impossible in integers. In the contrary case, take a particular ϵ, and determine uniquely integers $E_i, \epsilon_i, \lambda_i$ by means of the equations

$$EE_1 = B - \epsilon^2, \qquad E_1 E_2 = B - \epsilon_1^2, \qquad E_2 E_3 = B - \epsilon_2^2, \qquad \cdots,$$

$$\epsilon_1 = \lambda_1 E_1 - \epsilon, \qquad \epsilon_2 = \lambda_2 E_2 - \epsilon_1, \qquad \epsilon_3 = \lambda_3 E_3 - \epsilon_2, \qquad \cdots,$$

$$\frac{\sqrt{B} + \epsilon}{E_1} > \lambda_1 > \frac{\sqrt{B} + \epsilon}{E_1} - 1, \qquad \frac{\sqrt{B} + \epsilon_1}{E_2} > \lambda_2 > \frac{\sqrt{B} + \epsilon_1}{E_2} - 1, \qquad \cdots,$$

[75] Mém. Acad. Berlin, 23, année 1767, 1769, 242; Oeuvres, 2, 1868, 406–495. German transl. by E. Netto, Ostwald's Klassiker, No. 146, Leipzig, 1904.

where the effect of the inequalities is to insure that the λ's shall be positive integers making $0 < \epsilon_i < \sqrt{B}$. It is proved at length that, if the proposed equation is solvable, we will finally reach a least positive integer μ such that the term E_μ is identical with E and such that $E_{\mu+1} = E_1$, whence $E_{\mu+\nu} = E_\nu$, and also that $E_m = \pm 1$ for a certain m, $0 \leq m \leq \mu$. Then ϵ_{m-1} equals the greatest integer β which is $< \sqrt{B}$. For brevity, set

$$f_j = \frac{(\epsilon + \sqrt{B})(\epsilon_1 + \sqrt{B})(\epsilon_2 + \sqrt{B})\cdots(\epsilon_{j-1} + \sqrt{B})}{E_1 E_2 \cdots E_{j-1}},$$

$$f_m = R + S\sqrt{B}, \qquad f_\mu = X + Y\sqrt{B}.$$

Since $E_m = \pm 1$, $f_m \bar{f}_m$ gives $R^2 - BS^2 = \pm E$, and the general solution is given by

$$r + s\sqrt{B} = (R + S\sqrt{B})(X + Y\sqrt{B})^n.$$

By actually multiplying together the factors in f_m, it is shown that

$$R = \beta l_{m-1} + l_{m-2}, \qquad S = l_{m-1},$$

where the l's are derived from the relations (p. 448)

$$l = 1, \qquad l_1 = \lambda_1 l, \qquad l_2 = \lambda_2 l_1 + l, \qquad l_3 = \lambda_3 l_2 + l_1, \qquad l_4 = \lambda_4 l_3 + l_2,$$
$$l_5 = \lambda_5 l_4 + l_3, \qquad \cdots.$$

The notation is at fault if $m = 0$, when we have $R = 1$, $S = 0$, and if $m = 1$, when we have $R = \epsilon = \beta$, $S = 1$.

Application is made (pp. 454–94) to various numerical equations (13). For Pell's equation (pp. 494–5), we have $E = 1$, whence $\beta = \epsilon$, $m = 0$, $R = 1$, $S = 0$,

$$X = \beta l_{\mu-1} + l_{\mu-2}, \qquad Y = l_{\mu-1}, \qquad r + s\sqrt{B} = (X + Y\sqrt{B})^n,$$

where n is a positive integer such that $n\mu$ is even or odd according as $r^2 - Bs^2 = +1$ or -1. For the former, n is arbitrary if μ is even, but n must be even if μ is odd. Hence if B is any positive number not a square, $r^2 - Bs^2 = +1$ has positive integral solutions. Lagrange noted (pp. 457–461) that Euler's[65, 71] method to derive an infinitude of integral solutions of $ax^2 + bx + c = y^2$ from a given solution does not always lead to all integral solutions unless fractional values of the parameters be used or unless, in $y^2 - Bx^2 = A$, A is a prime.

Lagrange[75a] investigated the approximation of roots of algebraic equations by continued fractions and proved that the real roots of any quadratic equation with rational coefficients can be developed into a periodic continued fraction, and conversely.

Lagrange[76] derived his preceding formulas for the solution of

$$\pm E = r^2 - Bs^2$$

[75a] Mém. Acad. Berlin, 23, année 1767, 1769; 24, année 1768, 1770; Oeuvres, II, 560–652 (especially 603–15). Traité de la résolution des équations numériques, 1798; ed. 2, 1808, Ch. VI; Oeuvres, VIII, 41–50, 73–131.

[76] Mém. Acad. Berlin, 24, année 1768, 1770, 236; Oeuvres, II, 662–726. For simplification, see Lagrange.[85]

by a method first applied to equations of any degree n (see Lagrange[1] of Ch. XXIII).   His method for $t^2 - \Delta u^2 = A$, where Δ is positive and not a square, is as follows.   First, consider solutions with u prime to A.   Then we can determine integers θ and y such that $t = \theta u - Ay$, $\theta < \frac{1}{2}A$.   For this value of t, the initial equation becomes, after division by A,

$$E_1u^2 - 2\theta uy + Ay^2 = 1,$$

where $(\theta^2 - \Delta)/A = E_1$ is an integer.   Employ in turn each value of θ for which $\theta^2 \equiv \Delta \pmod{A}$ and solve the new equation by developing into a continued fraction either root of the corresponding quadratic

$$E_1 - 2\theta Y + AY^2 = 0.$$

Second, for solutions with $u = ru'$, $A = r^2A'$, whence $t = rt'$, with u', A' relatively prime, we have only to treat $t'^2 - \Delta u'^2 = A'$ as before.

The same method applies to $Bt^2 + Ctu + Du^2 = A$, $C^2 > 4BD$.   By the same substitution we now get $E_1u^2 - Quy + ABy^2 = 1$, where $E_1 = (B\theta^2 + C\theta + D)/A$, $Q = 2B\theta + C$.

He noted that a conjecture made by Euler[71] is false since $101 = x^2 - 79y^2$ has no integral solutions, although $101 = -4 \cdot 4 \cdot 79 + 38^2 - 79$.

He applied (p. 719–723) the method of his former paper to deduce the solution $u = 34$, $t = 123$, of $101 = t^2 - 13u^2$, chosen probably in view of his correspondence with Euler next mentioned.

Euler[77] stated he found trouble in applying Lagrange's[75] method of solv'ng (13) to the case $101 = p^2 - 13q^2$.   By that method we seek an integer $\alpha < 101/2$ such that $\alpha^2 - 13$ is divisible by 101.   This is true for $\alpha = 35$.   Then (14) becomes $A_1 = 12 = p_1^2 - 13q_1^2$.   Since 12 is divisible by the square 4, set the quotient 3 equal to $t^2 - 13u^2$.   Then $t = 4$, $u = 1$, whence $p_1 = 8$, $q_1 = 2$.   By Lagrange's method,

$$p = \frac{\alpha p_1 \mp Bq_1}{A_1} = \frac{35 \cdot 8 \mp 13 \cdot 2}{12}, \qquad q = \frac{\alpha q_1 \mp p_1}{A_1} = \frac{35 \cdot 2 \mp 8}{12}$$

As these are not integers, one should conclude that the problem is impossible. However, $p = 123$, $q = 34$ are solutions, which fact led Euler to believe that Lagrange's method is not sufficient.   He noted that this solution 123, 34 is given by $p_1 = 47$, $q_1 = 13$:

$$p = 123 = (35 \cdot 47 - 13 \cdot 13)/12, \qquad q = 34 = (35 \cdot 13 - 47)/12.$$

But what reason leads us to suppose that $p_1 = 47$, $q_1 = 13$?

To test whether $A = p^2 \pm Bq^2$ is possible or not, Euler gave for the case A a prime the following rule, of which he had no proof: Subtract from A any multiple of $4B$; if $A - 4nB$ is of the form ab^2, where a is a prime or unity, and if $a = p^2 \pm Bq^2$ is solvable, then the proposed equation is solvable.   Thus, $101 = p^2 - 13q^2$ is solvable since $101 - 4 \cdot 13 = 7^2$ and $1 = p^2 - 13q^2$ is solvable.

---

[77] Letter to Lagrange, Jan., 1770; Euler's Opera postuma, 1, 1862, 571–3; Lagrange's Oeuvres, XIV, 214–8.   See Lagrange,[85] end.

Lagrange's reply has not been preserved, but it convinced Euler[78] of the correctness of Lagrange's treatment of $101 = p^2 - 13q^2$, though, being then blind, Euler confessed he did not follow the real meaning of all the deductions, nor the significance of all the letters introduced.

Euler[79] noted that $ar^2 - 4 = s^2$ implies that $ax^2 + 1 = y^2$ holds for

$$x=\tfrac{1}{2}p^2(q^2-1), \qquad y=\tfrac{1}{2}q(q^2-3), \qquad p=rs, \qquad q=s^2+2.$$

Thus, if $a = 61$, we may take $r = 5$, $s = 39$ and deduce the large numbers x, y in his table.

E. Waring[80] quoted results due to Brouncker and Euler.

Euler[81] treated, essentially as had Brouncker,[49] $an^2 + 1 = y^2$, where a is positive and not a square. Thus, for $a = 5$, y is $> 2n$ and Euler set $y = 2n + p$, whence $n^2 = 4np + p^2 - 1$, $n = 2p + \sqrt{5p^2 - 1}$. The radical exceeds $2p$, whence $n > 4p$. Set $n = 4p + q$, whence $p^2 = 4pq + q^2 + 1$, $p = 2q + \sqrt{5q^2 + 1}$. Having now the initial radical, we may set $q = 0$ and obtain $p = 1$, $n = 4$, $y = 9$. For $a = e^2 \pm 2$ or $e^2 \pm 1$, we can give explicit solutions n, y:

$$(e^2\pm2)e^2+1\equiv(e^2\pm1)^2, \qquad (e^2\pm1)(2e)^2+1\equiv(2e^2\pm1)^2.$$

He repeated[82] his table[72] of the least positive solutions of $an^2 + 1 = m^2$, $a < 100$.

Euler[83] treated $f = a + bx + cx^2 = \square$ as had Diophantus when a or c is a square; also the case in which f is a product of two linear functions, l, m of x, by equating f to the square of lk, as well as the case in which f equals $l^2 + mn$. In Ch. V, Euler noted certain forms which are never equal to rational squares, as $3x^2 + 2$, $3t^2 + (3n + 2)u^2$, $5t^2 + (5n \pm 2)u^2$. In Ch. VI, he noted that, given $af^2 + bf + c = g^2$, we can find new solutions of $ax^2 + bx + c = y^2$. Subtract and factor each new member; thus we may set

$$p(x-f)=q(y-g), \qquad q(ax+af+b)=p(y+g).$$

Multiply the first by p and the second by q and subtract. Hence

$$x=ng-mf-\frac{b(m+1)}{2a}, \qquad y=mg-naf-\tfrac{1}{2}bn, \qquad m=\frac{aq^2+p^2}{aq^2-p^2}, \qquad n=\frac{2pq}{aq^2-p^2}.$$

To obtain integral solutions take $p^2 = aq^2 + 1$ and change the sign of g. Thus

$$x=2gpq+f(aq^2+p^2)+bq^2, \qquad y=g(aq^2+p^2)+2afpq+bpq, \qquad p^2-aq^2=1.$$

The method for $ax^2 + c = y^2$ is similar, but simpler, giving $x = qg + pf$, $y = pg + aqf$, and is derived a second way (§ 86) given earlier by Euler.[66]

[78] Opera postuma, I, 574; letter, March, 1770, to Lagrange, Oeuvres, XIV, 219.

[79] Ibid., 585; letter, Sept. 24, 1773, to Lagrange, Oeuvres, XIV, 239–40.

[80] Meditationes Algebraicae, 1770, 180–199; ed., 3, 1782, 308–337.

[81] Algebra, St. Petersburg, 2, 1770, Ch. 7, §§ 96–111; French transl., Lyon, 2, 1774, pp. 116–134; Opera Omnia, (1), I, 379–87.

[82] Also in Nova Acta Acad. Petrop., 10, ad annum 1792, 1797 (1777), 27; Comm. Arith., II, 185.

[83] Algebra, II, Chs. 4–6, §§ 38–95; French transl., 2, 1774, pp. 50–115; Opera Omnia, (1), I, 349–78.

Euler[84] solved $ax^2 + 1 = y^2$ for special types of numbers a. Given $p^2 = b^2 + c^2$, determine g, f so that $bg - cf = \pm 1$ and take $q = bf + cg$, $a = f^2 + g^2$; then $ap^2 - 1 = q^2$, $x = 2pq$, $y = 2q^2 + 1$. Next, if $ap^2 \mp 2 = q^2$, the divisor p^2 of $q^2 \pm 2$ must be of the form $b^2 \pm 2c^2$; hence take $a = f^2 \pm 2g^2$, $cf - bg = 1$ or -1, $q = bf \pm 2cg$. If $ap^2 \pm 4 = q^2$, the divisor p^2 of $q^2 \mp 4$ must be of the form $b^2 \mp c^2$; hence take $a = f^2 \mp g^2$, $cf - bg = 2$ or -2, $q = bf \mp cg$.

Lagrange[85] simplified his[76] method applicable to equations of any degree. Of two methods to solve $F \equiv Cy^2 - 2nyz + Bz^2 = 1$ in integers, one is to render F a minimum, and the other consists in applying transformations which replace $F = 1$ by $L\xi^2 - 2N\xi\psi + M\psi^2 = 1$, where $2 \mid N \mid$ exceeds neither $\mid L \mid$ nor $\mid M \mid$, while the determinants $N^2 - LM$ and $n^2 - CB = A$ are equal. By multiplication by M, we get $v^2 - A\xi^2 = M$ where $v = M\psi - N\xi$. If $A = -a$, where $a > 0$, it is proved that $\xi = 0$, $M = 1$. If $A > 0$, v/ξ is a convergent of the continued fraction for $\sqrt{A}$. Euler's[77] example, $101 = x^2 - 13y^2$ is now (pp. 614–620) transformed into $z^2 - 13w^2 = -1$ which is solved by use of the continued fraction for $\sqrt{13}$.

Euler[108] of Ch. XXII deduced an infinitude of solutions of $\alpha^2 - \lambda\beta^2 = 4$ from one solution.

Petri Paoli[86] treated $a + c^2x^2 = y^2$. Since a is a difference of two squares, set $y = cx + 1$, $cx + 2$, $\cdots$, in turn. Then $a = 2cx + 1$, $4cx + 4$, $6cx + 9$, $\cdots$. For a odd, use the first, third, $\cdots$ terms, so that x will be an integer chosen from the series $(a-1)/(2c)$, $(a-9)/(6c)$, $\cdots$. Similarly for a even. If a is positive, the terms of the series decrease and there is a finite number of trials. The case in which a is negative can be reduced to the preceding.

A. M. Legendre[87] obtained important conditions for the solvability of equations of degree 2 by use of Lagrange's[75] method for $x^2 - By^2 = A$, where A and B are integers with no square factor and $A > B > 0$. By that method,

$$(15) \quad \alpha^2 - B = AA'k^2, \; \alpha'^2 - B = A'A''k'^2, \cdots, \alpha \leqq A/2, \; \alpha' = \mu A' \pm \alpha \leqq A'/2, \cdots,$$

where A', $\cdots$ have no square factors, and $A^{(n)} < B$, so that the proposed equation depends upon

$$(16) \qquad x^2 - By^2 = A', \qquad x^2 - By^2 = A'', \cdots, \qquad x^2 - By^2 = A^{(n)}.$$

Legendre proved that, if for $x^2 - By^2 = A$ and the first transformed equation (16) there exist integers α, α', β, β' such that

$$\alpha^2 \equiv B \; (\text{mod } A), \quad \alpha'^2 \equiv B \; (\text{mod } A'), \quad \beta^2 \equiv A, \quad \beta'^2 \equiv A' \; (\text{mod } B),$$

the like conditions hold for the second transformed equation (16). Since $\alpha''^2 \equiv B \; (\text{mod } A'')$, by (15), it remains only to prove the existence of an integer β'' for which $\beta''^2 \equiv A'' \; (\text{mod } B)$. If θ is a prime factor of B, we

---

[84] Opusc. Anal., 1, 1783 (1773), 310; Comm. Arith. Coll., II, 35–43.

[85] Additions to Euler's Algebra, Lyon, 2, 1774, pp. 464–516, 561–635; Oeuvres de Lagrange, VII, 57–89, 118–164; Euler's Opera Omnia, (1), I, 548–573, 598–637.

[86] Opuscula analytica, Liburni, 1780, 122.

[87] Mém. Acad. Sc. Paris, 1785, 507–513. Cf. Legendre, Théorie des nombres, 1798, 43–50; ed. 2, 1808, 35–41; ed. 3, 1, 1830, 41–48; German transl. by Maser, I, 41–49. In his texts, Legendre introduced the factor z^2 in the right members of (16).

seek an integer λ for which $\lambda^2 \equiv A''$ (mod θ). First, let θ divide A'. Then by (15), θ divides α. Since k' has no divisor in common with B, which has no square factor, and hence is prime to θ, we can find integers n, p such that $k\beta = nk' - p\theta$. Hence

$$A''k'^2 = \frac{\alpha'^2 - B}{A'} = \frac{(\mu A' \pm \alpha)^2 - B}{A'} = \mu^2 A' \pm 2\mu\alpha + Ak^2 \equiv Ak^2 \ (\text{mod } \theta),$$

$$0 \equiv k^2(\beta^2 - A) \equiv k^2\beta^2 - A''k'^2 \equiv (n^2 - A'')k'^2, \qquad n^2 \equiv A'' \ (\text{mod } \theta).$$

Second, let θ be not a divisor of A' and hence not of β'. We may set $\alpha' = n\beta'k' - p\theta$. Then

$$0 \equiv A''k'^2(\beta'^2 - A') \equiv A''\beta'^2 k'^2 - \alpha'^2 \equiv \beta'^2 k'^2(A'' - n^2) \quad (\text{mod } \theta).$$

The preceding result leads to the theorem: The equation $x^2 - By^2 = A$ is solvable in integers if A and B are quadratic residues of each other, and if, in the first transformed equation $x^2 - By^2 = A'$, A' is a quadratic residue of B.

We readily deduce the more elegant theorem: If each of the positive numbers a, b, c has no square factor and if no two have a common factor and if there exist integers λ, μ, ν such that

$$\frac{a\lambda^2 + b}{c}, \qquad \frac{c\mu^2 - b}{a}, \qquad \frac{c\nu^2 - a}{b}$$

are all integers, then $ax^2 + by^2 = cz^2$ has integral solutions not all zero; if the three conditions are not all satisfied there are no integral solutions. Applying to $(cz)^2 - bcy^2 = acx^2$ the earlier theorem, we have the conditions $\alpha^2 \equiv bc$ (mod ac), $\beta^2 \equiv ac$ (mod bc), $\beta'^2 \equiv A'$ (mod bc). Set $\alpha = c\mu$, $\beta = c\nu$. Then the first two give $c\mu^2 \equiv b$ (mod a), $c\nu^2 \equiv a$ (mod b). By (15₁), $c\mu^2 - b = aA'k^2$, while ak^2 is prime to bc. Hence the third condition becomes $ak^2\beta'^2 \equiv c\mu^2 - b$ (mod bc). This will hold if $a\lambda^2 + b \equiv 0$ (mod c) is solvable. For, it is solvable for β' modulo b since $c\nu^2 k^2 \beta'^2 \equiv c\mu^2$ (mod b) is solvable for β'.

Legendre[88] proved that $x^2 - ay^2 = -1$ has integral solutions if a is a prime $4n+1$. Lagrange[74] had stated that he was not certain of a converse that, if a is a sum of two squares, every number $x^2 - ay^2$ is also of the form $ay_1^2 - x_1^2$; Legendre noted that this is true if a is a prime, but fails for $a = 2\cdot 17, 5\cdot 41, 13\cdot 17$. If a is a prime $8n+3$, $ax^2 - y^2 = 2$ is solvab'e. If a is a prime $8n-1$, $y^2 - ax^2 = 2$ is solvable. While each of the preceding three theorems was here treated separately, Legendre, in ed. 2, 1808, 54–60, first gave a preliminary discussion applicable to all the cases. Although he took A to be a prime, it suffices [Dirichlet[108]] to assume that A is positive and has no square factor. Let p, q be the least positive integral solutions of $p^2 - Aq^2 = 1$. The g.c.d. of $p-1$ and $p+1$ is $f = 1$ or 2. Hence

$$p+1 = fMg^2, \qquad p-1 = fNh^2,$$

where $MN = A$, $fgh = q$. By subtraction, $2 = fMg^2 - fNh^2$. We must take for M, N the various pairs of factors (including unity) of A. Let A be a prime. The case $2 = 2g^2 - 2Ah^2$ is excluded since $h < q$, $g < p$. Let A be a prime $4n+1$. Then in $2 = Ag^2 - h^2$ and $2 = g^2 - Ah^2$, g and h are not both

[88] Mém. Acad. Sc. Paris, 1785, 549–551; Théorie des nombres, 1798, 65–67; ed. 3, 1, 1830 64–71; Maser, I, 65–72.

even (since the right members would be multiples of 4), and hence both are odd, whence $g^2 \equiv h^2 \equiv 1$ (mod 8), and the right members would be multiples of 4. Hence the only possibility is the case $2 = 2Ag^2 - 2h^2$, so that $h^2 - Ag^2 = -1$ is solvable. Besides the remaining two theorems for primes $8n+3$, $8n-1$, cited above, Legendre proved that one of

$$Mx^2 - Ny^2 = \pm 1$$

is solvable if M and N are primes of the form $4n+3$. Given a positive integer A not a square, it is always possible to decompose it into two factors M, N, such that one of $Mx^2 - Ny^2 = \pm 1$, $Mx^2 - Ny^2 = \pm 2$ is solvable when the signs are suitably chosen. When $x^2 - Ay^2 = -1$ is solvable, A is a sum of two squares. Cf. Arndt.[124] In Table XII, he gave the least positive solutions of $m^2 - an^2 = -1$, when it is solvable, and of $m^2 - an^2 = +1$ in the contrary case, for $2 \leqq a \leqq 1003$, a not a square [errata, Cunningham,[259, 309] Richaud,[198] Whitford[4] (p. 97), Gérardin[311]], but with no indication as to which equation has the solution listed. It was reprinted (with fewer errata) as Table X in ed. 3, 1, 1830, and abridged to $a \leqq 135$ in ed. 2, 1808.

J. Tessanek[89] considered $(a^2+b)n^2+1 = \square$, say $(an+p)^2$. Set $n = p+q$. Then p satisfies a quadratic. Write $b - a = h$, $2a+1-b = g$. Then

$$gp = hq + \sqrt{(a^2+b)q^2+g}.$$

Replace p by $q+r$ and solve for q in terms of r. Thus

$$g'q = h'r + \sqrt{(a^2+b)r^2-g'},$$

where

$$h' = g - h = 3a - 2b + 1, \qquad g' = \frac{a^2+b-(g-h)^2}{g} = 2h-g+b = 4b-4a-1.$$

Replace q by $r+s$ and solve for r in terms of s. Thus

$$g''r = h''s + \sqrt{(a^2+b)s^2+g''},$$

where

$$h'' = g' - h' = 6b - 7a - 2,$$
$$g'' = \frac{a^2+b-(g'-h')^2}{g'} = 2h'-g'+g = 12a-9b+4.$$

Replace r by $s+t$. Then

$$g'''s = h'''t + \sqrt{(a^2+b)t^2-g'''},$$
$$h''' = g'' - h'', \qquad g''' = \frac{a^2+b-(g''-h'')^2}{g''}.$$

According to the method of Pell,[62-4] one ultimately obtains an equation in which the number g under the radical is $+1$. To find values of n for various a's, set $g = 1$ or $g'' = 1$, etc., whence $b = 2a$; or $3b = 4a+1$, $s = 0$, $r = q = 1$, $p = 2$, $n = 3$; etc. The terms free of a, b in 1, g, g'', $g^{(iv)}$, $\cdots$ are 1, 1, 4, 25, $\cdots$, i. e., the squares of 1, 1, 2, 5, 13, 34, $\cdots$, whose differences of second order give the same series. Thus the scale of relation is $u_{n+1} = 3u_n - u_{n-1}$, so that the general term is expressible in terms of the roots of $1 - 3z + z^2 = 0$; likewise for the coefficients of b, a.

---

[89] Abh. Böhmischen Gesell. Wiss., Prag, 2, 1786, 160–171.

John Leslie[90] treated $x^2+y^2+bxy=a^2$ by factoring a^2-y^2, solved

$$Ax^2+Bx+C=y^2$$

if A, C or B^2-4AC is a square, and derived a second solution of $ax^2+b=y^2$ from one solution.

P. Paoli[91] noted that, if $t=h$, $u=k$ give one set of rational solutions of $At^2+B=u^2$, all are given by

$$t=\frac{hr^2-2kr+Ah}{r^2-A}, \qquad u=k+r(t-h).$$

P. Cossali[92] discussed Euler's and Lagrange's methods to solve (13).

C. F. Gauss[93] showed how to find all solutions of $t^2-Du^2=m^2$, given two linear substitutions which transform any reduced form $AX^2+2BXY+CY^2$ of determinant D into the same quadratic form (see quadratic forms in Vol. III).

J. C. L. Hellwig[94] gave an exposition of Pell's and other equations of degree 2.

R. Adrain[95] reproduced the simpler proofs from Euler's[83] Algebra, II, Chs. 4–5.

F. Pezzi[96] employed the continued fraction

$$x=a+\frac{1}{a_1}+\frac{1}{a_2}+\cdots+\frac{1}{a_{n-1}}+\frac{1}{x_n}=\frac{x_nM_n+M_{n-1}}{x_nN_n+N_{n-1}},$$

where M_n/N_n is the convergent derived by deleting $1/x_n$. Take $x=\sqrt{A}$, $x_1=1/(\sqrt{A}-a)$, etc. Then $x_n=(\sqrt{A}+b_n)/c_n$, where

$$b_n=(-1)^n\{AN_nN_{n-1}-M_nM_{n-1}\}, \qquad c_n=(-1)^n\{M_n^2-AN_n^2\}.$$

By substituting this value of x_n and the corresponding value of x_{n+1} in $x_n=a_n+1/x_{n+1}$ and equating rationals and irrationals, and changing n to $n-1$, we get

$$b_n=a_{n-1}c_{n-1}-b_{n-1}, \qquad c_{n-1}c_n=A-b_n^2, \qquad x_n=c_{n-1}/(\sqrt{A}-b_n).$$

Since the a's do not exceed $2a$, the a's repeat after a certain number n of terms. Then $M_n^2=AN_n^2+(-1)^n$. Hence $x^2-Ay^2=1$ is solvable in an infinitude of ways, likewise $x^2-Ay^2=-1$ if and only if the period length n is odd. Consider any solutions of $M_m^2=AN_m^2+(-1)^m$. If N_m is even, M_m is odd and m even. If A is even and N_m odd, M_m is odd and $(-1)^n=(-1)^m$. If A and N_m are odd, N_m is even and $(-1)^n=(-1)^{m+1}$.

[90] Trans. Roy. Soc. Edinburgh, 2, 1790, 193–209. Reprinted in the Math. Repository (ed., Leybourn), London, 1, 1799, 364; 2, 1801, 17; Encycl. Britannica. Cf. Berkhan.[135]

[91] Elementi d'algebra, Pisa, 1, 1794, 165–6.

[92] Origine, trasporto in Italia . . . Algebra, Parma, 1, 1797, 146–155.

[93] Disquisitiones Arithmeticae, 1801, arts. 162, 198–202; Werke, 1, 1863, 129, 187; German transl. by Maser, 1889, 120, 177–87. Cf. Dirichlet.[133]

[94] Anfangsgründe der Unbest. Analytik, Braunschweig, 1803, 80–184.

[95] The Math. Correspondent, New York, 1, 1804, 212–222 (first American math. periodical).

[96] Memorie di Mat. e di Fisica Soc. Ital. Sc., Modena, 13, 1807, I, 342–365.

C. Kramp[97] treated periodic continued fractions and application to $Ay^2+1=\square$. The error (p. 283) on $11y^2+49=x^2$ was corrected in the second note.

P. Tédenat[98] stated that, if $y^2-Ax^2=B$ is solvable in integers, its solution reduces to the integration of the equation $y_{t+2}-2my_{t+1}+y_t=0$ in finite differences, the integral being $y=(r+s)/2$, $x=(r-s)/(2\sqrt{A})$, where

$$r=(Y+X\sqrt{A})(m+n\sqrt{A})^{z-1}, \qquad s=(Y-X\sqrt{A})(m-n\sqrt{A})^{z-1},$$

Y, X being the least integral solutions of $Y^2-AX^2=B$, and m, n being integral solutions of $m^2-An^2=1$. This is Euler's[72] result in changed notation.

P. Barlow[99] gave 15 theorems on $x^2-Ny^2=1$ and the fundamental solution for $N\leqq102$. He[100] gave general formulas for the solution of $x^2-Ny^2=\pm A$ or z^2.

C. F. Degen[101] gave in his introduction an account of $y^2=ax^2+1$ by the development of $\sqrt{a}$ into a continued fraction, and its solution by an artifice for certain a's, as $a=p^2\pm1$, $p^2\pm2$. His table I (pp. 3–109) gives, for $a\leqq1000$ and not a square, the solutions of $y^2=ax^2+1$ and the continued fraction for $\sqrt{a}$ [errata, Cunningham[259, 309]]. For example, in the entry

$$209\,[=a]\begin{vmatrix} 14 & 2 & 5 & 3 & (2) \\ 1 & 13 & 5 & 8 & (11) \\ 3220\,[=x] \\ 46551\,[=y] \end{vmatrix}$$

the first line gives the continued fraction

$$\sqrt{209}=14+\frac{1}{2+}\frac{1}{5+}\frac{1}{3+}\frac{1}{2+}\frac{1}{3+}\frac{1}{5+}\frac{1}{2+}\frac{1}{28+}\frac{1}{2+}\cdots.$$

The second line shows auxiliary numbers 1, 13, 5, 8, 11, 8, 5, 13, 1 arising in the process. Thus,

$$R=\sqrt{209}=14+\frac{1}{\alpha}, \qquad \alpha=\frac{1}{R-14}=\frac{R+14}{13}=2+\frac{1}{\beta},$$

$$\beta=\frac{13}{R-12}=\frac{R+12}{5}=5+\frac{1}{\gamma},\cdots.$$

Table II (pp. 109–112) gives the solutions of $y^2=ax^2-1$ when solvable [omitted when a is of the form t^2+1, when $y=t$, $x=1$ is a solution]. It is said to be solvable only for those values ($\neq2$, 5) of a which correspond in table I to a period with an even number of terms. For extensions of Degen's tables, see Bickmore,[219] and Whitford, p. 398 below.

<hr>

[97] Annales de Math. (ed., Gergonne), 1, 1810–11, 261–285, 319–320, 351–2.

[98] *Ibid.*, p. 349.

[99] Theory of numbers, London, 1811, 294. In $x^2-56587y^2=1$, the figure 7 is omitted; cf. A. Martin, Bull. Phil. Soc. Washington, 11, 1888, 592, and Martin.[168]

[100] New Mathematical Tables, London, 1814, 266.

[101] Canon Pellianus sive tabula simplicissimam aequationis celebratissimae $y^2=ax^2+1$ solutionem pro singulis numeri dati valoribus ab 1 usque ad 1000 in numeris rationalibus iisdemque integris exhibens. Havniae [Copenhagen], 1817.

P. N. C. Egen[102] gave the 121 values of $A<1000$ for which $x^2-Ay^2=-1$ is solvable.

J. L. Wezel[103] proved that if S is the denominator of a complete quotient $(\sqrt{A}+r)/S$ for the continued fraction for $\sqrt{A}$, and if p/q is a convergent, then $p^2-Aq^2=\pm S$. By the periodicity, we ultimately get an $S=1$. Thus $x^2-Ay^2=\pm 1$ is solvable for the plus sign, and for the minus sign only if the length of the period is odd. Also $x^2-Ay^2=\pm C$ is solvable if there occurs in the continued fraction for $\sqrt{A}$ a complete quotient of denominator C.

In the chapter on biquadratic residues in Vol. III will be given reports on the paper by G. L. Dirichlet (Jour. für Math., 3, 1828, 35–69) where he discussed $t^2\pm qu^2=ps^2$, p and q being primes and $p\equiv 1 \pmod 4$, and the related pamphlet of 1861 by H. R. Götting.

J. Baines[104] found values of n for which

$$25\cdot\frac{1^4+2^4+\cdots+n^4}{1^2+2^2+\cdots+n^2}\equiv 15n^2+15n-5=\square.$$

Set $n=m+1$. Then $15m^2+45m+25=(mr/s\pm 5)^2$ if $m=5s(9s\mp 2r)/D$, where $D=r^2-15s^2$. As by Euler, $D=1$ if $(s,r)=(1,4),(8,31),(63,244),(496,1924),\cdots$, whence $n=6,86,401,5361,\cdots$.

F. T. Poselger[105] treated $rx^2+1=\square$ by continued fractions.

C. G. J. Jacobi[106] stated that the solutions of $x^2-ay^2=1$ can be expressed in terms of the sine and cosine of $2m\pi/a$, and stated that he possessed a generalization to the case in which a is a product of several factors. If $a=bc$, we can find in an infinitude of ways four integers u,v,w,x such that the product of the four factors $u\pm v\sqrt{b}\pm w\sqrt{c}\pm x\sqrt{bc}$ is unity, where two or four of the signs are plus. The resulting relation can easily be given the three forms $y^2-bz^2=1$, $y_1^2-cz_1^2=1$, $y_2^2-az_2^2=1$. Hence the solutions $y,\cdots,z_2$ depend on u,v,w,x. The latter can be expressed by trigonometric functions.

T. L. Pistor[107] gave an exposition, illustrated by examples, of the methods of Gauss and Legendre to reduce the general quadratic equation in x,y to $v^2-Dy^2=N$, its solution by continued fractions if $D>0$ and by trial if $D=-d$, using $y=0,\pm 1,\pm 2,\cdots$, up to $\sqrt{N/d}$. On p. 44 is given a table of the least solution of Pell's equation $x^2-Dy^2=1$, $D=2,\cdots,200$.

G. L. Dirichlet[108] recalled Legendre's[88] result that if p,q are the least positive integral solutions of $p^2-Aq^2=1$, then $2=fMg^2-fNh^2$, where $f=1$ or 2, and $MN=A$ is a decomposition of A. Dirichlet proved that at most one of the latter equations, in addition to $1=g^2-Ah^2$, is solvable. Besides Legendre's theorems for primes $A=4n+1,8n+3,8n-1$, Dirichlet

---

[102] Handbuch der allgemeinen Arith., Berlin, 1819–20; ed. 2, I, 1833, 457; II, 1834, 467; ed. 3, I, 1846, 456; II, 1849, 468. Cf. Seeling.[148]

[103] Annales Acad. Leodiensis, Liège, 1821–2, 24–30.

[104] The Gentleman's Diary, or Math. Repository, London, 1831, 38, Quest. 1268.

[105] Abh. Akad. Wiss. Berlin (Math.), 1832, 1.

[106] Letter to Legendre, May 27, 1832; Werke, I, 458; Jour. für Math., 80, 1875, 276; Ann. de l'Ecole Normale Sup., 6, 1869, 176–7; Bull. Sc. Math. Astr., 9, 1875, 139. Cf. Koenig.[126]

[107] Über die Auflösung der unbest. Gl. 2. Grades in ganzen Zahlen, Progr., Hamm, 1833.

[108] Abh. Akad. Wiss. Berlin, 1834, 649–664; Werke, I, 219–236.

proved that, when $A=2a$, where a is a prime, $2t^2-au^2=+1$ is solvable for $a=8n+7$, $2t^2-au^2=-1$ for $a=8n+3$, and $t^2-2au^2=-1$ for $a=8n+5$. This method of exclusion yields no result when $a=8n+1$. But using also the quadratic reciprocity law, he proved that $t^2-2au^2=-1$ is solvable if a is a prime $16n+9$ such that $2^{(a-1)/4}\equiv-1 \pmod{a}$, though the conditions are not necessary. If a and b are both primes $4n+3$, $at^2-bu^2=(a/b)$ is solvable.* If a and b are both primes $4n+1$ and if $(a/b)=-1$, $t^2-abu^2=-1$ is solvable; but if $(a/b)=1$, and* $(a/b)_4=-1$, $(b/a)_4=-1$, $t^2-abu^2=-1$ is solvable, though the conditions are not necessary. He gave criteria for the solvability of $t^2-abcu^2=-1$, where a, b, c are primes $4n+1$. Finally, Dirichlet removed the initial hypothesis that p, q give the least solution of $p^2-Aq^2=1$.

M. A. Stern[109] developed the theory of continued fractions and in the final article (pp. 327–341) made application to $x^2-Ay^2=D$, in particular when $D=\pm1$, ±2. He tabulated 42 forms for A, like m^2n^2+2m and $(6n\pm1)^2+(8n\pm2)^2$, such that there is a small number of explicitly given partial denominators in the continued fraction for $\sqrt{A}$, whence one finds at once the least solution of $x^2-Ay^2=\pm1$.

B. Peirce and T. Strong[110] solved $376x^2+114x+34=y^2$ by setting $376x+57=x'$ and treating $376y^2-x'^2=9535$ by the theory of binary quadratic forms.

C. Gill[111] noted the solution $(1364557)^2-369(71036)^2=25$ and that in the least solution of $t^2-940751u^2=1$, u has 55 digits and t has 58 digits.

C. G. J. Jacobi[112] stated that, if p is a prime $4n+1$, and $x^2-py^2=-4$, then

$$\sqrt{p}\,(x+y\sqrt{p})=2^{(p+1)/2}\,\Pi\,\sin^2\frac{a\pi}{p},$$

where a ranges over the quadratic residues, between 0 and $p/2$, of p. If q is a prime $8n+3$, and $x^2-qy^2=-2$, then

$$x+y\sqrt{q}=\sqrt{2}\,\Pi\,\sin\left(\frac{a\pi}{q}+\frac{\pi}{4}\right).$$

If q and q' are primes $4n+3$, and q is a quadratic residue of q', then

$$2^{(q-1)/2\cdot(q'-1)/2}\,\Pi\,\sin\left(\frac{a\pi}{q}+\frac{a'\pi}{q'}\right)=\sqrt{q}\,x+\sqrt{q'}\,y,$$

where $qx^2-q'y^2=4$. Cubing $\frac{1}{2}(\sqrt{q}x+\sqrt{q'}y)$, we get solutions of $qu^2-q'v^2=1$.

---

* Legendre's symbol (a/b) denotes $+1$ or -1 according as $x^2\equiv a \pmod{b}$ is solvable or not. Let c be a prime $4n+1$, and k an integer not divisible by c for which $(k/c)=+1$, viz., $k^{(c-1)/2}\equiv+1 \pmod{c}$. According as $k^{(c-1)/4}\equiv+1$ or $-1 \pmod{c}$, Dirichlet wrote $(k/c)_4=+1$ or -1, respectively.

[109] Jour. für Math., 10, 1833, 1–22, 154–166, 241–274, 364–376; 11, 1834, 33–66, 142–168, 277–306, 311–350.

[110] Math. Miscellany, 1, 1836, 362–5; French transl., Sphinx-Oedipe, 8, 1913, 117–9.

[111] *Ibid.*, 230.

[112] Monatsber. Akad. Wiss. Berlin, 1837, 127; Jour. für Math., 30, 1845, 166; Werke, VI, 263–4; Opuscula Mathematica, 1, 1846, 324–5. Proof by Genocchi.[130]

G. L. Dirichlet[113] solved $t^2-pu^2=1$ by use of trigonometric functions and remarked that the method is not so well adapted to numerical calculation as that by continued fractions and does not give the least positive solutions. Let $a_1, \cdots, a_s$ be the $s=(p-1)/2$ quadratic residues of the odd prime p, and let $b_1, \cdots, b_s$ be the quadratic non-residues. Write $i=\sqrt{-1}$. In

$$Y+Z\sqrt{\pm p}=2\prod_{j=1}^{s}(x-e^{2\pi a_j i/p}), \qquad Y-Z\sqrt{\pm p}=2\prod_{j=1}^{s}(x-e^{2\pi b_j i/p}),$$

where the upper or lower sign is taken according as $p=4\mu+1$ or $4\mu+3$, Y and Z are polynomials in x whose coefficients (as shown by Gauss, Disq. Arith., art. 357) are integers. By multiplication, we get

$$Y^2\mp pZ^2=4X, \qquad X=\frac{x^p-1}{x-1}.$$

Let $p=4\mu+1$. For $x=1$, let Y, Z become the integers g, h. Then $g^2-ph^2=4p$. Hence $g=pk$, $h^2-pk^2=-4$. It remains to evaluate g and h. Since $a_1, \cdots, a_s$ have in some order the same remainders as $1^2, 2^2, \cdots, s^2$ when divided by p, we have

$$g+h\sqrt{p}=2\prod_{j=1}^{s}(1-e^{2\pi j^2 i/p})=2^{(p+1)/2}\prod_{j=1}^{s}\sin j^2\pi/p\equiv\alpha,$$

since

$$1-e^{2\pi j^2 i/p}=-2i\sin\frac{j^2\pi}{p}\cdot e^{\pi j^2 i/p}, \qquad 1+2^2+\cdots+s^2=\frac{p(p^2-1)}{24}\equiv(-1)^{(p-1)/4}$$

$$\text{(mod 2).}$$

In terms of the trigonometric product α, we evidently have

$$h=\frac{\alpha}{2}-\frac{2}{\alpha}, \qquad k=\frac{1}{\sqrt{p}}\left(\frac{\alpha}{2}+\frac{2}{\alpha}\right).$$

To pass from these solutions of $h^2-pk^2=-4$ to solutions of $t^2-pu^2=1$, let first $p=8\nu+1$; then h and k are both even, so that

$$\left(\frac{h}{2}\right)^2-p\left(\frac{k}{2}\right)^2=-1, \qquad t+u\sqrt{p}=\left(\frac{h}{2}+\frac{k}{2}\sqrt{p}\right)^2.$$

For $p=8\nu+5$, it is stated that h and k are both odd, whence solutions t, u are easily deduced. But R. Dedekind[114] noted that both h and k can be even, as for $p=37, 101$, etc. Finally, if $p=4\mu+3$, it is shown that, for $x=i$, Y and Z become $g(1\pm i)$ and $h(1\mp i)$, where g and h are real integers, and the upper or lower sign holds according as $p\equiv7$ or 3 (mod 8). Evidently X becomes i. Hence $Y^2+pZ^2=4X$ takes a form equivalent to $g^2-ph^2=\pm2$. From this solvable equation, we pass to $t^2-pu^2=1$ by setting $(g+h\sqrt{p})^2=2t+2u\sqrt{p}$. The expressions for g and h in terms of trigonometric functions can be found as before by use of $x=i$, but are not given.

---

[113] Jour. für Math., 17, 1837, 286–290; Werke, I, 343–350. Reproduced by P. Bachmann, Die Lehre . . . Kreistheilung, 1872, 294–9.
[114] Dirichlet's Werke, II, 418.

Dirichlet[115] noted that while $k=g/p$ is positive, the determination of the sign of h presents difficulties. He showed that h has the same sign as

$$\log\left(\frac{k\sqrt{p}+h}{k\sqrt{p}-h}\right)=-\sqrt{p}\,\Sigma\left(\frac{n}{p}\right)\frac{1}{n},$$

where n ranges over the positive integers not divisible by the prime p, and the symbol (n/p) is Legendre's.

C. d'Andrea[116] proved by use of continued fractions that $x^2-Du^2=1$ is solvable.

Dirichlet[117] noted that, if P is an integer >1 not necessarily a prime,

$$\tfrac{1}{2}(Y+Z\sqrt{P})=\Pi(x-e^{2b\pi i/P}),$$

where b ranges over the integers $<P$ and prime to P for which $(b/P)=-1$, and Y, Z are polynomials in x with integral coefficients. For $x=1$, let Y and Z become the integers Y_1, Z_1. Then, if $\epsilon=1$ or $\sqrt{P}$ according as the number of prime factors of P is >1 or $=1$,

$$(T+U\sqrt{P})^h=\left(\frac{Y_1+Z_1\sqrt{P}}{2\epsilon}\right)^e,\qquad e\equiv 4-2\left(\frac{2}{P}\right),$$

where h is the number of classes of binary quadratic forms of determinant P, and T, U give the least positive solutions of $t^2-Pu^2=1$. For example, if $P=17$,

$$Y=2x^8+x^7+5x^6+7x^5+4x^4+7x^3+5x^2+x+2,$$
$$Z=x^7+x^6+x^5+2x^4+x^3+x^2+x,$$

$Y_1=34$, $Z_1=8$, $e=2$, $T=33$, $U=8$, whence $h=1$.

G. W. Tenner[118] gave a convenient method to convert $\sqrt{a}$ into a continued fraction. Let α^2 be the largest square $<a$. Then proceed as for $a=113=10^2+13$.

I	II	III		IV	V	VI
10	×	10	=	113	—	13
1,	7,	3,		9,	104,	8
1,	5,	5,		25,	88,	11
1,	4,	6,		36,	77,	7
2,	2,	8,		64,	49,	7.

Divide $10+10$ by 13 and write the quotient 1 in column I and the remainder 7 in II in the second line. Subtract 7 from $\alpha=10$ and write the remainder 3 in III, and its square 9 in IV. Write the difference $a-9=104$ in V. Divide 104 by 13 (under VI in the preceding line) and write the quotient 8 in VI. Similarly, to form the third row divide $\alpha+3$ (3 of III) by 8 (of VI) and write the quotient 1 in I and the remainder 5 in II; subtract it from α and write the remainder 5 in III, its square in IV, $a-25=88$ in V, and its quotient 11 by 8 (of VI) in VI. Continue until we find in VI

<hr>

[115] Jour. für Math., 18, 1838, 270; Werke, I, 371–2.
[116] Trattato elementare di aritmetica e d'algebra, II, 1840, 671, Naples.
[117] Jour. für Math., 21, 1840, 153–5; Werke, I, 493–6.
[118] Einige Bemerkungen über die Gleichung $ax^2\pm 1=y^2$, Progr., Merseburg, 1841.

or II a number equal to the one above it. Then column I gives the denominators (quotients) and VI the complete quotients in the continued fraction; if the repeated number (7 in our example) occurs in VI, the last number 2 in I is the last term of the first half of the symmetrical period with an even number of terms;* but if the repeated number occurs in II, the last number in I is the middle term of the symmetrical period with an odd number of terms. If $y^2-ax^2=-1$ is solvable, let L/l, M/m be the last two convergents for $\sqrt{a}$, the second corresponding to the last quotient in the first half period; it is stated that $x=l^2+m^2$, $y=Ll+Mm$ [cf. Euler,[72] end]. For example, if $a=113$, $\alpha=10$, the half period is 10, 1, 1, 1, 2 and the convergents are 10/1, 11/1, 21/2, 32/3, 85/8, whence $x=3^2+8^2=73$, $y=3\cdot32+8\cdot85$. But if we use 1, 1, 1, 2 and the convergents 1/1, 1/2, 2/3, 5/8 to $\sqrt{a}-\alpha$, we have the same x, while $y=\alpha(l^2+m^2)+l\lambda+m\mu$. If there be an odd number of quotients $\alpha,\ \cdots,\ 2\alpha$, let K/k, L/l, M/m be the last three convergents for $\sqrt{a}$, the third corresponding to the middle quotient; it is stated that $x=(k+m)l$, $y=(K+M)l\pm1=(k+m)L\mp1$ [equivalent to Euler's $y=lM+kL$, since $kL-Kl=\pm1$]. Tenner continued Degen's table from 1001 to 1020.

Dirichlet[119] proved that, if D is a complex integer not a square, $t^2-Du^2=1$ is solvable in complex integers and deduced all solutions. It applies[120] without change to the case of real numbers. The proof rests on the lemma: if a is a given complex irrational number, we can find an infinitude of pairs of complex integers x, y $(y\neq0)$ such that $N(x-ay)<4/N(y)$, where $N(k+bi)=k^2+b^2$ for k and b real. Then, since the modulus of $r+s$ does not exceed the sum of the moduli of r and s,

$$\sqrt{N(x+ay)} \leqq \sqrt{N(x-ay)} + \sqrt{N(2ay)}.$$

Since $N(y)\geqq1$, the lemma gives

$$\sqrt{N(x^2-a^2y^2)} < 4\sqrt{N(a)} + \frac{4}{N(y)} < 4\sqrt{N(a)}+4.$$

Hence $N(x^2-a^2y^2)$ remains less than a fixed limit for an infinitude of pairs of complex integers. Now take $a=\sqrt{D}$. Hence x^2-Dy^2 takes the same value $l\neq0$ for an infinitude of pairs x, y, and hence for an infinitude of pairs for which the x's and y's differ by multiples of l:

$$x^2-Dy^2=x_1^2-Dy_1^2=l, \qquad x\equiv x_1, \qquad y\equiv y_1 \quad (\mathrm{mod}\ l).$$

By multiplication, $(xx_1-Dyy_1)^2-D(xy_1-yx_1)^2=l^2$. Since xy_1-yx_1 is divisible by l, also xx_1-Dyy_1 is divisible by l. Hence $t^2-Du^2=1$ is solvable in complex integers t, u $(u\neq0)$. All solutions are shown to be given without duplication by

$$t+u\sqrt{D}=\pm(T+U\sqrt{D})^n \qquad (n=0,\ \pm1,\ \pm2,\ \cdots),$$

---

* Note that $\sqrt{113} = 10 + \cfrac{1}{1} + \cfrac{1}{1} + \cfrac{1}{1} + \cfrac{1}{2} + \cfrac{1}{2} + \cfrac{1}{1} + \cfrac{1}{1} + \cfrac{1}{1} + \cfrac{1}{20} + t,\quad t = \sqrt{113} - 10.$

[119] Jour. für Math., 24, 1842, 328; Werke, I, 578–588.

[120] Abh. Akad. Wiss. Berlin, 1854, 113; Jour. de math., (2), 2, 1857, 370; Werke, II, 155, 176. Cf. Dedekind.[141]

where T, U is a fundamental solution, i. e., one for which $N(T+U\sqrt{D})$ is the minimum of all the $N(t+u\sqrt{D})>1$. If D is real and positive, t, u are both real or both pure imaginaries. Thus if the fundamental solution is real, all solutions are real. But if it be imaginary, only even values of n give real solutions. Since pure imaginary solutions give real solutions of $t^2-Du^2=-1$, the fundamental solution is imaginary or real according as the latter equation has real solutions or not, and the least positive solutions are T/i, U/i in the former case.

Du Hays[121] derived, for the case $b=0$, Euler's[65] recursion formulæ between consecutive sets of solutions of $ax^2+c=\square$, and gave the nth set.

Chabert[122] treated $ny^2+py+q=\square$ by equating it to $n(y-\beta)(y-\beta')$ and setting $(y-\beta)n/f=(y-\beta')f$. Use an irrational f if β, β' are irrational. While we cannot always get rational x, y, the process is said to be far simpler than Legendre's.

G. Eisenstein[123] proposed the problem to find a criterion to decide a priori if $p^2-Dq^2=4$ is solvable in odd integers, given that D is a positive integer $8n+5$, i. e., if the number of improperly primitive classes of quadratic forms of determinant D equals the number of properly primitive classes of determinant D or is three times the latter number.

F. Arndt[124] extended the work of Legendre[88] on $p^2-Aq^2=1$, who treated only the cases in which A is a prime or a product of two primes. Let p, q be the least positive solutions. First, let A be odd. Let θ_1 be the g.c.d. of $p+1$, q, and θ_2 that of $p-1$, q. Then

$$p+1=\tfrac{1}{2}\theta_1^2\rho_1,\ \ p-1=\tfrac{1}{2}\theta_2^2\rho_2,\ \ \theta_1\theta_2=2q,\ \ \rho_1\rho_2=A,\ \ 1=(\tfrac{1}{2}\theta_1)^2\rho_1-(\tfrac{1}{2}\theta_2^2)\rho_2\ \ (p\text{ odd});$$
$$p+1=\ \ \theta_1^2\sigma_1,\ \ p-1=\ \ \theta_2^2\sigma_2,\ \ \theta_1\theta_2=\ \ q,\ \ \sigma_1\sigma_2=A,\ \ 2=\theta_1^2\sigma_1-\theta_2^2\sigma_2\ \ \ \ \ \ (p\text{ even}).$$

If $A=4m+1$, only the first system of relations holds and ρ_1, ρ_2 are both $\equiv1\ (\mathrm{mod}\ 4)$ if $\tfrac{1}{2}\theta_1$ is odd, while both are $\equiv3\ (\mathrm{mod}\ 4)$ if $\tfrac{1}{2}\theta_1$ is even. If $A=4m+3$, either system may hold; if the first holds, $\rho_1\equiv1$, $\rho_2\equiv3\ (\mathrm{mod}\ 4)$. If A is an odd power of a prime $4m+1$, then $\rho_1=A$, $\rho_2=1$, and

$$-1=(\tfrac{1}{2}\theta_2)^2-(\tfrac{1}{2}\theta_1)^2A,$$

whence -1 is a quadratic residue of A, and the number k of terms in the period of the continued fraction for $\sqrt{A}$ is odd. Let $(\sqrt{A}+I_n)/B_n$ be any complete quotient; then if k is odd $A=B_s^2+I_s^2$, $s=(k+1)/2$. If, for $A=4m+1$, the number k of terms in the period is even, the denominator of the middle complete quotient is odd. If $A=2^n$, where n is odd and >3, it is proved that

$$p=2p_0^2-1,\qquad q=p_0q_0,\qquad p_0^2-2^{n-2}q_0^2=1.$$

If p_0, q_0 be the least solutions of the latter, then p, q give the least solutions of $p^2-2^nq^2=1$. Since $p_0=3$, $q_0=1$ when $A=8$, we can find the solutions step by step. Finally there is treated the case $A=2^nA'$, A' odd.

[121] Jour. de Math., 7, 1842, 325–30.
[122] Nouv. Ann. Math., 3, 1844, 250–3.
[123] Jour. für Math., 27, 1844, 86.
[124] Disquisitiones nonnullae de fractionibus continuis, Diss. Sundiae, 1845, 32 pp. Extract in Jour. für Math., 31, 1846, 343–358.

F. Arndt[125] simplified the solution of $x^2-Ay^2=\pm1$ when A has a square factor.

J. F. Koenig[126] stated that Jacobi had remarked to him that if

$$A=a+b\sqrt{f}+c\sqrt{g}+d\sqrt{f}\sqrt{g}, \qquad B=a-b\sqrt{f}-c\sqrt{g}+d\sqrt{f}\sqrt{g},$$

and C, D are derived from A, B by changing the sign of $\sqrt{g}$, then $AB\cdot CD$, $AC\cdot BD$, $AD\cdot BC$ equal, respectively

(α) $\qquad\qquad\qquad m^2-fgn^2, \qquad m'^2-fn'^2, \qquad m''^2-gn''^2,$

where

(β)
$$\begin{aligned}
\pm m &=a^2-fb^2-gc^2+fgd^2, & \pm n &=2(ad-bc),\\
\pm m' &=a^2+fb^2-gc^2-fgd^2, & \pm n' &=2(ab-gcd),\\
\pm m'' &=a^2-fb^2+gc^2-fgd^2, & \pm n'' &=2(ac-fbd).
\end{aligned}$$

Given values of m, $\cdots$, n'' for which the expressions (α) are unity, Jacobi desired solutions a, $\cdots$, d of (β). Koenig employed

$$z=a^2+fb^2+gc^2+fgd^2$$

and noted that a^2, $\cdots$, d^2 are linear functions of z, while, by computation,

$$z=mm'm''\pm fgnn'n''.$$

He gave a table of values of a, b, c, d for $f=2, 3, 5, 6, 7$, $g\leqq20$, and values of a, $\cdots$, d giving $x^2-fgy^2=-1$ for f, $g<100$, $f\cdot g<1000$.

J. B. Luce[127], to solve $x^2-ny^2=z^i$, set $n=a^2\pm b$, $\sqrt{n}=a\pm j$, whence

$$\sqrt{n}=a\pm\cfrac{1}{m\pm\cfrac{1}{2a\pm\cfrac{1}{m\pm\cdots}}} \qquad\qquad (m=2a/b).$$

In the resulting successive convergents, take the numerators and denominators as values of x, y. If $m=2a/b$ is integral, $p^2-nq^2=1$ for $p=am+1$, $q=m$. If not integral, seek a square whose product by n leads to an integral value of $2a/b$. He gave a table of such square multipliers for $n\leqq158$.

F. Arndt[128] was led by investigations on binary quadratic forms to $x^2-Dy^2=\pm4$, $D>0$. If $D\equiv0\pmod 4$, its roots are $x=2t$, $y=u$, where

$$t^2-\tfrac{1}{4}Du^2=\pm1.$$

If $D\equiv2$ or $3\pmod 4$ or $D\equiv1\pmod 8$, its roots are $x=2t$, $y=2u$, where $t^2-Du^2=\pm1$. For the remaining case $D\equiv5\pmod 8$, he tabulated the least solutions for those values <1005 of D for which the equation $x^2-Dy^2=\pm4$ has relatively prime solutions, the solutions being for $x^2-Dy^2=-4$ if it is solvable (such a D being marked $D*$). If x, y give the least positive solutions of the latter, $X=x^2+2$, $Y=xy$ give the least positive solutions of $X^2-DY^2=+4$. If the last is solvable in relatively prime numbers, its least solution is easily deduced from that for $x^2-Dy^2=1$.

[125] Archiv Math. Phys., 12, 1849, 239–243.

[126] Zerlegung der Gleichung $x^2-fgy^2=\pm1$ in Factoren, Progr., Königsberg, 1849, 23 pp. Extract in Archiv Math. Phys., 33, 1859, 1–13. Cf. Jacobi.[106]

[127] Amer. Jour. Sc. Arts (ed., Silliman), (2), 8, 1849, 55–60.

[128] Archiv Math. Phys., 15, 1850, 467–478.

C. Hermite[129] proved that if D is positive, $x^2 - Dy^2 = 1$ has an infinitude of integral solutions, and all are given by

$$x + y\sqrt{D} = (a + b\sqrt{D})^i \qquad (i = 0, \pm 1, \pm 2, \cdots)$$

where a, b are solutions such that $a + b\sqrt{D}$ is a minimum.

A. Genocchi[130] proved the results stated by Jacobi[112] by means of $Y^2 \mp pZ^2 = 4X$ [cf. Dirichlet[113]]. For $x = i = \sqrt{-1}$, let Y and Z become $y + y_1 i$ and $z + z_1 i$. According as the prime p is of the form $8k + 3$ or $8k + 7$, we have

$$y^2 - pz^2 = \mp 2, \qquad y \mp z\sqrt{p} = \pm(-1)^k K/\sqrt{2}, \qquad K = \pm 2^{(p+1)/2} \Pi \sin\left(\frac{\pi}{4} - \frac{r\pi}{p}\right),$$

where the product extends over the $(p-1)/2$ quadratic residues of p.

P. L. Tchebychef[131] proved that if α, β give the least positive solutions of $x^2 - Dy^2 = 1$, and if $x^2 - Dy^2 = \pm N$ is solvable, one solution has

$$0 \leq x \leq \sqrt{(\alpha \pm 1)N/2}, \qquad 0 \leq y \leq \sqrt{(\alpha \mp 1)N/(2D)}.$$

If a, b and a_1, b_1 are solutions x, y within these limits of $x^2 - Dy^2 = \pm N$, then $(ab_1 + a_1 b)(ab_1 - a_1 b)$ is a multiple of N, while neither factor is. Hence if $x^2 - Dy^2 = \pm N$ has two distinct sets of solutions within these limits, N is composite.

A. Göpel[132] proved, by use of continued fractions, that if A is a prime of the form $4A' + 3$ or the double of such a prime, $x^2 - Ay^2 = \pm 2$ is solvable, the sign being $+$ or $-$ according as A (or $\frac{1}{2}A$) is $\equiv 7$ or $3 \pmod 8$, and related theorems as to the values of A for which $x^2 - Ay^2 = 2$ is solvable, to be given in Vol. III under binary quadratic forms.

G. L. Dirichlet[133] noted that if $f = ax^2 + 2bxy + cy^2$ has for its determinant $D = b^2 - ac$ a number not a square, and if σ is the g.c.d. of a, $2b$, c, and if

$$x = \lambda x' + \mu y', \qquad y = \nu x' + \rho y', \qquad \lambda \rho - \mu \nu = 1,$$

is a transformation with integral coefficients of determinant unity which transforms f into itself, then

$$\lambda = (t - bu)/\sigma, \qquad \mu = -cu/\sigma, \qquad \nu = au/\sigma, \qquad \rho = (t + bu)/\sigma,$$

where t, u are integral solutions of $t^2 - Du^2 = \sigma^2$; and conversely, if t, u are integral solutions, the values of $\lambda, \cdots, \rho$ are integers which determine a transformation of f into itself. For the more difficult case in which D is positive, and f is a reduced form, obtain from the period of reduced forms defined by f all the transformations of f into itself and hence, by the above, deduce all solutions of $t^2 - Du^2 = \sigma^2$. This theory, which will be given under binary quadratic forms in Vol. III, is closely connected with the continued fraction for the positive root of $a + 2b\omega + c\omega^2 = 0$.

[129] Jour. für Math., 41, 1851, 209–211; Oeuvres, I, 185–7.

[130] Mém. Couronnés Acad. Sc. Belgique, 25, 1851–3, IX, X.

[131] Jour. de Math., 16, 1851, 257–265; Oeuvres, I, 73–80; Sphinx–Oedipe, 10, 1915, 4, 18.

[132] Jour. für Math., 45, 1853, 1–14. Summary, *ibid.*, 35, 1847, 313–8; Jour. de Math., 15, 1850, 357–362. Cf. Smith,[139] § 123, p. 783; Coll. Math. Papers, I, 284–8.

[133] Abh. Akad. Wiss. Berlin, 1854, 111–4; French transl., Jour. de Math., (2), 2, 1857, 370–3; Werke, II, 155–8, 175–8. Zahlentheorie, § 62, § 83, 1863; ed. 2, 1871; ed. 3, 1879; ed. 4, 1894. Cf. H. Minkowski, Geometrie der Zahlen, 1, 1896, 164–170.

Dirichlet[134] recalled the fact that if T, U are the least positive solutions of $t^2-Du^2=1$, where D is positive and not a square, all positive solutions are given by

$$l_n+u_n\sqrt{D}=(T+U\sqrt{D})^n \qquad (n=1, 2, \cdots).$$

An infinitude of values of u_n are divisible by any positive integer S. He proved that, if N is the least n for which u_n is divisible by S, the remaining n's are the successive multiples of N. If $D'=DS^2$ and if T', U' give the least positive solutions of $t^2-D'u^2=1$, then N is determined by

$$T'+U'\sqrt{D'}=(T+U\sqrt{D})^N.$$

For any prime factor p of S, let ν be the least index for which u_ν is divisible by p and let p^δ be the highest power of p dividing u_ν. Then, if e is arbitrary, the exponent of the highest power of p dividing $u_{\nu e}$ is $\delta+\epsilon$, where ϵ is the exponent of the highest power of p dividing e. Let ν_i, δ_i be the values corresponding to the general prime factor p_i of $S=\Pi p_i^{\alpha_i}$ and let N be the l.c.m. of $\nu_i p_i^{\alpha_i-\delta_i}$ $(i = 1, 2, \cdots)$. When $\alpha_1, \alpha_2, \cdots$ increase indefinitely, S/N approaches a limit. The application to quadratic forms will be given under that topic.

C. A. W. Berkhan[135] gave an exposition of the theory of $ax^2+1=y^2$ and a table of solutions for $a\leqq160$.

M. A. Stern[136] applied new theorems on continued fractions to shorten the work of forming an extended table of least solutions of $x^2-Ay^2=1$. Given the period for one number, we can find an infinitude of numbers the continued fraction for whose square root has a known period. He gave a table showing the manner in which the continued fractions for the square roots of 163 of the numbers <1000 can be derived from that for 2.

A. Cayley[137] gave for $D<1000$, $D\equiv5$ (mod 8), a table showing the least odd solutions of $x^2-Dy^2=-4$, when it is solvable, or, if not, of $x^2-Dy^2=+4$, when the latter is solvable. The computation was made by means of Degen's[101] table; if in the second line of the entry for D the number 4 does not occur, there is no solution of $x^2-Dy^2=4$; if the rank of the place in which 4 occurs is even, this equation and also $x^2-Dy^2=-4$ is solvable; if of odd rank, only $x^2-Dy^2=4$ is solvable. Also the least solution can be found by means of the series of quotients (in the first line of the entry) by stopping at the number preceding that above 4 and computing the continued fraction determined by this series. From the least solution of $\tau^2-D\nu^2=-4$ we get the least solution $x=\tau^2+2$, $y=\tau\nu$, of $x^2-Dy^2=+4$, and the least solution $X=(\tau^3+3\tau)/2$, $Y=(\tau^2+1)\nu/2$, of $X^2-DY^2=-1$. From the least solution of $T^2-DU^2=4$ we get the least solution $x=(T^3-3T)/2$, $y=(T^2-1)U/2$, of $x^2-Dy^2=1$.

---

[134] Monatsber. Akad. Wiss. Berlin, 1855, 493–5; Jour. de Math., (2), 1, 1856, 76–9; Jour. für Math., 53, 1857, 127–9; Werke, II, 183–194.

[135] Lehrbuch der Unbestimmten Analytik, Halle, 2, 1856, 121–193.

[136] Jour. für Math., 53, 1857, 1–102.

[137] Jour. für Math., 53, 1857, 369–371; Coll. Math. Papers, IV, 40. Reprinted, Sphinx-Oedipe, 5, 1910, 51–3. Errata, Cunningham,[309] p. 59. Extension by Whitford.[302]

G. C. Gerono[138] proved, following Lagrange,[85] that $x^2 - ny^2 = 1$ has an infinitude of integral solutions, if n is positive and not a square. If x, y are positive integral solutions, x/y is a convergent of even rank of the continued fraction for $\sqrt{n}$ and corresponds to the next to the last incomplete quotient of one of the periods.

H. J. S. Smith[139] stated the principal theorems relating to $t^2 - Du^2 = 1$ or 4 by use of Euler's[72] notation $(q_1, \cdots, q_n)$. He noted, as had Lagrange[75] and Gauss[93] (Art. 222), that the methods used by Euler[65, 71, 83] are incomplete because he always assumed that a first solution is known and merely deduced from it those solutions which belong to the same set, whereas there may exist solutions belonging to a different set, and lastly because he gave no method to distinguish between the integral and fractional values contained in his formulas for x, y.

L. Kronecker[140] noted that if T, U are the least solutions of $T^2 - PU^2 = 1$, $\log(T + U\sqrt{P})$ can be expressed in terms of special theta functions or elliptic functions, and the number of classes of binary quadratic forms of determinant P. He deduced approximate values for T, U; likewise, for the least solutions 4, 1 of $t^2 - 17u^2 = -1$, $4 + \sqrt{17}$ has the two approximations

$$\frac{2}{9} e^{(5/18)\pi\sqrt{17}}, \qquad \frac{1}{\sqrt{5}} e^{(1/10)\pi\sqrt{85}}.$$

R. Dedekind[141] proved the existence of integral solutions t, u $(u \neq 0)$ of $t^2 - Du^2 = 1$ by the method used by Dirichlet[120] for complex integers, but replacing his lemma by the following: There exist infinitely many pairs of integers x, y such that $x^2 - Dy^2$ is numerically $< 1 + 2\sqrt{D}$.

C. Richaud[142] stated that $x^2 - Ny^2 = -1$ is solvable for various types of values of N: If $A, \cdots, L$ are primes of the form $8n + 5$ and $N = 2A^a$, $2A^{2a+1}B^{2\beta+1}$ or $2A^{2a}B^{2\beta}\cdots L^{2\lambda}$. If $B, \cdots, L$ are not included among the linear divisors of $t^2 - 2Au^2$, and $N = 2A^aB^\beta$, $2A^aB^{2\beta+1}C^{2\gamma+1}$ or $2A^{2a+1}B^{2\beta}\cdots L^{2\lambda}$. If $a, b, \cdots, l$ are primes of the form $8n + 1$, and are not included among the linear divisors of $t^2 - 2Au^2$, and $N = 2A^{2m+1}a^a$, $2A^{2m+1}a^{2a+1}b^{2\beta+1}$ or $2A^{2m+1}a^{2a}\cdots l^{2\lambda}$. If $A, \cdots, L$ are primes not included among the linear divisors of $t^2 - \omega u^2$, where ω is a prime $4n + 1$, and $N = \omega^m A^{2a+1}$, $\omega^{2m+1}A^a$, $\omega^{2m+1}A^{2a+1}B^{2\beta+1}$, $\omega^{2m+1}A^{2a}\cdots L^{2\lambda}$. Also for 8 more such sets of N's. He[143] proved these results and similar ones by use of the continued fraction for $\sqrt{N}$ and the reciprocity law for quadratic residues.

Richaud[144] gave minimum integral values of x, y satisfying $x^2 - Ay^2 = 1$ for $A = a^2 \pm d$ (d a divisor > 1 of $2a$) and for many values of A such as $(9a+3)^2 \pm 9$, $(9a+6)^2 \pm 9$, $(25a+5)^2 - 25$. Likewise for $x^2 - Ay^2 = -1$.

[138] Nouv. Ann. Math., 18, 1859, 122–5, 153–8.

[139] Report British Assoc., 1861, §§ 96, 97, pp. 313–9; Coll. Math. Papers, I, 195–202.

[140] Monatsber. Akad. Wiss. Berlin, 1863, 44; French transl. in Annales sc. de l'école normale sup., 3, 1866, 302–8. Cf. Smith, Report British Assoc., 1865, § 138, p. 372; Coll. Math. Papers, I, 354–8.

[141] Dirichlet's Zahlentheorie, §§ 141–2, 1863; ed. 2, 1871; ed. 3, 1879; ed. 4, 1894.

[142] Jour. de Math., (2), 9, 1864, 384–8.

[143] *Ibid.*, (2), 10, 1865, 235–280; (2), 11, 1866, 145–176.

[144] Atti Accad. Pont. Nuovi Lincei, 19, 1866, 177–182.

M. A. Stern[145] proved (p. 27) that $x^2-Ay^2=d_n$ has one and but one solution in integers if d_n $(0<d_n<\sqrt{A})$ is the denominator of a complete quotient which belongs to a partial denominator a_{n+1} of the " negative " periodic continued fraction

$$[a, a_1, a_2, \cdots]=a-\frac{1}{a_1}-\frac{1}{a_2}-\cdots$$

for $\sqrt{A}$, and x, y are the numerator and denominator of the convergent $[a, a_1, \cdots, a_n]$. The first d_n which is unity leads to the solution of $x^2-Ay^2=1$ in least integers; this d_n is the denominator belonging to the final term of the first period. He found (pp. 30–43) the conditions for $d_m=2$. Finally there is a table giving for $N<100$ the partial denominators of the half period and the complete quotients for the negative continued fraction for $\sqrt{N}$. Lagrange[85] had shown by an example that Pell's equation cannot be solved by use of a continued fraction in which the partial denominators have signs chosen at will.

J. Frischauf[146] noted that Gauss[93] (Arts. 197–202) obtained the least solutions T, U of $t^2-Du^2=\sigma^2$ by use of a reduced quadratic form of determinant D. It is here shown that T, U are independent of the particular reduced form used.

N. de Khanikof[147] used a table showing the last two digits of the root of a square ending in 01, 04, $\cdots$, 96 to find the endings of possible integral solutions of $A+Bt^2=u^2$.

P. Seeling[148] treated the form of numbers the continued fractions for whose square roots have periods with a given number g of the terms, treated the cases $g = 2, \cdots, 7$ in detail, and tabulated the period of the continued fraction for $\sqrt{A}$, $2\leqq A\leqq 602$. He noted that Egen[102] omitted from his table all numbers of the form n^2+1, though they belong there. Egen stated that $x^2-Ay^2=-1$ is solvable only when the period of the continued fraction for $\sqrt{A}$ has an odd number of quotients. Seeling stated that it is possible in relatively prime integers x, y only when $A=4m+1$ or $4m+2$. Hence if the period for $\sqrt{A}$ has an odd number g of quotients, $A=4m+1$ or $4m+2$; this is proved for $g=1, 3, 5, 7$.

L. Öttinger[149] gave tables showing several solutions of $x^2-Ay^2=\pm b$ for $A=2, \cdots, 20$; $b=1, \cdots, 10, 3^k, 5^k, 7^k$ $(k=1, 2, 3, 4)$. If we have found by continued fractions the least solution of $p^2-Aq^2=\pm b$ and know a solution of $t^2-Au^2=1$ or -1, another solution of $x^2-Ay^2=\pm b$ is given by $\alpha=pt\pm Aqu, y=pu\pm qt$.

A. Meyer[150] proved by use of ternary forms that if D is a positive integer, 2^σ the highest power of 2 dividing D, $\sigma\leqq 4$, S^2 the greatest odd square dividing D, and $D=2^\sigma S^2D_1$, then there exist integers ξ, η, relatively prime

---

[145] Abhand. Gesell. Wiss. Göttingen (Math.), 12, 1866, 48 pp.

[146] Sitzungsber. Akad. Wiss. Wien (Math.), 55, II, 1867, 121.

[147] Comptes Rendus Paris, 69, 1869, 185–8.

[148] Archiv Math. Phys., 49, 1869, 4–44.

[149] *Ibid.*, 193–222.

[150] Diss., Zürich, 1871; Vierteljahrsschrift Naturf. Gesell. Zürich, 32, 1887, 363–382. Cf. Got.[299]

to $2D$, such that for all primes p and q satisfying

$$p \equiv \xi, \qquad q \equiv \eta \qquad (\mathrm{mod}\ 8SD_1),$$

the equation $t^2 - pqDu^2 = 1$ has a fundamental solution T, U for which neither $T+1$ nor $T-1$ is divisible by pq.

L. Lorenz[151] found the number of integral solutions of $m^2 + en^2 = N$, where $e = 1, 2, 3, 4$ or -1, and N is a given positive integer, by transforming the series

$$\sum_{m,\,n=-\infty}^{+\infty} q^{m^2+en^2} \qquad (q<1)$$

into a series of another form and finding the term q^N of the latter. For details when $e=1$ see Lorenz[94] of Ch. VI.

P. Seeling[152] noted that, if A is positive and not a square, and the continued fraction for $\sqrt{A}$ has the symmetric period n; $a, b, \cdots, b, a, 2n$, solutions x, y of $x^2 - Ay^2 = \pm 1$ are given by the numerator and denominator of the convergent belonging to the quotient $2n$. The sign is plus if the number of quotients in the period is even; while if it be odd, the sign is plus after $2, 4, 6, \cdots$, periods, minus after $1, 3, 5, \cdots$ periods. If $x^2 - Ay^2 = -1$ and the number of quotients in the period is odd, then $A = 4m+1$ or $4m+2$ and A has no factor $4m+3$; if A is a prime $4m+1$, the number of terms in the period for $\sqrt{A}$ is odd; if A is a product of two or more primes $4m+1$ or the double of such a product, no general rule has been found. Finally, he tabulated all numbers $A < 7000$ for which the period of $\sqrt{A}$ has an odd number of terms, so that $x^2 - Ay^2 = -1$ is solvable.

A. B. Evans and A. Martin[153] found the least solution of $rx^2 + 1 = \square$, where $r = 940751$, and noted that $rx^2 + 38 = \square$ has no integral solution.

Moret-Blanc[154] noted that if $x = h$, $y = k$ is a solution of $2x^2 - 1 = y^2$, then $x = hu + kv$, $y = 2hv + ku$ give a second solution, provided $u^2 - 2v^2 = 1$, as for $u = 3, v = 2$.

F. Didon stated and C. Moreau[155] proved that, if $D = (4n+2)^2 + 1$, where n is a positive integer, $t^2 - Du^2 = 4$ has no solution in odd integers, and the least positive solution is $t = 16(2n+1)^2 + 2$, $u = 8(2n+1)$.

O. Schlömilch[156] discussed the continued fraction for $\sqrt{\alpha^2/4 \pm \beta}$.

L. Matthiessen[157] noted that if $x = f$, $y = g$ give the least solution of $ax^2 - y^2 = 1$, all solutions are given by

$$a\left\{ a^n f^{2n+1} + \binom{2n+1}{2} a^{n-1} f^{2n-1} g^2 + \binom{2n+1}{4} a^{n-2} f^{2n-3} g^4 + \cdots \right\}^2 - (af^2 - g^2)^{2n+1}$$

$$= \left\{ \sum_{j=0}^{n} \binom{2n+1}{2j+1} a^{n-j} f^{2n-2j} g^{2j+1} \right\}^2.$$

---

[151] Tidsskrift for Math., (3), 1, 1871, 97. Cf. *J. Petersen, *ibid.*, p. 76.

[152] Archiv Math. Phys., 52, 1871, 40–9.

[153] Math. Quest. Educ. Times, 16, 1871, 34–6.

[154] Nouv. Ann. Math., (2), 11, 1872, 173–7.

[155] *Ibid.*, 48; (2), 12, 1873, 330–1.

[156] Zeitschrift Math. Phys., 17, 1872, 70–71.

[157] *Ibid.*, 18, 1873, 426.

If $x=f$, $y=g$ give the least solution of $ax^2-y^2=-1$, all solutions are given by the preceding and a similar formula.

D. S. Hart[158] stated that, if the fundamental set of solutions p_0, q_0 of $p^2-Nq^2=\pm1$ has been found, so that we have a set in addition to 1, 0, the simplest method to find successively all further sets of solutions is to use the relations $p=2p_0r+r'$, $q=2p_0s\mp s'$, where r, s are the last found values of p, q, and r', s' the next preceding values.

B. Minnigerode[159] modified the theory as presented by Dirichlet[133] by using a different definition of reduced forms and using the continued fraction

$$\omega=a_0-\cfrac{1}{a_1-}\cfrac{1}{a_2-}\cdots-\cfrac{1}{a_{r-1}-}\cfrac{1}{\omega},$$

with negative quotients (see the chapter on binary quadratic forms in Vol. III).

W. Schmidt[160] showed that all positive solutions of $t^2-Du^2=\pm4$ are given by the development into a continued fraction of a root of a certain reduced binary quadratic form of determinant D.

T. Muir[160a] treated the development into a continued fraction of the square root of any positive integer or fraction. In particular, he obtained (p. 19) in general form the results of Euler,[72] calling Euler's $(a, \cdots, l)$ a continuant $K(a, \cdots, l)$.

D. S. Hart and W. J. C. Miller[161] proved by use of $p^2-103n^2=1$ that $103(3x-2)^2+1=\square$ has no integral solution x and that 22421/3 is the least positive solution.

M. Collins and A. M. Nash[162] proved that $x^2+D^m=(N^2+D)y^2$ is solvable in rational numbers if $m=2n+1$ by taking

$$x+Ny=kD(y+D^n), \qquad x-Ny=(y-D^n)/k.$$

Several[163] solved $x^2-953y^2=\pm1$ by the continued fraction for $\sqrt{953}$.

S. Tebay[164] noted that, if p, q are the least solutions of $x^2-ny^2=1$, then

$$x=\tfrac{1}{2}(\eta^r+\eta^{-r}), \qquad y=\tfrac{1}{2}n^{-1/2}(\eta^r-\eta^{-r}), \qquad \eta=p+qn^{1/2}.$$

Let $na^2\pm k=m^2$. To solve $nt^2\pm k=\square$, set $t=a+\tau$. Then

$$m^2+2na\tau+n\tau^2=\square=(m+\tau x/y)^2, \qquad \text{if} \qquad \tau=2y(nay-mx)/(x^2-ny^2).$$

S. Bills[165] illustrated a "new, practical" method of solving $x^2-Ay^2=\pm1$ by taking $A=953$. Then $S=30$ is the root of the square just $<A$. From

---

[158] Math. Quest. Educ. Times, 20, 1874, 64.

[159] Göttingen Nachrichten, 1873, 619–652. Cf. A. Hurwitz.[205]

[160] Zeitschrift Math. Phys., 19, 1874, 92–94.

[160a] The Expression of a Quadratic Surd as a Continued Fraction, Glascow, 1874, 32 pp. Cf. R. E. Moritz, Ueber Continuanten und gewisse ihrer Anwendungen im Zahlentheoretischen Gebiete, Diss. Strassburg, Göttingen, 1902.

[161] Math. Quest. Educ. Times, 20, 1874, 66–7; 28, 1878, 65–66.

[162] Ibid., 22, 1875, 23–24.

[163] Ibid., 78–80; 23, 1875, 107.

[164] Ibid., 23, 1875, 30.

[165] Ibid., 98–99.

$0, 1, S$ as the initial triple and M, N, P as any triple, derive the next triple by

$$M_1 = NP - M, \qquad N_1 = \frac{A - M_1^2}{N}, \qquad P_1 = \left[\frac{S + M_1}{N_1}\right],$$

where $[k]$ is the largest integer $\leqq k$. When we reach a triple with $P = 2S$, we get a solution. Application is made to solve $nt^2 \pm k = \square$.

Several writers[166] proved that, if r is the least integer for which $Ar^2 - 1 = \square$ and if $AR^2 + 1 = \square$, then R is a multiple of r.

D. S. Hart[167] showed that 560 is the least z making $953z^2 + 87z + 1 = \square$.

A. Martin[168] noted that $x = 1284836351$ gives the least solution of $x^2 - 5658y^2 = 1$, whereas Barlow[99] gave erroneously a number of 48 digits. [Barlow solved $x^2 - 56587y^2 = 1$; the omission of 7 was a misprint.]

H. J. S. Smith[169] proved that, if T, U are the least integral solutions of $T^2 - DU^2 = (-1)^i$, then $T + U\sqrt{D}$ equals the product of the i complete quotients in a period in the development of $\sqrt{D}$ as a continued fraction. Also theorems on the number of different periods of complete quotients.

D. S. Hart[170] gave a " new " method to solve $x^2 - Ay^2 = 1$. Set $A = r^2 \pm m$. Then $(x + ry)(x - ry) = 1 \pm my^2$. Set $x - ry = 1$. Then

$$y = \pm 2r/m, \qquad x = 1 \pm 2r^2/m.$$

But the solutions are not in general integers. He and A. Martin[171] found positive integral solutions of $94x^2 + 57x + 34 = \square$.

A. Kunerth[172] required rational values of p for which $x = N/D$ is an integer, N and D being given quadratic functions of p with integral coefficients. Replacing p by a suitable linear function of q, we get*

$$x_1 \equiv ax - A = \frac{dq + f}{q^2 - g}, \qquad (f - dq)x_1 + d^2 = \frac{S}{q^2 - g},$$

where $S = f^2 - gd^2$ is known. Any common factor of d, f may be removed from each member of the second equation. Write v/w for the rational number q and equate each positive or negative factor of S in turn to $v^2 - gw^2$. Hence for g negative, there is only a finite number of trials. To apply to $y^2 = ax^2 + bx + c$ with the given solution x_1, y_1, set $y = p(x - x_1) + y_1$ in $y^2 = a(x^2 - x_1^2) + b(x - x_1) + y_1^2$. We get

$$x - x_1 = \frac{-2y_1 p + 2ax_1 + b}{p^2 - a}.$$

The case $b = 0$ is treated at length. The method is applied (pp. 24–32) to Pell's equation $y^2 = ax^2 + 1$; as $y = px + 1$, $x = -2p/(p^2 - a)$. He reproduced (pp. 56–8) Tenner's rule.[118]

---

[166] Math. Quest. Educ. Times, 23, 1875, 109–110; 24, 1876, 109–111.

[167] *Ibid.*, 25, 1876, 97.

[168] The Analyst, Des Moines, 2, 1875, 140–2; Math. Quest. Educ. Times, 26, 1876, 87; Math. Magazine, 2, 1890, 59.

[169] Proc. London Math. Soc., 7, 1875–6, 199–208; Collectanea Mathematica, Milan, 1881, 117; Coll. Math. Papers, 2, 1894, 148.

[170] Math. Quest. Educ. Times, 28, 1878, 29–30.

[171] *Ibid.*, 101–2; 24, 1876, 39–40.

[172] Sitzungsber. Akad. Wiss. Wien (Math.), 75, II, 1877, 7–58.

* There are five errors of signs on pp. 15–16. In the examples the signs are correct.

A. Martin[173] noted that, in the least solution of $x^2-9817y^2=1$, x has 97 digits.

D. S. Hart[174] noted that $(r^2+s^2)y^2-1=\square$ for $y=m^2+n^2$, if

$$ms=rn\pm\sqrt{(r^2+s^2)n^2}\pm s,$$

where one of r, s is odd and the other even, while n is to be found by trial.

Martin[174a] found the least solution of $x^2-9781y^2=1$.

S. Roberts[175] noted that if $t^2-Du^2=-1$ is solvable in integers, where $D=2^\mu\alpha^a\beta^b\cdots$, $\mu=0$ or 1 and α, β, $\cdots$ odd, then $t^2-D'u^2=-1$ is solvable, where $D'=2^\mu\alpha^{a+2p}\beta^{b+2q}\cdots$. Since any prime $4n+1$ is a D, any odd power of it is a D'. If $D=s^2d$, the solvability of $t^2-du^2=-1$ is a necessary, but not sufficient, condition for the solvability of $t^2-Du^2=-1$.

Roberts[176] proved that, if t, u are the least solutions of $t^2-Au^2=1$, there are values t_1, u_1, less than t, u, for which either $Mt_1^2-Nu_1^2=\pm1$, $MN=A$, or $Mt_1^2-Nu_1^2=\pm2$, $MN=A$, unless $M=1$. If the first of these equations is solvable and $M<N$, then M is the middle denominator of the period of the continued fraction for $\sqrt{A}$; but if the second holds, and not the first, $2M$ is the middle denominator.

H. Brocard[177] gave a bibliography and historical notes on Pell's equation.

K. E. Hoffmann[178] recalled that Lagrange proved that x_0, y_0 is a solution of $x^2-Ay^2=1$ if x_0/y_0 is the convergent corresponding to the first or first two periods of the continued fraction for $\sqrt{A}$. Other solutions follow from

$$x_n+y_n\sqrt{A}=(x_0+y_0\sqrt{A})^n.$$

While it is usually merely stated that x_n/y_n is a convergent to a later complete period, a direct proof is here given by use of the " closed form " of a periodic continued fraction (ibid., 62, 1878, 310–6).

A. Kunerth[179] gave a " practical " method of solving

$$(17)\qquad\qquad y^2=ax^2+bx+c.$$

If a rational solution is known, we may transform (17) into

$$(18)\qquad\qquad y^2=(\alpha x+\beta)^2+(\gamma x+\delta)(\epsilon x+\zeta).$$

Hence every such transformation yields two values $-\delta/\gamma$ and $-\zeta/\epsilon$ of x giving rational solutions. If $x=m/n$, $y=r/n$ is a solution of (17), take $\gamma=n$, $\delta=-m$. Then $r=m\alpha+n\beta$, from which we may determine α, β. Then ϵ, ζ may be found from (18). To proceed without a known solution, subtract $(\alpha x+\beta)^2$ from (17) and employ the condition that the difference be a product of two linear functions:

$$(19)\qquad\qquad (b-2\alpha\beta)^2-4(a-\alpha^2)(c-\beta^2)=\Delta^2.$$

[173] The Analyst, Des Moines, 4, 1877, 154–5.
[174] Ibid., 5, 1878, 118–9.
[174a] Math. Visitor, 1, 1878, 26–7.
[175] Proc. London Math. Soc., 9, 1877–8, 194.
[176] Ibid., 10, 1878–9, 30–32.
[177] Nouv. Corresp. Math., 4, 1878, 161–9, 193–200, 228–232, 337–343.
[178] Archiv Math. Phys., 64, 1879, 1–8.
[179] Sitzungsber. Akad. Wiss. Wien (Math.), 78, II, 1878, 327–37.

Set $D = b^2 - 4ac$, $\beta = (K + b\alpha)/(2a)$.   Then $K^2 = a\Delta^2 + D(\alpha^2 - a)$.   Hence we have to assign to Δ and α such values that the latter sum is a square.

To apply (pp. 338–346) this method to the congruence $y^2 \equiv c \pmod{b}$, where b is a prime, we have (17) for $a = 0$.   Then (19) holds for $\Delta = b + 2p\alpha$ if

$$\alpha = \frac{-bw(v + \beta w)}{v^2 - cw^2}, \qquad \frac{v}{w} = p.$$

The first denominator may be made equal to $\pm b$ if $\pm b$ is a quadratic residue of c.   Then $\alpha = \mp w_0(v_0 + \beta w_0)$.

Kunerth[180] continued the same subject.   Let α_1, β_1 be a solution of $r = m\alpha + n\beta$.   Then $\alpha = \alpha_1 - np$, $\beta = \beta_1 + mp$.   Substitute these in (18), with $\gamma = n$, $\delta = -m$.   After several reductions, we get

$$-\epsilon x - \zeta = (np^2 - 2\alpha_1 p - \epsilon)x - (mp^2 + 2\beta_1 p + \zeta).$$

Then (17) has an integral solution if and only if p can be chosen to make the value of x for which the preceding vanishes an integer.

A. B. Evans and others[181] proved that, if p_n/q_n is the last convergent in the first period of the continued fraction for $\sqrt{A}$, and r is the largest integer $\leqq \sqrt{A}$, then $p_n = rq_n - q_{n-1}$.   Hence we can derive x from y in a solution of $x^2 - Ay^2 = 1$.

J. de Virieu[182] used the final digits to show that xy is divisible by 5 in

(20) $$24x^2 + 1 = y^2.$$

E. Lionnet[183] stated and M. Rocchetti and F. Pisani[183] proved easily that three successive sets (x_i, y_i) of solutions of (20) or $2x^2 + 1 = 3y^2$ satisfy $x_{n+1} = 10x_n - x_{n-1}$, $y_{n+1} = 10y_n - y_{n-1}$, with $(x_1, y_1) = (0, 1)$ or $(1, 1)$, $(x_2, y_2) = (1, 5)$ or $(11, 9)$, respectively.   For solutions x of the second equation, $3x^2 + 2$ is of the form $360n + 5$ and is simultaneously a sum of three consecutive squares and a sum of two consecutive squares.   For $x^2 + 1 = 2y^2$, $x_n = 6x_{n-1} - x_{n-2}$, $y_n = 6y_{n-1} - y_{n-2}$, $(x_1, y_1) = (1, 1)$, $(x_2, y_2) = (7, 5)$.

S. Réalis[184] used $x^2 - ky^2 = (\alpha^2 - k\beta^2)(A^2 - kB^2)^2$, where

$$x = \alpha A^2 - 2k\beta AB + k\alpha B^2, \qquad y = -\beta A^2 + 2\alpha AB - k\beta B^2,$$

to derive a new solution of $x^2 - ky^2 = h$ from a given solution α, β and a solution of $A^2 - kB^2 = 1$.

H. Poincaré[185] noted that, if m is odd, and a, b give the least integral solutions of $a^2 - mb^2 = 1$ and c, d give the least odd integral solutions of $c^2 - md^2 = 4$, then

$$\left(\frac{c + d\sqrt{m}}{2}\right)^3 = a + b\sqrt{m}.$$

Several[186] proved easily that $x_{n+p} = 2x_p x_n - x_{n-p}$, $y_{n+p} = 2x_p y_n - y_{n-p}$, if x_n, y_n be the nth set of positive integral solutions of $x^2 - Ny^2 = 1 [x_0 = 1, y_0 = 0]$.

[180] Sitzungsber. Akad. Wiss. Wien (Math.), 82, II, 1880, 342–75.
[181] Math. Quest. Educ. Times, 30, 1879, 49.
[182] Nouv. Ann. Math., (2), 17, 1878, 476.
[183] *Ibid.*, (2), 18, 1879, 479, 528; (2), 20, 1881, 425–7, 373–4.   Cf. Pisani[5] of Ch. VII.
[184] Nouv. Corresp. Math., 6, 1880, 306–312, 342–350.
[185] Comptes Rendus Paris, 91, 1880, 846.
[186] Math. Quest. Educ. Times, 34, 1880, 114.

W. P. Durfee[187] stated that, if x_0, y_0; x_1, y_1; $\cdots$ be the integral solutions of $ax^2-y^2=-1$, arranged according to magnitude, then

$$x_n y_{n+t}-x_{n+t}y_n=-x_t, \qquad ax_n x_{n+t}-y_n y_{n+t}=-y_t.$$

S. Günther[138] noted that the solution of $2x^2-1=y^2$ was apparently known to Plato. Its complete solution implies that of $2x^2+1=y^2$, and conversely. To solve $(a^2+b^2)x^2-1=y^2$, seek the integral solutions of $\xi^2-(a^2+b^2)\eta^2=a^2\xi$ and test whether or not $a^2\xi$ divides $2(a^2+b^2)\eta^2\pm 2b\xi\eta$; if so we have a solution of the initial equation.

E. de Jonquières[189] found the period of the continued fraction for $\sqrt{A}$ for special types of numbers A, and treated periodic continued fractions whose numerators differ from unity.

D. S. Hart[190] stated that a process, simpler than Euler's and Lagrange's, to find integral solutions of $ax^2+bx+c=\square$ is to subtract such a square $(lx+m)^2$ that the difference will factor into two linear functions with integral coefficients. Then $L^2+MN=\square=(L-Mr/s)^2$ gives x; equate its denominator to unity.

E. Catalan[191] discussed $Ax^2=y^2+1$. Thus A is of the form a^2+b^2. If p, q give the least solution, x is divisible by p. Set $x=pz$; then

$$(q^2+1)z^2=y^2+1.$$

Hence consider $(a^2+1)x^2=y^2+1$. For its solutions,

$$x_n=2(2a^2+1)x_{n-1}-x_{n-2}, \qquad n\geqq 3.$$

It is shown that x_n is a sum of three squares if $n\geqq 3$. If $b>1$ in the initial equation, x_n is a sum of four squares. Every integer $y>1$, for which $(a^2+1)x^2=y^2-1$, is a sum of three squares. Cf. Catalan[63] of Ch. VII.

S. Roberts[192] proved that $q^2-Dr^2=1$ can be solved by using the nearest integral limits exclusively or superior limits exclusively as the partial quotients belonging to the continued fraction for $\sqrt{D}$, instead of using the customary inferior limits exclusively. But he admitted his results are due to Stern[145] and Minnigerode.[159]

G. de Longchamps[193] gave a bibliography of Pell's equation.

J. Perott[194] proved that there exists a positive integer λ such that, in

$$t_\lambda+u_\lambda \sqrt{d}=(t_1+u_1 \sqrt{d})^\lambda,$$

u_λ is divisible by a given odd prime, where t_1, u_1 give the least positive solutions of $t^2-du^2=1$. He repeated (pp. 342-3) Poincaré's[185] remark.

M. Weill[195] noted that $x^2-Ay^2=N^2$ has the solution $x=Au^2+t^2$, $y=2tu$, if t, u give a solution of $t^2-Au^2=N$. Taking $N=1$, consider $a, a_1, a_2, \cdots$,

[187] Johns Hopkins University Circular, 1, 1882, 178.
[188] Blätter für Bayer. Gymnasialschulwesen, 18, 1882, 19-24.
[189] Comptes Rendus Paris, 96, 1883, 568, 694, 832, 1020, 1129, 1210, 1297, 1351, 1420, 1490, 1571, 1721.
[190] Math. Magazine, 1, 1882-4, 40-1.
[191] Assoc. franç. av. sc., 12, 1883, 101; Atti Accad. Pont. Nuovi Lincei, 37, 1883-4, 84-95.
[192] Proc. London Math. Soc., 15, 1883-4, 247-268.
[193] Jour. de math. élém., 1884, 15 (1885, 171, on continued fractions).
[194] Jour. für Math., 96, 1884, 335-7.
[195] Nouv. Ann. Math., (3), 4, 1885, 189-193.

$a_k = 2a_{k-1}^2 - 1$, obtained from $a^2 - Au^2 = 1$, $y_1 = 2au$, $a_1 = Au^2 + a^2 = 2a^2 - 1$, $\cdots$. He gave an explicit expression for a_k and noted the connection with the formula for $\cos m\phi$ in terms of $\sin \phi$ and $\cos \phi$.

H. van Aubel[196] proved the statement by Brocard[177] that

$$x_{m+1} = 2px_m - x_{m-1}, \qquad y_{m+1} = 2py_m - y_{m-1},$$

give the relations between three consecutive sets of solutions of $x^2 - Ay^2 = 1$, where p, q give the least solutions. Also theorems giving p and q in terms of the convergents found near the middle of the period of the continued fraction for $\sqrt{A}$. If the period has an odd number of terms, A is a sum of two relatively prime squares, but not conversely. He treated values of A, b for which the solution $x = by + 1$, $y = 2b/(A - b^2)$ of $x^2 - Ay^2 = 1$ is integral. He noted cases when integral solutions can be derived from two sets of fractional solutions.

Several[197] solved the problem to find the polygons the number $x(x-3)/2$ of whose diagonals is a square, by treating $(2v-1)^2 - 8u^2 = 1$.

H. Richaud[198] found the least solution of $x^2 - Ny^2 = -1$ for $N = 1549$. He noted corrections to Legendre's[88] table for $N = 823$ and 809.

J. Vivante[199] treated $Dx^2 - 3 = y^2$ (cf. binary quadratic forms).

E. Lucas[200] gave periods of the continued fraction for $\sqrt{n}$, when n is a quadratic function.

Several[201] solved $x^2 - 19y^2 = 81$.

J. Perott[202] reviewed various classic papers on $t^2 - Du^2 = -1$ and proved that, if q is a prime of the form $16n + 9$, $t^2 - 2qu^2 = -1$ is solvable if and only if $2^{(q-1)/4} \equiv -1 \pmod{q}$; while, if q is a prime $16n + 1$, the condition $2^{(q-1)/4} \equiv 1 \pmod{q}$ is necessary, but not sufficient. If q is a prime $8n + 5$, $t^2 - 2q^2u^2 = -1$ is always solvable; but, if q is a prime $8n + 1$, a necessary condition is that, in the decomposition $q = c^2 + 2d^2$, d be divisible by 8. This condition is sufficient if q is of the form $16m + 9$, but not if $q = 16m + 1$.

F. Tano[203] proved that $x^2 - Ay^2 = -1$ is solvable if $A = a_1a_2 \cdots a_n$, where n is odd and $a_1, \cdots, a_n$ are distinct primes $\equiv 1 \pmod 4$ and if at most one of Legendre's symbols (a_i/a_j) is $+1$ for $i < j$. He gave theorems on the case $A = 2a_1 \cdots a_n$.

J. Knirr[204] gave in detail the Indian[30] cyclic method to solve $z^2 - cx^2 = 1$, claiming a simplification. This method is said to be much shorter than that by continued fractions. He tabulated the least solutions for $c \leqq 152$.

A. Hurwitz[205] developed any real number x_0 into a continued fraction by use of $x_0 = a_0 - 1/x_1$, $x_1 = a_1 - 1/x_2$, $\cdots$, where a_n is chosen so that $x_n - a_n$ lies between $-1/2$ and $+1/2$. Minnigerode[159] had shown that the de-

[196] Assoc. franç. av. sc., 14, II, 1885, 135–151.
[197] Mathesis, 6, 1886, 162.
[198] Jour. de Math. Elém., (3), 1, 1887, 181–3. Cf. Whitford,[4] p. 97.
[199] Zeitschr. Math. Phys., 32, 1887, 287–300.
[200] Jour. de math. spéciales, 1887, 1.
[201] Math. Quest. Educ. Times, 48, 1888, 48.
[202] Jour. für Math., 102, 1888, 185–223.
[203] Jour. für Math., 105, 1889, 160–9.
[204] Die Auflösung der Gleichung $z^2 - cx^2 = 1$, 18. Jahresberich Oberrealschule, 1889, 34 pp.
[205] Acta Math., 12, 1889, 367–405.

velopment is periodic if x_0 is a root of a quadratic equation with integral coefficients. The necessary and sufficient condition for the solvability of $x^2-Dy^2=-1$ is that

$$\sqrt{D}=(a_0;\ a_1,\ a_2,\ \cdots,\ a_r,\ -a_1,\ -a_2,\ \cdots,\ -a_r;\ a_1,\ a_2,\ \cdots).$$

G. Chrystal[206] gave an exposition of $x^2-Cy^2=\pm H$ convenient for English readers.

F. Tano[207] proved by developing $\sqrt{a^2\pm4}$ into a continued fraction that $x^2-(a^2+4)y^2=-1$ is solvable in integers when a is any odd integer, while $x^2-(a^2-4)y^2=-1$ is impossible except when $a=3$. There are infinitely many integral solutions of $x^2-kz^2=\pm a$ if a is any odd integer and k a sum of two squares. To prove that there are infinitely many integral solutions of

$$x^2+y^2+z^2=u^2+v^2+w^2+N,$$

where N is any integer, we add the two equations

$$x^2-(a^2+4)y^2=a,\qquad x_1^2-(a^2-4)y_1^2=-(2a-5)$$

if N is odd; but, if N is even, we first change the second members to $-a$, 4. By multiplying $x^2-a^2y^2-4y^2=\pm a$ by $u_i^2-a^2v_i^2+4v_i^2=1$ for $i=1,\ 2,\ \cdots$, in turn, we find that there is an infinitude of integral solutions of

$$\sum_{r=1}^{k}x_r^2-\sum_{r=1}^{k+1}y_r^2=\pm a\qquad\left(k=\frac{3^n-1}{2}\right).$$

G. Frattini[208] noted that, if x_0, y_0 is the fundamental solution of $x^2-(a^2+1)y^2=-N$, viz., a solution with $0<y_0\leqq\sqrt{N}$, then all its solutions are given by

$$x+y\sqrt{a^2+1}=(\pm x_0+y_0\sqrt{a^2+1})(a+\sqrt{a^2+1})^n,$$

where n ranges over the values 0, 2, 4, $\cdots$; while all solutions of $x^2-(a^2+1)y^2=+N$ are given by the same formula where n ranges over the positive odd integers.

Frattini[209] proved that, if K, H $(H<\sqrt{N})$ form a solution of $x^2-(a^2-1)y^2=N$, every solution in positive integers is given by

$$x+y\sqrt{a^2-1}=(K+H\sqrt{a^2-1})(a+\sqrt{a^2-1})^m,\qquad m=0,\ 1,\ 2,\ \cdots.$$

Let $\alpha^2-D\beta^2=1$, $\beta\neq0$. Multiplying $x^2-Dy^2=N$ by β^2, we get

$$(\beta x)^2-(\alpha^2-1)y^2=N\beta^2,$$

whose solutions are derived from one by the preceding formula, viz.,

$$x+y\sqrt{D}=(K+H\sqrt{D})(\alpha+\beta\sqrt{D})^m,\qquad m=0,\ 1,\ 2,\ \cdots.$$

When N is changed to $-N$, the same formulas hold if we replace K by $\pm K$, where, in the final formula, $H\leqq\sqrt{N(\alpha+1)/2D}$. Tchebychef's[131] first result is a corollary.

[206] Algebra, 2, 1889, 450–60; ed. 2, 2, 1900, 478–86.
[207] Bull. des Sc. Math., (2), 14, I, 1890, 215–8.
[208] Periodico di Mat., 6, 1891, 85–90.
[209] *Ibid.*, 169–180.

Frattini[210] reduced the solution of $x^2-Dy^2=N$ to the solution of one of the equations $x^2-Dy^2=N\rho^\lambda$, where $\rho=2m+1-n>0$, m^2 being the largest square $<D=m^2+n$. Let x, y be a solution of the given equation such that $y\geqq\sqrt{N/\rho}$. Then $x\leqq(m+1)y$. Let $x=(m+1)y-h$, whence $h\geqq0$. Then our equation becomes a quadratic for y; the radical in the root y must be an integer k. Thus

$$y=\frac{h(m+1)\pm k}{\rho}, \qquad k^2-Dh^2=N\rho.$$

The sign before k must be plus. Hence if $y\geqq\sqrt{N/\rho}$, and if k, h give positive integral solutions of $x^2-Dy^2=N\rho$, positive integral solutions of $x^2-Dy^2=N$ are given by

$$x+y\sqrt{D}=f(k+h\sqrt{D}), \qquad f=(m+1+\sqrt{D})/\rho.$$

Applying this result to the new equation, we conclude that, if $h\geqq\sqrt{N}$, positive integral solutions of the proposed equation are given by

$$x+y\sqrt{D}=f^2(k'+h'\sqrt{D}),$$

k', h' being positive integral solutions of $x^2-Dy^2=N\rho^2$. The reciprocal of f is $m+1-\sqrt{D}<1$. Thus we finally reach an equation $x^2-Dy^2=N\rho^\lambda$ with a solution y exceeding $\sqrt{N\rho^\lambda/\rho}$, and hence a solution of the proposed equation.

Frattini[211] used similarly $x^2-Dy^2=N(-n)^\lambda$, $\lambda=1$, 2, $\cdots$, to solve $x^2-Dy^2=N$, and applied the two methods to $x^2-Dy^2=-N$. He[212] deduced the theorem of Tchebychef.[131]

Frattini[213] supplemented and interpreted geometrically the theorem of Tchebychef. From Frattini[209] we derive the result: If 0, q_1, q_2, $\cdots$ are values of y in successive positive integral solutions of $x^2-Dy^2=1$, the series 0, $q_1\sqrt{N}$, $q_2\sqrt{N}$, $\cdots$ separate the positive integral solutions of $x^2-Dy^2=N$ in such a way that the number of solutions, in which y equals or exceeds any number of that series and is less than the following, is constant. The geometric interpretation is that the vectors of the successive solutions of $x^2-Dy^2=1$ divide the angle between the positive x-axis and the line of slope $1/\sqrt{D}$ into consecutive angles each of which contains an equal number of points with integral coördinates satisfying $x^2-Dy^2=N$. Again, if 1, p_1, p_2, $\cdots$ are the values of x, the series 0, $\sqrt{N(p_1+1)/2D}$, $\sqrt{N(p_2+1)/2D}$, $\cdots$ separate the solutions of $x^2-Dy^2=-N$ as before; for interpretation, use the y-axis instead of the x-axis.

C. A. Roberts[214] gave only the denominators of the continued fractions for $\sqrt{p}$, where p is a prime $4n+1\leqq10501$ (thus giving what corresponds only to the first line of each entry in the table by Degen,[101] and not the least solution of $x^2-py^2=\pm1$). The introduction to the table is by A. Martin.

[210] Periodico di Mat., 7, 1892, 7–15.
[211] *Ibid.*, 49–54, 88–92, 119–22.
[212] *Ibid.*, 123–124, 172–7.
[213] Atti Reale Accad. Lincei, Rendiconti, (5), 1, 1892, Sem. 1, 51–7; Sem. 2, 85–91.
[214] Math. Magazine, 2, 1892, 105–120.

E. Lemoine[215] proved that all positive solutions of $x^2+1=2y^2$ are given by $x_n=N_{2n-1}+N_{2n}$, $y_n=N_{2n}$, where $N_{n-1}a+N_nb$ is the nth term of the series $u_1=a$, $u_2=b$, $\cdots$, $u_n=2u_{n-1}+u_{n-2}$, so that

$$N_{2n-1}=2^{2n-3}+\binom{2n-4}{1}2^{2n-5}+\binom{2n-4}{2}2^{2n-7}+\cdots+\binom{2n-4}{2n-1}2,$$

$$N_{2n}\ =2^{2n-2}+\binom{2n-3}{1}2^{2n-4}+\binom{2n-3}{2}2^{2n-6}+\cdots+\binom{2n-3}{2n-2}2^2+1.$$

If x, y is a solution of $x^2+1=2y^2$, then $x+2y$, $x+y$ is a solution of $x^2-1=2y^2$ and the same holds if the equations are interchanged.

G. B. Mathews,[216] employing the fundamental solution $(T,\ U)$ of $t^2-Du^2=\sigma^2$, and the notation of hyperbolic functions, put

$$\phi=\cosh^{-1}(T/\sigma)=\sinh^{-1}(U\sqrt{D}/\sigma).$$

Then the general solution is $T_n=\sigma\cosh n\phi$, $U_n=(\sigma/\sqrt{D})\sinh n\phi$.

K. Schwering[217] started with Jacobi's elliptic function $x=\sin\operatorname{am}u$, the function inverse to

$$u=\int_0^x dx/\sqrt{1-x^4},$$

and an "odd" integral complex number $\eta=a+bi$, where a is odd and b even, so that $q=a^2+b^2$ is odd. Then

$$\sin\operatorname{am}(\eta u)=\pm\frac{x^q+a_1x^{q-4}+a_2x^{q-8}+\cdots+a_\nu x}{1+a_1x^4+a_2x^8+\cdots+a_\nu x^{q-1}}=\frac{x\phi(x^4)}{\chi(x^4)},\qquad \nu\equiv\frac{q-1}{4}.$$

If η is a complex prime of the form $4k+3+(4k'+2)i$, then $\phi(x^4)$, on which depends the division of the lemniscate by η, is factorable into

$$\phi(x^4)=Y^2-\eta Z^2.$$

Let g be a primitive root of the prime q, so that $g^\nu\equiv i\ (\operatorname{mod}\eta)$. Taking $x=1$, we get odd complex integral solutions t, u of $t^2-\eta u^2=2i(-1)^{\operatorname{ind}(1+i)}$ By squaring $t+\sqrt{\eta}u$ we get complex integral solutions of $T^2-\eta U^2=1$.

H. Weber[218] employed the modular equation (an algebraic equation in u and v of degree 24 in each) which holds between the two elliptic functions $u=f(\omega)$, $v=f(23\omega)$, to deduce the identity $X^2M-Y^2N=1$,

$$2X=(B-1)(B-4)(B^2-4B+2),\qquad M=B^3-5B^2+8B-5,$$
$$2Y=(B-3)(B^3-6B^2+10B-6),\qquad N=B^3-5B^2+4B-1.$$

Squaring $X\sqrt{M}+Y\sqrt{N}$, we get $x+y\sqrt{D}$, where $x^2-Dy^2=1$, $D=MN$.

C. E. Bickmore[219] computed (for a committee of which A. Cayley was chairman) a table, extending Degen's[101] and showing, for $1001\leqq a\leqq1500$, the least solutions of $y^2=ax^2-1$ when a is not of the form t^2+1 (in the contrary case, $y=t$, $x=1$, give a solution), and, when the latter is not solv-

[215] Jornal de Sc. Math e Astr. (ed., Teixeira), 11, 1892, 68–76, 115.

[216] Theory of Numbers, 1892, 93.

[217] Jour. für Math., 110, 1892, 63–4 (112, 1893, 37–8).

[218] Math. Annalen, 43, 1893, 185–196.

[219] Report British Assoc. for 1893, 1894, 73–120; Cayley's Coll. Math. Papers, 13, 1897, 430–467. Errata by Cunningham.[256, 309]

able, the least solutions of $y^2 = ax^2 + 1$. From a solution of the former we get the solution $y_1 = 2y^2 + 1$, $x_1 = 2xy$ of $y_1^2 = ax_1^2 + 1$.

A. Hurwitz[220] proved that the positive relatively prime solutions of $u^2 - Dv^2 = m$, where $|m| < 2\sqrt{D}$, are given by the fractions u/v approximating to $\sqrt{D}$, where u/v and r/s are said to form a pair of fractions approximating to x if x lies between them and if $su - rv = 1$.

A. H. Bell[221] found a special solution of $x^2 = Ny^2 + 1$ by setting

$$x = -1 + Nym/n,$$

whence $y = 2mn/(m^2N - n^2)$ and asked when the denominator is unity. He treated the case $N = 94$ and $x^2 - 61y^2 = -1$.

Emma Bortolotti[222] noted that a root of a quadratic equation with discriminant A and having as coefficients polynomials in x can be developed into a periodic continued fraction whose elements are linear functions of x if and only if $Au^2 - v^2 = 1$ is solvable in polynomials u, v in x. If A is of odd degree in x, the latter equation is evidently impossible.

A. Meyer[223] noted that if $t^2 - Du^2 = 1$ has a fundamental solution T, U, in which U is relatively prime to a divisor D_1 of D, it has solutions in which u is congruent modulo D_1 to an arbitrarily given number.

G. Speckmann[224] employed the identity

$$(na^2 \pm m)^2 - \left(\frac{n^2a^2 \pm 2nm}{x^2} \right)(ax)^2 = m^2,$$

for $m = 1$, and called the resulting solutions of Pell's equation regular if $x = 1$ and irregular if x^2 is a divisor > 1 of $n^2a^2 \pm 2nm$. To solve $x^2 - Dy^2 = M$ ($M \neq$ square), he sought a square η^2 such that $M + \eta^2$ is a square ξ^2; then a solution is $x = \xi + k\eta^2$, $y = \eta$, if $D = 1 + 2k\xi + k^2\eta^2$.

G. Frattini[225] discussed the solution of $x^2 - Ay^2 = 1$, where A is a polynomial in u, especially when A is of degree 2 or 4.

Ch. de la Vallée Poussin[226] indicated the advantage in using continued fractions in which all but the first quotient are negative integers.

G. Speckmann[227] noted that the fundamental solutions T, U of $t^2 - Du^2 = 1$ are $T = x + 2$, $U = 1$, if $D = x^2 + 4x + 3$; $T = 2x + 3$, $U = 2$, if $D = x^2 + 3x + 2$; etc. He noted identities like

$$(na^3 + m)^3 - (n^3a^6 + 3mn^2a^3 + 3m^2n)a^3 = m^3.$$

A. Palmström[228] gave many recursion formulas and relations between sets of solutions of $(a+2)x^2 - (a-2)y^2 = 4$. If x_1, y_1 are the least positive

[220] Math. Annalen, 44, 1894, 425-7.
[221] Amer. Math. Monthly, 1, 1894, 53-4, 92-4, 169, 239-240.
[222] Rendiconti Circolo Mat. Palermo, 9, 1895, 136-149.
[223] Jour. für Math., 114, 1895, 240.
[224] Ueber unbest. Gleichungen, Leipzig and Dresden, 1895.
[225] Giornale di Mat., 33, 1895, 371-8; 34, 1896, 98-109.
[226] Annales Soc. Sc. Bruxelles, 19, 1895, 111.
[227] Archiv Math. Phys., (2), 13, 1895, 327-333; 14, 1896, 443-5.
[228] Bergens Museums Aarbog for 1896, Bergen, 1897, No. 14, 11 pp. (French).

solutions of $x^2-Ay^2=-1$, then

$$(4x_1^2+4)\left(\frac{y}{y_1}\right)^2-4x_1^2\left(\frac{x}{x_1}\right)^2=4,$$

so that y/y_1, x/x_1 have the same properties as the above x, y, where now $a=4x_1^2+2$. If x_1 is the least positive integer for which $x^2-Ay^2=1$, we see that, by taking $a=4x_1^2-2$, the solutions of odd rank have the same properties as the solutions of $x^2-Ay^2=-1$.

C. Störmer[229] quoted the known result that, if a, b are the least positive solutions of $x^2-Ay^2=-1$, other solutions are given by

$$x_{2n+1}+y_{2n+1}\sqrt{A}=(a+b\sqrt{A})^{2n+1},$$

and solutions of $x^2-Ay^2=+1$ are given by $x_{2n}+y_{2n}\sqrt{A}=(a+b\sqrt{A})^{2n}$. He proved that

$$\alpha-\beta=2\tan^{-1}\frac{a}{x_{2n}},\qquad \alpha+\beta=2\tan^{-1}\frac{b}{y_{2n}},\qquad \alpha=\tan^{-1}\frac{1}{x_{2n-1}},\qquad \beta=\tan^{-1}\frac{1}{x_{2n+1}}.$$

Störmer[230] noted that if $x^2-Dy^2=\pm1$ $(D>0)$ has positive integral solutions and y_1 is the least y, either there is no solution y such that every prime divisor of y divides also D, or there is only one such solution, viz., y_1.

A. Thue[231] proved that in $x^2-Dy^2=m$ the least positive y is $\leqq v\sqrt{m}$, where v is a positive number for which $u^2-Dv^2=1$, provided D is not a square and $u>1$.

A. Boutin[232] tabulated the periods of continued fractions for $\sqrt{n}$, $n\leqq200$, and when n is one of 30 special quadratic functions of a parameter [cf. Stern[109]]. He[233] gave the complete solution of $y^2-(m^2-1)x^2=1$, with details when $m=2$.

H. Brocard[234] gave references to problems depending on $x^2-2y^2=\pm1$.

E. de Jonquières[235] proved by the use of binary quadratic forms that $(a^2-4)x^2-4y^2=\pm1$ is not solvable if $a\neq3$, that $(a^2-1)x^2-4y^2=\pm1$ is not solvable [error for -1], that $(a+1)x^2-ay^2=1$ $(a>0)$ has the least solutions $x=4a+1$, $y=4a+3$, that $(ma^2\pm1)x^2-my^2=\pm1$ has the least solutions $x=4ma^2\pm1$, $y=4ma^3\pm3a$, and gave long expressions for solutions of $(ma^2\pm4)x^2-my^2=\pm1$ (a and m odd). The method employed is similar to that of Gauss (Disq. Arith., art. 195), but with the variation (inspired by Legendre) that he omitted from the period of neighboring reduced forms those having the middle term zero. He applied (pp. 1077–81, 1177) Gauss' method of reduction to $(ma^2+4)x^2-my^2=1$. He gave (p. 1837) values of D for which $t^2-Du^2=-1$ is solvable in integers: $D=a^2(n^2+1)$, $D=4n^2+4n+5$, where a is a divisor of any term of odd rank in the recurring series having 0, 1, $2n$ as initial terms and having $2n$, 1 as the scale of relation. It is not solvable if $D=a^2(n^2+1)$, n a multiple of a.

[229] Nyt Tidsskrift for Math., 7, B, 1896, 49.
[230] Videnskabs-Selskabets Skrifter, Christiania, 1897, No. 2, 48 pp. Cf. Störmer.[274]
[231] Archiv for Math. og Naturvidenskab, 19, 1897, No. 4.
[232] Mathesis, (2), 7, 1897, 8–13.
[233] Ibid., (2), 8, 1898, 159–161.
[234] Ibid., 112–3.
[235] Comptes Rendus Paris, 126, 1898, 863–871, 991–7 (correction, 132, 1901, 750, and l'intermédiaire des math., 8, 1901, 108).

De Jonquières[236] noted that a solution of $t^2-Du^2=-1$ or $-m^2$ can be found from two similar transformations of a quadratic form (A, B, C) into (a, b, c) or its inverse $(-a, b, -c)$.

R. W. D. Christie[237] found the least solution of $x^2-103y^2=1$.

G. Ricalde[238] stated that if $x=1, x_1, x_2, \cdots; y=0, y_1, y_2, \cdots$ are the integral solutions of $x^2-Ay^2=1$ (A not a square), $2(x_{2n}+1)$ is a square t^2 and y_{2n} is a multiple of t; if $2(x_{2n+1}-1)$ is a square for one value of n, it is a square k^2 for every n, and y_{2n+1} is a multiple of k, and one has the solutions of $u^2-Av^2=-1$.   A. Palmström (pp. 210–11) noted that the first statement follows from

$$x_{2n}+y_{2n}\sqrt{A}\equiv(x_1+y_1\sqrt{A})^{2n}=(x_n+y_n\sqrt{A})^2, \qquad x_{2n}=x_n^2+Ay_n^2=2x_n^2-1,$$
$$y_{2n}=2x_ny_n.$$

As to the second statement, Palmström proved that $(x_{2n+1}\mp1)/(x_1\mp1)$ are squares, whence $2(x_{2n+1}-1)$ is a square for every n if for one n.   If u_1, v_1 are the least solutions of $u^2-Av^2=-1$,

$$x_{2n+1}+y_{2n+1}\sqrt{A}=(u_1+v_1\sqrt{A})^{4n+2}=(u_{n+1}+v_{n+1}\sqrt{A})^2$$
$$=2u_{n+1}^2+1+2u_{n+1}v_{n+1}\sqrt{A},$$

so that $2(x_{2n+1}-1)$ is a square.   But the latter may be true when
$$u^2-Av^2=-1$$
is impossible.

A. Goulard[239] proved that, if m is odd, $2(x_{mp}-1)$ is a square if and only if $2(x_p-1)$ is a square.   The latter is not a square if p is even, while, for p odd, it is a square if and only if $u^2-Av^2=-1$ is solvable.

A. Cunningham and R. W. D. Christie[240] each noted that $X^2-pY^2=1$ becomes $x^2-py^2=\mp2$ under the transformation $X=x^2\pm1$, $Y=xy$.   Then if p is a prime, it is of the form $8n+3$ or $8n-1$ according as the upper or lower sign holds.   By choosing values of x, y, we get solutions of the proposed equation.

C. de Polignac[241] proved that if t_1, u_1 are the least positive solutions of $t^2-Du^2=1$, where D is positive and not a square, and t_n, u_n any other solutions, there exists a linear substitution $x_1=(Q_1x+S_1)/(P_1x+R_1)$ whose nth power $x_n=(Q_nx+S_n)/(P_nx+R_n)$ gives $t_n=Q_n, u_n=P_n/u_1$.

G. Ricalde[242] gave the identities solving $x^2-Ay^2=1$:

$$(k^2n\pm1)^2-n(k^2n\pm2)k^2=1, \qquad (8n+25)^2-(4n^2+25n+39)4^2=1,$$
$$\{8[n^3+(n+1)^3]^2+1\}^2-[(2n+1)^2+4]\{4[n^3+(n+1)^3][n^2+(n+1)^2]\}^2=1,$$

as well as those due to Euler.[65]   He and others[243] made minor remarks on the linear relations between three successive solutions of $x^2-ay^2=\pm1$.

---

[236] Comptes Rendus Paris, 127, 1898, 596–601, 694–700.   Slightly different from Gauss.[93]

[237] Math. Quest. Educ. Times, 70, 1899, 51.

[238] L'intermédiaire des math., 6, 1899, 75.

[229] *Ibid.*, 7, 1900, 93.

[240] Math. Quest. Educ. Times, 73, 1900, 115–7.

[241] *Ibid.*, 75, 1901, 67–8.

[242] L'intermédiaire des math., 8, 1901, 256.   The third identity lacked the first exponent 2.

[243] *Ibid.*, 59, 286–7.

A. Boutin[244] noted that, if A is a properly chosen quadratic function of m, $x^2-Ay^2=\pm1$ are solved completely by an infinitude of polynomials in m, which satisfy certain differential equations of order two. Thus for

$$y^2-(m^2+1)x^2=1, \qquad y^2-(m^2+1)x^2=-1,$$

the recurring series

$$x_0=0,\ x_1=1,\ \cdots,\ x_n=2mx_{n-1}+x_{n-2};\ y_0=1,\ y_1=m,\ \cdots,\ y_n=2my_{n-1}+y_{n-2}$$

for even indices solve the first equation, and for odd indices the second. As functions of m, x_n and y_n satisfy the differential equations

$$(m^2+1)\frac{d^2x_n}{dm^2}+3m\frac{dx_n}{dm}-(n^2-1)x_n=0, \qquad (m^2+1)\frac{d^2y_n}{dm^2}+m\frac{dy_n}{dm}-n^2y_n=0.$$

Similar remarks are made for $A=25m^2-14m+2$ and for $x^2-Ay^2=1$, $A=m^2-1$, $m\alpha^2+2$, $m(m\alpha^2+1)$.

J. Romero[245] noted that $(ny^2\pm x)^2-(n^2y^2\pm2nx+A)y^2=\pm1$ if

$$x^2-Ay^2=\pm1.$$

A. S. Werebrusow noted that in $x^2-Ay^2=\pm1$, A may have the form $a^2m^2+2bm+c$ if $b^2-a^2c=\pm1$.

A. Holm[246] employed the $(n+1)$th divisor D_n when $\sqrt{C}$ is converted into a continued fraction the length of whose period is c. Let p_c, q_c be the fundamental solution of $x^2-Cy^2=1$. From one solution p_n, q_n of $x^2-Cy^2=(-1)^nD_n$ we get all the solutions by use of

$$x-y\sqrt{C}=\pm(p_n-q_n\sqrt{C})(p_c-q_c\sqrt{C})^m,$$

where, if c is even, m ranges over all integers, positive, negative or zero; while, if c is odd, m ranges over only the even integers.

H. Weber[247] treated $t^2-Du^2=\pm4$ from the standpoint of quadratic numbers $\frac{1}{2}(t+u\sqrt{D})$, where t and u are integers.

Necessary or sufficient conditions that $x^2-Dy^2=-1$ be solvable have been noted.[248]

E. B. Escott asked and A. S. Werebrusow[249] replied for what values of $a, b, \cdots$ is $[a, b, \cdots, a]/[b, c, \cdots, b]$ integral (cf. Dirichlet's Zahlentheorie p. 49).

P F. Teilhet[250] stated and several proved that if β is a root of $\gamma^2-3\beta^2=1$, and $\beta\neq0$, then $6\beta^2+1$ is not a square. Hence $n(n+1)(n+2)=3A^2$ is impossible.

P. von Schaewen[251] made $f\equiv Ax^2+Bx+C$ a square in the following cases (in which $D=B^2-4AC$): (i) $A=n^2A_1$, $D=m^2D_1$, $A_1+D_1=\square=q^2$, since

$$f=\frac{m^2}{4n^2}+A\left(x+\frac{B+mq}{2A}\right)\left(x+\frac{B-mq}{2A}\right)$$

[244] L'intermédiaire des math., 9, 1902, 60–62.

[245] Ibid., p. 182.

[246] Proc. Edinburgh Math. Soc., 21, 1902–3, 163–180.

[247] Archiv Math. Phys., (3), 4, 1903, 201; Algebra, I, 1895, 395–400; ed. 2, 1898, 438–443.

[248] L'intermédiaire des math., 10, 1903, 102, 224; 11, 1904, 156–8, 242; 12, 1905, 53–6, 249–250; 13, 1906, 243–7 (Werebrusow's results are erroneous).

[249] Ibid., 10, 1903, 98; 11, 1904, 154–6.

[250] Ibid., 11, 1904. 68–9, 182–4.

[251] Zeitschr. Math. Naturw. Unterricht, 34, 1903, 325–34. Progr. Gym. Glogau, 1906.

is of Euler's form P^2+QR. (ii) $C+D=\square$. (iii) $-AD=\square$. (iv) $-CD=\square$. (v) One of $A(1-D)$, $C(1-D)$, $D(1-A)$, $D(1-C)$ a square, and the generalizations to $D=m^2D_1$, $A(1-D_1)=\square=q^2$ (etc.), since then

$$f=\left(\frac{mq}{2A}\right)^2+A\left(x+\frac{B+m}{2A}\right)\left(x+\frac{B-m}{2A}\right).$$

R. W. D. Christie[252] noted that, if $ad-bc=\pm1$, $a^2+b^2=P$, $x^2+1=Py$ is satisfied by

$$x=nP+Q, \qquad y=n^2P+2nQ+c^2+d^2, \qquad \pm Q\equiv ac+bd.$$

The problem is now to choose a, b, c, d to make $y=\square$. He and others (p. 87) solved $x^2-149y^2=1$ without using continued fractions. He and E. B. Escott (p. 119) gave the identity

$$\{k(4n^2a^2\mp4na+4n^2+1)+(2na\mp a+2n)\}^2+1$$
$$=\{(2na\mp1)^2+(2n)^2\}\{(2kn+1)^2+(2kna\mp k+a)^2\}^2.$$

Christie[253] noted that if p_n/q_n is a convergent to $\sqrt{D}$, where D is a prime $4m+1$, then $q_{2n+1}=q_n^2+q_{n+1}^2$ [cf. Euler,[72] end].

G. Frattini,[254] employing a positive integer D and positive rational numbers E, F, defined the index of $E+F\sqrt{D}$ to be the maximum number of such factors into which it can be decomposed. If one solution α, β of $x^2-Dy^2=1$ is known, all solutions of $x^2-Dy^2=N$ are given by

$$x+y\sqrt{D}=(\alpha+\beta\sqrt{D})^k(x'\pm y'\sqrt{D}),$$

where the index of the particular solution x', y' does not exceed half the index of the solution of the Pell equation. But we may regard as known the solutions whose indices do not exceed a given limit (depending only on a finite number of trials).

Frattini[255] extended the preceding results to the algebraic case in which D, N, x, y are polynomials in a parameter a. Finally, he proved that, if D is a positive integer or a polynomial of even degree in a, $x^2-Dy^2=1$ is solvable if and only if $\sqrt{D}$ is developable into a simple periodic continued fraction such that

$$\sqrt{D}=(a_1, a_2, \cdots, a_n, c+\sqrt{D}),$$

where the a's and c are integers if D is integral, otherwise polynomials in a.

A. Cunningham[256] gave the least solutions of both $\tau^2-Dv^2=\pm1$, $D<100$, from Degen's[101] table, but checked by Legendre's[88]; also further (multiple) solutions for $D\leqq20$; also the least odd solutions of $\tau^2-Dv^2=\pm2$, ±8, ±16 for $D<500$, and $D=\pm4$ for $D<1000$ (computed from data in Degen's table). He noted three errors in the table by Bickmore.[219]

Cunningham and Christie[257] showed how to find an infinitude of integers

[252] Math. Quest. Educ. Times, (2), 6, 1904, 98–101.
[253] Educ. Times, 57, 1904, 41.
[254] Periodico di Mat., 19, 1904, 1–15.
[255] *Ibid.*, 57–73. Cf. Frattini,[283] H. E. Heine, Jour. für Math., 48, 1854, 256–8.
[256] Quadratic Partitions, 1904, 260–6.
[257] Math. Quest. Educ. Times, (2), 7, 1905, 79–80.

X_n having the same Y in $X_n^2-P_nY^2=-1$. They and A. H. Bell[258] solved $x^2-19y^2=-3$ without using the usual convergents.

Cunningham[259] used known solutions of $y^2-Dx^2=-1$ to factor numbers of the form y^2+1.

A. Aubry[260] give a history and exposition of the Pell equation.

J. Schröder[261] noted that if P_α/Q_α ($\alpha=1, 2, \cdots$) are the convergents to

$$1+\frac{1}{k}+\frac{1}{k}+\cdots,$$

$$(\sqrt{k}-1)^\alpha=(-1)^{\alpha-1}(\sqrt{k}\,Q_\alpha-P_\alpha)$$

holds only for $k=2$. P. Epstein (p. 310) noted that this result for $k=2$ is a case of the known relation between the general solution of $x^2-Dy^2=\pm1$ and its least solution. It is also a case of the following theorem. If $D=a^2+b$, and b is a divisor of $2a$, while Z_k/N_k are the convergents to $\sqrt{D}$, then

$$(\sqrt{D}-a)^k=(-1)^{k-1}b^{[k/2]}(N_k\sqrt{D}-Z_k).$$

Several writers[262] discussed the p's for which $x^2-(y^2-1)p^2=1$ is solvable.

A. H. Holmes[263] noted that 41 is the least prime y for which

$$7x^2-111=y^2.$$

A. Holm[264] noted that, if p, q give a particular solution of $x^2-Cy^2=\pm D$, and r, s one of $x^2-Cy^2=1$, all positive solutions of the former are given by

$$x-y\sqrt{C}=\pm(p-q\sqrt{C})(r-s\sqrt{C})^n, \qquad n=0, \pm1, \pm2, \cdots$$

R. W. D. Christie[265] noted that if we set $x=\cos\theta$, $y=\sin\theta$,

$$X_2=\cos 2\theta=2x^2-1, \qquad X_3=\cos 3\theta=4\cos^3\theta-3\cos\theta=4x^3-3x, \cdots,$$

$$Y_2=\sin 2\theta=2xy, \qquad Y_3=\sin 3\theta=(4\cos^2\theta-1)\sin\theta=4x^2y-y, \cdots,$$

which give the successive sets of solutions of $X_n^2-PY_n^2=1$ if $X_1=x$, $Y_1=y$ is the first set [cf. Wallis,[48] Euler[65]]. This was verified for any n.

Christie[266] proved that, if p_n, q_n are any convergents of $p_n^2-2q_n^2=\pm1$,

$$2\tan^-\frac{q_n}{q_{n+1}}\pm\tan^{-1}\frac{1}{p_{2n+1}}=\frac{\pi}{4}=2\tan^{-1}\frac{p_n}{p_{n+1}}\pm\tan^{-1}\frac{1}{p_{2n+1}}.$$

Christie[267] noted that successive solutions of $X^2-pY^2=1$ are given by

$$X_{n+1}=2xX_n-X_{n-1}, \qquad Y_{n+1}=2xY_n-Y_{n-1},$$

the initial solutions being 1, 0; x, y. From a solution of $x^2-601y^2=-1$, one of $X^2-601Y^2=1$ is found (pp. 54–5).

---

[258] Math. Quest. Educ. Times, (2), 8, 1905, 28–30, 58.

[259] Ibid., 83; Mess. Math., 35, 1905–6, 166–185. He noted (p. 183) eight errata in Degen's[101] table and various errata in Legendre's[83] tables of 1798 and 1830, including $A=397$ (cf. A. Gérardin, l'interméd. des math., 24, 1917, 57–8).

[260] Mathesis, (3), 5, 1905, 233.

[261] Archiv Math. Phys., (3), 9, 1905, 206–7.

[262] L'intermédiaire des math., 13, 1906, 93, 229–230; 14, 1907, 136.

[263] Amer. Math. Monthly, 13, 1906, 191 (148–9 for erroneous solution).

[264] Math. Quest. Educ. Times, (2), 10, 1906, 29.

[265] Ibid., (2), 9, 1906, 111.

[266] Ibid., 52–3.

[267] Ibid., (2), 11, 1907, 39. Cf. p. 96.

A. Auric[268] developed into a continued fraction the root of any quadratic equation of discriminant Δ; it is a question of factoring $t\pm2$, where t, u give the least solution of $t^2-\Delta u^2=4$.

B. Niewenglowski[269] noted that $x^2-ay^2=-1$ is solvable if and only if the least positive integral solutions of $x^2-ay^2=+1$ are of the form $x=1+2u^2$, $y=2uv$. The latter represents an hyperbola; if P and P_1 are points on it with integral coördinates, the line through P parallel to the tangent at P_1 cuts the hyperbola in a new point with integral coördinates.

A. Cunningham[270] gave tests for the divisibility of solutions of

$$\tau^2-Dv^2=\pm1$$

by a prime.

The existence of a fundamental solution of Pell's equation is a corollary to Dirichlet's theorem on the units in any algebraic field. For the case of a quadratic field, reference may be made to J. Sommer's[271] text.

"E. A. Majol"[272] gave eight values, 75, 78, 321, $\cdots$, of Δ for which there is a common prime divisor $4m+3$ of Δ and y in the fundamental solution of $x^2-\Delta y^2=1$.

A. Boutin[273] gave the period of the continued fraction for $\sqrt{A}$ for many forms of A, chiefly quadratic functions of a, and for various such A's listed the least solutions of $x^2-Ay^2=\pm1$. He listed the values of N, $0<N<1023$, for which $x^2-Ny^2=-1$ is solvable, a necessary and sufficient condition for which is that there be an odd number of terms in the period of incomplete quotients in the development of $\sqrt{N}$.

*C. Störmer[274] gave a simple proof of his[230] theorem and applied it to solve the following problem: Given the primes $p_1\ \cdots,\ p_n$, find all positive integers N for which $N(N+h)$ is divisible by no prime other than $p_1,\ \cdots,\ p_n$ when $h=1$ or 2. This is solved by the theorem that, if $a=1$ or 4, all positive integral solutions x of $x^2-1=ap_1^{z_1}\cdots p_n^{z_n}$ occur among the fundamental solutions of the equations $x^2-D_iy^2=1(i=1,\ \cdots,\ \nu)$, where $D_1,\ \cdots,\ D_\nu$ are all the values of $ap_1^{\epsilon_1}\cdots p_n^{\epsilon_n}$ when $\epsilon_1,\ \cdots,\ \epsilon_n$ take independently the values 1, 2.

G. Fontené[275] proved that, if a, b give the least positive solutions of $x^2-ky^2=1$, all solutions are given by $x+y\sqrt{k}=(a+b\sqrt{k})^n$; the proof is essentially the classic proof, but follows the proof by Mlle. J. Borry (*ibid.*, 13, 1907, 316).

A. Chatelet[276] proved by an elementary formulation of the classic method of solution by continued fractions that, if k is not a square, $x^2-ky^2=1$ is always solvable.

[268] Bull. Soc. Math. de France, 35, 1907, 121–5.
[269] *Ibid.*, 126–131; Wiadomosci Mat. Warsaw, 12, 1908, 1–26 (Polish).
[270] Report British Assoc. for 1907, 462–3.   Cf. Cunningham.[281]
[271] Vorlesungen über Zahlentheorie, 1907, 98–107, 113, 338–45, 355–8; French transl. of revised text by A. Lévy, 1911, 103–113, 119, 351–7, 370–3.
[272] L'intermédiaire des math., 15, 1908, 142–3.
[273] Assoc. franç. av. sc., 37, 1908, 18–26.
[274] Nyt Tidsskrift for Mat., 19, B, 1908, 1–7; Fortschritte der Math., 39, 1908, 246.
[275] Bull. math. élémentaires, 14, 1908–9, 209–212.
[276] *Ibid.*, 307–331.

R. W. D. Christie[277] expressed the solutions of $x^2-5y^2=\pm4$ in terms of fifth roots of unity. He and others[278] obtained a double infinitude of fractional solutions of $x^2-py^2=1$ from one integral solution:

$$\left(\frac{x}{z}\right)^2-p\left(\frac{py-y+2}{(p+1)z}\right)^2=1,\qquad z=\frac{2py-p+1}{p+1}.$$

E. B. Escott and A. Cunningham[279] factored u_{84} in $t_n^2-2u_n^2=(-1)^n$. Christie and Cunningham[280] proved that, if $p_n^2-2q_n^2=\pm1$,

$$(p_np_{n+1})^2+(2q_nq_{n+1})^2=q_{2n+1}^2,\qquad p_np_{n+1}-2q_nq_{n+1}=\pm1.$$

Cunningham[281] applied his[270] method to factor v_{66} in $\tau_n^2-2v_n^2=1$, and gave further examples of the factorization of solutions of Pell equations.

Cunningham[282] noted relations between the solutions of $x^2-3y^2=-2$, $z^2-3w^2=1$, in connection with the factorization of $(a^6+27b^6)/(a^2+3b^2)$.

G. Frattini[283] proved that if D and N are polynomials in a, and D is of even degree, and if $x^2-Dy^2=1$ has a known solution in polynomials in a, then all solutions of $x^2-Dy^2=N$ can be found from one.

G. Fontené[284] noted that, if a, b give the least positive solution of $x^2-ky^2=1$,

$$x_n=2ax_{n-1}-x_{n-2},\qquad y_n=2ay_{n-1}-y_{n-2}.$$

A. Lévy[285] gave another proof of the result proved by Fontené.[275]

A. Gérardin[286] noted that, if $t_n^2-du_n^2=1$,

$$u_{2n}=2u_nt_n=t_1u_{2n-1}+u_1t_{2n-1},\qquad t_n=2t_1t_{n-1}-t_{n-2},\qquad \frac{u_{n-1}}{u_1}\cdot\frac{u_{n+1}}{u_1}=\left(\frac{u_n}{u_1}\right)^2-1,$$

$$t_n=t_1t_{n-1}+du_1u_{n-1},\qquad u_n=t_1u_{n-1}+u_1t_{n-1},\qquad t_{2n}=t_n^2+du_n^2,$$
$$t_{n-1}t_{n+1}=t_n^2+du_1^2.$$

Each u_k/u_1 is a composite integer. For $f_n^2-dg_n^2=-1$,

$$f_n=(4f_0^2+2)f_{n-1}-f_{n-2}=(2f_0^2+1)f_{n-1}+2df_0g_0g_{n-1},$$
$$g_n=(2f_0^2+1)g_{n-1}+2f_0g_0f_{n-1}.$$

G. Ascoli[287] gave an elementary treatment of $ax^2+bx+c=y^2$.

F. Ferrari[288] cited known results leading to a practical method to find all integral solutions of $x^2-Dy^2=\pm1$ in the solvable cases.

W. Kluge[289] proved that for the integral solutions of

$$x_n^2-2kx_ny_n-y_n^2=(-1)^n\rho$$

[277] Math. Quest. Educ. Times, (2), 13, 1908, 35–6.
[278] Ibid., (2), 14, 1908, 56.
[279] Ibid., 105–6.
[280] Ibid., (2), 15, 1909, 74–75.
[281] Ibid., 95–6; (2), 17, 1910, 64–5.
[282] Ibid., (2), 17, 1910, 110–2.
[283] Atti del IV congresso internaz. dei mat., 2, 1909, 178–182. Cf. Frattini.[255]
[284] Bull. math. élémentaires, 15, 1909–10, 65.
[285] Ibid., 66.
[286] Sphinx-Oedipe, 5, 1910, 17–29.
[287] Suppl. al Periodico di Mat., 14, 1910–11, 33–8.
[288] Ibid., 69–75.
[289] Verhandlungen der Versammlung deutscher Philologen und Schulmänner, Leipsic, 51, 1911, 135–7. Unterrichtsblätter Math. Naturwiss., Berlin, 19, 1913, 9–11.

the relations $y_{n+1}=x_n$, $x_{n+1}=2kx_n+x_{n-1}$ hold. To apply to $t_n^2-Du_n^2=(-1)^n$, when $D=k^2+1$, make the substitution $t_n=ku_n+v_n$; then u_n, v_n satisfy the initial equation with $\rho=-1$. Hence $u_{n+1}=2ku_n+u_{n-1}$, $t_{n+1}=2kt_n+t_{n-1}$.

A. Cunningham[290] discussed the values of D for which

$$(ab\pm1)^2-(a\pm b)^2\equiv0 \quad (\mathrm{mod}\ (24D)^2),$$

where a, b are of the form $2Dn+1$ and $Dx^2\pm y^2=ab$.

H. B. Mathieu[291] asked if $(m^2-1)x^2+1=y^2$ has solutions not given by

$$x_1=0,\ x_2=1,\ \cdots,\ x_{n+1}=2mx_n-x_{n-1};\ \cdots,\ y_{n+1}=2my_n-y_{n-1}.$$

E. Dubouis[292] stated that there are no others in view of Fontené,[275] the exposition by Legendre being insufficient. All the solutions can be found[293] by applying Gauss, Disq. Arith., art. 200.

R. Fueter[294] noted that Dirichlet[108] gave sufficient, but not necessary, conditions that $x^2-my^2=-4$ be solvable for certain positive integers m not squares. When $m\equiv1\ (\mathrm{mod}\ 8)$, x and y are even and the problem reduces to $x^2-my^2=-1$; a necessary, but not sufficient, condition that it be solvable is that in the domain defined by $\sqrt{-m}$ there be an even number of classes in every genus.

A. Cunningham[295] wrote τ_x', v_x' and τ_x, v_x for the xth solutions of $\tau'^2-2v'^2=-1$, $\tau^2-2v^2=1$, and noted that E. Lucas (Ch. XVII of Vol. I of this History) proved that every prime p divides some v_x, where $x=(p-1)/n$ when $p=8\omega\pm1$, $x=(p+1)/n$ when $p=8\omega\pm3$, and $n=2m$. It is here proved that, if $n=4m$, $8m$, $16m$ or $32m$, then $p=8\omega+1=a^2+b^2=c^2+2d^2$ with $b=4\beta$, $d=2\delta$, and the number of factors 2 of n is given. If $n=6m$ either $p=8\omega\pm1=3\omega'+1$, $p=G^2+6H^2$, or $p=8\omega\pm3=3\omega'-1$, $p=2G^2+3H^2$.

St. Bohniček[296] proved that if π is a semiprimary prime in the domain R defined by a fourth root of unity and if the norm of π is $\equiv1\ (\mathrm{mod}\ 8)$, $\xi^2-\pi\eta^2=2$, $\xi_1^2-\pi\eta_1^2=1$ have the solutions

$$\xi=\frac{T^2-T_1^2}{(1-i)TT_1},\qquad \eta=\frac{T^2+T_1^2}{(1-i)TT_1\sqrt{\pi}},\qquad \xi_1=\frac{T^4+T_1^4}{2\sqrt{\pi}},\qquad \eta_1=\frac{T^4-T_1^4}{2\pi},$$

so that ξ, η are odd, ξ_1 and η_1 integral numbers in R. Here $T=\Pi S_{2s}$, $T_1=\Pi S_{2s+1}$, where S_r is the lemniscate function defined (p. 680) in terms of Jacobi's theta functions. But $\xi^2-\pi\eta^2=i$ or $2i$ is not solvable in integral numbers with ξ, η odd in the second case. If π is semiprimary, $\xi^2-\pi\eta^2=4$, $\xi_1^2-\pi\eta_1^2=4i$ have odd solutions ξ, η in R only if $\pi\equiv1\ (\mathrm{mod}\ \lambda^4)$, $\pi\not\equiv1\ (\mathrm{mod}\ \lambda^5)$, where $\lambda=1+i$. There are similar theorems for $\xi^2-\pi\eta^2=1$, 2, i or $2i$, when the norm of π is not $\equiv1\ (\mathrm{mod}\ 8)$. Application is made (pp. 719–725) to $x^2-py^2=\pm1$, -2, ±4, where p is a rational odd prime, use being made of cyclotomic functions.

E. E. Whitford[4] gave an extended history of Pell's equation and (pp. 98–112) extended the tables of Degen[101] and Bickmore[219] by listing for

[290] L'intermédiaire des math., 18, 1911, 166–7.
[291] *Ibid.*, 220.
[292] *Ibid.*, 19, 1912, 47.
[293] L'intermédiaire des math., 19, 1912, 72.
[294] Jahresber. d. Deutschen Math.-Vereinigung, 20, 1911, 45–46.
[295] British Assoc. Report for 1912, 412–3.
[296] Sitzungsber. Akad. Wiss. Wien. (Math.), 121, IIa, 1912, 701–7.

$1500 < A \leqq 1700$ the least solutions of $x^2-Ay^2=-1$, when solvable, and always those of $x^2-Ay^2=+1$. He noted (pp. 154–5) that the former is solvable for 38 of the 110 composite numbers $A=a^2+b^2$ between 1501 and 2000. Finally (pp. 162–190) he tabulated for $1500 < A \leqq 2012$ the period and auxiliary numbers for the continued fraction for $\sqrt{A}$ [corresponding to the first two lines in Degen's table].

R. Remak[297] modified Dedekind's[141] proof of the existence of solutions of $x^2-Dy^2=1$ and obtained upper limits on the least positive solutions:

$$x < (g+1)^{2g^3+1}, \qquad y \leqq (g+1)^{2g^3}, \qquad g \equiv [\sqrt{4D}].$$

Known methods of solving $y^2-2z^2=-1$ have been recalled.[298]

Th. Got[299] simplified the proofs by A. Meyer[150].

M. Simon[300] noted that Brahmegupta's first rule shows that he knew how to solve all equations (a) $4(\lambda^2 \mp 2)x^2+1=y^2$ and (b) $(\lambda^2 \pm 2)x^2+1=y^2$. The identity $(\lambda^2 \pm 2)\lambda^2 = (\lambda^2 \pm 1)^2 - 1$ gives $x=\lambda$, $y=\lambda^2 \pm 1$ for (b) and $x=\lambda/2$, $y=\lambda^2 \mp 1$ for (a). But if λ is odd, and a solution α, β of (a) is found, it becomes $(\beta^2-1)x^2/\alpha^2+1=y^2$, which is satisfied if $x=2\alpha\beta$, whence the solution is $x=\lambda(\lambda^2 \mp 1)$, $y=2(\lambda^2 \mp 1)^2 - 1$.

G. Métrod[301] noted that in $u^2-2v^2=1$, $v \neq 2^\alpha$, $\alpha > 1$, and $v \neq (2a)^e$. In $u^2-3v^2=1$, $v=2^t$ only for $t=0$, 2; v is not a power of an odd prime, and $v \neq (2a)^t$, where a is an odd prime. In $u^2-pv^2=1$, where p is an odd prime, cases are noted in which $v=2^t$ or a^t, where a is an odd prime $\neq p$.

E. E. Whitford[302] extended Cayley's[137] table from $D=1000$ to $D=1997$, but gave the solution of both $x^2-Dy^2=-4$ and $x^2-Dy^2=+4$ when they are solvable. He noted the application to finding the fundamental unit ϵ (least unit > 1) of the domain defined by $\sqrt{D}$; the least positive solutions of $x^2-Dy^2=1$ do not determine ϵ when one of the equations $x^2-Dy^2=-1$, 4 or -4 is solvable.

O. Perron[303] obtains by use of continued fractions the limits

$$x < 2(b+1)^4(\tfrac{2}{3}b+1)^l, \qquad y < 2(b+1)^3(\tfrac{2}{3}b+1)^l, \qquad l \equiv 2b(b+1)-4, \qquad b \equiv [\sqrt{D}],$$

for the least positive solutions of $x^2-Dy^2=1$. Remak[297] had given larger limits. Cf. Schmitz,[308] Schur.[314]

T. Ono[304] stated that, if $x^2-5y^2=4$,

$$\frac{x-y\sqrt{5}}{2} = \frac{1}{x} + \frac{1}{xx_1} + \frac{1}{xx_1x_2} + \cdots, \qquad x_1=x^2-2, \qquad x_2=x_1^2-2, \cdots.$$

Infinite series involving successive solutions of this and $x^2-Dy^2=p^2$ have been treated.[305]

"V. G. Tariste"[306] noted relations between successive x's or y's for which $mx^2+nx+p=y^2$.

[297] Jour. für Math., 143, 1913, 250–4. Cf. Kronecker,[110] Perron.[303]
[298] L'intermédiaire des math., 20, 1913, 254–6.
[299] Annales Fac. Sc. Toulouse, (3), 5, 1913, 94–8.
[300] Archiv Math. Phys., (3), 20, 1913, 280–1.
[301] Sphinx-Oedipe, 8, 1913, 137–8.
[302] Annals of Math., 15, 1913–4, 157–160.
[303] Jour. für Math., 144, 1914, 71–73.
[304] L'intermédiaire des math., 20, 1913, 224.
[305] Ibid., 21, 1914, 37–38, 47–48; 22, 1915, 21–23, 277–8.
[306] Ibid., 22, 1915, 125–6.

A. S. Werebrusow[307] stated erroneous conditions involving $N = a_i^2 + b_i^2$ for the solvability of $x^2 - Ny^2 = -1$.

Thekla Schmitz[308] proved that, for the least positive solutions of $x^2 - Dy^2 = 1$, $x + y\sqrt{D} < 2e^{4D}$, where e is the base of natural logarithms.

A. Cunningham[309] described and noted errata in various tables on the Pell equation: Euler,[72] Legendre,[88] Degen,[101] Cayley,[137] and Bickmore.[219]

Kiveliovitchi[310] gave an elementary method of solving $6x^2 + 1 = y^2$. We may take $x = 2u$, $y = 5u - v$, $2v = w$. Then $x = 5w + 2r$, $y = 12w + 5r$, $r^2 = 6w^2 + 1$. Hence if $(x_1 = 0, y_1 = 1)$, $\cdots$, (x_i, y_i), $\cdots$ are the solutions arranged in order of increasing magnitude, $x_{i+1} = 5x_i + 2y_i$, $y_{i+1} = 12x_i + 5y_i$. The same method is said to apply to $ax^2 + 1 = y^2$ if $a = 4h + 2$, $4a + 1 = \square$.

A. Gérardin[311] applied the remark of Hart[174] on $Ay^2 - 1 = \square$, $A = r^2 + s^2$. To treat similarly $x^2 - Ay^2 = 2$, set $A = a^2 - 2b^2$, $\pm y = \alpha^2 - 2\beta^2$, and solve the system of equations

$$(b\alpha - a\beta)^2 - A\beta^2 = \pm b, \qquad (2b\beta - a\alpha)^2 - A\alpha^2 = \pm 2b.$$

Thus, if $A = 151 = 13^2 - 2\cdot 3^2$, we get $\beta = 7$, $\alpha = 59$, $y = 3383$, which leads to Legendre's solution of $x^2 - 151y^2 = 1$. For $x^2 - Ay^2 = -4$, set $A = a^2 + b^2$, $y = z^2 + t^2$ and solve the system

$$(bz - at)^2 - At^2 = \pm 2b, \qquad (bt + az)^2 - Az^2 = \mp 2b.$$

Thus, if $A = 3^2 + 10^2$, we get the least solution $t = 3$, $z = 4$. An error for $A = 397$ in Legendre's[88] table is noted. He announced an extension in MS. to 3000 of the table by Whitford.[302]

M. Cassin[312] gave relations between successive solutions of $x^2 = 3y^2 + 1$.

Several[313] gave relations between successive solutions of $z^2 - Dx^2 = \pm 1$ or c, and of $ux^2 - vy^2 = w$.

*J. Schur[314] obtained closer limits than had Remak,[297] Perron,[303] and Schmitz.[308]

On $x^2 - 3y^2 = 1$, see papers 100 of Ch. I; 12, 24, 29, 33, 51 of Ch. V; 94 of Ch. VII; 230 of Ch. XXI. On $2x^2 \pm 1 = \square$, see papers 112–129 of Ch. IV; 92 of Ch. XXIII. For $ax^2 + by^2 = c$ or $ax^2 + bxy + cy^2 = k$, see Ch. XIII. On $5x^2 \pm 4 = \square$, see Wasteels[72] of Vol. I, p. 405. On the application to factoring, see Vol. I, p. 368. For "Pell equations of higher order," see papers 313–23 of Ch. XXI, 19–25 of Ch. XXIII, and Ch. XXVI. Pell equations occur incidentally in the following papers: 56, 70, 107, 152, 178, 185, 187, 189, 196, 202, 204, 210, 219, 223, 227 of Ch. I; 135 of Ch. IV; 41, 109 of Ch. V; 138, 193 of Ch. VI; 66 of Ch. XV; 55 of Ch. XVI; 21 of Ch. XVII; 270–4 of Ch. XXI; 111, 250 of Ch. XXII; 95, 99, 163 of Ch. XXIII.

<hr>

[307] L'intermédiaire des math., 22, 1915, 202–3; 23, 1916, 56 for admission of errors.
[308] Archiv Math. Phys., (3), 24, 1916, 87–9. Cf. Perron.[303]
[309] Mess. Math., 46, 1916, 49–69.
[310] Soc. Math. de France, Comptes Rendus Séances, 1916, 30–1.
[311] Sphinx-Oedipe, 12, June 15, 1917, 1–3; l'enseignement math., 19, 1917, 316–8; l'intermédiaire des math., 24, 1917, 57–58.
[312] L'intermédiaire des math., 25, 1918, 28, 93.
[313] Ibid., 83–87; 26, 1919, 51–54.
[314] Göttingen Nachrichten, 1918, 30–6

CHAPTER XIII.

FURTHER SINGLE EQUATIONS OF THE SECOND DEGREE.

EQUATION LINEAR IN ONE UNKNOWN.

Brahmegupta[1] (born 598 A.D.) solved $axy = bx + cy + d$. Let e be an arbitrary number and set $q = (ad + bc)/e$. To the greatest and least of e, q add the least and greatest of b, c, and divide the sums by a. We get the values of x, y (that of x on adding to c and vice versa). Thus, if $xy = 3x + 4y + 90$, take $e = 17$, whence $q = 6$, $y = 17 + 3$, $x = 6 + 4$. Another method is to give a special value to one of the unknowns.

Bháscara[2] (born 1114) gave a like rule for $a = 1$, but added e and q to (or subtracted them from) b and c in either order, and gave both geometric and algebraic proofs of the rule. Thus for $xy = 4x + 3y + 2$, take $e = 1$, whence $q = 14$; adding 4, 3 to 1, 14 in both orders, we get 17, 5 and 4, 18 as sets of values of x, y; taking $e = 2$, we get 5, 11 and 10, 6. The same example was treated in § 209, p. 269, by assigning any value as 5 to y and deducing $x = 17$.

On $axy + bx + cy + d = 0$ see Wezel[86], and papers 121–141 (on optic formula) of Ch. XXIII; also, Bervi[61] of Vol. I, p. 451; and *P. von Schaewen.[2a]

L. Euler[3] noted that $4mn - m - n$ is never a square since

$$a^2 + 1 = (4n - 1)(4m - 1)$$

is impossible; also $4pmn - m - n$ is not a square if m is of the form $4n^2q - n$.

Euler[4] proved that no number of the form $4mn - m - n$ or $8mn - 3m - 3n$ can be a square, and many such propositions.

Euler[5] stated without proof that $4mnz - m - n = \square$ is impossible. This arises from the fact that the divisors of $mx^2 + y^2$ are of the form $4mz + 1$, so that $d = 4mz - 1$ is not a divisor, whence $dn \neq m + y^2$. He[6] treated similarly the case $m = 1$, and proved that $4mn - m - n^a \neq \square$.

P. Bédos[7] erred in his proof that $4mn - m - 1 \neq \square$.

Several[8] proved that $4mn - m - n$ is never a square or triangular number.

S. Günther[9] solved $y^2 - ax^2 = bz$ by use of the continued fraction

$$K = \frac{a}{2u} - \frac{a}{2u} - \frac{a}{2u} - \cdots.$$

---

[1] Brahme-sphut'a-sidd'hánta, Ch. 18 (Algebra), §§ 61–64. Algebra, with arith. and mensuration, from the Sanscrit of Brahmegupta and Bháscara, transl. by Colebrooke, 1817, pp. 361–2.

[2] Víja-gańita, §§ 212–4; Colebrooke,[1] pp. 270–2.

[2a] Zeitschrift für d. Realschulwesen, 38, 1913, 141–6.

[3] Corresp. Math. Phys., (ed., Fuss), 1, 1843, 191, 202 (180, 259, 260); letters to Goldbach, Jan. 19, and Feb., 1743.

[4] Comm. Acad. Petrop., 14, 1744–6, 151; Comm. Arith., I, 48–49; Op. Om., (1), II, 220.

[5] Opera postuma, 1, 1862, 220 (about 1778).

[6] Corresp. Math. Phys., (ed., Fuss), 1, 1843, 114–7; letter to Goldbach, Mar. 6, 1742.

[7] Nouv. Ann. Math., 11, 1852, 278 (Euler's correct proof, p. 279).

[8] Math. Quest. Educ. Times, 70, 1899, 73.

[9] Jour. de Math., (3), 2, 1876, 331–340.

Let Q_i be the denominator of its ith convergent.   Then

$$Q_{2n} = (2uQ_{n-1} - aQ_{n-2})^2 - aQ_{n-1}^2, \qquad 2uQ_{n-1} - aQ_{n-2} = Q_n.$$

Hence a solution is $y = Q_n$, $x = Q_{n-1}$, $bz = Q_{2n}$, the last being used to determine u and n:

$$Q_{2n} = \binom{2n+1}{1} u^{2n} + \binom{2n+1}{3} u^{2n-2}(u^2 - a) + \cdots + \binom{2n+1}{2n+1}(u^2 - a)^n \equiv 0 \pmod{b}.$$

If b is odd, set $k = (2p-1)b$; then $\binom{k}{\rho}$ is divisible by b if $\rho < k$, and we may take $2n = k - 1$.   If b is even, divide x and y by a power of 2.

P. Mansion[10] gave a short proof of the preceding $Q_n^2 - aQ_{n-1}^2 = Q_{2n}$.

S. Réalis[11] noted that, if α, β, γ is one solution of $ax^2 + bxy + cy^2 = hz$, a second is given by $x = (h+a-c)\alpha + (b+2c)\beta$, $y = (2a+b)\alpha + (h-a+c)\beta$, since

$$ax^2 + bxy + cy^2 \equiv h(P\alpha^2 + Q\alpha\beta + R\beta^2) + (a+b+c)^2(a\alpha^2 + b\alpha\beta + c\beta^2),$$
$$P = ah + 2a(a+b-c) + b^2, \qquad Q = bh + 2(ab+4ac+bc),$$
$$R = ch + 2c(b+c-a) + b^2.$$

If, for $c = 1$, we solve the initial equation for y, the radical will be a rational number u if $u^2 - Dx^2 = 4hz$, $D = b^2 - 4a$, which was treated (*ibid.*, p. 111) and if $D > 0$ by Günther.[9]

A. H. Holmes[12] proved that $96x - 96y + 21 = \square$ is impossible in integers. On $ax^2 + bx + c = Ky$ see Desmarest.[87]

SOLUTION OF $x^2 - y^2 = g$.

Diophantus, II, 11, took $g = 60$, $x = y+3$, 3 being a number $\leqq \sqrt{60}$, whence $y = 17/2$.

Leonardo Pisano[13] took a square $a^2 < g$ and set $(x+a)^2 = x^2 + g$, which determines x.   He gave a second method.   Let g be odd, $g = 2n+1$.   Since $1 + 3 + \cdots + (2n-1) = n^2$, we may take $y = n$, whence $n^2 + g = (n+1)^2$.   He treated separately the cases $g = 2k$, $g = 4k$.

R. Descartes[14] noted that $6^2 - 3^2 = 3^3$, $118^2 - 10^2 = 24^3$; $(ax)^2 - x^2 = x^3$ if $x = a^2 - 1$.

J. L. Lagrange[15] concluded from his general theory of binary quadratic forms f that every integer is of the form $y^2 - z^2$.   This[16] is not true of the double of an odd prime, and Lagrange's argument is conclusive only when the discriminant of f is not a square.

S. Canterzani[17] treated $x^2 + A = \square$, by deciding whether or not A is a sum of differences of consecutive squares.   First, let A be even.   The sum of $2f$ consecutive differences $2h+1$, $2h+3$, $\cdots$ is $4fh + 4f^2$ and hence $\neq A$

---

[10] Jour. de Math., (3), 2, 1876, 341.

[11] Nouv. Corresp. Math., 6, 1880, 348–350.

[12] Amer. Math. Monthly, 18, 1911, 70.

[13] La Practica Geometriae, 1220.   Scritti di L. Pisano, Rome, 2, 1862, 216–8.

[14] Oeuvres, X, 302, posthumous MS.   Cf. papers 23–26 of Ch. XX.

[15] Nouv. mém. Acad. Sc. Berlin, année 1773; Oeuvres, III, 714.

[16] L'intermédiaire des math., 18, 1911, 33.

[17] Memorie dell'Istituto Nazionale Italiano, Classe di Fis. e Mat., Bologna, 2, II, 1810, 445–76.

if A is not a multiple of 4.   For $A=4B$, the sum equals A if $h=B/f-f$; then $x^2+A=(h+2f)^2$ for $x=h$.   Next, let $A=2B+1$.   The sum of $2f+1$ consecutive differences $2h+1$, $\cdots$ is $(2f+1)(2h+2f+1)$, which can be made equal to A by choice of h, whence $x^2-A=h^2$ if

$$x=\frac{B+f+1}{2f+1}+f.$$

T. Clowes[18] noted that the difference of the squares of $x+1$ and $x-1$ equals the difference of the squares of $a+b$ and $a-b$ if $x=ab$.

L. Poinsot[19] stated that any integer N, not the double of an odd integer, can be represented as a difference of two squares and in as many ways n as N can be expressed as a product of two factors both odd and relatively prime or both even and with no common factor > 2.   If N has k distinct prime factors, $n=2^{k-1}$.

P. Volpicelli[20] took $g=2^\mu h_1^\alpha\cdots h_k^\tau$, where the h's are distinct primes. As known, the number of decompositions of g into two factors is

$$\nu=\tfrac{1}{2}(\mu+1)(\alpha+1)\cdots(\tau+1)$$

or $\nu+\tfrac{1}{2}$ according as at least one of the exponents $\mu, \alpha, \cdots, \tau$ is odd or all are even.   Hence, in the respective cases, the number of decompositions into two distinct even factors, i. e., the number of solutions of $x^2-y^2=g$, is

$$\nu_1=\tfrac{1}{2}(\mu-1)(\alpha+1)\cdots(\tau+1)$$

or $\nu_1-\tfrac{1}{2}$, if $\mu>0$.   For $\mu=0$, the number of solutions is ν or $\nu-\tfrac{1}{2}$, respectively.

R. P. L. Claude[21] noted that any odd integer $\neq 1$ is a difference of two squares since ab is the difference of the squares of $(a\pm b)/2$, while the double of an odd integer is not.   Every integer which is a difference of two squares is such as many times as there are different combinations $2, 3, \cdots, n$ at a time of its n prime factors.

G. C. Gerono[22] stated only known results.

L. Lorenz[23] concluded from

$$\tfrac{1}{2}\sum_{m,\,n=-\infty}^{+\infty} q^{m^2-n^2}=\sum_{m=1}^{\infty}\sum_{n=1}^{\infty}\{q^{4mn}+q^{(2m-1)(2n-1)}\}$$

that the number of solutions of $m^2-n^2=N$ is double the number of divisors of N or $N/4$ according as N is odd or is divisible by 4; none if $N/2$ is odd.

G. H. Hopkins[24] noted that in $x^2-y^2=(2a_1\cdots a_n)^2$, where $a_1, \cdots, a_n$ are primes, x or y has $(3^n-1)/2$ integral values.

---

[18] The Ladies' and Gentlemen's Diary (ed., M. Nash), New York, 3, 1822, 53–4.

[19] Comptes Rendus Paris, 28, 1849, 582.

[20] Atti Accad. Pont. Nuovi Lincei, 6, 1852–3, 91–103; Annali di Sc. Mat. e Fis., 6, 1855, 120–8; Comptes Rendus Paris, 40, 1855, 1150; Nouv. Ann. Math., 14, 1855, 314.

[21] Nouv. Ann. Math., (2), 2, 1863, 88–90.

[22] Ibid., 90–92.

[23] Tidsskrift for Math., (3), 1, 1871, 113–4.

[24] Math. Quest. Educ. Times, 16, 1872, 46–7.

A. Sýkora[25] repeated Claude's[21] first remark.

L. P. da Motta Pegado,[26] A. Z. Candido,[26] T. H. Miller,[27] G. Bisconcini,[28] and H. E. Hansen[29] stated known results.

"H. Rifoctitlee"[30] noted that every integer N is the quotient of two differences of two squares. For, $N=2(a^2-b^2)$ or a^2-b^2 according as $N\equiv 2 \pmod 4$ or not. Then apply formula (11) of Euler,[66] Ch. XII, for $e=1$.

W. Sierpinski[31] proved that the number $\tau(n)$ of distinct representations of a positive integer n as a difference of two squares is twice the difference between the number of even and odd divisors of n. Also

$$\phi(x)\equiv\sum_{n>0}^{x}\tau(n)=2[\sqrt{x}]-2\left[\frac{x-1}{2}\right]\left[\frac{x+1}{2}\right]+4\sum_{n>0}^{(x-1)/2}[x+n^2],$$

where $[t]$ is the greatest integer $\leqq t$. If $\theta(n)$ is the number of divisors of n,

$$\phi(x)=2\sum_{k>0}^{x}\theta(k)-2\sum_{k>0}^{x/2}\theta(2k)+2\sum_{k>0}^{x/4}\theta(k),\qquad \lim_{m=\infty}\frac{1}{m}\sum_{k=1}^{m}\{\tau(k)-\theta(k)\}=0.$$

S. Guzel[32] proved that

$$\frac{1}{n}\left|\sum_{k=1}^{n}\{\tau(k)-\theta(k)\}\right|<\frac{4}{\sqrt{n}}.$$

*A. L. Bartelds[32a] discussed $x^2-y^2=g$.

For solutions of $x^2-1=g$, see Störmer[274] of Ch. XII. Cf. Gill.[34]

Solution of $ax^2+bxy+cy^2=dz^2$.

Diophantus, IV, 10, desired two cubes the ratio of whose sum to the sum of their sides is a square. Taking s and $2-s$ as the sides, we must have $4-6s+3s^2=\square$, say $(2-4s)^2$, whence $s=10/13$.

Diophantus, IV, 11, 12, solved $x^3\pm y^3=x\pm y$. Take $x=rz,\ y=sz$. Then $(r^3\pm s^3)/(r\pm s)$ is to be a square. For the upper signs he found (as in IV, 10) that $r=5,\ s=8,\ z=1/7$. For the lower signs, take $r=s+1$, so that $3s^2+3s+1=\square$, say $(1-2s)^2$, whence $s=7,\ z=1/13$.

In these three problems, Diophantus made no use of the fact that $(x^3\pm y^3)/(x\pm y)=x^2\mp xy+y^2$. But, in V, 7, he made x^2+x+1 the square of $x-2$ for $x=3/5$, whence $3^2+3\cdot5+5^2=\square$.

C. G. Bachet in his comments solved similarly $f=p^2$ or $3p^2$, where $f=x^2\pm xy+y^2$. Fermat (Oeuvres, III, 249) remarked that we can solve $f=a$, where a is the product of a square by one or more primes of the form $3n+1$ or 3.

L. Euler[33] proved that if $fx^2+gxy+hy^2=tz^2$ is solvable for $t=k$, it is solvable for $t=kl$, where $l=p^2+gpq+fhg^2$. We have only to multiply the

<hr>

[25] Archiv Math. Phys., 61, 1877, 446-7.
[26] Jornal de Sc. Math. e Ast., 1, 1878, 150-5, 171-2.
[27] Proc. Edinburgh Math. Soc., 9, 1890-1, 23-5.
[28] Periodico di Mat., 23, 1908, 21.
[29] L'enseignement math., 18, 1916, 48-55.
[30] L'intermédiaire des math., 11, 1904, 25-6. Proof, 8, 1901, 238-40, by continued fractions.
[31] Wiadomosci Matematyczne, Warsaw, 11, 1907, Suppl., 89-110.
[32] Ibid., 111-9.
[32a] Wiskundig Tijdschrift, 13, 1916-7, 207-9.
[33] Opera postuma, 1, 1862, 209-211 (about 1771).

given equation by l and note that the product of $fx^2+gxy+hy^2$ by l is of that same form.

C. Gill[34] solved $x^2-y^2=bc$ by setting $x+y=b\cot A/2$.   Next,

$$x^2+axy+by^2=z^2$$

is satisfied by

$$z+x=y\cot A/2, \qquad z-x=(ax+by)\tan A/2.$$

Eliminate z.   The resulting equation gives x/y, whence

$$y=t(\sin A+a\sin^2 A/2), \qquad x=t(\cos^2 A/2-b\sin^2 A/2).$$

Take $t=m^2+n^2$, $\sin A=2mn/t$.   Then

$$x=m^2-bn^2, \qquad y=2mn+an^2, \qquad z=m^2+amn+bn^2.$$

G. L. Dirichlet[35] proved that $Az^2+2Bzy+Cy^2=x^2$ is solvable in integers, with x prime to $2D$, if the left member is a form of determinant D of the principal genus.

J. Neuberg[36] noted that $x^2-xy+y^2=z^2$ holds if

$$x=2pq-q^2, \qquad y=p^2-q^2, \qquad z=p^2-pq+q^2.$$

T. Pepin[37] gave special methods to obtain a particular solution of $ax^2+2bxy+cy^2=z^2$.   Given one solution $x=\alpha$, $y=\beta$, $z=\gamma$, to find all, eliminate $D=b^2-ac$ between

$$az^2=(ax+by)^2-Dy^2, \qquad a\gamma^2=(a\alpha+b\beta)^2-D\beta^2,$$

and write p/q for the irreducible fraction equal to $(\beta z-\gamma y)/(\beta x-\alpha y)$.   Hence

$$q(\beta z-\gamma y)=p(\beta x-\alpha y), \qquad p(\beta z+\gamma y)=q(a\beta x+a\alpha y+2b\beta y).$$

Conversely, these imply the initial quadratic equation.   Hence μx, μy, μz equal quadratic functions of p, q.   It is shown that μ is a factor of $2D\beta^2$.

A. Desboves[38] noted that by specializing his[159] formulas we find that the complete solution in integers of $X^2+bY^2+dXY=Z^2$ is

$$X=q^2-bp^2, \qquad Y=dp^2+2pq, \qquad Z=q^2+bp^2+dpq,$$

where (as below) $\pm$ is to be inserted before the second members.   For the case $d=0$, the ordinary method is to factor Z^2-X^2 and get

$$X=\alpha q^2-\beta p^2, \qquad Y=2pq, \qquad Z=\alpha q^2+\beta p^2 \qquad (b=\alpha\beta).$$

For each pair of factors α, β of b, the latter equations give all the solutions. It is inexact to say with A. M. Legendre[39] and others that the general solution includes as many particular formulas as there are ways to decompose b into two relatively prime factors.   The complete solution in integers of $X^2+Y^2=cZ^2$ for $c=m^2+n^2$ (the only solvable case in view of a theorem

<hr>

[34] Application of the angular analysis to the solution of indeter. problems of the second degree, New York, 1848, 15–17.

[35] Zahlentheorie, § 155, § 158, 1863; ed. 2, 1871; ed. 3, 1879; ed. 4, 1894.

[36] Nouv. Corresp. Math., 1, 1874–5, 197–8. Cf. papers 112a, 124, 125 of Ch. V, and 72 of Ch. IV. Cf. J. Bertrand, Traité élém. d'algèbre, 1851, 222–4.

[37] Atti Accad. Pont. Nuovi Lincei, 32, 1878–9, 89–97.

[38] Nouv. Ann. Math., (2), 18, 1879, 269; proofs, (3), 5, 1886, 226–33.

[39] Théorie des nombres, ed. 2, 1808, 29.

of Legendre) is
$$X = (cq^2 - p^2)m, \qquad Y = p^2n - 2cpq + cnq^2, \qquad Z = p^2 - 2npq + cq^2,$$
obtained from $x = m$, $y = n$, $z = 1$.   The complete integral solution of
$$aX^2 + bY^2 + dXY = cZ^2 \qquad\qquad (c = a + b + d)$$
is found from $x = y = z = 1$ to be
$$X = -bp^2 + cq^2, \qquad Y = (b+d)p^2 + cq^2 - 2cpq, \qquad Z = -bp^2 - cq^2 + (d+2b)pq.$$
By changing the nctation of the parameters, this becomes
$$X = q^2 - bcp^2, \qquad Y = (q + cp)^2 - acp^2, \qquad Z = (q + bp)^2 + b(a+d)p^2 + dpq.$$

J. Neuberg and G. B. Mathews[40] proved that the general rational solution of $x^2 + xy + y^2 = z^2$ is $x = p^2 - q^2$, $y = 2pq + q^2$, $z = p^2 + pq + q^2$. A. Cunningham[41] deduced $\frac{1}{2}x + y = t^2 - 3u^2$, $\frac{1}{2}x = 2tu$ from $(\frac{1}{2}x + y)^2 + 3(\frac{1}{2}x)^2 = z^2$.

Ch. J. de la Vallée Poussin[42] proved that a necessary and sufficient condition for integral solutions of $ax^2 + 2bxy + cy^2 = mz^2$, where m is prime to $2(b^2 - ac)$, and the g. c. d. of a, $2b$, c is unity, is that m be representable by a form of determinant $b^2 - ac$ and of the same genus as $ax^2 + 2bxy + cy^2$.

E. Sós[43] found the complete solution of
$$x^2 + bxy + y^2 = z^2 \text{ or } y(bx + y) = z^2 - x^2$$
by setting $y = \lambda(z - x)$, $\lambda(bx + y) = z + x$.   Eliminating y, we get
$$z = lx, \qquad l = \frac{\lambda^2 - \lambda b + 1}{\lambda^2 - 1} = \frac{p}{q},$$
where p/q is a fraction in its lowest terms.   Hence
$$x = \mu q, \qquad z = \mu p, \qquad y = \lambda\mu(p - q).$$
The same method applies to $ax^2 + bxy + cy^2 = z^2$, a or c a square.

A. Gérardin[44] found a general solution of $aX^2 + bXY + cY^2 = hZ^2$, given one solut on α, β, γ, by setting $X = \alpha + mx$, $Y = \beta + my$, $Z = \gamma$. Then m is determined rationally and
$$X = c\alpha y^2 - 2c\beta xy - (a\alpha + b\beta)x^2, \qquad Y = a\beta x^2 - 2a\alpha xy - (b\alpha + c\beta)y^2,$$
$$Z = a\gamma x^2 + b\gamma xy + c\gamma y^2.$$

Gérardin[45] granted that $ah^2 + bh + c = m^2$, replaced h by $h + x$, m by $m + fx$, found x rationally, and hence obtained a solution of $ay^2 + byz + cz^2 = v^2$:
$$y = hf^2 + ah + b - 2mf, \qquad z = f^2 - a, \qquad v = mf^2 - (2ah + b)f + ma.$$

A. Aubry[46] solved $2d^2x^2 \mp 2dx + 1 - d^2 = y^2$ for d and made the radical rational by means of a Pell equation.   L. Valroff[47] made the substitution
$$x = \frac{R \pm S}{2Y}, \qquad y = \frac{X}{S},$$

[40] Math. Quest. Educ. Times, 46, 1887, 97. See papers 36, 171. Cf. papers 68, 69 of Ch. IV.
[41] Ibid., 75, 1901, 33-4.
[42] Mém. couronnés et autres mém. acad. Belgique, 53, 1895-6, No. 3, 43-54.
[43] Zeitschrift Math. Naturw. Unterricht, 37, 1906, 186-190.
[44] Bull. Soc. Philomathique, (10), 3, 1911, 218.
[45] Sphinx-Oedipe, 1907-8, 177-9.
[46] L'intermédiaire des math., 20, 1913, 144.
[47] Sphinx-Oedipe, 7, 1912, 74-6.

and noted that the resulting equation in d has real roots if

$$S^2\left(\frac{R\pm S}{2Y}\right)^2+(X^2-S^2)\left\{2\left(\frac{R\pm S}{2Y}\right)^2-1\right\}=\square=\left\{\frac{R(R\pm S)-2Y^2}{2Y}\right\}^2,$$

which is a consequence of $2X^2+2Y^2=R^2+S^2$.

SOLUTION OF $ax^2+by^2=c$.

L. Euler[48] noted that

$$(a\alpha p^2+b\beta q^2)(abr^2+\alpha\beta s^2)=ab(apr\pm\beta qs)^2+a\beta(\alpha ps\mp bqr)^2.$$

He[49] noted that, if $m^2=abn^2+1$, then $ax^2-by^2=af^2-bg^2$ for

$$x\sqrt{a}+y\sqrt{b}=(f\sqrt{a}+g\sqrt{b})(m+n\sqrt{ab})^\lambda.$$

C. F. Kausler[50] treated the solution of $m'x^2+n'y^2=N$, where $N=4A+1$, $m'=4m+1$, $n'=4n+2$.   Thus $x=2X+1$, $y=2Y$, whence

$$(4m+1)X(X+1)+2(2n+1)Y^2=A-m=2B.$$

Let $B>4m+1$ and set $B=(4m+1)D+E$.   Then

(1) $$\frac{X(X+1)}{2}=D-z,\qquad z\equiv\frac{(2n+1)Y^2-E}{4m+1}.$$

Since $(2n+1)t-E=(4m+1)z$ has the solutions

$$t=pE+\mu(4m+1),\qquad z=qE+\mu(2n+1),$$

the question is whether $t=\square=Y^2$.   If so, we test (1_1) by the table of pronic numbers $X(X+1)$ in Nova Acta, XIV, 253.   A similar treatment is given for the case $m'=4m-1$, $n'=4n+1$.

C. F. Gauss[51] solved $mx^2+ny^2=A$ by the method of exclusions.

F. Arndt[52] noted that, if f, h are given relatively prime integers, the least solutions of $fp^2-hq^2=\pm k$, $k=1$ or 2, can be found, without using continued fractions, by means of the least solutions of $x^2-fhy^2=1$, given in Table X of Legendre's Théorie des nombres (errata noted, p. 246).   We have only to take $x=\mp1+2fp^2/k$, $y=2pq/k$.   He gave a table of the least roots of $\rho\theta^2-\rho'\theta'^2=1$ or 2 for $3\leqq\rho\rho'\leqq1003$.

S. Réalis[53] solved $(n+4)x^2-ny^2=4$ by formulas simpler than those given by the usual method of employing a Pell equation.   If α, β give a solution, then

$$x=\tfrac{1}{2}[(n+2)\alpha+n\beta],\qquad y=\tfrac{1}{2}[(n+4)\alpha+(n+2)\beta]$$

give a second solution.   We thus get an infinitude of sets of solutions $(1,1)$, $(1+n,3+n)$, $\cdots$, which are said to give all.   Replacing x by $2u+1$, y by $2v+1$, we get $(n+4)(u^2+u)=n(v^2+v)$.   Hence the above work solves the problem to find an infinitude of pairs of triangular numbers whose ratio is $n:n+4$.

[48] Opera postuma, 1, 1862, 490 (about 1769).
[49] *Ibid.*, 215 (about 1774).
[50] Nova Acta Acad. Petrop., 15, ad annos 1799–1802, 164–9.
[51] Disquisitiones Arith., art. 323; Werke, I, 1863, 391; German transl. by Maser, 377–383.
[52] Archiv Math. Phys., 12, 1849, 211–276.
[53] Nouv. Ann. Math., (3), 2, 1883, 535–542.

D. Hilbert[54] remarked that the proof that a proposed diophantine equation is not solvable in rational numbers is often made by showing that the corresponding congruence with respect to a prime or prime power modulus is impossible. For the case of a quadratic equation in two variables it follows conversely that the possibility of solving the congruence for every prime power modulus implies the possibility of solving the equation. For, the known criterion for the solvability of a ternary quadratic diophantine equation leads to the result: If m, n are any integers, the equation $mx^2 + ny^2 = 1$ is solvable for rational numbers x, y, if the congruence $mx^2 + ny^2 \equiv 1 \pmod{p^e}$ is solvable in integers x, y for every prime p and positive integer e. There is no immediate extension to higher equations, since

$$y^2 + 7(x^2 + 1)(x^2 - 2)^2(x^2 + 2)^2 = 0$$

is irreducible and has no rational solution, while the corresponding congruence modulo p^e is solvable whatever be the prime p and positive integer e. Again, $t^4 + 13t^2 + 81$ is an irreducible function which becomes reducible modulo p^e for every prime p and integer e.

Several writers[55] found all solutions of $x(x+1)/2 = y(y+1)/3$ by means of $2u^2 - 3z^2 = -1$.

On $Mx^2 - Ny^2 = \pm 1$ or 4, see Legendre,[88] Jacobi,[112] Weber,[218] Palmström,[228] and de Jonquières[235] of Ch. XII.

On $x^2 + qy^2 = m$ see Cornacchia[4] of Ch. XXIII.

On $ax^2 + cy^2 = n$, see Euler[56] and Nasimoff.[68]

SOLUTION OF $ax^2 + bxy + cy^2 = k$.

L. Euler[56] noted that the problem to find the minimum of $Ax^2 + 2Bxy + Cy^2$ for integral values $\neq 0$ of x, y presents no difficulty if $B^2 - AC \leq 0$ and hence is here treated for $B^2 - AC$ positive and not a square. Then the proposed form may be reduced to $mx^2 - ny^2$, where m and n are positive integers whose ratio is not a square. If $m = 1$, it can be given the value unity by Pell's theorem. If $n = 1$, it can be given the value -1.

If $mx^2 - ny^2 = k$ for $x = a$, $y = b$, it has an infinitude of solutions. For, if $p^2 - mnq^2 = 1$ (in an infinitude of ways, since $mn \neq \square$), then

$$mx^2 - ny^2 = (ma^2 - nb^2)(p^2 - mnq^2)^\lambda.$$

This holds if

$$x\sqrt{m} \pm y\sqrt{n} = (a\sqrt{m} \pm b\sqrt{n})(p \pm q\sqrt{mn})^\lambda,$$

so that we get x, y as rational functions of a, b, p, q.

The problem to make $mx^2 - ny^2$ a minimum corresponds to finding the rational fraction x/y giving the closest approximation to $\sqrt{n/m}$. Develop the latter into a periodic continued fraction and take the convergent obtained by continuing to the largest quotient. Thus, for $7x^2 - 13y^2$, the con-

---

[54] Göttingen Nachrichten (Math.), 1897, 52–54.

[55] L'intermédiaire des math., 22, 1915, 239, 255–260.

[56] Novi Comm. Acad. Petrop., 18, 1773, 218; Comm. Arith. I, 570; Opera Omnia, (1), III, 310. On the incompleteness of Euler's methods, see Smith[139] of Ch. XII.

tinued fraction for $\sqrt{91}/7$ has the quotients 1, $\overset{\circ}{2}$, 1, 3, 9, 3, 1, 2, $\overset{\circ}{2}$ (with the period marked).   Since

$$1+\frac{1}{2+}\frac{1}{1+}\frac{1}{3}=\frac{15}{11},$$

$x=15$, $y=11$, give the minimum 2 of $7x^2-13y^2$.

Given (p. 577) the solution $x=a$, $y=b$ of

$$f\equiv Ax^2-2Bxy+Cy^2=c \qquad (k\equiv B^2-AC>0,\ k\ne \Box),$$

to find an infinitude of solutions, use the solution of

$$\phi\equiv p^2-2Bpq+ACq^2=1,$$

corresponding to the Pell problem $p=Bq+\sqrt{kq^2+1}$.   Now Af has the factors $Ax-By\pm y\sqrt{k}$, and ϕ the factors $p-Bq\pm q\sqrt{k}$.   Hence we use

(1) $$Ax-By+y\sqrt{k}=(Aa-Bb\pm b\sqrt{k})(p-Bq\pm q\sqrt{k})^n,$$

for any of the four combinations of signs.   To find the minimum of f for integral x, y, develop the root $(B\pm\sqrt{k})/A$ into a continued fraction and proceed as above.

G. L. Dirichlet[57] noted that in addition to the infinite set (1) of solutions there may exist further similar sets of solutions.   Given any positive number σ, we can find one and only one solution x, y of set (1) for which

$$\sigma<Ax+(\sqrt{k}-B)y\leqq\sigma(t+q\sqrt{k}),$$

where t, q give any positive solution of $t^2-kq^2=1$.   All solutions of these inequalities can be found by a finite number of trials.   Hence we find the initial solutions a, b defining the various sets (1).

A. M. Legendre[58] discussed the integral solutions of

(2) $$Ly^2+Myz+Nz^2=\pm H.$$

After preliminary transformations, we may assume that z is prime to y and H.   Distinguish the cases in which the roots of $Lt^2+Mt+N=0$ are imaginary, real or equal.   First, let $4LN-M^2=B>0$.   Set $x=2Ly+Mz$. Then $x^2+Bz^2=C=4LH$.   Give to z the successive values 0, 1, $\cdots$, $[\sqrt{C/B}]$ and see whether the resulting value of $C-Bz^2$ is a square x^2 and then whether the resulting x makes $Mz\mp x$ a multiple of $2L$.   Second, let $4LN-M^2=-B$, B positive and not a square.   If $H<\frac{1}{2}\sqrt{B}$, develop a root of $Lx^2+Mx+N=0$ into a continued fraction; if one of the complete quotients $(\frac{1}{2}\sqrt{B}+I)/D$ has $D=H$, at least one of the equations (2) is solvable. But if $H>\frac{1}{2}\sqrt{B}$, we may set $y=nz+Hu$ where $n\leqq\frac{1}{2}H$.   Thus if Ln^2+Mn+N is not a multiple fH of H for some value of n between $-\frac{1}{2}H$ and $\frac{1}{2}H$, (2) is impossible; while if such a multiple is found, the equation reduces to $fz^2+gzu+hu^2=\pm 1$ for $g=2nL+M$, $h=LH$.   See Lagrange[76, 85] of Ch. XII.

E. F. A. Minding[59] noted that, if $A\equiv b^2-ac$ is positive and not a square, and if $H<\frac{2}{3}\sqrt{A}$, we can decide whether or not $ax^2+2bxy+cy^2=\pm H$ is

<hr>

[57] Bericht Akad. Wiss. Berlin, 1841, 280; Werke, I, 628–9.

[58] Théorie des nombres, 1798, 99–122 (77–98); ed. 2, 1808, 88–110 (68–87); ed. 3, 1830, I, 104–129 (81–103); transl. by Maser, I, 105–131 (81–105).

[59] Jour. für Math., 7, 1831, 140–2.

solvable in integers by developing a root of $av^2+2bv+c=0$ into a continued fraction, admitting negative terms.

H. Scheffler[60] treated $ax^2-2bxy-cy^2=k$. We may take x, y relatively prime. Let $D=b^2+ac$ be positive and not a square. Set $a=Q_0$, $b=P_0$, $c=Q_{-1}$ Develop the root $x/y=K=(\sqrt{D}+P_0)/Q_0$ into a continued fraction and let the quotients be $a_0, a_1, \cdots$. Set

$$P_n=a_{n-1}Q_{n-1}-P_{n-1}, \qquad Q_n=\frac{D-P_n^2}{Q_{n-1}}.$$

Take $Q_0'=k$ and seek all integers P_0', numerically $\leqq k/2$, such that $D-P_0'^2$ s divisible by k. For each such existing P_0', develop $K'=(\sqrt{D}+P_0')/k$ into a continued fraction. There is no solution unless we can assign a common period $P_r=P_s'$, $Q_r=Q_s'$ ($r+s$ even) of the two developments. By use of such a common period or a repetition of that period, he obtained a process for finding all relatively prime solutions x, y.

C. L. A. Kunze[61] treated $x^3\pm y^3=x\pm y$ in four cases.

J. J. Nejedli[62] assumed that $D=b^2+ac>0$ in

$$ax^2=2bxy+cy^2+k.$$

Set $x=a_0y+y_1$. We get a similar equation, apart from the sign of k,

$$(3) \qquad Q_1y^2=2P_1yy_1+ay_1^2-k, \qquad P_1=aa_0-b, \qquad Q_1=c-aa_0^2+2a_0b.$$

Taking a_0 to be the greatest integer in $r=(b+\sqrt{D})/a$ and repeating the process on (3), we can solve the given equation. The process is equivalent to the development of r into a continued fraction.

S. Réalis[63] noted the identity $f(x, y)=f(\alpha, \beta)f^2(A, B)$, where

$$f(x, y)=ax^2+bxy+cy^2,$$
$$x=(a\alpha+b\beta)A^2+2c\beta AB-c\alpha B^2, \qquad y=-a\beta A^2+2a\alpha AB+(b\alpha+c\beta)B^2.$$

Given the solution $f(\alpha, \beta)=h$, we get another solution of $f(x, y)=h$ if (as is not always the case) solutions of $f(A, B)=\pm1$ can be found. In particular, from solutions $f(\alpha, \beta)=\pm1$, $f(A, B)=\pm1$, we get new solutions of $f(x, y)=\pm1$.

J. J. Sylvester[64] proved that $fy^2+2gxy-2fx^2=\pm1$ is solvable in integers if $A=2f^2+g^2$ is a prime and f is odd. Since $u^2-Av^2=1$ is solvable, set $u+1=\sigma p^2$, $u-1=A\sigma q^2$, where p, q are relatively prime. Then

$$p^2-Aq^2=2/\sigma=\mp1 \text{ or } \pm2,$$

the upper signs being excluded by the form $8n+3$ of A. If $p^2-Aq^2=1$, $v=2pq$, we write p, q for u, v and p_1, q_1 for p, q and see in like manner that $p_1^2-Aq_1^2=1$ or -2. Finally, we reach $\pi^2-A\phi^2=-2$, where π and ϕ are odd. Since every prime divisor of π^2+2 is known to have the form r^2+2s^2,

[60] Jour. für Math., 45, 1853, 349–369.
[61] Ueber einige Aufg. Dioph. Analysis, Weimar, 1862.
[62] Ein Beitrag zur Auflösung unbest. quad. Gl., Progr. Laibach, 1874.
[63] Nouv. Corresp. Math., 6, 1880, 342–350.
[64] Math. Quest. Educ. Times, 34, 1881, 21–2.

$\pi\pm\sqrt{-2}=(g+f\sqrt{-2})(y+x\sqrt{-2})^2$. By the coefficients of $\sqrt{-2}$,
$$\pm1=f(y^2-2x^2)+2gxy.$$

S. Roberts used reduced quadratic forms and results of A. Göpel.

E. Cesàro[65] proved that the number of sets of positive integral solutions of
$$Ax^2+Bxy+Cy^2=n \qquad\qquad (A>0,\ C>0)$$
is $\pi/(2\delta)-B/\delta^2$ in mean, where $\delta^2=4AC-B^2$.

S. Réalis[66] noted that if α, β is a solution of $x^2+nxy-ny^2=1$ then $x=(n+1)\alpha-n\beta$, $y=(n+2)\alpha-(n+1)\beta$ is a solution. From the evident solution 1, 0, we get the solution $n+1$, $n+2$. Using $y=n+2$, and solving the initial equation we get $x=n+1$ and the new value $x=-n^2-3n-1$. Applying the formula to the latter we get a fourth solution, etc. The ath set x_a, y_a of solutions of this series is given, as well as recursion formulæ.

Réalis[67] noted that $mx^2-(m+n\pm1)xy+ny^2=h$ has the solution

(4) $x=(m-n)\alpha-(m-n\pm1)\beta,\qquad y=(m-n\mp1)\alpha-(m-n)\beta,$

if α, β is one solution. Starting from this set (4), we get again the first set α, β. Evidently (4) hold also for an equation derived from the given one by increasing m and n by the same number; also for
$$(2m\mp1)x^2-2(m+n)xy+(2n\mp1)y^2=h.$$

For $x^2-(n+2)xy+ny^2=1$, the solution 1, 0 gives the solution $n-1$, n. For $y=n$, we have $x=n-1$, n^2+n+1, and hence find an infinitude of solutions. There is treated the equation obtained from the last by changing the sign of the constant term, and
$$x^2-2(n+1)xy+(2n-1)y^2=1 \text{ or } -2.$$

Recursion formulæ are given for the integral solutions of $x^2-Axy+By^2=h$ when $A-2$ is divisible by $A-B-1$.

*P. S. Nasimoff[68] gave an exposition of Jacobi's series for elliptic functions and application to the number of solutions of $ax^2+bxy+cy^2=n$, in particular for $x^2+16y^2=n$, $4x^2+4xy+3y^2=n$, $ax^2+cy^2=n$ (a, c odd).

F. J. Studnička[69] noted that if p_k and q_k are the numerator and denominator of the kth convergent for the continued fraction
$$\frac{1}{a}+\frac{1}{a}+\frac{1}{a}+\cdots, \qquad q_n=p_{n+1}=a^n+\binom{n-1}{1}a^{n-2}+\binom{n-2}{2}a^{n-4}+\cdots,$$
$$(-1)^n=p_{n-1}q_n-p_nq_{n-1}=q_{n-2}q_n-q_{n-1}^2.$$

Using $q_n=aq_{n-1}+q_{n-2}$, we get
$$aq_{n-2}q_{n-1}+q_{n-2}^2-q_{n-1}^2=(-1)^n$$
and hence the solutions of $axy+x^2-y^2=\pm1$. Cf. Kluge[289] of Ch. XII.

[65] Mém. Soc. R. Sc. de Liège, (2), 10, 1883, No. 6, 197–9.
[66] Nouv. Ann. Math., (3), 2, 1883, 494–7.
[67] Ibid., (3), 3, 1884, 305–15. Errata, p. 448.
[68] Application of elliptic functions to the theory of numbers, Moscow, 1885, Ch. 1. French résumé in Annales sc. de l'école normale supér., (3), 5, 1888, 23–31.
[69] Prag Sitzungsber. (Math. Nat.), 1888, 92–95.

* Ferval[70] gave an infinitude of solutions of each of the equations
$$(2a^2-2a-1)x^2-4(a^2-1)xy+(2a^2+2a-1)y^2=1,$$
$$(a^2-a-1)x^2-(2a^2-3)xy+(a^2+a+1)y^2=1.$$

A. Hurwitz[71] called r/s and u/v a pair of approximating fractions for a number between them if $us-vr=1$. If $0<m<2\sqrt{D}$ and if at least one of A, C is positive, and $D=B^2-AC>0$, every pair of integral solutions of $Au^2+2Buv+Cv^2=m$ is such that u/v is an approximating fraction to one of the roots of $Ax^2+2Bx+C=0$. If both A and C are negative, we get the same result by assuming also that $v^2>-A/(2\sqrt{D}-m)$.

H. Scheffler[72] made successive additions to get p, 2^2p, 3^3p, $\cdots$ and then a table of values for $pn^2+p_1n_1^2$. The aim is to solve $ax^2+bxy+cy^2=q$.

R. W. D. Christie[73] solved $x^2+xy-y^2=\pm1$ by use of continued fractions. Cf. J. Wasteels[72] of Vol. I, p. 405, of this History.

A. Cunningham and Christie[74] solved $y^2-avy-av^2=1$.

A. Lévy[75] recalled the special case of Dirichlet's theorem on the units of an algebraic field, that if (a, b) is the least positive solution $\neq(1, 0)$ of $x^2+xy-ky^2=1$, where k is a positive integer, every solution (u, v) is given by
$$u+v\omega=(a+b\omega)^n, \qquad \omega^2-\omega-k=0.$$

Several writers[76] solved $x^2+xy+y^2=1$.

C. Ruggeri[77] used the series with the recursion formula $z_{n+1}=z_n+z_{n-1}$ to solve $ax^2-bxy+cy^2=k$, when $b^2-4ac=5m^2$.

See papers 88, 89; also Leslie[90] of Ch. XII.

SOLUTION OF $Ax^2+2Bxy+Cy^2+2Dx+2Ey+F=0$.

L. Euler[78] noted that if $Ax^2+2Bxy+Cy^2+2Dx+2Ey+F=0$ has the set of solutions $x=a$, $y=b$, and if $\Delta=B^2-AC>0$, so that $p^2=\Delta q^2+1$ is solvable, a second set of solutions is
$$x=a(p+Bq)+bCq+Eq+(p-1)(BE-CD)/\Delta,$$
$$y=b(p-Bq)-aAq-Dq+(p-1)(BD-AE)/\Delta.$$

J. L. Lagrange[79] showed how to find the rational and integral solutions of
$$(1)\qquad\qquad \alpha x^2+\beta xy+\gamma y^2+\delta x+\epsilon y+\zeta=0.$$
Solving it algebraically for x in terms of y, we get
$$2\alpha x+\beta y+\delta=\pm t, \qquad t^2=By^2+2fy+g,$$

<hr>

[70] Jour. de math. spéc., 1889, 94, 141.

[71] Math. Annalen, 44, 1894, 425–7.

[72] Vermischte Math. Schriften, Part II, Die Quadratische Zerfällung der Zahlen durch Differenzreihen, Braunschweig, 1897, 28–59.

[73] Math. Quest. Educ. Times, 73, 1900, 71.

[74] *Ibid.*, (2), 10, 1906, 24–25.

[75] Bull. de math. élém., 15, 1909–10, 113–5. Cf. J. Sommer, Vorlesungen über Zahlentheorie, 1907, 100–7; French transl. by Lévy, 1911, 103–113.

[76] Amer. Math. Monthly, 15, 1908, 44.

[77] Periodico di Mat., 25, 1910, 266–276.

[78] Novi Comm. Acad. Petrop., 11, 1765 (1759), 28; Comm. Arith., I, 317; Op. Om., (1), III, 76.

[79] Mém. Acad. Berlin, 23, anneé 1767, 1769, 272; Oeuvres, II, 377–381, 509–522. Cf. his simplifications in his additions to Euler's Algebra, 2, 1774, 554, 595–607; Oeuvres de Lagrange, VII, 113, 140–7; Euler's Opera Omnia, (1), I, 593, 615–22. Cf. Smith.[88]

where $B=\beta^2-4\alpha\gamma$, $f=\beta\delta-2\alpha\epsilon$, $g=\delta^2-4\alpha\zeta$.   Set $A=f^2-Bg$.   Then

$$By+f=\pm u, \qquad u^2=A+Bt^2,$$

$$y=\frac{\pm u-f}{B}, \qquad x=\frac{\pm t-\delta}{2\alpha}-\frac{\beta(\pm u-f)}{2\alpha B}.$$

Hence the rational solutions of (1) follow from the rational solutions of $u^2=A+Bt^2$.   The latter depend on the integral solutions of $Ar^2=p^2-Bq^2$, discussed by Lagrange.[110]

To obtain the integral solutions of (1), it is necessary that not only u and t be integers, but also that $\pm u-f$ be a multiple mB of B, and that $\pm t-\delta-\beta m$ be a multiple of 2α.   If B is negative, $u^2-Bt^2=A$ has only a finite number of integral solutions, which can be found by trial.   This is not true when B is positive, as will be assumed henceforth.   We may set $u=\sigma p$, $t=\sigma q$, where p, q are relatively prime.   By Lagrange[75] of Ch. XII, the solutions of $p^2-Bq^2=A/\sigma^2$ are given by

$$p+q\sqrt{B}=(a+b\sqrt{B})J, \qquad J\equiv(X+Y\sqrt{B})^n=\xi+\psi\sqrt{B},$$

whence

$$p=a\xi+Bb\psi, \qquad q=a\psi+b\xi; \qquad 2\xi,\ 2\sqrt{B}\psi=(X+Y\sqrt{B})^n\pm(X-Y\sqrt{B})^n.$$

Here a, b, X, Y are given integers for which $X^2-BY^2=\pm1$.   We may restrict attention to the case $X^2-BY^2=+1$, to which the contrary case is easily reduced.   The problem is now to choose positive integral values of the exponent n for which the resulting values of x, y are integers, viz., for which $\pm\sigma p-f$ is a multiple of B, and $\pm\sigma q-\delta-\beta(\pm\sigma p-f)/B$ is a multiple of 2α.   These two questions are special cases of the general question of the divisibility of

$$(2) \qquad\qquad F+Gp+Hq\equiv F+P(X+Y\sqrt{B})^n+Q(X-Y\sqrt{B})^n$$

by $R=r^m r_1^{m_1}\cdots$, where r, r_1, $\cdots$ are distinct primes.   It is easily shown that $(X\pm Y\sqrt{B})^\rho-1$ is divisible by r, where $\rho=2r$ if B is divisible by r, $\rho=r\pm1$ if $B^{(r-1)/2}\pm1$ is divisible by r, and $\rho=r$ if $r=2$.   Then $(X\pm Y\sqrt{B})^e-1$ is divisible by r^m for $e=r^{m-1}\rho$.   Hence if $n=ke+N$, (2) is divisible by r^m if and only if r^m divides the function obtained from (2) by replacing n by N, so that we need only test the values $<e$ of n.   Similarly we need only test the divisibility of (2) by $r_1^{m_1}$ for $n<r_1^{m_1-1}\rho_1$.   Suppose that the test succeeds for $n=N$ and for $n=N_1$, etc., in the respective cases.   Then determine n so that it shall have the remainder N when divided by $r^{m-1}\rho$, the remainder N_1 when divided by $r_1^{m_1-1}\rho_1$, etc.   We saw that also a second expression $F_1+G_1p+H_1q$ had to be divisible by a certain number R_1.   The conditions on n are similar to those just stated.   Hence the method leads to all the (infinitude of) integral solutions of (2) when it is solvable.

Lagrange[80] multiplied (1) by 4α and set

$$u=2\alpha x+\beta y+\delta, \qquad a=\beta^2-4\alpha\gamma, \qquad b=\beta\delta-2\alpha\epsilon, \qquad c=\delta^2-4\alpha\zeta.$$

We get $u^2=ay^2+2by+c$.   Multiply by a and write $t=ay+b$, $R=b^2-ac$.   Hence $t^2-au^2=R$.   Assume that it has a known solution $t=P$, $u=Q$.

---

[80] Miscellanea Taurinensia, 4, 1766–9; Oeuvres, I, 725–31.

Then (1) has the solution

$$(3) \qquad y = \frac{P-b}{a}, \qquad x = \frac{Q-\delta}{2\alpha} - \frac{\beta y}{2\alpha}.$$

Since we may change the sign of P or Q, we get four solutions. If $R = A^m B^n \cdots$, where A, B, $\cdots$ are expressible in a single way in the form $P^2 - aQ^2$, it is known that R is expressible in this form in exactly $\frac{1}{2}\pi$ ways when $\pi = (m+1)(n+1)\cdots$ is even, and in $(\pi+1)/2$ ways when π is odd. If a is negative there is only a finite number of solutions of $t^2 - au^2 = R$, since $t^2 - au^2 = 1$ is not solvable, so that the number of factors A, B, $\cdots$ is limited. But if a is positive, let p, q be the least solution of $p^2 - aq^2 = 1$; then every solution is given by

$$p' = \frac{r^m + s^m}{2}, \qquad q' = \frac{r^m - s^m}{2\sqrt{a}} \qquad (r = p + q\sqrt{a}, \quad s = p - q\sqrt{a})$$

for $m = 1, 2, 3, \cdots$ Then

$$R = (P^2 - aQ^2)(p'^2 - aq'^2) = P_1^2 - aQ_1^2$$

if

$$(4) \qquad P_1 = Pp' \pm aQq', \qquad Q_1 = Pq' \pm Qp'.$$

If we employ as P, Q the various sets corresponding to the factors >1 of the form $t^2 - au^2$ of R and take $m = 1, 2, 3, \cdots$, we get by (4) ail the rational solutions of $P_1^2 - aQ_1^2 = R$. Returning to (3), Lagrange proved that, if the values (3) of x, y which correspond to the case $m = 0$ are integers, there is an infinitude of values of m (the multiples of an assigned number depending only on α and a) for which the solutions x, y are integers.

L. Euler[81] gave two methods of finding the general rational solution of

$$f(x, y) \equiv Ax^2 + 2Bxy + Cy^2 + 2Dx + 2Ey + F = 0,$$

given one solution $x = a$, $y = b$.   In $f(x, y) - f(a, b) = 0$, set

$$2(xy - ab) = (x-a)(y+b) + (x+a)(y-b), \qquad \frac{x-a}{y-b} = \frac{p}{q}.$$

We get

$$(x+a)(Ap+Bq) + (y+b)(Bp+Cq) + 2Dp + 2Eq = 0.$$

Eliminating y by the second of the preceding pair of equations, we get

$$\omega x = -at - 2b(Bp^2 + Cpq) - 2Dp^2 - 2Epq,$$
$$\omega y = bt - 2a(Bq^2 + Apq) - 2Dpq - 2Eq^2,$$
$$\omega = Ap^2 + 2Bpq + Cq^2, \qquad t = Ap^2 - Cq^2,$$

and hence obtain, when p, q are rational, the most general rational solution of the proposed equation. Integral solutions may be obtained from values of p, q making $\omega = \pm 1$ or ± 2.

For the second method, set

$$k = B^2 - AC, \qquad N = (BD - AE)/k, \qquad P = Ax + By + D, \qquad Q = y + N.$$

Then

$$Af(x, y) \equiv (P + Q\sqrt{k})(P - Q\sqrt{k}) - \delta, \qquad \delta \equiv D^2 - AF - N^2 k.$$

<hr>

[81] Novi Comm. Acad. Petrop., 18, 1773, 185; Comm. Arith., I, 549–55; Op. Om., (1), III, 297.

Let G and H be the values of P and Q for $x=a$, $y=b$. Then
$$(P+Q\sqrt{k})(P-Q\sqrt{k})=(G+H\sqrt{k})(G-H\sqrt{k}).$$
Equate the first factor on the left to the second factor on the right and vice versa. Thus
$$y=-b-2N, \qquad x=a+\frac{2B(b+N)}{A}.$$

Or, use the Pell equation $s^2-kr^2=1$, having an infinitude of solutions if k is neither negative nor a square, and set
$$P+Q\sqrt{k}=(G+H\sqrt{k})(s+r\sqrt{k})^n.$$

By equating the terms free of $\sqrt{k}$, we get rational expressions for x, y.

Euler[82] treated the solution in integers of

(5) $$\alpha x^2+\beta x+\gamma=\zeta y^2+\eta y+\theta,$$

given one solution $x=a$, $y=b$. Denote the roots of $z^2=2sz-1$ by
$$p=s+\sqrt{s^2-1}, \qquad q=s-\sqrt{s^2-1}.$$
Make the substitution

(6) $$x=\frac{f}{\sqrt{\alpha}}p^n+\frac{g}{\sqrt{\alpha}}q^n-\frac{\beta}{2\alpha}, \qquad y=\frac{f}{\sqrt{\zeta}}p^n-\frac{g}{\sqrt{\zeta}}q^n-\frac{\eta}{2\zeta}.$$

Since $pq=1$, the members of (5) equal respectively
$$f^2p^{2n}+g^2q^{2n}+2fg+\gamma-\frac{\beta^2}{4\alpha}, \qquad f^2p^{2n}+g^2q^{2n}-2fg+\theta-\frac{\eta^2}{4\zeta}.$$

These are equal if

(7) $$4fg=\frac{\beta^2}{4\alpha}-\frac{\eta^2}{4\zeta}+\theta-\gamma.$$

For $n=0$, let $x=a$, $y=b$. Then (6) gives

(8) $$f+g=\frac{2\alpha a+\beta}{2\sqrt{\alpha}}, \qquad f-g=\frac{2\zeta b+\eta}{2\sqrt{\zeta}},$$

and the resulting value of $(f+g)^2-(f-g)^2$ reduces to (7) since (5) holds for $x=a$, $y=b$. Hence the values of f, g from (8) lead to solutions (6) of (5) provided s, in the expressions for p and q, is such that the resulting x, y are rational. For $n=1$, the expressions for x, y become, in view of (8),
$$x=as+\frac{\beta(s-1)}{2\alpha}+b\zeta r+\frac{\eta r}{2}, \qquad y=bs+\frac{\eta(s-1)}{2\zeta}+a\alpha r+\frac{\beta r}{2}, \qquad r=\sqrt{\frac{s^2-1}{\zeta\alpha}}.$$

Then $s^2=1+\alpha\zeta r^2$, a solvable Pell equation if $\alpha\zeta$ is positive and not a square. Hence if the latter is solved and we set p, $q=s\pm r\sqrt{\alpha\zeta}$ and define f, g by (8), then, for any integer n, (6) gives a solution, which is proved rational as follows. Call x', y' the values obtained from (6) by changing n to $n+1$; x'', y'' those by changing n to $n+2$. Then
$$x''=2sx'-x+\frac{\beta}{\alpha}(s-1), \qquad y''=2sy'-y+\frac{\eta}{\zeta}(s-1).$$

---

[82] Mém. Acad. Sc. St. Petersb., 4, 1811 (1778), 3; Comm. Arith., II, 263.

Since the values given by $n=0$ and $n=1$ are rational, those given by any n are rational. Euler stated that if we employ only even values of n, we obtain integral values for x, y. Cayley[152] gave a generalization to several variables.

A. M. Legendre[83] reduced $ay^2+byz+cz^2+dy+fz+g=0$ to

$$(9)\quad ay_1^2+by_1z_1+cz_1^2=\Delta D,\qquad D=b^2-4ac>0,\qquad -\Delta=af^2-bdf+cd^2+gD,$$

by setting $y=(y_1+\alpha)/D$, $z=(z_1+\beta)/D$, $\alpha=2cd-fb$, $\beta=2af-bd$. If (9) has a solution, it has an infinitude of solutions given by

$$(10)\quad y_1=\gamma F+\delta G,\qquad z_1=\epsilon F+\zeta G,\qquad F+G\sqrt{D}=(\phi+\psi\sqrt{D})^n,$$

where ϕ, ψ give the least solution of $\phi^2-D\psi^2=1$. It is a question of the values of n for which y and z are integers. Since

$$F\equiv\phi^n,\qquad G\equiv n\phi^{n-1}\psi,\qquad \phi^2\equiv1\ \ (\mathrm{mod}\ D),$$

we see that the expressions for y, z are integers if and only if

$$(n=2m)\qquad (\alpha+\gamma)\phi+2\delta\psi m\equiv(\beta+\epsilon)\phi+2\zeta\psi m\equiv0\quad(\mathrm{mod}\ D),$$
$$(n=2m+1)\qquad \gamma\phi+\alpha+n\delta\psi\equiv\epsilon\phi+\beta+n\zeta\psi\equiv0\qquad(\mathrm{mod}\ D).$$

In either case the resulting values of n are said to be of the form $V+Dk$ [denied by Dujardin[84]], where k is arbitrary, so that there is an infinitude of values n. It remains to solve the problem: if F and G are given by (10_3) and if $\phi^2-D\psi^2=1$, find all values of n such that $\lambda F+\mu G+\nu$ is divisible by a prime not dividing $D\psi$. For this, the method of Lagrange[79] is given.

Dujardin[84] agreed with the statements in the preceding paper down to the erroneous one that the values of n are of the form $V+Dk$. But the quantities δ, ζ are divisible by D and the conditions marked $(n=2m)$ and $(n=2m+1)$ are satisfied only if the coefficients of the unknowns are relatively prime. Hence $\alpha+\gamma$, $\beta+\epsilon$ must be divisible by D if n is even, and $\gamma\phi+\alpha$, $\epsilon\phi+\beta$ if n is odd; then the conditions cited are satisfied for all values of m. The correct conclusion is therefore that n varies according to an arithmetical progression of difference 2 (not D). The latter result is said to follow also from the law of recurrence between three consecutive solutions of (9), which leads also to the following rule. Given two consecutive solutions y_i', z_i' $(i=1,2)$ of (9); then if no one of the systems $y_i'+\alpha$, $z_i'+\beta$ $(i=1,2)$ is divisible by D, there is no solution in integers; but if one of the latter systems is divisible by D, then to every system of the same parity as it there corresponds a solution of the proposed equation.

C. F. Gauss[85] treated the integral solutions of

$$(11)\qquad ax^2+2bxy+cy^2+2dx+2ey+f=0.$$

Set

$$\Delta=\begin{vmatrix} a & b & d \\ b & c & e \\ d & e & f \end{vmatrix},\qquad \alpha=b^2-ac,\qquad \beta=be-cd,\qquad \gamma=bd-ae.$$

---

[83] Théorie des nombres, 1798, 451–7; ed. 3, 1830, II, 105–112, No. 439.
[84] Comptes Rendus Paris, 119, 1894, 843, 934. Reprinted, Sphinx-Oedipe, 4, 1909, 45–7.
[85] Disquisitiones Arithmeticae, 1801, arts. 216–221; Werke, I, 1863, 215. German transl. by H. Maser, 1889, pp. 205–211.

By the substitution $p=\alpha x+\beta$, $q=\alpha y+\gamma$, we get $ap^2+2bpq+cq^2=\alpha\Delta$.   The theory of binary quadratic forms leads to all representations of $\alpha\Delta$ by the form (a, b, c).   From the resulting sets of values of x, y, discard those which are not integral [cf. Smith[88]].

To find (art. 300) the rational solutions of (11), set $x=t/v$, $y=u/v$, and find the integral solutions of the resulting equation which is of the form considered by Gauss.[147]

J. L. Wezel[86] reduced $ax^2+cxy+dx+ey+C=0$ to $x_1y_1=k$ by a linear substitution, and treated equations solvable rationally for one variable. For $ax^2+by^2+2cxy+C=0$, we solve (p. 40) for x and find no trouble unless $B\equiv c^2-ab$ is positive and $\neq\square$.   The latter case is treated elegantly by continued fractions.   Develop the root $r=(\sqrt{B}-c)/a$ of $az^2+2cz+b=0$. Let $Q=(\sqrt{B}+\pi)/C$ be the complete quotient with denominator C, and p_0/q_0, p/q the convergents immediately preceding this.   Then

$$z=\frac{pQ+p_0}{qQ+q_0}=r, \qquad (ap+cq)^2=q^2B+aC(pq_0-p_0q), \qquad ap^2+2cpq+bq^2=\pm C,$$

since $\sqrt{B}$ is irrational.   For $ax^2+by^2+cxy+dx+ey+C=0$, we set

$$x=(x'+2bd-ce)/D, \qquad y=(y'+2ae-cd)/D,$$

where $D=c^2-4ab$, and get an equation of the form last treated:

$$ax'^2+by'^2+cx'y'+(ae^2-cde+bd^2)D+CD^2=0.$$

E. Desmarest[87] noted that the substitution $X=x/a$ reduces the solution of $aX^2+bX+c=Ky$ to the problem to find multiples x of a satisfying an equation of type $f_x\equiv x^2+qx+r=Py$.   To solve a particular equation of the latter type, he would employ two auxiliary doubly-entry tables, a complicated method based upon the functions

$$_0P_{2N-2}=f_nN^2-f_n'N+1, \qquad _0P_{2N-1}=f_nN^2+f_n'N+1$$

and the fact that their products by f_n are also of the form f_x, where $x=f_nN-n-q$ and f_nN+n, respectively.   One of the auxiliary tables has the headings f_n, $_0P_1$, $_0P_2$, $\cdots$ and in the body of the table are entered the values for successive K's of the roots R and remainders ρ defined, for example, when $N=2K+1$, by use of

$$_0P_{2N}=R^2+\rho, \qquad R=(2K+2)n+qK-q-1, \qquad \rho=A(K+1)^2, \qquad A\equiv 4r-q^2.$$

Troublesome methods are indicated (pp. 42, 43) by means of which the square R^2 nearest to the given P enables us to find the entry in the body of the table which will yield the desired value of n such that the heading of the column of the entry will for this n be the value of y (or a known multiple of y).   The example $X^2+31X+241=PY$ is treated (pp. 24–25, 301–2) for all primes $P<1000$; but he knew (p. 104) that it can be transformed by $X=x-15$ into $x^2+x+1=Py$, which is proved to be solvable if and only if the prime P is 3 or $3q+1$.

[86] Annales Acad. Leodiensis, Liège, 1821–2, 1–48.

[87] Théorie des nombres.   Traité de l'analyse indéterminée du second degré à deux inconnues . . ., Paris, 1852, 4–126.

To solve (pp. 127–221) $F+2dX+2eY+f=0$, where $F\equiv aX^2+2bXY+cY^2$, $\Delta=b^2-ac\neq0$, it is transformed as usual into $F=M$, which is treated as usual by the theory of binary quadratic forms. If $\Delta=0$, it is transformed into $u^2+r=Py$, which is of the type first treated. In each case there is a discussion as to which of the solutions are integral.

H. J. S. Smith[88] noted that Euler's[81, 82] methods are incomplete for the reasons noted in Ch. XII, Smith.[139] He modified Gauss'[85] method by employing the g.c.d. δ of α, β, γ, and employing the new variables $X=p/\delta$, $Y=q/\delta$. Thus $aX^2+2bXY+cY^2=\alpha'\Delta'$, where $\alpha'=\alpha/\delta$, $\Delta'=\Delta/\delta$. Then if X_n, Y_n is any representation of $\alpha'\Delta'$ by (a, b, c), we separate the integral from the fractional solutions x, y by separating (by Lagrange's method) those values of X_n, Y_n which satisfy the congruences $X_n-\beta/\delta\equiv0$, $Y_n-\gamma/\delta\equiv0 \pmod{\alpha'}$ from those which do not, and obtain a finite number of formulas exhibiting all integral solutions.

G. Wertheim[89] treated (1) as had Lagrange,[79] and by reducing it to $ax^2+2bxy+cy^2=M$ and then applying the theory of binary quadratic forms.

C. de Comberousse[90] treated (1) for the case $\gamma=0$, whence $y=Q/L$, where Q is a quadratic and L a linear function of x. Thus L must divide a certain constant N, whence set $L=d$, d any divisor of N.

Rautenberg[91] reduced the solution of an equation of degree two in two variables to $Bx^2+Cx+D=\square$ and gave other known results.

R. Marcolongo,[92] G. B. Mathews,[93] P. Bachmann,[94] and E. Cahen[95] treated (1).

Focke[96] gave the usual application of quadratic forms to our problem.

E. de Jonquières[97] showed by detailed examples that the methods of Lagrange (continued fractions) and Gauss (period of reduced forms) for solving indeterminate equations of the second degree are less different than they seem, since they employ the same auxiliary quantities, and rest on the development of practically the same ideas.

G. Bisconcini[98] noted that $x=y=2$ is the only positive integral solution of $xy=x+y$, and $x=0, 1, y=0, 1$, the only integral solutions of $x^2+y^2=x+y$.

J. Westlund[99] proved that $x^2+y^2=(2x-1)/3$ is impossible in integers.

C. Ciamberlini[100] stated that $(x+y)(x+y+1)+2y=a$ has a single positive integral solution if a is a positive integer.

T. Pepin[101] used the method of Gauss.[85]

[88] British Assoc. Report, 1861, 313; Coll. Math. Papers, 1, 1894, 200–2.
[89] Elemente der Zahlentheorie, 1887, 226–236, 369–374.
[90] Algèbre supérieure, 1, 1887, 185–191.
[91] Ueber dioph. Gl. 2 Gr., Progr. K. Gymn., Marienburg, 1887.
[92] Giornale di Mat., 25, 1887, 161; 26, 1888, 65.
[93] Theory of Numbers, 1892, 257–261.
[94] Arith. der Quad. Formen, 1898, 224–231.
[95] Élém. de la théorie des nombres, 1900, 286–299.
[96] Über die Auflösung d. dioph. Gleich. mit Hilfe der Zahlentheorie, Progr. Magdeburg, 1895.
[97] Comptes Rendus Paris, 127, 1898, 694–700.
[98] Periodico di Mat., 22, 1907, 121–2.
[99] Amer. Math. Monthly, 14, 1907, 61.
[100] Suppl. al Periodico di Mat., 11, 1908, 104–5.
[101] Mem. Pont. Accad. Nuovi Lincei, 29, 1911, 319–327.

U. Fornari[102] treated $(x-1)(x-2)+y(2x+y-1)=2m$.

W. A. Wijthoff[103] solved $(x+y+1)^2=9xy$.

M. Rignaux[103a] stated a complete solution of (11), with $ac<b^2$, by recurring series.

For $\frac{1}{2}x(x+1)=\frac{1}{3}y(y+1)$ see paper 55. For $3x(x+1)=y(y+1)$, see Euler[79] of Ch. I. T. L. Pistor[107] of Ch. XII gave Gauss'[85] method. On $ax^2-a'x=by^2-b'y$, see Gill[107] of Ch. I.

$$ax^2+by^2+cz^2=0 \quad (\text{EXCEPT } x^2+y^2=2z^2).$$

For $x^2+y^2=2z^2$, here excluded, see squares in arithmetical progression (Ch. XIV).

Diophantus, II, 20, proposed to find three squares such that

(1) $$y^2-x^2 : z^2-y^2=a : b,$$

where $a : b$ is a given ratio. He took $a/b=1/3$, $y=x+1$, whence

$$z^2=x^2+8x+4.$$

Take $z=x+3$, whence $x=5/2$. In IV, 45, he took $a/b=3$, $x^2=4$, $y=t+2$, whence $\frac{9}{4}z^2=3t^2+12t+9=(3-5t)^2$, if $t=21/11$.

Alkarkhi[104] (beginning of eleventh century) solved $x^2-y^2=2(y^2-z^2)$ by taking $y=z+1$, $x=z+2$, whence $z=1/2$.

Leonardo Pisano[105] first treated (1) for several special cases. For $b=a+1$, take $x=2a-1$, $y=2a+1$, $z=2a+3$; then $y^2-x^2=8a$, $z^2-y^2=8b$. In general, if integers h, k, n can be found such that

$$\sum_{i=1}^{a}(h+i)=ka, \qquad \sum_{j=1}^{n}(h+a+j)=kb,$$

then $y^2-x^2=8ka$, $z^2-y^2=8kb$ for

$$x=2h+1, \qquad y=2h+2a+1, \qquad z=2h+2a+2n+1.$$

The conditions for the above sums are

$$h+1+h+a=2k, \qquad (h+a)n+n(n+1)/2=kb,$$

or

$$k=h+\frac{a+1}{2}=\frac{n(n+a)}{2(b-n)}.$$

These fractions must equal integers, as in the case for the values $a=11$, $b=43$, $n=16$, $k=8$, $h=2$, used by Leonardo. A. Genocchi[106] remarked that Leonardo's method consists essentially in separating a progression $h+1$, $h+2$, $\cdots$, $h+m+n$ into two parts such that the sum of the first m terms is ka and the sum of the last n terms is kb, whence

$$2k=\frac{mn(m+n)}{bm-an}, \qquad 2h+1=\frac{2amn+an^2-bm^2}{bm-an}.$$

[102] Il Pitagora, 19, 1913, 57–60.

[103] Wiskundige Opgaven, 11, 1912–4, 192–5.

[103a] L'intermédiaire des math., 26, 1919, 9.

[104] Extrait du Fakhri, French transl. by F. Woepcke, 1853, 116.

[105] Tre Scritti, 103–112. Scritti, 2, 1862, 275–9 (Opuscoli). Cf. Ch. XVI.

[106] Annali di Sc. Mat. e Fis., 6, 1855, 351–2 (misprint of sign before bm^2 in the fraction for $2h+1$).

Since a, b are relatively prime, we can make $bm-an=1$ in an infinitude of ways. Then $x=2h+1$, $y=2(h+m)+1$, $z=2(h+m+n)+1$.

F. Woepcke[107] gave an analogous interpretation of Leonardo's method and wondered why Leonardo preferred this ingenious method to the more natural one [of Diophantus] of substituting $x=y+m$, $z=y-n$, and thus finding x, y, z as rational functions of m, n, a, b. The last method and other simple ones were used by C. L. A. Kunze.[108]

For a different presentation of Leonardo's method and a proof of the equivalence of the problem with that of concordant forms, see Genocchi[87] of Ch. XVI.

Matsunago,[109] in the first half of the eighteenth century, noted that $rx^2+y^2=z^2$ has the solution $x=2mn$, $y=rm^2-n^2$, $z=rm^2+n^2$. If $k-l=t^2$, $kx^2-ly^2=z^2$ has the solution $y=\alpha+t\beta$, $z=l\beta-\alpha t$, provided $x^2=\alpha^2+l\beta^2$, which is of the preceding type. Again, $(k^2+l^2)x^2-y^2=z^2$ for

$$x=c, \qquad y=ka\pm lb, \qquad z=la\mp kb, \qquad a^2+b^2=c^2.$$

J. L. Lagrange[110] treated the solution of

$$(2) \qquad\qquad Ar^2=p^2-Bq^2$$

in integers. The cases $A=\square$, $B=\square$ are easily treated (pp. 381–2) by the methods of Diophantus. In (2) let p, q, r be integers, p and q relatively prime, while A and B are integers neither a square nor divisible by a square, and (as may be assumed) $|A|>|B|$. A necessary condition is that there exist an integer α such that α^2-B is divisible by A. This is shown by multiplying (2) by $p_1^2-Bq_1^2$, using

$$(3) \qquad\qquad (p^2-Bq^2)(p_1^2-Bq_1^2)=(pp_1\pm Bqq_1)^2-B(pq_1\pm qp_1)^2,$$

and taking $pq_1-qp_1=\pm1$, whence $Ar^2(p_1^2-Bq_1^2)=\alpha^2-B$. We may .also take $|\alpha|<|A|/2$, since also $(\mu A\pm\alpha)^2-B$ is divisible by A. When such an α exists, $AA_1=\alpha^2-B$, set $\alpha_1=\mu_1A_1\pm\alpha$, the integer μ_1 and the sign being chosen so that $|\alpha_1|<|A_1|/2$. Then α_1^2-B is divisible by A_1; call the quotient A_2. In this manner we get a series of decreasing integers $|A|$, $|A_1|$, $|A_2|$, $\cdots$, and hence get $|A_n|\leqq|B|$. It suffices to stop when A_n is of the form a^2C, where C has no square factor and $|C|\leqq|B|$. Multiply together the equations

$$AA_1=\alpha^2-B, \quad\cdots, \qquad A_{n-1}A_n=\alpha_{n-1}^2-B$$

and use (3). Hence $AA_1^2\cdots A_{n-1}^2A_n=P^2-BQ^2$. Multiply by (2). We get $Cq_1^2=p_1^2-Br_1^2$, where $q_1=AA_1\cdots A_{n-1}ar$. Hence

$$(4) \qquad\qquad Br_1^2=p_1^2-Cq_1^2.$$

Conversely if (4) is solvable, (2) is solvable. Treating (4) as we did (2), we get $Cr_2^2=p_2^2-Dq_2^2$, etc. Since $|A|$, $|B|$, $|C|$, $\cdots$ form a decreasing series, we finally get a term ±1. If it be -1, we proceed and get $+1$. The resulting equation $Vz^2=x^2-y^2$ is easily solved in integers. Let

[107] Jour. de Math., 20, 1855, 59.
[108] Ueber einige Aufg. Dioph. Analysis, Weimar, 1862, 14–15.
[109] Y. Mikami, Abh. Gesch. Math. Wiss., 30, 1912, 231–2.
[110] Mém. Acad. Berlin, 23, année 1767, 1769, 385–406; Oeuvres, II, 384–399.

M be the g.c.d. of V and $x+y$ and set $V=MN$, $x+y=M\rho$. Then $z^2=\rho\sigma$, $x-y=N\sigma$, where σ is an integer. If l is the g.c.d. of ρ and σ, we have $\rho=lm^2$, $\sigma=ln^2$, whence

$$z=lmn, \qquad x=l(Mm^2+Nn^2)/2, \qquad y=l(Mm^2-Nn^2)/2.$$

We may set $l=2$, since we may multiply x, y, z by $2/l$.

L. Euler[111] stated that the general solution of $\alpha x^2+\beta y^2=\gamma z^2$ is given by

$$x\sqrt{\alpha}\pm y\sqrt{-\beta}=(f\sqrt{\alpha}\pm g\sqrt{-\beta})(p\sqrt{\alpha n}\pm q\sqrt{-\beta n})^2,$$

if one solution $\alpha f^2+\beta g^2=\gamma h^2$ is given; the solution by taking $n=1$ is not general. Again, by taking $x=fp+\beta gq$, $y=gp-\alpha fq$, we get $\alpha x^2+\beta y^2=\gamma h^2 R$, where $R=p^2+\alpha\beta q^2$ is the square of $r^2+\alpha\beta s^2$ for $p=r^2-\alpha\beta s^2$, $q=2rs$. Again, if we multiply the initial equation by h^2 and $\alpha f^2+\beta g^2=\gamma h^2$ by z^2 and subtract, we get

$$\frac{\alpha(hx+fz)}{gz-hy}=\frac{\beta(gz+hy)}{hx-fz}.$$

Set each fraction equal to p/q and equate the two values of z; we get y/x. To obtain another solution, set $F=\alpha\beta q^2-p^2$, $G=2pq$, $H=\alpha\beta q^2+p^2$, whence $H^2=F^2+\alpha\beta G^2$. Multiply the latter by $\gamma h^2=\alpha f^2+\beta g^2$. The product of the right members leads to the solution

$$z=hH, \qquad x=fF+\beta gG, \qquad y=gF-\alpha fG.$$

A necessary condition for $fx^2+gy^2=hz^2$ is that $-fg$ be a quadratic residue of h.

Euler[111a] made ax^2+cy^2 a square by use of

$$x\sqrt{a}+y\sqrt{-c}=(p\sqrt{a}+q\sqrt{-c})^2.$$

Euler[112] considered the rational solutions of

$$(5) \qquad\qquad\qquad fx^2+gy^2=hz^2.$$

If, for f and g fixed, the equation is solvable when $h=h_1$, h_2 and h_3, then it is solvable when $h=h_1h_2h_3$. He stated (p. 558) the elegant empirical theorem that if (5) is solvable when $h=h_1$ it is solvable also when $h=h_1\pm 4nfg$, provided the latter is a prime.[113]

If (5) be solvable, then (p. 566) $-fg$ is a quadratic residue of h. For, since x, y may be taken relatively prime, we can determine p, q so that $py-qx=1$. Then

$$(fx^2+gy^2)(fp^2+gq^2)=t^2+fg \qquad (t=fpx+gqy)$$

is divisible by h.

---

[111] Opera postuma, 1, 1862, 205–211 (about 1769–1771).

[111a] Algebra, St. Petersburg, 2, 1770, §§ 181–7; Lyon, 2, 1774, pp. 219–26; Opera Omnia, (1), I, 425–9. Cf. Euler[6] and Lagrange[63] of Ch. XX.

[112] Opusc. anal., I, 1783 (1772), 211; Comm. Arith., I, 556–569.

[113] A. M. Legendre, Mém. Acad. Sc. Paris, 1785, 523, stated that this theorem is true, but omitted the proof (not easy) as it was necessary to separate cases. He stated the generalization: If $fx^2\pm gy^2=hz^2$ is solvable then $fx^2\pm gy^2=cz^2$ is solvable if $c=h+fgn$ is a prime and if n is such that the two members of the quadratic equation are congruent modulo 8.

If (5) be solvable, also $fx^2+gy^2=h_1z^2$ is solvable when h_1 is a certain integer $<h$. For, $k\equiv t^2+fg$ is divisible by h for some integer $t<h/2$; call the quotient h_1. Then

$$f(tx\pm gy)^2+g(ty\mp fx)^2=(t^2+fg)(fx^2+gy^2)=khz^2=h^2h_1z^2,$$

so that $fX^2+gY^2=h_1Z^2$ is solvable. It is not shown that $h_1<h$. In case a set of decreasing values h, h_1, h_2, $\cdots$ eventually contains f or g, we can determine x, y (p. 569, § 62).

A. M. Legendre[114] proved [Legendre[87] of Ch. XII] that, if each of the positive integers a, b, c has no square factor and if no two of them have a common factor, then $ax^2+by^2=cz^2$ has integral solutions not all zero if and only if there exist three integers λ, μ, ν such that

$$\frac{a\lambda^2+b}{c}, \qquad \frac{c\mu^2-b}{a}, \qquad \frac{c\nu^2-a}{b}$$

are all integers.

Legendre[115] explained the method of Lagrange[110] to solve (2), modified by use of a principle employed elsewhere by Lagrange[76] of Ch. XII. The present method is essentially due to Lagrange.[115a] We may take A and B positive, since otherwise

$$x^2-Ay^2=-Bz^2 \qquad \text{or} \qquad x^2+Ay^2=Bz^2 \qquad (A>0,\ B>0).$$

In the second write $Bz=z'$, $AB=A'$, whence $z'^2-A'y^2=Bx^2$. The second is obtained from the first by the transpositions of two terms. Consider therefore $x^2-By^2=Az^2$, $A>B>0$, where y is prime to A and x, while A and B have no square factors. Set $x=\alpha y-Ay'$. Then

$$\left(\frac{\alpha^2-B}{A}\right)y^2-2\alpha yy'+Ay'^2=z^2.$$

The first coefficient must be an integer, say $A'k^2$, where A' has no square factor. By changing α by a multiple of A, we may take α between $-A/2$ and $A/2$. Multiply the resulting equation by $A'k^2$ and set $kz=z'$, $A'k^2y-\alpha y'=x'$; we get $x'^2-By'^2=A'z'^2$, $A'<A$. If $A'>B$, we repeat the process. Finally we get a similar equation with one coefficient unity and hence easily solved. While this method is not the simplest one for solving the proposed equation, it is a very luminous one.

C. F. Gauss[116] proved by use of ternary quadratic forms the theorem of Legendre[114] that, if no two of a, b, c have a common factor and if each is neither zero nor divisible by a square, then $ax^2+by^2+cz^2=0$ has integral solutions not all zero if and only if $-bc$, $-ac$, $-ab$ are quadratic residues of a, b, c, respectively, and a, b, c are not all of the same sign. If a, b, c are

---

[114] Mém. Acad. Sc. Paris, 1785, 512–3; Théorie des nombres, 1798, 49; ed. 2, 1808, 41; ed. 3, 1830, I, 47; German transl. by Maser, I, 49.

[115] Théorie des nombres, 1798, 36–41; ed. 2, 1808, 28–32; ed. 3, 1830, I, 33–39. (Maser, I, 36–39.) For his remark on $x^2+by^2=z^2$ see Legendre.[39]

[115a] Addition V to Euler's Algebra, 2, 1774, 538–55; Euler's Opera Omnia, (1), I, 586–94; Oeuvres de Lagrange, VII, 102–14.

[116] Disquisitiones Arith., arts. 294–8; Werke, I, 1863, 349. German transl. by Maser, pp. 335–343.

arbitrary integers, let α^2, β^2, γ^2 be the largest squares dividing bc, ac, ab, respectively, and set $\alpha a=\beta\gamma A$, $\beta b=\alpha\gamma B$, $\gamma c=\alpha\beta C$; then the former equation is solvable if and only if $AX^2+BX^2+CZ^2=0$ is solvable, and the latter falls under the above theorem since A, B, C are relatively prime in pairs and have no square factors. For, $bc/\alpha^2=BC$ is an integer without square factor, so that B, C are relatively prime and without square factors.

E. F. A. Minding[117] considered $x^2=Ay^2+Bz^2$, where A, B are without square factors. Let f be the g.c.d. of $A=af$ and $B=bf$. The equation is solvable if and only if A, B, $-ab$ are quadratic residues of B, A, f, respectively.

A. Genocchi[118] treated the equation $az^2+bx^2=(a+b)y^2$, equivalent to (1), by the methods of Lagrange and Paoli[91] of Ch. XII.

G. L. Dirichlet[119] treated $ax^2+by^2+cz^2=0$, where a, b, c are relatively prime in pairs. If u, v, w are given relatively prime solutions, we can deduce all solutions. Since au, bv, cw are relatively prime and au, for example, is even, we can find integers l (even), m and n such that $aul+bvm+cwn=1$. Set $al^2+bm^2+cn^2=h$. Then $u'=2l-hu$, $v'=2m-hv$, $w'=2n-hw$ are solutions, congruent to u, v, w, respectively, modulo 2. Hence, in

$$2u''=vw'-wv', \qquad 2v''=wu'-uw', \qquad 2w''=uv'-vu',$$

u'', v'', w'' are integers. If x, y, z are any integers,

(6) $\quad t=au'x+bv'y+cw'z, \qquad t'=aux+bvy+cwz, \qquad t''=u''x+v''y+w''z$

are integers and $t\equiv t'$ (mod 2). It is shown that, conversely, if t, t', t'' are any integers for which $t-t'$ is even,

$$(6') \qquad \begin{aligned} 2x &=ut+u't'-2bcu''t'', \qquad 2y=vt+v't'-2cav''t'', \\ 2z &=wt+w't'-2abw''t'', \end{aligned}$$

so that x, y, z are integers. Multiply the latter equations by ax, by, cz, add, and use (6). We get

$$ax^2+by^2+cz^2=tt'-abct''^2.$$

Hence if x, y, z are solutions of the initial equation, then t, t', t'', defined by (6), are integers for which $t\equiv t'$ (mod 2) and $tt'=abct''^2$. Conversely, if t, t', t'' are integers satisfying the last two conditions, the values of x, y, z given by (6') are integral solutions. Further, by use of the above relations he proved the following extension of Legendre's[114] theorem: If no two of a, b, c have a common factor and are not zero, $ax^2+by^2+cz^2=0$ is solvable in relatively prime integers if and only if $-bc$, $-ca$, $-ab$ are quadratic residues of a, b, c, respectively, and the latter are not all of the same sign; further, if $-bc\equiv A^2$ (mod a), $-ca\equiv B^2$ (mod b), $-ab\equiv C^2$ (mod c), there exist relatively prime solutions for which

$$Az\equiv by \pmod{a}, \qquad Bx\equiv cz \pmod{b}, \qquad Cy\equiv ax \pmod{c}.$$

[117] Anfangsgründe der Hoheren Arith., 1832, 84.
[118] Annali di Sc. Mat. e Fis., 6, 1855, 186–194, 348.
[119] Zahlentheorie, §§ 156–7, 1863; ed. 2, 1871; ed. 3, 1879; ed. 4, 1894.

J. Plana[120] stated that all integral solutions of $x^2 - 79y^2 = 101z^2$ are given by

$$x = \alpha \cdot 927p^2 + \frac{1}{\alpha}\,4572q^2 + 3126pq,$$

$$y = \alpha \cdot\ 74p^2 + \frac{1}{\alpha}\ 414q^2 +\ 462pq,$$

$$z = \alpha \cdot\ 65p^2 - \frac{1}{\alpha}\ 270q^2 +\ 30pq,$$

for $\alpha = 2$ or 1, where p, q are arbitrary integers.

G. Cantor[121] considered the solution in integers of $F = 0$, where F is any ternary quadratic form. A formal solution (ϕ, ψ, χ) is one for which $F(\phi, \psi, \chi) \equiv 0$ identically in x, y, where $\phi, \cdots$ are binary quadratic forms in x, y. In particular, let F be

$$[aa'a''] \equiv aX^2 + a'Y^2 + a''Z^2.$$

Let the greatest common divisor of the three coefficients of ϕ, and those for ψ and χ be relatively prime in pairs; then the formal solution (ϕ, ψ, χ) is primitive, and we can find integers w's for which

$$w\psi \equiv a''\chi, \qquad w\chi \equiv -a'\psi \quad (\text{mod } a)$$
$$w'\chi \equiv a\phi, \qquad w'\phi \equiv -a''\chi \quad (\text{mod } a')$$
$$w''\phi \equiv a'\psi, \qquad w''\psi \equiv -a\phi \quad (\text{mod } a''),$$

identically in x, y. By the two congruences in the first line,

$$(w^2 + a'a'')\psi\chi \equiv 0 \quad (\text{mod } a).$$

Then $w^2 + a'a'' \equiv 0$ if a is odd, or when a is even if ψ, χ are properly primitive. The solution (ϕ, ψ, χ) is said to pertain to the combination $\{w, w', w''\}$ if

$$w^2 + a'a'' \equiv 0 \ (\text{mod } a), \quad w'^2 + aa'' \equiv 0 \ (\text{mod } a'), \quad w''^2 + aa' \equiv 0 \ (\text{mod } a'').$$

The number of possible sets of roots is $2^{\omega + \eta}$, where ω is the number of distinct odd prime factors of the determinant $D = -aa'a''$ of the primary form $[aa'a'']$, while $\eta = 0$, 1 or 2, according as $D/4$ is not integral, an odd or even integer. Then, if $-a'a''$, $-a''a$, $-aa'$ are quadratic residues of a, a', a'', respectively, there is a primitive solution (ϕ, ψ, χ) of $[aa'a''] = 0$ pertaining to any chosen one of the $2^{\omega + \eta}$ combinations $\{w, w', w''\}$, and $[aa'a''] = 0$ has exactly $\sigma \cdot 2^{\omega + \eta}$ systems of primitive solutions, where $\sigma = 2$ if $D \equiv 0$ (mod 4), while $\sigma = 4$ in all other cases.

L. Calzolari[122] treated (7) $u^2 = Ax^2 + By^2$ by setting (8) $u = Yx + Xy$. The discriminant of the resulting quadratic in x, y is to be a square, whence

$$(9) \qquad\qquad AX^2 + BY^2 - AB = U^2.$$

Eliminate X between the latter and (8), using (7). We get $U^2y^2 = (uY - Ax)^2$,

$$(10) \qquad\qquad Ax = Yu \pm Uy, \qquad By = Xu \mp Ux.$$

Thus from a set of solutions of (9), we get one of (7), viz., that given by (8)

---

[120] Memorie R. Accad. Torino, (2), 20, 1863, 107, footnote.
[121] De Aequat. secundi Gradus indet., Diss. Berlin, 1867.
[122] Giornale di Mat., 7, 1869, 177–192.

and

(11)
$$\frac{x}{y}=\frac{XY\pm U}{A-Y^2},$$

and conversely.   Expressed geometrically, (7) is a cone with the vertex at the origin, and (8) is a plane through the vertex.   The intersections are two lines whose projections on the xy-plane are given by (11).   If $X_0,\ Y_0,\ U_0$ is a particular solution of (9), and if $u_0,\ x_0,\ y_0$ are the values given by (8) and (10), the general solution is

$$X=X_0-x_0t, \qquad Y=Y_0+y_0t, \qquad U=U_0-u_0t.$$

Calzolari[123] stated a theorem, which not only decides like Legendre's[114] the possibility or impossibility of integral solutions of

(12)
$$u^2=Ax^2\pm By^2 \qquad (A,\ B \text{ without square factors}),$$

but determines the general solution without recourse to the process of Lagrange.   We may set $A=a_1^2+\cdots+a_m^2,\ B=b_1^2+\cdots+b_n^2\ (m\leqq4,\ n\leqq4)$. Set $x_i=a_ix,\ y_i=b_iy.$ Then

(13)
$$u^2=\sum_{i=1}^{m}x_i^2\pm\sum_{i=1}^{n}y_i^2.$$

Let $p_1,\ \cdots,\ p_m,\ q_1,\ \cdots,\ q_n$ be arbitrary integers.   We may set

$$x_i=u-p\mp q+p_i, \qquad y_i=u-p\mp q+q_i, \qquad p=\Sigma p_i, \qquad q=\Sigma q_i.$$

Then (13) becomes $u-\Sigma x_i\mp\Sigma y_i+K=0$, where

$$Ku=(p\pm q)(\Sigma x_i\pm\Sigma y_i)-\Sigma p_ix_i\mp\Sigma q_iy_i.$$

In the former give to $x_i,\ y_i$ their values.   Then

$$u=p\pm q+k, \qquad (m\pm n-1)k=K,$$

(14)
$$x_i=p_i+k, \qquad y_i=q_i+k, \qquad u=p\pm q+k.$$

Substitute these values (14) into the two expressions for K.   Thus

(15)
$$(p\pm q)^2-\Sigma p_i^2\mp\Sigma q_i^2=(m\pm n-1)k^2.$$

For $k=0$, values $p_i,\ q_i$ satisfying (15) give $x_i,\ y_i$ from (14) which satisfy (13). Set $a=\Sigma a_i,\ b=\Sigma b_i.$ Then, for $k=0,\ xa=\Sigma x_i=\Sigma p_i=p,\ by=q,\ u=ax\pm by.$ Substitute this u into (12); we get a quadratic for x/y.   Hence (12) is solvable if and only if $c\equiv Ab^2\pm Ba^2\mp AB=\square$, and the general solution is $x=ab\pm c,\ y=a^2-A,\ u=Ab\pm ac$, where the signs of $a,\ b$ are ambiguous.

S. Réalis[124] stated that, if $A,\ B,\ C$ are relatively prime and without square factors, and if $\alpha,\ \beta,\ \gamma$ give one solution of $Ax^2+By^2+Cz^2=0$, the general solution is

$$x=\alpha(-Aa^2+Bb^2+Cc^2)-2a(B\beta b+C\gamma c),$$
$$y=\beta(\ \ \ Aa^2-Bb^2+Cc^2)-2b(A\alpha a+C\gamma c),$$
$$z=\gamma(\ \ \ Aa^2+Bb^2-Cc^2)-2c(A\alpha a+B\beta b),$$

where $a,\ b,\ c$ are arbitrary.

---

[123] Giornale di Mat., 8, 1870, 28–34.
[124] Nouv. Corresp. Math., 4, 1878, 369–71.

S. Roberts[125] treated the solution of $x^2-2Py^2=-z^2$ or $\pm 2z^2$, when each prime factor of P is of the form $8m+1$. If P have one of the forms

$$(8\alpha\pm 1)^2+16(2\beta+1)^2, \qquad (8k\pm 3)^2+8(2l+1)^2, \qquad (8k-1)^2-8(2l)^2,$$
$$(8k-3)^2-8(2l+1)^2,$$

the equations

$$2Py^2=(8u\pm 3)^2+(8v\pm 3)^2, \quad 2Py^2=16u^2+2(8v\pm 3)^2, \quad 2Py^2=4u^2-2(8v\pm 1)^2,$$

are solvable. If, moreover, P is a prime of one of those forms or an odd power of it, $x^2-2Py^2=2$ is solvable. There are three more such triples of equations leading to analogous conclusions.

T. Pepin[126] proved Legendre's criterion as quoted by Gauss.[116]

G. Heppel[127] treated $d^2=2a^2+b^2$ by setting $b=d-2f$, whence

$$d=f+a^2/(2f).$$

Thus a is even, $a=2q$. Hence express $2q^2$ as a product fh and take $d=h+f$, $b=h-f$, $a=2q$.

F. Goldscheider[128] expressed in terms of one solution the general solution of $ax^2+by^2+cz^2=0$ which satisfies the final congruences of Dirichlet.[119] He proved that there exists such a solution for which also $kx+k'y+k''z$ is relatively prime to a given odd integer s, if k, k', k'' are given integers whose g.c.d. is prime to s.

G. de Longchamps[129] wrote $x^2=y^2+pz^2$ in the form

$$(x+y)/(pz)=z/(x-y)=t.$$

Hence t must divide z. Set $z=2\lambda t$. Thus $x=\lambda(pt^2+1)$, $y=\lambda(pt^2-1)$, where λ and t are arbitrary. For $nx^2=y^2+(n-1)z^2$, see de Longchamps.[162]

P. Bachmann[130] gave a clear exposition of our subject.

R. P. Paranjpye[131] proved that all integral solutions of $x^2-z^2=2y^2$ are

$$x=k(\lambda^2+2\mu^2), \qquad y=\pm 2k\lambda\mu, \qquad z=\pm k(\lambda^2-2\mu^2),$$

where λ, μ are relatively prime. Since y is even, $x\mp z=2k\lambda^2$, $x\pm z=4k\mu^2$.

A. S. Werebrusow[132] noted that, if $\alpha^2-D\beta^2=ma^2$, a second set of solutions of $X^2-DY^2=mZ^2$ is given by

$$X+Y\sqrt{D}=(\alpha+\beta\sqrt{D})\left(x+\frac{b+\sqrt{D}}{a}y\right)^2, \qquad D=b^2-ac.$$

A. Cunningham,[133] to solve $x^2+y^2=Az^2$, used the known solutions $Y=(t^2+Au^2)d$, $Z=2tu/d$, $x=(t^2-Au^2)/d$, of $Y^2-AZ^2=x^2$, where $d=1$ or 2, and solutions of $\tau^2-Av^2=-1$. Then $(Y^2-AZ^2)(\tau^2-Av^2)=-x^2$, whence $y=\tau Y\mp AvZ$, $z=\tau Z\mp vY$ give the general solutions. A. Holm (p. 70)

[125] Proc. London Math. Soc., 11, 1879–80, 83–87.
[126] Atti Accad. Pont. Nuovi Lincei, 32, 1878–9, 88.
[127] Math. Quest. Educ. Times, 38, 1883, 56.
[128] Das Reziprozitätsgesetz der achten Potenzreste, Progr., Berlin, 1889, 8.
[129] El Progreso Mat., 4, 1894, 46; Jour. de math. élém., 18, 1894, 5.
[130] Arith. der Quad. Formen, 1898, 198–224, 231.
[131] Math. Quest. Educ. Times, 75, 1901, 119. Cf. papers 109, 116 of Ch. XVI.
[132] Mem. Sc. Univ. Moscow, 23; l'intermédiaire des math., 9, 1902, 187.
[133] Math. Quest. Educ. Times, (2), 9, 1906, 69–70.

noted that $A=a^2+b^2$, whence $(x+bz)/(az+y)=(az-y)/(x-bz)=m/n$ (say), which determine $x:y:z$.

P. F. Teilhet[134] stated that $x^2+y^2=(m^2+n^2)z^2$ implies

$$z=K(A^2+B^2), \qquad x=mK(A^2-B^2)\pm2nKAB, \qquad y=nK(A^2-B^2)\mp2mKAB.$$

F. Ferrari[135] noted the solution, with $K=1$, given by Teilhet.[134]

A. Gérardin[136] stated that the general solution of $x^2+2y^2=\square$ is

$$x=2l^2-m^2-n^2+2m(3n-4l), \qquad y=4l^2+2n^2-2m^2-2l(3n-m).$$

Gérardin[137] noted the identities

$$\{(p-q)b^2-qy^2+2bqy\}^2+pq\{2b(y-b)\}^2=\{(p+q)b^2+qy^2-2bqy\}^2,$$
$$(m^2+n^2)(mnx^2-2z^2)^2+2mn\{mnx^2+2z^2-2xz(m+n)\}^2$$
$$=\{(m+n)(mnx^2+2z^2)-4mnxz\}^2,$$

another similar to the last and several for $x^2+8y^2=z^2$.

A. Thue[138] discussed the possibility of $Ax^2+By^2=Cz^2$, where x, y, z are relatively prime in pairs and $z\geqq y\geqq x>0$. We can determine integers p, q, r without a common factor such that $px+qy=rz$, with p^2, q^2, r^2 all $<3z$. Then

$$(Bp^2+Aq^2)x^2-2Bprxz+(Br^2-Cq^2)z^2=0,$$
$$(Bp^2+Aq^2)y^2-2Aqryz+(Ar^2-Cp^2)z^2=0.$$

Hence

$$ax=Cq^2-Br^2, \qquad by=Cp^2-Ar^2, \qquad cz=Bp^2+Aq^2,$$
$$az+2Bpr=cx, \qquad bz+2Aqr=cy,$$

where a, b, c are integers. Let U be the greatest of $|A|$, $|B|$, $|C|$. By the last five equations, $|c|<6U$, $|a|<12U$, $|b|<12U$. But a, b, $-c$ satisfy the initial linear and quadratic equation. Thus the possibility of the latter can be decided by a finite number of trials.

L. Aubry[139] proved that if $pA^2=B^2+rC^2$, where B and C are prime to A, then $pX^2=Y^2+rZ^2$ for $X\leqq2\sqrt{r/3}$ if $r>0$, and $X\leqq\sqrt{-r}$ if $r<0$; if B and C are prime to p, then also $Y\equiv Ba$, $Z\equiv Ca$ (mod p).

Several writers[140] solved $13x^2+17y^2=230z^2$.

C. Alasia[141] solved $x^2-79y^2=101z^2$ [Plana[120]] by several classic methods.

G. Bonfantini[142] noted that the evident sufficient condition for integral solutions of $x^2+y^2=mz^2$ is that m be a sum of two squares. To prove that the condition is necessary, consider integers k_i, ρ_i, q_i such that

$$k_1=1+\rho_1^2, \qquad 1+\rho_2^2=k_2k_1, \qquad \rho_2=q_1k_1+\rho_1, \qquad 1+\rho_3^2=k_3k_2, \qquad \rho_3=q_2k_2+\rho_2, \cdots.$$

By induction, $k_m=\phi_m^2+(\psi_m+\rho_1\phi_m)^2$, where

$$\phi_1=1, \quad \phi_2=\phi_1q_1, \cdots, \quad \phi_i=\phi_{i-1}q_{i-1}+\phi_{i-2},$$
$$\psi_1=0, \quad \psi_2=1, \cdots, \quad \psi_i=\psi_{i-1}q_{i-1}+\psi_{i-2} \quad (i=3, \cdots, m).$$

[134] L'intermédiaire des math., 12, 1905, 81.
[135] Suppl. al Periodico di Mat., 12, 1908–9, 34–5.
[136] Assoc. franç., 1908, 17. To make[109] $m=0$ replace l by $l+2m$, n by $n+3m$.
[137] Sphinx-Oedipe, 1907–8, 109–110.
[138] Skrifter Videnskapsselskapet, Kristiania, 1, 1911, No. 4, p. 18.
[139] Sphinx-Oedipe, 8, 1913, 150 (error in his 7, 1912, 81–2.
[140] Wiskundige Opgaven, 11, 1912–4, 281–6.
[141] Giornale di Mat., 53, 1915, 292–302.
[142] Suppl. al Periodico di Mat., 18, 1915, 81–6. For $m=2$, $ibid.$, 17, 1914, 84–5.

M. Weill[143] obtained solutions $x=a+\lambda\delta$, $y=b+\lambda'\delta$, $z=1+\delta$, of

$$x^2+y^2=(a^2+b^2)z^2$$

by finding δ rationally in terms of a, b, λ, λ'. Again, $ax^2+by^2=(a+b)z^2$ has solutions of the form $x=1+\lambda\delta$, $y=1+\lambda'\delta$, $z=1+\delta$. To one of these two is reduced the solution of $ax^2+by^2=z^2$ when $a+b$ or ab is a square.

E. Cahen[144] noted that Weill's formulas do not give all solutions and showed how to find all solutions of $x^2+y^2=5z^2$.

E. Turrière[145] noted that, if a, b, c are the sides of a triangle two of whose medians are perpendicular, then $a^2+b^2=5c^2$, whose solutions are expressed rationally in two parameters.

A. Desboves[38] gave all solutions of $x^2+y^2=(m^2+n^2)z^2$. Cf. papers 133–5, 142–5 above; Catalan[63] of Ch. VII; G. F. Malfatti[19] of Ch. VIII; papers 191, 252, 294, 307, 311 of Ch. XII; and 225 of Ch. XXII.

R. Hoppe[29] of Ch. V solved $p^2-3q^2=r^2$; Euler[109] of Ch. XXII solved $\alpha^2+3\beta^2=\square$.

FURTHER SINGLE QUADRATIC EQUATIONS IN THREE OR MORE UNKNOWNS.

Bháscara[146] (born 1114) found four distinct numbers whose sum equals the sum of their squares. Take as the numbers y, $2y$, $3y$, $4y$. Then $10y=30y^2$, $y=1/3$.

C. F. Gauss[147] considered the solution in integers of

(1) $$f\equiv ax^2+a_1x_1^2+a_2x_2^2+2bx_1x_2+2b_1xx_2+2b_2xx_1=0.$$

If $a=0$, x is determined rationally in terms of x_1, x_2; to obtain integral solutions, multiply the three x's by the denominator of x. Next, let $a\neq0$. We derive the equivalent equation

$$L^2-A_2x_1^2+2Bx_1x_2-A_1x_2^2=0, \qquad L=ax+b_2x_1+b_1x_2,$$
$$A_2=b_2^2-aa_1, \qquad B=ab-b_1b_2, \qquad A_1=b_1^2-aa_2.$$

If $A_2=0$, $B\neq0$, we can give arbitrary values to x_2 and L and determine x and x_1 rationally. If $A_2=B=0$, either A_1 is not a square and $x_2=L=0$ or $A_1=k^2$ and $L=\pm kx_2$. Finally, let $a_2\neq0$, $A_2\neq0$. Then

$$A_2L^2-(A_2x_1-Bx_2)^2+Dax_2^2=0,$$

where D is the determinant of f, whence $Da=B^2-A_1A_2$. If $D=0$, we have linear factors. If $D\neq0$, criteria for solvability were given by Gauss.[116]

Given one solution α, α_1, α_2 of $f=0$, we can transform f into a like form with $a=0$ (treated above). In fact, determine integers β, $\cdots$, γ_2 so that

$$\alpha(\beta_1\gamma_2-\beta_2\gamma_1)+\alpha_1(\beta_2\gamma-\beta\gamma_2)+\alpha_2(\beta\gamma_1-\beta_1\gamma)=1.$$

Then the desired transformation is

$$x=\alpha y+\beta y_1+\gamma y_2, \qquad x_1=\alpha_1y+\beta_1y_1+\gamma_1y_2, \qquad x_2=\alpha_2y+\beta_2y_1+\gamma_2y_2.$$

---

[143] Nouv. Ann. Math., (4), 16, 1916, 351–5.
[144] Ibid., (4), 17, 1917, 463–5.
[145] L'enseignement math., 18, 1916, 89–90.
[146] Vîja-gañita, § 119; Colebrooke,[1] p. 200.
[147] Disq. Arith., 1801, art. 299; Werke, I, 1863, 358; German transl., Maser, 344–6.

Aida Ammei,[148] just after 1807, noted that $x_1^2 + 2x_2^2 + \cdots + nx_n^2 = y^2$ has the solution

$$x_1 = -a_1^2 + \sum_{r=2}^{n} r a_r^2, \qquad x_r = 2a_1 a_r, \qquad y = \sum_{j=1}^{n} j a_j^2,$$

and that $x_1^2 + 3x_2^2 + 6x_3^2 + \cdots + \tfrac{1}{2}n(n+1)x_n^2 = y^2$ has the solution

$$x_1 = -a_1^2 + \sum_{r=2}^{n} \frac{r(r+1)}{2} a_r^2, \qquad x_r = 2a_1 a_r, \qquad y = \sum_{r=1}^{n} \frac{r(r+1)}{2} a_r^2.$$

G. Libri[149] noted that $ax^2 + by^2 + cz^2 + d = 0$ is solvable if $a'x^2 + b'y^2 + c'z^2 = 0$ is, where a', b', c' are any three of a, b, c, d. For example, if

$$an^2 + br^2 + cm^2 = 0,$$

we get a solution $x = np + q$, $y = rp + s$, $z = mp + t$, where p is found rationally in terms of the indeterminates q, s, t. If an^2, br^2, cm^2 are relatively prime integers and if no one of a, b, c is divisible by 4, we can assign the value ± 1 to the denominator of the fraction for p and hence get integral solutions x, y, z.

Every integer can be expressed in the form $F \equiv x^2 + 41u^2 - 113z^2$ since $F = 0$ is solvable. Likewise for $23x^2 + y^2 - 13z^2$ and $ax^2 + 5z^2 - 2y^2$, where a is a prime $\equiv 3, 13, 27, 37 \pmod{40}$.

A. Cauchy[150] treated the homogeneous equation $F(x, y, z) = 0$ of degree N, with the given set of integral solutions a, b, c. Let x, y, z be another set. The ratios of u, v, w are determined by $au + bv + cw = 0$, $xu + yv + zw = 0$. Then

$$F(wx, wy, -ux - vy) = 0, \qquad F(wa, wb, -ua - vb) = 0.$$

Set $y/x = p$, $b/a = P$. Then

$$F_1 \equiv F(w, wp, -u - vp) = 0, \qquad F_2 \equiv F(w, wP, -u - vP) = 0.$$

Let ϕ, χ, ψ be the partial derivatives of $F(x, y, z)$ with respect to x, y, z. Then

$$x\phi + y\chi + z\psi = NF(x, y, z), \qquad a\phi(a, b, c) + b\chi + z\psi = 0.$$

Thus $au + bv + cw = 0$ is satisfied by

$$(2) \qquad u = \phi(a, b, c) + br - cn, \qquad v = \chi + cm - ar, \qquad w = \psi + an - bm,$$

for m, n, r arbitrary integers. If the latter can be chosen to make $F_1 = F_2$ for a rational $p (p \neq P)$, we get the new solution $x : y : z = w : wp : -u - vp$ of $F = 0$.

To apply (pp. 292–301) this general method to

$$F(x, y, z) = Ax^2 + By^2 + Cz^2 + Dyz + Ezx + Fxy,$$

note that the condition $F_1 = F_2$ now gives $p = P$ or

$$p = -P + [(Ev + Du)w - Fw^2 - 2Cuv]/\alpha, \qquad \alpha \equiv Bw^2 - Dvw + Cv^2.$$

Replace P by its value b/a and use $au + bv + cw = 0$, $F(a, b, c) = 0$. Thus $y/x = p = a\beta/(b\alpha)$, where $\beta = Cu^2 - Ewu + Aw^2$, $\gamma = Av^2 - Fuv + Bu^2$. Then all

[148] Y. Mikami, Abh. Gesch. Math. Wiss., 30, 1912, 248. See papers 59, 66 of Ch. IX.
[149] Memoria sopra la teoria dei numeri, Firenze, 1820, 10–14.
[150] Exercices de mathématiques, Paris, 1826; Oeuvres, (2), VI, 286.

solutions of $F(x, y, z)=0$ are given by $ax/\alpha=by/\beta=cz/\gamma$, where α, β, γ have been defined and u, v, w are given by (2). In particular, $x=\alpha/a$, $y=\beta/b$, $z=\gamma/c$ are solutions.

To apply this method for $N=3$, we remove the factor $p-P$ from F_1-F_2 and have a quadratic in p, whose discriminant is to be made a perfect square if new rational solutions exist. To avoid treating this quadratic, Cauchy[287] of Ch. XXI gave a method independent of the above.

G. Poletti[151] treated the general equation of degree two in three unknowns. First, for solution in rational numbers, solve for one unknown u in terms of the other two v, w. Since the radical Z is to be rational, a quadratic function of v, w is to be a square Z^2. Solve the latter for v; a new radical Y is to be rational, whence

$$\alpha w^2+2\beta w+\gamma+rZ^2=Y^2.$$

Solving this for w, we see that a radical X is to be rational:

(F) $$X^2=AY^2+BZ^2+C.$$

Hence the rational solution of the initial equation is equivalent to that of (F), where A, B, C are given integers. This in turn is evidently equivalent to the solution in relatively prime integers of

(G) $$x^2=Ay^2+Bz^2+Ct^2.$$

Set $\pi=x^2-Ay^2$ and call ϕ the g.c.d. of x, y; ψ that of z, t. The quotient π_1 of π by $\phi^2\psi^2$ is an integer. Thus the problem reduces to

(H) $$x_1^2-Ay_1^2=\pi_1\psi^2, \qquad Bz_1^2+Ct_1^2=\pi_1\phi^2,$$

where $x_1=x/\phi$ and $y_1=y/\phi$ are relatively prime, and likewise also z_1, t_1. From Legendre's theory of the quadratic forms of divisors of $x_1^2-Ay_1^2$, we get π_1 as a quadratic function of two parameters y', z', and ψ as one of y_1', z_1'; then by the composition of quadratic forms, we get x_1, y_1 as functions of the four parameters y', z', y_1', z_1'. To get the linear forms $4A\xi+b_i$ of the divisors π_1, use Legendre's text. By (H_2) these must divide $\rho^2+BC\sigma^2$, whose divisors are of certain linear forms $4BC\xi_1+\beta_i$. Equate each of the latter to $4A\xi+b_i$ and solve for integers ξ, ξ_1. For each such set of solutions, we can tell by a theorem of Legendre whether or not (H_2) is solvable in integers. There is a similar discussion of the solution of (F) in integers.

A. Cayley[152] treated the generalization of Euler's[82] equation (5), viz.,

(3) $$\phi(x, y) \equiv \alpha x^2+\beta x+\gamma-\zeta y^2-\eta y-\theta=\phi(a, b).$$

This is a special case of

(4) $$(abcfgh)(x'y'z')^2=(abcfgh)(xyz)^2,$$

where the second member denotes $ax^2+by^2+cz^2+2fyz+2gxz+2hxy$. It is assumed that the latter has a linear automorph (transformation into itself), which may be taken to be such that $z'=z$. For $z'=z=1$, $h=0$, (4) becomes (3). We can find a solution of (4) by Hermite's method: set $x'=2\xi-x$,

---

[151] Memorie Accad. Sc. Torino, 31, 1827, 409–49. Cf. Atti della Società Ital. delle Scienze residente in Modena, Vol. 19.

[152] Nouv. Ann. Math., 16, 1857, 161–5; Coll. Math. Papers, III, 205–8.

$$y' = 2\eta - y, \quad z' = 2\zeta - z,$$

$$ax + hy + gz = a\xi + h\eta + g\zeta - qC\eta + qF\zeta, \qquad C = ab - h^2,$$
$$hx + by + fz = h\xi + b\eta + f\zeta + qC\xi - qG\zeta, \qquad F = gh - af,$$
$$gx + fy + cz = g\xi + f\eta + c\zeta - qF\xi + qG\eta, \qquad G = fh - bg,$$

where q is arbitrary. Multiply these by ξ, η, ζ and add. Thus

$$(abcfgh)(\xi\eta\zeta)(xyz) = (abcfgh)(\xi\eta\zeta)^2,$$

so that we get (4). Using the multipliers C, F, G, we get $z = \zeta$. Then the first two equations readily give

$$(1 + q^2 C)x' = (1 + 2qh - q^2 C)x + 2qby + 2(qf + q^2 G),$$
$$(1 + q^2 C)y' = (1 - 2qh - q^2 C)y - 2qax + 2(-qg + q^2 F),$$

which satisfy (4) identically with $z' = z = 1$. Taking $h = 0$, we get values making $ax^2 + 2gx + c + by^2 + 2fy$ identically equal to the same function of x', y'. To pass to formulas exactly equivalent to Euler's, set

$$(1 - q^2 ab)/(1 + q^2 ab) = s = \sqrt{1 - abr^2}.$$

H. J. S. Smith[153] stated criteria for the solvability of (1) in integers, whether the coefficients are real integers or complex integers $p + qi$. It suffices to consider the case in which the coefficients $a, \cdots, b_2$ of f have no common divisor, while f is an indefinite form of determinant $\neq 0$. Let Ω be the g.c.d. of the nine two-rowed minors of the determinant $\Omega^2 \Delta$ of f. Let ΩF be the contravariant $(b^2 - a_1 a_2)x^2 + \cdots$ of f. Let $\overline{\Omega}$, $\overline{\Delta}$, $\overline{\Omega\Delta}$ be the quotients of Ω, Δ, $\Omega\Delta$ by the greatest squares contained in them respectively. Let ω be any odd prime dividing $\overline{\Omega}$, but not $\overline{\Delta}$; δ one dividing $\overline{\Delta}$ but not $\overline{\Omega}$; θ one dividing both $\overline{\Omega}$, $\overline{\Delta}$. Then $f = 0$ is solvable in integers $\neq 0$ if and only if

$$\left(\frac{\overline{\Omega}}{\delta}\right) = \left(\frac{F}{\delta}\right), \qquad \left(\frac{\overline{\Delta}}{\omega}\right) = \left(\frac{f}{\omega}\right), \qquad \left(\frac{-\overline{\Omega\Delta}}{\theta}\right) = \left(\frac{f}{\theta}\right)\left(\frac{F}{\theta}\right),$$

where the symbols in the left members are those of Legendre, and those in the right members are generic characters of f (Eisenstein, Jour. für Math., 35, p. 125). This theorem is a generalization of the criteria for the solvability of $ax^2 + a_1 x_1^2 + a_2 x_2^2 = 0$.

A. Meyer[154] proved the preceding theorem for forms of odd determinant.

P. Bachmann[155] proved that, if F is a ternary quadratic form, all solutions of $p^2 - F(q, q', q'') = 2^h \delta$, in which p is divisible by δ, are obtained by a definite rule from any one solution and all solutions of

$$t^2 - F(u, u', u'') = 1.$$

The left member repeats under multiplication.

S. Réalis[156] stated that all integral solutions of $x^2 + ny^2 = u^2 + nv^2$ are given by (cf. Gérardin[167])

$$x = \alpha^2 + n\beta^2 - n\gamma^2, \qquad\qquad y = (\gamma - \alpha)^2 + n(\gamma - \beta)^2 - \gamma^2,$$
$$u = \alpha^2 + n(\alpha - \gamma)^2 - n(\alpha - \beta)^2 \qquad v = \beta^2 + n(\beta - \gamma)^2 - (\alpha - \beta)^2.$$

[153] Proc. Roy. Soc. London, 13, 1864, 110–1; Coll. Math. Papers, 1, 1894, 410–1.
[154] Jour. für Math., 98, 1885, 177–9.
[155] Jour. für Math., 71, 1870, 299–303.
[156] Nouv. Ann. Math., (2), 18, 1879, 508.

E. Cesàro[157] found various sets of solutions of

$$v^2 - v(x+y+z) + xy + yz + zx = 2w^2.$$

Réalis stated and Rochetti[158] proved that

$$2(xy+yz+zx) - (x^2+y^2+z^2) = 4n^2$$

has an infinitude of solutions. Solving for z, we are to make $xy - n^2 = \square$, whence $n^2 + c^2$ is to be expressed as a product of two factors. Or, choose p, q, r, s so that $n = pr \pm qs$; then four solutions are

$$x = p^2 + q^2, \qquad y = r^2 + s^2, \qquad z = (p \pm s)^2 + (q \pm r)^2.$$

A. Desboves[159] gave the complete solution in integers of the general homogeneous quadratic equation in n variables, when one solution x, y, $\cdots$ is given. Regard mx, my, $\cdots$ as the same solution as x, y, $\cdots$. First, let $n = 3$:

$$(5) \qquad aX^2 + bY^2 + cZ^2 + dXY + eXZ + fYZ = 0.$$

Let $X = \rho x$, $Y = \rho y + p$, $Z = \rho z + q$. Then (5) gives ρ as a rational function of p, q, x, y, z, so that

$$\begin{aligned}
X &= -(bp^2 + cq^2 + fpq)x, \\
(6) \qquad Y &= (dx + by + fz)p^2 - cyq^2 + (ex + 2cz)pq, \\
Z &= -bzp^2 + (ex + fy + cz)q^2 + (dx + 2by)pq.
\end{aligned}$$

This is the general solution of (5), since we can find p, q such that (6) becomes any assigned solution. A convenient modification (pp. 233–5) of the method of Gauss[147] leads to (6). Special cases of (6) have been noted above (Desboves[38]).

For any n, set $X = \rho x + r$, $Y = \rho y + p$, $\cdots$ in the proposed equation $F(X, Y, \cdots) = 0$. We get ρ and then

$$X = Mr - Nx, \qquad Y = Mp - Ny, \qquad Z = Mq - Nz, \cdots,$$

$$N = F(r, p, q, \cdots), \qquad M = x\frac{\partial F}{\partial r} + y\frac{\partial F}{\partial p} + \cdots.$$

The results are no more general than those for the case $r = 0$.

A. Meyer[160] gave criteria for the solvability in integers of

$$(7) \qquad ax^2 + by^2 + cz^2 + du^2 = 0,$$

where a, b, c, d are integers not zero without square factors and such that no three have a common factor. Write (a, b) for the positive g.c.d. of a, b, and set

$$\begin{aligned}
a &= (a, b)(a, c)(a, d)\alpha, & b &= (b, a)(b, c)(b, d)\beta, \\
c &= (c, a)(c, b)(c, d)\gamma, & d &= (d, a)(d, b)(d, c)\delta.
\end{aligned}$$

Then necessary conditions for the solvability of (7) in integers not all zero are (I) a, $\cdots$, d are not all of the same sign, and (II) $-(a, c)(a, d)(b, c)(b, d)\gamma\delta$ is a quadratic residue of (a, b), with five similar conditions derived by permuting the letters. Again, (7) is solvable if and only if conditions (I)

---

[157] Nouv. Corresp. Math., 6, 1880, 273.

[158] Mathesis, (1), 1, 1881, 165.

[159] Nouv. Ann. Math., (3), 3, 1884, 225–39.

[160] Vierteljahrsschrift Naturforsch. Gesell. Zürich, 29, 1884, 209–222.

and (II) hold and either $abcd \equiv 2, 3, 5, 6, 7 \pmod 8$; or $abcd \equiv 1$ and $a+b+c+d \equiv 0 \pmod 8$; or $abcd \equiv 4 \pmod 8$ and, if a and b are even and c and d odd, either $\frac{1}{4}abcd \equiv 3, 5, 7 \pmod 8$, or $\frac{1}{4}abcd \equiv 1 \pmod 8$ and

$$\frac{a}{2} + \frac{b}{2} + c + d \equiv \frac{(cd)^2 - 1}{2} \pmod 8.$$

He gave necessary and sufficient conditions for integral solutions of $f = 0$, where f is any quaternary quadratic form.   Finally,

$$ax^2 + by^2 + cz^2 + du^2 + ev^2 = 0$$

is solvable in integers not all zero if the coefficients are odd and not all of the same sign.

H. Minkowski[161] defined an invariant D in terms of the prime factors of the determinant of the quadratic form and proved that zero can be represented rationally by every indefinite quadratic form in 5 or more variables, by one in 4 variables if D is not divisible by the square of a prime, by one in 3 variables if $D = 1$, and by one in two variables if $D = -1$.

G. de Longchamps[162] would solve $x^2 \Sigma \lambda_i = \Sigma \lambda_i y_i^2$ by choosing integers $x, \alpha_1, \cdots, \alpha_n$ for which $\Sigma \lambda_i \alpha_i^2 = 2x \Sigma \lambda_i \alpha_i$ (for example, by taking $\alpha_1, \cdots, \alpha_{n-2}$ arbitrary even integers and choosing α_{n-1} so that $\Sigma_{i=1}^{i=n-1} \lambda_i \alpha_i - 2$ is divisible by λ_n, and taking α_n to be the quotient) and then finding $y_1, \cdots, y_n$ from $x - y_i = \alpha_i$.   Application is made to $nx^2 = y^2 + (n-1)z^2$ and to

$$x^2 - xy + y^2 = z^2.$$

The discriminant of the latter equation in x is $y^2 - 4(y^2 - z^2)$, which must be a square k^2; whence a solution is $z = 7, y = 5, k = 11, x = 8$.

P. Bachmann[163] proved Meyer's[160] theorems.

A. Meyer[164] discussed the solution of $p^2 - \Omega F(q, q', q'') = \epsilon$ [cf. Bachmann[155]].   For this and the next paper, see the chapter on quadratic forms.

G. Humbert[165] treated the integral solutions of $x^2 - 4yz - 4tu = A$.

Anonymous writers[166] stated that $x^2 + y^2 - z^2 = 2u^2$ has the solutions

$$x = 2ak(c^2 - ak), \quad y = c^2(c^2 - 4ab), \quad z = \{c^2 - 2a(a+b)\}^2 - 2a^2(a^2 - 2b^2),$$
$$u = 2ac(c^2 - ak), \quad k \equiv a + 2b.$$

Or we may compare the known solutions of $y^2 - z^2 = h^2$, $x^2 + h^2 = 2u^2$ and take $h = a^2 - b^2 = 2m^2 - l^2$; hence an infinitude of solutions can be found from one.

A. Gérardin[167] stated that $x^2 + hy^2 = z^2 + ht^2$ has the solutions (Réalis[156])

$$x = m^2 + n^2 + hp^2 - 2m(n+hp), \qquad y = n^2 + hp^2 - m^2,$$
$$z = m^2 + n^2 - hp^2 + 2n(hp - m), \qquad t = m^2 + hp^2 - n^2 + 2p(n-m);$$
$$x = n^2 + hp^2 - hm^2, \qquad y = n^2 + hp^2 + hm^2 - 2m(n+hp),$$
$$z = n^2 - hp^2 + hm^2 + 2hn(p - m), \qquad t = hp^2 + hm^2 - n^2 + 2p(n-hm).$$

[161] Jour. für Math., 106, 1890, 14. Gesamm. Abhandl., I, 227.
[162] El Progreso Mat., 4, 1894, 40-7; Jour. de math. élém., 18, 1894, 5.
[163] Arith. der Quad. Formen, 1898, 259-266, 553.
[164] Jour. für Math., 116, 1896, 321.
[165] Jour. de Math., (5), 9, 1903, 43.
[166] Sphinx-Oedipe, 1907-8, 30, 95-6.
[167] *Ibid.*, 107-9.

F. Ferrari[168] noted that the solution of $x_0^2 + A_1 x_1^2 + \cdots + A_n x_n^2 = x_{n+1}^2$ reduces to the solution of $\Sigma x_i^2 = x_{k+1}^2$.

O. Degel[169] noted that, if x, y, z, s are homogeneous coördinates, the surface

$$x^2 + my^2 + nz^2 + 2ayz + 2bxy + 2cxz = s^2$$

can be represented on a plane (Clebsch, Jour. für Math., 65, 1866, 380) by

$$\xi_1 = \rho x - \sigma, \qquad \xi_2 = \rho y, \qquad \xi_3 = \rho z, \qquad \xi_4 = \rho s - \sigma.$$

Take $\xi_4 = 0$ as the plane and set $\rho s = \sigma$ in the initial equation. We get σ rationally. Hence $\rho x, \cdots, \rho s$ are expressed as homogeneous quadratic functions of ξ_1, ξ_2, ξ_3. By the same method he[170] treated

$$x^2 + y^2 + z^2 - 2yz - 2zx - 2xy = s^2$$

and found

$$\rho x = (u+2)(u+v), \qquad \rho y = uv, \qquad \rho z = 2u, \qquad \rho s = 2v - u^2.$$

Several writers (pp. 164–6) gave solutions.

A. Gérardin[171] found m from $(1+ma)^2 + (1+mb)^2 - (mc)^2 = 2$, whence

$$(c^2 + 2ab + a^2 - b^2)^2 + (c^2 + 2ab + b^2 - a^2)^2 - \{2c(a+b)\}^2 = 2(a^2 + b^2 - c^2)^2.$$

He noted (p. 22) the identity

$$(g^2 - f^2)^2 + (g^2 - 2fg)^2 + (f^2 - 2fg)^2 = 2(f^2 - fg + g^2)^2.$$

Also $f^2 - fg + g^2 = k^2$ for $f = p^2 + 2pq - 3q^2$, $g = 4pq$, $k = p^2 + 3q^2$.

Gérardin[172] gave solutions of cases of

$$x^2 + 2(by + cz)x + my^2 + 2ayz + nz^2 = \square .$$

O. Degel[173] stated that all solutions of $11x^2 = y^2 - 3z^2 - w^2 + 2u^2 + 2s^2 + 10t^2$ are given by $\rho x = A + 2aB$, $\rho y = A + 2bB$, $\rho z = A + 2cB$, $\rho w = A + 2dB$, $\rho u = A + 2eB$, $\rho s = A + 2fB$, $\rho t = A$, where $a, \cdots, f$ are distinct and $\neq 0$, and

$$A = 11a^2 - b^2 + 3c^2 + d^2 - 2e^2 - 2f^2, \qquad B = -11a + b - 3c - d + 2e + 2f.$$

" V. G. Tariste "[174] noted that, if $\alpha_1, \cdots, \alpha_n$ is one set of integral solutions of $m_1 x_1^2 + \cdots + m_n x_n^2 = 0$, all solutions are given by

$$x_k = M\left\{ -\alpha_k \sum_{i=1}^{n} m_i \alpha_i'^2 + 2\alpha_k' \sum_{i=1}^{n} m_i \alpha_i \alpha_i' \right\},$$

where the α' are any rational numbers and M is such that the x's are integers. Gérardin (pp. 136–7) remarked that this result follows by taking $x_i = \alpha_i + m\alpha_i'$ $(i = 1, \cdots, k)$.

L. Aubry[175] discussed the integral solutions of $x_1 y_1 + \cdots + x_n y_n = 0$.

W. Mantel[176] treated $xy + xz + yz = N$.

For $x^2 + y^2 = 2a^2 + 2b^2$, see papers 83–87 of Ch. V.

On $\Sigma(x_i^2 + x_i) = g$, see Bachet[1a] of Ch. VIII. On $\Sigma x_i^2 - \Sigma y_i^2 = g$, see Tano[207] of Ch. XII. On $x^2 + 3y^2 = u^2 + 3v^2$, see papers 201 and 211 of Ch. XXII.

---

[168] Suppl. al Periodico di Mat., 11, 1908, 129–131.
[169] L'intermédiaire des math., 15, 1908, 151–2.
[170] Ibid., 16, 1909, 167.
[171] Sphinx-Oedipe, 6, 1911, 74–5.
[172] L'intermédiaire des math., 18, 1911, 202–3.
[173] Ibid., 20, 1913, 226.
[174] Ibid., 21, 1914, 49.
[175] Ibid., 23, 1916, 133–4.
[176] Wiskundige Opgaven, 11, 1914, 448–90.

CHAPTER XIV.

SQUARES IN ARITHMETICAL OR GEOMETRICAL PROGRESSION.

THREE SQUARES IN ARITHMETICAL PROGRESSION, $x^2+z^2=2y^2$.

This topic is closely connected with congruent numbers, Ch. XVI, especially papers 41, 67, 68, 120, 141. It may be stated in terms of triangular numbers (Ch. I[179]).

Diophantus, III, 9, used three special squares in A. P. (see Ch. XV).

Jordanus Nemorarius[1] in the thirteenth century found that

$$r=b^2-c^2/2, \qquad v=b^2+bc+c^2/2, \qquad q=b^2+2bc+c^2/2$$

make $v^2-r^2=q^2-v^2=2b^3c+3b^2c^2+bc^3$. Here b is any integer, c any even integer. In his notations, set $a=b+c$, $d=a+b$, $h=ac$, $k=bc$, $e=ad$, $f=bd$. Then $e=h+k+f$, and a solution is $v=(h+f)/2$, $r=f-v$, $q=e-v$.

Regiomontanus,[2] or Johann Müller (1436–1476), proposed in letters the problems: Find 3 squares in A. P., the sum of whose integral roots is 214; find 3 squares in A. P., the least being >20000; find 3 squares in harmonical progression.

F. Vieta[3] took A^2, $(A+B)^2$ and $(D-A)^2$ as the squares. Hence

$$(D-A)^2=A^2+4AB+2B^2; \qquad A=\frac{D^2-2B^2}{4B+2D}.$$

Hence we may take D^2-2B^2, D^2+2B^2+2BD and D^2+2B^2+4BD as the sides of the three squares.

Fermat[4] proposed to St. Croix, Sept., 1636, that he find three squares in A. P. the common difference being a square.

Fermat[5] knew a rule for finding three numbers whose squares are in A. P. Apparently the numbers were r^2-2s^2, $r^2+2rs+2s^2$, $r^2+4rs+2s^2$. Replacing r by $p-q$ and s by q, we obtain Frenicle's set $p^2-2pq-q^2$, p^2+q^2, $p^2+2pq-q^2$. To derive the latter, Frenicle[6] noted that the squares of $a-b$, c, $a+b$ are in A. P. if $a^2+b^2=c^2$, and took $a=p^2-q^2$, $b=2pq$, $c=p^2+q^2$.

L. Euler[7] deduced from the solution $y=1$ of $1+x^2=2y^4$ the solution $y=13$, etc. Cf. Cunningham[79] of Ch. XX.

To find three integers the sum of any two of which is a square and whose squares are in A. P., "Amicus"[8] took $2a^2b^2\pm(a^4-b^4)$, a^4+b^4 as the

---

[1] Elementa Arithmetica decem libris, demostrationibus Jacobi Fabri Stapulensis, Paris, 1496, 1514, Book 6, Theorem 12.

[2] C. T. de Murr, Memorabilia Bibl. publ. Norimbergensium, Pars I, 1786, 145, 159, 201. Cf. M. Cantor, Geschichte der Math., II, 1892, 241, 263.

[3] Zetetica, 1591, V, 2; Opera Math., 1646, 76. Same by J. Prestet, Elemens des Math., ou Principes . . ., Paris, 1675, 326.

[4] Oeuvres, II, 65; III, 287.

[5] Oeuvres, II, 234; letter from Frenicle to Fermat, Sept. 6, 1641 (tables by Frenicle, p. 237).

[6] Triangles rectangles en nombres, prop. XI. Full reference in Ch. IV.[52]

[7] Algebra, 2, 1770, Ch. 9, Art. 140; French transl., 2, 1774, p. 167; Opera Omnia, (1), I, 402.

[8] Ladies' Diary, 1795, 38, Quest. 974; Leybourn's Math. Quest. from L. D., 3, 1817, 297.

numbers. Their squares are in A. P. and their sums by twos will be squares if $2a^2+2b^2=\square$, which is known to hold if $a, b=2mn\pm(m^2-n^2)$. The same problem was treated by A. Cunningham and F. Phillips.[9] A. E. Jones[10] started with any three numbers

$$x=-m^2+2mn+n^2, \qquad y=m^2+n^2, \qquad z=m^2+2mn-n^2$$

whose squares are in A. P., and called P, Q, R the values obtained from them by replacing m by x^2, n by z^2. Then P, Q, R are the desired numbers since

$$P+Q=2z^2(x^2+z^2)=4z^2y^2, \qquad P+R=4x^2z^2, \qquad Q+R=2x^2(x^2+z^2)=4x^2y^2.$$

C. Campbell[11] treated the similar problem to find three numbers x, y, z the difference of any two of which is a square and whose squares are in A. P. Let $x-y=m^2$, $x-z=n^2$, $y-z=p^2$. Then $n^2-p^2=m^2$. Take $n+p=ms$, $n-p=m/s$. Since $x^2+z^2=2y^2$ gives x, we get y and z in terms of m, n.

J. Cunliffe[12] treated the problem to find 3 squares in A. P. such that the sum of each and its root shall be a square.

J. Wright[12a] found three squares x^2, y^2, z^2 in harmonical progression such that each exceeds its root by a square. If $a, b=2rs\pm(r^2-s^2), c=r^2+s^2$, $a^2+b^2=2c^2$ and the squares of $x=n^2/d, y=bn^2/(cd), z=bn^2/(ad)$ are in harmonical progression. For $d=m(2n-m)$, $x^2-x=\square$. Also, $y^2-y=\square$ if $b^2n^2-bcd=\square=(bn-pm)^2$, which holds if $m=2bn(c-p)/(bc-p^2)$. Then $z^2-z=\square$ if $(bc-p^2)^2-4ap(b-p)(c-p)=\square=(bc+2ap-p^2)^2$, which gives $p=2bc/(b+c-a)$.

J. Ivory[12b] found two sets a^2, b^2, c^2 and a_1^2, b_1^2, c_1^2 of three squares in A. P. having the same sum. The conditions are $a^2+c^2=2b^2=2b_1^2=a_1^2+c_1^2$, or $4b^2=\Sigma(a\pm c)^2=\Sigma(a_1\pm c_1)^2$. Hence we require a square which is a sum of two squares in two ways. The least numbers are obtained from

$$25^2=7^2+24^2=15^2+20^2.$$

To find three numbers whose sum is 117 and whose squares are in A. P., S. Jones[13] took $x, 5x, 7x$ as the numbers, whence $x=9$. S. Ryley took $2mn\pm(m^2-n^2)$, m^2+n^2 as the numbers. Then $(n+2m)^2=117+3m^2=\square$ for $m=3$; the resulting numbers 9, 45, 63 are said to give the only solution in positive integers.

R. Adrain[14] used the squares $u^2-y=(u-p)^2$, u^2, $u^2+y=(u+q)^2$, whence $2pu-y=p^2$, $y-2qu=q^2$. Solving, we get $u=(p^2+q^2)/\{2(p-q)\}$. There results Frenicle's[5] solution.

J. Surtees[15] noted that $(a-n)^2$, a^2+n^2, $(a+n)^2$ are in A. P. and $a^2+n^2=\square$ if $a=r^2-1$, $n=2r$.

J. R. Young[16] found three squares in A. P. such that the roots increased

[9] Math. Quest. Educ. Times, 24, 1913, 107.
[10] Math. Quest. and Solutions, 5, 1918, 62–3.
[11] The Gentleman's Diary, or Math. Repository, London, No. 65, 1805, 40–1, Quest. 873.
[12] Math. Repository (ed., Leybourn), London, 3, 1804, 97–106, Prob. 7.
[12a] New Series of Math. Repository (ed., T. Leybourn), 1, 1806, I, 99.
[12b] Ibid., 121–3.
[13] The Gentleman's Math. Companion, London, 2, No. 9, 1806, 15–17.
[14] The Math. Correspondent, New York, 2, 1807, 14.
[15] Ladies' Diary, 1811, 39, Quest. 1217; Leybourn's Math. Quest. L. D., 4, 1817, 139.
[16] Algebra, 1816; Amer. ed., 1832, 333–4 (329–31).

by 2 give squares, the sum of the first and third of which is also a square. Take $q=1$ in Frenicle's set; we get p^2+2p-1, p^2+1, p^2-2p-1. Hence the conditions are $p^2+3=\square$, $2p^2+2=\square$. Set $p=m+1$, and let the second equal $(nm+2)^2$, whence $m=4(1-n)/(n^2-2)$. Then

$$(p^2+3)(n^2-2)^2/4=n^4-2n^3+2n^2-4n+4=(n^2-n+\tfrac{1}{2})^2$$

if $n=5/4$, whence $p=23/7$.

To find three squares in A. P. such that any root plus unity is a square, H. Clay[17] took x^2, a^2x^2, b^2x^2 as the squares. Set $x+1=(r+1)^2$. Then $ax+1=(sr+1)^2$ determines r. Then $bx+1=\square$ if a certain quartic in s is a square, which is the case if $s=(2pq-4b)/(q^2-1)$. Finally, choose a and b so that 1, a^2, b^2 are in A. P. A. B. Evans[18] took $a=5$, $b=7$ and proceeded similarly. S. Bills[19] employed the numbers a, $b=2pq\pm(p^2-q^2)$; $c=p^2+q^2$, whose squares are in A. P., and took ax/y^2, bx/y^2, cx/y^2 as the roots of the required three squares. Then $ax+y^2=(r+y)^2$, $bx+y^2=(s+y)^2$ determine x, y. Then $cx+y^2=\square$ if a quartic in r is a square, which is the case if $r=s(a+b-c)/(2b)$. W. J. Miller[19] called the numbers x, y, z and set $x+1=m^2$, $y+1=n^2$, $z+1=p^2$, $m+n=r(n+p)$, $m-n=s(n-p)$, whence

$$\frac{m}{r-s+2rs}=\frac{n}{r+s}=\frac{p}{2-r+s}=\frac{1}{k}.$$

Then $x^2+z^2=2y^2$ reduces to $k^2=f(r,s)$, which is solved. D. T. Griffiths[20] took x^2-1, y^2-1, z^2-1 as the numbers. Their squares are in A. P. if $x^2+y^2-2=a(y^2+z^2-2)$, $y^2-z^2=a(x^2-y^2)$. Taking $a=1/2$ (the value when the squares are 1, 5^2, 7^2), and eliminating z, we get $5x^2-y^2=4$. This holds if $x=5$, $y=11$, whence $z=13$.

To find three squares in A. P. such that each less its root is a square, Smyth[21] took a^2x^2, b^2x^2, c^2x^2, $p=1/a$, etc. Then x^2-px, $\cdots$ are made squares in the usual way. " Epsilon " used the numbers $1/X$, a/X, b/X, where $X=2x-x^2$ and where 1, a^2, b^2 are in A. P. Now $1/X^2-1/X=\square$. Again, $t^2-t=\square$ if $t=(k+l)^2/(4kl)$. It is shown that a/X and b/X are of the latter form if

$$\frac{1}{X}=\frac{\{4ab-(ab-a-b)^2\}^2}{8ab(ab+b-a)(ab+a-b)(a+b-ab)}.$$

To find three squares in harmonical progression the sum of whose roots is a given biquadrate d^4, " Epsilon "[22] took a, $c=2mn\pm(m^2-n^2)$ and $b=m^2+n^2$. Then the squares of h/a, h/b, h/c are in harmonical progression; equating their sum to d^4, we get h.

A. Guibert[23] noted that the common difference of 3 squares in A. P. is a multiple of 24, and similar theorems. The general solution in positive relatively prime integers of $a^2+c^2=2b^2$ is stated to be

[17] The Gentleman's Math. Companion, London, 5, No. 25, 1822, 151-4.
[18] Math. Quest. Educ. Times, 16, 1872, 27-28.
[19] Ibid., 11, 1869, 88-91.
[20] Ibid., 63, 1895, 46-7.
[21] The Gentleman's Math. Companion, London, 5, No. 26, 1823, 214-8.
[22] The Gentleman's Math. Companion, London, 5, No. 28, 1825, 365-6.
[23] Nouv. Ann. Math., (2), 1, 1862, 213-9.

$$a = \pm(p^2 - q^2 - 2pq), \qquad b = p^2 + q^2, \qquad c = p^2 - q^2 + 2pq,$$

where p, q are relatively prime and one even. Extending the A. P., he proved that the nth term is a square if $q = 1$, $2p = n - 2$, or $q = 2$, $p = n - 2$.

"Civis"[24] proved that the common difference of three rational squares in A. P. is never 17. For, if so, $4ab(a^2 - b^2) = 17q^2$. Put $a = r^2$, $b = s^2$, $r^2 - s^2 = 8v^2$, whence $r = 2v^2 + 1$, $s = 2v^2 - 1$. Put $u = q/(8rsv)$. Then $17u^2 - 1 = 4v^4$, which is impossible in view of the formula for z in the known solution of $17u^2 - 1 = z^2$. A. Martin[25] noted that the theorem is evident for integers since a multiple of 4 cannot equal 17.

To find[26] three squares in A. P. such that each exceeds its root by a square, employ Frenicle's numbers (say l, m, n), and take lx, mx, nx as the roots of the required squares. Then $l^2x^2 - lx = \square$, etc., are solved as in Ch. XVIII.

D. André[27] noted that, if three squares are in A.P.,

$$2y^2 = x^2 + z^2, \qquad y^2 = \left(\frac{x+z}{2}\right)^2 + \left(\frac{x-z}{2}\right)^2 \equiv a^2 + c^2, \qquad x = a + c, \qquad z = a - c.$$

G. R. Perkins[28] treated the problems 1 [2]: Find three squares in A. P. such that each less [plus] its roots is a square. Take the numbers to be the squares of $\xi \pm \frac{1}{2}$, $\eta \pm \frac{1}{2}$, $\zeta \pm \frac{1}{2}$, where $4\xi = x + x^{-1}$, $4\eta = y + y^{-1}$, $4\zeta = z + z^{-1}$, and the signs are $+$ or $-$ according as the problem is 1 or 2. Then each square $\pm$ its root is a square. The squares are in A. P. if

$$\eta + \xi \pm 1 = m, \qquad \eta - \xi = n, \qquad \zeta + \eta \pm 1 = \frac{m(p+1)}{p}, \qquad \zeta - \eta = \frac{np}{p+1}.$$

These give ξ, η, ζ and $n = 2m(p+1)/N$, where $N \equiv 2p(2p+1)$. The desired numbers are the squares of $\xi \pm \frac{1}{2} = m/a$, m/b, m/c, where

$$(2p^2 - 1)a = (2p^2 + 2p + 1)b = (2p^2 + 4p + 1)c = N.$$

It remains to make x, y, z rational, using $4\xi = x + x^{-1}$, etc. This requires that $m^2 \mp tm$ be a square for $t = a, b, c$. Now

$$m^2 \mp am = (m \mp k)^2 \qquad \text{if} \qquad m = \frac{k^2}{\pm(2k - a)}.$$

Then $m^2 \mp bm = \square$ if $k^2 - b(2k - a) = \square$, say $(k - l)^2$, whence k is a rational function of l. Then $k^2 - c(2k - a) = \square$ if $l = (a + b - c)/2$. For Prob. 1, $p > 2$; if $p = 3$, m/a, $\cdots$, m/c are quotients of numbers of 14 digits [cf. Hart[5] of Ch. XVIII]. Three times as many digits are involved in the answer by D. Kirkwood[29] who started with x^2, $25x^2$, $49x^2$. For Prob. 2, the use of $p = 1$ gives the answers due to Williams[6] of Ch. XVIII.

[24] The Lady's and Gentleman's Diary, London, 1866, 56–7, Quest. 2041.
[25] Math. Quest. Educ. Times, 52, 1890, 87.
[26] Ibid., 14, 1871, 54. A Collection of Diophantine Problems by J. Matteson, pub. by A. Martin, Washington, D. C., 1888, § 10, pp. 14–16.
[27] Nouv. Ann. Math., (2), 10, 1871, 295–7.
[28] The Analyst, 1, 1874, 101–5.
[29] Stoddard and Henkle, University Algebra, N. Y., 1861, p. 494.

A. Cunningham[30] investigated the sets of three numbers <10000 whose squares are in A. P. the ratio of the greatest to the least being as great (or as small) as possible.

W. A. Whitworth[31] noted that if three squares without a common factor are in A. P., the middle one is $\equiv 1$, 25 or 49 (mod 120) and each of the others is $\equiv 1$ or 49 (mod 240).

J. Neuberg[32] and J. Déprez[33] investigated "automédian" triangles, viz., those whose medians are proportional to the sides a, b, c. If $a>b>c$, the condition is $a^2+c^2=2b^2$.

G. Bisconcini[34] noted that, if A is the common difference of three squares x_i^2 in A. P., then $x_2^2-x_1^2=A$, $x_3^2-x_1^2=2A$. By the latter, x_1, $x_3=(2A\mp\lambda^2)/(2\lambda)$. Thus $\lambda=2a_2$, $A=2a_1a_2$, $x_1=a_1-a_2$, $x_3=a_1+a_2$. By the first condition, $x_2^2=a_1^2+a_2^2$. It is stated [incorrectly[35]] that $a_1=r^2-s^2$, $a_2=2rs$, whence $A=4rs(r^2-s^2)$, which he called a number of Fibonacci.

C. Botto[35] noted the incompleteness of the solution by Bisconcini. To obtain all relatively prime solutions of $x^2+y^2=2z^2$, note that x and y are odd, and set $p=(x+y)/2$, $q=(x-y)/2$. Then $p^2+q^2=z^2$. Since p and q are relatively prime, p, $q=u^2-v^2$, $2uv$, and $z=u^2+v^2$. The same substitution reduces $x^2-y^2=2z^2$ to $2pq=z^2$, whence p, $q=a^2$, $2b^2$ and $z=2ab$.

G. Métrod[36] noted that $u^2-2v^2=-x^2$ has the solutions

$$u=u_n(a^2+2b^2)+4v_nab, \qquad v=2u_nab+v_n(a^2+2b^2), \qquad u_n^2-2v_n^2=-1.$$

E. Turrière[37] noted that the sides of an automédian triangle are

$$a=\lambda(1+2t-t^2), \qquad b=\lambda(1+t^2), \qquad c=\lambda(1-2t-t^2).$$

A. Gérardin[38] noted that for the automédian triangle with the sides 31, 41, 49, the sum of the sides is a square 121. J. Rose[39] noted that by Turrière's formula, $a+b+c=\lambda(3-t^2)$ becomes a square by choice of λ.

R. Goormaghtigh[40] restricted the last problem to relatively prime integral sides, whence these are the absolute values of $\alpha^2-\beta^2\pm2\alpha\beta$, $\alpha^2+\beta^2$. The perimeter is a square if $\alpha^2+\beta^2+4\alpha\beta=u^2$, whence $\alpha+2\beta=v$, $v^2=3\beta^2+u^2$. Thus $\beta=pq$, $v=(3p^2+q^2)/2$.

See papers 15, 35, 62 of Ch. XV; 20 of Ch. XVII; 5, 8, 16 of Ch. XVIII; 7, 8, 48, 49, 57, 114, 143 of Ch. XIX; 11 of Ch. XXII. On $x^2+1=2y^2$, see papers 112-129 of Ch. IV; 154, 188, 215, 234, 298 of Ch. XII; 92 of Ch. XXIII.

Papers without novelty on $x^2+z^2=2y^2$.

A. Boutin, Jour. de math. élém., (4), 4 [19], 1895, 12 [Vieta³].
Plakhowo, ibid., (5), 21, 1897, 95 [Frenicle⁵].

---

[30] Math. Quest. Educ. Times, 71, 1899, 56.
[31] Ibid., 72, 1900, 98.
[32] Mathesis, 9, 1889, 261-4; (3), 1, 1901, 280.
[33] Mathesis, (3), 3, 1903, 196-200, 226-30, 245-8.
[34] Periodico di Mat., 24, 1909, 157-70.
[35] Ibid., 232-4.
[36] Sphinx-Oedipe, 8, 1913, 130-1.
[37] L'enseignement math., 18, 1916, 87-8.
[38] L'intermédiaire des math., 23, 1916, 173.
[39] Ibid., 24, 1917, 20-22.
[40] Ibid., 88-90.

H. S. Vandiver, Amer. Math. Monthly, 9, 1902, 79–80; others, 7, 1900, 82–3, 112–3.
A. Gérardin, Sphinx-Oedipe, 1906–7, 95, 161–2 [Vieta,[3] bibliography].
F. Ferrari, Suppl. al Periodico di Mat., 11, 1908, 77–8 [Frenicle[5]].
A. Gérardin, Assoc. franç., 1908, 15–17 [bibliography].
A. Tafelmacher, l'intermédiaire des math., 15, 1908, 102, 259.
Welsch, *ibid.*, 16, 1909, 19, 156 [no novelty in authors cited].
A. Martin, Amer. Math. Monthly, 25, 1918, 124.
E. Bahier, Recherche . . . Triangles Rectangles en Nombres Entiers, 1916, 212–7.

Four squares in arithmetical progression.

Fermat[41] proposed the problem to Frenicle May (?), 1640 and stated (Fermat[11] of Ch. XV) that it is impossible. Euler[109] of Ch. XXII, P. Barlow,[42] and M. Collins[43] proved the problem is impossible.

B. Bronwin and J. Furnass[43a] took relatively prime squares x^2, y^2, z^2, w^2. By $y^2-x^2=z^2-y^2=w^2-z^2$, we must have $y+x=2ab$, $y-x=2cd$, $z+y=2ac$, $z-y=2bd$, $w+z=2bc$, $w-z=2ad$. By the two values of y and those of z, $(a+d)b=(a-d)c$, $(c+d)a=b(c-d)$. But the g.c.d. of the four numbers $a\pm d$, $c\pm d$ is 1 or 2. Hence $a+d=\delta c$, $a-d=\delta b$, $c+d=\epsilon b$, $c-d=\epsilon a$, $\delta=1$ or 2, $\epsilon=1$ or 2. These are inconsistent since a is prime to d.

A. Genocchi[44] proved the impossibility of 4 squares in A. P. and the following generalization (of the case $p=2$). The four expressions $x\mp(p+1)y$ and $x\mp(p-1)y$ are not all squares if p is a prime $8m\pm3$ such that $p+1$ and $p-1$ admit no prime divisor $4m+1$, and x,y are relatively prime.

Several writers[45] failed to find a solution.

L. Aubry[46] proved by descent the impossibility of 4 squares in A. P.

E. Turrière[47] gave a proof.

Numbers in arithmetical progression all but one being squares.

A. Guibert[48] noted that if A^2, B^2, C, D^2 (all but C being squares) are in A. P., they are the products by a square of a similar progression of odd integers relatively prime by twos. From the conditions $A^2+C=2B^2$, $B^2+D^2=2C$, eliminate C. Then $D^2=3B^2-2A^2$. The known method of solution gives

$$A=2p^2-2pq-q^2, \qquad b=2p^2+q^2, \qquad d=2p^2+4pq-q^2.$$

A. Cunningham[49] found five integers in A. P., four being squares. If v^2, w^2, X, y^2, z^2 are in A. P., $v^2+3z^2=(2y)^2$, $3v^2+z^2=(2w)^2$, which require that the five numbers be equal (Collins,[43] pp. 17–23). Next, let all but the fourth

---

[41] Oeuvres, II, 195.
[42] Theory of Numbers, 1811, 257.
[43] A Tract on the possible and impossible cases of quadratic duplicate equalities . . ., Dublin, 1858, 16. Abstract in British Assoc. Reports for 1855, 1856, Trans. of Sections, 4. The Ladies' and Gentleman's Diary, London, 1857, 92–6.
[43a] The Gentleman's Diary, or Math. Repository, London, No. 73, 1813, 42–43.
[44] Comptes Rendus Paris, 78, 1874, 433–5.
[45] Amer. Math. Monthly, 5, 1898, 180.
[46] Sphinx-Oedipe, 6, 1911, 1–2.
[47] L'enseignement math., 19, 1917, 240–1.
[48] Nouv. Ann. Math., (2), 1, 1862, 249–252. Cf. Pocklington[83] of Ch.
[49] Math. Quest. Educ. Times, (2), 9, 1906, 107–8.

be squares, the first three being v^2, w^2, x^2. As known, v, $x = t^2 - u^2 \mp 2tu$, $w = t^2 + u^2$. Since the common difference of these squares is $\delta = 4tu(t^2 - u^2)$, the fifth number is $w^2 + 3\delta = z^2$. This has an infinitude of solutions t, u, z derivable in succession from the minimum solution. From the solution 7^2, 13^2, 17^2, 409, 23^2, there are deduced two solutions in much larger integers.

Squares in geometrical progression.

Beha-Eddin[50] (1547–1622) included (as Prob. 6) among the 7 problems remaining unsolved from former times: Find 3 squares in G. P. whose sum is a square. Nesselmann noted that the problem is impossible since $x^2 + x^2y^2 + x^2y^4 = \square$ has no rational solution [Adrain,[113] Anderson,[114] Genocchi,[119] Pocklington[138] of Ch. XXII].

To find three squares in G. P. and three numbers in A. P. such that the three sums of corresponding terms are squares, W. Saint[51] took a^2, a^2x^2, a^2x^4 as the squares in G. P. and $2a+1$, ax^2+a+1, $2ax^2+1$ as the numbers in A. P. It suffices to make $a^2x^2 + ax^2 + a + 1 = \square = (ax + x/2)^2$, say, whence $a = \frac{1}{4}x^2 - 1$. Others took x^2, $4x^2$, $16x^2$ and either 1, $4x+1$, $8x+1$ or $2ax + a^2$, $8ax + 4a^2$, $14ax - 7a^2$.

W. Wright[52] found three squares x^2, a^2x^2, a^4x^2 in G. P. each plus its root being a square. Thus $x^2 + x = \square$, $x^2 + x/a = \square$, $x^2 + x/a^2 = \square$, which are satisfied in the usual way (Ch. XVIII).

To find three squares in G. P. each less its root being a square, J. Anderson[53] took x^2, xy, y^2 as the roots and $x^2 - 1 = (p-x)^2$, $y^2 - 1 = (q-y)^2$, which give x, y. Then $x^2y^2 - xy = \square$ leads to a quartic in p which is solved as usual. Isaac Newton (l. c.) took $\{r^2/(2r-1)\}^2$, r^2, $(2r-1)^2$ as the numbers. The first of the three conditions is satisfied identically. Take $r^2 - r = n^2r^2$, whence $r = 1/(1-n^2)$. Then $(2r-1)^2 - (2r-1) = \square$ if $2n^2 + 2 = \square$. Set $n = m+1$. Then $2n^2 + 2 = (sm+2)^2$ determines m.

S. Ward[54] found three squares x^2, $4x^2$, $16x^2$ in G. P., such that if any one of them is increased by its root, the sum is a square. Take $x^2 + x = p^2x^2$. The remaining two conditions become $2p^2 + 2 = \square$, $p^2 + 3 = \square$, which hold[16] if $p = 23/7$.

<hr>

[50] Essenz der Rechenkunst von Mohammed Beha-eddin ben Alhossain aus Amul, arabisch u. deutsch von G. H. F. Nesselmann, Berlin, 1843, p. 56. French transl. by A. Marre, Nouv. Ann. Math., 5, 1846, 313.

[51] The Diary Companion, Suppl. to Ladies' Diary, London, 1806, 36–37.

[52] The Gentleman's Math. Companion, 5, No. 24, 1821, 41–44.

[53] Ibid., 5, No. 27, 1824, 274–7.

[54] J. R. Young's Algebra, Amer. ed., 1832, 341.

CHAPTER XV.

TWO OR MORE LINEAR FUNCTIONS MADE SQUARES.

Diophantus, II, 12, solved $x+2=\square$, $x+3=\square$ (the first instance of a " double equality ") by resolving the difference of the two linear functions into two factors in a suitable manner; here he took 4 and 1/4. Take the square of half the difference of the two factors and equate it to the smaller expression, whence $225/64=x+2$. Or equate the square of half the sum of the actors to the greater expression. To solve without using a double equation, take $x=y^2-2$ and make $x+3=y^2+1$ a square, say by equating it to $(y-4)^2$, whence $y=15/8$.

Diophantus II, 13 relates to $9-x=\square$, $21-x=\square$; while II, 14 relates to $x-n=\square$, $x-m=\square$.

Diophantus, III, 5, 6, required three numbers such that their sum is a square and the sum of any pair exceeds the third by a square. Hence the sum of the three squares is a square, as for 4, 9, 36.

Diophantus, III, 7, 8, required three numbers whose sum and sums by pairs are squares. Let the sum of all three be $(x+1)^2$, the sum of the first and second be x^2, the sum of the second and third be $(x-1)^2$. Then the sum of the first and third is $6x+1$ and equals 121 if $x=20$.

Diophantus III, 9 relates to three numbers in arithmetical progression whose sums by pairs are squares. Since x^2, $(x+1)^2$, $(x-8)^2$ are in A.P. if $x=31/10$, we seek three numbers whose sums by twos are the numbers 961, 1681, 2401 just found.

Diophantus III, 10 relates to three numbers such that the sum of any pair of them added to a given number a gives a square, and such that the sum of the three added to a gives a square. For $a=3$, take the sum of the first two to be x^2+4x+1, the sum of the last two to be x^2+6x+6, and the sum of all three to be $x^2+8x+13$. Then the numbers are $2x+7$, x^2+2x-6, $4x+12$. The sum $6x+22$ of the first, third, and a, is the square 100 if $x=13$. In III, 11, a is negative.

Diophantus, III, 18 and IV, 35, noted that his method does not make $ax+b$ and $cx+d$ squares if $a:c$ is not the ratio of two squares.[1]

Diophantus, IV, 14, made $x+1$, $y+1$, $x+y+1$, $x-y+1$ squares.

Diophantus, IV, 22, found three numbers in G. P., the difference of any two being a square. In V, 1 [2], he found three numbers in G. P. such that each less [plus] the same given number is a square.

Diophantus, IV, 45, made $8x+4$ and $6x+4$ squares by subtraction.

Diophantus, V, 12, 14, treated the problems to divide unity into 2 or 3 parts such that, if the same given number is added to each part, the sums will be squares [see Chs. VI, VII].

---

[1] Cf. G. H. F. Nesselmann, Algebra der Griechen, 1842, 335–40. Cf. 86 of Ch. XIX.

Brahmegupta[2] (born 598 A.D.) made $ax+1$ and $bx+1$ both squares, viz., of $(3a+b)/(a-b)$ and $(a+3b)/(a-b)$, by taking $x=8(a+b)/(a-b)^2$.

He made (§§ 80–81, p. 369) $x+y$, $x-y$, $xy+1$ all squares by taking

$$x=\frac{2a^2}{b^4}(a^2+b^2), \qquad y=\frac{2a^2}{b^4}(a^2-b^2),$$

whence

$$x+y=\left(\frac{2a^2}{b^2}\right)^2, \qquad x-y=\left(\frac{2a}{b}\right)^2, \qquad xy+1=\frac{(2a^4-b^4)^2}{b^8}.$$

He made (§§ 82–85, pp. 370–1) $x+a$ and $x+b$ squares by taking

$$x=\left\{\frac{1}{2}\left(\frac{a-b}{e}+e\right)\right\}^2-a,$$

whence

$$x+b=\left\{\frac{1}{2}\left(\frac{a-b}{e}-e\right)\right\}^2.$$

To make $ax+b$ a square (§§ 86–87, pp. 371–2), put it equal to an arbitrarily assumed square and solve the equation for x.

Bháscara[3] (born 1114 A.D) made $3y+1$ and $5y+1$ squares by equating the first to $(3n+1)^2$, whence $5y+1=15n^2+10n+1=\square$ for $n=2$ or 18.

Alkarkhi[4] (beginning of eleventh century) solved $x+10=y^2$, $x+15=z^2$ by setting $z=y+2/3$ in $y^2+5=z^2$; also, $x+3=y^2$, $x+5=z^2$ by taking $z+y=4$, $-y=1/2$.

G. Gosselin[5] found three numbers (13/9, 133/9, 253/9) in A. P. which become squares when increased by 4; three numbers (1/9, 15/9, 48/9) whose sum is a square, the first a square, and the sum of the first and either of the other two is a square; four numbers (25, 16, 12, 11) whose sum is a square, while the excess of the first over the second, second over third, third over fourth are squares.

Rafael Bombelli[6] required three numbers, the sum of any two of which increased by 6 and the sum of all three increased by 6 are squares. He gave $38^4/_5$, $55^1/_5$, $14^{49}/_{100}$. He found (p. 458) a number which added to 4 and to 6 makes two squares.

F. Vieta[7] generalized the method of Diophantus III, 10 [11]. If the numbers are x, y, z, let

$$x+y=(A+B)^2-a, \qquad y+z=(A+D)^2-a, \qquad x+y+z=(A+G)^2-a.$$

Then

$$x=2AG+G^2-2AD-D^2, \qquad z=2AG+G^2-2AB-B^2,$$
$$x+z+a=4AG+2G^2-2AB-B^2-2AD-D^2=\square,$$

say F^2, by choice of a rational A.

[2] Brahme-sphut'a-sidd'hánta, Ch. 18 (Algebra), §§ 78–79. Algebra, with arith. and mensuration, from the Sanscrit of Brahmegupta and Bháscara, transl. by Colebrooke, 1817, pp. 368–9.

[3] Vija-gan'ita, § 197; Colebrooke,[2] p. 259.

[4] Extrait du Fakhri, French transl. by F. Woepcke, Paris, 1853, 86, 101.

[5] De Arte magna, seu de occulta parte numerorum, Paris, 1577, 74–5.

[6] L'algebra opera, Bologna, 1579, 496.

[7] Zetetica, 1591, V, 4[5], Francisci Vietae opera mathematica, ed. Francisci à Schooten, Lugd. Bat., 1646, p. 77.

C. G. Bachet[8] treated $ax+b=\Box$, $ax+c=\Box$, by finding two rational squares whose difference equals $b-c$. To solve $8x+4=\Box$, $6x+4=\Box$, take the double 4 of the side 2 of the common square 4, and the difference $2x$ of the left members, and one fourth of $2x$. Then the square of $\frac{1}{2}(\frac{1}{2}x+4)$ equals $8x+4$ and the square of $\frac{1}{2}(\frac{1}{2}x-4)$ equals $6x+4$. By either condition, $x=112$. Next, let the constant terms be distinct squares, as in $10x+9=\Box$, $5x+4=\Box$. Seek two numbers (5 and 1) whose sum is double the root 3 of the larger square and whose difference is double the root 2 of the smaller square. Take one of these numbers 1 and 5 as one of two factors whose product gives the difference $5x+5$ of the given functions. From $x+1$ and 5, we get

$$\left(\frac{x+6}{2}\right)^2 = 10x+9, \qquad \left(\frac{x-4}{2}\right)^2 = 5x+4, \qquad x=28.$$

But the factors $5x+5$, 1 give $\{\frac{1}{2}(5x+6)\}^2 > 10x+9$. Next, for $65-6x=\Box$, $65-24x=\Box$, multiply the first by 4 and we have a problem of the first type. For $16-x=\Box$, $16-5x=\Box$, seek two squares whose difference is the quadruple of x. Take $4-N$ as the side of the larger square. Then $(4-N)^2-4\{16-(4-N)^2\}=16-40N+5N^2$ is the smaller square, say $(4-7N)^2$, whence $N=4/11$. Thus the squares are $(40/11)^2$ and $(16/11)^2$, one fourth of whose difference gives $x=336/121$. [Bachet here used the same letter for x and N and put $4-6N$ erroneously for $4-7N$.]

Fermat,[9] commenting on Diophantus III, 10 and V, 30, desired four numbers such that the sum of any pair increased by a given number a gives a square. Let $a=15$. The three squares 9, 1/100, 529/225 are such that the sum of any pair increased by 15 gives a square (as found by Diophantus, V, 30, who took 9 as one square and solved $x^2+24=\Box$, $y^2+24=\Box$, $x^2+y^2+15=\Box$). Take as the four numbers

$$x^2-15, \qquad 6x+9, \qquad \tfrac{1}{5}x+\tfrac{1}{100}, \qquad \tfrac{46}{15}x+\tfrac{529}{225}$$

(the last three being of the form $2nx+n^2$). Then three of the conditions are satisfied identically. The remaining three conditions are

$$\tfrac{31}{5}x+(\tfrac{49}{10})^2=\Box, \qquad \tfrac{136}{15}x+(\tfrac{77}{15})^2=\Box, \qquad \tfrac{49}{15}x+(\tfrac{25}{6})^2=\Box,$$

a "triple equation" in which each constant term is a square. To treat[10] such a problem, $x+4=\Box$, $2x+4=\Box$, $5x+4=\Box$, replace x by an expression, like x^2+4x, which if increased by 4 gives a square. Then it remains only to solve the "double equation" $2x^2+8x+4=\Box$, $5x^2+20x+4=\Box$, from one solution $x=c$ of which we can deduce a second, by replacing x by $x+c$. Fermat[11] later explained this method in detail. It is stated (§§ 9–11) that the method fails for

$$(1) \qquad\qquad ax+1=\Box, \qquad bx+1=\Box, \qquad (a+b)x+1=\Box.$$

---

[8] Diophanti Alexandrini Arith., 1621, 435–9. Comment on Diop., VI, 24 (p. 177 above).

[9] Oeuvres, I, 292, 326–7; French transl., III, 242, 263–4.

[10] Oeuvres, I, 334–5; III, 269–270. Comment on Diophantus VI, 24. For further examples, Fermat[91, 100] of Ch. XIX.

[11] J. de Billy, Inventum Novum, Toulouse, 1670, Part II, §§ 1–28; German transl. by P. von Schaewen, Berlin, 1910; Oeuvres de Fermat, III, 360–374 (p. 329 for $2x+12=\Box$, $2x+5=\Box$).

Thus, if $a=2$, $b=3$, we substitute $2x^2+2x$ for x to satisfy the first identically; then the other two become $6x^2+6x+1=\square$, $10x^2+10x+1=\square$, one solution being $x=-1$; but this makes the unknown $2x^2+2x$ zero [von Schaewen[81]]. Although the method fails for $a=5$, $b=16$, $x=3$ is a solution. For $a=1$, $b=2$, there is no solution, whence four squares (the first being taken as unity) cannot be in A. P.

M. Petrus[12] found three squares A^2, B^2, C^2 such that the difference of any two is a square and the difference of the sides of any two is a square. He first gave a process to find four numbers p, s, t, q such that p^2+s^2, t^2+q^2 and $pstq$ are squares, while $p/s>t/q$, solutions being 112, 15, 35, 12 and 364, 27, 84, 13.   From the former he derived the answer to the first problem:

$$A=26633678, \qquad B=29316722, \qquad C=40606322.$$

In general, we have the answer[13]

$$\frac{B}{2}=(pt-sq)^2+(pq-st)^2, \qquad \frac{C}{2},\frac{A}{2}=(pt+sq)^2\pm(pq+st)^2,$$

since

$$C+A=4(pt+sq)^2, \qquad C-A=4(pq+st)^2, \qquad B+A=4(pt-sq)^2,$$
$$B-A=4(pq-st)^2, \qquad C-B=16ptsq, \qquad C+B=4(p^2+s^2)(t^2+q^2).$$

Renaldini[14] (1615–1698) treated Petrus'[12] initial problem (in Part II) and (in Sec. 1 of Part III) duplicate and triplicate equalities.

J. Prestet[15] treated the problem of Diophantus III, 7.   Let the sum of the first and third be x^2, that of the first and second y^2, that of all three z^2. Then the numbers are $x^2+y^2-z^2$, z^2-x^2, z^2-y^2.   The sum of the last two is not easily made a square.   Since $2=1/25+49/25$, set $x=z/5$.   Then the sum of the last two is $49z^2/25-y^2=(a-7z/5)^2$ if $z=5(a^2+y^2)/14$.   But the numbers obtained this way are larger than those of Diophantus and Vieta.

For Diophantus III, 9, he used (p. 326) z, $z+d$, $z+2d$, with $2z+d=y^2$, $2z+3d=x^2$, which give z and d.   To avoid fractions, multiply the numbers by 4.   Hence the numbers are $3y^2-x^2$, y^2+x^2, $-y^2+3x^2$.   It remains to make the sum $2y^2+2x^2$ of the first and third a square.   Express 2 as a sum of two squares, the smaller between 1/2 and 1.   By Diophantus II, 10, the root of the smaller is $(c^2-2c-1)/(c^2+1)$.   By trial, 9 is the first integer c giving a fraction $(31/41)>3/4$.   Thus $2(31^2+49^2)=82^2$.   Hence $x^2=2401$, $y^2=961$.   He gave also a less special solution.   He treated (p. 329) analogously Diophantus III, 10.

J. Ozanam[16] found two numbers, such that each when increased by a square (say unity) gives a square, and such that their sum and their difference increased by another square (say $t^2=x^2+2x+1$) shall give squares. The required numbers are taken to be $168t^2$ and $120t^2$.   Then the final

---

[12] Arithmeticae Rationalis Mengoli Petri, Bononiae, 1674, 1st Pref.   Cf. Euler.[28]

[13] Reconstructed from the author's inadequate notes on Petrus.

[14] Caroli Renaldinii Mathematum Analyticae Artis Pars Tertia, 1684; reviewed in Acta Eruditorum, 1685, p. 178.

[15] Elemens des Math. ou Principes Generaux . . ., Paris, 1675, 325.

[16] Letter, Oct., 13, 1676, to de Billy, Bull. Bibl. Storia Sc. Mat. e Fis., 12, 1879, 517.

conditions are satisfied since $168+120+1=17^2$, $168-120+1=7^2$. To make $168t^2+1$ and $120t^2+1$ squares, we have a double equality, satisfied by $x=-1648825564/1242622079$.

G. W. Leibniz[17] discussed the problem to find three numbers the sum and difference of every pair of which are squares.

M. Rolle[18] found four numbers the difference of any two of which is a square, and the sum of any two of the first three is a square:

$$A=y^{20}+21y^{16}z^4-6y^{12}z^8-6y^8z^{12}+21y^4z^{16}+z^{20},$$
$$B=10y^2z^{18}-24y^6z^{14}+60y^{10}z^{10}-24y^{14}z^6+10y^{18}z^2,$$
$$C=6y^2z^{18}+24y^6z^{14}-92y^{10}z^{10}+24y^{14}z^6+6y^{18}z^2, \qquad D=A+B+C.$$

For $y=1$, $z=2$, $A=2399057$, $B=2288168$, $C=1873432$, $D=6560657$.

T. F. de Lagny[19] solved $4x+6=y^2$, $9x+13=z^2$ by a "new method." Eliminating x, we have $9y^2/4-1/2=z^2$. Hence $9y^2-2=\square$, say the square of $3y-a$. Thus y is found in terms of a.

P. Halcke[20] divided 6 into two parts such that each increased by 6 gives a square, and made $6+x$, $12-x$ both squares.

Malézieux[21] proposed the first problem of Fermat.[9] It is a question of finding three equal sums of two squares.

The problem to find three numbers the sum and difference of any two of which are squares received at the time of its proposal no comment except the mere statement by C. Bumpkin[22] that 1873432, 2399057, 2288168 furnish an answer.

J. Landen[23] took as the numbers

$$x=\tfrac{1}{2}(f^4g^4+g^4+f^4+1); \qquad y, z=\tfrac{1}{2}(f^4g^4-g^4-f^4+1)\pm2f^2g^2.$$

Then $x\pm y$, $x\pm z$, $y-z$ are squares. It remains to make

$$E=f^4g^4-g^4-f^4+1=y+z$$

a square. Set $g=f+r$. Then $E=\square$ if

$$1+\frac{(f+r)^4-f^4}{f^4-1}=\left\{1+\frac{2f^3r}{f^4-1}+\frac{(f^6-3f^2)r^2}{(f^4-1)^2}\right\}^2,$$

which gives r and hence $g=f(f^8+6f^4-3)/(1+6f^4-3f^8)$. The case $f=2$ gives Bumpkin's[22] answer. Or we may take f^2g^2+1, f^2+g^2, $2fg$ as the numbers, whence $x\pm z$, $y\pm z$ are squares. For the preceding value of g it is verified that $f^2g^2\pm(g^2+f^2)+1$ are squares, whence their product E is a square. Or we may make $E=(f^4-1)(g^4-1)$ a square by equating it to $(f^4-1)^2(g^2+1)^2$, whence $g=f^2/\sqrt{2-f^4}$. Set $f=1-d$; then $2-f^4$ becomes a

[17] MS. dated Apr. 1, 1676, in Bibliothek Hannover. D. Mahnke, Bibliotheca Math., (3), 13, 1912-3, 39. Cf. Euler.[28]

[18] Journal des Savans, Aug. 31, 1682; Sphinx-Oedipe, 1906-7, 61-2. Cf. Coccoz,[74] Rignaux.[89]

[19] Nouv. Elemens d'Arith. et d'Algebre, Paris, 1697, 451-5.

[20] Deliciae Math., oder Math. Sinnen-Confect, Hamburg, 1719, 235.

[21] Eléments de Geométrie de M. le Duc de Bourgogne, par de Malézieux, 1722; Sphinx-Oedipe, 1906-7, 4-5, 45.

[22] Ladies' Diary, 1750, p. 21, Quest. 311. Cf. Euler.[28]

[23] C. Hutton's Diarian Miscellany, extracted from Ladies' Diary, 3, 1775, 398-401, Appendix. Leybourn's Math. Quest. proposed in Ladies' Diary, 2, 1817, 19-22. Cf. Euler.[28]

quartic in d which is the square of $1+2d-5d^2$ for $d=12/13$.   C. Wildbore's solution is the same as Landen's second with $f=a/b$, $g=x/y$.   C. Hutton took $4x$, $4+x^2$, $1+4x^2$ as the numbers.   Then $5x^2+5$ and $3x^2-3$ are to be squares.   The product $15x^4-15$ is a square for $x=2$, and for $x=z-2$ becomes a quartic in z which is made a square by the usual method.   He obtained Bumpkin's[22] answer.   T. Leybourn[24] took $x+y=u^2$, $x+z=v^2$, $y+z=w^2$; it remains to make u^2-v^2, v^2-w^2, u^2-w^2 squares, which is known[25] (Lowry[65a] of Ch. XIX) to be the case if

$$u=(m^2+n^2)(r^2+s^2), \qquad v=2mn(r^2-s^2)+2rs(m^2-n^2) \qquad w=2mn(r^2+s^2),$$
$$m=r^4+6r^2s^2+s^4, \qquad n=4rs(r^2-s^2).$$

P. Cheluccii[26] treated Diophantus III, 7.   From $x+y+z=r^2$, $x+y=s^2$, $x+z=t^2$, $y+z=v^2$ follow $x=t^2-r^2+s^2$, $y=r^2-t^2$, $z=r^2-s^2$, $2r^2-t^2-s^2=v^2$. Set $t=r-m$, $s=r-n$.   Then $r=(v^2+m^2+n^2)/(2m+2n)$.

L. Euler[27] treated the problem to make $x+a$, $x+b$, $x+c$ all squares. Set $x=z^2-a$, $z=p/q$, $b-a=m$, $c-a=n$.   Then p^2+mq^2 and p^2+nq^2 are to be squares.*   This is impossible if $m=-n=f^2$ or $2f^2$, and if $m=1$, $n=2$. Several solutions are found when $m=2$, $n=6$.   In §§ 213–8, pp. 264–271 (Opera, 446–9), he made $x+a$, $x+b$ squares, also $a+x$, $a-x$.

Euler[28] treated the problem to make $x\pm y$, $x\pm z$, $y\pm z$ all squares.   Let $y=x-p^2$, $z=x-q^2$, and $p^2+r^2=q^2$, whence $y-z=r^2$, $y+z=2x-p^2-q^2$. Equate the last sum to t^2, whence $2x=t^2+p^2+q^2$.   It remains to make $x+y=t^2+q^2$ and $x+z=t^2+p^2$ both squares.   To satisfy $p^2+r^2=q^2$, take $p=a^2-b^2$, $r=2ab$, $q=a^2+b^2$.   To make t^2+q^2 and t^2+p^2 squares, viz., $t^2+a^4+b^4\pm2a^2b^2=\square$, it suffices to make $t^2+a^4+b^4=c^2+d^2$, $2a^2b^2=2cd$, which are satisfied if $a=fh$, $b=gk$, $c=f^2g^2$, $d=h^2k^2$, and

(2) $$t^2=(f^4-k^4)(g^4-h^4).$$

By means of a table of values of m^4-n^4 for $m\leqq15$, $n\leqq9$, $n<m$, he found the solutions $520^2=(3^4-2^4)(9^4-7^4)$ and $975^2=(3^4-2^4)(11^4-2^4)$ of (2) and hence

$$x=434657, \qquad y=420968, \qquad z=150568,$$
$$x=2843458, \qquad y=2040642, \qquad z=1761858.$$

J. L. Lagrange[29] treated $a+bx=t^2$, $c+dx=u^2$ by eliminating x; thus

$$(dt)^2=dbu^2+(ad-bc)d,$$

the second member being made a square in the usual way.   To make

$$ax+by=t^2, \qquad cx+dy=u^2, \qquad hx+ky=s^2,$$

[24] Math. Quest. proposed in Ladies' Diary, 2, 1817, 19–22.

[25] New Series of Math. Repository (ed., T. Leybourn), 3, 1814, I, 163, Quest. 310.

[26] Institutiones analyticae, Viennae, 1761, 135.

[27] Algebra, St. Petersburg, 2, 1770, § 223; French transl., Lyon, 2, 1774, pp. 281–5.   Opera omnia, (1), I, 454–6.   Cf. Haentzschel[163] of Ch. XXII and paper 82 below.

* Euler's further discussion will be given under concordant forms, Ch. XVI.

[28] Algebra, 2, 1770, § 235; 2, 1774, pp. 314–9.   Opera Omnia, (1), I, 470–3.   Same problem in papers 12, 14, 17, 18, 22, 23, 24, 30, 33, 34, 57, 74, 85, 89.   See papers 40–45 of Ch. XIX.

[29] Addition VI, arts. 62–63, to Euler's Algebra, 2, 1774, 557–561.   Euler's Opera Omnia, (1), I, 595–7.   Oeuvres de Lagrange, VII, 115–7.

eliminate x and y, and choose $z = u/t$ so that

$$\frac{ak-bh}{ad-cb} z^2 - \frac{ck-dh}{ad-cb} = \square.$$

In the "Repository solution of the problem to find three numbers the sum and difference of any two of which are squares,"[30] $1 \pm 5x - 2x^2 \mp 2x^3 + 5x^4 \pm x^5$ are taken as the square roots of the sum and difference of the first and second numbers, while $1 \pm 3x + 6x^2 \mp 6x^3 - 3x^4 \mp x^5$ are taken as the square roots of the sum and difference of the first and third numbers. Hence the three numbers are determined. Here x is any square. Taking $x = 9$, we get numbers 4387539232, etc., of ten digits each.

C. Hutton[31] noted that $y + 1 = \square$ if $y = 4x^2 - 4x$. Then $\frac{1}{2}y + 1 = (2ax - 1)^2$ gives x.

Euler[32] solved the problem to make $z - a^2 v, \cdots, z - d^2 v$ squares, where $a^2, \cdots, d^2$ are four given squares, by investigating a quadrilateral the sines of whose angles p, q, r, s are $ax, \cdots, dx$, where $a, \cdots, d$ are given numbers. Let $A, \cdots, D$ be their cosines. Since $\sin (p+q) + \sin (r+s) = 0$, etc., we get $aB + bA + cD + dC = 0$ and two similar relations obtained by interchanging b, c, and B, C; or b, d and B, D. Hence we get the ratios of $A, \cdots, D$ as cubic functions $\alpha, \cdots, \delta$ of $a, \cdots, d$. Thus $A = \alpha y, \cdots,$ $D = \delta y$. Then $a^2 x^2 + \alpha^2 y^2 = 1$, $b^2 x^2 + \beta^2 y^2 = 1$, and we find that $x^2 = v/z$, $y^2 = 1/z$, where

$$v = (a+b+c+d)(a+b-c-d)(b-a+c-d)(a+c-b-d),$$
$$z = 4(bc-ad)(ac-bd)(ab-cd).$$

Hence

$$\sin p = a \sqrt{\frac{v}{z}}, \qquad \cos p = \frac{\alpha}{\sqrt{z}}, \qquad z - a^2 v = \alpha^2.$$

Euler[33] required three numbers x, y, z such that the sum and difference of any two are squares. Let $x > y > z$ and set

$$(3) \qquad x = p^2 + q^2 = r^2 + s^2, \qquad y = 2pq, \qquad z = 2rs.$$

Then $x \pm y = (p \pm q)^2$, $x \pm z = (r \pm s)^2$. Also $p^2 + q^2 = r^2 + s^2$ if

$$(4) \qquad p = ac + bd, \qquad q = ad - bc, \qquad r = ad + bc, \qquad s = ac - bd.$$

Thus $x = (a^2 + b^2)(c^2 + d^2)$. It remains to make

$$y + z = 4cd(a^2 - b^2), \qquad y - z = 4ab(d^2 - c^2)$$

both squares. Their product is a square if

$$cd(d^2 - c^2) = n^2 ab(a^2 - b^2).$$

Take $d = a$. Then $a^2 = (n^2 b^3 - c^3)/(n^2 b - c)$. Take $a = b \pm c$, and take b equal to the numerator of the resulting fraction for b/c. Thus

$$b = 2 \mp n^2, \qquad c = 2n^2 \mp 1, \qquad a = n^2 \pm 1.$$

[30] The Diarian Repository, or Math. Register . . . by a Society of Mathematicians, London, 1774, 522–3. Cf. Euler.[28]
[31] Miscellanea Math., London, 1775, 110.
[32] Mém. Acad. Sc. St. Petersb., 5, anno 1812, 1815 (1780), 73; Comm. Arith., II, 380–5.
[33] Ibid., 6, 1813–4 (1780), 54; Comm. Arith., II, 392–5. Cf. Euler.[28]

It remains to make $y-z$ a square. Since $d=a$,
$$ab(d^2-c^2)=3n^2(n^2\pm1)(2\mp n^2)^2.$$
Choose the lower signs. Then $3(n^2-1)$ is the square of $(n+1)f/g$ if
$$n=\frac{f^2+3g^2}{3g^2-f^2}.$$
Multiply the resulting values of a, b, c by $(3g^2-f^2)^2$; we get
$$a=d=4f^2g^2, \qquad b,\ c=f^4\mp2f^2g^2+9g^4,$$
$$p=8f^2g^2(f^4+9g^4), \qquad q=-(f^4-9g^4)^2,$$
$$r=f^8+30f^4g^4+81g^8, \qquad s=16f^4g^4.$$
For $f=g=1$, we get $p=q=5$, $r=7$, $s=1$, whence $x=y=50$, $z=14$. From one solution x, y, z, we get (§ 15) a second solution
$$(5) \qquad X=\frac{y^2+z^2-x^2}{2}, \qquad Y=\frac{x^2+z^2-y^2}{2}, \qquad Z=\frac{x^2+y^2-z^2}{2}.$$

In the "additamentum" (§ 16), Euler treated the problem to find three squares x^2, y^2, z^2 whose differences are squares. Using (3) and (4), we have
$$x^2-y^2=(p^2-q^2)^2, \qquad x^2-z^2=(r^2-s^2)^2, \qquad y^2-z^2=4(p^2q^2-r^2s^2),$$
the last being a square if $abcd(a^2-b^2)(d^2-c^2)=\square$. This is satisfied if
$$a=d=n^2\pm1, \qquad b=2n^2\mp1, \qquad c=n^2\mp2.$$
From one solution we get a second by (5).

E. Waring[34] noted that, in the problem to find three numbers the sum and difference of any two of which are squares, four of the conditions are satisfied if we employ either of Landen's[23] notations for the numbers or the notation $a^2x^2+b^2y^2$, $2abxy$, $a^2y^2+b^2x^2$, but gave no discussion. He recalled Rolle's[18] values A, B, C.

Euler[35] treated the problem to find four positive numbers in arithmetical progression such that the sum of any two is a square:
$$A+B=p^2, \qquad A+C=q^2, \qquad A+D=B+C=r^2, \qquad B+D=s^2, \qquad C+D=t^2.$$
Hence all are expressible in terms of p, q, r, subject to two conditions
$$2r^2=p^2+t^2=q^2+s^2.$$
Thus $r=\boxed{2}=x^2+y^2$. We get $2r^2=\boxed{2}$ and satisfy $2r^2=p^2+t^2$ by taking
$$p=\pm(x^2-y^2)-2xy, \qquad t=\pm(x^2-y^2)+2xy,$$
the first term being positive, whence $p<t$. Similarly, we satisfy $2r^2=q^2+s^2$ by taking $r=x_1^2+y_1^2$ and
$$q=\pm(x_1^2-y_1^2)-2x_1y_1, \qquad s=\pm(x_1^2-y_1^2)-2x_1y_1 \qquad (q<s).$$
Then $x^2+y^2=x_1^2+y_1^2$ is satisfied by taking
$$x=fz+1, \qquad x_1=fz-1, \qquad y=z-f, \qquad y_1=z+f,$$
as may be done without loss of generality by removing a common square

---

[34] Meditationes Algebraicae, ed. 3, 1782, 328.

[35] Posthumous paper, 1781, Comm. Arith., II, 617–25; Opera postuma, 1, 1862, 119–127. Reprinted, Sphinx-Oedipe, 4, 1909, 33–42.

factor from our numbers.   Then A, B, C, D are all positive if $p^2+q^2>r^2$, a condition expressed in terms of f and z and treated at length by Euler. For $z=4$, $f=7/2$, we drop the factor $1/2$ and get $x=30$, $x_1=26$, $y=1$, $y_1=15$, $p=839$, $q=329$, $r=901$; multiplying the resulting A, $\cdots$, D by 4, we get the integral solutions

$$722, \qquad 432\ 242, \qquad 2\ 814\ 962, \qquad 3\ 246\ 482.$$

J. Leslie[36] made $z+1=\square$, $v+1=\square$, $z+v+1=$ given $\square$ by setting $z=x^2-1$, $v=y^2-1$.

P. Cossali[37] made $F=hx+n^2$ and $F+fx$ squares by taking $F=(y+n)^2$,

$$F+fx=(y+n)^2+\frac{f}{h}(y^2+2yn)=(py-n)^2,$$

thus finding y.   Next, if $(ad-bc)/(a-c)$ is a square r^2, $ax+b$ and $cx+d$ are made squares.   Set $cx+d=(y+r)^2$; for the resulting x,

$$ax+b=\frac{a}{c}(y^2+2ry)+r^2=(py-r)^2.$$

If $(bc-ad)/c$ is a square q^2, set $cx+d=y^2$; then $ax+b=y^2a/c+q^2$ can be made the square of $q-ky$.   To make (pp. 145–6) $H+x=\square$, $H-x=\square$, according to L. Pisano, we have only to express $2H$ as a sum of two squares.

To find three numbers in geometrical progression the difference of any two of which is a square, R. Nicholson[38] took nx, n^2x, n^3x as the numbers.   Since the ratios of their differences are $1:n+1:n$, take $n=v^2$, $v^2+1=\square=(v+s)^2$.   For the resulting v, $n-1$ is a square.   Taking x to be a square, we get an answer.   J. Cunliffe took na^4, na^2b^2, nb^4 as the numbers, where $n=a^2-b^2$; the single condition $a^2+b^2=\square$ is satisfied if $b=r^2-s^2$, $a=2rs$.

To find three numbers in geometrical progression whose sum is a square, several[39] took x^2, nx^2, n^2x^2, $1+n+n^2=\square=(ne-1)^2$.

To find[40] three numbers the difference of any two being a square, take $x-y=16v^2$, $x-z=25v^2$, $y-z=9v^2$, where v and z are arbitrary; or take $5x^2$, x^2, b^2+x^2, where $4x^2-b^2=(2x-n)^2$ gives x; or take $(x+1)^2$, $2x+1$, $4x$, where $2x-1=\square$.

J. Cunliffe[41] made $x-y$, etc., and $x+y-z$, etc., squares.   Take

$$x+y-z=a^2, \qquad x+z-y=b^2, \qquad y+z-x=c^2.$$

Equate $x-y=\frac{1}{2}(b^2-c^2)$ to e^2, $x-z=\frac{1}{2}(a^2-c^2)$ to d^2.   Then

$$y-z=\frac{1}{2}(a^2-b^2)=d^2-e^2$$

must be a square, whence $d=2rs(m^2+n^2)$, $e=2rs(2mn)$.   Set

$$(a+b)r=2s(d+e), \qquad (a-b)s=r(d-e),$$

which give

$$a=(m^2+n^2)(r^2+2s^2)-2mn(r^2-2s^2), \qquad b=2mn(r^2+2s^2)-(m^2+n^2)(r^2-2s^2).$$

[36] Trans. Roy. Soc. Edinb., 2, 1790, 193, Prob. IV.
[37] Origine, Trasporto in Italia . . . Algebra, 1, 1797, 105–7.
[38] The Gentleman's Diary, or Math. Repository, 1798, No. 58; Davis' ed., 3, 1814, 290.
[39] The Gentleman's Math. Companion, London, 1, No. 2, 1799, 18.
[40] *Ibid.*, 21.
[41] The Gentleman's Diary, or Math. Repository, London, No. 61, 1801, 43, Quest. 806.

Take $c=2mn(r^2+2s^2)-(m^2-n^2)(r^2-2s^2)$. Then $c^2=b^2-2e^2$ gives
$$n : m = 12r^2s^2-r^4-4s^4 : 8s^4-2r^4.$$

Cunliffe[42] treated the last problem and Prob. 8: Divide n into four parts the difference of any two parts being a square. Also Prob. 9: Find four numbers whose sum and sums by twos are squares.

R. Adrain[43] made two or three linear functions rational squares as had Lagrange.[29]

Several[44] found two numbers such that if unity be added to each and to their sum and difference, the sums are squares. The numbers $x^2\pm2x$ answer the first two conditions. Then $4x+1=\square=p^2$, $2x^2+1=\square$. Take $p=r+1$. Then $16(2x^2+1)=(r^2+4)^2$ if $r=-8$, whence the numbers are 120, 168.

S. Johnson[45] found integers x, y, z, v such that their sum and the sum of any two are squares and $2(v+x+y)=\square$. Set $x+y+z+v=a^2$, $x+z=b^2$, $y+z=c^2$, $x+y=d^2$. Thus $2z=b^2+c^2-d^2$. Then $v+x=a^2-y-z=a^2-c^2$, $v+y=a^2-b^2$, $v+z=a^2-d^2$ must be squares. Set $a^2-c^2=e^2$, $a^2-d^2=f^2$, $c=rp-f$, $e=sp+d$. Then $c^2+e^2=d^2+f^2$ gives $p=(2rf-2sd)/(r^2+s^2)$. To obtain integers, take $f=(n^2-m^2)(r^2+s^2)$, $d=2mn(r^2+s^2)$. Then
$$e=(r^2-s^2)\cdot2mn+(n^2-m^2)\cdot2rs, \qquad c=(r^2-s^2)(n^2-m^2)-2nm\cdot2rs.$$
By $a^2=d^2+f^2$, $a=(n^2+m^2)(r^2+s^2)$. Thus $a^2-b^2=\square$ if $b=(n^2+m^2)\cdot2rs$. Finally,
$$2(v+x+y)=2a^2+d^2-b^2-c^2=n^4(r^2+s^2)^2+\cdots$$
$$=\left\{n^2(r^2+s^2)-nm\left(\frac{4r^3s-4rs^3}{r^2+s^2}\right)+m^2(r^2+s^2)\right\}^2$$
if $n : m = 2rs(r^2+s^2)^2 : s^6-r^2s^4+s^2r^4-r^6$.

Johnson[46] used the same methods to find x, y, z, v whose sum is a square and difference of any two is a square. J. Cunliffe took $v-x=c^2$, $v-y=b^2$, $v-z=a^2$, $v+x+y+z=n$; it remains to make $x-y=b^2-c^2$, $x-z=a^2-c^2$, $y-z=a^2-b^2$ squares. Hence we desire three squares a^2, b^2, c^2 the difference of any two of which is a square. This is stated to be true if $a^2=485809$, $b^2=451584$, $c^2=462400$.

The problem[46a] to find three numbers in A. P., the sum of any two of which exceeds the remaining one by a square, reduces to $x^2+z^2=2y^2$ (Ch. XIV).

J. Cunliffe[46b] found two rational numbers (x^2+n and y^2+n) such that each and their sum and their difference exceed a given number n by squares. The condition $x^2+y^2+n=\square=(x+v)^2$ gives x in terms of y, v, n. Then $x^2-y^2-n=\square=(n-v^2-y^2)^2/(4v^2)$ if $n^2-2nv^2=\square=(rv-n)^2$, which determines v.

[42] The Math. Repository (ed., Leybourn), London, 3, 1804, 97–106.
[43] The Math. Correspondent, New York, 1, 1804, 237–241; 2, 1807, 7–11.
[44] Ladies' Diary, 1804, pp. 38–9, Quest. 1111; Leybourn's Math. Quest. L. D., 4, 1817, 23.
[45] The Gentleman's Math. Companion, London, 2, No. 8, 1805, 46–8.
[46] *Ibid.*, 2, No. 9, 1806, 35–6.
[46a] New Series of Math. Repository (ed., Leybourn), 1, 1806, I, 7–10.
[46b] *Ibid.*, 2, 1809, I, 9–11.

C. F. Kausler[47] treated the problem to divide a given number a into n parts such that the sum of any $n-1$ parts shall be a square. [The treatment by Diophantus, V, 17, of the case $n=4$ was given in Ch. VIII.] The treatment for n is similar to that for his first case $n=5$. Then, by addition, the sum of the 5 squares $s_1^2, \cdots, s_5^2$ is $4a$. First, find a square P^2 approximately equal to $4a/5$, say

$$P^2 = \frac{4a}{5} + \frac{1}{25z^2}, \qquad 20az^2 + 1 = \square = (1 - mz)^2, \qquad z = \frac{2m}{m^2 - 20a}.$$

Since every number is a sum of 5 squares, set

$$4a = g_1^2 + \cdots + g_5^2, \qquad P = \frac{M}{N} = g_i + \frac{\alpha_i}{N}, \qquad s_i = g_i + \alpha_i x.$$

Thus $\alpha_i = M - g_i N$. Since $\Sigma s_i^2 = 4a$, we get $x = -2\Sigma g_i \alpha_i / \Sigma \alpha_i^2$. Thus, if $a = 21$, the nearest square root of $20a$ is $m = 21$, whence $z = 2$, $M = 41$, $N = 10$. Since $4a = 1 + 9 + 25 + 49$, $1 = (9 + 16)/25$, the g's are 3/5, 4/5, 3, 5, 7, the α's are 35, 33, 11, -9, -29, and $x = 1676/16785$.

To find[48] three numbers x, vx, v^2x in geometrical progression such that each increased by a given number n is a square. From $x + n = c^2$, $vx + n = (d + c)^2$, we get x, v. In the resulting value of $v^2x + n$, put $c^2 - n = r^2$; then

$$d^4 + 4d^3c + 2d^2(2c^2 + r^2) + 4r^2dc + r^2c^2 = \square = (d^2 - 2rd - rc)^2$$

gives d. The desired numbers are r^2, $\frac{1}{4}r^2 - n$, $(\frac{1}{4}r^2 - n)^2/r^2$, where $r = (n - s^2)/(2s)$ makes $r^2 + n = \square = c^2$.

Several[49] found four integers whose sum is a^2 and excess of the sum of any three over the fourth is a square b^2, c^2, d^2 or e^2. Hence $b^2 + c^2 + d^2 + e^2 = 2a^2$, which determines a rationally if we take $c = p - a$, $d = q - a$.

To find[49a] two numbers ($v^2 - n$ and $w^2 - n$) whose difference is a square and such that if each and their sum be increased by the same number n there result squares, we have to make $v^2 - w^2$ and $v^2 + w^2 - n$ squares and hence a certain quartic function a square.

J. Winward[50] found N integers whose sum is a square m^2 and sum of any $N-1$ of them is a square. Take $(2m-n)n$, $(2m-2n)(2n)$, $(2m-3n)(3n)$, $\cdots$, $\{2m-(N-1)n\}(N-1)n$ as the first $N-1$ numbers, and m^2 less their sum as the Nth. Then m^2 exceeds the jth number ($j < N$) by $(m - jn)^2$. Equating the excess of m^2 over the Nth number to $(nr)^2$, we get m in terms of n, r.

Several[51] solved $z + a^2 = \square$, $z/n + a^2 = \square$ by known methods.

To find[52] four integers whose differences are squares, let $x = 2lmn$, $y = l(m^2 - n^2)$. Then five of the differences of u, $u + x^2$, $u + x^2 + y^2$,

[47] Mém. Acad. Sc. St. Pétersbourg, 1, 1809, 271–282.

[48] The Gentleman's Math. Companion, London, 2, No. 13, 1810, 264–5.

[49] The Gentleman's Diary, or Math. Repository, London, No. 71, 1811, 35, Quest. 963. For 3 numbers, Gentleman's Math. Companion, 5, No. 29, 1826, 362–4.

[49a] New Series of Math. Repository (ed., Leybourn), 3, 1814, I, 105–8.

[50] The Gentleman's Math. Companion, London, 5, No. 25, 1822, 141–2.

[51] Ladies' Diary, 1823, 35–36, Quest. 1390.

[52] The Gentleman's Math. Companion, London, 5, No. 26, 1823, 202–4.

$u+(l^2m^2+n^2)^2$ are squares. It remains to make $(l^2m^2+n^2)^2-l^2(m^2+n^2)^2=\square$. Take $l=2$. Then $3(4m^4-n^4)=\square$. From the case $m=n=1$, we get the new solution $m=37$, $n=23$ by Euler's[67] method of Ch. XXII.

W. Wright[53] found three numbers v^2-1, x^2-1, y^2-1 whose sum is a square, each plus unity is a square, and the sum of the roots of the latter squares is a square. Take $v^2+x^2+y^2-3=(v+p)^2$, $v+x+y=q^2$.

To find three numbers such that the sum of the first and second and difference of first and third are squares, the sum of whose roots shall be a square and equal to the sum of the required three numbers, F. N. Benedict[54] took the latter to be $a^2x^2-x^2$, x^2, $(b^2+a^2-1)x^2$. Then $ax+bx=cx^2$ determines x, where $c=2a^2+b^2-1$. Finally, $c=\square=(b-m)^2$ gives b.

Several[55] found three numbers x^2-1, y^2-1, z^2-1 in arithmetical progression, whose sum is a square and each plus unity is a square. Use the known solution x, $z=\pm(m^2-n^2)+2mn$; $y=m^2+n^2$ of $x^2+z^2=2y^2$. To make $x^2+y^2+z^2-3=3(m^2+n^2)^2-3=\square$, take $n=1$ and solve $3m^2+6=\square$ as usual.

W. Wright[56] found three integers x, y, z, double the difference of any two being a square, also double the difference of the sum of any two and the third. First, solve $n(a^2-b^2)=p^2$, $n(c^2-b^2)=q^2$. Since $p^2-q^2=n(a^2-c^2)$, take $a+c=(p+q)t/(vn)$, $a-c=(p-q)v/t$, which give a, c. Then $p^2-q^2=\square$ if $p=2tvn(d^2+e^2)$, $q=2tvn\cdot2de$. For brevity, set $r=t^2+nv^2$, $s=t^2-nv^2$. Then $a=r(d^2+e^2)+2des$, $c=s(d^2+e^2)+2der$. Then $n(c^2-b^2)=q^2$ or $c^2-q^2/n=\square$, becomes a quartic in d, which is satisfied if $d=2rse/(4t^2v^2n-s^2)$. The case $n=1/2$ leads to a solution of the initial problem. Set $2(x+y-z)=a^2$, $2(y+z-x)=b^2$, $2(x+z-y)=c^2$, which give x, y, z. Then the init al three conditions require that $\frac{1}{2}(c^2-b^2)$, $\cdots$ be squares.

J. R. Young[57] treated Diophantus III, 7, 9 somewhat as had Prestet.[15] To make (pp. 347–51) $x\pm y$, $x\pm z$, $y\pm z$ all squares, take $x+y=u^2$, $x+z=v^2$, $y+z=w^2$. Then $x-y=v^2-w^2$, $x-z=u^2-w^2$ are squares if $u=ac+bd$, $v=ad+bc$, $w^2=4abcd$. Then $y-z=(a^2-b^2)(c^2-d^2)$. For $a=9$, $b=4$ $c=81$, $d=49$, we get Euler's[28] first answer, believed to give the smallest possible numbers. Or we may make $a^2-b^2=\square$ by taking $a=m^2+n^2$, $b=2mn$ and similarly for c^2-d^2. Other methods are based on the choice

$$u^2=(a^2+b^2)(c^2+d^2), \qquad v=ac\pm bd, \qquad w=ad\pm bc.$$

He (p. 345) treated Diophantus IV, 14.

F. T. Poselger[58] treated $A=\square$, $B=\square$ for the case in which $A-B$ is factorable into pq (cf. Diophantus II, 12). We may set

$$A, B=[(y^2p\pm q)/(2y)]^2$$

since

$$(y^2p+q)^2-(y^2p-q)^2=4y^2pq.$$

---

[53] The Gentleman's Math. Companion, London, 5, No. 28, 1825, 369–71.

[54] The Math. Diary, New York, 1, 1825, 27.

[55] The Gentleman's Math. Companion, London, 5, No. 29, 1826, 361–2.

[56] Ibid., 5, No. 30, 1827, 574–5.

[57] Algebra, 1816. American edition by S. Ward, 1832, 324–6, 335–6.

[58] Abh. Akad. Wiss. Berlin (Math.), 1832, 1.

S. Ryley[59] found three numbers whose sum, sum of any two, and difference of any two plus unity are squares. Take $x+y=a^2$, $x+z=b^2$, $y+z=1$. The remaining conditions reduce to

$$2a^2+2b^2+2=n^2, \qquad a^2-b^2+1=r^2.$$

Then $4b^2=n^2-2r^2=(n-rm)^2$ if $n=r(m^2+2)/(2m)$. Take $r=2m$. Then $4a^2=n^2+2r^2-4=\square$ if $m^2+12=\square=(s-m)^2$, say. Several used the numbers $2x^2+2y^2-\frac{1}{2}$, $2x^2-2y^2+\frac{1}{2}$, $2y^2-2x^2+\frac{1}{2}$, which satisfy five of the conditions. To satisfy $4x^2-4y^2+1=v^2$, take $x+y=v+1$, $4(x-y)=v-1$. For the resulting x, y, $16(2x^2+2y^2+\frac{1}{2})=17v^2+30v+25=(av-5)^2$, by choice of v.

Fr. Buchner[60] solved $x+1=p^2$, $x-1=q^2$ by setting $p+q=m$, $p-q=2/m$, and sim larly for $x+a=p^2$, $x-b=q^2$.

T. Baker[61] found four numbers p^2-s, q^2-s, r^2-s, s such that the sum of any two is a square, the difference of any two increased by a square r^2 (which is to be found) is a square, and the sum of all four diminished by r^2 is a square. Set $2s=r^2-t$. We need only make p^2+t, q^2+t, r^2+t, $A=p^2-q^2+r^2$, $B=p^2+q^2-r^2+t$ squares. Equate the first three to the squares of $p+t/x$, $q+t/y$, $r+t/z$ respectively, thus finding p, q, r. Then $A=\{p-v(q+r)\}^2$ determines t, and $B=\square$ holds if

$$v=\frac{x(y-z)}{x^2-yz}+\frac{(x^2+yz)^3}{2x(y+z)(x^4+y^2z^2)}.$$

S. Jones[62] found four positive integers x, y, z, $y+z-x$ half of whose sum is a square, the sum of any two is a square, the difference of any two increased by a given square e^2 is a square, and the sum of the four diminished by e^2 is a square. Take $x+y=a^2$, $x+z=b^2$, $y+z=c^2$, $2y+z-x=d^2$, $y+2z-x=e^2$, whence $a^2+e^2=b^2+d^2=2c^2$, which are satisfied if

$$a,\ e=\frac{\{2pv\mp(p^2-v^2)\}c}{p^2+v^2}; \qquad b,\ d=\frac{\{2p\pm(p^2-1)\}c}{p^2+1}.$$

Then all further conditions are satisfied if $b^2-c^2+e^2=\square$, i. e.,

$$f=m^2p^4+4n^2p^3v+2m^2p^2v^2-4n^2pv^3+m^2=\square, \qquad m=p^2+2p-1, \qquad n=p^2+1.$$

Now f is the square of $mp^2-2n^2pv/m+mv^2$ if $v=2m^2p/n^2$.

T. Baker[63] found five integers p^2-t, q^2-t, r^2-t, s^2-t, t the sum of any two of which is a square. Set $2t=p^2+q^2-m^2$. We need only make

$$r^2+m^2-q^2,\quad r^2+m^2-p^2,\quad s^2+m^2-q^2,\quad s^2+m^2-p^2,\quad A=r^2+s^2+m^2-p^2-q^2$$

squares. Equate the first four to the squares of

$$r+x(m-q), \qquad r+z(m-p), \qquad s+y(m-q), \qquad s+w(m-p),$$

respectively. The resulting relations serve to express r/m, s/m, p/m, q/m rationally in terms of x, y, z, w. The condition $A=\square$ is satisfied by making special assumptions.

[59] Ladies' Diary, 1836, 34–5, Quest. 1586.
[60] Beitrag zur Auflös. Unbest. Aufg. 2 Gr., Progr. Elbing, 1838.
[61] The Gentleman's Diary, or Math. Repository, London, 1838, 88–9, Quest. 1360.
[62] *Ibid.*, 86–8.
[63] *Ibid.*, 1839, 33–5, Quest. 1385.

C. Gill[64] found five numbers the sum of every three being a square. He used trigonometry.

To find three integers in geometrical progression, such that each plus unity is a square, Judge Scott[65] took x^2-1, $2x(x^2-1)$, $4x^2(x^2-1)$. It remains only to satisfy $2x(x^2-1)+1=\square=p^2$; take $2x+2=p\pm1$, $x^2-x=p\mp1$. A. Martin used x, xy, xy^2 and took $y=a^2x+2a$. Then $xy+1=\square$, $xy^2+1=(1+2a^2x)^2$ if $x=(4a-4)/a$, and $x+1=b^2$ gives a. D. S. Hart used x, xy, xy^2 with $x=m^2+2m$.

A. Emmerich,[66] to solve $4x+5=u^2$, $5x+4=v^2$, eliminated x to show that $u=3\alpha$, $v=3\beta$, $5\alpha^2-4\beta^2=1$, every solution of which is given by

$$2\beta\pm\alpha\sqrt{5}=(2\pm\sqrt{5})^{2n+1}.$$

To find[67] three integers in arithmetical progression such that the sum of every two is a square. To find[68] two numbers such that if unity be added to each of them or to their sum or to their difference, the resulting sums are all squares.

A. Martin[69] found three numbers the sum of any two of which is a square and the sum of the resulting three squares is a square. Set $x+y=p^2$, etc. The condition $p^2+q^2+r^2=w^2$ is satisfied if

$$p=2st(u^2+v^2), \quad q=2uv(s^2-t^2), \quad r=(s^2-t^2)(u^2-v^2), \quad w=(s^2+t^2)(u^2+v^2).$$

Several[70] solved $a^2+x=y^2$, $a^2+x/p=z^2$ by use of

$$y^2-pz^2=a^2-pa^2=(am\pm pan)^2-p(am\pm an)^2, \quad m^2-pn^2=1.$$

H. Brocard[71] discussed three numbers in geometrical progression, each plus unity a square.

P. W. Flood[72] found three numbers, the first two being squares, the sum of all and the sum of any two being squares. Take $16x^2$, $9x^2$, y^2-10xy. It remains to satisfy $9x^2-10xy+y^2=\square$, $16x^2-10xy+y^2=\square$; eliminate x^2.

R. W. D. Christie[73] solved $x+1=a^2$, $y+1=b^2$, $x+y+1=c^2$, $x-y+1=d^2$. Take $e=g^2-h^2$, $f=2gh$, $a=g^2+h^2$. Then

$$a^2=e^2+f^2, \quad b^2=2ef+1=\square=(1+2gh)^2$$

if $g=\frac{1}{2}(h\pm r)$, where $r^2=5h^2+4$ is solved by continued fractions.

Coccoz[74] noted that the sum and difference of any two of the three numbers 2399057, 2288168 and 1873432 are squares, and gave a general solution depending on a function of degree 20 [Rolle[18]].

---

[64] Application of the angular anal. to indeter. prob. degree 2, N. Y., 1848, p. 60.

[65] Math. Quest. Educ. Times, 14, 1871, 95–6.

[66] Mathesis, 10, 1890, 174–5.

[67] Amer. Math. Monthly, 1, 1894, 96, 136, 169.

[68] *Ibid.*, 280, 325.

[69] Math. Quest. Educ. Times, 61, 1894, 115–6.

[70] *Ibid.*, 65, 1896, 115.

[71] Nouv. Ann. Math., (3), 15, 1896, 288–290.

[72] Math. Quest. Educ. Times, 68, 1898, 53.

[73] *Ibid.*, 69, 1898, 38.

[74] L'illustration, July 20, 1901. Cf. Gérardin,[85] Euler.[28]

To find[75] three integers the difference of every two of which is a square. Likewise[76] for four integers. To make[77] $x+y+z$, $x+y$, $y+z$, $z+x$ all squares.

Several[78] solved $3x+1=\square$, $7x+1=\square$.

A. Cunningham[79] found integers $x_1, \cdots, x_r$ such that, if a given number N be added to their sum s or to the sum of any $r-1$ of them, the results are squares. From $s+N=\sigma^2$, $s-x_i+N=\sigma_i^2$, we get $x_i=\sigma^2-\sigma_i^2$ $(i=1, \cdots, r)$. Then the initial condition can be written

$$(r-1)\sigma^2+N-\sigma_5^2-\cdots-\sigma_r^2=\sigma_1^2+\cdots+\sigma_4^2.$$

We may assign any values to σ, σ_5, $\cdots$, σ_r such that the left member is positive and hence a sum of four squares.

A. Gérardin[80] treated the problem to find a number N which can be separated into four parts such that the sum of any two parts is a square. We need only use a number N which is a sum of two squares in three ways. Or we may employ the formula for $N=(a^2+b^2)(m^2+p^2)$ as a sum of two squares and take $m=f^2-g^2$, $n=2fg$, whence

$$N=\{a(f^2-g^2)\pm 2bfg\}^2+\{b(f^2-g^2)\mp 2afg\}^2=\{a(f^2+g^2)\}^2+\{b(f^2+g^2)\}^2.$$

P. von Schaewen[81] remarked that the triple equality (1) is not solvable by the method of Fermat or by any known method and proved that there is a solution $x \neq 0$ if and only if $a^2(z^2-1)^2+4b^2z^2=\square$ has a solution other than $z=0$, $z=1$. For de Billy's case $a=2$, $b=3$, the condition is $(z^2-1)^2+9z^2=\square$, which has no rational solutions other than $z=0$, $z=1$, as proved by Euler[144] of Ch. XXII. Thus the triple equation has only the solution $x=0$.

E. Haentzschel[82] treated the following problem. Given e_1, e_2, e_3, find a rational number s such that $s-e_1$, $s-e_2$, $s-e_3$ shall be rational squares. Their product $v^2/4$ must be a square. The relation

$$v^2=4(s-e_1)(s-e_2)(s-e_3)$$

is satisfied if s is Weierstrass' function $\wp(u)$ and $v=\wp'(u)$. Hence the problem is to find a rational value of $\wp(u)$ such that also $\wp'(u)$ is rational. The solution is effected by means of the relation between $\wp(2u)$ and $\wp(u)$, and shown to be equivalent to that by Euler[27] for his case of rational e_1, e_2, e_3 [cf. Haentzschel[156] of Ch. V]. Here is treated at length the case $e_1=-8$; e_2, $e_3=4\pm 3\sqrt{-3}$.

H. C. Pocklington[83] noted that the first, second, fifth and tenth terms of an arithmetical progression are not all squares, unless the first is zero or all are equal.

[75] Amer. Math. Monthly, 9, 1902, 113, 230.
[76] *Ibid.*, 10, 1903, 206–7.
[77] *Ibid.*, 141–3.
[78] Math. Quest. Educ. Times, 8, 1905, 79–80.
[79] *Ibid.*, (2), 9, 1906, 30–1.
[80] Sphinx-Oedipe, 1907–8, 10–12.
[81] Bibliotheca Math., (3), 9, 1908–9, 289–300.
[82] Jahresbericht d. Deutschen Math.-Vereinigung, 22, 1913, 278–284.
[83] Proc. Cambridge Phil. Soc., 17, 1914, 117.

E. Haentzschel, A. Korselt, and P. von Schaewen[84] treated the problem to find 3 numbers in arithmetical progression the sum of any two of which is a square (Diophantus III, 9).

A. Gérardin[85] noted further cases of Euler's relation (2):

$$13920^2 = (7^4 - 3^4)(17^4 - 1), \qquad 62985^2 = (14^4 - 5^4)(18^4 - 1),$$
$$3567^2 = (5^4 - 4^4)(21^4 - 20^4), \qquad 2040^2 = (2^4 - 1)(23^4 - 7^4),$$
$$7800^2 = (9^4 - 7^4)(11^4 - 2^4), \qquad 230880^2 = (17^4 - 9^4)(29^4 - 11^4).$$

He and A. Cunningham[86] noted solutions of

$$P(x+y) + Qx = \square, \qquad P(x+y) + Qy = \square.$$

E. Turrière[87] obtained a second solution from one of $ax + a' = \square$, $bx + b' = \square$.

H. R. Katnick[88] noted that $z \pm n$ can be made squares if n is even.

M. Rignaux[89] gave Rolle's[18] solution in factored form, and also

$$A = \Pi(81p^8 \pm 36p^6q^2 + 38p^4q^4 \pm 4p^2q^6 + q^8),$$
$$B = 16p^2q^2(9p^4 + q^4)(81p^8 - 2p^4q^4 + q^8), \qquad C = 32p^4q^4(27p^4 + q^4)(3p^4 + q^4).$$

In terms of any given solution are expressed two new solutions.

On linear functions made squares, see Genocchi[44] of Ch. XIV.

---

[84] Jahresber. d. Deutschen Math.-Vereinigung, 24, 1915, 467–471; 25, 1916, 138–9, 139–145, 351–9.

[85] L'intermédiaire des math., 22, 1915, 230–1 (50–1).

[86] *Ibid.*, 75, 233–5.

[87] L'enseignement math., 18, 1916, 423–4.

[88] Amer. Math. Monthly, 24, 1917, 339–40.

[89] L'intermédiaire des math., 25, 1918, 129.

CHAPTER XVI.

TWO QUADRATIC FUNCTIONS OF ONE OR TWO UNKNOWNS MADE SQUARES.

Congruent numbers k; $x^2 \pm k = \square$ both solvable.

Diophantus, III, 22, found solutions of $(x_1+x_2+x_3+x_4)^2 \pm x_i = \square$ [see Ch. VI] and, in V, 9, found solutions of $x_i^2 \pm (x_1+x_2+x_3) = \square$. In each case he began with the fact that in any right triangle having the hypotenuse h and legs a, b, the numbers $h^2 \pm 2ab$ are squares.

An anonymous Arab manuscript,[1] written before 972, contains the problem [of congruent numbers]: Given an integer k, to find a square x^2 such that $x^2 \pm k$ are both squares. The most convenient artifice to solve this problem is stated to be the theorem that if $x^2+y^2=z^2$, then $z^2 \pm 2xy = (x \pm y)^2$. [Hence $2xy$ is a congruent number if x, y are the legs of a right triangle.] It is stated that, if the triangle is primitive and if $x^2 \pm k$ are squares, the final digits of these squares are 1 or 9, with the express statement that the digit is not 5 [squares of odd numbers end in 1, 5 or 9]. An example is given: Using the primitive right triangle with the sides 3, 4, we get $2xy=24$, $5^2+24=7^2$, $5^2-24=1^2$. A table gives the expression of the odd numbers 3, $\cdots$, 19 in various ways as sums of two relatively prime parts a, b; also the sides $2ab$, $a^2 \pm b^2$ of a right triangle, and $k=2(2ab)(a^2-b^2)$; finally, u and v in $z^2+k=u^2$, $z^2-k=v^2$, where z is the corresponding hypotenuse. The table has 34 such k's. Woepcke noted that if we delete their square factors, we get the following 30 " primitive congruent numbers ":

5	34	210	429	2730
6	65	221	546	3570
14	70	231	1155	4290
15	110	286	1254	5610
21	154	330	1785	7854
30	190	390	1995	10374.

Woepcke remarked (p. 352) that there is no indication that the Arabs knew Diophantus prior to the translation by Aboul Wafâ († 998), but they may well have derived the problem of congruent numbers from the Hindus who were early acquainted with the indeterminate analysis of Diophantus.

Mohammed Ben Alhocain,[2] in an Arab manuscript of the tenth century, stated that the principal object of the theory of rational right triangles is to find a square which when increased or diminished by a certain number k becomes a square. He proved geometrically Diophantus' result that if $x^2+y^2=z^2$ then $z^2 \pm 2xy = (x \pm y)^2$, so that z^2 is the required square. Again,

---

[1] Imperial Library of Paris. French transl. by F. Woepcke, Atti Accad. Pont. Nuovi Lincei, 14, 1860–1, 250–9 (Recherches sur plusieurs ouv. Leonardo Pise, 1st part, III). Some of the results in the MS. were cited by Woepcke, Annali di Mat., 3, 1860, 206.

[2] French transl. by F. Woepcke, Atti Accad. Pont. Nuovi Lincei, 14, 1860–1, 350–3.

start with any two numbers a, b and take $k = ab(a+b)/(a-b)$.   Then

$$\left[\frac{a^2+b^2}{2(a-b)}\right]^2 \pm k = \left[\frac{a+b}{2} \pm \frac{ab}{a-b}\right]^2.$$

Or we may form the right triangle with the legs a^2-b^2, $2ab$ and take as k the double of their product.

Alkarkhi[3] (beginning of the eleventh century), to make $\xi+\xi^2$ and $\xi-\xi^2$ squares, began by solving the system $y+x^2 = \square$, $y-x^2 = \square$.   Set $y = 2x+1$, so that $y+x^2 = \square$.   Then $y-x^2 = 2x+1-x^2$ will be the square of $1-x$ if $x=2$.   Then $x^2 = 4$, $y = 5$ and $\xi = 4/5$ [since the initial system is satisfied if $\xi^2/\xi = x^2/y$].   The method is stated to be useful in the solution of $x^2+mx = \square$, $x^2-nx = \square$.   Although this problem does not belong directly to the present subject, it has been inserted here in view of the use by Leonardo of the same method.

Leonardo Pisano[4] mentioned about 1220 the problem, which had been proposed to him by Johann Panormitanus of Palermo, to find a square which when either increased or decreased by 5 gives a square.   He stated that the answer is the square of $3+\frac{1}{4}+\frac{1}{6}$ $[=\frac{41}{12}]$; for, its square increased by 5 gives the square of $4\frac{1}{12}$, and decreased by 5 gives the square of $2+\frac{1}{3}+\frac{1}{4}$ $[=\frac{31}{12}]$.   He said that he would treat such questions in a work to be entitled "liber quadratorum."   The latter,[5] dated 1225, opened with a bare mention of this special problem, but later[6] took up the general problem: To find a number which added to a square and subtracted from the same square gives squares; or, what is equivalent, to find three squares x_1^2, x_2^2, x_3^2 and a number (congruum) y such that

$$x_2^2 - y = x_1^2, \qquad x_2^2 + y = x_3^2.$$

Since any square is the sum of consecutive odd numbers $1, 3, \cdots$, beginning with unity, y must equal the sum of those odd numbers which enter the sum for x_2^2 and not in x_1^2, and again those in x_3^2 and not in x_2^2.   He proposed to determine y so that the number of consecutive odd numbers whose sum is $x_2^2 - x_1^2$ shall bear to the number making up $x_3^2 - x_2^2$ a given ratio a/b.   Let first

$$(1) \qquad \frac{a}{b} < \frac{a+b}{a-b}.$$

To treat together[7] the two cases separated by Leonardo, let s and t represent a and b when $a+b$ is even, but represent $2a$ and $2b$ when $a+b$ is odd.   Set

$$(2) \qquad \begin{aligned} m &= s(a-b), & n &= t(a-b), & u &= np, \\ p &= s(a+b), & q &= t(a+b), & v &= mq, \end{aligned}$$

---

[3] Extrait du Fakhrî, French transl. by F. Woepcke, Paris, 1853, (28), p. 85; same in (27), pp. 111-2.

[4] At the beginning of his Opuscoli, published by B. Boncompagni in Tre Scritti Inediti di L. Pisano, Rome, 1854, 2, and in Scritti di L. Pisano, Rome, 2, 1862, 227.

[5] Tre Scritti, 55, seq. Scritti, II, 253-283.   B. Boncompagni, Comptes Rendus Paris, 40, 1855, 779, and R. B. McClenon, Amer. Math. Monthly, 26, 1919, 1-8, gave a summary of the topics treated in the liber quadratorum.   Cf. O. Terquem, Annali di Sc. Mat. Fis., 7, 1856, 140-7; Nouv. Ann. Math., 15, 1856, Bull. Bibl. Hist., 63-71.   Xylander wrongly said that Leonardo borrowed from Diophantus (cf. Libri,[24] II, 41).

[6] *Invenire numerum*, Tre Scritti, p. 83; Scritti, II, 265.

[7] A. Genocchi, Note analitiche sopra Tre Scritti . . . , Annali di Scienze Mat. e Fis., 6, 1855, 275-8.   Cf. Leonardo[105] of Ch. XIII.

all of which are even. By (1), $s(a-b) < t(a+b)$, whence $m < q$. Now $v = mq$ is the sum of m consecutive odd numbers

(3) $q-(m-1), \cdots, \quad q-3, \quad q-1, \quad q+1, \quad q+3, \cdots, \quad q+(m-1)$.

Similarly, $u = np$ is the sum of n consecutive odd numbers equidistant by twos from p. Thus the numbers of terms in the sums for v and u have the ratio $m : n = s : t = a : b$. There are $(q-m)/2$ odd numbers $< q-m$; their sum z_1 is $(q-m)^2/4$. The sum z_2 of the odd numbers $< q+m$ is $(q+m)^2/4$. Between $q-m$ and $q+m$ lie the m consecutive odd numbers (3), so that their sum is v. But

$$m+n = (s+t)(a-b) = (a+b)(s-t) = p-q, \qquad q+m = p-n.$$

Thus the n odd numbers between $p-n$ and $p+n$, whose sum is u, are the n odd numbers which follow $q+m$. Finally, the sum z_3 of the odd numbers $< p+n$ is $(p+n)^2/4$. Hence $z_1+v = z_2$, $z_2+u = z_3$, while z_1, z_2, z_3 are squares; further,

$$v = mq = st(a-b)(a+b) = np = u.$$

Thus the proposed problem is solved by taking[8]

$$y = v = u, \qquad x_1^2 = z_1, \qquad x_2^2 = z_2, \qquad x_3^2 = z_3.$$

Next, if the inequality sign in (1) is reversed, we have only to interchange m and q in the definitions (2), which were used only to obtain $q+m = p-n$, $v = u$. As the latter hold also now, the preceding discussion holds for the present case also. The case $a : b = a+b : a-b$ is shown to be impossible in integers.[9]

Leonardo[10] gave several numerical examples. For $a=5$, $b=3$, then $y=240$, $x_1=7$, $x_2=17$, $x_3=23$. For $a=3$ or 2, $b=1$, then $y=24$, $x_1=1$, $x_2=5$, $x_3=7$. For $a=5$, $b=2$, then $y=840$, $x_1=1$, $x_2=29$, $x_3=41$. For $a=7$, $b=5$, then $y=840$, $x_1=23$, $x_2=37$, $x_3=47$. Note[11] that 24 is the least congruent number for which the three squares x_i^2 are integers; but with fractions, we can find smaller as shown later.

For, Leonardo[12] proved that if a and b are relatively prime and if $a+b$ is even then $cb(a+b)(a-b)$ is divisible by 24 and stated[13] that a similar proof holds if a and b are not relatively prime. He proved also that, if one of a and b is even and the other odd, $2a \cdot 2b(a+b)(a-b)$ is divisible by 24. Thus he was able to state[14] that any congruent number is a multiple of 24.

The product[15] of 24 by any square h^2 is a congruent number and the corresponding squares are the products of those for 24 by h^2. We also get congruent numbers by multiplying 24 by a sum of squares $1^2+2^2+3^2+\cdots$

[8] B. Boncompagni, Annali di Sc. Mat. e Fis., 6, 1855, 135, quoted Leonardo's solution to be $y = 4ab(a^2-b^2)$, $x_2 = a^2+b^2$, $x_1, x_3 = 2ab \pm (b^2-a^2)$. But this corresponds only to the case $s=2a$, $t=2b$.

[9] Tre Scritti, 96; Scritti, II, 271. Genocchi,[7] pp. 292-3.

[10] Tre Scritti, 88-92; Scritti, II, 268-70. Genocchi,[7] pp. 278-9.

[11] Tre Scritti, 90-93; Scritti, II, 269-270. Genocchi,[7] pp. 280-1.

[12] *Si duo numeri*, Tre Scritti, 80; Scritti, II, 264.

[13] Tre Scritti, 82; Scritti, II, 265.

[14] Tre Scritti, 92; Scritti, II, 270. Genocchi,[7] pp. 273-4.

[15] *Quotiens enim* 24, Tre Scritti, 93; Scritti, II, 270. Genocchi,[7] p. 283, p. 254.

or $1^2+3^2+5^2+\cdots$ or $h^2+(2h)^2+(3h)^2+\cdots$. For example,
$$24(1^2+3^2+5^2)=840.$$

To find[16] a congruent number whose fifth part is a square, take $a=5$ and determine b so that b, $a+b$, $a-b$ are all squares, say g^2, h^2, k^2, respectively. Then $5=g^2+k^2$. Either $g=1$, $k=2$, whereas $a+b=5+1$ is not a square, or $g=2$, $k=1$, whence $4ab(a^2-b^2)=720$ is the desired congruent number. Returning to the earlier problem to make $x^2\pm5$ both squares, and using the values $a=5$, $b=4$, just found, we have $s=10$, $t=8$, and, by (2), $m=10$, $q=72$, whence $z_2=(82/2)^2$, $x_2=41$. Since $720=5\cdot12^2$, we reduce the numbers in the ratio $1:12$ and get the solution $x=41/12$.

Leonardo[17] affirmed that no square can be a congruent number. This proposition is of special historical importance since it implies that the area of a rational right triangle is never a square and that the difference of two biquadrates is not a square. Leonardo stated without proof[18] the lemma that if a congruent number were a square there would exist integers a, b for which $a:b=a+b:a-b$ (proved impossible earlier).

Leonardo[19] noted that many numbers are not congruent; but any number is a congruent if the quotient of any congruent number by it is a square. A number is congruent if it equals one of the four numbers a, b, $a+b$, $a-b$, and if the remaining three are squares. For example, 16, 9, $16+9$ are squares, so that $16-9=7$ is a congruent number. To make $x^2\pm x$ both squares, let k be a congruent number and $g^2-k=f^2$, $g^2+k=h^2$; then we have the solution $x=g^2/k$ since
$$\left(\frac{g^2}{k}\right)^2-\frac{g^2}{k}=\left(\frac{fg}{k}\right)^2,\qquad\left(\frac{g^2}{k}\right)^2+\frac{g^2}{k}=\left(\frac{gh}{k}\right)^2.$$

To make $X^2\pm mX$ both squares, we set $X=mx$ and are led to the preceding problem, whence $X=mg^2/k$. Leonardo considered the example with $k=24$, $g=5$. Cf. Alkarkhi,[3] and Ch. XVIII.

Luca Paciuolo[20] reproduced part of Leonardo's Liber Quadratorum; he gave as the first five "congruente" numbers 24, 120, 336, 720, 1320, their corresponding squares ("congruo"[21]) being 5^2, 13^2, 25^2, 41^2 61^2. From n and $n+1$ he derived the congruent number $2n(n+1)\{2(n+n+1)\}$, the corresponding square being $\{n^2+(n+1)^2\}^2$. He made $x^2\pm b$ fractional squares for $b=5$, 7, 13; and solved $x^2+10=\square$, $x^2-11=\square$. He gave a table of 52 congruent numbers, of which only[22] 14 are primitive, the latter being all in the table in the Arab MS.[1] (viz., the first six and 65, 70, 154, 210, 231, 330, 390, 546); the Arab had the advantage of excluding values a, b not

---

[16] *Volo invenire*, Tre Scritti, 95; Scritti, II, 271. Genocchi,[7] p. 288.

[17] Tre Scritti, 98; Scritti, II, 272. Cf. Ch. XXII.

[18] For a proof, with a historical discussion, see Genocchi,[7] pp. 293–310 (pp. 131–2). Cf. F. Woepcke, Jour. de Math., 20, 1855, 56; extract in Comptes Rendus Paris, 40, 1855, 781.

[19] Tre Scritti, 98; Scritti, II, 272. Genocchi,[7] pp. 310–3, 345–6.

[20] Luce de Burgo, Summa de arithmetica geometria, Venice, 1494; ed. 2, Toscolano, 1523, ff. 14–18.

[21] Thus interchanging Leonardo's two terms. Cf. Bibl. Math., (3), 3, 1902, 144. Also noted by Boncompagni.[8]

[22] F. Woepcke, Annali di Mat., 3, 1860, 206; Atti Accad. Pont. Nuovi Lincei, 14, 1860–1, 259.

relatively prime. The dependence of the work of Paciuolo upon that of Leonardo was pointed out in detail by B. Boncompagni[23] and by G. Libri.[24]

F. Ghaligai[25] also borrowed [Libri,[24] III, 145] from Leonardo; he gave $5^2+24=7^2$, $5^2-24=1$, stating that 24 is the least congruent number. To find another, start with 1 and 3; add and double the sum, getting 8; multiply by $3-1$, getting 16; multiply by 1×3, getting 48; its double 96 is a congruent number; in fact, $1+3^2=10$ and $10^2-96=2^2$, $10^2+96=14^2$.

F. Feliciano[26] gave the same congruent numbers and rule as had Paciuolo.[20]   He gave $x^2=6\frac14$ as the solution of $x^2\pm6=\square$.

P. Forcadel de Beziers[27] employed right triangles with one leg less by unity than the hypotenuse h, citing $h=5, 13, 25, 41, 61$. Their squares are "congrus" numbers, the corresponding "congruens" being 24, 120, 336, 720, 1320 [double the products of the two legs]. He gave,[28] for $n=1, 2, 3, 4, 5$, the congrus $(4n^2+1)^2$ and corresponding congruens $8n(4n^2-1)$.

N. Tartaglia[29] quoted two rules of Leonardo, as given by Luca Paciuolo, for forming congruent numbers, one rule by use of two consecutive numbers, the other by use of[30] $(a^2+b^2)^2\pm4ab(a^2-b^2)=\square$.

G. Gosselin[30] treated (f. 75 verso) the problem: Given a square 100, to find the congruent number. Separate the double 20 of the side into two parts $2L$ and $20-2L$ whose product equals the product of two other numbers of difference 20, say L, $20+L$. Thus $L=4$ and $8\times12=4\times24$ is the required congruent number 96. Conversely, given a congruent number, to find the square (f. 77, verso). "This is the problem which Luca, Pisano, Tartaglia, Cardan and Forcadelus found so difficult, in investigating which they consumed not a little oil; nevertheless they did not succeed and it remained unsolved up to the present; let us now explain that difficult thing." Given the congruent number 96, to find the square Q such that $96+Q$ is the required square. Hence the sum $192+Q$ of the latter and 96 must be a square. Thus the difference of two squares is $96=4\cdot24=6\cdot16=8\cdot12$. But $\frac12(8+12)=10$ is excluded since $100\neq192+Q$, while $\frac12(4+24)=14$ and $14^2=192+Q$ gives $Q=4$, yielding the answer $96+Q=100$.

Beha-Eddin[32] (1547–1622) listed, among the seven problems remaining unsolved from former times, as Prob. 2 that to make x^2+10 and x^2-10 both squares. As noted by Nesselmann, it is impossible.

[23] Annali di Sc. Mat. e Fis., 6, 1855, 135–154.
[24] Hist. Sc. Math. en Italie, ed. 2, Halle, 1865, II, 39; III, 137–140, 265–271.
[25] Summa de Arithmetica, Florence, 1521, f. 60; Practica d'arithmetica, Florence, 1552, 1548, f. 61, left.
[26] Libro di Arithmetica & Geometria speculatiua & praticale: Francesco Feliciano . . . Intitulato Scala Grimaldelli, Venice, 1526, etc., Verona, 1563, etc., ff. 3–5 (unnumbered pages 7, 8).
[27] L'arithmeticqve, I, 1556, Paris, ff. 8, 9.
[28] The related right triangle has the sides $4n$, $4n^2-1$, $4n^2+1$.
[29] La Seconda Parte del General Trattato di numeri et misure, Venice, 1556, ff. 143–6.
[30] The final factor, given as $a+b$, was corrected by the translator, G. Gosselin, 1578, 91.
[31] De Arte magna, seu de occulta parte num., Paris, 1577.
[32] Essenz der Rechenkunst von Mohammed Beha-eddin ben Alhossain aus Amul, arabisch u. deutsch von G. H. F. Nesselmann, Berlin, 1843, p. 55. French transl. by Aristide Marre: Khelasat al Hissab, ou Essence du Calcul de Beha-eddin Mohammed ben al-Hosain al-Aamouli, Nouv. Ann. Math., 5, 1846, 313; ed. 2, corrected and with new notes, Rome, 1864.

Fermat[2] of Ch. XXII proved that the difference of two biquadrates is never a square. Hence no congruent number is a square.

L. Euler[33] noted (as had Leonardo) that $p^2 \pm 5q^2$ are both squares for $p=41$, $q=12$; $p^2 \pm 7q^2$ both squares for $p=337$, $q=120$. He made $p^2 \pm aq^2$ squares also for $a=6$, 14, 15, 30. The method is that used by him[76] for concordant numbers.

P. Cossali[34] undertook to reconstruct Leonardo's Liber Quadratorum, then believed to be lost. A sufficient (adverse) report will be found under Genocchi,[35] Woepcke[36] and Boncompagni.[37]

A. Genocchi[35] stated that Cossali[34] was wrong in believing that Leonardo's method of making $x^2 \pm a$ both squares is only special. While indirect, it is general and succeeds when the problem is solvable. In fact, it coincides exactly with the formulas obtained by Euler[76] after complicated calculations. This coincidence escaped Cossali, who filled many pages with useless calculations without discovering the general solution.

F. Woepcke[36] noted that of the [26 distinct] congruent numbers in the table of 29 lines by Cossali,[34] p. 126, only 12 are primitive, including all but 65 and 154 of those noted under Luca Paciuolo.[20]

B. Boncompagni[37] disagreed with the explanation by Cossali,[34] p. 132, of Leonardo's method. The latter had remarked that h will be a congruent number if its quotient by a given congruent number h_1 is a square q^2. According to Cossali's interpretation, q is rational only when $(h_1+2)(2h_1+2)(3h_1+4)$ is a rational square; while a more plausible interpretation leads always to a rational q.

"L. Pisanus"[38] made $n^2 \pm 13$ and n^2 all rational squares. Since

$$\{d^2+(d+1)^2\}\{(d+1)^2+(d+2)^2\} \pm 4(d+1)^2$$

are the squares of $2d^2+4d+3$ and $2d^2+4d+1$, take $d=2$ and we get $13 \cdot 25 \pm 36 = \square$. In $(a^2+b^2)^2 \pm 4ab(a^2-b^2) = (a^2 \pm 2ab - b^2)^2$, take $a=ct^2$, $b=s^2$. Then

$$(c^2t^4+s^4)^2 \pm 4ct^2s^2(ct^2+s^2)(ct^2-s^2) = \square.$$

Take $c=13$, $t^2=25$, $s^2=36$. But $(13 \cdot 25)^2 - 36^2$ is the product of the squares found before. Hence $(c^2t^4+s^4)^2/\{4t^2s^2(c^2t^4-s^4)\}$ is the required square n^2.

J. Hartley[39] took $x^2+13=(x+y)^2$, $x^2-13=(x-yz)^2$, and from the two rational values of x got $y^2=13(z-1)/\{z(z+1)\}$. The latter is a square for $z=(r^2+s^2)/(2rs)$ if $r^2+s^2=13$, $2rs=\square$. Take $r=-3-gt$, $s=2-t$. Then $r^2+s^2=13$ gives $t=(4-6g)/(g^2+1)$. Take $g=2$, whence $r=1/5$, $s=18/5$,

[33] Algebra, 2, 1770, § 226; French transl., Lyon, 2, 1774, p. 291; Opera Omnia, (1), I, 459.

[34] Origine, trasporto in Italia, primi progressi in essa dell'algebra, 1, 1797, 115–172. Cf. G. Libri, Histoire des Sc. Math. en Italie, ed. 2, III, 1865, 139, 140, 265.

[35] Comptes Rendus Paris, 40, 1855, 775–8.

[36] Atti Accad. Pont. Nuovi Lincei, 14, 1860–1, 259.

[37] Annali di Sc. Mat. e Fis., 6, 1855, 149–151.

[38] Ladies' Diary, 1803, p. 41, Quest. 1099; and Prize Prob. 1118, 1804, pp. 44–6; Leybourn's Math. Quest. L. D., 4, 1817, 10–11, 31–33. The Prize problem stated that there are rational squares x^2, y^2 such that $x^2 \pm 13$ are squares, and $13y^2$ is the area of a right triangle whose sides are integers; 13 is a sum of two squares, double the product of whose roots is a square, and if the latter square be added to and subtracted from 13 the results are squares.

[39] The Diary Companion, Supplement to the Ladies' Diary, London, 1803, 45.

and $2rs = \square$.   Thus $x = 106921/D$, $x^2 + 13 = (127729/D)^2$, $x^2 - 13 = (80929/D)^2$, $D = 19380$.

P. Barlow[40] proved by descent that 1 and 2 are not congruent numbers.

J. Cunliffe[41] noted that if, when n is given, a rational v can be found for which $n + v^2$ and $n - v^2$ are rational squares, we can deduce a rational x for which $x^2 + n$ and $x^2 - n$ are rational squares.   Take $(a+b)^2$ and $(a-b)^2$ as the latter.   Then $x^2 = a^2 + b^2$, $n = 2ab$.   To satisfy the former, take $a = (p^2 - q^2)/(2r)$, $b = pq/r$.   Then $(p^2 - q^2)pq = nr^2$.   Take $p = n$, $q = v^2$.   Then $n^2 - v^4 = \square$, which holds if $n \pm v^2$ are squares.   Application is made to the case $n = 13$ by expressing 13 as a sum of two rational squares in two ways.

" Umbra "[42] noted that $x^2 + n = a^2$, $x^2 - n = b^2$ can be solved if
$$n = (c^2 + d^2)/s^2, \qquad c^2 - d^2 = \square.$$
For, $2x^2 = a^2 + b^2$ is known to hold if a, $b = (2pq \pm p^2 \mp q^2)/r$, $x = (p^2 + q^2)/r$. Then $n = \tfrac{1}{2}(a^2 - b^2) = 4pq(p^2 - q^2)/r^2$.   Taking $p = c^2$, $q = d^2$, we have
$$r^2 = 4c^2 d^2 s^2 (c^2 - d^2),$$
whence r is rational since $c^2 - d^2 = \square$.   Similarly, $x^2 \pm n = \square$ are solvable if $n = (c^2 - d^2)/s^2$, $c^2 + d^2 = \square$, or if n is double the sum of two squares the double of whose difference is a square.

A. Genocchi[43] noted that the problem to make $x^2 \pm hq^2$ both squares is equivalent to the single equation $x^4 - h^2 q^4 = \square$.   By the direct, but laborious, method of Fermat (on Diophantus VI, 26), used by Lagrange (see papers 37–41, 54 of Ch. XXII), Genocchi treated the example $h = 5$ far enough to reach the special solution due to Leonardo.[4]   The direct solution of $x^2 \pm h = \square$ leads to $4mn(m^2 - n^2) = hg^2$ or the problem to form a rational right triangle with a given area.   The absence of a treatment of the latter leaves an evident lacuna in Diophantus VI, 6–11 (V, 8 deduced a new solution from one).   The method by Euler[33] is identical with that of Leonardo.

Genocchi (pp. 206–9) proved that an integer y is of the form $4mn(m^2 - n^2)$ in only a finite number of ways.   To two solutions x of $x^2 \pm y = \square$, each x a sum of two squares, correspond distinct values of y.   From one solution (pp. 251–3) of $x^2 \pm k = \square$, we readily get others.   Cf. Young[134] of Ch. XIX.

Genocchi[44] proved that $r^4 + 4s^4$, $2r^4 + 2s^4$, $r^4 - s^4$ are congruent numbers; also $r^4 + 6r^2 s^2 + s^4$ and $\pm(r^4 - 6r^2 s^2 + s^4)$ if one of the integers r, s is even and the other odd.   No prime $8m + 3$ is a congruent number.

Genocchi[45] proved that the double of a prime $8k + 5$ is not a congruent number.

Matthew Collins[46] proved that the only congruent numbers < 20 are 5, 6, 7, 13, 14, 15; that a prime $a = 4n + 3$ is not a congruent number if,

[40] Theory of Numbers, London, 1811, 109, 114.

[41] T. Leybourn's Math. Quest. from Ladies' Diary, 3, 1817, 368–71.

[42] The Gentleman's Math. Companion, London, 4, No. 21, 1818, 750–2.

[43] Annali di Sc. Mat. e Fis., 6, 1855, 129–134, 291–2.

[44] *Ibid.*, 313–7.   Cf. Genocchi.[58]

[45] Il Cimento, Rivista di Sc. Let ed Arti, Torino, 6, 1855, 677–9.   Genocchi,[7] p. 299 for the number 10.

[46] A Tract on the possible and impossible cases of quadratic duplicate equalities . . ., Dublin, 1858, 60 pp.   Abstr. in British Assoc. Reports for 1855, 1856, II, 2–5; and in The Lady's and Gentleman's Diary, London, 1857, 92–6.

for $m < a/2$, $m^2 - 2$ is not divisible by a (examples: $a = 11$, 19, 43). To treat $x^2 \pm 5y^2 = \square$, we add and subtract and get $2x^2 = z^2 + w^2$, $10y^2 = z^2 - w^2$. Set $z = z' + w'$, $w = z' - w'$, where z', w' are relatively prime. Thus

$$z'^2 + w'^2 = x^2,$$

whence $z' = m^2 - n^2$, $w' = 2mn$, $x = m^2 + n^2$, and $z'w' = 5y^2/2$, whence $y = 2y'$, $mn(m^2 - n^2) = 5y'^2$. If n is divisible by 5, $n = 5q^2$, $m = p^2$, $m + n = r^2$, $m - n = s^2$, leading to a pair like the initial equations, so that this case is excluded. If $m = 5p^2$, we get $n = q^2$, $m \pm n = r^2$, s^2, whence $5p^2 + q^2 = r^2$, $5p^2 - q^2 = s^2$. As the latter are satisfied by $p = 1$, $q = 2$, whence $m = 5$, $n = 4$, we get the solution [Leonardo's] $x = 41$, $y = 12$, $z = 49$, $w = 31$. In general, given a solution of

$$ax^2 + by^2 = nz^2, \qquad abx^2 - y^2 = \pm nw^2,$$

then $X = n(z^4 + w^4)/2$, $Y = 2xyzw$ make

$$4(X^2 \pm abY^2) = n^2(t \pm v)^2, \qquad t = z^4 - w^4, \qquad v = 2z'w^2,$$

and hence give a solution of $X^2 + abY^2 = \square$, $X^2 - abY^2 = \square$. For example, if $a = 5$, $b = n = 1$, we have $5x^2 \pm y^2 = \square$, holding for $x = 1$, $y = 2$, whence $X = 41$, $Y = 12$ satisfy $X^2 \pm 5Y^2 = \square$.

F. Woepcke[47] found 12 congruent numbers associated with the given one $2xy$, where $x^2 + y^2 = z^2$, viz., zx, zy, $x^2 - y^2$, $z^2 + x^2$, $z^2 + y^2$, $4xy(x^2 - y^2)$,

$$\left(\frac{z \pm x}{2}\right)^2 + (z \mp x)^2, \qquad \pm 2z^2 \mp (x + y \mp 2z)^2, \qquad (x - y \pm 2z)^2 - 2z^2.$$

In fact, $x = a^2 - b^2$, $y = 2ab$. In $2xy$ replace a by z and b by x and drop the square factor $4(z^2 - x^2) = 4y^2$; we get $xz = a^4 - b^4$. But if we replace b by y, we get yz. In $a^4 - b^4$, take $a = x$, $b = y$, and drop the square factor $x^2 + y^2 = z^2$; we get $x^2 - y^2$. Double the product of the latter by the congruent number $2xy$ is a congruent number; etc. He computed the above 12 functions for each right triangle in the Arab manuscript.[1]

Woepcke[48] treated the problem, proposed to him by Boncompagni: Given a congruent number k, to find a congruent number K such that the product kK of the two is another congruent number. If k is formed from a, b, where $2a^2 - b^2 = c^2$, then

$$ab(a^2 - b^2) \cdot ac(a^2 - c^2) = bc \cdot a^2(b^2c^2 - a^4).$$

If k is formed from two numbers of ratio r, where

$$r^4 - 2r^2 - 8r + 9 = w^2$$

and K is formed from two numbers of the ratio

$$\rho = \frac{-(r-1)^2 \pm w}{2(r-1)},$$

then kK is a congruent number formed from two numbers of the ratio $\sigma = (-r^2 + 3 \pm w)/2$. For, then

$$\left(r - \frac{1}{r}\right)\left(\rho - \frac{1}{\rho}\right) = \sigma - \frac{1}{\sigma}, \qquad \frac{p}{q} - \frac{q}{p} = \frac{1}{4p^2q^2} \cdot 4pq(p^2 - q^2).$$

---

[47] Annali di Mat., 3, 1860, 206-15. Same in Atti Accad. Pont. Nuovi Lincei, 14, 1860-1. 259-67.
[48] Annali di Mat., 4, 1861, 247-55.

Like results hold if we take (cf. Lucas[51])

$$9r^4 - 20r^3 - 2r^2 + 20r + 9 = w^2,$$

$$\rho = \frac{r^2 - 4r - 1 \pm w}{2(r-1)(2r+1)}, \qquad \sigma = \frac{2(2r+1)}{-3r^2 + 2r + 3 \pm w}.$$

If kK is a congruent number and hence equal to $4\alpha\beta(\alpha^2 - \beta^2)p^2/q^2$, we may set

$$k = 2\lambda(2\alpha\beta), \qquad K' = \lambda(\alpha^2 - \beta^2), \qquad K' \equiv K\lambda^2 q^2/p^2.$$

Thus if also K and hence K' is a congruent number, then k is the double of a leg of a right triangle whose second leg is a congruent number.

If $kK = K_1$ is a relation between three congruent numbers, the last formulas show that $\sigma = 2\lambda\beta$ and $\sigma_1 = 2\lambda\alpha$ are solutions of the system

$$\sigma^4 + \phi\sigma^2 = \psi^2, \qquad \sigma_1^4 - \phi\sigma_1^2 = \psi^2,$$

where $\phi = 4\lambda K'$, $\psi = \lambda k$. Conversely, if one of these equations can be solved, kK' and hence kK is a congruent number.

To find congruent numbers K, K_1 such that $kK = K_1$, where k is a given congruent number, take as K_1 in turn the 12 types in the earlier paper,[47] each type multiplied by an arbitrary rational square. Give K_1 the form $4\alpha\beta(\alpha^2 - \beta^2)p^2/q^2$ and equate the latter to kK. Hence

$$\frac{k}{4\alpha\beta}(\alpha^2 - \beta^2) = \frac{q^2 k^2}{16 p^2 \alpha^2 \beta^2} \cdot K, \qquad 2\left(\frac{k}{4\alpha\beta} \cdot 2\alpha\beta\right) = k,$$

so that the leg $\alpha^2 - \beta^2$ of a rational triangle is a congruent number and the other leg $2\alpha\beta$ is $k/2$. But this solves $kK = K_1$ for K.

G. Le Secq. Destournelles[49] proved the impossibility in integers of the pair

$$x^2 + y^2 = z^2, \qquad x^2 - y^2 = u^2.$$

The equation obtained by adding these may be written

$$x^2 = \left(\frac{z+u}{2}\right)^2 + \left(\frac{z-u}{2}\right)^2.$$

The terms on the right may be assumed relatively prime. Thus

$$\frac{z+u}{2} = \alpha\beta, \qquad \frac{z-u}{2} = \frac{\alpha^2 - \beta^2}{2},$$

or vice versa, where α, β are odd relatively prime integers. Substituting either set into $2y^2 = z^2 - u^2$, we get

$$y^2 = \alpha\beta(\alpha^2 - \beta^2), \qquad \alpha = m^2, \qquad \beta = n^2, \qquad m^4 - n^4 = \square.$$

Thus $m^2 \pm n^2 = 2k^2, 2l^2$, so that

$$k^2 + l^2 = m^2, \qquad k^2 - l^2 = n^2.$$

But these are like the initial equations with $k < x$, $l < y$.

A. Genocchi[50] stated that $x^2 \pm h$ are not both rational squares when h is a prime $8m+3$ or the product of two such primes, or the double of a prime $8m+5$, or the double of the product of two such primes.

---

[49] Congrès Sc. de France, Rodez, 40, I, 1874, 167–182; Jornal de Math. e Ast., 3, 1881.
[50] Comptes Rendus Paris, 78, 1874, 433–5. Reprinted, Sphinx-Oedipe, 4, 1909, 161–3.

E. Lucas[51] noted that a is a congruent number if and only if $x^4 - a^2y^4 = z^2$ is solvable; then $a \equiv 0, \pm 1 \pmod 5$ if xy is not divisible by 5.   A congruent number does not end in 2, 3, 7 or 8 when y is not divisible by 5.   We are led to congruent numbers by the problem to find three squares in arithmetical progression whose common difference is the product of a by a square.   The equations (pp. 184–6)

$$(4) \qquad\qquad x^2 - 5y^2 = u^2, \qquad x^2 + 5y^2 = v^2$$

were studied, but not completely solved, by Leonardo,[4, 16] Paciuolo,[20] Euler,[33] Collins,[46] and Genocchi,[7] p. 289.   We may assume that x, y, u, v are relatively prime, so that x and v are odd, y even.   Hence in view of the first equation we may set

$$x - u = 10r^2, \qquad x + u = 2s^2, \qquad y = 2rs \quad (r, s \text{ relatively prime}).$$

By the second equation (4), $(5r^2 + 3s^2)^2 - 8s^4 = v^2$, whence

$$5r^2 + 3s^2 \pm v = 2p^4, \qquad 5r^2 + 3s^2 \mp v = 4q^4, \qquad s = pq.$$

Adding the first two of these we get

$$(p^2 - q^2)(p^2 - 2q^2) = 5r^2.$$

Since the factors on the left are relatively prime, we find after considering residues modulo 5 that the only two admissible cases are $p^2 - q^2 = \pm 5g^2$, $p^2 - 2q^2 = \pm h^2$.   For the upper sign, the evident solution $p = 3$, $q = 2$, $g = h = 1$, leads to Leonardo's solution $x = 41$, $y = 12$, $u = 31$, $v = 49$ of (4).   For the lower sign, we get the system $q^2 - 5g^2 = p^2$, $q^2 + 5g^2 = h^2$, like (4); hence from one solution we get the second:

$$X = u^2x^2 + 5v^2y^2, \qquad U = u^2x^2 - 5v^2y^2, \qquad V = u^4 - 2x^4, \qquad Y = 2xyuv,$$

which differ only in form from the formulas by Genocchi.   Lucas solved (pp. 191–3) the equation to which Woepcke[48] was led:

$$9a^4 - 20a^3b - 2a^2b^2 + 20ab^3 + 9b^4 = c^2.$$

This may be written $d^2 + 44a^2b^2 = 9c^2$, where $d = 9a^2 - 10ab - 9b^2$.   Thus

$$3c \pm d = 2p^2, \qquad 3c \mp d = 22q^2, \qquad ab = pq.$$

Set $b = mq$, $p = ma$.   From $p^2 - 11q^2 = \pm d$ we get a quadratic for m with a rational root if $13a^2q^2 \pm (a^4 + 11q^4) = z^2$.   For the upper sign,

$$(2a^2 + 13q^2)^2 - 4z^2 = 125q^4.$$

If we take the factors of the left member to be r^4 and $125s^4$, and add, we get

$$(r^2 - 13s^2)^2 - 4a^2 = 44s^4.$$

Call the factors of the left member $\pm 2u^4$, $\mp 22v^4$; adding, we get

$$13u^2v^2 \pm (u^4 + 11v^4) = r^2,$$

which is like the initial quartic, but with smaller values of the unknowns.   A like result is proved in the remaining admissible cases.   The system[52] $x^2 \pm 6y^2 = \square$ is treated (pp. 180–4) by the method used for the generalization given in the next paper.   From Leonardo's solution $x = 5$, $y = 2$, is deduced $1201^2 \pm 6 \cdot 140^2 = 1249^2$, 1151^2.

[51] Bull. Bibl. Storia Sc. Mat., 10, 1877, 170–193.
[52] Also in Nouv. Ann. Math., (2), 15, 1876, 466–70.

Lucas[53] noted that if there exist relatively prime solutions of
$$x^2 - Ay^2 = u^2, \qquad x^2 + Ay^2 = v^2,$$
then A is of the form $\lambda\mu(\lambda^2 - \mu^2)$. For, by addition,
$$\left(\frac{u+v}{2}\right)^2 + \left(\frac{u-v}{2}\right)^2 = x^2, \qquad \frac{u+v}{2} = \lambda^2 - \mu^2, \qquad \frac{u-v}{2} = 2\lambda\mu, \qquad x = \lambda^2 + \mu^2,$$
where λ, μ are relatively prime and one is even. Hence
$$u, \ v = \lambda^2 - \mu^2 \pm 2\lambda\mu, \qquad y = 2, \qquad A = \lambda\mu(\lambda^2 - \mu^2).$$
He next showed how to derive a second solution from one, given that
A is a congruent number resolved into its prime factors. If α, β are two
integers whose product is A, the second equation gives
$$v + x = 2\alpha e^2, \qquad v - x = 2\beta f^2, \qquad y = 2ef.$$
Substitute the resulting v, x into the first given equation; then
$$(\alpha e^2 - 3\beta f^2)^2 - u^2 = 8\beta^2 f^4.$$
The two factors of the left member equal $\pm 2\beta_1^2 g^4$, $\pm 4\beta_2^2 h^4$, where $\beta_1\beta_2 = \beta$,
$gh = f$. For the upper sign, we add and get
$$(\beta_1 g^2 + \beta_2 h^2)(\beta_1 g^2 + 2\beta_2 h^2) = \alpha e^2.$$
The two factors equal $\alpha_1 p^2$ and $\alpha_2 q^2$, where $\alpha_1\alpha_2 = \alpha$, $pq = e$. Taking
$\alpha_1 = \alpha_2 = \beta_1 = 1$, we have a system like the proposed. Hence a solution
x, y, u, v leads to the second solution
$$X = u^2 x^2 - A v^2 y^2, \qquad Y = 2xyuv, \qquad V = u^2 x^2 + A v^2 y^2, \qquad U = u^4 - 2x^4.$$
For the lower sign above, we obtain a complicated set of formulas giving
a new solution from one. The formulas are said to give all solutions when
$A = 6$ and for the problem $x^2 \pm (x+2) = \square$ of Beha-Eddin.[32] By combining
Lucas' result (*ibid.*, p. 433) with the results of Fermat and Genocchi,[45]
Lucas concluded (p. 514) that $xy(x^2 - y^2) = Az^2$ has no rational solution if
$A = 1, 2, p, 2q$, where p and q are primes of the respective forms $8n+3$,
$8n+5$.

S. Günther[54] treated $x^2 + a = y^2$, $x^2 - a = z^2$ by setting
$$x - y = m(z - x), \qquad x + y = \frac{1}{m}(z + x),$$
which determine y and z in terms of x and m. Substituting these into one
of the proposed equations, we get x as a function of m:
$$x^2 = \frac{a(m^2 + 1)^2}{4m - 4m^3}.$$
Set $m = ap^2$. Then x is rational if $1 - a^2 p^4 = \square$. Hence we seek among the
rational solutions of $1 - a^2\xi^2 = \eta^2$ those values of ξ which are squares. If
such exist, a is a congruent number, otherwise not. We can not go further
with the general solution of the system since the character of a decides
whether or not such a biquadratic root of the Pell equation exists.

[53] Nouv. Ann. Math., (2), 17, 1878, 446.
[54] Prag Sitzungsberichte, 1878, 289–94.

S. Roberts[55] proved the known result that if $x^2 \pm Py^2$ are squares then $\pm Py^2$ is of the form $tab(a^2-b^2)$, where $t=4$ or 1 according as a, b are of different or like parity. He stated that the values of P which are primes or the doubles of primes are all obtained by the rule of Leonardo which makes three of a, b, $a \pm b$ squares, and carried further the analysis of Genocchi.[43] Inadmissible values of P are primes $8k+3$ or doubles of primes $8k+5$. He proved that various classes of primes P are excluded, all being such that $x^2-2Py^2=-1$ has no solution.

A. Desboves[56] started with a congruent number $\lambda\mu(\lambda^2-\mu^2)$, changed λ to x^2, μ to y^2, absorbed the factor x^2y^2 into the term Y^2 of $X^2 \mp aY^2 = \square$, and obtained the congruent number x^4-y^4. Since $a^4+d^4=b^4+c^4$ is solvable and hence also

$$(a^4-b^4)(d^4-b^4) = (ad)^4 - (bc)^4,$$

we can find an infinitude of numbers which are differences of two biquadrates and whose product is such a difference, and hence an infinitude of solutions of Boncompagni's[48] problem to find two congruent numbers whose product is a congruent number.

A. Genocchi[57] proved that the following numbers are not congruent: a prime $8k+3$ or the product of two such primes; the double of a prime $8k+5$ or the double of the product of two such primes.

Genocchi[58] stated his[44] results and that no congruent number is a product of a square by a prime $8m+3$, or by double a prime $8m+5$, or by a product of two primes $8m+3$, or by double the product of two primes $8m+5$.

G. Heppel[59] found a such that $101^2+a=(101+k)^2$, $101^2-a=(101-l)^2$ by taking $l=k+c$, whence $2k^2=202c-2kc-c^2$. Since c is a factor of $2k^2$, but not of k, $c=4$. Thus $k=18$, $a=3960$.

M. Jenkins[60] found an integer a for which $(m^2+n^2)^2 \pm a = (h \pm t)^2$. Then $a=2ht$, $(m^2+n^2)^2=h^2+t^2$. One solution of the latter is $h=m^2-n^2$, $t=2mn$.

G. B. Mathews[61] discussed $x^2 \pm a = \square$. From $x^2+a=(x+m)^2$, we get x. Then $x^2-a=N/(4m^2)$, where $N=a^2-6am^2+m^4$. Set $N=(a-m^2k/l)^2$. Then $m^2=fa$, $f=2l(k-3l)/(k^2-l^2)$. Take $a=fb^2$, where b is arbitrary. Then $m=fb$ and x is found.

R. Aiyar[62] noted that, if $A^2 \pm 4B$ are squares, A and B are expressible in one and but one way in the forms $A=\lambda(m^2+n^2)$, $B=\lambda^2mn(m^2-n^2)$, where m and n are relatively prime and one is even.

A. Cunningham and R. W. D. Christie[63] solved $x^2-y^2=y^2-w^2=cz^2$, where c is given, as $c=65$, by use of $x^2-2y^2=-w^2$. [Hence $y^2-cz^2=w^2$, $y^2+cz^2=x^2$.]

[55] Proc. Lond. Math. Soc., 11, 1879–80, 35–44.
[56] Assoc. franç., 9, 1880, 242.
[57] Memorie di Mat. e Fis. Soc. Ital. Sc., (3), 4, 1882, No. 3.
[58] Nouv. Ann. Math., (3), 2, 1883, 309–10.
[59] Math. Quest. Educ. Times, 40, 1884, 119.
[60] *Ibid.*, 41, 1884, 65–6.
[61] *Ibid.*, 107–8.
[62] Math. Quest. Educ. Times, 65, 1896, 100.
[63] Math. Quest. Educ. Times, (2), 13, 1908, 77–9.

G. Bisconcini[64] determined the numbers $A = 4rs(r^2 - s^2)$ which are products of powers of three primes. If 2 and 3 are the only prime factors of A, then $A = 24$.

R. D. Carmichael[65] proved that the system $q^2 + n^2 = m^2$, $m^2 + n^2 = p^2$ has no positive integral solutions [whence $m^2 + n^2 = p^2$, $m^2 - n^2 = q^2$].

H. B. Mathieu[66] asked if $x^2 + A = u^2$, $x^2 - A = v^2$ are completely solved by the identity

$$\{(a \pm b)^2 + b^2\}^2 \pm 4ab(a \pm b)(a \pm 2b) = (a^2 + 2b^2 \pm 4ab)^2.$$

L. Aubry[67] replied that all solutions of $2x^2 = u^2 + v^2$ are given by

$$u, v = l(r^2 \pm 2rs - s^2); \qquad x = l(r^2 + s^2); \qquad A = 4l^2rs(r^2 - s^2).$$

The case $l = 1$, $r = a \pm b$, $s = b$, gives the above identity, which with

$$\{(a \pm b)^2 + b^2\}^2 \mp 4ab(a \pm b)(a \pm 2b) = (a^2 - 2b^2)^2$$

give all relatively prime solutions. [Cf. Ch. XIV.]

G. Métrod[68] treated $x^2 + y = u^2$, $x^2 - y = v^2$. For $u_1 = a^2 - b^2$, etc.,

$$u_1^2 + v_1^2 = x^2.$$

Hence $(u_1 + v_1)^2 + (u_1 - v_1)^2 = 2x^2$; $u, v = a^2 - b^2 \pm 2ab$, $y = 4ab(a^2 - b^2)$.

J. Maurin and A. Cunningham[69] noted that from one solution of $x^2 - ny^2 = z^2$, $x^2 + ny^2 = t^2$, we get a second solution $X = x^4 + n^2y^4$, $Y = 2xyzt$.

A. Gérardin[70] listed the values < 1000 of h for which $x^2 \pm hy^2 = \square$ for $x < 3722$. He noted (pp. 57–9) the solutions

$$x = 16f^8 + 24f^4g^4 + g^8, \qquad y = 4fg(4f^4 - g^4), \qquad h = 4f^4 + g^4;$$
$$x = f^8 + 6f^4g^4 + g^8, \qquad y = 2fg(f^4 - g^4), \qquad h = 2f^4 + 2g^4.$$

L. Bastien[71] listed the 25 values < 100 of h for which $x^2 \pm hy^2$ are not squares, and stated (besides Genocchi's[50] results) the following cases of impossibility: h the double of a prime $16m + 9$; h a prime $8m + 1 = g^2 + k^2$, with $g + k$ a quadratic non-residue of h (as for 17, 73, 89, 97).

T. Ono[72] noted that $x^2 \pm 5y^2$ are squares for $x = 41$, $y = 12$ [Leonardo[4]] and $x = 3344161$, $y = 1494696$.

G. Candido[73] noted that, from two sets of solutions (x_i, y_i) of the system $x^2 \pm uy^2 = \square$, we get a third set by Euler's identity

$$(x_1x_2 \pm uy_1y_2)^2 + u(x_1y_2 \mp uy_1x_2)^2 = (x_1^2 + uy_1^2)(x_2^2 + uy_2^2).$$

E. Turrière[74] noted that if x, y, z are the rational coördinates of a point M on the quartic space curve $x^2 + a = y^2$, $x^2 + b = z^2$, the osculating plane at M is

$$(b - a)x^3X - by^3Y + az^3Z = ab(b - a),$$

<hr>

[64] Periodico di Mat., 24, 1909, 157–170.
[65] Amer. Math. Monthly, 20, 1913, 213–6.
[66] L'intermédiaire des math., 20, 1913, 2.
[67] *Ibid.*, 211–2. Practically same by Welsch, 212–3.
[68] Sphinx-Oedipe, 8, 1913, 130–1.
[69] L'intermédiaire des math., 21, 1914, 20–21, 176–8.
[70] *Ibid.*, 22, 1915, 52–3.
[71] *Ibid.*, 231–2.
[72] *Ibid.*, 117.
[73] *Ibid.*, 23, 1916, 111–2.
[74] L'enseignement math., 17, 1915, 315–324.

and meets the curve at a new point M_1 whose coördinates are rational and easily found. Thus if we employ Leonardo's solution $x=41/12$, $y=49/12$, $z=31/12$ when $a=5$, $b=-5$, we obtain in succession an infinitude of rational solutions. Or we may find the points with rational coördinates, on the hyperbola $y^2-x^2=a$, by setting $x+y=u$, $y-x=a/u$, and identifying their abscissas with those of the analogous points $x=(v^2-b)/(2v)$, $y=(v^2+b)/(2v)$ on $z^2-x^2=b$, obtaining the condition $uv(u-v)=av-bu$. The tangent at a rational point (u, v) meets the cubic at a new rational point. Finally, $x^2+a=y^2$, $x^2-a=z^2$ have the solutions y, $z=x$ $(\cos\theta\pm\sin\theta)$; $x^2=a/\sin 2\theta$; hence the rational solutions are given by those rational values of $\tan\theta/2$ for which $\sin 2\theta$ is a rational square; there are none if $a=1$ or 2.

For congruent numbers of order n, see papers 200–1, 210, 222 of Ch. XXIII, and 320 of Ch. XXII.

CONCORDANT FORMS: x^2+my^2, x^2+ny^2 BOTH SQUARES.
RELATED PROBLEMS.

Diophantus, II, 15, required x, m, n such that x^2+m and x^2+n are squares, given the sum $m+n$. He took $m=4x+4$, $n=6x+9$, $m+n=20$, whence $x=7/10$. In II, 16, $(x+2)^2-m$ and $(x+2)^2-n$ are squares if $m=4x+4$, $n=2x+3$; for $m+n=20$, $x=13/6$. The same problems occur in III, 23, 24.

Diophantus, II, 17, required x, m, n such that x^2+m and x^2+n are squares, given the ratio m/n. He took $x=3$, $m/n=3$, $n=(y+3)^2-9$. The condition is $3^2+3\{(y+3)^2-9\}=3y^2+18y+9=\square$, say $(2y-3)^2$, whence $y=30$.

Certain Arab writers[1,2] of the tenth century treated the special problem to make x^2+k and x^2-k both squares, taking k as given, unlike Diophantus.

Rafael Bombelli[75] divided 40 into two parts (30 and 10) such that if each be subtracted from the same square $(30\frac{1}{4})$ the remainders are squares.

L. Euler[7] treated the problem, equivalent to Diophantus II, 17: If a and b are given integers, find z, p, q, r, s such that

$$(1) \qquad p^2+azq^2=r^2, \qquad p^2+bzq^2=s^2.$$

Eliminating z, we get $p^2=(br^2-as^2)/(b-a)$. Since the latter is a square for $r=s$, set $r=s+(b-a)t$. Then $p^2=s^2+2bst+b(b-a)t^2$. Set $p=s+tx/y$. Equate to t the numerator of the resulting fraction for t/s. Thus

$$t=2xy-2by^2, \qquad s=b(b-a)y^2-x^2, \qquad p=(x-by)^2-aby^2,$$
$$r=(b-a)(2xy-by^2)-x^2, \qquad q^2z=4xy[(b-a)y-x](by-x).$$

Simplifications arise if we set $x=v+by$; then

$$p=v^2-aby^2, \qquad q^2z=4vy(v+ay)(v+by).$$

Thus we take v and y arbitrary, and choose q^2 to be the greatest square factor of the final expression. It is shown (§ 230) that p^2+q^2 and p^2+3q^2 are not both squares. Cf. Euler.[33]

<hr>

[75] L'Algebra Opera, Bologna, 1579, 461.
[76] Algebra, St. Petersburg, 2, 1770, §§ 225–230; French transl., Lyon, 2, 1774, pp. 286–302; Opera omnia, (1), I, 456–464. Cf. Euler[27] of Ch. XV.

Euler[77] called x^2+my^2 and x^2+ny^2 concordant forms if they can both be made squares by choice of integers x, y each not zero; otherwise, discordant forms. He treated the problem: Given an integer m, to find all integers n for which the two forms are concordant. Set $m=\mu\nu$, where one factor may be unity. Then $x^2+my^2=(\mu p^2+\nu q^2)^2$ if $x=\pm(\mu p^2-\nu q^2)$, $y=2pq$. To make $x^2+ny^2=w^2$, we must take $n=(w^2-x^2)/(4p^2q^2)$, where x has the preceding value. Then both factors $w\pm x$ must be even. Set $p^2q^2=r^2s^2$, where $2r^2$ divides $w+x$, and $2s^2$ divides $w-x$, the respective quotients being f and g. Then $n=fg$. Hence we consider x, r, s as known and seek f, g such that $fr^2-gs^2=x$. The latter is satisfied by $f=hs^2\pm\sigma x$, $g=hr^2\pm\rho x$, where ρ/σ is the convergent preceding r^2/s^2 for the continued fraction for the latter. A table (§ 10) gives the values of ρ, σ for $r^2\leqq 12^2$, $s^2\leqq r^2$. For $m=1$, Euler found (§ 12) those values numerically <100 of n which result from small values of r, s. It is known that $x^2\pm y^2$ are discordant; also x^2+y^2, x^2+2y^2. Proof is given (§§ 15–19) that x^2+y^2, x^2+3y^2 are discordant; also (§§ 20–23, § 31) that $x^2+y^2=z^2$, $x^2+4y^2=v^2$ are impossible. Hence

$$z^2-y^2=x^2, \qquad z^2+3y^2=v^2$$

and $v^2-4y^2=x^2$, $v^2-3y^2=z^2$ are discordant. But x^2+y^2, x^2+7y^2 are squares for $x=3$, $y=4$. In papers 109, 110, 113–4 of Ch. XVI it is noted that $x^2\pm y^2$ and $x^2\pm 4y^2$ are not both squares; also, $x^2\pm y^2$ and $x^2\mp 3y^2$.

Euler[78] satisfied $x^2+my^2=\square$ as in the last paper,[77] and noted that then $x^2+ny^2=(\mu p^2-\nu q^2+2Mp^2q^2)^2$ if $n=M^2p^2q^2+M(\mu p^2-\nu q^2)$, where M is arbitrary. If $M=N+\nu/p^2$, $n=(Np^2+\nu)(Nq^2+\mu)$.

Euler[79] noted that x^2+aby^2 is a square for $x=\zeta(ap^2-bq^2)$, $y=2\zeta pq$; that $x^2+cdy^2=\square$ for $x=\eta(cr^2-ds^2)$, $y=2\eta rs$. Hence set

$$\zeta pq=\eta r\varepsilon=\zeta\eta fghk, \qquad p=\eta fg, \qquad q=hk, \qquad r=\zeta fh, \qquad s=gk.$$

By the values of x,

$$\frac{g^2}{h^2}=\frac{\zeta}{\eta}\cdot\frac{\zeta\eta cf^2+bk^2}{\zeta\eta af^2+dk^2}.$$

Set $\theta=\zeta\eta$. Hence must $\theta(\theta cf^2+bk^2)(\theta af^2+dk^2)=\square$. But this condition was not discussed.

Euler[80] had previously treated the more special problem to find all integers N such that A^2+B^2 and A^2+NB^2 are both squares for $AB\neq 0$. Take $A=x^2-y^2$, $B=2xy$. Then

$$A^2+B^2=(x^2+y^2)^2, \qquad A^2+NB^2=(x^2-y^2)^2+4Nx^2y^2=z^2.$$

The last gives for N an expression which is an integer if $z=x^2+2\alpha x^2y^2\pm y^2$. According as the upper or lower sign is chosen, we have

$$N=(\alpha x^2+1)(\alpha y^2+1) \qquad\text{or}\qquad (\alpha x^2-1)(\alpha y^2+1)+1.$$

To investigate the rational α's for which the first N is integral, when x and y are integral, set $\alpha=a/(q^2s^2)$, $x=pq$, $y=rs$, where a is an integer, while p,

[77] Mém. Acad. Sc. St. Petersb., 8, 1817–8 (1780), 3; Comm. Arith., II, 406.
[78] Opera postuma, 1, 1862, 253 (about 1769).
[79] Ibid., 256 (about 1782).
[80] Nova Acta Acad. Petrop., 11, 1793 (1777), 78; Comm. Arith., II, 190–7.

etc., may become unity. Then, if $a=s=1$, we have $N=(p^2+1)(r^2+q^2)/q^2$. If $p=7$, $q=5$, q^2 divides p^2+1, and $N=2r^2+50$. If $a=-1$, $s=1$, then $N=(p^2-1)(r^2-q^2)/q^2$ and q^2 divides p^2-1 if $p=3$, $q=2$, etc. A list is given of the resulting N's numerically <100; those $\leqq 50$ and >0 are 7, 10, 11, 17, 20, 22, 23, 24, 27, 30, 31, 34, 41, 42, 45, 49, 50. But the problem is not proved impossible for the omitted values of N.

Euler[81] made $a^2x^2+b^2y^2$ and $a^2y^2+b^2x^2$ both squares by taking

$$\frac{ax}{by}=\frac{p^2-q^2}{2pq}, \qquad \frac{ay}{bx}=\frac{r^2-s^2}{2rs}.$$

By division, we get x^2/y^2. Hence it suffices to make the quotient of $pq(p^2-q^2)$ by $rs(r^2-s^2)$ a square, a problem* which had been frequently treated, but not completely solved. The first of three special methods is to take $s=q$, $r=p+q$; then we are to make $(p-q)/(p+2q)=\square$, which is the case if $p=u^2+2t^2$, $q=u^2-t^2$; the resulting solution is

$$a=3tu, \qquad b=2(u^2-t^2), \qquad x=t(2u^2+t^2), \qquad y=u(u^2+2t^2).$$

To obtain the general solution, we may take $s=q$ without loss of generality, since it is only a question of ratios. Then

$$n\equiv\frac{p(p^2-q^2)}{r(r^2-q^2)}=\square, \qquad q^2=\frac{p^3-nr^3}{p-nr}.$$

Set $p=rv$. Then $(v^3-n)/(v-n)=\square=(v-z)^2$ if

$$(n+2z)v^2-z(2n+z)v+n(z^2-1)=0.$$

From a given solution z, v, we get a second solution

$$v'=\frac{z(z+2n)}{2z+n}-v, \qquad z'=2v'-z.$$

Thus $v=0$, $z=1$ leads to the second solution

$$v'=\frac{1+2n}{2+n}, \qquad z'=\frac{3n}{2+n}.$$

Replace n by t^2/u^2; we get the above special solution. He investigated the third solution v'', z'', and also started with $v=0$, $z=-1$; $z=0$, $v=\pm 1$; $v=\infty$. Further, he treated the general condition for $n=4$ and $n=1/4$. In conclusion, he found $a, \cdots, d$ such that

$$a^2b^2+c^2d^2, \qquad a^2c^2+b^2d^2, \qquad a^2d^2+b^2c^2$$

are all squares. For $f=t^2-3u^2$, $g=2tu$, we have

$$f^2+3g^2=h^2, \qquad h=t^2+3u^2.$$

Then a solution is $a=2g$, $b=2h$, $c=f+g$, $d=f-g$, and the three quartics are the squares of f^2+7g^2, $2(f^2\mp fg+2g^2)$.

C. F. Degen[82] treated $x^2+my^2=p^2$, $x^2+ny^2=q^2$. We may set

$$p=\alpha(mt+nu), \qquad q=\alpha(nt+mu).$$

[81] Mém. Acad. Sc. St. Pétersbourg, 11, 1830 (1780), 12; Comm. Arith., II, 425–37.

* Euler,[67] seq., and Euler[77] of Ch. IV, Petrus[12] and Euler[33] of Ch. XV, Euler[18, 19] of Ch. XVIII, Euler[252] of Ch. XXII.

[82] Mém. présentés acad. sc. St. Pétersbourg par divers savans, 1, 1831 (1823), 29–38.

To avoid fractions set $\alpha=2$; then $y^2=4(m+n)(t^2-u^2)$. Set $m+n=fg$, $t^2-u^2=4fgz^2$. Hence $t=fz^2+g$, $u=fz^2-g$, $y=4fgz$, and we obtain " our fundamental solution "

$$\frac{x^2}{4g^2}+4mn=(f^2z^2-fg)^2.$$

Let k^2 be the maximum square dividing $mn=k^2L$, and set

$$x=4gk\xi, \qquad \pm A=\frac{f^2z^2-fg}{2k}.$$

Then $\xi^2+L=A^2$. Now $fg\pm2kA$ must be a square f^2z^2, say B^2. Thus $y=4gB$. Hence to exhibit the simplest solution, set[83] $m+n=fg=k^2L$, and let ϕ be any factor of L such that $\xi^2+L=A^2$ is satisfied [identically] by

$$2\xi=\frac{L}{\phi}-\phi, \qquad 2A=\frac{L}{\phi}+\phi.$$

Let $B=kC$. Then $fg\pm2kA=B^2$ becomes

$$L\pm\frac{1}{k}\left(\frac{L}{\phi}+\phi\right)=C^2.$$

If the latter can be solved, we have $x=(L/\phi-\phi)/2$, $y=C$, since $x:y=\xi:C$. It is proved (p. 33) that x^2+my^2 and x^2+ny^2 are concordant if

$$(m+1)(n+1)=\square=P^2,$$

since they equal $(mn+2m+1)^2$ and $(mn+2n+1)^2$ for $x=mn-1$, $y=2P$; also, if $m+n=2Q^2$, since they equal $(3m+n)^2$ and $(3n+m)^2$ for $x=m-n$, $y=4Q$.

M. Collins[46] proved that $x^2+y^2=\square$, $x^2+ay^2=\square$ are impossible for $1<a<20$, except for $a=7$, 10, 11, 17; also, $x^2-y^2=\square$, $x^2-ay^2=\square$ for $1<a<13$, except $a=7$, 11. If we know solutions of

$$x^2+ay^2=nz^2, \qquad y^2+bx^2=nw^2,$$

then $X=x^2w^2-y^2z^2$ and $Y=2xyzw$ are solutions of

$$X^2+Y^2=\square, \qquad X^2+abY^2=\square.$$

C. H. Brooks and S. Watson[84] found that x^2+y^2 and x^2+Ay^2 can be simultaneously squares only for the following 41 positive integers $A\leqq100$: 1, 7, 10, 11, 17, 20, 22, 23, 24, 27, 30, 31, 34, 41, 42, 45, 49, 50, 52, 57, 58, 59, 60, 61, 68, 71, 72, 74, 76, 77, 79, 82, 85, 86, 90, 92, 93, 94, 97, 99, 100. Set $x/y=v$, $v^2+1=(v+n)^2$, $v^2+A=(v-pn)^2$. The two rational values of v give $n^2=(A+p)/(p^2+p)$, where p may be any positive or negative integer or fraction for which A is integral.

S. Bills[85] gave a theorem said to include all the theorems by Collins.[46] The equations $x^2+Ay^2=\square$, $x^2+By^2=\square$, cited as (F), are satisfied if

$$x=mp^2-\frac{A}{m}q^2=nr^2-\frac{B}{n}s^2, \qquad y=2pq=2rs.$$

In view of the latter, take $p=fg$, $q=hk$, $r=fh$, $s=gk$. Then the former holds

---

[83] But the author had previously set $mn=k^2L$.

[84] The Lady's and Gentleman's Diary, London, 1857, 61–3, Quest. 1911.

[85] The Lady's and Gentleman's Diary, 1861, 82–4.

if (F_1) $mg^2-nh^{'}=Nk^2$, $m^{-1}Ah^2-n^{-1}Bg^2=Nf^2$, or if (F_2) $mf^2+n^{-1}Bk^2=Nh^2$, $nf^2+m^{-1}Ak^2=Ng^2$. Hence the solution of (F) can be derived from the solution of (F_1) or (F_2). Giving suitable values to m, n, N, A, B, we can readily derive all of Collins' formulas from (F_1) and (F_2).

A. Genocchi[86] stated that x^2+h and x^2+k are not both squares if (i) $h=1$, k a prime or square of a prime $8m\pm3$ provided the odd prime factors of $k-1$ are all of the form $4n+3$; (ii) $h=2$, k a prime $8m+3$ or double of a prime $8m+5$, provided the odd prime factors of $k-2$ are all of the form $8n+7$; (iii) h a prime $8m\pm3$, k a prime $8m+7$, provided the odd prime factors of $h-k$ are all of the form $4n+3$ and quadratic non-residues of k; (iv) h a prime $8m+3$, $k=h^2$, provided the odd prime factors of $h-1$ are all of the form $4n+3$; (v) h a prime, $k=hp$, where h and p are primes one of the form $8m+3$ and the other of the form $8m+7$, provided the prime factors of $p-1$ other than 2 and h are all of the form $4n+3$ and quadratic non-residues of h.

A. Genocchi[87] treated (1) by the method of Diophantus: Set $r=mx+p$, $s=nx-p$. Then $bzq^2=(r^2-p^2)b/a$, so that the second equation (1) becomes

$$[(mx+p)^2-p^2]\frac{b}{a}=(nx-p)^2-p^2, \qquad x=\frac{2p(an+bm)}{an^2-bm^2},$$

p being given in the problem of Diophantus II, 17. In Euler's[76] problem, p is unknown; the first of the preceding equations determines p in terms of m, n, x; then $azq^2=amnx^2(m+n)/(an+bm)$. By setting $n=bl$, $x=2(m+al)$, we get formulas derived from Euler's by changing p, l, m into $-p$, y, v. Genocchi noted that the present problem is equivalent to that of solving $y^2-x^2:z^2-y^2=a:b$, treated fully by Leonardo Pisano.[6] For, (1) gives $r^2-p^2:s^2-p^2=a:b$, and conversely, if we set $r^2-p^2=azq^2$. Genocchi proved (pp. 9–23) that the system $x^2+a=\square$, $x^2+b=\square$ is impossible in rational numbers for $a=1$ and b a prime $8k\pm3$ such that $b-1$ has no prime divisor $4t+1$ (as $b=3$, 5, 13, 19, 29, 37, 43); for $a=2$ and b a prime $8k+3$ such that every divisor of $b-2$ is of the form $8t+7$ (as $b=163$, 331, 449); $a=2$, $b=2A$, A a prime $8k+5$ such that $A-1$ has no odd prime divisor not of the form $8t+7$ (as $A=5$, 29, 197, 317); $a=A$, $b=AB$, where A, B are primes, one of the form $8k+7$ and the other $8k+3$, such that A is a quadratic residue of B when $A=8k+7$, $B-1$ has no odd prime divisor $4t+1$ not a quadratic residue of A, and, in case $A=8k+3$, $(B-1)/2$ is divisible by A if a quadratic residue of A (as $A=3$, $B=7$; $A=7$, $B=3$ or 19; $A=11$, $B=23$); a a prime $8k+3$, every odd prime divisor of $a-1$ being of the form $4t+3$, $b=a^2$; $a=1$, $b=\pm8$ or the negative of a prime $8k\pm3$ or the square of a prime $8k\pm3$, no prime divisor of $b-1$ being of the form $4t+1$ in the third case.

The following three papers relate to the system $t^2+u^2=2v^2$, $t^2+2u^2=3w^2$.

E. Lucas[88] treated the equivalent system $2v^2-u^2=t^2$, $2v^2+u^2=3w^2$ and showed how to get new solutions from one. [Cf. Pepin.[90]]

[86] Comptes Rendus Paris, 78, 1874, 433–5.
[87] Memorie di Mat. e Fis. Soc. Italiana Sc., (3), 4, 1882, No. 3.
[88] Nouv. Ann. Math., (2), 16, 1877, 409–416.

G. C. Gerono[89] took $t=1$, without loss of generality. Since u is odd, $u=2k+1$, the first condition gives $v^2=k^2+(k+1)^2$. He proved that also $v=m^2+(m+1)^2$. Using the [unproved] theorem of de Jonquières[26] of Ch. XVII, we get $v=5$ (excluded) or 1.

T. Pepin[90] noted that Lucas[88] did not treat all possible cases, whereas the omitted cases add new solutions. By using a somewhat different method, we get all solutions by a single set of formulas. We may limit to relatively prime solutions t, u and take v and w positive. By the first equation,

$$v^2=\left(\frac{t+u}{2}\right)^2+\left(\frac{t-u}{2}\right)^2, \qquad \frac{t+u}{2}=a^2-4b^2, \qquad \frac{t-u}{2}=4ab, \qquad v=a^2+4b^2,$$

for a, b relatively prime, a odd. By the second given condition,

$$t+u\sqrt{-2}=(1+\sqrt{-2})(c+d\sqrt{-2})^2, \qquad w=c^2+2d^2.$$

Comparing the values of t and of u, we get equations equivalent to

$$a^2+4ab-4b^2=c^2+2cd-2d^2, \qquad 4ab=3cd.$$

Thus $a=\alpha\lambda$, $b=3\beta\mu$, $c=\alpha\beta$, $d=4\lambda\mu$. Inserting these into the difference of the two preceding equations, we get a quadratic giving

$$\frac{\mu}{\alpha}=\frac{\beta\lambda\pm\sqrt{(3\beta^2-4\lambda^2)(2\lambda^2-3\beta^2)}}{2(9\beta^2-8\lambda^2)}.$$

Since the radical must be rational, $3\beta^2-4\lambda^2=\pm\gamma^2$, $2\lambda^2-3\beta^2=\pm\delta^2$. The upper signs are excluded modulo 3. Hence $2\lambda^2=\gamma^2+\delta^2$, $3\beta^2=\gamma^2+2\delta^2$, a pair like the given pair. Hence a solution v, w, t, u leads to a second solution

$$v_1=\alpha^2v^2+36\mu^2w^2, \qquad w_1=\alpha^2w^2+32\mu^2v^2, \qquad t_1, u_1=(\alpha v\mp6\mu w)^2-72\mu^2w^2,$$

where $\mu:\alpha=vw\pm tu:2(9w^2-8v^2)$. Starting from $v=w=t=u=1$, we get $\mu/\alpha=0$ or 1, the second giving $v_1=37$, $w_1=33$, $t_1=47$, $u_1=23$; etc. It is proved that we get all solutions in this way.

To find[91] two squares whose sum is double a square and difference is 10 times a square, take x, $y=2pq\pm(p^2-q^2)$. Then $x^2+y^2=2(p^2+q^2)^2$, $x^2-y^2=8pq(p^2-q^2)=10(12m^2)^2$ if $p=5m$, $q=4m$.

J. H. Drummond and W. F. King[92] proved that $2x^2-y^2=\square$, $2y^2-x^2=\square$ imply $x^2=y^2$.

A. Gérardin[93] noted that x^2+ny^2 and nx^2+y^2 are squares if

$$n=(\alpha^2+\beta^2)^2-1,$$

or $n=7$, $x=3$, $y=1$, or $n=17$, $x=8$, $y=1$.

Several writers[94] gave solutions of the last problem.

R. Goormaghtigh[95] made Sx^2+Py^2 and Sy^2+Px^2 both squares.

L. Aubry[95a] proved that $2y^2+u^2$ and $3y^2+u^2$ are not both squares.

[89] Nouv. Ann. Math., (2), 17, 1878, 381–3.
[90] Atti Accad. Pont. Nuovi Lincei, 32, 1878–9, 281–292.
[91] Math. Quest. Educ. Times, 63, 1895, 64.
[92] Amer. Math. Monthly, 6, 1899, 47–8, 151–5.
[93] L'intermédiaire des math., 22, 1915, 128.
[94] *Ibid.*, 23, 1916, 63–4, 205–7.
[95] *Ibid.*, 184–5.
[95a] *Ibid.*, 26, 1919, 84–5. For $u=1$, Rignaux.[131]

$$x^2+y \text{ AND } x+y^2 \text{ BOTH SQUARES.}$$

Diophantus, II, 21, took $y=2x+1$. Then $x^2+y=\square$. Let

$$x+y^2 \equiv 4x^2+5x+1$$

be the square of $2x-2$. Then $x=3/13$.

Alkarkhi[96] (beginning of eleventh century), after repeating this solution, added the condition $x^2+y^2=\square$, taking $x=3z$, $y=4z$.

Rafael Bombelli[97] set $y=4(x+1)$. Let $x+y^2 \equiv 16x^2+33x+16$ be the square of $4x-6$. Then $x=20/81$.

W. Emerson[98] treated the problem.

L. Euler[99] set $x^2+y=(p-x)^2$, $y^2+x=(q-y)^2$, whence

$$x=\frac{2qp^2-q^2}{4pq-1}, \qquad y=\frac{2pq^2-p^2}{4pq-1}.$$

Euler had first inserted the value of y from $x^2+y=p^2$ into y^2+x, obtaining $(p^2-x^2)^2+x=\square$, which he stated would be difficult to solve.

R. Adrain[100] noted that Euler's last condition is satisfied if we take $x=4p^2x^2$. Again, for $p+x=v$, it becomes

$$v^2(v-2x)^2+x=v^4-4v^3x+4v^2x^2+x=(v^2+2vx)^2 \qquad \text{if} \qquad x=8v^3x, \ v=\tfrac{1}{2}.$$

The equivalent problem $x^2-y=\square$, $y^2-x=\square$ was solved[101] as by Euler.[99]

J. W. West[102] noted that Euler's[99] condition is satisfied if $p^2-x^2=y$, $x=2y+1$. Solve the quadratic in x obtained by eliminating y.

C. A. Laisant,[103] after recalling Euler's[99] solution in rational numbers, remarked that the system is evidently impossible in positive integers, since in

$$y=(z-x)x+(z-x)z, \qquad x=(t-y)y+(t-y)t, \qquad z>x, \qquad t>y,$$

$y>x$ by the first equation and $x>y$ by the second. Similarly for negative solutions.

A. Auric[104] noted that Euler's solution is not general, since his problem is equivalent to the solution in integers of the homogeneous system $x^2+uy=z^2$, $ux+y^2=t^2$, which can be solved for x, y after giving arbitrary values to z, t, u (by factoring z^2-t^2).

L. Aubry[105] and G. Quijano[106] proved the impossibility of integral solutions.

---

[96] Extrait du Fakhrî, French transl. by F. Woepcke, Paris, 1853, 88–9.
[97] L'algebra opera, Bologna, 1579, 467.
[98] A Treatise of Algebra, London, 1764, 1808, p. 239.
[99] Algebra, 2, 1770, art. 239; French transl., Lyon, 2, 1774, 335–6. Opera Omnia, (1), I, 482.
[100] The Math. Correspondent, New York, 2, 1807, 11–13.
[101] The Ladies' and Gentlemen's Diary (ed., M. Nash), N. Y., 2, 1821, 45.
[102] Math. Quest. Educ. Times, 67, 1897, 64.
[103] Nouv. Ann. Math., (4), 15, 1915, 106–8.
[104] Ibid., 280–1.
[105] L'intermédiaire des math., 22, 1915, 67, simpler on p. 226.
[106] Ibid., 23, 1916, 87–8.

x^2+y^2-1 AND x^2-y^2-1 BOTH SQUARES.

Bháscara[107] (born 1114) gave sets of values of x, y for which $x^2\pm y^2-1$ are both squares:

$$y=\frac{8a^2-1}{2a},\quad x=\frac{y^2}{2}+1;\qquad x=\frac{1}{2a}+a,\ y=1;\qquad x=8a^4+1,\ y=8a^3.$$

Bháscara[108] treated this problem and the similar one on $x^2\pm y^2+1$, as being due to an ancient author. To find two squares whose sum and difference increased by unity are squares, call the desired squares $[y^2=]4k^2$ and $[x^2=]5k^2-1$, the latter being a square for $k=1$ or 37. For decrease by unity, use $4k^2$ and $5k^2+1$, a square for $k=4$ or 72.

Having chosen the coefficient 4, the other coefficient (5) is to be determined so that when 4 is added or subtracted we get a square. Thus $2\cdot4$ is the difference of two squares. Taking 2 as the difference of their roots, we get the roots to be 1 and 3, whence $5=4+1^2=3^2-4$. Similarly, taking 36 as the first coefficient, we must make 72 a difference of two squares. Taking 6 as the difference of their roots, we get 45 as the second coefficient; taking 4, we get 85.

J. Cunliffe[108a] solved $x^2=\frac{1}{2}(c^2+d^2)+1$, $y^2=\frac{1}{2}(c^2-d^2)$ by taking $c=d+n$, $y=rn$, whence $n=2d/(2r^2-1)$ by the second condition. Take $d=s(2r^2-1)$, $x=ts+1$. The first condition is satisfied if $(4r^4+1-t^2)s=2t$. For $t=2r^2$, we get Bháscara's final answer.

E. Clere[109] treated the same pair $x^2+y^2-1=z^2$, $x^2-y^2-1=u^2$. By subtraction, $2y^2=z^2-u^2$. Let $y=pq$ and set $z+u=2q^2$, $z-u=p^2$ [thus limiting to integral solutions]. Substituting the resulting values of z, u into the proposed first equation, we get $4x^2=4+4q^4+p^4$, which is a square if $p=q^2$. Thus we have the special solution

$$x=1+q^4/2,\qquad y=q^3,\qquad z=q^2+q^4/2,\qquad u=q^2-q^4/2.$$

A. Genocchi[110] proved that all rational solutions are given by

$$y=\frac{2gpq}{l},\qquad x=\frac{l+r}{l},\qquad l\equiv\frac{g^2(p^4+4q^4)}{2r}-\frac{r}{2},$$

where p, q are relatively prime integers, q odd; r an integral divisor of $g^2(p^4+4q^4)$ and $r\equiv g\ (\mathrm{mod}\ 2)$. We may give any rational values to g, p, q, r and, without loss of generality, replace r by gr. Then $y=4pqr/d$, $x=(p^4+4q^4+r^2)/d$, where $d=p^4+4q^4-r^2$. If we set $r=2q^2$, $p=1$, we get $y=8q^3$, $x=8q^4+1$; if we set $r=p^2-2q^2$, $p=-1/2$, we get also the first set by Bháscara.

---

[107] Lílávatí (arith.), §§ 59–61. Algebra, with arith. and mensuration, from the Sanscrit of Brahmegupta and Bháscara, transl. by Colebrooke, London, 1817, p. 27. Lilawati or a treatise on arith. and geom. by Bhascara Acharya, transl. by John Taylor, Bombay, 1816, 35.

[108] Víja-gañita (algebra), § 194; Colebrooke,[107] pp. 257–9.

[108a] New Series Math. Repository (ed., T. Leybourn), 2, 1809, I, 199.

[109] Nouv. Ann. Math., 9, 1850, 116–8.

[110] Ibid., 10, 1851, 80–85.

T. Pepin[111] found that all rational solutions are given by

$$dx = m^2p + n^2q, \qquad dy = 4mnst, \qquad dz = 2mn(s^2 + 2t^2), \qquad du = 2mn(s^2 - 2t^2),$$

where $d = m^2p - n^2q$, while m, n are relatively prime, also s, t. Further, $s^4 + 4t^4 = pq$. To obtain integral solutions, take

$$p = 1, \qquad d = \pm 1, \qquad m + \sqrt{5}\,n = (2 + \sqrt{5})^k.$$

Various writers[112] gave solutions.

J. H. Drummond[113] took $x^2 + y^2 + 1 = (m+1)^2$, $x^2 - y^2 + 1 = (m-1)^2$, $m = 2n^2$, whence $x = 2n^2$, $y = 2n$.

E. B. Escott[114] asked if $x^2 + y^2 - 1$, $x^2 - y^2 + 1$ are squares when $xy \neq 0$, $x \neq y$, for integral values other than $x = 13$, $y = 11$; in other words, if $4mn(m^2 - n^2) + 1 = \square$ ($mn \neq 0$, $m \neq n$) has a solution other than $m = 3$, $n = 2$. Several replies[115] show there is an infinitude of solutions.

R. P. Paranjpye[116] gave Bháscara's[107] third solution. Suppose in $2y^2 = z^2 - t^2$ that the common factor of y, z, t is a square. Since the difference of two squares is divisible by 8, we may set $z + t = 4\xi^2$, $z - t = 2\eta^2$, $y = 2\xi\eta$. Then $x^2 = 1 - y^2 + z^2 = 4\xi^4 + \eta^4 + 1$. Assume that $\eta^4 = 4\xi^2$, whence $\xi = 2p^2$, $\eta = 2p$.

$$x^2 + 2fxy + hy^2, \quad x^2 + 2gxy + ky^2 \text{ both squares.}$$

Beha-Eddin[32] listed as the last of seven problems remaining unsolved from former times that to make $x^2 \pm (x+2)$ both squares. His translator, Nesselmann (pp. 72–3), discussed the problem.

A. Marre[117] found only the solution $x = -17/16$ and concluded that the problem is impossible in positive integers.

A. Genocchi[118] called the squares $(p+q)^2$ and $(p-q)^2$, whence $x^2 = p^2 + q^2$, $x + 2 = 2pq$. By eliminating x, $(4p^2 - 1)q^2 - 8pq - (p^2 - 4) = 0$. By taking the first or third coefficient zero, we get $x = -2$, $-17/16$, $34/15$.

E. Lucas[119] solved completely the corresponding homogeneous equations

$$x^2 + xy + 2y^2 = u^2, \qquad x^2 - xy - 2y^2 = v^2,$$

where x, y, u, v may be assumed relatively prime. Adding, we see that the sum of the squares of $(u \pm v)/2$ is x^2, whence

$$\tfrac{1}{2}(u+v) = r^2 - s^2, \qquad \tfrac{1}{2}(u-v) = 2rs, \qquad x = r^2 + s^2.$$

Substitute the resulting values of u, v, x into the equation obtained by subtracting the proposed equations, we get $2y^2 + xy = 4rs(r^2 - s^2)$, whence $y = \tfrac{1}{4}(-x \pm t)$, where

$$(1) \qquad\qquad (r^2 + s^2)^2 + 32rs(r^2 - s^2) = t^2.$$

[111] Nouv. Ann. Math., (2), 14, 1875, 63.
[112] Math. Visitor, 2, 1887, 66–70.
[113] Amer. Math. Monthly, 9, 1902, 232.
[114] L'intermédiaire des math., 12, 1905, 76.
[115] *Ibid.*, 207–211; 13, 1906, 25. Cf. Zerr[50] of Ch. XIX.
[116] Jour. Indian Math. Club, Madras, 1, 1909, 188–9.
[117] Nouv. Ann. Math., 5, 1846, 323.
[118] Annali di Sc. Mat. e Fis., 6, 1855, 132, 303–4.
[119] Nouv. Ann. Math., (2), 15, 1876, 359–365. Same in Bull. Bibl. Storia Sc. Mat., 10, 1877, 186–191.

Hence the product of $r^2+16rs-s^2\pm t$ is $252r^2s^2$; call the factors $\pm14(3p)^2$, $\pm2q^2$ and add and subtract. Thus

$$r^2+16rs-s^2=\pm(63p^2+q^2),\qquad rs=pq.$$

For the upper sign, one solution of (1) leads to two new solutions:

$$R=m(r^2+s^2),\qquad S=nt,\qquad T=63n^2(r^2+s^2)-m^2t^2,$$
$$n\equiv4rs(r^2-s^2),\qquad m\equiv t(r^2+s^2)\pm(r^4+s^4-6r^2s^2),$$

so that the proposed pair has the solutions

$$x=r^2+s^2,\qquad 4y=-r^2-s^2\pm t,\qquad u,\ v=r^2-s^2\mp2rs.$$

For the lower sign, the problem is reduced to the earlier case.

A. Gérardin[120] used the known solution of $u^2+v^2=2x^2$:

$$u=2m^2-l^2,\qquad v=2m^2+l^2-4lm,\qquad x=2m^2+l^2-2lm.$$

It remains to make $8u^2-7x^2\equiv(x+4y)^2$ a square:

$$4m^4+l^4-88l^2m^2+56lm^3+28l^3m=\square=(2m^2-14ml-l^2)^2.$$

Then $x=34$, $y=15$, $u=46$, $v=14$ [Genocchi]. It is stated that we have also $y=-32$.

L. Euler[121] solved $x^2+2fxy+hy^2=P^2$, $x^2+2gxy+ky^2=Q^2$. Subtract and set $P-Q=(f-g)y$, whence $P+Q=2x+y(h-k)/(f-g)$. Squaring and adding, we get $2P^2+2Q^2$; equating to the value obtained by adding the proposed equations, we get

$$x:y=(f-g)^4-2(h+k)(f-g)^2+(h-k)^2:4(f-g)(f^2-g^2-h+k).$$

N. Fuss[122] made $f_1\equiv x^2+2axy+y^2$ and $f_2\equiv x^2+2bxy+y^2$ both squares, say p^2 and q^2. Then $p^2-q^2=2(a-b)xy$. Hence $x=4(a+b)$, $y=(a-b)^2-4$ is a particular solution since

$$f_1=[(a-b)(3a+b)-4]^2,\qquad f_2=[(a-b)(3b+a)-4]^2.$$

To find n such that $x^2\pm2nxy+y^2$ are squares, say $(p\pm q)^2$, we have

$$x^2+y^2=p^2+q^2,\qquad nxy=pq.$$

Set $p=\alpha xy$, $q=n/\alpha$. Then $n^2=\alpha^2(x^2+y^2)-\alpha^4x^2y^2$.

A. S. Werebrusow[123] reduced the system

$$\alpha x^2+2\alpha'xy+\alpha''y^2=u^2,\qquad \beta x^2+2\beta'xy+\beta''y^2=v^2$$

to $au^4+2bu^2v^2+cv^4=z^2$, $a=\beta'^2-\beta\beta''$, $b=\alpha\beta''+\alpha''\beta-2\alpha'\beta'$, $c=\alpha'^2-\alpha\alpha''$.

H. B. Mathieu[124] gave the solutions $x=15$, $y=-8$; $x=1768$, $y=2415$, of $x^2+y^2=\square$, $x^2+xy+y^2=\square$. L. Aubry[125] gave a general discussion.

Adrain,[113] Genocchi,[119] etc., of Ch. XXII proved that $x^2\pm xy+y^2$ are not both squares.

<hr>

[120] Sphinx-Oedipe, 1906–7, 162; Assoc. franç., 1908, 17.

[121] Opera postuma, I, 1862, 254 (about 1777). Nova Acta Acad. Petrop., 13, 1795–6 (1778), 45; Comm. Arith., II, 292.

[122] Mém. Acad. Sc. St. Pétersbourg, 9, 1824 (1820), 151–160.

[123] Math. Soc. Moscow, 26, 1098, 497–543; Fortschritte, 39, 1908, 259.

[124] L'intermédiaire des math., 17, 1910, 219.

[125] *Ibid.*, 283–5.

Two functions of one unknown made squares.

C. G. Bachet[126] treated the double equality

$$4N^2 + 3N - 1 = \square, \qquad 4N^2 + 4N - 1 = \square,$$

by factoring the difference N into $\frac{1}{4}$ and $4N$, where $4N$ is the double of $\sqrt{4N^2}$, and equating the squares of $\frac{1}{2}(4N \mp \frac{1}{4})$ to the given left members, whence, in either case, $\frac{7}{2}N = 65/64$. If the second equation is changed to $4N^2 - N - 1 = \square$, use the factors 1 and $4N$.

For $N^2 - 12 = \square$, $\frac{13}{2}N - 12 = \square$, use the factors N and $N - 13/2$ of the difference, so that their sum shall contain the double of $\sqrt{N^2}$.

For $4N^2 - N - 4 = \square$, $4N^2 + 15N = \square$, use the factors 4 and $4N + 1$ of the difference.

For $N^2 - 6144N + 1048576 = \square$, $N + 64 = \square$, first multiply the latter by 16384.

Fermat[127] treated many double and triple equalities.

J. L. Lagrange[128] considered briefly the system

$$(1) \qquad a + bx + cx^2 = \square, \qquad \alpha + \beta x + \gamma x^2 = \square.$$

If $a + bf + cf^2 = g^2$, the general solution of the first is

$$x = (fm^2 - 2gm + b + cf)/(m^2 - c).$$

Then the product of the second quadratic by $(m^2 - c)^2$ is a quartic function of m. There is no known rule to make the latter a square. If $a = \alpha = 0$, set $x = 1/y$; we are led to the simple problem $by + c = \square$, $\beta y + \gamma = \square$.

R. Adrain[129] treated $ax^2 + b = \square$, $cx^2 + d = \square$, given $ar^2 + b = e^2$, by setting $x = r + y$. Then $ax^2 + b = e^2 + 2ary + ay^2 = (zy - e)^2$ determines y rationally in z. For this value of y, the second condition becomes $Q = \square$, where Q is a quartic in z; but no treatment is given. Next, consider (1) for the case in which c and γ are squares; by multiplication by squares, we may assume that the coefficients of x^2 are equal and proceed as in the following example. For $x^2 - x + 7 = A^2$, $x^2 - 7x + 1 = B^2$, we have $6x + 6 = A^2 - B^2$. Take

$$2x + 2 = A + B, \qquad 3 = A - B.$$

Inserting $x + 5/2$ for A in the first given equation, we get $x = 1/8$.

Several[130] solved $1 - 8x = \square$, $x - 4x^2 + 4 = \square$ by inserting $x = (1 - a^2)/8$ into the second condition. Two answers are

$$x = 19740/177241, \qquad 72165/578888.$$

W. Welmin[130a] employed the elliptic function $\phi(\lambda)$ obtained by the inversion of the integral

$$(2) \qquad \lambda = \int_0^x \frac{dx}{\sqrt{(ax^2 + b)(cx^2 + d)}}.$$

---

[126] Diophanti Alexandrini Arith., 1621, 439–440. Comment on Diop. VI, 24 (p. 177 above).

[127] Oeuvres, III, 329–376, French transl. of J. de Billy's Inventum Novum. See de Billy[65] of Ch. IV; Fermat[9-11] and Ozanam[16] of Ch. XV; Fermat[373] of Ch. XXI; Fermat[40] of Ch. XXII.

[128] Additions to Euler's Algebra, 2, 1774, 557–9. Euler's Opera Omnia, (1), I, 596; Oeuvres de Lagrange, VII, 115–7. Extracts by Cossali,[17] 108–113.

[129] The Math. Correspondent, New York, 1, 1804, 238–240.

[130] Math. Miscellany, Flushing, N. Y., 1, 1836, 67–72.

[130a] Annales Univ. Warsaw, 1913, 1–17 (in Russian).

If λ be chosen so that $\phi(\lambda)$, $\{a\phi^2(\lambda)+b\}^{1/2}$ and $\{c\phi^2(\lambda)+d\}^{1/2}$ take rational values, rational solutions of the pair of equations $ax^2+b=\square$, $cx^2+d=\square$, are $x_1=\phi(\lambda)$, $x_2=\phi(2\lambda)$, $x_3=\phi(\lambda+2\lambda)$, $\cdots$.   In order that there be an infinity of solutions, it is necessary that the integral (2) have an irrational ratio to the same integral extended from x to ∞.

M. Rignaux[131] proved that $2y^2+1$ and $3y^2+1$ are both squares only when $y=0$.

Miscellaneous pairs of quadratic functions made squares.

Diophantus, II, 31, made $xy\pm(x+y)$ squares.   Since $2^2+3^2\pm2\cdot2\cdot3$ is a square, take $xy=(2^2+3^2)x^2$, $x+y=2\cdot2\cdot3x^2$, whence $y=13x$, $14x=12x^2$.

Paul Halcke[132] gave three ways of solving the problem.

L. Aubry, Welsch and E. Fauquembergue[133] proved that the problem is impossible in integers.

Diophantus, II, 26, found two numbers ($12x^2$ and $7x^2$) such that the square ($16x^2$) of their sum minus either number gives a square.   Hence $19x^2=4x$.

This problem was treated by J. H. Rahn and J. Pell,[134] and the latter treated (p. 102) the corresponding problem (Diophantus, III, 3) for three numbers.

Bháscara[135] made $7y^2+8z^2$ and $7y^2-8z^2+1$ both squares.   Treating the first by the method of the "affected square" (Ch. XII) with $8z^2$ as the additive quantity and $2z$ as the least root, we get $7(2z)^2+8z^2=(6z)^2$.   For $y=2z$, the second expression becomes $20z^2+1$ and is a square for $z=2$ or 36.

W. Emerson[136] made $xy+x$ and $xy+y$ squares.

Fr. Buchner[137] made $xy-x$ and $xy-y$ squares by taking $y=p^2x+1$. Then $xy-y=(px-m)^2$ if $x=(m^2+1)/(2mp-p^2+1)$.

S. Tebay[138] made $x^2+cxy+y^2\pm a$ squares.   Let $x^2+cxy+y^2+a=(y+p)^2$ determine y.   Then $x^2+cxy+y^2-a=\square$ if $x^4+\cdots=(x^2-cpx+q)^2$, which gives x.

Several[139] proved that $P+Q=R^2$, $P^2+Q^2=S^2$ imply that P^3+Q^3 is a sum of two squares:

$$P^3+Q^3=R^2\{\tfrac{1}{4}(S+P-Q)^2+\tfrac{1}{4}(S-P+Q)^2\}.$$

Also PQ is divisible by 12.   To find[140] all integral solutions P, Q, set $Q=Pq/p$.   Then $P+Q=R^2$ gives P, while $P^2+Q^2=s^2P^2$ if $p^2+q^2=p^2s^2$ and hence if $p=m^2-n^2$, $q=2mn$.

---

[131] L'intermédiaire des math., 25, 1918, 94–5.

[132] Deliciae Mathematicae, oder Math. Sinnen-Confect, Hamburg, 1719, 245–6.

[133] L'intermédiaire des math., 18, 1911, 71–2, 285–6; 20, 1913, 249.

[134] Rahn's Algebra, Zurich, 1659, 110.   An Introduction to Algebra, transl. by T. Brancker . . . augmented by D. P[ell], London, 1668, 100.

[135] Vija-gañita, § 187; Colebrooke,[107] p. 252.

[136] A Treatise of Algebra, London, 1764, 1808, p. 379.

[137] Beitrag zur Aufl. unbest. Aufg. 2 Gr., Progr. Elbing, 1838.

[138] Math. Quest. Educ. Times, 44, 1886, 62–3.

[139] Ibid., 54, 1891, 38.

[140] Ibid., 60, 1894, 128.   Cf. Teilhet[359] of Ch. XXI.

A. C. L. Wilkinson[141] made $2y^2-z^2$ and $2z^2-y^2+1$ squares x^2 and u^2. Put $x=a+b$, $z=a-b$. Then $x^2+z^2=2y^2$ gives $a^2+b^2=y^2$; a, $y=k(m^2\mp n^2)$; $b=2kmn$. Take $m=9$, $n=7$. Then $u^2=193(2k)^2+1$. By the continued fraction for $\sqrt{193}$, $u=6224323426849$, $k=224018302020$.

A. Gérardin and R. Goormaghtigh[142] treated

$$(y-x)^2+x=A^2, \qquad (y-x)^2-y=B^2.$$

Add and set $t=x-y$; then $2t^2+t=A^2+B^2$, an easy problem.

---

[141] Jour. Indian Math. Club, 2, 1910, 193.
[142] L'intermédiaire des math., 22, 1915, 193; 24, 1917, 84-5.

CHAPTER XVII.

SYSTEMS OF TWO EQUATIONS OF DEGREE TWO.

TWO QUADRATIC EQUATIONS IN TWO UNKNOWNS.

Beha-Eddin[1] (1547–1622) included (as Prob. 3) among the 7 problems remaining unsolved from former times that to make $x^2+y=10$, $y^2+x=5$. Nesselmann noted that there is no rational solution. Marre ′p. 323) noted that it leads to $x^4-20x^2+x+95=0$, having no rational solution.

Cataldi[2] required x, y when $x^2 + y^2$ and $xy/(x - y)^2$ have given values, and treated separately the case in which the values are 20 and 1.

Fermat[3] treated the problem to find in how many ways a given number m is the difference of two numbers whose product is a square. If $m=2^k p^a q^b$, where p and q are odd primes, the number of ways is $2ab+a+b$. If there is a third odd prime r with the exponent c, the number of ways is $4abc+2ab+2ac+2bc+a+b+c$; etc.

If[4] $x^2+y=y^2+x$, $x^2+y^2=\square$, then $x=y$ or $x=1-y$. For the latter, $x^2+y^2=(ry-1)^2$ gives y.

A. Martin[5] found the rational solutions of $x+y=x^2+y^2=\square$ by setting $x=az$, $y=bz$, $z=(a+b)/(a^2+b^2)$, where $a^2+b^2=\square$, the last being satisfied in the usual way. M. Brierley[6] took $y=rx$ and found $x=(r+1)/(r^2+1)$. Then take $r=3/4$, whence $r^2+1=\square$.

J. Hammond,[7] to divide a product N of two unknown primes x, y into two parts P, Q, each >1, such that $PQ\equiv-1 \pmod{N}$, tabulated for each m $(1<m\leqq15)$ all solutions n, n_1, P, Q, N, x, y of $P+Q=N$, $PQ+1=mN$, whence $P=m+n$, $Q=m+n_1$, $N=2m+n+n_1$, $nn_1=m^2-1$.

PROBLEMS OF HERON AND PLANUDE; GENERALIZATIONS.

Heron of Alexandria[8] (first century B.C.) treated the two problems:
(I) Find two rectangles such that the area of the first is three times the area of the second [and the perimeter of the second is three times the perimeter of the first]. It is stated that the sides of the first are $3^3 \cdot 2=54$ and $54-1=53$; those of the second, $3(53+54)-3=318$ and 3; the areas are 2862 and 954 [semi-perimeters 107, 321].

---

[1] Essenz der Rechenkunst von Mohammed Beha-eddin ben Alhossain aus Amul, arabisch u. deutsch von G. H. F. Nesselmann, Berlin, 1843, p. 55. French transl. by A. Marre, Nouv. Ann. Math., 5, 1846, 313. Cf. Genocchi, Annali di Sc. Mat. e Fis., 6, 1855, 297.

[2] Nuova Algebra Proportionale, Bologna, 1619, 51 pp. (chiefly on cubes and cube roots, pp. 42–43).

[3] Oeuvres, II, 216; letter from Fermat to Mersenne, Dec. 25, 1640.

[4] The Gentleman's Math. Companion, London, 4, No. 20, 1817, 643–4.

[5] Math. Quest. Educ. Times, 62, 1895, 70.

[6] *Ibid.*, 67, 1897, 72.

[7] Math. Quest. and Solutions, 1, 1916, 18–19.

[8] Liber Geeponicus (ed., F. Hultsch), 218–9. H. Schöne, Heronis Opera, III, Leipzig, 1903.

(II) Find two rectangles of equal perimeters such that the area of the second is 4 times the area of the first. The sum of two sides of the first rectangle is taken to be $4^3-1=63$, and one side $4-1=3$, so that the other is $63-3=60$. For the second rectangle one side is $4^2-1=15$ and the other $63-15=48$. The areas are 180, 720.

Pappus of Alexandria[9] (end of third century) discussed the simpler (determinate) problem: Given a parallelogram, find a second whose sides have a given ratio to the sides of the first, while the areas have a given ratio.

Maximus Planude[10] (about 1260–1310) discussed the problem to find two rectangles of equal perimeters such that their areas have a given ratio $b:1$. The solution is given in words; expressed algebraically, it states that the sides of one are $b-1$ and b^3-b, those of the other b^2-1 and b^3-b^2.

G. Valla [1] applied the last rule for $b=3, 4$.

M. Cantor[12] noted that the general [but see Zeuthen[15]] solution of Planude's problem is as follows: sides of first a and $b(b+1)a$, sides of second $(b+1)a$ and b^2a.

P. Tannery[13] discussed the generalization of Heron's two problems:

$$(1) \qquad\qquad a(x+y)=u+v, \qquad xy=buv,$$

and stated that the general solution, obtained by setting $a=pq$, $a^2b-1=rs$, $\beta b=mn$, is

$$u=\alpha\beta q, \qquad v=a(x+y)-u, \qquad x=abu+\alpha^2q^2ur, \qquad y=abu+\beta ms.$$

For $a=b$, Heron gave $x=2a^3$, $y=2a^3-1$, $u=a$, $v=2a(2a^3-1)$; for $a=1$,

$$(2) \qquad x=b^2-1, \qquad y=b^2(b-1), \qquad u=b-1, \qquad v=b(b^2-1).$$

* Ad. Steen[14] discussed the rational solutions of Planude's problem.

H. G. Zeuthen[15] noted that, to obtain Heron's solution of (1) for $a=b$, it suffices to assume that $u=a$, whence by the first equation v is a multiple za of a. Then $x+y=1+z$, $xy=a^3z$. Eliminating z, we get

$$(x-a^3)(y-a^3)=a^3(a^3-1),$$

which holds if $x-a^3=a^3-1$, $y-a^3=a^3$. Next, for $a=1$, try $v=bx$, $y=b^2u$. Then the first equation (1) gives $(b-1)x=(b^2-1)u$, which is satisfied by (2). If we replace the common factor $b-1$ in (2) by a, we get Cantor's solution, which is however not the general one. If we use $v/x=m$ in place of the earlier $v/x=b$, we get $y=mbu$ by (1_2) and find that the general solution of

[9] Sammlung, Buch III, Pappus ausgabe (ed., Hultsch), Berlin, 1875, 1877, 1878, 126. Cf. M. Cantor, Geschichte Math., ed. 3, 1, 1907, 454.

[10] Computation (Rechenbuch, Livre de Calcul). Greek text by C. I. Gerhardt, Halle, 1865, pp. 46, 47. German transl. (inadequate) by H. Waeschke, Halle, 1878. M. Cantor, Geschichte Math., ed. 3, 1, 1907, 513.

[11] De Expetendis et fugiendis rebus opus, Aldus, 1501, Liber IV (=Arithmeticae, III), Cap. 13,

[12] Die Römischen Agrimensoren, 1875, 62–3, 194–5.

[13] L'Arith. des Grecs dans Héron d'Alexandrie, Mém. soc. sc. phys. et nat. Bordeaux, (2), 4, 1882, 192.

[14] Tidsskrift for Math., (5), I1, 139–147.

[15] Bibliotheca Math., (3), 8, 1907–8, 118–120, 127–9.

(1) for $a=b$ is

$$\frac{x}{mb-1}=\frac{y}{mb(m-1)}=\frac{u}{m-1}=\frac{v}{m(mb-1)}.$$

G. Lemaire[16] and E. B. Escott[17] gave the solution

$$c,\quad \frac{cb^2}{1+b};\quad \frac{c}{1+b},\quad cb$$

of Planude's problem. It becomes (2) for $c=b^2-1$. Escott gave two particular solutions of the problem to find two parallelopipeds with equal sums of sides, equal surfaces, and with volumes in a given ratio q:

$$x+y+z=a+b+c,\qquad xy+yz+zx=ab+bc+ca,\qquad xyz=qabc.$$

See papers 438–440 of Ch. XXI.

U. Bini[18] gave two solutions in integers of the last problem and nine sets of solutions of Planude's problem, each involving a parameter. He rationalized the discriminant of the quadratic with the roots x, y, satisfying (1) for $a=1$.

System $x=2y^2-1$, $x^2=2z^2-1$.

Fermat[19] stated that $x=7$ is the only integral solution, excluding of course the evident solution $x=\pm1$. Cf. pp. 56, 57 of Vol. I of this History.

E. Lucas[20] wrote $x=2y^2-w^2$, $w=\pm1$. Then $x^2=(2y^2+w^2)^2-2(2yw)^2$. Multiply the latter by $-1=1^2-2\cdot1^2$. Thus $x^2=2r^2-s^2$, where

$$r=2y^2+w^2-2yw,\qquad s=2y^2+w^2-4yw.$$

In view of the proposed second condition, set $s=\pm1$, whence $x^2=2r^2-s^2$ becomes

$$r^2=\left(\frac{x+1}{2}\right)^2+\left(\frac{x-1}{2}\right)^2.$$

Also $r=(y\pm1)^2+y^2$, since $w=\pm1$. Thus r and r^2 are sums of squares of consecutive integers and hence $r=5$, $x=7$, by papers 26–30.

T. Pepin[21] treated $2y^2(y^2-1)=z^2-1$, obtained by eliminating x. For y odd, $y=\alpha\beta$, $z\pm1=2\alpha^2h$, $z\mp1=8\beta^2k$, whence $\alpha^2\beta^2-1=8hk$, $\pm1=\alpha^2h-4\beta^2k$, so that

$$(\alpha^2\pm4k)(\beta^2\mp h)=4hk.$$

Thus $\alpha^2\pm4k=mh$, $\beta^2\mp h=4nk$, where m, n are integers making $mn=1$. If $m=n=+1$, the lower sign is excluded and the upper gives $2h=\beta^2+\alpha^2$, $8k=\beta^2-\alpha^2$, whence $\alpha^2\beta^2-1=8hk$ becomes $\alpha^4-2\gamma^2=1$, $\gamma\equiv(\alpha^2-\beta^2)/2$. The case $m=n=-1$ leads to the same relation. This Pell equation has no integral solutions except $\alpha=\pm1$, $\gamma=0$. Next, let y be even, $y=2u$. Then

$$z^2=(2u)^4+(4u^2-1)^2,\qquad z=f_1^2+4g^2,\qquad \pm(4u^2-1)=f^2-4g^2,\qquad \pm4u^2=4fg,$$

<hr>

[16] L'intermédiaire des math., 14, 1907, 287.

[17] Ibid., 15, 1908, 11–13.

[18] Ibid., 15, 1908, 14–18.

[19] Oeuvres, II, 434, 441; letters to Carcavi, Aug., 1659, Sept., 1659. Cf. C. Henry, Bull. Bibl. Storia Sc. Mat. e Fis., 12, 1879, 700; 17, 1884, 342, 879, letter from Carcavi to Huygens, Sept. 13, 1659 (extract from letter from Fermat).

[20] Nouv. Ann. Math., (2), 18, 1879, 75–6. His u, x, y are replaced by x, y, w.

[21] Atti Accad. Pont. Nuovi Lincei, 36, 1882–3, 23–33.

f, g relatively prime integers; the upper sign is excluded by use of modulus 4. Restricting to positive integers, we have $u = \alpha\beta$, $f = \alpha^2$, $g = \beta^2$. Thus $(\alpha^2 + 2\beta^2)^2 = 1 + 2(2\beta^2)^2$, the discussion of which as a Pell equation leads to the condition $m^4 - 2n^4 = \pm 1$. The upper sign is excluded as it leads to $1 + c^4 = d^2$. For the lower sign, $m^2 = n^2 = 1$, as noted in his next paper (*ibid.*, p. 35).

A. Genocchi[22] treated $z^2 = y^4 + (y^2 - 1)^2$, obtained by eliminating x. For y odd, $y^2 - 1 = (f^2 - g^2)/2$, $y^2 = fg$, where f, g are odd and relatively prime. Thus $f = m^2$, $g = n^2$ and $2(m^4 + 1) = (m^2 + n^2)^2$, whence $m^2 = n^2 = 1$, $x = 1$, by known theorems on $2r^4 + 2s^4 = \square$. For y even, $y^2 - 1 = f^2 - g^2$, $y^2 = 2fg$, where f, g are relatively prime, and f is even. Thus $f = 2\alpha^2$, $g = \beta^2$ and $p^2 = 1 + 8\alpha^4$, where $p = 2\alpha^2 + \beta^2$. Hence $p \pm 1 = 2m^4$, $p \mp 1 = 4n^4$. Thus $m^4 \mp 1 = 2n^4$, which is impossible unless $m^2 = 1$, whence $x = -1$ or 7.

Genocchi[23] cited his[43] paper of Ch. XVI in which he proved that $2r^4 + 2s^4 \neq \square$, whence $\rho^4 + \sigma^4 \neq 2\square$, so that Pepin's[21] condition $m^4 + 1 = 2(n^2)^2$ requires $m^2 = 1$.

S. Réalis[24] gave a discussion quite similar to that by Genocchi.[22]

E. Turrière[25] treated the system $x = 2y^2 - 1$, $x^2 = 2z^2 - 1$.

A number and its square both sums of two consecutive squares.

E. de Jonquières[26] gave a proof, valid only when y is a prime, that y and y^2 are both sums of two consecutive squares only when $y = 5$. Like errors invalidate his[27] result that if a number and its square are both expressible in the form $x^2 + t(x+1)^2$, then $t = 1, 2, 4, 5, 7,$ or 9. Cf. Lucas,[20] and papers 89, 90 of Ch. XVI.

T. Pepin[28] reduced the problem to a certain quartic which he did not solve completely. If $y = P^2 + P_1^2$, all decompositions of y^2 into a sum of two relatively prime squares are given by $y^2 = (P^2 - P_1^2)^2 + (2PP_1)^2$. Taking $2PP_1$ and $\pm(P^2 - P_1^2)$ as consecutive integers, de Jonquières assumed that P and P_1 must be consecutive. While this condition is necessary if y is a power of a prime or the double of such a power, it is in general not necessary.

E. Catalan[29] asked for numbers $2x$ expressible as a sum of squares of two consecutive odd numbers, while $(2x)^2$ is a sum of squares of two consecutive even numbers, citing the case $2x = 10$. G. C. Gerono[30] proved that $2x = 10$ is the only solution of the equivalent system

$$x = 4y^2 + 1, \qquad x^2 = z^2 + (z+1)^2.$$

[22] Nouv. Ann. Math., (3), 2, 1883, 306–10.
[23] Bull. Bibl. Storia Sc. Mat., 16, 1883, 211–2.
[24] *Ibid.*, p. 213.   Reproduced, Sphinx-Oedipe, 4, 1909, 175–6.
[25] L'enseignement math., 19, 1917, 243–4.
[26] Nouv. Ann. Math., (2), 17, 1878, 219–20, 241–7, 289–310; (2), 18, 1879, 464–5. Cf. Meyl[30] of Ch. IV.
[27] *Ibid.*, (2), 17, 1878, 419–24, 433–46. Cf. Assoc. franç. av. sc., 7, 1878, 40–49.
[28] Atti Accad. Pont. Nuovi Lincei, 32, 1878–9, 295–8.
[29] Nouv. Ann. Math., (2), 17, 1878, 518.
[30] *Ibid.*, 521.

E. Lionnet[31] stated that 1 and 5 are the only sums of squares of two consecutive integers whose product is a sum of such squares; 1 and 5 are the only primes x, y, each a sum of squares of two consecutive integers, such that x^2 and y^2 are such sums of squares. Similarly, 1, 13 and their biquadrates are sums of squares of consecutive integers. Cf. Lionnet[314] of Ch. XXII.

Miscellaneous systems of two equations.

Bháscara[30] of Ch. XII gave a solution of the system $x^2+y^2+xy=z^2$, $(x+y)z+1=\square$. On systems of two equations involving sums of squares, see papers 108, 176 of Ch. VI; 97, 259 of Ch. VII.

"Umbra"[32] found numbers ax, bx, cx, $\cdots$ whose sum added to or subtracted from the sum of their squares gives a square. Set $s=a+b+c+\cdots$. Choose a, b, $\cdots$ so that the sum of their squares is a square q^2 (by setting $q=a+m$ and finding a). Hence $q^2x^2\pm sx$ are to be squares. Take $t=s/q^2$. Then $x^2\pm tx$ are to be squares. Determine x by $x^2+tx=(k-x)^2$. Then $x^2-tx=\square$ if $k^2-2kt-t^2=\square=(n-k)^2$, which gives k.

R. F. Muirhead,[33] to find pairs of quadratic equations $x^2-px+q=0$, $x^2-qx+p=0$, all of whose roots are integers $\geqq0$, found all integral solutions $\geqq0$ of $\alpha+\beta=\alpha'\beta'$, $\alpha'+\beta'=\alpha\beta$. Set $r=(\alpha-1)(\beta-1)$, $r'=(\alpha'-1)(\beta'-1)$, whence $r+r'=2$. It is shown that $\alpha\neq0$. Hence either $r=0$, $r'=2$, $\alpha'=2$, $\beta'=3$, $\alpha=1$, $\beta=5$, or $r=2$, $r'=0$, or $r=r'=1$, $\alpha=\alpha'=\beta=\beta'=2$. He solved also the pairs $\beta\pm\alpha=\alpha'\beta'$, $\beta'-\alpha'=\alpha\beta$.

A. Cunningham[34] solved $S_1=S_2=S_3$, where
$$S_i=500(N_i^2-N_{i+1}^2)+r(N_i-N_{i+1}),$$
by multiplying by $2\cdot10^3$ and setting $a_j=10^3N_j+r$. Thus
$$a_1^2-a_2^2=a_2^2-a_3^2=a_3^2-a_4^2.$$
But if four integral squares are in A. P., they are known to be equal.

M. Rignaux[35] gave integral solutions of the two systems
$$xy+zt=\square, \qquad xz-yt=\square; \qquad xy+zt=xz-yt=u^4.$$
A. Boutin[36] proved that $x^2-2y^2=1$, $y^2-3z^2=1$ imply $y^2=4$.

[31] Nouv. Ann. Math., (2), 20, 1881, 514.
[32] The Gentleman's Math. Companion, London, 4, No. 20, 1817, 673–5.
[33] Math. Quest. Educ. Times, 70, 1899, 84–6.
[34] Ibid., (2), 10, 1906, 29.
[35] L'intermédiaire des math., 25, 1918, 113–5.
[36] Ibid., 26, 1919, 123.

CHAPTER XVIII.

THREE OR MORE QUADRATIC FUNCTIONS OF ONE OR TWO UNKNOWNS MADE SQUARES.

a^2x^2+dx, b^2x^2+ex, $\cdots$ MADE SQUARES.

J. Cunliffe[1] took $v^2+mv=(d-v)^2$, whence $v=d^2/(2d+m)$. Then

$$v^2+nv=\square$$

if $d^2+2dn+mn=(q-d)^2$, which gives d. Then $v^2+pv=\square$ if

$$(q^2-mn)^2+4p(q+n)(q^2-mn)+4mp(q+n)^2=\square=(q^2-2pq-mn)^2,$$

whence $q=(p-m-n)/2$.

W. Wright[2] equated the numerators and denominators of the two values of d given by $d^2+2dn+mn=(q-d)^2$ and $d^2+2dp+mp=(t-d)^2$. Thus $q^2-mn=t^2-pm$, $n+q=p+t$. By division, $q+t=-m$. Hence

$$q=(p-m-n)/2.$$

A. B. Evans[3] made k^2x^2-kx a square for $k=a$, b, c. From

$$a^2x^2-ax=(a-m)^2x^2$$

we get x. Then $b^2x^2-bx=\square$, $c^2x^2-cx=\square$ if

$$a^2b^2c^2-abc^2d=y^2, \qquad a^2b^2c^2-ab^2cd=z^2, \qquad d=2am-m^2.$$

Subtract and set $bc(2a-m)=y+z$, $m(ab-ac)=y-z$. Substitute the resulting y into $a^2b^2c^2-abc^2d=y^2$; we get

$$m=\frac{4abc(ab-bc+ac)}{4ab^2c-(ab+bc-ac)^2}.$$

J. Matteson[4] solved $d^2+2dn+mn=A^2$, $d^2+2dp+mp=B^2$ by taking $2d+m=A+B$, $n-p=A-B$. Inserting the resulting value of A into the first of the initial equations, we get d rationally. An equal value of d is obtained by use of B. It is stated that if m, n, p be any three of the numbers 2016, 3000, 3696, 4056 (or any three of certain 13 numbers of 6 or 7 digits), the six expressions $v^2\pm mv$, $v^2\pm nv$, $v^2\pm pv$ are all squares when $v=65^2$.

D. S. Hart[5] found three squares such that each increased by its root shall be a square. Let ax, bx, cx be the roots. Take $a^2x^2+ax=m^2x^2$. For the resulting value of x, b^2x^2+bx and c^2x^2+cx are squares if $a^2b^2-a^3b+abm^2$ and $a^2c^2-a^3c+acm^2$ are squares. Since this is the case when $m=a$, set $m=a+n$. Multiply the resulting expressions by c^2 and b^2 respectively. Then shall

$$abc^2n^2+2a^2bc^2n+a^2b^2c^2=\square=A^2, \qquad ab^2cn^2+2a^2b^2cn+a^2b^2c^2=B^2.$$

---

[1] The Math. Repository (ed., Leybourn), London, 3, 1804, 97. The Gentleman's Math. Companion, London, 3, No. 14, 1811, 300–2. Same in Math. Quest. Educ. Times, 14, 1871, 54; 24, 1876, 28.

[2] The Gentleman's Math. Companion, 5, No. 24, 1821, 59–60; 5, No. 26, 1823, 214.

[3] Math. Quest. Educ. Times, 14, 1871, 55–6.

[4] The Analyst, Des Moines, 2, 1875, 46–9.

[5] *Ibid.*, 3, 1876, 81–3.

Factoring the difference, we set $a(c-b)n = A-B$, $2abc+bcn = A+B$. Insert the resulting value of A into the equation involving A^2. We find that

$$n = 4abc(ab+ac-bc)/\{(ac+bc-ab)^2 - 4abc^2\}.$$

$$x = \frac{\{(ac+bc-ab)^2 - 4abc^2\}^2}{8abc(ac+bc-ab)(ac+ab-bc)(ac-ab-bc)}.$$

The initial squares will be in A. P. if we take

$$a = 2rs - r^2 + s^2, \qquad b = r^2 + s^2, \qquad c = 2rs + r^2 - s^2;$$

whence $a=1$, $b=5$, $c=7$ if $r=2$, $s=1$. Then $x = 151321/7863240$, a result found by J. D. Williams[6] by starting with the squares x^2, $25x^2$, $49x^2$. For $r=4$, $s=3$, we get $a=17$, $b=25$, $c=31$, $x=-X$, where

$$X = (864571)^2/11011044931800,$$

and hence a solution of $a^2X^2 - aX = \square$, $\cdots$, $c^2X^2 - cX = \square$ [Perkins[28] of Ch. XIV].

Hart[7] made $k^2x^2 + kx = \square$ for $k = a'$, b', $\cdots$. Divide by k^2 and set $a = 1/a'$, $\cdots$. Then x^2+ax, x^2+bx, $\cdots$ are to be squares. Set $x = z^2$. Then z^2+a, z^2+b, $\cdots$ are to be squares. Suppose that z^2 is a sum of two squares in the required number of ways: $z^2 = m^2 + n^2 = p^2 + q^2 = \cdots$, and set $a = 2mn$, $b = 2pq$, $\cdots$. Then $z^2 + a = (m+n)^2$, $z^2 + b = (p+q)^2$, $\cdots$.

J. Matteson[8] gave the solutions by Hart[5, 7] with amplifications.

G. B. M. Zerr[9] solved the system $x^2+y^2 = z^2+w^2 = \square$, $x^2-w^2 = z^2-y^2 = \square$, also the system

$$(m^2+n^2)^2x^2 \pm (m^2+n^2)x = \square, \qquad (m^2-n^2)^2x^2 \pm (m^2-n^2)x = \square,$$

$$4m^2n^2x^2 \pm 2mnx = \square.$$

P. von Schaewen[10] made $4x^2 - 2x$, $4x^2 + 3x$, $4x^2 + 5x$ all squares. Setting $x = 1/(4x_1)$, we are to make $1 - 2x_1$, $1 + 3x_1$, $1 + 5x_1$ all squares [von Schaewen[81] of Ch. XV].

On three squares which increased or decreased by their roots give squares, see papers 12, 12a, 21, 26, 52–54 of Ch. XIV. For two squares, papers 3, 19 of Ch. XVI; 32 of Ch. XVII.

Three linear and quadratic functions of two unknowns made
squares.

Brahmegupta[2] of Ch. XV made $x+y$, $x-y$ and $xy+1$ all squares.

To find two numbers whose product is a square and product plus the square of either is a square, J. Hampson[11] took b^2a and a as the numbers. It remains to make $b^2+1 = \square = (b-c)^2$, say, which gives b. R. Mallock

[6] Algebra, Boston, 1840, 413.

[7] Math. Quest. Educ. Times, 39, 1883, 47–9.

[8] Collection of Diophantine Problems with Solutions (ed., A. Martin), Washington, D. C., 1888, pp. 10–20.

[9] Amer. Math. Monthly, 15, 1908, 17–18. Erroneous solution in J. D. Williams' Algebra, 1832, 419.

[10] Archiv Math. Phys., (3), 17, 1911, 249–250.

[11] Ladies Diary, 1763, p. 34, Quest. 491; Leybourn's Math. Quest. L. D., 2, 1817, 209.

took two perpendicular segments AC and CD; let CB be the altitude of triangle ACD.   Then AB and DB measure the required numbers.

T. Thompson[12] divided a given square a^2 into two parts

$$\frac{r^2-2rs^2}{4rs+1}, \qquad \frac{s^2+2r^2s}{4rs+1},$$

such that each plus the square of the other is a square.   Take $s=r+1$. Then the sum of fractions is a^2 if $2r+1=a^{-1}$, whence $r=(1-a)/(2a)$.

J. Whitley[13] took $x^2+y=(x+v)^2$, $y^2+x=(y+z)^2$, which give x, y in terms of v, z [Euler[99] of Ch. XVI].   Take $v=1-z$.   Then $x+y=a^2$ gives $z=(a-1)/(2a)$.

J. Cunliffe[14] found two numbers whose sum increased or decreased by their difference or difference of their squares give squares.   He took x and $1-x$ as the numbers.   Since either difference is $1-2x$, $2-2x$ and $2x$ are to be squares.   Take $2x=4n^2$, $n=s-1/2$.   Then

$$2-2x=1+4s-4s^2=\square=(2rs-1)^2$$

gives s.

W. Wright and Winward[15] took x and y as the numbers required in the last problem.   Then $2x$, $2y$, $x+y\pm(x^2-y^2)$ are to be squares.   Set $x+y=p$, $x-y=q$.   Then $p\pm q$ and $p\pm pq$ are to be squares.   Take $p+pq=n^2$. Then $p-pq=\square$ if $1-q^2=\square=(1-rq)^2$, whence $q=2r/(r^2+1)$.   Set

$$n=m(r+1)/(r^2+1).$$

Then $p\pm q=\square$ if $(r^2+1)(m^2\pm 2r)=\square$.   Now $r^2+1=\square$ if $r=(v^2-1)/(2v)$. Take $v=2$, whence $r=3/4$.   Take $m=P/2$.   Then $m^2\pm 2r=\square$ if $P^2\pm 6=\square$. Set $P^2+6=(3R-P)^2$, which gives P.   Set $R=t+2$.   Then $P^2-6=\square$ if $4+\cdots+9t^4=\square=(2+36t+3t^2)^2$, whence $t=47/6$.   B. Gompertz took $x+y=pk^2$, $1+x-y=1/p$ and by a long discussion obtained the preceding numerical answer.

"Jesuiticus"[16] imposed the further condition that $x+y=\square$.   Thus $x+y=r^2$, $2x=p^2$, $2y=q^2$, $1+x-y=m^2$, $1-x+y=n^2$, whence $p^2+q^2=2r^2$, $m^2+n^2=2$.   Take $p=m$, $q=n$, whence $r=1$.   Then $m^2+n^2=2$ if

$$m,\ n=(u^2-v^2\pm 2uv)/(u^2+v^2).$$

Several[17] solved easily the problem to find two positive rational numbers such that each and the sum s of their squares exceed their product by squares, and the problem when s is replaced by $\sqrt{s}$.

Four quadratic functions of two unknowns made squares.

L. Euler[18] made $AB\pm A$, $AB\pm B$ all squares.   Set $A=x/z$, $B=y/z$; then $xy\pm xz$, $xy\pm yz$ are to be squares.   Since $a^2+b^2\pm 2ab=\square$, set

$$xy=a^2+b^2=c^2+d^2, \qquad xz=2cd, \qquad yz=2ab.$$

[12] The Gentleman's Diary, or Math. Repository, No. 55, 1795.   A. Davis' ed., London, 3, 1814, 229–30.

[13] Ibid., No. 68, 1808, 36–7, Quest. 917.

[14] Ladies' Diary, 1810, p. 40, Quest. 1203; Leybourn's M. Quest. L. D., 4, 1817, 122–4.

[15] The Gentleman's Math. Companion, London, 3, No. 16, 1813, 421–4.

[16] Ladies' Diary, 1839, 41–42, Quest. 1638.

[17] Math. Quest. Educ. Times, 5, 1866, 60–1.

[18] Novi Comm. Acad. Petrop., 19, 1774, 112;   Comm. Arith., II, 53–63; Op. Om., (1), III, 338.

Then

$$z^2 = \frac{4abcd}{a^2+b^2}, \qquad \frac{x}{z} = \frac{2cd}{z^2} = \frac{a^2+b^2}{2ab}, \qquad \frac{y}{z} = \frac{c^2+d^2}{2cd},$$

The problem to choose $a, \cdots, d$ so that $a^2+b^2=c^2+d^2$ and so that the expression for z^2 shall be a square was treated by Euler in §§ 3–17. In § 18, he began by setting (in accord with the above)

$$A = \frac{a^2+b^2}{2ab}, \qquad B = \frac{c^2+d^2}{2cd}.$$

Then $A\pm 1 = (a\pm b)^2/(2ab)$, $B\pm 1 = (c\pm d)^2/(2cd)$. Hence the conditions are

$$\frac{c^2+d^2}{4abcd} = \square, \qquad \frac{a^2+b^2}{4abcd} = \square.$$

Make the numerators the squares of r^2+s^2 and p^2+q^2 by setting

$$a = p^2-q^2, \qquad b = 2pq; \qquad c = r^2-s^2, \qquad d = 2rs.$$

To make the common denominator a square, we have the condition

$$pq(p^2-q^2) \div rs(r^2-s^2) = \square,$$

which is satisfied if we have two rational right triangles the ratio of whose areas is a square [cf. Euler[81] of Ch. XVI]. The above ratio is α/β for $p=3\alpha$, $q=2\beta-\alpha$, $r=3\beta$, $s=2\alpha-\beta$ and for seven similar sets. The case $\alpha=9$, $\beta=4$ gives $p=27$, $q=-1$, $r=12$, $s=14$. By a table (p. 60) of values of $xy(x^2-y^2)$, we get right triangles of equal areas $2\cdot 3\cdot 5\cdot 7$ for $x=5$, $y=2$; $x=6$, $y=1$; $x=8$, $y=7$; also two of equal area for

$$r = p = m^2+mn+n^2, \qquad q = m^2-n^2, \qquad s = n^2+2mn.$$

Euler[19] made the four expressions $AB\pm A\pm B$ all squares. Set $A=x/z$, $B=y/z$. Then $xy\pm z(x+y)$ and $xy\pm z(x-y)$ shall be squares. This will be the case if

$$xy = a^2+b = c^2+d^2, \qquad z(x+y) = 2ab, \qquad z(x-y) = 2cd,$$

whence

$$x = \frac{ab+cd}{z}, \qquad y = \frac{ab-cd}{z}, \qquad z^2 = \frac{a^2b^2-c^2d^2}{a^2+b^2}.$$

Since xy shall be a sum of two squares in two ways, set

$$a = pr+qs, \qquad b = ps-qr, \qquad c = pr-qs, \qquad d = ps+qr.$$

Then

$$x = \frac{2rs(p^2-q^2)}{z}, \qquad y = \frac{2pq(r^2-s^2)}{z}, \qquad z^2 = \frac{4pqrs(p^2-q^2)(r^2-s^2)}{(p^2+q^2)(r^2+s^2)}.$$

---

[19] Novi Comm. Acad. Petrop., 15, 1770, 29; Mém., 11, 1830 (1780), 31; Comm. Arith., I, 414; II, 438. The simpler solution here reproduced is given in the second of these two papers, and is practically the same as that in Euler's posthumous paper, Comm. Arith., II, 586–7; Opera postuma, 1, 1862, 137–9. In two letters to Lagrange (Oeuvres, XIV, 214, 219), Jan. and March, 1770, Euler (Opera postuma, 1, 1862, 573–4) gave discussions occurring in the first and third of these papers. On the related problem to find p, q, r, s such that $\lambda pqrs(p^4-s^4)(q^4-r^4) = \square$, see Euler, Opera postuma, I, 487–490 (about 1766). The second letter is quoted in l'intermédiaire des math., 21, 1914, 129–131, and in Sphinx-Oedipe, 7, 1912, 57–8. First paper in Opera Omnia, (1), III, 148.

The final expression is a square if and only if
$$pq(p^4-q^4)\cdot rs(r^4-s^4)=\square.$$
By special assumptions, Euler was led to the values
$$p=(\alpha+\beta)(\alpha+2\beta),\qquad q=\beta(3\beta-\alpha),\qquad r=4\beta(\alpha+2\beta),\qquad s=\alpha^2+4\alpha\beta-\beta^2.$$
Then
$$p^2+q^2=(\alpha^2+\beta^2)v,\qquad r^2+s^2=(p+q)v,\qquad v\equiv\alpha^2+6\alpha\beta+13\beta^2,$$
$$r+s=(\alpha+\beta)(\alpha+7\beta),\qquad r-s=(3\beta+\alpha)(3\beta-\alpha),\qquad p-q=s.$$
The condition is thus $(\alpha+3\beta)(\alpha+7\beta)(\alpha^2+\beta^2)=\square$, which is treated by the usual methods.   From $\alpha=2$, $\beta=1$ and $\alpha=-17$, $\beta=7$, we get the solutions
$$A,\ B=\frac{13\cdot29^2}{8\cdot9^2},\qquad \frac{5\cdot29^2}{32\cdot11^2};\qquad \frac{13^2\cdot53^2}{3\cdot4\cdot7\cdot59^2},\qquad \frac{37\cdot13^2\cdot53^2}{3\cdot7\cdot4^2\cdot5^2\cdot19^2}.$$

Euler[20] made $x+y\pm x^2$, $x+y\pm y^2$ all squares.   Replace x, y by x/z, y/z. Then $(x+y)z\pm x^2$, $(x+y)z\pm y^2$ are to be squares.   This will be the case if
$$x^2=2AB,\qquad y^2=2CD,\qquad (x+y)z=A^2+B^2=C^2+D^2.$$
The final equality holds if
$$A=ac+bd,\qquad B=ad-bc,\qquad C=ad+bc,\qquad D=ac-bd.$$
The first two conditions hold if $x=Af$, $y=Cg$, $2B=Af^2$, $2D=Cg^2$.   By the latter,
$$\frac{a}{b}=\frac{2c+df^2}{2d-cf^2},\qquad \frac{a}{b}=\frac{2d+cg^2}{2c-dg^2},$$
which are equal if
$$\frac{d}{c}=\frac{2(f^2-g^2)\pm\sqrt{R}}{4+f^2g^2},\qquad R\equiv(4+f^4)(4+g^4).$$
Hence the problem will be solved if we make $R=\square$.   Set $g=1$.   Since $R=\square$ for $f=1$ (which makes $x=y$), we take $f=1+t$.   Then R is the square of $5+2t+13t^2/5$ if $t=60/11$.   Dropping the common factor 13 in x, y, we get
$$x=4\cdot11\cdot71,\qquad y=4\cdot37\cdot61,\qquad z=\frac{5\cdot37^2\cdot61^2}{2\cdot49\cdot31}.$$

<hr>

[20] Mém. Acad. Sc. St. Pétersbourg, 11, 1830 (1780), 46; Comm. Arith., II, 447.

CHAPTER XIX.

SYSTEMS OF THREE OR MORE EQUATIONS OF DEGREE TWO IN THREE OR MORE UNKNOWNS.

x^2+y^2, x^2+z^2, y^2+z^2 ALL SQUARES.

Paul Halcke[1] gave the solution $x=44$, $y=240$, $z=117$.

N. Saunderson[2] satisfied $x^2+z^2=\square$ by expressing z^2 as a product of two factors aw and $z^2/(aw)$ and taking half their difference as x. Similarly for $y^2+z^2=\square$. Take $a^2+b^2=c^2$. Then

$$x=\pm\frac{1}{2}\left(aw-\frac{z^2}{aw}\right), \qquad y=\pm\frac{1}{2}\left(bw-\frac{z^2}{bw}\right),$$

$$x^2+y^2=\frac{1}{4}c^2w^2-z^2+\frac{c^2z^4}{4a^2b^2w^2}.$$

Equate the sum of the last two terms to zero. Hence $w=cz/(2ab)$. To obtain integers, let $z=4abc$. Then $x=a(4b^2-c^2)$, $y=b(4a^2-c^2)$. For $a=3$, $b=4$, we get $x=117$, $y=44$, $z=240$.

L. Euler[3] made the last two sums squares by taking

$$\frac{x}{z}=\frac{p^2-1}{2p}, \qquad \frac{y}{z}=\frac{q^2-1}{2q}.$$

Then the first sum will be a square if

$$q^2(p^2-1)^2+p^2(q^2-1)^2=\square.$$

First, let $q-1=p+1$. Then must $2p^4+8p^3+6p^2-4p+4=\square$. Since 4 is a square, the condition is satisfied in the usual manner if $p=-24$. Next, $q-1=2(p+1)$ leads to the solution $p=48/31$, and $q-1=4(p-1)/3$ to $p=2/13$. For

$$q+1=(p+1)(t+1)/(p+t),$$

both $(p+1)^2$ and $(p-1)^2$ may be cancelled and the condition becomes

$$t^2p^4+2t(t^2+1)p^3+2t^2p^2+(t^2+1)^2)p^2+(t^2-1)^2p^2+2t(t^2+1)p+t^2=\square,$$

say the square of $tp^2+(t^2+1)p-t$. Hence $p=-4t/(t^2+1)$, where t is arbitrary. If $x=a$, $y=b$, $z=c$ is one solution of our problem, $x=ab$, $y=bc$, $z=ac$ is another.

Euler[4] made $S-A^2$, $S-B^2$, $\cdots$ squares, where $S=A^2+B^2+\cdots$. Thus S is to be expressed as a $\boxed{2}$ in several ways, the most general way being

$$S=B^2+\left[\frac{(f^2-1)x+2fy}{f^2+1}\right]^2, \qquad B=\frac{2fx-(f^2-1)y}{f^2+1},$$

if $S=x^2+y^2$ is one way. For three numbers $A=x$, B, C, take as C the

---

[1] Deliciae Mathematicae, Oder Math. Sinnen-Confect, Hamburg, 1719, 265.

[2] The Elements of Algebra, 2, 1740, 429–431.

[3] Algebra, 2, 1770, art. 238; French transl., 2, 1774, pp. 327–335. Opera Omnia, (1), I, 477–82. Cf. Fuss[95] and Schwering[150] of Ch. V.

[4] Novi Comm. Acad. Petrop., 17, 1772, 24; Comm. Arith., I, 467; Op. Omnia, (1), III, 201.

function derived from B by replacing f by $(f+1)/(f-1)$, viz.,

$$C = \frac{(f^2-1)x-2fy}{f^2+1}.$$

Then $x^2+B^2+C^2=S=x^2+y^2$ gives $x/y=8f(f^2-1)/(f^2+1)^2$. Take y equal to the denominator and multiply all the numbers by f^2+1. Hence

$$A=8f(f^4-1), \qquad B=(1-f^2)(f^4-14f^2+1), \qquad C=2f(3f^4-10f^2+3).$$

For $f=2$, we get 240, 117, 44. E. B. Escott[5] also gave the last solution.

Euler[6] set $x^2=4mnpq$, $y=mp-nq$, $z=np-mq$. Then

$$x^2+y^2=(mp+nq)^2, \qquad x^2+z^2=(np+mq)^2.$$

For $y=2(m^2-n^2)rs$, $z=(m^2-n^2)(r^2-s^2)$, we get

$$y^2+z^2=(m^2-n^2)^2(r^2+s^2)^2, \qquad p=2mrs-n(r^2-s^2), \qquad q=2nrs-m(r^2-s^2).$$

The resulting expression for $x^2/4$ is a quartic function of r which is the square of $mnr^2-(m^2+n^2)rs+mns^2$ if $r=4mn$, $s=m^2+n^2$. Then

$$x=2mn(3m^2-n^2)(3n^2-m^2), \qquad y=8mn(m^4-n^4),$$
$$z=(m^2-n^2)(m^2-4mn+n^2)(m^2+4mn+n^2),$$

which, apart from signs, equal the products of n^6 by Euler's[4] values when $f=m/n$. The simplest solution arises from $m=2$, $n=1$: $x=44$, $y=240$, $z=117$, whence x^2+y^2, etc., are the squares of 244, 125, 267.

From the sum of the roots of three squares, the sum of any two of which is a square, subtract the area of a right triangle; the remainder is a square which if decreased by the sides of the triangle yields remainders which are squares in arithmetical progression. L. Blakeley[7] took $44x$, $117x$, $240x$ as the roots of the required squares, the sum of 44^2, 117^2, 240^2 by twos being known to be squares; also let $3y$, $4y$, $5y$ be the sides of the right triangle of area $6y^2$. Then $401x-6y^2=\square=a^2$. Also, $a^2-3y=b^2$, $a^2-4y=c^2$, $a^2-5y=d^2$. Take $y=r^2-2dr$. Then

$$c^2=(d-r)^2, \qquad b^2=d^2-4dr+2r^2, \qquad a^2=d^2-10dr+5r^2.$$

The product of the last two is a square if $d=r/6$. Then $d^2=r^2/36$, $c^2=25r^2/36$, $b^2=49r^2/36$ are in A. P.

P. Barlow[8] noted that the first part of this question is satisfied if the roots of the squares are $575z/48$, $485z/44$ and z (from J. Bonnycastle's Algebra, p. 148). Next, we need a square which if diminished by each side of a right triangle the remainders are three squares in A. P., whence the sides of the triangle are in A. P., and hence proportional to 3, 4, 5. Let the squares be $(a^2+2ab-b^2)^2$, $(a^2+b^2)^2$, $(b^2+2ab-a^2)^2$, with the common difference $\delta=4ab(a^2-b^2)$. Thus let 3δ, 4δ, 5δ be the sides. Then

$$(b^2+2ab-a^2)^2+5\delta=\square=(b^2-4ab-a^2)^2$$

if $8a^3b-8b^3a=12a^2b^2$, i. e., $(2a-2b)(a+b)=3ab$, which holds if $a=2b$.

---

[5] L'intermédiaire des math., 8, 1901, 103-4.
[6] Posth. paper, Comm. Arith., 2, 1849, 650; Opera postuma, 1, 1862, 103-4.
[7] Ladies' Diary, 1805, p. 43, Quest. 1131; Leybourn's Math. Quest. L. D., 4, 1817, 45-6.
[8] The Diary Companion, Supplement to Ladies' Diary, London, 1805, 45-6.

J. Cunliffe[9] set $x^2+y^2=(x+y-a)^2$, $x^2+z^2=(x+z-b)^2$. From the resulting two values of x we get

$$z=\frac{(a^2-d^2)y-ad(a-d)}{2dy+a^2-2da}, \qquad d=a-b.$$

Then $y^2+z^2=\square$ if $4d^2y^4+4d(a^2-2ad)y^3+\cdots+a^2d^2(a-d)^2=\square$. Let it be the square of $2dy^2+(a^2-2ad)y-ad(a-d)$. Then

$$y=\frac{2ad(2ad-d^2)}{(a+d)^2(a-d)+4ad^2}=\frac{2a(a-b)(a^2-b^2)}{4a^3+b^3-4a^2b}.$$

"Calculator"[10] first solved $x^2+y^2=a^2$, $x^2+z^2=b^2$. Take $b=rv-a$, $z=y-sv$; then $a^2-y^2=b^2-z^2$ gives $v=(2ra-2sy)/(r^2-s^2)$, whence b, z are known. To satisfy $x^2+y^2=a^2$ take

$$a=(r^2-s^2)(m^2+n^2), \qquad y=(r^2-s^2)(2mn), \qquad x=(r^2-s^2)(m^2-n^2).$$

Then $z=(r^2+s^2)\cdot2mn-2rs(m^2+n^2)$. Then y^2+z^2 becomes a quartic in m which is equated to the square of $m^2-mn(r^2+s^2)/(rs)-n^2$, whence $m:n=4rs:r^2+s^2$. Taking $n=r^2+s^2$, we have

$$x=(s^2-r^2)(r^4-14r^2s^2+s^4), \qquad y=8rs(r^4-s^4), \qquad z=2rs(3r^2-s^2)(r^2-3s^2),$$

which equal the products of s^6 by Euler's[4] values for $f=r/s$. Cf. Euler.[6]

S. Ward[11] took $x^2+y^2=a^2$, $x^2+z^2=(m+n)^2$, $y^2+z^2=(m-n)^2$. Then

$$4x^2=2a^2+8mn, \qquad 4y^2=2a^2-8mn, \qquad 4z^2=4m^2+4n^2-2a^2.$$

Let $2a^2=m^2+16n^2$. Then the first two expressions are squares and the third becomes $3m^2-12n^2=\square$. Take $m=np$, $3p^2-12=f^2(p-2)^2$. Thus

$$p=\frac{2(f^2+3)}{f^2-3}, \qquad (f^2-3)^2a^2/n^2=10f^4-36f^2+90.$$

Set $f=1+q$. The quartic is the square of $8-2q+\frac{5}{4}q^2$ if $q=-16/3$. Then $x=240$, $y=44$, $z=117$, which appear to be the least numbers.

W. Lenhart[12] took $x=(p^2-1)/(2p)$, $y=2q/(q^2-1)$, $z=1$. Then

$$x^2+y^2=\square$$

if

$$(p^2-1)^2(q^2-1)^2+16p^2q^2=\square=\{(p^2-1)(q^2-1)+8\}^2,$$

provided $p^2+q^2=5=1+4$. As usual,

$$(s^2+1)p=s^2+4s-1, \qquad (s^2+1)q=2(s^2-s-1),$$

$s\neq1$ or 3. For $s=2$, $p=11/5$, $q=2/5$.

C. Gill[13] obtained Euler's[6] result by setting

$$t=a\cos A+z\sin A, \qquad y=z\cos A-a\sin A$$

and c, x to be the analogous functions of B. Then $a^2+z^2=b^2+y^2=c^2+x^2$.

---

[9] New Series Math. Repository (ed., Leybourn), London, 1, 1806, II, 39. Also in Math. Repository, 3, 1804, 5.

[10] The Gentleman's Math. Companion, London, 4, No. 19, 1816, 626–7. Same with altered lettering, S. Bills, The Mathematician, London, 3, 1850, 200–1.

[11] J. R. Young's Algebra, Amer. ed., 1832, 338–9.

[12] Math. Miscellany, 2, 1839, 132. Reproduced in Math. Magazine, 2, 1898, 215–6; Sphinx-Oedipe, 8, 1913, 84.

[13] Application of Angular Analysis . . . , N. Y., 1848; Reproduced in Math. Quest. Educ. Times, 17, 1872, 82–3.

Take $A+B=90°$. Then $x^2+y^2=a^2$ if $z=2a\sin 2A$. Take $\cot\tfrac{1}{2}A=r/s$, $a=(r^2+s^2)^3$.

C. L. A. Kunze[14] set $x=2mn$, $y=m^2-n^2$, $z=m^2a/b-n^2b/a$. Then

$$x^2+y^2=(m^2+n^2)^2, \qquad x^2+z^2=\left(\frac{a}{b}m^2+\frac{b}{a}n^2\right)^2.$$

Take a, b to be legs of a rational right triangle with hypotenuse h, and set $n=mh/(2b)$. Then $y^2+z^2=h^2n^4/a^2$. Multiplying the resulting x, y, z by $4ab^2/m^2$, we get

$$x=4abh, \qquad y=a(4b^2-h^2), \qquad z=b(4a^2-h^2).$$

The last solution was obtained also by taking

$$x=2mn, \qquad y=mn^2-m, \qquad z=nm^2-n.$$

Then the first two conditions are satisfied, while

$$y^2+z^2=m^2n^2(m^2+n^2-4)+m^2+n^2=\square$$

if $m^2+n^2=4$. Take $m=2a/h$, $n=2b/h$, $a^2+b^2=h^2$, and multiply x, y, z by $h^3/2$. We get the former solution.

Judge Scott[15] took $x^2+y^2=(y-m)^2$, $x^2+z^2=(z-n)^2$, which determine y, z. Take $m/s=(p^2-q^2)/(p^2+q^2)$, $n/s=2pq/(p^2+q^2)$, whence $m^2+n^2=s^2$. Then $y^2+z^2=\square$ if $s^2x^4-4m^2n^2x^2+m^2n^2s^2=\square=s^2x^4$, say, whence $x=s/2$. Take $s=16pq(p^4-q^4)$. We get Euler's[6] answer. The latter was obtained also by A. Martin (*ibid.*), who satisfied $u^2-y^2=w^2-z^2$ by taking $u=a(r^2-s^2)$, $y=b(r^2-s^2)$, $w=a(r^2+s^2)-2brs$, $z=b(r^2+s^2)-2ars$. Then $u^2-y^2=\square=x^2$ if $a=p^2+q^2$, $b=2pq$. There remains the condition $y^2+z^2=\square$. Divide by $4r^2s^2$ and take $m=(r^2+s^2)/(rs)$. Then a quartic in p is to be a square, say $(p^2-mpq-q^2)^2$, whence $p/q=4/m$.

C. Chabanel[16] used the devices of Diophantus for a similar problem. Set

$$\gamma=\alpha^2-\beta^2, \qquad \delta=2\alpha\beta, \qquad \gamma_1=\gamma\delta, \qquad \delta_1=\gamma^2, \qquad z^2=8\alpha\beta\gamma^2,$$
$$x=\gamma/t-\gamma_1 t, \qquad y=\delta/t-\delta_1 t.$$

Then $x^2+z^2=(\gamma/t+\gamma_1 t)^2$, $y^2+z^2=(\delta/t+\delta_1 t)^2$. Since $4\gamma\gamma_1=4\delta\delta_1=z^2$,

$$x^2+y^2=\frac{\gamma^2+\delta^2}{t^2}+(\gamma_1^2+\delta_1^2)t^2-z^2,$$

which is a square for $t=z/(\alpha^4-\beta^4)$ since $\gamma^2+\delta^2=(\alpha^2+\beta^2)^2$ and

$$\gamma_1^2+\delta_1^2=(\alpha^4-\beta^4)^2.$$

Multiplying the initial x, y, z by $2(\alpha^2+\beta^2)\sqrt{2\alpha\beta}$, we get

$$X^2+Y^2=p^2, \qquad Y^2+Z^2=q^2, \qquad Z^2+X^2=r^2$$

for

$$X, \ q=(\alpha^2-\beta^2)[(\alpha^2+\beta^2)^2\mp 16\alpha^2\beta^2], \qquad Z=8\alpha\beta(\alpha^4-\beta^4),$$
$$Y, \ r=2\alpha\beta[(\alpha^2+\beta^2)^2\mp 4(\alpha^2-\beta^2)^2], \qquad p=(\alpha^2+\beta^2)^3,$$

where the upper signs give X, Y. For $\alpha=2$, $\beta=1$, we get Halcke's[1] solution.

---

[14] Ueber einige Aufg. Dioph. analysis, Weimar, 1862, pp. 7–9.
[15] Math. Quest. Educ. Times, 17, 1872, 82–3. Cf. Martin[20].
[16] Nouv. Ann. Math., (2), 13, 1874, 289–292.

J. Neuberg[17] satisfied Euler's[3] condition $p^2+q^2-4+1/p^2+1/q^2=w^2$ by the special values $w=2/(pq)$, $p^2+q^2=4$. The latter holds if

$$p=\frac{4rs}{r^2+s^2}, \qquad q=\frac{2(r^2-s^2)}{r^2+s^2}.$$

Hence $x^2+y^2=\zeta^2$, $y^2+z^2=\xi^2$, $z^2+x^2=\eta^2$ for

$$x=8rs(r^4-s^4); \qquad \xi=(r^2+s^2)^3; \qquad y,\ \zeta=(r^2-s^2)\{(r^2+s^2)^2\mp16r^2s^2\};$$
$$z,\ \eta=2rs\{(r^2+s^2)^2\mp4(r^2-s^2)^2\}.$$

C. Leudesdorf[18] solved the equivalent system $2(u^2+v^2-w^2)=x^2$, $2(u^2+w^2-v^2)=y^2$, $2(v^2+w^2-u^2)=z^2$ by use of trigonometric functions (cf. Gill[13]). G. Heppel repeated Neuberg's[17] solution.

J. Matteson[19] obtained Euler's[4] result by the method of Euler.[3]

A. Martin[20] varied Scott's[15] method by making the first two terms of the quartic in x cancel (giving $x=2mn/s$), instead of the last two.

K. Schwering[21] proceeded as had Neuberg[17] with λ, μ in place of p, q. To connect the result with elliptic functions, set $R(p)=p^4+1+p^2\rho$, $\rho=q^2+1/q^2-4$,

$$u=\int_0^p \frac{dp}{\sqrt{R(p)}}, \qquad p=\psi(u).$$

Then p is a well-known elliptic function. By the addition theorem,

$$\psi(u+v)=\frac{\psi(u)\psi'(v)+\psi(v)\psi'(u)}{1-\psi^2(u)\psi^2(v)}.$$

Hence if $\psi(u)$ and $\psi'(u)$ are rational, also $\psi(2u)$, $\psi(3u)$, $\cdots$, $\psi'(2u)$, $\cdots$ are rational. Thus one solution p, q yields an infinitude of solutions. The relation of the same problem to Abel's theorem is considered on p. 11.

Several writers[22] gave solutions.

*F. Ferrari[23] gave an infinitude of solutions.

R. F. Davis[24] gave Neuberg's[17] solution.

A. Martin[24a] gave another derivation of Euler's[6] result.

H. Olson[24b] proved that, if $x^2+y^2=u^2$, $x^2+z^2=v^2$, $y^2+z^2=w^2$, the product $xyzuvw$ is divisible by $3^4\cdot4^4\cdot5^2$.

M. Rignaux[24c] stated that all solutions of $x^2+y^2=\square$, etc., are given by

$$x=2mnpq, \qquad y=mn(p^2-q^2), \qquad z=pq(m^2-n^2), \qquad y^2+z^2=\square,$$

and noted four solutions, involving parameters, of the final condition.

[17] Nouv. Corresp. Math., 1, 1874–5, 199–202.

[18] Math. Quest. Educ. Times, 34, 1881, 95–6.

[19] Collection of Diophantine Problems . . ., ed., Martin, Washington, D. C., 1888, 21.

[20] Math. Magazine, 2, 1898, 214–5.

[21] Geom. Aufgaben mit rationalen Lösungen, Progr., Düren, 1898, 9.

[22] Amer. Math. Monthly, 6, 1899, 123–5; Math. Quest. Educ. Times, 68, 1898, 104; (2), 11, 1907, 26–7.

[23] Suppl. al Periodico di Mat., 14, 1910–11, 138–140.

[24] Math. Quests. and Solutions, 2, 1916, 24–25; Math. Mag., 2, 1898, 215.

[24a] Amer. Math. Monthly, 25, 1918, 305–6.

[24b] Ibid., 304–5.

[24c] L'intermédiaire des math., 25, 1918, 127.

The preceding problem is evidently equivalent to that of finding a rectangular parallelopiped whose edges and diagonals of faces are all rational. If we add the condition that also a diagonal of the solid shall be rational, we have a problem which H. Brocard[25] attempted to prove impossible by means of the terminal digits. P. Tannery[26] noted that the proof is insufficient since it supposes that the numbers in question are relatively prime in pairs.

V. M. Spunar[27] noted that the last problem is impossible.

A. Mukhopâdhyây[28] proved it impossible [if the edges be relatively prime integers]. The solutions of $x^2+y^2=\square$ are known to be $x=2k$, $y=k^2-1$. Similarly, $y=2l$, $z=l^2-1$; $z=2m$, $x=m^2-1$. Then

$$x^2+y^2+z^2=x^2+(l^2+1)^2=\square$$

requires $x=2n$, $l^2+1=n^2-1$, whereas $n^2-l^2=2$ has no integral solutions.

M. Rignaux[29] remarked that the problem is difficult and not yet solved. He satisfied three of the conditions, but not the fourth.

A. Transon[30] stated falsely that a tetrahedron with six integral edges cannot have among its solid angles a tri-rectangular trieder, and stated that one can find, in an infinitude of ways a tretahedron $OABC$ with integral values of the three edges meeting at O, and of the areas of the four faces, while the three face angles at O are right angles. C. Chabanel[31] and C. Moreau[31] gave the solution

$$OA=4xyz, \qquad OB=2y(x^2+y^2-z^2), \qquad OC=2x(x^2+y^2-z^2),$$
$$\text{area } ABC=2xy(x^2+y^2-z^2)(x^2+y^2+z^2).$$

FOUR SQUARES WHOSE SUMS BY THREES ARE SQUARES.

L. Euler[32] applied his[4] method to $A=x$, B, C and the following D, but was led to a condition difficult to treat and abandoned that method. Next, take $A=y$, B and C as in Euler,[4] and $D=\{2px-(p^2-1)y\}/(p^2+1)$. Then

$$B^2+C^2=x^2+y^2-2gxy, \qquad g=\frac{4f(f^2-1)}{(f^2+1)^2}.$$

Since $S=x^2+y^2$, the condition $S=y^2+B^2+C^2+D^2$ gives $y^2+D^2-2gxy=0$. Inserting the value of D, we get

$$4p^2x^2=2g(p^2+1)^2xy-(p^2-1)^2y^2-(p^2+1)^2y^2+4p(p^2-1)xy,$$
$$4p^2x/y=g(p^2+1)^2+2p(p^2-1)\pm(p^2+1)R, \qquad R^2=g^2(p^2+1)^2+4gp(p^2-1)-4p^2.$$

Take $R=gp^2+2p+g$. Then $p=-g$, $4gx/y=2(g^4+1)$ or 4. Using the

<hr>

[25] L'intermédiaire des math., 2, 1895, 174–5.

[26] *Ibid.*, 3, 1896, 227.

[27] Amer. Math. Monthly, 24, 1917, 393.

[28] Math. Quest. Educ. Times, 41, 1884, 60.

[29] L'intermédiaire des math., 26, 1919, 55–57.

[30] Nouv. Ann. Math., (2), 13, 1874, 64; correction, 200.

[31] *Ibid.*, 340–3.

[32] Novi Comm. Acad. Petrop., 17, 1772, 24; Comm. Arith., I, 467–72; Opera Omnia, (1), III, 203. Second method reproduced by Martin, Math. Mag., 2, 1898, 217–8.

latter value 4 and dropping the common factor x, we get

$$A = g, \qquad B = \frac{2f - g(f^2 - 1)}{f^2 + 1}, \qquad C = \frac{f^2 - 1 - 2fg}{f^2 + 1}, \qquad D = -g, \qquad g = \frac{4f(f^2 - 1)}{(f^2 + 1)^2}.$$

Using the former value and taking $y = 2g$, $g = m/n$, and multiplying $A, \cdots,$ D by $(f^2 + 1)n^4$, we get

$$A = 2mn^3(f^2 + 1), \qquad\qquad B = 2f(m^4 + n^4) - 2mn^3(f^2 - 1),$$
$$C = (f^2 - 1)(m^4 + n^4) - 4fmn^3, \qquad D = 2m^3n(f^2 + 1).$$

In his second method (§§ 56–60), Euler denoted the squares by v^2, x^2, y^2, z^2. Let a, α be two numbers for which $a^2 + \alpha^2 = A^2$. Let

$$v^2 + y^2 + z^2 = \left(\frac{Av + \alpha x}{a}\right)^2, \qquad x^2 + y^2 + z^2 = \left(\frac{Ax + \alpha v}{a}\right)^2,$$
$$y^2 + v^2 + x^2 = \left(\frac{Ay - az}{\alpha}\right)^2, \qquad z^2 + v^2 + x^2 = \left(\frac{Az - ay}{\alpha}\right)^2.$$

The first two lead to a single condition and the last two to a single one:

$$a^2(y^2 + z^2) = \alpha^2(v^2 + x^2) + 2\alpha A vx, \qquad \alpha^2(v^2 + x^2) = a^2(y^2 + z^2) - 2aAyz.$$

By adding these two equations, we get $z = \alpha vx/(ay)$. The first of the two becomes

$$\alpha^2 x^2(v^2 - y^2) = 2\alpha A vxy^2 + \alpha^2 v^2 y^2 - a^2 y^4.$$
$$\frac{\alpha x}{y} = \frac{Avy \pm \sqrt{R}}{v^2 - y^2}, \qquad R = \alpha^2 v^4 + a^2 y^4.$$

To make $\sqrt{R}$ rational, set $v = y(1 + s)$, $\sqrt{R} = y^2(A + 2\alpha^2 s/A + \alpha s^2)$. Of the resulting two solutions, one is complicated, while the other (given by $x/y = 1$) is

$$v = a(A^2 - 2\alpha^2), \qquad x = y = 2\alpha aA, \qquad z = \alpha(A^2 - 2\alpha^2).$$

To obtain a simpler solution in which the numbers are distinct, take two numbers b, β such that $b^2 + \beta^2 = B^2$, and set $\alpha v^2 = \beta M$, $ay^2 = bM$. Then $\sqrt{R} = BM$. But $a\beta/(\alpha b)$ must be the square of v/y; take it to be m^2/n^2. Thus

$$\frac{v}{y} = \frac{m}{n}, \qquad \frac{x}{y} = \frac{Abm \pm aBn}{a\beta n - \alpha bn}, \qquad \frac{z}{y} = \frac{\alpha m}{an} \cdot \frac{x}{y}.$$

Taking $a = 21$, $\alpha = 20$, $b = 35$, $\beta = 12$, we get $A = 29$, $B = 37$, $m = 3$, $n = 5$. For the lower sign, $x/y = 3/8$. Hence $v = 168$, $x = 105$, $y = 280$, $z = 60$. Finally, he noted the solution

$$v = 4fg(f + g)(3f - g)k, \qquad y = 4fg(f - g)(3f + g)k, \qquad x = lk, \qquad z = 2fgl,$$
$$k = 3f^2 + g^2, \qquad l = (f^2 - g^2)(9f^2 - g^2).$$

M. S. O'Riordan[33] developed the idea underlying Euler's first solution. Let $S = A^2 + B^2 + C^2 + D^2$, $S - A^2 = \alpha^2$, $\cdots$, $S - D^2 = \delta^2$. To obtain a number

[33] The Gentleman's Math. Companion, London, 2, No. 12, 1809, 185–7; Math. Repository (ed., Leybourn), New Series, 6, II, 1835, 1–4. Reproduced in Math. Magazine, 2, 1898, 218–9.

S which is a sum of two squares in four ways, employ

$$T = (a^2+b^2)(c^2+d^2) = E_{\pm}^2 + F_{\pm}^2, \qquad E_{\pm} = ac \pm bd, \qquad F_{\pm} = ad \mp bc,$$
$$S = T(e^2+f^2) = (eE_{\pm} + kfF_{\pm})^2 + (fF_{\pm} - keF_{\pm})^2, \qquad k^2 = 1.$$

Hence $S = A^2 + \alpha^2 = B^2 + \beta^2 = C^2 + \gamma^2 = D^2 + \delta^2$ if

$$A = eE_+ - fF_+, \qquad B = eF_- - fE_-, \qquad C = eE_- - fF_-, \qquad D = eE_+ + fF_+,$$
$$\alpha = eF_+ + fE_+, \qquad \beta = eE_- + fF_-, \qquad \gamma = eF_- + fE_-, \qquad \delta = eF_+ - fE_+.$$

It remains to satisfy the condition $S = \Sigma A^2$ or, if we prefer, $A^2 + D^2 = \gamma^2 - B^2$, viz.,

$$(A+D)^2 + (A-D)^2 = 2(\gamma+B)(\gamma-B), \qquad (eE_+)^2 + (fF_+)^2 = 2efE_-F_-.$$

Divide by f^2 and set $=fw$. Thus

$$w^2(ac+bd)^2 + (ad-bc)^2 = 2w(ac-bd)(ad+bc).$$

The roots w are rational if the discriminant

$$(ac-bd)^2(ad+bc)^2 - (ac+bd)^2(ad-bc)^2 = 4abcd(a^2-b^2)(c^2-d^2)$$

is a square. Take $a = mb$, $c = nd$, $mn = r(n-1)$. Then shall

$$r(n+1)(rn+n-r)(rn-n-r) = \square.$$

Take $n = 2r$. Then shall $2r^2 - 3r = \square$, as is the case for $r = 3s^2/(2s^2-1)$. For $s = 1$, we get Euler's solution 168, 105, 280, 60. Removing the restriction $n = 2r$, let $(nr+n-r)k = (nr-n-r)l$. Then shall $n(n+1)(n-1)e = \square$, $e = (l+k)/(l-k)$. Take $n = e+x$. There results the answer $a = l+k$, $b = l-k$, $c = (l^2+k^2)^2$, $d = 4lk(l^2-k^2)$.

B. Gompertz[34] employed x^2, y^2, z^2, w^2,

$$x = (y^2+z^2-p^2)/(2p), \qquad w = (y^2+z^2-q^2)/(2q).$$

Then $x^2+y^2+z^2$ and $w^2+y^2+z^2$ are squares. Also, $x^2+w^2+z^2$ and $x^2+w^2+y^2$ are squares if

$$f_j \equiv (y^2+z^2)^2(p^2+q^2) + (p^2+q^2-4j^2)p^2q^2 = \square$$

for $j = y$ and $j = z$. Take $p = (q^2-r^2)/(2r)$, $y = (q^2+r^2)/(4r)$. Then

$$p^2+q^2 = 4y^2$$

and $f_y = \square$. Set $z = ty$, $pq/y^2 = b$. Then $f_z = \square$ if $(1+t^2)^2 + b^2(1-t^2) = \square$. Set $t = 1+v$. The condition becomes $v^4 + \cdots = \square = (2+Av \pm v^2)^2$ and holds if $A = 2 - b^2/2$, $v = \pm b^2/4 - 1$. For $q = 2$, $r = 1$, we get Scott's[38] solution.

C. Gill[35] treated the problem to find n squares the sum of any $n-1$ of which is a square. He[36] gave elsewhere his solution for $n = 5$ and remarked that the smallest numbers given by his formulas are so very large as to discourage any attempt to compute them. For $n = 3$, see Gill.[13] The method was adapted to the case $n = 4$ by S. Bills.[37] If z^2, y^2, x^2, w^2 are the required squares, their sum shall equal

$$a^2 + z^2 = b^2 + y^2 = c^2 + x^2 = d^2 + w^2.$$

[34] The Gentleman's Math. Companion, 2, No. 12, 1809, 182–4. Reproduced (essentially) by A. Martin, Math. Mag., 2, 1898, 216.

[35] Application of the angular analysis . . ., New York, 1848, 69–76.

[36] The Lady's and Gentleman's Diary, London, 1850, 53–5, Quest. 1797.

[37] Math. Quest. Educ. Times, 16, 1872, 108–110.

Take
$$b = a \cos A + z \sin A, \qquad y = a \sin A - z \cos A,$$
and c, x; d, w corresponding functions of angles B, C. It remains only to satisfy $y^2 + x^2 + w^2 = a^2$, viz.,
$$x^2(\Sigma \sin^2 A - 1) - az\Sigma \sin 2A + z^2\Sigma \cos^2 A = 0.$$
The discriminant must be a square, whence
$$k^2 = 2\Sigma \cos 2A + 2\Sigma \cos (A - B).$$
Take $C = A + B - 90°$. Then $k^2 = \sin 2A \cdot \sin 2B$. Take $\sin 2A = \tan B/2$. Then $k = \sin 4A/(1 + \sin 2A)$. The case $\cot A/2 = 2$ leads to the solution [due to Euler,[32] § 58]:
$$z = 186120, \qquad y = 23838, \qquad x = 102120, \qquad w = 32571.$$
Bills gave also 280, 105, 60, 168 and 1120, 3465, 1980, 672.

Judge Scott[38] found 639604, 3456000, 3750000, 832797 [due to Euler,[32] § 55].

S. Tebay[39] gave the solution x^2, $\cdots$, u^2, where
$$x = (s^2 - 1)(s^2 - 9)(s^2 + 3), \qquad y = 4s(s - 1)(s + 3)(s^2 + 3),$$
$$z = 4s(s + 1)(s - 3)(s^2 + 3), \qquad u = 2s(s^2 - 1)(s^2 - 9).$$
A. Martin[38a] gave a complete solution by the method of Tebay.[39]

Three squares whose differences are squares.

Under Euler[28] of Ch. XV are cited various papers on the related problem to make $x \pm y$, $x \pm z$, $y \pm z$ all squares.

L. Euler[40] made the differences of x^2, y^2, z^2 squares by taking
$$\frac{x}{z} = \frac{p^2 + 1}{p^2 - 1}, \qquad \frac{y}{z} = \frac{q^2 + 1}{q^2 - 1},$$
whence $x^2 - z^2$ and $y^2 - z^2$ are squares. Also $x^2 - y^2 = \square$ if
$$P = (p^2q^2 - 1)(q^2 - p^2) = \square.$$
Each factor will be a square if
$$pq = \frac{a^2 + b^2}{2ab}, \qquad \frac{q}{p} = \frac{c^2 + d^2}{2cd}.$$
The product of the latter must be a square q^2. Take a, $b = f \pm g$; c, $d = h \pm k$. Then must $(f^4 - g^4)(h^4 - k^4) = \square$ [cf. Euler[28] of Ch. XV.]

J. Cunliffe[40a] treated the problem.

"Calculator"[41] took
$$x = (r^2 + s^2)(m^2 + n^2), \qquad y = (r^2 + s^2)(m^2 - n^2), \qquad z = 4rsmn - (r^2 - s^2)(m^2 - n^2).$$
Then $x^2 - y^2$ and $x^2 - z^2$ are the squares of $2mn(r^2 + s^2)$ and
$$2rs(m^2 - n^2) + 2mn(r^2 - s^2).$$

---

[38] Math. Quest. Educ. Times, 16, 1872, p. 108.
[39] Ibid., 68, 1898, 103–4.
[39a] Ibid., 24, 1913, 81–2.
[40] Algebra, 2, 1770, §§ 236–7; 2, 1774, pp. 320–7; Opera Omnia, (1), I, 473–7.
[40a] The Math. Repository (ed., Leybourn), London, 3, 1804, 5–10.
[41] The Gentleman's Math. Companion, London, 3, No. 14, 1811, 334–6.

For $q = (r^2 - s^2)/(rs)$,

$$\frac{y^2 - z^2}{4r^2s^2} = m^4 + 2qm^3n - 6m^2n^2 - 2qmn^3 + n^4 = (m^2 - qmn + n^2)^2,$$

if $m/n = (q^2 + 8)/(4q)$. Or we may use

$$\frac{z^2 - y^2}{(z+y)^2} = \frac{z - y}{z + y} = \frac{A}{B}, \qquad A = r^2n^2 - r^2m^2 + 2rsmn, \qquad B = s^2m^2 - s^2n^2 + 2rsmn.$$

Take $B = (tn - sm)^2$ to get m. Then $A = r^2n^2$ if $r = 4ts^2/(t^2 - 3s^2)$. He[42] later used the same x, but took $z = 2mn(r^2 + s^2)$, $a = (r^2 + s^2)(m^2 - n^2)$, whence $x^2 - z^2 = a^2$. Set $b = a - rv$, $y = z + sv$; then $a^2 + z^2 = b^2 + y^2$ gives v in terms of a, r, s, z. Finally, $y^2 - z^2 = \square$ if a quartic in m is the square of (say) $m^2 - mn(r^2 - s^2)/(rs) + n^2$, whence

$$m : n = r^4 + 6r^2s^2 + s^4 : 4rs(r^2 - s^2).$$

J. Cunliffe[43] obtained Calculator's[41] first result by the same method.

S. Ward[44] discussed Euler's[40] final condition. Set $f = f'g$, $h = h'k$,

$$(f'^4 - 1)(h'^4 - 1) = (f'^4 - 1)^2(h'^2 - 1)^2,$$

which reduces to $f'^4/h'^2 = f'^4 - 2$. The latter is a square if $f'^2 = (r^2 + 2s^2)/(2rs)$, and $r^2 + 2s^2 = \square$ if $r = t^2 - 2$, $s = 2t$. The value for f'^2 is a square if $t(t^2 - 2) = \square$. Taking $t = 2$, we get $x/z = -41/9$, $y/z = 185/153$. Or we may treat $P = \square$ by setting $q = mp$ and treating $(m^2p^4 - 1)(m^2 - 1) = \square$ by the usual method for quartics, one solution $p = 1$ being known.

W. Lenhart[45] took the roots of the three squares to be

$$\frac{x^2 + y^2}{x^2 - y^2}, \qquad \frac{v^2 + w^2}{v^2 - w^2}, \qquad 1.$$

The square of either the first or the second exceeds unity by a square. Hence it remains only to make the difference of their squares a square, viz., $(vx + wy)(vx - wy)(vy + wx)(vy - wx) = \square$. Take $v = ty + x$, $w = tx - y$, whence $vy + wx = t(vx - wy)$. Then shall $t(vx + wy)(vy - wx) = \square$, which holds if

$$x^2 - y^2 + 2txy = \square, \qquad y^2 - x^2 + 2xy/t = \square.$$

The second condition is satisfied if $x = 2y/t$. Then the first becomes $4 + 3t^2 = \square = (2 - pt)^2$, say, whence we get t and $x = p^2 - 3$, $y = 2p$, $v = (p^2 + 1)^2 + 8$, $w = 2(p^2 - 3)p$. Or we may take $x^2 - y^2 + 2txy = (x - py)^2$, whence $x = p^2 + 1$, $y = 2(p + t)$. Then $t^2(y^2 - x^2 + 2xy/t) = (ty - r)^2$ if

$$-t^2x^2 + 4t^2x = -4rt^2, \qquad 4ptx = r^2 - 4ptr.$$

Then

$$r = \frac{x^2 - 4x}{4}, \qquad t = \frac{r^2}{4p(r+x)} = \frac{r^2}{px^2} = \frac{(p^2 - 3)^2}{16p}.$$

Dividing the values of x and y by $d = (p^2 + 1)/(8p)$ and those of v and w

---

[42] The Gentleman's Math. Companion, London, 4, No. 19, 1816, 628–31.

[43] Ibid., 5, No. 26, 1823, 262–4.

[44] J. R. Young's Algebra, Amer. ed., 1832, 339–341.

[45] Math. Miscellany, 2, 1839, 129–132; French transl., Sphinx-Oedipe, 8, 1913, 83–4.

by $d/2$, we have

$$x = 8p, \qquad y = p^2 + 9, \qquad w = 8p(p^2 - 9), \qquad v = p^4 + 2p^2 + 81 = \Pi(p^2 \pm 4p + 9).$$

Three squares, sum of any two less third a square.

L. Euler[46] gave four methods to solve

$$(1) \qquad y^2 + z^2 - x^2 = p^2, \qquad x^2 + z^2 - y^2 = q^2, \qquad x^2 + y^2 - z^2 = r^2.$$

(i) Let $s = x^2 + y^2 + z^2$. Since $s = p^2 + 2x^2$, etc., s must be expressible in three ways in the form $a^2 + 2b^2$, whence s must have at least three prime factors of that form. Take $m = ac \pm 2bd$, $n = bc \mp ad$, $u = mf \pm 2ng$, $v = nf \mp mg$. Then

$$m^2 + 2n^2 = (a^2 + 2b^2)(c^2 + 2d^2), \qquad (m^2 + 2n^2)(f^2 + 2g^2) = u^2 + 2v^2.$$

Take $u^2 + 2v^2 = s$. By using the four combinations of signs, we get four sets of values of u, v. As we need only three sets, omit that given by both lower signs. Set

$$(2) \qquad \begin{aligned} &p,\, q = f(ac + 2bd) \pm 2g(bc - ad), \qquad r = f(ac - 2bd) + 2g(bc + ad), \\ &x,\, y = f(bc - ad) \mp g(ac + 2bd), \qquad z = f(bc + ad) - g(ac - 2bd), \end{aligned}$$

where the upper signs give p and x. Compute $x^2 + y^2 + z^2$ and compare with the earlier expression $u^2 + 2v^2$ for s; we get

$$(3) \qquad \begin{aligned} &Ff^2 + Gg^2 + 2Cfg = 0, \qquad F = (b^2 - a^2)c^2 + (a^2 - 4b^2)d^2 - 2abcd, \\ &G = (a^2 - 4b^2)c^2 + 4(b^2 - a^2)d^2 + 4abcd, \qquad C = -(bc + ad)(ac - 2bd). \end{aligned}$$

Taking $F = 0$, we get $c : d = 2b - a : b - a$ or $-a - 2b : b + a$, and also $f : g = -G : 2C$. The same solution results also from $G = 0$.

(ii) By (3), $f : g = -(C \pm V) : F$, where

$$V^2 = C^2 - FG = (a^2 - 2b^2)^2 Q,$$

$$Q = c^4 + 8mc^3d - 4c^2d^2 - 16mcd^3 + 4d^4, \qquad m = \frac{ab}{a^2 - 2b^2}.$$

Let Q be the square of $c^2 - 4mcd + 2d^2$. Then $c : d = 2m^2 + 1 : 2m$.

(iii) Use p, q, x, y given by (2), but take

$$r = f(\alpha c - 2\beta d) + 2g(\beta c + \alpha d), \qquad z = f(\beta c + \alpha d) - g(\alpha c - 2\beta d),$$

where α, β are such that $\alpha^2 + 2\beta^2 = a^2 + 2b^2$. Hence we now get new values for F, G, C in (3). For $F = 0$, we get

$$c : d = -\alpha - 2b : \beta + a \qquad \text{or} \qquad -\alpha + 2b : \beta - a.$$

He deduced the following simple solution of the problem: Start with any two integers m and n, m odd, and set

$$s = m^2 + 2n^2, \qquad t = m^2 - 2n^2, \qquad u = 2mn,$$

or take s, t, u such that $s^2 = t^2 + 2u^2$; we have the solution

$$\begin{aligned} &x = s(s + u)\rho - 2t^2\sigma, \qquad y = s(s + u)\rho + 2t^2\sigma, \qquad z = st\rho + 2t\sigma^2, \\ &p = st\rho + 4t(s + u)\sigma, \qquad q = st\rho + 4t(s - u)\sigma, \qquad r = s\sigma\rho - 4t^2\sigma, \end{aligned}$$

where $\rho = 3s + 4u$, $\sigma = s + 2u$.

---

[46] Posth. paper, Comm. Arith., II, 603–16; Opera postuma, 1, 1862, 105–118. French transl. in Sphinx-Oedipe, 1906–7, 163–83.

(iv) The first two equations (1) are satisfied if $z=mn(A-B)$, $y+x=2m^2A$, $y-x=2n^2B$, $p=mn(A+B)$, $q=mn(a^2-2ab-b^2)$, where $A=a(a+b)$, and $B=b(a-b)$. The third equation (1) becomes

$$2m^4A^2+2n^4B^2-m^2n^2(A-B)^2=r^2.$$

Set $m=f+g$, $n=f-g$, $r=(A+B)f^2+4(A-B)fg-(A+B)g^2$. Then

$$f:g=B^2-A^2:2AB.$$

A. M. Legendre[47] noted that the last two conditions (1) are evidently satisfied if

$$x=r^2+s^2, \qquad y=r^2+rs-s^2, \qquad z=r^2-rs-s^2.$$

Then the first condition becomes $r^4-4r^2s^2+s^4=\square$. Set $r=s(2+\phi)$ and make the quartic function of ϕ the square of $1+8\phi+\alpha\phi^2$. The case $\alpha=1$ gives $\phi=-23/4$, $r=15$, $s=4$, whence $x=241$, $y=269$, $z=149$, which is apparently the least solution.

J. Cunliffe[48] noted that (1) give $x^2=\frac{1}{2}(q^2+r^2)$, etc., whence

$$r^2=2x^2-q^2=2y^2-p^2.$$

Hence, if we set $x=y+\rho v$, $q=p+\sigma v$, we get $v=(2\sigma p-4\rho y)/(2\rho^2-\sigma^2)$. To satisfy $2y^2-p^2=r^2$, set

$$y=D(m^2+n^2), \quad p=D(n^2-m^2+2mn), \quad r=D(m^2-n^2+2mn), \quad D=2\rho^2-\sigma^2.$$

The resulting value of $\frac{1}{2}(p^2+q^2)$ will equal the square of

$$z=m^2A-\frac{2mn}{A}(4\rho^4+4\rho^3\sigma+2\rho\sigma^3+\sigma^4)-n^2(2\rho^2-2\rho\sigma+\sigma^2),$$

where $A=2\rho^2+2\rho\sigma+\sigma^2$, if

$$m:n=4\rho^2\sigma^2+4\rho^3\sigma+2\rho\sigma^3:4\rho^4+4\rho^3\sigma+2\rho\sigma^3+2\rho^2\sigma^2+\sigma^4.$$

Taking $\rho=\sigma=1$, he obtained, as his least answer, $x=149$, $y=269$, $z=241$.

D. S. Hart[49] noted that (1) are equivalent to $2r^2+2q^2=\square$, $2r^2+2p^2=\square$, $2q^2+2p^2=\square$. The first is satisfied if $r=\rho^2-2\sigma^2$, $q=\rho^2+4\rho\sigma+2\sigma^2$. Set $p=l+r$, $a=\rho^2+2\rho\sigma+2\sigma^2$. Then the last two conditions of the problem become $2l^2+4rl+4r^2=\square$, $2l^2+4rl+4a^2=\square$. Equating the latter to $(2a-lt)^2$, we get l in terms of t, r, a. Then the former becomes a quartic in t. S. Bills satisfied the first two of Hart's conditions by taking

$$q=\frac{P^2-Q^2+2PQ}{P^2-Q^2-2PQ}\cdot r, \qquad p=\frac{R^2-S^2+2RS}{R^2-S^2-2RS}\cdot r.$$

The third condition leads to a quartic.

G. B. M. Zerr[50] took x^2z^2, y^2z^2 and z^2 as the squares and set

(A) $$x^2+y^2-1=(t+u)^2, \qquad x^2-y^2+1=(t-u)^2.$$

Since $x^2=t^2+u^2$, take $t=n(p^2-q^2)$, $u=2npq$, whence $x=n(p^2+q^2)$. Take

[47] Théorie des nombres, 1798, 461-2; ed. 2, 1808, 434; ed. 3, 1830, II, 127; German transl. by Maser, 2, 1893, 124.

[48] The Gentleman's Diary, London, No. 62, 1802, 41-2, Quest. 823. Math. Repository (ed., Leybourn), 3, 1804, 97.

[49] Math. Quest. Educ. Times, 20, 1874, 84-6.

[50] Amer. Math. Monthly, 10, 1903, 207-8. Cf. papers 114-5 of Ch. XVI.

$y = 2mn - 1$.   Then the first condition (A) is satisfied if

$$n = m/z, \qquad z = m^2 - pq(p^2 - q^2).$$

There remains the condition

$$(y^2 - x^2 + 1)z^2 = 2m^4 + 2p^2q^2(p^2 - q^2)^2 - m^2(p^2 + q^2)^2 = \square,$$

which is satisfied if $m = p^2 - q^2$, $p^4 + q^4 - 4p^2q^2 = (p^2 - 2rq^2)^2$, whence

$$p^2 = \tfrac{1}{4}q^2(4r^2 - 1)/(r - 1).$$

For $r = 13$, $p = 15q/4$ and the numbers are proportional to Legendre's.

FURTHER SETS OF THREE OR MORE LINEAR FUNCTIONS OF THREE OR MORE
SQUARES MADE SQUARES.

Leonardo Pisano,[51] to make $x^2 + y^2$, $x^2 + y^2 + z^2$, $x^2 + y^2 + z^2 + w^2$, $\cdots$ all squares, took the first square x^2 to be 9.   Then the second, y^2, is the sum 16 of all odd numbers 1, 3, 5, 7 preceding 9, whence $9 + 16 = \square = 25$.   As the third square take the sum 144 of all odd numbers < 25 whence $144 + 25 = \square = 169$.   As the fourth square take $1 + 3 + \cdots + 167 = 7056$ whence $7056 + 169 = \square = 7225$.   As the fifth square take

$$1 + 3 + \cdots + 7223 = 13046444.$$

Leonardo noted (p. 279) that, since 7225 is the square of 85, not a prime, we can get several values for the fifth square.   Besides that given above we may take the sum of all odd numbers $\leq 7225/5 - 5 - 1$ and get the square 720^2, or the sum of all odd $\leq 7225/25 - 25 - 1$ and get 132^2.   A. Genocchi[52] noted that a fourth solution was omitted, viz., the sum 204^2 of all odd $\leq 7225/17 - 17 - 1$.

F. Feliciano[53] gave only 9, 16, 144.

N. Tartaglia[54] obtained 25, 144, 7056 by Leonardo's method.

J. de Billy[55] found the squares 9, 1/100, $(23/15)^2$ such that if 15 is added to the sum of any two of them there results a square.   [Due to Diophantus, V, 30; cf. Fermat[9] of Ch. XV.]

L. Euler,[46] p. 604, stated that it is not possible to find four squares such that if each be subtracted from the sum of the remaining three the difference is always a square.

H. Faure[56] proved the last theorem by use of the lemma that $2x^2 + 2y^2 + 2xy = z^2$ is impossible in integers.

Euler[57] noted five sets of solutions, like $p = 89$, $q = 191$, $r = 329$, of

$$p^2 + q^2 = 2z^2, \qquad p^2 + r^2 = 2y^2, \qquad q^2 + r^2 = 2x^2.$$

---

[51] Scritti, II, 254, note on margin; 279.   Tre Scritti, 57, 112.
[52] Annali di Sc. Mat. e Fis., 6, 1855, 355-6.
[53] Libro di Arith . . . Scala Grimaldelli, Venice, 1526, f. 5.
[54] La Seconda Parte Gen. Trattato Numeri et Misure, Venice, 1556, f. 142 left.
[55] Diophantvs Geometria, Paris, 1660, 117-8.
[56] Nouv. Ann. Math., 16, 1857, 342-4.
[57] Opera postuma, 1, 1862, 259-60 (about 1782).

To make $x^2+y^2+2z^2$, $x^2+z^2+2y^2$, $y^2+z^2+2x^2$ all squares, A. M. Legendre[58] set $y=x+2p$, $z=x+2q$. Then $x^2+y^2+2z^2=4(x+f)^2$ for $(2f-p-2q)x=p^2+2q^2-f^2$. Equating this to the value found similarly from $x^2+z^2+2y^2=\square$, he was led to the values

$$x=7p^2-30pq+7q^2, \qquad y=23p^2-14pq+7q^2, \qquad z=7p^2-14pq+23q^2.$$

Substitute these into $y^2+z^2+2x^2$ and set $p/q=1+\theta$. Then shall

$$1+2\theta+2\theta^2+\theta^3+\frac{169}{256}\theta^4=\square.$$

The particular solution $\theta=208$ gives $x=18719$, $y=62609$, $z=18929$.

T. Pepin[59] noted that also $\theta=-1$ and -2 (whence $x=y=7$, $z=23$; $x=y=z=1$) and applied his first formulas (Ch. XXII[157]) with $x_1=0$, $x_2=-1$, $x_3=-2$ and found $\theta=-8/15$, whence $x:y:z=77:77:253$.

C. Gill and W. Wright[60] made $x^2+y^2+z^2+v^2$, $x^2+y^2-z^2+v^2$, $x^2-y^2+z^2+v^2$, $y^2+z^2-x^2+v^2$ squares. To satisfy the second and third conditions, take $2vx=y^2-z^2$, say $2v=y+z$, $x=y-z$. The fourth condition holds if $y^2+10yz+z^2=\square=(y-p)^2$, which gives y. Clearing of denominators, we now have

$$y=2p^2-2z^2, \qquad v=9z^2+2pz+p^2, \qquad x=2p^2-4pz-22z^2.$$

Then the first condition leads to a quartic in p; equating it to $(3p^2-2pz+d)^2$, we get $d=-23z^2/3$.

To find four squares the double of whose sum is a square, and double the difference between the sum of any three and the fourth is a square, they took $(x+y)^2$, $(x-y)^2$, v^2, z^2. Then two conditions are satisfied if $v+z=4x$, $v-z=y$, and the solution follows readily.

The solutions of the system $2x^2+2y^2-3z^2=\square$, etc., and the system $x^2+2(y^2-z^2)=\square$, etc., offer no special interest.

To find three numbers such that the square of each plus the product of the same number and the sum or difference of the remaining two gives a square, several[61] used the numbers a^2, b^2, c^2. Then the conditions reduce to $a^2+b^2+c^2=\square$, $a^2+c^2-b^2=\square$, $a^2+b^2-c^2=\square$. To satisfy the first two, take $b^2=2ac$. Equate the third to $(cn-a)^2$. Take $n=-3/4$.

A. Gérardin[62] treated the system $N=Ph^2-k^2$ ($P=n+1$, $n+2$, $\cdots$, $n+\alpha$).

E. Fauquembergue[63] made the four functions $x^2\pm hy^2$, $u^2\pm hy^2$ squares. H. Holden[63a] showed that

$$A\equiv\alpha x^2+\beta y^2+\gamma z^2, \qquad B\equiv\alpha y^2+\beta z^2-\gamma x^2, \qquad C\equiv\alpha z^2+\beta x^2-\gamma y^2$$

[58] Théorie des nombres, 1798, 460–1; ed. 2, 1808, 433–4; ed. 3, 1830, II, 125; German transl. by Maser, II, 122. J. Cunliffe, New Series of Math. Repository (ed., T. Leybourn), 1, 1806, I, 189–191, used the same method with $2p-2q$, $-2p$ replaced by m, n, and obtained an equivalent result.

[59] Atti Accad. Pont. Nuovi Lincei, 30, 1876–7, 219–20.

[60] The Gentleman's Math. Companion, London, 5, No. 30, 1827, 579–83.

[61] Ladies' Diary, 1833, 38–39, Quest. 1547.

[62] L'intermédiaire des math., 23, 1916, 88–93. He gave 139 examples.

[63] *Ibid.*, 24, 1917, 38–9.

[63a] Messenger of Math., 48, 1918, 77–87, 166–179.

can usually be made squares by values of x, y, z which are rational functions of a parameter k whenever the auxiliary equation $\beta\gamma p^2 + \gamma\alpha q^2 = \alpha\beta r^2$ can be similarly solved. For, if rational functions p_1, q_1, r_1 of k satisfy the latter, a linear relation $p_1 x + q_1 y + r_1 z = 0$ implies $A = \square$. Similarly, if rational functions p_2, q_2, r_2 of m satisfy the auxiliary equation, then $r_2 x + p_2 y + q_2 z = 0$ implies $B = \square$. Solving the two linear equations, we obtain x, y, z as quadratic functions of m and k. For these values, C becomes a quartic function of m whose first and last coefficients are squares of functions of k, so that C can be made a square. For $\alpha = \beta = 2$, $\gamma = 1$, we have the problem of a rational triangle with rational medians. Euler's[46] equations (1) are treated by this method and by a related method. In the second paper he used the method to make $px^2 + q^2 y^2 - pz^2$, $py^2 + q^2 z^2 - pw^2$, $pz^2 + q^2 w^2 - px^2$, $pw^2 + q^2 x^2 - py^2$ all squares.

On $2x^2 + 2y^2 - z^2 = \square$, etc., see triangles with rational medians, Ch. V.

Quadratic forms in x and y, x and z, y and z, made squares.

J. Cunliffe[64] found rational numbers x, y, z such that

(4) $$x^2 - xy + y^2, \qquad x^2 - xz + z^2, \qquad y^2 - yz + z^2$$

are squares, by equating the first and second to the squares of $4a - x$, $4b - x$, whence

$$x = \frac{16a^2 - y^2}{8a - y} = \frac{16b^2 - z^2}{8b - z}.$$

Equate the denominators. Thus $y = 5a - 3b$, $z = 5b - 3a$. Then

$$y^2 - yz + z^2 = (7a - nb)^2$$

if $a : b = n^2 - 49 : 14n - 94$. J. Whitley equated the first two functions (4) to the squares of $x - ny$ and $x - mz$; hence take

$$x = (n^2 - 1)(m^2 - 1), \qquad y = (m^2 - 1)(2n - 1), \qquad z = (n^2 - 1)(2m - 1).$$

Set $p = 2n - 1$, $q = n^2 - 1$, $v = n^2 - n + 1$. Then $v^2 = p^2 - pq + q^2$. Equating

$$y^2 - yz + z^2 = p^2 m^4 - 2pqm^3 + (4q^2 + pq - 2p^2)m^2 + (2pq - 4q^2)m + v^2$$

to the square of $pm^2 - qm + v$, we get m rationally.

To find rational numbers such that[65]

(5) $$x^2 + xy + y^2, \qquad x^2 + xz + z^2, \qquad y^2 + yz + z^2$$

are squares, equate the first and second to the squares of $x + y - m$ and $x + z - n$. We get x and z in terms of y. The third condition leads to a quartic in y, which is made a square as usual.

Lowry[65a] made $\alpha \equiv x^2 + axy + by^2$, $\beta \equiv x^2 + a_1 xz + b_1 z^2$, $\gamma = y^2 + a_2 yz + b_2 z^2$ squares. Set $r = n(a_1 n + 2m)$, $s = m^2 - b_1 n^2$, $\rho = u(au + 2v)$, $\sigma = v^2 - bu^2$. Take $y/x = \rho/\sigma$, $z/x = r/s$. Then $\alpha\sigma^2/x^2 = (v^2 + auv + bu^2)^2$. Similarly, $\beta = \square$. Since

[64] The Gentleman's Math. Companion, London, 3, No. 14, 1811, 310–11.

[65] Ibid., 4, No. 21, 1818, 757–60; J. Cunliffe, Leybourn's Math. Repository, New Ser., 2, 1809, I, 93–5. Cf. Ch. V.[113]

[65a] New Series of Math. Repository (ed., T. Leybourn), 3, 1814, I, 153–164.

$z/y = r\sigma/(s\rho)$, $\gamma = \square$ if

$$(6) \qquad s^2\rho^2 + a_2 rs\rho\sigma + b_2 r^2 \sigma^2 = \square,$$

or $b_2 r^2 v^4 + \cdots + ku^4 = \square$, $k \equiv b_2 r^2 - aa_2 brs + a^2 s^2$. This quartic is made a square in a special way for special values of m and n for which $k = \square$. For the case $b_1/b = \square = d^2$, make $\sigma = s$ by taking $u = dn$, $v = m$. Then (1) becomes $\rho^2 + a_2 r\rho + b_2 r^2 = \square = (\rho - re/t)^2$, if $2et + a_2 t^2 = r/n$, $e^2 - b_2 t^2 = \rho/n$, which are linear in m and n.

An anonymous writer[65b] gave an elegant solution for the case $b = b_1 = b_2 = 1$. Take $x = nR$, $y = m^2 - n^2$, $z = nS$, where $R = an \pm 2m$, $S = a_2 n + 2m$. Then

$$\alpha = (m^2 \pm amn + n^2)^2, \qquad \gamma = (m^2 + a_2 mn + n^2)^2, \qquad \beta = n^2(R^2 + a_1 RS + S^2).$$

Also, $\beta = n^2(p^2 + a_1 pq + q^2)^2$ if $R = p^2 - q^2$, $S = a_1 q^2 + 2pq$. Comparing the two expressions for R and the two for S, we get m and n as fractions whose common denominator is $2(a_2 \mp a)$, which may be omitted since x, y, z are homogeneous in m and n. For $a = a_2$, use the lower sign.

J. Whitley and W. Rutherford[66] equated $p^2 x^2 + xy + y^2$ and $p^2 z^2 + xz + x^2$ to the squares of $px + y - a$ and $pz + x - b$, finding x and y in terms of z. Then $p^2 y^2 + yz + z^2 = \square$ if a quartic in z is a square.

W. Lenhart[67] took $x = abc$, $y = bdf$, $z = cfn$ in (5). By Lagrange's Addition, § 90, to Euler's Algebra (Lagrange[63] of Ch. XX), the resulting functions are squares if

$$p^2 - q^2 = ac, \qquad p_1^2 - q_1^2 = ab, \qquad p_2^2 - q_2^2 = bd,$$
$$2pq + q^2 = df, \qquad 2p_1 q_1 + q_1^2 = fn, \qquad 2p_2 q_2 + q_2^2 = cn.$$

To solve the equations in the first column, set $p + q = a$, $p - q = c$, $2p + q = d/r$, $q = rf$. From the two values for $2p$ and the two for q, we get $c = a - 2rf$, $d = r(2a - rf)$. Similarly, by the equations in the second column, $b = a - 2sf$, $n = s(2a - sf)$. By the two in the third column, $2p_2 = d + b = tc - n/t$, $2q_2 = d - b = 2n/t$. Eliminating t, we get $(3d+b)(d-b) = 4nc$. Inserting the earlier values of c, d, b, n, we get

$$\{(6r+1)a - (2s+3r^2)f\}\{(2r-1)a + (2s-r^2)f\} = 8s(a - 2rf)(a - \tfrac{1}{2}sf).$$

The final factor will occur also on the left if $2s + 3r^2 = \tfrac{1}{2}s(6r+1)$. Then

$$a = 12r^3(5 - r) + 5r^2, \qquad f = 12r^2 + (3 - 2r) - 2r - 1.$$

Next, for (4), equate the last two functions to A^2 and B^2. Their differences are equal if $A + B = 2(x + y - z)$, $A - B = \tfrac{1}{2}(x - y)$. Insert the resulting value of B into $y^2 - yz + z^2 = B^2$. Thus $z = (3x^2 + 10xy + 3y^2)/\{8(x+y)\}$. Finally, $x^2 - xy + y^2 = \square$ if $x = p^2 - q^2$, $y = 2pq - q^2$.

T. Strong (p. 301) equated $(x+y)^2 - Axy$, $(x+z)^2 - Bxz$, $(y+z)^2 - Dyz$ to the squares of $x + y - a$, $x + z - b$, $y + z - c$. By the first two we get y and z in terms of x. Then the third condition states that two quadratic functions of x are equal. We may equate the constant terms or the coefficients of x^2 and get x rationally.

[65b] New Series of the Math. Repository (ed., T. Leybourn), 3, 1814, I, 151–3. Slightly modified solution by A. Martin, The Analyst, Des Moines, 5, 1878, 124–5.

[66] Ladies' Diary, 1834, 37–8, Quest. 1560.

[67] Math. Miscellany, Flushing, N. Y., 1, 1836, 299–301.

N. Vernon (p. 302) equated the first and second functions (5) to the squares of $(r^2-xy)/(2r)$, and $(s^2-xz)/(2s)$. Then $x+y=(r^2+xy)/(2r)$, etc., which give x, y in terms of z. Then the third function becomes a quartic in z which is made a square as usual.

D. S. Hart[68] noted that $x^2+xy+y^2=\square$ if $x=m^2-n^2$, $y=2mn+n^2$. Then $x^2+xz+z^2=\square$ if $z=x(2pq+q^2)/(p^2-q^2)$. Take $m=2$, $n=1$, $p=r+\frac{1}{2}q$. Then $y^2+yz+z^2=\square$ if $r=7q/4$, $p=9q/4$. Hence an answer is 195, 325, 264.

A. Martin and A. B. Evans[69] took $x^2+axy+y^2=(mx-y)^2$ to get x/y. Then $x^2+axz+z^2$ and $y^2+ayz+z^2$ are made squares by known methods.

Several writers[70] made the functions (5) squares. R. F. Davis[71] noted the solutions 7, 8, -15 and 435, 4669, 1656.

N. G. S. Aiyar[72] solved $x^2+xy+y^2=c^2$, etc., by geometry, algebra and trigonometry, without attention to rational values.

A. Gérardin[73] assumed that a solution of $\alpha^2+\alpha\beta+\beta^2=A^2$ is known and sought a solution of

$$x^2+\alpha x+\alpha^2=B^2, \qquad x^2+\beta x+\beta^2=C^2,$$

by setting $B=x+u$ or $B=\alpha-xp/q$, or $x=t-\alpha-\beta$, obtaining a quartic function of t which is made a square in three ways. There is found a solution in positive integers by functions of the sixth degree.

E. Turrière[74] considered the system

$$Ax^2+Bxy+Cy^2=\square, \qquad Dy^2+Eyz+Fz^2=\square, \qquad Gz^2+Hzx+Ix^2=\square,$$

under the assumption that each has a set of rational solutions, say x_0, y_0 for the first. Solving the first with $y-y_0=Z(x-x_0)$, where Z is a parameter, we get x and y rationally in terms of Z. Similarly, z/y is rational in $X=(z-z_1)/(y-y_1)$, and x/z in $Y=(x-x_2)/(y-y_2)$. The condition that the product of the values of y/x, z/y, x/z be unity is of the sixth degree in X, Y, Z. The problem is thus reduced to finding the rational points on a certain sextic surface.

M. Rignaux,[74a] to treat the last system, would use a solution $x=x_0$, $y=y_0$ of the first equation, where x_0, y_0 are quadratic functions of two parameters m, n; likewise a solution $x=x_1$, $z=z_1$ of the third equation in terms of parameters p, q. Hence take $x=x_0x_1$, $y=y_0x_1$, $z=z_0x_1$. The given second equation becomes a quartic in m, n and is solvable in known special cases.

$xy+a$, $xz+a$, $yz+a$ all squares.

Diophantus, III, 12, 13 and IV, 20, asked for three numbers such that the product of any two increased by a given number a shall be a square. For $a=12$, he found 2, 2, 1/8; for $a=-10$, complicated fractions; for $a=1$, x, $x+2$, $4x+4$. In V, 27, the numbers themselves are to be squares.

---

[68] Math. Quest. Educ. Times, 20, 1874, 59–60.

[69] *Ibid.*, 21, 1874, 45–6.

[70] The Math. Visitor, 1, 1880, 105–6, 129–30; Amer. Math. Monthly, 1, 1894, 208 for (4).

[71] Math. Quest. Educ. Times, 11, 1907, 25.

[72] Jour. of Indian Math. Club, 2, 1910, 24–25.

[73] Nouv. Ann. Math., (4), 16, 1916, 62–74.

[74] *Ibid.*, (4), 18, 1918, 43–49. For such a system, see Ch. V, p. 223.

[74a] L'intermédiaire des math., 25, 1918, 132–3.

F. Vieta[75] generalized the method of Diophantus III, 12 [13]. Let A be the second number. Then the first is $(B^2-a)/A$, and the third is $(D^2-a)/A$. Hence must

$$\frac{B^2-a}{A}\cdot\frac{D^2-a}{A}+a=\square.$$

We can make $B^2-a=F^2$, $D^2-a=G^2$ in an infinitude of ways. Then $F^2G^2+aA^2$ is to be a square, say $(FG-HA)^2$. Hence $A=2HFG/(H^2-a)$.

C. G. Bachet[76], who doubted that Diophantus had a general solution, used the canon: Subtract the given number from each of two squares and divide the remainders by the difference of the roots of these squares; then the quotients and the difference of the roots are three numbers giving a solution. For $a=6$, take $N+3$ and $2N+3$ as the roots of the squares; then N, $N+6+3/N$ and $4N+12+3/N$ give a solution.

De Sluse[77] took an arbitrary square b^2 and set $d=b^2-a$, $xy=x^2+2xb+d$, whence $xy+a=(x+b)^2$. Similarly, we can set $z=xc^2/e^2+2bc/e+d/x$, whence $xz+a=(xc/e+b)^2$. Let $yz+a$ be the square of $(cx+cb)/e+b+d/x$. Thus

$$\frac{2b^2c}{e}+\frac{dc^2}{e^2}=\frac{b^2c^2}{e^2}+\frac{2dc}{e}.$$

When b^2 is replaced by $d+a$, this reduces to $2=c/e$, so that the required numbers are x, $x+2b+d/x$, $4x+4b+d/x$. For a negative, $a=-A$, call the numbers x, $y=x+A/x$, $z=xb^2/c^2+A/x$. Then $xy-A=x^2$, $xz-A=x^2b^2/c^2$, $yz-A=(xb/c+A/x)^2$ if $b/c=2$.

N. Saunderson[78] (blind from infancy) gave the solution

$$x=\frac{r^2-a}{r-s}, \qquad y=\frac{s^2-a}{r-s}, \qquad z=r-s \quad \text{or} \quad 2x+2y-(r-s),$$

where r and s exceed $\sqrt{a}$ and $r>s$. For $a=1$, a solution is

$$x, \qquad y=\alpha^2x+2\alpha, \qquad z=\beta^2x+2\beta, \qquad \alpha-\beta=\pm 1.$$

V. Ricatti[79] treated the problem.

L. Euler[79a] set $xy+a=p^2$, $z=x+y\pm 2p$, whence $xz+a=(x\pm p)^2$, $yz+a=(y\pm p)^2$. For $a=12$, $p=4$, then $x=y=2$, $z=12$. For $a=12$, $p=5$, then $x=1$, $y=13$, $z=4$ or 24. In art. 231, he noted that for $a=1$ the general solution is

$$x=(p^2-1)/z, \qquad y=(q^2-1)/z, \qquad z=\{(p^2-1)(q^2-1)-r^2\}/(2r).$$

Euler[80] treated $AB-1=p^2$, $AC-1=q^2$, $BC-1=r^2$. Thus

$$A^2B^2C^2=l(r^2+1), \qquad l=(p^2+1)(q^2+1)=m^2+n^2, \qquad m=pq\pm 1, \qquad n=p\mp q,$$
$$A^2B^2C^2=(mr+n)^2+(nr-m)^2.$$

[75] Zetetica, 1591, V, 7[8], Francisci Vietae Opera mathematica, ed. Francisci à Schooten, Lugd. Bat., 1646, 78.

[76] Diophanti Alex., 1621, 149, 215.

[77] Renati Francisci Slusii, Mesolabum, accessit pars altera de analysi et miscellanea, Leodii Eburonum, 1668, 177-8.

[78] The Elements of Algebra, 2, 1740, 390-5.

[79] Institutiones analyticae a Vincentio Riccato, Bononiae, I, 1765, 64.

[79a] Algebra, 2, 1770, art. 232 (end of art. 233); 2, 1774, p. 305 (pp. 310-1); Opera Omnia, (1), I, 465 (468).

[80] Posth. paper, Comm. Arith., II, 577-9; Opera postuma, 1, 1862, 129-131.

Set $-ABC=mr+n+t(nr-m)$.   Then, for $d=n(t^2-1)+2mt$,

$$dr=m(t^2-1)-2nt, \quad d^2(r^2+1)=(m^2+n^2)(t^2+1)^2, \quad dABC=(m^2+n^2)(t^2+1)$$

$$A=d/(t^2+1), \quad dB=(p^2+1)(t^2+1), \quad dC=(q^2+1)(t^2+1).$$

To obtain integral solutions, set $B=(p^2+1)/A$, $C=(q^2+1)/A$.   Then

$$BC-1=(m^2+n^2-A^2)/A^2$$

is a square if $A=n=p-q$.   Then $B=A+C+2q$.   It remains to make q^2+1 divisible by A, which requires that $A=\boxed{2}$.   If $A=5$, $q=5u\pm2$, then $C=5u^2\pm4u+1$, $B=5u^2+14u+10$ or $5u^2+6u+2$.   Among other ways of obtaining integral solutions, take $AB=1+p^2$, $(AC-1)(BC-1)=(mC+1)^2$, whence $C=(A+B+2m)/Q$, where $Q=AB-m^2$.   Then

$$AC-1=(A+m)^2/Q, \quad BC-1=(B+m)^2/Q.$$

Hence we set $Q=n^2$, whence $m^2+n^2=p^2+1$.   Take $m=ap+\alpha$, $n=\alpha p-a$, where $a^2+\alpha^2=1$; for example, $a=(f^2-g^2)/(f^2+g^2)$, $\alpha=2fg/(f^2+g^2)$.   Then

$$C=\{A+B\pm2(ap+\alpha)\}/(\alpha p-a)^2.$$

For $f=1$, $g=0$, $C=A+B\pm2p$.   For $f=2p$, $g=1$, $C=(A+B)f_1^2\pm2pf_1(f_1+2)$ where $f_1=4p^2+1$.   Next, we take $f=f_1$, $g=2p$.   In this way Euler obtained $C=(A+B)M^2\pm2pMN$, where $(M,\ N)=(1,\ 1)$, $(4p^2+1,\ 4p^2+3),\cdots$ are given by a recurring series with the scale of relation $4p^2+2$, -1; he gave the general terms.

J. Leslie[81] made $xy+1$, $xz+1$, $yz+1$ squares by factoring (cf. Buchner[83]).

P. Cossali[82] gave the result due to Saunderson.[78]

Fr. Buchner[83] treated $xy+1=p^2$, $xz+1=q^2$, $yz+1=r^2$.   Then

$$x=\frac{p+1}{m}=\frac{q+1}{n}, \quad y=m(p-1)=l(r-1), \quad z=n(q-1)=\frac{r+1}{l}.$$

Thus p, q, r and hence also x, y, z are functions of m, n, l.

A. B. Evans[84] to make $xy-1$, etc., squares, took $x=a^2+b^2$, $y=c^2+d^2$, $z=e^2+f^2$, $E=bc-ad$, $F=be-af$, $G=de-cf$.   Then $xy-E^2$, $xz-F^2$, $yz-G^2$ are squares.   Take $e=a+c$, $f=b+d$.   Then $F=E$, $G=-E$.   It remains only to make $E=\pm1$.

E. Bahier[85] noted the answer $a-1$, a, $4a-1$ and gave de Sluse's[77] values with $x=1$ and Saunderson's[78] with the second z.

PROBLEMS RELATED TO THE LAST ONE.

Diophantus, III, 17, 18 [19], treated the problem (which evidently reduces to the last one): to find three numbers such that the product of any two increased [diminished] by the sum of those two gives a square.[86]

<hr>

[81] Trans. Roy. Soc. Edinb., 2, 1790, 209, Prob. XII.

[82] Origine, Trasporto in Italia . . . Algebra, 1, 1797, 102.

[83] Beitrag zur Auflös. unbest. Aufg. 2 Gr., Prog. Elbing, 1838, p. 9.

[84] Math. Quest. Educ. Times, 14, 1871, 75–6; 29, 1878, 90–1.

[85] Recherche Méthodique et Propriétés des Triangles Rectangles en Nombres Entiers, Paris, 1916, 198–9.

[86] In Diophantus IV, 38, 40, the results are to be given numbers, instead of squares.   His condition that each number must be 1 less than a square is not necessary, as noted by Stevin, Les Oeuvres math. de Simon Stevin . . . par A. Girard, 1625, 589; 1634, 148. Thus if the numbers are 14, 23, 39, an answer is 4, 2, 7.

Take y and $4y+3$ as two of the numbers, which each increased by unity have a ratio which is a square $1/4$.   From

$$y(4y+3)+y+4y+3 = \square = (2y-3)^2,$$

we get $y=3/10$.   For the numbers $3/10$, $42/10$, x, the conditions are

$$\tfrac{13}{10}x+\tfrac{3}{10} = \square, \qquad \tfrac{26}{5}x+\tfrac{21}{5} = \square.$$

By the usual method (Ch. XV), $x=7/10$.   Cf. Nesselmann,[95a] pp. 142–4.

N. Saunderson[87] found three numbers a, b, c such that the product of any two increased by t times their sum is a square.   Since $(a+t)(b+t)=n^2+t^2$, express n^2+t^2 as a product of two factors, say $n+r$, $n-s$, each $>t$.   Then $c=a+b+t\pm 2n$ and

$$a=\frac{r^2+t^2}{r-s}-t, \qquad b=\frac{s^2+t^2}{r-s}-t, \qquad c=r-s-t \text{ or } 2a+2b+2t-(r-s-t).$$

When $t=1$, take $r-s=1$, whence $a=r^2$, $b=s^2$, $c=0$ or $2a+2b+2$.

The same numbers are such that the product of any two increased by t times the third is a square (Diophantus, III, 14).

P. Cossali[82] noted that if the product of any two of x^2, z^2, $2\{x^2+z^2+(z-x)^2\}$ be increased by $(z-x)^2$ times the sum of the two or by $(z-x)^2$ times the third, we get a square.   On adding $2(z-x)^2$ to each of these three numbers, we get three numbers such that the product of any two diminished by $(z-x)^2$ times either their sum or the third gives a square.

Diophantus[88], V, 3 [4], required three numbers such that any one of them or the product of any two of them increased [diminished] by a given number a is a square.   He quoted from his *Porisms* that if $x+a=m^2$, $y+a=n^2$, $xy+a=\square$, then m and n are consecutive numbers.[89]   Thus if $a=5$ we take $x=(z+3)^2-5$, $y=(z+4)^2-5$ as two of the required numbers, and $2(x+y)-1=4z^2+28z+29$ as the third.   We are to make

$$4z^2+28z+34 = \square,$$

say $(2z-6)^2$.   Hence $z=1/26$.

For V, 4, Diophantus took $a=6$, $x=z^2+6$ and $y=(z+1)^2+6$ as two of the numbers, and $2(x+y)-1$ as the third.   The latter less 6 is

$$4z^2+4z+19 = (2z-6)^2$$

if $z=17/28$.

Diophantus' method shows that $xy+a$, $xz+a$, $yz+a$, $x+a$, $y+a$ are all squares if $x=m^2-a$, $y=(m+1)^2-a$, $z=(2m+1)^2-4a$.   To make also $z+a=\square$, say $(2m-r)^2$, we have $m=(r^2+3a-1)/\{4(1+r)\}$.

Fermat (Oeuvres, III, 250) gave a solution for the case $a=1$.   In

$$y=\tfrac{169}{5184}x+\tfrac{13}{36}, \qquad z=\tfrac{7225}{5184}x+\tfrac{85}{36},$$

the constant terms increased by unity give squares; further, $xy+1$, $xz+1$, $yz+1$ are squares.   The "triple equation" $x+1=\square$, $y+1=\square$, $z+1=\square$ is readily solved since the constant terms are squares (Ch. XV).

---

[87] The Elements of Algebra, 2, 1740, 399–405; French transl., Sphinx-Oedipe, 1908–9, 3–9.

[89] But this is incorrect; $m-n=\pm 1$ is a sufficient but not necessary condition for $xy+a=\square$. In fact, by eliminating x, y, we get $m^2n^2-a(m^2+n^2-1)+a^2=\square$. While this is satisfied if $m^2+n^2-1=2mn$, whence $m=n\pm 1$, it can be satisfied as usual by setting $m=n\pm 1+\mu$.

PRODUCT OF ANY TWO OF FOUR OR FIVE NUMBERS INCREASED BY UNITY A SQUARE.

Diophantus, IV, 21, required four numbers such that the product of any two increased by unity is a square. He took x, $x+2$, $4x+4$ as the first three (by IV, 20), and $(3x+1)^2-1$ as the product of the first and fourth. Thus the fourth is $9x+6$. The product of the second and fourth, increased by unity, is $9x^2+24x+13$; let it equal $(3x-4)^2$, whence $x=1/16$. The remaining conditions are now satisfied.

Rafael Bombelli[90] treated the problem for four numbers.

Fermat[91] took 1, 3, 8 as the first three numbers. The conditions on the fourth number x are $x+1=\square$, $3x+1=\square$, $8x+1=\square$. His method (Fermat[10, 11] of Ch. XV) of solving a "triple equation" gives $x=120$.

L. Euler[92] gave the solution a, b, $c=a+b+2l$, $d=4l(l+a)(l+b)$, where $ab+1=l^2$, and noted the cases 3, 8, 1, 120 and 3, 8, 21, 2080. He extended the question to five numbers, by seeking z such that $1+az$, $\cdots$, $1+dz$ are all squares. Denote the product of these four sums by $P=1+pz+qz^2+rz^3+sz^4$, where therefore $p=a+b+c+d$, $\cdots$, $s=abcd$. Let P be the square of $1+\frac{1}{2}pz+gz^2$, where $g=q/2-p^2/8$. Then

$$r+sz=pg+g^2z, \qquad z=(r-pg)/(g^2-s).$$

For brevity set $a+b+l=f$, $d/4=k$. Then

$$k=fl^2+lab, \qquad c=f+l, \qquad p=2f+4k,$$
$$q=(a+b+c)d+(a+b)c+ab=8fk+f^2-1, \qquad s=4abk(f+l).$$

Now $k=f(ab+1)+lab$, $4k^2=4kf+4kab(f+l)$. Hence

$$1+q+s=(2k+f)^2=\tfrac{1}{4}p^2, \qquad g=-\tfrac{1}{2}(1+s).$$

The denominator g^2-s of z is fortunately the square of $(s-1)/2$. Thus

$$z=\frac{4r+2p(1+s)}{(s-1)^2}$$

and P is a square. Euler stated that each factor $1+az$, etc., is then a square. Taking $a=1$, $b=3$, we have $l=2$, $c=8$, $d=120$, $p=132$, $q=1475$, $r=4224$, $s=2880$, $z=777480/2879^2$, and the ten expressions $ab+1$, $\cdots$, $dz+1$ are the squares of

$$2,\ 3,\ 11,\ 5,\ 19,\ 31,\ \tfrac{3011}{2879},\ \tfrac{3259}{2879},\ \tfrac{3809}{2879},\ \tfrac{10079}{2879}.$$

To obtain smaller (but fractional) numbers, set $a=1/2$, $b=5/2$. Then

$$c=6, \qquad d=48, \qquad z=44880/128881.$$

A. M. Legendre[93] verified Euler's preceding assertion that $1+az$, etc., are squares by noting that a, b, c, d are the roots of

$$\xi^4-p\xi^3+q\xi^2-r\xi+s=0$$

---

[90] L'algebra opera, Bologna, 1579, p. 543.

[91] Oeuvres, III, 251.

[92] Opusc. anal., 1, 1783, 329; Comm. Arith., II, 45. Results stated in a letter to Lagrange, Sept. 24, 1773 (Oeuvres, XIV, 235–40); Euler's Opera postuma, 1, 1862, 584–5.

[93] Théorie des nombres, ed. 3, 2, 1830, 142–4; Maser's transl., 2, 1893, 138.

and showing that when $r\xi$ is replaced by its value from the preceding equation, $(s-1)^2(\xi z+1)$ becomes $(2\xi^2-p\xi-s-1)^2$.

C. O. Boije af Gennäs[94] gave the solution

$$r, \quad s(rs+2), \quad (s+1)(rs+r+2), \quad 4(rs+1)(rs+r+1)(rs^2+rs+2s+1).$$

For $r=1$, $s=2$, we get 1, 8, 15, 528.

J. Knirr[95] took as the four numbers

$$n, \quad a^2n+2a, \quad b^2n+2b, \quad p^2n+2p.$$

The product of the second and third, increased by unity, is

$$\{abn+(a+b)\}^2+\{1+4ab-(a+b)^2\}$$

and is a square if the final part is zero, whence $b=a\pm1$. The product of the second and fourth, increased by unity, is then the square of $1+pq$ if

$$p(q^2-a^2n^2-2an)=2a^2n+4a-2q.$$

The coefficient of p is unity if $q=an+1$. G. H. F. Nesselmann[95a] took $b=a+1$, $p=a+2$.

C. C. Cross[96] gave the set due to Boije[94] with r, s replaced by m, $n-1$. He and others failed to find five such numbers. He[97] later took the fifth number to equal the first one m, the only new condition being $m^2+1=\square$, for example, $m=(k^2-1)/(2k)$.

M. A. Gruber[98] noted a special case of Euler's[92] five numbers.

A. Gérardin[99] obtained special solutions by recurring series.

Fermat[100] treated the problem to find four numbers such that the product of any two increased by the sum of those two gives a square. He made use of three squares such that the product of any two increased by the sum of the same two gives a square. Stating that there is an infinitude of such sets of three squares, he cited 4, $3504384/d$, $2019241/d$, where $d=203401$. However, he actually used the squares $25/9$, $64/9$, $196/9$, of Diophantus V, 5, which have the additional property that the product of any two increased by the third gives a square. Taking these three squares as three of our numbers and x as the fourth, we are to satisfy

$$\tfrac{34}{9}x+\tfrac{25}{9}=\square, \qquad \tfrac{73}{9}x+\tfrac{64}{9}=\square, \qquad \tfrac{205}{9}x+\tfrac{196}{9}=\square.$$

This "triple equation" with squares as constant terms is readily solved. T. L. Heath[101] found x to be the ratio of two numbers each of 21 digits.

L. Euler[102] gave a more general treatment of the latter problem. Let A, B, C, D denote the numbers increased by unity. Then $AB-1$, $\cdots$, $CD-1$ are to be squares. Take $AB=p^2+1$,

$$C=\frac{A+B+2(ap+\alpha)}{(\alpha p-a)^2}, \qquad D=\frac{A+B+2(bp+\beta)}{(\beta p-b)^2}, \qquad a^2+\alpha^2=b^2+\beta^2=1.$$

[94] Nouv. Ann. Math., (2), 19, 1880, 278–9; E. Lucas, Théorie des nombres, 1891, 129.
[95] Die Auflösung der Gleichung $z^2-cx^2=1$, 18. Jahresbericht Oberrealschule, 1889, 31.
[95a] Zeitschr. Math. Phys., Hist.-lit. Abt., 37, 1892, 167.
[96] Amer. Math. Monthly, 5, 1898, 301–2.
[97] *Ibid.*, 6, 1899, 85–87.
[98] *Ibid.*, 122–3.
[99] L'intermédiaire des math., 23, 1916, 14–15.
[100] Oeuvres, III, 242–3. A special case of our main problem since $xy+x+y=(x+1)(y+1)-1$.
[101] Diophantus of Alexandria, ed. 2, 1910, p. 163.
[102] Posth. paper, Comm. Arith., II, 579–582; Opera postuma, 1, 1862, 131–4.

Then five of the conditions are satisfied. There remains $CD-1=\square$. Replacing $A+B$ by its value $(A^2+p^2+1)/A$, we see that the condition becomes $A^4+2A^3k+\cdots+(p^2+1)^2=\square$, where $k=(a+b)p+\alpha+\beta$. The quartic is the square of A^2+Ak-p^2-1 if

$$A\{k^2-4(p^2+1)-4(ap+\alpha)(bp+\beta)+(\alpha p-a)^2(\beta p-b)^2\}=4k(p^2+1).$$

This solution is of course not general. For instance, if $\alpha=\beta=0$, $a=1$, $b=-1$, then the preceding A is zero, whereas we may obtain solutions as follows. We have, in this case, $C=A+B+2p$, $D=A+B-2p$. Then

$$CD-1=(A+B)^2-4p^2-1=\square=q^2, \qquad (A-B)^2=q^2-3=\square=(q-r)^2.$$

Thus $q=(r^2+3)/(2r)$. Also $A+B=2p+s$ if $p=(q^2+1-s^2)/(4s)$. Set $r=2$, $s=15/4$. Then $q=7/4$, $p=-2/3$, $A=C=13/12$, $B=4/3$, $D=15/4$. For $r=2$, $s=7/2$, we get

$$A=\tfrac{289}{224}, \qquad B=\tfrac{233}{224}, \qquad C=\tfrac{65}{56}, \qquad D=\tfrac{7}{2}.$$

For $b=-a$, $\beta=-\alpha$, Euler found $C,D=\{\alpha(A+B)\pm(4a+2)\}/(4\alpha)$ and noted that all the resulting solutions are fractional. He cited the solution $A=D=1$, $B=2$, $C=5$, and asked if there are other integral solutions.

PRODUCT OF ANY TWO OF FOUR NUMBERS INCREASED BY n A SQUARE.

C. G. Bachet[103] proposed the problem and took $n=3$. From $(N+2)^2$ and $(N+6)^2$ subtract 3 and divide the remainders by the difference 4 of the roots of the squares; we get

$$a=\tfrac14N^2+N+\tfrac14, \qquad b=\tfrac14N^2+3N+\tfrac{33}{4}.$$

As the third number, he took

$$c=2(a+b)-4=N^2+8N+13.$$

Hence by a general canon, $ab+3$, $ac+3$, $bc+3$ are squares. Take the fourth number to be $d=4$. Then $ad+3$ and $bd+3$ are squares. Finally,

$$cd+3=(2N-10)^2 \text{ if } N=5/8.$$

He gave also a second method of solution.

Fermat[104] remarked that it is easy to deduce a solution from Diophantus[88] V, 3. As three of the numbers take solutions x_1, x_2, x_3 of the latter problem. As the fourth number, take $x+1$. We then have a "triple equation" $x_ix+x_i+n=\square$, whose constant terms x_i+n are squares, and hence easily solved (Ch. XV).

P. Iacobo de Billy[105] took $n=4$, R as the first number, and $R+2$, $2R+2$, $3R+2$ as the roots of the squares obtained when R is one factor. Thus the remaining three numbers are $R+4$, $4R+8$, $9R+12$. Then $(R+4)(9R+12)+4$ is the square of $3R-8$ if $R=1/8$. The other two conditions are seen to be satisfied.

N. Saunderson[37] (p. 398) took any number $a>\sqrt{n}$, subtracted $4a^2-3n$ from any larger square b^2, and called d the quotient obtained on dividing

---

[103] Dioph. Alex., 1621, 150.
[104] Oeuvres, III, 254.
[105] Diophantvs Geometria, Paris, 1660, 100.

the remainder by $4a+2b$. Then an answer is given by

$$d, \qquad e=(a^2-n)/d, \qquad f=d+e+2a, \qquad g=3e+f+2a.$$

Thus for $n=3$, take $a=2$, $b=3$. Then $d=1/7$, $e=7$, $f=78/7$, $g=253/7$.

L. Euler[106] called the numbers A, B, C, D. Set $AB=p^2-n$. Equate the product of $AC+n$ and $BC+n$ to $(Cx+n)^2$; then

$$C=\frac{n(A+B-2x)}{x^2-AB}, \qquad AC+n=\frac{n(A-x)^2}{x^2-AB}.$$

Hence $(x^2-AB)/n$ is to be a square y^2, whence

$$C=(A+B-2x)/y^2, \qquad x^2-ny^2=p^2-n.$$

Similarly,

$$D=(A+B-2v)/z^2, \qquad v^2-nz^2=p^2-n.$$

In $CD+n=\square$, replace $A+B$ by $(A^2+p^2-n)/A$. Hence

$$A^4-2A^3(x+v)+2A^2(p^2-n)+A^2ny^2z^2+4A^2xv-2A(p^2-n)(x+v)+(p^2-n)^2$$

is to be a square. It can be made the square of $A^2-A(x+v)-(p^2-n)$ by choice of a rational A. To simplify the formulae, Euler took $v=-x$, $z=y$. Then the condition becomes

$$(A^2-p^2+n)^2+nA^2y^2(y^2-4)=\square$$

and is satisfied if $y=2$. It remains only to satisfy $p^2=x^2-3n$. Set $p=x-t$. Then $x=(t^2+3n)/(2t)$, $p=(3n-t^2)/(2t)$. To secure homogeneity, set $x, p=(3nu^2\pm t^2)/(2tu)$. Then

$$AB=(nu^2-t^2)(9nu^2-t^2)/(4t^2u^2),$$

$$A=\frac{f(nu^2-t^2)}{2gtu}, \qquad B=\frac{g(9nu^2-t^2)}{2ftu}, \qquad C, D=\frac{n(f\pm3g)^2u^2-(f\mp g)^2t^2}{8fgtu}.$$

To find four numbers such that the product of any two increased by the sum of the four is a square, we have only to take $mA, \cdots, mD$, where $m=(A+B+C+D)/n$, while $A, \cdots, D$, n are the numbers given by the preceding solution. Euler gave two solutions in integers: 15, 175, 310, 475 and 36, 96, 264, 504. Since n may be negative, we obtain four numbers the product of any two of which decreased by the sum of the four is a square. A solution in integers is 8, 24, 44, 80.

E. Bahier,[85] pp. 199–208, employed the numbers of Saunderson,[78] taking his two values of z as two of the four numbers. There remains only the condition $(r+s)^2-3a=\square$, which is satisfied by expressing $3a$ as a difference of two squares.

Other products of numbers in pairs increased by linear functions made squares.

J. Collins[107] made the six functions $xy\pm v$, $xz\pm v$, $yz\pm v$ squares, where $v=x+y+z$. Take $xy\pm v=(t\pm s)^2$, $xz\pm v=(r\pm q)^2$, $yz\pm v=(p\pm n)^2$, and (1) $\frac{1}{2}v=ts=rq=pn$. Then $xy=t^2+s^2$, $xz=r^2+q^2$, $yz=p^2+n^2$. Take $t=(a^2-b^2)g$,

---

[106] Comm. Arith., II, 582–5 (posth. paper); Opera postuma, 1, 1862, 134–7; Algebra, 2, 1770, arts. 233–4; 2, 1774, pp. 306–14; Opera Omnia, (1), I, 465–9.

[107] The Gentleman's Math. Companion, London, 2, No. 10, 1807, 66–7.

$s = 2abg$, $r = (a^2 - c^2)g$, $q = 2acg$, $p = (d^2 - a^2)g$, $n = 2adg$. Then

$$x = \frac{g(a^2+b^2)(a^2+c^2)}{a^2+d^2}, \qquad y = \frac{g(a^2+d^2)(a^2+b^2)}{a^2+c^2}, \qquad z = \frac{g(a^2+c^2)(a^2+d^2)}{a^2+b^2}.$$

To satisfy (1), take $a = f^2 + fh + h^2$, $b = f^2 - h^2$, $c = 2fh + h^2$, $d = f^2 + 2fh$. For four numbers, see Euler.[106]

J. Cunliffe[108] made $xy + z$, etc., squares by taking $y - x = 2n$, $z = n^2$. Then $xy + z = (x+n)^2$, while $xz + y$ and $yz + x$ are linear functions of x and may be equated to squares. S. Jones took $y = x - 1$, $z = x - 4$. "J. B." took $y = t^2 x - v^2$, $z = v^2 x$, whence $xy + z = t^2 x^2$. Then $xz + y = (vx - r)^2$ gives x. From $yz + x = \square$, we get a quartic in r which is solved as usual.

W. Wright[109] took $xy - a = p^2$, $yz - b = q^2$ and made $p^2 + a$ and $q^2 + b$ squares. Then $xz - c = \square$ if $\square/y^2 - c = \square$, which is easily satisfied.

Cunliffe[110] took $xy + z = A^2$, $xz + y = B^2$. Thus $(y+z)(x+1) = A^2 + B^2$. Hence set $y + z = a^2 + b^2$, $x + 1 = c^2 + d^2$, $A = ac + bd$, $B = ad - bc$. Also, $(y-z)(x-1) = A^2 - B^2$. Hence take $y - z = A - B$, $x - 1 = A + B$. By the two values of x, we get b in terms of a, c, d. To get integral values of b, equate the denominator $c - d$ to unity.

D. S. Hart[111] made $xy + z$, etc., and $xy + x + y$, etc., all squares by taking $x = n^2$, $y = (n+1)^2$, $z = 4(n^2 + n + 1)$.

E. N. Barisien[112] treated the system

$$xz - y = t^2, \qquad (z+a)x - y = u^2, \qquad (z+b)x - y = v^2.$$

Subtract the first from the other two. Thus

$$ax = u^2 - t^2, \qquad bx = v^2 - t^2, \qquad av^2 - bu^2 = (a-b)t^2.$$

Set $v = t + h$, $u = t + l$. Discarding the denominator $2ha - 2lb$, we have

$$t = bl^2 - ah^2, \quad u = bl^2 + ah^2 - 2alh, \quad v = ah^2 + bl^2 - 2blh, \quad x = 4lh(h - l)(bl - ah).$$

Then y, z can be found from $Az - y = B$. Set $B = Ap + r$, $z = q + p$; then $y = Aq - r$. [Take $g = -ah/l$, $f = -l^2/a$]. Then

$$x = 4f^2 g(a+g)(b+g), \qquad u = f(g^2 + 2ag + ab), \qquad v = f(g^2 + 2bg + ab),$$
$$t = f(g^2 - ab), \qquad z = g, \qquad y = f^2\{3g^4 + 4g^3(a+b) + 6abg^2 - a^2b^2\}.$$

He[113] elsewhere merely stated the latter solution.

"V. G. Tariste"[114] treated the case $a = 1$, $b = 2$ of the last problem. Then $v^2 + t^2 = 2u^2$, whose general solution is $u = \lambda(A^2 + B^2)$; v, $t = \lambda(A^2 - B^2 \pm 2AB)$.

Several writers[115] made $xy + z$, $yz + x$, $xz + y$ all squares (Diophantus III, 14).

E. Bahier,[85] pp. 208–212, made $xy - v$, $xz - v$, $yz - v$ squares the sum of two of which equals the third.

[108] The Gentleman's Math. Companion, London, 3, No. 17, 1814, 463–6.

[109] Ibid., 467–8.

[110] Ibid., 5, No. 27, 1824, 349–53.

[111] Math. Quest. Educ. Times, 28, 1878, 67–8.

[112] Sphinx-Oedipe, 1907–8, 180–1.

[113] Mathesis, (3), 9, 1909, 154–5.

[114] L'intermédiaire des math., 19, 1912, 38–9.

[115] Zeitschr. Math. Phys., Hist.-lit. Abt., 37, 1892, 138; Math. Quest. Educ. Times, 25, 1914, 40, 102–4; Amer. Math. Monthly, 24, 1917, 88–89, 294.

Further equations whose quadratic terms are sums of products.

Bháscara[116] (born 1114) treated the problem to make $w+2$, $x+2$, $y+2$, $z+2$ the squares of numbers in A. P., and $wx+18$, $xy+18$, $yz+18$ all squares, such that the sum of the roots of the seven squares when increased by 11 gives 13^2. Since $18/2$ is the square of 3, the roots of the first four squares are y, $y+3$, $y+6$, $y+9$. Then the roots of $wx+18$, etc., are found to be y^2+3y-2, $y^2+9y+16$, $y^2+15y+52$. The sum of the roots plus 11 gives $3y^2+31y+95=13^2$, $y=2$.

Diophantus, IV, 16, solved $z(x+y)=a$, $y(x+z)=b$, $x(y+z)=c$, when $a=35$, $b=32$, $c=27$, by assuming that $x=15/z$, $y=20/z$, whence $z=5$.

Rallier des Ourmes[117] obtained $2xz=a+c-b$, etc., by elimination from Diophantus' equations. From $yz=m$, $xz=n$, $xy=p$ follows $y=\sqrt{pm/n}$, etc. For $a=24$, $b=45$, $c=49$, we get $m=10$, $n=14$, $p=35$, whence $x=7$, $y=5$, $z=2$. He gave also a solution by listing the pairs of complementary factors of the smallest two, 24 and 45, of the three given numbers:

$$24=1\cdot24=2\cdot12=3\cdot8=4\cdot6, \qquad 45=1\cdot45=3\cdot15=5\cdot9.$$

From each list select a pair of factors with a common sum, as $2\cdot12$, $5\cdot9$, and select by trial one of a pair as one unknown and the cofactor as the sum of the other two unknowns.

To find n numbers, given the product of each by the sum of all the others, list the pairs of cofactors of each of the smallest $n-1$ of the n given numbers and select those pairs, one from each list, which have the same sum (the sum of the unknowns). The smallest cofactor of each pair is one of the smallest $n-1$ of the unknowns and their sum subtracted from the total sum gives the largest unknown. For $n=5$, use $180=4\cdot45$, $294=7\cdot42$, $418=11\cdot38$, $444=12\cdot37$; the unknowns are 4, 7, 11, 12,

$$15=49-(4+7+11+12).$$

S. Jones[118] took $x(y+z)=a^2x^2$, $y(x+z)=b^2$, $z=ax+b$, which give x, y, z. Then $z(x+y)=\square$ if $a^2+2a-1=\square=(a-n)^2$ and $a^2-2a+3=\square$. The latter becomes a quartic in n which is a square if $n=-2/3$.

L. Euler[119] developed a method to make various functions simultaneously equal to squares. The method will be explained for his problem (§§ 31–34): Given an integer n, find integers x, y, z such that $xy+n$, $xz+n$, $yz+n$, $xy+xz+yz+n$ are all squares. For any set of solutions of

$$f \equiv x^2+y^2+z^2-2xy-2xz-2yz-4n=0$$

and for any function P, P^2-f is a square. Taking $P=x+y-z$, we find that $4(xy+n)$ is a square. Taking $P=x-y+z$, we find that $4(xz+n)$ is a square. Similarly, $yz+n$ is a square. Taking $P=x+y+z$, we find that $4(xy+xz+yz+n)$ is a square. Now $f=0$ if $z=x+y+2v$, where $v^2=xy+n$. To satisfy the latter take any integer for v and take x and y to be any pair of

[116] Vija-gañita, §§ 143–4. Algebra with arith . . . from Sanskrit . . . of Bháscara, transl. by Colebrooke, 1817, 218–9.

[117] Mém. de Mathématique et de Physique, Paris, 5, 1768, 479–84.

[118] The Gentleman's Math. Companion, London, 3, No. 15, 1812, 348–9.

[119] Novi Comm. Acad. Petrop., 6, 1756–7, 85–114; Comm. Arith., I, 245–259; Opera Omnia, (1), II, 399–427. French transl., Sphinx-Oedipe, 8, 1913, 97–109.

integers whose product is $n-v^2$. Then
$$xy+n=v^2, \quad xz+n=(x+v)^2, \quad yz+n=(y+v)^2, \quad xy+xz+yz+n=(x+y+v)^2,$$
the right members being the reduced values of $P^2/4$, for the respective P's.

To solve an interesting related problem (§§ 35–39), take
$$f=x^2+y^2+z^2-2xy-2yz-2xz-2a(x+y+z)-b=0$$
and $P=x+y\pm z\pm a$ for the four combinations of signs. Then
$$4(xy+xz+yz)+4a(x+y+z)+a^2+b, \qquad 4(xy+xz+yz)+a^2+b,$$
$$F=4xy+4a(x+y)+a^2+b, \qquad 4xy+4az+a^2+b,$$
and the expression obtained from the last two by permuting the variables, are all squares. Now $f=0$ if $z=x+y+a\pm v$, provided x and y make $F=v^2$. The latter is the case if $x+a$ and $y+a$ are two numbers whose product is $(v^2-b+3a^2)/4$. In particular, if $a=1$, $b=-1$, we see how to find three numbers x, y, z such that
$$xy+z, \qquad xz+y, \qquad yz+x, \qquad xy+x+y, \qquad xz+x+z, \qquad yz+y+z,$$
$$\sigma=xy+xz+yz, \qquad \sigma+x+y+z$$
are all squares. The simplest solution is $x=1$, $y=4$, $z=12$. Solutions in which also the numbers themselves are squares are
$$\tfrac{9}{64}, \ \tfrac{25}{64}, \ \tfrac{40}{16}; \quad \tfrac{25}{9}, \ \tfrac{64}{9}, \ \tfrac{196}{9}.$$

Euler[120] asked for numbers p, q, r, $\cdots$ such that the product of each by the sum of the remaining numbers is a square. Hence if S be their sum, $p(S-p)$, $q(S-q)$, $\cdots$ are to be squares. Take $p(S-p)=f^2p^2$, etc. Hence the desired numbers are
$$\frac{S}{1+f^2}, \qquad \frac{S}{1+g^2}, \qquad \cdots, \qquad \left(\frac{1}{1+f^2}+\frac{1}{1+g^2}+\cdots=1\right).$$
Take $f=a/\alpha$, etc. For three numbers, let them be
$$\frac{a^2}{a^2+\alpha^2}, \qquad \frac{(ab-\alpha\beta)^2}{d}, \qquad \frac{(a\beta-\alpha b)^2}{d} \qquad [d=(a^2+\alpha^2)(b^2+\beta^2)].$$
The sum of the last two is $1-4a\alpha b\beta/d$. The sum of all three is therefore unity if $a^2(b^2+\beta^2)=4a\alpha b\beta$, whence $a:\alpha=4b\beta:b^2+\beta^2$. Taking $a=4b\beta$ and multiplying the initial numbers by d, we get the solution
$$16b^2\beta^2(b^2+\beta^2), \qquad \beta^2(3b^2-\beta^2)^2, \qquad b^2(3\beta^2-b^2)^2.$$
For four numbers, Euler gave the solutions (1, 2, 2, 5), (1, 10, 34, 125), (5, 9, 26, 90), (5, 32, 61, 512) and solutions involving two parameters. For five numbers, he gave 2, 40, 45, 58, 145.

Euler[121] gave a special method of treating the last problem. Select any number, like $S=130$, which is in several ways a sum of two parts whose product is a square, viz.,

$p=$	2,	5,	13,	26,	32,	40,	49,	65,
$S-p=$	128,	125,	117,	104,	98,	90,	81,	65.

---

[120] Novi Comm. Acad. Petrop., 17, 1772, 24; Comm. Arith., I, 459–66; Op. Om., (1), III, 188.
[121] Opera postuma, 1, 1862, 260 (about 1769).

Selected values of p give an answer if their sum is 130, as for 2, 5, 26, 32, 65, and 2, 13, 26, 40, 49.

Euler[122] found a, b, c, d so that $ab-cd$, $ac-bd$, $bc-ad$ are squares. Call the first two expressions x^2, y^2, and solve for b, c. Take $2x=a+d+v$, $2y=a+d-v$. Then

$$b, c = \frac{(a+d)^2 \pm 2(a-d)v+v^2}{4(a-d)}, \qquad bc-ad = \left[\frac{a^2-6ad+d^2-v^2}{4(a-d)}\right]^2.$$

For $v=d=8$, $a=24$, we get $b=21$, $c=13$.

S. Tebay[123] found four positive integers $a_1, \cdots, a_4$ such that $a_1a_2+a_3a_4$, $a_1a_3+a_2a_4$, $a_1a_4+a_2a_3$, $\Sigma a_i a_j$ are squares.

A. Gérardin[123a] made $xy+zt$ and $xz-yt$ squares by several methods.

SQUARES INCREASED BY LINEAR FUNCTIONS MADE SQUARES.

Let $\sigma=x_1+x_2+x_3$. Diophantus, II, 35, and Bombelli[124] made $x_i^2+\sigma$ a square for $i=1, 2, 3$. Diophantus, II, 36, made each $x_i^2-\sigma$ a square. Diophantus, V, 9, made each $x_i^2 \pm \sigma$ a square. Diophantus, III, 1, made each $\sigma-x_i^2$ a square by taking $x_1=x$, $x_2=2x$, $\sigma=5x^2$, $5=(2/5)^2+(11/5)^2$, $x_3=2x/5$, whence $x=17/25$. J. Whitley[125] took $x_1=x$, $x_2=nx$, $x_3=mx$, $\sigma-x_1^2=a^2x^2$, which gives x. Then $1+a^2-n^2$ and $1+a^2-m^2$ are to be squares, which is the case if $\frac{1}{2}n^2=a=m$.

Diophantus, IV, 17, made $x_1+x_2+x_3$, $x_1^2+x_2$, $x_2^2+x_3$, $x_3^2+x_1$ all squares by taking $x_2=4x$, $x_1=x-1$, $16x^2+x_3=(4x+1)^2$, whence $x_3=8x+1$. Then

$$x_1^2+x_2=(x+1)^2, \qquad x_1+x_2+x_3=13x=\square=169y^2,$$
$$x_3^2+x_1=13^2 \cdot 8^2 y^4+13 \cdot 17y^2=\square=(13 \cdot 8y+1)^2,$$

whence $y=55/52$, $x=13y^2$.

Fermat[126] suggested that a more elegant solution is obtained by setting $x_1=x$, $x_2=2x+1$, $x_3=4x+3$, whence

$$x_1+x_2+x_3=7x+4=\square, \qquad x_3^2+x_1=16x^2+25x+9=\square,$$

a "double equation" with squares as constant terms. He stated that a similar device will solve the analogous problem in four or a greater number of unknowns.

J. Anderson[127] took $x_1^2+x_2=(p-x_1)^2$, $x_2^2+x_3=(q-x_2)^2$, $x_3^2+x_1=(r-x_3)^2$, which give x_1, x_2, x_3. In Σx_1, equate the coefficient of r^2 to zero, whence $q=1/4$. Other writers gave essentially Diophantus' solution.

S. Ward[128] took $x_2=1-2x_1$, $(1-2x_1)^2+x_3=A^2$, $1-x_1+x_3=B^2$. Then $A^2-B^2=4x_1^2-3x_1$. Take $A+B=2x_1$, $A-B=2x_1-3/2$, whence $B=3/4$, $x_1=x_3+7/16$. Then $16(x_3^2+x_1)=(4x_3-p/q)^2$ determines x_3.

[122] Mém. Acad. Sc. St. Petersb., 5, anno 1812, 1815 (1780), 73 (§ 21); Comm. Arith., II, 385-91.
[123] Math. Quest. Educ. Times, 52, 1890, 117.
[123a] L'intermédiaire des math., 26, 1919, 17-18.
[124] L'algebra opera di R. Bombelli, Bologna, 1579, 485.
[125] Ladies' Diary, 1807, 37, Q. 1155; Leybourn's Math. Quest. L. D., 4, 1817, 72-3.
[126] Oeuvres, I, 301; French transl., III, 249.
[127] The Gentleman's Math. Companion, London, 5, No. 26, 1823, 204-7.
[128] J. R. Young's Algebra, Amer. ed., 1832, 337-8.

Diophantus, II, 34, made x^2-y, y^2-z, z^2-x squares. In IV, 18, these and $x+y+z$ are made squares.

T. Strong[129] made x^2-y, x^2-z, y^2-x, y^2-z all squares. Take

$$x^2-y=(x-ay)^2, \qquad x^2-z=(x-bz)^2, \qquad y^2-x=(y-cx)^2.$$

Hence x, y, z are rational functions of a, b, c. Equate the resulting expression for y^2-z to $(e-1/b)^2$. We get b rationally in terms of e, a, c. For $a=1$, $c=e=2$, we get $x=5/4$, $y=3/2$, $z=14/9$.

Ricatti[130] found three numbers such that if the square of each be added to the remaining two the sums are squares. He used the numbers x, $2x$, 1.

R. Adrain[131] took

$$x^2+y+z=(m-x)^2, \qquad y^2+x+z=(n-y)^2, \qquad z^2+x+y=(r-z)^2,$$

and solved the resulting system of three linear equations for x, y, z.

To make $s+x^2$, $s+y^2$, $s+z^2$ squares, where $s=x+y+z$, "A.B.L." [132] equated them to $(x+v)^2$, $(y+t)^2$, $(z+k)^2$ and solved algebraically the resulting linear equations. "Epsilon" took $y+z=1/4$. Then

$$x+\tfrac{1}{4}+(\tfrac{1}{4}-y)^2=(\tfrac{1}{4}-y+p)^2$$

gives x, and $x+\tfrac{1}{4}+y^2=\square$ if $\tfrac{1}{2}p=q^2+2pq-2qy$, which gives y. W. Wright took $(v-1)r$, $(x-1)r$ and $(y-1)r$ as the numbers, and r^2 as their sum, whence $r=v+x+y-3$. The conditions become $v^2-2v+2=\square=(p-v)^2$, etc., which determine v, x, y.

H. J. Anderson[133] found n numbers whose sum s exceeds the square of each by a square. Express $s=x^2+y^2$ as a sum of two squares $x'^2+y'^2$, $x''^2+y''^2$, $\cdots$, in n ways (Euler, Algebra, II, § 219) by taking $x'=a'y-b'x$, $y'=a'x+b'y$, $x''=a''y-b''x$, $y''=a''x+b''y$, $\cdots$, where

$$a'=\frac{2mn}{m^2+n^2}, \qquad b'=\frac{m^2-n^2}{m^2+n^2}, \qquad a''=\frac{2pq}{p^2+q^2}, \qquad b''=\frac{p^2-q^2}{p^2+q^2}, \cdots.$$

Take x, x', x'', $\cdots$ as the required numbers. Their sum s is of the form $Ax+By$. Thus $s=x^2+y^2$ if $4By-4y^2+A^2=\square$. For $n=4$, C. Farquhar used the numbers w, wx, wy, wz. Set $\sigma=1+x+y+z$. Then $w\sigma-w^2=\square=x^2w^2$ gives σ. Then take

$$2x=y^2, \qquad x^2+1-z^2=\{1+p(x-z)\}^2,$$

which determines z.

J. R. Young[134] found three squares x_i^2 and a number a such that $x_i^2\pm a$ are all squares. Take $x_i^2=m_i^2+n_i^2$, $a=2m_in_i$, $m_i=r_i^2-s_i^2$, $n_i=2r_is_i$, whence $x_i=r_i^2+s_i^2$. It remains to make the values $4r_is_i(r_i^2-s_i^2)$ of a equal. Take $r_1=r_2=r_3=r$. Thus $s_i(r^2-s_i^2)$ are to be equal. The values for $i=1$ and 2 are equal if $r^2=s_1^2+s_1s_2+s_2^2$. Thus $4r^2-3s_2^2$ is to be a square. Hence take

---

[129] Amer. Jour. Sc. and Arts (ed., Silliman), 1, 1818, 426–7.

[130] Institutiones analyticae a Vincentio Riccato, Bononiae, 1, 1765, 64.

[131] The Math. Correspondent, New York, 2, 1807, 13–14.

[132] The Gentleman's Math. Companion, London, 5, No. 25, 1822, 125–30.

[133] Math. Diary, New York, 1, 1825, 151–4.

[134] Algebra, 1816. S. Ward's Amer. ed., 1832, 346–7. A like discussion for two squares had been given by J. Cunliffe, New Series of Math. Repository (ed., T. Leybourn), 1, 1806, I, 221–2.

$r=f^2+3g^2$, $s_2=4fg$. For $f=2$, $g=1$, $s_1=-5$ or -3, $r=7$, $s_2=8$. We may take as s_3 the second value -3, whence $a=3360$, $x_1=74$, $x_2=113$, $x_3=58$.

A. B. Evans[135] found n numbers a_i such that $a_i^2+a_{i+1}=\square$ ($i=1$, $\cdots$, $n-1$), $a_n^2+a_1=\square$. All but the last condition are satisfied if $a_r=m^2+2ma_{r-1}$ ($r=2$, $\cdots$, n), whence

$$a_n=A+2^{n-1}m^{n-1}a_1, \qquad A=m^2+2m^3+2^2m^4+\cdots+2^{n-2}m^n.$$

Then $a_n^2+a_1=(2^{n-1}m^{n-1}a_1+p)^2$ gives a_1. D. S. Hart took $m=1$.

SQUARE OF EACH OF THREE NUMBERS PLUS PRODUCT OF REMAINING TWO
A SQUARE.

L. Euler[136] found solutions of $x^2+yz=p^2$, $y^2+xz=q^2$, $z^2+xy=\square$. Then $p^2-q^2=(x-y)(x+y-z)$. Set $p-q=x-y$, $p+q=x+y-z$, whence

$$p=x-\tfrac{1}{2}z.$$

Then $x^2+yz=p^2$ gives $z=4(x+y)$. The third condition becomes

$$16(x+y)^2+xy=\square,$$

say[136a] $(4x+4y+s)^2$. Then $(x-8s)(y-8s)=65s^2$. Hence set $x-8s=5ts/u$, $y-8s=13us/t$, and to avoid fractions take $s=tu$. Thus $x=8tu+5t^2$, $y=8tu+13u^2$. He stated that the same solution is found if we start by taking $x=(yz-s^2)/(2s)$, the resulting numbers being $s(8t+s)$, $t(t-8s)$, $4(s^2+t^2)$.

To give another method, set $x=a^2+2b$, $y=b^2+2a$, $z=ab(ab-4)$. The first two conditions are satisfied and the third becomes

$$a^4b^4-(8a^3-2)b^3+17a^2b^2+4ab+2a^3=\square,$$

which is not discussed. But he noted the solutions $x, y, z=33, 185, 608$ and $297, 377, 320$. Nesselmann,[95a] p. 141, treated this quartic with $a=-1/p$.

J. Lynn[137] took 1, $x-1$, $4x$ as the numbers. Then two of the conditions are satisfied and the third is $(4x)^2+x-1=\square=(4x\pm a)^2$, say, which determines x.

S. Ward[138] took $x=mz$, $y=nz$, $m+n=1/4$. Then the first two expressions are squares. The third is a square if $1+\tfrac{1}{4}n-n^2=\square$, say $(1-cn)^2$, which gives n.

J. H. Drummond[139] took w^2, mw^2, nw^2 as the numbers. Then $1+mn$, m^2+n, $m+n^2$ are to be squares. Taking $n=\tfrac{1}{4}-m$, it remains to make $1+mn=\square$, say $(1-pm)^2$, which gives m.

W. Wright[140] made $\alpha=x^2+4yz$, $\beta=y^2+4zx$, $\gamma=z^2+4xy$ and $x+y+z$ squares. Take $x=y+z$. Then β and γ are squares. Take

$$\Sigma x=2y+2z=4u^2.$$

[135] Math. Quest. Educ. Times, 20, 1874, 86–7.
[136] Opera postuma, 1, 1862, 258–9 (about 1782).
[136a] J. Cunliffe, New Series of Math. Repository (ed., T. Leybourn), 2, 1809, I, 172–3, chose it equal to $(4ry-4x)^2$ to obtain x rationally in terms of y, r. We may give any desired value to $x+y+z$.
[137] C. Hutton's Miscellanea Mathematica, London, 1775, 236–7.
[138] J. R. Young's Algebra, Amer. ed., 1832, 336.
[139] Amer. Math. Monthly, 9, 1902, 232. Misprint of m^2x^2 for mx^2.
[140] The Gentleman's Math. Companion, London, 3, No. 15, 1812, 346–7.

Then $\frac{1}{4}\alpha = u^4 + 2u^2z - z^2 = (mz - u^2)^2$ if $z = 2u^2(m+1)/(m^2+1)$.   S. Jones took $\beta = (2z - y)^2$ and found $2(n^2 - an)k$, $2(a+n)k$, $2k^2$, where $k = a^2 + n^2$.

W. Wallace[140a] made $\alpha \equiv xy + z^2$, $\beta \equiv xz + y^2$, $\gamma \equiv yz + x^2$, and $\alpha^{1/2} + \beta^{1/2} + \gamma^{1/2}$ squares by taking $\alpha = (2y+z)^2$, $\beta = (2z+y)^2$, whence $x = 4(y+z)$.   Then $\gamma = r^2$ if $yz = \Pi\{r \pm 4(y+z)\}$.   Equate the factors to ym/n and zn/m.   We get y, z and hence x as rational functions of m, n, r.   Omitting the common denominator, we have $x = 4(m^2+n^2)r$, $y = (8mn+n^2)r$, $z = (m^2-8mn)r$.   Then α, β, γ equal the squares of $(m^2+8mn+2n^2)r$, $(2m^2-8mn+n^2)r$, $(4m^2+mn-4n^2)r$.   The sum $(7m^2+mn-n^2)r$ of these is a square if r equals the first factor or the quotient of it by any square.

Miscellaneous Systems of Equations of Degree Two.

Diophantus, III, 2, made $s^2 + x_i$ $(i=1, 2, 3)$ rational squares, where $s = x_1 + x_2 + x_3$.   In Diophantus, III, 3, $s^2 - x_i$ $(i=1, 2, 3)$ are made squares. T. Brancker[141] treated the latter problem.   A. Gérardin[142] gave several integral solutions of the last two problems.

Diophantus, III, 4, made $x_i - s^2$ $(i=1, 2, 3)$ rational squares.

To find x_1, x_2, $\cdots$ such that

$$s^2 + x_i = p_i^2, \qquad s^2 - x_i = q_i^2,$$

where $s = \Sigma x_i$, "Comes"[143] noted that since p_i^2, s^2, q_i^2 are squares in arithmetical progression we may use the known values

$$p_i = s(m_i^2 - n_i^2 + 2m_in_i)/(m_i^2+n_i^2), \qquad q_i = s(n_i^2 - m_i^2 + 2m_in_i)/(m_i^2+n_i^2).$$

Then $s = \Sigma x_i$ gives s.   For Diophantus' solution, see the first page of Ch. VI.

A. Gérardin and R. Goormaghtigh[144] made $s^2 - x_i^2$ $(i=1, 2, 3)$ squares; also $s^2 - (s - x_i)$; also $s^2 - (s - x_i)$ $(i=1, 2, 3, 4)$, where $s = x_1 + \cdots + x_4$. The latter[145] made $s^2 + x_i$ $(i=1, \cdots, n)$ squares, also $s^2 - (s - x_i)$, where $s = x_1 + \cdots + x_n$.

Leonardo Pisano[146] treated cases of $x_1^2 + x_1 + \cdots + x_n = y_1^2$, $y_1^2 + x_2^2 = y_2^2$, $y_2^2 + x_3^2 = y_3^2$, $\cdots$, $y_{n-1}^2 + x_n^2 = y_n^2$.

J. Cunliffe[147] made $\sigma + x_i$ $(i=1, 2, 3)$ squares, where $\sigma = x_1^2 + x_2^2 + x_3^2$.

S. Ryley[148] made $\alpha = x^2 + yz + y^2$, $\beta = x^2 + yz + z^2$, $\gamma = y^2 + yz + z^2$ squares. Take $\alpha = a^2$, $\beta = b^2$.   Then $y^2 - z^2 = a^2 - b^2$.   Hence take $(a+b)r = (y+z)s$, $(a-b)s = (y-z)r$, which give a, b in terms of y, z.   Now $\gamma = \square$ if

$$y = 2rs(m^2 + 2mn), \qquad z = 2rs(n^2 - m^2).$$

Then $a^2 - yz - y^2$ becomes a function of r, s, m, n of degree 4 in n, which will

[140a] New Series of Math. Repository (ed., T. Leybourn), 3, 1814, I, 21–23.

[141] An Introduction to Algebra, transl. out of the High-Dutch by T. Brancker, much altered and augmented by D. P[ell], London, 1668, 102–4.

[142] L'intermédiaire des math., 22, 1915, 197–8.

[143] The Gentleman's Math. Companion, London, 4, No. 21, 1818, 752–7.

[144] L'intermédiaire des math., 22, 1915, 220–1, 244; 23, 1916, 136–141, 155–7, 209–11; 24, 1917, 13–14.

[145] Nouv. Ann. Math., (4), 16, 1916, 401–26.

[146] Scritti di L. Pisano, 2, 1862, 279–83.   Cf. F. Woepcke, Jour. de Math., 20, 1855, 61–62; A. Genocchi, Annali Sc. Mat. Fis., 6, 1855, 193–205, 357–9.

[147] Math. Repository (ed., Leybourn), London, 3, 1804, 97–106.

[148] The Gentleman's Math. Companion, London, 1, No. 8, 1805, 42–4.

equal the square of
$$x = n^2(r^2-s^2) + nm(2r^4-4s^2r^2-2s^4)/(r^2-s^2) - 2s^2m^2$$
if $m : n = s^2+r^2 : 2s^2-2r^2$.

To make $\alpha = x^2+y^2+s$, $\beta = x^2+z^2+s$, $\gamma = y^2+z^2+s$ squares, where $s = xy+xz+yz$, S. Ryley[149] took $y = 1$, $z = 3$. Then
$$\alpha = \square, \qquad \beta = x^2+4x+12 = (x+n)^2$$
if $x = (12-n^2)/(2n-4)$, and $\gamma(2n-4)^2 = (4-14n)^2$ if $n = -16$, whence
$$x : y : z = 61 : 9 : 27.$$
J. Cunliffe took $x = 3z$, $y = n-z$. Then $\gamma = (n+z)^2$. Make $\beta = a^2$, by choice of n. Then $16z^2\alpha = a^4-10a^2z^2+153z^4 = \square$ if $a = 19z/3$. Or take $\alpha = (rn-3z)^2$, $z = r^2-1$, whence $n = 2(3r+1)$. Then $\beta = \square$ if $r = 5/3$, whence
$$x : y : z = 4 : 32 : 12.$$
"Limenus" took $\alpha = a^2$, $\beta = b^2$, $\gamma = c^2$. Then $x^2+c^2 = y^2+b^2 = z^2+a^2$. Hence take a number $(m^2+m_1^2)(n^2+n_1^2)(q^2+q_1^2)$ which is a sum of two squares in three ways, whence
$$x = mn_1q+mnq_1-m_1nq+m_1nq_1,$$
while y (or z) is the similar expression with only the second (or first) term negative. Set $v = m/m_1$, $r = n/n_1$, $s = q/q_1$. Then $(x+y+z)^2+x^2 = a^2+b^2$ becomes $fv^2-4(r+s)v = f+4rs+4$, where $f = (r^2-1)(s^2-1)$. Thus the square of $fv-2rs-2s$ is known; equate the root to $f+2rs+2+C$ and take $C = -2$ to cancel the terms in s^4, s^3. Hence $2rs = -1$, $v = (2r^2-3r-1)/(2r^2+r-1)$. Take $q = -n_1$, $q_1 = 2n$, $m = 2n^2-3nn_1-n_1^2$. Then
$$x = 4n^4+n_1^4-n^2n_1^2, \qquad y = 4n^4-4n^3n_1+n^2n_1^2-4nn_1^3-n_1^4,$$
$$z = -4n^4+8n^3n_1+n^2n_1^2+2nn_1^3+n_1^4.$$
The least positive numbers found are 19, 13, 2.

To make $\xi^2+\eta^2+\zeta^2+2\xi\eta-2\xi\zeta+2\eta\zeta$, etc., squares, W. Wright[150] put $\xi = x+y$, $\eta = x+z$, $\zeta = y+z$ and noted that the problem is reduced to the preceding one, for which he took $y = px$, $z = 3x$, and found p so that $p^2+4p+12 = (p-r)^2$; finally, $4p+13 = \square$ if $r = 16$. Others equated the first function $(\xi+\eta+\zeta)^2-4\xi\zeta$ to $(\xi+\eta)^2$, whence $\zeta = 2\xi-2\eta$, or to $(2\xi-\zeta/2)^2$, whence $\xi = \eta+\zeta/2$. Then the difference of the other two initial functions factors.

J. Cunliffe[151] made $x^2+y^2+a(x+y)$, $x^2+z^2+b(x+z)$, $y^2+z^2+c(y+z)$ squares by taking $x = rv$, $y = sv$, $z = tv$, where $r^2+s^2 = e^2$, $r^2+t^2 = f^2$, $s^2+t^2 = g^2$. Take $m = a(r+s)/e^2$, $n = b(r+t)/f^2$, $p = c(s+t)/g^2$. Then the quotients of the initial functions by e^2, f^2, g^2 are v^2+mv, v^2+nv, v^2+pv, which are made squares (Cunliffe[1] of Ch. XVIII).

D. S. Hart[152] equated the same initial functions to the squares of $x+y$, $x+z$, $y+z$. Then $a(x+y) = 2xy$, etc., determine x, y, z rationally in terms of a, b, c.

---

[149] The Gentleman's Math. Companion, London, 2, No. 9, 1806, 31–35.

[150] Ibid., 5, No. 29, 1826, 502–6.

[151] Ibid., 3, No. 14, 1811, 300–2. Same by J. Matteson, The Analyst, Des Moines, 2, 1875, 46–9.

[152] Math. Quest. Educ. Times, 17, 1872, 37.

W. Wright[153] noted that $sx-yz=m^2$, $sy-xz=n^2$, $sz-xy=r^2$, $s=x+y+z$, lead to the problem $(x+y)^2=m^2+n^2$, $(x+z)^2=m^2+r^2$, $(y+z)^2=n^2+r^2$ at the beginning of this Chapter.

S. Jones[154] made $\alpha=sx+yz$, $\beta=sz+yx$, $\gamma=sy+zx$ squares, where $s=x+y+z$, by taking $y=a-x$, $\alpha=b^2$, $\gamma=c^2$, whence $x=(a^2+b^2-c^2)/(2a)$, etc., and $\beta=\square$.

J. R. Young[155] found four numbers whose sum is a square and such that if unity be added to the product of the sum by any one of them there results a square. Let the numbers be $x\pm1$, $x\pm y$. It suffices to make $4x$, $4x^2\pm4xy+1$ squares. Take $x=4$ and set $65-16y=m^2$, $65+16y=n^2$. Then $m^2+n^2=130$, which holds if $m=3$, $n=11$.

W. Wright and others found[156] four numbers v, x, y, z whose sum is a square n^2 and such that $vn^2+1=\square$, etc. Equate vn^2+1, xn^2+1, yn^2+1, zn^2+1 to the squares of $1+s$, $1+r$, $1+q$, $1+p$. By addition, $s^2+2s+1=n^4$ if $r^2+q^2+p^2+2r+2q+2p=1$. The latter is solved for r after taking $q=m-1$, $p=lm-1$. Several solvers used the numbers $x\pm1$, $x\pm y$.

To make x^2+y^2+S, x^2+z^2+S, y^2+z^2+S squares, where

$$S=2xy+2xz+2yz,$$

W. Wright[157] noted that the functions factor, being $a(b+c)$, $b(a+c)$, $c(a+b)$, where $a=x+y$, $b=x+z$, $c=y+z$. Take $b=na$, $c=ma$, $n(m+1)=n^2\xi^2$, $m(n+1)=(n\xi-p)^2$. We get m and n. Then $m+n=N/D$, where N and D are quadratic in ξ. Take $N=(p\xi+q)^2$ to get ξ. Then $D=\square$ becomes a quartic in q. C. Holt noted that one condition is satisfied if the numbers are $5n-m$, $m-4n$, $4n$. Baines wrote $s=x+y+z$; thus s^2-z^2, s^2-y^2, s^2-x^2 are to be squares, say of $(s+z)/m$, $(s+y)/n$, $(s+x)/r$. We get x, y, z. To satisfy $\Sigma x=s$, take $r=3$, $n=-37/36$, $m=25/21$.

To find three numbers double the sum of any two less the third being a square, double the sum of any two of their squares less the square of the third being a square, while the last three squares have the same property, W. Wright[158] used the numbers x, y, $x+y$. Then all but the first three conditions are satisfied. Take $x+y=a^2$, $4x+y=(2a-p)^2$. For the resulting x, a, $4y+x=\square$ if $p^4+54p^2y+9y^2=\square=(3y-v)^2$, which gives y.

To make[159] $2(v+x+y+z)=\square=4a^2$, $\alpha=2(x+y+z)^2-2v^2=\square$, etc., note that $\alpha=4a^2(x+y+z-v)$. Hence take $x+y+z-v=4b^2$, etc. The condition $a^2=b^2+c^2+d^2+e^2$ is satisfied by taking $a=e+r$ and finding e.

Several[160] discussed the problem to make $a+b$, $b+c$, $b-c$, $a+2b+c+d$ and $a^2+bc+bd+cd$ squares, $(a+b)(b-c)=b+c$, and $b^2-c^2=1$.

[153] The Gentleman's Math. Companion, London, 3, No. 17, 1814, 462–4.
[154] Ibid., 3, No. 18, 1815, 317–8.
[155] Algebra, 1816. Amer. ed. by S. Ward, 1832, 331.
[156] The Gentleman's Math. Companion, London, 5, No. 26, 1823, 240–2.
[157] Ibid., 5, No. 29, 1826, 500–2.
[158] Ibid., 5, No. 30, 1827, 575–6.
[159] Ibid., 558.
[160] Math. Miscellany, Flushing, N. Y., 1, 1836, 154–5.

J. Matteson[161] found four squares such that fifteen linear or quadratic functions of the squares or their roots shall be squares.

A. Martin and H. W. Draughon[162] found three integers such that the square of the sum of any two less the square of the third is a square.

A. Gérardin[163] treated $x^2-(y-z)^2=a$, $y^2-(x-z)^2=b$, $z^2-(x-y)^2=c$. Set $y=z+u$, $x=z+u+w$, $z=w+h$. Then $c=hr$, $a=rs$, $b=hs$, where $r=h+2w$, $s=r+2u$.

RATIONAL ORTHOGONAL SUBSTITUTIONS.

L. Euler[164] stated that he had a general solution of the problem to find 16 integers arranged in a square such that the sum of the squares of the numbers in each row or column or either diagonal are all equal, while the sum of the products of corresponding numbers in any two rows or columns is zero. The example given is the following:

$$
\begin{array}{rrrr}
68 & -29 & 41 & -37 \\
-17 & 31 & 79 & 32 \\
59 & 28 & -23 & 61 \\
-11 & -77 & 8 & 49.
\end{array}
$$

Euler[165] treated orthogonal substitutions on $n=3$, 4, 5 variables, i. e., linear substitutions leaving unaltered the sum of the squares of the variables. He expressed the coefficients in terms of trigonometric functions. For $n=3$, he noted the rational solution

$$
\begin{array}{ccc}
p^2+q^2-r^2-s^2 & 2qr+2ps & 2qs-2pr \\
2qr-2ps & p^2-q^2+r^2-s^2 & 2pq+2rs \\
2qs+2pr & 2rs-2pq & p^2-q^2-r^2+s^2,
\end{array}
$$

each entry being divided by $p^2+q^2+r^2+s^2$. For $n=4$ he gave two similar rational solutions of which the second is

$$
\begin{array}{cccc}
ap+bq+cr+ds & ar-bs-cp+dq & -as-br+cq+dp & aq-bp+cs-dr \\
-aq+bp+cs-dr & as+br+cq+dp & ar-bs+cp-dq & ap+bq-cr-ds \\
ar+bs-cp-dq & -ap+bq-cr+ds & aq+bp+cs+dr & as-br-cq+dp \\
-as+br-cq+dp & -aq-bp+cs+dr & -ap+bq+cr-ds & ar+bs+cp+dq,
\end{array}
$$

in which the sum of the products of corresponding numbers in any two rows or columns is zero, while the sum of the squares of the numbers in any row or column is $\sigma=(a^2+b^2+c^2+d^2)(p^2+q^2+r^2+s^2)$. For his[164] former problem, we require also that the sum of the squares of the numbers in

---

[161] Math. Quest. Educ. Times, 18, 1873, 35–7. Same in his Collection of Diophantine Problems with Solutions (ed., A. Martin), Washington, D. C., 1888, 22–4.

[162] Amer. Math. Monthly, 1, 1894, 361–2.

[163] Sphinx-Oedipe, 8, 1913, 30–1.

[164] Opera postuma, 1, 1862, 576–7, letter to Lagrange, Mar. 20, 1770. Quoted by Legendre, Théorie des nombres, 2, 1830, 144; Maser's German transl., II, 140.

[165] Novi Comm. Acad. Petrop., 15, 1770, 75; Comm. Arith., I, 427–443.

either diagonal shall be σ, viz.,

$$(ac+bd)(pr+qs)=0, \qquad (ab+cd)(pq+rs)+(ad+bc)(ps+qr)=0.$$

He gave two special cases, one of which is his[161] above solution.

G. R. Perkins[166] employed as the numbers of the first row of his square

$$
\begin{array}{llll}
pp'+qq'+rr'+ss', & pr'+qs'-rp'-sq', & ps'-qr'+rq'-sp', & pq'-qp'-rs'+sr', \\
-pq'+qp'-rs'+sr', & -ps'+qr'+rq'-sp', & pr'+qs'+rp'+sq', & pp'+qq'-rr'-ss', \\
-pr'+qs'+rp'-sq', & pp'-qq'+rr'-ss', & -pq'-qp'+rs'+sr', & ps'+qr'+rq'+sp', \\
ps'+qr'-rq'-sp', & -pq'-qp'-rs'-sr', & -pp'+qq'+rr'-ss', & pr'-qs'+rp'-sq'
\end{array}
$$

those whose sum of squares equals $(p^2+q^2+r^2+s^2)(p'^2+\cdots)$. By writing in reverse order the functions of the first row and changing the signs of r, s in the first two terms and the signs of p, q in the last two terms, we get the entries in the second row. We derive the third row from the first, and fourth from the second, by moving each term one place to the right or left without crossing the middle vertical column, and changing the signs of q, s or those of p, r according as the term is moved to the right or left. Two of the various possible such squares are given. Of the conditions required by Euler,[164] all are now satisfied except those relating to the two diagonals. Take $s=0$. The latter conditions become

$$p'r'=q's', \qquad p(p'q'-r's')=r(p's'-q'r').$$

By further specializations, he obtained the solution

$$
\begin{array}{llll}
42+2q & -11+4q & 24-\ q & 2-8q \\
-18+8q & -16+\ q & 24+4q & 38+2q \\
11+4q & 42-2q & -\ 2-8q & 24+\ q \\
16+\ q & -18-8q & -38+2q & 21-4q.
\end{array}
$$

C. Avery[167] proceeded as had Perkins, without describing the process to choose the signs, and obtained the solution

$$
\begin{array}{llll}
48+4q & -44+3q & 51-2q & -7-6q \\
-47+6q & 21+2q & 64+3q & 12+4q \\
44+3q & 48-4q & 7-6q & 51+2q \\
-21+2q & -47+6q & -12+4q & 64-3q.
\end{array}
$$

The case $q=5$ yields Euler's[164] answer.

V. A. Lebesgue[168] gave orthogonal substitutions in 3 variables in trigonometric form. He[169] quoted Euler's[164, 5] solution of the problem of 16 integers.

L. Bastien[170] took four integers α, β, γ, δ such that $\alpha\beta/(\gamma\delta)$ is the square of r/s, where r, s are relatively prime integers. Write

$$x=r(\delta^2-\gamma^2), \qquad y=s(\alpha\gamma-\beta\delta), \qquad t=r(\alpha\gamma-\beta\delta), \qquad u=s(\beta^2-\alpha^2).$$

[166] Math. Miscellany, New York, 2, 1839, 102–5.
[167] *Ibid.*, 101.
[168] Nouv. Ann. Math., 9, 1850, 46–51.
[169] *Ibid.*, 15, 1856, 403–7.
[170] Sphinx-Oedipe, 7, 1912, 12.

Then a solution of Euler's[164] problem is

$$\begin{array}{cccc}
\alpha x+\beta y & -\beta x+\alpha y-2\delta t & 2\alpha y+\gamma u+\delta t & \gamma t-\delta u \\
-2\beta y-\gamma t-\delta u & \gamma u-\delta t & \beta x+\alpha y & -\alpha x+\beta y-2\gamma t \\
\beta x+\alpha y+2\delta t & \alpha x-\beta y & \gamma t+\delta u & 2\alpha y+\gamma u-\delta t \\
-\gamma u-\delta t & -2\beta y+\gamma t-\delta u & \alpha x+\beta y+2\gamma t & \beta x-\alpha y,
\end{array}$$

the sum of the squares of the numbers in any row, column or diagonal being $(\alpha^2+\beta^2)(x^2+y^2)+(\gamma^2+\delta^2)(t^2+u^2)$.

Fuss[95] of Ch. V made p^2+s^2, q^2+s^2, r^2+s^2 squares with $pq+pr+qr=s^2$.

PAPERS NOT AVAILABLE FOR REPORT.

S. Günther, Ziele u. Resultate d. neueren Math. Hist. Forschung, Erlangen, 1876, 50–53.

J. Favaro, Notize storico-critiche sulla costruzione delle equazioni, Modène, 1878.

G. de Longchamps, Jour. de math. élém., 1882, 192; 28, 1894, 5.

Ferrent, *ibid.*, (2), 3, 1884, 121, 155, 169, 193, 217, 241; 1885, 3, 170–1.

J. Novák, Ueber unbest. Gl. 2 Grades mit 2 Unbek., Progr. Budweis, 1890.

CHAPTER XX.

QUADRATIC FORM MADE AN NTH POWER.

Binary Quadratic Form Made a Cube.

Diophantus, VI, 19, to find a right triangle the sum of whose area x and hypotenuse h is a square and perimeter is a cube, took 2 and x as the legs and $h+x=25$, noting that the square 25 when increased by 2 becomes the cube of 3. Then $h^2=x^2+2^2$ gives $x=621/50$.

Jordanus[35] of Ch. XII noted that $x(x+1)$ is never a cube.

C. G. Bachet[1] noted that from $5^2+2=3^3$ we can find other [rational] numbers x making x^2+2 a cube. Let $x=5-N$. To make $27-10N+N^2$ the cube of $3-z$ equate the second term $-27z$ of its cube to $-10N$, whence $z=10N/27$. We now get N. In VI, 20, we have $17=2^3+3^2$ and seek a cube which increased by 17 gives a square; take $N-2$ and $3+t$ as the sides of the cube and square, and equate the second terms $12N$ and $6t$ of the expansions, whence $N=10$, $t=20$.

Fermat[2] stated that he could give a rigorous proof that 25 is the only integral square which is less than a cube by 2.

Fermat[3] stated elsewhere this result on 25 and the fact that 4 and 121 are the only integral squares which when increased by 4 give a cube.

L. Euler[4] proved that $x^3+1=\square$ has no [positive] rational solution except $x=2$. To show that, for a and b relatively prime, $a^3b+b^4=\square$ only when $a=2$, $b=1$, set $a+b=c$. The condition becomes $bcg=\square$, $g\equiv c^2-3bc+3b^2$. First, let c be not divisible by 3. Then b, c, g are relatively prime and hence each is a square. Set $g=(bm/n-c)^2$ and solve for b/c. If m is not divisible by 3, $c=\pm(m^2-3n^2)$, $b=\pm(2mn-3n^2)$. For the lower sign, c is not a square. Hence $c=m^2-3n^2=\square=(m-np/q)^2$, $m/n=(p^2+3q^2)/(2pq)$. Then $b/n^2=G/(pq)$, $G=p^2-3pq+3q^2$. Thus $pqG=\square$, so that the method of descent applies. Next, for $m=3k$, $b:c=n^2-2kn:n^2-3k^2$. As before,

$$c=n^2-3k^2=(n-kp/q)^2, \qquad b/n^2=(p^2+3q^2-4pq)/(3q^2+p^2).$$

Hence $(3q^2+p^2)(p-q)(p-3q)=\square$. Let $p-q=t$, $p-3q=u$. Then

$$tu(3t^2-3tu+u^2)=\square$$

and the method of descent applies. Finally, let $c=3d$. Then

$$bd(b^2-3bd+3d^2)=\square$$

is of the initial type with the former b, c replaced by d, b. Since b is prime to 3, the descent applies. It is stated that a like proof shows that $x^3-1\neq\square$.

---

[1] Diophanti Alex. Arith., 1621, 423–5.

[2] Oeuvres, I, 333–4; French transl., III, 269.

[3] Oeuvres, II, 345, 434, letters to Digby, Aug., 1657, and to Carcavi, Aug., 1659. E. Brassinne, Précis des Oeuvres math. de P. Fermat et de l'Arith. de Diophante, Mém. Acad. Sc. Toulouse, (4), 3, 1853, 122, 164.

[4] Comm. Acad. Petrop., 10, 1738, 145; Comm. Arith. Coll., I, 33–34; Opera Omnia, (1), II, 56–58. Proof republished by E. Waring, Medit. Algebr., ed. 3, 1782, 374–5.

Euler[5] applied to $x^3+1=\square$ his[144] method of Ch. XXII to make a cubic or a quartic a square, finding no solutions except 0, -1, 2, and stated that there are no others.   Cf. Euler[157] of Ch. XXI.

Euler,[6] to make ax^2+cy^2 a cube, assumed that

$$x\sqrt{a}+y\sqrt{-c}=(p\sqrt{a}+q\sqrt{-c})^3,$$

whence $x=ap^3-3cpq^2$, $y=3ap^2q-cq^3$.   For Fermat's case x^2+2, we have (Art. 193) $a=1$, $c=2$, $y=\pm1$, whence $q(3p^2-2q^2)=\pm1$, and q divides unity. Taking $q=1$, we have $3p^2-2=\pm1$, whence $p^2=1$, $x^2=25$.   A like proof is given (Art. 192) of Fermat's result that 4 and 121 are the only integral squares which when increased by 4 give a cube.   But (Arts. 195–6) for $2x^2-5$ the method leads to no solution, whereas the solution $x=4$ exists and the above assumption is shown to fail.

A. M. Legendre[7] treated Fermat's problems as had Euler.[6]

V. A. Lebesgue[8] proved that $x^2-7=y^3$ is impossible.   For, if y is even, x is odd and $x^2=8n+1\ne(2v)^3+7$; while, if y is odd, $x^2+1=(y+2)Q$ is impossible since the prime divisors of $Q=(y-1)^2+3$ are of the form $4n+3$.

L. Öttinger[9] noted that $x^2-y^2=\mp z^3$ has the general solution

$$\{4m^3\pm3mr(2m\pm r)\}^2-\{(m\pm r)(4m^2\pm2mr+r^2)\}^2=\mp(2mr\pm r^2)^3.$$

T. Pepin[10] criticized Euler's[6] proofs, noting that there may exist sets of formulas for x and y other than the set deduced by Euler's assumption. He proved Fermat's[2, 3] assertions.   He studied the solution of $x^2+cn^{2a}=z^3$ for $c=1, 2, 3, 4, 7$, n being 1 or an odd prime, and z being odd if $c=7$, and proved that the following are not cubes: x^2+1 $(x>0)$; x^2+3; $4x^2+7$; x^2+9; x^2+n^2 if $n=108l+k(k=23, 35, 59, 71, 95)$, or $n=83, 263, 407$, or if n is a prime $12l+7$ with $7<n<1350$; x^2+2n^2 if n is a prime $24l+5$ or $24l+7$; x^2+3n^2 if n is a prime $6l+5$ or its square; x^2+5.   Also, $x^2+9^2=z^3$ only for $x=\pm46$, $x^2+7^2=z^3$ only for $x=\pm524$; $x^2+11^2=z^3$ only for $x^2=4$.

H. Brocard[11] and others gave various solutions of $x^3+17=y^2$.

G. C. Gerono[12] proved that $x^3=y^2+17$ is impossible in integers by use of $(x+2)\{(x-1)^2+3\}=y^2+5^2$ and the divisors of a sum of two squares.

E. de Jonquières[13] proved that $x^3+a=y^2$ is impossible in integers if $a=c^3-4$, $|c|\equiv1, 3, 7$ (mod 8), or $a=c^3-4^t$, $|c|\equiv3, 5, 7$ (mod 8), $t>1$, or $a=c^3-1$, $c=2(2d+1)$, and hence if $a=-3, -5, 7, -9, 11, -17, 23, -43, 61$; also for $a=4, 6, 14, 16$ if $x\ne0$.

F. Proth[14] stated and E. Lucas[14] proved that $x^2+3=y^3$ is impossible since $y=r^2+3s^2$, while $x^2-3=y^3$ holds only for $x=2$, $y=1$.

---

[5] Algebra, 2, 1770, Ch. 8, Art. 121; French transl., 2, 1774, pp. 135–152; Opera Omnia, (1), I, 392.   Mém. Acad. Sc. St. Pétersbourg, 11, 1830 (1780), 69; Comm. Arith. Coll., II, 478.

[6] Algebra. II, Ch. 12, Arts. 187–196.   Opera Omnia, (1), I, 429–434.

[7] Théorie des nombres, ed. 3, II, 1830, Art. 336, p. 12.

[8] Nouv. Ann. Math., (2), 8, 1869, 452–6, 559.

[9] Archiv Math. Phys., 49, 1869, 211.

[10] Jour. de Math., (3), 1, 1875, 318–9, 345–358.   Details in Pepin.[72]

[11] Nouv. Corresp. Math., 3, 1877, 25, 49; 4, 1878, 50.   Cf. Escott,[27] Brocard.[56]

[12] Nouv. Ann. Math., (2), 16, 1877, 325–6.

[13] Ibid., (2), 17, 1878, 374–380, 514–5.

[14] Nouv. Corresp. Math., 4, 1878, 121, 224.

T. Pepin[15] applied de Jonquières'[13] method to obtain the generalization that $x^3 + a = y^2$ is impossible if a is of the form $c^3 - 4^a b^2$, where b and c are odd, while b has no divisor $4l + 3$, and $c \equiv 1, 3, 7$ (mod 8) if $\alpha = 1$, $c \equiv 3, 5, 7$ (mod 8) if $\alpha > 1$. Also if $a = 8(2d + 1)^3 - b^2$, and b is prime to 3 and does not have two factors (equal or distinct) of the form $4l + 3$; for example, $a = -17$ or 47. Also, if $a = 8c^3 - 2b^2$, where $c = 4k + 1$ and b is an odd number not having two equal or distinct prime factors of the forms $8l + 5$ or $8l + 7$; for example, $a = 6, -10, 118, -58$. Also, if $a = 8c^3 + 2b^2$, $c = 4k + 3$, and b is odd and without two prime factors $8l + 3$ or $8l + 5$. Also in several analogous cases.

E. Catalan[16] noted that some, but not all, solutions of $x^2 + 3y^2 = z^3$ are

$$x = \tfrac{1}{2}(\alpha + \beta)(\alpha - 2\beta)(\beta - 2\alpha), \qquad y = \tfrac{3}{2}\alpha\beta(\alpha - \beta), \qquad z = \alpha^2 - \alpha\beta + \beta^2.$$

S. Réalis[17] gave identities showing solutions of $x^3 + k = y^2$ if $k = b^2(8b - 3a^2)$, $b^2(b - 3a^2)$, $b(3a^2 + b)^2$, $4a^2(a^2 + 1)$. Given one solution $\alpha^3 + k = \beta^2$, another follows from the identity, obtained by Euler's[5] process,

$$\left(\frac{9\alpha^4 - 8\alpha\beta^2}{4\beta^2} \right)^3 + \beta^2 - \alpha^3 = \left(\frac{27\alpha^6 - 36\alpha^3\beta^2 + 8\beta^4}{8\beta^3} \right)^2.$$

Réalis[18] stated that, if $z^2 - 3\alpha z - \alpha^3 + \beta^2 = 0$ has integral roots,

$$x^3 + \{(\alpha + 1)^3 - (\beta + 1)^2\}z = y^2$$

has integral solutions $x = \alpha - z$, $y = \beta - z$; for example, if $\alpha = a^2$, $\beta = \pm a^3$; $\alpha = 2$, $\beta = \pm 1$; $\alpha = 32$, $\beta = \pm 64$. If

$$8\beta - 3\alpha^2 - 6\alpha + 1 = \square,$$

$x^3 - \alpha^3 + \beta^2 = y^2$ has integral solutions other than $x = \alpha$, $y = \beta$. Cf. Ch. XXI.[346]

T. Pepin[19] proved there is one and only one square which becomes an odd [Pepin[33]] cube on adding 2, 13, 47, 49, 74, 121, 146, 191, 193, 301, 506, 589, 767, 769, 866 or 868. No square > 0 added to 1, 3, 5, 27, 50, 171, or 475 becomes an odd cube. The only solutions of $x^2 + 11 = y^3$ are $x = 4$, 58; the only solution of $x^2 + 19 = y^3$ is $y = 7$. If a is one of the primes 11, 17, 29, 37, 47, 83, 96, 107, 181, 197, 233, 359, 421, 569, 757, 827, there is a single square which becomes an odd cube on adding $11a^2$. If $a < 1000$ and a is of one of the linear forms $38l + 3, 13, 15, 21, 27, 29, 31, 33, 37$ and $a \neq 29$, 89, 173, 281, 331, 569, 953, no square increased by $19a^2$ is an odd cube. Also, similar theorems.

Pepin[20] gave sixteen special theorems on $x^2 + g = z^3$, proved only under the assumption that x is even and z odd.

Pepin[21] proved that $x^2 + n \neq z^3$ if $n = 5, 6, 10, 12, 14, \cdots, 98$; $4x^2 + n \neq z^3$ if $n = 7, 15, 39, 47, 55, 63, 71, 79$; $x^2 + 44 = z^3$ only for $x^2 = 81$; and gave several theorems on $x^2 + 11y^2 = z^3$ [all provided z is odd, Pepin[33]].

[15] Annales Soc. Sc. Bruxelles, 6, 1881–2, 86–100.
[16] Mém. Soc. Sc. Liège, (2), 10, 1883, No. 1, p. 10.
[17] Nouv. Ann. Math., (3), 2 ,1883, 289–297.
[18] *Ibid.*, 334–5. Proof of first by E. Fauquembergue, (3), 4, 1885, 379; of second by H. Brocard, (3), 10, 1891, p. 7* of Exercices.
[19] Mem. Pont. Accad. Nuovi Lincei, 8, 1892, 41–72; Extract, Sphinx-Oedipe, 1908–9, 188–9. Cf. Pepin.[75]
[20] Comptes Rendus Paris, 119, 1894, 397–9; corrections, 120, 1895, 494 [Pepin[33]].
[21] *Ibid.*, 120, 1895, 1254–6.

E. Fauquembergue[22] gave an insufficient proof that $x^3+2 \neq y^2$ if $x \neq -1$.

C. Störmer[23] solved $x^2-y^2=z^3$ by means of the identity

$$\{x(x^2+3y^2)\}^2 - \{y(y^2+3x^2)\}^2 = (x^2-y^2)^3.$$

A. Goulard[24] proved that $x^2-1=z^3$ only for $x^2=9$, since $x^2-1=8w^3$ has no solution except when $w=0$ or 1 [Legendre,[81] of Ch. I]. T. Pepin (pp. 283–5) reduced the question to $u^3+x^3=2y^3$ which holds only for $u=x$ [Legendre, Théorie des nombres, ed. 2, 1808, 347].

E. de Jonquières[25] treated $x^2-a^2=y^3$. For $a=3$, E. B. Escott[26] noted the solutions $y=-2, 0, 3, 6, 40$ and stated that there are no others <1155.

Concerning Fermat's assertion that 25 is the only square which increased by 2 gives a cube, H. Delannoy[27] remarked that Euler's[6] proof is incomplete since if applied to $x^2+47=z^3$ it yields $x=500$ but not the solution $x=13$. P. Tannery[28] replied that the proof as given by Legendre[7] depends on the fact that every divisor of x^2+2 is of the form p^2+2q^2, while not every divisor of x^2+47 is of the form p^2+47q^2. I. Ivanoff (p. 47) explained the difference by the fact that in the domain $R(\sqrt{-2})$ of the complex integers depending on $\sqrt{-2}$ the introduction of ideals is superfluous, but not for $R(\sqrt{-47})$. E. Landau[29] supplemented Ivanoff's remark by noting that a second circumstance is necessary to justify Euler's conclusion that $(x+\sqrt{-2})(x-\sqrt{-2})=t^3$ implies that $x \pm \sqrt{-2}$ are cubes in $R(\sqrt{-2})$, viz., that, in $R(\sqrt{-2})$, ± 1 are the only units (complex integers dividing unity). From the superfluity of the introduction of ideals, we can conclude only that, if a product of two relatively prime complex integers is a cube, each of the two factors is a product of a cube by a unit. For $R(\sqrt{2})$, the introduction of ideals is unnecessary, but $(x+\sqrt{2})(x-\sqrt{2})=t^3$ does not imply that $x \pm \sqrt{2}$ are cubes of integers $\alpha+\beta\sqrt{2}$. Cf. Euler[183] of Ch. XXI.

A. Boutin[30] stated that $x^3-7y^2=1$ for $x=1, 2, 4, 22$, but for no other values <196. Other writers[31] stated that any new solution has at least 1400 digits in y.

E. Fauquembergue[32] noted that $px^2+mxy+qy^2=z^3$ for

$$x=p(f^3-3pqfg^2-mpqg^3), \qquad y=p^2g\{3f^2+3mfg+(m^2-pq)g^2\},$$
$$z=p(f^2+mfg+pqg^2).$$

T. Pepin[33] remarked that all the theorems in his[19–21] papers on insolvable equations $x^2+cy^2=z^3$ were subject to the restriction that z is odd. The enunciation of this restriction is necessary if $c=8l$ or $8l+7$ since in these

---

[22] Mathesis, (2), 6, 1896, 191. Criticized by L. Aubry, l'interméd. des math., 18, 1911, 204.

[23] L'intermédiaire des math., 2, 1895, 309.

[24] Ibid., 3, 1896, 135.

[25] Ibid., 6, 1899, 91–5; 5, 1898, 257 ($a=3$). Cf. Descartes,[14] Ch. XIII; Tait,[25] Ch. XXI.

[26] Ibid., 7, 1900, 135.

[27] Ibid., 5, 1898, 221–2.

[28] Ibid., 6, 1899, 48.

[29] Ibid., 8, 1901, 145–7.

[30] Ibid., 8, 1901, 278.

[31] Ibid., 11, 1904, 44 (9, 1902, 109, 183–5).

[32] Ibid., 9, 1902, 311–2.

[33] Ann. Soc. Sc. Bruxelles, 27, II, 1902–3, 121–170. Extract in Sphinx-Oedipe, 5, 1910, 10–13 (of numéro spécial), 42–6.

two cases z can be even without x and y being even.   That the solution of the equation is effected by different formulas according as z is even or odd is shown by the case $c = 47$.   Then all relatively prime solutions in which z is odd are

$$z = f^2 + 47g^2, \qquad x = f(f^2 - 141g^2), \qquad y = g(3f^2 - 47g^2),$$

where f and g are relatively prime and one even.   All solutions of $x + 47y^2 = (2u)^3$, where u is odd, are

$$x = 13f^3 + 60f^2g - 168fg^2 - 144g^3,$$
$$y = f^3 - 12f^2g - 24fg^2 + 16g^3, \qquad u = 3f^2 + 2fg + 16g^2,$$

with similar expressions when $z = 4u$, $8u$, $16u$, etc.   The cases $c = 35$, $c = 499$ are treated (p. 142, p. 155)

Pepin[34] noted that $2x^3 = 3y^2 - 1$ has the solution $x = 61$, $y = 389$, but left undecided the question of an infinitude of solutions.   One of two methods is based on the theorem that all relatively prime solutions of $2x^3 = 3y^2 - z^2$ are given by

$$x = f^2 - 3g^2, \qquad y = fA + 3gB \text{ or } 3fA - 15gB, \qquad z = fA + 9gB \text{ or } -5fA + 27gB,$$

where $A = f^2 + 9g^2$, $B = f^2 + g^2$.   It remains to find f, g such that $z = \pm 1$.

G. de Longchamps[35] stated that $px^2 + qy^2 = z^3$ always has integral solutions.   [In fact, a solution is $x = \alpha t$, $y = \beta t$, $z = t \equiv p\alpha^2 + q\beta^2$.]

H. Brocard[36] listed the known values of a for which $x^3 - y^2 = a$ is impossible and the values for which there is a single solution.

E. B. Escott,[37] A. Cunningham and R. F. Davis[38] treated $x^2 - 17 = y^3$.

A. S. Werebrusow[39] expressed Euler's[6] solution of $x^2 + cy^2 = z^3$ in terms of $\alpha = -2p$, $\beta = -p^2 + 3cq^2$.

Several[40] solved $x^2 + 3y^2 = 4z^3$ completely by use of identities.

U. Bini[41] gave a solution of $x^2 + 3y^2 = z^3$ involving two parameters.

An anonymous writer[42] noted that $17y^2 - 1 = 2x^3$ has no solution with $1 < y \leqq 55$.

A. Cunningham[43] gave a tentative method to solve $x^3 = y^2 + a$.   Choose a modulus m, preferably 10^3 or 10^4, and find the values $< m$ of x for which $x^3 - a$ is a quadratic residue of m.   By use of various m's we finally get the possible linear forms of x.   Application is made to $a = -17$, $a = -127$.

Several[44] solved $x^2 + x \pm 1 = y^3$.

Welsch[45] applied the theory of binary quadratic forms to justify Legendre's[7] determination by use of $\sqrt{-3}$ of all solutions of $x^2 + 3y^2 = z^3$.

[34] Nouv. Ann. Math., (4), 3, 1903, 422–8.

[35] L'intermédiaire des math., 9, 1902, 115.

[36] *Ibid.*, 10, 1903, 284.

[37] *Ibid.*, 12, 1905, 43–45.   Amer. Math. Monthly, 26, 1919, 239–41.   Cf. Brocard.[11, 55]

[38] Math. Quest. Educ. Times, (2), 8, 1905, 53–4.   Cf. Cunningham.[43]

[39] L'intermédiaire des math., 10, 1903, 152.   Cf. E. B. Escott, 11, 1904, 101–2.

[40] *Ibid.*, 14, 1907, 168; 18, 1911, 279.

[41] *Ibid.*, 14, 1907, 192.

[42] Sphinx-Oedipe, 1906–7, 79.

[43] Math. Quest. Educ. Times, (2), 14, 1908, 106–8.

[44] L'intermédiaire des math., 15, 1908, 244; 16, 1909, 201; 17, 1910, 126; 23, 1916, 4.

[45] *Ibid.*, 17, 1910, 179–180.

E. B. Escott[46] noted that, if $y^3 = 2x^2 - 1$ is solvable, $y = 24n^2 - 1$ or $2n^2 - 1$.

L. Aubry[47] proved that $x^2 + 1 + 2^{2k} = y^3$ is impossible. If x is odd, $x^2 + 2^{2k}$ is a sum of two relatively prime squares, so that the factors of $y^3 - 1$ are $\equiv 1 \pmod 4$. Thus $y - 1 \equiv 1$, which gives $y^2 + y + 1 \equiv 3 \pmod 4$. If $x = 2^n z$, where z is odd,

$$2^{2n}\{(2^{k-n})^2 + z^2\} = (y-1)(y^2 + y + 1).$$

Since $y^2 + y + 1$ is odd, its prime factors are of the form $4t + 1$. Thus $y - 1$ is divisible by 2^{2n} and hence by 4. Again, $y^2 + y + 1 \equiv 3 \pmod 4$.

L. Aubry and E. Fauquembergue[48] proved that $2x^2 - 1 = y^3$ has no solutions other than $x = 0$, $y = -1$; $x = \pm 1$, $y = 1$; $x = \pm 78$, $y = 23$.

A. Gérardin[49], to make $G \equiv x^2 + xy + y^2$ a cube, assumed that

$$(1 + mx)^2 + (1 + mx)(my) + (my)^2 = (1 + mf)^3,$$
$$f^3 m^2 + (3f^2 - G)m = -3f + 2x + y,$$

and took $-3f + 2x + y = 0$. Then f and m are expressed in terms of x, y. To make the result symmetrical, set $y = q/3$, $x = p + q/3$. Hence

$$X^2 + XY + Y^2 = Z^3$$

for

$$X = q^3 + 3pq^2 - p^3, \qquad Y = -3pq(p+q), \qquad Z = p^2 + pq + q^2,$$

a result obtained otherwise by A. Desboves.[50]

Gérardin[51] treated $aX^2 + bXY + cY^2 = hZ^3$, given one solution α, β, γ. After substituting $X = \alpha + mx$, $Y = \beta + my$, $Z = \gamma + mf$, equate the coefficients of the first powers of m (by choice of f); thus m is determined rationally.

L. Aubry[52] proved that 25 is the only square which increased by 2 gives a cube [Fermat[2]]. He[53] proved that $x^2 + a = y^3$ is impossible for $a = 4A^2 + B^3$ if $B \equiv 1 \pmod 4$ and A is not divisible by the square of a prime $4n - 1$ dividing B, or by 3 if B is not divisible by 3, or by 3^3 if B is divisible by 3. Hence it is impossible for $a = 17$.

E. Landau[54] proved that $x^3 + 2 = y^2$ has only a finite number of solutions by means of Thue's result that $\alpha^3 + 3\alpha^2\beta + 6\alpha\beta^2 + 2\beta^3 = 1$ has only a finite number of solutions (Thue[9] of Ch. XXIII), and Landau's[29] discussion above.

H. Brocard[55] gave eight sets of solutions of $x^2 - y^3 = 17$.

L. J. Mordell[56] investigated $y^2 - k = x^3$ by elementary methods, by the theory of ideals, and by the arithmetical theory of binary cubic forms. In particular, he listed the values of k between -100 and 100 for which he believed there is an infinitude of solutions.

[46] Amer. Math. Monthly, 16, 1909, 96.

[47] Sphinx-Oedipe, 6, 1911, 26–27; stated by F. Proth, Nouv. Corresp. Math., 4, 1878, 64, 223.

[48] Sphinx-Oedipe, 6, 1911, 103–4; 8, 1913, 170–1 (122–3 for E. B. Escott's proof that a solution $y > 23$ has more than 256 digits).

[49] Assoc. franç. av. sc., 40, 1911, 10–12.

[50] Nouv. Ann. Math., (2), 18, 1879, 269, formula (8) with $a = b = 1$.

[51] Bull. Soc. Philomathique, (10), 3, 1911, 222–5.

[52] Sphinx-Oedipe, 7, 1912, 84.

[53] L'intermédiaire des math., 19, 1912, 231–3.

[54] Ibid., 20, 1913, 154.

[55] Ibid., 62–3. Cf. Brocard.[11]

[56] Proc. London Math. Soc., (2), 13, 1914, 60–80.

A. Gérardin[57] summarized the known results on $x^3 - k = z^2$; he noted the solutions $2^3 - 4 = 2^2$, $5^3 - 4 = 11^2$, contrary to de Jonquières'[13] assertion that only one solution exists. Given one solution x_0, z_0, Gérardin deduced (*ibid.*, 163-5) the second solution [Réalis[17]]

$$x = \{3x_0^2/(2z_0)\}^2 - 2x_0.$$

Set $x_0 = 2p$, where p is a prime. Then $z_0 = p^j$, $2p^j$, $3p^j$, $6p^j$ ($j = 0, 1, 2$). There result twelve integral values of k for which the given equation is solvable. For $k = (2p - 1)^2(9p^2 - 2p + 1)$, the solutions include $x = 2p$, $2p - 1$, $2 - 4p$, $4p^2 - 2p$, $(12p^2 - 6p + 1)^2 - 4p + 1$.

L. Bastien[58] listed the values $3, 5, 6, 9, 10, 12, 14, 16, 17, \cdots, 99$ of $n \leqq 100$ for which $q^3 - k^2 = n$ is impossible, the values $n = 1, 2, 8, 13, 29, \cdots$, 81 for which there is a single solution, and the values for which there are more than one solution.

Crussol[59] noted cases when $x^3 + k = y^2$ has $7, 9, 34$ and 41 solutions. A. Gérardin (*ibid.*, p. 16) noted cases when it has 21 solutions.

A. Gérardin[60] proved that all solutions of $x^2 + 3y^2 = z^3$ are given by

$$(\alpha^3 - 9\alpha\beta^2)^2 + 3(3\alpha^2\beta - 3\beta^3)^2 = (\alpha^2 + 3\beta^2)^3.$$

T. Hayashi[61] proved that $y^2 + 1 \neq z^3$ for $y \neq 0$; $y^2 - 1 \neq z^3$ for $y^2 \neq 0, 1, 9$.

A. Cunningham[62] proved that, if p is prime, $x^3 - p^2 = 2 \cdot 10^6$ has the single solution $x = 129$, $p = 383$.

L. J. Mordell[62a] noted that no equation $x^2 + a = y^3$ has an infinitude of integral solutions.

For $2x^2 + 2x + 13 = y^3$, see paper 161 of Ch. I. On $27b^2 + 1 = 4c^3$, see Kronecker[23] of Ch. XXI.

Binary Quadratic Form Made an nth Power.

J. L. Lagrange[63] noted that the mth power of $f = x^2 + axy + by^2$ can be expressed in the same form $F = X^2 + aXY + bY^2$ by employing the factors $x + \alpha y$, $x + \beta y$ of f and taking $X + \alpha Y$ to be the expansion of $(x + \alpha y)^m$. The resulting values of X, Y make F an mth power.

L. Euler[64] stated that he used this method for $f = x^2 + ny^2$ in the first edition of his algebra.[6]

Euler[65] noted that, if $N = a^2 + nb^2$, N^λ is of the form $x^2 + ny^2$, and asked for the least $x \neq 0$ or least $y \neq 0$ for which $N^\lambda = x^2 + ny^2$. Let

$$(a + b\sqrt{-n})^\lambda = A + B\sqrt{-n}, \qquad a = \sqrt{N}\cos\phi, \qquad b\sqrt{n} = \sqrt{N}\sin\phi.$$

[57] Sphinx-Oedipe, 8, 1913, 145-9.

[58] *Ibid.*, 9, 1914, 15-16.

[59] *Ibid.*, 43-44.

[60] L'intermédiaire des math., 21, 1914, 129.

[61] Nouv. Ann. Math., (4), 16, 1916, 150-5.

[62] Math. Quest. and Solutions (3), 3, 1917, 74.

[62a] London Math. Soc. Records of Proceedings, Nov. 14, 1918.

[63] Addition IX to Euler's Algebra, Lyon, 2, 1774, 636-644; Euler's Opera Omnia, (1), 1, 1911, 638-643; Oeuvres de Lagrange, VII, 164-170. For $f = x^2 - By^2$, Lagrange, Mém. Acad. Sc. Berlin, 23, année 1767; Oeuvres, II, 522-4.

[64] Opera postuma, I, 1862, 571-3, letter to Lagrange, Jan., 1770; Oeuvres de Lagrange, XIV, 214.

[65] Nova Acta Acad. Petrop., 9, 1791 (1777), 3; Comm. Arith., II, 174-182.

Then
$$a+b\sqrt{-n}=\sqrt{N}(\cos\phi+i\sin\phi),\qquad A=N^{\lambda/2}\cos\lambda\phi,\qquad B=\{N^{\lambda/2}\sin\lambda\phi\}/\sqrt{n}.$$
Hence B is a minimum $\neq 0$ for a rational value of λ approximately equal to $\pi k/\phi$, where k is an integer.

Euler[66] made x^2+7 a biquadrate. For $x=(7p^2-q^2)/(2pq)$, it is the square of $(q^2+7p^2)/(2pq)$. To make the latter a square, take $q=pz$, whence we are to make $2z(7+z^2)=\square$. Since an evident solution is $z=1$, set $z=1+y$. We get $16+20y+6y^2+2y^3$, which is the square of $4+5y/2$ for $y=1/8$.

A. M. Legendre[67] treated $Ly^2+Myz+Nz^2=bP$, where P is a product of powers of several variables, in particular, x^k.

G. L. Dirichlet[68] recalled that if l is an odd prime not dividing a and if $\delta^2-a\epsilon^2=l$ it is known that $d^2-ae^2=l^n$ holds for the numbers d, e given by $d+e\sqrt{a}=(\delta+\epsilon\sqrt{a})^n$. It is proved that d, e are relatively prime. If also $d_1^2-ae_1^2=kl^n$, where d_1, e_1 are relatively prime, and k is odd and prime to al, we can find solutions of $t^2-au^2=k$ such that
$$(d\pm e\sqrt{a})(t\pm u\sqrt{a})=d_1+e_1\sqrt{a}$$
for a suitable choice of signs. Application is made to show that, if P, Q are relatively prime, the most general manner of making P^2-5Q^2 a fifth power, odd and not divisible by 5, is to set
$$P+Q\sqrt{5}=(M\pm N\sqrt{5})^5(t\pm u\sqrt{5}),\qquad t^2-5u^2=1,$$
where M, N are relatively prime, one even and M not divisible by 5. If P, Q are relatively prime, both odd, and Q is divisible by 5, the most general way to make $P^2-5Q^2=4z^5$ is to set
$$P+Q\sqrt{5}=(\phi+\psi\sqrt{5})^5/16,$$
where ϕ, ψ are relatively prime, both odd, and ϕ is prime to 5.

Cauchy's papers on the representation of p^k or $4p^k$, where p is a prime, by x^2+ny^2 will be considered under binary quadratic forms. Luce[127] of Ch. XII discussed $x^2-ny^2=z^i$.

F. Landry[69] obtained a new kind of continued fraction from
$$A=a^2+r,\qquad \sqrt{A}=a+\frac{r}{\sqrt{A}+a}=a+\frac{r}{2a+}\frac{r}{2a+}\cdots.$$
If m/n is a convergent of order u, $m^2-An^2=(-1)^u r^u$. Hence to solve $x^2-Ay^2=z^m$, take as z any integer for which $A=a^2-z$.

V. A. Lebesgue[70] recalled the fact that, if a is an odd prime and A is an odd integer dividing t^2+a, but not divisible by a, $A^u=x^2+ay^2$ holds for an infinitude of values μ, when x, y are relatively prime. The least μ is

[66] Algebra, St. Petersburg, 2, 1770, § 160; French transl., Lyon, 2, 1774, pp. 191–3; Opera Omnia, (1), I, 413.

[67] Théorie des nombres, 1798, 435–40; ed. 2, 1808, 374–9; ed. 3, 1830, II, 43–49; German transl. by Maser, II, 43–50.

[68] Jour. für Math., 3, 1828, 354; Werke, I, 21.

[69] Cinquième mémoire sur la théorie des nombres, Paris, July, 1856.

[70] Jour. de Math., (2), 6, 1861, 239–240.

said to be even if A is a quadratic non-residue of a or if $A = 4n+3$, $a = 4k+1$. When $\mu = 2\nu$, y is odd. Then $A'-x = p^2$, $A'+x = aq^2$, where p^2 and aq^2 are relatively prime. Hence $2A' = p^2 + aq^2$ and ν is a minimum.

L. Öttinger[71] tabulated solutions of $x^2 - y^2 = z^n$, $n = 2, 3, 4$, and gave the identity

$$\{(4m^2 \pm 2mr + r^2)^2 - 8m^4\}^2 - \{4m(m \pm r)(2m^2 \pm 2mr + r^2)\}^2 = (2mr \pm r^2)^4.$$

T. Pepin[72] proved that, if c is positive and such that there is a single class of positive odd quadratic forms of determinant $-c$ (as for $c = 1, 2, 3, 4, 7$), the most general manner of solving $x^2 + cy^2 = z^m$, where x, y are to be relatively prime integers and z odd, is to set

$$(p + q\sqrt{-c})^m = P + Q\sqrt{-c}, \qquad x = \pm P, \qquad y = \pm Q, \qquad z = p^2 + cq^2,$$

where p, q are any relatively prime integers for which z is odd. Hence we can justify the method of Euler for $c = 1$ or 2. Next (pp. 333–8), let n be a positive integer such that all the quadratic forms of determinant $-n$ are distributed into various genera each composed of a single class; then all relatively prime solutions of $x^2 + ny^2 = z^{2m+1}$, with z odd, are obtained from

$$(1) \qquad\qquad \pm x \pm y\sqrt{-n} = (p + q\sqrt{-n})^{2m+1},$$

where p, q are relatively prime. But for $x^2 + ny^2 = z^{2m}$, z odd, we use (1) with the exponent m, and employ the complete solution

$$z = \frac{af^2 + bg^2}{k}, \qquad p = \frac{af^2 - bg^2}{k}, \qquad q = \frac{2fg}{k} \quad (k = 1, 2)$$

of $p^2 + nq^2 = z^2$, where for a, b are to be taken all the decompositions of n into two relatively prime factors, except that when $k = 2$, $n = 8t$, the two factors shall have 2 as their g.c.d. For $ax^2 + cy^2 = z^m$, $a > 1$, $c > 1$ (pp. 339–343), when ac is one of the numbers for which the number of classes of quadratic forms of determinant $-ac$ equals the number of genera, there is no solution in integers $\neq 0$ if m is even; while if m is odd we get all relatively prime solutions with z odd from

$$\pm \sqrt{a}\,x \pm \sqrt{-c}\,y = (\sqrt{a}\,p + \sqrt{-c}\,q)^m, \qquad z = ap^2 + cq^2,$$

where p, q are relatively prime. Thus $2x^2 + 3$ and $2 + 3y^2$ can not equal cubes.

Pepin[15] proved that $x^5 + a$ is not a square if $a = 32(2d+1)^5 - 5b^2$, where b is prime to 10 and has no prime factor $20l+11$. If $d = 0$, $b = 1, 3$, then $a = 27, -13$.

M. d'Ocagne[73] solved $x^2 - ky^2 = z^n$ in positive integers by use of

$$\phi(\alpha, \beta, n) = \sum_{i=0}^{[(n-1)/2]} \binom{n-i-1}{i} \alpha^{n-2i-1}\beta^i.$$

[71] Archiv Math. Phys., 49, 1869, 211–222.
[72] Jour. de Math., (3), 1, 1875, 325. Results for $m = 3$ cited under Pepin.[10]
[73] Comptes Rendus Paris, 99, 1884, 1112.

A solution involving an arbitrary positive integer a is

$$x = a\phi(2a, k-a^2, n) + (k-a^2)\phi(2a, k-a^2, n-1), \qquad y = \phi(2a, k-a^2, n),$$

$z = \pm(k-a^2)$ for n even; $z = -(k-a^2)$, $a > \sqrt{k}$, for n odd.

M. Weill[74] repeated Euler's[6] method for $ax^2 + cy^2 = z^n$.

T. Pepin[75] proved that, if the number of classes of quadratic forms of determinant $-c$ is relatively prime to n, all relatively prime integral solutions of $x^2 + cy^2 = z^n$ are given by

$$\pm x \pm y\sqrt{-c} = (p + q\sqrt{-c})^n, \qquad z = p^2 + cq^2.$$

For $n = 3$, the solvability depends upon whether or not the triplication of a quadratic form gives the principal class.

H. S. Vandiver[76] found an infinitude of, but not all, solutions of $x^2 + bxy + cy^2 = z^n$.

G. Candido[77] employed Lucas' functions U_n, V_n:

$$V_n = (p + \sqrt{p^2 - q})^n + (p - \sqrt{p^2 - q})^n, \qquad (\tfrac{1}{2}V_n)^2 - (p^2 - q)U_n^2 = q^n.$$

Change q to $p^2 - q$. Thus $x^2 - qy^2 = z^n$ has the solution $x = \tfrac{1}{2}V_n$, $y = U_n$, $z = p^2 - q$.

A. Cunningham[78] noted that $y^2 + y + 1 = x^n$ is impossible if $n > 3$, $x^n < 2\cdot 10^8$. R. W. D. Christie stated that x must be of the form $a^2 + a + 1$ and inferred that $n \neq 4$, 5, $n \neq 3$ unless $a = 2$.

Cunningham[79] noted that the only solution of $\tfrac{1}{2}(q^2 + 1) = p^4$ with $q < 1600000$ is $q = 239$, $p = 13$. Christie obtained this solution by making special assumptions. Cf. Störmer[137-9] of Ch. VI; Euler[7] of Ch. XIV; Euler[53] and Pepin[58] of Ch. XXII.

U. Bini[80] stated that the method of Desboves[142] of Ch. XXIII [cf. Lagrange[63]] does not lead to the determination of the form of the solutions of $x^2 + axy + by^2 = z^n$ for every integer n.

A. S. Werebrusow[81] gave polynomials X, Y of degree n in x, y making

$$AX^2 + 2BXY + CY^2 = (ax^2 + 2bxy + cy^2)^n.$$

E. B. Escott[82] noted that solutions of $X^2 - DY^2 = 4Z^n$ are given by

$$(\alpha^n + \beta^n)^2 - D\left\{\frac{\alpha^n - \beta^n}{\sqrt{D}}\right\}^2 = 4(x^2 - Dy^2)^n, \qquad \alpha, \beta = x \pm y\sqrt{D}.$$

But not all solutions are so obtained.[83]

O. Degel[84] treated the homogeneous equation obtained from the last one by replacing X, Y, Z by x_i/x_4 ($i = 1, 2, 3$). The section C by $x_2 = 0$

---

[74] Nouv. Ann. Math., (3), 4, 1885, 189.

[75] Mem. Accad. Pont. Nuovi Lincei, 8, 1892, 41–72.

[76] Amer. Math. Monthly, 9, 1902, 112.

[77] Giornale di Mat., 43, 1905, 93–6. Cf. Candido.[87]

[78] Math. Quest. Educ. Times, (2), 8, 1905, 69–70.

[79] Ibid., 9, 1906, 23–24; 14, 1908, 77.

[80] L'intermédiaire des math., 14, 1907, 246.

[81] Ibid., 15, 1908, 153; Mat. Sbornik, 22.

[82] L'intermédiaire des math., 15, 1908, 153.

[83] Ibid., 17, 1910, 2, 137–8, 229–30.

[84] Ibid., 17, 1910, 253–5.

lies on the cone $x_1^2x_4^{n-2}-4x_3^n=0$, which every plane $x_3=\mu x_4$ cuts in two lines having $x_1=\pm2\sqrt{\mu^n}\,x_4$. We get rational coördinates of the general point P on C by taking μ to be a square if n is odd. For example, let $n=2m$. The general point on the line joining $P=(2\mu^m,\,0,\,\mu,\,1)$ and $(p,\,1,\,0,\,0)$ is $(2\mu^m+\delta p,\,\delta,\,\mu,\,1)=(x)$, which is on the surface if $\delta=4p\mu^m/(D-p^2)$ and gives rational solutions x_i. The same problem was treated by others.[85]

F. Ferrari[86] made $f\equiv x^2+axy+by^2$ an nth power. Let $f=(x+\alpha y)(x+\beta y)$. A sufficient condition is $x+\alpha y=(r+\alpha s)^n$. The latter becomes linear in α by use of $\alpha^2-a\alpha+b=0$. Hence we get $x,\,y$ as polynomials in $r,\,s,\,a,\,b$.

G. Candido[87] used Lucas' $u_k,\,v_k$ satisfying

$$(\tfrac{1}{2}v_k)^2-\left(\frac{p^2}{4}-q\right)u_k^2=q^k$$

to show that for $p=2\lambda+a\mu$, $q=\lambda^2+a\mu\lambda+b\mu^2$, an infinitude of solutions of $x^2+axy+by^2=z^k$ is given by $2x=v_k-a\mu u_k$, $y=\mu u_k$, $z=q$. The explicit formulas are given in the cases $k=2,\,3,\,4$ and for $a=0$ or $b=0$.

F. Ferrari[88] used, as had Lagrange, the expansion of $(a_1+ia_2\sqrt{\alpha}\,)^n$ to find A's such that $A_{k1}^2+\alpha A_{k2}^2=(a_1^2+\alpha a_2^2)^k$.

E. Swift[89] proved that the number $n(n-3)/2$ of diagonals of an n-gon is not a biquadrate.

By Thue[211] of Ch. XXVI, $x^2-h^2=ky^n$ $(n>2)$ has only a finite number of solutions. On $1+y^2\neq x^n$, see Lebesgue[68] of Ch. VI. On $1+2y^2=3^k$, see Fauquembergue[158] of Ch. XXIII. On $1\pm4x^n=\square$, see papers 7, 8, 169 of Ch. XXVI.

$$a_1x_1^2+\cdots+a_nx_n^2\ \text{Made a Cube or Higher Power.}$$

S. Réalis[90] noted that $u_1^2u_2=\alpha(u_1x_2-vx_1)^2+\beta(u_1y_2-vy_1)^2+\gamma(u_1z_2-vz_1)^2$ if

$$u_i=\alpha x_i^2+\beta y_i^2+\gamma z_i^2,\qquad v=2(\alpha x_1x_2+\beta y_1y_2+\gamma z_1z_2).$$

J. Neuberg[91] took $x_2=x_1$, $y_2=y_1$, $z_2=-z_1$ in the preceding result to get

$$\alpha x^2+\beta y^2+\gamma z^2=(\alpha x_1^2+\beta y_1^2+\gamma z_1^2)^3,\qquad \frac{x}{x_1}=\frac{y}{y_1}=\alpha x_1^2+\beta y_1^2-3\gamma z_1^2,$$

$$\frac{z}{z_1}=3\alpha x_1^2+3\beta y_1^2-\gamma z_1^2.$$

E. N. Barisien[92] noted that any sixth power is the sum of two squares diminished by a third:

$$n^6\equiv\{(n+2)(n^2-2n-2)\}^2+\{4n(n+1)\}^2-\{2(n+1)(n+2)\}^2.$$

[85] L'intermédiaire des math., 18, 1911, 35.
[86] Periodico di Mat., 25, 1909-10, 59-66. Cf. Lagrange.[63]
[87] *Ibid.*, 27, 1912, 265-8. Cf. Candido.[77]
[88] *Ibid.*, 28, 1913, 71-8.
[89] Amer. Math. Monthly, 23, 1916, 261-2.
[90] Nouv. Corresp. Math., 4, 1878, 325.
[91] Mathesis, 1, 1881, 74.
[92] Le matematiche pure ed applicate, 2, 1902, 35-36.

An identity (p. 253) shows that 4 times the cube of any even integer is a ▣ less a ▣.

G. de Longchamps[93] noted that $\alpha x^2 + \beta y^2 + \gamma z^2 + \delta t^2 = u^3$ for

$$\frac{x}{f} = \frac{y}{g} = \alpha f^2 + \beta g^2 - 3\gamma i^2 - 3\delta k^2, \qquad \frac{z}{i} = \frac{t}{k} = 3\alpha f^2 + 3\beta g^2 - \gamma i^2 - \delta k^2,$$

$$u = \alpha f^2 + \beta g^2 + \gamma i^2 - \delta k^2.$$

The case $\delta = t = 0$ gives Neuberg's result.

An anonymous writer[94] noted the solution $x = 3$, $y = 12$, $z = 11$, $u = 2$ of $x^2 + y^2 - z^2 = u^5$.

J. Rose[95] noted the solution $x = 4v^2$, $y = 4v^3$, $z = 4v^2(v-1)$, $u = 2v$, and a solution with $y = z + 1$.   Mehmed-Nadir gave the solution

$$x = b(a^2 + b^2)(a^2 - b^2); \qquad z,\, y = \tfrac{1}{2}\{(a^2 \pm 1)(a^2 + b^2)^2 \pm 4b^4\}; \qquad u = a^2 + b^2;$$

and noted that the same x, u, with $Y = a(a^2 + b^2)^2$, $Z = 2ab^2(a^2 + b^2)$, satisfy $x^2 + Y^2 + Z^2 = u^5$.

"V. G. Tariste"[96] stated that all sets of solutions of $x^2 + y^2 - z^2 = u^5$ are given by seven sets of formulas like $u = 4a$, $x = 2b$, $u^5 - x^2 = 4\alpha\beta$; $y,\, z = \alpha \pm \beta$.

F. L. Griffin and G. B. M. Zerr[96a] discussed $x_1^2 + \cdots + x_n^2 = y^4$.

W. H. L. Janssen van Raay[97] solved $x^3 = x^2 + y^2 + z^2$.

G. Candido[98] found a solution of $\Sigma x_i^2 = y^p$ by expanding $\Pi(\alpha_j^2 + \beta_j^2)$.

R. D. Carmichael[99] gave a four-parameter solution of $x^2 + ay^2 + bz^2 = w^4$.

[93] L'intermédiaire des math., 10, 1903, 111–2.
[94] *Ibid.*, 14, 1907, 244.
[95] *Ibid.*, 15, 1908, 46.
[96] *Ibid.*, 19, 1912, 38.
[96a] Amer. Math. Monthly, 17, 1910, 147–8.
[97] Wiskundige Opgaven, 12, 1915, 209–11.
[98] Periodico di Mat., 30, 1915, 45–47.
[99] Diophantine Analysis, New York, 1915, 46.

CHAPTER XXI.

EQUATIONS OF DEGREE THREE.

Impossibility of $x^3+y^3=z^3$.

According to Ben Alhocain a defective proof was proposed before 972 by the Arab Alkhodjandi.[1]

The Arab Beha-Eddin[2] (1547–1622) listed among the problems remaining unsolved from former times that to divide a cube into two cubes.

Fermat[3] stated that it is impossible to decompose a cube into two cubes.

Fermat proposed the problem to find two cubes whose sum is a cube to Sainte-Croix Sept., 1636 (Oeuvres de Fermat, II, 65; III, 287), to Frenicle May(?), 1640 (Oeuvres, II, 195), to the mathematicians of England and Holland Aug. 15, 1657 (Oeuvres, II, 346; III, 313). Oddly enough, Frans van Schooten[4] proposed Feb. 17, 1657, the same problem to Fermat. Fermat[5] insisted that the problem is impossible.

Frenicle[6] proposed the equivalent problem to find r central hexagons, with consecutive sides, whose sum is a cube. By a central hexagon of n sides he meant the number

$$H_n = 1+6+2\cdot6+3\cdot6+\cdots+(n-1)6 = n^3-(n-1)^3.$$

The sum of $H_n, H_{n-1}, \cdots, H_{n-r+1}$ is thus a cube z^3 if and only if

$$n^3 = (n-r)^3+z^3.$$

J. Kersey[6a] stated that J. Wallis proved that no rational cube equals a sum of two rational cubes, but gave no reference.

L. Euler[7] stated Aug. 4, 1753 that he had proved the problem impossible.

Euler[8] gave the following proof, incomplete at one point. We may assume that x and y are relatively prime and both odd. Set $x+y=2p$, $x-y=2q$. Then we are to prove that $2p(p^2+3q^2)$ is not a cube. Suppose that it is a cube. First, let p be not divisible by 3. Then $p/4$ and p^2+3q^2

---

[1] F. Woepcke, Atti Accad. Pont. Nuovi Lincei, 14, 1860–1, 301.

[2] Essenz der Rechenkunst von Mohammed Beha-eddin ben Alhossain aus Amul, arabisch u. deutsch von G. H. F. Nesselmann, Berlin, 1843, p. 55. French transl. by A. Marre, Nouv. Ann. Math., 5, 1846, 313, Prob. 4; ed. 2, Rome, 1864. Cf. A. Genocchi, Annali di Sc. Mat. e Fis., 6, 1855, 301, 304.

[3] Observation 2 on Diophantus (quoted in full in Ch. XXVI on Fermat's last theorem). Oeuvres de Fermat, I, 291; III, 241. The problem was sent (1637?) by Fermat to Mersenne to be proposed to St. Croix; cf. P. Tannery, Bull. des sc. math., (2), 7, 1883, 8, 121–3.

[4] Correspondance of Huygens, No. 378, Oeuvres complètes de Chr. Huygens, 2, 1889, 17; Oeuvres de Fermat, 3, p. 558.

[5] Oeuvres, II, 376, 433, letter to Digby, Apr. 7, 1658, to Carcavi, Aug. 1659.

[6] Solutio duorum problematum . . . 1657 [lost]; Oeuvres de Fermat, III, 605, 608.

[6a] The Elements of Algebra, London, Book III, 1674, 73.

[7] Corresp. Math. Phys. (ed., Fuss), 1, 1843, 618. Also stated in Novi Comm. Acad. Petrop., 8, 1760–1, 105; Comm. Arith. Coll., I, 287, 296; Opera Omnia, (1), II, 557, 574.

[8] Algebra, 2, 1770, Ch. 15, art. 243, pp. 509–16; French transl., 2, 1774, pp. 343–51; Opera Omnia, (1), I, 484–9. Reproduced by A. M. Legendre, Théorie des nombres, 1798, 407–8; ed. 3, 1830, II, 7; transl. by Maser, II, 9.

are relatively prime integers, so that each is a cube. Since p^2+3q^2 is a cube, he stated without rigorous proof (cf. Ch. 12, Arts. 188–191) that it is the cube of a number t^2+3u^2 of like form and that $p+q\sqrt{-3}$ is the cube of $t+u\sqrt{-3}$. [Cf. papers 6, 10, 27–29, 72 of Ch. XX; also 30, 36 and 183 below.]

Hence $p=t(t^2-9u^2)$, $q=3u(t^2-u^2)$. But also $p/4$ shall be a cube. The same is true of the product $2p$ of $2t$, $t+3u$, $t-3u$, which are relatively prime since p and hence t is not divisible by 3. Thus the last two are cubes, f^3 and g^3, whence $2t=f^3+g^3$. Thus we have two cubes f^3, g^3, much smaller than x^3, y^3, whose sum is a cube $2t$. A similar method of descent is used in the remaining case $p=3r$, when the product of the relatively prime numbers $9r/4$ and $3r^2+q^2$ is a cube. As before, $r=3u(t^2-u^2)$. Since

$$\frac{8}{27}\cdot\frac{9r}{4}=\frac{2r}{3}=2u(t+u)(t-u)$$

is a cube and is the product of three relatively prime factors, each factor is a cube: $t+u=f^3$, $t-u=g^3$, so that f^3-g^3 is a cube $2u$.

J. A. Euler[9] noted that, if $p^3+q^3+r^3=0$ is possible, $x=p^2q$, $y=q^2r$, $z=r^2p$ satisfy $x/y+y/z+z/x=0$ or $x^2z+y^2x+z^2y=0$. In attempting to prove the latter impossible, he stated that yx is divisible by z, but admitted in a note that one can only conclude that the denominator of the irreducible fraction equal to y/z is a divisor of xy. For $v=xy/z$, we get $x/y+v/x+y/v=0$, $v<x$. Continuing, we get solutions in smaller integers.

L. Euler[10] noted that $p^3+q^3=r^3$ implies $AB(A+B)=1$ for $A=p^2/(qr)$, $B=q^2/(pr)$. Set $A=\alpha B$. Then $B^3\alpha(\alpha+1)=1$, whereas $\alpha(\alpha+1)$ is not a cube.

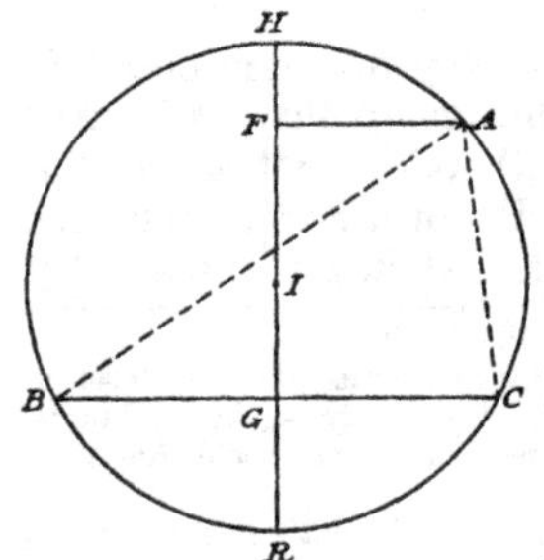

N. Fuss I[11] noted that $a^3=b^3+c^3$ implies that $a^6-4b^3c^3=(b^3-c^3)^2$. Conversely, $a^6-4d^3=\square$ implies $a^3=p^2+pq^3$ (since the square root of A^2-dB^2 is of the form p^2-dq^2), whence $p=r^3$, $p+q^3=$ cube.

J. Glenie[12] constructed on a given right line BC as base a triangle ABC such that $AB^3+AC^3=BC^3$. Through the mid point G of BC draw a perpendicular GH to it and take

$$GH=BC\,\frac{3\sqrt{5}}{2\sqrt{31}},\qquad GF=BC\,\frac{\sqrt{5\cdot31}}{24}.$$

Draw the circle HBC; let it cut the parallel FA to BC at A. Without proof he stated that ABC is the required triangle.

To make $AB^3+AC^3=2BC^3$ or $3BC^3$ (Probs. 2, 3), take

$$GH=BC\sqrt{\tfrac{15}{11}},\qquad GF=BC\cdot\tfrac{4}{9}\sqrt{\tfrac{11}{15}},\qquad\text{or}\qquad GH=\tfrac{3}{2}BC,\qquad GF=\tfrac{1}{2}BC.$$

He treated the corresponding three problems on the difference of cubes.

---

[9] L. Euler's Opera postuma, I, 1862, 230–1 (about 1767).

[10] Ibid., 236–7 (about 1769).

[11] Ibid., 242 (about 1778).

[12] The Antecedental Calculus . . . and the Constructions of Some Problems, London, 1793, 16 pp., p. 13.

A. G. Kästner[13] checked the construction by use of trigonometric functions and logarithmic tables.

I. K. Hagner[14] set $a=BC$, $b=GH$, $c=GF$. Then

$$GR=\frac{a^2}{4b},\qquad FA^2=FR\cdot HF=(b-c)\frac{(a^2+4bc)}{4b}.$$

Having GA^2, we see that BA and AC are

$$\tfrac{1}{2}\sqrt{4bc+a^2}\pm\tfrac{1}{2}a\sqrt{\frac{b-c}{b}},$$

$$BA^3+AC^3=\left\{\frac{(4b^2-3a^2)c+4a^2b}{4b}\right\}\sqrt{4bc+a^2}.$$

Equating this to a^3, and writing $4bc+a^2=(a+2f)^2$, which gives $c=(a+f)f/b$, we get

$$b^2=\frac{3a^2(a+f)(a+2f)}{4\{2a^2+(a+f)(a+2f)\}}.$$

By the expression for BA, we must have $b>c$, whence $f<a(0.29\cdots)$. The value $f=a/4$ gives Glenie's solution. Taking $f=(k-3/2)a$, we see that the expression for b^2 is the square of $\{3-6k+5k^2/2\}a/(4k^2-6k+6)$ if $k=24/23$, whence $b=\pm5a/38$. If in Euler's[8] equation $2p(p^2+3q^2)=z^3$, we set $2p=rz$, we obtain q, whence

$$x,\,y=\frac{rz}{2}\pm\frac{z\sqrt{4-r^3}}{2\sqrt{3r}}$$

and see why the cubic equation is solved by use of a curve of order 2. For $r=3/2$, we get Glenie's case.

C. F. Hauber[15] proved Glenie's construction and solved

$$x^3+y^3=\frac{p}{q}\,a^3,\qquad x+y=\frac{m}{n}\,a$$

for x, y and discussed their geometric constructibility, but made no discussion as to rationality.

J. W. Becker[16] gave a construction simpler than Glenie's, as he avoided irrationals. Take a circle of radius $IR=152$, lay off $RG=124$, $RF=279$, draw perpendiculars FA and BC to IR to cut the circle at the vertices A, B, C of the required triangle (see above figure). For Prob. 2, take $IR=639$, $RG=198$, $RF=550$. For Prob. 3, take $IR=5$, $RG=1$, $RF=4$. In general, let the sum of the cubes of the sides equal e times the cube of the base a. Denote the sum of the sides by as, the difference by ad. Thus

$$a^3(\tfrac{1}{2}s+\tfrac{1}{2}d)^3+a^3(\tfrac{1}{2}s-\tfrac{1}{2}d)^3=a^3e,\qquad d=\sqrt{\frac{4e-s^3}{3s}}.$$

He asked if s can be chosen to make d rational, stating it to be impossible

[13] Archiv der reinen u. angewandten Math. (ed., Hindenburg), 1, 1795, 352–6, 481–7.
[14] Ibid., 2, 1798, 448–457.
[15] Ibid., pp. 458–70.
[16] Ibid., 471–80.

if $e=1$. For $e=2$, take $s=2$, whence $d=0$ and the triangle is equilateral. No general discussion was given.

C. F. Kausler[17] gave a complex and inconclusive argument to show that $x^3 \pm y^3$ is not a cube. His first theorem is that $x-y$ and x^2+xy+y^2 are not both cubes; the proof rests on Euler's[8] lemma about p^2+3q^2 a cube.

C. F. Gauss[18] proved by descent that $x^3+y^3+z^3=0$ is impossible in integers, using an imaginary cube root of unity.

P. Barlow[19] gave an erroneous proof [Barlow[15] of Ch. XXVI].

A. M. Legendre[20] proved that the even one of x, y, z is divisible by 3 and then by descent that $x^3+y^3=(2^m 3^n u)^3$ is impossible, where u is not divisible by 2 or 3.

Schopis[21] undertook a proof of the impossibility of

$$(x+y)^3 - x^3 = \text{cube},$$

in integers. If the equation holds, then $y^3 Q = \text{cube}$, $Q=z^3$, where

$$Q = \frac{3x^2}{y^2} + \frac{3x}{y} + 1.$$

Solving for y, we get

$$y = \frac{3x \pm x\sqrt{12z^3 - 3}}{2(z^3 - 1)}.$$

Thus $12z^3 - 3 = w^2$. The quotient of w^2+3 by 12 must be an integer, whence $w=6n+3$, and

$$z^3 = 3n^2 + 3n + 1.$$

He stated that the second member is a cube only when $n=0$ or -1, whence $z=1$, and the denominator of y would be zero.

L. Calzolari[22] attempted to prove the equation impossible.

L. Kronecker[23] noted that the theorem that $r^3+s^3=1$ has no rational solutions with $rs \neq 0$ is equivalent to the fact that $4a^3+27b^2=-1$ has no rational solutions other than $a=-1$, $b=\pm 1/3$. The latter are the only values of the coefficients of a cubic $x^3+ax+b=0$ with rational coefficients and discriminant unity.

G. Lamé[24] noted that, if x and y are relatively prime, x^3+y^3 is the product of two relatively prime factors δ, q, where δ is $D=x+y$ or $3D$ according as D is not or is divisible by 3, and q is of the form A^2+3B^2. Then if a sum of two cubes is a cube, we transpose the single even cube and get $x^3+y^3=(2z)^3$,

[17] Nova Acta Acad. Petrop., 13, ad annos 1795-6 (1802), 245-54.

[18] Werke, Il, 1863, 387-390, posthumous MS. Quoted, Nouv. Corresp. Math., 4, 1878, 136.

[19] Theory of Numbers, London, 1811, 132-140.

[20] Mém. Acad. Roy. Sc. de l'Institut de France, 6, année 1823, 41, §49 (=Suppl. 2 to Théorie des nombres, ed. 2, 1808). Théorie des nombres, ed. 3, 1830, art. 653, pp. 357-60; German transl. by Maser, II, 348.

[21] Einige Sätze aus der unbestim. Analytik, Progr. Gumbinnen, 1825. Repeated in Zeitschr. Math. Naturw. Unterricht, 23, 1892, 269-270.

[22] Tentativo per dimostrare il teorema di Fermat . . . , Ferrara, 1855; Extract by D. Gambioli, Periodico di Mat., 16, 1901, 155-8.

[23] Jour. für Math., 56, 1859, 188; Werke, I, 121.

[24] Comptes Rendus Paris, 61, 1865, 921-4, 961-5. Extract in Sphinx-Oedipe, 4, 1909, 43-4.

whence δ and q must be cubes.   It is stated that (cf. Euler[8])

$$q=(a^2+3b^2)^3=A^2+3B^2,\qquad A=a(a^2-9b^2),\qquad B=3b(b^2-a^2).$$

In

$$A^2+3B^2=\frac{(A+B)^3+(A-B)^3}{(A+B)+(A-B)}=\frac{(3B+A)^3+(3B-A)^3}{18B},$$

$\delta=2A$ or $18B$, according as $x+y$ is not or is divisible by 3.   But a and $3b$ are relatively prime and not both odd.   Hence δ is a cube only if $a=4k^3$, $a-3b=i^3$, $a+3b=j^3$; or $b=4k^3$, $b-a=i^3$, $b+a=j^3$, in the respective cases. In either case, $j^3+i^3=(2k)^3$ and i, j, k are smaller than x, y, z.   He noted numerical results like

$$(7^3+2^3)(8^3-7^3)=39^3,\qquad (43^3-36^3)(54^3-5^3)=(12^3+1)^3=(10^3+9^3)^3.$$

P. G. Tait[25] noted that $x^3+y^3=z^3$ implies

$$(x^3+z^3)^3y^3+(x^3-y^3)^3z^3=(z^3+y^3)^3x^3$$

and said that this leads easily to a proof of the impossibility of integral solutions of the former equation.   Every cube is a difference of two squares of which one is divisible by 9 since

$$x^3=\left[\frac{x(x+1)}{2}\right]^2-\left[\frac{x(x-1)}{2}\right]^2.$$

T. Pepin[26] proved the impossibility of $x^3+y^3=z^3$.

S. Günther[27] showed how the square root occurring in the solution x, y of $x^3+y^3=a^3$, $x+y=z$, can be replaced by a cube root which is " absolutely irreducible."

J. J. Sylvester[28] gave a proof of the impossibility.

R. Perrin[29] showed how one (hypothetical) set of integral solutions of $a^3+b^3+c^3=0$ leads to a new set of integral solutions.

Schuhmacher[30] stated that Euler[8] erred in affirming that $p+q\sqrt{-3}$ must be the cube of $t+u\sqrt{-3}$, since it might be $\alpha^\lambda(t+\alpha u)^3$, where $\alpha^3=1$. He argued that the first of Euler's two cases may be dispensed with.

J. Sommer[31] proved Kummer's[63] result (Ch. XXVI) that $x^3+y^3=z^3$ is not solvable in integral numbers of the domain defined by a cube root of unity.

H. Krey[32] made the impossibility proof by use of the theory of quadratic forms.   Set $f(x, y)=x^2-xy+y^2$.   Then $2f$ is an improperly primitive form of determinant -3 and of class number 1.   We can represent properly by f any positive odd number, not divisible by 3, all of whose prime factors p have -3 as a quadratic residue.   If (u, v) is a representation of m, and (u', v') of m', then

$$(uu'+vv'-uv',\quad uu'+vv'-vu')$$

[25] Proc. Roy. Soc. Edinburgh, 7, 1869–70, 144 (in full).
[26] Jour. de Math., (2), 15, 1870, 225–6.
[27] Sitzungsber. Böhm. Ges. Wiss., Prag, 1878, 112–9.
[28] Amer. Jour. Math., 2, 1879, 393; Coll. Math. Papers, 3, 1909, 350.
[29] Bull. Math. Soc. France, 13, 1884–5, 194–7.   Reprinted, Sphinx-Oedipe, 4, 1909, 187–9.
[30] Zeitschrift Math. Naturw. Unterricht, 25, 1894, 350.
[31] Vorlesungen über Zahlentheorie, 1907, 184–7.
[32] Math. Naturwiss. Blätter, 6, 1909, 179–180.

is a representation of mm'. Taking $u'=v$, $v'=u$, we get $m^2=f(2uv-u^2, 2uv-v^2)$. First, if $x+y$ is not divisible by 3, it is relatively prime to $f=(x+y)^2-3xy$, so that it and f are cubes. By the above,

$$m^3=f(u^3-u^2v+uv^2, \quad v^3-uv^2+u^2v).$$

When this is taken as f, the sum u^3+v^3 of the arguments is a cube (corresponding to $x+y$). Thus the method of descent applies. The case in which $x+y$ is a multiple of 3 leads by a like argument to a descent.

P. Bachmann[33] amplified the proofs by Euler[8] and Legendre.[20]

R. Fueter[34] proved that if $\xi^3+\eta^3+\zeta^3=0$ is solvable by numbers $\neq0$ of an imaginary quadratic domain $k(\sqrt{m})$, where $m<0$, $m\equiv2$ (mod 3), then the class number of k is divisible by 3. It is solvable in the real domain $k(\sqrt{-3m})$ if and only if solvable in $k(\sqrt{m})$. In particular, Kummer's result that it is not solvable in $k(\sqrt{-3})$ is a consequence of the fact that it is not solvable in rational numbers. To give a direct proof, let $\alpha^3+\beta^3=z^3$, α, $\beta=\frac{1}{2}(x\pm y\sqrt{-3})$, where x, y, z are integers distinct from 0, and set α^3, $\beta^3=\frac{1}{2}(X\pm Y\sqrt{-3})$, $z^3=X$. Then

$$\left(\frac{X+Y}{2}\right)^3+\left(\frac{X-Y}{2}\right)^3=\left(z\cdot\frac{x^2+3y^2}{4}\right)^3.$$

If m and n are integers prime to 3, the domain defined by a cube root of m^3+27n^3 has its class number a multiple of 3, and $\Sigma\xi^3=0$ is solvable.

W. Burnside[35] discussed the solution of $x^3+y^3+z^3=0$ in quadratic domains.

R. D. Carmichael[36] gave a series of lemmas leading to a proof of the fact, stated by Euler,[8] that $p^2+3q^2=s^3$ (p, q relatively prime, s odd) implies that s is of the form t^2+3u^2, etc.

Further proofs by Holden[80]; also Korneck,[149] Stockhaus[231], and Rychlik[232] of Ch. XXVI.

TWO EQUAL SUMS OF TWO CUBES.

Diophantus, V, 19, mentioned without details the theorem in the *Porisms* that the difference of two cubes is always a sum of two cubes (cf. p. 607).

P. Bungus[37] remarked that while a square is often the sum of two squares, a cube is first composed of three cubes, citing $6^3=3^3+4^3+5^3$.

F. Vieta[38] required two cubes whose sum equals the difference B^3-D^3 of two given cubes ($B>D$). Call $B-A$ the side of the first required cube and B^2A/D^2-D the side of the second. Thus $(B^3+D^3)A=3D^3B$ and hence

$$(1) \qquad x^3+y^3=B^3-D^3, \qquad x=\frac{B(B^3-2D^3)}{B^3+D^3}, \qquad y=\frac{D(2B^3-D^3)}{B^3+D^3}.$$

<hr>

[33] Niedere Zahlentheorie, 2, 1910, 454–8.

[34] Sitzungsber. Akad. Wiss. Heidelberg (Math.), 4, A, 1913, No. 25.

[35] Proc. London Math. Soc., (2), 14, 1914, 1.

[36] Diophantine Analysis, 1915, 67–70.

[37] Numerorum Mysteria, 1591, 1618, 463; Pars Altera, 65.

[38] Zetetica, 1591, IV, 18–20; Opera Mathematica, ed. by Frans van Schooten, Lugd. Batav., 1646, 74–75. A wrong sign in (2) is corrected on p. 554.

Using the same sides for (2); sides $A-D$, $D^2 A/B^2 - B$ for (3), he got

$$(2) \qquad x^3 - y^3 = B^3 + D^3, \qquad x = \frac{B(B^3 + 2D^3)}{B^3 - D^3}, \qquad y = \frac{D(2B^3 + D^3)}{B^3 - D^3};$$

$$(3) \qquad x^3 - y^3 = B^3 - D^3, \qquad x = \frac{D(2B^3 - D^3)}{B^3 + D^3}, \qquad y = \frac{B(2D^3 - B^3)}{B^3 + D^3}.$$

C. G. Bachet,[39] in his commentary on Diophantus IV, 2 (to solve $x - y = g$, $x^3 - y^3 = h$), gave Vieta's results (1)–(3). He was able to express the difference of two given cubes as a sum of two positive cubes only when the greater of the given cubes exceeds the double of the smaller.

A. Girard[39a] noted that, if $D^3 > \frac{1}{2} B^3$ in (1), we first apply (3) repeatedly until we obtain two cubes the smaller of which is less than one-half the larger, and then use (1).

Fermat[40] noted that in the case $B^3 < 2D^3$, expressly excluded by Bachet, we can make $B^3 - D^3$ a sum of two positive cubes. Let, for example, $B = 5$, $D = 4$. By Vieta's formula (3), we get

$$5^3 - 4^3 = \left(\tfrac{248}{63}\right)^3 - \left(\tfrac{5}{63}\right)^3.$$

Of the new cubes, the first exceeds the double of the second. Hence their difference is a sum of two cubes by (1). Thus 5^3–4^3 is the sum of two positive cubes, "which would doubtless astonish Bachet." Further, if we employ the three formulas in succession, and repeat the operations indefinitely, we obtain an infinitude of pairs of cubes satisfying the same conditions; for, from the two cubes whose sum equals the difference of the given cubes, we can find by (2) two new cubes whose difference equals the sum of our two cubes and hence equals the difference of the two original cubes; from this new difference of two cubes we pass to a sum of two cubes, and so on indefinitely. The condition $B^3 < 2D^3$ imposed by Bachet on (3) is not necessary; being given the cubes 8 and 1, we can find two new cubes with the same difference. Bachet would doubtless say that this is impossible. Nevertheless I have found that[41]

$$\left(\tfrac{1265}{183}\right)^3 - \left(\tfrac{1256}{183}\right)^3 = 8 - 1.$$

Further, after what precedes, I solve happily the problem (not known by Bachet): To separate the sum of two cubes into two new cubes, and indeed in an infinitude of ways. Thus to find two cubes whose sum is $8 + 1$, I first seek by (2) two cubes 8000/343 and 4913/343 whose difference is $8 + 1$. As the double of the smaller exceeds the larger, we apply (3) and afterwards (1) and obtain the solution. If we wish a second solution, we apply (2), etc."

Fermat[42] proposed as a new problem to Brouncker, Wallis and Frenicle: Given a number composed of two cubes, to divide it into two other cubes.

---

[39] Diophanti Alex. Arith., 1621, 179–182, 324.

[39a] L'arith. de Simon Stevin . . . annotations par A. Girard, Leide, 1625, 635; les Oeuvres Math. de Simon Stevin de Bruges par A. Girard, 1634, 159.

[40] Oeuvres, I, 297–9; French transl., III, 246–8.

[41] By (1), $8 - 1 = (4/3)^3 + (5/3)^3$. Then apply (2) for $B = 5/3$, $D = 4/3$.

[42] Oeuvres, II, 344, 376; letters from Fermat to Digby, Aug. 15, 1657; Apr. 7, 1658.

He would be content if Brouncker would divide $8+1$ into two other rational cubes.

Without indicating his method, Frenicle[43] gave the solutions

$$9^3+10^3=1^3+12^3, \qquad 9^3+15^3=2^3+16^3, \qquad 15^3+33^3=2^3+34^3,$$
$$16^3+33^3=9^3+34^3, \qquad 19^3+24^3=10^3+27^3.$$

J. Wallis[44] gave 22 additional solutions

$$27^3+30^3=3^3+36^3, \qquad (4\tfrac{1}{2})^3+(7\tfrac{1}{2})^3=1^3+8^3, \ \cdots.$$

" If these do not suffice, I will furnish as many as he wishes; and so easily that in an hour I would promise a hundred . . . ." Letter XXVI contains Frenicle's reply; he points out that all of Wallis' solutions were obtained from the known solutions by simple multiplication or division. " You should therefore not be astonished that he agrees so readily to furnish a hundred such combinations in an hour; what is easier than to multiply or divide small numbers? Indeed, it would be still easier to indicate the divisions, not making the reductions, unless he wished to disguise more his artificial solutions." Frenicle added that it would have been easy to give essentially new solutions and then cited 13 such (Oeuvres de Fermat, III, 535). Wallis (p. 538, letter XXVIII) claimed that Frenicle had been guilty of the same fault.

Wallis (p. 599, letter XLIV, June 30, 1658) was not more fortunate[45] in regard to Fermat's problem to express 9 as the sum of two positive cubes; he expressed 9 as the difference of the cubes of 20/7 and 17/7, and said that the method to employ to express 9 as the sum of two cubes would be to find in a table of cubes two whose sum is 9 times a cube! Vieta and Bachet had found no difficulty in expressing B^3+D^3 as a difference of two cubes, but had not attacked the more difficult problem $x^3+y^2=B^3+D^3$.

J. Prestet[46] treated the problem to find two cubes whose sum equals the difference of two given cubes (even when the smaller exceeds one-half the greater), using first (3) and then (1). To find two cubes whose difference is the sum B^3+D^3 of two given cubes, solve (2), then $z^3+v^3=x^3-y^3$, and then $t^3-f^3=z^3+v^3$. To find two cubes whose difference is B^3-D^3, solve (1) and then $z^3-v^3=x^3+y^3$.

L. Euler[47] noted that there exist integral solutions of

$$(4) \qquad\qquad\qquad A^3+B^3+C^3=D^3.$$

Euler[48] derived Vieta's formula (2) and noted that it does not give all the solutions. For $B=4$, $D=3$, we have $37y=465$, $37x=472$, whereas

[43] Commercium Epistolicum de Wallis, letter X, Brouncker to Wallis, Oct. 13, 1657; French transl. in Oeuvres de Fermat, III, 419–420.
[44] Commercium, letter XVI, Wallis to Digby, Nov., 1657. Oeuvres de Fermat, III, 436.
[45] Cf. Frenicle, letter to Digby, Oeuvres de Fermat, III, 605, 609.
[46] Nouveaux elemens des math., Paris, 2, 1689, 260–1.
[47] Corresp. Math. Phys. (ed., Fuss), 1, 1843, 618, Aug. 4, 1753.
[48] Novi Comm. Acad. Petrop., 6, 1756–7, 155; Comm. Arith., I, 193; Op. Om., (1), II, 428. Reproduced without reference by E. Waring, Meditationes Algebr., ed. 3, 1782, 325.

there exists the simpler solution $x=6$, $y=5$. To treat (4), he set

(5) $\qquad\qquad A=p+q, \qquad B=p-q, \qquad C=r-s, \qquad D=r+s.$

Thus

(6) $\qquad\qquad\qquad\qquad p(p^2+3q^2)=s(s^2+3r^2).$

Taking

$$p=ax+3by, \qquad q=bx-ay, \qquad s=3cy-dx, \qquad r=dy+cx,$$

we have

$$p^2+3q^2=\beta(x^2+3y^2), \qquad s^2+3r^2=\gamma(x^2+3y^2), \qquad \beta=a^2+3b^2, \qquad \gamma=d^2+3c^2.$$

Hence our equation becomes $\beta(ax+3by)=\gamma(3cy-dx)$, whence

$$x=-3nb\beta+3nc\gamma, \qquad y=na\beta+nd\gamma.$$

Writing λ, $\mu=3ac\pm3bc\mp ad+3bd$, we get

$$A=n\lambda\gamma-n\beta^2, \qquad B=n\mu\gamma+n\beta^2, \qquad C=n\gamma^2-n\lambda\beta, \qquad D=n\gamma^2+n\mu\beta.$$

The abbreviators β, γ, λ, μ were not used by Euler; but their introduction[49] enables us to point out the identity which underlies his solution. In

$$A^3+B^3+C^3-D^3=n^3(\gamma^3-\beta^3)\{\lambda^3+\mu^3-3\beta\gamma(\lambda+\mu)\}$$
$$=n^3(\gamma^3-\beta^3)(\lambda+\mu)(\lambda^2-\lambda\mu+\mu^2-3\beta\gamma),$$

it is the final factor which vanishes, and this in view of the identity

$$\beta\gamma\equiv(3bc-ad)^2+3(ac+bd)^2=\left(\frac{\lambda-\mu}{2}\right)^2+3\left(\frac{\lambda+\mu}{6}\right)^2,$$

which in turn follows from

$$(c+b\sqrt{-3})(d+c\sqrt{-3})=ad-3bc+(ac+bd)\sqrt{-3}.$$

Euler noted (p. 206) that we may solve similarly $l\pi=\lambda\rho$, where $\pi=mp^2+nq^2$, $\rho=mr^2+ns^2$, while l, λ are any linear functions of p, q, r, s, by setting

$$p=nfx+gy, \qquad q=mfy-gx, \qquad r=nhx+ky, \qquad s=mhy-kx.$$

Then

$$\pi=(g^2+mnf^2)(nx^2+my^2), \qquad \rho=(k^2+mnh^2)(nx^2+my^2).$$

Hence x/y is ratonal.

Euler[50] treated (4) by setting, without loss of generality,

$$A=(m-n)p+q^2, \qquad B=(m+n)p-q^2,$$
$$C=p^2-(m+n)q, \qquad D=p^2+(m-n)q.$$

Then $(A+B)(A^2-AB+B^2)=(D-C)(D^2+DC+C^2)$ becomes, after division by $2m(p^3-q^3)$, $m^2+3n^2=3pq$. Thus $m=3k$, where $pq=n^2+3k^2$. But he had proved in the same paper that every divisor of n^2+3k^2, in which n and k are relatively prime, is of like form. Thus

$$p=a^2+3b^2, \qquad q=c^2+3d^2, \qquad m=3(bc\pm ad),$$

while n is $ac\mp3bd$ or its negative.

[49] L. E. Dickson, Amer. Math. Monthly, 18, 1911, 110–111.
[50] Novi Comm. Acad. Petrop., 8, annees 1760–1, 1763, 105; Comm. Arith., I, 287; Opera Omnia, (1), II, 556.

Euler[51] deduced Vieta's formula (2) and noted that in (6) the second factors have a common divisor of like form t^2+3u^2. From

$$(7) \qquad p^2+3q^2 = (f^2+3g^2)(t^2+3u^2), \qquad s^2+3r^2 = (h^2+3k^2)(t^2+3u^2),$$

he concluded that

$$(8) \qquad p=ft+3gu, \qquad q=gt-fu, \qquad s=ht+3ku, \qquad r=kt-hu.$$

Inserting the values of p, s and (7) into (6) and deleting the common factor t^2+3u^2, we obtain t/u rationally. To avoid fractions, take u equal to the denominator. Thus

$$(9) \qquad u=f(f^2+3g^2)-h(h^2+3k^2), \qquad t=3k(h^2+3k^2)-3g(f^2+3g^2).$$

For f, g, h, k arbitrary, formulæ (5), (8), (9) give the general solution of (4). Special cases are

$$7^3+14^3+17^3=20^3, \qquad 11^3+15^3+27^3=29^3, \qquad 1^3+6^3+8^3=9^3, \qquad 3^3+4^3+5^3=6^3.$$

W. Emerson[52] repeated Vieta's discussion and treated the problem to find three cubes whose sum is both a cube and a square. Cf. Hill[84] of Ch. XXIII.

J. P. Grüson[53] gave (1).

S. Jones[54] deduced (1) and (2).

J. R. Young[55] passed from (4) to (6) as had Euler. Set $p=m^2$, $s=n^2$. Then (6) becomes

$$3n^6+9r^2n^2-3m^6=9m^2q^2=(c-3rn)^2, \text{ if } r=\frac{c^2-3n^6+3m^6}{6cn}.$$

Take $m=1$, $n=2$, $c=3d$ and drop the common denominator $4d$. Hence

$$(d^2+16d-21)^3+(16d-d^2+21)^3+(2d^2-4d+42)^3=(2d^2+4d+42)^3.$$

He also solved (4) by taking[56] $A=m-1$, $B=n^2-p$, $C=n^2+p$, $D=m+1$, whence $9m^2=3n^2p^2+3(n^6-1)=(q-3np)^2$, say. Hence

$$np, \quad m=\{q^2\mp3(n^6-1)\}/(6q).$$

Multiplying the resulting values of A, $\cdots$, D by $6nq$, we get

$$A, D=n\{q^2\mp6q+3(n^6-1)\}; \qquad B, C=\mp q^2+6n^3q\pm3(n^6-1).$$

F. T. Poselger[57] treated the transformation of a sum or difference of two cubes into a difference or sum of two positive cubes.

J. P. M. Binet[58] expressed Euler's[48] solution of

$$(10) \qquad x^3+y^3=z^3+u^3$$

[51] Algebra, 2, 1770, arts. 245, 248; French transl., 2, 1774, pp. 351, 360. Opera Omnia, (1), I, 490–7.

[52] A Treatise of Algebra, London, 1764, 1808, 382–4.

[53] Enthüllte Zaubereyen und Geheimnisse der Arith., Berlin, 1796, 125–8, and Zusatz at end of Theil I.

[54] The Gentleman's Diary, or Math. Repository, London, No. 90, 1830, 38–9.

[55] Algebra, 1816, S. Ward's edition, 1832, 351–2. Reproduced, Math. Mag., 2, 1895, 154–5.

[56] Reproduced, Math. Mag., 2, 1898, 254.

[57] Akad. Wiss. Berlin Math. Abhandl., 1832, 27–31.

[58] Comptes Rendus Paris, 12, 1841, 248–50. Reprinted, Sphinx-Oedipe, 4, 1909, 29–30.

in the explicit form

$$x = \rho^2 - \sigma\rho', \qquad y = \sigma'\rho' - \rho^2, \qquad z = \rho\sigma' - \rho'^2, \qquad u = \rho'^2 - \rho\sigma,$$
$$\rho = f^2 + 3g^2, \qquad \rho' = f'^2 + 3g'^2, \qquad \sigma,\ \sigma' = ff' + 3gg' \pm (3fg' - 3f'g).$$

He stated that we may set $f' = 1$, $g' = 0$ without loss of generality and hence express the general solution of (10) in the form

$$(11) \qquad x = k^2 - l, \qquad y = -k^2 + m, \qquad z = km - 1, \qquad u = -kl + 1,$$

where $k = a^2 + 3b^2$, $l = a - 3b$, $m = a + 3b$. We may take $\alpha = m/3$, $\beta = -l/3$ as new parameters in place of a, b, and get

$$x = 3\beta + 9t^2, \qquad y = 3\alpha - 9t^2, \qquad z = 9\alpha t - 1, \qquad u = 9\beta t + 1,$$

where $t = k/3 = \alpha^2 + \beta^2 - \alpha\beta$. The case $\alpha = \beta = 1$ gives $3^3 + 4^3 + 5^3 = 6^3$.

*V. Bouniakowsky[59] treated (4).

C. Richaud[60] noted that in $(x+1)^3 - x^3 = y^3 + z^3$, $y + z$ is of the form $t^2 + 3u^2$, whence $2x = -1$, $2y = s + v$, $2z = s - v$, where

$$t^2 - sv^2 = \frac{s^3 - 1}{3}.$$

From one solution of the last equation we get the second solution

$$t' = \frac{(s+1)t + 2sv}{s-1}, \qquad v' = \frac{2t + (s+1)v}{s-1}.$$

Hence from one solution a, $b = d - 1$, c, d of (4), by replacing x, y, z by $d - 1$, c, a, and hence t, s, v by $2d - 1$, $c + a$, $c - a$, respectively, we get another solution:

$$A = \frac{a(a+c) - c - 2d + 1}{a + c - 1}, \qquad B = \frac{(a+c)(c+d-a-1) + d}{a+c-1} = D - 1,$$

$$C = \frac{c(a+c) - a + 2d - 1}{a + c - 1}, \qquad D = \frac{(a+c)(c+d-a) + d - 1}{a+c-1},$$

since $A = \frac{1}{2}(s - v')$, $C = \frac{1}{2}(s + v')$, $D = \frac{1}{2}(t' + 1)$. Thus the solution 3, 5, 4; 6 leads to 1, 8, 6; 9 and -8, 50, 29; 53.

H. Grassmann[61] reduced (10) to

$$\tfrac{1}{3}(a^3 - b^3) = bd^2 - ac^2,$$

by setting $x = a + c$, $y = a - c$, $z = b - d$, $u = b + d$, and stated that a/b must be a square, whence $a = m\alpha^2$, $b = m\beta^2$,

$$\tfrac{1}{3}m^2(\alpha^6 - \beta^6) = (\beta d + \alpha c)(\beta d - \alpha c).$$

Giving artibrary integral values to α, β, m, and expressing the left member as a product pq, we get d, c from $\beta d \pm \alpha c = p$, q.

C. Hermite[62] derived Binet's solution (11) of (10) from a general property of cubic surfaces. Let ω be an imaginary cube root of unity. The lines

[59] Memoirs Imper. Acad. Sc., St. Petersburg, 6, 1865, 142 (In Russian).
[60] Atti Accad. Pont. Nuovi Lincei, 19, 1865-6, 183-6.
[61] Archiv Math. Phys., 49, 1869, 49; Werke, 2, pt. I, 1904, 242-3. Error indicated by *A. Hurwitz, Jahresber. d. Deutschen Math.-Vereinigung, 27, 1918, 55-56.
[62] Nouv. Ann. Math., (2), 11, 1872, 5-8; Oeuvres, III, 115-7.

$x=\omega$, $y=\omega^2 z$ and $x=\omega^2$, $y=\omega z$ lie on the surface (10) with $u=1$. Each of these generators meets the line

$$x=az+b, \qquad y=pz+q$$

if

$$\frac{\omega-b}{a}=\frac{q}{\omega^2-p}, \qquad \frac{\omega^2-b}{a}=\frac{q}{\omega-p},$$

whence $p=b$, $q=(1+b+b^2)/a$, and the z-coördinates of the points of intersection are respectively

$$z_1=\frac{\omega-b}{a}, \qquad z_2=\frac{\omega^2-b}{a}.$$

The third root of $(az+b)^3+(pz+q)^3=z^3+1$ is

$$z=\frac{(1+b+b^2)^2-a^3(1-b)}{a(1-a^3-b^3)}.$$

Then also x and y are rational in a, b. To obtain simpler formulas, replace a by $1/a$, b by b/a. Then

(12) $\qquad sx=r(a+2b)-1, \qquad sy=r^2-a-2b, \qquad sz=r^2-a+b,$

where $r=a^2+ab+b^2$, $s=a^3-b^3-1$. Passing to the homogeneous equation (10) and changing b to $2b$, a to $a-b$, we get (11) with x, y, z, u replaced by z, $-y$, x, $-u$.

Several[63] expressed $8+27$ and $1+8$ as sums of two new rational cubes.

G. Korneck[64] stated that all integral solutions are obtained by taking positive and negative integers m, t, f in

$$x=6m^3tf+t(t\pm m)r+3t(t\mp m)f^2, \qquad y=6m^3tf-t(t\pm m)r-3t(t\mp m)f^2,$$
$$z=-6t^3mf+m(m\pm t)r+3m(m\mp t)f^2, \qquad u=6t^3mf+m(m\pm t)r+3m(m\mp t)f^2,$$

where $r=m^4+m^2t^2+t^4$.

E. Catalan[65] noted that (4) is satisfied identically by

$$A=(2x-1)(2x^3-6x^2-1), \qquad B=(x+1)(5x^3-9x^2+3x-1),$$
$$C=3x(x+1)(x^2-x+1), \qquad D=3x(2x-1)(x^2-x+1).$$

S. Réalis[66] proposed a problem which was solved by P. Sondat;[67] if α, β, γ, δ is one set of solutions of $x^3+y^3+u^3+v^3=0$, another set is

$$u=\alpha A-B, \qquad v=\beta A-B, \qquad x=\gamma A+B, \qquad y=\delta A+B,$$
$$A=\alpha+\beta+\gamma+\delta, \qquad B=\alpha^2+\beta^2-\gamma^2-\delta^2.$$

The new set yields similarly the given set, apart from a common factor.

G. Brunel[68] treated, for n an odd prime, the equation

(13) $$x_1^n+x_2^n=\begin{vmatrix} y_1 & y_2 & \cdots & y_{n-1} & 0 \\ 0 & y_1 & \cdots & y_{n-2} & y_{n-1} \\ \cdot & \cdot & \cdots & \cdot & \cdot \\ y_2 & y_3 & \cdots & 0 & y_1 \end{vmatrix}\equiv f(y_1, \cdots, y_{n-1}),$$

[63] Math. Quest. Educ. Times, 16, 1872, 95–6; 17, 1872, 84.
[64] Auflösung $x^3+y^3+z^3=u^3$ in ganzen Z., Progr. Kempen, 1873.
[65] Nouv. Corresp. Math., 4, 1878, 352–4, 371–3. Cf. Catalan.[123]
[66] Nouv. Ann. Math., (2), 17, 1878, 526; Nouv. Corresp. Math., 4, 1878, 350.
[67] Nouv. Ann. Math., (2), 18, 1879, 378.
[68] Mém. Soc. Sc. Phys. et Nat. de Bordeaux, (3), 2, 1886, 129–141.

the determinant being $y_1^3 + y_2^3$ if $n = 3$, and y_1^2 if $n = 2$. Proceeding as had Hermite[62] and considering the intersections of (13) with the general line in space of n dimensions

$$y_i = a_i x_1 + b_i x_2 \quad (i = 1, \cdots, n-1),$$

it is shown that the coördinates of any point on (13) are expressed rationally as functions of $n-1$ parameters $a_1, \cdots, a_{n-1}$:

$$x_1 = 1 - B, \qquad x_2 = A - 1, \qquad y_i = a_i(1 - B) + (a_{n-1} - a_{i-1})(1 - A)$$
$$(i = 1, \cdots, n-1),$$

where $a_0 = 0$, $b_1 = -a_{n-1}$, $b_i = a_{i-1} - a_{n-1}$ $(i = 2, \cdots, n-1)$,

$$A = f(a_1, \cdots, a_{n-1}), \qquad B = f(b_1, \cdots, b_{n-1}).$$

V. Schlegel[69] treated $a_1^3 + a_2^3 + a_3^3 = a_4^3$ by setting

$$a_1 + a_2 = m^2(a_4 - a_3), \qquad m^2 a_1 a_2 + a_4 a_3 = p^2 - q^2,$$
$$a_4 + a_3 + m(a_1 - a_2) = n(p - q), \qquad a_4 + a_3 - m(a_1 - a_2) = \frac{p+q}{n}.$$

These become, for $a_1 + a_2 = x$, $a_1 - a_2 = y$, $a_4 + a_3 = u$, $a_4 - a_3 = v$,

$$x = m^2 v, \qquad m^2(x^2 - y^2) + u^2 - v^2 = 4(p^2 - q^2),$$
$$u + my = n(p - q), \qquad u - my = \frac{p+q}{n}.$$

The last two give u, y; the second of the four becomes

$$\frac{3(p+q)}{mx - v} = \frac{mx + v}{p - q}.$$

Equate each member to r. We thus get x and v in terms of p, q, r, m. By $x = m^2 v$,

$$r^2 = \frac{3(p+q)(m^3 + 1)}{(p - q)(m^3 - 1)}.$$

For any m, we can choose $p \pm q$ to make r rational; then the a_i are rational.

A. Martin[70] gave Vieta's derivation of (1) with $B = r$, $D = -s$, and with $B = p$, $D = q$.

C. Moreau[71] gave the ten numbers $< 100,000$ which are sums of two positive cubes in two ways.

A. S. Werebrusow[72] gave the formula

$$(M\psi \mp \omega\phi^2)^3 + (-N\psi \pm \omega\phi^2)^3 = (M\phi \mp \omega\psi^2)^3 + (-N\phi \pm \omega\psi^2)^3,$$

where $M^2 + MN + N^2 = 3\omega^2 \phi\psi$, $\omega^3 = 1$ [Teilhet[78]].

K. Schwering[73] stated that the general solution of (10) is

$$(14) \qquad x = m\alpha - n^2, \qquad y = -m\beta + n^2, \qquad z = n\alpha - m^2, \qquad u = -n\beta + m^2,$$

where

$$(15) \qquad\qquad\qquad \alpha^2 + \alpha\beta + \beta^2 = 3mn.$$

[69] El Progreso Mat., 4, 1894, 169–171.
[70] Math. Magazine, 2, 1895, 153–4; Amer. Math. Monthly, 9, 1902, 79.
[71] L'intermédiaire des math., 5, 1898, 66 [253; 4, 1897, 286].
[72] Ibid., 9, 1902, 164–5; 11, 1904, 96, 289. Math. Soc. Moscow, 25, 1905, 417–437.
[73] Archiv Math. Phys., (3), 2, 1902, 280–4.

To get Binet's[58] solution, set $m=1$, $n=a^2+3b^2$, α, $\beta=a\mp3b$. By (14),

$$x^3+y^3-z^3-u^3=(m^3-n^3)(\alpha-\beta)(\alpha^2+\alpha\beta+\beta^2-3mn).$$

H. Kühne[74] expressed the preceding solution in terms of three independent parameters by replacing α by $3pr$, β by $3qr$, m by p^2+pq+q^2, n by $3r^2$, whence (15) is satisfied identically. Thus

$$x=3spr-9r^4, \qquad y=-3sqr+9r^4, \qquad z=9pr^3-s^2, \qquad u=-9qr^3+s^2,$$

where $s\equiv p^2+pq+q^2$, satisfy (10) identically. Not only do any p, q, r lead to a solution α, β, m, n of (15), but conversely, by multiplying them by a common factor, we can make $n/3$ a square, necessarily r^2, and then $p=\alpha/(3r)$, $q=\beta/(3r)$.

D. Mirimanoff[75] wrote (10), with $u=1$, in the form

$$(x-1)(x-\omega)(x-\omega^2)+y^3=z^3.$$

Set $y=u(x-\omega)+v(x-\omega^2)$, $z=u\omega^2(x-\omega)+v\omega(x-\omega^2)$, and divide by $(x-\omega)(x-\omega^2)$. We get

$$Dx=1+3(\omega^2-1)uv^2+3(\omega-1)u^2v, \qquad D=1+3(1-\omega)uv^2+3(1-\omega^2)u^2v.$$

Hence we get all solutions (except $x=\omega$, ω^2) by giving all values to u, v. Real solutions result if and only if $u+v$, $\omega^2u+\omega v$, $\omega u+\omega^2v$ are real, i. e., if u and v are conjugate. Writing b, a, $-a-b$ for these three sums, we obtain Hermite's solution (12).

A. Holm[76] derived (2) by the tangent method. Set $x=X+B$, $y=Y-D$ and take $Y=XB^2/D^2$. Then $X=0$ or $3BD^3/(B^3-D^3)$. The latter gives (2).

H. Kühne[77] discussed diophantine equations such that the n variables are expressible rationally in $n-1$ parameters. His[74] solution of (10) is an example of the method.

P. F. Teilhet[78] remarked that the solution by Werebrusow[72] is not the general one and stated that all solutions of (10) with $4(x-u)=3(z-y)$ are obtained by equating the two expressions

$$\left(\frac{21m^2+n^2\pm2mn}{2}\right)^3+\left(\frac{21m^2-n^2\mp16mn}{4}\right)^3,$$

or by equating the two

$$\left(\frac{3m^2+7n^2\pm2mn}{2}\right)^3+\left(\frac{3m^2-7n^2\mp16mn}{4}\right)^3,$$

where m, n are both even or both odd.

A. Gérardin[79] derived (2) from

$$\frac{x-B}{y+D}=\frac{y^2-Dy+D^2}{x^2+Bx+B^2}=m$$

by setting $x=B+mh$, $y=h-D$, and equating to zero the constant term of the quadratic for h. Thus $m=D^2/B^2$, $h=3B^3D/(B^3-D^3)$. Similarly for (1).

---

[74] Archiv Math. Phys., (3), 4, 1903, 180. Cf. Fujiwara.[85]

[75] Nouv. Ann. Math., (4), 3, 1903, 17–21.

[76] Proc. Edinburgh Math. Soc., 22, 1903–4, 43.

[77] Math. Naturwiss. Blätter, 1, 1904, 16–20, 29–33, 45–58. Cf. Kühne[169], Ch. XXIII.

[78] L'intermédiaire des math., 11, 1904, 31.

[79] Sphinx-Oedipe, 1906–7, 90–93, (52); l'intermédiaire des math., 16, 1909, 85.

H. Holden[80] obtained all integral solutions a, b, c, d of

$$a(a^2+pb^2)=c(c^2+pd^2),$$

for such values of p that any factor of a or c, not of the form l^2+pm^2, is a factor of both. This is true if there is a single properly primitive class of quadratic forms of determinant $-p$ and if, when there are improperly primitive classes, the highest power of 2 which divides l^2+pm^2 has an even exponent. The conditions hold for $p=1$, ±2, 3, -5, -13, -29, -53, -61. For $p=3$, we have the equivalent equation

$$(a+b)^3+(a-b)^3=(c+d)^3+(c-d)^3$$

and hence the complete solution of (10).   He proved that there is no integral solution of the initial equation with $a=b$ and hence none of $x^3=y^3+z^3$.

J. Jandasek[81] gave the identity

$$(3u^3+3u^2v+2uv^2+v^3)^3\equiv(3u^2v+2uv^2+v^3)^3+(uv^2)^3+(3u^3+3u^2v+2uv^2)^3.$$

K. Petr[82] noted that Euler's[48] solution of $x^3+y^3+z^3=u^3$ may be written in the form

$x:y:u:-z$

$$=A^2E+2BC-BD:-A^2E+BC+BD:B^2E+2AC-AD:-B^2E+AC+AD,$$

where C, D are arbitrary and $ABE^2=C^2-CD+D^2$. It is thus not essentially different from Binet's solution.

Binet's[58] solution is claimed[83] to be not general.

R. Norrie[84] treated (4) by taking $A=rx_1+\lambda$, $B=rx_2+\mu$, $C=rx_3-\mu$, $D=rx_0+\lambda$. Thus $\alpha r^3+3\beta r^2+3\gamma r=0$, where $\alpha=x_0^3-x_1^3-x_2^3-x_3^3$,

$$\beta=\lambda x_0^2-\lambda x_1^2-\mu x_2^2+\mu x_3^2, \qquad \gamma=\lambda^2 x_0-\lambda^2 x_1-\mu^2 x_2-\mu^2 x_3.$$

We may make $\gamma=0$ by choice of x_0. Then $\alpha r^3+3\beta r^2=0$ for $r=-3\beta/\alpha$. The resulting values of A, B, C, D in terms of $x_1, x_2, x_3, \lambda, \mu$ are of high degree and much more complicated than the complete solution by Euler[48] and Binet.[58]

M. Fujiwara[85] showed that the formulas by Schwering[73] and Kühne[74] can be deduced by simple substitutions from formula (11) of Euler and Binet.

A. Gérardin[86] gave the identities

$$(g^4\pm9f^3g)^3+(3f^2)^6\equiv(9f^4\pm3fg^3)^3+(g^2)^6,$$

$$(7\alpha^2-16\alpha\beta-3\beta^2)^3+(14\alpha^2+4\alpha\beta+6\beta^2)^3$$

$$=(14\alpha^2-4\alpha\beta+6\beta^2)^3+(7\alpha^2+16\alpha\beta-3\beta^2)^3,$$

and one similar to the latter.

---

[80] Messenger Math., 36, 1906–7, 189–192.

[81] Casopis, Prag, 39, 1910, 94–5.

[82] *Ibid.*, 40, 1911, 99–102.   In the Fortschritte report the sign before AD in u is wrong.

[83] L'intermédiaire des math., 18, 1911, 265–6; 19, 1912, 116.

[84] University of St. Andrews 500th Anniversary, Mem. Vol., Edinburgh, 1911, 50–1.

[85] Tôhoku Math. Jour., 1, 1911, 77–8; Archiv Math. Phys., (3), 19, 1912, 369.

[86] L'intermédiaire des math., 19, 1912, 7.   Cf. pp. 116–8 for references.   He gave the first in Assoc. franç. av. sc., 40, 1911, 12.

G. Osborn[87] gave Young's[55] identity and

$$(x^2-7xy+63y^2)^3+(8x^2-20xy-42y^2)^3+(6x^2+20xy-56y^2)^3$$
$$=(9x^2-7xy+7y^2)^3.$$

J. W. Nicholson,[88] using one solution of $m^3=n^3+p^3+r^3$, found that

$$(my-bx)^3=(ny-bx)^3+(py-ax)^3+(ry+ax)^3$$

holds if $x:y=m^2b-n^2b-p^2a+r^2a:mb^2-nb^2-pa^2-ra^2$.

J. E. A. Steggall,[89] to solve $x^3-u^3=y^3-v^3$, took $x-u=p$, $x+u=q$, $y-v=s$, $y+v=r$. Then (6) implies $p^2+3q^2=\mu s$, $s^2+3r^2=\mu p$, whence

$$(3qr)^2=(\mu s-p^2)(\mu p-s^2)=(ps-\mu k)^2,$$
$$\mu=\frac{p^3+s^3-2kps}{ps-k^2}, \qquad 3q^2=\frac{(s^2-kp)^2}{ps-k^2}.$$

Since $ps-k^2=3t^2$, we get $p+q=\{s^2+p(3t-k)\}/(3t)$, etc. Hence

$$x=\frac{L^2+p^3(3t-k)}{6tp^2}, \qquad y=\frac{p^4+pL(3t-k)}{6tp^2},$$
$$u=\frac{L^2-p^3(3t+k)}{6tp^2}, \qquad v=\frac{p^4-pL(3t+k)}{6tp^2},$$

where $L=k^2+3t^2$, is the most general rational solution.

R. D. Carmichael[90] obtained a rational solution, involving four parameters, of

$$x^3+y^3+z^3-3xyz=u^3+v^3+w^3-3uvw,$$

by employing the factor $x+y+z$ of the left member. Taking $z=w=0$, he deduced formulæ (11) of Euler and Binet, which he proved to give the general solution.

T. Hayashi[91] noted that C. Shiraishi published in his book of 1826 the solutions[91a] (attributed to Gokai Ampon) of $x^3+y^3+z^3=u^3$:

$$u=y+1, \quad z=3a^2, \quad x=6a^2\pm3a+1, \quad y=9a^3+6a^2+3a \quad \text{or} \quad 9a^3-6a^2+3a-1.$$

Replacing a by α/β and passing to the homogeneous form, we get

$$x=6\alpha^2\beta+3\alpha\beta^2+\beta^3, \qquad y=9\alpha^3+6\alpha^2\beta+3\alpha\beta^2, \qquad z=3\alpha^2\beta, \qquad u=y+\beta^3;$$

and in like manner

$$x=6\alpha^2\beta-3\alpha\beta^2+\beta^3, \qquad u=9\alpha^3-6\alpha^2\beta+3\alpha\beta^2, \qquad z=3\alpha\beta^2, \qquad y=u-\beta^3.$$

Further, S. Baba, Mathematics, vol. 2, 1830, gave the solution

$$x=(a^6-4)a, \qquad y=6a^3+a^6-4, \qquad z=a^6-6a^3-4, \qquad u=(a^6+8)a$$

of (10); S. Kaneko, Mathematics, vol. 2, 1845, gave the first solution of Frenicle.[43] Kawakita, in Algebraic Solutions, vol. 2, compiled from a

[87] Math. Gazette, 7, 1913–4, 361.
[88] Amer. Math. Monthly, 22, 1915, 224–5.
[89] Proc. Edinburgh Math. Soc., 34, 1915–6, 11–17.
[90] Diophantine Analysis, New York, 1915, 63–65.
[91] Tôhoku Math. Jour., 10, 1916, 15–27 (in Japanese).
[91a] For a briefer account, see D. E. Smith and Y. Mikami, A History of Japanese Mathematics, Chicago, 1914, 233–5.

manuscript by Baba, solved (10) by setting

$$x=a+b, \quad y=a-b, \quad z=bc, \quad u=d-bc, \quad 2a^3+6ab^2-d^3+3bcd^2-3b^2c^2d=0.$$

Take $a=c^2d/2$. Then $12bc=d(4-c^6)$. Take $c^3=\alpha$, $\alpha^2-4=\beta$, and multiply the resulting values of x, y, z, u by $12c/d$; we get

$$x=6\alpha-\beta, \qquad y=6\alpha+\beta, \qquad z=-\beta c, \qquad u=12c+\beta c \qquad (\alpha=c^3, \ \beta=\alpha^2-4).$$

M. Weill[92] noted that if x_i, y_i, z_i, u_i give two solutions of (10), we can evidently find δ rationally so that $x_1+\delta x_2$, $\cdots$, $u_1+\delta u_2$ is a solution. Given only one solution, we obtain a new solution $x_1+\rho t$, $y_1+\lambda t$, $z_1+\mu t$, $u_1+\nu t$, if $At^2+3Bt+3C=0$, where

$$A=\rho^3+\lambda^3-\mu^3-\nu^3, \qquad B=\rho^2 x_1+\lambda^2 y_1-\mu^2 z_1-\nu^2 u_1, \qquad C=\rho x_1^2+\lambda y_1^2-\mu z_1^2-\nu u_1^2.$$

We may choose λ, $\cdots$, ρ to make $C=0$ or $A=0$ and get t rationally.

For three consecutive cubes whose sum is a cube, see papers 245–267.

For minor results on our subject, see Schier[67] of Ch. XXIII.

Three equal sums of two cubes.

Fermat's[40] method of solution was given above.

W. Lenhart[93] found four integers the sum of any two of which is a cube. Three of the conditions are satisfied if x, m^3-x, n^3-x, r^3-x be taken as the numbers. The remaining conditions require that m^3+n^3-2x, m^3+r^3-2x, n^3+r^3-2x be cubes, say s^3, a^3, b^3. Eliminating x, we have

$$(1) \qquad\qquad r^3+s^3=a^3+n^3=b^3+m^3.$$

By his[186] table of numbers expressible as a sum of two cubes,

$$46969=\left(\tfrac{95}{7}\right)^3+\left(\tfrac{248}{7}\right)^3=\left(\tfrac{149}{12}\right)^3+\left(\tfrac{427}{12}\right)^3=\left(\tfrac{341899}{30291}\right)^3+\left(\tfrac{1081640}{30291}\right)^3.$$

Rejecting the common denominator, we get integers (one of 24 digits and three of 22 digits) solving the initial problem.

A. B. Evans[94] obtained the last result otherwise. By Euler,[51] for $f=7$, $g=k=14$, $h=16$,

$$1043^3+2989^3=1140^3+2976^3=7^3\cdot3^3\cdot2^6\cdot13\cdot3613.$$

Now $13\cdot3613=41^3-28^3$ can be expressed as a sum of two cubes by the usual method. The final answer involves numbers of 22 and 24 digits.

J. Matteson[95] obtained Lenhart's result by the method of Evans.

H. Brocard[96] noted that the sum of any two of the numbers $20012\tfrac{1}{2}$, $-15916\tfrac{1}{2}$, $19291\tfrac{1}{2}$, $-20020\tfrac{1}{2}$ is a cube. E. B. Escott[97] noted that 6044, 7780, -1948, -6052 have this property.

E. Fauquembergue[98] gave an erroneous solution of (1) with 5 parameters.

[92] Nouv. Ann. Math., (4), 17, 1917, 41–46.
[93] Math. Miscellany, New York, 1, 1836, 155–6.
[94] Math. Quest. Educ. Times, 15, 1871, 91–2.   His factor 2^3 should be 2^6.
[95] Collection Dioph. Problems, pub. by A. Martin, Washington, D. C., 1888, 1–4.
[96] L'intermédiaire des math., 8, 1901, 183–4.
[97] *Ibid.*, 9, 1902, 16.
[98] *Ibid.*, 9, 1902, 155; 10, 1903, 82 (Sphinx-Oedipe, 1906–7, 80, 125).

A. S. Werebrusow[99] gave the solution[72]

$$[(M+N)\psi\pm\omega\phi^2]^3+[-(M+N)\phi\mp\omega\psi^2]^3$$
$$=(-M\psi\pm\omega\phi^2)^3+(M\phi\mp\omega\psi^2)^3=(-N\psi\pm\omega\phi^2)^3+(N\phi\mp\omega\psi^2)^3,$$

in which $M^2+MN+N^2=3\omega^2\phi\psi$, $\omega^3=1$. He[100] noted that

$$(2) \qquad x^3+y^3=x_1^3+y_1^3=x_2^3+y_2^3$$

holds for

$$x_2=\frac{x_1^2y-x^2y_1}{xy-x_1y_1}, \qquad y_2=\frac{xy_1^2-x_1y^2}{xy-x_1y_1},$$

and the values derived from the latter by interchanging x_1, y_1. He[101] used this result to get the general solution of (2).

Fauquembergue[102] remarked that the last formula follows from the identity

$$(y_1^3-y^3)(x_1^2y-x^2y_1)^3+(x^3-x_1^3)(y_1^2x-y^2x_1)^3=(x^3y_1^3-y^3x_1^3)(xy-x_1y_1)^3,$$

due to A. Desboves[103], by taking $x^3+y^3=x_1^3+y_1^3$ and dividing the result by the product of $(xy-x_1y_1)^3$ by $x^3-x_1^3=y_1^3-y^3$.

A. Gérardin[104] stated that the least solution of (2) in integers >1 is probably $x=560$, $y=70$, $x_1=552$, $y_1=198$, $x_2=525$, $y_2=315$.

Fauquembergue[105] noted that if Cauchy's[287] formulas are applied to $x^3+y^3=19z^3$, which has the solution $x=3$, $y=-2$, $z=1$, we get

$$19=(\tfrac{8}{3})^3+(\tfrac{1}{3})^3=(\tfrac{5}{2})^3+(\tfrac{3}{2})^3=(\tfrac{92}{35})^3+(\tfrac{33}{35})^3=(\tfrac{27323}{10386})^3+(\tfrac{9613}{10386})^3=\cdots,$$

so that $19\cdot363510^3$ is a sum of two positive integral cubes in various ways.

SOLUTION OF $2(x^3+z^3)=y^3+t^3$.

R. Amsler[106] noted the solution $x=u_{n+1}$, $z=v_n$, $y=u_n+u_{n+1}$, $t=v_n+v_{n+1}$, where u_n and v_n are the nth coefficients of the developments of

$$(1-3x-3x^2-x^3)^{-1}, \qquad (1+3x+3x^2-x^3)^{-1}.$$

A. Gérardin[107] noted the identities

$$(a^3+3b^3)^3+(a^3-3b^3)^3=2\{(a^3)^3+(3ab^2)^3\},$$
$$(\alpha^2+4\alpha\beta-\beta^2)^3+(\beta^2+4\alpha\beta-\alpha^2)^3=2\{(\alpha+\beta)^6-(\alpha-\beta)^6\}.$$

Gérardin[108] gave several solutions, as

$$x=2a(a^3-c^3), \quad y=c(c^3-4a^3), \quad z=b(2a^3+c^3), \quad t=d(2a^3+c^3), \quad 2(a^3+b^3)=c^3+d^3.$$

[99] L'intermédiaire des math., 9, 1902, 164; 11, 1904, 288; Matem. Sborn. (Math. Soc. Moscow), 25, 1905, 417–37.
[100] L'intermédiaire des math., 12, 1905, 268; 25, 1918, 139, for numerical examples in which x_2 and y_2 are integers.
[101] Matem. Sborn. (Math. Soc. Moscow), 27, 1909, 146–169.
[102] L'intermédiaire des math., 14, 1907, 69.
[103] Nouv. Ann. Math., (2), 18, 1879, 407. Special case of Desboves.[302]
[104] L'intermédiaire des math., 15, 1908, 182; Sphinx-Oedipe, 1906–7, 80, 128.
[105] Sphinx-Oedipe, 1906–7, 125.
[106] Nouv. Ann. Math., (4), 7, 1907, 335. Proof by L. Chanzy, (4), 16, 1916, 282–5; same in Sphinx-Oedipe, 9, 1914, 93–4.
[107] Sphinx-Oedipe, 1910, 179.
[108] *Ibid.*, 9, 1914, 143–4; Nouv. Ann. Math., (4), 16, 1916, 285–7, where Y, Z should be interchanged.

Relations between five or more cubes.

To divide a given cube k^3 into n $(n>2)$ positive cubes, J. Whitley[109] took a, $k-v$, vk^2/a^2-a, dv, ev, $\cdots$ as the roots of the required cubes. Then

$$v=\frac{3ka^3(k^3-a^3)}{k^6+a^6(d^3+e^3+\cdots-1)}.$$

S. Ryley took a, $v-a$, $k-a^2v/k^2$, dv, ev, $\cdots$ as the roots; then

$$3k^3a(k^3-a^3)=v\{k^6(1+d^3+e^3+\cdots)-a^6\}.$$

F. Elefanti[110] noted that

$$9^3=1+6^3+8^3,\qquad 13^3=1+5^3+7^3+12^3,\qquad 16^3=4^3+6^3+7^3+9^3+14^3,$$

and that 28^3 is a sum of 9 cubes, also of 11 cubes; etc. For the second relation see Bouniakowsky[54] of Ch. VIII.

Y. Hirano[111] noted that

$$(a^3+36c^3)^3+(36c^3\pm b^3)^3+(a^3\pm b^3)^3+(\pm 6abc)^3$$
$$=(36c^3)^3+(a^3)^3+(b^3)^3+(a^3\pm b^3+36c^3)^3.$$

A. Martin[112] noted that the sum of the cubes of rm, $q-rm$, sm, p_1q, $\cdots$, $p_{n-3}q$ will equal the cube of $sm+qr^2/s^2$ by choice of m/q. Also,

$$1^3+2^3+4^3+12^3+24^3=25^3,\qquad 1^3+2^3+52^3+216^3=217^3.$$

S. Réalis[113] noted that $z_1^3+\cdots+z_4^3=z^3$ if

$$z_1,\ z_3=\pm 3\alpha\beta(\alpha-\beta)+\gamma^3;\qquad z_2,\ z_4=\pm 3\alpha\beta(\alpha-3\beta)\pm 6\beta^3-\gamma^3.$$

This is not the general solution since $\Sigma z_i=0$.

E. Catalan[114] noted that $x^3=6(x-1)^2+(x-2)^3+2$ gives

$$x^3(x^3-2)^3+(2-x)^3(x^3+1)^3+(2x^3-1)^3-(x^3+1)^3=6(x-1)^2(x^3+1)^3.$$

Taking $x=7/4$ or $x=1+6(a/b)^3$, we get a solution of $X^3+Y^3+Z^3=S^3+T^3$ in positive integers. If we multiply each term by $27(x^6-x^3+1)^3x^9$, combine the third and fourth terms and replace x^3 by x, we get

$$(2x-1)^3(2x^3-6x^2-1)^3+(5x^3-9x^2+3x-1)^3(x+1)^3+27x^3(x^2-x+1)^3(x+1)^3$$
$$=27x^3(2x-1)^3(x^2-x+1)^3.$$

D. S. Hart[115] found cubes whose sum is a cube by taking $1^3+\cdots+n^3=S$ and seeking by trial to make $S-(s+m)^3+s^3$ a sum of cubes.

S. Tebay[116] noted that, if $x=aa_1$, $y=aa_2$, $z=aa_3$, $2u^3=n$,

$$(1)\qquad\qquad x^3+y^3+z^3=2u^3$$

becomes $a^{-3}=n^{-1}\Sigma a_1^3$. First, solve $a_1^3+a_2^3=nr^3+s^3$ by setting

$$2u^3r^3+s^3=(ur+t)^3+(ur-t)^3=2u^3r^3+6urt^2,$$

---

[109] Ladies' Diary, 1832, 41–2, Quest. 1536.

[110] Quar. Jour. Math., 4, 1861, 339.

[111] Easy Solution of Math. Problems, 1863. Cf. Hayashi, Tôhoku Math. Jour., 10, 1916, 18.

[112] Math. Quest. Educ. Times, 21, 1874, 104.

[113] Nouv. Corresp. Math., 4, 1878, 350–2.

[114] *Ibid.*, 352–4, 371–3.

[115] Math. Quest. Educ. Times, 23, 1875, 82–3; Math. Magazine, 1, 1882–4, 173–6.

[116] Math. Quest. Educ. Times, 38, 1883, 101–3.

whence $s^3 = 6urt^2$.   Take $t = 3n^3$, $r = 4u^2m^3$, whence $s = 6umn^2$.   Hence solutions are a_1, $a_2 = 4u^3m^3 \pm 3n^3$.   Next, for $a_3 = p - s$, our initial equation becomes

$$a^{-3} = r^3 + \frac{p^3}{n} - \frac{3p^2s}{n} + \frac{3ps^2}{n} = \left(r + \frac{ps^2}{nr^2}\right)^3, \qquad \text{if} \qquad p = \frac{3nr^3s}{nr^3 - s^3}.$$

Special sets of five cubes whose sum is a cube have been noted.[117]

A. Martin[118] noted that the sum of the cubes of $p+q$, $p-q$, $r-p$, s is the cube of $r+p$ if $p = \frac{1}{6}s^3/(r^2 - q^2)$; that of $a+b-c$, $a+c-b$, $b+c-a$, y is the cube of $a+b+c$ if $y^3 = 24abc$, whence take $a = 3p^3$, $b = 3q^3$, $c = r^3$ or take $y = 2a$, $c = a^2/(3b)$; the sum of the cubes of $pa+nt$, $qa-nt$, $ra-nt$, nt is of the form $sa^3 + R$ and is a cube if $s \equiv p^3 + q^3 + r^3$ is a cube and if $R = 0$, which determines t.   Next, he gave Whitley's[109] result.

Finally, given that $p_1^3 + \cdots + p_n^3$ is a cube, to find $n+1$ cubes whose sum is a cube.   If n is odd, take x, $p_1 - x$, $p_2 - x$, $p_3 + x$, $p_4 - x$, $p_5 + x$, $\cdots$, $p_n + x$ as the roots of the desired cubes, where

$$x = (p_1^2 + p_2^2 - p_3^2 + p_4^2 - \cdots - p_n^2)/(p_1 + \cdots + p_n).$$

If n is even, take x, $p_1 + x$, $p_2 - x$, $p_3 + x$, $p_4 - x$, $\cdots$, $p_{n-1} + x$, $p_n - x$ as the roots, and $(t+x)^3$ as the sum of their cubes, where

$$x = (t^2 - p_1^2 + p_2^2 - p_3^2 + p_4^2 - \cdots + p_n^2)/(p_1 + \cdots + p_n - t).$$

Martin[119] found cubes whose sum is a cube b^3 by selecting b^3 between n^3 and $S = 1^3 + \cdots + n^3$ and seeking by trial to express $S - b^3$ as a sum of distinct cubes $\leqq n^3$.   Also by seeking to express $p^3 - q^3$ as a sum of distinct cubes $\neq q^3$.   He tabulated the values of S for $n \leqq 342$.

R. W. D. Christie[120] gave 14 cases like $4^3 = 1 + 1 + 2^3 + 3^3 + 3^3$ of a cube equal to a sum of five cubes.

Ed. Collignon[121] noted that there is no positive integral solution of

$$x^p + (x-1)^p + \cdots + (x-k)^p = (x+1)^p + \cdots + (x+k)^p \qquad (p = 3 \text{ or } 4).$$

A. Gérardin[122] gave numerical examples of equal sums of three cubes.

A. S. Werebrusow[123] noted that (1) holds if

$$x = u+v, \qquad y = u-v, \qquad u = a^2m^3, \qquad v = bn^3, \qquad z = -6mn^2, \qquad ab = 6.$$

From two sets of solutions a third set is derived.

A. Gérardin[124] gave, besides two more complicated identities of like type,

$$(6\alpha\beta)^3 + (9\alpha^2 + \beta^2 - \alpha\beta)^3 + (9\alpha^2 - \beta^2 + \alpha\beta)^3 = (9\alpha^2 - \beta^2 - \alpha\beta)^3 + (9\alpha^2 + \beta^2 + \alpha\beta)^3.$$

Gérardin[125] discussed $a^3 + b^3 + hc^3 = (a+b)^3 + hd^3$.   For $a = pm$, $c = d+m$,

[117] Amer. Math. Monthly, 2, 1895, 329–331.
[118] Math. Magazine, 2, 1895, 156–160.
[119] *Ibid.*, 185–190.   Two examples, Martin[68] of Ch. XXIII.
[120] Math. Quest. Educ. Times, (2), 4, 1903, 71.
[121] Sphinx-Oedipe, 1906–7, 129–133.
[122] *Ibid.*, 120–4.
[123] Math. Soc. Moscow, 26, 1908, 622–4.
[124] Assoc. franç., 38, 1909, 143–5.
[125] Sphinx-Oedipe, 5, 1910, 178.

it becomes
$$hm^2+3(dh-bp^2)m+3(hd^2-pb^2)=0.$$
To make the constant term zero, set $h=b^2$, $p=d^2$; then, for $b=x^3$,
$$(3d^6-3d^3x^3)^3+(x^6)^3+(3d^4x^2-2dx^5)^3=(3d^6-3d^3x^3+x^6)^3+(dx^5)^3.$$
By annulling the coefficient of m, he obtained
$$(3p)^3+(p^2+3)^3+p(p^2+3)(p+3)^3=(p^2+3p+3)^3+p(p^2+3)p^3.$$
Again,
$$(x^2-6y^2)^3+(6x^2-17xy)^3+(8x^2-36xy+54y^2)^3$$
$$=(9x^2-36xy+48y^2)^3+(36y^2-17xy)^3.$$

E. Barbette[126] employed the first method of Martin[119] to show that
$$3^3+4^3+5^3=6^3,\qquad 1+6^3+8^3=9^3=1+3^3+4^3+5^3+8^3,$$
$$3^3+4^3+5^3+8^3+10^3=12^3=6^3+8^3+10^3,$$
$$1+5^3+6^3+7^3+8^3+10^3=13^3=5^3+7^3+9^3+10^3,$$
$$2^3+3^3+5^3+7^3+8^3+9^3+10^3=14^3$$
are the only sets of distinct cubes $\leqq 10^3$ whose sum is a cube.

R. Norrie[84] would find n cubes whose sum is a cube by taking
$$(rx_1+\lambda)^3+(rx_2-\lambda)^3+(rx_3+\mu)^3+(rx_4-\mu)^3+\cdots$$
$$+(rx_{n-1}+\rho)^3+(rx_n-\rho)^3=(rx_0)^3,$$
$$(rx_1+\lambda)^3+(rx_2+\mu)^3+(rx_3-\mu)^3+\cdots+(rx_{n-1}+\rho)^3+(rx_n-\rho)^3=(rx_0+\lambda)^3,$$
according as n is even or odd.

A. Gérardin[127] noted that the sum of the cubes of $x-1$, x, $x+1$, $2f-1$, $2f$, $2f+1$ is of the form $3t(t^2-2q)$ if $t=x+2f$, $q=3fx-1$.

R. D. Carmichael[128] noted that (1) has the special solution
$$x=\rho^3\pm6\sigma^3,\qquad y=\rho^3\mp6\sigma^3,\qquad z=-6\rho\sigma^2,\qquad u=\rho^3,$$
and obtained a set of solutions of $x^3+y^3+z^3+u^3=3t^3$ involving five parameters. A special solution of $x^3+2y^3+3z^3=t^3$ is x, $t=2n^3\mp m^3$, $y=m^3$, $z=2mn^2$.

The double of a cube may be a sum of four cubes.[129]

A. Gérardin[130] derived a solution of $x^3+y^3+z^3=hv^3$ from a given solution, and deduced a solution of
$$A+B+C=X+Y+Z,\qquad A^3+B^3+C^3=X^3+Y^3+Z^3.$$

M. Weill[131] derived a third solution $x=x_1+\lambda(x_2-x_1)$, $\cdots$ from two given solutions of $x^3=y^3+z^3+t^3+u^3$; likewise for $ax^3+by^3+cz^3+dt^3=0$.

E. Fauquembergue[132] treated $x^3+y^3+z^3=4u^3$ by setting $x=2a$, $y=4b+1$, $z=4c-1$, $2b-2c+1=f$, $b+c=g$. Then $2a^3+3f^2g+4g^3=u^3$, which is satis-

[126] Les sommes de p-ièmes puissances distinctes égales à une p-ième puissance, Liège, 1910, 105–132.
[127] L'intermédiaire des math., 19, 1912, 136.
[128] Amer. Math. Monthly, 20, 1913, 304–6.
[129] L'intermédiaire des math., 21, 1914, 144, 188–190; 22, 1915, 60.
[130] *Ibid.*, 22, 1915, 130–2 (error for $h=2$); 23, 1916, 107–110.
[131] Nouv. Ann. Math., (4), 17, 1917, 46, 51–53.
[132] L'intermédiaire des math., 24, 1917, 40.

fied if $a=6$, $f=1$, $g=9$, $u=15$, giving $12^3+17^3+19^3=4\cdot15^3$. This contradicts the statement by E. Turrière[133] that $x^3+y^3+z^3=nt^3$ is impossible if $n\equiv4$ or 5 (mod 9).

A. S. Werebrusow[134] gave two equal sums of four cubes.

SUM OF THREE CUBES MADE A SQUARE.

V. Bouniakowsky[135] used $\int x(x+b)dx$ to get the identity
$$(x+b)^2(2x-b)+b^3\equiv x^2(2x+3b).$$
Set $2x-b=(x+b)\lambda^3$, $2x+3b=\mu^2$. Then
$$X^3+Y^3=Z^2, \qquad X=\frac{3\lambda}{8-\lambda^3}, \qquad Y=\frac{2-\lambda^3}{8-\lambda^3}, \qquad Z=\frac{\lambda^3+1}{8-\lambda^3}.$$
Multiply by $(8-\lambda^3)^3$. Thus
$$(3\lambda)^3+(2-\lambda^3)^3+(\lambda^3+1)^3=[3(\lambda^3+1)]^2.$$

E. Catalan,[136] by use of the toroid, obtained the identity
$$(a^4+2ab^3)^3+(b^4+2a^3b)^3+(3a^2b^2)^3=(a^6+7a^3b^3+b^6)^2,$$
which gives an infinitude of, but not all, solutions of $x^3+y^3+z^3=u^2$.

E. Lucas[137] deduced from formulas of Cauchy[287] the generalization
$$A(Aa^4+2Bab^3)^3+B(Bb^4+2Aa^3b)^3+A^2B^2(3a^2b^2)^3=(A^2a^6+7ABa^3b^3+B^2b^6)^2$$
of Catalan's[136] identity.

A. Desboves[138] gave a new proof of the last identity.

A. S. Werebrusow[139] derived from one solution a, b, c, d the second solution
$$(a+\alpha x)^3+(b-\alpha x)^3+(c+x)^3=(d+\delta x)^2,$$
$$2d\delta=3(a^2-b^2)\alpha+3c^2, \qquad x=\delta^2-3(a+b)\alpha^2-3c.$$
We may start from the solution $(n^2)^3=(n^3)^2$.

A. Gérardin[140] gave the identities
$$(9x^4+8u^3x)^3+(4u^4)^3+(4u^3x)^3=(8u^6+36u^3x^3+27x^6)^2,$$
$$\{a^4-8ab^3(c^3+d^3)\}^3+(ct)^3+(dt)^3=\{a^6+20a^3b^3(c^3+d^3)-8b^6(c^3+d^3)^2\}^2,$$
where $t=4a^3b+4b^4(c^3+d^3)$.

Gérardin[141] tabulated solutions of $x^3+y^3+z^3=u^2$.

BINARY CUBIC FORM MADE A CUBE.

Fermat[142] solved $Ax^3+Bx^2+Cx+D=z^3$ if $D=d^3$ by setting
$$z=d+Cx/(3d^2),$$
or if $A=a^3$ by setting $z=ax+B/(3a^2)$, while if both $D=d^3$ and $A=a^3$ there

---

[133] L'enseignement math., 18, 1916, 421.

[134] L'intermédiaire des math., 25, 1918, 75-6.

[135] Bull. Ac. Sc. St. Pétersbourg, Phys. Math., 11, 1853, 72.

[136] Bull. Acad. Roy. de Belgique, (2), 22, 1866, 29; Mélanges Math., 1868, 58; Nouv. Corresp. Math., 1, 1874-5, 153, foot-note.

[137] Bull. Bibl. Storia Sc. Mat. Fis., 10, 1877, 176.

[138] Nouv. Ann. Math., (2), 18, 1879, 409.

[139] L'intermédiaire des math., 15, 1908, 136-7.

[140] Sphinx-Oedipe, 8, 1913, 29.

[141] L'intermédiaire des math., 23, 1916, 9-10.

[142] J. de Billy's Inventum novum, III, §§ 27-30, Oeuvres de Fermat, III, 386-8.

are three ways cf solving. Thus, for $x^3+2x^2+4x+1=z^3$, $z=x+1$ gives $x=1$, $z=x+2/3$ gives $x=-19/72$, $z=1+\frac{4}{3}x$ gives $x=-90/37$, and each of these primitive solutions furnishes new solutions as above. Cases when the preceding methods fail are noted in § 30; there is no rational solution $x\neq0$ of $1+3x+3x^2+4x^3=z^3$ or of $x^3-3x^2\pm3x\pm1=z^3$; for

$$x^3+2x^2+3x+1=z^3,$$

$z=1+x$ gives $x=0$, while $z=x+2/3$ gives the only primitive solution [von Schaewen[150] noted the additional primitive solutions $x=-1$, $x=-1/2$].

L. Euler,[143] after reproducing (§§ 147–151) essentially Fermat's methods, treated the new case in which a particular solution $x=h$, $z=k$, is known. Taking $x=h+y$ we get a cubic whose constant term is a cube. Since $4+x^2=z^3$ for $x=2$ or $x=11$, we may apply the last method, or set $x=(2+2y)/(1-y)$ and get $(8+8y^2)(1-y)=w^3$ or set $x=(2+11y)/(1\pm y)$.

L. Euler[144] proved that $py^3\pm p^2x^3=z^3$ is impossible if p is a prime. For, $z=pA$, whence $p^2A^3\mp px^3=y^3$. Then $y=pB$, whence $p^2B^3=pA^3\mp x^3$. Then $x=pC$, etc., and x, y, z are divisible by an indefinitely large power of p.

W. L. Krafft[145] would make x^3+ny^3 the cube of $p^3+nq^3+n^2r^3-3npqr$ by setting

$$x+y\sqrt[3]{n}=(p+q\sqrt[3]{n}+r\sqrt[3]{n^2})^3,$$

which determines x, y, subject to the condition $p^2r+pq^2+nqr^2=0$, whence

$$p=\frac{1}{2r}\{-q^2+\sqrt{q^4-4nqr^3}\}.$$

To make the radical rational, set $q=s^2$, $s^6-4nr^3=t^2$, whence take $s^3+t=2f^3$, $s^3-t=2ng^3$. Then $s^3=f^3+ng^3$, which is like the initial equation, but in smaller numbers.

P. Paoli[146] treated $a+b^3x^3=y^3$ by setting $y=bx+m$, solving the quadratic in x and making the radical rational. Thus $12am-3m^4$ is to be a square, which he accomplished by trying values of $m<\sqrt[3]{4a}$. A like method was stated to apply to $a+bx+c^3x^3=y^3$.

D. M. Sensenig[147] treated without novelty $ax^3+bx^2+cx+d=y^3$, when a or d is a cube.

A. Desboves[148] stated that if $T=cZ^3$ and $F=cZ^2$, where T and F are binary forms of the third and fourth degrees in X and Y, are such that $T=0$ and $F=0$ are solvable in integers, one can determine a solution (X, Y, Z) of cne of the equations knowing a solution (x, y, z) of an equation of the same degree by formulas giving X, Y, Z as cubic functions of x, y, z, in case of $T=cZ^3$, and, in case of $F=cZ^2$, by functions of degree four in x, y and of degree eight in z.

[143] Algebra, St. Petersburg, 2, 1770, Ch. 10, §§ 147–161; French transl., Lyon, 2, 1774, pp. 177–195; Opera Omnia, (1), I, 406–414.

[144] Opera postuma, I, 1862, 217 (about 1775).

[145] Ibid., 234.

[146] Opuscula analytica, Liburni, 1780, 128–130.

[147] The Analyst, Des Moines, 3, 1876, 104.

[148] Comptes Rendus Paris, 90, 1880, 1069. Cf. Desboves[159] of Ch. XXII.

E. Landau, A. Boutin, P. Tannery, and A. S. Werebrusow[149] considered $x^3+3x^2y+6xy^2+2y^3=1$ or z^3.

P. von Schaewen[150] treated $Ax^3+Bx^2y+Cxy^2+Dy^3=z^3$. If $A=a^3$, $B=0$, we have

$$(z-ax)(z^2+axz+a^2x^2)=y^2(Cx+Dy),$$

which is satisfied if $m(z-ax)=ny$, $n(z^2+\cdots)=m(Cx+Dy)y$. Eliminating z, we get

$$\frac{x}{y}=\frac{1}{6a^2mn}\{Cm^2-3an^2\pm E^{\frac{1}{2}}\}, \qquad E=C^2m^4+12a^2Dm^3n-6aCm^2n^2-3a^2n^4.$$

We can always make E a square. Next, if $A=a^3$, $B\neq0$, we replace $ax+By/(3a^2)$ by x_1 and y by $3a^2y_1$ and are led to the first case. Finally, if neither A nor D is a cube, but $x=p$, $y=q$, $z=r$ is a known solution, set $qx=py+s$ to obtain a cubic in which the coefficient of y^3 is r^3. For Fermat's example, $x^3+2x^2y+3xy^2+y^3=z^3$, set $X=x+y$, $x=Y$. Then

$$X^3-XY^2+Y^3=z^3, \qquad E=m^4+12m^3n+6m^2n^2-3n^4.$$

Many solutions are found: $(x, y, z)=(1, -1, 1)$, $(3, -7, -1)$, $(1, -2, 1)$, $(6, -13, 5)$, etc., whereas Fermat's method gave the primitive solution $x=19$, $y=-45$.

J. von Sz. Nagy[151] noted that a principle of Poincaré's[15] of Ch. XXIII enables us to transform the cubic curve $f\equiv a^3x^3+pxy^2+qy^3-z^3=0$ without double points, treated by von Schaewen, by the birational transformation

$$x=pm^2-3an^2\pm rm, \qquad y=6a^2mn, \qquad z=a(pm^2+3an^2\pm rm)$$

into the quartic curve $p^2m^4+12a^2qm^3n-6apm^2n^2-3a^2n^4-r^2m^2=0$, and conversely the last into $f=0$ by

$$m=y^2, \qquad n=y(z-ax), \qquad \pm r=3a(z-ax)^2+6a^2(z-ax)x-py^2.$$

To pass to the non-homogeneous form, use x/y, z/y, n/m, r/m.

E. Haentzschel,[152] starting from a given solution $x=h$, $y=k$, of

$$y^3=a_0x^3+3a_1x^2+3a_2x+a_3\equiv f(x),$$

derived a second solution by applying the substitution

$$x=(ht-a_1h^2-2a_2h-a_3)/\tau, \qquad \tau\equiv t+a_0h^2+2a_1h+a_2,$$

giving

$$y^3=\{t^3+3C_2(h)t+C_3(h)\}f(h)/\tau^3,$$

where C_2 and C_3 are the quadratic and cubic covariants of $f(x)$, and choosing t so that $3C_2(h)t+C_3(h)=0$. We may begin with the identity

$$4C_2^3(x)+C_3^2(x)=Df^2(x),$$

where D is the discriminant of f, set $v=-C_3/f$, $v^2=4s^3+D$; then

$$f(x)=(\sqrt{-C_2(x)/s})^3.$$

[149] L'intermédiaire des math., 8, 1901, 147, 309; 9, 1902, 111, 283; 10, 1903, 108; 13, 1906, 196-7.

[150] Jahresbericht d. Deutschen Math.-Vereinigung, 18, 1909, 7-14.

[151] *Ibid.*, 401-2.

[152] *Ibid.*, 22, 1913, 319-29.

Given a pair of values v, s satisfying $v^2 = 4s^3 + D$, we can find new pairs by use of the addition theorem for the elliptic function $\wp(u)$. Only such a value v is useful for which the cubic equation[153] $v = -C_3/f$ has a rational root x. The simplest case $D = \square$ is treated at length and illustrated for $19y^3 = x^3 + z^3$.

L. Holzer treated[154] $(x+y)(x^2+y^2) = 4Cz^3$. J. de Billy[155] (p. 41) treated $(x+y)(x^2+y^2) = z^3$.

Candido[179] of Ch. XXIII made the product of a linear and a quadratic factor a cube.

Binary Cubic Form made a Square.

J. de Billy[155] treated many problems $f = \square$, where f is a cubic or quartic in one or more variables with numerical coefficients.

Fermat[156] treated $20x^3 + 5x^2 + 40x + 16 = z^2$. For $z = 4 + 5x$, $x = 1$. To deduce a second solution, set $x = 1 + y$. Then

$$20y^3 + 65y^2 + 110y + 81 = \left(9 + \frac{55y}{9}\right)^2 \qquad \text{for} \qquad y = \frac{-112}{81}.$$

From the latter, we get a third solution.

L. Euler[157] made $F \equiv f^2 + bx + cx^2 + dx^3 = \square$ by setting $F = (f + px)^2$, where $2fp = b$, whence $x = (p^2 - c)/d$, or by setting $F = (f + px + qx^2)^2$ and choosing p and q to make the terms in x and x^2 cancel, whence

$$p = b/(2f), \qquad q = (c - p^2)/(2f), \qquad x = (d - 2pq)/q^2.$$

But it often happens that neither of these two methods leads to a value $\neq \pm f$ of x, as for example for $f^2 + dx^3$, and then we resort to trial. For $3 + x^3 = \square$, set $x = 1 + y$ to obtain $4 + \cdots + y^3$. But for $1 + x^3$, $x = 2 + y$ gives $9 + 12y + 6y^2 + y^3$ and neither of the two methods leads to a value of x other than 0, 2, -1; in fact, $1 + x^3 = \square$ only when $x = 0$, 2, -1.

Euler[144] of Ch. XXII applied to cubics his method to make a quartic a square.

W. L. Krafft,[158] given $ma^3 + n = b^2$, made $mx^3 + n = z^2$ by setting $x = a + y$, $z = b + 3ma^2y/(2b) \equiv z_1$ or $z = z_1 + py^2$ and in the latter case requiring that the terms y^2 shall cancel. A. J. Lexell treated the case $n = k^2$ by setting $x = ay$, whence $(b^2 - k^2)y^3 = z^2 - k^2$, and taking $(b \pm k)y^2 = z \pm k$, $(b \mp k)y = z \mp k$.

L. Euler[159] noted that $1 + z - z^3 = \square$ for $z = 11/9$.

Krafft[160] made $x^3 + ny^3$ a square for relatively prime integers x, y, by setting

$$x + y\alpha^\gamma \sqrt[3]{n} = (p + \alpha^\gamma q \sqrt[3]{n} + \alpha^{2\gamma} r \sqrt[3]{n^2})^2 \qquad (\gamma = 0, 1, 2; \ \alpha^3 = 1).$$

[153] Treated by Haentzschel, Sitzungsber. Berlin Math. Gesell., 10, 1910, 20.
[154] Monatshefte Math. Phys., 26, 1915, 289.
[155] Diophanti Redivivi, Lvgdvni, 1670, Pars Posterior.
[156] J. de Billy's Inventum novum . . ., Oeuvres de Fermat, III, 385.
[157] Algebra, St. Petersburg, 2, 1770, Ch. 8, §§ 112–127; French transl., Lyon, 2, 1774, pp. 135–152; Opera Omnia, (1), 1, 1911, 388–396. Reproduced, Sphinx-Oedipe, 1908–9, 49–57.
[158] Euler's Opera postuma, 1, 1862, 211–2 (about 1770).
[159] *Ibid.*, 217.
[160] *Ibid.*, 232–4.

Thus $x = p^2 + 2nqr$, $y = 2pq + nr^2$, $0 = 2pr + q^2$, which holds if $p = 2a^2$, $r = -b^2$, $q = 2ab$. The product of the three factors is the square of $p^3 + nq^3 + n^2 r^3 - 3npqr$.

J. L. Lagrange[161] proved that $r^3 - As^3 = q^2$ for

$$(1) \qquad r = 4t(t^3 - Au^3), \quad s = -u(8t^3 + Au^3), \quad q = 8t^6 + 20At^3u^3 - A^2u^6.$$

He took a cube root a of unity and set

$$p = t + ua\sqrt[3]{A} + xa^2\sqrt[3]{A^2}, \qquad p^2 = T + Ua\sqrt[3]{A} + Xa^2\sqrt[3]{A^2}.$$

Thus

$$T = t^2 + 2Aux, \qquad U = Ax^2 + 2tu, \qquad X = u^2 + 2tx.$$

Then the factor $r - as\sqrt[3]{A}$ of the given cubic function will be of the form p^2 if $r = T$, $s = -U$, $X = 0$. Substituting the value $x = -u^2/(2t)$ from $X = 0$ into the first two conditions, we get

$$r = t^2 - \frac{Au^3}{t}, \qquad s = -\frac{Au^4}{4t^2} - 2tu.$$

In the product $P = t^3 + Au^3 - 3Atux + A^2x^3$ of the expressions p in which a takes its three values, we insert the above value of x and obtain q. To avoid fractions multiply r and s by $4t^2$, and q by $8t^3$.

Euler[162] noted that this product P may be made equal to any power.

Lagrange[163] extended the method from $a^3 = 1$ to $\alpha^3 - a\alpha^2 + b\alpha - c = 0$, with the roots α_1, α_2, α_3. Then

$$F(x, y, z) \equiv \prod_{i=1}^{3} (x + \alpha_i y + \alpha_i^2 z) = x^3 + ax^2y + (a^2 - 2b)x^2z + bxy^2 + (ab - 3c)xyz$$
$$+ (b^2 - 2ac)xz^2 + cy^3 + acy^2z + bcyz^2 + c^2z^3$$

is such that its product by $F(x_1, y_1, z_1)$ is $F(X, Y, Z)$, where

$$X + \alpha Y + \alpha^2 Z = (x + \alpha y + \alpha^2 z)(x_1 + \alpha y_1 + \alpha^2 z_1).$$

In particular, the square of $F(x, y, z)$ is $F(X, Y, Z)$, where

$$X = x^2 + 2cyz + acz^2, \qquad Y = 2xy - 2byz + (c - ab)z^2,$$
$$Z = 2xz + y^2 + 2ayz + (a^2 - b)z^2.$$

We may make $Z = 0$ by choice of x rational in y, z. Hence

$$X^3 + aX^2Y + bXY^2 + cY^3 = V^2$$

has solutions involving the parameters y, z, with $V = F(x, y, z)$. The same method leads to solutions of $F(X, Y, Z) = V^m$.

A. M. Legendre[164] made $Z = 0$ by taking $y = (u - a)z$, $2x = (b - u^2)z$. Then replacing u by u/v, we see that X, Y, V are proportional to

$$X = u^4 - 2bu^2v^2 + 8cuv^3 + (b^2 - 4ac)v^4, \qquad Y = -4v(u^3 - au^2v + buv^2 - cv^3),$$
$$V = u^6 - 2au^5v + 5bu^4v^2 - 20cu^3v^3 - 5(b^2 - 4ac)u^2v^4$$
$$- (8a^2c - 2ab^2 - 4bc)uv^5 - (b^3 - 4abc + 8c^2)v^6.$$

[161] Mém. Acad. R. Sc. Berlin, 23, année 1767, 1769; Oeuvres, II, 532.

[162] Opera postuma, 1, 1862, 571–3; letter to Lagrange, Jan., 1770, Oeuvres, XIV, 216.

[163] Addition IX to Euler's Algebra, 2, 1774, 644–9 [misprint of sign in X, § 92]. Oeuvres de Lagrange, VII, 170–9. Euler's Opera Omnia, (1), I, 643–50.

[164] Théorie des nombres, ed. 3, II, 1830, § 465, p. 139. German transl. by Maser, 2, 1893, 133.

A. Desboves[165] gave for $a=b=0$ this result with v replaced by $v/2$.

He[166] reduced $ax^3+by^3=cz^2$ to Lagrange's[161] case by multiplication by a^2c^3.

H. Brocard[167] noted that $x^3+(2a+1)(x-1)=y^2$ has the special solution

$$x=(a+1)^2+2(a+1)-1, \qquad y=(a+1)^3+3(a+1)^2-1.$$

R. F. Davis[168] made $8x^3-8x+16$ the square of px^2+x-4, obtaining a quadratic for x with rational roots if $8p^3-8p+16=\square$. Hence solutions like $p=0$, ±1, 2 lead to new solutions x.

G. de Rocquigny[169] proposed for solution $x^3-x\pm1=y^2$. H. Brocard[170] noted that for the upper sign it has solutions $x=0$, 1, 3, 5. E. B. Escott[171] noted that for the lower sign it is impossible as shown by use of modulus 3.

L. C. Walker[172] reproduced Lagrange's[163] work, applying it to $x^3+ay^3=z^2$.

The least positive integral solution[173] of $x^3-66y^3=\square$ has $x=25$.

L. Aubry[174] found restrictions on possible solutions of $x^3+x^2+2x+1=\square$.

A. Gérardin[175] assumed that x_0, y_0, z_0 is a known solution of

$$ax^3+bx^2y+cxy^2+dy^3=z^2$$

and took $x=x_0+mf$, $y=y_0+mg$, $z=z_0+mh$. There results a quadratic equation $Am^2+Bm+C=0$. He took in turn

$$A=0, \qquad B=0, \qquad C=0, \qquad B^2-4AC=\square.$$

L. J. Mordell[176] wrote the proposed cubic in the form

$$(2) \qquad g^2=4h^3-g_2ha^2-g_3a^3,$$

which is the syzygy connecting the seminvariants a,

$$h=b^2-ac, \qquad g_2=ae-4bd+3c^2, \qquad g_3=ace+2bcd-ad^2-b^2e-c^3,$$

and $g=a^2d-b^3+3bh$ of the quartic

$$f=ax^4+4bx^3y+6cx^2y^2+4dxy^3+ey^4.$$

Given integral solutions of (2) in which a is odd and prime to h, we can find integers a, $\cdots$, e such that f has the invariants g_2 and g_3, and b is prime to a. Conversely, every such quartic yields a solution of (2) with a odd and prime to h. Hence to find all solutions (with y odd and prime to x) of

$$(3) \qquad z^2=4x^3-g_2xy^2-g_3y^3,$$

take a representative f of each class of binary quartics with the invariants g_2, g_3; apply to f a suitable linear substitution $\binom{p\ r}{q\ s}$ of determinant unity to obtain a quartic f' having a' odd and prime to b'; then $x=h'$, $y=a'$,

[165] Comptes Rendus Paris, 87, 1878, 161.
[166] Nouv. Ann. Math., (2), 18, 1879, 398.
[167] Nouv. Corresp. Math., 3, 1877, 23–24.
[168] Proc. Edinb. Math. Soc., 13, 1894–5, 179–80.
[169] L'intermédiaire des math., 9, 1902, 203.
[170] Ibid., 10, 1903, 131.
[171] Ibid., 132.
[172] Amer. Math. Monthly, 10, 1903, 49–50.
[173] Math. Quest. Educ. Times, (2), 14, 1908, 29.
[174] L'intermédiaire des math., 18, 1911, 276–7.
[175] Sphinx-Oedipe, 8, 1913, 161.
[176] Quar. Jour. Math., 45, 1913–4, 170–186.

viz., $y=f(p, q)$, $x=H(p, q)$, H being the Hessian of f. Thus the complete solution of (3), in relatively prime integers x, y, is given by a finite number of pairs of quartic forms in two parameters p, q. In particular, five such pairs of quartics give all solutions of $z^2=x^3+y^3$ in which y is odd and prime to x.

R. F. Davis[177] noted that if $x=p$ is a solution of $ax^3+bx+c^2=\square$, two further solutions are the rational roots of $(apx-b)^2=4ac^2(x+p)$.

E. Fauquembergue[178] proved that $x^2=(y+1)(y^2+4)$ has no integral solutions except $(x, y)=(2, 0)$ and $(10, 4)$, since $p^2q^2-1=p^4-q^4$ implies $p=q=1$.

A. Gérardin[179] proposed that special cubics be made squares. He and L. Aubry[180] gave a partial solution for $2x^3+x^2+1=\square$.

E. Haentzschel[180a] made use of Weierstrass' $\wp$-function to study

$$\prod_{i=1}^{3}(h_i^2 x+1)=\square, \qquad h_1=h_2+h_3,$$

where h_2 and h_3 are rational or conjugate complex numbers. As an example he treated Euler's[157] problem $x^3+1=\square$.

For $x^3+x^2+x+1=\square$ see pp. 54–58 of Vol. I of this History.

For $f=\square$, where f is a certain cubic, see papers 154–6 of Ch. V, 82 of Ch. XV, and 163 of Ch. XXII.

NUMBERS THE SUM OF TWO RATIONAL CUBES: $x^3+y^3=Az^3$.

Fermat[40] indicated a process to get an infinitude of solutions from one. J. Prestet[181] employed Fermat's process to get the solution

$$X=x(2y^3+x^3), \qquad Y=-y(2x^3+y^3), \qquad Z=z(x^3-y^3).$$

J. L. Lagrange[161] reduced the problem, by means of his theory of polynomials which repeat under multiplication, to the solution of $tu^2+t^2v=Auv^2$. Setting $u=ft$, $v=fgt$, and dividing by f^2gt^3, we get

$$h\equiv\frac{1}{f}+\frac{1}{g}=Afg.$$

Set $l=1/f-1/g$. Then $h(h^2-l^2)=4A$. Set $l=kh$. Then $4A/(1-k^2)$ is h^3, so that $2A^2(1-k^2)$ is the cube of $2A/h$. But he did not complete the discussion.

L. Euler[182] proved that $y=x$ if $A=2$.

L. Euler[183] proved the impossibility of $x^3+y^3=4z^3$ and that the problem is equivalent to the impossibility of $1+2x^3=\square$ in rational numbers, $x\neq0$. To discuss $x^3+y^3=nz^3$, set $x=a+b$, $y=a-b$, $z=2v$. Then $a(a^2+3b^2)=4nv^3$.

---

[177] Math. Quest. Educ. Times, (2), 24, 1913, 67–8.

[178] L'intermédiaire des math., 21, 1914, 81–3.

[179] *Ibid.*, 22, 1915, 104, 128.

[180] *Ibid.*, 23, 1916, 132–3.

[180a] Sitzungsber. Berlin Math. Gesell., 16, 1917, 85–92.

[181] Nouveaux elemens des Math., Paris, 2, 1689, 260–1. Cf. Lucas, Amer. Jour. Math., 2, 1879, 178; Cauchy,[287] end.

[182] Algebra, 2, 1770, Art. 247; French transl., 2, 1774, pp. 355–60; Opera Omnia, (1), I, 491.

[183] Opera postuma, 1, 1862, 243–4 (about 1782).

Take

$$a = p(p^2 - 9q^2), \qquad b = 3q(p^2 - q^2), \qquad v = r(p^2 + 3q^2).$$

Then $a^2 + 3b^2 = (p^2 + 3q^2)^3$, $a = 4nr^3$. Hence take $p = \alpha f^3$, $p + 3q = 2\beta g^3$, $p - 3q = 2\gamma h^3$, $\alpha\beta\gamma = n$, $fgh = r$. Substituting the resulting values of p, q into $p = \alpha f^3$, we get $\alpha f^3 = \beta g^3 + \gamma h^3$. If the latter be solvable, the proposed equation is solvable. He noted (pp. 244–5) that $16^2 - 3 \cdot 23^2 = (1 - 3 \cdot 2^2)^3$, whereas $16 + 23\sqrt{3} \neq (1 + 2\sqrt{3})^3$. In general, $x^2 - ny^2 = (p^2 - nq^2)^3$ implies

$$x \pm y\sqrt{n} = (f \pm g\sqrt{n})(p \pm q\sqrt{n})^3, \qquad f^2 - ng^2 = 1,$$

but not the relation with the first factor omitted.

A. M. Legendre[184] proved that, for $A = 2$, every set of integral solutions has $x = \pm y$, while for $A = 2^m$, $m > 1$, $x = -y$, and observed that, for $A \equiv \pm 3$ or ± 4 (mod 9), z must be divisible by 3. He stated that the equation is impossible for $A = 3, 5, 6$, whereas for $A = 6$ it has the solutions[185] $x = 37$, $y = 17$, $z = 21$.

On geometrical aspects of the problem, see Glenie,[12] Becker.[16]

Wm. Lenhart[186] gave a table of 11 pages expressing 2581 integers < 100000 as a sum of the cubes of two positive rational numbers. Formulas used in the construction of the table were deduced as follows from

$$x^3 + y^3 = (x + y)Q, \qquad Q = x^2 - xy + y^2.$$

First, let $x + y = a^3$, $x > y$, where a is even. For $j = 1, 2, 3, \cdots$, take $x = s + j$, $y = s - j$, $2s = a^3$. Then

$$\text{(A)} \qquad \left(\frac{s+j}{a}\right)^3 + \left(\frac{s-j}{a}\right)^3 = s^2 + 3j^2,$$

the successive values of $3j^2$ being computed by their differences. For a odd, take $x = s + j$, $y = s - (j - 1)$; the new right member is $s^2 + s + 3j^2 - 3j + 1$. Similarly for $x + y = a'a^3$ or $9a'a^3$. Next, let $Q = m^3$. Then

$$x + y = (x/m)^3 + (y/m)^3,$$

whence

$$\left(\frac{nx+y}{m}\right)^3 + \left(\frac{(n+1)x - ny}{m}\right)^3 = (n^2 + n + 1)\{(2n+1)x - (n-1)y\},$$

with three similar formulas. Euler's[6] solution (Ch. XX) of $Q = m^3$ is quoted. Finally, let $Q = m'm^3$; then

$$\left(\frac{am^3 + a'x}{m}\right)^3 + \left(\frac{am^3 + a'y}{m}\right)^3 = \{2am^3 + a'(x+y)\}F,$$

$$F = a^2m^3 + aa'(x+y) + a'^2m',$$

from which is derived four similar formulas whose right members have

<hr>

[184] Théorie des nombres, Paris, 1798, 409; Mém. Acad. R. Sc. de l'Institut de France, 6, année 1823, 1827, § 51, p. 47 (=pp. 29–31 of Suppl. 2 to ed. 2, 1808, of Théorie des nombres). This Supplément is reproduced in Sphinx-Oedipe, 4, 1909, 97–128; errata, 5, 1910, 112. Théorie des nombres, ed. 3, 2, 1830, 9.

[185] G. Lamé, Comptes Rendus Paris, 61, 1865, 924.

[186] Math. Miscellany, Flushing, N. Y., 1, 1836, 114–128, Suppl. 1–16 (tables).

the factor F. In the continuation (pp. 330–6), it is noted that

$$\left(\frac{s'x+r'm^3}{m}\right)^3+\left(\frac{s'y-r'm^3}{m}\right)^3=s'(x+y)\{3r'^2m^3+3s'r'(x-y)+s'^2m'\}$$

if $Q=m'm^3$. If also $x+y=a^3$, we may simplify this formula. To apply to (A), divide each member by a^3 and set $(s^2+3j^2)/m^3=m'$; hence

$$\left(\frac{s'(s+j)+r'm^3}{am}\right)^3+\left(\frac{s'(s-j)-r'm^3}{am}\right)^3=s'(3r'^2m^3+6r's'j+s'^2m').$$

G. L. Dirichlet[187] proved by descent the impossibility of $x^3\pm y^3=4z^3$. Hence $x^3\pm y^3=2^nz^3$ is impossible, having been proved by Euler for $n=0$, $n=1$.

J. P. Kulik[188] tabulated the odd numbers to 12097 (to 18907) which are differences (sums) of two cubes, and gave the cubes.

J. J. Sylvester[189] stated that there are no solutions for $A=2, 3$. He[190] proposed the question: If p and q are primes of the respective forms $18l+5$ and $18l+11$, it is impossible to decompose p, q^2, $2p$, $4q$, $4p^2$, $2q^2$ into a sum of two rational cubes.

C. A. Laisant[191] proved that $a^3-b^3=10^{n_1}+\cdots+10^{n_k}$ is impossible if $k=3, 4$ or 5.

Moret-Blanc[192] stated that $a^3-b^3=h\cdot10^n$ is impossible if $h=1, 2$ or 8.

T. Pepin[193] proved that, if p and q are primes of the respective forms $18l+5$ and $18l+11$, the equation is impossible when $A=p$, p^2, q, q^2, $2p$, $2q^2$, $4p^2$, $4q$, $9p$, $9q$, $9p^2$, $9q^2$, $5p^2$, $5q$, $25p$, $25q^2$. If the sum or difference of two numbers is a cube, their product is expressible algebraically as the sum of two cubes. Hence the double of a triangular number is a sum of two rational cubes. Since a prime $6m+1$ is of the form A^2+3B^2, it is a sum of two rational cubes if one of the three numbers $2A$, $3B\pm A$ is a cube, or if $2B$ or $A\pm B$ is the triple of a cube.

Pepin[194] proved that Euler's and Legendre's use of numbers $a+b\sqrt{-3}$ is legitimate and hence showed that the equation is impossible for $A=14$, $21, 38, 39, 57, 76, 196$, and stated that it is impossible for $31, 93, 95, 190$.

E. Lucas[195] noted that a solution x, y, z yields the solution

$$(1) \quad \begin{aligned} X=x^9-y^9+3x^3y^3(2x^3+y^3), \qquad Y&=y^9-x^9+3x^3y^3(2y^3+x^3), \\ Z=3xyz(x^6+x^3y^3+y^6). \end{aligned}$$

For $A=9$, we get 919, -271, 438, and in general all solutions with z even (not given by Prestet, Euler, Legendre). For $A=7$, we get[196] 73,

[187] Werke, II, Anhang, 352–3.

[188] Tafeln der Quadrat- und Kubik-Zahlen aller Zahlen bis Hundert Tausend . . ., Leipzig, 1848.

[189] Annali di Sc. Mat. e Fis., 7, 1856, 398; Math. Papers, II, 63.

[190] Nouv. Ann. Math., (2), 6, 1867, p. 96.

[191] Ibid., (2), 8, 1869, 315. J. Joffroy stated that $a^3-b^3=k\cdot10^n$ is impossible.

[192] Ibid., (2), 9, 1870, 480.

[193] Jour. de Math., (2), 15, 1870, 217–236; Extract, Sphinx-Oedipe, 4, 1909, 27–8. Proof for p, p^2, q, q^2 by Hurwitz,[312] p. 220.

[194] Jour. de Math., (3), 1, 1875, 363–372.

[195] Bull. Bibl. Storia Sc. Mat., 10, 1877, 174–6. Nouv. Corresp. Math., 2, 1876, 222.

[196] Stated by Lucas, Nouv. Ann. Math., (2), 15, 1876, 83.

-17, 38, and all solutions with z even. This solution is simpler than Fermat's[41] 1265, -1256, 183.

S. Réalis[197] noted that, from the solution 1, 2, 1 of $x^3+y^3=9z^3$, Prestet's formulas give the solution 17, -20, -7, from which the new formulas

$$X=2x^2-4xy+9yz-9z^2, \qquad Y=2y^2-xy+9xz-18z^2, \qquad Z=2x^2-4xz-yz+z^2$$

give $3\cdot919$, $-3\cdot271$, $3\cdot438$ and hence the solution by Lucas.[195] For $A=7$, an analogous second set of formulas was given by Réalis.

Lucas[198] noted that integral solutions exist if and only if A is of the form $ab(a+b)/c^3$, where a, b, c are integers. For, if x, y, z are solutions, $a=x^3$, $b=y^3$ give $ab(a+b)=A(xyz)^3$. The converse is true by the identity

$$(2) \quad [x^3-y^3+6x^2y+3xy^2]^3+[y^3-x^3+6y^2x+3yx^2]^3$$
$$=xy(x+y)\cdot3^3(x^2+xy+y^2)^3.$$

For $x=1$, $y=2$, we get $17^3+37^3=6\cdot21^3$, contrary to Legendre.[184]

Lucas[199] proved Sylvester's theorem that the equation is impossible for $A=p$, $2p$, $4q$, q^2, $4p^2$, $2q^2$, where p and q are primes $18l+5$, $18l+11$, respectively. Combining this result with that of Lucas,[198] we see that $xy(x+y)=Az^3$ is impossible in rational numbers (excluding zero and equal values) if $A=p$, $2p$, $4q$, $4p^2$, q^2, $2q^2$, 1, 2, 3, 4, 18, 36.

A. Desboves[200] derived the identity (2) by Lucas from Lagrange's[163] theory of polynomials which repeat under multiplication.

J. J. Sylvester[201] proved that pq, p^2q^2, pp_1^2, qq_1^2 are not sums of two rational cubes if p, p_1 are primes $18l+5$ and q, q_1 primes $18l+11$. These with p, q, p^2, q^2, their products by 9, and $2p$, $4q$, $4p^2$, $2q^2$, give all known types not resolvable into a sum or difference of two rational cubes. He announced the theorem that if ρ, ψ, ϕ are primes of the respective forms $18n+1$, $+7$, $+13$, while each is not of the form f^2+27g^2 and hence does not have 2 as a cubic residue, then no one of the numbers 2ρ, 4ρ, $2\rho^2$, $4\rho^2$, 2ψ, $4\psi^2$, 4ϕ, $2\phi^2$ is a sum of two rational cubes. If v is a prime $6n+1$ not having 3 as a cubic residue, then neither $3v$ nor $3v^2$ is a sum of two cubes. By all of these results, we know whether or not any number $\leqq100$ (except perhaps 66) is a sum of two rational cubes. Proofs of the above theorems rest on the linear form of the divisors of x^3-3x+1. He stated the empirical theorem that every prime $18n\pm1$ or else its triple is expressible in the form[202] $x^3-3xy^2\pm y^3$.

A. Desboves[203] gave two proofs of Lucas' identity (2) and noted that the replacement of x by x^3 and y by y^3 yields Lucas' (1). He showed that

[197] Nouv. Ann. Math., (2), 17, 1878, 454–7.

[198] Ibid., 425–6. Cf. Candido[179] of Ch. XXIII.

[199] Ibid., 507–14. This and his[198] preceding paper are duplicated in Amer. Jour. Math., 2, 1879, 182–4.

[200] Comptes Rendus Paris, 87, 1878, 159.

[201] Comptes Rendus Paris, 90, 1880, 289, 1105 (correction); Amer. Jour. Math., 2, 1879, 280, 389–393. Coll. Math. Papers, 3, 1909, 430, 437; 312, 347–9.

[202] A. M. Sawin, Annals of Math., 1, 1884–5, 58–63, noted that x and y are relatively prime integers if and only if n is an integer.

[203] Nouv. Ann. Math., (2), 18, 1879, 400, 491; (3), 5, 1886, 577.

$x^3+y^3=Az^3$ has integral solutions if $A=xy(x+y)$, x^3+y^3, $2x^6+6y^2$, $x(y^3-x)$, or $x^3-y^3-3xy(x+2y)$, and hence if $A=6$, 7, 9, 12, 15, 17, 19, 20, 22, 26, 28, 30, 37.

E. Catalan[204] noted that $xy(x+y)=z^3$ is impossible in view of the identity (2) and the impossibility of $r^3+s^3=t^3$. Lucas'[198] paper implies this result.

E. Lucas[205] proved certain and stated others of the preceding theorems by Sylvester[201] and Pepin,[193] and remarked that, if $x^3-3xy^2+y^3=3Az^3$ has solutions, then[206]

$$[2x^3-3x^2y-3xy^2+2y^3]^3+[x^3+3x^2y-6xy^2+y^3]^3=A[3z(x^2-xy+y^2)]^3,$$

the divisors of the resulting A's being of the form $18n\pm1$. In the third paper he cited cases (A a prime $18n+13$, A a square of a prime $18n+7$, etc.) in which $x^3+y^3=Az^3$ can be completely solved by the method of tangents and secants, citing Sylvester's theory of residuation.

T. Pepin[207] proved (p. 110) Sylvester's[201] theorem on 2ρ, 4ρ, 2ψ, etc., and remarked (p. 75) that the first three are covered by the method used by Pepin[194] for $2\cdot7$, $2\cdot19$, $4\cdot19$. He proved (p. 109) the results stated by Sylvester[201] on the 16 types pq, $\cdots$, $2q^2$, as well as the theorem (pp. 113–4): If

$$\rho=(9m+4)^2+3(9n\pm4)^2, \qquad \psi=(9m+2)^2+3(9m\pm2)^2,$$
$$\phi=(9m+1)^2+3(9n\pm1)^2, \qquad \zeta=m^2+27(3n\pm1)^2$$

are primes, no one of the numbers

$$18(\rho,\ \psi,\ \zeta,\ \phi^2,\ \psi^2,\ \zeta^2), \qquad 36(\rho,\ \phi,\ \zeta,\ \rho^2,\ \psi^2,\ \zeta^2)$$

is a sum of two rational cubes.

C. Henry[208] proved that any number of the form $A=f^{12}-9g^{12}$ and its double are expressible as sums of two cubes:

$$2A=\left[\frac{Af^6+3g^6B}{f^2C}\right]^3+\left[\frac{Af^6-3g^6B}{f^2C}\right]^3,$$

if $B=f^{12}-g^{12}$, $C=f^{12}+3g^{12}$.

H. Delannoy[209] proved by descent that $x^3+y^3=4z^3$ is impossible. The problem $x^3+y^3=20^3\cdot105489$ has been treated.[210]

T. R. Bendz[211] misquoted Lucas' (2), whence his criticism is invalid.

K. Schwering[212] put the equation into the form

$$1+\left(-\frac{z}{x}\sqrt[3]{A}\right)^3=\left(\frac{-y}{x}\right)^3,$$

[204] Nouv. Corresp. Math., 5, 1879, 91.

[205] Bull. Soc. Math. France, 8, 1879–80, 173–182; Comptes Rendus Paris, 90, 1880, 855–7; Nouv. Ann. Math., (2), 19, 1880, 206–11. Related results from these papers are quoted under Lucas[70] of Ch. XXV.

[206] Sylvester, Comptes Rendus Paris, 90, 1880, 347 (Coll. Math. Papers, III, 432), had stated that there exist solutions in functions of degree 9.

[207] Atti Accad. Pont. Nuovi Lincei, 34, 1880–1, 73–131.

[208] Nouv. Ann. Math., (2), 20, 1881, 418–20. The right member of his formula (3) is A, in error for $2A$.

[209] Jour. math. élémentaires, (5), 1 (année 21), 1897, 58–9.

[210] Amer. Math. Monthly, 5, 1898, 181.

[211] Öfver diophantiska ekvationen $x^n+y^n=z^n$, Diss., Upsala, 1901, 15–18.

[212] Archiv Math. Phys., (3), 2, 1902, 285.

and found an infinity of solutions from one by treating

$$1+z^3-(mx+n)^3\equiv(1-m^3)(x-\alpha)(x-\beta)(x-\gamma)$$

by his method[238] for $x^3+y^3=z^2$ to obtain $(\gamma^3+1)^{\frac{1}{2}}$ and γ as functions of $\alpha=\beta$.

A. S. Werebrusow[213] discussed the form of numbers A expressible as the sum of two rational cubes. Elsewhere he[214] took

$$x+y=A_0z_0^3,\qquad x^2-xy+y^2=A_1z_1^3,\qquad A=A_0A_1,\qquad z=z_0z_1,$$

whence A_1 is of the form $(s,\ t)\equiv s^2+st+t^2$, and $z_1=(a,\ b)$. Then

$$z_1^3=(M,\ N),\qquad M=a^3+3a^2b-b^3,\qquad N=-a^3+3ab^2+b^3,$$
$$A_1z_1^3=(s,\ t)(M,\ N),\qquad x=(s+t)M+sN,\qquad y=tM+(s+t)N,$$

with similar formulas derived by interchanging s and t or M and N. Further treatment was given for $z_1=1$, $A_1=1$, 3 or 7.

A. Cunningham[215] discussed $x^3-y^3=17z^3$, obtaining integral solutions with $z=7$. From the solution $x=18$, $y=-1$, $z=7$ of $x^3+y^3=17z^3$, Prestet's formula leads to positive integral solutions smaller than those given by Lucas' (1).

R. W. D. Christie[216] noted results due to Desboves.[203]

Christie[217] noted that, if $p=a^3-6ab^2-3a^2b-b^3$, $X^3-pY^3=1$ has the solution

$$x=\frac{a^3-3ab^2-b^3}{3ab(a+b)},\qquad y=\frac{a^2+ab+b^2}{3ab(a+b)},$$

and hence also $X=1/x$, $Y=-y/x$.

A. Cunningham[218] treated $x^3+y^3=Cz^3$ for $x,\ y$ relatively prime by setting

$$x+y=X,\qquad x^2-xy+y^2=Y,\qquad z=\zeta Z.$$

The g. c. d. of X, Y is 1 or 3. Let C be prime to 3. Then $XY=C\zeta^3Z^3$,

$$X=C\zeta^3,\ Y=Z^3;\quad\text{or}\quad \zeta=3\zeta',\ X=9C\zeta'^3,\ Y=3Z^3.$$

Since Z^3 is a factor of Y and is prime to 3, $Z=A^2+3B^2$. Hence $Z^3=A_1^2+3B_1^2$. But, for y even, $Y=(x-\tfrac{1}{2}y)^2+3(\tfrac{1}{2}y)^2$. Hence, if $Y=Z^3$, $x-\tfrac{1}{2}y=\pm A_1$, $\tfrac{1}{2}y=\pm B_1$. If $Y=3Z^3$, $x-\tfrac{1}{2}y=\pm 3B_1$, $\tfrac{1}{2}y=\pm A_1$. For y odd,

$$Y=\left(\frac{x+y}{2}\right)^2+3\left(\frac{x-y}{2}\right)^2.$$

There is treated also the case $C\equiv0\pmod 3$.

T. Hayashi[219] concluded from the impossibility of rational solutions of $x^3+y^3=3z^3$ that $4\alpha(\alpha+\beta)(\alpha+2\beta)/6$ is never a cube.

R. D. Carmichael[220] noted that, if $A=2^m$, we may take $x,\ y,\ z$ odd and proved that one of the variables must be zero, except for the trivial solution $x=y=z$ which occurs if $m=1$.

[213] Matem. Sborn. (Math. Soc. Moscow), 23, 1902, 761–3.
[214] L'intermédiaire des math., 9, 1902, 300–3.
[215] Math. Quest. Educ. Times, (2), 2, 1902, 38 [48], 73.
[216] *Ibid.*, (2), 3, 1903, 109–110.
[217] *Ibid.*, (2), 13, 1908, 90. Cf. Desboves.[203]
[218] *Ibid.*, 27–30.
[219] Nouv. Ann. Math., (4), 10, 1910, 83–6.
[220] Diophantine Analysis, 1915, 70–72.

J. G. van der Corput[221] applied quadratic forms to prove the impossibility of $x^3 \pm y^3 = p^m z^3$ if p is a prime $\equiv 2$ or $5 \pmod 9$.

B. Delaunay[222] stated that, if ρ is an integer not a cube, $\rho x^3 + y^3 = 1$ has no integral solutions if the fundamental unit u of the domain defined by $r = \sqrt[3]{\rho}$ is not of the form $Br + C$, but has the single solution $x = B, y = C$, if it be of that form. Here u is Dirichlet's $ar^2 + br + c$, where a, b, c are integers not of like sign, whose powers, with positive and negative exponents, give all the units $\alpha r^2 + \beta r + \gamma$, where α, β, γ are integers.

M. Weill[223] used the identity

$$\Sigma \{ u^3 - 9uv^2 \pm (3v^3 - 3u^2v) \}^3 = 2(u^3 - 9uv^2)(u^2 + 3v^2)^3$$

to show that, if one solution of $x^3 + y^3 = Az^3$ is known, a second is

$$X = \beta^3 + 6\alpha\beta^2 + 3\alpha^2\beta - \alpha^3, \quad Y = \alpha^3 + 6\alpha^2\beta + 3\alpha\beta^2 - \beta^3, \quad Z = 3xyz(\alpha^2 + \alpha\beta + \beta^2),$$

where $\alpha = x^3$, $\beta = y^3$, and to obtain solutions when $A = 3c^2 + 3c + 1$.

W. S. Baer[224] proved that n can be represented in the form $n = \phi(u) + \phi(v)$, where $\phi(x) = \alpha x^3 + \gamma x$, with u, v, α, γ integers and $u > \xi, v > \xi$, if and only if n is a product of two integers: $n = kl$, where $k > 2\xi$, $l = \alpha l' + \gamma$, $l' < k^2 - 3k\xi + 3\xi^2$, l' being integral and $4l' - k^2$ the triple of a square. Then u and v will be relatively prime if and only if the g. c. d. of k and l' is 1 or 3, and in the latter case l' is not divisible by 3^2. The theorem can be extended to cubics $\Phi = AX^3 + BX^2 + CX + D$, where $A, \cdots, D$ are integers and B is divisible by $3A$, since $X = x - B/(3A)$ transforms $6(\Phi - \delta)$ into ϕ. In particular, let $\alpha = 1$, $\gamma = 0$, $\xi = 0$. Then n is representable as a sum of two positive cubes if and only if n is a product of two positive integers k and l such that $l < k^2$ and $4l - k^2$ is the triple of a square; the cubes will be relatively prime if and only if the g. c. d. of k and l is 1 or 3, and in the latter case l is not not divisible by 3^2.

If h is a positive integer, and p is a prime or unity, $u^3 + v^3 = hp^v$ has only a limited number of relatively prime positive solutions, and the remaining solutions are readily deduced. But $u^3 + v^3 = w^2$ has an infinitude of positive solutions of which u and v are relatively prime.

L. Varchon[224a] proved that $x^3 - y^3 = 2^a 5^b$ is impossible in integers $\neq 0$; Moret-Blanc's[192] result is a corollary.

M. Rignaux[224b] derived (1), (2) and analogous identities from a common source.

Sum or difference of two cubes a square.

L. Euler[225] noted that $x^3 + y^3 = \square$ for $x = pz/r$, $y = qz/r$, $z = r^3/(p^3 + q^3)$. To obtain integers, set $r = n(p^3 + q^3)$; then

$$x = n^2 p(p^3 + q^3), \qquad y = n^2 q(p^3 + q^3).$$

<hr>

[221] Nieuw Archief voor Wiskunde, (2), 11, 1915, 64–8.
[222] Comptes Rendus Paris, 162, 1916, 150–1.
[223] Nouv. Ann. Math., (4), 17, 1917, 54–9.
[224] Tôhoku Math. Jour., 12, 1917, 181–9.
[224a] Nouv. Ann. Math., (4), 18, 1918, 356–8.
[224b] L'intermédiaire des math., 25, 1918, 140–2.
[225] Novi Comm. Acad. Petrop., 6, ad annos 1756–7, 1761, 181; Comm. Arith. Coll., 1, 1849, 207; Opera Omnir, (1), II, 454.

To obtain relatively prime integers x, y, when p, q are integers, we must employ fractional values for n. To obviate this, Euler gave a second method. The factors $x+y$, x^2-xy+y^2 have the g.c.d. 1 or 3. In the first case, he put the second factor equal to the square of p^2-pq+q^2 and stated that $\pm x = p^2-2pq$, $\pm y = p^2-q^2$. The upper sign is excluded since

$$x+y = 3p^2-(p+q)^2 \neq \square.$$

For the lower sign, $x+y = (p+q)^2-3p^2 = \square$ if

$$p = 2mn, \qquad q = 3m^2-2mn+n^2,$$
$$x = 4mn(3m^2-3mn+n^2), \qquad y = (m-n)(3m-n)(3m^2+n^2).$$

In the second case, $x^2-xy+y^2 = 3(p^2-pq+q^2)^2$, $(x+y)/3 = \square$. As the three subcases lead to equivalent results, consider the case

$$x = 2p^2-2pq-q^2, \qquad y = p^2-4pq+q^2, \qquad (x+y)/3 = p^2-2pq = \square.$$

The last condition is satisfied if $p = 2m^2$, $q = m^2-n^2$, whence

$$x = 3m^4+6m^2n^2-n^4, \qquad y = -3m^4+6m^2n^2+n^4.$$

Euler[226] noted the examples $1+2^3 = 3^2$, $8^3-7^3 = 13^2$, $37^3+11^3 = 228^2$, $65^3+56^3 = 671^2$, $71^3-23^3 = 588^2$, $74^3-47^3 = 549^2$.

Several[227] found that the difference of 7^3 and 8^3 is a square by considering x^3, $(x+1)^3$, and, by use of tables of cubes, found that this pair and 7^3, 14^3 give the least solutions.

C. H. Fuchs[228] discussed $x^3+y^3 = az^2$. Let x, y, z have no common factor, a no square factor. If x or y is even, set $x+y = p$, $x-y = q$. Then $p(p^2+3q^2) = 4az^2$. If p is not a multiple of 3,

$$p = \alpha t^2, \qquad p^2+3q^2 = 4\beta u^2, \qquad \alpha\beta = a.$$

Since β is a divisor of p^2+3q^2, it is of that form. Thus $4\beta = \mu^2+3\nu^2$. Also $u = \xi^2+3\eta^2$. By use of $\sqrt{-3}$, he got

(1) $$p = \mu(\xi^2-3\eta^2)-6\nu\xi\eta, \qquad q = \nu(\xi^2-3\eta^2)+2\mu\xi\eta.$$

The case $p = 3P$ is similar. For xy odd, set $2p = x+y$, $2q = x-y$. One of the three cases has $p = 2p'$, a odd. Then $p = 2\alpha t^2$, $p^2+3q^2 = \beta u^2$. He again got (1).

R. Hoppe[229] obtained the general solution of $x^3+y^3 = z^2$ in relatively prime integers by setting $pq = z^2$, $p = x+y$, $q = (x+y)(x-2y)+3y^2$, where p and q have the greatest common factor 1 or 3. In the first case all solutions are given by

$$\theta^2 x = a(a^3-8b^3), \qquad \theta^2 y = 4b(a^3+b^3), \qquad \theta^3 z = a^6+20a^3b^3-8b^6,$$

where a is odd, and $\theta = 3$ or 1 according as 3 is or is not a divisor of $a+b$. Second, if p, q have the factor 3, the solutions are [Euler[225]]

$$\eta^2 x = a^4+6a^2b^2-3b^4, \qquad \eta^2 y = 3b^4+6a^2b^2-a^4, \qquad \eta^3 z = 6ab(a^4+3b^4),$$

[226] Opera postuma, 1, 1862, 241.
[227] Ladies' Diary, 1812, 35, Quest. 1227; Leybourn's M. Quest. L. D., 4, 1817, 149.
[228] De Formula $x^3+y^3 = az^2$, Diss. Vratislaviae, 1847, 33 pp.
[229] Zeitschrift Math. Phys., 4, 1859, 304-5.

where a is not divisible by 3, while $\eta=2$ or 1 according as a, b are both odd or not both odd.

C. Richaud[230] solved $(x+1)^3-x^3=y^2$ for x and made the radical rational. Thus $(2y)^2-1=3r^2$, whence $x=0$, $y=1$; $x=7$, $y=13$; $x=104$, $y=181$; etc. The same solutions were given by Moret-Blanc,[231] who remarked that $x^3+(x+1)^3=y^2$ only for $x=0$, 1 (cf. E. Lucas, Mathesis, 1887, 200).

W. J. Greenfield[232] gave numerical solutions of $x^3-y^3=\square$.

M. Weill[233] noted that $(-3\alpha^2)^3+(\alpha^3+4)^3=(1+\alpha^3)(\alpha^3-8)^2$.

P. F. Teilhet[234] gave the solutions 65, 56, 671; 5985, 5896, 647569.

E. Fauquembergue[235] reproduced Euler's[225] formulas with $p=n$, $q=m$. Replacing p by $\beta-\alpha$, q by $-\alpha$, we obtain the formulas of Axel Thue[236] for $x^3+y^3=z^2$, who noted that, if z is not divisible by 3, then $x^2-xy+y^2=B^2$. Thus, for relatively prime p and q, $px=q(B+x-y)$, $qy=p(B-x+y)$, since the product of the second factors is xy. Eliminating B, we get

$$x/y = (q^2-2pq)/(p^2-2pq).$$

In case the numerator and denominator have a common factor, it is 3, and $p-2q=3p_1$; set $q_1=q+2p_1$; we get

$$x : y = q_1^2-2p_1q_1 : p_1^2-2p_1q_1.$$

Hence in every case we may set

$$x=\pm(q^2-2pq), \qquad y=\pm(p^2-2pq), \qquad B=\mp(p^2-pq+q^2).$$

Now $x+y$ must be a square, A^2. Hence $(q-2p)^2-3p^2=\pm A^2$, so that the lower sign is excluded. From $2pq=(p-q)^2-A^2$, we get

$$2p\alpha=\beta(p-q+A), \qquad q\beta=\alpha(p-q-A), \qquad \frac{p}{q}=\frac{\beta^2+2\alpha\beta}{2\alpha\beta-2\alpha^2},$$

where α, β are relatively prime. Any common factor of the numerator and denominator divides 6. If it be 3, we reduce to a like fraction as above. If it be 2, then β and hence p and y are even; but we may assume that if either x or y is even, x is even. Thus in every case we may set

$$p=\pm(\beta^2+2\alpha\beta), \qquad q=\pm(2\alpha\beta-2\alpha^2),$$
$$x=4\alpha(\alpha^3-\beta^3), \qquad y=\beta(\beta^3+8\alpha^3).$$

It follows that $X^6+Y^3=z^2$ is impossible in integers if z is not divisible by 3. For, if the preceding x or y be a square, $\alpha=k^2$, $\alpha^3-\beta^3=h^2$, or $\beta=k_1^2$, $\beta^3+8\alpha^3=h_1^2$, respectively; in either case, $X_1^6+Y_1^3=z_1^2$ in smaller integers.

Multiplying x and α by $\sqrt[3]{A}$, y and β by $\sqrt[3]{B}$, we see that $Ax^3+By^3=z^2$ has the integral solutions

$$x=4\alpha(A\alpha^3-B\beta^3), \qquad y=\beta(B\beta^3+8A\alpha^3), \qquad z=B^2\beta^6-20AB\alpha^3\beta^3-8A^2\alpha^6.$$

---

[230] Atti Ac. Pont. Nuovi Lincei, 19, 1865–6, 185.

[231] Nouv. Ann. Math., (3), 1, 1882, 364; cf. (2), 20, 1881, 515; l'intermédiaire des math., 9, 1902, 329; 10, 1903, 133.

[232] Math. Quest. Educ. Times, 23, 1875, 85–6.

[233] Nouv. Ann. Math., (3), 4, 1885, 184. Cf. Gérardin.[242]

[234] L'intermédiaire des math., 3, 1896, 246.

[235] Ibid., 4, 1897, 110–12. Cf. the remarks, 112–15.

[236] Ibid., 5, 1898, 95; Det Kgl. Norske Videnskabers Selskabs Skrifter, 1896, No. 7.

"Alauda"[237] noted that $nx^2 = y^3 + z^3$ if $x = 3n$, $y = 2n$, $z = n$. E. Fauquembergue (*ibid.*, 6, 1899, 131) gave [Euler[225]]

$$ab\{6(a^2 + 3b^2)\}^2 \equiv (6ab + a^2 - 3b^2)^3 + (6ab - a^2 + 3b^2)^3.$$

K. Schwering[238] obtained an infinity of solutions by means of the relation between Abel's theorem and certain diophantine equations, first indicated by Jacobi[148] of Ch. XXII. Set

$$x^3 + 1 - (mx + n)^2 \equiv (x - \alpha_1)(x - \alpha_2)(x - \alpha_3).$$

By the coefficients of x^2 and x,

$$\frac{-m}{2n} = \frac{\alpha_1 + \alpha_2 + \alpha_3}{\alpha_1\alpha_2 + \alpha_1\alpha_3 + \alpha_2\alpha_3}.$$

Substitute for m, n their values from $m\alpha_i + n = (\alpha_i^3 + 1)^{\frac{1}{2}}$ for $i = 1, 2$. Thus

$$\alpha_3 = \frac{\alpha_1^2\alpha_2^2 - 4(\alpha_1 + \alpha_2)}{\alpha_1\alpha_2(\alpha_1 + \alpha_2) + 2 + 2\sqrt{(\alpha_1^3 + 1)(\alpha_2^3 + 1)}}.$$

Hence we get $m\alpha_3 + n$ and thus $(\alpha_3^3 + 1)^{\frac{1}{2}}$. Take $\alpha_1 = \alpha_2 = \alpha$. Then

$$\alpha_3 = \frac{\alpha^4 - 8\alpha}{4\alpha^3 + 4}, \qquad -\sqrt{\alpha_3^3 + 1} = \frac{\alpha^6 + 20\alpha^3 - 8}{8(\alpha^3 + 1)\sqrt{\alpha^3 + 1}}.$$

By eliminating α_3, we get the desired solution

$$(\alpha^3 - 8)^3\alpha^3 + 64(\alpha^3 + 1)^3 = (\alpha^6 + 20\alpha^3 - 8)^2.$$

The corresponding Abel theorem is here $\Sigma d\alpha_i / \sqrt[3]{(\alpha_i^3 + 1)^2} = 0$.

A. S. Werebrusow[239] gave Euler's[225] final solution.

F. de Helguero[240] solved $(x - y)t = z^2$, where $t = x^2 + xy + y^2$. Set $d = 3$ or 1, according as t is or is not divisible by 3. Then $x - y = d\alpha^2$, $t = d\beta^2$. Thus $d\beta^2$ has one of the three representations by $x^2 + xy + y^2$. It remains to make $d(x - y) = \square$. According as $d = 3$ or 1, this reduces to $u^2 - v^2 = w^2$ or $u^2 - 3v^2 = 1$.

F. Pegorier[241] discussed $(x + 1)^3 - x^3 = \square$.

A. Gérardin[242] noted that one solution of $\alpha^3 + \beta^3 = \gamma^2$ implies a second since

$$(\alpha^3 + 4\beta^3)^3 - (3\alpha^2\beta)^3 \equiv (\alpha^3 + \beta^3)(\alpha^3 - 8\beta^3)^2.$$

W. H. L. Janssen van Raay[243] discussed the solution of $x^3 + y^3 = z^2$. Cashmore[244] gave the first solution due to Hoppe.[229]

See Bouniakowsky,[135] Mordell,[176] and Baer[224]; also Catalan[122a] and Tafelmacher[160] of Ch. XXVI.

[237] L'intermédiaire des math., 5, 1898, 75–6.
[238] Archiv Math. Phys., (3), 2, 1902, 285–8.
[239] L'intermédiaire des math., 11, 1904, 153.
[240] Giornale di Mat., 47, 1909, 362–4.
[241] Bull. de math. élém., 14, 1908–9, 51–52.
[242] L'intermédiaire des math., 18, 1911, 201–2. Cf. Weill.[233]
[243] Wiskundige Opgaven, 12, 1915, 67–71 (Dutch).
[244] L'intermédiaire des math., 23, 1916, 224.

SUM OF CUBES OF NUMBERS IN ARITHMETICAL PROGRESSION A CUBE.

L. Euler[245] treated the problem to find three consecutive numbers $x-1$, x, $x+1$, the sum $3x^3+6x$ of whose cubes is a cube. Since $x=4$ gives a solution, set $x=4+y$. Then $6^3+150y+36y^2+3y^3$ is to be the cube of a number, say $6+fy$. The coefficients of y are equal if $108f=150$, and then $1871y=-7452$, $x=32/1871$. Or we may take $3x^3+6x=27x^3z^3$, whence $x^2(18z^3-2)=4$, and $18z^3-2$ is to be a square. Since this is the case for $z=1$, set $z=1+v$; the cubic in v is the square of $4+27v/4$ if $v=-15/32$.

J. R. Young[246] required that the sum of the cubes of $a-a/x$, a, $a+a/x$ be a cube. Hence, as by Euler, $3+6/x^2$ is to be a cube. To make $x^2=2n^3$, take $x=2nq$, whence $n=2q^2$. Then $3n^3+3=24q^6+3$ is to be a cube, which is true if $q=1$.

C. Pagliani[247] treated the problem to find 1000 consecutive numbers the sum of whose cubes is a cube. The sum of the cubes of $x+1$, $\cdots$, $x+m$ is $s=m(y+1)(y^2+2y+m^2)/8$ for $y=2x+m$. Let $m=8n^3$. Then s will be the cube of $n(y+4n^2)$ if $y=0$ or

$$3(4n^2-1)y=2(32n^6-24n^4+1).$$

Writing v for $2n$, we see that this is equivalent to saying that

$$(x+1)^3+(x+2)^3+\cdots+(x+v^3)^3=\{vx+\tfrac{1}{2}v^3(v+1)\}^3$$

if $6x=(v^2-1)^2-3(v^3+1)$. Then x is integral if v is not divisible by 3. The cases $v=2$, 4, 10 give

$$(1)\quad 3^3+4^3+5^3=6^3,\quad 6^3+7^3+\cdots+69^3=180^3,\quad 1134^3+\cdots+2133^3=16830^3.$$

W. Lenhart[248] treated the problem of m consecutive cubes whose sum is a cube. First, let $m=2n$. The sum of the cubes of $s+1$, $\cdots$, $s+n$, s, $s-1$, $\cdots$, $s-n+1$ is $\sigma=(2s+1)(ns^2+ns+n^3)$. Set $n=4n_1^3$ and divide σ by $(2n_1)^3$; we get

$$s^3+\tfrac{3}{2}s^2+\tfrac{1}{2}s(32n_1^6+1)+8n_1^6=(s+2n_1^2)^3,$$

if $3s=8n_1^4-4n_1^2-1$. To make s an integer >1, take n_1 prime to 3. For $n_1=1$, the roots of the 8 cubes are 2, 1, 3, 0, 4, -1, 5, -2, leading to (1_1). For $n_1=2$ or 5 we get (1_2), (1_3). Again, we can equate σ to the cube of $n+s(2n^2+1)/(3n)$ by choice of s in terms of n. Second, let $m=2n+1$. Then

$$\Sigma=\sigma+(s-n)^3=ms^3+\tfrac{1}{4}sm(m^2-1).$$

Since Σ is a cube for $s=1/2$, set $s=1/2+t$ and take $m=m_1^3$. Thus

$$\frac{\Sigma}{m_1^3}=\frac{1}{8}m_1^6+\frac{1}{4}(m_1^6+2)t+\frac{3}{2}t^2+t^3=(\tfrac{1}{2}m_1^2+t)^3,$$

if $t=(m_1^4-2m_1^2-2)/6$, whence $s=(m_1^2-1)^2/6$. Again, let $\Sigma=p^3m^3s^3$. Then,

[245] Algebra, 2, 1770, art. 249; French transl., 2, 1774, p. 365. Opera Omnia, (1), I, 497-8.
[246] Algebra, 1816; Amer. ed., 1832, 332.
[247] Annales de math. (ed., Gergonne), 20, 1829-30, 382-4.
[248] Math. Miscellany, New York, 2, 1839, 127-132; French transl., Sphinx-Oedipe, 8, 1913,

for $p = 1 + r$,

$$\frac{1}{4s^2} = \frac{p^3 m^2 - 1}{m^2 - 1} = \left\{ 1 + \frac{3m^2 r}{2(m^2 - 1)} \right\}^2$$

if $r = \frac{3}{4}(4 - m^2)/(m^2 - 1)$, whence $s = 4(m^2 - 1)^2/\{18m^2 + 9 - (m^2 - 1)^2\}$.

V. A. Lebesgue[249] stated that, if x and r are positive integers,

(2) $x^3 + (x+r)^3 + (x+2r)^3 + \cdots + [x + (n-1)r]^3 = (x+nr)^3$

is impossible except for $n = 3$, $x = 3r$. If we write

(3) $s = 2x + (n-1)r, \qquad \sigma = s^2 + (n^2 - 1)r^2,$

we obtain for the left member of (2) the expression $ns\sigma/8$. He considered it a difficult problem to make the latter a cube, and remarked that it was impossible for $n = 2$ by Euler's[3] theorem.

A. Genocchi[250] treated the last problem $ns\sigma/8 = y^3$. Set $s = rt$, $2y = rz$. Then $nt(t^2 + n^2 - 1) = z^3$. Following Fermat's method,[143] set $t = 1 + u$, $z = n + pu$, and equate the terms of the first degree in u. Hence

(4) $$p = \frac{n^2 + 2}{3n}, \qquad u = \frac{3n(1 - p^2)}{p^3 - n}.$$

The cases $n = 3$, $r = 107$; $n = 4$, $r = 1$; $n = 5$, $r = 13$, give respectively

(5) $149^3 + 256^3 + 363^3 = 408^3, \qquad 11^3 + 12^3 + 13^3 + 14^3 = 20^3,$
(6) $230^3 + 243^3 + 256^3 + 269^3 + 282^3 = 440^3.$

B. Boncompagni[251] proposed for solution the same problem (2) and

(7) $x^3 + (x+r)^3 + \cdots + [x + (n-1)r]^3 = v^3.$

V. Bouniakowsky[252] noted the particular solution $r_0 = 2$, $x_0 = -n + 2$, $v_0 = n$, of (7), and that this solution leads to the second solution

$$r = r_0 = 2, \qquad x = x_0 + u, \qquad v = v_0 + pu,$$

where p and u are given by (4), and thus derived (5), etc. Starting from the latter, we obtain new solutions. For $n = 3$, $ns\sigma/8$ is the cube of $v_1 v_2$ if

$$3(x+r) = v_1^3, \qquad (x+r)^2 + 2r^2 = v_2^3.$$

The general solution of the second equation is known to be

$$x + r = \pm(p^3 - 6pq^2), \qquad r = \pm(3p^2 q - 2q^3), \qquad v_2 = p^2 + 2q^2.$$

Taking the upper signs, we see by the first condition that

$$p = 3p', \qquad v_1 = 3w, \qquad 3p'^3 - 2p'q^2 = w^3.$$

From the evident solution $p' = q = w = 1$, we get $p = v_1 = 3$, $q = 1$, etc. In (2), he set $r = \lambda x$ and noted that the rational cubic for λ has no rational root when $n < 8$ except for $n = 3$, and stated that (1_1) is the only solution in positive cubes.

[249] Annali di Mat., (1), 5, 1862, 328.
[250] *Ibid.*, 329.
[251] Nouv. Ann. Math., (2), 3, 1864, 176; Zeitschr. Math. Phys., 9, 1864, 284.
[252] Bull. Acad. Sc. St. Pétersbourg, 8, 1865, 163–170.

A. Genocchi[253] treated (7), i. e., to make $ns\sigma$ a cube. Set

$$m=n^2-1, \qquad s=n^2s'^3, \qquad s+r\sqrt{-m}=(p+q\sqrt{-m})^3.$$

Then

$$r=q(3p^2-mq^2), \qquad n^2s'^3=p(p^2-3mq^2).$$

Set

$$np=8v'^3, \qquad p+q\sqrt{3m}=(s''+\tfrac{1}{3}r''\sqrt{3m})^3.$$

From the resulting rational expressions for p, q we get

$$ns''[s''^2+(n^2-1)r''^2]=8v'^3,$$

which is of the same form as the initial equation $ns\sigma=8v^3$. Hence one solution r'', s'', v' leads to a second solution r, s, v, etc. But not all solutions are so obtained. More convenient formulæ are obtained by setting $r=g+z$, $2v=h+pz$, where $r=g$, $2v=h$ is one set of solutions.

L. Matthiessen[254] noted the particular solutions of (7):

$$n=2p+3, \qquad x=-2p-1, \qquad r=2, \qquad v=2p+3;$$
$$n=2p+4, \qquad x=-\ p-1, \qquad r=1, \qquad v=\ p+2.$$

Also that 351120^3 is a sum of k positive cubes for $k=3, 4, 5, 6, 7, 8$.

A. B. Evans[255] noted that the sum of the cubes of the first n^3 integers is a cube only if $n=1$, since $(n^3+1)/2$ is not a cube if $n>1$ [Euler[182] on $x^3\pm y^3=2z^3$].

D. S. Hart[256] took $2n-1$ consecutive integers, x being the middle one. The sum of their cubes is $(2n-1)x^3+(2n^3-3n^2+n)x$. For $2n-1=p^3$, the sum is a cube if $s=x^3+\tfrac{1}{4}(p^6-1)x$ is a cube. Take $x=\tfrac{1}{2}+y$, $8s=(2y+p^2)^3$; we get y and $x=(p^2-1)^2/6$. For $2n$ cubes, add the term $(x+n)^3$. The answer is now $x=\{(p^2-1)^2-3\}/6$.

A. Martin[257] noted that the sum of the cubes of x, $x+1$, $\cdots$, $x+n^3-1$ is a cube if $x=(n^4-3n^3-2n^2+4)/6$.

Hart[258] expressed the difference of $1^3+\cdots+n^3$ and $(S+m)^3-S^3$ as a sum of cubes by trial.

S. Réalis[259] stated that $z_1^3+\cdots+z_n^3=(5n+3)z^3$ has a solution with $z_1, \cdots, z_n$ in arithmetical progression, and solutions with $z=1$, $n\neq2$.

A. Martin[260] proved that $1^3+2^3+\cdots+n^3$ is not a cube if $n>1$, since $n(n+1)/2\neq p^3$. For, $(2n+1)^2=8p^3+1$ is of the form $x^3+1=\Box$, which holds (Euler[157]) only if $x=0$, -1, 2. He listed (p. 188) sets of 20, 25 and 64 consecutive cubes whose sum is a cube, besides known cases.

[253] Annali di Mat., 7, 1865, 151–8; Atti Accad. Pont. Nuovi Lincei, 19, 1865–6, 43–50. French transl., Jour. de Math., (2), 11, 1866, 179; Sphinx-Oedipe, 4, 1909, 73–8. Account by M. Cantor, Zeitschr. Math. Phys., 11, 1866, 248–251.
[254] Zeitschr. Math. Phys., 13, 1868, 348–350.
[255] Math. Quest. Educ. Times, 14, 1871, 32–33.
[256] Ibid., 15, 1871, 24–6 (Math. Magazine, 1, 1884, 173–6).
[257] Ibid, p. 26. Same by J. Matteson, Collection of Dioph. Problems, 1888, Probs. 4, 5.
[258] Math. Quest. Educ. Times, 23, 1875, 82–83.
[259] Nouv. Corresp. Math., 6, 1880, 525–6.
[260] Math. Magazine, 2, 1895, 159.

E. B. Escott[261] proved that, for $2 \leqq n \leqq 5$,
$$k^n + (k+1)^n + \cdots + (k+m)^n = (k+m+1)^n$$
has only the following integral solutions: (1_1) and
$$3^2 + 4^2 = 5^2, \qquad (-2)^3 + (-1)^3 + 0^3 + 1^3 + \cdots + 5^3 = 6^3.$$

L. Matthiessen[262] noted that if fractional values of x, v are allowed in (7), we may set $r=1$. Write $u=2x+n-2$, $v=pu+n/2$. The usual form of (7) becomes a quadratic in u:
$$(n-8p^3)u^2 + 3n(1-4p^2)u + n^3 + 2n - 6n^2p = 0.$$
Evident solutions are obtained by equating to zero the first or third coefficient. In the second case, integers x are found only for $n=2$, $n=4$.

F. Hromádko[263] noted that $x=3$ is the only positive integral solution of $x^3 + (x+1)^3 + (x+2)^3 = (x+3)^3$ [Lebesgue[249]].

E. Grigorief[264] obtained the special solutions
$$15^2 + 52^3 + 89^3 + \cdots + 348^3 = 495^3, \qquad 76^3 + 477^3 + 878^3 + \cdots + 2883^3 = 3016^3,$$
$$435^3 + 506^3 + 577^3 + 648^3 + 719^3 + 790^3 = 1155^3.$$

" L. N. Machaut "[265] treated (2) by setting $x/r=u$ and obtaining a cubic for u with a real positive root $(u=3)$ only for $n=3$, leading to (1_1).

J. N. Vischers[266] proved Lebesgue's[249] first result when $n=3$.

L. Aubry[267] proved that 3, 4, 5 are the only three consecutive integers the sum of whose cubes is a cube.

Sum of Cubes of Numbers in Arithmetical Progression a Square.

To find five integers in A. P. the sum of whose cubes is a square (or sum of squares is a cube), J. Stevenson[268] used $nx-2x$, $nx-x$, nx, $nx+x$, $nx+2x$, the sum of whose cubes $5n^3x^3 + 30nx^3$ will equal m^2x^2 by choice of x (or sum of squares $5n^2x^2 + 10x^2 = m^3x^3$ by choice of x). Several solved both questions simultaneously by using x^2, $2x^2$, $3x^2$, $4x^2$, $5x^2$, whose sum of cubes is $(15x^3)^2$ and sum of squares is $55x^4 = a^3x^3$, if $x=a^3/55$; take $a=55$.

Several[269] made the sum $n^2(2n^2-1)$ of the cubes of the first n odd integers a square by using Euler's[83] solutions (Ch. XII) $n=1, 5, 29, \cdots$ of $2n^2-1 = \square$.

A. Genocchi[253] discussed the rational solutions of
$$(1) \qquad x^3 + (x+r)^3 + (x+2r)^3 + \cdots + (x+nr-r)^3 = y^2.$$
In view of (3) of Lebesgue[249] the problem is $ns\sigma = 8y^2$. Set $2y=nst$. Solving $\sigma = 2nst^2$ for s, we see that $n^2t^4 - (n^2-1)r^2 = \square = (nt^2-rp)^2$. Hence
$$dr = 2npt^2, \qquad ds = 2n(n^2-1)t^2 \text{ or } 2np^2t^2,$$

[261] L'intermédiaire des math., 5, 1898, 254–6; 7, 1900, 141.
[262] Zeitsch. Math. Naturw. Unterricht, 33, 1902, 372–5.
[263] *Ibid.*, 34, 1903, 258.
[264] L'intermédiaire des math., 9, 1902, 319.
[265] *Ibid.*, 15, 1908, 163–4.
[266] Wiskundig Tijdschrift, 5, 1908, 65.
[267] Sphinx-Oedipe, 6, 1911, 142–3.
[268] The Gentleman's Diary, or Math. Repository, London, 1814, 36–7, Quest. 1010.
[269] Ladies' Diary, 1832, 36, Quest. 1529.

where $d = n^2 - 1 + p^2$. The general solution thus involves the rational parameters p, t.

E. Catalan[270] stated that, if $r = 1$, integral solutions of (1) are

$$n = kb^2\gamma, \qquad x = \frac{(a^2 - kb^2)\gamma + 1}{2}, \qquad y = \frac{abu\gamma}{2}, \qquad (a^4 + k^2 b^4)\gamma^2 - \frac{2}{k}u^2 = 1,$$

where $k = 1$ or 2, while a, b are relatively prime integers. For example, if $a = 5$, $b = 1$, we may take $\gamma = 313$, $u = 7850$ (in place of 1850 in Table X in Legendre's Théorie des nombres), whence $n = 626$, $x = 3600$.

Catalan,[271] in treating the integral solutions of (1) for $r = 1$, wrote $\alpha = 2ns$, $\beta = \sigma$, where s, σ are given by Lebesgue's[249] (3) for $r = 1$. The problem is then to make $\alpha\beta$ the square $16y^2$ of an integer. Since sn is even, y will then be an integer. But his separation into two cases lacks generality and his solution is incomplete. His[272] later discussion leads to the following result: Take any two relatively prime integers p, q, one even, and express $pq/2$ as a product of a square u'^2 by a number θ without a square factor; then if

$$(p^2 + q^2)\gamma^2 - 4\theta v^2 = 1$$

has integral solutions γ, v, we have

$$2x = (q - p)\gamma + 1, \qquad 2(x + n - 1) = (q + p)\gamma - 1, \qquad y = (u'v\theta\gamma)^2.$$

M. Cantor[273] reported on Catalan's[271, 2] discussion of the preceding equation $\alpha\beta = 16y^2$, where α and β are integers divisible by 4 for which $\beta \pm \alpha + 1$ are squares, and obtained two sets of solutions, in which p and q are relatively prime integers, one an odd square and the other either half of an even square or an even square. In the first case, $(p^2 + q^2)\gamma^2 - u^2 = 1$ yields integers γ, u, and then $y^2 = 2pq(\gamma u/4)^2$. In the second case, if $(p^2 + q^2)\gamma^2 - 2w^2 = 1$ has integral solutions γ, w, then $y^2 = pq(\gamma w/2)^2$. In each case, $n = p\gamma$,

$$2x = (q - p)\gamma + 1.$$

C. Richaud[274] treated (1) for $r = 1$, viz., $l^2 - k^2 = y^2$, where $2k = x(x - 1)$, $2l = (x + n)(x + n - 1)$. Certain, but not all, solutions arise from $l = a^2 + b^2$; k, $y = 2ab$, $a^2 - b^2$. Eliminating x, y, we get a quartic equation. For $k = 2ab$, it becomes

$$m^2 - (4t^4 + 1)n^2 = -1, \qquad m = 2(a + b), \qquad nt = a - b,$$

with an infinitude of solutions $m = 2t^2$, $n = 1$; $m = 32t^6 + 6t^2$, $n = 16t^4 + 1$; etc. Note that the sum of the numbers x, $x + 1$, $\cdots$, $x + n - 1$ is a square, $(a - b)^2$. For a general r, (1) becomes $ns\sigma = 8y^2$ by Lebesgue's[249] (3). For[275] $ns/2 = \alpha b^2$, $\sigma/4 = \alpha a^2$, $y = \alpha ab$, he eliminated s and discussed at length the resulting

[270] Bull. Acad. Roy. de Belgique, (2), 22, 1866, 339–40.

[271] Atti Accad. Pont. Nuovi Lincei, 20, 1866–7, 1–4; Nouv. Ann. Math., (2), 6, 1867, 63–67; Mélanges Math., 1868, 99–103.

[272] Atti Accad. Pont. Nuovi Lincei, 20, 1866–7, 77; Nouv. Ann. Math., (2), 6, 1867, 276–8; Mélanges Math., 1868, 248–251.

[273] Zeitschr. Math. Phys., 12, 1867, 170–2.

[274] Atti Accad. Pont. Nuovi Lincei, 20, 1866–7, 91–110.

[275] In the alternative case $ns/4 = \alpha b^2$, $\sigma/2 = \alpha a^2$, $y = \alpha ab$, not treated, there are two misprints for 4.

equation, for the case $\alpha=1$, whence the sum of x, $x+r$, $\cdots$, $x+(n-1)r$ is b^2. In the most interesting case $\alpha=1$, $r=2$, the eliminant becomes $a^2-(t^4+1)n^2=-1$ for $b=nt$. It has an infinitude of solutions $(a, n)=(t^2, 1)$, $(4t^6+3t^2, 4t^4+1)$, etc. Taking $t=1$, we have $x=1$ and the following result: While the sum of n consecutive odd numbers 1, 3, $\cdots$ is always a square n^2, the sum of the cubes of the same n odd numbers will be the square of an when $a^2-2n^2=-1$. Examples are $(a, n)=(1, 1)$, $(7, 5)$, $(41, 29)$, $(239, 169)$.

E. Lucas[275a] stated that the sum of the cubes of five consecutive integers is a square only when the middle number is 2, 3, 98 or 120. The sum of two consecutive cubes is a square only for the cubes 1 and 8.

G. R Perkins[276] solution of (1) differs only in notation from Genocchi's[253].

E. Lucas[277] asked when the sum of 7 consecutive cubes is a square.

Several[278] found that the sum of the cubes of the first n odd integers is a square if $2n^2-1=\square$, $n=1, 5, 29, \cdots$.

M. A. Gruber[279] attempted to show that a sum of cubes of n consecutive integers is a square only for $1^3+2^3+\cdots+n^3=(1+\cdots+n)^2$.

A. Cunningham[280] desired a sum of successive odd cubes equal to a square. The sum S_r of the successive odd cubes 1, 3^3, $\cdots$, $(2r-1)^3$ is $r^2(2r^2-1)$ and is a square if $r=5$. Next,

$$(2\rho+1)^3+\cdots+(2r-1)^3=S_r-S_\rho=(r^2-\rho^2)(2r^2+2\rho^2-1)$$

is a square z^2 if, upon setting $x=2r^2$, $y=2\rho^2$,

$$(2x-1)^2-(2y-1)^2=2(2z)^2.$$

Solutions are found by making special assumptions.

W. A. Whitworth[281] expressed $\sqrt{2}$ as a continued fraction, took a convergent N/D, with D odd, and got

$$1^3+3^3+\cdots+(2D-1)^3=N^2D^2.$$

Cunningham[282] asked for a sum of successive cubes

$$S_{m,\,n}=(n+1)^3+(n+2)^3+\cdots+m^3$$

equal to the product of a square by q. Since

$$S_{m,\,0}=1^3+2^3+\cdots+m^3=T_m^2, \qquad T_m=\tfrac{1}{2}m(m+1), \qquad S_{m,\,n}=S_{m,\,0}-S_{n,\,0},$$

we set $T_m=\xi T_n$ and see that $S_{m,\,n}\div q$ is a square if $(\xi^2-1)/q$ is a square. For each such ξ, we test $T_m=\xi T_n$ by a table of triangular numbers (de Joncourt's, 1772) and find suitable pairs m, n. Solutions are found for $q=2, \cdots, 11$.

M. A. Gruber[283] noted that $n=1$ and $n=5$ are the only cases in which

$$1^3+3^3+5^3+\cdots+(2n-1)^3=\square, \qquad (2n-1)^3=\square.$$

[275a] Recherches sur l'analyse indéterminée, Moulins, 1873, 92. Extract from Bull. Soc. d'Emulation du Département de l'Allier, 12, 1873, 532.

[276] The Analyst, Des Moines, 1, 1874, 40.

[277] Nouv. Corresp. Math., 2, 1876, 95.

[278] Math. Quest. Educ. Times, 53, 1890, 55. Cf. Brocard[92] of Ch. XXIII.

[279] Amer. Math. Monthly, 2, 1895, 197-8.

[280] Math. Quest. Educ. Times, 72, 1900, 45-46 (error); 73, 1900, 132-3.

[281] Ibid., 72, 1900, 46.

[282] Ibid., 75, 1901, 87-88.

[283] Amer. Math. Monthly, 7, 1900, 176.

L. Matthiessen[284] discussed (1) in three ways. One way is to multiply (7), the corresponding equation with the right number v^3, by z^3, where $v^3z^3=y^2$. Thus, for $11^3+12^3+13^3+14^3=20^3$, take $z=5$, whence $y=1000$.

H. Brocard, "E. A. Majol," and F. Ferrari[285] discussed a sum of three consecutive cubes equal to a sum of two squares.

L. Aubry[286] treated $(y-k)^3+y^3+(y+k)^3\equiv 3y(y^2+2k^2)=u^2$. First, let $y=2a^2$, $y^2+2k^2=6b^2$, $u=6ab$. Then $2a^4=3b^2-k^2$, which is satisfied if

$$a=q^2-3p^2, \qquad b=q^4+4pq^3+18p^2q^2+12p^3q+9p^4,$$
$$k=q^4+12pq^3+18p^2q^2+36p^3q+9p^4.$$

Second, let $y=6a^2$, $y^2+2k^2=2b^2$, $u=6ab$. Then $18a^4=b^2-k^2$, which holds if

$$a=2pq, \qquad b=rp^4+sq^4, \qquad k=rp^4-sq^4,$$

$(r,\ s)=(72,\ 1)$ or $(9,\ 8)$. For $p=q=1$, the second set gives

$$23^3+24^3+25^3=204^2,$$

which occurs in a manuscript of Lucas'. Or we may set $y=3a^2$ or a^2.

Homogeneous cubic equation $F(x, y, z)=0$.

A. Cauchy[287] derived a second solution from a given solution a, b, c. Let $\phi(x, y, z)$, χ, ψ be the first partial derivatives of $F(x, y, z)$ with respect to x, y, z, respectively. Then $F=0$ for

(1) $$x : y : z = as-t\alpha : bs-t\beta : cs-t\gamma,$$

where, if $u=\phi(a, b, c)$, $v=\chi(a, b, c)$, $w=\psi(a, b, c)$, the parameters α, β, γ satisfy $u\alpha+v\beta+w\gamma=0$, while

$$s=F(\alpha, \beta, \gamma), \qquad t=a\phi(\alpha\ \beta, \gamma)+b\chi(\alpha, \beta, \gamma)+c\psi(\alpha, \beta, \gamma).$$

We may take α, β, $\gamma=0$, w, $-v$; $-w$, 0, u; or v, $-u$, 0. In each case one of the terms (1) is very simple. He showed that we may take such a simple value and obtain the following solution

(2) $$\frac{a^2x}{F(0, w, -v)}=\frac{b^2y}{F(-w, 0, u)}=\frac{c^2z}{F(v, -u, 0)}.$$

These become

(3) $$\frac{x}{a(Bb^3-Cc^3)}=\frac{y}{b(Cc^3-Aa^3)}=\frac{z}{c(Aa^3-Bb^3)}$$

for the case

(4) $$F\equiv Ax^3+By^3+Cz^3+Kxyz=0.$$

If a, b, c and a', b', c' are two given sets of solutions of $F=0$, where F is any ternary cubic form, Cauchy obtained a third set by expanding

$$F(as-ta', bs-tb', cs-tc')=0$$

[284] Zeitschr. Math. Naturw. Unterricht, 37, 1906, 190-3.

[285] L'intermédiaire des math., 15, 1908, 41-43.

[286] Sphinx-Oedipe, 8, 1913, 28-9. Cf. Lucas[88a] of Ch. XXIII.

[287] Exercices de mathématiques, Paris, 1826, 233-260; Oeuvres de Cauchy, (2), 6, 1887, 302. For a less effective method, see Cauchy[150] of Ch. XIII.

and obtaining $stL = 0$, where L is a linear function of s, t, which is zero for
$$s = a\phi(a'\ b',\ c') + b\chi(a',\ b',\ c') + c\psi(a',\ b',\ c'),$$
$$t = a'\phi(a,\ b,\ c) + b'\chi(a,\ b,\ c) + c'\psi(a,\ b,\ c).$$
Then the resulting third set of solutions of $F = 0$ is

(5)
$$x : y : z = as - ta' : bs - tb' : cs - tc'.$$

By (3) for $A = B = 1$, $C = -a^3 - b^3$, $K = 0$, $c = 1$, we see that
$$x = a(a^3 + 2b^3), \qquad y = -b(b^3 + 2a^3), \qquad z = a^3 - b^3$$
satisfy $x^3 + y^3 = (a^3 + b^3)z^3$ [Prestet[181]].

For geometrical interpretations of Cauchy's results, see Lucas.[296]

A. M. Legendre[288] deduced from one solution of $x^3 + ay^3 = bz^3$ the second solution
$$X = x(x^3 + 2ay^3) \qquad Y = -y(2x^3 + ay^3), \qquad Z = z(x^3 - ay^3).$$

Given X, Y, Z, the determination of x, y, z depends on a quartic equation.

J. J. Sylvester[289] stated that (4) can be transformed into
$$A'u^3 + B'v^3 + C'w^3 + Kuvw \qquad (A'B'C' = ABC),$$
where uvw is a factor of z, provided (i) the ratio of two of the coefficients A, B, C is a cube, (ii) the "determinant" $27ABC + K^3$ has no positive prime factor $6l + 1$, and (iii) if 2^m and 2^n are the highest powers of 2 dividing ABC and K, respectively, then either m is of the form $3k \pm 1$ or, if not, m exceeds $3n$. If α, β, γ give one solution of (4) and if we set

(6)
$$F = A\alpha^3, \qquad G = B\beta^3, \qquad H = C\gamma^3, \qquad x = F^2G + G^2H + H^2F - 3FGH,$$
$$y = FG^2 + GH^2 + HF^2 - 3FGH, \qquad z = \alpha\beta\gamma(F^2 + G^2 + H^2 - FG - FH - GH),$$

then $x^3 + y^3 + ABCz^3 + Kxyz = 0$. For the case $A = B = 1$, C a prime, and $27C + K^3$ positive and not divisible by a prime $6k + 1$, he[290] gave a process to obtain all integral solutions of (4) from one initial solution $P = (e, g, i)$. The process is to apply to P repetitions of transformation (6) and the transformation, depending also upon P, from one system l, m, n to the system
$$\lambda = 3gm(gl - em) + 3Cin(il - en) + K(gil^2 - e^2lm),$$
$$\mu = 3Cin(im - gl) + 3el(em - gl) + K(eim^2 - g^2lm),$$
$$\nu = 3el(en - il) + 3gm(gn - im) + K(egn^2 - i^2lm),$$
or to the system obtained by interchanging e and g.

Sylvester[291] stated that $F \equiv x^3 + y^3 + z^3 + 6xyz = 0$ is not solvable in integers; likewise for $2F = 27nxyz$ when $27n^2 - 8n + 4$ is a prime; and for $4F = 27nxyz$ when $27n^2 - 36n + 16$ is a prime. Set $M^3 - 27A = \Delta^3\Delta_1$ where Δ_1 has no cubic factor. If Δ_1 is even and contains no factor of the form $f^2 + 3g^2$, and if A is a prime, $x^3 + y^3 + Az^3 = Mxyz$ has no integral solution

[288] Théorie des nombres, ed. 3, 2, 1830, 113-7; Maser's transl., 2, 1893, 110-4.
[289] London, Edinburgh, Dublin Phil. Mag., 31, 1847, 189-191, 293-6 for corrected theorems; Coll. Math. Papers, 1, 1904, 107-13.
[290] Phil. Mag., 31, 1847, 467-471; Coll. Math. Papers, I, 114-8.
[291] Annali di Sc. Mat. e Fis., 7, 1856, 398-400; Math. Papers, II, 63-4.

except when $-M/A$ is the square of an integer. Likewise if A is of the form $p^{3w\pm1}$, where p is a prime. Also without the assumption that Δ_1 is even, provided it has no factor f^2+3g^2, while $A=2^{3w\pm1}$; or $A/2$ is a prime $qi\pm4$, and $M/9$ is an integer; or $A/4$ is a prime $qi\pm2$, and $M/18$ is an integer; or A is a prime and A, B are of the respective forms $qn\pm2$, $qn\pm6$, or $qn\pm4$, $qn\pm3$, or $qn\pm3$, qn.

E. Lucas[292] stated Cauchy's results on the cubic (4) as follows: (i) If a, b, c is one set of integral solutions, another set x, y, z is given by

$$\frac{x}{a}+\frac{y}{b}+\frac{z}{c}=0, \qquad Aa^2x+Bb^2y+Cc^2z=0.$$

(ii) If a, b, c and a', b', c' are two distinct sets of solutions, then

$$\begin{vmatrix} x & y & z \\ a & b & c \\ a' & b' & c' \end{vmatrix}=0, \qquad Aaa'x+Bbb'y+Ccc'z=0$$

give a third set. But (i) and (ii) do not yield all solutions. Lucas[293] had stated as exercises these results without relation to Cauchy. They were verified by Moret-Blanc,[294] and restated by A. Gérardin.[295]

Lucas[296] stated the generalizations to any homogeneous cubic $F(x, y, z)=0$. 1°. The tangent at a point m_1 with rational coördinates x_1, y_1, z_1, and on $F=0$, cuts the cubic at a rational point m, i. e.,

$$F=0, \qquad x\frac{\partial F}{\partial x_1}+y\frac{\partial F}{\partial y_1}+z\frac{\partial F}{\partial z_1}=0$$

determine x, y, z rationally. The point m is distinct from m_1 unless the tangent is parallel to an asymptote or passes through a point of inflexion. 2°. The secant m_1m_2 through two rational points on the cubic cuts the cubic in a rational point (in general distinct from m_1, m_2). 3°. The conic through five rational points on a cubic cuts it in a sixth rational point.

S. Réalis[297] obtained a second solution (quadratic in α, β, γ) of $x^3+2y^3+3z^3=6xyz$ from one solution α, β, γ.

Réalis[298] noted that all integral solutions except $x=y=z$ of

$$x^3+y^3+z^3=3xyz$$

are given by

$$x=(a-b)^3+(a-c)^3, \qquad y=(b-c)^3+(b-a)^3, \qquad z=(c-a)^3+(c-b)^3.$$

If α, β, γ is one set of solutions of

$$Ax^3+By^3+Cz^3=(A+B+C)xyz,$$

another set is given by

$$x=(A+B+C)(\alpha^2-\beta\gamma)+3(B\beta^2+C\gamma^2)-3\alpha(B\beta+C\gamma),$$

[292] Bull. Bibl. Storia Sc. Mat., 10, 1877, 175; Amer. Jour. Math., 2, 1879, 178.
[293] Nouv. Ann. Math., (2), 14, 1875, 526.
[294] Ibid., (2), 20, 1881, 201.
[295] Sphinx-Oedipe, 5, 1910, 90.
[296] Nouv. Ann. Math., (2), 17, 1878, 507–9; Amer. Jour. Math., 2, 1879, 180.
[297] Nouv. Corresp. Math., 4, 1878, 346–52.
[298] Ibid., 5, 1879, 8–11.

and values of y, z derived by permuting the triples of letters cyclically. All solutions of $x^3 + y^3 + z^3 = x^2y + y^2z + z^2x$ are given.

A. Desboves[299] proved that if x, y, z is one set of solutions of $Ax^3 + By^3 + Cz^3 = 0$, a second set of solutions is given by

$$X = x(Ax^3 + 2By^3), \qquad Y = -y(2Ax^3 + By^3), \qquad Z = z(Ax^3 - By^3).$$

For $A = 1$ this result is due to Legendre.[288]

J. J. Sylvester[300] called the intersection of the tangent at a point P on a cubic with the cubic the tangential of P. He proved for $A = B = C = 1$ that (3) gives the tangential to (4) at the point (a, b, c) and that the point on the cubic collinear with (a, b, c) and (a', b', c') has the coördinates

$$(7) \qquad bca'^2 - b'c'a^2, \qquad cab'^2 - c'a'b^2, \qquad abc'^2 - a'b'c^2.$$

A. Desboves[301] noted that Cauchy's formula (5) becomes, for (4),

$$x = 3Bbb'(ab' - ba') + 3Ccc'(ac' - ca') - K(a^2b'c' - a'^2bc),$$

with similar expressions for y, z. Since a, b, c and a', b', c' satisfy (4), we can express A, B as linear functions of C, K. Substitute the resulting value of B into x, etc. We get (7). This result, which is simpler than, but equivalent to, Cauchy's (5), had been found otherwise by Sylvester,[300] whose published announcement without proof was limited to the case $A = B = C = 1$, and, for $K = 0$, but A, B, C arbitrary, by Desboves[302] and by P. Sondat.[303] From the fact that (7) satisfy $Ax^3 + By^3 + Cz^3 = 0$, we have the identity

$$(b^3c'^3 - b'^3c^3)(a^2b'c' - a'^2bc)^3 + (c^3a'^3 - c'^3a^3)(b^2a'c' - b'^2ac)^3$$
$$+ (a^3b'^3 - a'^3b^3)(c^2a'b' - c'^2ab)^3 = 0.$$

This leads to solutions of the system of equations [cf. Bini[438]]

$$x^3 + y^3 + z^3 = x_1^3 + y_1^3 + z_1^3, \qquad xyz = x_1y_1z_1 \quad \text{or} \quad x + y + z = x_1 + y_1 + z_1.$$

Desboves[304] simplified Cauchy's proofs of (2) and (5), gave also a direct proof of (2), and showed that a^2 divides $F(0, w, -v)$, etc., a fact seemingly overlooked by Cauchy. Hence we may take $x = F(0, w, -v)/a^2$, etc., obtaining polynomials of degree 4 for x, y, z. As new results, he proved that if one solution of $F = 0$ is given we can reduce its complete solution to that of a biquadratic equation. He sought an F such that the latter is $A\xi^4 + B\eta^4 = C\zeta^2$, where $C = A + B$, the only biquadratic hitherto solved completely. The resulting F is

$$AC(x + y)z^2 + 2Cy^2z - (x - y)(x^2 + y^2).$$

He obtained the solution of $f(x, y) + cz^3 = 0$, with coefficients of special type, given solutions m, n of the cubic $f(x, y) = 0$.

A. Holm[305] noted that the tangent to a cubic at a rational point, not an inflexion point, cuts the cubic in a new rational point. In case there is a

[299] Nouv. Ann. Math., (2), 18, 1879, 404. Same by R. Norrie.[84]
[300] Amer. Jour. Math., 3, 1880, 61–6; Coll. Papers, 3, 1909, 354–7.
[301] Nouv. Ann. Math., (2), 20, 1881, 173–5; (3), 5, 1886, 563–5.
[302] Ibid., (2), 18, 1879, 407–8.
[303] Ibid., (2), 19, 1880, 459.
[304] Ibid., (3), 5, 1886, 545–579.
[305] Proc. Edinburgh Math. Soc., 22, 1903–4, 40.

rational asymptote, the line parallel to it and through a rational point cuts it again in a rational point.

A. S. Werebrusow[306] obtained solutions of (4) with $K=0$ from one solution.

B. Levi[307] considered a cubic equation with rational coefficients which corresponds to a cubic curve C of genus unity (transformable birationally into a straight line) and determined points on C by use of an elliptic parameter. By a configuration of rational points on C is meant the set of all rational points deduced from one or more rational points by the operations of finding the tangential point to a given point and finding the third intersection with C of the secant joining two points of the set. There are theorems on the number of points in a finite configuration of such rational points (cf. Hurwitz[312]). There is a discussion of the cubic

$$xz^2 - y(y-x)(y-kx) = 0$$

into which any cubic with a rational point can be transformed birationally.

A. Thue[308] considered $Ax^3 + By^3 = Cz^3$ in which x, y, z are relatively prime in pairs and $z \geqq y \geqq x > 0$. We can find integers p, q, r, without a common factor and numerically $< \sqrt{3z}$, such that $px + qy = rz$. Hence

$$ax = Cq^3 - Br^3, \qquad by = Ar^3 - Cp^3, \qquad cz = Aq^3 - Bp^3,$$

where a, b, c are integers. Hence $Aax + Bby = Ccz$. From this and the former linear relation we get the ratios of x, y, z. He introduced further numbers and deduced many relations with the aim to obtain limits for a, b, c, etc.

L. Chanzy[309] applied Lucas'[296] three methods to the equation

$$y^3 + px^2 + qx + ry + s = 0.$$

The tangent at (x_1, y_1) meets the cubic at the point with the ordinate

$$-p\left(\frac{3y_1^2 + r}{2px_1 + q}\right)^2 - 2y_1.$$

The line joining the known points (x_1, y_1), (x_2, y_2) meets the cubic in the point with the ordinate

$$y_3 = -p\left(\frac{x_2 - x_1}{y_2 - y_1}\right)^2 - y_1 - y_2,$$

while x_3 follows from $(x_3 - x_1)(y_2 - y_1) = (y_3 - y_1)(x_2 - x_1)$.

L. J. Mordell[310] considered a ternary cubic form $F(x, y, z)$. Given one set of solutions, we can find a linear unitary substitution which transforms $F = 0$ into $S_1\xi^2 + 2S_2\xi + S_3 = 0$, where S_j is a function of degree j of η, ζ. Its discriminant $f = S_2^2 - S_1S_3$ is a binary quartic whose invariants are

---

[306] Matem. Sborn. (Math. Soc. Moscow), 27, 1909, 211–227.

[307] Atti IV Congresso Internaz. Mat., Roma, 2, 1909, 173–7. Supplement to his four papers, Atti R. Accad. Sc. Torino, 41, 1906, 739–64; 43, 1908, 99–120, 413–434, 672–681.

[308] Skrifter Videnskapsselsk. Kristiania (Math.), 1, 1911, No. 4, pp. 19–21; 2, 1911, No. 15, 7 pp. The related No. 20 is considered under Thue[178] of Ch. XXIII.

[309] Sphinx-Oedipe, 8, 1913, 166–7.

[310] Quar. Jour. Math., 45, 1913–4, 181–6.

numerical multiples of the invariants S and T of F. If $S_1=b\eta+c\zeta$, f is a square for $\eta=c$, $\zeta=-b$. Thus (Mordell[162] of Ch. XXII) if we find[176] all rational solutions of

$$t^2=4s^3+108Ss-27T,$$

we can deduce all the rational solutions of $F=0$. The method is applied in detail to the canonical cubic $x^3+y^3+z^3+6mxyz=0$.

W. H. L. Janssen van Raay[311] solved $y/z+z/x+x/y=3$ in integers by reducing it to $a^3+b^3+c^3=3abc$.

A. Hurwitz[312] proved (p. 226) that a curve (4) with integral coefficients has either no rational point or an infinity of rational points if A, B, C are not zero and relatively prime in pairs, while no one of them is divisible by a square of a prime, and at most one of them is ±1. Next, if $A=B=1$, $C\neq\pm1$, and C is not divisible by a square of a prime, the curve has either 1, 2 or an infinity of rational points. Finally, if $A=B=C=1$, $K\neq1,-3$, -5, the curve has 3 or an infinity of rational points. There is a discussion of cubic curves without a double point (genus 1), the coefficients of whose equation belong to an algebraic field. A rational point is one whose coordinates are proportional to three numbers of the field. By use of an elliptic parameter, there are found all complete sets of a finite number of rational points, such that the line joining any two (distinct or identical) meets the curve in a point of the set. The most general cubic curves with exactly one or exactly four rational points are determined. Cf. Levi.[307]

M. Weill,[312a] starting with one solution a, b, c of $Ax^3+By^3+Cz^3=0$, wrote $x=a+\lambda\delta$, $y=b+\lambda'\delta$, $z=c+\delta$, and equated to zero the coefficient $A\lambda a^2+B\lambda'b^2+Cc^2$ of 3δ, and hence found δ rationally, thus obtaining the second solution (3) due to Cauchy. Given two sets of solutions a, b, c and a', b', c', he wrote $x=a+\delta a'$, etc., found δ rationally, and obtained Desboves'[301] special case of Cauchy's (5).

Ternary Cubic Form made a Constant.

J. L. Lagrange[163] determined cubic forms $F(x,\,y,\,z)$ whose product by $F(X,\,Y,\,Z)$ is of that form. Cf. Libri[64, 65] of Ch. XXV.

G. L. Dirichlet[313] employed the roots α, β, γ of a cubic equation with integral coefficients and without rational roots. Let $F(x,\,y,\,z)$ denote the product of $x+\alpha y+\alpha^2z$ by the similar functions of β and γ. First, let a single root α be real. If T, U, V form a fundamental solution of $F(T,\,U,\,V)=1$, and X, Y, Z form one solution of $F(x,\,y,\,z)=m$, an infinite set of solutions of the latter is given by the development of

$$x+\alpha y+\alpha^2z=(X+\alpha Y+\alpha^2Z)(T+\alpha U+\alpha^2V)^n.$$

One solution of any set can be found by a finite number of trials. But if all three roots are real, it is stated that there exist two fundamental solutions from which all can be found by multiplication and powering.

<hr>

[311] Wiskundige Opgaven, 12, 1915, 206–8.
[312] Vierteljarhschrift d. Naturfor. Gesell. Zürich, 62, 1917, 207–29.
[312a] Nouv. Ann. Math., (4), 17, 1917, 47–51.
[313] Bericht Akad. Wiss. Berlin, 1841, 280–5; Werke, I, 625–32.

G. Eisenstein[314] proved that, if p is a prime $3m+1$, $27(x^{p-1}+\cdots+x+1)$ can be expressed in the form

$$\Phi = u^3 + pp_1y^3 + pp_2z^3 - 3puyz,$$

where $y=v+w\rho$, $z=v+w\rho^2$, and u, v, w are polynomials in x with real coefficients, while $\rho^2+\rho+1=0$, and p_1, p_2 are the primary complex prime factors of p. The product of two forms Φ is of like form. When $\Phi=1$ has real integral solutions other than $u=1$, $y=z=0$, an infinitude of solutions can be derived from one, as by Pell's equation.

C. Souillart[315] and E. Mathieu[315] proved that the product of two forms

$$C \equiv x^3+y^3+z^3-3xyz = - \begin{vmatrix} x & y & z \\ y & z & x \\ z & x & y \end{vmatrix}$$

is of the same form and stated that a like theorem holds for cyclic determinants of order n. This was proved for C by J. Petersen.[316]

E. Meissel[317] wrote (x, y, z) for $x^3+Ay^3+A^2z^3-3Axyz$, where A is positive and not a cube. Let $\theta^3=1$ and x, y, z be integral solutions of $(x, y, z)=1$. Let

$$(x+\theta y\rho+\theta^2 z\rho^2)(a+\theta b\rho+\theta^2 c\rho^2)=1, \qquad \rho=\sqrt[3]{A}.$$

By the product of this for the three values of θ, we get $(x, y, z)(a, b, c)=1$. By the three equations which follow from the above,

$$a=x^2-Ayz, \qquad b=Az^2-xy, \qquad c=y^2-xz,$$

which give a second solution of $(x, y, z)=1$. An nth solution follows from $(x+\theta y\rho+\theta^2 z\rho^2)^n$. Solutions of $(x, y, z)=1$ are found for each $A<82$.

G. B. Mathews[318] proved that if the integer m can be represented by

$$F(x, y, z) = x^3+ny^3+n^2z^3-3nxyz,$$

it can be represented in an infinity of ways. $F(x, y, z)=1$ has integral solutions and all solutions can be derived from a single fundamental solution ξ, η, ζ by use of

$$\xi_k+\eta_k t+\zeta_k t^2 = (\xi+\eta t+\zeta t^2)^k, \qquad t=\sqrt[3]{n}.$$

H. W. Lloyd Tanner[319] wrote $\phi(x, y, z)$ for the norm of $x+y\theta+z\theta^2$, where $\theta^3+3k\theta-b=0$, and called $u+v\theta+w\theta^2$ a unit if $\phi(u, v, w)=1$. He obtained a correspondence between the units and the proper automorphs of ϕ, i. e., linear transformations of ϕ into itself, and investigated improper and associated automorphs.

H. S. Vandiver[320] noted that the circulant (cyclic determinant) of order n is a product of n linear factors

$$a_1+\omega^k a_2+\omega^{2k} a_3+\cdots+\omega^{n-k} a_n \qquad (k=0, 1, \cdots, n-1),$$

[314] Jour. für Math., 28, 1844, 289–303.

[315] Nouv. Ann. Math., 17, 1858, 192–4; 19, 1860, 320–2. Cf. Math. Quest. Educ. Times, 63, 1895, 35–6.

[316] Tidsskrift for Math., 1872, 57.

[317] Beitrag zur Pell'schen Gleichung höherer Grade, Progr., Kiel, 1891.

[318] Proc. London Math. Soc., 21, 1891, 280–7. On $F=0$, see Maillet[160] of Ch. XXIII.

[319] Ibid., 27, 1895–6, 187–199.

[320] Amer. Math. Monthly, 9, 1902, 96–8.

where ω is a primitive nth root of unity.   The product of two circulants of order n is such a circulant.   This is used to prove that

$$x^3 + ay^3 + a^2z^3 - 3axyz = v^n$$

has an infinitude of integral solutions for every pair of integers n, a.

R. D. Carmichael[321] proved that every prime $\neq 3$ is representable in one and but one way by $f = x^3 + y^3 + z^3 - 3xyz$, where x, y, z are ≥ 0.   All positive integers are representable by f with x, y, z each ≥ 0, with the sole exception of the integers divisible by 3, but not by 9.   A prime $6n+1$ can be represented in one and but one way by f with at least one variable negative.

A. Cunningham[322] considered primes of the preceding form f.   Take $y = x + \beta$, $z = x + \gamma$.   Then $f = AB$, $A = 3x + \beta + \gamma$, $B = \beta^2 - \beta\gamma + \gamma^2$.   If $B = 1$, then $\beta = \gamma = \pm 1$, $f = 3x \pm 2$.   Since any prime $p > 3$ is of the last form, we get positive integers x, y such that f represents p.   Next, let $A = 1$; if B is prime it is of the forms $6\omega + 1 = k^2 + 3l^2$.

E. Turrière[323] noted that the above form f represents the rational number n when $x = n$, $y = n + 1/3$, $z = n - 1/3$.   If $n \equiv 1 \pmod 3$, it represents n when $x = y = (n-1)/3$, $z = (n+2)/3$.

Miscellaneous Single Diophantine equations of degree three.

Bháscara[323a] noted that the sum of the cubes of y, $2y$, $3y$, $4y$ equals the sum of their squares if $100y^3 = 30y^2$, whence $y = 3/10$.

T. Robinson[323b] found two cubes x^3, v^3x^3 and a square m^2x^2 in arithmetical progression, since $2v^3x^3 = x^3 + m^2x^2$ determines x rationally.

A. J. Lexell[323c] noted that, if a cubic equation has rational roots, its discriminant is a square.

J. L. Lagrange[324] employed the " tangent method " to determine new solutions of the cubic equation $f(x, y) = 0$ from one set of solutions p, q.   Set $x = p + t$, $y = q + u$, and take

$$t\frac{\partial A}{\partial p} + u\frac{\partial A}{\partial q} = 0, \qquad A \equiv f(p, q).$$

Substituting the resulting expression for u into $f(p + t, q + u) = 0$, we may delete the factor t^2 and thus express t, and hence u, as a rational function of the partial derivatives of A.   Cf. Lagrange[252] of Ch. XXII.

To express $1^2 + 2^3$ as a sum of another square and cube, J. Cunliffe[325] took $9 = v^2 + (2 - x)^3$, $v = 21x^2 - 6x - 1$, whence $x = 253/441$.   J. Whitley took $9 = x^3 + (3 - nx)^2$, whence $2x + n^2 = \sqrt{24n + n^4}$, which equals $5 + pq - q^2$ if

[321] Bull. Amer. Math. Soc., 22, 1915, 111–7.   Cf. Carmichael.[90]

[322] Math. Quest. and Solutions, 1, 1916, 14–15.

[323] L'enseignement math., 18, 1916, 417–20.

[323a] Víja-gańita, § 119.   Algebra . . . from Sanscrit of Brahmegupta and Bháscara, transl. by Colebrooke, 1817, 200.

[323b] The Gentleman's Diary, or Math. Repository, London, No. 25, 1765; Davis' ed., 2, 1814, 98.

[323c] Euler's Opera postuma, 1, 1862, 504–6 (about 1770).

[324] Nouv. mém. acad. Berlin, année 1777, 1779, 153; Oeuvres, IV, 396.

[325] The Gentleman's Math. Companion, London, 2, No. 13, 1810, 220–1.

$n=1+q$ and $25+10pq+(p^2-10)q^2-2pq^3+q^4=(5+pq-q^2)^2$. The last holds if $10p=28$, $q=-51/60$, whence $x=15/16$. Cf. Gérardin.[346]

W. Lenhart[326] discussed $\Sigma(x_i^3+x_i)=\Sigma(y_i^3+y_i)$, where $i=1,\ \cdots,\ n$. Assign any values to x_i, y_i $(i=3,\ \cdots,\ n)$. Then seek numbers (in his[205] table of sums of two cubes) $t=x_1^3+x_2^3$, $t'=y_1^3+y_2^3$, such that

$$y_1+y_2-x_1-x_2+\zeta=t-t',$$

where ζ depends on the chosen values of x_i, y_i, $i\geqq3$. For $n=2$, he found $x_1=5$, $x_2=6$, $y_1=7$, $y_2=1$. For $n>2$ he took $t=t'$ and found

$$(12,\ 5,\ 1;\ 11,\ 8,\ 2), \qquad (14,\ 13,\ 11,\ 8;\ 17,\ 12,\ 5,\ 3),$$
$$(21,\ 14,\ 10,\ 4,\ 1;\ 20,\ 17,\ 5,\ 3,\ 2).$$

B. Peirce (*ibid.*) took $x_i=a_ix+b_i$, $y_i=a_ix+b_{n-i+1}$ and found that the condition gives

$$x=\{\Sigma a_i(b_{n-i+1}^2-b_i^2)\}/\{\Sigma a_i^2(b_i-b_{n-i+1})\}.$$

R. Hoppe[327] considered the rational solutions of $x^3+y^3=x-y$. Set $y=x(1-u)/(1+u)$. Then x and y are rational in u if $u/(1+3u^2)=\square$. If u is a solution,

$$w=\frac{u}{1+3u^2}\left\{\frac{2(1+3u^2)}{1-3u^2}\right\}^2$$

is a second solution, etc. The nth such solution is found.

C. Hermite[328] noted the solution $x=a(ab-c^2)$, $y=a^3-b^2c$, $z=b(c^2-ab)$, $u=a^2c-b^3$ of

$$(1)\qquad\qquad x^2y+y^2z+z^2u+u^2x=0.$$

J. Joffroy[329] stated that $a^2-b^3=7\cdot10^n$ is impossible. A. Morel gave an erroneous extension to $a^2-b^3\neq10^{n_1}+\cdots+10^n$.

S. Réalis[330] gave long cubic functions x, y, z, w of α, β, γ for which

$$x^3+y^3+z^3\equiv(\alpha^3+\beta^3+\gamma^3)w^2.$$

Réalis[331] obtained as solutions of (1):

$$x=3(\alpha^2-\alpha\beta+\beta^2), \qquad y=-\alpha^2+3\alpha\beta-5\beta^2,$$
$$z=-3\alpha^2+9\alpha\beta-9\beta^2, \qquad u=\alpha^2-\alpha\beta+3\beta^2,$$

as well as formulas of the third and fourth degrees.

T. Pepin[332] noted that a surface of degree m is osculated at an arbitrary point of a given surface only when there is a positive integer n satisfying

$$m^3+6m^2+11m=3(n+1)(n+2),$$

and proved that 1, 5, 20 are the only integral values <675 of m. E. de Jonquières[333] used the discriminant of the quadratic in n to show that

[326] Math. Miscellany, New York, 2, 1839, 96–7; Extract, Sphinx-Oedipe, 8, 1913, 93–4.
[327] Zeitschr. Math. Phys., 4, 1859, 359–61.
[328] Nouv. Ann. Math., (2), 6, 1867, 95.
[329] Nouv. Ann. Math., (2), 10, 1871, 95–6, 288.
[330] Nouv. Corresp. Math., 4, 1878, 346–52.
[331] Nouv. Ann. Math., (2), 18, 1879, 301–4.
[332] Jour. de Math., (3), 7, 1881, 71–108.
[333] Atti Accad. Pont. Nuovi Lincei, 37, 1883–4, 183–8.

either $m=5t$, whence $m=5$, $n=9$, or $m=20$, $n=58$, if $m<300$; or $m=25k+1$, whence $m=1$, $n=1$, if $m<1000$.

Réalis[333a] noted that the double of any square, as well as the triple of the square of any even number >2, equals the excess of a sum of two squares over a sum of two cubes.

M. Weill[334] noted that (1) has the solution $x=pA$, $y=hp^3-1$, $z=px$, $u=-hpy$, where $A=ph^2+1$; also, $x=HA^2$, $y=-AB$, $z=H^2A$, $u=hHB$, where $H=h^3-p$, $B=p^3h+3ph^2-h^5+1$. The last solution is based on the identity $A^3-hH^3=(1+h^5)B$.

H. S. Vandiver and W. F. King[335] proved the impossibility of
$$x^2y+xz^2=y^2z.$$

G. Bisconcini[336] noted that $x^3-y^3=(x+y)^2$ has the single solution $x=1$, $y=0$, in integers; $x^3+y^3=x^2+y^2$ has only the solutions $x=1$, $y=0$ or 1; $(x-y)^3=xy$ or x^2+y^2 has various solutions.

References[337] on cubic equations with integral roots are in place.

A. Cunningham[338] noted that one method of solving $x^3+y^3=z^2+u^2$ is to make $x+y$ and x^2-xy+y^2 both sums of two squares.

A. Gérardin[339] satisfied $\Sigma x^3=\Sigma y^2$ by taking
$$(1+mx)^3+(my)^3+(mz)^3-(m\alpha)^2-(m\beta)^2=(1+\tfrac{3}{2}xm+gm^2)^2$$
and equating the coefficients of m^2 (thus determining g), so that m is found rationally. Another method is to take $g=0$.

R. Norrie[84] noted that from one set $a_1,\ \cdots,\ a_n$ of solutions not all zero of a homogeneous cubic equation in $X_1,\ \cdots,\ X_n$ we can in general deduce further sets by substituting $X_i=rx_i+a_i$ $(i=1,\ \cdots,\ n)$, thus deriving $\alpha r^3+\beta r^2+\gamma r=0$. Since γ is linear, we can make $\gamma=0$ by choice say of x_n in terms of $x_1,\ \cdots,\ x_{n-1}$. Then take $r=-\beta/\alpha$. The method is applied to $bx(x^2-b^2)=u^2+2v^2$ and to
$$\mu_1V_1^3+\cdots+\mu_nV_n^3+xy(x-y)=\lambda z^3.$$

As to this method see Lagrange[324] and the related method of Cauchy[287] and Lucas.[296]

A. Cunningham and E. B. Escott[340] made $xy(x+y)+l$ a cube, where $l=x-y$ or $2x+2y$; also $xy\pm2(x+y)$ is made a cube.

Welsch[341] noted that 1, 2, 3 are the only three positive integers whose sum equals their product. For n integers see papers 150–2 of Ch. XXIII.

A solution[342] of $\Sigma x_i^2-\Sigma y_i^2=\Sigma u_i^3$ is x_i, $y_i=\tfrac{1}{2}(u_i^2\pm u_i)$. This and other solutions are found by decompositions of $u^3=x^2-y^2$.

[333a] Nouv. Ann. Math., (3), 2, 1883, 295–6.
[334] Nouv. Ann. Math., (3), 4, 1885, 184–8.
[335] Amer. Math. Monthly, 9, 1902, 293–4; 10, 1903, 22. Cf. Euler[9]; also Hurwitz[212] of Ch. XXVI.
[336] Periodico di Mat., 22, 1907, 125–9.
[287] L'intermédiaire des math., 15, 1908, 47–8, 152, 239; 16, 1909, 208.
[338] Ibid., 18, 1911, 210–3.
[339] Bull. Soc. Philomathique, (10), 3, 1911, 226–233. Cf. paper 285 above.
[340] L'intermédiaire des math., 19, 1912, 164–5, 273.
[341] Ibid., 69.
[342] Ibid., 20, 1913, 190, 239–40.

L. Aubry[343] noted solutions, involving two parameters, of
$$xyz-(x^2+y^2+z^2)w+4w^3=0.$$
Special solutions (p. 207) are given for $b=7$, 61, 2281, 99905 of
$$b^2x+by+z=(x+y+z)^3.$$

L. Aubry[344] treated $x^3+x+y^3+y=z^3+z$ by setting $x+y=2u$, $x-y=2v$, $z=pu$, whence $2(u^2+3v^2+1)=p(p^2u^2+1)$, which is solved as a Pell equation in u, v.

E. B. Escott[345] treated the preceding problem by setting $y=x+d$, $z=x+b$, $x=k(b-d)$, and found eight sets of solutions. Next, for
$$x^3+x+y^3+y+z^3+z=a^3+a,$$
set $y=x+d$, $z=e-x$, $a=x+b$, $d=b+ke$. The discriminant of the resulting equation for x must be a square, $9s^2$. Thus $k=3n-1$. For $n=0$,
$$2x=e-b, \qquad 2y=b-e, \qquad 2z=2a=b+e.$$
For $n=1$, we get the solution (in which ρ is a rational parameter)
$$4x=2e-R-\rho, \qquad 8y=-16e-R+\rho, \qquad 4z=2e+R+\rho,$$
$$8a=16e-R+\rho, \qquad R\equiv(21e^2+4)/\rho.$$

He gave (pp. 126-7) solutions of each of the equations $x^3\pm xy+y^3=z^2$, $x^3\pm x^2y^2+y^3=z^2$. L. Aubry (p. 47) reduced $x^3-xy+y^3=z^2$ to a Pell equation by setting $\tfrac{1}{2}(x\pm y)=v$, u.

A. Gérardin[346] noted that, if $a^3-b^3=f^2-g^2$, then
$$(1+ma)^3-(mb)^3=(1+mf)^2-(mg)^2$$
becomes a quadratic equation for m. By equating to zero one of the three coefficients, we find new solutions of $x^3-y^3=F^2-G^2$. Cf. Cunliffe[325]; also Réalis[17, 18] of Ch. XX.

P. Bachmann[347] solved $k^3-(p_1^2+p_2^2+p_3^2)k=2p_1p_2p_3$ in positive integers. We may assume that $p_i=h_ik_i$ ($i=1$, 2, 3), $k=fk_1k_2k_3$, where $f=1$ or 2. Multiplying the given equation by fk_3^2, we get
$$(f^2k_1^2k_3^2-h_2^2)(f^2k_2^2k_3^2-h_1^2)=(fh_3k_3^2+h_1h_2)^2.$$
The factors on the left are equated to ns_1^2 and ns_2^2 respectively, by use of solutions of $x^2-h^2=ns^2$.

Cashmore[348] stated erroneously that $x^3+y^3=u^2+v^2$ for
$$x,\ y=2(a^2+b^2\pm2eh\pm2fg), \qquad u=4(a^3-ab^2+2beg+6bfh),$$
$$v=4(b^3-a^2b+2aeg+6afh).$$

R. Goormaghtigh[349] solved $x^3+2x+y^3=\square$.

T. Hayashi[350] proved that $x^2y+y^2z+z^3=0$ is impossible in integers $\neq0$.

<hr>

[343] L'intermédiaire des math., 20, 1913, 95.
[344] Sphinx-Oedipe, 8, 1913, 46-7. Cf. Lenhart.[326]
[345] *Ibid.*, 123-4.
[346] *Ibid.*, 14.
[347] Archiv Math. Phys., (3), 24, 1915, 89-90.
[348] L'intermédiaire des math., 23, 1916, 224.
[349] *Ibid.*, 200-1.
[350] Nouv. Ann. Math., (4), 16, 1916, 161-5.

E. Maillet[350a] discussed $y^3 - y = c^3(x^3 - x)$, where c is rational. For each value of c there is only a finite number of integral solutions.

Solutions[350b] have been found for the equation in binomial coefficients

$$\binom{u+1}{3} + \binom{v+1}{3} = \binom{w+1}{3}, \qquad u^3 - u + v^3 - v = w^3 - w.$$

The sum of the first n odd cubes can[350c] be expressed as a sum of seven squares $\neq 0$. Special solutions of $x^3 + y^3 + z^3 = k(x+y+z)$ are noted (p. 155).

On $l(mp^2 + nq^2) = \lambda(mr^2 + ns^2)$, where l, λ are linear functions of p, q, r, s, see papers 48, 51, 55, 80, 89. On $tu^2 + t^2v = Auv^2$, see Lagrange, p. 572. On $\phi(u) + \phi(v) = g$, see Baer.[224]

On $x(1-x^2) = Ay^2$, see Tweedie[74] of Ch. IV.

On $x + a/x = y^2$, see Leibniz[64] and Terquem[70] of Ch. XXII. On equations of degree three involving products of consecutive numbers, see papers 28, 32, 56, 58, 59, and 63 of Ch. XXIII. On $xy(x+y) = Az^3$, see Euler[10], Lucas[199], Catalan,[204] and Hayashi[219]; also Lucas[150] of Ch. I. Chuquet[34] of Ch. XII expressed 20 as a sum of three positive rational cubes; on the general topic, see papers 404–29; also Ch. XXV, end. On $x^2 + y^2 + z^2 = kxyz$, see the papers cited under Hurwitz[174] of Ch. XXIII.

Systems of equations of degree three in two unknowns.

Diophantus, IV, 29, 30, made $xy \pm (x+y)$ cubes. Take $y = x^2 - x$. Then the condition with the upper sign is satisfied and that for the lower sign requires $x^3 - 2x^2 = \text{cube} = (\tfrac{1}{2}x)^3$, say, whence $x = 16/7$.

Bombelli[351] treated the same problem.

Bháscara[352] noted that the sum and difference of $4y^2$ and $5y^2$ are squares and their product $20y^4$ is a cube, $(10y)^3$, if $y = 50$. The sum of the cubes of y^2 and $2y^2$ is $9y^6$, a square, and the sum $5y^4$ of their squares is a cube, $(5y)^3$, if $y = 25$. Under Bháscara[30] of Ch. XII is given his solution of $x - y = \square$, $x^2 + y^2 = z^3$, and of $y^2 + z^3 = \square$, $y + z = \square$.

L. Euler[353] discussed $x + y = \square$, $x^2 + y^2 = p^3$. Hence take $p = a^2 + b^2$, $x = a(a^2 - 3b^2)$, $y = b(3a^2 - b^2)$. Then $x + y = (a-b)Q$, $Q = a^2 + 4ab + b^2$. Set $a - b = c^2$. Then $Q = 6b^2 + 6bc^2 + c^4 = (c^2 + 3bf/g)^2$ if $b/c^2 = 2g(g-f)/(3f^2 - 2g^2)$. Then x and y will be positive if $b = 2g(g-f)$, $c^2 = 3f^2 - 2g^2$. The latter is satisfied if $f = 11$, $g = 1$, $c = 19$, or if $f = -3$, $g = 1$, $c = 5$, whence $b = 8$, $a = 33$, $x = 29601$, $y = 25624$. For three numbers he gave only results:

$$35 + 9 + 5 = 7^2, \qquad 35^2 + 9^2 + 5^2 = 11^3; \qquad 67 + 9 + 5 = 9^2, \qquad 67^2 + 9^2 + 5^2 = 19^3.$$

[But the last sum equals $5 \cdot 919 \neq 19^3$.]

W. Spicer,[354] to find two squares whose sum is a square and difference a cube, took $a = \tfrac{1}{2}x^2 + \tfrac{1}{2}x^3$ and $b = \tfrac{1}{2}x^2 - \tfrac{1}{2}x^3$ as the squares with the sum x^2

[350a] Nouv. Ann. Math., (4), 18, 1918, 289–292.
[350b] Zeitschrift Math. Naturw. Unterricht, 50, 1919, 95–6.
[350c] L'intermédiaire des math., 26, 1919, 77–8, 109–10.
[351] L'algebra opera di Rafael Bombelli, Bologna, 1579, 553.
[352] Vija-gaṅita, §§ 121–2. Colebrooke,[323a] 201–2.
[353] Opera postuma, 1, 1862, 255–6 (about 1782).
[354] Ladies' Diary, 1766, 33–4, Quest. 536; C. Hutton's Diarian Miscellany, 3, 1775, 220; Leybourn's M. Quest. L. D., 2, 1817, 251.

and difference x^3. Choose the squares

$$c = \frac{4n^2}{(1+n^2)^2}, \qquad d = \frac{(1-n^2)^2}{(1+n^2)^2}$$

with the sum 1 and set $a = cx^2$, $b = dx^2$, either of which gives x.

J. Leslie[355] made $x+y$ and x^3+y^3 squares, by division.

W. Cole[356] made $x-y$, x^2-y^2, x^3-y^3 all squares by taking $x-y=a^2$, $x+y=m^2a^2$, whence $x^3-y^3=\square$ if $3m^4+1=\square$, which holds if $m=2$. J. Young took $m=2$ initially.

J. Saul[357] made $x+y=s^2$, $x^2+y^2=v^2$ and x^3+y^3 a square. By elimination of y from the first two equations, $s^4-2s^2x+2x^2=v^2$. Let $v=s^2-rx$. Then $x=s^2(2-2r)/(2-r^2)$. Then $x^2-xy+y^2=\square$ if $r^4-6r^3+14r^2-12r+4=\square$, say $(r^2-3r+5/2)^2$, whence $r=3/4$.

To divide[358] a given square a^2 into two parts such that the difference of their squares and the difference of their cubes are both squares, an anonymous solver called b^2 the difference of the parts, whence the difference of their squares is $(ab)^2$. The quotient of the difference of their cubes by b^2 is to be a square, whence $3a^4+b^4=\square$. Put $a=bx$, $x=2-z$. Then $3x^4+1=49+\cdots+3z^4$ is the square of $7-48z/7+12\cdot51z^2/49$ by choice of z.

J. Whitley[359] found two positive fractions such that each plus the square of the other is a square, while the difference of their squares or their cubes is a square. Let the fractions be $(1\pm4v^2)/8$, whose sum and difference are squares. The difference of their cubes is a square if $3+16v^4=\square=a^2$. Let $v=\frac{1}{2}-z$, $\frac{1}{2}a=1-z+2z^2$. Hence $z=\frac{1}{4}=v$. B. Gompertz took $x=az$ and $y=tz$ as the fractions, where $a=(1+t^2)/2$. Then $x^2-y^2=\square$. Take $x+y^2=p^2z^2$, $y+x^2=q^2z^2$. We get two values of z which are equal if $as(q-a)=t(p-t)$, $q+a=s(p+t)$. These give p and q. Then $x^3-y^3=(rz)^2$ gives z, which equals the earlier value of z if $cs\,(a^2s-t^2)(a-st)=\square$, where $c=1/(a^3-t^3)$. Take $t=3$, $s=1$. Hence $x=5/32$, $y=3/32$.

S. Jones[360] made $x+y=a^2$, $x^2+y^2=\square=(bx-y)^2$ by choice of x, y. Then $x^3+y^3=\square$ if $b^4-2b^3+2b^2+2b+1=\square=(b^2-b+\frac{1}{2})^2$, whence $b=-\frac{1}{4}$. W. Wright took $a=1$, proceeded similarly, and found y from

$$(1-y)^2+y^2=\square=(1-my)^2.$$

Then $1-3y+3y^2=\square$ if $m^4-6m^3+14m^2-12m+4=\square=(m^2-3m-2)^2$, whence $m=8/3$, $y=15/23$.

Lowry[360a] eliminated $x=a^2-y$ from x^2+y^2 and x^2-xy+y^2 and equated the resulting expressions to the squares of a^2-yr/s and $a^2-ye/(sw)$; the conditions hold if $w=1$, $4r=3s$, $e=5s/4$. J. Cunliffe took $x=R^2-S^2$, $y=2RS$, $x^2-xy+y^2=(R^2-RS+S^2)^2$, whence $R=4S$; then the desired numbers are $a^2x/(x+y)$, $a^2y/(x+y)$.

<hr>

[355] Trans. Roy. Soc. Edinburgh, 2, 1790, 211.

[356] Ladies' Diary, 1787, 36–7, Quest. 853; Leybourn's M. Quest. L. D., 3, 1817, 155–6.

[357] The Gentleman's Diary, or Math. Repository, No. 55, 1795; Davis' ed., 3, 1814, 235.

[358] Ibid., No. 56, 1796; Davis' ed., 3, 1814, 249.

[359] The Gentleman's Math. Companion, London, 2, No. 12, 1809, 169–71.

[360] Ibid., 3, No. 18, 1815, 323–4.

[360a] New Series of Math. Repository (ed., T. Leybourn), 3, 1814, I, 169–172.

To find two integers the difference of whose squares is a cube and the difference of whose cubes is a square, J. R. Ambler[361] took x^3+2 and x^3-2 as the numbers, the difference of whose squares is $(2x)^3$. The difference of the cubes is a square if $3x^6+4 = \square = (2x^3-2)^2$, $x=2$. J. Davey used the numbers x, y and set $x^2-y^2=z^3$, $x+y=n^2z$, which give x, y in terms of z. Then $x^3-y^3=\square$ if $3n^8+z^2=\square=(rn^4-z)^2$, which gives z.

W. Snip[362] made x^2+y^2 and x^3+y^3 squares by t king $x=(m^2-n^2)v$, $y=2mnv$. Then $x^3+y^3=a^2b^2v^2$ determines v rationally.

J. Anderson[363] made $x+y$ a square and $x-y$, x^2+y^2 cubes by setting $x\pm yi=(p\pm qi)^3$. Then $x-y=p^3-3p^2q-3pq^2+q^3=(q-p)^3$ if $p=3q$. Hence $x=18q^3$, $y=26q^3$, $x+y=\square$ if $q=11$. Ashcroft used the numbers $(x^4\pm x^3)/2$ whose sum is x^4 and difference is x^3. Their sum of squares is $(4x^8+4x^6)/8$, which is a cube if $4x^2+4=5^3$, $x=11/2$.

S. Ward[363a] took $y=x+Y$, $Y=8r^3$, $x=Yz$. Then $(x^2+y^2)/Y^2$ equals $2z^2+2z+1$, which is the cube of $1+2z/3$ if $z=9/4$. Then $x+y=44r^3=\square$ if $r=11$.

Several[364] found two integers whose sum is a square and difference a cube, while if each number be doubled the new sum is a cube and difference a square. Take $x+y=4a^6$, $x-y=8b^6$.

To make $x-y$, x^2-y^2, x^3-y^3 rational squares [Cole[356]], J. Whitley[365] used the numbers $x=2z^2+2v^2$, $y=2z^2-2v^2$; then shall

$$x^2+xy+y^2=4(3z^4+v^4)=\square,$$

which is true if $z=v$ or $z=2v$. H. Godfray took $x=m^2+n^2$, $y=2mn$; then $x^2+xy+y^2=(m^2+mn+5n^2/2)^2$ if $n=-4m/7$.

Several[366] solved $x+y=\square$, $x^2+y^2=\square$, $x^2+y^3=x^3+y^2$.

Several[367] found two numbers the difference of whose squares is a cube and difference of cubes a square.

H. W. Curjel[368] found two numbers x, y whose sum and difference are squares, sum of squares a cube, and sum of cubes a square. By the first and last conditions, $x^2-xy+y^2=\square$, which holds if $x=z(2mn-n^2)$, $y=z(m^2-n^2)$. Then $y\pm x$ are squares if $m=9$, $n=4$, $z=\square$, whence $x=56z$, $y=65z$. Then $x^2+y^2=7361z^2$. Thus take $z=7361^4$.

P. F. Teilhet[369] stated that all pairs of numbers whose sum and sum of squares are squares are $(A^2-B^2)M^2N$ and $2ABM^2N$, where A and B are relatively prime and not both even, $N\equiv A^2-B^2+2AB$, and where M^2N is an integer. He asked when also the sum of their cubes is a square, as for 345,184.

<hr>

[361] Ladies' Diary, 1816, 38–9, Quest. 1291; Leybourn's M. Quest. L. D., 4, 1817, 221–3.
[362] The Gentleman's Math. Companion, London, 4, No. 20, 1817, 659–60.
[363] Ibid., 4, No. 21, 1818, 719–21.
[363a] Young's Algebra, Amer. ed., 1832, 342–3.
[364] Ladies' Diary, 1821, 32–5, Quest. 1362.
[365] The Lady's and Gentleman's Diary, London, 1849, 49–50, Quest. 1779.
[366] Math. Visitor, 1, 1880, 100–1, 126.
[367] Amer. Math. Monthly, 1, 1894, 95–6, 325.
[368] Math. Quest. Educ. Times, 62, 1895, 51–2.
[369] L'intermédiaire des math., 10, 1903, 124. Cf. papers 139–40 of Ch. XVI.

A. S. Werebrusow[370] found an infinitude of solutions of the last question. Teilhet[371] gave a more general treatment of the problem.

A. Gérardin[372] treated the system $x^3+hy^3=a^3+hb^3$, $x+hy=a+hb$, and found many solutions, such as

$$x,\ a=(9m^2-1)\alpha^2\mp 18m\alpha\beta-3\beta^2;\qquad y,\ b=(9m^2-1)\alpha^2\pm 6\alpha\beta+3\beta^2;\qquad h=3m.$$

In l'intermédiaire des mathématiciens are discussed the problems:

$$P(x+y)+Qx=z^3,\qquad P(x+y)+Qy=w^3,\qquad 22,\ 1915,\ 145\text{--}6.$$
$$(y-x)^2+x=z^p,\qquad (y-x)^2-y=w^p,\qquad p\geqq 3,\qquad 196;\ 23,\ 1916,\ 68\text{--}9;\ 24,\ 1917,\ 85\text{--}6.$$
$$(x+y)^3+x=a^2,\qquad (x+y)^3+y=b^3,\qquad 23,\ 1916,\ 141\text{--}2.$$
$$x^3-hy^3=\square,\qquad x^3+hy^3=\square,\qquad 22,\ 1915,\ 53,\ 232;\ 24,\ 1917,\ 39.$$
$$x^3+y^3=a^3-b^3,\qquad x^3-y^3=c^3+d^3,\qquad 26,\ 1919,\ 145.$$

SYSTEMS OF EQUATIONS OF DEGREE THREE IN THREE UNKNOWNS.

Diophantus, IV, 6 found that x^2+z^2 is a square and y^3+z^2 a cube for $y=16/7$, $x=3y$, $z=4y$. In IV, 7, 8, he found that x^2+z^2 is a cube and y^3+z^2 a square for $x=5$, $y=5$, $z=10$ and for $x=40$, $y=20$, $z=80$.

J. de Billy proposd the problem to find three numbers such that if their product is subtracted from any one of the numbers or from the difference of any two or from the product of the second by the first or third or from the square of the second, there results always a square. He expressed his belief that $3/8$, 1, $5/8$ is the only solution.

Fermat[373] replied that [if the numbers are denoted by A, 1, $1-A$] the problem reduces to the double equal ty

$$A^2-A+1=\square,\qquad A^2-3A+1=\square,$$

which has an infinitude of solutions. In addition to de Billy's solution $A=3/8$, Fermat gave $A=10416/51865$.

Malézieux[374] proposed the problem to find three rational numbers in A. P. such that one obtains a square by adding to their product either the difference of the s uares of any two of them or the sum of the three differences of the three numbers.

E. Fauquembergue[375] gave the solution $1/31$, $25/589$, $1/19$.

J. Ozanam[376] asked for three numbers in G. P. such that one obtains squares by adding to their product the square of each number, and such that if these fractional squares are reduced to their simplest forms the sums by twos of the square roots of the numerators are three cubes in G. P.

"J. Hob"[377] solved the first part, saying the entire problem is impossible.

---

[370] L'intermédiaire des math., 10, 1903, 319–20.
[371] Ibid., 11, 1904, 167–70.
[372] Sphinx-Oedipe, 5, 1910, 1–12.
[373] Oeuvres, II, 437, letter to de Billy, Aug. 26, 1659.
[374] Unedited letter to de Billy, Sept. 6, 1675. Cf. P. Tannery, l'intermédiaire des math., 3, 1896, 37. Éléments de Geométrie de M. le Duc de Bourgogne, par de Malézieux, 1722.
[375] L'intermédiaire des math., 6, 1899, 115–6.
[376] Unedited letter to de Billy, June 25, 1676. Cf. P. Tannery, l'intermédiaire des math., 3, 1896, 57; C. Henry, Bull. Bibl. Storia Sc. Mat. e Fis., 12, 1879, 517.
[377] L'intermédiaire des math., 4, 1897, 253.

E. Fauquembergue[378] called the numbers x/y, x, xy.   Then

(1) $$x^3+x^2y^2, \qquad x^3+x^2, \qquad x^3+x^2/y^2$$

are made squares by removing the factors x^2, and making the product $(x+y^2)(x+1)(x+1/y^2)$ a square by Fermat's[156] method.   Setting $y=\alpha/\beta$, we get $x=N/(4\alpha^4\beta^4)$, where

$$N=(\alpha^2+\alpha\beta+\beta^2)(\alpha^2+\alpha\beta-\beta^2)(\alpha^2-\alpha\beta+\beta^2)(-\alpha^2+\alpha\beta+\beta^2).$$

Then (1) are the squares of

$$\frac{N(\alpha^4+\alpha^2\beta^2-\beta^4)}{8\alpha^6\beta^6}, \qquad \frac{N(\alpha^4-\alpha^2\beta^2+\beta^4)}{8\alpha^6\beta^6}, \qquad \frac{N(-\alpha^4+\alpha^2\beta^2+\beta^4)}{8\alpha^6\beta^6}.$$

These fractions are said to be arithmetically irreducible.   The sums by twos of the numerators are $2N\alpha^4$, $2N\alpha^2\beta^2$, $2N\beta^4$, which are in G. P., but are not made cubes as required.

L. Euler[379] desired three rational numbers whose sum, product and sum of products by twos are all squares.   Denote the numbers by nx, ny, nz. Then

$$xyz(x+y+z)=\square=v^2(x+y+z)^2, \qquad z=v^2(x+y)/(xy-v^2).$$

Then $n^3xyz=\square$ requires $n=m^2xy(x+y)(xy-v^2)$.   By the sum of products by twos,

$$xy+\frac{v^2(x+y)^2}{xy-v^2}=\square.$$

Set $xy-v^2=u^2$, $x=tv$.   Then the preceding condition becomes

$$v^4(t^2+1)^2+u^2v^2(3t^2+2)+u^4(t^2+1)=\square=[v^2(t^2+1)+su^2]^2,$$

$$\frac{v^2}{u^2}=\frac{t^2+1-s^2}{2s(t^2+1)-3t^2-2}.$$

Set $s=t-r$ and multiply numerator and denominator by t^2+1-s^2.   Thus

$$4rt^4-2(3r^2+3r-1)t^3+(2r^3+3r^2+2r-3)t^2-2(3r-1)(r+1)t$$
$$+2(r-1)(r+1)^2=Q^2,$$

$$\frac{v}{u}=\frac{2rt-r^2+1}{Q}, \qquad x=tv, \qquad y=\frac{u^2+v^2}{tv}, \qquad z=\frac{v^2(x+y)}{u^2}.$$

Rational values of t are found from $r=1$, $3/2$, 3, 9.   The simplest numbers derived from $r=3/2$, $t=60/19$, are $705600/d$, $196/4157$, $361/557$, where $d=2315449$.   The corresponding integral solutions are $705600d$, $109172d$, $1500677d$.   Euler[380] had expressed his belief that these give the least integers.

E. Fauquembergue[381] used a simpler method and obtained

$$4a^2b^4(a^2+b^2), \qquad (a^4-b^4)^2, \qquad 4a^4b^2(a^2+b^2),$$

whose product is a square, sum is $(a^2+b^2)^4$, and sum of products by twos is $4a^2b^2(a^2+b^2)^2(a^4+b^4)^2$.   For $a=2$, $b=1$, we get 80, 225, 320.

[378] L'intermédiaire des math., 5, 1898, 86-7.
[379] Novi Comm. Acad. Petrop., 8, 1760-1, 64; Comm. Arith., I, 239; Op. Om., (1), II, 519.
[380] Corresp. Math. Phys. (ed., Fuss), 1, 1843, 631, Aug. 23, 1755.
[381] L'intermédiaire des math., 6, 1899, 95-96.

To find[381a] three integers whose sum and sums by twos are cubes, take

$$x+y+z=(b+n)^3, \qquad x+y=b^3, \qquad x+z=c^3.$$

Then

$$y+z=2(b+n)^3-b^3-c^3=(b+2n)^3 \qquad \text{if} \qquad b=-(c^3+6n^3)/(6n^2).$$

J. Cunliffe[382] noted that $x+y-z=a^3$, $x+z-y=b^3$, $y+z-x=c^3$ imply $x+y+z=a^3+b^3+c^3$, which has been made a cube by many writers.

Several[382a] found three squares in A. P. the sum of whose square roots is a cube by using the know expressions $v(2pq\pm q^2\mp p^2)$, $v(p^2+q^2)$ for the roots, and equating their sum vk, where $k=p^2+4pq+q^2$, to s^3k^3.

J. Anderson[383] made $xyz+1$, $xy+1$, $xz+1$ and $yz+1$ all squares by equating the last two to $(pz-1)^2$ and $(qz-1)^2$, whence $x=p^2z-2p$, $y=q^2z-2q$. Then the first two will be the squares of $1+2pqz$ and $pqz-p-q$ if $z=4+2(p+q)/(pq)$ and $p=q+1$, respectively.

To find three numbers whose sum is a cube and the sum of any two a square, J. Foster[384] took $x+y=m^2a^2$, $x+z=m^2b^2$, $y+z=m^2c^2$, $x+y+z=d^3m^3$, whence $m=(a^2+b^2+c^2)/(2d^3)$. Many solvers used the numbers $2(x^2+y^2-z^2)$, $2(x^2+z^2-y^2)$, $2(z^2+y^2-x^2)$, whose sums by twos are squares. To satisfy $2(x^2+y^2+z^2)=8p^6$, take $y^2=4p^6-n^2=(cn-2p^3)^2$, which determines n, and take $z=2mn/(m^2+1)$, whence $n^2-z^2=\square=x^2$.

W. Lenhart[385] found that the sum, and sum of any two of, 1982015, 2759617 and 44286264 are cubes; they are the excesses of 366^3 over the cubes in

$$168^3+359^3+361^3=2\cdot366^3.$$

S. Bills[386] obtained the same result from

$$x^3+(z+1)^3+(z-1)^3=2(z+v)^3, \qquad z^2+\frac{v^2-1}{v}z=\frac{x^3-2v^3}{6v}.$$

The root z involves the square root of $6vx^3-3v^4-18v^2+9$ which is equated to $6v(x+2a)(x-a)^2$. For the resulting value of x, $6v(x+2a)=\square$ if

$$v^4+6v^2+16a^3v-3=\square=(v^2+3)^2,$$

whence $v=3/(4a^3)$. Take $a=\frac{1}{2}$. Several writers[387] solved the same problem.

To find three integers in arithmetical progression whose common difference is a cube, the sum of any two less the third is a square, and the sum of the roots of the resulting squares is a square, S. Bills[388] took x^2-y^3, x^2, x^2+y^3 as the numbers. To make $x^2\pm2y^3$ squares, take $x=uy$, whence $u^2\pm2y$ are known to be squares if $u=t(p^2+q^2)$, $y=2pql^2(p^2-q^2)$. It remains to make $2t(p^2+4pq+q^2)pq(p^2-q^2)$ a square, say of $2pq(p^2+4pq+q^2)(p^2-q^2)r$, thus finding t. Other solvers used the numbers $x^2\mp xy+y^2$, x^2+y^2 in A. P.

[381a] New Series of Math. Repository (ed., T. Leybourn), 2, 1809, I, 31–33.
[382] The Gentleman's Math. Companion, London, 3, No. 14, 1811, 282–3.
[382a] New Series of Math. Repository (ed., T. Leybourn), 3, 1814, I, 111–5.
[383] The Gentleman's Math. Companion, London, 5, No. 26, 1823, 238–9.
[384] Ladies' Diary, 1826, 35–6, Quest. 1434.
[385] Math. Miscellany, New York, 1, 1836, 123.
[386] Math. Quest. Educ. Times, 12, 1869, 80.
[387] Math. Visitor, 2, 1887, 84–8.
[388] Math. Quest. Educ. Times, 12, 1869, 91–2.

and took $x=2mn$, $y=m^2-n^2$. Then shall $m^2+4mn+n^2=\square$, say $(m+pn)^2$, which gives m/n. Take $p=3/2$. Then xy is a cube if $n=300$.

J. Matteson[389] solved the last problem and that with the final condition replaced by the following: The sum of the roots of the squares is an eighth power, the squares being a seventh, a fifth and a fourth power, and the arithmetical mean of the required numbers a square.

To find three positive integers whose sum, sum of squares, and sum of cubes, are squares, A. B. Evans[390] took ax, ay, az, where $a=x+y+z$, and set $a(x^3+y^3+z^3)=a^2(x-y+z)^2$, whence $y^2+y(x+z)-3xz=0$. The radical in y is rational if $x^2+14xz+z^2=\square=(zm/n-x)^2$, which holds if $x=m^2-n^2$, $z=2mn+14n^2$. In the resulting expression for Σx^2, set $m=p-8n$ and equate to the square of $p^2-16pn-83n^2$. Thus $p/n=1332/83$. Then $ax=412095790665$, etc. Several solvers used $15mx$, $15my$, $8m(x+y)$, whose sum of cubes is divisible by their sum. Thus a linear and two quadratic functions are to be squares, which is true if $m=d^2/\{23(x+y)\}$.

D. S. Hart[391] divided unity into three positive parts whose sum of squares and sum of cubes are squares by taking x/s, y/s, z/s as the parts, where $s=x+y+z$. The conditions are satisfied if $s=\square$, $\Sigma x^2=\square$, $\Sigma x^3=\square$, which is the preceding problem. He[392] found three numbers whose sum and sum of squares are cubes, and sum of cubes a square. Let ax^3, bx^3, cx^3 be the numbers. Their sum of cubes will equal $(x^5)^2$ if $x=\Sigma a^3$. To make Σa and Σa^2 cubes, equate their product to $(a+b-c)^3$; the roots of the resulting quadratic for a are rational if $b^4+2b^3c-9b^2c^2+6bc^3-7c^4=\square$. Set $b=2c+d$. The new quartic is a square if $d=35c/9$ or $116c/315$.

To find three integers whose sum, product and sum of squares are all squares, S. Tebay[393] used the numbers xy, $x(x+y)$, $y(x+y)$, while A. B. Evans used xa^2, ya^2, xya^2, with $x=y+1$.

D. S. Hart[394] found three numbers, say ax, bx, cx, such that if the sum of their cubes be added to or subtracted from the square of each, the sums and remainders are squares. Set $d=a^3+b^3+c^3$. Then $a^2x^2+dx^3=\square=e^2x^2$, $a^2x^2-dx^3=\square=f^2x^2$ give $x=(e^2-a^2)/d=(a^2-f^2)/d$. Similarly, $b^2x^2\pm dx^3=g^2x^2$, h^2x^2 give $x=(g^2-b^2)/d=(b^2-h^2)/d$, while $c^2x^2\pm dx^3=k^2x^2$, l^2x^2 give

$$x=(k^2-c^2)/d=(c^2-l^2)/d.$$

By the numerators of x, $e^2=2a^2-f^2$, $g^2=2b^2-h^2$, $k^2=2c^2-l^2$. The first is satisfied if $a=P^2+Q^2$, $f=2PQ-P^2+Q^2$. As in Diophantus V, 8, take three right triangles of equal area, with the hypotenuses $49+9$, $49+25$, $49+64$. For $P=7$, $Q=3$, we get $a=58$, $f=2$. Similarly, $P=7$, $Q=5$ give $b=74$, $h=46$; $P=8$, $Q=7$ give $c=113$, $l=97$. Hence we get ax, etc.

Problems solved in the American Mathematical Monthly: Three numbers the sum of whose cubes is a square and sum of squares a cube (1, 1894, 363). Three integers the sum of any two of which is a cube (p. 208, p. 279).

---

[389] Collection of Diophantine Problems, Washington (ed., Martin), 1888, pp. 5–7.
[390] Math Quest. Educ. Times, 17, 1872, 30–1.
[391] Ibid., 21, 1874, 100–1.
[392] Ibid., 26, 1876, 102.
[393] Ibid., 23, 1875, 31.
[394] Math. Visitor 2, 1882, 17–18.

Three integers whose sum is a cube and sum of any two less the third a cube (2, 1895, 86–7). Three positive integers the product of the first by the sum of the other two a square and sum of their cubes a square[395] (p. 196). Four positive integers each less double the cube of their sum a cube (7, 1900, 49–50). Three positive integers whose sum, sum of squares, and sum of cubes, are all squares (9, 1902, 145–6), or all cubes (24, 1917, 240).

R. F. Davis and others[396] made $X^3+Y^2+Z^2$, $Y^3+X^2+Z^2$, $Z^3+X^2+Y^2$ all squares. Take $X=2(1-y)$, $Y=2(1+y)$, $Z=2(1-y^2)$. Then the first two equal the squares of $2(2\mp y+y^2)$. The third is a square if $y=\pm 4/3$.

A. Martin[397] solved $A^2+B^2+C^2=\square$, $A^3+B^3+C^3=D^3$. As the solutions of the latter he employed the products of a by the values given by Young.[56] Take $n^2+2=(n-r)^2$. Then $\Sigma A^2=\square$ becomes a quartic for q whose solution, found as usual, is a very long expression for q. Take $r=3$, whence $n=7/6$. Then $q=\alpha/\beta$, where $\alpha=81420385$, $\beta=11290752$. Take $a=6\beta^2$. Then A, B, C are integers each of 17 digits:

$$A=11868013975030087,\quad B=16269106368215226,\quad C=88837226814909894.$$

M. Rignaux[398] noted a solution of the last problem involving parameters m, n, g such that $m^2+2n^2=\square$; in his numerical example, A and C are negative, while A, B, C contain only 6 or 7 digits.

P. Tannery and H. Brocard[399] noted that 3, 4, 5 yield by multiplication

$$54+72+90=6^3,\qquad 54^3+72^3+90^3=108^3.$$

E. B. Escott[400] gave numbers without common factor:

$$3+4-6=1^3,\qquad 3^3+4^3-6^3=-5^3;$$
$$36+37-46=3^3,\qquad 36^3+37^3-46^3=-3^3.$$

H. Brocard[401] gave $9+15-16=2^3$, $9^3+15^3-16^3=2^3$ and $24+2-18=2^3$, $24^3+2^3-18^3=20^3$.

A. Gérardin[402] noted that, if $x+y+z$, Σx^2 and Σx^3 are all cubes, x, y, z are in neither geometrical nor arithmetical progression. He and others[403] noted special sets of integral solutions of $x+y+z=c^2$, $x^2+y^2+z^2=b^3$; also values making s^3-x-y, s^3-y-z, s^3-x-z all squares or all cubes, where $s=x+y+z$ (ibid., 23, 1916, 5–6); s^3-x, s^3-y, s^3-z all squares (pp. 157–9); $xyz+x^2$, $xyz+y^2$, $xyz+z^2$ all squares (24, 1917, 37–8); s^2-x-y, s^2-y-z, s^2-x-z all cubes (22, 1915, 220).

[395] Also, Math. Quest. Educ. Times, 24, 1913, 63–4.
[396] Math. Quest. Educ. Times, 64, 1896, 26.
[397] Math. Magazine, 2, 1898, 254–5.
[398] L'intermédiaire des math., 24, 1917, 79–80. He corrected a misprint in a citation of Martin's solution, correctly quoted in 7, 1900, 162.
[399] L'intermédiaire des math., 6, 1899, 190.
[400] Ibid., 7, 1900, 141.
[401] Ibid., 10, 1903, 14.
[402] Sphinx-Oedipe, 9, 1914, 38–9.
[403] L'intermédiaire des math., 22, 1915, 172; 23, 1916, 93.

To find n numbers the cube of whose sum increased (or diminished) by any one of them gives a cube.

Diophantus, V, 18 [19], required three numbers x_i such that, if s denotes their sum, $s^3 + x_i$ $[s^3 - x_i]$ are cubes. Set $x_i = (a_i^3 - 1)s^3$ $[x_i = (1 - a_i^3)s^3]$. Since $\Sigma x_i = s$, we have $(\Sigma a_i^3 - 3)s^2 = 1$ $[= -1]$. For the first problem, $\Sigma a_i^3 - 3 = \square$, take $a_1 = m+1$, $a_2 = 2-m$, $a_3 = 2$; then

$$\Sigma a_i^3 - 3 = 9m^2 - 9m + 14 = (3m-4)^2,$$

if $m = 2/15$; thus $s = 5/18$. For the second problem, $3 - \Sigma a_i^3 = \square$, $a_i^3 < 1$, Diophantus took the square to be $2\frac{1}{4}$, whence $\Sigma a_i^3 = \frac{3}{4} = 162/216$. Hence we have to express 162 as the sum of three cubes. Now $162 = 125 + 64 - 27$. By the theorem in the " Porisms," the difference of two cubes is always a sum of two cubes. Having thus the three cubes [not given by Diophantus] and $2\frac{1}{4}s^2 = 1$, whence $s = 2/3$, we obtain the numbers x_i. Cf. Bachet.[404]

Diophantus, V, 20, required three numbers x_i of sum s, such that $x_i - s^3$ are cubes. Set $x_i = (a_i^3 + 1)s^3$. Then $\Sigma a_i^3 + 3$ is to be a square $1/s^2$. Let $a_1 = m$, $a_2 = 3 - m$, $a_3 = 1$. Then $9m^2 - 27m + 31 = \square = (3m-7)^2$, say, whence $m = 6/5$, $s = 5/17$.

C. G. Bachet[404] believed that Diophantus had found by accident the square $2\frac{1}{4}$ which 3 exceeds by a number expressible as a sum of three cubes <1, and stated that he could not solve the problem if $2\frac{1}{4}$ be replaced by $2\frac{7}{9}$. He completed the computation omitted by Diophantus. [By Vieta's[38] formula (1)], $64 - 37$ is the sum of the cubes of $40/91$ and $303/91$. Thus $162/216$ is the sum of the cubes $125/216$, $20^3/(91^3 \cdot 27)$, $101^3/(91^3 \cdot 8)$. Subtracting them from unity and multiplying the remainders by $s^3 = (2/3)^3$, we obtain the answers $91/27^2$, etc., which Bachet expressed as fractions with a common denominator, but with the common factor 27 in all terms.[415] The reduced denominator is $549353259 = 91^3 \cdot 27^2$.

A. Girard[39c] noted that we may employ Bachet's value $2\frac{7}{9}$ since $3 - 2\frac{7}{9} = 162/9^3$ is a sum of three cubes. Or we may employ $2\frac{14}{25}$ which 3 exceeds by the sum $440/1000$ of the cubes $216/1000$, $216/1000$ and $8/1000$; the resulting solution of Diophantus V, 19 is $49/256$, $49/256$, $62/256$ [since $s^3 - x_1 = (3/8)^3$, $s^3 - x_3 = (1/8)^3$].

Fermat[405] would not admit that Diophantus was led to $2\frac{1}{4}$ by chance and remarked that it is not difficult to rediscover his method. " Take $x-1$ as the side of the required square between 2 and 3. Then $3 - (x-1)^2$ is to be the sum of three cubes. Take as sides of two of the cubes linear functions of x such that, if the sum of their cubes be subtracted from $2 + 2x - x^2$, the result contains only two terms in x of consecutive degrees. This can be done in an infinitude of ways. Take $1 - x/3$ and $1 + x$ as the sides of two of the cubes; then the result mentioned is

$$\frac{-13}{3}x^2 - \frac{26}{27}x^3.$$

Equating this to $-c^3x^3$, we have $x = 117/(27c^3 - 26)$. We are to choose

---

[404] Diophanti Alex. Arith., 1621, 324.
[405] Oeuvres, III, 258–9.

c so that $1-x/3>0$. Since the third cube is negative, we apply[406] the Porism. Here Bachet was again embarrassed; he confessed that he could express the difference of two given cubes as a sum of two cubes only when the greater of the given cubes exceeded the double of the smaller."

Ludolph van Ceulen[407] (1540–1610), at the end of his Dutch work on the circle, proposed 100 problems the 68th and 69th of which are to find three and four numbers such that if each be subtracted from the cube of their sum the remainders are cubes. For three numbers his solution, communicated in letters, is the one published by van Schooten,[408] his successor at the University of Leyden. After learning van Ceulen's method, N. Huberti obtained answers for four numbers, quoted by van Schooten:

$$867160/C, \qquad 787400/C, \qquad 13527640/D, \qquad 14087528/D$$
$$(C=4657463, \quad D=125751501);$$
$$12172736/k, \qquad 11296152/k, \qquad 9112168/k, \qquad 4724776/k \quad (k=64481201);$$

and the further answer for three numbers:

$$15817815000/G, \qquad 9568152000/G, \qquad 8925120000/G \quad (G=86526834967).$$

Frans van Schooten[408] first found three cubes such that on subtracting them from 4^3, the sum of the remainders is a square. Let the roots of the cubes be $N-1$, $4-N$, 2. Subtracting their cubes from 64, we get $65-3N+3N^2-N^3$, $48N-12N^2+N^3$ and 56. Equate their sum

$$121+45N-9N^2$$

to $(11+N)^2$; hence $N=23/10$. Thus the above remainders equal

$$a=61803/1000, \qquad b=59087/1000, \qquad 56.$$

Now let the three desired numbers be an^3, bn^3, $56n^3$ and their sum $4n$, whence $n=20/133$. Hence the answer is

$$494424/D, \qquad 472696/D, \qquad 448000/D \qquad (D=2352637).$$

Their sum is 80/133, whose cube diminished by the three numbers gives as remainders the cubes of 26/133, 34/133, 40/133.

J. H. Rahn and J. Pell[409] treated Diophantus V, 18, 19, 20 at length. Pell's solution differs little from van Schooten's, except in using $(11+mN)^2$ in place of $(11+N)^2$, and is given in Wallis' Algebra, Engl. ed., 1685, p. 219, with van Schooten's answer.

J. Kersey[410] employed, for Diophantus V, 19, $a_1=53/144$, $a_2=27/144$, $a_3=16/144$, whence $3-\Sigma a_i^3=(247/144)^2$, $s=144/247$; thus the desired num-

---

[406] To secure Diophantus' value $(x-1)^2=2\frac{1}{4}$, we must take $x=5/2$ or $-1/2$, whence $27c^3=364/5$ or $-8\cdot26$, so that c is irrational. Hence Fermat's process is not general, although it leads to a solution by setting $c=5/3$, whence $x=13/11$, and the sides of the cubes are 20/33, 72/33, $-65/33$, as noted by Heath, "Diophantus," ed. 2, 214.

[407] Van den Circkel, 1596, 1615. Latin transl. by W. Snellius, 1619. Cf. Bull. Bibl. Storia Sc. Mat. e Fis., 1, 1868, 141–156.

[408] Exercitationvm Math., 1657, Liber V, Sect. 13, 434–6. Reproduced by C. Hutton, The Diarian Miscellany, London, 1, 1775, 138–9.

[409] Rahn's Algebra, Zurich, 1659. An Introduction to Algebra, transl. out of the High-Dutch by T. Brancker, much altered and augmented by D. P[ell], London, 1668, 105–131. Cf. Wallis' Algebra, Ch. 59.

[410] The Elements of Algebra, London, Book III, 1674, 111–4, 104–5.

bers are the ratios of 2837107, 2966301, 2981888 to 15069223.   Or we may take $a_1 = 103/(9 \cdot 23)$, $a_2 = 12/(9 \cdot 23)$, $a_3 = 1/9$, whence $s = 9 \cdot 23/1053$, giving $x_1 = 7777016/43243551$, etc.   Or, $a_1 = 41/64$, $a_2 = 39/64$, $a_3 = 3/63$,

$$s = 8 \cdot 16/185, \qquad x_1 = 1545784/6331625.$$

Or, $a_1 = 67/88$, $a_2 = 87/88$, $a_3 = 22/88$, $s = 176/221$, $x_1 = 3045672/10793861$. For four numbers, take $\alpha_1, \cdots, \alpha_4$ to be the ratios of 4684, 4836, 3485, 3315 to 1360.   Then $\Sigma(64 - \alpha_i^3) = t^2$, $t = 16027/1360$; the desired numbers are $x_i = (16 - \alpha_i^3)s^3$, $s = \Sigma x_i = 1/t$.   His[411] solution of V, 18 is the same as in Diophantus.

The answer of van Schooten[408] was given without details in the Ladies' Diary, 1717, Question 51.   J. Hampson[412] gave without details the smaller answer to Diophantus, V, 19: $13851/D$, $19467/D$, $18954/D$, where $D = 85184$. He[413] also stated two answers to Diophantus V, 18: ratios of 23625, 1538 and 18577 to 157464; ratios of 18954, 4184 and 271 to 132651.

J. Landen[414] took zy, zx, zv as the numbers in Diophantus V, 20, and p^2z as their sum, and zs, zr, zq as the roots of the cubes, finding the answer $341/D$, $854/D$, $250/D$, where $D = 4913$; no details were given.

The " Repository solution "[415] is a repetition of that by Diophantus as completed by Bachet,[404] leading to $162707336/d$, $134953209/d$, $68574961/d$, where $d = 549353259$.   It is also noted that 37 is the sum of the cubes of $18/7$ and $19/7$, whence $s = 2/3$ and a new answer is $68256/k$, $67229/k$, $31213/k$, where $k = 250047$.

For Diophantus V, 18, J. Bennett[416] took nx as one number and s as the sum of the three.   Let $s^3 + nx = (s + x)^3$, whence $x = \frac{1}{2}\sqrt{4n - 3s^2} - 3s/2$. Taking $n/s^2 = 21$, 31, 57, we get $nx = 63s^3$, $124s^3$, $342s^3$, which will be the desired numbers if their sum is s, i. e., if $s = 1/23$.   J. Ryley[417] used the numbers x, y, $a - x - y$; then $a^3 + x = a^3s^3$, $a^3 + y = a^3n^3$ give x, y.   Let $a^3 + a - x - y = m^3a^3$.   Then $a^2f = 1$, $f = m^3 + n^3 + s^3 - 3$.   Take $n = 2 - r$, $s = 1 + r$, $f = (2vm - 3r)^2$, which gives r in terms of m, v.

T. Leybourn[418] noted that Diophantus V, 18 is satisfied by taking $(a^3 - u^6)v^3$, $(b^3 - u^6)v^3$, $(c^3 - u^6)v^3$ as the numbers, if u^2v is their sum.   The latter requires $F = a^3 + b^3 + c^3 - 3u^6 = \square$.   Take $a = p + q$, $b = r - p$, $c = s$. Then $F = 3(q + r)p^2 + 3(q^2 - r^2)p + q^3 + r^3 + s^3 - 3u^6$.   Take $3(q + r) = n^2$; then $F = (m - np)^2$ determines p rationally.   By trial, he found that $F = (23)^2$ if $a = 4$, $b = 5$, $c = 7$, $u = 1$.   For Diophantus V, 20, he[419] took $(a^3 + u^6)v^3$, $\cdots$, $(c^3 + u^6)v^3$ as the numbers and u^2v as their sum.   Then

$$G = a^3 + b^3 + c^3 + 3u^6 = \square.$$

---

[411] The Elements of Algebra, London, Book III, 1674, 101.

[412] Ladies' Diary, 1747, 27, Quest. 275.

[413] Ibid., 1748, 27, Quest. 288.

[414] Ladies' Diary, 1749, 26, Quest. 304; C. Hutton's Diarian Miscellany, 2, 1775, 270; Leybourn's Math. Quest. proposed in Ladies' Diary, 2, 1817, 7–9.

[415] The Diarian Repository; or, Math. Register . . . Collection of Math. Quest. from Ladies' Diary, by a Society of Mathematicians, London, 1774, 81–2.

[416] Ladies' Diary, 1805, 43–4, Quest. 1132; Leybourn's M. Quest. L. D., 4, 1817, 46–7.

[417] The Diary Companion, Supplement to Ladies' Diary, London, 1805, 46–7.

[418] Leybourn's Math. Quest. proposed in Ladies' Diary, 1, 1817, 405–7.

[419] Ibid., 2, 1817, 7–9.

By trial, $G=37^2$ if $a=5$, $b=8$, $c=9$, $u=1$. Setting $b=3q^2-a$, we see that G becomes a quadratic in a, and $G=(m-3qa)^2$ determines a rationally.

M. Noble[420] gave a note on the history of Diophantus V, 19, citing papers reported on above. He noted that one solution leads to an infinitude. For, if $3-\Sigma a_i^3=a^2$, then $3-\Sigma(a_i+g_ix)^3=(a+fx)^2$, provided

$$-3Ax-3Bx^2-Cx^3=2afx+f^2x^2, \qquad A=\Sigma a_i^2 g_i, \qquad B=\Sigma a_i g_i^2, \qquad C=\Sigma g_i^3.$$

We may take $3A+2af=0$, $x=(-f^2-3B)/C$. He also gave the following solution. Let x, y, z be the desired numbers and s their sum. Then $x=s^3-a^3$, $y=s^3-b^3$, $z=s^3-c^3$. Thus

$$(1) \qquad s=3s^3-a^3-b^3-c^3.$$

Take $s=u^2v$, $a=(p+q)v$, $b=(u^2-q)v$, $c=(u^2-p)v$. Then (1) requires that

$$u^2=v^2F, \qquad F=Eu^2+3(u^4-p^2)q-3(p+u^2)q^2=\Box, \qquad E=u^4+3pu^2-3p^2.$$

Set $E=(u^2+pm/n)^2$. We get p and hence

$$E=(ku^2)^2, \qquad k=(3n^2+3mn-m^2)/(3n^2+m^2).$$

Equating F to the square of u^3k+qr/e, we get q. Then evidently

$$v=\frac{eu}{eku^3+rq}, \qquad x=\{u^6-(p+q)^3\}v^3, \qquad \cdots, \qquad z=\{u^6-(u^2-p)^3\}v^3.$$

Wm. Lenhart[421] found n numbers x_i such that if each be added to the cube of their sum s the sum shall be a cube α_i^3. Thus $s+ns^3=\Sigma\alpha_i^3$. Take $s=1/r$. Then $r^2+n=\Sigma(r\alpha_i)^3$. But in another paper, Lenhart[62] of Ch. XXV, he showed how to express a number (here r^2+n) as a sum of cubes. Again, to find n numbers x_i such that if each be subtracted from the cube of their sum s the remainder shall be a cube β_i^3, we have $ns^3-s=\Sigma\beta_i^3$. Take $s=r/t$. Then $r(nr^2-t^2)=\Sigma(t\beta_i)^3$. If $r=t$, the problem is to find n cubes, each <1, whose sum is $n-1$. It was discussed in the paper cited. Here let $t>r$, t being prime to r. The following tentative process was used. From nr^2 subtract in turn the terms of a decreasing series of squares prime to r and beginning with the first square $<nr^2$ and ending with the square just $>r^2$; Multiply each remainder by r and seek (as in the paper cited) a separation of the product into cubes $(t\beta_i)^3$. For $n=4$, take $r=12$, $t=19$; $2580=a+b$, $a=1241=9^3+8^3$, $b=1339=2^3+11^3$. Using his table of sums of two cubes, he found various answers for $n=3$ and one for $n=5$.

S. Bills,[422] to find $x_i=(1-a_i^3)s^3$, $s=\Sigma x_i$, would solve $n-a_1^3-\cdots-a_n^3=k^2$ by taking arbitrary values for k, a_4, $\cdots$, a_n and using the theorem that any number is a sum of three rational squares. Similarly,[423] to find $x_i=(a_i^3-1)s^3$, we have $\Sigma a_i^3-n=1/s^2$; set $K=1/s$ and assign arbitrary values to a_3, $\cdots$, a_n and solve $a_1^3+a_2^3+d=K^2$, where $d=a_3^3+\cdots+a_n^3-n$. Put

$$a_1=\tfrac{3}{2}+v, \qquad a_2=\tfrac{3}{2}-v;$$

then $K^2-9v^2=27/4+d=fg$. Taking $K+3v=f$, $K-3v=g$, we get K and v.

[420] Leybourn's Math. Quest. proposed in Ladies' Diary, 1, 1817, 52–62.
[421] Math. Miscellany, New York, 1, 1836, 263–7.
[422] Math. Quest. Educ. Times, 22, 1875, 71.
[423] Ibid., 24, 1876, 52–3.

A. B. Evans[424] found three positive numbers whose sum is unity such that each plus unity is a cube. Take $a_i^3 x^3 - 1$ as the numbers. Then $x^3 \Sigma a_i^3 = 4$. Set $p = s + a_3$, $4r^3 = a_1^3 + a_2^3 - s^3$. Eliminating the a_i^3, we get

$$\frac{1}{x^3} = \frac{1}{4}(p^3 - 3p^2 s + 3ps^2 + 4r^3) = \left(r + \frac{ps^2}{4r^2}\right)^3, \qquad \text{if } p = \frac{12r^3 s}{4r^3 - s^3}.$$

Then $x = (4r^3 - s^3)/(4r^4 + 2rs^3)$. For $r = 9$, $s = 5$, the condition $a_1^3 + a_2^3 = 3041$ is satisfied if $a_1 = 1404/133$, $a_2 = 1637/133$. Cf. papers 426, 428.

D. S. Hart[425] found N $(N \leqq 5)$ numbers such that if each be subtracted from the cube of their sum s the remainder is a cube. For 3 numbers x, y, z, let $s^3 - x = m^3$, $s^3 - y = n^3$, $s^3 - z = p^3$. Then $m^3 + n^3 + p^3 = 3s^3 - s$, which is satisfied if

$$m = \tfrac{1}{28}, \qquad n = \tfrac{2}{28}, \qquad p = \tfrac{15}{28}, \qquad s = \tfrac{9}{14}, \qquad x = \tfrac{17}{64}, \qquad y = \tfrac{13}{49}, \qquad z = \tfrac{351}{3136}.$$

These give answers involving the least numbers found to date. For $N = 4$, $m^3 + n^3 + p^3 + q^3 = 4s^3 - s$, which is satisfied if $s = 5/9$, $m = 3/27$, $n = 5/27$, $p = 6/27$, $q = 13/27$; the desired numbers are the ratios of 3348, 3250, 3159, 1178 to 27^3. For $N = 5$, take $s = \tfrac{1}{2}$, and m, $\cdots$, r to be the ratios of 1, 3, 4, 5, 8 to 18.

R. Davis[426] divided unity into three parts such that each increased by unity is a cube. He and D. S. Hart (p. 133) treated Diophantus V, 20.

S. Tebay,[427] to make $a^3 - x_i$ a cube where $a = \Sigma x_i$, and hence $na^3 - a$ a sum of n cubes, would express $n - a^{-2}$ as a sum of n cubes, the roots of three of which are $m - s$, $m - t$, $s + t - m$. Let $H - n + m^3$ be the negative of the sum of the remaining cubes. Then $a^{-2} = H + 3st(2m - s - t)$. Equate the last product to $9r^2 s^2 t^2$, thus determining t. Then

$$a^{-2}(3r^2 s + 1)^2 = 9r^2(s^2 - 2ms + \tfrac{1}{2}Hr^2)^2, \qquad 24r^2(3mr^2 + 1)s = 9Hr^6 - 4.$$

Hence s and t are found rationally in terms of r, m, H. He[428] expressed 2 as a sum of three rational cubes. But $3s^3 - s = 2$ if $s = 1$. Hence, as by Hart,[425] we have three numbers whose sum is unity and such that unity exceeds each by a cube. He tested eleven sums of three cubes by the method of Hart,[425] but found no answer in quite so small numbers as Hart's, his smallest answer being 13/49, 17/64, 351/(49·64), with the sum 9/14.

A. Holm[429] treated Diophantus V, 19 by starting with Diophantus' formula

$$3 - \left(\frac{5}{6}\right)^3 - \left(\frac{2}{3}\right)^3 - \left(\frac{-1}{2}\right)^3 = \left(\frac{3}{2}\right)^2.$$

To find positive solutions of $3 - \Sigma a_i^3 = \square$, take $a_1 = 5/6$, $a_2 = \tfrac{2}{3} - x$, $a_3 = -\tfrac{1}{2} + x$. Then

$$\frac{9}{4} + \frac{7}{12}x - \frac{1}{2}x^2 = \square = \left(-\frac{3}{2} + rx\right)^2, \qquad \text{if} \qquad x = \frac{36r + 7}{12r^2 + 6}.$$

[424] Math. Quest. Educ. Times, 25, 1876, 31. Cf. Strong[61] and Lenhart[62] of Ch. XXV.
[425] Ibid., 26, 1876, 66-8.
[426] Math. Visitor, 1, 1880, 107.
[427] Math. Quest. Educ. Times, 38, 1883, 81-2.
[428] Ibid., 101-3.
[429] Math. Quest. Educ. Times, (2), 9, 1906, 98.

To make the cubes positive take $\frac{1}{2}<x<\frac{2}{3}$. This is the case if $r=11/2$, whence $x=5/9$. Thus the ratios of $351, 832, 833$ to 3136 answer Diophantus' problem.

A. Gérardin, R. Goormaghtigh and others discussed in l'intermédiaire des mathématiciens the following problems in which s is the sum of the unknowns:

s^3-x and s^3-y cubes, 22, 1915, 222; 23, 1916, 142–4, 210–1.

s^3-x, s^3-y, s^3-z all cubes or all biquadrates, 22, 1915, 245; 23, 1916, 4–5.

s^3-x, s^3-y, s^3-z, s^3-t all cubes or all squares, 23, 1916, 28–9, 52–3.

s^3-x_1, $\cdots$, s^3-x_5 all cubes or all squares, 100–1.

s^3-x_1, $\cdots$, s^3-x_n all cubes or all squares, n odd $\geqq 5$, 24, 1917, 114–5.

SYSTEMS OF EQUATIONS OF DEGREE THREE IN FOUR OR MORE UNKNOWNS.

Alkarkhi[430] (beginning of eleventh century) solved $x^2-y^3=z^2$, $x^2+y^3=t^2$ by setting $x=2y$, $z=my$, $t=ny$, whence $y=4-m^2=n^2-4$, $m^2+n^2=8$; take $m^2=4/25$, $n^2=196/25$. He treated various similar problems.

J. Ozanam[431] asked for four numbers such that one obtains a square by adding to the product of the first three the product of any two of the four.

W. Wright[432] found four numbers the product of any three added to unity being a square. Substitute the value of z from $xyz+1=(pz+1)^2$ into $vyz+1$ and $vxz+1$. The results are squares if

$$F=p^2-2vyp+vxy^2=(p-q)^2,$$

which determines p, and $G=p^2-2vxp+vx^2y=\square$. The latter leads to a quartic in q which is equated to the square of $q^2-2vxq+2v^2xy-vxy^2+2vx^2y-2v^2x^2$, thus determining q. Then $vxy+1=n^2$ determines x. J. Baines took $wxy+1=a^2$, $wxz+1=b^2$, $wyz+1=c^2$, which determine w, x, y in terms of z. Take $(b^2-1)(c^2-1)=1=z$. Then $xyz+1=(a^2-1)^2+1=(41/9)^2$ if $a=7/3$.

J. Anderson[433] found n numbers whose sum is a square such that the square of each exceeds the cube of their sum by a square. Let the numbers be s^2x_i, where $\Sigma x_i=1$. Then shall $x_i^2-s^2=\square=(sp_i-x_i)^2$, say, whence $x_i=s(p_i^2+1)/(2p_i)$. W. Watson used the numbers x_is^3 with the sum s^2. Then $x_i^2-1=\square=(x_i-m_i)^2$ gives x_i. The condition on the sum gives s.

Several[434] found four numbers x, $x+y$, $x+2y$, $x+3y$ in arithmetical progression whose sum s of squares is a square and sum p of the product of the extremes and the product of the means is a cube. Take $y=vx$. Then $s=\square$ if $4+12v+14v^2=(rv-2)^2$, which gives v. Take $r=4$. Then $v=14$, $p=478x^2$, which is a cube if $x=478$.

S. Ward[435] found four numbers a, b, c, x such that the product of any three added to unity shall be a square. Set $m=ab$, $n=ac$, $p=bc$, and let $mx+1=(1-rx)^2$, whence $x=(2r+m)/r^2$. Then shall

$$r^2(nx+1)=r^2+2rn+mn=A^2, \qquad r^2(px+1)=r^2+2rp+mp=B^2.$$

[430] Extrait du Fakhri, French transl. by F. Woepcke, Paris, 1853, 134.

[431] Letter to de Billy, May 9, 1676; Bull. Bibl. Storia Sc. Mat., 12, 1879, 517.

[432] The Gentleman's Math. Companion, London, 5, No. 24, 1821, 47–8.

[433] Ibid., 5, No. 27, 1824, 266–8.

[434] The Math. Diary, New York, 1, 1825, 55–6.

[435] Amer. edition of J. R. Young's Algebra, 1832, 343–5.

Thus $A^2-B^2=(2r+m)(n-p)$. Take $A+B=2r+m$, $A-B=n-p$, which give A. Hence $r^2+2rn+mn=A^2$ gives

$$r=\frac{a^2bc-\frac{1}{4}(ab+ac-bc)^2}{ab-ac-bc}.$$

Taking any values of a, b, c which satisfy $abc+1=\square$, we get an answer. For example, $a=\frac{1}{2}$, $b=2$, $c=3$ give Young's answer $x=16016/25$.

On four integers the sum of any two of which is a cube, see Lenhart,[93] etc.

A. Genocchi[436] noted that early arithmeticians knew that $x=3$, $y=4$, $z=5$, $s=6$ satisfy $xy=2s$, $x^2+y^2=z^2$, $x^3+y^3+z^3=s^3$, and proved that this is the only integral solution. If the third condition is replaced by $x^3+y^3+z^3=s^nt$, where $n>1$ and t is an unknown integer, he proved that either $b=1$, $a=3$, $n=2$, $m=1$, $t=3$ or $m=3$, $t=1$, or $b=1$, $a=2$, $n=3$, $m=t=1$, or $b=1$, $a=2$, $n=2$, $m=2$, $t=3$ or $m=1$, $t=6$.

P. W. Flcod[437] noted that the six cubes

$$(\tfrac{1}{4})^3,\ (\tfrac{1}{3})^3,\ (\tfrac{5}{12})^3,\ (\tfrac{1}{6})^3,\ (\tfrac{53}{168})^3,\ (\tfrac{75}{168})^3,$$

of which the sum of the first three is $1/8$ and the sum of the last three is $1/8$, are such that on adding any one to the square of the sum of the remaining five we obtain a square.

U. Bini[438] considered $xyz=uvw$ with $\Sigma x^2=\Sigma u^2$ or $\Sigma x^3=\Sigma u^3$, and $\Sigma x=\Sigma u$, $\Sigma x^3=\Sigma u^3$, the second pair being equivalent to $\Sigma x=\Sigma x'$, $xyz=x'y'z'$.

L. E. Dickson[439] showed how to obtain all sets of integral solutions of the last pair of equations, as well as of the pair[440]

$$xyz=x'y'z',\qquad xy+xz+yz=x'y'+x'z'+y'z',$$

which express the condition that two rectangular parallelopipeds shall have integral edges, equal volumes and equal surfaces. Cf. papers 16–18 of Ch. XVII.

A. Gérardin[441] noted that d^3-x^2, d^3-y^2, d^3-z^2, d^3-t^2 are all squares, where $d=x+y-z-t$, if $x=65$, $y=488$, $z=481$, $t=7$.

Gérardin[442] gave three sets of solutions of $x^3+y^3+z^3=t^3+u^3+v^3$, $xyz=tuv$, including the solution

$$\frac{x}{p^2}=\frac{t}{pq}=p^3+2q^3,\qquad \frac{y}{pq}=\frac{u}{q^2}=-q^3-2p^3,\qquad \frac{z}{q^2}=\frac{v}{p^2}=p^3-q^3.$$

The same pair of equations and $\Sigma x=\Sigma t$ have the solution

$$\frac{x}{p}=\frac{t}{q}=pq-r^2,\qquad \frac{y}{q}=\frac{u}{r}=qr-p^2,\qquad \frac{z}{r}=\frac{v}{p}=pr-q^2.$$

[436] Atti Accad. Pont. Nuovi Lincei, 19, 1865–6, 49; Annali di Mat., 7, 1865, 157; French transl., Jour. de Math., (2), 11, 1866, 185–7.

[437] Math. Quest. Educ. Times, 70, 1899, 52.

[438] L'intermédiaire des math., 16, 1909, 41–3, 112. Cf. Desboves[302]; also Sphinx-Oedipe, 8, 1913, 140, and Ch. XXIV.

[439] Messenger Math., 39, 1909–10, 86–7.

[440] *Ibid.*, and Amer. Math. Monthly, 16, 1909, 107–114.

[441] L'intermédiaire des math., 23, 1916, 76.

[442] Nouv. Ann. Math., (4), 15, 1915, 564–6.

CHAPTER XXII.

EQUATIONS OF DEGREE FOUR.

Sum or difference of two biquadrates never a square; area of a rational right triangle never a square.

Leonardo Pisano[1] recognized the fact, but gave an incomplete proof, that no square is a congruent number (i. e., x^2+y^2 and x^2-y^2 are not both squares), while the latter is the area of a rational right triangle. Four centuries later, Fermat[2] stated and proved the result thus implied by Leonardo: *no right triangle with rational sides equals a square with a rational side.* The occasion was the twentieth of Bachet's problems inserted at the end of Book VI of Diophantus: to find a right triangle whose area is a given number A. The necessary and sufficient condition given by Bachet was that $(2A)^2+K^4=\square$ for a suitable K. For, this condition implies that $2A/K$ and K are legs of a right triangle of area A; while, conversely, if K and H are legs of a right triangle of area A, they are proportional to K^2 and $2A$, which are therefore legs of a right triangle. He quoted two conditions given by F. Vieta, Zetetica, 1591, IV, 16, of which the first is that the area increased by some biquadrate should be a biquadrate, and expressed doubt as to the necessity of the conditions.

Fermat's proof is of especial interest as it illustrates in detail his method of infinite descent and as it presents the only instance of a detailed proof left by him. In the left column is given a translation of Fermat's account and in the right column proofs[3] of the statements.

<table>
<tr><td>

" If the area of a right triangle were a square, there would be two biquadrates whose difference is a square, and hence two squares whose sum and difference are squares. Thus there would be a square equal to the sum of a square and the double of a square, such that the sum of the two component squares

</td><td>

If the sides have a common factor, the area has a square factor which may be removed. Since we may therefore assume that the sides x, y, z are relatively prime, we may apply the rule of Diophantus and set $x=2mn$, $y=m^2-n^2$, where m and n are relatively prime integers not both odd. Then $mn(m^2-n^2)$ shall

</td></tr>
</table>

---

[1] Tre Scritti, 1854, 98; Scritti, 2, 1862, 272. See Leonardo[17] of Ch. XVI.

[2] Fermat's marginal notes in his copy of Bachet's edition of Diophantus' Arithmetica; Oeuvres de Fermat, Paris, 1, 1891, 340; 3, 1896, 271.

[3] Cf. H. G. Zeuthen, Geschichte der Math. in XVI and XVII Jahrhundert, 1903, 163. In the elaboration of Fermat's proof by A. M. Legendre, Théorie des nombres, 1798, 401–4; ed. 2, 1808, 340–3, use is made of the theory of quadratic forms to show that $\xi=r^2+2s^2$; while P. Bachmann, Niedere Zahlentheorie, 2, 1910, 451–4, employed the uniqueness of factorization of the integral algebraic numbers $a+b\sqrt{-2}$. Both completed the final step in the proof by comparing the areas of the initial and new triangles. H. Dutordoir, Annales de la Société Sc. de Bruxelles, 17, 1892–3, I, 49, announced in eight lines that he could fill in an elementary manner the gaps left in this proof by Fermat. For the elaboration used in the text, see L. E. Dickson, Bull. Amer. Math. Soc., (2), 17, 1911, 531–2.

is a square. But if a square is the sum of a square and the double of a square, its root is likewise the sum of a square and the double of a square, which I can easily prove. It follows that this root is the sum of the two legs of a right triangle, one of the squares forming the base and the double of the other square the height. This right triangle will therefore be formed from two squares whose sum and difference are squares. But[4] both of these squares can be shown to be smaller than the squares of which it was assumed that the sum and difference are squares. Similarly, we would have smaller and smaller integers satisfying the same conditions. But this is impossible, since there is not an infinitude of positive integers smaller than a given one. The margin is too narrow for the complete demonstration and all its developments."

be a square, whence $m=a^2$, $n=b^2$, $a^4-b^4=\square$, where a and b are relatively prime, one even and the other odd. Thus a^2+b^2 and a^2-b^2 are relatively prime. Hence $a^2+b^2=\xi^2$, $a^2-b^2=\eta^2$, ξ and η being odd integers. Also $\xi^2=\eta^2+2b^2$. Set
$$e=(\xi+\eta)/2, \qquad f=(\xi-\eta)/2.$$
Then e and f are integers and $ef=b^2/2$. A common factor of e and f would divide ξ, η, b^2 and a^2. Hence e and f are relatively prime. We may take e odd (changing if necessary the sign of η). Thus $e=r^2$, $f=2s^2$, $2rs=b$, where r and s are integers. Hence $\xi=e+f=r^2+2s^2$, $\eta=r^2-2s^2$. Also $a^2=b^2+\eta^2=r^4+4s^4$. The right triangle with the legs r^2 and $2s^2$ has the area r^2s^2. It is therefore formed (in the sense of Diophantus, as above) from two squares $m_1=a_1^2$ and $n_1=b_1^2$, its sides being $2m_1n_1$ and $m_1^2\pm n_1^2$. Thus $2m_1n_1=2s^2$, $m_1^2-n_1^2=r^2$. By $m_1n_1=s^2$, we get $a_1b_1=s$, a factor of $b=2rs$. Hence[5] a_1 and b_1 are each less than b and hence less than a.

Fermat's[6] observations on Diophantus II, 8 and V, 32 includes the statements that the sum of two biquadrates is never a biquadrate or a square.

Fermat had proposed, Sept., 1636, to Sainte-Croix that he find a right triangle whose area is a square (Oeuvres, II, 65; III, 287); to Frenicle, May (?), 1640, (Oeuvres, II, 195); to Wallis, Apr. 7, 1658 (Oeuvres, II, 376). Fermat stated that the problem is impossible in a letter to Pascal, Sept. 25, 1654 (Oeuvres, II, 313). The attempted[7] proof by J. Wallis, June 30, 1658 (Oeuvres de Fermat, III, 599) goes no further than a proof of the rule of Diophantus for the sides of a right triangle. Fermat referred in a letter to Carcavi, Aug., 1659 (Oeuvres, II, 431–6, see 436) to proofs by the " descente indéfinie " which he had sent to Carcavi and Frenicle con-

---

[4] As translated by Heath, Diophantus of Alex., ed. 2, 1910, 293–5. Tannery (Oeuvres de Fermat, III, 272) gave the incorrect reading: But the sum of these two squares can be shown to be smaller than that of the first two of which it was assumed that the sum and difference are squares.

[5] Or, by $a_1^2+b_1^2\leqq a_1^4+b_1^4=a$.

[6] Oeuvres, I, 291, 327; III, 241, 264.

[7] Criticized by Frenicle, Oeuvres de Fermat, III, 606, 609.

cerning negative theorems, and cited in the same letter the theorem under discussion.

Frenicle de Bessy[8] (†1765) gave a proof, published posthumously, the principle of which is doubtless due to Fermat in view of the letters just cited. It suffices to prove it for a primitive right triangle. Denote the sides by $2mn$, $m^2 \pm n^2$. If the area is a square, the odd leg $m^2 - n^2$ is a square l^2 and the even leg $2mn$ the double of a square. Thus we have a second primitive triangle whose hypotenuse is m, odd leg l and even leg n. Since mn is a square and m is relatively prime to n, m and n are both squares. Denote the sides of the second triangle by $2ef$, $e^2 \pm f^2$, where e and f are relatively prime. Since $n = 2ef$ is a square, one of the numbers e, f is an odd square and the other the double of a square. Let $e = r^2$, $f = 2s^2$. Also $e^2 + f^2 = m$ is a square a^2. Thus a, e, f are the sides of a third primitive right triangle whose area is the square $r^2 s^2$. Its sides are less than the corresponding sides of the second triangle:

$$a < a^2 = m, \qquad f < 2ef = n, \qquad e < (e+f)(e-f) = l.$$

The sides of the second are less than the corresponding sides of the first: $m < m^2 + n^2$, $n < 2mn$, $l < l^2 = m^2 - n^2$. Hence from the first primitive triangle with a square area we have derived another primitive triangle (the third[9]) with a square area and with smaller sides.

G. Wertheim[10] reproduced the last proof in slightly modified form.

Frenicle proved in like manner (p. 175) that no right triangle has each leg a square and hence the area of a right triangle is never the double of a square. He concluded (p. 178) that no square is the sum of two biquadrates and that $x^4 - 4z^4 = y^2$ is impossible in integers.

Fermat had proposed to St. Croix Sept., 1636 that he find two biquadrates whose sum is a biquadrate (Oeuvres, II, 65; III, 287), to Frenicle, May (?), 1640 (II, 195).

G. W. Leibniz[11] proved, in a manuscript dated Dec. 29, 1678, that the area of a primitive right triangle with integral sides is not a square. The sides are $x^2 \pm y^2$, $2xy$, one being even. Then if $x^2 - y^2$ and xy are both squares, x and y are both squares; also $x+y$ and $x-y$ (since a common factor 2 would make $x^2 - y^2$ even, contrary to the above). But y, $x-y$, x, $x+y$ are not all squares. For, if so, the last three give squares in arithmetical progression whose common difference is a square, "which is absurd." Further, if $x^2 - y^2 = xy$, then $(y+x)y = \square$, $(x-y)x = \square$, and each of the four factors would be a square, just disproved. He noted several corollaries. In view of the triangle formed from x and 1, $(x-1)x(x+1)$ is not a square. The difference of two biquadrates is not a square. For, if $v^4 - w^4 = \square$, the

---

[8] Traité des Triangles Rectangles en Nombres, Paris, 1676, 101–6; Mém. Acad. Sc. Paris, 5, 1666–1699; éd. Paris, 5, 1729, 174; Recuil de plusieurs traitez de mathématique de l'Acad. Roy. Sc. Paris, 1676.

[9] Identical with Fermat's second triangle.

[10] Zeitschrift Math. Phys., 44, 1899, Hist. Lit. Abt., 4–7.

[11] Math. Schriften (ed., C. I. Gerhardt), 7, 1863, 120–5. In a fragment, dated July, 1679, Leibniz merely stated that the problem is impossible; see L. Couturat, Opuscules et fragments inéditis de Leibniz, Paris, 1903, 578.

area of the triangle formed from v^2 and w^2 would be a square. Again,

$$x/y - y/x \neq \square \qquad \text{by} \qquad (x^2 - y^2)xy \neq \square.$$

J. Ozanam[12] stated that $x^4 \pm y^4 \neq z^4$. For, $a^4 - b^4$ is the area of the right triangle whose sides are the ratios of $2a^2b^2$, $a^4 - b^4$, $a^4 + b^4$ to ab, and is not a square " as proved by Messieurs de l'Acad. Roy. Sc. and also by R. P. de Billy."

L. Euler[13] proved that $a^4 + b^4 \neq \square$ if $ab \neq 0$. For, if $(a^2)^2 + (b^2)^2 = \square$, where a and b are relatively prime, then $a^2 = p^2 - q^2$, $b^2 = 2pq$, where p and q are relatively prime, one even and the other odd. By $p^2 - q^2 = \square$, p is odd, whence q is even. By $p(2q) = b^2$, p and $2q$ are squares. By $p^2 = a^2 + q^2$, we get $p = m^2 + n^2$, $q = 2mn$, m and n relatively prime. Since $2q = \square$, $mn = \square$ and $m = x^2$, $n = y^2$. Thus $x^4 + y^4$ is a square p, and x, y are less than a, b. By a similar proof, $a^4 - b^4 \neq \square$ unless $b = 0$ or $b = a$.

E. Waring[14] and A. M. Legendre[15] reproduced literally these proofs by Euler.

C. F. Kausler[16] treated $x^4 + y^4 = z^4$ by use of the lemma that $x^2 \pm y^2$ are not both squares. Equating $x = 2PQ$, $y = P^2 - Q^2$ (from $x^2 + y^2 = \square$, x, y relatively prime) to x, $y = p^2 + q^2$, $p^2 - q^2$ or to $(p^2 + q^2)/2$, $(p^2 - q^2)/2$, in either order, where p and q are relatively prime, and odd in the latter case, we are led to a contradiction. Now $x^4 = z^4 - y^4$ requires $z^2 + y^2$, $z^2 - y^2 = m^4n^4$, 1 or m^3n^4, m, $\cdots$ (19 cases); 7 cases are excluded by the lemma, others by $z^2 + y^2 > z^2 - y^2$ or $(z^2 - y^2)^3 > z^2 + y^2$. Finally, if $z^2 + y^2 = m^3$, $z^2 - y^2 = mn^4$, then $2z^2 = m(m^2 + n^4)$, while $m = 2$ is easily excluded. Thus [a prime factor of] m is a factor of z and hence of y.

P. Barlow[17] noted that, if the area $xy/2$ of a right triangle (x, y, z) were a square w^2, then $z^2 \pm 4w^2 = (x \pm y)^2$, whereas it was proved by descent (p. 109) that $x^2 + y^2$ and $x^2 - y^2$ are not both squares. Also (p. 119), $x^4 + y^4 \neq \square$.

J. Horner[18] noted that if $x/y \pm y/x = \square$, where x, y are relatively prime, then $x = m^2$, $y = n^2$, $m^4 \pm n^4 = \square$, contrary to a known result.

Schopis[19] proved $x^4 + y^4 = z^2$ impossible, using the impossibility of $x^4 - y^4 = 2z^2$. Next (pp. 6-10), $x^4 + y^4 = 2z^2$ is impossible; likewise (p. 11) $x^4 - y^4 = z^2$.

A. M. Legendre[20] stated that the above[3] proof that the area of a right triangle is not a square shows that $a^4 - b^4 \neq \square$ if $a \neq b$, $b \neq 0$. [But in

[12] Journal des Sçavans, 1680, p. 85.

[13] Comm. Acad. Petrop., 10, 1747 (1738), 125-34; Comm. Arith., I, 24-34; Opera omnia, (1), II, 38. Same proofs in Euler's Algebra, 2, Ch. 13, arts. 202-8, St. Petersburg, 1770, p. 418; French transl., Lyon, 2, 1774, pp. 242-54; Opera omnia, (1), 1, 1911, 437; Sphinx-Oedipe, 1908-9, 59-64.

[14] Meditationes Algebraicae, Cambridge, ed. 3, 1782, 371-2.

[15] Théorie des nombres, Paris, 1798, 404; ed. 2, 1808, 343; ed. 3, 1830, II, 5; German transl. by Maser, 2, 1893, 5.

[16] Nova Acta Acad. Petrop., 13, ad annos 1795-6 (1827), Mem., 237-44.

[17] Theory of Numbers, 1811, 121 (cf. 144).

[18] The Gentleman's Diary, or the Math. Repository, London, No. 80, 1820, 37.

[19] Einige Sätze aus der unbestimmten Analytik, Progr. Gumbinnen, 1825.

[20] Théorie des nombres, ed. 3, 2, 1830, § 325, p. 4, Cor. (Maser, II, p. 4).

that proof it was known that a and b are not both odd, a criticism due to A. Genocchi[21]].

J. A. Grunert[22] reproduced Euler's proof that $a^4 + b^4 \neq c^2$.

O. Terquem[23] proved by descent that $x^4 \pm y^4 = z^2$ is impossible.

J. Bertrand[23a] proved that $x^4 + y^4 \neq z^2$ somewhat as had Euler.

P. Volpicelli[24] proved that no congruent number is a square. For, if $pq(p^2 - q^2) = a^2$, then $h^2 \equiv (p^3 q + pq^3)^2 = a^4 + 4p^4 q^4$, $(a^4 - 4p^4 q^4)^2 = h^4 - (2apq)^4$, whereas a difference of two biquadrates is not a square.

V. A. Lebesgue[25] proved the impossibility of $x^4 + y^4 = z^2$ by descent. It suffices to treat $(2^a p)^4 + y^4 = z^2$, where p, y, z are all odd, and y, z are relatively prime. The factors $z \pm y^2$ of $(2^a p)^4$ have no common factor other than 2. Hence

$$z \pm y^2 = 2t^4, \qquad z \mp y^2 = 2^{4a-1} u^4, \qquad p = tu, \qquad \pm y^2 = t^4 - 2^{4a-2} u^4.$$

The lower sign is inadmissible. Hence $t^4 - y^2 = 2^{4a-2} u^4$. Thus

$$t^2 \pm y = 2v^4, \qquad t^2 \mp y = 2^{4a-3} z^4, \qquad vz = u, \qquad t^2 = v^4 + (2^{a-1} z)^4.$$

T. Pepin[26] proved the impossibility of $x^4 - y^4 = z^2$ in integers $\neq 0$.

W. L. A. Tafelmacher[27] proved the impossibility of $x^4 + y^4 = z^4$.

D. Gambioli[28] proved that $x^4 - y^4 = z^2$ is impossible in integers $\neq 0$.

T. R. Bendz[29] proved by descent from $x^4 + 4y^4 = z^2$ that the area of a right triangle is not a square.

L. Kronecker[30] amplified Euler's[13] proof.

G. B. M. Zerr[31] employed unproved assumptions in an attempt to prove that the area of no right triangle is a square.

A. Bang[32] noted that relatively prime solutions of $x^4 - z^4 = y^4$ imply

$$x^2 + z^2 = 2y_1^4, \qquad x \pm z = 2y_2^4, \qquad x \mp z = 2^2 y_3^4, \qquad y = 2y_1 y_2 y_3.$$

Thus $y_1^4 - y_2^8 = 4y_3^8$, so that

$$y_1^2 + y_2^4 = 2u_1^8, \qquad y_1 \pm y_2^2 = 2u_2^8, \qquad y_1 \mp y_2^2 = 2^8 u_3^8, \qquad y_3 = 2u_1 u_2 u_3.$$

Hence $u_1^8 - u_2^{16} = 2^{14} u_3^{16}$, so that

$$u_1^4 + u_2^8 = 2v_1^{16}, \qquad u_1^2 + u_2^4 = 2v_2^{16}, \qquad u_1 \pm u_2^2 = 2v_3^{16}, \qquad u_1 \mp u_2^2 = 2^{11} v_4^{16},$$

and $u_3 = v_1 v_2 v_3 v_4$. By the third and fourth, $u_1^2 + u_2^4 = 2v_3^{32} + 2^{21} v_4^{32}$. Then by the second,

$$(v_2^4)^4 - (v_3^8)^4 = (2^5 v_4^8)^4,$$

<hr>

[21] Annali di Sc. Mat. e Fis., 6, 1855, 316, foot-note. His like criticism of the proof by Terquem[23] is not valid.

[22] Klügel's Math. Wörterbuch, 5, 1831, 1143.

[23] Nouv. Ann. Math., 5, 1846, 71-4.

[23a] Traité élém. d'algèbre, 1851, 224-7.

[24] Atti Accad. Pont. Nuovi Lincei, 6, 1852-3, 89-90.

[25] Exercices d'analyse numér., 1859, 83-4; Introd. à la théorie des nombres, 1862, 71-3.

[26] Atti Accad. Pont. Nuovi Lincei, 36, 1882-3, 35-36.

[27] Anales de la Universidad de Chile, 84, 1893, 307-320.

[28] Periodico di Mat., 16, 1901, 149-150.

[29] Öfver diophantiska ekvationen $x^n + y^n = z^n$, Diss. Upsala, 1901, 5-9.

[30] Vorlesungen über Zahlentheorie, 1, 1901, 35-8.

[31] Amer. Math. Monthly, 9, 1902, 202.

[32] Nyt Tidsskrift for Matematik, 16, B, 1905, 35-36.

like the proposed equation, but with $v_2^4 < \sqrt{x}$. Rychlik[232] of Ch. XXVI gave a proof.

J. Sommer[33] reproduced Euler's[13] proof of the impossibility of $x^4 + y^4 = z^2$ in integers and Hilbert's[153] proof (Ch. XXVI) of its impossibility in complex integers $a + bi$.

A. Bottari[34] proved $x^4 + y^4 = z^2$ impossible by use of an unnecessarily complicated set of solutions of $x^2 + y^2 = z^2$.

F. Nutzhorn[35] gave a complicated proof of the impossibility of $x^4 + y^4 = z^4$.

R. D. Carmichael[36] gave a new proof that neither of the equations $m^4 - 4n^4 = \pm t^2$ is possible in integers each $\neq 0$. Hence the system $p^2 - 2q^2 = km^2$, $p^2 + 2q^2 = \pm kn^2$ is impossible in integers each $\neq 0$. Thus the area of a right triangle is not the double of a square. Hence $m^4 + n^4 = \alpha^2$ is impossible in integers each $\neq 0$.

SOLUTION OF $2x^4 - y^4 = \square$; RIGHT TRIANGLE WHOSE HYPOTENUSE AND SUM
     OF LEGS ARE SQUARES; $x^2 + y^2 = B^4$, $x + y = A^2$. ALSO, $x^4 - 2y^4 = \square$,
$$z^4 + 8w^4 = \square.$$

Fermat[37] proposed to St. Martin and Frenicle, May 31, 1643, the problem to find a rational right triangle whose hypotenuse and the sum of whose legs are squares. Fermat[38] affirmed that the smallest such triangle with rational sides is that with the sides[39]

(1)   4 687 298 610 289,   4 565 486 027 761,   1 061 652 293 520.

Fermat's[40] method consists in forming the right triangle from $x+1$, x; its sides are $2x^2 + 2x + 1$, $2x + 1$, $2x^2 + 2x$. The first and the sum $2x^2 + 4x + 1$ of the last two shall be squares. By the usual method of Diophantus, we get $x = -12/7$. The triangle is therefore formed from $-5/7$, $-12/7$. Employing 5, 12 instead, we get[41] (169, -119, 120). When a negative result is obtained it is in accord with a general procedure of Fermat to repeat the operation and to form the triangle from $x+5$, 12. Its sides are $(x+5)^2 \pm 12^2$ and $24(x+5)$. Hence $x^2 + 10x + 169$ and $x^2 + 34x + 1$ are to be squares, say a^2 and $b^2/169$. Then $b^2 - a^2 = 168x^2 + 5736x$. Taking

$$b - a = 14x, \qquad b + a = 12x + 2868/7,$$

we get $a = -x + 1434/7$. Comparing its square with the earlier a^2, we get

$$x = \frac{1343 \cdot 1525}{7 \cdot 2938} = \frac{2048075}{20566}.$$

---

[33] Vorlesungen über Zahlentheorie, 1907, 176–193. French transl. by A. Lévy, 1911, 184–199.

[34] Periodico di Mat., 23, 1908, 109.

[35] Nyt Tidsskrift for Mat., 23, B, 1912, 33–38.

[36] Amer. Math. Monthly, 20, 1913, 213–21.

[37] Oeuvres, II, 259–63.

[38] Oeuvres, I, 336; III, 270, observation on Bachet's comment on Diophantus VI, 24. Also, Oeuvres, II, 261 (259, 263), letter to Mersenne, Aug. 1, 1643.

[39] Cited by Frenicle, Mém. Acad. Sc., 5, 1666–99; éd. Paris, 1729, 56–71. Since his numerical search was fruitless, he doubtless learned of Fermat's solution from Mersenne.

[40] Inventum Novum, I, 25, 45; III, 32; Oeuvres, III, 340, 353, 388.

[41] Whence the hypotenuse and leg difference of (169, 119, 120) are squares.

The ratio of $x+5$ to 12 is that of 2150905 to 246792. The triangle formed from these is (1). He noted that the problem is equivalent to that to find two numbers whose sum is a square and sum of squares is a biquadrate.

Fermat[42] noted that in the right triangle (156, 1517, 1525) the square of the difference of the legs exceeds the double of the square of the least leg by a square. Without giving details he added that this triangle serves to find a right triangle whose hypotenuse is a square and whose least side differs from the other two by squares.

Frenicle[43] gave details on the last problem. An analysis followed by numerical trials led him to the triangle, formed from $b=156$ and $a=1517$, having the sides

$$2ab=473304, \qquad a^2-b^2=2276953, \qquad a^2+b^2=2325625=1525^2.$$

The least side differs from the other two by the squares of 1343 and 1361. As remarked by A. Genocchi[44] these results imply that $2x^4-y^4=\square$ has the solution $x=1525$, $y=1343$ [Lagrange,[54] Euler[55] (third memoir), Lebesgue[56]].

E. Torricelli[45] proposed the problem to find a right triangle with integral sides whose hypotenuse, sum of legs and sum of hypotenuse and larger leg are all squares. E. Lucas[46] stated that this problem was proposed by Fermat and that its solution depends on $x^4-2y^4=z^2$. In fact, Fermat[46a] proposed the problem to Torricelli. An attempt to trace its origin has been made by E. Turrière.[46b] Cf. *M. Cipolla.[46c]

J. Ozanam[47] treated the problem of Fermat[37] by the method essentially the same as employed by L. Euler.[48] If the legs are x, y, then $x+y$ is to be a square and x^2+y^2 a biquadrate. In this form the problem was proposed by Leibniz. Euler made x^2+y^2 a square $(p^2+q^2)^2$ by taking $x=p^2-q^2$, $y=2pq$. Then p^2+q^2 is a square for $p=r^2-s^2$, $q=2rs$, whence

$$x^2+y^2=(r^2+s^2)^4.$$

It remains to make

$$x+y=r^4+4r^3s-6r^2s^2-4rs^3+s^4$$

a square. It will be the square of $r^2-2rs+s^2$ if $r=3s/2$. Taking $r=3$, $s=2$, we obtain a negative value -119 for x. Setting $r=3s/2+t$, we get

$$16(x+y)=s^4+37\cdot8s^3t+51\cdot8s^2t^2+160st^3+16t^4,$$

which is the square of $s^2+148st-4t^2$ if $s/t=84/1343$. Taking $s=84$, we get $r=1469$ and x, y as in (1).

<hr>

[42] Oeuvres, II, 265–6, letter to Carcavi, 1644.

[43] Methode pour trouver la solution des problèmes par les exclusions, Ouvrages de math., Paris, 1693; Mém. Acad. R. Sc. Paris, 5, 1666–99 [1676]; éd. 1729, 81–5.

[44] Atti R. Accad. Sc. Torino, 11, 1876, 811–29.

[45] G. Loria, l'intermédiaire des math., 24, 1917, 97–8. Cf. 25, 1918, 83.

[46] Bull. Bibl. Storia Sc. Mat. Fis., 10, 1877, 289.

[46a] Letter from Mersenne to Torricelli, Dec. 25, 1643, Bull. Bibl. Storia Sc. Mat. Fis., 8, 1875, 411; Oeuvres de Fermat, 4, 1912, 82–3 (cf. p. 88).

[46b] L'enseignement math., 20, 1919, 245–268.

[46c] Atti Accad. Gioenia sc. nat. Catania, (5), 11, 1919, No. 11.

[47] Nouv. elemens algebre, Amsterdam, 2, 1749, 480–1.

[48] Algebra, 2, 1770, art. 240, pp. 503–5; French transl., 2, 1774, p. 336; Opera Omnia, (1), I, 483–4.

Euler[49] noted that $x^4-2y^4=(p^2-2q^2)^2$ for $y^2=2pq$, $x^2=p^2+2q^2$. The latter holds if $\pm p=r^2-2s^2$, $q=2rs$. Then $2pq=y^2=\pm 4rs(r^2-2s^2)$. Set $r=t^2$, $s=u^2$. For the upper sign, $t^4-2u^4=\square$, whereas t and u are smaller than x, y. Hence take the lower sign. Thus a solution of $2u^4-t^4=\square$ yields a solution of $x^4-2y^4=\square$. For $t=u=1$, we get $x=3$, $y=2$. Then for $t=3$, $u=2$, we get $x=113$, $y=84$. Again, $u=13$, $t=1$ gives $x=57123$, $y=6214$. Lebesgue[56] (end) noted that this solution is incomplete.

Euler[50] treated $x+y=\square$, $x^2+y^2=(z^2+1)^4$ by taking $x=z^4-6z^2+1$, $y=4z^3-4z$. Then $x+y$ is the product of the two factors $z^2+(2\pm 2\sqrt{2})z-1$, which he equated to $(z+p\pm q\sqrt{2})^2$. By the rational and the irrational parts, we get

$$z=\frac{pq}{1-q}, \qquad p=\frac{q\pm\sqrt{2q^4-1}}{1+q}.$$

Thus $q=13$ gives $p=18$ or $-113/7$, $q=-13$ gives $p=21$ or $113/6$.

Euler[51] reduced (2) to (7) by setting $v=2x^4+y^4$, whence $z^4+8(xy)^4=v^2$. Conversely, let $q^4+8p^4=r^2$; then $8p^4=(r+q^2)(r-q^2)$, so that q and r are odd. First, let $r+q^2=2\alpha$, $r-q^2=4\beta$, where α is odd. Then $p^4=\alpha\beta$, and α, β are relatively prime; whence $\alpha=s^4$, $\beta=t^4$, $p=st$. By subtraction, and cancellation of 2, $q^2=s^4-2t^4$. Second, let $r-q^2=2\alpha$, $r+q^2=4\beta$, where α is odd. Proceeding as before, we get $q^2=2t^4-s^4$. While in the second case only we obtained (2), the reduction can always be made since $f^4+8g^4=h^2$ implies $2x^4-y^4=z^2$ for

$$x=f^3+2fg^2-gh, \qquad y=f^3-4fg^2+gh, \qquad z=f^6+f^4g^2-6f^3gh+24f^2g^4-8g^6.$$

In quoting this solution, Lebesgue,[56] p. 74, gave f^2g incorrectly for fg^2 in x.

Euler[52] noted that $x+y=B^2$, $x^2+y^2=A^4$ imply $(x-y)^2=2A^4-B^4$. The latter is the square of $\eta^2+2\xi\eta-\xi^2$ if $A^2=\xi^2+\eta^2$, $B^2=(\xi+\eta)^2-2\eta^2$. Taking $\eta=2abcd$, we have $A=a^2b^2+c^2d^2$ if $\xi=a^2b^2-c^2d^2$, and $B=a^2c^2-2b^2d^2$ if $\xi+\eta=a^2c^2+2b^2d^2$. The two values of $\xi+\eta$ are equal if

$$\frac{a}{d}=\frac{bc\pm r}{c^2-b^2} \qquad\text{or}\qquad \frac{d}{a}=\frac{bc\mp r}{2b^2+c^2}, \qquad r^2=2b^4-c^4.$$

Hence $2A^4-B^4$ is a rational square if $2b^4-c^4$ is. Taking $b=c=1$, we have $a=3$, $d=2$, $\xi=5$, $\eta=12$, $A=13$, $B=1$, $2\cdot 13^4-1=239^2$; since $B<A$, x and y are not both positive. Taking $b=13$, $c=1$, we have $r=239$, $a/d=-3/2$ or $113/84$. For $a=3$, $d=-2$, then $\xi=1517$, $\eta=-156$, $A=1525$, $B=-1343$, which[53] do not yield positive x, y. For $a=113$, $d=84$, then $A=2165017$, $B=-2372159$, and we obtain very large solutions. [In fact, Fermat's (1). Since $x-y=\eta^2+2\xi\eta-\xi^2$, $x+y=B^2$, we have $x=2\xi\eta$, $y=\xi^2-\eta^2$. Thus x, y are the legs of the right triangle formed from ξ, η. Here $\xi=2150905$, $\eta=246792$, as in Fermat's solution.]

<hr>

[49] Algebra, II, art. 211; French transl., pp. 260–3; Opera Omnia, (1), I, 444–5.
[50] Opera postuma, 1, 1862, 491 (about 1774).
[51] Opera postuma, 1, 1862, 221–2 (about 1780).
[52] Opusc. anal., 1, 1783 (1773), 329; Comm. Arith., II, 47.
[53] The method of Euler, Algebra, 2, art. 140, to make $2x^4-1$ a square does not give all solutions since 1525/1343 is omitted (remarked by Lebesgue[56]). E. Fauquembergue, l'intermédiaire des math., 5, 1898, 94, claimed to prove that $x=1$, $x=13$ are the only integral solutions.

J. L. Lagrange[54] discussed Fermat's[37] problem at length. From $p+q=y^2$, $p^2+q^2=x^4$, he derived, after setting $z=p-q$,

$$(2) \qquad 2x^4 - y^4 = z^2.$$

The problem reduces to the solution of (2) since we have

$$(3) \qquad p = \tfrac{1}{2}(y^2+z), \qquad q = \tfrac{1}{2}(y^2-z).$$

Lagrange was evidently not acquainted with Euler's[52] paper of 1773 in which he derived (2) and obtained four sets of solutions $A=x$, $B=y$; indeed, Lagrange omitted the set 1525, 1343, in his citation of Euler. Given any integers x, y for which $2x^4 - y^4 = \square$, Lagrange gave a method to obtain smaller integral solutions; then by reversing the process and starting with $x=y=1$, he concluded that all pairs of larger solutions can be found in the order of their magnitude. While Euler's simpler procedure appears to give all the solutions in this manner, he did not prove that this is the case.

We may assume that x and y are relatively prime. A simple argument shows that x, y, z are all odd. By (2),

$$(z+y^2)^2 = (2x^2)^2 - (z-y^2)^2 = (2x^2+z-y^2)(2x^2-z+y^2).$$

Denote these (even) factors by $2mp$, $2mq$, where p and q are relatively prime. Then pq must be a square. Hence, replacing p, q by p^2, q^2,

$$2x^2+z-y^2 = 2mp^2, \qquad 2x^2-z+y^2 = 2mq^2, \qquad z+y^2 = 2mpq.$$

Eliminating z from the first two, by means of the third, we get

$$x^2 - y^2 = mp(p-q), \qquad x^2+y^2 = mq(p+q).$$

Thus $m=1$ or 2, since m is a divisor of $2x^2$ and $2y^2$. If $m=2$, set $p+q=q'$, $q-p=p'$. Whether $m=1$ or 2, we obtain equations of the form

$$(4) \qquad x^2 - y^2 = p(p-q), \qquad x^2+y^2 = q(p+q).$$

Thus p is odd. Set $(x+y)/p = 2m/n$, where n is odd and prime to m. Then $x+y=2ms$, $p=ns$, where s is an integer. By (4_1), $x-y=2nt$, $p-q=4mt$, where t is an integer prime to s. Thus

$$(5) \qquad x = ms+nt, \qquad y = ms-nt, \qquad p=ns, \qquad q=ns-4mt.$$

Then the product of (4_2) by $(s^2-8t^2)/n^2$ gives

$$s^4+8t^4 = u^2, \qquad u \equiv 3st + \frac{m}{n}(s^2-8t^2).$$

Since m and n are relatively prime we therefore have

$$(6) \qquad m = (u-3st)/l, \qquad n = (s^2-8t^2)/l \qquad (l \text{ an integer}).$$

If $m=0$, then $s/t = \pm 1$, $n^2=1$, $x^2=y^2=1$. Hence if (2) has a set of relatively prime solutions x, y not both of numerical value unity, then by (5) the greater of x, y exceeds the greater of the corresponding solutions s, t of

$$(7) \qquad s^4+8t^4 = u^2,$$

and s, t are relatively prime and not both of absolute value unity. Conversely, from relatively prime solutions s, t, we obtain by (6) and (5) relatively prime solutions x, y of (2).

---

[54] Nouv. Mém. Acad. Sc. Berlin, année 1777 [1779], 140; Oeuvres, 4, 1869, 377-98.

Let s, t be relatively prime solutions of (7).   Then s is odd and

$$u+s^2=2\mu\omega, \qquad u-s^2=2\mu\rho, \qquad 8t^4=4\mu^2\omega\rho,$$

where ω and ρ are relatively prime.   Thus μ divides t^2.   Also $s^2=\mu(\omega-\rho)$. Hence $\mu=1$, $\omega=2q^4$, $\rho=r^4$, or $\omega=q^4$, $\rho=2r^4$, whence

$$u=2q^4+r^4, \qquad s^2=2q^4-r^4; \qquad \text{or} \qquad u=q^4+2r^4, \qquad s^2=q^4-2r^4.$$

Conversely, if $2q^4-r^4=s^2$ or $q^4-2r^4=s^2$ and we set $t=qr$, we have solutions s, t of (7).   If s and t are relatively prime and numerically distinct from unity, the same is true of q and r, while the greater of s, t exceeds the greater of q, r.   The first of these two equations is of type (2).

Applying to the second, $q^4-2r^4=s^2$, a discussion entirely similar to that just used, Lagrange obtained

$$s=8n^4-p^4, \qquad q^2=8n^4+p^4; \qquad \text{or} \qquad s=n^4-8p^4, \qquad q^2=n^4+8p^4.$$

The former becomes the latter if we interchange n with p and change the sign of s.   The solution of $q^4-2r^4=s^2$ is therefore reduced to that of $q^2=n^4+8p^4$, of type (7), by setting $r=2pn$, $s=n^4-8p^4$.   Further, q and r exceed n and p.

The method leads to all solutions not only of (2) but also of (7) and of $q^4-2r^4=\square$.   Starting with the evident solutions $s=t=1$, $u=\pm3$ of (7), we deduce the solutions $r=2st=2$, $q=u=\pm3$, $k=7$, of $q^4-2r^4=k^2$; and, by (6), (5), solutions of (2): $m=0$, $n=-1$, $l=7$, $x=y=z=1$, or $m=-6$, $n=-7$, $l=1$, $x=13$, $y=1$, $z=239$.   For $r=2$, $q=3$, $s=7$, we deduce the solutions $s=7$, $t=qr=6$, $u=113$ of (7); from 13, 1, 239, we get the solutions $s=239$, $t=13$, $u=57123$ of (7).   Starting again with one of the latter sets, we obtain new sets of solutions of (2) and $q^4-2r^4=\square$.   In this manner, the sets of solutions of (2) in order of magnitude are $(x, y, z)=(1, 1, 1)$, (13, 1, 239), (1525, 1343, 2750257), (2165017, 2372159, 1560590745759), $\cdots$.   The corresponding sets (3) are p, $q=1$, 0; 120, -119; 2276953, -473304; and the last two numbers (1).   Lagrange therefore proved Fermat's assertion that (1) gives the sides of the least right triangle whose hypotenuse and sum of legs are squares.   But Lagrange evidently merely transcribed the statement by Fermat, without making a numerical verification, as the value $15\cdots9$ of z given by Lagrange (pp. 142, 150, 151; Oeuvres, 380, 393–4) is erroneous [Genocchi[44]], the correct value being the difference $350\cdots1$ of the last two numbers (1).

Three of Euler's[55] posthumous papers of 1780 relate to Fermat's[37] problem.   In the first paper we find a slight modification of his[48] discussion. Taking $s=2$, $r=3+v$, we get

$$x+y=1+148v+102v^2+20v^3+v^4=(1+74v-v^2)^2,$$

if $v=1343/42$.   Thus $p=1385\cdot1553$, $q=168\cdot1469$, yielding Fermat's solution (1).

Euler, in the second paper, employed his[48] notations, and obtained $x+y=A^2-2B^2$, where $A=r^2+2rs-s^2$, $B=2rs$.   Taking $A=t^2+2u^2$, $B=2tu$,

[55] Mém. Acad. Sc. St. Pétersbourg, 9, 1819–20, 3; 10, 1821–22, 3; 11, 1830, 1; [Comm. Arith., II, 397, 403, 421.

we have $A^2 - 2B^2 = (t^2 - 2u^2)^2$.   Noting that a solution involving fractions may be replaced by an integral solution, he took $s=1$, whence $r=tu$. Equating the two expressions for A, we get

$$t^2 u^2 + 2tu - 1 = t^2 + 2u^2.$$

For $u=1$, $t=3/2$. The latter leads to the second value $u = -13$, which in turn gives $t = -113/84$. Then $u = 301993/1343$, etc. Euler stated that it is easy to see that the pairs of adjacent values of u, t give all sets of rational solutions. From the formulas for the sum of the roots of a quadratic equation, we see that

$$u' + u = \frac{2t}{2 - t^2}, \qquad t' + t = \frac{2u'}{1 - u'^2},$$

if u, t, u', t' are consecutive terms of the series.

Euler, in the third paper, set $A/B = (1+x)/(1-x)$ in $2A^4 - B^4 = \square$.   Thus

$$1 + 12x + 6x^2 + 12x^3 + x^4 \equiv (1 + 6x + x^2)^2 - 32x^2 = \square.$$

In accord with his[143] general method, he set $1 + 6x + x^2 = \lambda(p^2 + 8q^2)$, $x = \lambda pq$. Cf. Euler,[143] end.

V. A. Lebesgue[56] gave a method simpler than Lagrange's (whose article he had apparently not seen) to obtain from given solutions of (2) a smaller set of solutions.   Since $p^2 + q^2 = x^4$, we may set $p = 2mn$, $q = m^2 - n^2$, $x^2 = m^2 + n^2$, where n is even since $p+q$ is a square y^2.   By the third relation, $m = r^2 - s^2$, $n = 2rs$, $x = r^2 + s^2$, where one of the integers r, s is even and the other odd.   Changing the sign of y if necessary, we may assume that, of the factors $r^2 + 2rs - s^2 \pm y$ of $8r^2 s^2$ (in view of $p + q = y^2$), the one with the upper sign is divisible by 2 but not by 4.   For r odd we may therefore set

$$r^2 + 2rs - s^2 + y = 2\frac{t}{u} r^2 \qquad r^2 + 2rs - s^2 - y = 4\frac{u}{t} s^2,$$

where u, t are odd and relatively prime.   Multiplying the sum by $\frac{1}{2}ut$, we get

(8) $$r^2(t^2 - ut) - 2rsut + s^2(2u^2 + ut) = 0.$$

For s odd, the right members are obtained by interchanging r, s, and the new sum is derived from (8) by replacing r/s by $-s/r$, and t by $-t$.   By (8)

$$\frac{r}{s}(t^2 - ut) = ut \pm \sqrt{ut(2u^2 - t^2)}.$$

Since ut and $2u^2 - t^2$ are relatively prime, each is a square or the negative of a square.   But t and u are odd, and $t^2 - 2u^2$ is of the form $8k - 1$ and not a square.   Hence, taking u and t positive, we may set $u = f^2$, $t = g^2$, $2f^4 - g^4 = h^2$. Then

$$\frac{r}{s} = \frac{f}{g}\frac{A}{B}, \qquad A = 2f^2 + g^2, \qquad B = fg \mp h.$$

If x, y do not have a common square factor, r, s are relatively prime and $\sigma r = fA$, $\sigma s = gB$, where σ is prime to f and g.   Then $y = r^2 t/u - 2s^2 u/t$ and

$$\sigma^2 y = g^2 A^2 - 2f^2 B^2, \qquad \sigma^2 x^2 = f^2 A^2 + g^2 B^2, \qquad \sigma^4 z = C^2 - 2(f^2 A^2 - g^2 B^2)^2,$$

---

[56] Jour. de Math., 18, 1853, 73–86.   Reprinted, Sphinx-Oedipe, 6, 1911, 133–8.

where $C=f^2A^2+2fgAB-g^2B^2=g^2A^2+2f^2B^2$. Now f divides r, g divides s, while r and s are $<\sqrt{x}$. Hence each set of integral solutions of (2) with $x^2\neq1$ leads to a set of smaller solutions. For $f=13$, $g=1$, $h=\pm239$, we get $A=3\cdot113$, $B=-2\cdot113$ or $3\cdot84$; for the first, $\sigma=113$, $r=39$, $s=-2$, $x=1525$, $y=-1343$; for the second, $\sigma=3$, $r=13\cdot113$, $s=84$, $x=2165017$, $y=-2372159$.

Lebesgue noted that $x^4\pm2^my^4=z^2$ has integral solutions only when $m=4n\pm3$ and then may be made to depend upon (2); likewise, $2^mx^4-y^4=z^2$ only when $m=4n+1$. But $x^4\pm y^4=2^mz^2$ is impossible in integers. All of these cases except $x^4\pm8y^4=z^2$ and $8x^4-y^4=z^2$ had been treated by Euler, Algebra 2, Ch. 13, whose[49] solution of $x^4-2y^4=z^2$ is incomplete (Art. 211).

E. Lucas[57] gave a complete solution of $x^4-2y^4=\pm z^2$ and $x^4+8y^4=z^2$, based on the complete solution of $u^2+v^2=y^4$. He[46] obtained the usual results concerning Fermat's[37] problem.

T. Pepin[58] treated $2x^4-1=\square$ by his[157] final method. He[59] remarked that Lebesgue[56] merely stated, but did not prove, that his formulas lead to all solutions of (2) under a given limit. Pepin obtained the same solutions by a simpler method proved complete. If x, y, z are relatively prime by pairs,

$$x=p^2+q^2, \qquad \pm z\pm y^2i=(1+i)(p+qi)^4,$$

where p, q are relatively prime and q may be taken even. Then

$$\pm y^2=(p^2-q^2+2pq)^2-8p^2q^2, \qquad \pm z=p^4-\cdots,$$

the lower sign being excluded by use of modulus 8. Thus

$$\pm(p^2-q^2+2pq)\pm y=2r^2, \qquad \pm(p^2-q^2+2pq)\mp y=4s^2, \qquad rs=pq,$$

r, s being relatively prime. By the last, $p=\lambda\mu$, $q=hk$, $r=\lambda h$, $s=\mu k$, where λ, μ, h, k are integers relatively prime by pairs, k alone being even. From $p^2-q^2+2pq=r^2+2s^2$ (the lower sign having been excluded by modulus 4), $k^2(2\mu^2+h^2)-2\lambda\mu hk+\lambda^2(h^2-\mu^2)=0$, whence

$$\frac{k}{\lambda}=\frac{\mu h\pm\sqrt{2\mu^4-h^4}}{2\mu^2+h^2}, \qquad \frac{h}{\mu}=\frac{\lambda k\pm\sqrt{\lambda^4-2k^4}}{\lambda^2+k^2}.$$

Thus μ, h form a solution of (2), while $\lambda^4-2k^2=\square$. The above is valid if $x>1$, whence $q\neq0$. Thus any solution except $x=y=z=1$ leads to a solution $x'=\mu$, $y'=h$, $z'=\sqrt{2\mu^4-h^4}$, in smaller numbers, and given by

$$x=\lambda^2\mu^2+h^2k^2, \qquad \pm y=\lambda^2h^2-2\mu^2k^2, \qquad \pm z=y^2-8\lambda hk(\lambda^2\mu^2-h^2k^2),$$

where $2\mu^4-h^4=t^2$, $k/\lambda=(\mu h\pm t)/(2\mu^2+h^2)$, from whose numerator and denominator common factors are to be suppressed. We can therefore compute the successive sets of solutions of (2) starting with $x=y=z=1$.

[57] Recherches sur l'analyse indéterminée, Moulins, 1873, 25–32. Extract from Bull. Société d'Émulation Dept. de l'Allier, 12, 1873, 467–72. Same in Bull. Bibl. Storia Sc. Mat. Fis., 10, 1877, 239–45.

[58] Atti Accad. Pont. Nuovi Lincei, 30, 1876–7, 220–2.

[59] Ibid., 36, 1882–3, 37–40.

S. Réalis[60] noted that if $\alpha^4 - 2\beta^4 = \gamma^2$, then $x^4 - 2y^4 = z^2$ for

$$x = 3(339\alpha^3 + 392\beta^3) + 8\alpha\beta(216\alpha + 211\beta) + 7\gamma(113\alpha + 96\beta),$$
$$y = 4(147\alpha^3 - 226\beta^3) - 27\alpha\beta(5\alpha + 64\beta) + 7\gamma(108\alpha + 113\beta).$$

For $\alpha = \gamma = 1$, $\beta = 0$, $x : y : z = 113 : 84 : 7967$.   For $\alpha = 3$, $\beta = 2$, $\gamma = 7$,

$$x = 57123, \qquad y = 6214, \qquad z = 3262580153.$$

A. Gérardin[61] treated the last problem, assuming that also a second solution $A^4 - 2B^4 = C^2$ is known.   Set

$$(\alpha + Au)^4 - 2(\beta + Bu)^4 = (\gamma - Su + Cu^2)^2.$$

Then

$$\{4(A^3\alpha - 2B^3\beta) + 2CS\}u^2 + \{6(A^2\alpha^2 - 2B^2\beta^2) - S^2 - 2\gamma C\}u$$
$$+ 4A\alpha^3 - 8B\beta^3 + 2\gamma S = 0.$$

Equating to zero the coefficient of u^2, we get S and u.   Taking $A = 3$, $B = -2$, $C = -7$, we obtain Réalis' solution.   Starting with $3^4 - 2\cdot 2^4 = 7^2$, set

$$(3 + mx)^4 - 2(2 + my)^4 = \{7 + \tfrac{2}{7}(27x - 16y)m + gm^2\}^2$$

and annul the coefficient of m^2; we get g and m in terms of x, y and hence a solution of the sixth degree.   Modifying the last method, we again get Réalis' solution.

A. Cunningham[62] noted that the solution of (2) by Lebesgue[56] and Lucas[57] appears to be complete and to indicate that the only integral solutions of $x^2 - 2y^4 = -1$ are (1, 1) and (239, 13).   But Euler's[53] solution of (2) yields only half the solutions.

L. C. Walker[63] quoted Fermat's last two integers (1), whose sum is a square and sum of squares is a biquadrate.

$ax^4 + by^4$ MADE A SQUARE OR MULTIPLE OF A SQUARE.

The cases $x^4 \pm y^4$, $2x^4 - y^4$, $x^4 - 2y^4$, $x^4 + 8y^4$ have been treated above. For $x^4 - h^2y^4$, see Congruent Numbers in Ch. XVI, especially papers 43, 54.

G. W. Leibniz[64] treated before 1678 the problem to find an integer x such that $x + a/x = y^2$, where a is a given integer and y is to be rational.   Set $a = bc$, $x = bz$, where c and z are relatively prime integers.   Set $y = v/w$, a fraction in its lowest terms.   Then $bz^2 + c = zv^2/w^2$, so that z is divisible by w^2. Similarly, since cw^2/z is an integer $v^2 - bzw^2$, w^2 is divisible by z.   Hence $z = w^2$ and $bw^4 + c = v^2$.   Since c is the product of $v \pm w^2\sqrt{b}$, it exceeds each of the factors and hence their difference, whence $c^2 > 4bw^4$.   The resulting tentative process to solve $x + a/x = y^2$ is to express a as a product bc of two integers, choose an integer w such that $4bw^4 < c^2$ and test the value $x = bw^2$ (or what is equivalent, see if $bw^4 + c$ is a square).

[60] Nouv. Corresp. Math., 6, 1880, 478–9.
[61] Sphinx-Oedipe, 6, 1911, 87–8.
[62] Math. Quest. Educ. Times, (2), 14, 1908, 76–8.
[63] Amer. Math. Monthly, 11, 1904, 39.
[64] Math. Schriften (ed., C. I. Gerhardt), 7, 1863, 114–9.

L. Euler[65] proved that $2a^4 \pm 2b^4$ is not a square if $a \neq b$ by means of the fact that $x^4 \mp y^4$ is not a square. Likewise for $4x^4 \pm y^4$, $x^4 - 4y^4$, $\pm(4x^4 - 2y^4)$. [Cf. Frenicle,[9] Bendz,[29] Carmichael.[36]] He proved that neither $ma^4 \pm m^3 b^4$ nor its double is a square. Also[66] that $a^4 + 2b^4 \neq \square$ if $b \neq 0$.

Euler[67] treated $a + ex^4 = \square$, supposing known one solution: $a + eh^4 = k^2$. Set $x = h + y$. Then

$$a + ex^4 = k^2 + 4eh^3 y + 6eh^2 y^2 + 4ehy^3 + ey^4$$

will be the square of $k + 2eh^3 y/k + eh^2(k^2 + 2a)y^2/k^3$ if

$$y = 4hk^2(2a - k^2)/(3k^4 - 4a^2).$$

By use of the substitution $x = h(1+y)/(1-y)$, $a + ex^4$ becomes a quartic having both the constant term and the coefficient of y^4 squares, and hence is more readily made a square.

J. L. Lagrange[68] proved that if $s^4 + at^4 = u^2$ a second set of solutions of $x^4 + ay^4 = z^2$ is given by

$$x = s^4 - at^4, \qquad y = 2stu, \qquad z = u^4 + 4as^4 t^4.$$

To deduce this result, Lagrange made assumptions which he recognized were not necessary ones. Assume that $z = m^2 + an^2$. Then the given equation is satisfied if $y^2 = 2mn$, $x^2 = m^2 - an^2$. The latter holds if $m = p^2 + aq^2$, $n = 2pq$, $x = p^2 - aq^2$. The resulting expression for y^2 is a square if $p = s^2$, $q = t^2$, $p^2 + aq^2 = u^2$. From the second solution, one deduces similarly a third, etc. But not all sets are necessarily obtained in this way. He remarked that the simplest and most general method for such equations is perhaps that by factors in his Addition IX to Euler's Algebra (Lagrange[163] of Ch. XXI).

A. E. Kramer[69] treated $px^4 - y^4 = z^2$, where p is an odd prime, and x, y are relatively prime. Let $p = n^2 + m^2$. Then

$$(y^2 + mx^2)(y^2 - mx^2) = (nx^2 + z)(nx^2 - z).$$

He took $m = r^2$. First, let one of y, r be odd and the other even, so that x is even. Set $y^2 + r^2 x^2 = ab$, $nx^2 + z = ac$, where b, c are relatively prime. Then the long equation gives $y^2 - r^2 x^2 = dc$, $nx^2 - z = db$. Then a, b, c, d are odd and a, d are relatively prime. Since $\alpha \equiv na - r^2 d$, $\beta \equiv r^2 a + nd$ have no common factor except possibly p, while $b\alpha = c\beta$, we have $\alpha = sc$, $\beta = sb$, where $s = \pm 1$ or $\pm p$. Let e be the g. c. d. of d and $y + rx$; h that of α/s and $y - rx$. Since $y^2 - r^2 x^2 = d\alpha/s$, we get $y + rx = ef$, $y - rx = gh$, $d = eg$, $\alpha/s = fh$, where f, g are relatively prime, as also e, h. Substituting the values of y, rx,

<hr>

[65] Comm. Acad. Petrop., 10, 1747 (1738), 125; Comm. Arith., I, 28; Opera Omnia, (1), II, 47. Algebra, St. Petersburg, 2, 1770, arts. 209–10; French transl., Lyon, 2, 1774, 254–263. Opera Omnia, (1), 1, 442–3.

[66] The proof in his Algebra is the shorter. The latter was reproduced by A. M. Legendre, Théorie des nombres, 1798, p. 405; Maser, II, 7; E. Waring, Medit. Algebr., ed. 3, 1782, 373–4.

[67] Algebra, St. Petersburg, 2, 1770, Arts. 138–9; French transl., 2, 1774, pp. 162–7; Opera Omnia, (1), I, 400–2.

[68] Nouv. Mém. Acad. Sc. Berlin, année 1777, 1779, 151; Oeuvres, IV, 395. Reproduced by E. Waring, Meditationes Algebraicae, ed. 3, 1782, 371.

[69] De quibusdam aequationibus indeter. quarti gradus, Diss., Berlin, 1839.

d, a, given by the last four equations, into $y^2+r^2x^2=a\beta/s$, we get

$$2\left\{\frac{\dfrac{e}{h}\left(n^2f^2-2\dfrac{p}{s}r^2g^2\right)-(p+r^4)fg}{2nr}\right\}^2=sf^4+\frac{p}{s}g^4.$$

Denote the quantity in brackets by A. Evidently s is not negative. According as s is unity or the prime p, we get $2A^2=f^4+pg^4$ or g^4+pf^4. Conversely, any solution of one of the latter equations leads to a solution of the proposed equation with $p=n^2+r^4$, since f, g, A determine x, y, z.

Next, let y, r be both even or both odd. The only modification needed in the above case is to divide $y^2\pm r^2x^2$, $nx^2\pm z$, $y\pm rx$ by 2, and use $d/2=eg$. The result is $B^2=sf^4+4g^4p/s$, where

$$nrB\equiv\frac{e}{h}\left(n^2f^2-4\frac{p}{s}r^2g^2\right)-(p+r^4)fg.$$

For $s=1$, we have $B^2=f^4+4pg^4$, which implies

$$\frac{B\mp f^2}{2}=pb^4,\qquad\frac{B\pm f^2}{2}=c^4,\qquad g=bc.$$

Hence the initial equation is reduced to a similar one $pb^4-c^4=\pm d^2$, where c, b are relatively prime. It thus remains to consider $c^4-pb^4=d^2$. First, let one of c, d be even and the other odd. Then $c^2\pm d=pe^4$, $c^2\mp d=h^4$, $b=eh$, whence $h^4+pe^4=2c^2$. Next, let c and d be both odd or both even. Then $(c^2\pm d)/2=4pv^4$ or pv^4, $(c^2\mp d)/2=u^4$ or $4u^4$. Then $c^2=u^4+4pv^4$ or $c^2=4u^4+pv^4$, which is reduced to the former type by multiplication by 4.

O. Terquem[70] proved that neither x^4+2y^4 nor $x^4\pm4y^4$ nor x^4-8y^4 is a square if $y\neq0$, and that $z\pm1/z$ is not a square.

 * J. Bertrand[71] treated $ax^4+by^4=\square$.

C. G. Sucksdorff[72] treated $2^mx^4\pm2^ny^4=2^pz^2$ for x, y, z odd, positive and relatively prime. It suffices to treat eight cases having $n=p=0$, $m=4\mu+0$, 1, 2, 3; four with the minus sign having $m=p=0$, $n=4\mu+0$, 1, 2, 3; four having $m=n=0$, $p=2\mu+0$, 1. First, $2^{4\mu}x^4+y^4=z^2$. The factors $z\pm2^{2\mu}x^2$ must be α^4, β^4, where $\alpha\beta=y$. By subtraction, $2^{2\mu+1}x^2=\alpha^4-\beta^4$. Hence

$$\alpha+\beta=2u^2,\qquad\alpha-\beta=2v^2,\qquad\alpha^2+\beta^2=2t^2.$$

Eliminating α, β, we get $u^4+v^4=t^2$, of the given type. A like method of descent applies to $2^{4\mu+1}x^4+y^4=z^2$, whence

$$z\pm y^2=2\alpha^4,\qquad z\mp y^2=2^4\beta^4,\qquad\pm y^2=\alpha^4-8\beta^4$$

(lower sign excluded since the sum of two odd squares is not divisible by 8); thus $8\beta^4=\alpha^4-y^2$, $\alpha^2\pm y=2\gamma^4$, $\alpha^2\mp y=4\delta^4$, whence $\alpha^2=\gamma^4+2\delta^4$. For

$$2^{4\mu+1}x^4-y^4=z^2$$

reference is made to Euler's[67] treatment of $a+ex^4=\square$, where $-y^4$ is taken as a; various solutions result. The impossibility of $2^{4\mu+2}x^4+y^4=z^2$ follows

<hr>

[70] Nouv. Ann. Math., 5, 1846, 75–78.

[71] Traité élém. d'algèbre, Paris, 1850, 244.

[72] Disquisitio au et quatenus aequatio $2^mx^4\pm2^ny^4=2^pz^2$ solutione gaudeat in integris. . . . Helsingfors, 1851, 16 pp.

as for the first equation. Next, $x^4 - 2^{4\mu+1}y^4 = z^2$ implies

$$x^2 \pm z = 2\alpha^4, \qquad x^2 \mp z = 2^4\beta^4,$$

whence $x^2 = \alpha^4 + 8\beta^4$, $x \pm \alpha^2 = 2\gamma^4$, $x \mp \alpha^2 = 4\delta^4$, $\pm \alpha^2 = \gamma^4 - 2\delta^4$. For the upper sign we have an equation like the proposed. For the lower sign, there are solutions, as $\alpha = \gamma = \delta = 1$. The impossibility of $x^4 - 2^{4\mu+t}y^4 = z^2$ $(t = 0$ or $2)$ follows from $x^2 \pm z = 2\alpha^4$, $x^2 \mp z = 2^{3+t}\beta^4$, $x^2 = \alpha^4 + 2^{2+t}\beta^4$. The impossibility of $x^4 + y^4 = 2^{2\mu+1}z^2$, $x \neq y$, follows from $(x^4 - y^4)^2 = 2^{4\mu+2}z^4 - 4x^4y^4$.

Lebesgue's[56] results concerning the equations in the last paper have been quoted. Cf. Schopis[19] on $x^4 + y^4 \neq 2z^2$.

E. Lucas[73] listed and treated the solvable equations

$$(1) \qquad\qquad ax^4 + by^4 = cz^2,$$

in which 2 and 3 are the only primes dividing a, b or c, viz., $(a, b, c) =$ $(1, -1, 24)$, $(1, -2, \pm 1)$, $(1, 2, 3)$, $(1, 3, 1)$, $(1, -6, 1)$, $(1, 8, 1)$, $(1, 9, 1)$, $(1, -12, 1)$, $(1, 18, 1)$, $(1, 24, 1)$, $(1, \pm 36, 1)$, $(1, -54, 1)$, $(1, -72, 1)$, $(1, 216, 1)$, $(3, -1, 2)$, $(3, -2, 1)$, $(4, -1, 3)$, $(4, -3, 1)$, $(9, -1, 8)$, $(9, -8, 1)$, $(27, -2, 1)$.

T. Pepin[74] stated that there is no rational solution of $px^4 - 36y^4 = z^2$ if p is a prime of the form $a^2 + 9b^2$, and many such theorems with 36 replaced by new numbers, usually by the discriminant of the quadratic form for p.

Lucas stated and Moret-Blanc[75] proved that $x = 1$, $y = 0$ and $x = 3$, $y = 2$ are the only integral solutions $\geqq 0$ of $x^4 - 5y^4 = 1$.

Lucas[75a] proved that either of $4v^4 - u^4 = 3w^4$, $9v^4 - u^4 = 8w^4$ implies

$$u^4 = v^4 = w^4.$$

Pepin[76] noted that necessary conditions for relatively prime integral solutions of $Au^2 = Bx^4 + Cy^4$ are that AB, AC and $-BC$ be quadratic residues of C, B, A, respectively, and that $-BC^3$ be a biquadratic residue of A. He proved that $u^2 = 3y^4 - 2x^4$ is completely solved by the repeated application of

$$x = \lambda^2\mu^2 - 3f^2g^2, \qquad y = \lambda^2f^2 + 2\mu^2g^2, \qquad u = x^2 - 12\lambda\mu fg(\lambda^2f^2 - 2\mu^2g^2),$$

where λ, μ, f, g are integers relatively prime in pairs such that

$$g : \lambda = f\mu \pm \sqrt{3f^4 - 2\mu^4} : 3f^2 + 2\mu^2.$$

The same analysis gives the complete solution of $x^4 - 6y^4 = z^2$ and

$$x^4 + 24y^4 = z^2.$$

He treated other rare cases in which the complete solution is found: $x^4 + 7y^4 = 8u^2$ and $7x^4 - 2y^4 = 5u^2$, with the respective auxiliaries $x^4 + 28y^4 = z^2$ and $x^4 - 350y^4 = z^2$.

---

[73] Recherches sur l'analyse indéterminée, Moulins, 1873; extract from Bull. Soc. d'Emulation du Département de l' Allier, 12, 1873, 441–532. Bull. Bibl. Storia Sc. Mat. Fis., 10, 1877, 239–58.

[74] Comptes Rendus Paris, 78, 1874, 144–8; 88, 1879, 1255; 91, 1880, 100 (reprinted, Sphinx-Oedipe, 5, 1910, 56–7); 94, 1882, 122–4.

[75] Nouv. Ann. Math., (2), 14, 1875, 526; 20, 1881, 203–5.

[75a] Nouv. Ann. Math. (2), 16, 1877, 415.

[76] Atti Accad. Pont. Nuovi Lincei, 31, 1877–8, 397–427.

A. Desboves[77] employed the identity

$$(2) \qquad (y^2 + 2yx - x^2)^4 + (2x+y)x^2y(2y+2x)^4 = (x^4 + y^4 + 10x^2y^2 + 4xy^3 + 12x^3y)^2,$$

and that obtained by changing x to x^2 and y to y^4, to show that

$$(3) \qquad x^4 + ay^4 = z^2$$

is solvable in integers if a is of the form $(2x+y)x^2y$ or $2x^2+y^4$. By changing x to $x+y$ in (2) and making other simple transformations, he[78] proved that (3) is solvable in integers if $a = -x^2(x^2+y^2)$, $\pm y^2 - x^4$, $-x(x+1)$, $y(y\pm 2x^2)$, $x^2(2x+y^4)$, $y^4 - 2x^2$, $-2xy(x^2-y^2)(x^4+y^4-6x^2y^2)$ with $z = \square$ in the last case; and, by other identities, if $a = -8(x^8+y^8)$, $-x(x^2+4)$, $-x^8-4$. If (1) has solutions x, y, z, then Fermat's method conveniently applied leads to the new solution[79]

$$(4) \quad X = x(4a^2x^8 - 3c^2z^4), \qquad Y = y(4b^2y^8 - 3c^2z^4), \qquad Z = z[4c^4z^8 - 3(ax^4 - by^4)^4],$$

of different type from Lagrange's solution when $a = c = 1$. For the examples of Lucas[73] not under Lagrange's case and for which (4) do not give all solutions, we have $(a+b)c$ a square, say v^2. Using fractional values, we may set $y = 1$. Then $ac(x^4 - 1) + v^2 = c^2z^2$. Setting $x = (t+1)/(t-1)$, we get an equation for which Fermat's method is applicable. If x, y, z is a solution of (1), then[80]

$$x_1 = 2ax^4 - cz^2, \qquad y_1 = 2xyz, \qquad z_1 = c^2z^4 + 4ax^4(cz^2 - ax^4)$$

is a solution of $x^4 + abc^2y^4 = z^2$. The latter becomes $x^4 + u(v^2 - u)y^4 = z^2$ for $ac = u$, $(a+b)c = v^2$. Hence (1) is solvable if $a = c = 1$, $b = u(v^2 - u)$, as shown also by the identity

$$(5) \qquad (2u - v^2)^4 + u(v^2 - u)(2v)^4 = (v^4 - 4u^2 + 4uv^2)^2.$$

E. Lucas[81] obtained from one solution of $\lambda x^4 + \mu y^4 = (\lambda + \mu)z^2$ the two solutions

$$X = 4\mu\rho n^2 x^2 y^2 z^2 - m^2v^2, \qquad Y = 4\lambda\rho m^2 x^2 y^2 z^2 - n^2 v^2,$$
$$Z = (4\mu\rho n^2 x^2 y^2 z^2 + m^2 v^2 + 4\mu mnxyzv)^2 + 16\lambda\mu m^2 n^2 x^2 y^2 z^2 v^2,$$

where $\rho = \lambda + \mu$, $v = \lambda x^4 - \mu y^4$, $m = \pm 4\lambda\mu x^4 y^4 \pm \rho^2 z^4 - 2\rho xyzv$, $n = v^2 - 4\lambda\rho x^2 y^2 z^2$. Since the proposed equation is satisfied if $x = y = z = \pm 1$, we obtain two new solutions. Thus $3x^4 - 2y^4 = z^2$ has the solutions

$$33, 13, 1871; \quad 28577, 8843, 1410140689.$$

If (1) has the solution (x_0, y_0, z_0), it may be written in the form

$$\lambda\left(\frac{x}{x_0}\right)^4 + \mu\left(\frac{y}{y_0}\right)^4 = (\lambda + \mu)\left(\frac{z}{z_0}\right)^2, \qquad \lambda = ax_0^4, \qquad \mu = by_0^4, \qquad \lambda + \mu = cz_0^2.$$

He stated that his formulas above solve completely twenty equations of

[77] Comptes Rendus Paris, 87, 1878, 159–161. Reproduced, with pp. 321–2, 522, 598, in Sphinx-Oedipe, 4, 1909, 163–8.

[78] Comptes Rendus Paris, 87, 1878, 321–2.

[79] Ibid., 522; correction, 599. Reproduced in Desboves' Questions d'algèbre, ed. 4, 1892. Cf. Desboves.[123]

[80] Ibid., 598.

[81] Nouv. Ann. Math., (2), 18, 1879, 67–74. In Lucas' expression for Z the coefficient 4 of the final term should be 16. If we adopt his change of signs in m, we must alter a sign in his Z.

type (1) in which a, b, c contain only the prime factors 2 and 3 [erroneous for $4x^4 - 3y^4 = z^2$, Desboves[91]].

Desboves[82] again gave (2) and, by replacing y by $v-x$ and then x^2 by u, deduced (5).   He noted that (3) is solvable in the further cases $a = x(y^2 - x)$, $-xy^2(x+y)$, $-x(x+y^2)$, $-x^2y^2(x^2-y^2)^2$.   He again (*ibid.*, p. 440) gave (4). He noted (*ibid.*, p. 436–7) that (1) has the solutions

$$X = 3ax^4 - by^4, \qquad Y = 4ax^3y, \qquad Z = ax^4 + by^4$$

if $c = 81a^3x^6 - 14a^2bx^4y^4 + ab^2y^8$, and gave a simpler derivation of Lagrange's solution of (3).   For $ax^4 + by^4 = cz^4$, see Desboves.[262]

Solutions of $x^4 + y^4 = 17z^2$ are 1, 2, 1 and 13, 2, 41, neither of which can be obtained (*ibid.*, p. 495) from a solution x, y, z by the formulas (4).

T. Pepin[83] gave the complete solution of $7x^4 - 5y^4 = 2z^2$ in integers. Then[84] $X = z$, $Y = xy$, $Z = (7x^4 + 5y^4)/2$ give all the solutions of $X^4 + 35Y^4 = Z^2$ in which Y is odd; while those with Y even are all obtained by the method of descent.

S. Réalis[85] noted that $x^4 - 3y^4 = 13z^2$ has the solution

$$x = 76\alpha^3 + 96\alpha^2\beta + 135\alpha\beta^2 + 156\beta^3 + 13\gamma(19\alpha + 12\beta),$$
$$y = 52\alpha^3 + 28\alpha^2\beta - 96\alpha\beta^2 - 57\beta^3 + 13\gamma(16\alpha + 19\beta),$$

if $\alpha^4 - 3\beta^4 = 13\gamma^2$, and asked for the value of z.

Pepin[86] noted that in Euler's[144] method of making a quartic $V = P^2 + QR$ a square, not only a rational root of $R = 0$ or $Q = 0$ or $S = 0$ or $T = 0$ leads to an infinity of solutions of $V = \square$, but this may be true of further roots. The latter happens for $11x^4 - 7y^4 = z^2$, whence $V = 11 - 7\xi^4 = P^2 + QR$, $P = 2\xi$, $Q = 11 + 7\xi^2$, $R = 1 - \xi^2$.   The complete solution is obtained by descent to two irreducible solutions 1, 1, 2 and 2, 1, 13 by four sets of formulas, among them being an infinity of solutions which escape the methods of Fermat and Euler.   To obtain (pp. 42–48) the complete solution of $x^4 + 20y^4 = z^2$, that of $5n^4 - m^4 = 4t^2$ is found by descent.   From one set of solutions x, y, z of (1) for $c = a + b$ is derived,[87] by special assumptions, the new solutions

$$X = \lambda^2 x^2 - bc\mu^2 y^2, \qquad Y = \lambda^2 y^2 - ac\mu^2 x^2, \qquad Z = Y^2 - 4a\lambda\mu xy(\lambda^2 x^2 + bc\mu^2 y^2),$$

where $\mu : \lambda = xy \pm z : ax^2 - bz^2$.

Pepin[88] obtained by descent all solutions of $13x^4 - 11y^4 = 2z^2$ and all of $8x^4 - 3y^4 = 5z^2$, whereas Euler's[144] method to make $40\xi^4 - 15 = \square$ does not give all solutions.

A. Desboves[89] proved that, if (x, y, z) and (x', y', z') are solutions of (1), a new solution is given by

(6)
$$x'X = x^2\lambda^2 - bcy^2\mu^2, \qquad y'Y = y^2\lambda^2 - acx^2\mu^2,$$
$$x'^4z'^2Z = [(x^2\lambda^2 + bcy^2\mu^2)z' + 2bxyy'^2\lambda\mu]^2 + 4abx^2y^2x'^4\lambda^2\mu^2,$$

<hr>

[82] Nouv. Ann. Math., (2), 18, 1879, 434.

[83] Jour. de Math., (3), 5, 1879, 405–24.

[84] *Ibid.*, (5), 1, 1895, 351–8.

[85] Nouv. Corresp. Math., 6, 1880, 479.

[86] Atti Accad. Pont. Nuovi Lincei, 36, 1882–3, 49–67.

[87] *Ibid.*, 67–70. Cf. Lucas.[81]

[88] *Ibid.*, 38, 1884–5, 20–42.

[89] Comptes Rendus Paris, 104, 1887, 846–7.

where $\lambda=ax^2x'^2-by^2y'^2$, $\mu=xyz'+zx'y'$.   For $a+b=c$, we may set

$$x'=y'=z'=1$$

and deduce his[159] and Pepin's[87] formulas.   For $a=c=1$, $x'=z'=1$, $y'=0$, we get Lagrange's formula.   He announced the empirical result that the complete solution of (1) in integers is given by as many systems (6) as (1) has primitive solutions $(x',\ y',\ z')$.   For $8x^4-3y^4=5z^2$, Pepin's ten systems reduce to the two systems (6) with $(x',\ y',\ z')=(1,\ 1,\ 1)$, $(2,\ 1,\ 5)$. For[90] the case $c=a+b$, set $x=y=z=1$ in (6) and drop the accents; we get

$$X=a(a-b)x^3-b(3a+b)xy^2-2bcyz,$$

while Y is derived from X by interchanging a, b and x, y.   He gave another set of formulas of like degree.   By finding a relation

$$EX^2x^2+GY^2y^2-2LXYxy-H(X^2y^2+Y^2x^2)=0$$

such that Y/X is a function of y, x, involving only the irrationality $(ax^4+by^4)^{1/2}$, he obtained the quadratic formulas

$$X=-(a-b)^2x^2+4bcy^2,\qquad Y=[2c^2-(a-b)^2]xy+2c(a-b)z,$$

$$Z=4b(a-b)xy[4acx^2+(a-b)^2y^2]+[2c^2-(a-b)^2][(a-b)^2x^2+4bcy^2]z,$$

and stated that a like discussion may be made for

$$ax^4+by^4+dx^2y^2=cz^2,\qquad c=a+b+d.$$

Desboves[91] noted that, if (1) is solved completely by (6) when $(x',\ y',\ z')$ is replaced by $(x_i',\ y_i',\ z_i')$ for $i=1,\ 2,\ 3$, in succession, then any one of these solutions is called primitive if one does not get it when one determines all solutions given by the other two and continues the calculations with them.

Desboves[92] stated that we can find, by a single system of formulas (not given), the complete solution of $ax^4-by^4=2z^2$ when a and b are consecutive primes $8n+7$ and $8n+5$ or $8n+5$ and $8n+3$.

T. Pepin[93] treated $x^4+2^k\cdot7y^4=z^2$ for $k=2\alpha$ and $4\alpha+3$.   He[94] gave a detailed discussion of

$$5x^4-3y^4=2z^2,\qquad 5x^4-2y^4=3z^2,\qquad 3x^4+5y^4=8z^2,\qquad 8x^4-5y^4=3z^2.$$

E. B. Escott[95] noted that if in $x^4+y^4=az^2$ we set $x=zk/l$ we obtain a quadratic for z^2 which will be rational if $(al)^4-(2a)^2(ky)^4=(aml^2)^2$, so that the problem reduces to the pair of equations $p^2\pm2aq^2=\square$ (Ch. XVI).

Axel Thue[96] proved that $x^4-2^my^4=1$ has no integral solutions.

Escott[97] solved $4A^4+1=B^2C$ by noting that the left member has the factors $2A^2\pm2A+1$, whence $(2A\pm1)^2+1\equiv0$ (mod B^2).

A. Gérardin[98] noted that if $(\alpha,\ \beta,\ \gamma)$ and $(A,\ B,\ C)$ are two solutions of (1), $x=\alpha+Au$, $y=\beta+Bu$, $z=\gamma+Su+Cu^2$ give a new solution provided a certain

[90] Comptes Rendus Paris, 104, 1887, p. 1832.
[91] *Ibid.*, 1602-3.
[92] Assoc. franç. av. sc., 16, 1887, I, 175 (in full).
[93] Mem. Acc. Pont. Nuovi Lincei, 4, 1888, 227.
[94] *Ibid.*, 9, I, 1893, 247-284.
[95] L'intermédiaire des math., 7, 1900, 199 (reply to 3, 1896, 130).
[96] Archiv for Math. og Naturvidenskab, 25, 1903, No. 3.
[97] L'intermédiaire des math., 12, 1905, 155-6.
[98] Bull. Soc. Philomathique, (10), 3, 1911, 234-6; Sphinx-Oedipe, 6, 1911, 101-2.

quadratic equation in u is satisfied. Equating to zero the coefficient of u^2 by choice of S, we get u rationally. He deduced Réalis'[85] result.

A. Cunningham[99] listed all $a^4 + b^4 = mc^2 < 10^7$, $1 + y^4 = mc^2$, $y < 1000$.

E. Fauquembergue[100] proved Lucas'[73] result that 3, 1, 2 is the only set of solutions of $x^4 - y^4 = 5z^4$.

W. Mantel[101] proved by descent that $x^4 + 2^n y^4 \neq z^2$ unless $n \equiv 3 \pmod 4$.

H. C. Pocklington[102] proved by descent the impossibility of

$$x^4 - py^4 = z^2, \qquad x^4 - p^2 y^4 = z^2, \qquad x^4 - y^4 = pz^2, \qquad x^4 + 2y^4 = z^2,$$

where p is a prime $8m + 3$, and indicated (p. 119) the solution of

$$2x^4 - y^4 = \pm z^2.$$

R. D. Carmichael[103] treated $x^4 + my^4 = nz^2$. If there is a solution, there is an integer ρ such that $n\rho^2 = s^4 + mt^4$. Hence we are led to the equation

$$(7) \qquad x^4 + my^4 = (s^4 + mt^4)z^2.$$

A special solution, other than the evident one $x = s$, $y = t$, $z = 1$, is obtained by setting $z = p^2 + mq^2$. Then (7) is satisfied if

$$x^2 = s^2(p^2 - mq^2) + 2mt^2 pq, \qquad y^2 = t^2(p^2 - mq^2) - 2s^2 pq.$$

A solution of this double equation is found by the method of Fermat:

$$x = sp - 2s(s^8 - m^2 t^8), \qquad y = tp + 2t(s^8 - m^2 t^8), \qquad z = p^2 + 16ms^4 t^4(s^4 - mt^4)^2,$$
$$p = (s^4 + mt^4)^2 + 4ms^4 t^4.$$

By the method of infinite descent, he proved (pp. 19–21) that there is no set of integers, all different from zero, satisfying either of the equations $x^4 - 4y^4 = \pm z^2$. Hence the area of no rational right triangle is the double of a square; this implies that $x^4 + y^4 = z^2$ has no integral solutions all different from zero.

A. Gérardin[104] explained three methods to obtain the complete solution of $ax^4 + by^4 = cz^2$, given one solution.

A. Auric[105] solved $ax^4 + by^4 = cd^2 z^2$ by eliminating z between it and the auxiliary equation $mx^2 + ny^2 = cdz$ and making the discriminant of the eliminant a square.

M. Rignaux[106] obtained an infinitude of solutions of $x^4 - y^4 = az^2$, given one solution. *E. Haentzschel[106a] discussed (1).

$ax^4 + by^4 + dx^2 y^2$ MADE A SQUARE.

L. Euler[107] noted that in making $F \equiv x^4 + kx^2 y^2 + y^4$ a square there is a lack of generality in assuming that F is the square of $x^2 + y^2 p/q$ or

<hr>

[99] L'intermédiaire des math., 18, 1911, 45–6.
[100] L'intermédiaire des math., 19, 1912, 281–3.
[101] Wiskundige Opgaven, 11, 1912–4, 491–5.
[102] Proc. Cambridge Phil. Soc., 17, 1914, 110.
[103] Diophantine Analysis, 1915, 77–79.
[104] L'intermédiaire des math., 22, 1915, 149–161.
[105] *Ibid.*, 23, 1916, 7–8.
[106] *Ibid.*, 24, 1917, 14.
[106a] Sitzungsber. Berlin Math. Gesell., 16, 1917, 9–16.
[107] Nova Acta Acad. Petrop., 10, ad annum 1792, 1797 (1777), 27; Comm. Arith., II, 183.

$x^2+xyp/q\pm y^2$. By a certain device he was led to the case $k=fx^2+2\sqrt{1+fy^2}$ in which F is the square of $y^2+x^2\sqrt{1+fy^2}$. For $1<f<100$, he gave the least integer y for which the radical is rational. For half of the positive values of $k<100$ and for 30 negative values numerically <100, tables show values of $x:y$ for which F is a square.

Euler[108] resumed the solution of $x^4+mx^2y^2+y^4=z^2$. The resulting fraction for m can be given an integral form by use of a rational number a for which $z=ax^2y^2-(x^2\pm y^2)$. Then $m\pm2=(ax^2\mp2)(ay^2-2)$. We may set $x=pq$, $y=rs$, $a=b/(p^2r^2)$, where p, q are relatively prime, likewise r, s. Then

$$m\pm2=(bq^2\mp2r^2)(bs^2-2p^2)/(p^2r^2).$$

Set $bs^2-2p^2=cr^2$, $bs^2+cr^2=2n$, $bc=\lambda$. Then $n^2-p^4=\lambda y^2$, where y^2 is the largest square dividing n^2-p^4. Thus $m=(\lambda q^2\mp2n)/p^2$. Conversely, for assigned values of p, n, q, the integer $x=pq$ and the largest square y^2 dividing n^2-p^4 are solutions of the proposed equation with the preceding value of m. In fact,

$$x^4=q^4(n^2-\lambda y^2), \qquad mx^2y^2=q^2y^2(\lambda q^2\mp2n), \qquad z^2=(y^2\mp q^2n)^2.$$

Euler gave tables of solutions with a slightly changed notation. In conclusion (p. 498), he gave a more elegant method for the case $m=\lambda\zeta^2\pm\alpha$, where $\alpha^2-4=\lambda\beta^2$. Then $x=\beta$, $y=2\zeta$, $z=\beta^2\pm2\alpha\zeta^2$ are solutions. Starting with two sets of solutions α, β and 2, 0 of the Pell equation, he derived the solution

$$A=g^n+h^n, \qquad B=(g^n-h^n)/\sqrt{\lambda}, \qquad g\equiv\frac{\alpha+\beta\sqrt{\lambda}}{2}, \qquad h\equiv\frac{\alpha-\beta\sqrt{\lambda}}{2}.$$

Since $gh=1$, $A^2-\lambda B^2=4$. Thus for $m=\lambda f^2\pm A$ (f arbitrary), we get the solutions $x=B$, $y=2f$ of the quartic equation.

Euler[109] proved that $m^4+14m^2n^2+n^4$ is not a square if m and n are relatively prime and m is even and n odd (excluding $m=0$, $n=1$), or if m and n are both odd (excluding $m=n=1$). The question was reduced to one on $\alpha^2+3\beta^2=\square$. By setting $x=m^2-n^2$, $y=2mn$, we see that x^2+y^2 and x^2+4y^2 are not both squares for x odd, y even $\neq0$. Another corollary is that $p(p+q)(p+2q)(p+3q)\neq\square$, so that four squares cannot be in arithmetical progression. Another corollary is $p^4-p^2q^2+q^4\neq\square$ if $p^2\neq q^2\neq0$, and is derived by setting $p=m+n$, $q=m-n$ for p and q odd, and $p+q=m$, $p-q=n$ when one of p, q is even and the other odd.

Euler[110] elsewhere stated that $x^4-x^2+1\neq\square$ if $x^2\neq1$ or 0. This was proved by the editor of the 1810 English edition, p. 112, by showing in the Appendix that p^2-q^2 and p^2+3q^2 are not both squares.

C. F. Kausler[111] wrote $z=x/y$ in Euler's[107] quartic F. The problem is now to make $z^4+kz^2+1=\square$, or as a generalization $f^2+bZ+eZ^2=P^2$, $Z=z^2$.

[108] Mém. Acad. Sc. St. Petersb., 7, années 1815–6, 1820 (1782), p. 10; Comm. Arith., II, 492. For misprints and errata see Cunningham.[136]

[109] Mém. Acad. Sc. St. Pétersbourg, 8, années 1817–18 (1780), 3; Comm. Arith., II, 411–13. Same results by V. A. Lebesgue, Nouv. Ann. Math., (2), 2, 1863, 68–77.

[110] Algebra, 2, 1770, art. 142; 2, 1774, p. 169; Opera Omnia, (1), I, 403.

[111] Nova Acta Acad. Petrop., 13, ad annos 1795–6, Mém., pp. 205–36.

Thus $Z(b+eZ)=P^2-f^2$. For a suitably chosen rational A, we may set
$$b+eZ=A(P+f), \qquad Z=(P-f)/A.$$
Eliminating P, we get $Z=(2fA-b)/(e-A^2)$. In our case, $e=f=1$, $b=k$, whence $Z=(k-2A)/(A^2-1)$ is to be a square z^2. Thus $k-2A=mp^2$, $A^2-1=mq^2$. Of the solutions of the latter Pell equation, those are to be selected which satisfy the first equation (a "solution" which he admitted was imperfect). By eliminating m and setting $2A=\alpha$, $p/q=2n$, we get $k=\alpha+(\alpha^2-4)n^2$, the case treated by Euler[108] at the end of his second paper. Kausler treated at length (pp. 219–236) the problem to make k integral by choice of rational values of α, n.

N. Fuss[112] required integers m such that $x^4+mx^2y^2+y^4=z^2$. Set
$$m-2=\alpha\beta, \qquad m+2=\gamma\delta.$$
Then $z^2-(x^2+y^2)^2=\alpha\beta x^2y^2$, $z^2-(x^2-y^2)^2=\gamma\delta x^2y^2$. For $x=pq$, $y=rs$, we have
$$z+x^2+y^2=\alpha q^2s^2, \qquad z-x^2-y^2=\beta p^2r^2,$$
$$z+x^2-y^2=\gamma p^2s^2, \qquad z-x^2+y^2=\delta q^2r^2.$$
Eliminating z and replacing x, y by their values, we get three linear equations between α, β, γ, δ, which give
$$\gamma=\frac{\alpha q^2-2r^2}{p^2}, \qquad \delta=\frac{\alpha s^2-2p^2}{r^2}, \qquad \beta=\frac{\alpha q^2s^2-2p^2q^2-2r^2s^2}{p^2r^2},$$
of which the last may be replaced by $\gamma\delta=\alpha\beta+4$. If $p=r=1$, then $\gamma\delta=(\alpha q^2-2)(\alpha s^2-2)$, and α, q, s may be given any values; as the values of $m<100$ we get 2, 8, 12, 16, 17, 22, 23, 26, 31, $\cdots$, 94.

R. Adrain[113] proved by descent that $x^4+x^2y^2+y^4 \neq \square$. He and T. Strong (p. 151) also noted that $(x^2+y^2)^2-x^2y^2=a^2$ requires that $a^2+x^2y^2=\square$ and $a^2-3x^2y^2=\square=(x^2-y^2)^2$, whereas a^2+q^2 and a^2-3q^2 are not both squares (Euler's Algebra, Second English transl., II, 481). H. J. Anderson[114] noted that we may take x and y positive and relatively prime. If x and y are both odd, $x^4+x^2y^2+y^4=8n+3 \neq \square$. Hence we may take x even, y odd. Thus $(x^2+y^2)^2-x^2y^2$ is an odd square, whence $x^2+y^2=p^2+q^2$, $xy=2pq$. By an argument like that in Euler's Algebra, II, Art. 230, we conclude that r^2-s^2 and r^2-4s^2 are odd squares, where s is even, and r, s are divisors of x, y, and similarly that t^2-u^2 and t^2-4u^2 are odd squares, where u is even, and t, u are divisors of r, s. Finally, we would reach odd squares v^2-w^2 and v^2-4w^2, where $\frac{1}{2}w$ no longer has divisors. Hence the problem is impossible.

A. M. Legendre[115] found only two solutions of $m^4-4m^2n^2+n^4=p^2$, viz., $(m, n, p)=(15, 4, 191)$, $(442, 161, 364807)$. The complete solution, including $(2, 1, 1)$, was given by E. Lucas.[116]

[112] Mém. Acad. Sc. St. Pétersbourg, 9, 1824 (1820), 159.
[113] The Math. Diary, New York, 1, 1825, 147–150. Cf. Genocchi[119] and Pocklington[128]; also Beha-Eddin[50] of Ch. XIV and Kausler[10] of Ch. XXVI.
[114] *Ibid.*, 150–1.
[115] Théorie des nombres, ed. 3, 2, 1830, 127; Maser, II, 124. See Legendre[47] of Ch. XIX.
[116] Recherches sur l'analyse indéterminée, Moulins sur Allier, 1873, p. 67. Bull. Bibl. Storia Sc. Mat. Fis., 10, 1877, 291–2.

V. A. Lebesgue[117] noted that if $x^4+ax^2y^2+by^4=z^2$ has the solution $x=r$, $y=s$, $z=p$, it has also the solution

$$x=r^4-bs^4, \qquad y=2prs, \qquad z=p^4-(a^2-4b)r^4s^4.$$

A. Desboves[118] remarked that this generalization of the result by Lagrange[68] for $a=0$ is insignificant since it is made by replacing his initial identity (the following for $d=0$) by

$$(u^2-bv^2)^2+d(u^2-bv^2)(2uv+dv^2)+b(2uv+dv^2)^2=(u^2+duv+bv^2)^2,$$

which Lagrange gave in his addition IX to Euler's algebra (French transl., 2, 1774, 640).

A. Genocchi[119] proved by descent that $x^4+x^2y^2+y^4 \neq \square$.

T. Pepin[120] treated $x^4+8x^2+1=\square$.

E. Lucas[121] deduced two solutions (X, Y, Z) from a given solution (x, y, z) of

$$x^4-2(a+2f^2)x^2y^2+(a^2+b^2)y^4=z^2.$$

For brevity, set

$$\Delta=4f^4+4af^2-b^2, \qquad\qquad n=z^2+4f^2x^2y^2,$$
$$m=-bxyz\pm f[x^4-(a^2+b^2)y^4], \qquad \alpha=(\Delta n^2x^2y^2+m^2z^2)/f,$$
$$\beta=4m^2x^2y^2-n^2z^2, \qquad\qquad \gamma=4m^2x^2y^2+n^2z^2.$$

Then

$$X=16amnxyz\beta+b(16m^2x^2y^2z^2-\beta^2)^2, \qquad Y=2\gamma\alpha, \qquad Z=\Delta\gamma^4-4\alpha^4.$$

A. Desboves[122] noted that if x, y, z satisfy $ax^4+by^4+dx^2y^2=cz^2$, then

$$X=ax^4-by^4, \qquad Y=2xyz, \qquad Z=c^2z^4+(4ab-d^2)x^4y^4$$

satisfy $X^4+abc^2Y^4+cdX^2Y^2=Z^2$; while[123]

$$X=x(4bcy^4z^2-q^2), \qquad Y=y(4acx^4z^2-q^2), \qquad Z=z\{4fx^4y^4q^2-(c^2z^4-fx^4y^4)^2\}$$

satisfy the initial equation if $q=ax^4-by^4$, $f=d^2-4ab$. Cf. Desboves.[90]

T. Pepin[124] treated $ax^4+2bx^2y^2+cy^4=n^2$, a necessary condition for which is that the quadratic form (a, b, c) represent n^2. If one such representation is known, all are given by quadratic functions of two parameters. But in returning to our quartic we are led again to the problem to make a quartic a square.

Moret-Blanc[125] found solutions of $x^4-5x^2y^2+5y^4=\square$ and

$$(x^5+y^5)/(x+y)=\square$$

by Euler's method.

S. Réalis[126] proved that $2y^4-2y^2+1=\square$ only for $y=0, 2$.

[117] Jour. de Math., 18, 1853, 84; Nouv. Ann. Math., (2), 11, 1872, 83-6.
[118] Comptes Rendus Paris, 87, 1878, 925.
[119] Annali di Sc. Mat. e Fis., 6, 1855, 302. Cf. Adrain.[113]
[120] Atti Accad. Pont. Nuovi Lincei, 30, 1876-7, 222-4. Cf. Euler, Algebra 2, Ch. 9, Art. 144.
[121] Nouv. Ann. Math., (2), 18, 1879, 73.
[122] Ibid., (2), 18, 1879, 384; for $a=c=1$, p. 437. Verification, (2), 19, 1880, 461-2.
[123] Ibid., (2), 18, 1879, 440; implied, Comptes Rendus Paris, 87, 1878, 522.
[124] Atti Accad. Pont. Nuovi Lincei, 32, 1878-9, 79-128.
[125] Nouv. Ann. Math., (2), 20, 1881, 150-5.
[126] Bull. Bibl. Storia Sc. Mat. Fis., 16, 1883, 213; reproduced, Sphinx-Oedipe, 4, 1909, 175-6.
    See papers 19-25 of Ch. XVII.

E. Fauquembergue[127] gave the general solution of $(x^2+y^2)(2x^2-y^2)=2z^2$.
A. Gérardin[128] gave x, y, $z=3f$, $4f$, $5f^2$ and $h/2$, $2h/3$, $5h^2/36$.

$x^4+4x^2+1=y^2$ is impossible in rational numbers.[129]　Cf. Pietrocola.[131]

T. Pepin[130] treated $x^4-8x^2y^2+8y^4=z^2$ by the method of descent applicable only if y is even; then $x=X^8-8Y^8$, $y=2XYZ$, $z=Z^4-32X^4Y^4$. For y odd the equation is reduced to the pair $pq=rs$, $p^2-4q^2+4pq+8s^2-r^2=0$, to which the method of descent is applicable. There exist only six sets of solutions x, y, z, each $\neq 0$, with $y<10^{10}$.

C. Pietrocola[131] discussed the equivalent equations

$$x^4+4hx^2y^2+(2h-1)^2y^4=z^2, \qquad (x^2+2hy^2+z)(x^2+2hy^2-z)=(4h-1)y^4.$$

From one solution he derived another and proved the equation impossible if $h=1$. The last result had been proposed as a problem by P. Tannery.[132]

A. S. Werebrusow[133] listed many values of m between -100 and $+100$ for which $x^4+mx^2y^2+y^4=z^2$ is impossible, and stated that it is impossible for m positive or for $m=8k+3$ negative if $m+2$ and $m-2$ are primes.

A. Gérardin[134] noted that the last statement fails for $m=99$.

Gleizes and H. B. Mathieu[135] gave special expressions for m for which the equation is solvable.

A. Cunningham[136] noted that the equation is solvable for $m=60$, 99, -72, -96, contrary to Werebrusow,[133] and for $m=91$, -90, contrary to Euler[108] (p. 495, p. 498); and corrected various misprints on pp. 496–8 of Euler's paper.

L. Aubry[137] stated that Werebrusow's[133] theorem is true for a positive $m\equiv 1$, 5 or 7 (mod 8), and a negative $m=-(8k+5)$, but false for a positive $m=8k+3$. Aubry (pp. 57–9) treated $x^4+bx^2y^2+cy^4=dz^2$, given

$$d=p^4+bp^2q^2+cq^4,$$

by setting $x^2=p^2u-cq^2v$, $y^2=q^2u+(bq^2+p^2)v$ and deducing an equation of the initial form, whence one solution leads to two new solutions.

H. C. Pocklington[138] proved that $x^4-x^2y^2+y^4$, $x^4+14x^2y^2+y^4$ are neither squares if $x\neq y$. If N is not of the form $8n\pm 3$ and is not divisible by any prime $4n+1$, and at the same time $N\mp 4$ is an odd power of an odd prime (including unity), then $(x^2+y^2)^2\mp Nx^2y^2=z^2$ is impossible in integers. For $N=1$ and the upper signs, we see that $x^4+x^2y^2+y^4=z^2$ is impossible. Also $x^4-14x^2y^2+y^4=z^2$ is impossible. There is a list of values of $n<100$ for which $x^4\pm nx^2y^2+y^4=z^2$ is impossible. The complete solution is given of

$$x^4-4x^2y^2+y^4=z^2.$$

[127] L'intermédiaire des math., 4, 1897, 70.
[128] Ibid., 16, 1909, 175.
[129] Ibid., 1897, 20, 83, 203, 229; 1898, 89, 128; 1900, 87–90; 1903, 158; 1905, 109.
[130] Mem. Accad. Pont. Nuovi Lincei, 14, 1898, 71–85.
[131] Giornale di mat., 36, 1898, 77–80.
[132] L'intermédiaire des math., 1897, 20, 30, 203.
[133] Ibid., 15, 1908, 52, 282 (corrections); Mat. Sbornik, Moscow, 26, 1908, 599–617.
[134] L'intermédiaire des math., 16, 1909, 154.
[135] Ibid., 15, 1908, 159.
[136] Ibid., 17, 1910, 201.
[137] Ibid., 18, 1911, 203.
[138] Proc. Cambridge Phil. Soc., 17, 1914, 111–118.

Cases in which $ax^4+dx^2y^2+by^4=\square$ is impossible were noted by Lebesgue,[30, 37] Genocchi[85, 93] and Pepin[98] of Ch. XXVI; by Desboves[188] of this Chapter. Solvable cases by Pepin[132] of Ch. I, Haentzschel[143] of Ch. V.

QUARTIC FUNCTION MADE A SQUARE.

Fermat[139] sought rational values of x for which

$$(1) \qquad f(x) \equiv a+bx+cx^2+dx^3+ex^4$$

shall equal the square of a rational number, where $a, \cdots, e$ are integers. The case in which a or e is the square of an integer is quite simple. For $a=\alpha^2$, the first three terms of $f(x)$ are identical with those of the square of

$$\alpha+\frac{b}{2\alpha}x+\frac{1}{2\alpha}\left(c-\frac{b^2}{4\alpha^2}\right)x^2.$$

Comparing the terms with the factor x^3, we obtain

$$x=-\frac{8\alpha^2[b(4\alpha^2c-b^2)-8\alpha^4d]}{64\alpha^6e-(4\alpha^2c-b^2)^2}.$$

Hence from a particular solution $f(\xi)=\alpha^2$, we may obtain new solutions since $f(\xi+x)=\alpha^2+bx+\cdots+ex^4$ falls under the last case.

The same special cases were treated similarly by L. Euler[140] and A. M. Legendre.[141]

T. F. de Lagny[142] made $x^4-10x^3+26x^2-7x+9$ the square of x^2-5x+3 for $x=23/5$.

L. Euler[143] treated in a posthumous paper the equation

$$a^2x^4+2abx^3y+cx^2y^2+2bdxy^3+d^2y^4=\square.$$

Set $c-b^2-2ad=mn$. Then

$$(2) \qquad (ax^2+bxy+dy^2)^2+mnx^2y^2=z^2.$$

This is satisfied if

$$ax^2+bxy+dy^2=\lambda(mp^2-nq^2), \qquad xy=2\lambda pq, \qquad z=\lambda(mp^2+nq^2).$$

Admitting fractional solutions, we may set $y=1$. Then

$$4\lambda^2ap^2q^2+2b\lambda pq+d=\lambda(mp^2-nq^2).$$

For a fixed λ, let p and q be given solutions. Let p' be the second root of this quadratic in p, whence

$$p'=-p-2bq/(4\lambda aq^2-m).$$

Then p', q' are corresponding values if

$$q'=-q-2bp'/(4\lambda ap'^2+n).$$

[139] Diophanti Alexandrini Arith. Libri Sex . . . Doctrinae Analyticae Inventum Novum; Collectum à J. de Billy ex varijs Epistolis quas ad eum . . . misit P. de Fermat, p. 30. French transl., Oeuvres de Fermat, 3, 1896, 377–388 (the term x^4 is omitted on p. 388, § 31).

[140] Algebra, 2, 1770, Ch. 9, Nos. 128–137; French transl., Lyon, 2, 1774, pp. 153–162. Opera Omnia, (1), 1, 1911, 396–400. Sphinx-Oedipe, 1908–9, 67–78.

[141] Théorie des nombres, 1798, 458–9; ed. 3, 2, 1830, 123; Maser, II, 120.

[142] Nouv. Elemens d'Arith. et d'Alg., Paris, 1697, 496.

[143] Mém. Acad. Sc. St. Petersb., 11, 1830 [1780], 1; Comm. Arith., II, 418.

From p', q', we get p'', q'', etc. Any two consecutive terms of p, q, p', q', p'', $\cdots$ yield a solution with $y=1$. Proceeding in the reverse order, we obtain a sequence q, p, q_1, p_1, q_2, $\cdots$, any two consecutive terms of which yield a solution.

To obtain an initial pair of solutions, set $y=1$ and let the quartic be the square of ax^2+bx-d or of ax^2-bx-d; then

$$x=\frac{4bd}{b^2-2ad-c} \quad \text{or} \quad x=\frac{b^2-2ad-c}{4ab}.$$

To treat $\alpha C^4 \pm \beta = \square$, where $\alpha \pm \beta$ is a square a^2, set $C=(1+x)/(1-x)$. Then

$$a^2x^4+4(\alpha \mp \beta)x^3+6a^2x^2+4(\alpha \mp \beta)x+a^2 = \square,$$

which is of the above type. Euler treated in detail the cases

$$2A^4-B^4=\square, \qquad 3A^4+B^4=\square, \qquad \tfrac{3}{2}A^4-\tfrac{1}{2}B^4=\square.$$

Euler[144] treated $V \equiv A+Bx+Cx^2+Dx^3+Ex^4 = \square$. If V can be given the form P^2+QR, where

$$P=a+bx+cx^2, \qquad Q=d+ex+fx^2, \qquad R=g+hx+ix^2,$$

then $V=(P+Qy)^2$, where $2Py+Qy^2-R=0$. The latter is also quadratic in x, viz., $Sx^2+Tx+U=0$. From initial solutions x, y, we obtain[145] $x'=-x-T/S$; then from x' we get $y'=-y-2P'/Q'$, etc. As in the preceding paper, we thus obtain two series of solutions of $V=\square$.

If, for $E=0$, $V=f^2$ for $x=a$, we may take

$$P=f, \qquad Q=x-a, \qquad R=B+C(x+a)+D(x^2+ax+a^2).$$

For a general V, let $V=f^2$ for $x=a$. When x is replaced by $a+t$, let V become $f^2+\alpha t+\beta t^2+\gamma t^3+\delta t^4$. Then $V=P^2+QR$ for

$$P=f+\frac{\alpha t}{2f}, \qquad Q=t^2, \qquad R=\beta-\frac{\alpha^2}{4f^2}+\gamma t+\delta t^2.$$

Euler[146] gave ten values of x for which

$$a^2+2abx+(b^2+d^2-f^2)x^2+2dex^3+e^2x^4=z^2,$$

including

$$x=(-d\pm f)/e, \qquad z=a+bx; \qquad x=-a/(b\pm f), \qquad z=x(ex+d).$$

G. Libri[147] treated $a^2x^4+bx^3+cx^2+dx+e=z^2$ with all coefficients positive (since we may replace x by x_1+h). Multiply by $4a^2$ and set

$$2az=2a^2x^2+bx+v.$$

Thus

(3) $$\qquad (4a^2v+b^2-4a^2c)x^2+(2bv-4a^2d)x+v^2-4a^2e=0.$$

[144] Mém. Acad. Sc. St. Petersb., 11, 1830 [1780], 69; Comm. Arith., II, 474. Cf. Pepin.[86]

[145] This method of solving any equation quadratic in x and in y was given by Euler also in Mém. Acad. Sc. St. Petersb., 11, 1830, 59; Comm. Arith., II, 467. For applications to rational quadrilaterals, see Kummer,[133] and Schwering[150] of Ch. V. Cf. papers 55, 143, 148, 155; also Pepin[140] of Ch. IV, Güntsche[91, 152] of Ch. V. On the relation of elliptic functions to an equation quadratic in x and in y, see G. Frobenius, Jour. für Math., 106, 1890, 125–188.

[146] Opera postuma, 1, 1862, 266 (about 1782).

[147] Jour. für Math., 9, 1832, 282.

A positive v cannot surpass a certain number L which makes every coefficient in (3) positive; hence we have only to try $v=0, 1, \cdots, L-1$. If $v=-t$, where $0<t<x$, let s be the least t for which $4a^2(t+c)>b^2$ and substitute $s+w$ for $-v$ in (3), we get an equation like $Ax^2+Bx+4a^2e=(s+w)^2$ with all coefficients positive, whereas $x^2>t^2=(s+w)^2$; hence the only cases to try are $v=-1, \cdots, -(s-1)$. Finally, if $v=-u, 0<u>x$, let r be the remainder $<x$ on dividing u by x and n the quotient. Set

$$4a^2z^2 = [2a^2x^2 + (b-n)x - r]^2.$$

By $z^2>a^2x^4$, we have $b>n$ and need only try $n=1, \cdots, b-1$.

C. G. J. Jacobi[148] stated that the analysis of Euler[144-5] to find an infinitude of rational values of x, given one, making the quartic $f(x)$ a square is the same as that of Euler's[149] (earlier) solution of the transcendental equation

$$(4) \qquad\qquad \Pi(y) = n\Pi(x), \qquad \Pi(x) \equiv \int_0^x \frac{dx}{\sqrt{f(x)}}.$$

For the latter, Euler used a chain of n equations $f(p, q)=0$, $f(q, r)=0$, $f(r, s)=0, \cdots$, where

$$f(p, q) = \alpha + 2\beta(p+q) + \gamma(p^2+q^2) + 2\delta pq + 2\epsilon pq(p+q) + \zeta p^2q^2$$

is symmetrical in p and q, whereas in the diophantine problem Euler's canonical equation $Qy^2+2Py-R=Sx^2+Tx+U=0$ is not symmetric in x, y, as pointed out by L. Schlesinger,[150] who discussed at length Jacobi's above remark. The latter had been discussed by T. Pepin.[151] Jacobi observed that the analysis of the multiplication of elliptic integrals (4) gives an infinitude of rational y's for which also $\sqrt{f(y)}$ is rational, if a rational x makes $\sqrt{f(x)}$ rational, and drew from the theory of abelian integrals the conclusion[152]: If $f(x)$ is of the fifth or sixth degree and if one rational value of x makes $\sqrt{f(z)}$ rational, there exist an infinitude of x's of the form $a+b\sqrt{c}$, with a, b, c rational, for which $\sqrt{f(x)}=a'+b'\sqrt{c}$, with a', b' rational; and the extension to $f(x)$ of degree $2n+1$ or $2n+2$ and x's satisfying an equation of degree n with rational coefficients. J. Ptaszycki[153] remarked that the last theorem follows at once from the representation of a rational function by means of polynomials which enter in the development into a continued fraction of the square root of this function. The generalization of Jacobi's theorem has been considered by J. von Sz. Nagy.[154]

The problem to make a quartic a rational square was proposed in 1865 as a prize subject by the Accad. Nuovi Lincei of Rome.

L. Calzolari[155] wrote $a+bv+cv^2+dv^3+ev^4=w^2$ in the form

$$4ew^2 = a' + 2b'v + c'v^2 + Q^2, \qquad Q \equiv 2ev^2 + dv + k,$$
$$a' = 4ae - k^2, \qquad b' = 2be - dk, \qquad c' = 4ce - 4ek - d^2.$$

[148] Jour. für Math., 13, 1834, 353–5; Werke, II, 51–5.
[149] Institutiones Calculi Integralis, 1, 1763, Ch. 6, Prob. 83, § 642.
[150] Jahresber. d. Deutschen Math.-Vereinig., 17, 1908, 63 (with history of $f(x) = \square$).
[151] Atti Accad. Pont. Nuovi Lincei, 30, 1876–7, 224–37.
[152] Cf. Jacobi, Jour. für Math., 32, 1846, 220; Werke, II, 135; Schwering[238] of Ch. XXI.
[153] Jahresber. d. Deutschen Math.-Vereinig., 18, 1909, 1–3.
[154] Ibid., 4–7. Cf. Nagy[16] of Ch. XXIII.
[155] Giornale di Mat., 7, 1869, 317–50.

Set $Q=y/x$, $2w=z/x$, $c'v+b'=u/x$. Then
$$u^2=Ax^2+By^2+Cz^2, \qquad A=b'^2-a'c', \qquad B=-c', \qquad C=c'e,$$
which can be given the form $A_1(u^2-x^2)=B_1(y^2-z^2)$ by choice of k. The solutions for u, x, y, z are evident. Substitute these in the quadratic in u, x, y obtained by eliminating v between $Q=y/x$, $c'v+b'=u/x$. For example, if $v^4-2=w^2$, then $w^2=(v^2+k^2)-2kv^2-k^2-2$. Set $v^2+k=u/x$, $v=y/x$, $w=z/x$. Then
$$u^2=(k^2+2)x^2+2ky^2+z^2, \qquad u^2-z^2=2(9x^2-4y^2)$$
for $k=-4$. It has the solutions
$$u, z=\tfrac{1}{2}(\alpha\gamma\pm\beta\delta); \qquad 12x, 8y=2\beta\gamma\pm\alpha\delta.$$
Substitute these in $ux=kx^2+y^2$ (obtained by eliminating v). Thus
$$L\alpha^2+M\alpha\beta+N\beta^2=0,$$
with coefficients quadratic in γ, δ. Taking $L=0$ we get four sets of solutions α, β, γ, δ; likewise four from $N=0$.

S. Bills[156] made $f\equiv x^4+4x^3+8x^2+7x+6=\square$ by noting that $f=4$ for $x=-1$ and setting $f=Q^2$, $Q=x^2+2x+k$, where k is chosen so that $Q=\pm2$ for $x=-1$.

T. Pepin[157] made use of the notations of Euler,[140] viz., (1) and
$$\theta(x)=f+gx+hx^2, \qquad F(z)=f(z)-\theta^2(z)=a-f^2+\cdots+(e-h^2)z^4.$$
Pepin took x_1, x_2, x_3 arbitrary but distinct, and determined f, g, h, x by
$$(5) \qquad \theta(x_i)=\epsilon_i\sqrt{f(x_i)}, \qquad \epsilon_i^2=1, \qquad x=\frac{2gh-d}{e-h^2}-x_1-x_2-x_3 \quad (i=1, 2, 3).$$

Then x_1, x_2, x_3, x are the roots of $F(z)=0$. Hence if x_1, x_2, x_3 are three solutions of $f(x)=\square$, then f, g, h are rational and x is a new solution. Next, let $x_3=x_1$; then $F'(x_1)=0$, and (5) for $i=3$ is to be replaced by the derivative of (5) for $i=1$. Finally, for $x_1=x_2=x_3$, we use (5) for $i=1$ and its first and second derivatives, and so obtain a second solution from a first. Then the preceding case gives a third solution and (5) a fourth solution.

Pepin[158] noted that if a quartic $f(x)$ can be transformed into a square by replacing x by a rational function, then $F\equiv y^2-f(x)=0$ is a unicursal curve and hence has three double points, whence the partial derivatives of F with respect to x and y vanish, showing that f has a double root. The problem is then to make the remaining quadratic factor a square. The problem to make a product of two binary quadratic forms a square is treated by means of a congruence. Conditions are given in order that a reciprocal quartic shall never be a rational square for a rational value of the variable.

A. Desboves[159] noted that if x, y, z is a set of solutions of
$$aX^4+bY^4+dX^2Y^2+fX^3Y+gXY^3=cZ^2,$$

[156] Math. Quest. Educ. Times, 22, 1875, 91-2.
[157] Atti Accad. Pont. Nuovi Lincei, 30, 1876-7, 211-37.
[158] Atti Accad. Pont. Nouvi Lincei, 32, 1878-9, 166-202.
[159] Comptes Rendus Paris, 88, 1879, 638-40, 762 (correction). Cf. Desboves[148] of Ch. XXI.

formulas can be found giving in general four sets of solutions.   In

$$ax^4\left(\frac{X}{x}\right)^4+by^4\left(\frac{Y}{y}\right)^4+\cdots=cz^2\left(\frac{Z}{z}\right)^2$$

consider ax^4, etc., as coefficients; we thus have an equation of the first type having now $a+\cdots+g=c$ (an artifice due to Lucas[81] for $d=f=g=0$). After dividing such an equation by c and setting $X=(\rho+x)/(\rho+1)$, we get an equation in ρ to which Fermat's method applies.   The explicit formulas for the two sets of solutions are very long (each furnishing two sets by changing the sign of z).

F. Romero[160] proved that $x^4+x^3+x^2+x-1=y^2$ has no positive integral solutions.   For, y is odd and the equation becomes

$$x(x+1)(x^2+1)=2\{m^2+(m+1)^2\}.$$

Thus $x=4n+2$, and $4n+3$ would divide the sum of the squares of two relatively prime integers.

E. Lucas[161] discussed $f(x)=y^2$, where $f(x)$ is a quartic with rational coefficients.   Set $y\phi(x)=F(x)$, where $\phi=x^p+a_1x^{p-1}+\cdots$ with rational a's, while F is of degree $p+2$.   Then $F^2=f\phi^2$ is an equation of degree $2p+4$ in which enter $2p+3$ unknowns besides x.   If we know $2p+3$ sets of rational solutions x_i, y_i of $y^2=f(x)$, no two of which differ merely in the sign of y, and determine the coefficients in $y\phi=F$ so that it shall be satisfied by these $2p+3$ sets, these coefficients will be rational.   Then $F^2=f\phi^2$ will furnish a new rational x which leads to a new set of rational solutions of $y^2=f(x)$. We may take two or more of the x_i equal; if $x_2=x_1$, we replace

$$F^2(x_2)=f(x_2)\phi^2(x_2)$$

by the derivative of $\pm\sqrt{f(x_1)}=F(x_1)/\phi(x_1)$.   Taking all of the x_i equal, we see that one solution of $f(x)=y^2$ leads to an infinite sequence of solutions. (Cf. Pepin.[158])   If $f(x)$ has a rational root α, we may take

$$F=(x-\alpha)\phi_{p+1}(x).$$

If f has a rational quadratic factor $q(x)$, we may take $F=q\psi_p$ and apply the above method to $2p+1$ sets of solutions.

L. J. Mordell[162] assumed that we have one solution of $f=z^2$, where f is a binary quartic with the invariants g_2, g_3.   Then we can transform f into a quartic with leading coefficient z^2.   The syzygy between its seminvariants (cf. Mordell[176] of Ch. XXI) is $g^2=4h^3-g_2hz^4-g_3z^6$.   Thus g/z^3, h/z^2 give rational solutions of

$$t^2=4s^3-g_2s-g_3.$$

It is shown that the knowledge of all rational solutions of the latter leads to all rational solutions of $f=z^2$.

E. Haentzschel[163] treated $y^2=f(x)=a_0x^4+\cdots+a_4$.   First, let $f(x)=0$ have a rational root r and apply the substitution

$$x=r+\tfrac{1}{4}f'(r)/(s-t),\qquad t=f''(r)/24.$$

<hr>

[160] Nouv. Ann. Math., (2), 18, 1879, 328.
[161] Nouv. Corresp. Math., 5, 1879, 183–6.
[162] Quar. Jour. Math., 45, 1913–4, 178–181.
[163] Jour. für Math., 144, 1914, 275–283.

We obtain Weierstrass' normal form

$$(6) \qquad v^2 = 4s^3 - g_2 s - g_3 = 4(s-e_1)(s-e_2)(s-e_3),$$

where g_2, g_3 are the invariants of f; also $y = \pm\frac{1}{4}f'(r)v/(s-t)^2$. Euler[27] of Ch. XV discussed the problem to find s such that $s-e_i$ are squares for $i=1, 2, 3$ (whence their product gives $v^2/4$), but evidently restricted attention to the case in which each e_i is rational. Haentzschel showed how, from three primitive solutions of (6), to find four infinite sets of solutions by means of Weierstrass $\wp$-function.

Removing the assumption of a rational rcot r, but assuming one solution x_0, y_0 of $f = y^2$, he applied a certain linear fractional transformation giving a quartic whose leading coefficient is a square.

G. Humbert[164] stated that all the methods which have been proposed to deduce rational solutions of $ax^4 + \cdots + e = z^2$ from one or more initial solutions are identical at bottom, and gave the method in geometrical and analytic form.

On $x^4 \pm x^3 y + x^2 y^2 \pm x y^3 + y^4 = \square$, see papers 63–66 of Ch. II, Vol. I. On $xy(x^2-y^2) = Az^2$, see papers 11, 18; also Congruent Numbers in Ch. XVI.

For other special quartics made squares, see papers 101 of Ch. I; 21, 92–4, 96–7, 109, 138–40 of Ch. IV; and 9, 72, 73, 77, 92, 133 of Ch. V; and various papers of Chs. XIV–XX.

$$A^4 + B^4 = C^4 + D^4.$$

L. Euler[165] took $A = p+q$, $D = p-q$, $C = r+s$, $B = r-s$ and derived

$$(1) \qquad pq(p^2+q^2) = rs(r^2+s^2).$$

Set $p = ax$, $q = by$, $r = kx$, $s = y$. Then

$$y^2/x^2 = (k^3 - a^3 b)/(ab^3 - k).$$

If $k = ab$, $x = 1$, then $y = \pm a$, $C = \pm A$, $B = \mp D$. Set therefore $k = ab(1+z)$. Then $y^2/x^2 = a^2 Q/(b^2-1-z)^2$, where

$$Q = (b^2-1)^2 + (b^2-1)(3b^2-1)z + 3b^2(b^2-2)z^2 + b^2(b^2-4)z^3 - b^2 z^4.$$

Let Q be the square of $b^2 - 1 + fz + gz^2$ and choose f, g to make the terms in z, z^2 agree. Thus

$$f = \frac{3b^2-1}{2}, \qquad g = \frac{3b^4 - 18b^2 - 1}{8(b^2-1)}, \qquad z = \frac{b^2(b^2-4) - 2fg}{b^2 + g^2}.$$

Then $x : y = b^2 - 1 - z : a(b^2 - 1 + fz + gz^2)$. As examples, Euler took $b=2$, $b=3$, and found the solution

$$A = 2219449, \qquad B = -555617, \qquad C = 1584749, \qquad D = 2061283,$$

and an erroneous[166] one replaced in his next paper by

$$A = 12231, \qquad B = 2903, \qquad C = 10381, \qquad D = 10203.$$

<hr>

[164] L'intermédiaire des math., 25, 1918, 18–20.
[165] Novi Comm. Acad. Petrop., 17, 1772, 64; Comm. Arith., I, 473; Op. Om., (1), III, 211.
[166] This error was also noted in l'intermédiaire des math., 2, 1895, 6, 394; 7, 1900, 86; Mathesis, 1889, 241–2.

Euler[167] treated $a^4-b^4=c^4-d^4$ by setting

$$(a^2+b^2)p=(c^2-d^2)q, \qquad (a^2-b^2)q=(c^2+d^2)p.$$

Multiply the first by p, the second by q, add and subtract. Let $q^2-p^2=s^2$. Then

(2) $$b^2s^2=c^2(p^2+q^2)-2c^2pq, \qquad 2d^2pq=a^2s^2-b^2(p^2+q^2).$$

In (2_1) take $bs=a(q-p)+2p(a-c)x$, whence

$$a:c=2px^2+q:2px^2+2(q-p)x-q.$$

Taking a and c equal to these expressions, and multiplying (2_2) by $s^2/(2pq)$, we find that

$$d^2s^2=q^2(q-p)^2-4q(q-p)(q^2+p^2)x+2(q^2-p^2)^2x^2$$
$$+2(q^2-6pq+p^2)(p^2+q^2)x^2+8p(q-p)(p^2+q^2)x^3+4p^2(q-p)^2x^4,$$

which is readily made a square since the first and last coefficients are squares. For $p=3$, $q=5$, we have $s=4$ and

(3) $$d^2=\tfrac{25}{4}-85x-206x^2+102x^3+9x^4.$$

If we seek to make three terms of d^2 identical with those of the square of $5/2-17x+\alpha x^2$ or of $\alpha+17x+3x^2$, we find that $c^4=a^4$. But

$$\alpha^2+2\alpha\beta x+\gamma x^2+2\delta\epsilon x^3+\epsilon^2x^4=z^2, \qquad \beta^2+\delta^2-\gamma=\square=\zeta^2,$$

for $z=\alpha+\beta x$, $x=-(\delta\pm\zeta)/\epsilon$; also for $z=x(\epsilon x+\delta)$, $x=-\alpha/(\beta\pm\zeta)$. For the special form (3) we therefore get $x=-15$, $11/3$, $1/18$ or $5/22$, each leading to a permutation of the same values

$$a=542, \qquad b=359, \qquad c=514, \qquad d=103.$$

Euler[168] treated the following generalization of (1):

$$pq(mp^2+nq^2)=rs(mr^2+ns^2).$$

Set $q=ra$, $s=pb$. Then $p^2:r^2=na^3-mb:nb^3-ma$. Set

$$a=b(1+z), \qquad \alpha=nb^2/(nb^2-m), \qquad \beta=\alpha-1.$$

Then

$$p^2:r^2=C:1-\beta z, \qquad C\equiv1+3\alpha z+3\alpha z^2+\alpha z^3.$$

We may make $C(1-\beta z)=\square$ by the usual methods for quartics. But we obtain much simpler solutions by making $C/(1-\beta z)=(1+dz)^2$, viz.,

$$3\alpha-2d+\beta+(3\alpha+2\beta d-d^2)z+(\alpha+\beta d^2)z^2=0.$$

Taking $2d=3\alpha+\beta$, we get $z=-3/(4\alpha+4\beta d^2)$, $p/r=1+dz$.

For $m=n=1$, $b=3$, we get $\alpha=9/8$, $\beta=1/8$, $d=7/4$, $z=-96/193$, $p/r=25/193$, and obtain the solution $p=25$, $r=193$, $q=291$, $s=75$ of (1),

<hr>

[167] Mém. Acad. Sc. St. Petersb., 11, 1830 (1780), 49; Comm. Arith., II, 450. Euler wrote c^2+d^2 in his second equation and c^2-d^2 in his third. His further formulas require that d^2 be replaced by $-d^2$, which would invalidate the conclusions. In the present report, d^2 has been replaced by $-d^2$ at the outset, so that the remaining developments become correct as they stood.

[168] Nova Acta Acad. Petrop., 13, ad annos 1795–6, 1802 (1778), 45; Comm. Arith., II, 281. To conform with the notations of Euler's first paper, the interchange of a with p, b with q, c with r, d with s has been made. Also, Opera postuma, 1, 1862, 246–9 (about 1777).

whence
$$(4) \qquad 158^4 + 59^4 = 133^4 + 134^4.$$

For $m = n = 1$, $b = f/g$, we get $\alpha = f^2/(f^2 - g^2)$. In the resulting fraction for p/r, take p to be the product of the numerator by g. We obtain the solution of (1).

$$(5) \quad \begin{aligned} p &= g(f^2 + g^2)(-f^4 + 18f^2g^2 - g^4), \qquad r = 2g(4f^6 + f^4g^2 + 10f^2g^4 + g^6), \\ q &= 2f(f^6 + 10f^4g^2 + f^2g^4 + 4g^6), \qquad s = f(f^2 + g^2)(-f^4 + 18f^2g^2 - g^4). \end{aligned}$$

The case $f = 2$, $g = 1$ gives $p = 275$, $q = 928$, $r = 626$, $s = 550$, whence
$$2379^4 + 27^4 = 729^4 + 577^4.$$

From one set of solutions of (1) we obtain the second set

$$p' = p + q + r + s, \qquad q' = p + q - r - s, \qquad r' = p - q + r - s, \qquad s' = p - q - r + s.$$

A. Desboves[169] noted that $1203^4 + 76^4 = 1176^4 + 653^4$.

Desboves[170] wrote $s/q = m$ in (1) and obtained $p^3 + pq^2 - m^3q^2r - mr^3 = 0$. Regard m as a parameter. From the solution $p = m$, $q = r = 1$, we derive by Cauchy's formula the new solution

$$p = 2m(m^6 + 10m^4 + m^2 + 4), \qquad q = (m^2 + 1)(-m^4 + 18m^2 - 1),$$
$$r = 2(4m^6 + m^4 + 10m^2 + 1).$$

Replace m by f/g. The resulting solution is not new, as supposed by Desboves,[171] but[172] is Euler's (5). For $f = 1$, $g = 3$, we get (4). For $f = 1$, $g = 2$, we get Desboves'[169] numbers.

A. Cunningham[173] discussed the solution of the problem and proved the impossibility of $x^4 + y^4 = \xi^4 + 4\eta^4$.

R. Norrie,[174] starting with an evident solution of (1) took $p = \rho x_1 - s$, $r = \rho x_2 - q$; thus

$$(qx_1^3 - sx_2^3)\rho^3 + 3qs(x_2^2 - x_1^2)\rho^2 + \{(q^2 + 3s^2)qx_1 - (3q^2 + s^2)sx_2\}\rho = 0.$$

After making the coefficient of ρ zero by choice of x_2/x_1, we have only to take $-\rho$ equal to the ratio of the coefficient of ρ^2 to that of ρ^3. After reductions, we obtain Euler's (5). The same method applies also to

$$\lambda(\rho x_1 + a)^4 + \mu(\rho x_2 + b)^4 = \lambda(\rho x_1 + c)^4 + \mu(\rho x_2 + d)^4, \qquad \lambda a^4 + \mu b^4 = \lambda c^4 + \mu d^4.$$

A. S. Werebrusow[175] gave $239^4 + 7^4 = 227^4 + 157^4$ and Euler's solution (4).

T. Hayashi[176] reduced the problem to the solution of $3u^4 + v^4 = w^2$, from one solution of which we obtain an infinitude (Desboves[77]).

F. Ferrari[177] expressed $(4^2 + 5^2)(7^2 + 8^2)(4^2 + 15^2)(13^2 + 20^2)$ as a sum of two squares in eight ways and noted that the squares are biquadrates in two cases, giving Euler's (4).

[169] Nouv. Corresp. Math., 5, 1879, 279.

[170] Assoc. franç., 9, 1880, 239–242.

[171] Nouv. Corresp. Math., 6, 1880, 32.

[172] Noted by E. Fauquembergue, Mathesis, 9, 1889, 241–2; reproduced in Sphinx-Oedipe, 5, 1910, 93–4.

[173] Messenger Math., 38, 1908–9, 83–9.

[174] University of St. Andrews 500th Anniversary Mem. Vol., Edinburgh, 1911, 60–1.

[175] L'intermédiaire des math., 20, 1913, 197; 19, 1912, 205.

[176] The Tôhoku Math. Jour., 1, 1912, 143–5.

[177] Periodico di Mat., 28, 1913, 78.

E. Fauquembergue[178] gave the identity

$$(2\alpha^2-15\alpha\beta-4\beta^2)^4+(4\alpha^2+15\alpha\beta-2\beta^2)^4=(4\alpha^2+9\alpha\beta+4\beta^2)^4+s^2,$$
$$s=4\alpha^4+132\alpha^3\beta+17\alpha^2\beta^2+132\alpha\beta^3+4\beta^4,$$

while by Fermat's method we may make $s=\square$ in an infinitude of ways,
e. g., $\alpha=8$, $\beta=25$.

A. S. Werebrusow[179] gave $292^4+193^4=256^4+257^4$.

J. E. A. Steggall[180] treated $x^n-u^n=y^n-v^n$ by setting

$$\lambda x=1+ab, \qquad \lambda y=1+ac, \qquad \lambda u=a^{n-1}+b, \qquad \lambda v=a^{n-1}+c,$$

which determine a, b, c, λ in terms of x, y, u, v. He discussed only the case
$n=4$, whence

$$4a(1+a^4)+6(b+c)a^2=(b+c)(b^2+c^2).$$

This is satisfied if $b+c=2a(1+t)$, and

$$4\{(1+a^4)(1+t)+a^2(1+t)^2(2-2t-t^2)\}=(1+t)^2(b-c)^2.$$

A particular value making the left member a square is

$$t=\frac{8(1+a^2)^2(1-18a^2+a^4)}{(1+14a^2+a^4)^2+64a^2(1+a^2)^2},$$

whence we derive one of Euler's tentative solutions. The smallest set of
integral solutions is said to be (4).

M. Rignaux[181] recalled [Euler[168]] that (1) is unaltered by the substitution

$$p=P+Q+R+S, \quad q=P+Q-R-S, \quad r=P-Q+R-S, \quad s=P-Q-R+S.$$

He obtained (p. 128, pp. 133–4) various solutions of (1).

A. Gérardin[181a] noted that (1) has the solution

$$p=a^7+a^5-2a^3+a, \qquad q=3a^2, \qquad r=a^6-2a^4+a^2+1, \qquad s=3a^5,$$

which is simpler than Euler's solution (5).

$$A^4+hB^4=C^4+hD^4.$$

E. Grigorief[182] noted that

$$19^4+5\cdot281^4=417^4+5\cdot117^4, \qquad 74^4+5\cdot101^4=147^4+5\cdot63^4,$$

the latter being erroneous. He[183] found an infinitude of solutions when
$h=2$, the least having eleven digits (from $u=33$, $v=13$), by making special
assumptions leading to the condition $3u^4-2v^4=w^2$.

A. S. Werebrusow[184] gave $139^4+2\cdot34^4=61^4+2\cdot116^4$.

A. Gérardin[185] treated $a^4+hb^4=c^4+hd^4$ by setting $a-c=m$, $d-b=x$,

[178] L'intermédiaire des math., 21, 1914, 17 (18–19, bibliography).
[179] *Ibid.*, 18.
[180] Proc. Edinburgh Math. Soc., 34, 1915–6, 15–17.
[181] L'intermédiaire des math., 25, 1918, 27–28.
[181a] *Ibid.*, 24, 1917, 51.
[182] L'intermédiaire des math., 9, 1902, 322; 10, 1903, 245.
[183] *Ibid.*, 14, 1907, 184–6.
[184] *Ibid.*, 17, 1910, 127.
[185] Sphinx-Oedipe, 6, 1911, 6–7, 11–13. Cf. Norrie.[174]

$a+c=p(d+b)$; thus

$$2(2mp^3-hx)b^2+2x(2mp^3-hx)b+(mp^3x^2-hx^3-2c^2pm-2cm^2p)=0.$$

Equate to zero the coefficient of b^2. Then that of b is zero, and we obtain m and h rationally in terms of p, c, x. In the special cases $p=cx$ and $c=x=1$, the resulting identities are simple. He gave solutions of the systems formed by $x^4+mx^2y^2+y^4=a^2$ and various other quartics.

Gérardin[186] gave solutions of $a^4+hb^4=c^4+hd^4$ for 26 numerical values of h, and noted the solution $a=2p^2$, $c=2p$; b, $d=p\mp1$; $h=2p^3(p^2-1)$.

SUM OF THREE BIQUADRATES NEVER A BIQUADRATE.

L. Euler[165, 167] stated that this theorem was hardly to be doubted, though not yet proved. Again he[168] stated " It has seemed to many Geometers that this theorem $(x^n+y^n\neq z^n,\ n>2)$ may be generalized. Just as there do not exist two cubes whose sum or difference is a cube, it is certain that it is impossible to exhibit three biquadrates whose sum is a biquadrate, but that at least four biquadrates are needed if their sum is to be a biquadrate, although no one has been able up to the present to assign four such biquadrates. In the same manner it would seem to be impossible to exhibit four fifth powers whose sum is a fifth power, and similarly for higher powers."

Euler[187] noted that $abc(a+b+c)=1$ has the rational solutions 4, 1/3, 1/6, and $abcd(a+b+c+d)=1$ the solutions 4/3, 3/2, $-1/3$, $-3/2$. Hence we cannot infer the impossibility of $p^4+q^4+r^4=s^4$ by setting $a=p^3/qrs$, $b=q^3/prs$, $c=r^3/pqs$; nor that of $p^5+q^5+r^5+s^5=t^5$ by setting $a=p^4/qrst$, $\cdots$, $d=s^4/pqrt$.

A. Desboves[188] expressed doubt as to the theorem and proved the impossibility of $p^4\pm6p^2q^2-7q^4=\square$ in connection with a study of

$$X^4+Y^4-Z^4=2T^2,$$

which has the solutions

$$X=x^2\mp y^2,\qquad Y=x^2\pm y^2,\qquad Z=2xy,\qquad T=x^4-y^4.$$

L. Aubry[189] proved that the fourth power of an integer $\leqq1040$ is not a sum of three biquadrates.

A. S. Werebrusow[189a] showed that no solution can be found by making each term a biquadrate in Euler's identity

$$(a^2+b^2+c^2+d^2)^2=(a^2+b^2-c^2-d^2)^2+(2ac+2bd)^2+(2ad-2bc)^2.$$

SUM OF FOUR OR MORE BIQUADRATES A BIQUADRATE.

L. Euler[190] remarked that it seemed possible to assign four biquadrates whose sum is a biquadrate, but that he had found no example, whereas he

[186] Sphinx-Oedipe, 8, 1913, 13.
[187] Opera postuma, 1, 1862, 235–7 (about 1769). Cf. Euler.[249]
[188] Nouv. Corresp. Math., 6, 1880, 32. Cf. Sphinx-Oedipe, 8, 1913, 27.
[189] Sphinx-Oedipe, 7, 1912, 45–6. Stated, l'interméd. des math., 19, 1912, 48.
[189a] L'intermédiaire des math. 21, 1914, 161.
[190] Corresp. Math. Phys. (ed., Fuss), 1, 1843, 618 (623), Aug. 4, 1753. See preceding topic.

could give five biquadrates with a biquadrate as sum. He[167] again remarked that he was trying to find four such biquadrates.

Euler[191] gave an incomplete discussion of the " difficult " problem to find four biquadrates whose sum is a biquadrate. Evidently

$$A^4+B^4+C^4+D^4=E^4$$

for

$$A^2=(p^2+q^2+r^2-s^2)/n, \qquad B^2=2ps/n, \qquad C^2=2qs/n,$$
$$D^2=2rs/n, \qquad E^2=(p^2+q^2+r^2+s^2)/n.$$

These five functions are to be made squares. This will be true of the first and last if

(1) $$\qquad (p^2+q^2+r^2)/n=a^2+b^2, \qquad s^2/n=2ab.$$

Then $s^2=2abn=\square$ if $2n=\alpha\beta$, $a=\alpha f^2$, $b=\beta g^2$, whence $s=\alpha\beta fg$. Next,

$$\frac{2ps}{n}=4pfg=4x^2, \qquad \frac{2qs}{n}=4qfg=4y^2, \qquad \frac{2rs}{n}=4rfg=4z^2,$$

if $p=x^2/(fg), q=y^2/(fg), r=z^2/(fg)$. Substitute these values into (1_1); we get

$$x^4+y^4+z^4=\tfrac{1}{2}ab(a^2+b^2).$$

But no discussion of this final condition is given.

D. S. Hart[192] employed the sum

$$\sigma=\tfrac{1}{5}n^5+\tfrac{1}{2}n^4+\tfrac{1}{3}n^3-\tfrac{1}{30}n$$

of n consecutive biquadrates $1^4, \cdots, n^4$, and

$$(s+m)^4=s^4+\sigma+t-\sigma, \qquad t\equiv(s+m)^4-s^4.$$

Thus $(s+m)^4$ can be expressed as a sum of biquadrates if $\sigma-t$ is. Evidently $n>8$. For $n=9$, $s=14$, $m=1$, $\sigma-t=3124=1^4+2^4+3^4+5^4+7^4$, yielding

(2) $$\qquad 4^4+6^4+8^4+9^4+14^4=15^4.$$

For $n=20$, $s=30$, $m=4$, 34^4 is the sum of the fourth powers of 1, 3, 4, 5, 9, 10, 11, 12, 14, 15, 16, 17, 18, 19, 30.

A. Martin[193] gave (2).

A. Martin[194] started with the identity

$$(1+4m^4)^4=1^4+(2m)^4+96(m^2)^4+(4m^3)^4+(4m^4)^4.$$

But $96=3^4+2^4-1^4$. Hence the new right member has six positive biquadrates and the term $-(m^2)^4$. For $m=2$, the latter cancels $(2m)^4$ and we get

$$1^4+8^4+12^4+32^4+64^4=65^4,$$

which was communicated to him by D. S. Hart. For $m=3$,

$$325^4=A+108^4+324^4,$$

where

$$A=1+6^4+18^4+27^4-9^4=28^4+10^4+8^4+7^4=26^4+20^4+10^4+8^4+3^4,$$

[191] Opera postuma, 1, 1862, 216–7 (about 1772).
[192] Math. Quest. Educ. Times, 14, 1871, 86–7.
[193] Ibid., 20, 1873, 55. L'intermédiaire des math., 1, 1894, 26.
[194] Math. Magazine, 2, 1896, 173–184.

so that we get 6 or 7 biquadrates whose sum is a biquadrate. Multiplying (2) by 2^4 and by 5^4 and eliminating 30^4, we see that 75^4 is the sum of the fourth powers of 8, 12, 16, 18, 20, 28, 40, 45, 70. Finally, he tabulated the values of $S = 1^4 + \cdots + n^4$ for $n \leqq 285$ to use in seeking by trial to express $S - b^4$ as a sum of distinct biquadrates $\leqq n^4$. Example in Martin,[68] Ch. XXIII.

E. Fauquembergue[195] gave the identity

$$(4x^4 + y^4)^4 = (4x^4 - y^4)^4 + (4x^3 y)^4 + (4x^3 y)^4 + (2xy^3)^4 + (2xy^3)^4,$$

which becomes $5^4 = 3^4 + 4^4 + 4^4 + 2^4 + 2^4$ for $x = y = 1$.

C. B. Haldeman[196] noted that $a^4 + b^4 + (a+b)^4 \equiv 2(a^2 + ab + b^2)^2$ [Proth[227]]. Hence on adding $d^4 + e^4$, the sum will be a biquadrate if $a^2 + ab + b^2 = de$ and $d^2 + e^2 = \square$. To satisfy the latter, take $e = (d^4 - 4z^4)/(4dz^2)$; then the former condition gives

$$a = \frac{-bz \pm t}{2z}, \qquad t = \sqrt{d^4 - 4z^4 - 3b^2 z^2}.$$

Take $t = d^2 - z^2$, whence $d^2 = \frac{1}{2}(3b^2 + 5z^2)$. Since $b = z$ makes d rational, set $b = y + z$, and take $d = 2z + sy/t$, whence we find y and then b, d. Or we may take $d = 2$, $z = 1$, whence $t = \sqrt{12 - 3b^2}$; set $b = v + 1$, $t = sv/t + 3$, whence we get v and

$$(3) \qquad \Sigma(2s^2 \pm 12st - 6t^2)^4 + \Sigma(4s^2 \mp 12t^2)^4 + (3s^2 + 9t^2)^4 = (5s^2 + 15t^2)^4.$$

Or, finally, take $d = 9$, $e = 4$, $a^2 + ab + b^2 = 4 \cdot 37$ since $2(4 \cdot 37)^2 + 9^4 + 4^4 = 15^4$. Since $b = 6$ gives a rational value for a, set $b = 6 + r$. Then

$$(2a + b)^2 = 592 - 3b^2 = -3r^2 - 36r + 484 = \left(\frac{rs}{t} + 22\right)^2,$$

by choice of r rationally in s, t. Hence the sum of the fourth powers of $8s^2 + 40st - 24t^2$, $6s^2 - 44st - 18t^2$, $14s^2 - 4st - 42t^2$, $9s^2 + 27t^2$, $4s^2 + 12t^2$ equals $(15s^2 + 45t^2)^4$. For $s = 1$, $t = 0$, we get (2), which is believed to be the solution in least integers.

For six biquadrates, add $e^4 + f^4$ to each member of his[239] identity (1) and take $3(3a^2 + t^2)^2 = ef$. It remains to make $e^2 + f^2 = \square$, say the square of $1201(3a^2 + t^2)/140$, whence $e = 7(3a^2 + t^2)/20$. Or we may take the sum of three of the six to be

$$(4) \qquad Q_{a,b} = (2a)^4 + (a+b)^4 + (a-b)^4 = 2(3a^2 + b^2)^2$$

and the others to be the fourth powers of 6, 12, 13 or 26, 27, 42 and the sum of the six to be 15^4 or 45^4.

For seven biquadrates, take $Q_{a,b} + d^4 + e^4 + (2g)^4 + g^4 = (3g)^4$, $3a^2 + b^2 = de$, whence $d^2 + e^2 = 8g^2$, which holds if $e = +7d$, $g = -5d/2$. Take $d = y + a$, $b = ry/t + 2a$. Then $y = 2a(7t^2 - 2rt)/(r^2 - 7t^2)$ and we have an answer. Or use $Q_{a,b} + Q_{d,e} + 3^4 = 5^4$, which is satisfied if $3a^2 + b^2 = 4$, $3d^2 + e^2 = 16$; taking $b = 2 - as/t$, $e = 4 - dv/z$, we get a, b in terms of s, t, and d, e in terms of v, z. Next, $Q_{a,b} + Q_{d,e} + 2^4 + 1^4 = 3^4$ if $3a^2 + b^2 = 4 = 3d^2 + e^2$ (like preceding case).

---

[195] L'intermédiaire des math., 5, 1898, 33.

[196] Math. Magazine, 2, 1904, 288–296. The editor Martin noted (p. 349 and in his 1900 paper below) that this MS. had been long in the editor's hands.

To find a sum of n biquadrates equal to a biquadrate for $n = 9, 10, 11, 12$, multiply (3) by a suitable biquadrate and eliminate one biquadrate by use of one of the earlier results. Finally, given

$$2^4 + 6^4 + 8^4 + 2^4 + 7^4 + 12^4 = 13^4, \qquad 2 + 6 = 8,$$

we can find a, b so that $2^4 + 6^4 + 8^4 = a^4 + b^4 + (a+b)^4 = 2(a^2 + ab + b^2)^2$. Thus $a^2 + ab + b^2 = 2^2 + 2 \cdot 6 + 6^2 = 52$, $a = \frac{1}{2}(-b - \sqrt{208 - 3b^2})$. Set $b = y + 6$,

$$208 - 3b^2 = -3y^2 - 36y + 100 = \left(10 + \frac{sy}{t}\right)^2,$$

whence we get y, b, a. Take $s = 2$, $t = 1$. Then $7y = -76$, $7b = -34$, $7a = 58$ and

$$\left(\tfrac{58}{7}\right)^4 + \left(\tfrac{34}{7}\right)^4 + \left(\tfrac{24}{7}\right)^4 + 2^4 + 7^4 + 12^4 = 13^4.$$

A. Martin[197] employed methods admittedly similar to Haldeman's, whose manuscript was in his hands, but found many new sets of biquadrates whose sum is a biquadrate. For 5 biquadrates, take

$$Q_{a,\,b} + y^4 + \left(\frac{y^2 - e^2}{2e}\right)^4 = \left(\frac{y^2 + e^2}{2e}\right)^4,$$

which reduces to $2e(3a^2 + b^2) = y(y^2 - e^2)$. First, take $y = 2e$; then

$$b^2 = 3e^2 - 3a^2 = \left\{\frac{s}{t}(e-a)\right\}^2, \qquad \text{if} \qquad a = \frac{s^2 - 3t^2}{s^2 + 3t^2} \cdot e,$$

which for $e = 2(s^2 + 3t^2)$ leads to Haldeman's (3). For $y = 3e$, we get a result equivalent to the last. The next solvable case is $y = 8e$, giving

$$(12s^2 + 120st - 36t^2)^4 + (36s^2 + 24st - 108t^2)^4 + (16s^2 + 48t^2)^4$$
$$+ (24s^2 - 96st - 72t^2)^4 + (63s^2 + 189t^2)^4 = (65s^2 + 195t^2)^4.$$

For $y = 13e$, we get a similar formula. Next, let

$$Q_{x,\,y} + w^4 + z^4 = s^4, \qquad 3x^2 + y^2 = wz.$$

The first becomes $w^2 + z^2 = s^2$, whence take $z = 2pq$, $w = p^2 - q^2$, $s = p^2 + q^2$. The case $p = 2$, $q = 1$, leads to (3). Omitting the discussions found to be unfruitful, let $p = r + 2q$, $x = t + 2q^2$. Then

$$y^2 = wz - 3x^2 = 2qr^3 + 12q^2r^2 - 3t^2 + A, \qquad A = 22q^3r - 12q^2t.$$

Take $A = 0$, whence $t = 11qr/6$. Set $y = qrm/n$. We get q in terms of m, n, whence

$$(88n^2\alpha + 2304n^4)^4 + \Sigma\{(44n^2 \pm 24mn)\alpha + 1152n^4\}^4 + (48n^2\beta)^4$$
$$+ (\beta^2 - 576n^4)^4 = (\beta^2 + 576n^4)^4, \qquad \alpha = 12m^2 - 23n^2, \qquad \beta = 12m^2 + 25n^2.$$

In the Congress paper, on the contrary, he took $t = -4q$ and found the special solution $2^4 + 13^4 + 32^4 + 34^4 + 84^4 = 85^4$. For n biquadrates, $n = 6, 7, 8, 11$, he took

$$Q_{a,\,b} + 2^4 + 7^4 + 12^4 = 13^4, \qquad\qquad Q_{a,\,b} + 2^4 + 4^4 + 5^4 + 8^4 = 9^4,$$
$$Q_{a,\,b} + Q_{c,\,d} + 5^4 + 6^4 = 9^4, \qquad Q_{a,\,b} + Q_{c,\,d} + Q_{e,\,f} + 7^4 + 14^4 = 21^4,$$

and found other sets by combination.

---

[197] Deux. Congrès Internat. Math., 1900, Paris, 1902, 239–248. Reproduced with additions in Math. Mag., 2, 1910, 324–352.

E. Barbette[198] used the final method of Martin[194] to show that (2) is the only sum of distinct biquadrates $\leqq 14^4$ equal to a biquadrate, and that

$$4^5+5^5+6^5+7^5+9^5+11^5=12^5$$

is the only sum of distinct fifth powers $\leqq 11^5$ equal to a fifth power.

R. Norrie[199] found (in confirmation of Euler's[190] conjecture)

$$(5) \qquad 353^4=30^4+120^4+272^4+315^4,$$

by a series of special assumptions which lead to this single result. Next (p. 77),

$$(u^2+v^2)^4 = (u^2-v^2)^4+(2uv)^4+(x+y)^4+(x-y)^4+(2y)'$$

provided [see (4)]

$$2uv(u^2-v^2)=x^2+3y^2.$$

To solve the latter, set $u=rx_1+2$, $v=1$, $x=rx_2+3$, $y=rx_3+1$. Hence

$$2x_1^3r^3+(12x_1^2-x_2^2-3x_3^2)r^2+(22x_1-6x_2-6x_3)r=0.$$

Equate the coefficient of r to zero. Then the equation gives r. For 6 biquadrates (p. 80), use

$$(X^4+Y^4)^4\equiv(X^4-Y^4)^4+(2XY^3)^4+8X^4Y^4(X^8-Y^8),$$

$$X^8-Y^8\equiv2(2xy)^4(x^8+16y^8)(X^4+Y^4), \qquad X=x^4+4y^4, \qquad Y=x^4-4y^4.$$

From the latter,

$$X^{2r+3}-Y^{2r+3}=2(2xy)^4(x^8+16y^8)(X^4+Y^4)(X^8+Y^8)\cdots(X^{2r+2}+Y^{2r+2}),$$

the second member being double the sum of 2^{r+2} biquadrates. Hence $(X^{2r+2}+Y^{2r+2})^4$ equals a sum of $2^{r+2}+2$ biquadrates. Returning to the 6 biquadrate case, take $x=u^3$, $y=2v^3$, whence x^8+16y^8 equals the value of

$$b^{12}+c^{12}\equiv\frac{\{b^3(b^4-3c^4)\}^4+\{c^3(c^4-3b^4)\}^4+\{2bc(b^4-c^4)\}^4(b^4+c^4)}{(b^4+c^4)^4}$$

for $b=u^2$, $c=2v^2$. Thus we get a biquadrate expressible simultaneously as a sum of 6, 8 or 10 biquadrates. The sum of two of these biquadrates has the factor u^8+16v^8, which as before can be replaced by the sum of four rational biquadrates. In this way we can assign a biquadrate which is a sum of any even number >4 of biquadrates.

For 7 biquadrates (p. 84), take $t=(x^4+y^4+z^4)/8$ in

$$(t+1)^4\equiv(t-1)^4+8t+8t^3$$

and set $x=p^2-q^2$, $y=2pq+q^2$, $z=2pq+p^2$. We get a relation between biquadrates, one with the coefficient 2, for which we substitute the sum of three rational biquadrates given by Gerardin's[208] (3). Again,

$$\{(x^2+3y^2)^2+4z^4\}^4\equiv\{(x^2+3y^2)^2-4z^4\}^4+(2z)^4\{(x^2+3y^2)^4+(2z^2)^4\}\{T+(2y)^4\},$$

where $T=(x+y)^4+(x-y)^4$. But for any $r\gtreqqless2$, we can express T (in one of its two occurrences) as a sum of r biquadrates and hence obtain a bi-

<hr>

[198] Les sommes de p-ièmes puissances distinctes égales à une p-ième puissance, Liège, 1910, 133–146.

[199] University of St. Andrews 500th Anniversary Memorial Vol., Edinburgh, 1911, 89.

quadrate expressed as a sum of $r+5$ biquadrates.   In fact,

$$\{bc^3(\tau+2\Sigma)\}^4+(2c^4\Sigma-b^4\tau)^4\equiv\{bc^3(\tau-2\Sigma)\}^4+(2c^4\Sigma+b^4\tau)^4+(2bc\tau)^4\Sigma,$$

where $\tau=c^8-b^8$, $\Sigma=x_1^4+x_2^4+\cdots+x_n^4$.

Equal sums of biquadrates.

A. Martin[200] tabulated various sets of numbers having equal sums of fourth powers, as 1, 2, 9 and 3, 7, 8; 1, 9, 10 and 5, 6, 11; 1, 11, 12 and 4, 9, 13; 1, 5, 8, 10 and 3, 11.

C. B. Haldeman[201] noted that the sums $Q_{a,\,b}$ and $Q_{d,\,e}$ of three biquadrates are equal if $3a^2+b^2=3d^2+e^2$.   Taking $e=b-v$, we get b, e rationally in terms of a, d, v and see that

$$(4av)^4+(3a^2-3d^2-2av-v^2)^4+(3a^2-3d^2+2av-v^2)^4$$

is unaltered by the interchange of a and d.   For $a=1$, $d=v=2$, we get

$$8^4+9^4+17^4=3^4+13^4+16^4.$$

Next, let

$$Q_{a,\,b}+d^4=\left(\frac{d^2+s^2}{2s}\right)^4+\left(\frac{d^2-s^2}{2s}\right)^4,$$

whence $b^2=N^2/(4s^2)$, $N^2=d^4-s^4-12a^2s^2$.   Taking

$$N=d^2-2p^2s^2/(3q^2),\qquad d=v+3aq/p,$$

we get a and d rationally.   Or take $N=d^2-s^2$, whence $d^2=6a^2+s^2$, set $a=2s+y$ and solve as usual.   Again,

$$Q_{a,\,b}+1^4=Q_{d,\,e}+3^4\qquad\text{if}\qquad 3a^2+b^2=7,\qquad 3d^2+e^2=3.$$

Take $a=1+x$, $d=\frac{1}{2}+y$ and solve as usual.   Finally, to find a sum of four biquadrates equal to a sum of three, employ his[239] identity (1) and equate the left member to $Q_{m,\,n}$.   The resulting condition, $3(3a^2+t^2)^2=3m^2+n^2$ is satisfied if

$$m=\left(\frac{z^2-3r^2}{z^2+3r^2}\right)f,\qquad n=\left(\frac{6rz}{z^2+3r^2}\right)f,\qquad f=3a^2+t^2.$$

A. Cunningham[202] noted that $X^4+Y^4+x^4+z^4=X_1^4+x_1^4+y_1^4+z_1^4$ follows by combining a solution of each of $X^4+Y^4=X_1^4+Y_1^4$, $x^4+Y_1^4+z^4=x_1^4+y_1^4+z_1^4$. Again, $x^4+y^4+2u_1^4=x_1^4+y_1^4+2u^4$ follows from

$$x^4+y^4+z^4=2u^4,\qquad x_1^4+y_1^4+z_1^4=2u_1^4$$

(solved, Cunningham[240]) with $u=A^2+3B^2$, $u_1=A_1^2+3B_1^2$, $AB=A_1B_1$, whence $z=z_1$.

A. S. Werebrusow[203] gave an incorrect proof of the impossibility of $x^4+y^4+z^4=3u^4$ in relatively prime integers.

[200] Math. Magazine, 2, 1896, 183.
[201] *Ibid.*, 2, 1904, 286–8.   For the notation Q, see Haldeman[196] (4).
[202] Messenger Math., 38, 1908–9, 103–4.
[203] L'intermédiaire des math., 15, 1908, 281.   Cf. 16, 1909, 55, 208; 17, 1910, 279.

F. Ferrari[204] noted the identity

$$(a^2+2ac-2bc-b^2)^4+(b^2-2ba-2ac-c^2)^4+(c^2+2ab+2bc-a^2)^4$$
$$=2(a^2+b^2+c^2-ab+ac+bc)^4.$$

while U. Bini (*ibid.*) gave the identity

$$[a(d+c)-b(c-3d)]^4+[2(bc-ad)]^4+[a(d-c)-b(c-3d)]^4$$
$$=[a(d-c)\pm b(c+3d)]^4+[2(bc+ad)]^4+[a(d+c)+b(c-3d)]^4,$$

with the plus sign. A. Gérardin (*ibid.*, 19, 1912, 254) stated that the sign should be minus and gave other such identities. Welsch (*ibid.*, 132, 184) gave another method of correcting the signs: retain the plus sign, but change the final term of the first member to $-b(c+3d)$.

A. Cunningham[205] found numbers expressible in several ways in the form $x^4+y^4+z^4$ by use of $x^4+y^4\equiv 2u^2-z^4$, $u=x^2+xy+y^2$, $z=x+y$, and expressing this u in the form A^2+3B^2 in several ways.

E. Miot[206] stated that [the case $b=c$ of Ferrari's[204] identity]

$$(1) \qquad (4pq)^4+(3p^2+2pq-q^2)^4+(3p^2-2pq-q^2)^4=2(3p^2+q^2)^4$$

and noted cases when a sum of three squares equals a sum of three biquadrates and a sum of three eighth powers. Welsch[207] stated that Miot's solution is erroneous and noted that

$$2a^2=(x^2-y^2)^2+(x^2-z^2)^2+(y^2-z^2)^2=(u^4-v^4)^2+(u^4-w^4)^2+(v^4-w^4)^2$$

always implies that

$$2a^4=\Sigma(x^2-y^2)^4=\Sigma(u^4-v^4)^4.$$

A. Gérardin[208] noted cases of two equal sums of three biquadrates and gave four methods of finding particular solutions of

$$(2) \qquad x^4+y^4=z^4+u^4+v^4,$$

the fourth leading to the solution

$$x=128p^9+pq^8, \qquad y,\ z=64p^8q\mp 12p^4q^5-q^9, \qquad u=3pq^8, \qquad v=128p^9-2pq^8.$$

[It is expressed by the next identity with $h=1$, $l=q$, and p replaced by $2p$.] He gave 16 identities which follow by a change of variable from

$$(p^9-4ph^2l^8)^4+(6ph^2l^8)^4+h(p^8l+3hp^4l^5-4h^2l^9)^4$$
$$=(p^9+2ph^2l^8)^4+h(p^8l-3hp^4l^5-4h^2l^9)^4.$$

In conclusion, he gave

$$(3) \qquad (p^2-q^2)^4+(2pq+q^2)^4+(2pq+p^2)^4=2(p^2+pq+q^2)^4.$$

A. Martin[209] gave (1) and (3).

E. Miot[210] noted the solution 37, 17; 35, 26, 3 of (2).

[204] L'intermédiaire des math., 16, 1909, 83.
[205] Math. Quest. Educ. Times, (2), 14, 1908, 83–4. Same in Mess. Math., 38, 1908–9, 101–2.
[206] L'intermédiaire des math., 17, 1910, 214.
[207] *Ibid.*, 18, 1911, 64.
[208] Assoc. franç., 39, 1910, I, 44–55. Same in Sphinx-Oedipe, 5, 1910, 180–6; 6, 1911, 3–6; 8, 1913, 119.
[209] Math. Magazine, 2, 1910, 351.
[210] L'intermédiaire des math., 18, 1911, 27–28.

R. Norrie[211] gave several methods to solve

$$(4) \qquad\qquad x^4+y^4+z^4=u^4+v^4+w^4.$$

First, take $x=rx_1+a$, $y=rx_2+b$, $z=rx_3+c$, $u=rx_1-a$, $v=rx_2+c$, $w=rx_3+b$. We obtain a cubic in r whose constant term is zero. The coefficient of r will be zero if $x_3=x_2+2x_1a^3/(b^3-c^3)$. Then $-r$ is the ratio of the coefficient of r^2 to that of r^3. Second, he noted that

$$\{x_2y_2^3(x_1^4+2y_1^4)\}^4+\{x_1y_1^3(x_2^4-2y_2^4)\}^4+\{2x_1y_1^3x_2^3y_2\}^4$$

equals identically the sum derived by interchanging the subscripts 1, 2. Replacing x_1, y_1, x_2, y_2 by their reciprocals and multiplying each root by $(x_1y_1x_2y_2)^4$, we obtain a new integral function which is added to the former. Hence

$$\{x_2y_2^3(x_1^4+2y_1^4)\}^4+\{x_1y_1^3(x_2^4-2y_2^4)\}^4+\{x_2^3y_2(y_1^4+2x_1^4)\}^4+\{x_1^3y_1(y_2^4-2x_2^4)\}^4$$

is unaltered by the interchange of the subscripts 1, 2. Multiplying

$$(x_1^4+2y_1^4)^4-(x_1^4-2y_1^4)^4-(2x_1^3y_1)^4\equiv 4(2x_1y_1^3)^4$$

by the identity derived by interchanging the subscripts, we get two equal sums of five biquadrates. The third method is really Haldeman's[201] remark that $Q_{v,\,x}=Q_{v,\,u}$ if $3y^2+x^2=3v^2+u^2$. The general solution of the latter is stated to be

$$x,\ u=\{(3\lambda^2\pm1)v+(3\lambda^2\mp1)y\}/(2\lambda),$$

where λ is arbitrary. Again, $x^4+y^4+(x+y)^4$ is unaltered when x is replaced by $(3x-5y)/7$ and y by $(5x+8y)/7$. Changing the sign of y and subtracting the new identity from the former, we get

$$(7x+7y)^4+(3x+5y)^4+(8x-3y)^4+(5x-8y)^4$$
$$=(7x-7y)^4+(3x-5y)^4+(8x+3y)^4+(5x+8y)^4.$$

Finally there is given the identity, in which $\tau=\mu^2c^8-\lambda^2b^8$,

$$\lambda\{bc^3(\lambda\mu\tau+2\mu^3\nu x^4)\}^4+\mu(2\mu^3\nu c^4x^4-\lambda^2b^4\tau)^4$$
$$=\lambda\{bc^3(\lambda\mu\tau-2\mu^3\nu x^4)\}^4+\mu(2\mu^3\nu c^4x^4+\lambda^2b^4\tau)^4+\nu(2\lambda\mu bc\tau x)^4.$$

If we replace νx^4 by $\Sigma_{i=1}^{i=r}\nu_ix_i^4-\Sigma_{i=1}^{i=s}\kappa_iy_i^4$, we get a solution of

$$\sum_{i=1}^{r+2}\lambda_iu_i^4=\sum_{i=1}^{s+2}\mu_iv_i^4 \qquad (\lambda_1=\mu_1,\ \lambda_2=\mu_2).$$

In the last, Norrie made the restrictions that $s=r$, $\kappa_i=\nu_i$, whence $\lambda_i=\mu_i$.

A. Gérardin[212] noted the identity

$$(x^4-2y^4)^4+(2x^3y)^4+(3xy^3)^4=(x^4+2y^4)^4+(2xy^3)^4+(xy^3)^4.$$

E. N. Barisien[213] noted the identity (1).

Gérardin[214] quoted his[208] solutions of (2) involving two parameters with $x=z+u$ and noted that (3) is simpler than Ferrari's[204] formula, which follows by taking $a+c=p$, $b+c=-q$.

[211] University of St. Andrews 500th Anniversary, Edinburgh, 1911, 62–75.
[212] Bull. Soc. Philomathique, (10), 3, 1911, 236.
[213] Nouv. Ann. Math., (4), 11, 1911, 280–2.
[214] L'intermédiaire des math., 18, 1911, 200–1, 287–8.

"V. G. Tariste"[215] noted that (3) is derived from Bini's[204] formula by equating to zero one of the six biquadrates.

O. Birck[216] stated that (3), viz.;

$$x = -y = p^2 + pq + q^2, \qquad z = p^2 - q^2, \qquad u = q^2 + 2pq, \qquad v = -p^2 - 2pq,$$

gives the most general solution of $x + y = z + u + v = 0$ with either (2) or $x^2 + y^2 = z^2 + u^2 + v^2$. He noted that

$$7^4 + 28^4 = 3^4 + 20^4 + 26^4, \qquad 51^4 + 76^4 = 5^4 + 42^4 + 78^4.$$

A. S. Werebrusow[217] gave equal sums of three biquadrates involving many parameters and derived Gérardin's[204] formulas by specialization. He[218] gave $37^4 + 38^4 = 26^4 + 42^4 + 25^4$ and eight more such sets.

E. Fauquembergue[219] gave the identity

$$[2(\alpha^2 - \beta^2)]^4 + [\beta(4\alpha - 5\beta)]^4 + (2\alpha^2 - 5\alpha\beta + 2\beta^2)^4 = (2\alpha^2 - 4\alpha\beta + 3\beta^2)^4 + v^2,$$

where $v = 4\alpha^4 - 4\alpha^3\beta + 13\alpha^2\beta^2 - 36\alpha\beta^3 + 24\beta^4$, and found five sets making $v = \square$, all giving trivial solutions of (2). A. Tafelmacher[220] drew the same conclusion from a complete study of the identity derived by replacing α by $\beta + \gamma$.

L. Bastien[221] stated a solution of $x_1^4 + \cdots + x_n^4 = y_1^4 + \cdots + y_m^4$, $n \geqq 2$, $m \geqq 3$:

$$x_1 = \rho^3(\nu^4\rho^4\sigma - 8\tau\mu^4), \qquad x_2 = \nu^3(\rho^8\sigma + 8\tau\mu^4), \qquad x_i = 8\nu\rho^2\mu^3\tau\alpha_i \quad (i = 3, \cdots, n),$$
$$y_1 = \rho^3(\nu^4\rho^4\sigma + 8\tau\mu^4), \qquad y_2 = \nu^3(\rho^8\sigma - 8\tau\mu^4), \qquad y_i = 8\nu\rho^2\mu^3\tau\beta_i \quad (i = 3, \cdots, n),$$
$$\tau = \nu^8 - \rho^8, \qquad \sigma = \beta_3^4 + \cdots + \beta_m^4 - \alpha_3^4 - \cdots - \alpha_n^4.$$

R. D. Carmichael[222] noted that $x^4 + y^4 + 4z^4 = t^4$ has the special solution x, $t = \rho^4 \mp 2\sigma^4$, $y = 2\rho^3\sigma$, $z = 2\rho\sigma^3$. Solutions involving two parameters are given for $x^4 + ay^4 + az^4 = t^4$ and $x^4 + y^4 + az^4 = at^4$, if $a = 2$ or 8. Also,

$$(k^2 - 2k)^4 + (2k - 1)^4 + (k^2 - 1)^4 = 2(k^2 - k + 1)^4,$$

the case $p = k$, $q = -1$, of (3). By Cunningham,[173] $x^4 + y^4 - 4z^4 \neq t^4$.

A. S. Werebrusow[223] tabulated all solutions, each $\leqq 50$, of (4).

E. Miot[224] gave a solution of (4) involving a parameter; likewise for two equal sums of 4 or 5 biquadrates.

Werebrusow[225] noted that

$$(a + x)^4 + (b + x)^4 + (c - x)^4 = (a - x)^4 + (b - x)^4 + (c + x)^4$$

for

$$a = pv + (s + 3t)U, \qquad b = (3s^2t + 18st^2 + 18t^3)v + 3tU,$$
$$c = (p + 18t^3)v + (s + 3t)U, \qquad x = 3tV,$$

[215] L'intermédiaire des math., 19, 1912, 183–4.
[216] Ibid., 255.
[217] Ibid., 20, 1913, 105–6.
[218] Ibid., 58; error in fourth set, p. 301.
[219] Ibid., 245.
[220] Ibid., 21, 1914, 59–62.
[221] Sphinx-Oedipe, 8, 1913, 154–5.
[222] Amer. Math. Monthly, 20, 1913, 306–7.
[223] L'intermédiaire des math., 21, 1914, 153–5.
[224] Ibid., 155–6.
[225] Ibid., 23, 1916, 223. Math. Sbornik.

where
$$p = s^3 + 9s^2t + 18st^2, \quad s^3 + 12s^2t + 3bst^2 + 3bt^3 = P^2 + Q^2, \quad (P^2 + Q^2)v^2 = U^2 + V^2.$$

Relations involving both biquadrates and squares.

Diophantus, V, 32, treated $x^4 + y^4 + z^4 = v^2$ by setting $v = x^2 - k$. Then $x^2 = (k^2 - y^4 - z^4)/(2k)$. Take $k = y^2 + z^2$. Then $x^2 = y^2z^2/(y^2 + z^2)$. Hence $y^2 + z^2$ equals a square w^2. For $y = 3$, $z = 4$, we get $k = 25$, $x = 12/5$. Diophantus' method thus leads to the identity (cf. Fauquembergue[235])
$$(yz)^4 + (yw)^4 + (zw)^4 \equiv (w^4 - y^2z^2)^2, \qquad w^2 = y^2 + z^2.$$

Taking $y = ab$, $z = bc$, $w = ac$, we get [Norrie,[211] p. 91]
$$a^4 + b^4 + c^4 \equiv (a^2 - b^2 + c^2)^2, \qquad a^2b^2 + b^2c^2 = a^2c^2.$$

E. Waring[226] reproduced Diophantus' argument with k eliminated.

F. Proth[227] recalled that any prime N of the form $6x + 1$ is expressible in the form $N = a^2 + b^2 + ab$. Thus $2N = a^2 + b^2 + (a + b)^2$. By multiplication, $2N^2 = a^4 + b^4 + (a + b)^4$, whence
$$2(a^2 + ab + b^2)^2 \equiv a^4 + b^4 + (a + b)^4.$$

It is stated that if N is of the form $6x + 1$, whether prime or not, $2N^2$ is a sum of three biquadrates [incorrect, Kempner[42] of Ch.XXV, Diss., p. 44]. If N is expressible in two ways in the form $a^2 + b^2 + ab$, as
$$91 = 5^2 + 6^2 + 5 \cdot 6 = 1 + 9^2 + 1 \cdot 9,$$
we get a number expressible as a sum of three biquadrates in two ways:
$$2 \cdot 91^2 = 5^4 + 6^4 + 11^4 = 1^4 + 9^4 + 10^4.$$

S. Réalis[228] noted that $z_1^4 + z_2^4 + z_3^4 = 3z^2$ if
$$z_1 = 5s + 2\alpha\beta(2\alpha^2 + 5\beta^2) + 9\alpha^2\beta^2, \qquad z_2 = 5s + 2\alpha\beta(5\alpha^2 + 2\beta^2) + 9\alpha^2\beta^2,$$
$$z_3 = 5s + 16\alpha\beta(\alpha^2 + \beta^2) + 27\alpha^2\beta^2, \qquad z = t\{25t^3 + 72\alpha^2\beta^2(\alpha + \beta)^2\},$$
where $s = \alpha^4 + \beta^4$, $t = \alpha^2 + \alpha\beta + \beta^2$.

G. Dostor[229] gave the identity
$$(a + b + c - d)^4 + (a + b - c + d)^4 + (a - b + c + d)^4 + (-a + b + c + d)^4$$
$$= 4(a^2 + b^2 + c^2 + d^2)^2 + 16[(ab - cd)^2 + (ac - bd)^2 + (ad - bc)^2].$$

S. Réalis[230] noted that $v^4 + x^4 + y^4 = 2z^2$ is satisfied if
$$x = 2057\alpha^3 - 2541\alpha^2\beta + 2787\alpha\beta^2 - 391\beta^3,$$
$$y = 391\alpha^3 - 2787\alpha^2\beta + 2541\alpha\beta^2 - 2057\beta^3,$$
$$v = (2\alpha + 2\beta)(391\alpha^2 - 730\alpha\beta + 391\beta^2),$$
whence for $\alpha = 1$, $\beta = 0$ or 1,
$$46^4 + 121^4 + 23^4 = 2 \cdot 10467^2, \qquad 26^4 + 239^4 + 239^4 = 2 \cdot 57123^2.$$

[226] Meditationes Algebraicae, 1770, 194; ed. 3, 1782, 325.
[227] Nouv. Corresp. Math., 4, 1878, 179–181.
[228] Ibid., 350.
[229] Archiv Math. Phys., 60, 1877, 445.
[230] Nouv. Corresp. Math., 6, 1880, 238–9. Misquoted, C. A. Laisant, Algèbre, 1895, 221–2.

From a given solution is deduced a second by long formulas, whence

$$1^4+3^4+10^4=2\cdot71^2, \qquad 7^4+7^4+12^4=2\cdot113^2, \qquad 1^4+1^4+2^4=2\cdot3^2.$$

A. Martin[231] gave 9 biquadrates, $720^4, \cdots, 3120^4$, whose sum is a square.

Martin[232], assuming that the sum of the fourth powers of x, $x-ay$, $x-by$, $x-cy$, is a square, obtained $x/y=\alpha/\beta$, where α and β are polynomials in a, b, c, and took $x=\alpha$, $y=\beta$. By the same method, he[233] elsewhere found $199^4+271^4+343^4+559^4=344162^2$.

Martin and R. J. Adcock[234] repeated the solution by Diophantus and stated that Diophantus' result $12^4+15^4+20^4=481^2$ gives the least solution in integers.

E. Fauquembergue[235] noted that, if $\alpha^2+\beta^2=\gamma^2$,

$$(\alpha\beta)^4+(\beta\gamma)^4+(\gamma\alpha)^4=(\alpha^4+\alpha^2\beta^2+\beta^4)^2,$$

$$(2\alpha^2\beta\gamma^3)^4+(2\alpha\beta^2\gamma^3)^4+[(\alpha^2-\beta^2)\gamma^4]^4+[2\alpha\beta(\alpha^4+\gamma^4)]^4=[\gamma^{12}-4\alpha^2\beta^2(\alpha^4+\beta^4)^2]^2.$$

These two formulas were given also by A. Martin.[236] To find n biquadrates whose sum is a square, the latter took their roots to be x, $x-ay$, $x-by$, $x-cy$, p_1y, $\cdots$, $p_{n-4}y$. Then shall

$$4x^4-4(a+b+c)x^3y+6(\Sigma a^2)x^2y^2-4(\Sigma a^3)xy^3+(\Sigma a^4+\Sigma p_1^4)y^4=\square,$$

say the square of $2x^2-\Sigma a\cdot xy+\frac{1}{4}\{6\Sigma a^2-(\Sigma a)^2\}y^2$. Thus x/y is determined.

E. B. Escott[237] noted that

$$(m^2+mn+n^2)^4-(mn)^4-(mn+n^2)^4=[m(m+n)(m^2+mn+2n^2)]^2.$$

E. Fauquembergue[238] gave identities including

$$(a^4+2b^4)^4=(a^4-2b^4)^4+(2a^3b)^4+(8a^2b^6)^2$$
$$=(2a^2b^2)^4+(2a^3b)^4+(a^8-4a^4b^4-4b^8)^2.$$

C. B. Haldeman[239] found four biquadrates whose sum is a square:

$$(2a)^4+(a+b)^4+(a-b)^4+d^4=2(3a^2+b^2)^2+d^4=s^2.$$

Take $s=d^2+v$, $3a^2+b^2=vg$. Then v, b^2, s are determined rationally in terms of d, g, a. Take $g=2$, $a=3/7$. Then $b^2=4d^2/7-27/49$. Since b is rational for $d=1$, take $d=y+1$ and equate b to $ry/t+1/7$, thus determining y. Then

$$b=-(7r^2-56rt+4t^2)/(7k), \qquad d=(7r^2-2rt+4t^2)/k, \qquad k=7r^2-4t^2.$$

For $r=1$, $t=0$, we get $2^4+4^4+6^4+7^4=63^2$. Next, let the sum of the initial biquadrates equal $2s^2$. The condition is evidently satisfied if

$$s=\frac{d^4+2v^2}{4v}, \qquad 3a^2+b^2=\frac{d^4-2v^2}{4v}.$$

[231] Annals of Math., 5, 1889–90, 112–3.
[232] *Ibid.*, 6, 1891–2, 73.
[233] Amer. Math. Monthly, 1, 1894, 401–2.
[234] *Ibid.*, 279–80.
[235] L'intermédiaire des math., 1, 1894, 167 [6, 1899, 186].
[236] Math. Magazine, 2, 1898, 210–1.
[237] L'intermédiaire des math., 6, 1899, 51.
[238] *Ibid.*, 7, 1900, 412.
[239] Math. Magazine, 2, 1904, 285–6.

Take $d^2=2v$, $3a^2+b^2=(t+b)^2$. Thus b, d, v are found rationally in terms of a, t, whence

(1) $\quad (4at)^4+(3a^2+2at-t^2)^4+(3a^2-2at-t^2)^4+(6a^2+2t^2)^4=2\{3(3a^2+t^2)^2\}^2$.

For $a=1$, $t=2$, we get $3^4+5^4+8^4+14^4=2\cdot147^2$.

A. Cunningham[240], to solve $x^4+y^4+z^4=2u^{2n}$, took as u any number of the form $\alpha^2+3\beta^2$, whence u^{2n} is of the form A^2+3B^2 and a solution is $x=B-A$, $y=B+A$, $z=2B$.

A. Gérardin[241] noted that $(1+mx)^4+(my)^4+(mz)^4=(1+2mx)^2$ if
$$m^2(x^4+y^4+z^4)+4mx^3+2x^2=0.$$

Its discriminant must be a square, say $(2Sx)^2$, whence $x^4-y^4-z^4=2S^2$. Set $S=zU$, $y^2+kz^2=x^2$. Then $ky^2+\frac{1}{2}(k^2-1)z^2=U^2$. Hence the problem reduces to a " double equation," that of making the two binary quadratics squares.

E. N. Barisien[242] noted the identity
$$(2x^2+a^2)^4+(2x^2-a^2)^4+(4ax)^4=(4x^4+12a^2x^2+a^4)^2+(4x^4-12a^2x^2+a^4)^2.$$

Mehmed-Nadir[243] gave two special sets of solutions of
$$\tfrac{1}{2}(x^4+y^4+z^4)=u^2+v^2+w^2=\rho^2.$$

A. Cunningham and E. Miot[244] obtained solutions by use of the identity
$$x^4+y^4+(x+y)^4=2(x^2+xy+y^2)^2.$$

A. Gérardin[245] solved $X^4+Y^4+Z^4=A^2+B^2$ by use of the identity
$$(pa+qb)^2+(qa)^2+(2pb)^2=(pa-qb)^2+(qa+2bp)^2,$$

setting $q=af^2$, $p=2bg^2$. It remains to solve $ab(f^2+2g^2)=X^2$. For $a=b=1$, we may take $f=m^2-2n^2$, $X=m^2+2n^2$, $g=2mn$. He noted (ibid., p. 90) that
$$(\alpha^2+\beta^2)^4-(\alpha^2-\beta^2)^4-(2\alpha\beta)^4\equiv2\{2\alpha\beta(\alpha^2-\beta^2)\}^2.$$

R. Norrie,[199] pp. 90–92, would derive a second solution of
$$X_1^4+\cdots+X_n^4=X^2$$

from one solution $a_1^4+\cdots+a_n^4=a^2$ by setting $X_i=rx_i+a_i$, $X=r^2y+rx+a$, and making the coefficients of r and r^2 zero by choice of y, x. To obtain an explicit solution when $n>4$, take $t=x^2+xy+y^2$ in $(t^2+z^4)^2\equiv t^4+(z^2)^4+2t^2z^4$, whence $2t^2=x^4+y^4+(x+y)^4$. But x^4+y^4 can be expressed as a sum of r biquadrates P_i if $r>2$ [Norrie,[199] end]. Hence
$$\{(x^2+xy+y^2)^2+z^4\}^2=(x^2+xy+y^2)^4+(z^2)^4+\{z(x+y)\}^4+\sum_{i=1}^{r}(zP_i)^4.$$

E. N. Barisien[245a] wrote Proth's[227] identity in the form
$$a^4+b^4+(a+b)^4\equiv(a^2+ab+b^2)^2+a^2b^2+a^2(a+b)^2+b^2(a+b)^2.$$

[240] Messenger Math., 38, 1908–9, 101, 103.
[241] Bull. Soc. Philomathique, (10), 3, 1911, 239–240.
[242] Nouv. Ann. Math., (4), 11, 1911, 280–2.
[243] L'intermédiaire des math., 18, 1911, 217.
[244] Ibid., 19, 1912, 70–71.
[245] Sphinx-Oedipe, 6, 1911, 21–22.
[245a] Mathesis, (4), 4, 1914, 13.

R. D. Carmichael[246] showed that one solution of $x^4+ay^4+bz^4=\square$ leads to a second.

E. N. Barisien[247] noted that $N=(a^2+b^2)(c^2+d^2)(a^2c^2+b^2d^2)$ equals

$$\{ab(c^2\mp d^2)\}^2+\{cd(a^2\pm b^2)\}^2+(a^2c^2+b^2d^2)^2.$$

Let N' be derived from N by interchanging c and d. Then NN' is a sum of nine squares in four ways, in two of which two of the nine squares are biquadrates.

See papers 178, 188, 206–7, 219–20, 287–8, 292; also Gérardin, p. 38; Lucas[88a] of Ch. XXIII.

<h2 style="text-align:center">Miscellaneous Single Equations of Degree Four.</h2>

C. Wolf[248] treated $x^2y^2+x^2+y^2=\square$. First, make $x^2y^2+x^2=\square$, i.e., $y^2+1=v^2=(t-y)^2$, whence $y=(t^2-1)/2t$. Since $x^2y^2+x^2=x^2v^2$, it remains to make $x^2v^2+y^2=\square$, say $(z-vx)^2$; we thus obtain x.

L. Euler[249] made $P=(p^2-q^2)(q^2-r^2)$ a biquadrate by setting $p=a+b+2c$, $q=a+b$, $r=a-b$, whence $P=16abc(a+b+c)$. Consider therefore

$$xyz(x+y+z)=s^4.$$

Take $s^4=(x+y+z)^2p^2$. Thus

$$Dz=(x+y)p^2, \quad D(x+y+z)=xy(x+y), \quad Ds^2=xyp(x+y), \quad D=xy-p^2.$$

Set $x=nq^2$, $y=nr^2$, $nqr-p=k(q^2+r^2)$ and eliminate p. Thus

$$\frac{s^2}{n^2q^2r^2}=\frac{n\{-nqr+k(q^2+r^2)\}}{k\{-2nqr+k(q^2+r^2)\}}\equiv F.$$

For $n=2k$, $F=2(q-r)^2/(q^2+r^2-4qr)$. As Euler omitted the factor 2, it is not sufficient to make the denominator a square. Next, let $n=k$. Then $F=(q^2+r^2-qr)/(q-r)^2$. Equate the numerator to the square of $q+rf/g$. Thus $q:r=g^2-f^2:g^2+2fg$. Or we may begin by taking $p=2xy/(x+y)$, whence $s^2=2xy(x+y)^2/(x-y)^2$; take $x=2q^2$, $q=r^2$ to make $2xy=\square$.

Euler[250] treated $(p^2+1)^2+(q^2+1)^2=\square$ by setting

$$p^2+1=x^2-y^2, \quad q^2+1=2xy, \quad p=x-z.$$

Thus $2zx=z^2+y^2+1$. Take $y=2z$. Then $q^2=10z^2+1$, which is satisfied by $(z,q)=(2/3, 7/3), (2/9, 11/9), (6, 19)$.

Euler[251] treated $Ll=\square$, where $L=A+Bz+Cz^2$, $l=a+bz+cz^2$. Take $Ll=p^2l^2$. Then $L=p^2l$, which can be solved if one solution is known.

J. L. Lagrange[252] treated the more general problem to solve

$$F(x, y)\equiv f(x)+s(x)y+cy^2=0,$$

where f is of the fourth degree and s of the second. If $F(p, q)=0$, set

[246] Diophantine Analysis, 1915, 44.
[247] Nouv. Ann. Math., (4), 16, 1916, 390–1.
[248] Elementa Matheseos Universae, Halae, 1, 1742, 380.
[249] Opera postuma, 1, 1862, 239 (about 1769). Extract in Bull. Soc. Philomathique, (10), 3, 1911, 240–3. Cf. Euler,[187] Gérardin,[266] Kommerell.[270]
[250] Opera postuma, 1, 1862, 215–6 (about 1774).
[251] Ibid., 218–9 (about 1777).
[252] Nouv. Mém. Acad. Sc. Berlin, année 1777, 1779; Oeuvres, IV, 397.

$x=p+t$, $y=q+tz$.   After dividing by t, we obtain $B+Cz+tQ=0$, where Q is quadratic in t and z, while B and C are constants.   From the solution $t=0$, $z=-B/C$ of this cubic, we obtain a second by the tangent method.

Euler[253] treated as two separate problems the solution of

$$V_\pm \equiv x^4+y^4+z^4+v^4-2x^2y^2-2x^2z^2-2y^2z^2\pm2(x^2v^2+y^2v^2+z^2v^2)=0.$$

Then

$$x^2y^2\mp z^2v^2=\tfrac{1}{4}(x^2+y^2-z^2\pm v^2)^2, \qquad x^2z^2\mp y^2v^2=\tfrac{1}{4}(x^2+z^2-y^2\pm v^2)^2.$$

The left members will be squares if

$$(1) \qquad \frac{xy}{zv}=\frac{p^2\pm r^2}{2pr}, \qquad \frac{xz}{yv}=\frac{q^2\pm s^2}{2qs},$$

whence

$$(2) \qquad \frac{x^2+y^2-z^2\pm v^2}{2}=\frac{zv(p^2\mp r^2)}{2pr}, \qquad \frac{x^2+z^2-y^2\pm v^2}{2}=\frac{yv(q^2\mp s^2)}{2qs}.$$

From (1) we obtain x^2/v^2 and y^2/z^2 by multiplication and division.   Hence we have values a, b, c, d for which $x=at$, $v=bt$, $y=cu$, $z=du$.   Then (2) give

$$(a^2\pm b^2)t^2+(c^2-d^2)u^2=2mbdtu, \qquad (a^2\pm b^2)t^2-(c^2-d^2)u^2=2nbctu,$$
$$m=\frac{p^2\mp r^2}{2pr}, \qquad n=\frac{q^2\mp s^2}{2qs}.$$

By subtraction, we get t/u.   Hence we take

$$t=c^2-d^2, \qquad u=b(md-nc)$$

and obtain x, y, v, z.   Changing the sign of n, we obtain a second set of solutions.   Rational solutions result only when the product of the right members of (1) is a rational square.   For the upper signs, take $p=2fg$, $r=f^2-g^2$, $q=2hk$, $s=h^2-k^2$.   Then the condition is

$$fg(f^2-g^2)\cdot hk(h^2-k^2)=\square.$$

It is the square of $3mnfg(f-g)$ for

$$h=g, \qquad k=f-g, \qquad f=2m^2-n^2, \qquad g=m^2+n^2.$$

See Euler[81] of Ch. XVI.

Euler[254] used the preceding $V_+\equiv F$ to find x^2, $\cdots$, v^2 such that

$$\alpha\equiv x^2y^2-z^2v^2, \qquad \beta\equiv z^2x^2-y^2v^2, \qquad \gamma\equiv y^2z^2-x^2v^2$$

shall be squares.   We have

$$F+4\alpha=(x^2+y^2-z^2+v^2)^2, \qquad F+4\beta=(x^2+z^2-y^2+v^2)^2,$$
$$F+4\gamma=(y^2+z^2-x^2+v^2)^2.$$

Hence we seek solutions of $F=0$.   Solving the latter for x^2 we get

$$x^2=y^2+z^2-v^2+2T, \qquad T^2=y^2(z^2-v^2)-z^2v^2.$$

Now $z^2-v^2=\square$ for $z=5$, $v=3$, whence $T^2=16y^2-225=(4y-t)^2$ if

$$y=(225+t^2)/(8t).$$

---

[253] Acta Acad. Petrop., 2, II, 1781 (1778), 85; Comm. Arith., II, 366; Op. Om., (1), III, 429.
[254] Opera postuma, 1, 1862, 257–8 (about 1782).   For sums, instead of differences, see Euler[81] of Ch. XVI.

Taking $t=5$, we get $y=25/4$, $T=20$, $x=39/4$. Multiplying the unknowns by 4, we get the solution $x=39$, $y=25$, $z=20$, $v=12$. Or we may solve $F=0$ for v^2 and get $v^2=2S-x^2-y^2-z^2$, $S^2=x^2y^2+x^2z^2+y^2z^2$. Set $S=xy+tz$. Then

$$z=2txy/k, \qquad S=xy(x^2+y^2+t^2)/k, \qquad k=x^2+y^2-t^2.$$

Then v^2 is a complicated function of degree 6 and was not treated. A solution is said to result from $t=185/153$. For $t=13/3$, $x=5$, $y=4$, we get the above solution $x=39$, etc.

C. F. Kausler[255] treated the problem to find all rational numbers x, y for which $N\equiv(x^2-1)(y^2-1)$ is an integer. Set $y=p/q$, where p and q are relatively prime integers. The numerator and denominator of the resulting fraction for x^2 are $(N-1)q^2+p^2=mP^2$ and $p^2-q^2=mQ^2$. For $m=1$, the latter gives $p=(A^2+B^2)/d$, $q=(A^2-B^2)/d$, where A, B are relatively prime, one even or both odd according as $d=1$ or 2. The first condition then gives N which is an integer for $d=1$ if $P\pm2AB$ is divisible by $(A^2-B^2)^2$. For $m>1$, $p+q=mQ$, m or Q^2, the last two yielding (as far as numbers <100) only the same values of N as above. For $p+q=mQ$, then $p-q=Q$ and, dropping the common factor $Q/2$ in p, q, we have $p=m+1$, $q=m-1$, m even, $N=m(P^2-4)/(m-1)^2$. Then $P\mp2=R(m-1)^2$, whence

$$N=mR[(m-1)^2R\pm4].$$

G. Eisenstein[256] considered a binary cubic whose coefficients are variables. Its discriminant D is a quartic in these four variables. Given one solution of $D=$ constant, we can find an infinitude of solutions by means of the formulas for the coefficients of the cubic obtained by a linear transformation of determinant unity.

V. A. Lebesgue[257] noted that

$$a^2t^4+b^2u^4+c^2v^4-2bcu^2v^2-2acv^2t^2-2abt^2u^2=s^2$$

is satisfied identically by

$$t=x(by^2-cz^2), \qquad u=y(cz^2-ax^2), \qquad v=z(ax^2-by^2),$$

with s the product of the binomials, and by

$$t=x(cy^2-bz^2), \qquad u=y(az^2-cx^2), \qquad v=z(bx^2-ay^2).$$

Several[258] found two numbers whose sum equals the difference of their fourth powers. Let the numbers be $(n\pm1)x$. Then $x=(4n^2+4)^{-1/3}$ is rational if $n=\pm1$. Hence set $n=m+1$. Then $x=N^{-1/3}$, $N=(pm+2)^3$ if $p=2/3$, $m=9/2$.

E. Lucas[259] stated that the difference of two consecutive cubes is never a biquadrate. Moret-Blanc[75] noted that $3x^2+3x+1\neq z^4$ since $4z^4-1\neq3t^2$.

D. S. Hart[260] found rational numbers a, b, x for which

$$4x^4+4ax^3+4bx+ab=0.$$

[255] Nova Acta Acad. Petrop., 15, ad annos 1799–1802, 1806, 116–45.
[256] Jour. für Math., 27, 1844, 76.
[257] Comptes Rendus Paris, 59, 1864, 1069.
[258] Math. Quest. Educ. Times, 2, 1865, 77; cf. (2), 4, 1903, 68–9.
[259] Recherches sur l'analyse indéterminée[73], 1873, 92; extract in Mathesis, 8, 1888, 21.
[260] Math. Quest. Educ. Times, 24, 1876, 35–36.

Take $(2x^2+ax)^2=(ax-b)^2$. We get x rationally and a condition on a, b, which is solved for a. Take $b=-m^2/2$, whence a follows rationally.

A. Desboves[261] gave identities yielding an infinitude of solutions of $ax^3+by^3=cv^4$ for certain values of c. He[262] noted that $aX^4+bY^4\equiv cZ^3$ for
$$X=x(3ax^4-5by^4), \qquad Y=y(5ax^4-3by^4), \qquad Z=ax^4+by^4,$$
$$c=81a^2x^8-158abx^4y^4+81b^2y^8,$$
and gave long formulas yielding solutions of $aX^4+bY^4=cZ^4$ when c is represented by a certain form of degree 20. Further, $X^4-Y^4=cZ^4$ is solvable when c is of one of the forms
$$xy(x^2+4y^2), \qquad x^8+4y^8, \qquad 2xy(x^2-y^2)(x^4+y^4-6x^2y^2).$$

S. Réalis[263] gave various quartic equations not having a rational root, as
$$x^4-2\alpha^2x^2+4\alpha\beta x+\alpha^4+\beta^2=0, \qquad \beta\neq0, \qquad \beta\neq\pm4\alpha^2;$$
$$(x^2+2\alpha x+2\beta^2)^2+2\beta^2x^2=5(\alpha x^3+\beta^2x^2+2\alpha\beta^2x-\beta\gamma^3), \qquad \beta\not\equiv0 \pmod 5.$$
Several[264] solved $x^3+y^3=(x-y)^4$. Set $x+y=u$, $x-y=z$. Then
$$4z^4-3uz^2=u^3, \qquad 8z^2=u(3+r), \qquad r^2=16u+9.$$
Set $r^2=(8t\pm3)^2$. Hence there are two types of solutions.

R. W. D. Christie[265] made $12abc(a+b+c)$ a square, but not a biquadrate as claimed. A. Gérardin[266] noted that it is a biquadrate for $(a, b, c)=(1, 2, 6)$, $(3, 4, 9)$, etc.

E. Grigorief[267] noted that $11^4=12^3+17^3+20^3$. P. F. Teilhet[268] gave cases of $x^4=y_1^3+y_2^3+y_3^3$ for $x=3, 10, 17, 20, 29, 36, 43, 55, 62$. He[269] noted that
$$7^4=12^3+12^2+23^2=8^3+40^2+17^2=5^3+40^2+26^2,$$
$$8^4=14^3+34^2+14^2=12^3+48^2+8^2=\text{three such sums}.$$

K. Kommerell[270] gave as the positive integral solutions of
$$xyz(x+y-z)=t^2,$$
$$x=\tfrac{1}{2}T-\tfrac{1}{2}a(d^2y_1-e^2z_1), \qquad y=ad^2y_1, \qquad z=ae^2z_1, \qquad t=adey_1z_1U,$$
where y_1, z_1 are without square factors, d^2y_1 is relatively prime to e^2z_1, and $T^2-4y_1z_1U^2=a^2(d^2y_1-e^2z_1)^2$.

A. Hurwitz[271] proved that $x^3y+y^3z+z^3x=0$ is impossible since
$$u^7+v^7+w^7=0$$
is impossible.

<hr>

[261] Nouv. Ann. Math., (2), 18, 1879, 408.

[262] *Ibid.*, 440-4.

[263] *Ibid.*, (3), 2, 1883, 370; 4, 1885, 376, 427-31; Mathesis, 7, 1887, 96; Jour. de math. spéc., 1888, 90 (and questions 66, 67). Reprinted, C. A. Laisant's Algèbre, 1895, 224-6.

[264] Zeitschr. Math. Naturw. Unterricht, 20, 1889, 264-5.

[265] Educ. Times, 49, 1896.

[266] Bull. Soc. Philomathique, (10), 3, 1911, 244.

[267] L'intermédiaire des math., 9, 1902, 319.

[268] *Ibid.*, 10, 1903, 170-1.

[269] *Ibid.*, 11, 1904, 18.

[270] Math. Naturw. Mitteilungen, Stuttgart, (2), 7, 1905, 74-8. Cf. Brehm,[285] Euler[249]; also papers 12, 22 of Ch. V.

[271] Math. Annalen, 65, 1908, 428-30. Generalization, Hurwitz,[312] Ch. XXVI.

F. L. Griffin and G. B. M. Zerr[272] made a sum of n squares a biquadrate.

A. Gérardin[273] noted that s_4 is divisible by s_3, where

$$s_n = (9f^4)^n + (9f^3+1)^n - (9f^4+3f)^n.$$

For $f=1$, the quotient is -4175. E. Fauquembergue noted also that

$$5^4+3^4-6^4=59(5^3+3^3-6^3), \qquad 5^4+6^4-7^4=240(5^3+6^3-7^3).$$

A. Cunningham[274] expressed numbers in the form $(x^6+y^6)/(x^2+y^2)$ in several ways.

A. Cunningham[275] found certain types of solutions of

$$f(x, y)+f(x', y')=s(\xi, \eta)+s(\xi', \eta'), \qquad f(x, y) \equiv \frac{x^5 \mp y^5}{x \mp y}, \qquad s(\xi, \eta) = \frac{\xi^6+\eta^6}{\xi^2+\eta^2};$$

and (pp. 111–2) of $s(x, y)=s(x, z)$, $6xy=\square$, $y \neq z$. He[276] gave various criteria for the solvability of $\pm N = x_1^4 - 2y_1^2 = x_2^2 - 2y_2^4$, $N \equiv \pm 1 \pmod 8$. He discussed (p. 108) $q_1 q_2 q_3 = \square$, where $q_r = x_r^4 + y_r^4$. He[277] proved the existence of an infinitude of integral solutions of $F(x, y) = F(x', y')$ for each a/k, where

$$F(x, y) = ax^4 + 4ax^3y + kx^2y^2 + 4axy^3 + ay^4.$$

If $(k+10a)/(k-6a)$ is a rational square, $F(x, y)$ is a product of two factors. If (pp. 94–95) either of $(2a \mp 2b+c)(12a-2c)$ is of the form $-\alpha^2-\beta^2$,

$$\phi(x, y) \equiv ax^4 + bx^3y + cx^2y^2 + bxy^3 + ay^4 = \phi(x', y')$$

is usually solvable in integers. Certain numbers (pp. 39–40) can be expressed simultaneously in the forms

$$N_1 = \frac{x_1^3-y_1^3}{x_1-y_1}, \qquad N_2 = \frac{z_2^3+x_2^3}{z_2+x_2}, \qquad N_3 = \frac{z_3^3+y_3^3}{z_3+y_3}, \qquad N_4 = \frac{x_4^4+y_4^4+z_4^4}{x_4^2+y_4^2+z_4^2},$$

and $N_1'/3$, $N_2'/3$, $N_3'/3$, $N_4'/3$, where $N_1' = (x_1'^3 - y_1'^3)/(x_1'-y_1')$, etc. He[278] considered numbers expressible in two or four of the forms $\pm(x^4-2y^2)$, $\pm(x^2-2y^4)$. He[279] showed that certain binary quartic functions of four pairs of variables are equal for an infinite of set of values, by use of the above[275] $s(\xi, \eta)$.

He[280] solved $N_1+N_2=N_3+N_4$, where $N_r = (x_r^5-y_r^5)/(x_r-y_r)$.

He[281] gave a method to solve $x^3y-xy^3=a$.

H. B. Mathieu[282] noted that each triangular number which is a square yields a solution of $x^3+y^2=z^4$. Thus, $\Delta_{49} = 35^2$ gives

$$\Delta_{49}^2 - \Delta_{48}^2 = 49^3, \qquad 49^3+1176^2=35^4.$$

[272] Amer. Math. Monthly, 17, 1910, 147–8.
[273] Sphinx-Oedipe, 1906–7, 159–160.
[274] Mess. Math., 39, 1909–10, 97–128; 40, 1910–11, 1–36.
[275] Math. Quest. Educ. Times, (2), 16, 1909, 75.
[276] Ibid., (2), 17, 1910, 66–7.
[277] Ibid., (2), 19, 1911, 27–28.
[278] Ibid., (2), 22, 1912, 40–41, 107–9; 23, 1913, 62–6.
[279] Ibid., (2), 21, 1912, 89–90, 103–4.
[280] Ibid., (2), 26, 1914, 60.
[281] Ibid., (2), 27, 1915, 74–5.
[282] L'intermédiaire des math., 19, 1912, 129.

L. Aubry and H. Brocard[283] solved $2x^2y^2+1=x^2+y^2+z^2$ for $y=4$. Aubry[284] gave a solution involving three parameters of

$$y_2^2y_3^2+y_3^2y_1^2+y_1^2y_2^2-y_1y_2y_3y_4=0.$$

Brehm[285] solved $xyz(x+y-z)=t^2$ in integers.   Set $tq=xyp$, where p and q are relatively prime integers.   Then the equation gives $s(x+y-z)=rp^2x$, $ys=rq^2z$, where r and s are relatively prime integers.   Hence x, y, t are expressed in terms of t.

E. Swift[286] proved that $x^4-y^4=z^3$ is impossible for x prime to y.

R. D. Carmichael[287] noted that if x_0, y_0, u_0, v_0 give a solution of

$$x^4+ay^4=u^2+bv^2,$$

we can deduce a second solution [after performing the operations]:

$$x=x_0^4-ay_0^4+bv_0^2, \qquad y=2x_0y_0u_0, \qquad u=u_0^4+4x_0^4(ay_0^4-bv_0^2), \qquad v=4x_0^2u_0^2v_0.$$

F. L. Carmichael[288] obtained the solution

$$x=u_2^2+bv_2^2-ab^2v_2^2-ab^3v_2^2, \qquad y=2bv_2\{m^2+2mn-(b+ab^2+ab^3)n^2\},$$
$$u=u_1^2-bv_1^2+ay_1^2-abw_1^2, \qquad v=2u_1v_1+2ay_1w_1,$$

where

$$u_1=u_2^2-bv_2^2-ab^2v_2^2-ab^3v_2^2, \qquad v_1=2u_2v_2, \qquad y_1=2b^2v_2^2, \qquad w_1=2bv_2^2,$$
$$u_2=m^2+(b+ab^2+ab^3)n^2, \qquad v_2=2mn+2n^2;$$

also two simpler solutions, as well as solutions when $a/b=\square$, $a=0$ or $b=0$.

L. Bastien and L. Aubry[289] found the general solution of

$$x^2=(y^2-w)(z^2+w).$$

Several[290] treated $x^4-y^4=a^3+b^3$.

A. Gérardin[291] discussed $x^2+y^2+z^2=kxyz^2$.

A. Cunningham[292] treated $a^2+b^2=c^4+2d^2$, for b and c given.

L. Aubry[292a] solved $(x^2-y^2)(x^2+2y^2)=x^2-2y^2$.

Gérardin[157] of Ch. IV solved $x^4+6x^2y^2+y^4=\alpha^4+6\alpha^2\beta^2+\beta^4$.   On equations quadratic in x and in y, see note 145.   On $pq(mp^2+nq^2)=rs(mr^2+ns^2)$, see papers 168, 170, 174, 181.

To find n numbers whose sum is a square and sum of squares is a biquadrate.

For the case $n=2$, see papers 37–63.

G. W. Leibniz[293] considered the case $n=3$.

[283] L'intermédiaire des math., 19, 1912, 157–9, 3 (for special solutions).

[284] Ibid., 20, 1913, 95.

[285] Math. Naturw. Mitt., (2), 15, 1913, 20–21.   Cf. Kommerell.[270]

[286] Amer. Math. Monthly, 22, 1915, 70–1.

[287] Diophantine Analysis, 1915, 46–8.

[288] Amer. Math. Monthly, 23, 1916, 321–9.

[289] L'intermédiaire des math., 23, 1916, 36–8.

[290] Ibid., 123–4; 24, 1917, 66, 88, 133–4.

[291] Ibid., 24, 1917, 32.

[292] Ibid., 143–4.

[292a] Ibid., 23, 1919, 150–2.

[293] MS. in Bibliothek Hannover, about 1676.   Cf. D. Mahnke, Bibliotheca Math., (3), 13, 1912–3, 39.   J. Wallis, Opera Math., 3, 1699, 618, quoted a letter from Leibniz to Oldenburg, Oct. 26, 1674, in which this problem is mentioned (Bull. Bibl. Storia Sc. Mat. e Fis., 12, 1879, 519).

L. Euler[294] required four positive integers whose sum and sum of squares are biquadrates. He took them to be $x=a^2+b^2+c^2-d^2$, $y=2ad$, $z=2bd$, $v=2cd$. Then $\Sigma x^2 = (\Sigma a^2)^2$. Set $a=p^2+q^2+r^2-s^2$, $b=2ps$, $c=2qs$, $d=2rs$. Then $\Sigma a^2 = (\Sigma p^2)^2$. It remains to make $\Sigma x = \square^2$. Take $p=s-q+\frac{3}{2}r$. Then

$$\sqrt{\Sigma x} = 2q^2-3qr-2qs+\tfrac{13}{4}r^2+5rs+2s^2,$$

which for $q=r+t$ will be the square of $3r/2-u$ if

$$(t+3s+3u)r = u^2-2t^2+2ts-2s^2.$$

For $q=r=2$, $s=9$, $p=10$, we get $x=409$, $y=24$, $z=160$, $v=32$, $\Sigma x=5^4$, $\Sigma x^2=21^4$. Euler gave a similar treatment of the problem in five integers.

Euler[55] (first paper of 1780) treated the problem for $n=3, 4, 5$ and obtained the sets 8, 49, 64; 320, 400, 961; 16, 48, 104, 193; 32, 32, 88, 137; 16, 16, 32, 72, 89; 64, 152, 409; 17424, 108864, 580993, the last two sets having also the sum a biquadrate.

J. Cunliffe[294a] took x^2, $2xy$, $2y^2$ as the $n=3$ numbers, the sum of their squares being $(x^2+2y^2)^2$. Their sum is the square of $ry-x$ if $y=2r+2$, $x=r^2-2$. For $r=v-3$, $x^2+2y^2=(v^2-6v-9)^2$ if $v=28/5$.

Walmond and Mason[295] wrote x^4 for the biquadrate. Take $r=\sqrt{4x-5}$, $2x-1$ and x^2-2 as the $n=3$ numbers, their sum being $(x+1)^2$ if $r-3=1$, whence $x=21/4$. For $n=4$, take $r=\sqrt{6x-6}$, $x-2$, $x-1$, x^2-1, whose sum $=(x+1)^2$ if $r-4=1$, $x=31/6$. For $n=5$, take $r=\sqrt{4x-12}$, $x+1$, $x-1$, $2x-1$, x^2-3, whose sum $=(x+2)^2$ if $r-4=4$, $x=19$.

S. Bills[296] employed the identity of Aida[59] of Ch. IX:

$$u_1^2+\cdots+u_n^2 = (v_1^2+\cdots+v_n^2)^2, \qquad u_1=v_1^2+\cdots+v_{n-1}^2-v_n^2,$$
$$u_i=2v_n v_{i-1} \quad (i=2,\cdots,n),$$
$$v_1^2+\cdots+v_n^2 = (x_1^2+\cdots+x_n^2)^2, \qquad v_1=x_1^2+\cdots+x_{n-1}^2-x_n^2,$$
$$v_i=2x_n x_{i-1} \quad (i=2,\cdots,n).$$

The remaining condition $u_1+\cdots+u_n=\square$ becomes a quartic in x_n which is equated to the square of $x_n^2+2x_{n-1}x_n+x_1^2+\cdots+x_{n-1}^2$. Hence

$$x_n = r-\tfrac{3}{2}x_{n-1},$$

where $r=x_1+\cdots+x_{n-2}$.

A. B. Evans[297] used the numbers x, $a_1 y$, $\cdots$, $a_{n-1}y$ and wrote

$$m=a_2+\cdots+a_{n-1}, \qquad v=a_2^2+\cdots+a_{n-1}^2.$$

Take $x=a^2-py$. Then $x^2+(a_1^2+v)y^2=a^4$ determines y rationally. Hence $a^{-2}(a_1^2+v+p^2)^2[x+(a_1+m)y]=(a_1^2+pa_1+b)^2$, $b=pm+v-\frac{1}{2}p^2$, determines a_1 rationally.

D. S. Hart[298] used the numbers px^2-ax, px^2+ax, $\cdots$, Nx^2-Zx, Nx^2+Zx and, if n is odd, Sx^2. Equating the sum of their squares to $(xm/n)^4$, we get x^2.

---

[294] Opera postuma, 1, 1862, 255 (about 1782).
[294a] New Series of Math. Repository (ed., T. Leybourn, 3, 1814, I, 79–80.
[295] Ladies' Diary, 1827, 36–7, Quest. 1452. Reference was made to *Férussac, Bull. des Sc. Math., III, 276.
[296] Math. Quest. Educ. Times, 18, 1873, 104–5.
[297] *Ibid.*, 22, 1875, 69–71.
[298] *Ibid.*, 24, 1876, 55–57.

Examples for $n=4, \cdots, n=7$ are deduced. To proceed otherwise when $n=3$, employ the numbers $2mp$, $2rp$, $m^2+r^2-p^2$. Their sum is the square of $m-r+p$ if $m=(p^2-2rp)/r$. Then the sum of their squares equals $(m^2+r^2+p^2)^2$ and is a biquadrate if

$$r^2(m^2+r^2+p^2)=p^4+\cdots=(p^2-2rp+r^2)^2,$$

whence $p=4r$, $m=8r$, and the desired numbers are $64r^2$, $8r^2$, $49r^2$. A. Martin employed $2a_i s(i=1, \cdots, n-1)$, $a_1^2+\cdots+a_{n-1}^2-s^2$ as the n numbers and wrote $m=a_2+\cdots+a_{n-1}$. Then shall

$$2a_1 s+2ms+a_1^2+\cdots+a_{n-1}^2-s^2=A^2, \qquad a_1^2+\cdots+a_{n-1}^2+s^2=B^2.$$

Take $s=A-B$, $2a_1+2m-2s=A+B$. Then either of the preceding equations gives a_1.

R. Goormaghtigh[299] discussed $(x+y+z)^2=x^2+y^2+z^2=M^4$.

MISCELLANEOUS SYSTEMS OF EQUATIONS OF DEGREE FOUR.

Diophantus, V, 5, found three squares such that the product of any two added either to the sum of the same two or to the remaining one gives a square (cf. Fermat[100] of Ch. XIX).

J. Prestet[300] found three squares such that the product of any two added to the product of a given square a^2 by either the sum of those two or the remaining one gives a square. For $a=3$, he found 25, 64, 196.

Beha-Eddin[301] (1547–1622) included, among seven problems remaining unsolved from former times,

Prob. 1: $x-y=10$, $(x+x^{1/2})(y+y^{1/2})=$ given;

Prob. 5: $x+y=10$, $\dfrac{x}{y}+\dfrac{y}{x}=x$.

Fermat[302] noted that x^4-y^4 is a cube and $x-y=1$ if $x=13/22$, $y=-9/22$, while positive solutions can be found by setting $x=z+13/22$, $y=z-9/22$.

L. Euler[303] required three numbers x, y, z such that $k\equiv x^2y^2+x^2+y^2$, $x^2z^2+x^2+z^2$, $y^2z^2+y^2+z^2$, $x^2y^2+z^2$, $x^2z^2+y^2$, $y^2z^2+x^2$, $s\equiv x^2y^2+x^2z^2+y^2z^2$, $s+x^2+y^2+z^2$ shall be all squares. He took $z^2=x^2+y^2+1+2\sqrt{k}$. For $y=x+1$ we have $k=w^2$, $z^2=4w$, where $w=x^2+x+1$. Now $w=\square=(t-x)^2$ for $x=(t^2-1)/(2t+1)$. Then the solutions are

$$x=\frac{t^2-1}{2t+1}, \qquad y=\frac{t^2+2t}{2t+1}, \qquad z=\frac{2t^2+2t+2}{2t+1}.$$

Euler[304] treated the three problems to make (i) AB and AC squares; (ii) BC a square; (iii) B and C squares, where

$$A=x^2+y^2, \qquad B=t^2x^2+u^2y^2, \qquad C=u^2x^2+t^2y^2.$$

[299] L'intermédiaire des math., 25, 1918, 17-18.

[300] Elemens des Math., Paris, 1675, 331.

[301] Essenz der Rechenkunst von Mohammed Beha-eddin ben Alhossain aus Amul, arabisch u. deutsch von G. H. F. Nesselmann, Berlin, 1843, 55–6. French transl. by A. Marre, Nouv. Ann. Math., 5, 1846, 313. Cf. A. Genocchi, Annali di Sc. Mat. e Fis., 6, 1855, 297.

[302] Oeuvres, I, 300–1; French transl., III, 248–9. Observation on Diophantus, IV, 12.

[303] Novi Comm. Acad. Petrop., 6, 1756, 85; Comm. Arith., I, 258; Op. Om., (1), II, 426.

[304] Novi Comm. Acad. Petrop., 20, 1775 (1771), 48; Comm. Arith., I, 444; Op. Om., (1), III, 405.

It suffices to treat the case in which x and y are relatively prime, also t and u. For problem[305] (i), AB is the square of $Axy(p^2+q^2)$ if

$$t=xy(p^2-q^2)+2y^2pq, \qquad v=xy(p^2-q^2)-2x^2pq.$$

Then C is found to have the factor A, so that $AC=\square$ if

$$4p^2q^2x^4-4pq(p^2-q^2)x^3y+(p^4-6p^2q^2+q^4)x^2y^2+4pq(p^2-q^2)xy^3+4p^2q^2y^4=Q^2.$$

Taking $Q=2pqx^2-(p^2-q^2)xy-2pqy^2+\alpha y^2$, we have

$$\alpha(\alpha-4pq)y^2-2\alpha(p^2-q^2)xy+4pq(\alpha-pq)x^2=0.$$

For $\alpha=4pq$, we obtain the solution

$$x=2(p^2-q^2), \qquad y=3pq, \qquad t=3(p^4+p^2q^2+q^4), \qquad u=(p^2-q^2)^2.$$

For $\alpha=pq$, we obtain a similar solution. For $\alpha=\mp2p^2$, we get

$$x=p(p\pm2q), \qquad t=p(2p\pm q)(p^2\pm2pq+3q^2),$$
$$y=q(2p\pm q), \qquad u=q(p\pm2q)(q^2\pm2pq+3p^2).$$

For problem (ii), BC is a square if $x=3$, $y=5$, $t=11$, $u=45$, or if

$$x=3n^4+6m^2n^2-m^4, \qquad y=3m^4+6m^2n^2-n^4, \qquad t=mx, \qquad u=ny.$$

For problem (iii), we apply the last solution with $m^2+n^2=\square$.

Euler[306] required four numbers the four elementary symmetric functions of which are squares. For the numbers Mab, Mbc, Mcd, Mda the conditions reduce to

$$abcd=\square, \qquad bd(a^2+c^2)+ac(b+d)^2=\square, \qquad M=(ab+bc+cd+da)/f^2.$$

Finding the second condition impossible if $b/d=2$ or 3, Euler took $b/d=p^2/q^2$. Then must $p^2q^2(a^2+c^2)+ac(p^2+q^2)^2$ be a square, say that of $pqa+cm/n$. Thus a/c is found, and we readily form the condition that ac and hence $abcd$ shall be a square. By trial Euler found the two solutions[307] $a=64$, $b=9$, $d=4$, $c=49$ or 289, $M=1469$ or 4589; also one of another type: $a=16$, $b=5$, $c=5$, $d=4$, $f=3$, $M=21$. He discussed at length the problem to find b, d such that the initial second condition can be satisfied by choice of a, c.

Euler[308] treated $x^2+y^2+z^2=\square$, $x^2y^2+x^2z^2+y^2z^2=\square$. The first is satisfied if $x=p^2+q^2-r^2$, $y=2pr$, $z=2qr$. The second then becomes

$$(1) \qquad\qquad (p^2+q^2)(p^2+q^2-r^2)^2+4p^2q^2r^2=\square.$$

Set $n=(p-r)/q$ and eliminate r. Then shall

$$(p^2+q^2)\{2np+(1-n^2)q\}^2+4p^2(p-nq)^2=\square=R^2.$$

Set $R=(1-n^2)q^2+2npq+\alpha p^2$. The terms in p^2q^2 cancel if

$$2(1-n^2)\alpha=1+2n^2+n^4.$$

---

[305] F. van Schooten had proposed to find rational sides of a triangle given the base a, altitude b and ratio $m:n$ of the other sides (mz, nz). Thus $b=2mnxy$, $a=(m^2-n^2)(x^2+y^2)$, $z^2=(x^2+y^2)[(m\pm n)^2x^2+(m\mp n)^2y^2]$, falling under problem (i). The simplest solution is $x=3$, $y=5$, $m=28$, $n=17$, $a=33$, $b=28$.

[306] Novi Comm. Acad. Petrop., 17, 1772, 24; Comm. Arith., I, 450; Op. Om., (1), III, 172.

[307] Reproduced by A. Gérardin, l'intermédiaire des math., 16, 1909, 105–6.

[308] Acta Acad. Petrop., 3, I, 1782 (1779), 30; Comm. Arith., II, 457; Op. Om., (1), III, 453.

From the linear relation between p^4 and p^3q, we get

$$p : q = 8n(1-n^2) : 5 - 10n^2 + n^4.$$

J. A. Euler[309] treated his father's[308] problem.   Multiply

$$(p^2-1)^2 + 4p^2 \equiv (p^2+1)^2$$

by $4q^2$ and the like identity in q by $(p^2+1)^2$ and add.   Thus

$$(q^2-1)^2(p^2+1)^2 + 4q^2(p^2-1)^2 + 16p^2q^2 = (p^2+1)^2(q^2+1)^2.$$

Hence we have three squares whose sum is a square.   The sum of their products by twos is $4q^2$ times

$$(q^2-1)^2(p^2+1)^4 + 16p^2q^2(p^2-1)^2.$$

This is to be made a square.   Set $A = (p^2+1)^2$, $B = 4p(p^2-1)$.   Then $(q^2-1)^2A^2 + q^2B^2$ is to be a square, say $(Aq^2+v)^2$.   Then $q^2 = (A^2-v^2)/d$, where $d \equiv 2A^2 - B^2 + 2Av$.   Take $v^2 = A^2 - B^2$.   Then d is the square of $A+v$.   Now $A^2 - B^2 = (p^4 - 6p^2 + 1)^2$.   Hence

$$q = \frac{B}{A+v} = \frac{4p(p^2-1)}{2p^4 - 4p^2 + 2} = \frac{2p}{p^2-1}.$$

Hence, after multiplication by $(p^2-1)^2$, we have the solution

$$x = (6p^2 - p^4 - 1)(p^2+1), \qquad y = 4p(p^2-1)^2, \qquad z = 8p^2(p^2-1),$$
$$\Sigma x^2 = (p^2+1)^6, \qquad \Sigma x^2y^2 = 16p^2(p^2-1)^2[(p^2-1)^4 + 16p^4]^2.$$

For $p=2$, we get 35, 72, 96.   Next (p. 47) let $x = am$, $y = bm$, $z = cn$, where $a^2 + b^2 = c^2$, $m = 2pq$, $n = p^2 - q^2$.   Then

$$\Sigma x^2 = c^2(p^2+q^2)^2, \qquad \Sigma x^2y^2 = m^2[4a^2b^2p^2q^2 + c^4(p^2-q^2)^2].$$

The latter is the square of $m(a^4+b^4)$ if $p=a$, $q=b$.   Then

$$x = 2a^2b, \qquad y = 2ab^2, \qquad z = c(a^2-b^2) \qquad (a^2+b^2 = c^2)$$

is a solution.   It may be obtained by using his father's notations and assuming that $p^2 + q^2 = c^2$.   Then the condition (1) becomes

$$c^2(c^2 - r^2)^2 + 4p^2q^2r^2 = \square,$$

which is satisfied if $r = cp/q$, since the left member becomes $c^2(p^4+q^4)^2/q^4$.

The problem[309a] to find four integers whose sum is a biquadrate and sum of any two a square reduces to finding a biquadrate n^4 which is a sum of two squares in three ways.   Take as n a product of two or more primes $4k+1$.

J. Cunliffe[309b] noted that the problem to find three positive integers whose sum is a square and sums by twos are biquadrates is evidently equivalent to that to find three biquadrates half of whose sum is a square and the sum of any two exceeds the remaining one.   Half the sum of the fourth powers of $m+n+sv$, $m+rv$, $n+v(r+s)$ is $A^2 + 2Bv + \cdots + \alpha^2v^4$, where

<hr>

[309] Acta Acad. Petrop., pro anno 1779, I, 1782, Mém., pp. 40–48.

[309a] New Series of Math. Repository (ed., T. Leybourn), 1, 1806, I, 59–61.

[309b] *Ibid.*, 2, 1809, I, 178–9.   If we wave the condition that the numbers be positive, we may use the biquadrates m^4, n^4, $(m+n)^4$, half of whose sum is $(m^2+mn+n^2)^2$.

$A=m^2+mn+n^2$, $B=s(m+n)^3+m^3r+n^3(r+s)$, $\alpha=r^2+rs+s^2$. Equate it to the square of $A+vB/A\pm\alpha v^2$ to get v rationally.

Several[310] found 7 numbers in arithmetical progression the sum of whose cubes is a biquadrate. Let $nx-3x$, $nx-2x$, $\cdots$, $nx+3x$ be the numbers. Equating the sum of their cubes $7n^3x^3+84nx^3$ to m^4x^4, we get x. Or use x, $\cdots$, $7x$, the sum of whose cubes is $784x^3$.

To find a rectangular parallelopiped whose edges, sum of edges, and sum of faces, are rational squares, several[311] took x^2, y^2, z^2 as the adjacent edges, and $x^2+y^2+z^2=(x+y-z)^2$, whence $z=xy/(x+y)$. Then

$$S\equiv 2x^2y^2+2x^2z^2+2y^2z^2=\square$$

if $x^2+xy+y^2=\square=(rx-y)^2$, which gives x/y. C. Wilder took $S=4m^2y^2z^2$ [printed $S=4m^2$], and $x^2=myz(2-a^2)/(2a)$. Then

$$\Sigma x^2=x^2+(2m^2-1)y^2z^2/x^2=\square$$

if

$$\left(\frac{2+a^2}{2a}\right)^2 m^2-1=\square=\left\{b-\left(\frac{2+a^2}{2a}\right)m\right\}^2, \qquad m=\frac{a(b^2+1)}{b(2+a^2)}.$$

Eliminating m from the assumed expression for x^2, we get y in terms of x, z, a, b, which are arbitrary. [The solution is false as it satisfies neither of the proposed equations, but only the combination of them which was employed.]

To find three positive integers the sum of any two of which is a square and double the sum of all three is a biquadrate, R. Maffett and D. Robarts[312] took a^2, b^2, c^2 as the sums by pairs. Then shall $a^2+b^2+c^2$ be a biquadrate. Take $a=3(p^2+r^2)$, $b=4(p^2-r^2)$, $c=8pr$. Then $\Sigma a^2=(5p^2+5r^2)^2$, which equals $(25r^2)^2$ for $p=2r$.

To find two integers whose sum, sum of squares, and sum of cubes, are all squares, and sum of biquadrates is a cube, J. Whitley[313] used the numbers $x=2rs$, $y=r^2-s^2$, whence $x^2-xy+y^2=\square$ if $r=4s$. Call X, Y the products of x, y by $23=8+15$. Then $X=23\cdot8s^2$, $Y=15\cdot23s^2$ satisfy the first three conditions. Also $X^4+Y^4=23^3ts^8$, where $t=23(8^4+15^4)$, will be a cube if $s=t$. C. Gill used $x=b\sin A$, $y=b\cos A$ with the sum a^2. Then

$$x^3+y^3=a^2b^2(1-\sin A\cos A)=c^2$$

if $c=ab(1-\tfrac{1}{2}\sin A)$, $\cot\tfrac{1}{2}A=4$, whence $x=8b/17$, $y=15b/17$. By their sum, $b=17a^2/23$. The fourth condition is satisfied if $a=23^2\cdot54721$.

E. Lucas[313a] proved that $2v^2-u^2=w^4$, $2v^2+u^2=3z^2$ imply

$$u^2=v^2=w^2=z^2=1.$$

E. Lionnet[314] desired a number N which, as well as its biquadrate, is the sum of the squares of two consecutive integers. J. Lissençon wrote

[310] The Gentleman's Diary, or Math. Repository, London, No. 76, 1816, 39, Quest. 1043.
[311] The Math. Diary, New York, 1, 1825, 125-7.
[312] Ladies' Diary, 1833, 35, Quest. 1542.
[313] The Lady's and Gentleman's Diary, London, 1854, 52-3, Quest. 1857.
[313a] Nouv. Ann. Math., (2), 16, 1877, 414.
[314] Nouv. Ann. Math., (2), 19, 1880, 472-3. Repeated in Zeitschr. Math. Naturw. Unterricht, 12, 1881, 268.

$N = a^2 + (a+1)^2$, whence

$$N^4 = A^2 + B^2, \qquad A = -4a^4 - 8a^3 + 4a + 1, \qquad B = -8a^3 - 12a^2 - 4a.$$

Then $1 = A - B$ gives $a(a+1)^2(a-2) = 0$. The only answer, given by $a = 2$, is $N = 13$, $13^4 = 119^2 + 120^2$.

L. Bastien[315] solved the system $y^2 + z^2 + t^2 = 2x^2$, $y^4 + z^4 + t^4 = 2x^4$ by eliminating x. Thus $y^2 + z^2 - t^2 = \pm 2yz$, $y \mp z = \pm t$. Let $y = z + t$. Substitute this value of y in the first equation. We get $zt = (z+t+x)(z+t-x)$. Hence set $z = ab$, $t = cd$, $z+t+x = ac$, $z+t-x = bd$, $2b - c = hd$, $b - 2c = ha$. The solution is now evident.

A. Gérardin[316] gave special cases in which $s^4 - x$, $s^4 - y$, $s^4 - z$ are all squares or all cubes, where $s = x + y + z$.

L. Aubry[317] proved the impossibility of the system

$$g^4 + 9f^3 g = \square, \qquad 9f^4 + 3fg^3 = \square.$$

Gérardin[318] solved the system $x^4 + x^2 y^2 = a^2$, $y^4 + x^2 y^2 = b^2$, $x + y = c^2$. M. Rignaux[319] noted that $a = \alpha x$, $b = \beta y$, whence the system reduces to

$$x^2 + y^2 = \alpha^2 = \beta^2, \qquad x + y = c^2$$

and is easily solved.

E. Fauquembergue[320] discussed the system $x^4 - hy^4 = \square$, $x^4 + hy^4 = \square$.

A. Gérardin[321] discussed the system $\Sigma P^4 = \Sigma U^4$, $PQR = UVW$.

A. Cunningham[322] solved $X^4 - Z = A^2$, $X^4 + Z = B^2$ by taking any odd integer α and any even integer β and setting $X = \alpha^2 + \beta^2$.

Euler,[254] and Euler[81] of Ch. XVI, made $x^2 y^2 \mp z^2 v^2$, $x^2 z^2 \mp y^2 v^2$, $y^2 z^2 \mp x^2 v^2$ squares. Petrus[12] of Ch. XV made $p^2 + s^2$, $t^2 + q^2$, $pstq$ squares. Woepcke[48] of Ch. XVI treated $\sigma^4 + \phi \sigma^2 = \sigma_1^4 + \phi \sigma_1^2 = \square$. Gérardin[185] of Ch. XXII treated $x^4 + mx^2 y^2 + y^4 = a^2$ with other quartics.

---

[315] Sphinx-Oedipe, 8, 1913, 173.

[316] L'intermédiaire des math., 23, 1916, 150, 169. R. Goormaghtigh and A. Colucci gave solutions, 24, 1917, 134–5.

[317] *Ibid.*, 23, 1916, 129–131.

[318] *Ibid.*, 122–3.

[319] *Ibid.*, 24, 1917, 65–6.

[320] *Ibid.*, 39.

[321] *Ibid.*, 100–1.

[322] Math. Quest. and Sol., 4, 1917, 4–5.

CHAPTER XXIII.

EQUATIONS OF DEGREE n.

SOLUTION OF $f=\text{CONST.}$, WHERE f IS A BINARY FORM.

J. L. Lagrange[1] noted that, in seeking integral solutions of

$$A = Bt^n + Ct^{n-1}u + \cdots + Ku^n,$$

where $A, \cdots, K$ are given integers, we may take u relatively prime to A, and thus find integers θ, y such that $t = u\theta - Ay$. Inserting the value of t, we see that $B\theta^n + C\theta^{n-1} + \cdots + K$ must be divisible by A. If such an integer θ exists, the proposed equation reduces, after division by A, to

$$F(u, y) \equiv Pu^n + Qu^{n-1}y + \cdots + Vy^n = 1,$$

where $P, \cdots, V$ are given integers. Set $u/y = x$, $F(x, 1) = z$. Then $1/y^n = z$. The problem of solving $F = 1$ in integers reduces to the examination of the real values a of x for which z is zero or a minimum (whence $dz/dx = 0$). For such an a, Lagrange employed the continued fraction for a and two series of convergents and proved that u/y must equal one of these convergents l/L, whence $u = \pm l$, $y = \pm L$. While a root of $z = 0$ may lead to an infinitude of solutions, a root of $dz/dx = 0$ furnishes only a limited number.

A. M. Legendre[2] reproduced this method of Lagrange's, developing into a continued fraction each real root of $F(x, 1) = 0$ and also the real part of each imaginary root and forming their various convergents p/q. The least of the $F(p, q)$ is the minimum of $F(u, y)$ for integral values u, y. In case the minimum is ± 1, we have a solution of $F(u, y) = \pm 1$ and hence a solution of the initial equation $A = Bt^n + \cdots$.

H. Poincaré[3] noted that the problem reduces to the case of the representation of a number N by a form in which the leading coefficient is unity: $x^m + Ax^{m-1}y + \cdots$. We first solve the congruence $\xi^m - A\xi^{m-1} + \cdots \equiv 0 \pmod{N}$ and then determine by Hermite's method whether or not two decomposable forms in m variables are equivalent under m-ary linear transformation.

G. Cornacchia[4] gave a method of solving in integers

$$(1) \qquad \sum_{h=0}^{n} C_h x^{n-h} y^h = P,$$

when C_0 and C_n are positive and a root $x_0 > P/2$ of the corresponding congruence $\Sigma C_h x^{n-h} \equiv 0 \pmod{P}$ is known. Take y_0 such that $x_0 y_0 \equiv \pm 1 \pmod{P}$. Apply the g.c.d. process to P, x_0, and let $x_1, x_2, \cdots, x_m = 1$ be the remainders. Let $y_1, \cdots, y_m = 1$ be the corresponding remainders from P, y_0. Then if (1) has relatively prime integral solutions a, b such that $2ab < P$, this solution is one of the above pairs x_i, y_{m+1-i} or is a pair obtained

[1] Mém. Acad. Berlin, 24, année 1768, 1770, 236; Oeuvres, II, 662, 675. For $n = 2$, Lagrange[76] of Ch. XII.

[2] Théorie des nombres, 1798, 169–180; ed. 3, 1830, I, 179; German transl., Maser, I, 179.

[3] Comptes Rendus Paris, 92, 1881, 777. Cf. Poincaré.[24]

[4] Giornale di Mat., 46, 1908, 33–90.

similarly from another root of the congruence. The process is simplified, applied to $x^2 + qy^2 = m$ and compared with the method of binary quadratic forms.

CONDITIONS FOR AN INFINITUDE OF SOLUTIONS OF $f(x, y) = 0$.

C. Runge[5] considered an irreducible polynomial $f(x, y)$ with integral coefficients (i. e., not a product of such polynomials), and the algebraic function y defined by $f(x, y) = 0$. By one system of conjugate developments of y according to descending powers of x is meant those obtained from a single development by replacing the single algebraic number, in terms of which all the coefficients are expressed rationally, by its conjugate values and the fractional power of x by all its values. He proved that if the various developments of y form more than one system of conjugates there is only a finite number of integral values of x for which $f(x, y) = 0$ is satisfied by rational values of y. Also that $f = 0$ has an infinitude of pairs of integral solutions x, y only when x, y become infinite simultaneously and when the developments according to descending powers of one of these variables form a single system of conjugate developments. Hence necessary (but not sufficient) conditions for an infinitude of pairs of integral solutions x, y of $f(x, y) = 0$ are: (i) If f is of degree m in x and n in y, the coefficients of x^m and y^n are constants a, b. (ii) The algebraic function y defined by $f(x, y) = 0$ becomes infinite with x with the order of $x^{m/n}$. If $cx^\rho y^\sigma$ is a term of f, then $n\rho + m\sigma \leqq mn$. (iii) The sum of the terms for which

$$n\rho + m\sigma = mn$$

must be expressible in the form

$$b\Pi_\beta (y^\lambda - d_\beta x^\mu) \qquad (\beta = 1, 2, \cdots, n/\lambda),$$

where $\Pi(u - d_\beta)$ is a power of an irreducible function of u.

A. Boutin[6] raised the question as to the types of equations such that, if x_i, y_i $(i = n-1, n-2)$ are two sets of integral solutions,

$$(1) \qquad x_n = \alpha x_{n-1} + \beta x_{n-2}, \qquad y_n = \alpha y_{n-1} + \beta y_{n-2}$$

are also solutions. E. Maillet[7] treated the properties of one or two recurring series $x_{n+p} = \alpha_1 x_{n+p-1} + \cdots + \alpha_p x_n$ with rational (or integral) coefficients and proved that the only equations $F(x, y) = 0$, where F is without a rational divisor, with an infinitude of integral solutions given by a formula of recurrence (1) of the second order are either linear, quadratic

$$Ax^2 + Bxy + Cy^2 \pm H = 0,$$

or $\qquad (tv'y - t'vx)^p - (vu'x - uv'y)^q (tu' - ut')^{p-q} = 0,$

where p, q are relatively prime integers. If we consider rational solutions, we obtain an analogous result.

E. Maillet[8] proved theorems concerning arithmetically irreducible equations

$$(2) \qquad F(x, y) = \phi_n(x, y) + \phi_{n-1}(x, y) + \cdots + \phi_0 = 0,$$

---

[5] Jour. für Math., 100, 1887, 425–35.

[6] L'intermédiaire des math., 1, 1894, 20–21.

[7] Mém. Acad. Sc. Toulouse, (9), 7, 1895, 182–213.

[8] Comptes Rendus Paris, 128, 1899, 1383; Jour. de Math., (5), 6, 1900, 261–77.

where ϕ_j is homogeneous and of degree j. (I) Let $\phi_n(x, y)$ be arithmetically reducible; let c_1 be a simple real root of $\phi_n(1, c) = 0$ of degree λ; let $\psi_k(1, c)$ be an irreducible factor of ϕ_n, of degree k ($k < n$) and with the root c_1. Then $F = 0$ has, on the infinite branch whose asymptote has c_1 as angular coefficient, an infinitude of solutions only if one of the $\phi_i(1, c_1)$, $i = n-1$, $\cdots$, $n - k$, is not zero. (II) There exists no irreducible equation $F(x, y) = 0$ with integral coefficients having an infinitude of integral solutions on an infinite branch of $F = 0$ such that the angular coefficient of the asymptote is rational and not zero, if this coefficient is a simple root of $\phi_n(1, c) = 0$. If the real angular coefficients of the asymptotes of $F = 0$ are all rational, not zero and distinct, then $F = 0$ has only a finite number of integral solutions. By amplifying the case $k = 2$, he obtains a complicated third theorem; also one on $F(x, y, z) = 0$.

A. Thue[9] proved that, if $U(x, y)$ is an irreducible homogeneous polynomial with integral coefficients and c is a given constant, $U(p, q) = c$ has only a finite number of positive integral solutions p, q, when the degree of U exceeds 2.

A. Thue[10] considered homogeneous integral functions $P(x, y)$, $Q(x, y)$, $R(x, y)$ of degrees p, q, r, with integral coefficients, $P(x, y)$ being irreducible. If $p > q$, $p > 2$, $P = Q$ does not have an infinitude of pairs of integral solutions x, y. If $p > q > r$, $p < q + r$, $P + Q + R = 0$ is not satisfied by an infinitude of pairs of relatively prime integers x, y.

E. Maillet[11] completed a lacuna in the proof by Thue[9] and gave the following generalization of his theorem. Let ϕ_i be a homogeneous polynomial of degree i in x, y. While the coefficients of ϕ_0, $\cdots$, ϕ_s need not be rational, let ϕ_r ($r > s$) have integral coefficients and contain a term in x^r and one in y^r. If

$$\phi_r(x, y) - \phi_s(x, y) - \phi_{s-1}(x, y) - \cdots - \phi_0 = 0$$

is irreducible, it has an infinitude of integral solutions x, y only when s exceeds a specified quantity depending on the reducibility of $\phi_r = 0$. When ϕ_r is irreducible, this quantity is $r_1 - 2$ or $r_1 - 1$, according as $r = 2r_1$ or $r = 2r_1 + 1$.

Maillet[11a] gave a practical method to find an upper limit to the absolute values of the integral solutions x, y of an equation of type (2), subject to certain conditions on ϕ_n which imply that (2) has only a finite number of integral solutions.

RATIONAL POINTS ON THE PLANE CURVE $f(x, y, z) = 0$.

D. Hilbert and A. Hurwitz[12] treated homogeneous polynomials $f(x_1, x_2, x_3)$ of degree n with integral coefficients such that the curve $f = 0$ is of genus (or deficiency, geschlecht) zero. In view of results by M. Noether,[13] we

[9] Jour. für Math., 135, 1909, 303–4. Cf. Maillet.[11]
[10] Skrifter Videnskaps. Kristiania (Math.), 1, 1911, No. 3 (German).
[11] Nouv. Ann. Math., (4), 16, 1916, 338–345.
[11a] Ibid., (4), 18, 1918, 281–92.
[12] Acta Math., 14, 1890–1, 217–24.
[13] Math. Annalen, 23, 1884, 311–358.

can decide by rational operations whether or not $f=0$ is of genus zero and if so we can find by rational operations $n-1$ linearly independent ternary forms ϕ_i of degree $n-2$ with integral coefficients such that for arbitrary parameters λ_i the curve $f=0$ is cut by the curve

$$\lambda_1\phi_1+\cdots+\lambda_{n-1}\phi_{n-1}=0$$

in $n-2$ points varying with the parameters λ_i. Set

$$\Phi_i=\lambda_{i1}\phi_1+\cdots+\lambda_{i\,n-1}\phi_{n-1} \qquad (i=1,\,2,\,3),$$

where the λ_{ij} are arbitrary parameters. Transform $f=0$ by

$$y_1 : y_2 : y_3 = \Phi_1 : \Phi_2 : \Phi_3.$$

The result is $g(y_1,\,y_2,\,y_3)=0$, where g is an irreducible form of degree $n-2$ in the y's with integral coefficients. Now give to the parameters λ_{ij} such integral values that g remains irreducible. Since our transformation is birational, every rational point on $f=0$ corresponds to a rational point on $g=0$ and conversely. Hence the initial problem is reduced to the equation $g=0$ also of genus zero, but of lower degree by two units. Ultimately we reach an equation of degree 1 or 2. For a linear equation $l(u_1,\,u_2,\,u_3)=0$, we can evidently find three linear functions ω_i of the homogeneous parameter t_1/t_2 such that $u_1 : u_2 : u_3 = \omega_1 : \omega_2 : \omega_3$ gives all rational solutions of $l=0$ when t_1, t_2 take all integral values. By applying the inverses of our transformations, we get the initial $f=0$ and solutions $x_1 : x_2 : x_3 = \rho_1 : \rho_2 : \rho_3$, where the ρ_i are forms of degree n in t_1, t_2. The only missing solutions are those, finite in number and found rationally, which correspond to rational singular points of $f=0$, where our transformations cease to be birational. Second, if we reached a quadratic equation, it can be transformed rationally into $a_1u_1^2+a_2u_2^2+a_3u_3^2=0$, the a's without square factors and relatively prime in pairs. It has integral solutions if and only if the a's are not all of like sign and if $-a_2a_3$, $-a_3a_1$, $-a_1a_2$ are quadratic residues of a_1, a_2, a_3, respectively (papers 114, 116, 119 of Ch. XIII). When these conditions are satisfied, the conic has rational points and can be transformed birationally into a straight line; we proceed as before.

M. Noether[14] had earlier proved that a rational curve can be transformed birationally into a straight line or conic; a curve of order $2n$ with a $(2n-1)$-fold point is counted as curve of odd order.

H. Poincaré[15] proved the above result that any unicursal curve with rational coefficients is equivalent to a conic or a straight line, two curves being called equivalent if one can be transformed into the other by a birational transformation with rational coefficients. A curve $f=0$ of genus 1 (bicursal curve) with rational coefficients is equivalent to a curve of order p $(p\geqq3)$ if and only if $f=0$ has a rational group of p points, i. e., a set of p points such that every elementary symmetric function of their coordinates is rational.

---

[14] Math. Annalen, 3, 1871, 170.

[15] Jour. de Math., (5), 7, 1901, 161–233. For a special case, von Sz. Nagy[151] of Ch. XXI.

J. von Sz. Nagy[16] proved that any curve of genus 2 with rational coefficients is equivalent in general to a quartic curve and contains an infinitude of rational groups of two points.

J. von Sz. Nagy[17] cited the known fact that a curve C_n^p of order n and genus $p>1$ has in general no birational automorphs besides identity, and never more than $84(p-1)$, and concluded that we can derive at most a finite number of rational points from one. The birational automorphs of non-hyperelliptic and hyperelliptic curves are discussed. An example shows that from a rational point we do not in general obtain all other rational points by means of the birational automorphs of the curve.

J. von Sz. Nagy[18] wrote Q_n for the g.c.d. of n and $2p-2$ and proved that a curve C_n^p of order n and genus p contains infinitely many rational groups of hQ_n points if h is an integer for which $hQ_n>p-1$; it is equivalent to a curve C_m^p for $m>p+1$ if and only if it contains a rational group of m non-singular points. In particular, they are equivalent if m is a multiple of Q_n, and hence if $m=2p-2$, $p>2$, and the curves are not hyperelliptic.

E. Maillet[18a] considered a polynomial $f(x, y)$ of degree $n>2$, irreducible, with integral coefficients, and such that the curve $f=0$ is unicursal (of genus 0). If there are at least $n-3$ simple rational points, there is an infinitude corresponding to the rational values of a parameter t, and $x=f_2(t)/f_1(t)$, $y=f_3(t)/f_1(t)$, where the f_i are polynomials with integral coefficients having no common divisor, of degrees $n_i\leqq n$, one being of degree n (cf. papers 12, 15). The curve has an infinite number of points with *integral* coordinates only when f_1 is a constant or of one of the forms $\alpha(Mt+N)^n$, with α, M, N integers, or $\alpha(Mt^2+Nt+P)^{n/2}$, where n is even and N^2-4MP is positive and not a square, while α, M, N, P are integers. There are extensions to certain equations $f(x, y)=0$ of genus >0 and to certain unicursal surfaces.

For cubic curves of genus unity, see Levi[307] and Hurwitz[312] of Ch. XXI.

Equations formed from linear functions.

For related papers, see Lagrange,[142] Rados[194a]; papers 313–23 of Ch. XXI; and Ch. XX.

G. L. Dirichlet[19] stated a theorem, which he regarded as remarkable for its simplicity and importance: if an equation

(1) $$s^n+as^{n-1}+\cdots+gs+h=0$$

with integral coefficients has no rational divisor and if at least one of its roots $\alpha, \beta, \cdots, \omega$ is real, and if we set

$$\phi(\alpha)=x+\alpha y+\cdots+\alpha^{n-1}z,$$

then the indeterminate equation

(2) $$F(x, y, \cdots, z)\equiv\phi(\alpha)\phi(\beta)\cdots\phi(\omega)=1$$

[16] Math. Naturw. Berichte aus Ungarn, 26, 1908 (1913), 186 (168–195).
[17] Jahresbericht d. Deutschen Math.-Vereinigung, 21, 1912, 183–191.
[18] Math. Annalen, 73, 1913, 230–240, 600.
[18a] Comptes Rendus Paris, 168, 1919, 217–20; Jour. Ecole Polyt., (2), 20, 1919, 115–56.
[19] Comptes Rendus Paris, 10, 1840, 285–8; Werke, I, 619–623.

has an infinity of integral solutions. Application is made to functions considered by Lagrange[142] which repeat under multiplication. If such a function can take a given value, it takes the same value for an infinitude of sets of values of $x, \cdots, z$, under the assumption that the algebraic equation to which the function owes its origin has no rational divisor, but has at least one real root.

G. Libri[20] stated that the conditions imposed on (1) that there be a real root and no rational factor are not necessary, it sufficing to have $h = \pm 1$.

J. Liouville[21] proved Libri's theorem false. For, if (1) is $s^2 + 1 = 0$, then (2) is $(x+yi)(x-yi) = x^2 + y^2 = 1$, with only a finite number of integral solutions.

Dirichlet[22] noted that his theorem remains true if (1) has only imaginary roots, provided $n > 2$. The problem is that of the units of an algebraic domain.

P. Bachmann[23] treated the solution of $N = 1$, where N is the norm of the general algebraic number determined by a root of an equation of degree n.

H. Poincaré[24] noted that, for F defined by (2) by means of any equation (1), the problem to find integers β_i such that $F(\beta_1, \cdots, \beta_n)$ shall equal any given integer N reduces to the problem to form all complex ideals of norm N. In the solution of the latter one considers the congruences $s^n + a s^{n-1} + \cdots \equiv 0$ (mod μ), μ any divisor of N.

E. Meissel[25] considered the product, extended over the roots of $\theta^5 = 1$,

$$V = (x, y, z, u, v) = \Pi(x + \theta y \rho + \theta^2 z \rho^2 + \theta^3 u \rho^3 + \theta^4 v \rho^4), \qquad \rho = \sqrt[5]{A}.$$

By the reciprocal solution of $V = 1$ is meant $1/V = (a, b, c, d, e) = 1$, where

$$5a = \frac{\partial V}{\partial x}, \qquad 5Ae = \frac{\partial V}{\partial y}, \qquad 5Ad = \frac{\partial V}{\partial z}, \qquad 5Ac = \frac{\partial V}{\partial u}, \qquad 5Ab = \frac{\partial V}{\partial v}.$$

For $2 \leqq A \leqq 7$, he gave two primary solutions $V_1 = 1$, $V_2 = 1$, accompanied by their reciprocal solutions. He stated that two primary solutions always exist and deduced the solutions $V_1^m V_2^n$. He conjectured that, if p is a prime, the corresponding Pell equation of degree p has $\frac{1}{2}(p-1)$ primary solutions.

A. Thue[26] considered a homogeneous polynomial $F(x_1, \cdots, x_n)$ of degree $n - 1$ such that $F = 0$ can be given the form

$$(3) \qquad\qquad P_1 P_2 \cdots P_{n-1} = Q_1 Q_2 \cdots Q_{n-1},$$

where P_i, Q_i are linear functions of $x_1, \cdots, x_n$ with integral coefficients. Set

$$(4) \qquad a_1 P_1 = a_2 Q_1, \qquad a_2 P_2 = a_3 Q_2, \cdots, a_{n-1} P_{n-1} = a_1 Q_{n-1},$$

where the a's are any integers without common divisor. Then (4) if independent give $x_i = k\Delta_i$ $(i = 1, \cdots, n)$, where Δ_i is a homogeneous poly-

[20] Comptes Rendus Paris, 10, 1840, 311-4, 383.
[21] *Ibid.*, 381-2. Bull. des Sc. Math., (2), 32, I, 1908, 48-55.
[22] Bericht Akad. Wiss. Berlin, 1842, 95; 1846, 103-7; Werke, I, 638-644.
[23] De unitatum complexarum theoria., Diss., Berlin, 1864.
[24] Comptes Rendus Paris, 92, 1881, 777-9; Bull. Soc. Math. France, 13, 1885, 162-194.
[25] Beitrag zur Pell'schen Gleichung höherer Grade, Progr., Kiel, 1891.
[26] Det Kgl. Norske Videnskabers Selskabs Skrifter, 1896, No. 7 (German).

nomial of degree $n-1$ in $a_1, \cdots, a_{n-1}$. Finally we choose k to make these x's integers.

If $F=0$ can be given the form (3), every set of integral solutions of $P_i=0$, $Q_j=0$ $(i, j=1, \cdots, n-1)$ is evidently a solution of $F=0$. Conversely, if a certain number of integral solutions of $P_i=Q_j=0$ satisfy $F=0$, then $F=0$ can be given the form (3). In fact, if a polynomial $F(x_1, \cdots, x_n)$ of degree m always vanishes simultaneously with the products $U=P_1\cdots P_p$, $V=Q_1\cdots Q_q$ of linear functions of $x_1, \cdots, x_n$, such that not all the values for which any two are zero make a third zero, then $F\equiv AU+BV$, where A and B are polynomials in $x_1, \cdots, x_n$.

A. Palmström[27] extended the preceding method to the equation

$$(5) \qquad \begin{vmatrix} P_{11} & P_{12} & \cdots & P_{1\,n-1} \\ P_{21} & P_{22} & \cdots & P_{2\,n-1} \\ \cdot & \cdot & \cdots\cdots & \cdot \\ P_{n-1\,1} & P_{n-1\,2} & \cdots & P_{n-1\,n-1} \end{vmatrix} = 0,$$

where the P's are linear homogeneous functions of $x_1, \cdots, x_n$. For every set of integral x's satisfying (5) there exist $n-1$ relatively prime integers $a_1, \cdots, a_{n-1}$ satisfying

$$(6) \qquad a_1P_{i1}+a_2P_{i2}+\cdots+a_{n-1}P_{i\,n-1}=0 \quad (i=1, \cdots, n-1),$$

and conversely. From the latter, $x_i/x_n=\Delta_i/\Delta_n$, so that we may set $x_j=k\Delta_j$ $(j=1, \cdots, n)$ and choose k to make the x's integral. Here the a's have any values for which $\Delta_1, \cdots, \Delta_n$ are not all zero. In case the Δ's are all identically zero, so that only p of the equations (6) are independent, we can assign arbitrary values to $n-p-1$ of the x's and determine the remaining x's by p linear equations. He[130] gave a detailed example.

G. Métrod[27a] found the number of ways to decompose a given number into a product of n factors (including unity).

Product P_n of n consecutive integers not an exact power.

Chr. Goldbach[28] argued that a P_3 is not a square since its root would be a multiple of m and a divisor of $(m+1)(m+2)$, whence $m=1$ or 2.

J. Liouville[29] proved by use of Bertrand's postulate [Vol. I, Ch. XVIII] that $m(m+1)\cdots(m+n-1)$ is not a square or higher power if at least one factor $m, \cdots, m+n-1$ is a prime, or if $n>m-5$. The latter was proved similarly by E. Mathieu,[30] who verified the theorem for any n when $m\leqq 100$. In particular, $m!$ is not an exact power, a fact proved in the same way by W. E. Heal.[31]

Mlle. A. D.[32] proved that a P_3 is not an exact power.

[27] Skrifter Udgivne af Videnskabsselskabet, Christiania, 1900 (1899), Math.-Naturw. Kl., No. 7 (German).

[27a] L'intermédiaire des math., 26, 1919, 153–4. Cf. Minetola[192–3] of Ch. III, and Cesàro[30] of Ch. IX; also Index to Vol. I (under "Number," including $n=x^ay^b$).

[28] Corresp. Math. Phys. (ed., Fuss), 2, 1843, 210, letter to D. Bernoulli, July 23, 1724.

[29] Jour. de Math., (2), 2, 1857, 277. Cf. Moreau.[50]

[30] Nouv. Ann. Math., 17, 1858, 235–6.

[31] Math. Magazine, 1, 1882–4, 208–9.

[32] Nouv. Ann. Math., 16, 1857, 288–290. Proposed by Faure, p. 183.

G. C. Gerono[33] proved that $P_4 \neq \square$ by setting $(m+1)(m+4)=2p$, whence $(m+2)(m+3)=2(p+1)$, while $p(p+1) \neq \square$. "P. A. G."[34] gave a proof by use of

$$m(m+1)(m+2)(m+3)+1 = \{m(m+3)+1\}^2.$$

Gerono[35] proved that P_5, P_6 or P_7 is not a square.

V. A. Lebesgue[36] proved that P_5 is not a square or cube.

A. Guibert[37] proved that, if $8 \leqq n \leqq 17$, $P_n \neq \square$, while P_6 or P_9 or a product of any three integers in arithmetical progression is not a cube.

A. B. Evans[38] and G. W. Hill[39] proved that $P_6 \neq \square$.

D. André[40] proved that, if $n > 1$, $P_n \neq y^n$ or $y^n \pm 1$.

A. B. Evans[41] proved that P_5, P_6 or P_7 is not a square.

H. Bourget[42] proved that $P_5 \neq \square$.

R. Bricard[43] proved that $P_8 \neq \square$ by use of a Pell equation.

L. Aubry[44] proved that P_4 is not a cube by treating the case in which a single one of the four numbers is divisible by 3 and the case in which two are divisible by 3, necessarily the first and fourth, and examining in the second case the residues modulo 9 of the four numbers.

T. Hayashi[45] proved that P_2 or P_4 is not a square or cube, $P_3 \neq x^n$, $n \geqq 2$. Also (p. 166), $y(y+1)(2y+1) \neq x^n$, $n \geqq 2$.

S. Narumi[46] proved that $x(x+1) \cdots (x+n) = \square \neq 0$ is impossible if $n \leqq 202$.

T. Hayashi[47] proved that $P_5 \neq \square$.

FURTHER PROPERTIES OF PRODUCTS OF CONSECUTIVE INTEGERS.

J. Liouville[48] proved that, if p is a prime >5,

$$(p-1)!+1 \neq p^m, \qquad \left\{\left(\frac{p-1}{2}\right)!\right\}^2 + 1 \neq p^m.$$

Berton[49] verified that $P \equiv a(a+h)(a+2h)(a+3h) \neq p^4$ since

$$P = (a^2+3ah+h^2)^2 - h^4, \qquad p^4+h^4 \neq \square.$$

Hence the area $\sqrt{P}$ of an inscriptible quadrilateral whose sides are in arithmetical progression is not a square.

[33] Nouv. Ann. Math., 16, 1857, 393-4.
[34] *Ibid.*, 17, 1858, 98.
[35] *Ibid.*, 19, 1860, 38-42.
[36] *Ibid.*, 112-5, 135-6.
[37] *Ibid.*, 213 [400]; (2), 1, 1862, 102-9.
[38] The Lady's and Gentleman's Diary, London, 1870, 88-9, Quest. 2106.
[39] The Analyst, Des Moines, Iowa, 1, 1874, 28-29.
[40] Nouv. Ann. Math., (2), 10, 1871, 207-8.
[41] Math. Quest. Educ. Times, 27, 1877, 30; 44, 1886, 65-9.
[42] Jour. de math. élém., 1881, 66.
[43] L'intermédiaire des math., 17, 1910, 139-40.
[44] Sphinx-Oedipe, 8, 1913, 136.
[45] Nouv. Ann. Math., (4), 16, 1916, 155-8.
[46] Tôhoku Math. Jour., 11, 1917, 128-142.
[47] Nouv. Ann. Math., (4), 18, 1918, 18-21.
[48] Jour. de Math., (2), 1, 1856, 351.
[49] Nouv. Ann. Math., 18, 1859, 191.

C. Moreau[50] repeated the first remark by Liouville.[29]

H. Brocard[51] asked for values of x making $1+x!$ a square.   He[52] suggested that the only solutions are 4, 5, 7.

E. Lucas[53] noted that the product P of the first n primes is not of the form $a^p \pm b^p$, where a and b are positive integers and $p>1$, $P>2$.

E. Lionnet[54] stated that no product $1 \cdot 3 \cdot 5 \cdots$ of consecutive odd numbers is a square or higher power.   Moret-Blanc[55] proved the last statement by Bertrand's postulate.

Moret-Blanc[56] solved $y(y+1)(y+2)=x(x+1)$, proposed by Lionnet. Adding 1 to the product by 4, we are to make $4y^3+12y^2+8y+1=\square$, say $(my-1)^2$.   The discriminant of the quadratic in y is to be rational.   Thus $m=2n$, $n^4-6n^2-4n+1=\square$, which holds for $n=3$.   Thus solutions are $1 \cdot 2 \cdot 3 = 2 \cdot 3$, $5 \cdot 6 \cdot 7 = 14 \cdot 15$.   G. C. Gerono (p. 432) noted that, since $2x+1=2ny-1$, the initial equation becomes $y^2-(n^2-3)y+n+2=0$ and proved that $n=3$.

E. Lionnet proposed and Moret-Blanc[57] solved the problem to find N such that both N and $N/2$ are products of two consecutive integers, the smaller factor of $N/2$ being a product $x(x+1)$ of two consecutive integers. Thus

$$2(x^2+x)(x^2+x+1)=y^2+y, \qquad 8x^4+16x^3+16x^2+8x+1=(2y+1)^2.$$

Euler's process to deduce new solutions from $x=1$ leads only to $x=0$ or fractional values.

E. Lemoine[58] asked if the product of three consecutive numbers (besides 2, 3, 4) is of the form px^3, where p is a prime.   H. Brocard (p. 304) noted that the problem reduces to $y^3-y=px^3$, took $y=p$ and concluded that $x=2$, $y=3$.   Several replies (p. 369) show readily that 2, 3, 4 is the only solution.

E. B. Escott[59] proved that $x(x+4)(x+6) \neq \square$.

G. de Rocquigny[60] proposed for solution

$$x(x+1) \cdots (x+5) = y(y+1)(y+2).$$

E. B. Escott[61] noted the solutions $x=1$ or -6, $y=8$, besides the evident solutions $x=0$, -1, $\cdots$, -5.   P. F. Teilhet[62] proved that these are the only solutions by noting that the left member becomes $(z-4)z(z+2)$ for $z=(x+1)(x+4)$.

---

[50] Nouv. Ann. Math., (2), 11, 1872, 172.

[51] Nouv. Corresp. Math., 2, 1876, 287; Nouv. Ann. Math., (3), 4, 1885, 391.

[52] Mathesis, 7, 1887, 280.

[53] Nouv. Corresp. Math., 4, 1878, 123; Théorie des nombres, 1891, 351, Ex. 4.   Proof by P. Bachmann, Niedere Zahlentheorie, I, 1902, 44-6.

[54] Nouv. Ann. Math., (2), 20, 1881, 515.

[55] Ibid., (3), 1, 1882, 362.   Invalid objection by G. C. Gerono, p. 520.

[56] Nouv. Ann. Math., (2), 20, 1881, 431-2.   Same, Zeitschr. Math. Naturw. Unterricht, 13, 1882, 451.

[57] Nouv. Ann. Math., (2), 20, 1881, 375.

[58] L'intermédiaire des math., 2, 1895, 15.

[59] Ibid., 7, 1900, 211-3.

[60] Ibid., 9, 1902, 203.

[61] Ibid., 10, 1903, 132.

[62] Ibid., 12, 1905, 116-8.

P. F. Teilhet[63] stated for $m=3$ and several proved that, if m is a prime, $n(n+1)(n+2)=mA^2$ is impossible.

A. Gérardin[64] remarked that if $1+x!=y^2$ has solutions other than $x=4, 5, 7; y=5, 11, 71$, then y has at least 20 digits.

SUM OF nTH POWERS AN nTH POWER.

Euler (Ch. XXII, paper 187 and the one preceding it) expressed his belief that no sum of four fifth powers is a fifth power.

E. Collins[65] noted that if $N=1+n+n^2+\cdots+n^{k-1}$ is divisible by a prime p, then $p\equiv1$ (mod k), since $n^k=(n-1)N+1$. Henceforth, let this N be a prime. Then, if A is any integer not divisible by N, A^q is congruent to a power of n modulo N, where $q=(N-1)/k$, since A^q is a root of $x^k\equiv1$ (mod N), and its roots are powers of n. Hence if $a_1^q+\cdots+a_n^q=A^q$, while $a_1, \cdots, a_n$ are not divisible by the prime N, the difference of some two of the a_i^q is divisible by N. For example, if $n=2, k=3$, then $N=7, q=2$, whence if a sum of two squares (each prime to 7) is a square, their difference is divisible by 7. Again, let $N=1+5+5^2=31$; then $q=10$ and, if a sum of five tenth powers (not divisible by 31) be a tenth power, a difference of two of the powers is divisible by 31. He verified that $q>n$ except when $k=2$, or $k=3, n=2$. He conjectured that a sum of n numbers each an eth power is not an eth power if $n<e$.

F. Paulet[66] announced that no nth power is a sum of nth powers if $n>2$. A committee reported adversely, citing the known formula $6^3=3^3+4^3+5^3$.

O. Schier[67] made an erroneous discussion of $x^n+y^n+z^n=u^n$. First, let n be an odd prime. Then $x+y+z=u+nd$. Subtract its nth power from the given equation. The new left member has the factor $y+z$ which is said to be divisible by the factor n of the new right member. This admitted, the given equation would be impossible for n a prime >3 and hence for any $n>3$. Only special sets of solutions are found for $n=3$ and $n=2$.

A. Martin[68] found by tentative methods (Hart[115] and Martin[119] of Ch. XXI)

$$4^5+5^5+6^5+7^5+9^5+11^5=12^5, \qquad 5^5+10^5+11^5+16^5+19^5+29^5=30^5,$$

$$\sum_{k=1}^{100} k^3-1^3-6^3-11^3-21^3-43^3=294^3, \qquad 1^3+3^3+4^3+5^3+8^3=9^3,$$

$$\sum_{k=1}^{100} k^4-1^4-2^4-3^4-4^4-8^4-10^4-14^4-24^4-42^4-72^4=212^4.$$

Barbette[198] of Ch. XXII noted that the first result is the only one involving fifth powers each $\leq12^5$.

Martin[69] would by trial express $1^n+2^n+\cdots+x^n-b^n$ as a sum of distinct nth powers each $\leq x^n$. For $n=5, x=11, b=12$, we get his[68] first result.

[63] L'intermédiaire des math., 11, 1904, 68, 182–4.
[64] Nouv. Ann. Math., (4), 6, 1906, 223.
[65] Mém. Acad. Sc. St. Pétersbourg, 8, années 1817 et 1818, 1822, 242–6.
[66] Comptes Rendus Paris, 12, 1841, 120, 211.
[67] Sitzungsber. Akad. Wiss. Wien (Math.), 82, II, 1881, 883–892.
[68] Bull. Phil. Soc. Wash., 10, 1887, 107; in Smithsonian Miscel. Coll., 33, 1888.
[69] Math. Quest. Educ. Times, 50, 1889, 74–5.

Martin and G. B. M. Zerr[70] multiplied the numbers 4, 5, $\cdots$, 12 in the formula just cited by 42^4 and obtained six numbers whose sum is a fifth power 42^5 and sum of fifth powers is a fifth power.

Martin[71] multiplied his[68] first formula by 2^5 and replaced the new third term 12^5 by its value to get a formula for 24^5. There is an analogous longer formula for 50^5. Again,

$$1^6+2^6+4^6+5^6+6^6+7^6+9^6+12^6$$
$$+13^6+15^6+16^6+18^6+20^6+21^6+22^6+23^6=28^6.$$

Martin[72] found sets of fifth powers whose sum is a fifth power.

G. de Rocquigny[73] proposed for solution $(x-r)^m+x^m+(x+r)^m=y^m$. H. Brocard[74] noted $x=4$, $r=1$, $y=6$ $[m=3]$, and E. B. Escott[74] noted $x=1$, $r=2$, $y=3$, for m any odd number. Cf. Gelin,[93] also Escott[261] of Ch. XXI, and Bottari[190] of Ch. XXV.

A. Martin[75] found sixth powers whose sum is a sixth power by the tentative method of expressing p^6-q^6 as a sum of distinct sixth powers $\neq q^6$, or $S-b^6$ as a sum of sixth powers $\leqq n^6$, where $S=1^6+\cdots+n^6$. By each method he found his[71] example, also that the sum of the sixth powers of 1, 1, 2, 5, 9, 11, 12, 13, 15, 18, 21, 22, 23, 24 is 29^6 [false] and that of 1, 2, 2, 4, 5, 6, 8, 9, 10, 12, 14, 15, 18, 19, 27, 33, 49 is 50^6 (each with one repeated term). By combining these, he found eleven new sets of 29, 31 (seven), 32, 46, 47. He tabulated the values of n^6 and $1^6+\cdots+n^6$ for $n\leqq228$.

C. Bianca[76] noted that $s=a_1^p+\cdots+a_{n+1}^p$ is a pth power if

$$a_1 : a_2 : \cdots : a_{n+1} = b^n : b^{n-1}c : b^{n-2}cd : b^{n-3}cd^2 : \cdots : bcd^{n-2} : cd^{n-1},$$

where $b^p+c^p=d^p$. For, if $a_1=kb^n$, $\cdots$, then $s=(kd^n)^p$.

A. Martin[77] reported on sums of nth powers equal to an nth power.

* N. Agronomof[78] proved that $x_1^{2m+1}+\cdots+x_k^{2m+1}=0$ is solvable in integers if $k=4^n+1$ and $n\geqq m$. He proved the identity

$$\Sigma_1-\Sigma_2+\cdots+(-1)^{2m+1}\Sigma_{2m+2}=0,$$

where Σ_j denotes the sum of the $(2m+1)$-th powers of all the sums of $2m+2$ parameters taken j at a time. A. Filippov[78a] gave an account in French of this paper, with details for the case $m=2$.

---

[70] Math. Quest. Educ. Times, 55, 1891, 118.

[71] Quar. Jour. Math., 26, 1893, 225–7.

[72] Math. Papers Internat. Congress of 1893 at Chicago, 1896, 168–174. Republished, Math. Mag., 2, 1898, 201–8, with the following corrections: In Ex. 18, p. 173, insert 16^5; on p. 169, fourth line up, delete one 3^5; on p. 174, delete the final equation. In Part III (combining earlier sets) he added a new set of n fifth powers for $n=17$, 21, 24, 26, 28, 36, 42, 48, 52, 63, 67, 72 and three sets for $n=33$.

[73] L'intermédiaire des math., 9, 1902, 203.

[74] Ibid., 10, 1903, 131–3.

[75] Math. Mag., 2, 1904, 265–271.

[76] Il Pitagora Palermo, 13, 1906–7, 65–6.

[77] Proc. Fifth Intern. Congress of Math., 1912, I, 431–7.

[78] Izv. Fis. Mat. Obs. Kazan (Bull. Soc. Phys. Math. Kasan), 1914, 1915.

[78a] Tôhoku Math. Jour., 15, 1919, 135–40.

TWO EQUAL SUMS OF nTH POWERS.

A. Desboves[79] noted that $u^5+v^5=s^5+w^5$ has the complex solution

$$u,\ v=2xy\pm(x^2-2y^2);\qquad s,\ w=2xy\pm(x^2+2y^2)\sqrt{-1}.$$

J. W. Nicholson[80] recalled that, if $s=a_1+\cdots+a_m$,

$$s^n=\Sigma(s-a_1)^n-\Sigma(s-a_1-a_2)^n+\cdots-(-1)^m\Sigma(a_1+a_2)^n+(-1)^m\Sigma a_1^n.$$

Thus $11^n=9^n+8^n+5^n-6^n-3^n-2^n$ for $n=2$ or 1 [Euler[2], Ch. XXIV]; etc.

Several writers[81] determined the signs so that

$$1^n\pm2^n\pm3^n\pm\cdots\pm(2^{n+1})^n=1^n\pm3^n\pm5^n\pm\cdots\pm(2^{n+2}-1)^n.$$

A. de Farkas[82] proved it is impossible to find two different sets x_i and y_i such that for a and q are arbitrary

$$(x_1+a)^n+(x_2+aq)^n+(x_3+aq^2)^n+\cdots=(y_1+a)^n+(y_2+aq)^n+\cdots.$$

N. Agronomof[83] argued the existence of integral solutions of the equation

$$x_1^\rho+\cdots+x_h^\rho=y_1^\rho+\cdots+y_g^\rho,\qquad h\geqq2^{\rho-3},\qquad g\geqq2^{\rho-3},\qquad \rho>4.$$

But, as shown by Filippov[78a] for the case $\rho=5$, $h=g=4$, the method leads only to the trivial solution $x_1==-x_3$, $x_2=-x_4$, $y_1=-y_3$, $y_2=-y_4$.

C. B. Haldeman[83a] gave special rational solutions of $s_3=s_4$ and $s_8=s_n$, where s_n denotes a sum of n fifth powers.

On $x^n+v^n=y^n+u^n$, see Steggall[180] of Ch. XXII.

MISCELLANEOUS RESULTS ON SUMS OF LIKE POWERS.

J. Hill[84] noted that the sum of the cubes of $x^2/2$, $2x^2/3$, $5x^2/6$ is a sixth power x^6. Cf. Emerson[52] of Ch. XXI.

L. Euler[85] stated that no sum of three biquadrates is divisible by 5 or 29, which alone are exceptional. Cf. Gegenbauer[126] of Ch. XXVI.

R. Elliott[86] noted that $1^5+\cdots+n^5=\square$ if $F=\frac{1}{3}(2n^2+2n-1)=\square$ and took $n=x+1$. Then $9F=6x^2+18x+9=\square=(ax-3)^2$ determines x. The anonymous proposer solved $F=\alpha^2$ for n; the radical must be a rational number $3c$. Take $\alpha=p+q$, $c=p-q$. Then $p^2-10pq+q^2=1$, whence $24q^2+1=\square$, whose solution is known.

G. Libri[87] expressed as a trigonometric sum the number of sets of solutions of $x_1^a+\cdots+x_k^a+1\equiv0\pmod p$, where p is a prime $an+1$ [Libri[147]]. Cf. pp. 224–5 of Vol. I of this History.

---

[79] Assoc. franç., 9, 1880, 242–4.
[80] Amer. Math. Monthly, 9, 1902, 187, 211.
[81] Math. Quest. Educat. Times, (2), 13, 1908, 110–111.
[82] L'intermédiaire des math., 20, 1913, 79–80.
[83] Tôhoku Math. Jour., 10, 1916, 211.
[83a] Amer. Math. Monthly, 25, 1918, 399–402.
[84] Ladies' Diary, 1737, Quest. 192; Leybourn's Math. Quest. L. D., 1, 1817, 254–5. Cf. Math. Quest. Educ. Times, 66, 1897, 120.
[85] Opera postuma, I, 1862, 186 (between 1775 and 1779).
[86] Ladies' Diary, 1796, 40–1, Quest. 992; Leybourn's M. Quest. L. D., 3, 1817, 296–7.
[87] Mém. divers savants acad. sc. de l'Institut de France (math.), 5, 1838, 61–63.

V. Bouniakowsky[88] obtained the identity

$$(10\lambda^2+x)^5+(10\lambda^2-x)^5+8(10\lambda^2)^5=(10^3\lambda^5+10\lambda x^2)^2$$

from $\int\{(x+a)^4-(x-a)^4\}dx$ by setting $a=10\lambda^2$.

E. Lucas[88a] stated that the sum of the cubes of the first n (odd) integers is never a cube, fifth or eighth power (cube, fourth or fifth power). The sum of the cubes of three consecutive integers is never a square, cube or fifth power, except for $1^3+2^3+3^3=6^2$, $3^3+4^3+5^3=6^3$ [correction, Aubry[286] of Ch. XXI]. The sum of the first n biquadrates is never a square, cube or fifth power. The sum of the first n fifth powers is never a cube, fourth or fifth power.

E. Lucas[89] asked for what values of n the sum of the fifth powers of the first n odd numbers is a square. The problem reduces to

$$x^4-5x^2y^2+7y^4=3z^2,$$

whose complete solution was given by L. Aubry.[90]

Lucas[91] asked for what n's the sum of the fifth or seventh powers of 1, $\cdots$, n is a square. H. Brocard[91] noted that the sum of the fifth powers is $\frac{1}{4}n^2(n+1)^2t$, where $t=(2n^2+2n-1)/3$. To make $t=y^2$, we have

$$(2n+1)^2=6y^2+3,$$

which must have 9 as its final digit, whence $y=10m\pm1$. He noted the special solutions $y=n=1$; $y=11$, $n=13$. Cf. Moret-Blanc,[95] Fortey.[99]

H. Brocard[92] noted that the sum $n^2(2n^2-1)$ of the cubes of the first n odd numbers is a square for $n=1,5,29,169,985,\cdots$. As to Lucas'[88a] theorem that the sum s of the squares of the first n odd numbers is not a square, cube or fifth power, he stated that this is evident since $s=(2n-1)(2n)(2n+1)/6$. Lucas (p. 247–8) noted that this proof would require extensive developments; if p is a product of three consecutive numbers, $p/6$ is not a square if the first of the three numbers is odd, and also if it be even except for $2\cdot3\cdot4/6=2^2$, $43\cdot49\cdot50/6=140^2$.

Abbé Gelin[93] proved that $(x-1)^{2n}+x^{2n}+(x+1)^{2n}=y^{2n}$ is impossible and that the sum of like even powers of 9 or 12 consecutive integers is never an exact power (stated for 9 by Lucas, p. 248). The proof is by use of various properties of $\Sigma(N)$, obtained by adding the digits of N, then adding the digits of this sum, etc., until there results a sum with a single digit.

E. Lucas stated and H. Brocard, Radicke and E. Cesàro[94] proved that

$$\{1^5-3^5+5^5-\cdots-(4x-1)^5\}/\{1-3+5-\cdots-(4x-1)\}$$

[88] Bull. Acad. Sc. St. Pétersbourg (Phys.-Math.), 11, 1853, 65–74. Extract in Sphinx-Oedipe, 5, 1910, 14–16.

[88a] Recherches sur l'analyse indéterminée, Moulins, 1873, 91–2. Extract from Bull. Soc. d'Emulation du Département de l'Allier, 12, 1873, 531–2.

[89] Nouv. Corresp. Math., 2, 1876, 95.

[90] L'intermédiaire des math., 18, 1911, 60–62. Cf. 16, 1909, 283.

[91] Nouv. Corresp. Math., 3, 1877, 119–120. Cf. 4, 1878, 167.

[92] Ibid., 3, 1877, 166–7.

[93] Ibid., 388–390 (extract from Les Mondes, July 14, 1877).

[94] Ibid., 5, 1879, 112, 213–5; 6, 1880, 467.

is always a square, but never a biquadrate.

Moret-Blanc[95] found the x's for which (Lucas[91])

$$1^5 + \cdots + x^5 = \left\{ \frac{x(x+1)}{2} \right\}^2 \left\{ \frac{(2x+1)^2 - 3}{6} \right\} = \square.$$

Hence $(3u^2 - 1)/2 = v^2$ or $(3u - 2v)^2 - 6(v - u)^2 = 1$, whose solutions are given by the convergents of odd rank in the continued fraction for $\sqrt{6}$.

E. Catalan[96] noted that, if p is an odd prime and j is an odd integer $\leqq p - 1$, the sum of the $\frac{1}{2}(p-1)$th powers of j integers relatively prime to p is not divisible by p.

A. Berger[97] proved that, if s, m, n, g_1, $\cdots$, g_s are positive integers, and $\psi(n)$ is the number of positive integral solutions of $g_1 x_1^m + \cdots + g_s x_s^m = n$,

$$\lim_{n = \infty} \frac{\psi(1) + \cdots + \psi(n)}{n^{s/m}} = (g_1 \cdots g_s)^{-1/m} \frac{\Gamma(1 + 1/m)^s}{\Gamma(1 + s/m)}.$$

L. Gegenbauer[98] proved a generalization of Catalan's[96] theorem. If λ is one of the numbers 2, 3, 4, and if p is a prime $\equiv 1 \pmod{\lambda}$, and r an integer prime to λ and $< p^{1/t}$, where t is the largest integer $\leqq (\lambda + 1)/2$, then the sum of the $(p-1)/\lambda$th powers of r integers relatively prime to p is not divisible by p.

H. Fortey[99] found that $1^5 + \cdots + n^5 = \square$ for $n = 1$, 13, 133, 1321, $\cdots$, by use of $3y^2 - 2x^2 = 1$. Cf. Moret-Blanc.[95]

E. Lemoine[100] said that A is decomposed into maximum nth powers if $A = a_1^n + \cdots + a_p^n$, where a_1^n, a_2^n, a_3^n, $\cdots$ are the largest nth powers $\leqq A$, $A - a_1^n$, $A - a_1^n - a_2^n$, $\cdots$, respectively. Similarly, consider the decomposition $A = \alpha_1^n - \alpha_2^n + \alpha_3^n - \cdots \pm \alpha_p^n$, where α_1 is the least integer $\geqq \sqrt[n]{A}$ and R_1 the remainder $\alpha_1^n - A$, α_2 is the least integer $\geqq \sqrt[n]{R_1}$ and R_2 the remainder, α_3 the least integer $\geqq \sqrt[n]{R_2}$, etc., and call γ_p the least number requiring p powers. Then, for $n = 2$, $\gamma_1 = 1$, $\gamma_2 = 3$, $\gamma_3 = 6 = 3^2 - 2^2 + 1^2$, $\gamma_{p+1} = \frac{1}{4}\gamma_p^2 + 1$. For $n = 3$, he[101] gave elsewhere the possible forms of the final power a_p^3.

L. Aubry[102] proved that $-1^3 + 3^3 - 5^3 + \cdots + (4n - 1)^3$ is never a square, cube or biquadrate.

Welsch and E. Miot[103] noted cases in which $a^n + (a+1)^n + \cdots + (a+k)^n$ is of the form $l^2 - m^2$ and hence is a sum of consecutive odd numbers of which the least is $2m + 1$.

C. Bisman[104] noted that a sum of like even powers of $n^2 + 4$ numbers can be expressed as the algebraic sum of $n^2 + 5$ squares of which only one is taken negatively.

[95] Nouv. Ann. Math., (2), 20, 1881, 212.

[96] Mém. Soc. R. Sc. de Liège, (2), 13, 1880, 291. Cf. Gegenbauer.[98]

[97] Öfversigt K. Vetenskaps-Akad. Förhand., Stockholm, 43, 1886, 355–66.

[98] Sitzungsber. Akad. Wiss. Wien (Math.), 95, II, 1887, 838–842.

[99] Math. Quest. Educ. Times, 48, 1888, 30–31.

[100] Assoc. franç., 25, 1896, II, 73–7. For $n = 2$, see papers 20, 21 of Ch. IX.

[101] L'intermédiaire des math., 1, 1894, 232.

[102] Sphinx-Oedipe, 6, 1911, 38–9. E. Lucas, Nouv. Corresp. Math., 5, 1879, 112, had asked
    for solutions.

[103] L'intermédiaire des math., 20, 1913, 47–48.

[104] Mathesis, (4), 3, 1913, 257–9.

T. Suzuki[105] noted that there are at least $(p-2)(p-1)^{n-2}$ solutions of

$$a_1^{x_1}+\cdots+a_n^{x_n}\equiv 0 \quad (\mathrm{mod}\ p),$$

if two of the a's are primitive roots of the prime p.   Also there are solutions if a_1 is a primitive root and if not every $a_i\equiv 1$ (mod p) for $i=2,\cdots,n$.

Rational solutions of $x^y=y^x$.

L. Euler[106] set $y=tx$ and deduced $x^{t-1}=t$.   The graph is composed of $y=x$, a branch asymptotic to the positive x and y axes, and an infinity of isolated points.   Among the rational solutions are $(x,\ y)=(2,\ 4)$, $(3^2/2^2$, $3^3/2^3)$, $(4^3/3^3,\ 4^4/3^4)$.

D. Bernoulli[107] noted that, for $x\neq y$, the only integral solution is 2, 4; but that there is an infinitude of rational solutions.

J. van Hengel[108] remarked that $r^{r+n}>(r+n)^r$ if r and n are positive integers either one ≥ 3.   Thus if $a^b=b^a$, it remains to treat the cases $a=1$ or 2.   If $a=2$, $b>4$, whence $b=2+n$, we apply the above remark.

* C. Herbst[109] noted that 2, 4 give the only solution in integers.

* A. Flechsenhaar[110] and R. Schimmack[111] discussed the rational solutions.

A. M. Nesbitt[112] and E. J. Moulton[113] discussed the graph of $x^y=y^x$.

A. Tanturri[114] proved that 2, 4 give the only solution in integers.

Product of factors $(x+1)/x$ equal to such a fraction.

Fermat[115] proposed the problem to find in how many ways $(n+1)/n$ can be expressed as a product of k such fractions, citing the case $n=8$, $k=10$, as suitable to be proposed to all mathematicians of his time.   Tannery noted that of the decompositions of 9/8 the difference of the factors is least and greatest in respectively

$$\frac{90}{89}\cdot\frac{89}{88}\cdot\frac{88}{87}\cdot\frac{87}{86}\cdot\frac{86}{85}\cdot\frac{85}{84}\cdot\frac{84}{83}\cdot\frac{83}{82}\cdot\frac{82}{81}\cdot\frac{81}{80},\qquad \frac{9+1}{9}\cdot\frac{9^2+1}{9^2}\cdot\frac{9^4+1}{9^4}\cdots\frac{9^{256}+1}{9^{256}}\cdot\frac{9^{512}}{9^{512}-1}.$$

V. Bouniakcwksy[116] noted that an irreducible fraction a/b less than unity can be expressed in an infinitude of ways as a product of fractions of the form $x/(x+1)$.   We may often find fewer than the $b-a$ fractions

[105] Tôhoku Math. Jour., 5, 1914, 48–53.   Cf. papers 265–6 of Ch. XXVI.

[106] Introductio in analysin infin., lib. 2, cap. 21, § 519; French transl. by J. B. Labey, 2, 1797 and 1835, 297.

[107] Corresp. Math. Phys. (ed., Fuss), 2, 1843, 262; letter to Goldbach, June 29, 1728.

[108] Beweis des Satzes, das unter allen reellen positiven ganzen Zahlen nur das Zahlen Paar 4 und 2 für a und b der Gleichung $a^a=b^b$ genügt, Progr. Emmerich, 1888.

[109] Unterrichtsbl. für Math., 15, 1909, 62–3.

[110] *Ibid.*, 17, 1911, 70–3.

[111] *Ibid.*, 18, 1912, 34–5.

[112] Math. Quest. Educ. Times, (2), 23, 1913, 77–8.

[113] Amer. Math. Monthly, 23, 1916, 233.

[114] Periodico di Mat., 30, 1915, 186–7.

[115] Oeuvres, I, 397.   Quoted by Tannery, l'intermédiaire des math., 9, 1902, 170–1.

[116] Mém. Acad. Sc. St. Pétersbourg (Sc. Math. Phys.), (6), 3, 1844, 1–16.

used in

$$\frac{a}{b} = \frac{a}{a+1} \cdot \frac{a+1}{a+2} \cdots \frac{b-1}{b}.$$

Set

$$\frac{a}{b} = \frac{p}{q} \cdot \frac{u}{u+1},$$

whence

$$u = \frac{aq}{bp-aq}.$$

Consider the case $bp-aq=1$ and let $p=\alpha$, $q=\beta$ be the least solutions. Then

$$\frac{a}{b} = \frac{\alpha}{\beta} \cdot \frac{\alpha\beta}{\alpha\beta+1}.$$

Proceed similarly with α/β. Many numerical examples are given.

A. Padoa[117] noted the equivalence of

$$\frac{n+1}{n} = \frac{x+1}{x} \cdot \frac{y+1}{y}, \qquad (x-n)(y-n) = n(n+1).$$

Hence if n is given we obtain all couples x, y by finding all pairs of positive integers whose product is $n(n+1)$, and adding n to each factor.

J. E. A. Steggall[118] found positive integral solutions of

$$(1) \qquad \frac{x+1}{x} \cdot \frac{y+1}{y} = \frac{z+1}{z},$$

by noting that xy must be divisible by $x+y+1=a$, and hence $x(x+1)$ by a. Hence for any integer x, determine a factor $a>x+1$ of $x(x+1)$; then $y=a-x-1$, while $z=x-b$ where $b=x(x+1)/a$. T. W. Chaundy (pp. 74-5) deduced $(x-z)(y-z)=z(z+1)$ and set $z=pq$, $x-z=p_1q$, where p, p_1 are relatively prime. Hence $y-z=pq_1$, $p_1q_1=pq+1$.

G. Ascoli and P. Niewenglowski[119] gave solutions of (1).

A. M. Legendre[120] evaluated, up to $w=1229$,

$$\frac{2}{3} \cdot \frac{4}{5} \cdot \frac{6}{7} \cdot \frac{10}{11} \cdots \frac{w-1}{w}.$$

Optic formula $\dfrac{1}{x}+\dfrac{1}{y}=\dfrac{1}{a}$; generalization.

An anonymous writer[121] noted that, if three regular polygons of x, y, z sides fill the space about a point, then $1/x+1/y+1/z=1/2$. If there are four regular polygons of x, y, z, z sides, then $1/x+1/y+2/z=1$. The number of solutions is found, also for 5 or 6 polygons.

---

[117] L'intermédiaire des math., 10, 1903, 30–31.
[118] Math. Quest. Educ. Times, (2), 20, 1911, 50–1.
[119] Supplem. al Periodico di Mat., 14, 1911, 101–4, 116–7.
[120] Théorie des nombres, ed. 2, 1808; ed. 3, 1830. Table IX.
[121] Ladies' Diary, 1785, 40–1, Quest. 829; Leybourn's M. Quest. L. D., 2, 1817, 132–3.

D. André[122] deduced $x - a = d$, $y - a = e$, where $de = a^2$, the pair of divisors $d = e = -a$ of a^2 being excluded. Züge[123] gave $x = a + p^2$, $y = a + q^2$, where $pq = a$. F. Schilling[124] noted that Züge's solution is incomplete and gave that due to André with a geometrical interpretation of the optic formula.

A. Thorin[125] asked if $1/a = 1/a_1 + 1/a_2$ has integral solutions besides

$$a = mn, \qquad a_1 = m(n+1), \qquad a_2 = mn(n+1).$$

A. Palmström, J. Sadier, and C. Moreau[126] each gave the solution

$$a = \lambda mn, \qquad a_1 = \lambda m(m+n), \qquad a_2 = \lambda n(m+n),$$

and noted that

$$(1) \qquad \frac{1}{a} = \frac{1}{a_1} + \cdots + \frac{1}{a_n}$$

has the special solution

$$a = \lambda \alpha_1 \cdots \alpha_n, \qquad a_1 = \lambda s \alpha_1, \cdots, a_n = \lambda s \alpha_n, \qquad s \equiv \sum_{i=1}^{n} \frac{\alpha_1 \cdots \alpha_n}{\alpha_i}.$$

Dujardin[127] stated that, if $n = 2$, all solutions are given by

$$a_2 = a + \lambda, \qquad a_1 = a + \frac{a^2}{\lambda} \quad (\lambda \text{ a divisor of } a^2),$$

while (1) may be written $A a_n = a(B a_n + C)$, where $A = a_1 \cdots a_{n-1}$, and B, C are integral functions of $a_1, \cdots, a_{n-1}$ [with $C = A$]. Then $Ba = A - AC/\lambda$, where $\lambda = B a_n + C$. Hence give to $a_1, \cdots, a_{n-1}$ any values and choose a divisor λ of AC. Take as B and a two integers whose product is $A - AC/\lambda$. If $\lambda - C$ is divisible by B, we get a solution.

M. Lagoutinsky[128] stated that if $n = 3$ the complete solution of (1) is given by formulas involving 13 parameters.

V. V. Bobynin[129] discussed the expressing of fractions in the form $\Sigma 1/x_i$ in the papyrus of Akhmim (Achmîm), about the seventh century, and in the Liber Abbaci of Leonardo Pisano.

A. Palmström[130] treated, as an example of a more general type,[27]

$$\frac{1}{x_1} = \frac{1}{x_2} + \cdots + \frac{1}{x_n},$$

which may be written in the form

$$\begin{vmatrix} -x_2 & x_3 & 0 & 0 & \cdots & 0 \\ -x_2 & 0 & x_4 & 0 & \cdots & 0 \\ -x_2 & 0 & 0 & x_5 & \cdots & 0 \\ \cdot & \cdot & \cdot & \cdot & \cdot & \cdot \\ -x_2 & 0 & 0 & 0 & \cdots & x_n \\ x_1 - x_2 & x_1 & x_1 & x_1 & \cdots & x_1 \end{vmatrix} = 0.$$

[122] Nouv. Ann. Math., (2), 10, 1871, 298.
[123] Zeitschrift Math. Naturw. Unterricht, 26, 1895, 15–16.
[124] Ibid., 491–3.
[125] L'intermédiaire des math., 2, 1895, p. 3.
[126] Ibid., 299–302.
[127] Ibid., 3, 1896, 14.
[128] Ibid., 4, 1897, 175.
[129] Abh. Geschichte Math., IX, 1–13 (Suppl. Zeitsch. Math. Phys., 44, 1899).
[130] Skrifter Udgivne af Videnskabsselskabet, Christiania, 1900 (1899), Math.-Naturw. Kl., No. 7 (German). L'intermédiaire des math., 5, 1898, 81–3.

For integral solutions x_i there exist relatively prime integers a_i satisfying

$$-a_1x_2+a_ix_{i+1}=0 \ (i=2, \cdots, n-1), \qquad a_1(x_1-x_2)+a_2x_1+\cdots+a_{n-1}x_1=0,$$

and conversely. Hence

$$x_1=ka_1\cdots a_{n-1}, \qquad x_j=ka_1\cdots a_{n-1}(a_1+\cdots+a_{n-1})/a_{j-1},$$

k being chosen to make the x's integers.

M. Lagoutinsky[131] treated (1) for the case in which a, a_1, $\cdots$ have no common divisor. Call their l.c.m. A, and set $A/a=k$, $A/a_i=k_i$. Thus $k=\Sigma k_i$. Hence we take $k_1, \cdots, k_n$ to be any integers without a common divisor and find the l.c.m. A of these k_i's and $k=\Sigma k_i$. Then the solution is $a=A/k$, $a_i=A/k_i$.

Züge[132] solved $axy+bx+cy+d=0$ by multiplying by a. Thus $ax+c=P$, $ay+b=Q$, where $bc-ad=PQ$. For integral solutions, select the factors P, Q so that $P\equiv c$, $Q\equiv b$ (mod a). For the special case $xy=a(x+y)$, the result by André[122] follows.

P. Whitworth[133] noted that each divisor of $N^2=(x-N)(y-N)$ yields a solution of $1/x+1/y=1/N$.

P. Zühlke[134] gave, for $1/x+1/y=2/m$, $2x-m=p$, $2y-m=q$, $pq=m^2$. If m is odd the resulting x, y are integers.

E. Sós[135] noted that the general solution of $1/x=1/x_1+1/x_2$ is

$$x=ky_1y_2, \qquad x_1=ky_1(y_1+y_2), \qquad x_2=ky_2(y_1+y_2),$$

where y_1, y_2 are any relatively prime integers. Calling such a solution irreducible if $k=1$, and setting $x=p_1^{a_1}\cdots p_\nu^{a_\nu}$, where $p_1, \cdots, p_\nu$ are distinct primes, we find that there are $2^{\nu-1}$ essentially distinct irreducible solutions belonging to a given x, with x_2, x_1 counted the same as x_1, x_2; in all,

$$\frac{1}{2}\left\{\prod_{k=1}^{\nu}(1+2a_k)+1\right\}$$

essentially distinct solutions belonging to x. For the complete solution of

$$(2) \qquad \frac{1}{x}=\frac{1}{x_1}+\cdots+\frac{1}{x_n},$$

2^n-1 parameters y_i are introduced.

Sós[136] noted that, if the a's are given integers,

$$(3) \qquad \frac{a}{z}=\frac{a_1}{z_1}+\cdots+\frac{a_n}{z_n}$$

has (not the only) solutions $z=ax$, $z_i=a_ix_i$, if (2) holds. The complete solution in positive integers, with g.c.d. unity, is obtained for (3). The method is similar to that for the case $n=2$. Set $z_1=ZZ_1$, $z_2=ZZ_2$, where Z_1, Z_2 are relatively prime. Then

$$z=fZ, \qquad f=\frac{aZ_1Z_2}{a_1Z_2+a_2Z_1}.$$

[131] L'intermédiaire des math., 7, 1900, 198.
[132] Archiv Math. Phys., (2), 17, 1900, 329–32.
[133] Math. Quest. Educ. Times, 75, 1901, 85.
[134] Archiv Math. Phys., (3), 8, 1905, 88.
[135] Zeitschrift Math. Naturw. Unterricht, 36, 1905, 97.
[136] *Ibid.*, 37, 1906, 186–190.

Let $f=p/q$, where p, q are relatively prime.   Thus Z is a multiple z^1q of q and $z=z^1p$, $z_1=z^1qZ_1$, $z_2=z^1qZ_2$.

A. Flechsenhaar[137] and E. Schulte discussed $1/a+1/b=1/c$.   E. Sós (p. 113) treated (2).   W. Hofmann[138] discussed the integral solutions of

$$\frac{1}{a}+\frac{1}{b}=\frac{1}{c}, \qquad \frac{1}{a}-\frac{1}{b}=\frac{1}{b}-\frac{1}{c}.$$

G. Lemaire[139] transformed given decompositions $\Sigma 1/f$ of $9/10$ into others.

R. Janculescu[140] noted that in $1/x+1/y=1/z$, z will be integral only when the g.c.d. d of x and y is a multiple of $x/d+y/d$.

D. Biddle[141] solved each of $1/(a\pm b)+1/(c\pm a)=1/a$.

MISCELLANEOUS SINGLE EQUATIONS OF DEGREE $n>4$.

J. L. Lagrange[142] noted that, if a is a fixed nth root of unity, the product of two functions of the type

$$p\equiv t+ua\sqrt[n]{A}+xa^2\sqrt[n]{A^2}+\cdots+za^{n-1}\sqrt[n]{A^{n-1}}$$

is of like form.   Hence if we replace a by the different nth roots of unity and form the product of the functions so obtained from p, we obtain a rational function P of t, u, $\cdots$, z, A such that the product of two functions of type P is a third function of type P.   We can find P by eliminating ω between

$$\omega^n-A=0, \qquad t+u\omega+x\omega^2+\cdots+z\omega^{n-1}=l;$$

then P is the term free of l in the eliminant.   For example, if $n=2$, $P=t^2-Au^2$.   An application is to the solution of

$$(1) \qquad\qquad\qquad r^n-As^n=q^m.$$

We seek to express each factor $r-asA^{1/n}$ as an mth power p^m, where $a^n=1$, and p is the above linear function.   Then

$$p^m=T+Ua\sqrt[n]{A}+Xa^2\sqrt[n]{A^2}+\cdots+Za^{n-1}\sqrt[n]{A^{n-1}}.$$

Hence $r=T$, $s=-U$, $X=0$, $\cdots$, $Z=0$.   Thus (1) is solvable by this method if $X=0$, $\cdots$, $Z=0$ are solvable.   Although only $n-2$ equations in n variables, they do not always have rational solutions.   For details on the case $n=3$, $m=2$, and Lagrange's extension of the method in his addition IX to Euler's Algebra where $a^n=1$ is replaced by any equation of degree n, see papers 161-6 of Ch. XXI; also Ch. XX.

Lagrange[143] treated the problem to make $y=p/q$ an integer when $p=a+bx+\cdots$, $q=a^1+b^1x+\cdots$ are polynomials in x.   By eliminating x,

[137] Unterrichtsblätter Math., 16, 1910, 41, 41-2.

[138] *Ibid.*, 17, 1911, 14-15.

[139] L'intermédiaire des math., 18, 1911, 214-6.

[140] Mathesis, (4), 3, 1913, 119-120.

[141] Math. Quest. Educat. Times, (2), 25, 1914, 61-3.

[142] Mém. Acad R. Sc. Berlin, 23, année 1767, 1769; Oeuvres, II, 527-532.   Exposition by A. Desboves, Nouv. Ann. Math., (2), 18, 1879, 265-79; applications, 398-410, 433-444, 481-499; also by R. D. Carmichael, Diophantine Analysis, New York, 1915, 35-63.   Cf. Dirichlet[19]; also Libri[64, 65] of Ch. XXV.

[143] Addition IV to Euler's Algebra, 2, 1774, 527-533.   Oeuvres de Lagrange, VII, 95-8.   Euler's Opera Omnia, (1), I, 579.

we get $0 = A + Bp + Cq + Dp^2 + \cdots$. Replacing p by qy, we see that A must be divisible by q. Hence we take for q the various factors of A in turn and solve $q = a^1 + b^1 x + \cdots$ for rational x's. A special treatment is necessary when q reduces to the constant a^1. G. Libri[144] eliminated x between the congruences $p \equiv 0$, $q \equiv 0 \pmod{q}$ and obtained $D \equiv 0 \pmod{q}$, where D is a function of the coefficients of p, q. Next, seek the integral solutions of $q = d$ for each divisor d of D in turn, and then solve $y = p/q$. As another method he suggested (p. 317) the use of series.

A. J. Lexell[144a] found values of p, q, r, s for which

$$\frac{\lambda(p^2 + s^2)(q^2 + r^2)}{pqrs(p^2 - s^2)(q^2 - r^2)} = \square.$$

L. Euler[145] treated $v^2 z^2 r^2 + \Delta x^2 y^2 s^2 = \square$, where

$$r = ax^2 + 2bxy + cy^2, \qquad s = av^2 + 2bvz + cz^2.$$

To make s have the factor r, set

$$z = agx + (f + bg)y, \qquad v = (f - bg)x - cgy.$$

Then $s/r = f^2 + (ac - b^2)g^2 \equiv t$. The proposed equation becomes

$$v^2 z^2 + \Delta t x^2 y^2 = \square,$$

which is of type (2) of Euler[143], Ch. XXII. The case $b = 0$ was treated in more detail.

G. Libri[146] treated $a^n x^n + b x^{n-1} + \cdots + q = z^n$ with all coefficients positive. Set $z = ax + e$, whence $x^{n-1}(na^{n-1}e - b) + \cdots + (e^n - q) = 0$. Seek the least e for which all the coefficients are positive and the greatest e for which they are all negative. For each integer e within these limits, seek the positive integral solutions x. If the coefficients in the given equation are not all positive, set $x = A + y$ and choose A so that the coefficients of the resulting equation will all be positive.

Libri[147] investigated the integral solutions ≥ 0 of $\phi(x, y, \cdots) = 0$ for which $x < a$, $y < b$, $\cdots$, where a, b, $\cdots$ are given positive integers. Set

$$X = x(x-1)(x-2)\cdots(x-a+1), \qquad Y = y(y-1)\cdots(y-b+1), \cdots.$$

Let $F = 0$ be the result of eliminating x, y, $\cdots$ between $\phi = 0$, $X = 0$, $Y = 0$, $\cdots$. If the equation of condition $F = 0$ is satisfied, take the equation, say $X_1(x) = 0$, in one variable, preceding the final stage of elimination. Then if X_2 is the g.c.d. of X_1 and X, all possible integral values of x occur among the roots of $X_2 = 0$; similarly for the other variables. The same method applies to a congruence $\phi \equiv 0 \pmod{a}$. For a a prime p, $X \equiv x^p - x \pmod{p}$, $Y \equiv y^p - y \pmod{p}$. Since

$$\frac{1}{m}\sum_{k=0}^{m-1}\left(\cos\frac{2k\pi}{m} + i\sin\frac{2k\pi}{m}\right)^n = 1 \text{ or } 0,$$

[144] Jour. für Math., 9, 1832, 74–75.
[144a] Euler's Opera postuma, 1, 1862, 487–90 (about 1766).
[145] Mém. Acad. Sc. St. Petersb., 9, 1819 [1780], 14; Comm. Arith., II, 414.
[146] Memoria sopra la teoria dei numeri, Firenze, 1820, 24 pp.
[147] Mémorie sur la théorie des nombres, Mém. divers Savants Acad. Sc. de l'Institut de France (Math. Phys.), 5, 1838 (presented 1825), 1–75.

according as n is divisible by m or not, the number of roots of $\phi\equiv0\ (\mathrm{mod}\ m)$ is

$$\frac{1}{m}\sum_{x,\,y,\,\ldots=0}^{m}\sum_{k=0}^{m-1}\cos\frac{2k\phi(x,\ y,\ \cdots)\pi}{m}.$$

When applied to $\phi=x^2+c$, this formula leads to Gauss' results on trigonometric sums. Again, $x^2+Ay^2+B\equiv0\ (\mathrm{mod}\ p)$ has $p\pm1$ sets of solutions.

Libri[148] noted that the number of sets of positive integral solutions of $\phi(x,\ y,\ \cdots)=0$ and the number of sets in which $x,\ y,\ \cdots$ take the values $1,\ \cdots,\ n-1$ are approximately

$$\sum_{x,\,y,\,\ldots=1}^{\infty}e^{-10s\phi^2},\qquad\sum_{x,\,y,\,\ldots=1}^{n}e^{-10s\phi^2}\qquad(s=x+y+\cdots),$$

respectively. To apply the method of the preceding paper to the linear congruence $\phi=Ax-1\equiv0\ (\mathrm{mod}\ p)$, A not divisible by p, we use $x^{p-1}-1\equiv0$, or $(Ax)^{p-1}-1$. Since the division of the latter by ϕ is exact, we get $x=A^{p-2}$. Next, for $\phi=x^2+qx+r\equiv0\ (\mathrm{mod}\ 2p+1=\text{prime})$, we divide $x^{2p}-1$ by ϕ and require that the remainder be divisible by $2p+1$. Thus the conditions for two roots $\alpha,\ \beta$ [neither zero] are

$$\frac{\beta^{2p}-\alpha^{2p}}{\beta-\alpha}\equiv0,\qquad \alpha\beta\left(\frac{\beta^{2p-1}-\alpha^{2p-1}}{\beta-\alpha}\right)+1\equiv0\qquad(\mathrm{mod}\ 2p+1),$$

which by use of symmetric functions can be expressed in terms of q and r. For the case $x^2-s\equiv0\ (\mathrm{mod}\ 2p+1)$, the first condition is satisfied and the second reduces to $s^p-1\equiv0$. For $x^2+x+1\equiv0\ (\mathrm{mod}\ 6p+1)$, the first condition is equivalent to $(-3)^{3p}\equiv1$.

V. Bouniakowsky[149] noted that there is an infinitude of solutions of

$$x^mX^n+y^mY^n=z^mZ^n,$$

where $m,\ n$ are relatively prime. Determine $\alpha,\ \beta$ so that $m\alpha-n\beta=1$. Let a and b be arbitrary and $c=a+b$. Then a solution is

$$x=a^\alpha,\qquad y=b^\alpha,\qquad z=c^\alpha,\qquad X=b^\beta c^\beta,\qquad Y=a^\beta c^\beta,\qquad Z=a^\beta b^\beta.$$

New solutions follow from the integral form of

$$a^{m\alpha}/a^{n\beta}+b^{m\alpha'}/b^{n\beta'}=c^{m\alpha''}/c^{n\beta''}.$$

Similarly, if $p,\ q,\ r,\ \cdots$ are without a common factor, we may solve

$$\sum_{i=1}^{n}A_i x_i^p y_i^q z_i^r\cdots=0$$

by use of $\Sigma A_i a_i=0$, $p\alpha\pm q\beta\pm\cdots=1$, replacing a_i by $a_i^{p\alpha\pm q\beta\pm\cdots}$, throwing negative powers into the denominator and clearing of fractions.

G. C. Gerono[150] noted that if r is the radius of the circle inscribed in a triangle with sides $a,\ b,\ c$ and area Δ and if $x=a/r,\ y=b/r,\ z=c/r$, Heron's formula for Δ, and $\Delta=\frac{1}{2}pr$, where p is the perimeter, give

$$(y+z-x)(x+z-y)(x+y-z)=4(x+y+z).$$

Call the factors $2X,\ 2Y,\ 2Z$, respectively. Let $x,\ y,\ z$ be positive integers.

[148] Mem. Accad. Sc. di Torino, 28, 1824, 272-9; Jour. für Math., 9, 1832, 59.
[149] Bull. Acad. Sc. St. Pétersbourg, 6, 1848, 200-2. Cf. Hurwitz[212] of Ch. XXVI.
[150] Nouv. Ann. Math., 17, 1858, 360.

Then X, Y, Z are positive integers for which $XYZ = X+Y+Z$.   If X is the largest of X, Y, Z, then $XYZ < 3X$, $YZ = 2$ or 1.   We may take $Y=2$, $Z=1$. Then $z=5$, $y=4$, $x=3$.   See the next two papers, and 341 of Ch. XXI.

Housel[151] proved that the sum of n distinct positive integers equals their product only when the integers are 1, 2, 3.

J. Murent[152] discussed the positive integral solutions $(a_1, \cdots, a_n)$ of

$$x_1 + x_2 + \cdots + x_n = x_1 x_2 \cdots x_n \qquad (n>1).$$

One solution is $(n, 2, 1, \cdots, 1)$.   Always at least two a's exceed unity.   If $n>2$, at least one a is unity; call i the index of a solution $(a_1, \cdots, a_i, 1, \cdots, 1)$ with $a_1 > 1, \cdots, a_i > 1$.   Then $2^i - i \leqq n$; if $=n$, then $a_1 = \cdots = a_i = 2$. If $n = 5 = 2^3 - 3$, there is a single solution $(2, 2, 2, 1, 1)$ of index 3, while the only remaining solutions are $(3, 3, 1, 1, 1)$ and $(5, 2, 1, 1, 1)$ of index 2.

P. di San Robert[153] noted that $F(x, y, z) = 0$ can be solved by use of the slide rule only if reducible to $X(x) + Y(y) = Z(z)$, a necessary and sufficient condition for which is

$$\frac{d^2 \log R}{dx\,dy} = 0, \qquad R \equiv \frac{\partial F}{\partial x} \div \frac{\partial F}{\partial y}.$$

S. Réalis[154] noted that

$$Q = \frac{(a^2+a)\big[(a+1)^{m-1} - a^{m-1}\big]}{m-1}$$

is not an mth power, being between a^m and $(a+1)^m$, and that mQ is not divisible by $(a+1)^m - a^m$.

E. Lucas[155] noted that $x^k + x + k = y^2$ is impossible if k is odd.

S. Réalis[156] noted that, if $xy \neq 0$, $6xy(3x^4 + y^4) \neq z^3$ or $4z^3$.   The impossibility (p. 524−5) of

$$x^3 + y^6 = 9z+7 \text{ or } 7z+5, \qquad \sum_{i=1}^{7} x_i^6 = 9x+8$$

is easily verified by use of remainders modulo 9 or 7.   M. Rochetti[156a] expressed

$$3(\alpha^3 + \beta^3 + \gamma^3)^2 \{(\alpha+\beta)^3 + (\beta+\gamma)^3 + (\gamma+\alpha)^3\}$$

as a sum of three cubes.

A. Markoff[157] gave complicated formulas for all positive integral solutions of $x^2 + y^2 + z^2 = 3xyz$.

E. Fauquembergue[158] proved that $1 + 3 + 3^2 + \cdots + 3^n = y^2$ only when $n = 0, 1, 4$, by using the powers of $a + b\sqrt{-2}$ to treat $3^{n+1} = 1 + 2y^2$.

---

[151] Nouv. Ann. Math., (2), 1, 1862, 67–69.
[152] Ibid., (2), 4, 1865, 116–20.
[153] Atti della R. Accad. Sc. Torino, 2, 1866–7, 454–5.
[154] Nouv. Ann. Math., (2), 12, 1873, 450–1.
[155] Nouv. Corresp. Math., 4, 1878, 122, 224.
[156] Nouv. Ann. Math., (2), 17, 1878, 468.
[156a] Ibid., (2), 19, 1880, 459.
[157] Math. Annalen, 17, 1880, 396. Cf. Hurwitz.[174]
[158] Mathesis, (2), 4, 1894, 169–170.

G. Cordone[159] investigated polynomials U, V in x which satisfy
$$P_0(x)U^n + P_1(x)U^{n-1}V + \cdots = R(x)$$
identically in x, where the $P_i(x)$ are polynomials in x.

E. Maillet[160] considered recurring series u_0, u_1, $\cdots$ of rational terms with the generating equation $f(x) = x^q + a_1 x^{q-1} + \cdots + a_q = 0$ and law of recurrence

(2) $$u_{n+q} + a_1 u_{n+q-1} + \cdots + a_q u_n = 0,$$

where $a_1, \cdots, a_q$ are rational. An algebraic equation with rational coefficients is irreducible if and only if all the recurring series of rational terms having the equation as their generating equation admit the corresponding law of recurrence as an irreducible law. To apply this to diophantine equations, let

$$\Delta_q(n) = \begin{vmatrix} u_{n+q-1} & u_{n+q-2} & \cdots & u_n \\ u_{n+q} & u_{n+q-1} & \cdots & u_{n+1} \\ \cdot & \cdot & \cdots & \cdot \\ u_{n+2q-2} & u_{n+2q-1} & \cdots & u_{n+q-1} \end{vmatrix}$$

become $F(u_n, u_{n+1}, \cdots, u_{n+q-1})$ when $u_{n+2q-2}, \cdots, u_{n+q}$ are expressed in terms of $u_{n+q-1}, \cdots, u_n$ by means of (2). It is known that the law (2) is reducible if and only if $\Delta_q(0) = 0$. Hence $F(u_0, \cdots, u_{q-1}) = 0$ has rational solutions if and only if $f(x) = 0$ is reducible. If $u_0, \cdots, u_{q-1}$ give a rational solution, the same argument shows that $u_n, \cdots, u_{n+q-1}$ give a rational solution for n arbitrary. We get all the rational solutions by taking in turn all the maximum divisors $\chi(x) = x^t + \cdots + c_t$, with rational coefficients, of $f(x)$, i. e., a divisor not dividing any other divisor of $f(x)$, and forming all the recurring series of rational terms having $\chi(x) = 0$ as generating equation and any rational numbers as the first t terms $u_0, u_1, \cdots, u_{t-1}$. Among the recurring series which together give all the rational solutions of $F = 0$, those which give only a finite number of solutions are the ones whose generating functions are divisors $\theta(x)$, with rational coefficients, of $f(x)$, such that $\theta(x) = 0$ has as its roots only distinct roots of unity. For example, let $q = 3$ and $f(x) = x^3 - \gamma$. Then

$$F(u_0, u_1, u_2) = \gamma^2 u_0^3 + \gamma u_1^3 + u_2^3 - 3\gamma u_0 u_1 u_2.$$

Let γ be the cube of a rational number δ, so that f is reducible. The maximum divisors are $x - \delta$ and $x^2 + \delta x + \delta^2$. To the first correspond the solutions u_0, δu_0, $\delta^2 u_0$, where u_0 is any rational number. To the second correspond u_0, u_1, $-\delta(u_1 + \delta u_0)$, where u_0 and u_1 are any rational numbers. If γ is not the cube of a rational number, there is no rational solution of $F = 0$. Let (2) be an irreducible law for u_0, u_1, $\cdots$ and let $a_q = \pm 1$. Then $\Delta_q(0) = g \neq 0$, $F(u_n, \cdots, u_{n+q-1}) = \pm g$, so that we have rational solutions of the latter. There are similar results for integral solutions when the a's are integral.

D. Hilbert[161] treated the diophantine equation $D = \pm 1$, where

$$D = x_0^{2n-2}\Pi(t_i - t_k)^2 \quad (i = 1, \cdots, n; \; k = i+1, \cdots, n)$$

[159] Giornale di Mat., 33, 1895, 106, 218.
[160] Assoc. franç. av. sc., 24, II, 1895, 233–42.
[161] Göttingen Nachrichten (Math.), 1897, 48–52. Cf. Eisenstein[256] of Ch. XXII for $n=3$.

is the discriminant of $x_0 t^n + x_1 t^{n-1} + \cdots + x_n = 0$, with undetermined coefficients, and roots $t_1, \cdots, t_n$. By use of $x_1 = 0, \cdots, x_{n-2} = 0$, it is readily proved that $D = \pm 1$ has rational solutions. The main theorem is: For $n > 3$, $D = \pm 1$ is not solvable in integers; the only equations with integral coefficients and with the discriminant ± 1 are $Q \equiv (ut + v)(u^1 t + v^1) = 0$ and the cubic $Q[(u + u^1)t + v + v^1] = 0$, where u, u^1, v, v^1 are any integers for which $uv^1 - u^1 v = \pm 1$. The proof employs the theorem[162] that the discriminant of an algebraic domain is always distinct from ± 1 and the lemma (here proved by use of ideals): If an equation with integral coefficients is irreducible in the domain of rational numbers, its discriminant is an integer divisible by the discriminant of the domain determined by a root of the equation.

C. Störmer[163] noted that, if A, B, M_i, N_j are positive integers,

$$AM_1^{x_1} \cdots M_m^{x_m} - BN_1^{y_1} \cdots N_n^{y_n} = \pm 1 \text{ or } \pm 2$$

has only a finite number of sets (if any) of integral solutions x_i, y_j, and that these can be found by solving a finite number of Pell equations.

E. Fauquembergue[164] noted that $3x^2 = 4y^3 - z^6$ has no solutions with y, z relatively prime, since $(x + z^3)^3 - (x - z^3)^3 = (2yz)^3$ gives $x = z^3$, $y = z^2$. On $x^2 = z^6 - 4y^3$, see Fuss[11] of Ch. XXI.

G. B. Mathews[165] noted that $xy(x + y) = z^n$ has no solution if $n = 3m$, while if $n = 3m \pm 1$ the general solution is $(\lambda^n \xi, \lambda^n \eta, \lambda^3 \zeta)$, where (ξ, η, ζ) is the unique solution in which x/y equals a given irreducible fraction, and the g.c.d. of x and y is not divisible by an nth power.

A. Cunningham[166] solved in integers $N_1 N_3 = N_2 N_4$, where $N_r = x_r^4 + 4y_r^4$; also

$$\frac{N_1 N_3 N_5 \cdots N_{2r+1}}{N_0 N_2 N_4 \cdots N_{2r}} = \frac{N_a}{N_b}.$$

He solved $M_1 M_3 = M_2 M_4$, where $M_r = (x_r^6 + 3^3 y_r^6)/(x_r^2 + 3y_r^2)$.

S. O. Šatunovsky[167] discussed the solution in integers of

$$ax^{mn} + a_1 x^{mn-1} + \cdots + a_{mn} = by^n, \qquad b = \pm a/c^m.$$

P. F. Teilhet[168] gave, for $m = 1$, recurring series leading to all (an infinitude of) solutions of $x^{2m} - y^{2m} = x^m y^m - 1$ and asked if there are solutions when $m > 1$ other than $x = y = 1$.

* H. Kühne[169] noted that if the system of n functions $x_i = \phi_i(\xi_0, \cdots, \xi_{n-1})$ is equivalent to the system of n functions $\xi_i = f_i(x_0, \cdots, x_{n-1})$, the coefficients of the ϕ's and f's belonging to the same domain, there exists between the x's and the ξ's a connection (Verknüpfung) and these connections have the group property. This concept leads to a process of solving all

[162] Minkowski, Geometrie der Zahlen, 1896, 130.
[163] Comptes Rendus Paris, 127, 1898, 752.
[164] L'intermédiaire des math., 5, 1898, 106–7.
[165] Math. Quest. Educ. Times, 73, 1900, 37. For $z = 1$, Euler[10] of Ch. XXI.
[166] Ibid., 75, 1901, 43; (2), 1, 1902, 26–7, 38–9.
[167] Zap. mat. otd. obsc., Odessa, 20, 1902, 1–21 (Russian).
[168] L'intermédiaire des math., 9, 1902, 318.
[169] Math. Naturw. Blätter, 1, 1904, 16–20, 29–33, 45–58.

diophantine equations in n unknowns such that all the unknowns are expressible rationally in $n-1$ parameters. An instance is the method of solving $x^3+y^3+z^3+u^3=0$ used by Schwering[73] and Kühne[74] of Ch. XXI.

A. Cunningham[170] found solutions of

$$(3) \qquad (x^3+y^3)(X^3+Y^3)=\xi^3+\eta^3$$

by expressing $n=(x^3+y^3)/(x+y)$ in the form t^2+3u^2 in one of the three ways: $(\tfrac{1}{2}x-y)^2+3(\tfrac{1}{2}x)^2$ for x even, $(x-\tfrac{1}{2}y)^2+3(\tfrac{1}{2}y)^2$ for y even,

$$\left(\frac{x+y}{2}\right)^2+3\left(\frac{x-y}{2}\right)^2 \text{ for } x, y \text{ both odd,}$$

and by expressing $N=(X^3+Y^3)/(X+Y)$ in the form T^2+3U^2. Then

$$nN=A^2+3B^2, \qquad A=tT\mp3uU, \qquad B=tU\pm uT.$$

But A^2+3B^2 is expressible in the form $(\xi^3+\eta^3)/(\xi+\eta)$ in one of three ways. Hence (3) is reduced to $(x+y)(X+Y)=\xi+\eta$. R. W. D. Christie[171] noted the special solution

$$(1+n^3)\{(2n-1)^3+(n-2)^3\}=(n^2+2n-2)^3+(2n^2-2n-1)^3.$$

He[172] noted that $10^3+30^2=(3^3+7^2)(3^2+4^2)$. Cunningham noted that

$$A^3+B^2=(a^3+b^2)(c^2+d^2)$$

is satisfied if $A=A_1^2$, $a=\alpha^2$, $A_1^3=\alpha^3c\mp bd$, $B=\alpha^3d\pm bc$.

*P. S. Frolov[173] found the least solution of (4) for $x=1$.

A. Hurwitz[174] discussed the positive integral solutions $x_1, \cdots, x_n$ of

$$(4) \qquad x_1^2+\cdots+x_n^2=xx_1x_2\cdots x_n, \qquad n \geqq 3,$$

where x is an integer. If $\xi=(x, x_1, \cdots, x_n)$ is a solution, then evidently $\xi'=(x, x_1', x_2, \cdots, x_n)$ is a solution when $x_1'+x_1=xx_2\cdots x_n$. Similarly, $\xi''=(x, x_1, x_2', x_3, \cdots, x_n)$ is a solution when $x_2'+x_2=xx_1x_3\cdots x_n$. Call these solutions $\xi', \xi'', \cdots, \xi^{(n)}$ " neighbors " to ξ. Build the neighbors to each of these, etc. Then all such solutions are said to be " derived " from ξ. Call ξ a " fundamental " solution if no one of its n neighbors has a smaller sum $x_1+\cdots+x_n$. It is proved that ξ is a fundamental solution if and only if $2x_i^2 \leqq xx_1\cdots x_n$ for $i=1, \cdots, n$; that every solution is either a fundamental solution or can be derived from another one; that there is no positive integral solution of (4) when x is a given integer $>n$; that all positive integral solutions with $x=n$ can be derived from $x_1=\cdots=x_n=1$ (the case $n=3$ being due to Markoff[157]). If $n \geqq 5$ and if $x, x_1, \cdots, x_n$ form a fundamental solution of (4) with $x_1 \geqq x_2 \geqq \cdots \geqq x_n$, the last $n-2-k$ of the x_i's have the value unity, where k is determined by $2^k \leqq n < 2^{k+1}$.

E. B. Escott[175] cited two numerical equations $x^7+rx^5+sx^3+tx+k=0$ with rational roots [see Ch. XXIV[63]]. " Charbonier " (18, 1911, 62–3) employed the roots $a, b, -a-b, c, d, e, -c-d-e$.

[170] Math. Quest. Educ. Times, (2), 5, 1904, 76. [Cf. 27, 1915, 17–18.]
[171] Ibid., 100.
[172] Ibid., (2), 6, 1904, 115.
[173] Vest. opytn. fiziki (Spacinski's Bote Math.), Odessa, 1906, Nos. 419–20, pp. 243–55.
[174] Archiv Math. Phys., (3), 11, 1907, 185–96. Cf. papers 173, 186, 194, 195a.
[175] L'intermédiaire des math., 16, 1909, 242.

E. N. Barisien[176] noted that $x=f(n)$, $y=\phi(n)$ give solutions (but not necessarily all solutions) of the equation $F(x, y)=0$ obtained by eliminating n. Similarly when x, y, z are functions of n, m. A. Cunningham[177] gave the least solution 3, 4, 5, and the general solution of

$$(x^4+y^4+z^4)^2=2(x^8+y^8+z^8).$$

E. B. Escott[177a] noted that, if $X=x^2+1$,

$$(x^3+x^2+2x+1)(x^3-x^2+2x-1)=X^3-X-1.$$

A. Thue[178] considered solutions x, y, z, relatively prime in pairs, of

$$Ax^n+By^n+Cz^n-xyzU(x, y, z)=0,$$

where U is a homogeneous polynomial of degree $n-3$ whose coefficients, as well as A, B, C, are integers. Let n be odd. Let p, q, r be integers, not all zero, such that $px+qy+rz=0$. Then

$$(Ar^n-Cp^n)x^n+(Br^n-Cq^n)y^n=xyE_1, \quad E_1=r^nzU-\frac{C}{xy}\{(px)^n+(qy)^n+(rz)^n\},$$

with two similar equations derived by permuting x, y, z and p, q, r. Then

$$ax=Br^n-Cq^n, \quad by=Cp^n-Ar^n, \quad cz=Aq^n-Bp^n.$$

Hence $Aax+Bby+Ccz=0$, so that we have a second linear relation. Also $ay^{n-1}-bx^{n-1}=E_1$, with two similar equations. Let u be the greatest of x, y, z numerically; λ the greatest of p, q, r; l of A, B, C; m the greatest of the coefficients of U, and $\delta=\frac{1}{2}(n-2)(n-1)m+(2^{n-1}+1)l$. He proved the following theorems. If $ABC \neq 0$, $n \geqq 3$, and if p, q, r can be found such that $\lambda^{n-1}<u/(l\delta)$, then $a=b=c=0$. If our given function of degree n is irreducible, we can determine a function $K \geqq l\delta$ of A, B, C and the coefficients of U, such that no numbers p, q, r exist for which $\lambda^{n-1}<u/K$. If

$$Ax^n+By^n+Cz^n=0$$

has relatively prime solutions and if n is odd and >1, there do not exist solutions p, q, r not all zero of $px+qy+rz=0$ for which $\lambda^{n-1}<u/\{(2^{n-1}+1)l^2\}$.

G. Candido[179] considered a polynomial $f(x, y)$ with the factors $L=x+\alpha y$ and $\phi(x, y)$, where α is rational. Set $x+\alpha y=z^n$, $\phi=A$. Then $f(x, y)=Az^n$ has the solutions

$$x=\tfrac{1}{2}v_n(p, q), \quad y=\tfrac{1}{2}\mu u_n(p, q), \quad z=\lambda+\alpha\mu, \quad p\equiv 2\lambda+\alpha\mu, \quad q\equiv\lambda^2+\alpha\lambda\mu,$$

where u_k, v_k satisfy $(\tfrac{1}{2}v_k)^2-(\tfrac{1}{4}p^2-q)u_k^2=q^k$. Similarly, if f has the factor $Q=x^2+\beta xy+\gamma z^2$, where β, γ are rational, take it as z^n. Each method is applied in detail to solve $LQ=Az^3$; in the particular case $x^3+y^3=Az^3$, the solutions are those obtained by Lucas[198] of Ch. XXI.

A. Cunningham[180] proved that if $4x^3-y^3=3x^2yz^2$ in positive integers, then $x=y$, $z=1$. He discussed (p. 28) $x^5+y^5=t^2+u^2$, a necessary and

[176] Sphinx-Oedipe, 5, 1910, 76–77.
[177] Math. Quest. Educ. Times, (2), 15, 1909, 49; (2), 18, 1910, 101–2.
[177a] Ibid., (2), 17, 1910, 57.
[178] Skrifter Videnskapsselsk. Kristiania (Math.), 2, 1911, No. 20.
[179] Periodico di Mat., 27, 1912, 265–273.
[180] Math. Quest. Educat. Times, (2), 22, 1912, 69–70.

sufficient condition[181] being that $x+y$ and $N=(x^5+y^5)/(x+y)$ be $\boxtimes$.
Since $N \equiv (x^2-3xy+y^2)^2+5xy(x-y)^2$, set $x=\xi^2$, $y=5\eta^2$ and make $x+y=\boxtimes$.
E. Miot[182] took $x^5+y^5=2^k pqr^2$, where p is a prime $4n+1$, whence $2^k p=s^2+t^2$,
and multiplied the initial equation by q^5.   L. Aubry obtained an infinitude
of solutions by setting

$$x-1=an, \qquad y-1=bn, \qquad t-1=cn+dn^2, \qquad u-1=en+fn^2.$$

"V. G. Tariste"[183] noted that, if x, y, z are <10,

$$x^n+y^n+z^n+xyz=100x+10y+z$$

holds only for $n=3$ and then x, y, z are the digits of 370, 407 or 952.   A. H.
Holmes[184] obtained special solutions with $n=1$ or 2 by assuming that
$yz=100$ or $xz=10$.

A. Cunningham[185] noted that every prime $p=X^n-Y^n$, with $n=12m+7$,
can be expressed in the forms $(x^3 \pm y^3)/(x \pm y)$.   Cf. Cunningham.[187]

G. Frobenius[186] proved that $x^2+y^2+z^2=kxyz$ is solvable in positive
integers only for $k=3$ and $k=1$, while the latter case reduces to the former
by the substitution $x=3X$, $y=3Y$, $z=3Z$.   Cf. Hurwitz.[174]

Cunningham[187] noted that, if $n>3$, $X^n-Y^n=x^2+xy+y^2$ has an infinitude
of positive integral solutions.   He noted (24, 1913, 85–6) cases when
x^3-y^3 or x^7-y^7 is expressible in the form Q^2+1.   He expressed (26, 1914, 50)
the product of two numbers of type x^2+x+1 and (27, 1915, 102) the
product of three such factors in the form A^2+3B^2 in several ways.

T. Kojima[188] proved that if a rational function of several variables
with integral coefficients equals an nth power for all integral values of the
variables, it is an exact nth power.

H. Brocard[189] stated that $x=y=1$ is the only integral solution of
$x^x+y^y=x+y$, and that $x^x+y^y=xy$ has no positive integral solution.   These
problems were proposed by G. W. Leibniz.[190]

A. Cunningham[191] gave several solutions of $\Pi(x_i^2+x_i+1)=z^3$.

E. Fauquembergue[192] noted that the only solutions of $(4x^4-1)(4x-1)=y^2$
in integers are $x=0, 1, 2$; $\pm y=1, 3, 21$.

M. Rignaux[193] gave two identities $x^6+y^6=z^6+w^2$.

W. Mantel[194] proved that $x^2+y^2+z^2=x^2y^2t^2$ is impossible in integers;
that, if $n=2, 6, 9, 11, 12$, $x_1^2+\cdots+x_n^2=x_1x_2\cdots x_n$ has no positive integral
solutions, and gave the least solutions for $n=3$ (3, 3, 3), $n=4$ (2, 2, 2, 2),

[181] Republished, l'intermédiaire des math., 19, 1912, 227–8.
[182] *Ibid.*, 119–120.
[183] *Ibid.*, 133.
[184] Amer. Math. Monthly, 18, 1911, 69–70.
[185] L'intermédiaire des math., 20, 1913, 3.   Proof by Aubry, p. 120; by Welsch, p. 184.
[186] Sitzungsber. Akad. Wiss. Berlin, 1913, 458–87.
[187] Math. Quest. Educ. Times, 23, 1913, 31–32.
[188] Tôhoku Math. Jour., 8, 1915, 24.
[189] L'intermédiaire des math., 22, 1915, 61–2; 21, 1914, 101.
[190] Opera omnia (ed., L. Dutens), III, 85–6; letter to Oldenbourg, June 21, 1677.
[191] L'intermédiaire des math., 23, 1916, 41–2.
[192] *Ibid.*, 24, 1917, 41–42.
[193] *Ibid.*, 25, 1918, 7.   For $x^2+y^6=z^3+w^6$, see Gérardin[86] of Ch. XXI.
[194] Wiskundige Opgaven, 12, 1917, 305–9.

$n=5$ (1, 1, 3, 3, 4), $n=7$, 8, 10.   He stated and L. de Jong proved that the g.c.d. of solutions x, y, z of $x^2+y^2+z^2=xyz$ is 3, and listed seven sets of solutions.   Cf. Hurwitz[174].

G. Rados[194a] proved that if a polynomial $F(x)$ of degree n with integral coefficients decomposes with respect to every prime modulus into n linear factors with integral coefficients, then $F(x)$ decomposes algebraically into n linear factors with integral coefficients.

A. Korselt[194b] argued that, if $f(x, y)$ is a homogeneous function of degree $d>1$ with no multiple root, $f(x, y)=z^n$ is solvable in relatively prime integral rational functions x, y, z of any parameters if and only if $d=2$, n any, or $d=3$, $n=2$.

"V. G. Tariste" stated and R. Goormaghtigh[195] proved that $x^y-y^x=x-y$ has only the integral solutions $x=y+1=1$, 2, 3.

M. Rignaux[195a] proved by the theory of quadratic forms that
$$a^2+b^2+c^2=Kabc$$
holds, when $c=1$, only for $K=3$.   Cf. Hurwitz[174].

F. Irwin[195b] gave a method to find the integral solutions of
$$ax^r-bxy+y-c=0.$$

For $(x^n-1)/(x-1)=\square$, see Landau, p. 57 of Vol. I of this History.

On $pr(p^2-r^2):qs(q^2-s^2)$, see papers 67–77 of Ch. IV, Euler[81] of Ch. XVI, Euler[18, 19] of Ch. XVIII and Euler[253] of Ch. XXII.

For $k^2+4k\mu v=\square$, where $k=(\mu^2+1)(v^2+1)$, see Haentzschel[144] of Ch. V.

By Hilbert[54] of Ch. XIII an equation $f=0$ may have no rational solution, while $f\equiv0$ (mod p^e) is solvable when p is any prime.   From one solution of $F(x, y, z)=0$, Cauchy[150] of Ch. XIII found another.   For $(f^4-k^4)(g^4-h^4)=\square$, see Euler[28] and Gérardin[85] of Ch. XV, Ward[44] of Ch. XIX.   On $f(x)=\square$ see Jacobi,[152] etc., of Ch. XXII.   Brunel[68] of Ch. XXI solved $x_1^n+x_2^n=F$, where F is a cyclic determinant of order n.   Euler[187] of Ch. XXII noted rational solutions of $abcd(a+b+c+d)=1$.

MISCELLANEOUS SYSTEMS OF EQUATIONS OF DEGREE $n>4$.

C. Gill and T. Beverley[196] found numbers whose sum is a $4n$th power and such that if the square of each be added to their sum there results a square.   Take px^{2n}, qx^{2n}, $\cdots$ as the numbers and x^{4n} as their sum.   The final conditions give $p^2+1=\square$, $q^2+1=\square$, $r^2+1=\square$, $\cdots$, which hold if
$$p=\frac{y^2-x^{2n}}{2yx^n}, \qquad q=\frac{ax^{2n}-y^2/a}{2yx^n}, \qquad r=\frac{bx^{2n}-y^2/b}{2yx^n}, \cdots.$$
To make $p+q+\cdots=x^{2n}$, take $y=(a+b+\cdots-1)/(2x^n)$, $1/a+1/b+\cdots=1$.

J. Liouville[197] stated that, if there be a finite number of sets of positive

---

[194a] Math. és termés. értesitö (Hungarian Acad. of Sc.), 35, 1917, 20–30.
[194b] Archiv Math. Phys., 27, 1918, 181–3.
[195] L'intermédiaire des math., 25, 1918, 30, 95.
[195a] Ibid., 131–2.
[195b] Amer. Math. Monthly, 26, 1919, 270–1.
[196] The Gentleman's Math. Companion, London, 5, No. 28, 1825, 367–9.
[197] Jour. de Math., (2), 4, 1859, 271–2.   Cf. Gegenbauer.[202]

integral solutions of $f(x_1, \cdots, x_\mu)=0, \cdots, F(x_1, \cdots, x_\mu)=0$, and we set $x_i=d_i\delta_i$ in all possible ways and write $\eta=+1$ or -1 according as $d_1\cdots d_\mu$ is a product of an even or odd number of primes (equal or distinct), then $\Sigma\eta$ is the number of sets of solutions of the given equations in which each x_i is a square.

H. Delorme[198] noted that the system $x^{2m}=ay^{2n}+1$, $x^{2p+1}=by^{2q+1}+c$ is insolvable if $a+1$ and c are divisible by 3, while b is not [since impossible modulo 3].

A. B. Evans[199] found four integers $(ax^5, \cdots, dx^5)$ whose sum is a sixth power and the sum of any three a fifth power. Take $a+b+c+d=x$. Then the conditions are $x-a=p^5, \cdots, x-d=s^5$. Thus $x=\frac{1}{3}(p^5+q^5+r^5+s^5)$ is an integer if $p=3m$, $q=3m+1$, $r=3m+2$, $s=3m+3$, and then a, b, c, d are also integers.

A. Desboves[200] called a a congruent number of order m if the system
$$x^{2m}+ay^{2m}=u^2, \qquad x^{2m}-ay^{2m}=v^2$$
has integral solutions. For $m=2$, the quotient of the expression found for a by 16 is $xy(x^2-y^2)(x^4-6x^2y^2+y^4)/2$. Taking $x=2$, $y=1$, the latter becomes -21. The least congruent number of order 2 is 21. A. Gérardin[201] remarked that it seems more logical to call a a congruent number of order m if $x^m\pm ay^m=\square$ hold simultaneously. Cf. papers 210, 222, and Ch. XVI.

L. Gegenbauer[202] considered a set of positive integral solutions $x_1^0, \cdots, x_\mu^0$ of the system of equations $f_1(x_1, \cdots, x_\mu)=0, \cdots, f_r(x_1, \cdots, x_\mu)=0$, and any divisor δ_λ^0 of x_λ^0, and called the product $\delta_1^0\cdots\delta_\mu^0$ a divisor-product belonging to the set $x_1^0, \cdots, x_\mu^0$. Let $\chi(x)$ be a function for which $\chi(xy)=\chi(x)\chi(y)$ for all values x, y satisfying a definite condition. Let $X(n)=\Sigma\chi(d)$, where d ranges over all divisors of n. Then
$$\Sigma X(x_1^0)\cdots X(x_\mu^0)=\Sigma\chi(\delta_1^0\cdots\delta_\mu^0),$$
where on the left the summation extends over those sets of solutions x_1^0, $\cdots, x_\mu^0$ which satisfy the condition mentioned, while on the right the summation extends over all the divisor-products belonging to these sets of solutions. If we take $\chi(x)=+1$ or -1, according as x is a product of an even or odd number of primes (equal or distinct) and note that $\Sigma\chi(d)=+1$ or 0, according as n is a square or not, we obtain the theorem stated by Liouville.[197] Other special cases are obtained by taking $\chi(x)$ to be the number $\phi_k(x)$ of sets of k integers $<x$ and prime to x, or $\mu(x)$ of Vol. I, Ch. 19, and noting that $\Sigma\phi_k(d)=n^k$, $\Sigma\mu(d)=0$ if $n>1$.

Several writers[203] found two integers whose sum, difference and difference of squares are all twelfth powers (square, cube and biquadrate). Elsewhere[204] was added the condition that the product of the nine roots of these powers shall be a square, cube and biquadrate.

[198] Nouv. Ann. Math., (2), 1, 1862, 455–7.
[199] Math. Quest. Educ. Times, 25, 1876, 76.
[200] Nouv. Ann. Math., (2), 18, 1879, 490.
[201] L'intermédiaire des math., 22, 1915, 101.
[202] Sitzungsber. Akad. Wiss. Wien (Math.), 95, II, 1887, 606–9.
[203] Amer. Math. Monthly, 2, 1895, 128–9.
[204] Math. Quest. Educ. Times, 60, 1894, 37–38.

G. B. M. Zerr[205] found six positive integers x_i such that each diminished by $\frac{5}{2}(x_1+\cdots+x_6)^5$ becomes a fifth power.

Several[206] found three numbers in arithmetical progression whose sum is a sixth power.

E. Swift[207] proved that $x=0$, $y=1250a^6$ give the only integral solution of

$$x^2+y^2=\square, \qquad \tfrac{5}{4}(x^2+y^2)=z^3, \qquad xy=2x^3, \qquad 2(x+y)+\frac{xy}{x+y}=\square,$$

$$(x^4+y^4)(x^2+y^2)-(x^5+y^5)\sqrt{x^2+y^2}=\square.$$

A. Cunningham[208] discussed $x^{2n}+y^{2n}+z^{2n}=u^{4n}+v^{4n}+w^{4n}$ ($n=1,\ 2$) by use of the identity $a^4+b^4+(a+b)^4=2(a^2+ab+b^2)^2$. Employ the usual solution of $u^2+v^2=w^2$, and set $x=u^2-v^2-uv$, $y=2uv$, $z=x+y$. Then

$$u^8+v^8+w^8=2C^2, \qquad C=u^4+u^2v^2+v^4, \qquad 2C^2=x^4+y^4+z^4,$$

$$u^4+v^4+w^4=2C=x^2+y^2+z^2.$$

He[209] expressed two special sextics and two octics in the form Y^2-qxZ^2, where Y, Z are functions of x, and $q=17,\ 13,\ 19,\ 2$.

A. Gérardin[210] discussed the solution of $x^m+Ay^p=f^2$, $x^m-Ay^p=g^2$. Thus $2x^m=f^2+g^2$, so that x is a sum of two squares.

Gérardin[211] treated the system $x^6-1=4yz$, $8y^{3n}-1=xt$, by taking as x, t the factors $2y^n-1$, $4y^{2n}+2y^n+1$ in either order, or $t=1$, or, for $y=2$, $x=2^k-1$ or $2^{2k}+2^k+1$ where $n=k-1$.

E. N. Barisien[212] noted that $x^{12}=r^3+s^3-t^3=u^2-v^2-w^2$ for $r=9y^4$, $s=x^4+9xy^3$, $t=3x^3y+9y^4$, where u, v, w are sextic functions of x, y.

A. Cunningham[213] noted that if $N_m=x^m-y^m$, and m, n are primes both of the form $4k\pm1$, we can set $N_m=t_m^2\mp nu_m^2$, $N_n=t_n^2\mp mu_n^2$, simultaneously. He and R. F. Davis[214] proved that we can express $(x^{14}+x^7+1)/(x^2+x+1)$ in the forms A^2+3B^2 and C^2+7D^2.

Cunningham[215] investigated $N=\phi(x,\ y)=\phi(x',\ y')=\cdots$, where

$$\phi(x,\ y)=x^\beta y^n\pm x^a y^m$$

and x, y are relatively prime integers.

A. Gérardin[216] gave solutions of the system

$$2(x^3+y^3)=z^3+u^3+v^3, \qquad 2(x^2+y^2)^4=(v^2-z^2)^4+(v^2-u^2)^4+(u^2-z^2)^4.$$

L. Aubry[217] made $P(x+y)+Qx$ and $P(x+y)+Qy$ both nth powers.

[205] Amer. Math. Monthly, 5, 1898, 114.
[206] Ibid., 8, 1901, 48–9.
[207] Ibid., 15, 1908, 110–1. Problem proposed by J. D. Williams in 1832.
[208] Math. Quest. Educ. Times, (2), 14, 1908, 66–7 (reprinted, Mess. Math., 38, 1908–9, 102–3).
[209] Ibid., (2), 16, 1909, 105–6.
[210] Assoc. franc. av. sc., 37, 1908, 15–17.
[211] Sphinx-Oedipe, 6, 1911, 141–2.
[212] L'intermédiaire des math., 19, 1912, 194. Cf. Gérardin[86] of Ch. XXI.
[213] Math. Quest. Educ. Times, (2), 23, 1913, 21–22.
[214] Ibid., (2), 23, 1913, 86–8.
[215] Mess. Math., 44, 1914–5, 37–47.
[216] L'intermédiaire des math., 21, 1914, 143–4; 24, 1917, 111–2.
[217] Ibid., 23, 1916, 33–4. Cf. Sphinx-Oedipe, 10, 1915, 26–27.

A. Gérardin[218] noted cases in which s^5-x, s^5-y, s^5-z are squares, where $s=x+y+z$; s^n-x, $\cdots$, s^n-t are all cubes, where $s=x+y+z+t$, for

$$x,\ z=\mp(27p^9+144p^3)-108p^6-63, \qquad y=216p^6+1, \qquad t=126 \qquad (s=1).$$

He noted (pp. 197-8) cases when $Ps^n+Qx^m+Ry^m+\cdots$, $Ps^n+Qx^m+Rz^m+\cdots$, $\cdots$ are all pth powers, where $s=x+y+\cdots$.

Gérardin[219] gave the general solution of his problem to make $s\pm x$, $s\pm y$, $\cdots$, $s\pm\alpha$ all pth powers, where $s=x+y+\cdots+\alpha$.

R. Goormaghtigh[220] gave solutions of $x+y+z=s$, $s^2-x^2=A^p$, $s^2-y^2=B^p$, $s^2-z^2=C^p$, where p is 2 or any odd integer. He[221] stated that, for $A<1000000$, $A=1+x+\cdots+x^m=1+y+\cdots+y^n$ holds only for

$$31=1+5+5^2=1+2+2^2+2^3+2^4, \qquad 8191=1+2+\cdots+2^{12}=1+90+90^2,$$

in addition to evident solutions if x or y is negative.

Despujols[222] took $\phi_1^2+\phi_2^2=(a^2+b^2)^{n-1}$ in the identity

$$(a^2+b^2)(\phi_1^2+\phi_2^2)\pm h=\{(a\pm b)\phi_1+(b\mp a)\phi_2\}^2, \qquad h\equiv 2(a\phi_1+b\phi_2)(b\phi_1-a\phi_2),$$

to obtain a congruent number[201] h of order n. He stated that every congruent number of order n is of the form $2\theta^2\lambda\mu$, where $\theta^2(\lambda^2+\mu^2)=x^n$, and conversely.

On $x_1^2x_2^2x_3^2\pm x_i^2=\square$ $(i=1,2,3)$ see p. 174, p. 186.

Papers not available for report.

P. Lackerbauer, Lehrsätze und Aufgaben über Gleichheiten als Beitrag zur höheren unbestimmetn Analysis, Progr. Münnerstadt, 1834.

G. A. Longoni, Sui problemi di analisi indeterminata, Monza, 1840.

C. F. Meyer, Ein diophantische Problem, Progr. Potsdam, 1867.

Poeschko, Auflösungsmethode unbestimmter Gl., Progr. St. Pölten, 1869.

J. Slavik, Solution of indeter. equations (Czech), Progr. Königgrätz, 1877.

F. M. Costa Lobo, Résolution des équations indéterminées, Coimbra, 1885.

C. Alasia, Elementi della teoria generale delle equazioni . . . e delle equazioni indeterminante, Napoli, 1891.

A. Zinna, L'analisi diofantea, Trapani, 1900.

H. Zuschlag, Diophantische Gleich., Berlin, 1908.

J. Edaljii, Note on indeterminate equations, Jour. Indian Math. Soc., 3, 1911, 115.

H. Verhagen, An equation in three unknowns, Nieuw Tijdschrift voor Wiskunde, 3, 1915-6, 307-14.

---

[218] L'intermédiaire des math., 23, 1916, 169-170.
[219] Ibid., 207-8.
[220] Ibid., 24, 1917, 23-24.
[221] Ibid., 88 (p. 153, correction).
[222] Ibid., 26, 1919, 14-15.

CHAPTER XXIV.

SETS OF INTEGERS WITH EQUAL SUMS OF LIKE POWERS.

If $t=\frac{2}{3}(a+b+c)$, a, b, c and $t-a$, $t-b$, $t-c$ have the same sum and same sum of squares; this double property shall be denoted by

$$(1) \qquad a,\ b,\ c \overset{2}{=} t-a,\ t-b,\ t-c, \qquad t=\tfrac{2}{3}(a+b+c).$$

The separation of two sets of numbers by the symbol $\overset{n}{=}$ shall denote that they have the same sum of kth powers for $k=1, \cdots, n$.

Chr. Goldbach[1] noted that

$$\alpha+\beta+\delta,\ \alpha+\gamma+\delta,\ \beta+\gamma+\delta,\ \delta \overset{2}{=} \alpha+\delta,\ \beta+\delta,\ \gamma+\delta,\ \alpha+\beta+\gamma+\delta.$$

L. Euler[2] remarked that a, b, c, $a+b+c \overset{2}{=} a+b$, $a+c$, $b+c$. This is the case $\delta=0$ of Goldbach's result, but it implies the latter since (Frolov[7]) each number may be increased by any constant δ.

If[2a] N be chosen so that N, $N-a_1$, $\cdots$, $N-a_t$ have the same sum as n, $n+a_1$, $\cdots$, $n+a_t$, then the sum of the squares of the former numbers equals that of the latter.

E. Prouhet[3] noted that $1, \cdots, 27$ can be separated into three sets, two of which are 1, 6, 8, 12, 14, 16, 20, 22, 27 and 2, 4, 9, 10, 15, 17, 21, 23, 25, such that the sum and sum of squares of the numbers in any set are the same as for the other sets. As a generalization, it is stated that there are n^m numbers separable into n sets each of n^{m-1} terms such that the sum of the kth powers of the terms is the same for all the sets when $k<m$.

F. Pollock[4] noted the fact, equivalent to (1), that

$$p,\ p+a,\ p+2a+3n \overset{2}{=} p-n,\ p+a+2n,\ p+2a+2n.$$

F. Proth[5] noted that

$$a^2+ab+b^2, \qquad c^2+cd+d^2, \qquad (a+c)^2+(a+c)(b+d)+(b+d)^2$$

and the numbers derived by interchanging b and c have the same sum and sum of squares.

E. Cesàro[6] proved that if $a, \cdots, k$ form a rearrangement of $1, \cdots, 9$ and

$$a,\ b,\ c,\ d \overset{2}{=} d,\ e,\ f,\ g \overset{2}{=} g,\ h,\ k,\ a,$$

then $a=2$, $b=4$, $c=9$, $d=5$, $e=1$, $f=6$, $g=8$, $h=3$, $k=7$. Note that the three sets of four numbers each may be placed on the sides of a triangle, with a, d, g at the vertices.

---

[1] Corresp. Math. Phys. (ed., Fuss), 1, 1843, 526, letter to Euler, July 18, 1750.
[2] *Ibid.*, 549, letter to Goldbach, Sept. 4, 1751. Special case by Nicholson[80] of Ch. XXIII.
[2a] New Series of Math. Repository (ed., T. Leybourn), 3, 1814, I, 75–77.
[3] Comptes Rendus Paris, 33, 1851, 225.
[4] Phil. Trans. Roy. Soc. London, 151, 1861, 414.
[5] Nouv. Corresp. Math., 4, 1878, 377–8.
[6] *Ibid.*, 293–5. Question by F. Proth.

M. Frolov[7] noted that $\Sigma a^k = \Sigma b^k$, $\Sigma a_1^k = \Sigma b_1^k$, $k = 1, \cdots, n$, imply
$$\Sigma(a+h)^k = \Sigma(b+h)^k, \qquad \Sigma(a+a_1)^k = \Sigma(b+b_1)^k.$$

For $n = 2$ there must be at least 3 terms a; for $n = 3$, at least 4. For $n = 3$, the least terms are stated incorrectly[7a] to be 1, 5, 8, 12 and 2, 3, 10, 11. For $n = 3$, there are examples when the a's and b's together give 1, 2, $\cdots$, $2m$.

J. W. Nicholson[8] noted the identities
$$3a+3b, \ 2a+4b, \ a, \ b \overset{3}{=} 3a+4b, \ a+3b, \ 2a+b;$$
$$5a+10b, \ 4a+11b, \ 3a+5b, \ 2a+8b, \ 3a+3b, \ 2a+6b, \ a, \ b$$
$$\overset{5}{=} 5a+11b, \ 4a+6b, \ 3a+10b, \ 3a+8b, \ a+5b, \ 2a+3b, \ 2a+b,$$

there being one more term on the left than on the right. But for $n = 1$, $\cdots$, 5, the sum of the nth powers of the ten numbers $a\pm32$, $a\pm24$, $a\pm18$, $a\pm10$, $a\pm4$ equals the sum of the nth powers of the ten $a\pm30$, $a\pm28$, $a\pm16$, $a\pm8$, $a\pm6$.

A. Martin[9] noted the special case of (1):
$$a, \ b, \ 2a+2b \overset{2}{=} a+2b, \ 2a+b.$$

Also,
$$p, \ q, \ 2p+2q, \ 3p+3q \overset{2}{=} 3p+2q, \ 2p+3q, \ p+q;$$
$$a+b+c, \ a+b-c, \ a-b+c, \ -a+b+c \overset{2}{=} 2a, \ 2b, \ 2c.$$

R. W. D. Christie[10] noted that, if $t = e+f+g+h$,
$$s+e, \ s+f, \ s+g, \ s+h, \ s-t \overset{2}{=} s-e, \ s-f, \ s-g, \ s-h, \ s+t.$$

[Since we may reduce each term by s, we obtain an evident identity.]

A. Cunningham[11] noted that $x+y$, b, $c \overset{2}{=} x$, y, $b+c$ if $xy = bc$. Next, if a, b, $c \overset{2}{=} x$, y, z, then
$$a, \ b, \ c+kz, \ kc \overset{2}{=} x, \ y, \ z+kc, \ kz.$$

Similarly a solution in two sets of n numbers yields one in two sets of $n+1$ numbers. J. H. Taylor noted that if $a_1+a_3+\cdots+a_{2r-1} = a_2+a_4+\cdots+a_{2r}$, then
$$a_1+1, \ a_2, \ a_3+1, \ a_4, \ \cdots, \ a_{2r} \overset{2}{=} a_1, \ a_2+1, \ a_3, \ a_4+1, \ \cdots, \ a_{2r}+1.$$
If $b_1+\cdots+b_{2r} = 2r(n-r)-r$, then
$$b_1, \ \cdots, \ b_{2r}, \ n \overset{2}{=} b_1+1, \ \cdots, \ b_{2r}+1, \ n-2r.$$

H. M. Taylor noted the generalization of (1):
$$a_1, \ \cdots, \ a_n \overset{2}{=} t-a_1, \ \cdots, \ t-a_n, \qquad t = \frac{2}{n}(a_1+\cdots+a_n).$$

R. W. D. Christie noted that $ab+cd$, bc, $ad \overset{2}{=} bc+ad$, ab, cd, and
$$n-1, \ n-2, \ n+3, \ n-4, \ n+5, \ n+6, \ n-7$$
$$\overset{2}{=} n+1, \ n+2, \ n-3, \ n+4, \ n-5, \ n-6, \ n+7.$$

---

[7] Bull. Soc. Math. France, 17, 1888–9, 69–83; 20, 1892, 69–84. The second was reprinted in Sphinx-Oedipe, 4, 1909, 81–89.

[7a] On the proof-sheets Escott noted that 5, 1, 4, 8 $\overset{2}{=}$ 2, 2, 7, 7 has smaller terms. It is derived from 3, −1, 2, 6 $\overset{2}{=}$ 0, 0, 5, 5 of Escott[63] by increasing each term by 2.

[8] Amer. Math. Monthly, 1, 1894, 187.

[9] Math. Magazine, 2, 1898, 212–3, 220.

[10] Math. Quest. Educ. Times, (2), 2, 1902, 40. His condition $s = a+b+c+d$ is unnecessary.

[11] Ibid., (2), 4, 1903, 98–100.

A. Gérardin[12] noted that $x^3+y^3+z^3=(x+1)^3+(y-2)^3+(z+1)^3$ is equivalent to $\Delta_x+\Delta_z=(y-1)^2$, where $\Delta_x=x(x+1)/2$. He took $\Delta_x=1, 3, 6, 10, 15, \cdots$ in turn and found the possible z's $\leqq 100$ by use of a table of triangular numbers. He found 13 solutions like

$$1^3+15^3+12^3=2^3+10^3+16^3, \qquad 1+15+12=2+10+16.$$

The sum of the squares of 1, 15, 12 exceeds that of 2, 10, 16 by 10. Consider two of our 13 solutions for which the ratio of the excesses mentioned is a square m^2; multiply the numbers of the first solution by m and add to the second solution; in this way we get

$$2, 4, 20, 22, 33 \overset{3}{=} 1, 6, 16, 26, 32;$$
$$1, 4, 12, 13, 20 \overset{3}{=} 2, 3, 10, 16, 19;$$
$$3, 4, 15, 20, 23, 26 \overset{3}{=} 2, 5, 17, 18, 22, 27;$$
$$2, 6, 30, 46, 53, 73 \overset{3}{=} 3, 4, 34, 44, 51, 74;$$
$$2, 6, 44, 58, 63, 91 \overset{3}{=} 1, 8, 40, 60, 65, 90.$$

Others follow by adding two of these. From $x+y+z=x+2+y-4+z+2$, he got

$$1, 19, 23, 24, 32, 48 \overset{3}{=} 3, 15, 20, 25, 40, 44.$$

Gérardin[13] noted that $14, 23, 25, 138 \overset{2}{=} 7, 26, 30, 137$,

$$1, g+3, 3g+2, 4g+4 \overset{3}{=} 2, g+1, 3g+4, 4g+3,$$
$$2, 12, 15, 35, 38, 48 \overset{5}{=} 3, 8, 20, 30, 42, 47,$$

while $x+h, y+p, z \overset{3}{=} x, y, z+h+p$ is impossible. [The last fact is a case of Bastien's[48] evident theorem.]

H. B. Mathieu[14] noted that

$$l, l-m-an, l+(a-1)m-n \overset{2}{=} l-m-n, l-an, l+(a-1)m.$$

U. Bini[15] gave $a+b, c, d \overset{2}{=} c+d, a, b$ if $ab=cd$. [Cunningham.[11]]

E. B. Escott[16] showed how to find all solutions of

$$(2) \qquad \sum_{i=1}^{n} x_i = \Sigma y_i, \qquad \sum_{i=1}^{n} x_i^2 = \Sigma y_i^2,$$

for $n=3$. Set $x_i=X_i+S$, $y_i=Y_i+S$, where $3S=x_1+x_2+x_3$. But, if Σx_i is not divisible by 3, take $S=\Sigma x_i$, $3x_i=X_i+S$, $3y_i=Y_i+S$. Thus

$$\Sigma X_i=0=\Sigma Y_i.$$

Using these to eliminate X_3 and Y_3 from $\Sigma X_1 X_2 = \Sigma Y_1 Y_2$, we get

$$(3) \qquad X_1^2+X_1X_2+X_2^2=Y_1^2+Y_1Y_2+Y_2^2.$$

Hence the problem reduces to solving (3). To find all its solutions, let N be any number all of whose prime factors are of the form $6n+1$ or 3, besides square factors common to X_1, X_2, Y_1, Y_2. Then represent N in all ways in the form x^2+xy+y^2.

---

[12] Sphinx-Oedipe, 1906–7, 120–4.

[13] *Ibid.*, 1907–8, 27, 94–5. Also, a case of (1).

[14] L'intermédiaire des math., 14, 1907, 201. All the solutions, *ibid.*, 50, 200–3, by the other writers are special cases of (1).

[15] *Ibid.*, 227. His other solution is equivalent to (1).

[16] *Ibid.*, 15, 1908, 109–111.

A. Gérardin[17] noted that

$1, m+3, 2m-2, 4m+2, 5m-3, 6m-1$
$$\overset{3}{=} 2,\ m-1,\ 2m+3,\ 4m-3,\ 5m+1,\ 6m-2,$$

$x,\ x+3,\ x+5,\ x+6,\ x+9,\ x+10,\ x+12,\ x+15$
$$\overset{5}{=} x+1,\ x+2,\ x+4,\ x+7,\ x+8,\ x+11,\ x+13,\ x+14,$$

also the result due to G. Tarry:

$c,\ a+3b,\ 2a-b-c,\ 4a+5b-3c,\ 5a+b-4c,\ 6a+4b-5c$
$$\overset{5}{=} b+c,\ a-b,\ 2a+4b-c,\ 4a-3c,\ 5a+5b-4c,\ 6a+3b-5c.$$

Gérardin[18] noted that b^2+ab-a^2, $a^2+2ab-4b^2$, $4b^2$ and $4b^2$ have the same sum and sum of cubes as a^2+ab-b^2, $4b^2+2ab-a^2$, b^2 and b^2.

G. Tarry[19] gave

$b,\ a-3b+2c,\ 2a+2b-5c,\ 2a+4b-7c,\ 3a-6b+c,\ 3a-4b-c,\ 4a-b-6c,$
$4a+4b-11c,\ 5a-9b,\ 6a+5b-16c,\ 8a-11b-4c,\ 9a+3b-20c,\ 10a-10b-9c,$
$10a-5b-14c,\ 11a-2b-19c,\ 11a-21c,\ 12a-10b-13c,\ 12a-8b-15c,$
$13a-3b-22c,\ 14a-7b-20c$
$$\overset{9}{=} c,\ a+3b-4c,\ 2a-5b+2c,\ 2a-3b,\ 3a+2b-7c,\ 3a+4b-9c,\ 4a-7b,$$
$$4a-2b-5c,\ 5a+5b-14c,\ 6a-10b-c,\ 8a+4b-19c,\ 9a-11b-6c,$$
$$10a-4b-15c,\ 10a+b-20c,\ 11a-10b-11c,\ 11a-8b-13c,\ 12a-3b-20c,$$
$$12a-b-22c,\ 13a-9b-16c,\ 14a-6b-21c.$$

Welsch[20] stated that the general solution of (2) is

$$x_{n-1}=\tfrac{1}{2}(a-X+\lambda),\ \ x_n=\tfrac{1}{2}(a-X-\lambda),\ \ y_{n-1}=\tfrac{1}{2}(a-Y+\mu),\ \ y_n=\tfrac{1}{2}(a-Y-\mu),$$

with $x_i,\ y_i$ $(i=1,\ \cdots,\ n-2)$ arbitrary, where

$$X=\sum_{i=1}^{n-2} x_i,\qquad Y=\sum_{i=1}^{n-2} y_i,\qquad \lambda^2-\mu^2=(2a-X-Y)(X-Y)-2\sum_{i=1}^{n-2} x_i^2+2\sum_{i=1}^{n-2} y_i^2,$$

and $\lambda,\ \mu$ are of the same parity as $a-X$, $a-Y$. E. B. Escott (pp. 213–4) noted that one can proceed as he[16] had done for $n=3$.

H. B. Mathieu[21] asked if the general solution is

$$2su-uv+st,\ st+tv,\ su-2uv+tv \overset{2}{=} st-uv,\ 2su-2uv+st+tv,\ su+tv.$$

Numerical solutions not of this type were cited in reply.[22]

A. Gérardin[23] noted three cases of (1) in which $c=2a+2b=t$ [Martin[9]], and that $4p^2-3mp$, $3m^2+4mp-4p^2$ have the same sum and sum of cubes as $6m^2-3mp$, $2p^2+4mp-6m^2$, $3m^2-2p^2$.

U. Bini[24] set $y_s=x_s+r_s$ in (2), whence $\Sigma r_s=0$. By the latter, r_m is eliminated from the quadratic equation, which is then treated as a quadratic for r_1. Next, let

$$(4) \qquad\qquad x^n+y^n+z^n=u^n+v^n+w^n \qquad\qquad (n=1,\,2,\,4),$$

<hr>

[17] Sphinx-Oedipe, 1908–9, 96; errata, 144.

[18] *Ibid.*, 4, 1909, 44.

[19] *Ibid.*, 176.

[20] L'intermédiaire des math., 16, 1909, 89–90. For $n=3$, *ibid.*, 15, 1908, 280–1.

[21] *Ibid.*, 16, 1909, 219–220.

[22] *Ibid.*, 17, 1910, 72, 165.

[23] Assoc. franç. av. sc., 38, 1909, 143–5.

[24] Mathesis, (3), 9, 1909, 113–8; same method in Periodico di Mat., 25, 1910, 119–128.

where x, y, z are not a permutation of u, v, w. Then $x+y+z=0$ and the equation given by $n=4$ is a consequence of the others. Replacing z by $-x-y$ and w by $-u-v$, we get

$$x^2+xy+y^2=u^2+uv+v^2.$$

Let x_1, y_1, u_1, v_1 be one solution; the general solution is

$$x=P_1x_1+P_2x_2, \qquad y=P_1y_1+P_2y_2, \qquad u=P_1u_1+P_2u_2, \qquad v=P_1v_1+P_2v_2,$$
$$P_1=u_2^2+u_2v_2+v_2^2-x_2^2-x_2y_2-y_2^2,$$
$$P_2=2x_1x_2+2y_1y_2-2u_1u_2-2v_1v_2+x_1y_2+x_2y_1-u_1v_2-u_2v_1,$$

where x_2, y_2, u_2, v_2 are arbitrary. Various special solutions of (4) are given.

A. Gérardin[25] noted that

$$(f-2g)^k+(4f-g)^k+(3g-5f)^k=(4f-3g)^k+(2g-5f)^k+(f+g)^k \qquad (k=1, 2, 4).$$

He[26] gave $2d+3x$, $4d+2x$, $d \overset{2}{=} d+2x$, $4d+3x$, $2d$.

Welsch[27] stated that the general solution of (2) is

$$x_{n-2}=-\sum_{i=1}^{n-3} x_i+t+BD-AC, \qquad y_{n-2}=-\sum_{i=1}^{n-3} y_i+t+AB-CD,$$
$$x_{n-1}=t+AB, \qquad x_n=t-CD, \qquad y_{n-1}=t+BD, \qquad y_n=t-AC,$$

with x_i, y_i $(i=1, \cdots, n-3)$ arbitrary [false if $n>3$, since in $\Sigma x_i^2=\Sigma y_i^2$ only the terms free of the x's and y's cancel].

E. N. Barisien[28] gave the relations involving $1, \cdots, 32$:

$$1, 8, 10, 15, 20, 21, 27, 30 \overset{2}{=} 4, 5, 11, 14, 17, 24, 26, 31$$
$$\overset{2}{=} 2, 7, 9, 16, 19, 22, 28, 29 \overset{2}{=} 3, 6, 12, 13, 18, 23, 25, 32.$$

C. Bisman[29] gave six relations like the last, a numerical example of $\Sigma a^k=\Sigma b^k$ $(k=1, \cdots, n)$ for each $n \leqq 9$, and three identities of the type

$$a-b, a-2c, a+b+c, a+2b-c \overset{3}{=} a+2b, a+c, a-b-c, a+b-2c.$$

L. Aubry[30] treated $\Sigma x_i=\Sigma u_i$, $\Sigma x_i^3=\Sigma u_i^3$ $(i=1, 2, 3)$ by setting $x_i=1+y_in$, $u_i=1+v_in$, whence $\Sigma y_i=\Sigma v_i$. The cubic equation holds if

$$n=3(\Sigma v_i^2-\Sigma y_i^2)/(\Sigma y_i^3-\Sigma v_i^3).$$

E. B. Escott[31] applied his[16] method to the last problem.

A. de Farkas[32] noted that, if Σx, Σx^2, Σx^3 and $x_3+3x_4+\cdots+(m-1)x_m$ equal the analogous sums involving y's, then x_1+a, x_2+a+d, $\cdots$, $x_m+a+(m-1)d$ have the same sum and sum of cubes as y_1+a, $\cdots$ [false].

G. Tarry[33] stated that the first $2^n(2a+1)$ integers can be separated into two sets each of $2^{n-1}(2a+1)$ integers having the same sum of tth powers for $t=1, \cdots, n$. For $a=1$, $n=3$, the first set is $1, 3, 7, 8, 9, 11, 14, 16, 17, 18, 22, 24$.

[25] Assoc. franç. av. sc., 39, I, 1910, 44; Sphinx-Oedipe, 5, 1910, 182.
[26] Sphinx-Oedipe, 5, 1910, 177.
[27] L'intermédiaire des math., 18, 1911, 60 (for $n=3$), 205.
[28] Mathesis, (4), 1, 1911, 69.
[29] Ibid., 205-8, 264.
[30] L'intermédiaire des math., 19, 1912, 156-7. E. Miot (p. 3) gave two numerical solutions.
[31] Ibid., 263-4.
[32] Ibid., 182. His remark on p. 131 is the case $n=2$ of Frolov's[7] first result.
[33] Ibid., 200.

Tarry[34] gave (1) and noted that, for x arbitrary,

$$a, b, \cdots, h \overset{n}{=} p, q, \cdots, t$$

imply

$$a, \cdots, h, p+x, \cdots, \cdots, t+x \overset{n+1}{=} p, \cdots, t, a+x, \cdots, h+x.$$

By use of this lemma he found

$$6a-3b-8c, 5a-9c, 4a-4b-3c, 2a+2b-5c, a-2b+c, b$$
$$\overset{5}{=} 6a-2b-9c, 5a-4b-5c, 4a+b-8c, 2a-3b, a+2b-3c, c.$$

H. B. Mathieu[35] gave as the general solution of (2), for $n=3$,

$$l\pm(ab+ac), \qquad l(1-bd)+qab\mp ac, \qquad l(cd+1)\mp ab-qac.$$

L. Aubry (p. 234) noted that $x+y+z \overset{3}{=} u+v+w$ implies $xyz=uvw$.

O. Birck[36] noted that, if $x+y+z=0$,

$$(ix-ky)^n+(iy-kz)^n+(iz-kx)^n=(iy-kx)^n+(iz-ky)^n+(ix-kz)^n,$$
$$n=0, 1, 2, 4.$$

" V. G. Tariste "[37] noted that

$$(23n+57l)^e+(40n-6l)^e+(17n-63l)^e$$
$$=(23n-57l)^e+(40n+6l)^e+(17n+63l)^e,$$
$$e=2, 4.$$

Further such cases were given by E. B. Escott and A. Gérardin.[38]

E. Miot[39] stated that any $2^n(2a+1)$ numbers in arithmetical progression can be separated into two equal sets having the same sum of tth powers for $t=1, \cdots, n$, if $a>0$, $n>1$; while $t=1, \cdots, n-1$ if $a=0$. Hence, if in Tarry's[33] example we replace x by $a+(x-1)r$, we get

$$a, a+2r, a+6r, \cdots, a+23r \overset{3}{=} a+r, a+3r, \cdots, a+22r.$$

Tarry[40] noted that the number of terms in each member of the equations deduced in his[34] lemma is $2k-d$, if k is the number of terms in each member of the given equations, while x is expressible in d ways as a difference of two numbers belonging to the same member.   Given

$$1, 5, 10, 16, 27, 28, 38, 39 \overset{6}{=} 2, 3, 13, 14, 25, 31, 36, 40,$$

take

$$x=11=16-5=27-16=38-27=39-28=13-2=14-3=25-14=36-25.$$

Thus $d=8$,

$$1, 5, 10, 24, 28, 42, 47, 51 \overset{7}{=} 2, 3, 12, 21, 31, 40, 49, 50.$$

E. Miot (p. 85) noted that

$$1+n, 2+n, 10+n, 12+n, 20+n, 21+n$$
$$\overset{5}{=} n, 5+n, 6+n, 16+n, 17+n, 22+n.$$

[34] L'intermédiaire des math., 19, 1912, 219–221. Cf. Tarry.[46]

[35] Ibid., 225.

[36] Ibid., 19, 1912, 252–5. Cf. Birck[216] of Ch. XXII.

[37] Ibid., 129; cf. 201, 250.

[38] Ibid., 21, 1914, 126–9.

[39] Ibid., 20, 1913, 64–5. Generalization of Tarry.[33]

[40] Ibid., 68–70.

O. Birck (p. 182) took $x+y+z=0$ and
$$\xi=ix-ky, \quad \eta=iy-kz, \quad \zeta=iz-kx, \quad \pi=iy-kx, \quad \kappa=iz-ky, \quad \rho=ix-kz.$$
Then
$$n+\xi, \ n-\xi, \ n+\eta, \ n-\eta, \ n+\zeta, \ n-\zeta \overset{5}{=} n+\pi, \ n-\pi, \ n+\kappa, \ n-\kappa, \ n+\rho, \ n-\rho.$$

O. Birck[41] noted that
$$\xi^4+\eta^4+\zeta^4=\pi^4+\kappa^4+\rho^4, \qquad \xi+\eta+\zeta=\pi+\kappa+\rho, \qquad \eta-\xi=\kappa-\pi\neq0$$
for
$$\xi, \ \eta=i-\tfrac12(x\pm y); \qquad \kappa, \ \pi=i+\tfrac12(x\pm y); \qquad \zeta, \ \rho=k\pm x,$$

subject to the condition $k^3-i^3+(k-\tfrac14 i)x^2-\tfrac34 iy^2=0$. From one solution (i, k, x, y) of the latter he derived two or more new solutions.

A. Gérardin[42] noted that
$$p(p+a+b), \ p^2+2p(a+b)+2ab, \ p(a+b)+2ab$$
$$\overset{3}{=} ap, \ bp, \ p^2+p(a+2b)+2ab, \ p^2+p(2a+b)+2ab.$$

E. B. Escott[43] noted that (4), for $n=2, 4$, has the solutions
$$x=m^2+mn+3n^2, \qquad y=\ \ 2m^2-4mn-n^2, \qquad z=\ \ 3m^2-2n^2,$$
$$u=3m^2-mn+n^2, \qquad v=-m^2+4mn+2n^2, \qquad w=-2m^2+3n^2,$$

where m, n are odd, and gave two analogous solutions. Gérardin gave (*ibid.*) a process to obtain solutions.

Crussol[44] treated the last problem with the restriction $y+z=v+w$. The equations can be written in the form
$$(x+pn)^k+(y+pm)^k+(z-pm)^k=(x-pn)^k+(y-pm)^k+(z+pm)^k, \quad k=2, 4,$$
where m, n are relatively prime. Thus
$$xn=m(z-y), \qquad 4p^2n^2(n^2-m^2)=3n^2(z+y)^2+(n^2-4m^2)(z-y)^2.$$
Set $s=3\alpha^2-\beta^2(n^2-4m^2)$. Then the solution is
$$p=3\alpha^2+\beta^2(n^2-4m^2), \qquad z+y=ns+2\alpha\beta(n^2-4m^2), \qquad z-y=ns-6\alpha\beta n^2.$$

Crussol[45] noted that the system
$$(x+a)^k+(x-a)^k+(y+b)^k+(y-b)^k=(z+a)^k+(z-a)^k+(t+b)^k+(t-b)^k,$$
$$k=2, 4, 6,$$
is equivalent to $x^2+y^2=z^2+t^2$ and
$$6(a^2-b^2)=y^2+t^2-x^2-z^2, \qquad 10(a^2+b^2)=y^2+t^2+x^2+z^2.$$
Set $x=\alpha q-\beta p, \ y=\alpha p+\beta q, \ z=\alpha q+\beta p, \ t=\alpha p-\beta q$. Thus
$$3(a^2-b^2)=(\alpha^2-\beta^2)(p^2-q^2), \qquad 5(a^2+b^2)=(\alpha^2+\beta^2)(p^2+q^2),$$
$$\alpha^2+\beta^2=5(\gamma^2+\delta^2), \qquad \alpha=2\delta+\gamma, \qquad \beta=2\gamma-\delta, \qquad a=\gamma p+\delta q, \qquad b=\gamma q-\delta p,$$
$$3(\gamma^2-\delta^2)(p^2-q^2)=2\gamma\delta(2p+q)(p-2q).$$
The discriminant of this quadratic in γ, δ must be a square. The first of

[41] L'intermédiaire des math., 20, 1913, 273–7.
[42] Sphinx-Oedipe, 8, 1913, 134; correction, 157.
[43] *Ibid.*, 141–2. Cf. papers 206–7 of Ch. XXII.
[44] *Ibid.*, 175–6.
[45] *Ibid.*, 189.

three special solutions is 2, 16, 21, 25; 5, 14, 23, 24, given by $p=\delta=3$, $q=2$, $\gamma=5$.

G. Tarry[46] republished his[34] results and noted that

$$A_1, \cdots, A_k \overset{2n}{=} B_1, \cdots, B_k, \qquad A_i+A_{k-i}=2h=B_i+B_{k-i} \ (i=1, \cdots, k)$$

imply $A_1, \cdots, A_k \overset{2n+1}{=} B_1, \cdots, B_k$, as shown by subtracting h from every term of the given equations.   A. Aubry concluded that

$$A, B, C, -A, -B, -C \overset{5}{=} A', B', C', -A', -B', -C'$$

if

$$A = ab+a\beta+b\alpha-3\alpha\beta, \qquad B = -ab+a\beta+\alpha b+3\alpha\beta, \qquad C = 2a\beta+2\alpha b,$$
$$A' = ab+a\beta-b\alpha+3\alpha\beta, \qquad B' = -ab+a\beta-\alpha b-3\alpha\beta, \qquad C' = 2a\beta-2\alpha b,$$

since $\Sigma A^2 = \Sigma A'^2$, $\Sigma A^4 = \Sigma A'^4$.   Take $a=1$, $\alpha=2$, $b=3$, $\beta=4$ and add 32 to every term; thus

$$1, 12, 21, 43, 52, 63 \overset{5}{=} 3, 7, 28, 36, 57, 61.$$

Aubry noted that A_1+x, $B_1+y \overset{2}{=} A_1, B_1, x+y$ if $A_1x+B_1y=xy$.   Hence set $A_1=ab$, $B_1=cd$, $x=c\alpha$, $y=b\alpha$, $\alpha=a+d$.   Thus, if A, $B \overset{2}{=} \xi$, η, ζ, then $A=ab+bd+cd$, $B=ab+ac+cd$, whence

$$A^2-AB+B^2 = (a^2+ad+d^2)(b^2+bc+c^2)$$

But a^2+ad+d^2 has besides 3 only prime factors of the form $6k+1$.   If A^2-AB+B^2 is divisible by 3, $A+B=3h$ and A, $B \overset{2}{=} A-h$, $B-h$, $2h$. Hence A, $B \overset{2}{=} \xi$, η, ζ is solvable if and only if A^2-AB+B^2 is a multiple of 3 or has at least two prime factors $6k+1$.

Crussol[47] solved $a, b, c, d \overset{3}{=} a_1, b_1, c_1, d_1$.   After adding a suitable constant to each term we have $a+b+c+d=0$.   Set

$$A=a+b=-c-d, \qquad A_1=a_1+b_1=-c_1-d_1,$$
$$2B=a-b, \qquad 2B_1=a_1-b_1, \qquad 2C=c-d, \qquad 2C_1=c_1-d_1.$$

Then

$$A^2+(B+C)^2+(B-C)^2 = A_1^2+(B_1+C_1)^2+(B_1-C_1)^2,$$
$$A(B+C)(B-C) = A_1(B_1+C_1)(B_1-C_1).$$

The general solution of the latter is $A=\lambda px$, $B+C=\mu qy$, $B-C=vrz$, $A_1=\mu rx$, $B_1+C_1=vpy$, $B_1-C_1=\lambda qz$.   Then the former condition becomes $ex^2=fy^2+gz^2$, where $e=\mu^2r^2-\lambda^2p^2$, $f=\mu^2q^2-v^2p^2$, $g=v^2r^2-\lambda^2q^2$.   From the evident solutions $(x, y, z)=(v, \lambda, \mu)$ and (q, r, p), we get the general solution

$$x=v(\alpha^2f+\beta^2g), \qquad y=\lambda(\alpha^2f-\beta^2g)+2\mu\alpha\beta g, \qquad z=\mu(\alpha^2f-\beta^2g)-2\lambda\alpha\beta f.$$

L. Bastien[48] proved the impossibility of $x_1, \cdots, x_n \overset{n}{=} y_1, \cdots, y_n$ when the x's do not form a permutation of the y's.   For, the elementary symmetric functions of the x's equal those of the y's, so that the x's are the roots of the same equation of degree n as the y's.

[46] Sphinx-Oedipe, numéro spécial, June, 1913, 18–23; l'enseignement math., 16, 1914, 18–27 (prepared for press by Aubry after Tarry's death).
[47] Sphinx-Oedipe, 8, 1913, 156–7; special case $\lambda=\mu=v=1$, p. 134.
[48] Ibid., 171–2.

E. N. Barisien[49] noted that 1, 5, 9, 11, 15, 16 and 3, 4, 8, 10, 14, 18 and 2, 6, 7, 10, 14, 18 and 1, 5, 9, 12, 13, 17 have the same sum and sum of squares; also that

$$3, 4, 8, 11, 15, 16 \overset{2}{=} 2, 6, 7, 12, 13, 17.$$

A. Aubry[50] gave known and new solutions of $\Sigma a = \Sigma \alpha$, $\Sigma a^2 = \Sigma \alpha^2$, and proved the impossibility of $x, y \overset{3}{=} t, u, v$.

N. Agronomof[51] noted the case $a + c + 3 = 2b$ of (1).

A. Gérardin[52] gave a solution of $\Sigma A = \Sigma X$, $\Sigma A^3 = \Sigma X^3$:

$$A = 2p^2 - 9pq + 6q^2, \qquad B = 2pq, \qquad C = pq, \qquad X = -p^2 + 9pq - 12q^2,$$
$$Y = 2p^2 - 10pq + 12q^2, \qquad Z = p^2 - 5pq + 6q^2.$$

N. Agronomof[53] gave an 8 parameter solution of

$$\sum_{i=1}^{4} x_i^k = \sum_{i=1}^{4} y_i^k \qquad (k = 1, 2, 3).$$

For any solution of this system, we have

$$\sum_{i=1}^{4} (x_i + z)^k + \sum_{i=1}^{4} y_i^k = \sum_{i=1}^{4} (y_i + z)^k + \sum_{i=1}^{4} x_i^k \qquad (k = 1, 2, 3, 4),$$

z being arbitrary. Proceeding similarly, we can solve

$$\sum_{i=1}^{\nu} x_i^k = \sum_{i=1}^{\nu} y_i^k \qquad (\nu = 2^{n-1}; \; k = 1, \cdots, n).$$

By specializing the solution first cited, he obtained solutions of

$$\sum_{i=1}^{s} x_i^k = \sum_{i=1}^{4} y_i^k \qquad (k = 1, 2, 3; \; s = 1 \text{ or } 2 \text{ or } 3).$$

A. Filippov[53a] stated that the specialized solutions just mentioned are trivial since they reduce to $x_i = y_i$ or $y_i = 0$.

A. Gérardin[54] noted that $\Sigma x = \Sigma a$, $\Sigma x^2 = \Sigma a^2$ if $a = 3$, $b = 2$, $c = 1$,

$$x = (\; u^2 + 2uv + 3v^2)/D, \qquad y = (3u^2 + 8uv + 6v^2)/D,$$
$$z = (2u^2 + 8uv + 9v^2)/D, \qquad D = u^2 + 3uv + 3v^2.$$

R. Goormaghtigh[55] solved the same system by setting

$$x = Pg + Qp, \qquad y = Ph + Qq, \qquad z = P(k + l + m - g - h) + Qr,$$
$$a = Pk + Qp, \qquad b = Pl + Qq, \qquad c = Pm + Qr.$$

Then the equation obtained by eliminating z between the proposed equations determines P/Q as follows:

$$P = p(k - g) + q(l - h) + r(g + h - k - l),$$
$$Q = g^2 + h^2 + gh + kl + lm + mk - (g + h)(k + l + m).$$

<hr>

[49] Mathesis, (4), 3, 1913, 69.
[50] Annaes Sc. Acad. Polyt. do Porto, 9, 1914, 141–151.
[51] Suppl. al Periodico di Mat., 19, 1915, 20.
[52] Nouv. Ann. Math., (4), 15, 1915, 564; l'intermédiaire des math., 22, 1915, 130–2 (correction for $h = 2$); 23, 1916, 107–10. Cf. papers 130, 302, 438–40, 442 of Ch. XXI.
[53] Tôhoku Math. Jour., 10, 1916, 207–14.
[53a] Ibid., 15, 1919, 143.
[54] L'intermédiaire des math., 24, 1917, 55 (correction, p. 153).
[55] Ibid., 25, 1918, 20–21.

AN EQUIVALENT PROBLEM IN THE THEORY OF LOGARITHMS.

The system of equations $\Sigma a_i^k = \Sigma b_i^k$ $(k=1, \cdots, n)$ which we have been considering is equivalent to the system $\Sigma a_1 = \Sigma b_1$, $\Sigma a_1 a_2 = \Sigma b_1 b_2$, $\cdots$, $\Sigma a_1 a_2 \cdots a_n = \Sigma b_1 b_2 \cdots b_n$. Consider the equation having the roots a_1, a_2, $\cdots$ and that having the roots b_1, b_2, $\cdots$. Thus our problem is equivalent to the following: Find two equations of the same degree each having all its roots integral and the first n coefficients of the one equal to the corresponding coefficients in the other.

The latter problem occurs in the investigation of rapidly converging series convenient for the computation of logarithms. In the familiar series

$$\log\frac{m}{n} = 2M(k + \tfrac{1}{3}k^3 + \tfrac{1}{5}k^5 + \cdots), \qquad k = \frac{m-n}{m+n},$$

take, for example, $m = x^2$, $n = (x-1)(x+1)$. Then $\log(x+1)$ differs from $2\log x - \log(x-1)$ by a series in $k = 1/(2x^2-1)$. In general, we desire that m and n shall be polynomials in x whose roots are all integers such that k becomes a fraction whose numerator is a constant. We may remove the second terms of the polynomials by a linear substitution.

J. B. J. Delambre[56] took $m = x^3 + px + q$, $n = x^3 + px - q$, and assumed that $m = 0$ has the roots a, b, $-a-b$, and $n = 0$ the roots $-a$, $-b$, $a+b$, whence $p = -a^2 - ab - b^2$, $q = a^2 b + ab^2$. For $a = b = 1$, we have the formulas m, $n = x^3 - 3x \pm 2$, ascribed to Borda.

J. E. T. Lavernède[57] gave an extensive treatment of such polynomials, chiefly of degrees 3 and 4, and noted the examples

$$m = x^2(x+5)^2 = x^4 + 10x^3 + 25x^2, \qquad n = (x-1)(x+2)(x+3)(x+6) = m - 36;$$
$$m = x^2(x-7)^2(x+7)^2, n = (x-3)(x+3)(x-5)(x+5)(x-8)(x+8) = m - 14400;$$
$$m, n = (x\pm2)(x\pm4)(x\pm10)(x\mp7)(x\mp9) = x^5 - 125x^3 + 3004x \pm 5040.$$

S. F. Lacroix[58] quoted the preceding results and the following, attributed to Haros:

$$m = x^2(x-5)(x+5), \qquad n = (x-3)(x+3)(x-4)(x+4) = m + 144.$$

John Muller[59] had made only the following contribution to our subject:

$$\log(d+1)^2 = \log d + \log(d+2) + \log\frac{d^2+2d+1}{d^2+2d},$$

$$\log(d+3)^2 = \log(d+1)^2 + \log(d+4) - \log d - \log q, \qquad q = \frac{d^3+6d^2+9d+4}{d^3+6d^2+9d}.$$

The latter is applied when $d = 14$ to find $\log 17$, knowing $\log 15$, $\log 18$ and $\log 14$. Then $q = 2025/2023$. Taking $a = 2024$, $x = 1$, we have $q = (a+x)/(a-x)$, a series for the logarithm of which is found by subtracting the

---

[56] J. C. de Borda's Tables trigonométriques décimales ou Tables des logarithmes . . . revues, augmentées et publiées par Delambre, Paris, an IX (1800–1). Introduction.
[57] Notice des travaux de l'Acad. du Gard, 1807, 179–192; Annales de Math. (ed., Gergonne), 1, 1810–11, 18–51, 78–100. See Allman.[60]
[58] Traité du Calcul Diff. . . . Int., ed. 2, I, 1810, 49–52.
[59] Traité analytique des sections coniques, fluxions et fluentes . . ., Paris, 1760, 112. This topic does not occur in the earlier English edition, A Math. Treatise: containing a System of Conic Sections; with the Doctrine of Fluxions and Fluents . . ., London, 1736.

series for $\log (1-x/a)$ from that for $\log (1+x/a)$.   [If in the second formula we take $d=x-2$, we obtain Borda's[56] result.   If in the first we take $d=x-1$, we obtain the example $m=x^2$, $n=x^2-1$ given before the report on Delambre.[56]]

W. Allman[60] gave the result quoted under Delambre[56] and the first two results cited under Lavernède.

T. Knight[61] started with $x\equiv(x+n)\{x/(x+n)\}$, changed x into $x+n'$ in the fraction and multiplied by such a fraction as will restore equality:

$$x\equiv(x+n)\cdot\frac{x+n'}{x+n+n'}\cdot\frac{x(x+n+n')}{(x+n)(x+n')}.$$

In the final fraction change x into $x+n''$ and restore equality by annexing the new factor

$$\frac{x(x+n+n')(x+n+n'')(x+n'+n'')}{(x+n)(x+n')(x+n'')(x+n+n'+n'')}.$$

The expanded numerator has its first three terms the same as the corresponding terms of the expanded denominator, and also the fourth terms alike if $n''=n+n'$.   The rest of the paper is on the case $n=n'=n''=\cdots=-1$, and gives the general factor explicitly.

Secrétan[62] noted that

$$(x\mp1)(x\mp5)(x\pm7)(x\pm8)(x\mp9)=x^5-110x^3+2629x\mp2520.$$

E. B. Escott[63] spoke of $a_0x^n+a_1x^{n-1}+\cdots$ and $a_0'x^n+\cdots$ as having exactly their first r terms alike if $a_0=a_0',\ \cdots,\ a_{r-1}=a_{r-1}',\ a_r\neq a_r'$.   He readily proved theorem (I): If f and g are two polynomials in x having exactly their first r terms alike, then $f(x)\cdot g(x+d)$ and $g(x)\cdot f(x+d)$ have exactly their first $r+1$ terms alike.   Starting with $f=x-a$, $g=x$, and taking $d=-b$, we see that $(x-a)(x-b)$ and $x(x-a-b)$ have two terms alike. Taking the latter as f and g, and $d=-c$, we see that (Knight)

$$(x-a)(x-b)(x-c)(x-a-b-c),\qquad x(x-a-b)(x-a-c)(x-b-c)$$

have three terms alike.   Proceeding similarly, we obtain theorem (II): If we form the equation whose roots are the sums of $a_1,\ \cdots,\ a_n$ taken 1, 3, 5, $\cdots$ at a time, and that whose roots are the sums of the a's taken 2, 4, 6, $\cdots$ at a time, we obtain two functions of degree 2^{n-1} having exactly their first n terms alike.   For special a's common factors occur and may be removed.   Thus, if $n=4$ and if the a's are a, b, $a+b$, $a+2b$, four of the eight roots will be common and the remaining ones are 0, $a+3b$, $2a+b$, $3a+4b$, and a, b, $2a+4b$, $3a+3b$.   If in (I) we take $g=P(x)\equiv x(x+d)(x+2d)\cdots\{x+(n-1)d\}$ and $f=P+c$, and remove the common factor P/x, we obtain two functions $(P+c)(x+nd)$ and $(x+nd)P+cx$ of degree $n+1$ with exactly their first $n+1$ terms alike.   Again, taking $g=P(x)\cdot P(x+a)$ and $f=g+c$ in (I), and removing the common factor $g/\{x(x+a)\}$, we get

$$(x+nd)(x+a+nd)(g+c),\qquad (x+nd)(x+a+nd)g+cx(x+a),$$

<hr>

[60] Trans. Roy. Irish Acad., 6, 1797, 391–434.
[61] Phil. Trans. Roy. Soc. London, 1817, 217–33.
[62] Comptes Rendus Paris, 44, 1857, 1276–9.
[63] Quar. Jour. Math., 41, 1910, 141–167.

having all terms alike except the last two in each. Taking $n=2$ or 3 and making suitable assumptions, we find that these functions have two common linear factors (pp. 148–50, with changed notations). Besides employing roots in three or more arithmetical progressions, leading to a solution of degree 7 (p. 152), various special methods are used.

Escott, after reading the proof-sheets of this chapter, pointed out its relation to the derivation of formulas for the computation of π:

$$\tan^{-1}\frac{a}{x+\alpha}+\tan^{-1}\frac{b}{x+\beta}+\cdots \equiv \tan^{-1}\frac{p}{X},$$

where X is a real polynomial in x whose degree equals the number of fractions in the left member. Since

$$\tan^{-1}\frac{a}{y}=\frac{1}{2i}\log\frac{y+ai}{y-ai},$$

it suffices to have $(x+\alpha+ai)(x+\beta+bi)\cdots \equiv X+pi$. Of the polynomials m, n in the above problem on logarithms, we may employ here those containing only odd powers of x and a constant term. If in Delambre's[56] example we replace a by $-ai$ and b by $-bi$, we have

$$(x+ai)(x+bi)(x-ai-bi)\equiv x^3+(a^2+ab+b^2)x+ab(a+b)i,$$

$$\tan^{-1}\frac{a}{x}+\tan^{-1}\frac{b}{x}-\tan^{-1}\frac{a+b}{x}\equiv \tan^{-1}\frac{ab(a+b)}{x^3+(a^2+ab+b^2)x}.$$

By the former we have a product of factors like x^2+a^2 expressed as a sum of two squares (cf. note 13, p. 382 of Vol. I of this History). Escott noted that his[63] general results include as special cases Goldbach's[1] and Euler's[2] formulas, the first identity by Nicholson[8], the two formulas by J. H. Taylor,[11] as well as the following (after reducing each term by such a constant that the sum of the terms in either member becomes zero[16]): Gérardin's[13] 2, $\cdots$, 47, Gérardin,[17, 42] Tarry,[17] Miot,[40] and Aubry.[46]

In Sphinx-Oedipe, 10, 1915, 30, occur two examples of two sets of five numbers having equal sums of kth powers for $k=1, \cdots, 4$, the numbers being functions of six parameters.

CHAPTER XXV.

WARING'S PROBLEM AND RELATED RESULTS.[*]

WARING'S PROBLEM.

E. Waring[1] stated that every integer is a sum of at most 9 [positive integral] cubes, also a sum of at most 19 biquadrates, etc. Every integer N of the proper form is a sum of a finite number of terms $t = ax^m + bx^n + cx^r + \cdots$ (N being a multiple of 3 if $t = 3x^4 + 6x^3 + 24$). Cf. Maillet.[14]

J. A. Euler[2] stated that, to express every positive integer as a sum of positive nth powers, at least $T = \nu + 2^n - 2$ terms are necessary, where ν is the largest integer $< (3/2)^n$. For $n = 2, 3, 4, 5, 6, 7, 8$, $T = 4, 9, 19, 37, 73, 143, 279$ [cf. Vacca[18]].

A. R. Zornow,[3] at the suggestion of C. G. J. Jacobi, constructed a table of the least number of positive cubes composing each number $\leqq 3000$. The number of cubes was stated to be $\leqq 8$ except for 23, $\leqq 7$ for numbers > 454, $\leqq 6$ for numbers > 2183. The final statement and the second for 239 (which requires 9 cubes) are erroneous. Corrections were made by Z. Dase, who computed a table extending to 12000 and communicated it to Jacobi.[4] The largest number within the limits for which 7 cubes are required is 8042; for 8 cubes, 454. Jacobi considered the problem to find all the decompositions of a given number into the least number of cubes. He tabulated the numbers < 12000 which are sums of two cubes and those which are sums of three cubes.

C. A. Bretschneider[5] constructed at Jacobi's suggestion, a table giving all the decompositions of numbers $\leqq 4100$ into a sum of biquadrates, and a companion table showing the numbers which equal the sum of a given number of biquadrates but not fewer. For 79, 159, 239, 319, 399, 379 and 559, it is necessary to use 19 biquadrates; for the remaining numbers, at most 18. As far as $4096 = 4^6$, he verified that 37 fifth powers are needed, and 73 sixth powers. He repeated Euler's[2] statement.

J. Liouville[6] was the first to prove that every positive integer is the sum of a fixed number N_4 of biquadrates, in fact, of at most 53. He first proved that the product of any square by 6 is a sum of 12 biquadrates, in view of

$$6n^2 = \sum_4 x^4 + \sum_8 \{\tfrac{1}{2}(x \pm y \pm z \pm t)\}^4, \qquad 2n = \Sigma x^2.$$

---

[*] A. J. Kempner read critically the reports in this chapter and compared them with the original papers except for 2, 6, 38b, 44a, 54, 60–62, 64, 69, 72, which were not accessible to him. The statements concerning incorrect results in papers 6a, 13 and 17 are made on his authority.

[1] Meditationes algebraicae, Cambridge, 1770, 204–5; ed. 3, 1782, 349–350.

[2] L. Euler's Opera postuma, 1, 1862, 203–4 (about 1772).

[3] Jour. für Math., 14, 1835, 276–280.

[4] Jour. für Math., 42, 1851, 41–69; Jacobi, Werke, VI, 322–354, and 429–431 for corrections of the Journal article.

[5] Jour. für Math., 46, 1853, 1–28.

[6] In his lectures at the Collège de France; printed in V. A. Lebesgue's Exercices d'Analyse Numérique, Paris, 1859, 112–5. Cf. E. Maillet, Bull. Soc. Math. France, 23, 1895, bottom of p. 45.

But any number is of the form $6p+r$, $r=0$, $\cdots$, 5, while p is a sum $n_1^2+\cdots+n_4^2$ of four squares. By the earlier remark, $6p$ is a sum of 48 biquadrates. Hence $N_4 \leqq 48+5$.

E. Lucas[6a] gave the identity

$$(1) \quad 6(x_1^2+x_2^2+x_3^2+x_4^2)^2 = \Sigma(x_i+x_j)^4+\Sigma(x_i-x_j)^4 \quad (i,j=1,\cdots,4; \, i<j).$$

[It becomes Liouville's[6] identity for $x_1=x+y$, $x_2=x-y$, $x_3=z+t$, $x_4=z-t$]. Lucas also gave the incorrect identity

$$10(x_1^2+x_2^2+x_3^2+x_4^2)^3 = \sum_{12}(x_1\pm x_2)^6.$$

Assuming that every integer is a sum of nine cubes, he stated incorrectly that it follows that every integer is a sum of at most 26 sixth powers.

Lucas[7] noted the identities

$$24(x^2+y^2+z^2)^2 = 2(x+y+z)^4+2\sum_3(x+y-z)^4+\sum_3(2x)^4,$$

$$10(x^2+y^2+z^2+u^2)^3 = \sum_6(x+y)^6+\sum_6(x-y)^6+4\sum_4 x^6,$$

the second being erroneous [Fleck[23]], since the left member exceeds the right by $60(x^2y^2z^2+x^2y^2u^2+x^2z^2u^2+y^2z^2u^2)$.

S. Réalis[8] proved that 47 biquadrates are sufficient by using the result that any integer is a sum of 4 squares, one of which is arbitrary (under certain restrictions) and hence may be chosen a biquadrate.

E. Lucas[9] reduced the number to 45 as follows. Let $k=6p+r$. If $p=8h+j$ ($j=1$, 2, 3, 5 or 6), p is a ▣, and, by (1), k a sum of $3\cdot12+5$ biquadrates. If $p=8h$ or $8h+4$, $p-27$ is a ▣; then

$$k=6n_1^2+6n_2^2+6n_3^2+2\cdot3^4+r,$$

so that at most $3\cdot12+2+5$ biquadrates are needed. Finally, if $p=8h+7$, $p-14$ is a ▣, so that

$$k=6n_1^2+6n_2^2+6n_3^2+3^4+3+r, \qquad N_4\leqq 3\cdot12+4+5.$$

Lucas[10] obtained the lower value $N_4 \leqq 41$. Since $8h+j$ ($j=1$, 2, 3, 5, or 6) is a ▣, $48h+6j$ is a sum of 36 biquadrates. By subtracting at most five of the biquadrates 1^4, 2^4, 3^4 from any given number, we obtain one of these numbers $48h+t$ ($t=6$, 12, 18, 30, 36). By the tables our theorem is true for numbers $\leqq 5\cdot3^4$.

E. Maillet[11] proved that every positive integer is a sum of 21 or fewer cubes $\geqq 0$, five or more of which are 0 or 1. He employed the identity

$$\sum_{j=1}^3 (\{(\alpha+x_j)^3+(\alpha-x_j)^3\}) = 6\alpha(\alpha^2+x_1^2+x_2^2+x_3^2)$$

to conclude that $6\alpha(\alpha^2+m)$ is a sum of at most six positive cubes if $0\leqq m\leqq\alpha^2$ and if m is a sum of three squares, i. e., if $m \neq 4^h(8n+7)$. Under the similar conditions on m', $6A=6\alpha(\alpha^2+m)+6\alpha'(\alpha'^2+m')$ is a sum of at most twelve

---

[6a] Nouv. Corresp. Math., 2, 1876, 101.

[7] Jour. de math. élém. et spéc., 1, 1877, 126–7, Probs. 38, 39. Quoted by C. A. Laisant, Recueil de problèmes de math., algèbre, 1895, 125.

[8] Nouv. Corresp. Math., 4, 1878, 209–210.

[9] *Ibid.*, 323–5.

[10] Nouv. Ann. Math., (2), 17, 1878, 536–7.

[11] Assoc. franç. av. sc., 24, II, 1895, 242–7.

positive cubes. For α and α' odd and relatively prime and for every A' such that $\alpha < \alpha' < \alpha^2/8$, $8\alpha\alpha' \leqq A' \leqq \alpha'^3$, it is shown that there exist positive integers m and m' satisfying the earlier conditions and also $\alpha m + \alpha' m' = A'$. Hence every integral multiple $6A$ of 6, for which

$$6(\alpha^3 + \alpha'^3) + 48\alpha\alpha' \leqq 6A \leqq 6(\alpha^3 + \alpha'^3) + 6\alpha'^3, \qquad \alpha < \alpha' < \alpha^2/8,$$

with α, α' odd and relatively prime, is a sum of at most twelve positive cubes. Taking $\alpha = \gamma - 2$, $\alpha' = \gamma$, we see that the intervals obtained by varying γ overlap if γ exceeds a finite limit and is odd. Hence every multiple of 6 exceeding a certain finite limit is a sum of at most twelve positive cubes, whence $N_3 \leqq 12 - 5$ (at least five cubes being 0 or 1).

G. Oltramare[12] proved that any positive cube is the sum of 9 smaller cubes $\geqq 0$. Any number N is the sum $a^2 + b^2 + c^2 + d^2$ of four squares. Then $8x^3 + 6xN$ is the sum s of the cubes of $x \pm a$, $x \pm b$, $x \pm c$, $x \pm d$. For N odd, $N = 2x + 1$, we have $N^3 = 1^3 + s$. For $N_1 = 2^k N$, where N is odd, we multiply the last formula by 2^{3k}.

G. B. Mathews[13] argued that there is a considerable probability that all sufficiently large integers are expressible as sums of $p + 1$ pth powers, at least for some positive integers p. According to Kempner[42], this is not true when p is 6 or any power of 2.

E. Maillet[14] proved that if $\phi(x) = ax^5 + a_1 x^4 + \cdots + a_5$ equals a positive integer for every integer $x \geqq \mu$, then every integer n exceeding a certain function of $a, \cdots, a_5$ is the sum of a limited number N of positive numbers $\phi(x)$ and a limited number of units, where N is at most 6, 12, 96, 192 when ϕ is of degree 2, 3, 4, 5, respectively. For each function $\phi(x)$, the number of representations of n obtained increases indefinitely with n.

E. Lemoine[15] stated that every integer equals $p + s$, where s is a cube or a sum of distinct cubes, while p is one of the 24 numbers $0 - 6, 8 - 17, 27 - 33$.

L. Ripert[16] proved this statement.

R. D. von Sterneck[17] gave a table showing the number of cubes needed for the representation of all numbers $\leqq 40000$. From 8042 on, six cubes suffice. He stated incorrectly [Fleck[20]] that $3k^3$ is not the sum of three cubes unless they are equal. He conjectured incorrectly [Kempner[42]] that always about ten of any thousand consecutive numbers are sums of two cubes.

G. Vacca,[18] after citing Euler's statement, noted that $2^n \cdot \nu - 1$ is the sum of $\nu - 1$ numbers each 2^n and $2^n - 1$ units. [Thus, for $n = 2$, 7 is the sum of 4, 1, 1, 1, but not a sum of fewer than 4 squares; for $n = 3$, 23 is the sum of 8, 8 and seven units, but not a sum of fewer than 9 positive cubes; for $n = 4$, 79 is the sum of 16, 16, 16, 16 and 15 units, but not a sum of fewer than 19 biquadrates.]

[12] L'intermédiaire des math., 2, 1895, 30.
[13] Messenger of Math., 25, 1895–6, 69.
[14] Jour. de Math., (5), 2, 1896, 363–380; Bull. Soc. Math. France, 23, 1895, 40–49. Cf. papers 68, 72, 73, 117, 181–2 of Ch. I.
[15] Nouv. Ann. Math., (3), 17, 1898, 196.
[16] *Ibid.*, (3), 19, 1900, 335–6.
[17] Sitzungsber. Akad. Wiss. Wien (Math.), 112, IIa, 1903, 1627–66.
[18] L'intermédiaire des math., 11, 1904, 292–3.

E. Maillet[19] erroneously concluded that there is an infinitude of integers not a sum of fewer than 128 eighth powers >0.

A. Fleck[20] noted that in the proof by Lucas[10] it suffices to subtract at most three biquadrates unless the given number is $48m+t$, $t=10$, 11, 26, 27, 42, 43. For $t=10$, subtract 1^4+3^4; we get $6N$, where $N=4(2m-3)$ is a ▣ unless $2m-3\equiv7$ (mod 8), i. e., $m=1+4\mu$. In the latter case,

$$48m+10-5^4-3^4=6\cdot4(8\mu-27)$$

is a sum of 36 biquadrates since $4(8\mu-27)$ is a ▣. Treating similarly the remaining t's, he concluded that $N_4\leqq39$. He found that $N_3\leqq13$ by employing Maillet's[11] result and the formula, following from $r^3\equiv r$ (mod 6),

$$6N+r=6N+r^3-6k=r^3+6\mu=r^3+\sum_{12}x^3.$$

E. Landau[21] proved that every definite integral rational function of x of degree n with rational coefficients is a sum of 8 squares of integral rational functions with rational coefficients, and gave references to related problems.

A. Fleck[22] proved that the square (cube) of every definite integral rational function of x with rational coefficients is a sum of a finite determinable maximum number, independent of the degree and coefficients of the function, of fourth powers (sixth powers) of integral rational functions of degree $\leqq1$ with rational coefficients, i. e., linear functions and constants.

Fleck[23] remarked that Maillet's[14] limit 192 for N_5 can easily be reduced by about 36, but that the new limit is still far above the ideal limit 37 suggested by tables. To show that N_6 is finite, he used the identity

$$60(a^2+b^2+c^2+d^2)^3=\sum_4(a+b+c)^6+\sum_{12}(a+b-c)^6+2\sum_6(a+b)^6$$

$$+2\sum_6(a-b)^6+36\sum_4a^6.$$

Hence $60n^3$ is a sum of 184 sixth powers. Thus if m is any integer, $60m$ is the sum of at most $184N_3$ sixth powers. Since any integer is of the form $60m+r$, $r=0$, 1, $\cdots$, 59, we have $N_6\leqq184N_3+59$.

E. Landau[24] lowered the limit for N_4 to 38. Setting $x_4=x_3$ in (1), we see that $6n^2$ is a sum of 11 biquadrates if n is representable in the form $x_1^2+x_2^2+2x_3^2$, which is true if n is any odd number m. Hence $6m^2$ and $6\cdot16\ m^2$ are sums of 11 biquadrates. As above, $8k+j$ ($j=1$, 2, 3, 5 or 6) is a sum of three squares at least one of which is odd. Hence 6 times such a number is a sum of $11+12+12$ biquadrates. By arguments of the type used by Fleck,[20] we get $N_4\leqq38$. Except for numbers $48n+t$, $t=11$, 27, 43, he proved that 37 biquadrates suffice. For these cases, A. Wieferich[25] showed that 37 suffice. Hence $N_4\leqq37$.

[19] Annali di Mat., (3), 12, 1905, 173, note. Error admitted in l'intermédiaire des math., 20, 1913, 202.

[20] Sitzungsber. Berlin Math. Gesell., 5, 1906, 2–9.

[21] Math. Annalen, 62, 1906, 272–281.

[22] *Ibid.*, 64, 1907, 567–572.

[23] *Ibid.*, 561–6. To $N=192$ must be added the number of units.

[24] Rendiconti Circolo Mat. Palermo, 23, 1907, 91–6.

[25] Math. Annalen, 66, 1909, 106–8.

E. Maillet[26] proposed the following generalization of Waring's problem: Can k be taken sufficiently large that there shall be integral solutions of

$$\sum_{j=1}^{k} x_j^{n_1} = N_1, \qquad \sum_{j=1}^{k} x_j^{n_2} = N_2, \qquad \cdots, \qquad \sum_{j=1}^{k} x_j^{n_a} = N_a,$$

where $n_1, \cdots, n_a$ have given values, and $N_1, \cdots, N_a$ any values satisfying suitable conditions? For $a=2$, $n_1=2$, $n_2=1$, $k=4$, there is always a solution (Cauchy, Ch. VIII, p. 284) if N_1 is odd and N_2 is odd and

$$\sqrt{3N_1-2}-1 < N_2 < \sqrt{4N_1}.$$

E. Maillet[27] proved Waring's theorem for eighth powers, but gave no explicit limit for N_8. He proved in an elementary way that there is an infinitude of numbers each not a sum of n or fewer nth powers.

A. Hurwitz[28] proved that every integer is the sum of at most

$$37(6 \cdot 4 + 60 \cdot 12 + 48 + 6 \cdot 8) + 5039 = 36119$$

8th powers, in view of $N_4 \leqq 37$ and the identity

$$5040(a^2+b^2+c^2+d^2)^4 \equiv 6\sum_4 (2a)^8 + 60\sum_{12}(a\pm b)^8$$
$$+ \sum_{48}(2a\pm b\pm c)^8 + 6\sum_8 (a\pm b\pm c\pm d)^8.$$

In general, if there exists an identity (in a, b, c, d)

$$p(a^2+b^2+c^2+d^2)^n = \sum_{i=1}^{r} p_i(\alpha_i a + \beta_i b + \gamma_i c + \delta_i d)^{2n},$$

where $p, p_1, \cdots, p_i$ are positive integers and $\alpha_1, \cdots, \delta_r$ are integers, then

$$N_{2n} \leqq N_n(p_1 + \cdots + p_r) + p - 1,$$

so that N_{2n} would be finite if N_n is. He proved by use of the gamma function that there is an infinitude of positive integers each not the sum of n or fewer nth powers.

J. Schur[29] found the identity which proves N_{10} finite:

$$22680(a^2+b^2+c^2+d^2)^5 = 9\sum_4 (2a)^{10} + 180\sum_{12}(a\pm b)^{10}$$
$$+ \sum_{48}(2a\pm b\pm c)^{10} + 9\sum_8 (a\pm b\pm c\pm d)^{10}.$$

A. Wieferich[30] proved that $N_3 \leqq 9$ [except for a limited set of integers arising from a case[31] overlooked]. The proof consists in showing that any positive integer is the sum of three cubes together with $k = 6a^3 + 6am$, where $0 < A$ and $m = x_1^2 + x_2^2 + x_3^2 < A^2$. For, by Maillet,[11] k is then a sum of 6 positive cubes.

---

[26] L'intermédiaire des math., 15, 1908, 196, and Maillet.[27]

[27] Bull. Soc. Math. de France, 36, 1908, 69–77; Comptes Rendus Paris, 145, 1907, 1399.

[28] Math. Annalen, 65, 1908, 424–7.

[29] Math. Annalen, 66, 1909, 105 (in a paper published by Landau.)

[30] Math. Annalen, 66, 1909, 95–101.

[31] The case $\nu=4$ in $10648 < (0.4)5^{2\nu-\epsilon}$. Attention was called to this gap in the proof by P. Bachmann, Niedere Zahlentheorie, 2, 1910, 344, who indicated in his Zusätze, pp. 477–8, a long method of treating the omitted case, but himself made certain errors. The latter were incorporated in the unsuccessful attempt by E. Lejneek (Math. Ann., 70, 1911, 454–6) to fill the gap. The gap in Wieferich's proof was filled by Kempner.[42]

E. Landau[32] proved that every integer z exceeding a fixed value is the sum of at most 8 positive cubes. He proved that there exists a prime p not dividing z such that $8p^9 \leqq z < 12p^9$ and such that $p^2(p-1)$ is not divisible by 3. Hence $\beta^3 \equiv z \pmod{p^3}$ has positive integral solutions $\beta < p^3$. In $z = \beta^3 + p^3 M$, set $M = 6p^6 + M_1$. Then

$$7p^9 < p^3 M < 12p^6, \qquad p^6 < M_1 < 6p^6.$$

By the paper of Wieferich,[30] we can find an integer γ, $0 \leqq \gamma < 96$, such that $M_1 - \gamma^3 = 6m$, where $m = x_1^2 + x_2^2 + x_3^2$. For z sufficiently large, $m < p^6$, so that $0 \leqq x_i < p^3$, and

$$z = \beta^3 + p^3(6p^6 + M_1) = \beta^3 + (p\gamma)^3 + 6p^3(p^6 + x_1^2 + x_2^2 + x_3^2),$$

$$z = \beta^3 + (p\gamma)^3 + \sum_{i=1}^{3} \{(p^3 + x_i)^3 + (p^3 - x_i)^3\}.$$

A. Wieferich[33] proved that $N_5 \leqq 59$, $N_7 \leqq 3806$. He gave a table showing the least number of fifth powers required to represent each number $1, \cdots,$ 3011.

D. Hilbert[34] proved Waring's assertion that every positive integer z is the sum of at most N_m positive mth powers, where N_m is a finite number, not determined, depending upon m but not upon z. He first proved, by use of a five-fold integral (a 25-fold integral in the first paper,) the lemma (stated by Hurwitz,[28] who was unable to prove it) that there exists for every m (and $r = 5$) an identity in the x's

$$(x_1^2 + \cdots + x_r^2)^m = \sum_h \rho_h (a_{1h} x_1 + \cdots + a_{rh} x_r)^{2m},$$

where the a_{ih} are integers and the ρ_h are positive rational numbers. It is a simple step to prove Waring's theorem for powers whose exponents are 2^k, $k \geqq 2$. The case of any exponent is derived from this by an elementary, but long, discussion (not using calculus).

F. Hausdorff[35] proved Hilbert's lemma by use of integrals involving exponentials, the method being more suitable for computing the a's and ρ's.

E. Stridsberg[36] proved easily that Waring's theorem for μth powers would follow if it were shown that, if B is any real number, every positive integer $\geqq B$ can be written as $\Sigma \rho_\lambda P_\lambda^\mu$, where the P's are integers $\geqq 0$ and ρ_λ is a positive rational number depending only on μ. He noted that Hausdorff's elegant modification of Hilbert's proof can be reduced to an elementary study of binomial coefficients. Using symbolic powers of h, let $h^{2\mu}$ denote $(2\mu)!/\mu!$ for all even integers $2\mu \geqq 0$, and $h^{2\mu+1} = 0$ for all odd integers $2\mu + 1 \geqq 1$. A theorem of Hausdorff's becomes the simple one that, if $f(x)$ is any polynomial which is never negative for a real value of x, then

---

[32] Math. Annalen, 66, 1909, 102–5; Landau, Handbuch der Lehre von der Verteilung der Primzahlen, 1, 1909, 555–9. Cf. Landau.[39]

[33] Math. Annalen, 67, 1909, 61–75.

[34] Göttingen Nachr., 1909, 17–36; Math. Ann., 67, 1909, 281–300.

[35] Math. Annalen, 67, 1909, 301–5. Cf. Hurwitz.[44]

[36] Arkiv för Mat., Astr., Fysik, 6, 1910–11, No. 32, No. 39. French résumé in Math. Annalen, 72, 1912, 145–152.

$f(x+h)>0$ for x real [cf. Hurwitz[44]], since

$$f(x+h)] = \frac{1}{\Gamma(1/2)} \int_{-\infty}^{\infty} e^{-\alpha^2/4} f(x+\alpha)\,d\alpha,$$

being true for $f(x+h)=h'$. Hilbert's lemma is proved by use of

$$(h_1 x_1 + \cdots + h_r x_r)^m = h^m (x_1^2 + \cdots + x_r^2)^{m/2},$$

whence follows Hurwitz's theorem that Waring's theorem is true for $n=2m$ if true for $n=m$. Finally, he simplified the second (elementary) part of Hilbert's proof of Waring's theorem.

A. Boutin[37] gave the identities

$$\sum_8 \pm (x \pm y \pm z \pm u)^4 = 192xyzu, \qquad \sum_{2^n} \pm(\pm x_1 \pm \cdots \pm x_n)^n = n!\,2^n x_1 \cdots x_n,$$

the exterior sign being the product of the n interior signs.

P. Bachmann[38] gave an exposition of several of the preceding papers.

A. Fleck[38a] and W. Wolff[38b] proved that every definite quartic function of x with rational coefficients is a sum of five squares of rational integral functions with rational coefficients.

E. Landau[39] gave a new elementary proof that all numbers exceeding a certain limit and prime to 10 (or to the product of any two primes of the form $3m + 2$) are sums of at most 8 positive cubes. He here avoided the theory of the distribution of primes used in his[32] former proof.

J. Kürschák[40] generalized Liouville's[6] identity (1) to give

$$\Sigma\,(a_0 \pm a_1 \pm \cdots \pm a_k)^4 = 2^k \binom{3k}{k} (a_0^2 + \cdots + a_{3k}^2)^2,$$

where on the left occur all possible combinations of signs and all sets of $k+1$ of the $3k+1$ variables $a_0, \cdots, a_{3k}$. For $m \geq 3$, there is no identity

$$\Sigma\,(a_0 \pm a_1 \pm \cdots \pm a_k)^{2m} = C(a_0^2 + \cdots + a_n^2)^m.$$

A. Gérardin[41] noted that $(x^3+9y^3)^3$ is the sum of the cubes of x^3, y^3, $6y^3$, $8y^3$, $3x^2y$, $3xy^2$, $6xy^2$. Also $(x^3+3y^3)^3$ is the sum of the cubes of x^3, $3y^3$, $3xy^2$, $2x^2y$, x^2y. L. Rouve remarked that the former is the sum of the cubes of x^3, $3x^2y$, $9y^3$, $3xy^2$, $6xy^2$.

A. J. Kempner[42] considered the number $C(k, n)$ of the positive integers $\leq k$ which are sums of n or fewer positive nth powers, and the superior limit S of $C(k, n)/k$ for $k = \infty$. He proved that $S<1/n!$, whereas Hurwitz[28] and Maillet[27] had proved merely that $S<1$. It follows that there is an infinitude of positive integers of each of the forms $9l$, $9l+1$, $\cdots$, $9l+8$, such that each is not a sum of fewer than four positive cubes. There is an infinitude of positive integers each not a sum of fewer than nine sixth

[27] L'intermédiaire des math., 17, 1910, 122-3, 236-7. See papers 66-68 below.
[38] Niedere Zahlentheorie, 2, 1910, 328-48.
[38a] Archiv Math. Phys., (3), 10, 1906, 23-38; (3), 16, 1910, 275-6.
[38b] Vierteljahrsschrift Naturf. Gesell. Zürich, 56, 1911, 110-24.
[39] Archiv Math. Phys., (3), 18, 1911, 248-252.
[40] Ibid., 242-3.
[41] Sphinx-Oedipe, 6, 1911, 19, 95.
[42] Über das Waringsche Problem und einige Verallgemeinerungen, Diss., Göttingen, 1912. Extract in Math. Annalen, 72, 1912, 387.

powers, and an infinitude each not a sum of fewer than 2^{q+2} powers, with the exponent 2^q, for $q>1$. He lowered the known limit for N_6 to 970 by use of the identity

$$120(a^2+b^2+c^2+d^2)^3 = \sum_8 (a \pm b \pm c \pm d)^6 + 8 \sum_{12} (a \pm b)^6 + \sum_4 (2a)^6,$$

for $c=d$ and for $d=0$, and the fact that every number is of one of the forms $a^2+b^2+kc^2$ for $k=1, 2$. For the determination of upper limits for N_{12} and N_{14} from known limits for N_6 and N_7, he gave identities expressing $l(a^2+b^2+c^2)^n$ as a sum of $(2n)$th powers, for $n=6$ and 7, where l is a suitably chosen integer.

R. Remak[43] noted that Stridsberg[36] used integrals in a single place and applied the result proved by them only for the special case in which $f(\alpha) = g^2(\alpha)$. For this case Remak gave an elementary proof by use of the fact that a quadratic form in n variables is definite if the determinant of the part involving the first ν variables (suitably chosen) is positive for $\nu = 1, 2,$ $\cdots, n$. Hence the proof of Waring's theorem is reduced to algebraic processes.

A. Hurwitz[44] gave a new elementary proof of the theorem, used by Hausdorff,[35] Stridsberg[36] and Remak,[43] that if the real polynomial

$$f(x) = c_0 + c_1 x + \cdots + c_{2n} x^{2n},$$

not identically zero, is $\geqq 0$ for every real x, then

$$f(x) + \frac{1}{1!} f''(x) + \frac{1}{2!} f^{(4)}(x) + \cdots + \frac{1}{n!} f^{(2n)}(x)$$

is positive for every real x; likewise for $f(x) + f'(x) + \cdots + f^{(2n)}(x)$.

L. Orlando[44a] amplified Hurwitz's[44] proof.

G. Frobenius[45] also gave an algebraic proof of Waring's theorem by altering Stridsberg's proof at the point where he had used integrals.

E. Schmidt[46] used Minkowski's convex point sets in space of q dimensions to give a more luminous exposition of Hilbert's first lemma.

G. Loria[47] remarked that if Waring's minimum 19 for N_4 could be lowered to 16 [overlooking the facts noted by J. A. Euler], one would hope for a proof that every number is a sum of n^2 exact nth powers.

E. Landau[48] pointed out errors in the same journal on sums of cubes.

W. S. Baer[49] proved that every integer $\geqq 23 \cdot 10^{14}$ is a sum of 8 or fewer positive cubes, likewise every odd number $>175\ 396\ 368\ 704$, and every number $\equiv 8 \pmod{16}$. The following numbers are sums of 7 or fewer positive cubes: every number $2744s$ (s odd), all sufficiently large multiples of 16 or 27, all sufficiently large numbers $\equiv 0, 8, 16, 24, 28, 36, 44, 48, 56, 64$

---

[43] Math. Annalen, 72, 1912, 153–6.

[44] *Ibid.*, 73, 1912, 173–6. Cf. Orlando[44a]. For a generalization see G. Pólya, Jour. für Math., 145, 1915, 233.

[44a] Atti della R. Accad. Lincei, Rendiconti, 22, I, 1913, 213–5.

[45] Sitzungsber. Akad. Wiss. Berlin, 1912, 666–70.

[46] Math. Annalen, 74, 1913, 271–4.

[47] L'enseignement math., 15, 1913, 200–1.

[48] L'intermédiaire des math., 20, 1913, 177, 179.

[49] Beiträge zum Waringschen Problem, Diss., Göttingen, 1913, 74 pp.

(mod 72). He reduced the limit for N_6 to 478, that for N_5 to 58 and gave a simpler proof that $N_4 \leq 37$. For $k = 2744$, it is shown by elementary methods that every number $\equiv k$ (mod $2k$) is a sum of 7 or fewer positive cubes; hence if $C_7(x)$ denotes the number of positive integers $\leq x$ which are decomposable into 7 or fewer positive cubes,

$$(2) \qquad \frac{1}{2k} < \frac{C_7(x)}{x} \leq 1 \quad \text{for all sufficiently large } x.$$

His transcendental methods enabled him to replace $1/(2k)$ by $13/72$. He[50] later gave a direct elementary proof of the last result (2) for $k = 4096$ by noting that the integers ku, where u is positive and odd, can be decomposed into 7 positive cubes all of whose 7 bases exceed any assigned positive number g for every u exceeding a limit depending upon g.

E. Stridsberg[51] gave a brief elementary proof of Hurwitz's lemma [Hilbert[34]] without the use of integrals (Remak,[43] Frobenius[45]) or the gamma function. The proof is admitted to be otherwise essentially the same as his[36] former proof.

G. H. Hardy and S. Ramanujan[52] proved that the logarithm of the number of ways n is a sum of rth powers of positive integers (rearrangements of the same powers not being counted as distinct) is asymptotic to

$$(r+1) \left\{ \frac{1}{r} \, \Gamma\left(\frac{1}{r} + 1 \right) \cdot \zeta\left(\frac{1}{r} + 1 \right) \right\}^{r/(r+1)} n^{1/(r+1)},$$

where ζ denotes the Riemann zeta function, and Γ the gamma function.

Hardy and J. E. Littlewood[52a] made use of the theory of analytic functions (cf. Ch. III[221]) to prove that every positive integer, which exceeds a certain number depending on k alone, is a sum of at most $k \cdot 2^{k-1} + 1$ positive kth powers; for example, a sum of at most 33 biquadrates. The transcendental method leads not only to a proof of the existence of representations, but also to asymptotic formulas for their number. They since communicated to the author the improved result that at most $(k-2)2^{k-1} + 5$ positive kth powers are necessary; this gives 9 cubes, 21 biquadrates, 53 fifth powers, 133 sixth powers, etc.

Numbers expressible as sums of unlike powers.

D. André[53] proved that every even integer is the sum of a cube $\neq 0$ and three squares (since every $8n+3$ is a $\boxed{3}$). In general, if s is odd, every even integer $> 7^s$ is the sum of an sth power $\neq 0$ and three squares each $\neq 0$.

G. de Rocquigny[54] noted that every integer except 1, 2, 3, 4, 5, 7, 8, 10, 11, 18 is a sum of three cubes and three squares. He[55] stated many

[50] Math. Annalen, 74, 1913, 511–4.

[51] Arkiv för Mat., Astr., Fysik, 11, 1916–7, No. 25, pp. 35–9. His second paper with the same title, ibid., 13, 1919, No. 25, deals at length (pp. 31–70) with definite and semi-definite polynomials in x and incidentally with their occurrence in the literature on Waring's problem.

[52] Proc. London Math. Soc., (2), 16, 1917, 130.

[52a] Quar. Jour. Math., 48, 1919, 272 seq.

[53] Nouv. Ann. Math., (2), 10, 1871, 185–7.

[54] Travaux Sc. de l'Univ. Rennes, 3, 1904, 42.

[55] L'intermédiaire des math., 10, 1903, 109, 212; 11, 1904, 31, 56, 81, 99, 149, 171, 214.

theorems like the following: Every integer >36 is a sum of four squares and four biquadrates each $\neq 0$; every integer >14 is a sum of four squares and four cubes $\neq 0$.

P. F. Teilhet[56] verified that every integer up to 600, except 23, is a sum of two squares and two positive or zero cubes.

G. Lemaire[57] noted that 3, 6, 7, 11, 15, 19, 22, 23 are not sums of any number of powers of distinct numbers.

G. Rabinovitch[58] proved that every number >23 is expressible in one of the forms a^m+b^n, $a^m+b^n+c^p$, $\cdots$, where a, b, $\cdots$ are distinct, and m, n, p, $\cdots$ exceed unity.

A. Gérardin[59] proved the theorems due to André.[53]

EVERY NUMBER A SUM OF THREE RATIONAL CUBES.

S. Ryley[60] solved $a=x^3+y^3+z^3$ by taking $x=p+q$, $y=p-q$, $z=m-2p$. Then
$$36p^2q^2 = 6ap - 6pm^3 + 36p^2(m-p)^2$$
will, for $p=av^2/6$, equal the square of $av-av^2(m-av^2/6)$ if $m^3=2av(m-av^2/6)$. Let $m=dv$. Then $v=6ad/D$, where $D=3d^3+a^2$. Hence
$$x=\frac{(9d^6-30a^2d^3+a^4)D+72a^4d^3}{6adD^2}, \qquad y=\frac{30a^2d^3-9d^6-a^4}{6adD}, \qquad z=\frac{6ad^2D-12a^3d^2}{D^2}.$$
Reference is made to a less simple method in Leed's Correspondent, Quest. 211.

T. Strong[61] showed how to express any number a as sum of three or more rational cubes. Take x, $p-x$, $m-p$, r, s, $\cdots$ as the roots of the cubes. Thus
$$(3p^2-6px)^2 = 9p^2(p-2m)^2 + 12ap - 12p(m^3+r^3+s^3+\cdots).$$
The right member will be the square of $3p(p-2m)+2c$ if
$$p=c^2/(3a), \qquad c(2m-p)=m^3+r^3+s^3+\cdots.$$
Set $c=mn$, $r=mr'$, $s=ms'$, $\cdots$. The second condition gives
$$m=\frac{6an}{3a^2+n^3+r'^3+s'^3+\cdots}.$$
Hence giving any rational values to n, r', s', $\cdots$, we get rational values for $x=m-c/(3p)$, m, p, r, s, $\cdots$. Since we can in particular express 4 as a sum of three positive cubes, we can divide unity into three positive parts such that if each be increased by unity the sum is a cube [Evans,[424] Davis,[426] and Tebay[428] of Ch. XXI].

Wm. Lenhart,[62] to express A as a sum of three cubes, selected any cube r^3 and from Ar^3 subtracted a cube s^3 chosen by trial such that the difference

[56] L'intermédiaire des math., 11, 1904, 16–17.
[57] Ibid., 19, 1912, 218.
[58] Ibid., 20, 1913, 157.
[59] Ibid., 22, 1915, 207.
[60] Ladies' Diary, 1825, 35, Quest. 1420.
[61] Amer. Jour. Arts, Sc. (ed., Silliman), 31, 1837, 156–8.
[62] Math. Miscellany, Flushing, N. Y., 1, 1836, 122–8.

is a number t found in his[186] table (Ch. XXI) of numbers expressible as a sum of two positive rational cubes. Or, let $Ar^3+s^3=t=a^3+b^3$. Then A is the sum of the cubes of ax, bx, cx if $c=p-s$ and

$$\frac{1}{x^3}=\frac{t+c^3}{A}=\frac{p^3-3p^2s+3ps^2}{A}+r^3=\left(r+\frac{ps^2}{r^2A}\right)^3, \qquad p=\frac{3r^3As}{r^3A-s^3}.$$

Hence

$$ax=\frac{a(r^3A-s^3)}{rd}, \qquad bx=\frac{b(r^3A-s^3)}{rd}, \qquad cx=\frac{s(2r^3A+s^3)}{rd}, \qquad d=r^3A+2s^3.$$

As an application, 2 and 4 are expressed as sums of three positive rational cubes. The same table is used tentatively to express $n+1$ or $n-1$ as a sum of n cubes each >1 or each <1, with examples when $n=4$, 5, 6.

Several[63] expressed any number n as the sum of three rational cubes. Let their roots be $(1\pm z)/(2x)$, $(ax^2-1)/x$. The sum of their cubes is n if

$$z^2=1-4ax^2+4a^2x^4-\tfrac{4}{3}a^3x^6+\tfrac{4}{3}nx^3.$$

Assuming that $z=1-2ax^2+\tfrac{2}{3}nx^3$, we get $x=6an/(n^2+3a^3)$.

EVERY POSITIVE NUMBER A SUM OF FOUR POSITIVE RATIONAL CUBES, ETC.

G. Libri[64] noted that if m, n, r are solutions of $ax^3+by^3+cz^3=0$, then $aX^3+bY^3+cZ^3=d$ is solvable for d arbitrary. Set $X=mp+q$, $Y=np+s$, $Z=rp+t$. The new equation lacks p^3 and will lack p^2 and hence determine p rationally in terms of s, t, if we take $q=-(bn^2s+cr^2t)/(am^2)$.

If A is a multiple of 24, it is a sum of four cubes [not necessarily positive]:

$$A=(q-p)^3+(-p-3q)^3+2(p+q)^3, \qquad q=\pm 1, \qquad p=q-\frac{A}{24q^2}.$$

Next, let $A=24x+b$, $0<b<24$. If b is one of the numbers 1, 3, 5, 7, 8, 9, 11, 13, 15, 16, 17, 19, 21, 23, b^3-b is a multiple $24u$ of 24, whence $A=b^3+s$, where $s=24(x-u)$ is a sum of four cubes, so that A is a sum of five cubes. If b is not one of the above numbers, $b\pm 1$ is one of them. Hence every integer is a sum of six cubes one of which is 0 or 1. If

$$f=x^3+y^3+z^3+u^3, \qquad F=A^3+B^3+C^3+D^3,$$

we have the identity in r, s, t,

$$fF\equiv(g-r-s-t)^3+(g+r)^3+(s-g)^3+(t-g)^3,$$

(1)
$$g=\frac{fF+(r+s+t)^3-r^3-s^3-t^3}{3\{(r+s+t)^2+r^2-s^2-t^2\}}.$$

Every integer is the algebraic sum of 17 biquadrates, taken positively or negatively. The proof, similar to the above for cubes, follows from

$$A=3\left(p+\frac{r}{2}\right)^4+\left(p-\frac{r}{2}\right)^4-(p+r)^4-3p^4, \qquad p=\frac{-4A+3r^4}{12r^3}.$$

---

[63] Math. Quest. Educ. Times, 13, 1870, 63-4.
[64] Memoria sopra la teoria dei numeri, Firenze, 1820, 17-23.

Again, if $p = -1 - B/480$,

$$B = 30(p+2)^4 + 2(p-2)^4 - 20(p+1)^4 - 12(p+3)^4.$$

These two quartic forms repeat under multiplication.

Libri[65] proved that any positive rational number m equals the sum of four positive rational cubes. In the identity

$$(2) \qquad m = \left(\frac{m+6q^3}{6q^2}\right)^3 + \left(\frac{m-6q^3}{6q^2}\right)^3 - \left(\frac{m}{6q^2}\right)^3 - \left(\frac{m}{6q^2}\right)^3,$$

we can reduce the right member to a sum of four positive cubes. In

$$(3) \qquad a^3 - b^3 = a^3\left(\frac{a^3-2b^3}{a^3+b^3}\right)^3 + b^3\left(\frac{2a^3-b^3}{a^3+b^3}\right)^3,$$

take $a = (m+6q^3)/(6q^2)$, $b = m/(6q^2)$. Then the sum of the first and third terms in (2) is a sum $\alpha^3 + \beta^3$ of two positive cubes if $(m+6q^3)^3 > 2m^3$, where

$$\alpha = \frac{m+6q^3}{6q^2} \cdot \frac{\{(m+6q^3)^3 - 2m^3\}}{\{(m+6q^3)^3 + m^3\}}.$$

Now use (3) for $a = \alpha$, $b = m/(6q^2)$. Then $\alpha^3 - \{m/(6q^2)\}^3$ is a sum of two cubes each positive if

$$(m+6q^3)^3\{(m+6q^3)^3 - 2m^3\}^3 > 2m^3\{(m+6q^3)^3 + m^3\}^3,$$

which implies the preceding inequality and can be satisfied. Formula (1) is here repeated. It is stated that $3x^4 + y^4 - z^4 - 3u^4$ represents all rational numbers.

P. Tardy[66] gave the generalization to n factors of $4ab = (a+b)^2 - (a-b)^2$ and

$$24abc = (a+b+c)^3 - (a+b-c)^3 - (a-b+c)^3 + (a-b-c)^3.$$

This formula had been given by C. F. Gauss.[67]

E. Rebout[68] noted that, in this formula, also $24abc$ is a cube if $a=3$, $b=4$, $c=6$.

V. A. Lebesgue[69] remarked that every positive rational number is a sum of four positive rational cubes:

$$(4) \qquad n = \left(\frac{n}{6m^2}\right)^3 \{(2-a)^3 + a^3(b-1)^3 + b^3(c-1)^3 + c^3\},$$

where m^3 is a rational cube lying between $n/6$ and $n/12$, while

$$a = 1 + 6m^3/n, \qquad b = 2 - 3/(a^3+1), \qquad c = 2 - 3/(b^3+1).$$

---

[65] Jour. für Math., 9, 1832, 288–292; Mém. présentés pars divers Savants Acad. R. Sc. l'Institut de France (Math. Phys.), 5, 1838, 71–5. In Comptes Rendus Paris, 10, 1840, 313, Libri stated he had proved the theorem in his book, *Mémoires de Math. et de Phys., Florence, 1829, 152–168.

[66] Annali di Sc. Mat. Fis., 2, 1851, 287; cf. Nouv. Ann. Math., 2, 1843, 454. Cf. Boutin.[37]

[67] Werke, II, 1863, 387. Cf. H. Brocard, Nouv. Corresp. Math., 4, 1878, 136–8.

[68] Nouv. Ann. Math., (2), 16, 1877, 272–3.

[69] Exercices d'analyse numérique, Paris, 1859, 147–151.

E. Lucas[70] remarked that Lebesgue[69] appears not to have guessed what seems to have led Euler to obtain formula (4), viz., the problem to express a number as a sum of two cubes. Any positive rational number N is expressible in an infinitude of ways as a product or quotient of two sums of two positive rational cubes. To prove the former (which corresponds to Euler's theorem), employ the identities

$$(6LM+L^2-3M^2)^3+(6LM-L^2+3M^2)^3=2^23^2LM(L^2+3M^2)^2,$$
$$(L+M)^3+(L-M)^3=2L(L^2+3M^2).$$

Divide their product (member by member) by $(L^2+3M^2)^3$. Hence $2^33^2L^2M$ is expressed as a product of two sums of two cubes. Take $L=Bb^3$, $M=2^{\lambda-3}3^{\mu-2}Aa^3$. We get a decomposition of $N=2^\lambda3^\mu AB^2$, and we can choose a^3/b^3 to make all the cubes positive. As a corollary, E. Fauquembergue[70a] proved that the quadruple and square of $4p^6+27q^6$ are sums of two cubes, a problem proposed by Lucas.

G. Oltramare[71] noted that every integer is a sum of five integral cubes. R. Norrie[72] gave the identity

$$x=\left(\frac{c^2(d+2x)}{2d}\right)^4-\left(\frac{c^2(d-2x)}{2d}\right)^4+\left(\frac{2c^4x-b^4d}{2bcd}\right)^4-\left(\frac{2c^4x+b^4d}{2bcd}\right)^4,\qquad d=c^8-b^8.$$

He expressed (p. 58), 5, 17 and 41 as sums of five integral cubes, not all positive. Other solutions had been given by A. Cunningham.[73]

---

[70] Nouv. Ann. Math., (2), 19, 1880, 89–91; Bull. Soc. Math. France, 8, 1879–80, 180–2. No reference is made to Euler's writings. The author of this History has found no formula like (2) or (4) in Euler's papers or books. Nor did Libri[65] or Lebesgue[69] imply that such a formula is due to Euler. The fact that Lebesgue spoke of (3) as the transformation of Euler may have led Lucas to infer too hastily that also (2) is due to Euler.

[70a] Nouv. Ann. Math., (2), 19, 1880, 430.

[71] L'intermédiaire des math., 1, 1894, 25. Cf. 165–6, 244; 2, 1895, 325.

[72] University of St. Andrews 500th Anniversary Mem. Vol., 1911, 68.

[73] Math. Quest. Educ. Times, (2), 4, 1903, 49.

CHAPTER XXVI.

FERMAT'S LAST THEOREM, $ax^r + by^s = cz^t$, AND THE CONGRUENCE $x^n + y^n \equiv z^n$ (MOD p).[*]

For proofs of the impossibility of $x^n + y^n = z^n$ for $n = 3, 4$, see Chs. XXI, XXII.

Leo Hebreus,[1] or Lewi ben Gerson (1288–1344), proved that $3^m \pm 1 \neq 2^n$ if $m > 2$, by showing that $3^m \pm 1$ has an odd prime factor. The problem had been proposed to him by Philipp von Vitry in the following form: All powers of 2 and 3 differ by more than unity except the pairs 1 and 2, 2 and 3, 3 and 4, 8 and 9.

Fermat,[2] commenting about 1637 on Diophantus II, 8 (to solve $x^2 + y^2 = a^2$), stated that "it is impossible to separate a cube into two cubes, or a biquadrate into two biquadrates, or in general any power higher than the second into two powers of like degree; I have discovered a truly remarkable proof which this margin is too small to contain." This theorem is known as Fermat's last theorem.

Claude Jaquemet[3] (1651–1729), in a manuscript in the Bibliothéque Nationale de Paris and first attributed to Nicolas Malebranche[4] (1638–1715), attempted to prove Fermat's last theorem. In $a^z = x^z + y^z$ we may suppose x, y relatively prime. The quotient of $x^z + y^z$ by $x + y$ is

$$Q = x^{z-1} - yx^{z-2} + y^2 x^{z-3} - \cdots \pm y^{z-1}.$$

Then $x + y$ and Q have no common divisor d other than factors of z. For, it would divide

$$Q - (x^{z-1} + yx^{z-2}) = -2yx^{z-2} + y^2 x^{z-3} - \cdots \pm y^{z-1}.$$

Adding $2yx^{z-2} + 2y^2 x^{z-3}$, we get $3y^2 x^{z-3} - \cdots$. Finally, we get zy^{z-1}. But y is not divisible by d since x, y are relatively prime; hence z is. Similarly, $x - y$ and $(x^z - y^z)/(x - y)$ have no common divisor not a factor of z.

Suppose that a, x, y are relatively prime integers for which $a^z = x^z + y^z$, z odd. As just proved, at most one of the powers is divisible by z. First let x^z and y^z be not divisible by z. Let $x^z = p^z q^z$, $y^z = r^z s^z$, where r and s are relatively prime, also p and q. Then $a - pq = r^z$, $a - rs = p^z$. Thus the divisor $p - r$ of $p^z - r^z$ divides $pq - rs$. Dividing the latter by $p - r$, we get the remainder $pq - ps$ or $rq - rs$, neither zero, and "by continuing this process to infinity, we get no new remainders, so that $p - r$ is not a divisor of $pq - rs$." As pointed out by E. Lucas[4a] the last conclusion is wrong;

---

[*] H. S. Vandiver read critically the proof-sheets of this chapter and believes that the reports are accurate. Both he and the author compared the reports with the original papers when available.

[1] Cf. J. Carlebach, Diss. Heidelberg, Berlin, 1909, 62–4.

[2] Oeuvres, I, 291; French transl., III, 241. Diophanti Alexandrini Arith. libri sex, ed., S. Fermat, Tolosae, 1670, 61. Précis des Oeuvres math. de P. Fermat, par E. Brassinne, Mém. Acad. Sc. Toulouse, (4), 3, 1853, 53.

[3] Cf. A. Marre, Bull. Bibl. Storia Sc. Mat. Fis., 12, 1879, 886–894.

[4] Cf. C. Henry, ibid., 565–8.

[4a] Ibid., 568. Since he omitted the factor p before $q - s$, take k to be a multiple of p.

take any integer k and set $p(q-s)=k(p-r)$. Then $pq-rs=(p-r)(s+k)$. The second case in which a^z and x^z are not divisible by z differs from the preceding only as to signs.

L. Euler's[5] theorems on the linear forms of the divisors of $a^m\pm b^m$ are cited under Euler[5,6] of Ch. XVI of Vol. I of this History.

Lagrange's[142] method for $r^n-As^n=q^m$ is given in Ch. XXIII.

A. J. Lexell[6] considered $a^5+b^5=c^5$. Set $x+y=a^5$, $x-y=b^5$. Then

$$\frac{x^2-y^2}{4x^2}=\left(\frac{z}{x}\right)^5\equiv\frac{a^5b^5}{c^{10}}, \qquad x^6-4xz^5=x^4y^2=\Box.$$

Since the factors are relatively prime, $x=p^2$, $x^5-4z^5=q^2$. Hence

$$p^{10}-q^2=4r^5s^5, \qquad p^5+q=2r^5, \qquad p^5-q=2s^5, \qquad p^5=r^5+s^5.$$

N. Fuss I[7] noted that, if $1\pm4x^n=\Box$ is possible in rational numbers, $r^n+p^n=q^n$ would be possible in integers. To reduce the former to integers, set $x=pq/r^2$; then $r^{2n}\pm4p^nq^n=\Box$, say the square of r^n+2v, where v is prime to r. Then $\pm p^nq^n=v(r^n+v)$, whence $v=p^n$, $r^n+v=q^n$.

L. Euler[8] multiplied $a^n+b^n=c^n$ by $4a^n$ and added b^{2n}. Thus

$$(2a^n+b^n)^2=4a^nc^n+b^{2n}=\Box.$$

Euler[9] noted that he had failed in attempts to prove $x^n+y^n=z^n$ impossible if $n>2$.

C. F. Kausler[10] proved that $x^6+y^6=z^6$ is impossible in integers. For, if possible, set $x=mn$, where m is a prime. Of the forty cases, all are immediately excluded except two:

$$z^4+z^2y^2+y^4=m^6n^6 \text{ or } mn^6, \qquad z^2-y^2=1 \text{ or } m^5.$$

For the second alternatives, eliminate z^2. Then

$$3y^4+3y^2m^5+m^{10}=mn^6,$$

and m is a factor of $3y^4$. If y is divisible by m, z is, and x, y, z have a common factor. There remains the case $m=3$; then $z+y$, $z-y$ are 3^5, 1 or 3^4, 3 or 3^3, 3^2, cases readily excluded. The first alternative is excluded by the lemma: There are no integers y, z for which

$$z^4+z^2y^2+y^4\equiv(z^2-y^2)^2+3z^2y^2=\Box.$$

Sophie Germain[11] (1776–1831) stated in her first letter to Gauss, Nov. 21, 1804, that she could prove that $x^n+y^n=z^n$ is impossible if $n=p-1$,

<hr>

[5] Comm. Arith., I, 50–6, 269; II, 533–5.

[6] Euler's Opera postuma, 1, 1862, 231–2 (about 1768).

[7] *Ibid.*, 241 (about 1778). Cf. Euler.[8]

[8] *Ibid.*, 242 (about 1782).

[9] *Ibid.*, 587; letter to Lagrange, March 23, 1775. Corresp. Math. Phys. (ed., P. H. Fuss), 1, 1843, 618, 623, letters to Goldbach, Aug. 4, 1753, May 17, 1755. Novi Comm. Acad. Petrop., 8, 1760–1, 105; Comm. Arith. Coll., I, 296.

[10] Nova Acta Acad. Sc. Petrop., 15, 1806, ad annos 1799–1802, 146–155.

[11] The first and third letters were published in Oeuvres philosophiques de S. Germain, Paris, 1879, 298. Cinq lettres de Sophie Germain à C. F. Gauss, publiées par B. Boncompagni, Berlin, 1880, 24 pp. Reproduced in Archiv Math. Phys., 65, 1880, Litt. Bericht 259, pp. 27–31; 66, 1881, Litt. Bericht 261, pp. 3–10. Reviewed, with Gauss,[13] by S. Günther, Zeitschr. Math. Phys., 26, 1881, Hist.-Lit. Abt., pp. 19–26; Italian transl., Bull. Bibl. Storia Sc. Mat. e Fis., 15, 1882, 174–9.

where p is a prime $8k+7$. In her[12] fourth letter, Feb. 20, 1807, she stated that if the sum of the nth powers of any two numbers is of the form h^2+nf^2, the sum of these two numbers is of that form. Gauss[13] replied, April 30, 1807, that this is false, as shown by $15^{11}+8^{11}=h^2+11f^2$, whereas

$$15+8 \neq x^2+11y^2.$$

C. F. Gauss[14] gave a sketch of a proof of the impossibility of $a^5+b^5+c^5=0$ and noted that the method is not applicable to seventh powers.

P. Barlow[15] proved that if n is a prime and $x^n-y^n=z^n$ is solvable in integers prime in pairs, then one of the four sets of conditions

$$
\begin{array}{c|c|c|c}
x-y=r^n & n^{n-1}r^n & r^n & r^n \\
x-z=s^n & s^n & n^{n-1}s^n & s^n \\
y+z=t^n & t^n & t^n & n^{n-1}t^n
\end{array}
$$

must hold. For, $(x^n-y^n)/(x-y)$ is not divisible by a factor $\neq n$ of $x-y$, and if divisible by n, the quotient is prime to $x-y$ and to n. Hence z^n is divisible by $x-y$, and, if n is a factor of $x-y$, by $n(x-y)$, while the quotient is prime to n and to $x-y$. In the first case, $x-y=r^n$. In the second case, $n(x-y)=r^n=n^n r_1^n$, $x-y=n^{n-1}r_1^n$.

His attempt to prove $x^n-y^n=z^n$ impossible if $n>2$ involves the error (cf. Smith,[79] Talbot[84]) that a sum of fractions in their lowest terms is not an integer if the denominator of each fraction has a factor not dividing all the remaining denominators.

N. H. Abel[16] stated that, if n is a prime >2, $a^n=b^n+c^n$ is impossible in integers when one or more of the numbers a, b, c, $a+b$, $a+c$, $b-c$, $a^{1/m}$, $b^{1/m}$, $c^{1/m}$ are primes [cf. Talbot,[77] de Jonquières[117]]. If the equation is possible, then a, b, c have factors x, y, z, respectively, such that either [cf. Barlow[15]]

$$2a=x^n+y^n+z^n, \qquad 2b=x^n+y^n-z^n, \qquad 2c=x^n+z^n-y^n;$$
$$2a=n^{n-1}x^n+y^n+z^n, \qquad 2b=n^{n-1}x^n+y^n-z^n, \qquad 2c=n^{n-1}x^n+z^n-y^n;$$
$$2a=n^{n-1}(x^n+y^n)+z^n, \qquad 2b=n^{n-1}(x^n+y^n)-z^n, \qquad 2c=n^{n-1}(x^n-y^n)+z^n;$$

or values derived from the second set by permuting a, b, and x, y, and changing the signs of c and z; or values derived from the third set by replacing a by b, b by $-c$, c by a, x by y, y by $-z$, and z by x. Thus $2a$ must have one of the three forms listed, where x, y, z have no common factor. Finally, $2a \geqq 9^n+5^n+4^n$; the least one of a, b, c cannot be less than $(9^n-5^n+4^n)/2$. The editor, L. Sylow, remarked p. 338 that these theorems appear to contain some inaccuracies.

[12] Published by E. Schering, Abh. Gesell. Wiss. Göttingen, 22, 1877, 31–32.
[13] Lettera inedita di C. F. Gauss a Sofia Germain, publicata da B. Boncompagni, Firenze, 1879. Reproduced in Archiv Math. Phys., 65, 1880, Litt. Bericht 257, pp. 5–9.
[14] Werke, II, 1863, 390–1, posth. paper.
[15] Appendix to English transl. of Euler's Algebra. Proof "completed" by Barlow in Jour. Nat. Phil. Chem. and Arts (ed., Nicholson), 27, 1810, 193, and reproduced in Barlow's Theory of Numbers, London, 1811, 160–9.
[16] Oeuvres, 1839, 264–5; nouv. éd., 2, 1881, 254–5; letter to Holmboe, Aug. 3, 1823.

A. M. Legendre[17] remarked that the French Academy of Sciences had offered one of its prizes for a proof of Fermat's last theorem, but without awarding the prize. He considered $x^n+y^n+z^n=0$ for n a prime >2 and for relatively prime integers x, y, z each $\neq 0$. He noted (§§ 3, 4) that $x+y+z$ is divisible by n, and its nth power by $(x+y)(y+z)(z+x)$, by a proof criticized and completed by Catalan.[91] Let

$$\phi(y,\, z) = y^{n-1} - y^{n-2}z + y^{n-3}z^2 - \cdots + z^{n-1}$$

be the quotient of y^n+z^n by $y+z$. Then (§ 7) $y+z$ and ϕ have the g.c.d. n or are relatively prime according as x is or is not divisible by n.

First, let x be divisible by n. Then (§§ 8, 10)

$$
(1) \qquad
\begin{aligned}
y+z &= \frac{1}{n}a^n, & \phi(y,\, z) &= n\alpha^n, & x &= -a\alpha, \\
z+x &= b^n, & \phi(z,\, x) &= \beta^n, & y &= -b\beta, \\
x+y &= c^n, & \phi(x,\, y) &= \gamma^n, & z &= -c\gamma,
\end{aligned}
$$

where a is an integer divisible by n, and each prime factor of α, β or γ is of the form $2kn+1$. Each prime factor of α is of the form $2tn^2+1$ (§ 11), and x, assumed divisible by n, is divisible by n^2 (§ 13), both results being credited to Sophie Germain in the foot-note to § 22.

Second, let no one of the numbers x, y, z be divisible by n. Methods applicable only in the special cases $n=3$, 5, 7, 11, but not to $n=13$, etc., are given in §§ 14–20. To Sophie Germain is credited the proof (§§ 21–22) that, if n is an odd prime <100,

$$(2) \qquad\qquad x^n+y^n+z^n=0$$

has no integral solutions each prime to n. This proof is called "very ingenious, quite simple, and of an almost absolute generality." As noted above, $y+z$ is prime to $\phi(y,\, z)$, and their product equals $(-x)^n$; hence we may set

$$
(3) \qquad
\begin{aligned}
y+z &= a^n, & \phi(y,\, z) &= \alpha^n, & x &= -a\alpha, \\
z+x &= b^n, & \phi(z,\, x) &= \beta^n, & y &= -b\beta, \\
x+y &= c^n, & \phi(x,\, y) &= \gamma^n, & z &= -c\gamma,
\end{aligned}
$$

whence

$$(4) \qquad 2x = b^n+c^n-a^n, \qquad 2y = a^n+c^n-b^n, \qquad 2z = a^n+b^n-c^n.$$

THEOREM. If there exists an odd prime p such that

$$(5) \qquad\qquad \xi^n+\eta^n+\zeta^n \equiv 0 \quad (\mathrm{mod}\ p)$$

has no set of integral solutions ξ, η, ζ, each not divisible by p, and such that n is not the residue of the nth power of any integer modulo p, then (2) has no integral solutions each prime to n.

For, if x, y, z are integers satisfying (2), they satisfy congruence (5), so that one of them, say x is divisible by p. Then, by (4),

$$b^n+c^n+(-a)^n \equiv 0 \quad (\mathrm{mod}\ p).$$

---

[17] Sur quelques objets d'analyse indéterminée et particulièrement sur le théorème de Fermat, Mém. Acad. R. Sc. de l'Institut de France, 6, année 1823, Paris, 1827, 1–60. Same, except as to paging, Théorie des nombres, ed. 2, 1808, second supplément, Sept., 1825, 1–40 (reproduced in Sphinx-Oedipe, 4, 1909, 97–128; errata, 5, 1910, 112).

Hence a, b, or c is divisible by p. But if b were divisible by p, then, by (3), $y=-b\beta$ would be divisible by p, and hence by (2) also z would be divisible by p, whereas x, y, z have no common factor. Similarly, c is not divisible by p. Hence

$$a\equiv 0, \qquad x\equiv 0, \qquad z\equiv -y, \qquad \phi(x, y)\equiv y^{n-1}, \qquad \phi(y, z)\equiv ny^{n-1} \qquad (\bmod\ p).$$

Thus, by (3), $\gamma^n\equiv y^{n-1}$, $\alpha^n\equiv ny^{n-1}$. Hence $n\gamma^n\equiv\alpha^n$ (mod p). By the final equation (3), γ is prime to p. Hence we can determine an integer γ_1 such that $\gamma\gamma_1\equiv 1$ (mod p). Thus $n\equiv(\alpha\gamma_1)^n$ (mod p), contrary to hypothesis.

The theorem applies if $n=7$, $p=29$, since the residues of the seventh powers modulo 29 are ±1, ±12, no two of which differ by unity, and no one of which is congruent to 7. Similarly, for each odd prime $n<100$, S. Germain gave a p for which the theorem applies.

The condition that n shall not be a residue of an nth power requires that p be of the form $mn+1$, where evidently m is even. Legendre proved (§§ 23–28) that m must be prime to 3 and that both conditions in the theorem hold if $p=mn+1$ is a prime and $m=2$, 4, 8, 10, 14, 16 (but overlooked the exceptional character of $n=3$ when $m=14$, 16; cf. Dickson[195]). He concluded that (1) has no solutions prime to n for n an odd prime <197.

He proved[18] (§§ 38–47) that $x^5+y^5+z^5=0$ has no integral solutions and that if solutions of (2) exist for $n=7$, 11, 13 or 17, they involve a great number of digits (§§ 29–37).

Schopis[19] argued that, if $x^5-y^5=w^5$, where xyw is prime to 5, then

$$x-y=u^5,$$

and

$$x^4+x^3y+\cdots+y^4=u^{20}+5u^{15}y+10u^{10}y^2+10u^5y^3+5y^4$$

is a fifth power, say $(u^4+z)^5$. Thus

$$5yA=z(5u^{16}+10u^{12}z+10u^8z^2+5u^4z^3+z^4), \qquad A=u^{15}+2u^{10}y+2u^5y^2+y^3.$$

Thus z is divisible by 5 and the second member by 25. Thus A is divisible by 5, which is seen to be impossible.

G. L. Dirichlet[20] proved that there are no relatively prime integers x, y such that $x^5\pm y^5=2^m5^nAz^5$, m and n being positive integers, $n\neq 2$, and A not divisible by 2, 5 or a prime $10k+1$. With the same restrictions on A, the theorem holds also if $n=0$, $m\geqq 0$, and $2^mA\equiv 3$, 4, 9, 12, 13, 16, 21, or 22 (mod 25). If $n>0$, $n\neq 2$, and if A is not divisible by 2, 5 or a prime $10k+1$, there exist no relatively prime integers x, y such that $x^5\pm y^5=5^nAz^5$. The last shows that $x^5\pm y^5=z^5$ is impossible in integers (since one of the unknowns, say z, must be divisible by 5); the proof is analogous in the two cases z even and z odd, whereas Legendre[18] employed two methods.

---

[18] This proof was reproduced in Legendre's Théorie des nombres, ed. 3, II, 1830, arts. 654–663, pp. 361–8; German transl. by H. Maser, 1893, 2, pp. 352–9. If z is the unknown divisible by 5, the proof for the case z even is like Dirichlet's,[20] while that for z odd is by a special analysis.

[19] Einige Sätze Unbest. Analytik, Progr. Gumbinnen, 1825, 12–15.

[20] Jour. für Math., 3, 1828, 354–375; Werke I, 21–46. Read at the Paris Acad. Sc., July 11 and Nov. 14, 1825 and printed privately, Werke, I, 1–20. Cf. Lebesgue.[37]

A. M. Legendre[21] stated that the discussion of (2), at least for special exponents n, can be facilitated by a consideration of the cubic equation whose roots are x, y, z; for integral roots, the discriminant must be a perfect square. He was not entitled to conclude that $x+y+z$ and xyz are divisible by n^2, as he had not proved that one of the unknowns is divisible by n.

V. Bouniakowsky[22] argued that if $x^m+y^m+z^m=0$, where m is a prime and x, y, z are integers with no common factor, and if N is chosen so that $m=\phi(N)-1$ (which is possible for each prime $m<31$, except $m=13$), then $xyz(xy+xz+yz)$ is divisible by N. But he used Euler's theorem $x^{\phi(N)}\equiv1 \pmod{N}$ which is valid only when x is prime to N.

Dirichlet[23] proved by descent that (2) is impossible in integers for $n=14$, also the impossibility of

$$t^{14}-u^{14}=2^m\cdot7^{1+n}w^{14}.$$

G. Libri[24] considered the number N_2 of sets of positive solutions $<n$ of $x^3+y^3+1\equiv0 \pmod{n}$, for a prime $n=3p+1$. The equation for the three periods of nth roots of unity is found in the form

$$z^3+z^2-\tfrac{1}{3}(n-1)z-\tfrac{1}{27}[nN_2+3-(n+2)^2+9n]=0.$$

Comparing this with the known cubic, we get $N_2=n\pm a-2$, where

$$4n=a^2+27b^2$$

[Pepin[109]]. Since a is comprised between zero and $r=(4n-27)^{1/2}$, we have $N_2\leqq n-r-2$. Hence N_2 increases indefinitely with n, and from a certain limit on, $x^3+y^3+1\equiv0 \pmod{n}$ is always solvable with neither x nor y divisible by n. Having N_2, we can find the number of positive solutions $<n$ of $x^3+y^3+u^3+1\equiv0 \pmod{n}$.

If n is a prime $8m+1$, so that $n=a^2+16b^2$ in a single way, the same method of proof shows that the number of solutions of $x^4+y^4+1\equiv0 \pmod{n}$ is $n\pm6a-3$, which increases with n. It is stated that one can prove [Pellet,[128, 244] Dickson[199], Cornacchia,[217] Mantel[277]] that a limit to the prime p can be assigned such that, after passing it, the number of solutions of $x^n+y^n+1\equiv0 \pmod{p}$ will always increase. Hence it is futile to try to prove $u^n+v^n=z^n$ impossible by trying to show that one of the unknowns is divisible by an infinitude of primes.

E. E. Kummer[25] considered $x^{2\lambda}+y^{2\lambda}=z^{2\lambda}$, where λ is a prime, and x, y, z are relatively prime by pairs. We may take y even. The third of four possible cases is

$$z+y=u^{2\lambda}, \qquad z-y=w^{2\lambda}, \qquad z\pm x=2p^{2\lambda}, \qquad z\mp x=2^{2\lambda\nu-1}\lambda^{2\lambda\mu-1}q^{2\lambda},$$

This is the only possibility if $\lambda=8n+1$, or if $2\lambda+1$ is a prime. If the initial equation is solvable in integers, so is $r^{2\lambda}+s^{2\lambda}=2q^{2\lambda}$. As auxiliary to the

[21] Théorie des nombres, ed. 3, II, 1830, art. 451, pp. 120-2; German transl., Maser, II, pp. 118-120.

[22] Mém. Acad. Sc. St. Pétersbourg (Math.), (6), 1, 1831, 150-2.

[23] Jour. für Math., 9, 1832, 390-3; Werke, I, 189-194. Reproduced by Gambioli,[171] pp. 164-7.

[24] Jour. für Math., 9, 1832, 270-5.

[25] Jour. für Math., 17, 1837, 203-9.

proofs, it is shown[26] that if

$$\frac{a^n \pm b^n}{a \pm b} = (a \pm b)^{n-1} \mp n(a \pm b)^{n-3}ab + \frac{n(n-3)}{2}(a \pm b)^{n-5}a^2b^2 \mp \cdots$$

and $a \pm b$ have a common factor, it divides the last term $\pm n(ab)^{(n-1)/2}$, and hence is the prime n if a and b are relatively prime.   Since the coefficients n, $n(n-3)/2$, $\cdots$ are divisible by n, the exponent of the highest power of n dividing $a^n \pm b^n$ exceeds by unity that in $a \pm b$.

F. Paulet[27] attempted to prove Fermat's last theorem, but concluded without proof that $\alpha = \beta$ in $\alpha cr = \beta s$, where

$$\alpha = bmx^2 - (p-q)a, \qquad \beta = ar + (p-q)c + s.$$

In his second proof he equated corresponding summands of equal sums.

G. Lamé[28] proved that $x^7 + y^7 + z^7 = 0$ is impossible in relatively prime integers.   One of the unknowns, say x, is divisible by 7 (Legendre[17]). It is shown that $x + y + z = 7AP$, $P = \mu\nu\rho$, where μ, ν, ρ, 7 are relatively prime integers such that

$$z + y = 7^6\mu^7 = a, \qquad z + x = \nu^7 = b, \qquad x + y = \rho^7 = c.$$

He made use of the lemma (pp. 197–8) that [Bouniakowsky[34]]

$$(x+y+z)/\sqrt[7]{7(x+y)(z+x)(z+y)} = A = \square.$$

Thus A must be a square B^2.   Then

$$\Sigma a = 27B^2P, \qquad \Sigma a^2 + \Sigma ab = BD, \qquad abc = 7^6P^7, \qquad 3\Sigma a^4 + 10\Sigma a^2b^2 = 2^4B^{14}.$$

Eliminating a, b, c, we get an equation whose solution is shown to depend upon the impossible equation

$$U^8 - 3 \cdot 7^4U^4V^4 + 2^4 7^5V^8 = W^4.$$

For simplifications of this proof, see Lebesgue[30] and Genocchi.[85]

A. Cauchy[29] reported on Lamé's preceding paper and stated that his lemma is obtained by taking $n = 7$ in the generalization that $(x+y)^n - x^n - y^n$ is algebraically divisible not only by $nxy(x+y)$ but also (if $n > 3$) by $q = x^2 + xy + y^2$, and if $n = 6k+1$ by q^2.

V. A. Lebesgue[30] simplified Lamé's[28] proof by use of the lemma that

$$p^2 = q^4 - 2^{2a}3 \cdot 7^4q^2r^2 + 2^{4a+4}7^7r^4$$

is impossible in odd integers p, q, r, relatively prime in pairs, $r \neq 0$, if a is a positive integer.

Lebesgue[31] proved that if $X^n + Y^n = Z^n$ is impossible in integers, then $x^{2n} + y^{2n} = z^2$ is impossible.

---

[26] Also in Nouv. Ann. Math., 7, 1848, 239, 307–8.

[27] Corresp. Math. (ed., A. Quetelet), 11, 1839, 307–313.

[28] Comptes Rendus Paris, 9, 1839, 45–6; Jour. de Math., 5, 1840, 195–211.   Mém. présentés divers savants Acad. Sc. de l'Institut de France, 8, 1843, 421–437.

[29] Comptes Rendus Paris, 9, 1839, 359–363; Jour. de Math., 5, 1840, 211–5.   Oeuvres de Cauchy, (1), IV, 499–504.

[30] Jour. de Math., 5, 1840, 276–9, 348–9 (removal of obscurity in proof of lemma).

[31] Ibid., 184–5.

J. Liouville[32] noted that if $u^n+v^n=w^n$ is impossible in integers not zero, then $z^{2n}-y^{2n}=2x^n$ is impossible.

Cauchy[33] expressed $(x+y)^n-x^n-y^n$ in terms of x^2+xy+y^2 and $xy(x+y)$ for n odd $\leqq 13$.

V. Bouniakowsky[34] proved for $m=2, 3, 4, 5, 6, 7$ that

$$\sqrt[m]{A}+\sqrt[m]{B}=R$$

is impossible if R is rational and the radicals irrational. For $m=7$ set $C=(AB)^{1/7}$. We get $R^7-A-B=7RC(R^2-C)^2$, which implies the lemma of Lamé[28] (Cauchy[29]). For, by setting $A=a^7, B=b^7, R=a+b, C=ab$, we get

$$(a+b)^7-a^7-b^7=7ab(a+b)(a^2+ab+b^2)^2.$$

E. E. Kummer[35] submitted to Dirichlet about 1843 the manuscript giving what he then believed to be a complete proof of Fermat's last theorem. Dirichlet declared that the proof would be correct if it were shown not only that every number $a_0+a_1\alpha+\cdots+a_{\lambda-1}\alpha^{\lambda-1}$ (where α is a primitive λth root of unity and the a's are ordinary integers) is always a product of indecomposable numbers of that form, as shown by Kummer, but also that this were possible in only one way, which is unfortunately apparently not the case.

Frizon[36] announced a uniform process applicable to prime exponents $\leqq 31$.

V. A. Lebesgue[37] supplemented Dirichlet's[20] results by proving that, if A has no prime factor $10m+1$ and no factor which is a fifth power, $x^5+y^5=AB^5u^5$ is impossible in integers if A is a multiple of 5, or if $A\equiv\pm 2$, $\pm 3, \pm 4, \pm 6, \pm 8, \pm 9, \pm 11$, or ± 12 (mod 25). A like treatment is apparently not applicable to the remaining cases $A\equiv\pm 1, \pm 7$ (mod 25). The equation $x^{10}\pm y^{10}=Az^5$ is impossible if A has no prime factor $10m+1$. As auxiliary propositions, $a^2=b^4+50b^2c^2+125c^4$ is impossible, while

$$a^2=b^4+10b^2c^2+5c^4,$$

which can be reduced by descent to the case in which b and c are odd, is impossible if $c=5\cdot 2^i\cdot h^2$.

E. Catalan[38] expressed his belief that $x^m-y^n=1$ holds only for $3^2-2^3=1$.

S. M. Drach[39] argued that $x^n+y^n=z^n$ is impossible in integers if $n=2m+1>1$. For, by Euler's Algebra, 2, Ch. 12,

$$Y=c^mq^n+\Sigma A_i q^{n-2i}p^{2i}c^{m-i}a^i, \qquad Z=a^mp^n+\Sigma A_i p^{n-2i}q^{2i}a^{m-i}c^i$$

satisfy $aZ^2-cY^2=(ap^2-cq^2)^n$ if $A_i=\binom{n}{2i}$. Take $a=z, Z=z^m, c=y, Y=y^m$.

[32] Jour. de Math., 5, 1840, 360.

[33] Exercices d'analyse et de phys. math., 2, 1841, 137–144; Oeuvres, (2), XII, 157–166.

[34] Mém. Acad. Sc. St. Pétersbourg (Math.), (6), 2, 1841, 471–492. Extract in Bull. St. Péters., VIII, 1–2.

[35] K. Hensel, Gedächtnisrede auf E. E. Kummer, Abh. Gesch. Math. Wiss., 29, 1910, 22. [Cf. the less technical address by Hensel, E. E. Kummer und der grosse Fermatsche Satz, Marburger Akademische Reden, 1910, No. 23.]

[36] Comptes Rendus Paris, 16, 1843, 501–2.

[37] Jour. de Math., 8, 1843, 49–70.

[38] Jour. für Math., 27, 1844, 192. Nouv. Ann. Math., 1, 1842, 520; (2), 7, 1868, 240 (repeated by E. Lionnet). For $n=2$, Lebesgue[68] of Ch. VI.

[39] London, Edinburgh, Dublin Phil. Mag., 27, 1845, 286–9.

Then

$$z^n - y^n = x^n = (zp^2 - yq^2)^n, \qquad x = zp^2 - yq^2.$$

Then Z/z^m and Y/y^m give

$$1 = p^n \left[1 + \Sigma A_i \left(\frac{q^2 y}{p^2 z} \right)^i \right] = q^n \left[1 + \Sigma A_i \left(\frac{p^2 z}{q^2 y} \right)^i \right],$$

$$2z^{n/2}, \ 2y^{n/2} = (p\sqrt{z} + q\sqrt{y})^n \pm (p\sqrt{z} - q\sqrt{y})^n.$$

From the sum and difference of the resulting values of $p\sqrt{z} \pm q\sqrt{y}$,

$$\frac{p\sqrt{z}}{q\sqrt{y}} \{ (z^{n/2} + y^{n/2})^{1/n} - (z^{n/2} - y^{n/2})^{1/n} \} = \{(\ ) + (\ )\}.$$

Developing the difference of the two members by the binomial theorem, we get a series in y/z with every coefficient negative if $n > 1$. Next, the case $n = 2m$ is treated at length.

C. G. J. Jacobi[40] gave a table of the values of m' for which $1 + g^m \equiv g^{m'}$ (mod p), where p is a prime ≤ 103, $0 \leq m \leq 102$, and q is a primitive root of p.

O. Terquem[41] proved the theorem of Lebesgue[31] and the corollary of Liouville[32].

A. Vachette[42] noted that $x^m - y^n = (xy)^p$ is impossible in integers. For $p = mn$, set $z = (xy)^n$ and take $n = m$. Thus $x^m - y^m = z^m$ is impossible if z is a power of xy.

J. Mention[43] proved the formula [cf. Kummer[25]]:

$$(6) \qquad a^n + b^n = (a+b)^n - nab(a+b)^{n-2} + \frac{n(n-3)}{2} a^2 b^2 (a+b)^{n-4} - \cdots.$$

V. A. Lebesgue[44] obtained (6) by applying Waring's formula to the quadratic equation with roots a, b. Applying it to the cubic with the roots α, β, γ, we get $(\alpha + \beta + \gamma)^n$. For $n = 7$, the latter result is said to have been employed [in papers 28–30] to prove the impossibility of $x^7 + y^7 = z^7$ by a method simpler than that for exponents 3 and 5.

G. Lamé[45] claimed to have proved that, if n is an odd prime, $x^n + y^n = z^n$ is not satisfied by complex integers

$$(7) \qquad a_0 + a_1 r + \cdots + a_{n-1} r^{n-1},$$

where r is an imaginary nth root of unity and the a's are integers.

J. Liouville[46] pointed out the lacuna in Lamé's proof that he had not shown that a complex integer is decomposable into complex primes in a single manner.

Lamé (p. 352) admitted the lacuna and believed (on the basis of extensive tables of factorizations) that it could be filled; he affirmed (pp. 569–572) that the ordinary laws for integers hold for complex integers when $n = 5$.

---

[40] Jour. für Math., 30, 1846, 181–2; Werke, VI, 272–4.
[41] Nouv. Ann. Math., 5, 1846, 70–73.
[42] Ibid., 68–70.
[43] Nouv. Ann. Math., 6, 1847, 399 (proposed, 2, 1843, 327; 18, 1859, 172, 249).
[44] Ibid., 427–431.
[45] Comptes Rendus Paris, 24, 1847, 310–5.
[46] Ibid., 315–6.

Lamé stated (p. 888) that Fermat's equation is impossible for a series of exponents including $n=5, 11, 13$.

Lamé[47] presented his arguments in two long memoirs.

O. Terquem[48] suggested a subscription to Lamé for his[45] proof (!) declaring it the greatest discovery of the century in the mathematical world.

E. E. Kummer[49] pointed out the falsity of Lamé's[45] assumption that every complex integer can be decomposed into primes in a single way.

L. Wantzel[50] proved that Euclid's g.c.d. process holds for complex integers $a+b\sqrt{-1}$ [already proved by C. F. Gauss[51]] and for complex integers formed from an imaginary cube root of unity, and stated that a like result holds for complex integers (7), with n arbitrary, since the norm (or modulus) of (7) is <1 when $a_0, \cdots, a_{n-1}$ are between 0 and 1 [erroneous, Cauchy[52]].

A. Cauchy[52] showed that the final statement by Wantzel[50] is false for $n=7$ and for any prime $n=4m+1\geqq17$. He pointed out lacunæ in the proposed proof by Lamé[45] of Fermat's last theorem. He defined the factorial of (7) to be its product by the complex members obtained from it by replacing r by the remaining primitive nth roots of unity, and obtained upper limits for such factorials [norms]. He[53] proved that any common factor of $M_h=Ar^h+B$ and M_k divides M_0 if A and B are relatively prime.

Cauchy[54] attempted to prove the false theorem that the norm of the remainder obtained on dividing one complex number (7) by another can always be made less than the norm of the divisor. He concluded (falsely) that a product of complex integers (7) can be decomposed into complex primes in a single manner, and that the other laws of divisibility of integers hold for these complex integers.

Cauchy[55] noted (erroneous) conclusions which follow from the assumption that his preceding theorems hold for a given number n; in particular, errors relating to the factors $A+r^iB$ of A^n+B^n. He promised to discuss later the objections which can be raised against proofs in his preceding paper.

Cauchy[56] further developed the subject and admitted at the end of his final paper that his[54] basal theorem is false, failing for $n=23$.

Cauchy[57] obtained results most of which are included in Kummer's general theory. In the fifth paper, p. 181 (Oeuvres, p. 364), he stated that $a^n+b^n+c^n=0$ is impossible in relatively prime integers not divisible by

[47] Jour. de Math., 12, 1847, 137–171, 172–184.

[48] Nouv. Ann. Math., 6, 1847, 132–4.

[49] Comptes Rendus Paris, 24, 1847, 899–900; Jour. de Math., 12, 1847, 136.

[50] Comptes Rendus Paris, 24, 1847, 430–4.

[51] Comm. Soc. Sc. Gotting. Recentiores, 7, 1832, § 46; Werke, II, 1863, 117.   German transl. by H. Maser, Gauss' Untersuchungen über höhere Arith., 1889, 556.

[52] Comptes Rendus Paris, 24, 1847, 469–481; Oeuvres, (1), X, 240–254.

[53] Ibid., 347–8; Oeuvres, (1), X, 224–6.

[54] Ibid., 516–528; Oeuvres, (1), X, 254–268.

[55] Ibid., 578–584; Oeuvres, (1), X, 268–275.

[56] Ibid., 633–6, 661–6, 996–9, 1022–30; Oeuvres, (1), X, 276–285, 296–308.

[57] Ibid., 25, 1847, 37, 46, 93, 132, 177, 242, 285; Oeuvres, (1), X, 324–351, 354–371.

the odd prime n if

$$1+2^{n-4}+3^{n-4}+\cdots+\left(\frac{n-1}{2}\right)^{n-4}$$

is not divisible by n [i. e., if the Bernoullian number $B_{(n-3)/2}$ is not divisible by n], or if a certain number ω (p. 359) is prime to n. Cf. Genocchi,[64] Kummer.[65]

E. E. Kummer[58] proved that $x^\lambda-y^\lambda=z^\lambda$ is impossible for the series[59] of real primes λ for which (A) the number of non-equivalent ideal complex numbers formed from an imaginary λth root α of unity is not divisible by λ and (B) every complex unit $E(\alpha)$, which is congruent modulo λ to a rational integer, equals the λth power of another complex unit. These two conditions are satisfied if $\lambda=3, 5, 7$, but probably not for $\lambda=37$.

G. L. Dirichlet[60] noted that Kummer's condition (A) relates to a theory closely analogous to the fact that a number m for which D is a quadratic residue is not always represented by x^2-Dy^2, but by one of several quadratic forms, and similarly for the forms in $\lambda-1$ variables defined by norms of complex integers based on α.

Kummer[61] proved that, for the domain defined by an imaginary λth root α of unity, where λ is an odd prime, the number of classes of ideals is the product $H=h_1 h_2$ of the two integers

$$h_1=\frac{P}{(2\lambda)^{\mu-1}}, \qquad h_2=\frac{D}{\Delta},$$

where $\mu=(\lambda-1)/2$, and P, D, Δ are defined as follows. Let β be a primitive root of $\beta^{\lambda-1}=1$, and g a primitive root of λ. Then

$$P=\prod_{j=1}^{\mu}\phi(\beta^{2j-1}), \qquad \phi(\beta)=1+g_1\beta+g_2\beta^2+\cdots+g_{\lambda-2}\beta^{\lambda-2},$$

where g_i is the least positive residue of g^i modulo λ. Next,

$$e(\alpha)=\sqrt{\frac{(1-\alpha^g)(1-\alpha^{-g})}{(1-\alpha)(1-\alpha^{-1})}}$$

is a complex unit (a divisor of 1). Then, if lx denotes the real part of $\log x$,

$$D=\begin{vmatrix} le(\alpha) & le(\alpha^g) & \cdots & le(\alpha^{g^{\mu-2}}) \\ le(\alpha^g) & le(\alpha^{g^2}) & \cdots & le(\alpha^{g^{\mu-1}}) \\ \cdot & \cdot \cdot \cdot \cdot \cdot \cdot & \cdot & \cdot \\ le(\alpha^{g^{\mu-2}}) & le(\alpha^{g^{\mu-1}}) & \cdots & le(\alpha^{g^{2\mu-4}}) \end{vmatrix}.$$

[58] Berichte Akad. Wiss. Berlin, 1847, 132–9.

[59] "I prove that it is impossible for an infinitude of primes λ, but do not know for just which λ's the assumptions hold." That these λ's are infinite in number was believed, but not proved, by Kummer. He called the remaining primes exceptional (as 37, etc.). The same statements were made in 1847 in letters to Kronecker (Kummer,[35] pp. 75, 84). In his Vorlesungen über Zahlentheorie, 1, 1901, 23, Kronecker stated that Kummer proved the impossibility of $x^\lambda+y^\lambda=z^\lambda$ for an infinitude of primes λ and at first believed that his proof applied to nearly all λ's, but later believed the contrary. Kummer,[35] p. 32, is elsewhere quoted as believing it probable that there are approximately as many regular primes as irregular (exceptional) primes. A. Wieferich, Taschenbuch für Mathematiker u. Physiker, Leipzig, 2, 1911, 108–111, stated that Kummer proved Fermat's last theorem for an infinite series of exponents.

[60] Berichte Akad. Wiss. Berlin, 1847, 139–141; Werke, II, 254–5.

[61] Berichte Akad. Wiss. Berlin, 1847, 305–319. Same in Jour. für Math., 40, 1850, 93–138; Jour. de Math., 16, 1851, 454–498.

Let $\epsilon_1(\alpha), \cdots, \epsilon_{\mu-1}(\alpha)$ be units such that products of powers of them multiplied by $\pm \alpha^m$ give all the units. Then

$$\Delta = \begin{vmatrix} l\epsilon_1(\alpha) & \cdots & l\epsilon_{\mu-1}(\alpha) \\ \cdot & \cdot \cdot \cdot \cdot \cdot \cdot \cdot & \cdot \\ l\epsilon_1(\alpha^{g^{\mu-2}}) & \cdots & l\epsilon_{\mu-1}(\alpha^{g^{\mu-2}}) \end{vmatrix}.$$

It is shown that h_1 is divisible by λ if and only if λ divides the numerator of one of the first $(\lambda-3)/2$ Bernoullian numbers $B_1 = 1/6$, $B_2 = 1/30$, $\cdots$; while if h_2 is divisible by λ also h_1 is, but not conversely. He proved that if λ is not a divisor of H, condition (B) of Kummer[58] is satisfied. Hence if λ is an odd prime not dividing the numerator of any one of the first $(\lambda-3)/2$ Bernoullian numbers, $x^\lambda + y^\lambda = z^\lambda$ is impossible in integers.

The French Academy of Sciences[62] offered as a prize a gold medal of value 3000 francs for a proof of Fermat's last theorem. After several postponements of the date fixed for the award, the prize was finally (C. R., 44, 1857, 158) awarded to Kummer for his investigations on complex numbers, though he had not been a competitor.

Kummer[63] proved by use of prime ideals that, if λ is an odd prime not dividing the numerator of any one of the first $(\lambda-3)/2$ Bernoullian numbers, $u^\lambda + v^\lambda + w^\lambda = 0$ has no solution in integers, nor in complex integers

$$a_0 + a_1\alpha + a_2\alpha^2 + \cdots + a_{\lambda-2}\alpha^{\lambda-2},$$

where α is an imaginary λ-th root of unity. Thus there is no solution for $\lambda < 100$, except perhaps for $\lambda = 37, 59, 67$. This proof has been given in modern form, by use of Dedekind's ideals, by Hilbert.[153]

A. Genocchi[64] proved that, if n is an odd prime,

$$-2B_{(n-3)/2} \equiv 1 + 2^{n-4} + \cdots + \left(\frac{n-1}{2}\right)^{n-4} \quad (\bmod\ n)$$

and noted that this, in connection with a statement by Cauchy,[57] shows that $x^n + y^n + z^n = 0$ is impossible in integers not divisible by the odd prime n when n is not a divisor of the numerator of the Bernoullian number $B_{(n-3)/2}$, the last one of the Bernoullian numbers in Kummer's condition.

Kummer[65] noted that his assumption that B_n is not divisible by λ for $n = (\lambda-3)/2$ (as well as for smaller n's) corresponds to Cauchy's[57] condition

$$1^{\lambda-4} + 2^{\lambda-4} + \cdots + \left(\frac{\lambda-1}{2}\right)^{\lambda-4} \not\equiv 0 \quad (\bmod\ \lambda).$$

If not both $B_{(\lambda-3)/2}$ and $B_{(\lambda-5)/2}$ are divisible by λ, one of the solutions x, y, z of $x^\lambda + y^\lambda = z^\lambda$ must be divisible by λ. Proof by Kummer,[76] pp. 61-5.

---

[62] Comptes Rendus Paris, 29, 1849, 23; 30, 1850, 263-4; 35, 1852, 919-20. There were five competing memoirs for the prize proposed for 1850 and eleven for the postponed prize for 1853; but none were deemed worthy of the prize. Cf. Nouv. Ann. Math., 8, 1849, 362-3 and, for bibliography, 363-4; 9, 1850, 386-7.

[63] Jour. für Math., 40, 1850, 130-8 (93); Jour. de Math., 16, 1851, 488-98. Reproduced by Gambioli,[171] pp. 169-176.

[64] Annali di Sc. Mat. e Fis., 3, 1852, 400-1. Summary in Jour. für Math., 99, 1886, 316. This congruence is a special case of one proved by Cauchy, Mém. Acad. Sc. Paris, 17, 1840, 265; Oeuvres, (1), III, p. 17.

[65] Letter to L. Kronecker, Jan. 2, 1852, Kummer,[35] p. 91.

H. Wronski[66] pretended that the impossibility of $x^n+y^n=z^n$, $n>2$, follows from his[67] results on $z^n-Nv^n=Mu^n$.

F. Landry[68] proved Legendre's[17] statement for $p=mn+1$, $m=10$ and 14, when $n>3$, noting that $(14^7\pm1)/(14\pm1)$ are primes.

Landry[69] employed two primes ϕ and $\theta=2t\phi+1$, and an integer ϵ belonging to the exponent ϕ modulo θ. The congruence $1+\epsilon^z\pm\epsilon^v\equiv0$ (mod θ) can be reduced to $1+\epsilon\pm\epsilon^z\equiv0$ unless $x=\phi$ or $x=0$, whence $2\phi\equiv\pm1$. By use of the substitutions $\epsilon=\epsilon_1^{-1}$, $\epsilon=\epsilon_1^{1/z}$, etc., we can reduce $1+\epsilon+\epsilon^z\equiv0$ to a similar congruence with z replaced by the integral residues modulo ϕ of

$$z, \quad 1-z, \quad \frac{1}{z}, \quad \frac{z-1}{z}, \quad \frac{1}{1-z}, \quad \frac{z}{z-1}.$$

Excluding $z=1$ or 2, these six expressions are incongruent modulo ϕ unless ϕ is of the form $6l+1$ and then they reduce to two for two special values of z. If all three relations $1+\epsilon-\epsilon^z\equiv0$, $1-\epsilon+\epsilon^z\equiv0$, $1-\epsilon-\epsilon^z\equiv0$ are impossible for a single one of the above six values, then $1+\epsilon-\epsilon^z\equiv0$ is impossible for all six.

For Landry's third memoir (on primitive roots), see Vol. I, p. 119, p. 190 of this History; for his fifth memoir (on continued fractions), see Landry[69] of Ch. XX above.

Landry[70] recurred to the exception arising if $2^\phi\equiv\pm1$ (mod θ), where θ is a prime $2k\phi n+1$, n a prime >2. For $\phi=5$, 7, 11, 13, 17, 19, he found all the cases in which $2^\phi\mp1$ has such a factor θ. For example, if $\phi=11$, only when $n=31$, $\theta=683$. Aside from these exceptions, $1+\epsilon\pm\epsilon^z\equiv0$ does not hold for $z=\phi$ or $z=0$ when $\phi\leqq19$; nor for $z=2$, $\frac{1}{2}$, -1, or $z=3$, $1-3$, $\frac{1}{3}$, etc., except for a few special values of θ.

Landry[71] proved that, if θ is a prime $2k\phi n+1(n>3)$, $1+\epsilon\pm\epsilon^z\equiv0$ (mod θ) are each impossible for $\phi=5$, 7, 11, 13, 17, 19, aside from the exceptions for $\phi=11$, 13, 17 noted by Landry,[70] and the new exceptions, arising for $\phi=19$: $\theta=761$, $n=5$, $k=4$; $\theta=647$, $n=17$, $k=1$; $\theta=419$, $n=11$, $k=1$.

H. E. Heine[71a] considered $P^m-DQ^m=1$, where P, Q, D are polynomials in x.

L. Calzolari[72] noted that any given numbers x, y, z can be expressed in the form $x=v+w$, $y=u+w$, $z=u+v+w$ [since we may take $u=z-x$,

[66] Véritable science nautique des marées, Paris, 1853, 23. Quoted in l'intermédiaire des math., 23, 1916, 231–4, and by Guimaräes.[273]

[67] Réforme du savoir humain, 1847, 242. See p. 210 of Vol. 1 of this History.

[68] Premier mémoire sur la théorie des nombres. Demonstration d'un principe de Legendre relatif au théorème de Fermat, Paris, Feb. 1853, 10 pp.

[69] Deuxième mémorie sur la théorie des nombres. Théorème de Fermat, Paris, July, 1853, 16 pp.

[70] Quatrième mémorie sur la théorie des nombres. Théorème de Fermat, Paris, Feb. 1855, 27 pp.

[71] Sixième mémorie sur la théorie des nombres. Théorème de Fermat, 3ᵉ livre, Paris, Nov. 1856, 24 pp.

[71a] Jour. für Math., 48, 1854, 256–9.

[72] Tentativo per dimostrare il teorema di Fermat ... $x^n+y^n=z^n$, Ferrara, 1855. Extract by D. Gambioli,[171] 158–161.

$v = z - y$, $w = x + y - z$]. Let $x^n + y^n = z^n$, and set $x = z - u$, $y = z - v$. Then

$$z^n - n(u+v)z^{n-1} + \binom{n}{2}(u^2+v^2)z^{n-2} - \cdots + (-1)^n(u^n+v^n) = 0.$$

Hence $u^n + v^n$ is divisible by z. Similarly, $\alpha = u^n + (v-u)^n$ is divisible by x, and $\beta = v^n + (u-v)^n$ by y. His argument that Fermat's equation is impossible if n is odd and > 3 is unsatisfactory. By Cotes' theorem,

$$u^n + v^n = (u+v)\Pi(u^2 - 2uv \cos \lambda\pi/n + v^2),$$

where $\lambda = 1, 3, 5, \cdots, n-2$. The λth quadratic function has the factors

$$u + v \pm 2\sqrt{uv} \cos \lambda\pi/(2n).$$

But $u^n + v^n$ has the factor $z = u + v + w$, whence

$$w = 2\sqrt{uv} \cos \lambda\pi/(2n).$$

Similarly, since α is divisible by $x = u + (v-u) + w$, and β by $y = v + (u-v) + w$,

$$w = 2\sqrt{u(v-u)} \cos \frac{\lambda'\pi}{2n}, \qquad w = 2\sqrt{v(u-v)} \cos \frac{\lambda''\pi}{2n},$$

whereas the one is real and the other imaginary. He also claimed that the first w is symmetrical in u, v, while the third w is not. He made also the error of assuming that an even factor of a product of an odd by an even number must divide the latter.

J. A. Grunert[73] proved that, if $n > 1$, there are no positive integral values satisfying $x^n + y^n = z^n$ unless $x > n$, $y > n$, simultaneously. Set $z = x + u$ and apply the binomial theorem; hence $y^n > nx^{n-1}u$.

L. Calzolari[74] considered a triangle whose sides are integral solutions of $x^n + y^n = z^n$, n odd > 1. Thus $z^2 = x^2 - axy + y^2 \equiv P_2$ for a suitable value of a. It is stated that the polynomial $P_n \equiv x^n + y^n$ is divisible by P_2, the polynomial quotient P_{n-2} is divisible by P_2, etc., and finally the symmetric quotient $P_1 = x + y$ equals z, which is impossible. If $n = 2m$, $P_2^m \equiv P_n$, $a = 0$, $m = 1$.

G. C. Gerono[75] considered the integers x, y for which $a^x - b^y = 1$ for primes a, b. If $a > 2$, then $b = 2$, $a = 2^n + 1$, and $x = 1$, $y = n$ when $n > 1$, with also $x = 2$, $y = 3$ when $n = 1$. If $a = 2$, then $b = 2^n - 1$, $x = n$, $y = 1$.

E. E. Kummer[76] proved that for any relatively prime integral solutions of $x^\lambda + y^\lambda = z^\lambda$, where λ is any odd prime, and xyz is prime to λ,

$$(8) \qquad B_{(\lambda-i)/2}P_i(x, y) \equiv 0 \quad (\text{mod } \lambda) \qquad (i = 3, 5, \cdots, \lambda-2),$$

where B_j is the jth Bernoullian number and $P_i(x, y)$ is the homogeneous polynomial of degree i for which

$$\left(\frac{d^i \log (x + e^v y)}{dv^i}\right)_{v=0} = \frac{P_i(x, y)}{(x+y)^i}.$$

He proved that Fermat's equation is impossible in integers for odd prime exponents which satisfy the following three conditions:

[73] Archiv Math. Phys., 26, 1856, 119–120. Wrong reference by Lind,[241] p. 54.
[74] Annali di Sc. Mat. e Fis., 8, 1857, 339–345.
[75] Nouv. Ann. Math., 16, 1857, 394–8.
[76] Abh. Akad. Wiss. Berlin (Math.), for 1857, 1858, 41–74. Extract in Monatsb. Akad. Wiss. Berlin, 1857, 275–82.

(i) The factor h_1 of the class number H is divisible by λ, but not by λ^2.

(ii) For μ, g, $e(\alpha)$ defined as by Kummer[61], and for the integer $\nu < (\lambda-1)/2$ such that $B_\nu \equiv 0 \pmod{\lambda}$, there exists an ideal with respect to which as modulus the unit

$$E_\nu(\alpha) = \prod_{k=0}^{\mu-1} e(\alpha^{g^k})^{g-2k\nu}$$

is not congruent to a λth power, whence the second factor h_2 of H is not divisible by λ.

(iii) The Bernoullian number $B_{\nu\lambda}$ is not divisible by λ^3.

All three conditions are satisfied when $\lambda = 37$, 59, 67, the values < 100 for which he had not previously proved Fermat's theorem. [But Kummer (pp. 46–50) repeatedly used an earlier[76a] congruence involving logarithms which is not true in all cases, as noted by F. Mertens.[76b] The remark that this error vitiates also the present paper, and two further criticisms were made by H. S. Vandiver.[76c] First, Kummer (p. 42, bottom) relied on his paper in Jour. de Math., 16, 1851, 473, where he reduced h_1 modulo λ, but not modulo λ^n, $n > 1$, as now needed. Second, Kummer (p. 53) employed a decomposition of $\Psi_r(\alpha)$ which holds only when it contains only ideals of the first degree. Although the theorem on p. 61 is really subject to this restriction, it is applied (p. 67) to ideals $\Theta_r(\alpha)$ which are not proved to be of the first degree. Kummer,[76a] p. 120, had given the different decomposition when there occur ideals not all of the first degree.]

H. F. Talbot[77] proved (I) If n is odd > 1, $a^n = b^n + c^n$ is impossible in integers if a is a prime [Abel[16]]; (II) If n is any integer > 1, and if $a^n = b^n - c^n$ is possible when a is a prime, then $b - c = 1$. For (I), $(b+c)^n > b^n + c^n = a^n$, $b+c > a$; $b < a$, $c < a$, $b+c < 2a$. Hence $b+c$ is not divisible by the prime a, contrary to the given equation. Similarly for (II). Generalizations are given. If a is a prime and $m < n$, $a^m = b^n + c^n$ is impossible if n is odd, while $a^m = b^n - c^n$ is impossible if $b - c > 1$.

K. Thomas[78] attempted to prove Fermat's last theorem.

H. J. S. Smith[79] gave numerous references on Fermat's last theorem, noted that Barlow's[15] proof was erroneous, and reproduced the proof by Kummer[63] for regular primes.

A. Vachette[80] proved (6) and concluded that, if a, b are integers and n is a prime > 2, $(a+b)^n - a^n - b^n$ is divisible by $nab(a+b)$, and gave several expressions for the quotient. Set

$$A_k = (x+1/x)^k - x^k - 1/x^k, \qquad a = x+1/x.$$

Then A_{6n+7} is proved divisible by $(a^2-1)^2$ [Cauchy[29]]. There are proofs (pp. 264–5) of (6) by induction on n and by Waring's formula.

F. Paulet[81] gave an erroneous proof of Fermat's last theorem.

[76a] Kummer, Jour. für Math., 44, 1852, 134 (error, p. 133).

[76b] Sitzungsber. Akad. Wiss. Wien (Math.), 126, 1917, IIa, 1337–43.

[76c] Proc. National Acad. Sc., April, 1920.

[77] Trans. Roy. Soc. Edinburgh, 21, 1857, 403–6.

[78] Das Pythagoräische Dreieck und die Ungerade Zahl, Berlin, 1859, Ch. 10.

[79] Report British Assoc. for 1860, 148–152; Coll. Math. Papers, I, 131–7.

[80] Nouv. Ann. Math., 20, 1861, 160–6.

[81] Cosmos, 22, 1863, 385–9. Correction, p. 407, by R. Radau.

L. Calzolari[82] attempted a proof, starting as before.[74]

P. G. Tait[83] stated that if $x^m = y^m + z^m$ has integral solutions when m is an odd prime, then $x \equiv y \equiv 1$, $z \equiv 0 \pmod{m}$.

H. F. Talbot[84] noted that Barlow[15] made the same error in his proof of the impossibility of $x^n - y^n = z^n$ as for the case $n = 3$ (p. 139), where he stated that, if r, s, t are relatively prime in pairs,

$$\frac{t^2}{sr} - \frac{s^2}{tr} - \frac{9r^2}{st} \neq 6$$

since each fraction is in its lowest terms and each denominator has a factor not common with the other denominators, and hence the algebraic sum of the fractions is not an integer (by the false Cor. 2, Art. 13). On the contrary, we have

$$\frac{7}{2\cdot 3} + \frac{8}{3\cdot 5} + \frac{3}{2\cdot 5} = 2.$$

A. Genocchi[85] abbreviated Lamé's[28] proof for $n = 7$. Let x, y, z be roots of $v^3 - pv^2 + qv - pq + r = 0$. Then $x^7 + y^7 + z^7 = 0$ is equivalent to

$$p^7 - 7r(p^4 - p^2 q + q^2) + 7pr^2 = 0.$$

After excluding the trivial case $p = 0$, we may change q to $p^2 q$, r to $p^3 r$, and get $7r^2 - 7r(1 - q + q^2) = -1$. The radical in the expression for the root r must be rational. Thus $(1 - q + q^2)^2 / 4 - 1/7$ is a square. Set $2q - 1 = s/t$. Then

$$7^2(s^4 + 6s^2 t^2) - 7t^4 = (7u)^2.$$

Proof of the impossibility of the latter is not given.

Gaudin[86] attempted to prove that, if n is an odd prime, $(x + h)^n - x^n = z^n$ is impossible in rational numbers. Treating x/h as a new variable, we are led to the case $h = 1$. To avoid the author's complicated formulas, take $n = 5$. Then

$$(x+1)^5 - x^5 = 5x(x+1)\{x(x+1)+1\} + 1$$

is of the form $10t + 1$. Since z^5 is of that form, $z = 10s + 1$ and

$$z^5 = 5 \cdot 10s\{10s[10s(10s \cdot \overline{2s+1} + 2) + 2] + 1\} + 1,$$

which is said never to equal the first expression. His remaining two arguments are trivial.

I. Todhunter[87] proved Cauchy's[29] theorem and that, if $q = x^2 + xy + y^2$, $b = xy(x+y)$,

$$\frac{(x+y)^{2m} + x^{2m} + y^{2m}}{2m} = \frac{q^m}{m} + \Sigma \frac{(m-r-1)(m-r-2)\cdots(m-3r+1)}{(2r)!} q^{m-3r} b^{2r},$$

<hr>

[82] Annali di Mat., 6, 1864, 280–6.
[83] Proc. Roy. Soc. Edinburgh, 5, 1863–4, 181.
[84] Trans. Roy. Soc. Edinburgh, 23, 1864, 45–52.
[85] Annali di Mat., 6, 1864, 287–8.
[86] Comptes Rendus Paris, 59, 1864, 1036–8.
[87] Theory of Equations, 1861, 173–6; ed. 2, 1867, 189; 1888, 185, 188–9.

$$\frac{(x+y)^{2m+1}-x^{2m+1}-y^{2m+1}}{2m+1}$$

$$=q^{m-1}b+\Sigma\,\frac{(m-r-1)(m-r-2)\cdots(m-3r)}{(2r+1)!}\,q^{m-3r-1}b^{2r+1},$$

summed for $r=1,2,\cdots$. The first formula had been given earlier.[88]

Housel[89] proved Catalan's[38] empirical theorem that two consecutive integers, other than 8 and 9, can not be exact powers [with exponents >1].

E. Catalan[90] stated this theorem and those given under Catalan.[122a]

Catalan[91] set $p=x+y+z$, $P=p^n-x^n-y^n-z^n$ and proved that the quotient Q of P by $(x+y)(y+z)(z+x)$ is (for n odd >3)

$$p^{n-3}+H_1p^{n-4}+H_2p^{n-5}+\cdots+H_{n-3}$$
$$+\,y^{n-3}+H_1(x^2,z^2)y^{n-5}+H_2(x^2,z^2)y^{n-7}+\cdots+H_{(n-3)/2}(x^2,z^2)$$
$$+\,x^{n-3}+H_1(y^2,z^2)x^{n-5}+H_2(y^2,z^2)x^{n-7}+\cdots+H_{(n-3)/2}(y^2,z^2),$$

where $H_1=p$, $H_2=\Sigma x^2+\Sigma xy$, $H_3=\Sigma x^3+\Sigma x^2y+xyz$, $\cdots$,

$$H_q(x,z)=x^q+zx^{q-1}+z^2x^{q-2}+\cdots+z^q.$$

If n is a prime the coefficients of P and Q are divisible by n. Also,

$$Q-\frac{n(x^{n-1}-z^{n-1})}{x^2-z^2}\equiv n(y+z)(x+y)\phi,$$

where ϕ is a polynomial in x, y, z with integral coefficients.

G. C. Gerono[92] proved that, if x or y is a prime, $x^m=y^n+1$ holds in positive integers >1 only when $x=n=3$, $y=m=2$. See Carmichael.[226]

A. Genocchi[93] stated that $x^4+6x^2y^2-y^4/7=z^2$ is impossible in integers. Hence $x^7+y^7+z^7=0$ is not satisfied by values of x, y, z which are roots of a cubic equation with rational coefficients, a generalization of Lamé's[28] theorem.

E. Laporte[94] would deduce Fermat's last theorem from the fact that the series of powers higher than the second are formed by the summation of terms of arithmetical progressions preceded by extraneous terms.

Moret-Blanc[95] proved that the only positive integral solutions of

$$x^y=y^x+1$$

are $y=0$; $y=1$, $x=2$; $y=2$, $x=3$. A. J. F. Meyl[96] showed that the only positive integral solutions of $(x+1)^y=x^{y+1}+1$ are $x=0$, $x=y=1$, $x=y=2$.

[88] N. M. Ferrers and J. S. Jackson, Solutions of the Cambridge Senate-House Problems for 1848–1851, pp. 83–85.

[89] Catalan's Mélanges Math., Liège, ed. 1, 1868, 42–48, 348–9.

[90] *Ibid.*, 40–1; Revue de l'instruction publique en Belgique, 17, 1870, 137; Nouv. Corresp. Math., 3, 1877, 434. Proofs by Soons.[172]

[91] Mélanges Math., ed. 1, 1868, No. 47, 196–202; Mém. Soc. Sc. Liège, (2), 12, 1885, 179–185, 403. (Cited in Bull. des sc. math. astr., (2), 6, I, 1882, 224.)

[92] Nouv. Ann. Math., (2), 9, 1870, 469–471; 10, 1871, 204–6.

[93] Comptes Rendus Paris, 78, 1874, 435. Proof, 82, 1876, 910–3.

[94] Petit essai sur quelques méthodes probables de Fermat, Bordeaux, 1874. Reprinted in Sphinx-Oedipe, 4, 1909, 49–70.

[95] Nouv. Ann. Math., (2), 15, 1876, 44–6.

[96] *Ibid.*, 545–7.

F. Lukas[97] set $y=x-a$, $z=x-b$, $a<b$, in $y^n+z^n=x^n$, $n>2$.   Hence

$$x^n-\binom{n}{1}(a+b)x^{n-1}+\binom{n}{2}(a^2+b^2)x^{n-2}-\cdots+(-1)^n(a^n+b^n)=0.$$

Let $w_1, \cdots, w_n$ be its roots, all positive.   Then

$$\Sigma w_1=n(a+b), \qquad \frac{1}{n}\,\Sigma w_1^2=a^2+b^2+2nab=\text{integer},$$

which are said to be impossible if $n>2$.   This error was noted in Jahrbuch Fortschritte der Math., 7, 1875, 100.

T. Pepin[98] proved that $x^7+y^7+z^7=0$ is not satisfied by integers not divisible by 7, by use of the fact that $u^2=x^4+7^3y^4$ has no integral solutions with $y\neq0$ (proved by descent).   He proved (pp. 743–7) that the first equation has no solution in which one of the unknowns is divisible by 7.

J. W. L. Glaisher[99] expressed Cauchy's[29] theorem in a new form.   Let n be odd and set $x=c-b$, $y=a-c$.   Then

$$(x+y)^n-x^n-y^n=(b-c)^n+(c-a)^n+(a-b)^n\equiv E_n.$$

Then E_n is algebraically divisible by $E_3=3xy(x+y)$.   If $n=6m\pm1$, E_n is divisible by $E_2=2(x^2+xy+y^2)$.   If $n=6m+1$, E_n is divisible by $E_4=\frac{1}{2}E_2^2$.

Glaisher[100] expressed $(x+y)^n-x^n-y^n$, for n odd ≤13, in terms of $\beta=x^2+xy+y^2$ and $\gamma=xy(x+y)$.   [Earlier by Cauchy.[33]]

T. Muir[101] noted that x, y, $-x-y$ are the roots of $w^3-\beta w+\gamma=0$. Hence by Waring's formula for the sum of like powers of the roots,

$$\frac{(x+y)^{2m+1}-x^{2m+1}-y^{2m+1}}{2m+1}=\beta^{m-1}\gamma+\frac{(m-2)(m-3)}{1\cdot2\cdot3}\beta^{m-4}\gamma^3$$
$$+\frac{(m-3)\cdots(m-6)}{1\cdot2\cdot3\cdot4\cdot5}\beta^{m-7}\gamma^5+\cdots.$$

He gave a similar formula for $(x+y)^{2m}+x^{2m}+y^{2m}$.   For three variables, set

$$\beta=\Sigma x^2+\Sigma xy, \qquad \gamma=\Sigma x^2y+2xyz, \qquad \delta=xyz(x+y+z).$$

Then x, y, z, $-x-y-z$ are the roots of $w^4-\beta w^2+\gamma w-\delta=0$.   Thus

$$(x+y+z)^{2m+1}-x^{2m+1}-y^{2m+1}-z^{2m+1}$$
$$=\Sigma(-1)^{r+s+t-1}\frac{(2m+1)\cdot(r+s+t-1)!}{r!s!}(-\beta)^r\gamma^s\delta^t,$$

summed for all integral solutions ≥0 of $2r+3s+4t=2m+1$.   Since $s>0$, the sum has the factor $\gamma=\frac{1}{3}\{(x+y+z)^3-x^3-y^3-z^3\}$.

Glaisher[102] noted that Newton's identities give a recursion formula for $x_1^n+\cdots+x_m^n$, extended Cauchy's theorem to negative exponents, and gave recursion formulas for and factors of the sum of the pth powers of all the quantities $\pm a_1\pm\cdots\pm a_n$ in which r of the signs are negative.

[97] Archiv Math. Phys., 58, 1876, 109–112.
[98] Comptes Rendus Paris, 82, 1876, 676–9.
[99] Quar. Jour. Math., 15, 1878, 365–6.
[100] Messenger Math., 8, 1878–9, 47, 53.
[101] Quar. Jour. Math., 16, 1879, 9–14.
[102] Ibid., 89–98.

A. Desboves[103] noted that $aX^m + bY^m = cZ^n$ has integral solutions if and only if c is of the form $ax^m + by^m$; we can find a function c of a, b and as many parameters as one pleases such that integral solutions exist. Next, let $n = m$. Then we can find a, b, c so that there are two sets of solutions and these determine $a : b : c$. There exists such an equation with three sets of solutions if and only if

$$P^m + Q^m + R^m = U^m + V^m + T^m, \qquad PQR = UVT$$

have integral solutions $\neq 0$. We can solve $X^{4m} - a^2 Y^{4m} = Z^2$ if

$$a = \frac{(x+yi)^{4m} - (x-yi)^{4m}}{2i},$$

viz., by $X = x^2 + y^2$, $Y = 1$.

A. E. Pellet[104] considered, for p a prime, the congruence

$$At^m + Bu^n + C \equiv 0 \pmod{p}, \qquad ABC \not\equiv 0 \pmod{p}.$$

Let d be the g.c.d. of m, $p-1$; d_1 that of n, $p-1$. Set $x \equiv t^m$. Then x must satisfy the two congruences

$$x(x^{(p-1)/d} - 1) \equiv 0, \qquad (Ax + C)\left[\left(\frac{-Ax - C}{B}\right)^{(p-1)/d_1} - 1\right] \equiv 0 \pmod{p}.$$

Conversely, to each of the μ common roots of the latter two congruences correspond dd_1 sets of solutions of the proposed congruence, which therefore has μdd_1 sets of solutions. For $m = n = 2$, the two congruences have at least one common root, since the second is not $x^{(p-1)/2} + 1 \equiv 0$, being of higher degree. Hence $At^2 + Bu^2 + C \equiv 0 \pmod{p}$ is solvable (Lagrange,[9] etc., of Ch. VIII).

R. Liouville[105] claimed that $X^n + Y^n = Z^n$ is impossible if $n > 1$ and X, Y, Z are polynomials in a variable t. Set $\alpha = X/Z$. Then

$$U = \int \frac{\alpha^{n-1} d\alpha}{\sqrt[n]{1 - \alpha^n}} = \int \frac{Z}{Y}\left(\frac{X}{Z}\right)^{n-1} d\left(\frac{X}{Z}\right)$$

is a polynomial in $\sqrt[n]{1 - \alpha^n} = Y/Z$. Since dU/dt is the argument of the second integral,

$$Z^2 \frac{d}{dt}\left(\frac{X}{Z}\right) = -Z^2 \left(\frac{Y}{X}\right)^{n-1} \frac{d}{dt}\left(\frac{Y}{X}\right)$$

must be the product of Y by a polynomial A. Hence

$$A + \frac{Z^2 Y^{n-2}}{X^{n-1}} \frac{d}{dt}\left(\frac{Y}{Z}\right) = 0.$$

Thus X^{n-1} divides $Z^2 d(Y/Z)$. Call the quotient B and set $P = Y/Z$. Then

$$\frac{dP}{dt} = \frac{B}{Z^2} X^{n-1}, \qquad \frac{dP}{dt} \div (1 - P^n)^{(n-1)/n} = BZ^{n-3}.$$

But in the latter, the left member is infinite for a root of $P^n = 1$, while the

[103] Nouv. Ann. Math., (2), 18, 1879, 481–9.
[104] Comptes Rendus Paris, 88, 1879, 417–8.
[105] Comptes Rendus Paris, 89, 1879, 1108–10.

right member remains finite. This argument was called insufficient by E. Netto.[106]

A. Korkine[107] modified the last proof. Let Z be a polynomial in t whose degree m is not less than the degrees of X and Y. Then one of the latter is of degree m, say Y. Let $m-\lambda$ ($\lambda \geqq 0$) be the degree of X. Differentiate $(Y/X)^n+(Z/X)^n+1=0$ with respect to t. Then, since Y, Z have no common factor,

$$\frac{XY'-YX'}{Z^{n-1}}=\frac{ZX'-XZ'}{Y^{n-1}}$$

is an integral function. As the degrees of the numerators are $\leqq 2m-\lambda-1$ and that of the denominators is $m(n-1)$, we have

$$2m-\lambda-1-m(n-1)\geqq 0, \qquad m(3-n)\geqq\lambda+1, \qquad n<3.$$

A. Lefébure[108] separated into two classes the primes $p=2kn+1$. Into the first class, put the p's such that the algebraic sum of any three residues of nth powers modulo p cannot be a multiple of p. Into the second class, put the p's for which the algebraic sum of three residues is a multiple of p. It is claimed that all the p's in the first class are divisors of one of the integers satisfying $x^n+y^n=z^n$, so that every p is a divisor of x, y or z, or is in the second class. Hence if the first class is infinite, the equation is impossible. But the first class is not finite when the second is infinite [correction by Pepin[109]].

T. Pepin[109] noted that Libri[24] long ago pronounced judgment on an attempted proof like Lefébure's.[108] To prove Libri's assertion on

$$x^3+y^3+1\equiv 0 \ (\text{mod } p=3h+1),$$

Pepin showed (by use of Gauss, Disq. Arith., art. 338, on the equation for the three periods of roots of unity) that the number of sets of solutions of the congruence in positive integers $<p$ is $p+L-8$, where L is determined by $L^2+27M^2=4p$ and $L\equiv 1 \ (\text{mod } 3)$. Hence 7 and 13 are the only primes $3h+1$ which cannot divide a sum of three cubes without dividing one of them.

O. Schier[110] claimed to prove $x^n+y^n=z^n$ impossible in relatively prime integers if n is an odd prime. We have $x+y\equiv z \ (\text{mod } n)$. Expand by the binomial theorem

$$(x+y)^n=(z+nk)^n,$$

cancel x^n+y^n with z^n, and divide by the factor n. Thus

$$xy(x^{n-2}+y^{n-2})+\frac{n-1}{2}x^2y^2(x^{n-4}+y^{n-4})+\cdots=z^{n-1}nk+\cdots+n^{n-1}k^n.$$

Hence also the left member must be divisible by n. It is stated that this divisibility depends on that of the factors xy and $x+y$ occurring in every

[106] Jahrbuch Fortschritte Math., 11, 1879, 138.
[107] Comptes Rendus Paris, 90, 1880, 303–4 (Math. Soc., Moscow, 10, 1882, 54–6).
[108] *Ibid.*, 90, 1880, 1406–7.
[109] *Ibid.*, 91, 1880, 366–8. Reprinted, Sphinx-Oedipe, 4, 1909, 30–32.
[110] Sitzungsber. Akad. Wiss. Wien (Math.), 81, II, 1880, 392–8.

term.   Hence n divides x or y.   For, if $x+y$ and hence z is divisible by n, set $x=z+nk-y$ in the initial equation; the result is said to hold only if y is a multiple of n.

F. Fabre[111] proposed the question of the divisibility of $(x+y)^n-x^n-y^n$ by x^2+xy+y^2 and M. Dupuy proved (*ibid.*, 1881, 282-3) that n must be of the form $6a\pm1$.

If[112] $(\Sigma a)^{2n+1}=\Sigma a^{2n+1}$ is true for $n=1$ it is true for any n, since

$$(a+b)(a+c)(b+c)=0.$$

, A. E. Pellet stated and Moret-Blanc[113] proved that $At^3+Bu^2+C\equiv0$ (mod 7) is solvable if ABC is prime to 7.

E. Cesàro[114] proved that if $\psi(n)$ is the number of sets of positive integral solutions of $Ax^\alpha+By^\beta=n$, where A and B are positive integers,

$$\psi(1)+\cdots+\psi(n)=\frac{n^{1/\alpha+1/\beta}}{A^{1/\alpha}B^{1/\beta}}\int_0^1\sqrt[\alpha]{1-x^\beta}\,dx.$$

The ratio of $\psi(n)$ to the number of solutions of $x^\alpha+y^\beta=n$ is $A^{-1/\alpha}B^{-1/\beta}$, in mean.   Hence, for $\alpha=\beta=1$, $\psi(n)=n/(AB)$, in mean.   For $\alpha=\beta=2$, $\psi(n)=\pi/(4\sqrt{AB})$, in mean.   The mean of the sum of the pth powers of all the positive integral values which x can take in $x^k+y^k=n$ is found (p. 229).

C. M. Piuma[115] noted that, if no one of the coefficients A, B, C is divisible by the prime $m=pq+1$, then $Ax^p+By^q+C\equiv0$ (mod m) has integral solutions if and only if $Az+Bz_1+C\equiv0$ (mod m) has solutions for which $z\equiv x^p$, $z_1\equiv y^q$ are solvable for x, y, i. e., if

$$z(z^q-1)\equiv0,\qquad z_1(z_1^p-1)\equiv0\ (\text{mod } m)$$

are solvable.   Thus the initial congruence has solutions if and only if $P\equiv0$ (mod m), where P is the resultant of the equations corresponding to the last two and $Az+Bz_1+C=0$, so that P is a product of $(p+1)(q+1)$ linear factors.

For $q=2$, there are solutions if $C+A$ or $C-A$ is divisible by m, or if any one of the products $-BC$, $-B(C+A)$, $-B(C-A)$ is a quadratic residue of m.   In particular, $Ax^3+By^2+C\equiv0$ (mod 7) is solvable if no one of the coefficients is divisible by 7.   Cf. Pellet.[113]

E. Catalan, P. Mansion and de Tilly[116] gave adverse reports on two manuscripts submitted for the prize offered for 1883 by the Belgian Academy (p. 101) for a proof of Fermat's last theorem.

E. de Jonquières[117] proved that in $a^n+b^n=c^n$, $n>1$, the greater of a, b is composite.   Set $c=a+k$, $b>a$.   Then, by the binomial theorem,

[111] Jour. de math. élémentaire de Longchamps et de Bourget, 1880, No. 273, p. 528.

[112] Math. Quest. Educ. Times, 36, 1881, 105.

[113] Nouv. Ann. Math., (3), 1, 1882, 335, 475-6.

[114] Mém. Soc. R. Sc. de Liège, (2), 10, 1883, No. 6, 195-7, 224.

[115] Annali di Mat., (2), 11, 1882-3, 237-245.

[116] Bull. Acad. R. Belgique, (3), 6, année 52, 1883, 814-9, 820-3, 823-32.

[117] Atti Accad. Pont. Nuovi Lincei, 37, 1883-4, 146-9.   Reprinted in Sphinx-Oedipe, 5, 1910, 29-32.   Proof by S. Roberts, Math. Quest. Educ. Times, 47, 1887, 56-58; H. W. Curjel, 71, 1899, 100.

$b^n = (a+k)^n - a^n$ is divisible by k. But if $k \geqq b$, $c^n \geqq (a+b)^n > a^n + b^n$. Hence b^n is divisible by an integer k, $k > 1$, $k < b$. Similarly, if a is a prime $< b$, then $c - b = 1$. He[118] stated that if $a^n + b^n = c^n$ and a or b is a prime, the least of the two is a prime and the greater is composite and differs from c by unity.

G. Heppel[119] proved that, if n is a prime > 3, $(x+y)^n - x^n - y^n$ is divisible by $nxy(x+y)(x^2 + xy + y^2)$ and found the coefficients of the general term of the quotient.

P. A. MacMahon[120] employed his generalization of Waring's formula in Proc. Lond. Math. Soc., 15, 1883–4, p. 20, to prove the identity

$$S(x, y) + S(y, x) = \Sigma(-1)^{b+1} \frac{(a+b-1)!\,(a+3b)}{a!\,b!} (x^2 + xy + y^2)^a \{xy(x+y)\}^b,$$

summed for the integral solutions of $2a + 3b = n$, where

$$S(x, y) = \frac{(x+2y)x^n + (-1)^{n+1}(x-y)(x+y)^n}{(x-y)(x+2y)(2x+y)} \{2y(x+y) - x^2\}.$$

He gave a similar identity for three variables. The right member of the initial identity becomes $5xy(x+y)(x^2 + xy + y^2)^2$ if $n = 7$ [cf. Cauchy[29]].

E. Catalan[121] stated that if p is an odd prime,

$$(x+y)^p - x^p - y^p \equiv pxy(x+y)P^2,$$

where P is a polynomial with integral coefficients, holds only if $p = 7$ and $P = x^2 + xy + y^2$. He[122] proved this by taking $x = y = 1$. Thus $2^{p-1} - 1 = pN^2$, where N is an integer. Set $t = (p-1)/2$. Since $2^t - 1$ and $2^t + 1$ are relatively prime, having the difference 2, one of them is a square. The first is of the form $4n + 3$ and is not a square. Hence $2^t + 1 = M^2$. Thus the factors $M+1$, $M-1$ of 2^t are powers of 2 and their difference is 2. Hence $M - 1 = 2$, so that $p = 7$, $N = 3$ or $p = 3$, $N = 1$.

Catalan[122a] stated the empirical theorems: (I) $(x+1)^x - x^x = 1$ is impossible in integers except for $x = 0$ or 1. (II) $x^y - y^x = 1$ is impossible except for $x = 1$, $y = 0$ or $x = 3$, $y = 2$. (III) $x^p - 1 = P$, where p and P are primes, is satisfied only by $x = 2$, $p = 3$, $P = 7$. (IV) $x^n - 1 = P^2$ is impossible if P is a prime. (V) $x^2 - 1 = p^m$, for p a prime, is satisfied only by $x = 3$, $p = 2$, $m = 3$; $x = 2$, $p = 3$, $m = 1$. (VI) $x^p - q^y = 1$, where p and q are primes, is impossible except when $x = y = 3$, $p = q = 2$. (VII) $x^3 + y^3 = p^2$, where p is a prime, is impossible except when $x = 2$, $y = 1$, $p = 3$. (VIII) $x^n = \{(2^{n-2} - 1)^n + 1\}/2^{n-2}$ is impossible except when $n = 3$, $x = 1$. Cf. Gegenbauer.[133]

G. B. Mathews[123] proved for special primes p that $x^p + y^p = z^p$ is impossible if no one of x, y, z is a multiple of p. The method was suggested by

[118] Comptes Rendus Paris, 98, 1884, 863–4. Extract in Oeuvres de Fermat, IV, 154–5.
[119] Math. Quest. Educ. Times, 40, 1884, 124.
[120] Messenger Math., 14, 1884–5, 8–11.
[121] Nouv. Ann. Math., (3), 3, 1884, 351 (Jour. de math., spéc., 1883, 240).
[122] Ibid., (3), 4, 1885, 520–4.
[122a] Mém. Soc. R. Sc. Liège, (2), 12, 1885, 42–3 (earlier in Catalan[90]).
[123] Messenger Math., 15, 1885–6, 68–74.

Gauss' remarks for $p=3$ (Werke, 2, 1863, 387–391).   Since $z \equiv x+y \pmod{p}$,

$$D = (x+y)^p - x^p - y^p \equiv 0 \pmod{p^2}, \qquad D = pxy(x+y)\phi(x, y).$$

The equivalent congruence $xyz\phi(x, y) \equiv 0 \pmod{p}$ is proved insolvable for $p=3, 5, 11, 17$ unless at least one of the three unknowns is divisible by p. The method leaves in doubt the case $p=3n+1$ since the factor x^2+xy+y^2 of ϕ has real roots.

E. Catalan[124] stated 16 theorems on $a^n+b^n=c^n$, n a prime >3.   If a is a prime, then $a \equiv 1 \pmod{n}$; $a^n \equiv 1 \pmod{nb}$; every prime factor of $c-a$ divides $a-1$; $a+b$ and $c-a$ are relatively prime; also $2a-1$ and $2b+1$;

$$nb^{n-1} \leqq a^n \leqq n(b+1)^{n-1};$$

a and b exceed n; a^n-1 is divisible by $nb(b+1)(b^2+b+1)$.   Next, no one of $a+b$, $c-a$, $c-b$ is a prime.   If $a+b=c_1^n$, $c-a=b_1^n$, $c-b=a_1^n$, then c is divisible by n.   The ϕ [of Mathews[123]] is

$$H_1 x^{p-3} + H_2 x^{p-4} y + \cdots + H_1 y^{p-3}, \qquad H_k = \frac{1}{p}\left[\binom{p-1}{k} \pm 1 \right],$$

the sign being plus if k is even.

Catalan[125] stated the same theorems.   Also, if $a^n+b^n=c^n$, where a, b, c are relatively prime in pairs, and $a+b$ is divisible by n, it is divisible by n^{n-1}; if $a+b$ is divisible by a prime $p \neq n$, it is divisible by p^n; if $a+b$ is divisible by a power $>n^{n-1}$ of n, it is divisible by n^{2n-1}; if $a+b$ is divisible by a power $>p^n$ of a prime $p \neq n$, it is divisible by p^{2n}.

L. Gegenbauer[126] proved that 17, 29 and 41 are the only primes $p=4\mu+1$ not dividing a sum of three biquadrates prime to p.   Cf. Euler[83] of Ch. XXIII.

C. de Polignac[127] proved that $a^n-2^k = \pm 1$ is impossible unless $a=3$, $n=1$ or 2.

A. E. Pellet[128] found by use of inequalities in the theory of roots of unity that $x^q+y^q \equiv z^q \pmod{p}$, where p is a prime $q\omega+1$, has solutions x, y, z each not divisible by p for every ω exceeding a certain limit (not specified) for which $q\omega+1$ is a prime [Libri[24]].

P. Mansion[129] considered $x^n+y^n=z^n$, where x, y, z are relatively prime, $x<y<z$, n an odd prime.   By de Jonquières,[117] y is composite.   It is proved here that z is composite.   The proof that x is composite is erroneous, as later admitted.

M. Martone[130] attempted to prove Fermat's last theorem.

[124] Bull. Acad. Roy. Sc. Belgique, (3), 12, 1886, 498–500.   Reproduced in Oeuvres de Fermat, IV, 156–7.

[125] Mém. Soc. R. Sc. Liège, (2), 13, 1886, 387–397 (=Mélanges Math., 2, 1887, 387–397). Proofs of some of these theorems by Lind,[241] pp. 30–31, 41–43, and by S. Roberts, Math. Quest. Educ. Times, 47, 1887, 56–8.

[126] Sitzungsber. Akad. Wiss. Wien (Math.), 95, II, 1887, 842.

[127] Math. Quest. Educ. Times, 46, 1887, 109–110.

[128] Bull. Soc. Math. de France, 15, 1886–7, 80–93.

[129] Bull. Acad. Roy. Sc. Belgique, (3), 13, 1887, 16–17 (correction, p. 225).

[130] Dimostrazione di un celebre teorema del Fermat, Catanzaro, 1887, 21 pp.   Napoli, 1888. Nota ad una dimostr. . . ., Napoli, 1888 (attempt to complete the proof in the former paper).

F. Borletti[131] proved that, if n is a prime >2, $x^n + y^n = z^n$ has no positive integral solutions if z is a prime, and $x^{2n} - y^{2n} = z^{2n}$ has no integral solution if one of the unknowns is a prime; $x^n \pm y^n = 2^{an}$ is impossible if $n > 1$, and x, y are odd and relatively prime.

E. Lucas[132] proved Cauchy's[29] result. Set $q = a^2 + ab + b^2$,

$$r = ab(a+b), \qquad S_n = (a+b)^n + (-a)^n + (-b)^n.$$

Then $S_{n+3} = qS_{n+1} + rS_n$. Hence, by Waring's formula, S_n is divisible by q^2r if $n = 6m+1$; by q, and not by r, if $n = 6m+2$; by r, and not by q, if $n = 6m+3$; by q^2, and not by r, if $n = 6m+4$; by qr if $n = 6m+5$; by neither q nor r if $n = 6m$. As a generalization, if p is a prime,

$$(1 + x + x^2 + \cdots + x^{p-2})^n - 1 - x^n - x^{2n} - \cdots - x^{(p-2)n}$$

is divisible by $Q = 1 + x + \cdots + x^{p-1}$ if n is odd and prime to p, and by Q^2 if $n = 2p+1$. For p arbitrary, let $\phi(x) = 0$ be the equation for the primitive pth roots of unity. Then without details it is stated that

$$\{\phi(x) - x^\lambda\}^n - \phi(x^n)$$

is divisible by $\phi(x)$ for n odd and prime to p. [Apparently the term $x^{\lambda n}$ should be added, and λ taken to be the degree of $\phi(x)$, which degree is the number of integers $<p$ and prime to p.]

L. Gegenbauer[133] proved that, if α is a positive integer with at least one odd factor >1, and q is a prime, $x^\alpha + y^\alpha = q^n$ has positive integral solutions only when $q = 2$, $n = a\alpha + 1$, $x = y = 2^a$, or $\alpha = q = 3$, $n = 2 + 3a$, $x = 2 \cdot 3^a$, $y = 3^a$. Hence 3^2 is the only power of an odd prime representable as a sum of the αth powers of two relatively prime integers. A special case of this gives the seventh empirical theorem of Catalan.[122a] It is proved that if q is a prime, $x^{a+1} - q^n = 1$ is possible only for $x = 2$, $n = 1$, $a+1$ a prime, or $x = 3$, $a = 1$, $q = 2$, $n = 3$. Hence a prime other than $2^n - 1$ is not followed by a power, while 3^2 is the only power followed by a power of a prime. These imply the third, fourth, fifth and sixth empirical theorems of Catalan.

A. Rieke[134] attempted to prove $x^p + y^p = z^p$ impossible if p is an odd prime >3. He proved and used (6). From an equation of degree $t = (p-1)/2$ in a quantity m admitted to be doubtless irrational, he drew (p. 241) the meaningless conclusion "that m^t has the factor p, and m the factor $p^{1/t}$, and indeed for all values of m."

D. Varisco[135] failed to prove Fermat's last theorem since he concluded (p. 375) that there is a unique set of solutions $\sigma_1 = 0$, etc., of

$$\lambda_1 - \sigma_1 = 2ud, \qquad \lambda_1 d_1 - \sigma d = \eta, \qquad \sigma - \lambda = 2ud_1, \qquad \sigma_1 d_1 - \lambda d = \eta,$$

whereas the four equations are linearly dependent and have further sets of solutions. The fault seemed irreparable to O. Landsberg.[136]

[131] Reale Ist. Lombardo, Rendiconti, (2), 20, 1887, 222–4.

[132] Assoc. franç. av. sc., 1888, II, 29–31; Théorie des nombres, 1891, 276.

[133] Sitzungsber. Akad. Wiss. Wien (Math.), 97, IIa, 1888, 271–6.

[134] Zeitschrift Math. Phys., 34, 1889, 238–248. Errors noted by a "reader," 37, 1892, 57, and Rothholz.[140]

[135] Giornale di Mat., 27, 1889, 371–380.

[136] Ibid., 28, 1890, 52.

A. Rieke[137] again attempted to prove $x^p + y^p = z^p$ impossible, but again confused (pp. 251–2) algebraic and arithmetical divisibility, even for $p = 3$ (p. 253).

E. Lucas[138] proved (p. 267, p. 275) the theorem of Cauchy,[29] and (p. 370–1) the formulas (1), (3), (4) of Legendre[17], with the aim to show that, when x, y, z are relatively prime in pairs, no one of them is a prime or a power of a prime [cf. Markoff[157]]. He proved (p. 341) the first result due to Jaquemet.[3]

D. Mirimanoff[139] found in terms of the units a necessary and sufficient condition that the second factor [Kummer[61]] of the class number be divisible by λ. He treated in detail the case $\lambda = 37$.

J. Rothholz[140] used the theorem of Kummer[25] on the divisors of $a^n \pm b^n$ to show (?) that $x^{2n} \pm y^{2n} = z^{2n}$ has no integral solutions if n is a prime $4k + 3$ or if one of the numbers x, y, z is a prime and n is an odd prime; $x^n + y^n = z^n$ is impossible if x, y or z is a power of a prime, the prime not being $\equiv 1$ (mod n), while n is an odd prime; $x^n + y^n = (2p)^n$ is impossible if n and p are odd primes; $x^n \pm y^n = z^n$ is impossible if x, y or z has one of the values $1, \cdots, 202$. The history of the theorem is discussed at length. On p. 29 are pointed out two errors in the proof by Rieke.[134]

*W. L. A. Tafelmacher[141] proved Abel's formulas and congruencial corollaries from them. In the second paper he proved that Fermat's equation is impossible for $n = 3$, 5, 11, 17, 23, 29 and, in case $x + y - z \equiv 0$ (mod n^4) for $n = 7$, 13, 19, 31 [but with proofs valid only when no one of x, y, z is divisible by n, since the argument pp. 273–8 does not suffice to exclude the case in which one of these numbers is divisible by n].

H. Teege[142] proved that $x^5 + y^5 = 1$ has no rational solutions by setting $x + y = p/q$, $x/y = t$, $t + 1/t = z$, $(q/p)^5 = s$. Then

$$x^4 - x^3 y + \cdots + y^4 = s(x + y)^4, \qquad (s-1)z^2 + (4s+1)z + 4s + 1 = 0.$$

Since z is rational, $(4s+1)^2 - 4(s-1)(4s+1) = m^2$. Set $m = 5\mu$. Then $4s + 1 = 5\mu^2$. Let $\mu = b/a$, where a and b are relatively prime. Thus

$$4q^5 + p^5 = 5p^5 b^2 / a^2.$$

Hence a^2 divides $5p^5$. The impossibility of this equation is proved by considering the cases a divisible or not divisible by 5.

H. W. Curjel[143] proved that if $x^s - y^t = 1$ and x, y are primes, then z is a prime, t is a power of 2, and x or y equals 2.

Several[144] proved by use of cube roots of unity the known result that, if n is odd and not a multiple of 3, $(x+y)^n - x^n - y^n$ is divisible by $x^2 + xy + y^2$.

S. Levänen[145] discussed $x^5 + y^5 = 2^m z^5$, for x, y, z without common factor,

<hr>

[137] Zeitschr. Math. Phys., 36, 1891, 249–254. Error indicated in 37, 1892, 57, 64.

[138] Théorie des nombres, 1891. References in Introduction, p. xxix, where it is stated falsely that Kummer proved Fermat's theorem for all even exponents.

[139] Jour. für Math., 109, 1892, 82–88.

[140] Beiträge zum Fermatschen Lehrsatz. Diss. (Giessen), Berlin, 1892.

[141] Anales de la Universidad de Chile, Santiago, 82, 1892, 271–300, 415–37. Report from Lind,[241] p. 50.

[142] Zeitschr. Math. Naturw. Unterricht, 24, 1893, 272–3.

[143] Math. Quest. Educ. Times, 58, 1893, 25 (quest. by J. J. Sylvester).

[144] Ibid., 112.

[145] Öfversigt af Finska Vetenskaps-Soc. Förhandlingar, Helsingfors, 35, 1892–3, 69–78.

and m not divisible by 5 (since $x^5+y^5=z_1^5$ is impossible by Legendre[18]). By the residues of z^5, x^5+y^5 modulo 25, we see that m is not in the set $2, 4, 7, 9, 12, \cdots, 2n+[(n-1)/2]$. For z divisible by 5, we have $z=5tr$, $x+y=2^m5^4t^5$. Proceeding as did Legendre, we find that the equation is impossible.

D. Mirimanoff[146] proved by use of ideals that $x^{37}+y^{37}+z^{37}=0$ is impossible in integers.

H. Dutordoir[147] expressed his belief that $a^n+b^n=c^n$ is impossible in integers if n is a rational number other than 1 and 2. The fact that it is impossible when $n=1/2$ and one of a, b, c is not a perfect square is a case of the impossibility of

$$\sqrt{a}+\sqrt{b}=\sqrt{c}+\sqrt{d},$$

when c is different from a and b, and one of the four numbers $a, \cdots, d$ is not a square (Euclid, Elements, X, 42).

A. S. Bang[148] pointed out errors in various elementary proofs of special cases of Fermat's last theorem.

G. Korneck[149] claimed to prove Fermat's last theorem by means of the Lemma: If n and k are relatively prime (n odd) and divisible by no square >1, then in every solution in integers of $nx^2+ky^2=z^n$, x is divisible by n. E. Picard and H. Poincaré[150] pointed out the falsity of this Lemma by citing the examples $n=3$, $k=1$, $x=y=z=4$, and $n=5$, $k=3$, $x=1$, $y=3$, $z=2$. The Jahrbuch Fortschritte der Math., 25, 1893, 296, pointed out that § 3 of Korneck's paper shows a lack of knowledge of the nature of algebraic numbers.

Malvy[151] noted that, if a is a primitive root of a prime $p=2^n+1$, and if in $a^{2\mu+1}+1\equiv a^h$ (mod p) we give to μ the values $1, 2, \cdots, 2^{n-1}$, we obtain for h as many even as odd values. If in $a^{4\mu+2}+1\equiv a^h$ we give to μ the values $1, \cdots, 2^{n-2}$, we obtain α even and β odd values for h, while if $p=17$, $a=3$ or $p=257$, $a=5$, we have $\alpha=\beta$.

E. Wendt[152] proved that if n and $p=mn+1$ are odd primes,

$$r^n+s^n+t^n\equiv 0 \ (\text{mod } p)$$

has only solutions in which r, s or t is divisible by p if and only if p is not a divisor of

$$D_m=\begin{vmatrix} 1 & \binom{m}{1} & \binom{m}{2} & \cdots & \binom{m}{m-1} \\ \binom{m}{m-1} & 1 & \binom{m}{1} & \cdots & \binom{m}{m-2} \\ \cdot & \cdot & \cdot & \cdots & \cdot \\ \binom{m}{1} & \binom{m}{2} & \binom{m}{3} & \cdots & 1 \end{vmatrix},$$

[146] Jour. für Math., 111, 1893, 26–30.
[147] Ann. Soc. Sc. Bruxelles, 17, I, 1893, 81. Cf. Maillet.[285]
[148] Nyt Tidsskrift for Math., 4, 1893, 105–7.
[149] Archiv Math. Phys., (2), 13, 1894 (1895); 1–9. He noted, pp. 263–7, that the Lemma fails for $n=3$, $k=1$, and so gave a separate proof of the impossibility of $x^3+y^3=z^3$.
[150] Comptes Rendus Paris, 118, 1894, 841.
[151] L'intermédiaire des math., 1, 1894, 152; 7, 1900, 193 (repeated).
[152] Jour. für Math., 113, 1894, 335–347.

which is the resultant of $x^m=1$, $(x+1)^m=1$.   For, if we multiply the congruence by ω^n, where $\omega t \equiv 1$, we obtain a congruence of the form $x+1 \equiv y$ (mod p), where x and y are nth powers, so that their mth powers are congruent to unity.

He proved Legendre's[17] result concerning the cases $m=2, 4, 8, 16$.   If $m=2^\nu n^k$ can be chosen so that $mn+1$ is a prime not dividing D_m, where ν is not divisible by the prime n, then $a^n=b^n+c^n$ $(n>2)$ is not solvable in integers all prime to n.   If $mn+1$ is a prime dividing neither D_m nor n^m-1, the same conclusion holds.   [This result differs only in form from that by Sophie Germain[17]].

D. Hilbert[153] gave a simplification of Kummer's[63] proof of Fermat's theorem for regular prime exponents, and a proof that $\alpha^4+\beta^4=\gamma^2$ is impossible in complex integers $a+bi$.

G. B. Mathews[154] noted that, if p is an odd prime, and x, y, z are solutions of $x^p+y^p+z^p=0$, it is possible to choose k in an infinitude of ways such that $kp+1=q$ is a prime not a factor of x, y, z, or y^p-z^p, etc., and such that k is not divisible by 3.   Then, since x^p, y^p, z^p are distinct roots of $t^k \equiv 1$ (mod q), their sum is divisible by q.   Let $r=e^{2\pi i/k}$ and $P_k=\Pi(r^\alpha+r^\beta+r^\gamma)$, where the product extends over all triples of roots $r^\alpha, r^\beta, r^\gamma$ of $x^k=1$.   Then $P_k=\pm u_k^k$, where u_k is a positive integer.   Thus $u_k \equiv 0$ (mod q) if and only if three roots of $x^k \equiv 1$ (mod q) have a sum divisible by q.   Hence if it could be proved that for a given p there is an infinitude of primes $kp+1$ for which $u_k \equiv 0$ (mod q) is not satisfied, Fermat's theorem would follow [Libri[24]].

E. de Jonquières[155] noted that, if $n>2$, it is not possible to express c and b as algebraic functions of p, q such that c^n-b^n becomes $(pq)^n$ identically, and stated that this does not imply the impossibility of integral solutions.

G. Speckmann[156] discussed $T^x-DU^x=m^x$.

V. Markoff[157] noted that Lucas'[138] proof of Abel's[16] theorem that $a^n=b^n+c^n$ (n an odd prime) is impossible when a, b or c is a prime is incomplete as the case $a=b+1$ is not treated.   He asked if $(x+1)^n=x^n+y^n$ is impossible.

P. Worms de Romilly[158] stated that $a^p+b^p=c^p$, p a prime >2, implies

$$c=x+y+z, \qquad b=x+z, \qquad a=x+y,$$
$$x=\tfrac{1}{2}M(P+Q)p^{v+1}q^{u+1}, \qquad y=P=p^{p(v+1)-1}, \qquad z=Q=q^{p(u+1)},$$
$$Mp^{v+1}q^{u+1}=2^{\mu\theta}a^\theta-1, \qquad 2^{\mu\alpha}a^\alpha=P+Q,$$

p and q odd and relatively prime, $q>1$, and $u, v, \theta, \mu, \alpha$ integers $\geqq 0$.   [Since $c-b=y$ is a power of p, Fermat's equation is impossible by Abel's[16] result.]

[153] Jahresbericht d. Deutschen Math.-Vereinigung, 4, 1894-5, 517-25.   French transl., Annales Fac. Sc. Toulouse, [(3), 1, 1909;] (3), 2, 1910, 448; (3), 3, 1911, for errata, table of contents, and notes by Th. Got on the literature concerning Fermat's last theorem.
[154] Messenger Math., 24, 1894-5, 97-99.   Reprinted, Oeuvres de Fermat, IV, 159-61.
[155] Comptes Rendus Paris, 120, 1895, 1139-43 (minor error, 1236).
[156] Ueber unbestimmte Gleichungen, 1895.
[157] L'intermédiaire des math., 2, 1895, 23; repeated, 8, 1901, 305-6.
[158] Ibid., 2, 1895, 281-2; repeated, 11, 1904, 185-6.

If m is a prime $6k+1$, $(\alpha+1)^{m-1}\equiv 1$, $\alpha^{m-1}\equiv 1 \pmod{m^2}$ do not hold simultaneously. If m is a prime, the integers u, not divisible by m, which satisfy

$$(u^m+1)^m - u^{m^2} \equiv 1 \pmod{m^2}$$

are of the form $u=am-1$.

P. F. Teilhet[159] found A for which $x^n - Ay^n = 1$ by taking $x=y^n+1$, or, when n is even, $x=y^n-1$. H. Brocard (pp. 116–7) found special solutions when $n=3$, $n=5$. T. Pepin (pp. 281–3) noted that we may apply to $x^n - Ay^n$ the method of Lagrange in his Additions to Euler's Algebra to find the minima of any homogeneous polynomial in x, y.

W. L. A. Tafelmacher[160] treated $x^3+y^3=z^2$ and proved $x^6+y^6=z^6$ to be impossible.

H. Tarry[161] mentioned a mechanical device of double-entry tables for solving indeterminate equations, in particular, $x^m+y^m=z^n$.

F. Lucas[162] used Cauchy's[29] theorem to prove that, if x, y are relatively prime and m is an odd prime, when $x+y$ is prime to m it is prime to

$$Q = (x^m+y^m)/(x+y),$$

but when $x+y$ is divisible by m, $m(x+y)$ is prime to Q/m. From this he deduced Legendre's formulas (1) and (3).

Axel Thue[163] noted that, if L, M, N are functions of x such that $L^n - M^n = N^n$ for all values of x, where $n>2$, then $aL=bM=cN$, where a, b, c are constants. If $A^n - B^n = C^n$, then

$$(A^n+\alpha B^n)^3 - (\alpha A^n+B^n)^3 = (\alpha-1)^3(ABC)^n, \qquad \alpha^3=1.$$

If $p^n - q^n = r^n$, then $x^3 - y^3 = z^3(pqr)^n$ for

$$x = p^{3n}+3p^{2n}q^n-6p^nq^{2n}+q^{3n},$$
$$y = p^{3n}-6p^{2n}q^n+3p^nq^{2n}+q^{3n}, \qquad z=3(p^{2n}-p^nq^n+q^{2n}).$$

E. Maillet[164] considered, for a, b, c, x, y, z integers not divisible by the odd prime λ, the equation

$$ax^{\lambda^t}+by^{\lambda^t}=cz^{\lambda^t}.$$

A necessary condition for solutions is that the congruence

$$a+b\eta^{\lambda^t}\equiv c(\alpha+\beta\eta)^{\lambda^t} \pmod{\lambda^{t+1}}$$

have a solution η such that $0<\eta<\lambda$, $\alpha+\beta\eta\not\equiv 0 \pmod{\lambda}$, where $\alpha c\equiv a$, $\beta c\equiv b \pmod{\lambda}$. This is applied to show that $x^\lambda+y^\lambda=z^\lambda$ is impossible for $\lambda=197$, hence extending Legendre's limit to $\lambda<223$. By the method of Kummer it is shown that, if λ is a prime >3,

$$x^{\lambda^t}+y^{\lambda^t}+z^{\lambda^t}=0$$

is impossible in complex integers, formed from a λth root of unity, relatively prime by twos and prime to λ, if λ^{t-1} is the highest power of λ dividing the

[159] L'intermédiaire des math., 3, 1896, 116.
[160] Anales de la Universidad de Chile, 97, 1897, 63–80.
[161] Assoc. franç. av. sc., 26, 1897, I, 177 (five lines).
[162] Bull. Soc. Math. France, 25, 1897, 33–35. Extract in Sphinx-Oedipe, 4, 1909, 190.
[163] Archiv for Math. og Natur., Kristiania, 19, 1897, No. 4, pp. 9–15.
[164] Assoc. franç. av. sc., 26, 1897, II, 156–168.

number of classes of these complex integers, and hence for a value of t exceeding a certain limit depending on λ. He[165] later proposed the problem that the last theorem be proved without the restriction that x, y, z are prime to λ.

I. P. Gram's[166] paper was not available for report.

E. Maillet[167] applied Kummer's methods to $x^\lambda + y^\lambda = cz^\lambda$, where λ is a regular prime. The equation is impossible in integers if $c = \lambda$. It is impossible in real relatively prime integers not divisible by λ if $c = A\lambda^s$, $s = k\lambda + \beta \geqq 1$, $\beta = 0$ or 1, when $A = 1$ or $r_1^{b_1} \cdots r_i^{b_i}$, where $r_1, \cdots, r_i$ are distinct primes $\neq \lambda$, belonging to exponents $f_1, \cdots, f_i$ modulo λ such that

$$\frac{1}{f_1} + \cdots + \frac{1}{f_i} \leqq \frac{\lambda - 3}{\lambda - 1};$$

in particular, if $A = r_1^{b_1}$, $r_1 \not\equiv 1 \pmod{\lambda}$. For r a prime and $b < \lambda$, the equation with $c = r^b$ is impossible in real integers if $r^b \equiv -1 + t\lambda \pmod{\lambda^2}$, where t has at least one of the values $1, \cdots, \lambda - 1$; or if $\lambda = 5, 7, 17$, $r^b \equiv 4 \pmod{\lambda^2}$; or if $\lambda = 11$, $r^b \equiv 5$ or $47 \pmod{11^2}$; or if $\lambda = 13$, $r^b \equiv 17 \pmod{13^2}$. Finally, $x^7 + y^7 = cz^7$ is impossible in real integers for c a prime of one of the forms $49k \pm 3, \pm 4, \pm 5, 6, -8, \pm 9, \pm 10, -15, \pm 16, -22, \pm 23$ or ± 24.

H. J. Woodall[168] noted that $x^m + y^m - 1$ is divisible by xy if $y = x^m - 1$ (m even) or if $x = 2$, $y = 2^m - 1$ (m odd).

T. R. Bendz[169] stated that $x^n + y^n = z^n$ has integral solutions if and only if $\alpha^2 = 4\beta^n + 1$ has rational solutions [Euler[8]], as follows from

$$\left(\frac{2y^n + x^n}{x^n}\right)^2 = 4\left(\frac{yz}{x^2}\right)^n + 1.$$

He proved Abel's[16] formulas, also $x + y \equiv z \pmod 3$ and (p. 30)

$$(x+y)^n - x^n - y^n \equiv 0 \pmod{n^3},$$

when no one of x, y, z is divisible by n.

F. Lindemann[170] attempted to prove that $x^n = y^n + z^n$ is impossible if n is an odd prime. He later (p. 495) recognized the error in the computation, but stated that his work gives the first proof of Abel's[16] statement that if x, y, z are $\neq 0$ and relatively prime in pairs

$$2x = p^n + q^n + r^n, \qquad 2y = p^n + q^n - r^n, \qquad 2z = p^n - q^n + r^n$$

if no one of x, y, z is divisible by n, while, if z is divisible by n,

$$2x, \ 2y = p^n + q^n \pm n^{n-1}r^n, \qquad 2z = p^n - q^n + n^{n-1}r^n.$$

If $x + y + z$ is divisible by n^λ, then, in (2), $\alpha \equiv \beta \equiv \gamma \equiv 1 \pmod{n^{\lambda - 1}}$.

D. Gambioli[171] proved de Jonquières'[117] theorems, and the fact that in $x^n + y^n = z^n$ ($n > 1$), z is composite if n has an odd factor, or if x and y are

<hr>

[165] Congrès internat. des math., 1900, Paris, 1902, 426–7.
[166] Förhandlingar Skandinaviska Naturforskare, Götheborg, 1898, 182.
[167] Comptes Rendus Paris, 129, 1899, 198–9. Proofs in Acta Math., 24, 1901, 247–256.
[168] Math. Quest. Educ. Times, 73, 1900, 67.
[169] Öfver diophantiska ekvationen $x^n + y^n = z^n$, Diss., Upsala, 1901, 34 pp.
[170] Sitzungsber. Akad. Wiss. München (Math.), 31, 1901, 185–202.
[171] Periodico di Mat., 16, 1901, 145–192.

composite; but erred in his proof that the least unknown is composite. He gave abstracts of the papers by Calzolari,[72] Dirichlet,[23] Kummer,[63] and Legendre,[17] a list (191–2) of references on Bernoullian numbers and ideal complex numbers, and (189–191) a short proof of the impossibility of $x^5+y^5=z^5$. In an appendix (*ibid.*, 17, 1902, 48–50) he quoted Kummer[49] and Liouville[46] on the insufficiency of the proofs by Lamé,[45] and Cauchy.[54–56]

Soons[172] proved theorems stated by Catalan.[90]

P. Stäckel[173] proved Abel's theorem as given by Lindemann.[170]

G. Candido[174] proved a theorem of Catalan.[121]

* D. Gambioli's[175] paper was not available for report.

P. Whitworth[176] noted that if $\Sigma 1/x=0$, $\Sigma x=1$, then $\Sigma x^n = x^n+y^n+z^n$ equals a series in xyz.

P. V. Velmine[177] (W. P. Welmin) proved that, if m, n, k are integers >1, there exist rational integral functions u, v, w of a variable which satisfy $u^m+v^n=w^k$ only for the cases $u^m\pm v^2=w^2$, $u^3+v^3=w^2$, $\pm u^4+v^3=w^2$ (when the solution is easy), and $u^5+v^3=w^2$, the complicated formulas for whose solution are not proved to give all solutions. Cf. Korselt.[282]

D. Mirimanoff[178] studied $P(x)=(x+1)^l-x^l-1$ where l is a prime >3. Since it is unaltered when x is replaced by $-1-x$, a root α of $P(x)=0$ implies the roots

$$(9) \qquad \alpha,\ 1/\alpha,\ -1-\alpha,\ -1/(1+\alpha),\ -1-1/\alpha,\ -\alpha/(1+\alpha),$$

all of which are distinct unless $\alpha=0$ or -1 or $\alpha^2+\alpha+1=0$. Now P has the factors $x(x+1)$ and x^2+x+1. Set

$$E(x)=\frac{P(x)}{lx(x+1)(x^2+x+1)^\epsilon},$$

where $\epsilon=1$ if $l\not\equiv 1\ (\mathrm{mod}\ 3)$, $\epsilon=2$ if $l\equiv 1\ (\mathrm{mod}\ 3)$. Then $E(x)=0$ has only distinct imaginary roots which fall into sets of six. Thus $E(x)=\Pi e_j(x)$, where each $e_j(x)$ is of the form $x^6+1+3(x^5+x)+t(x^4+x^2)+(2t-5)x^3$, where t is real. If $E(x)$ has a factor which is irreducible in the domain of rational numbers, the factor is a product of certain of the $e_j(x)$.

A. S. Werebrusow[179] denoted u^2+uv-v^2 by $(u,\ v)$. Then $x^5+y^5=Az^5$ becomes

$$(x+y)(x^2-xy+y^2,\ x^2-2xy+y^2)=Az^5.$$

This decomposes into two equations, one being the second factor equated to $A_1z_1^5$, the other being $x+y=A_0z_0^5$, where $A_0A_1=A$, $z_0z_1=z$, and z_1 is a product of primes $5n+1$. Multiplying $(u,\ v)$ by $1=9^2-5\cdot4^2$ and its powers,

---

[172] Mathesis, (3), 2, 1902, 109.

[173] Acta Math., 27, 1903, 125–8.

[174] La formula di Waring e sue notevoli applicazioni, Lecce, 1903, 20.

[175] Il Pitagora, 10, 1903–4, 11–13, 41–43.

[176] Math. Quest. Educ. Times, (2), 4, 1903, 43.

[177] Mat. Sbornik (Math. Soc. Moscow), 24, 1903–4, 633–61, in answer to problem proposed by V. P. Ermakov, 20, 1898, 293–8. Cf. Jahrbuch Fortschritte Math., 29, 1898, 139; 35, 1904, 217.

[178] Nouv. Ann. Math., (4), 3, 1903, 385–97.

[179] L'intermédiaire des math., 11, 1904, 95–96; Math. Soc. Moscow (Mat. Sbornik), 25, 1905, 466–473 (Russian). Cf. Jahrbuch Fortschritte Math., 36, 1905, 277–8.

we conclude that for each power we get six representations of a prime by (u, v); but only three representations of 5. A composite number has 2^p representations if p is the number of its distinct prime factors $5n\pm1$.

Take $z_1 = (a, b)$. We get u, v such that $z_1^5 = (u, v)$ by using
$$(a, b)(\sigma, \tau) = (a\sigma+b\tau, b\sigma+a\tau+b\tau).$$

Then

(10) $\qquad (x-y)^2 = vs+(u+v)t, \qquad (x+y)^2 = (4u-v)s+(v-3u)t.$

The product of the square root of the last sum by (s, t) gives Az_0^5, so that we have the general form of A. Taking $x+y$ arbitrary, we get $x-y$ and then s, t by (10).

Mirimanoff[180] considered

(11) $\qquad\qquad\qquad x^\lambda+y^\lambda+z^\lambda=0$

for the case in which no one of the integral solutions x, y, z is divisible by the odd prime λ. By use of Kummer's congruences (8), he proved that (11) is impossible in integers prime to λ if at least one of the Bernoullian numbers* $B_{\nu-1}$, $B_{\nu-2}$, $B_{\nu-3}$, $B_{\nu-4}$ is not divisible by λ, where
$$\nu = (\lambda-1)/2;$$
also, for every $\lambda < 257$. In terms of Kummer's $P_i(t) = P_i(1, t)$, he defined the polynomials

(12) $\qquad \phi_i(t) = (1+t)^{\lambda-i}P_i(t) \equiv \sum_{k=1}^{\lambda-1} (-1)^{k-1}k^{i-1}t^k \quad (i=2, 3, \cdots, \lambda-1)$

modulo λ. Thus Kummer's criterion (8) is equivalent to the following. If (11) has solutions prime to λ, each of the six ratios $t = x/y, \cdots, z/x$ satisfies the congruences

(13) $\qquad \phi_{\lambda-1}(t) \equiv 0, \qquad B_{(\lambda-i)/2}\phi_i(t) \equiv 0 \pmod{\lambda} \quad (i=3, 5, \cdots, \lambda-2).$

An equivalent criterion not involving Bernoullian numbers is that each of the six ratios satisfies the congruences

(14) $\qquad \phi_{\lambda-1}(t) \equiv 0, \qquad \phi_{\lambda-i}(t)\phi_i(t) \equiv 0 \pmod{\lambda} \quad (i=2, 3, \cdots, \nu).$

E. Maillet[181] proved by Kummer's methods that $x^a+y^a=az^a$ $(a>2)$ has no real integral solutions $\neq 0$ if a is divisible by 4; or if a is even and divisible by a prime $4n+3$; or if $2<a\leqq100$, $a\neq37$, 59, 67, 74; or if a has no prime factor >17. Likewise for $x^a+y^a=baz^a$ if a is divisible by 4 and b is not; or if a is of the form $4n+2$ and has a prime factor $\lambda=4h+3$ such that b is not divisible by $\lambda^{\lambda-1}$; or if $a=p^i$, $b<p$, p being a prime $\geqq5$ not exceptional in the sense of Kummer; or if $a=3^i$, $b=2$ or 4, $i\geqq2$. Probably the second equation is impossible in integers $\neq0$ if $b=1$ or 2, $a>2$ or $a>3$, respectively.

R. Sauer[182] proved that $x^n=y^n+z^n$, $n>2$, does not hold if x or y or z is a power of a prime.

U. Bini[183] noted that, if $x+y+z=0$ and $k=2m+1$, $s=x^k+y^k+z^k$ is divisible by xyz. If $1/x+1/y+1/z=0$ and $k=3h+2$, s is divisible by

[180] Jour. für Math., 128, 1905, 45–68.
* If $B_{\nu-1}$ or $B_{\nu-2}$ is not divisible by λ, the conclusion was drawn by Kummer.[76]
[181] Annali di mat., (3), 12, 1906, 145–178. Abstracts in Comptes Rendus Paris, 140, 1905, 1229; Mém. Acad. Sc. Inscr. Toulouse, (10), 5, 1905, 132–3.
[182] Eine polynomische Verallgemeinerung des Fermatschen Satzes, Diss., Giessen, 1905.
[183] Periodico di Mat., 22, 1906–7, 180–3.

$x+y+z$, and $x^n y^n + x^n z^n + y^n z^n$ is divisible by $(xyz)^3$ if $n \geqq 5$. Proofs[184] have been given of the first result and the fact that, if $x+y+z=0$, s is a function of xyz and $xy+xz+yz$.

* G. Cornacchia[185] treated the congruence $x^n + y^n \equiv z^n \pmod{p}$.

P. A. MacMahon[186] noted that the integral solutions of $x^n - ay^n = z$ may be obtained by the development of $a^{1/n}$ into a continued fraction.

F. Lindemann[187] again[170] proved Abel's formulas and, after treating at great length each of the three cases, concluded that Fermat's equation is impossible in integers. A. Fleck[188] pointed out a serious error and various minor errors. I. I. Iwanov[189] noted errors, also in Lindemann's[170] first proof, where in (67) the modulus n^6 should be n^5.

A. Bottari[190] proved that if x, y, z are positive integers in arithmetical progression such that $x^n + y^n = z^n$, then either $n=1$ and $x = y/2 = z/3$ or $n=2$ and $x/3 = y/4 = z/5$. If x, y, z, t are positive integers in arithmetical progression such that $x^n + y^n + z^n = t^n$, then $n=3$, $x/3 = y/4 = z/5 = t/6$. Cf. Cattaneo.[192]

J. Sommer[191] omitted the restriction that n is a regular prime in stating that Kummer proved that $x^n + y^n = z^n$, for $n > 2$, is not solvable in complex integers based on an nth root of unity. He gave the proof for $n=3$ and $n=4$.

P. Cattaneo[192] gave a brief proof of the results of Bottari,[190] but included the false solution $n=1$, $x = y/2 = z/3 = t/4$.

A. S. Werebrusow[193] failed in his proof of Fermat's last theorem, the error being indicated by L. E. Dickson and others (*ibid.*, pp. 174–7).

Werebrusow[194] stated that $(x+y+z)^n - x^n - y^n - z^n$ has, for n odd, the factor $n(x+y)(x+z)(y+z)$. While this is true for n an odd prime, it fails for $n=9$, $x=y=z=1$ (*ibid.*, 16, 1909, 79–80).

L. E. Dickson[195] noted that, if α is a common root of the congruences

$$(15) \qquad\qquad z^m \equiv 1, \qquad (z+1)^m \equiv 1 \quad \pmod{p}$$

of Wendt,[152] the numbers (9) are common roots and are distinct if $2^m - 1$ is not divisible by p. They are the roots of a sextic in z which is unaltered when z is replaced by $1/z$ or by $-1-z$. The sextic must divide $z^m - 1$ modulo p. Set $x = z + 1/z$, $m = 2\mu$. The sextic becomes

$$C(x) = x^3 + 3x^2 + \beta x + 2\beta - 5.$$

From $z^\mu - 1/z^\mu = 0$ we get $f(x^2) = 0$, where $f(\omega)$ is of degree $\frac{1}{2}\mu - 1$ or $(\mu-1)/2$

---

[184] L'intermédiaire des math., 13, 1906, 142; 14, 1907, 22–23, 36–39, 92–95, 258.

[185] Sulla Congruenza $x^n + y^n = z^n \pmod{p}$, Tempio (Tortu), 1907, 18 pp.

[186] Proc. London Math. Soc., (2), 5, 1907, 45–58. For $z = \pm 1$, G. Cornacchia, Rivista di fisica, mat. sc. nat., Pavia, 8, II, 1907, 221–230.

[187] Sitzungsber. Akad. Wiss. München (Math.), 37, 1907, 287–352.

[188] Archiv Math. Phys., (3), 15, 1909, 108–111.

[189] Kagans Bote, 1910, No. 507, 69–70.

[190] Periodico di Mat., 22, 1907, 156–168.

[191] Vorlesungen über Zahlentheorie, 1907, 184. Revised French ed. by A. Lévy, 1911, 192.

[192] Periodico di Mat., 23, 1908, 219–20.

[193] L'intermédiaire des math., 15, 1908, 79–81.

[194] *Ibid.*, p. 125. Case $n=3$, in l'éducation math., 1889, p. 16.

[195] Messenger of Math., (2), 38, 1908, 14–32.

according as μ is even or odd.   Thus $f(x^2)$ must be divisible by

$$S(x) = C(x)C(-x) = x^6 + (2\beta - 9)x^4 + (\beta^2 - 12\beta + 30)x^2 - (2\beta - 5)^2.$$

Hence $\mu \geqq 7$.   For $\mu = 7$, $f(x^2) = x^6 - 5x^4 + 6x^2 - 1$ must be congruent to $S(x)$, whence $p = 2$.   For $\mu = 8$, $f(x^2) = x^6 - 6x^4 + 10x^2 - 4$, whence $p = 17$, contrary to $n > 1$.   The cases $\mu = 10, 11, 13$ are readily treated.   The conclusion is that, if n and $p = mn + 1$ are odd primes, m being prime to 3 and $m \leqq 26$, the congruence $\xi^n + \eta^n + \zeta^n \equiv 0 \pmod{p}$ has no integral solutions each prime to p, except for $n = 3$, $m = 10, 14, 20, 22, 26$; $n = 5$, $m = 26$; $n = 31$, $m = 22$.   A direct examination of (15) was made for $m = 28, 32, 40, 56, 64$. By use of these results and the theorem of S. Germain,[17] it was shown that Fermat's equation is impossible in integers prime to n for every odd prime exponent $n < 1700$.

Dickson[196] proved the last theorem for $n < 7000$ by extending the range of the m's to include all values < 74, as well as 76 and 128.

Dickson[197] factored certain numbers $m^m - 1$ for use in the last paper.

Dickson[198] discussed the following problem: Given an odd prime n, to find the odd prime moduli p for which $x^n + y^n + z^n \equiv 0 \pmod{p}$ has no solutions each prime to p.   We may take $p = mn + 1$, where m is not divisible by 3, since otherwise such solutions are evident.   The general results are applied to the cases $n = 3, 5, 7$.   For $n = 3$, the only values of p are 7 and 13 [cf. Pepin[109]].   For $n = 5$, $p = 11, 41, 71, 101$ [verified up to 1000 by Legendre[17]].   For $n = 7$, $p = 29, 71, 113, 491$.

Dickson[199] proved, by use of Jacobi's functions of roots of unity, that if e and p are odd primes such that

$$p \geqq (e-1)^2(e-2)^2 + 6e - 2,$$

then $x^e + y^e + z^e \equiv 0 \pmod{p}$ has integral solutions x, y, z, each prime to p. In particular this establishes the conjecture by Libri.[24]   Also, $x^4 + y^4 \equiv z^4$ (mod p) has solutions prime to p for every prime $p = 4f + 1$ exceeding 17 [and different[200] from 41].

P. Wolfskehl[201] bequeathed to the K. Gesellschaft der Wissenschaften zu Göttingen one hundred thousand marks to be offered as a prize for a complete proof of Fermat's last theorem.   It may be noted that Wolfskehl[202] was the author of a paper on the related subject of the class number for complex numbers formed of eleventh or thirteenth roots of unity.

[196] Quar. Jour. Math., 40, 1908, 27–45.   The omitted value $n = 6857$ was later shown in MS. to be excluded.

[197] Amer. Math. Monthly, 15, 1908, 217–222.   See p. 370 of Vol. I of this History; also, A. Cunningham, Messenger of Math., 45, 1915, 49–75.

[198] Jour. für Math., 135, 1909, 134–141.

[199] Ibid., 135, 1909, 181–8.   Cf. Pellet,[123, 244] Hurwitz,[213] Cornacchia,[217] and Schur.[283]

[200] On p. 188, line 11, it is stated that for f even and < 14, $p = 4f + 1$ is a prime only when $f = 4$, $p = 17$, thus overlooking $f = 10$, $p = 41$.   The fact that $x^4 + y^4 \equiv 1 \pmod{41}$ has no solutions each prime to 41 was communicated to the author by A. L. Dixon.

[201] Göttingen Nachrichten, 1908, Geschäftliche Mitt., 103.   Cf. Jahresbericht d. Deutschen Math.-Vereinigung, 17, 1908, Mitteilungen u. Nachrichten, 111–3.   Fermat's Oeuvres, IV, 166.   Math. Annalen, 66, 1909, 143.

[202] Jour. für Math., 99, 1886, 173–8.

No mention will be made here of numerous[203] recent false proofs[204] of Fermat's last theorem, published mostly as pamphlets. Errors in some of these have been noted by A. Fleck,[205] B. Lind[241] (p. 48), J. Neuberg,[206] and D. Mirimanoff.[207]

E. Schönbaum[208] gave a historical introduction to and exposition of the elements of the theory of algebraic numbers; also Kummer's proof, in simplified form, of Fermat's last theorem for the case of regular primes.

* A. Turtschaninov[209] proved and slightly generalized Abel's[16] theorem

* F. Ferrari[210] discussed the infinitude of solutions of each of

$$x^n \pm y^n = z^{n+1}, \qquad x^{2n+1} \pm y^{2n+1} = z^{2n}.$$

A. Thue[211] stated that there are no [not an infinite number of] integral solutions of any of the equations, with $n > 2$, h and k given positive integers,

$$x^n + (x+k)^n = y^n, \qquad x^2 - h^2 = ky^n, \qquad (x+h)^3 + x^3 = ky^n, \qquad (x+h)^4 - x^4 = ky^n.$$

These results are consequences of the theorem (pp. 27–30) that, if $r > 2$ and a, b, c are any positive integers, $c \neq 0$, there is not an infinitude of pairs of positive integral solutions p, q of $bp^r - aq^r = c$.

A. Hurwitz[212] proved that, if m and n are positive integers not both even, $x^m y^n + y^m z^n + z^m x^n = 0$ has integral solutions $\neq 0$ if and only if $u^t + v^t + w^t = 0$ has such solutions, where $t = m^2 - mn + n^2$. Cf. Bouniakowsky,[149] Ch. XXIII.

Hurwitz,[213] after citing Dickson's[199] proof by cyclotomy, gave an elementary, but long, proof that, if a, b, c are integers $\neq 0$ and e is an odd prime,

$$ax^e + by^e + cz^e \equiv 0 \pmod{p}$$

has A sets of solutions x, y, z each not divisible by the prime p, where

$$\frac{A}{p-1} > p + 1 - (e-1)(e-2)\sqrt{p} - \eta^e \qquad (\eta = 0, 1 \text{ or } 3).$$

Hence $A > 0$ when p exceeds a limit depending on e.

A. Wieferich[214] proved that if $x^p + y^p + z^p = 0$ is possible in integers prime to p, where p is an odd prime, then $2^{p-1} - 1$ is divisible by p^2. He deduced this criterion from the conditions (13) derived by Mirimanoff[180]

[203] According to W. Lietzmann, Der Pythagoreische Lehrsatz, mit einem Ausblick auf das Fermatsche Problem, Leipzig, 1912, 63, more than a thousand false proofs were published during the first three years after the announcement of the large prize.

[204] Titles in Jahrbuch Fortschritte Math., 39, 1908, 261–2; 40, 1909, 258–261; 41, 1910, 248–250; 42, 1911, 237–9; 43, 1912, 254, 274–7; 44, 1913, 248–50.

[205] Archiv. Math. Phys., (3), 14, 1909, 284–6, 370–3; 15, 1909, 108–111; 16, 1910, 105–9, 372–5; 17, 1911, 108–9, 370–4; 18, 1911, 105–9, 204–6; 25, 1916–7, 267–8.

[206] Mathesis, (3), 8, 1908, 243.

[207] Comptes Rendus Paris, 157, 1913, 491; error of E. Fabry, 156, 1913, 1814–6. L'enseignement math., 11, 1909, 126–9.

[208] Casopis, Prag, 37, 1908, 384–506 (Bohemian).

[209] Spaczinski Bote, 1908, No. 454, 194–200 (Russian).

[210] Suppl. al Periodico di Mat., 11, 1908, 40–2.

[211] Skrifter Videnskabs-Selskabet Christiania (Math.), 1908, No. 3, p. 33.

[212] Math. Annalen, 65, 1908, 428–30. Case $m = 2$, $n = 1$ by Euler[9] and Vandiver,[225] Ch. XXI.

[213] Jour. für Math., 136, 1909, 272–292.

[214] Ibid., 293–302. For outline of proof, see Dickson,[233] 182–3.

from Kummer's criterion. Shorter proofs have since been given by Mirimanoff[223] and Frobenius.[228]

P. Mulder[215] noted that if n is an odd prime and a^n+b^n is divisible by n, it is divisible by n^2. Proof as by Kummer.[25]

Chr. Ries[216] argued that $a^{2n}+b^{2n}=c^{2n}$ $(n>1)$ is impossible in integers by considering the two factors of a^{2n} whose difference is $2b^n$, but assumed that every prime factor of $2b^n$ divides b.

G. Cornacchia[217] employed the theory of roots of unity to investigate the number of sets of solutions of $x^n+y^n\equiv1$ (mod p), where p is a prime of the form $nk+1$. There are proper solutions for $n=3$ if $p\neq7$, 13; for $n=4$, if $p\neq5$, 13, 17, 41; for $n=6$, if $p\neq7$, 13, 19, 43, 61, 97, 157, 277; for $n=8$, if $\neq17$, 41, 113; for n any odd prime if $p>(n-2)^2n(n-1)+2(n+3)$. For p a prime $nk+1$, $x^n+y^n+z^n\equiv0$ (mod p) has proper solutions for $n=4$ if $p\neq5$, 17, 29, 41 [Gegenbauer[126]]; for $n=6$, if $p\neq13$, 61, 97, 157, 277, 31, 223, 7, 67, 79, 139; for $n=8$ if $p\neq17$, 41, 113, 89, 233, 137, 761. He proved a theorem like that of Dickson,[199] but with a limit

$$p>(e-2)^2e(e-1)+2(e+3)$$

which is larger than Dickson's if $e>3$.

A. Flechsenhaar[218] considered, for n a prime >3,

$$(16)\qquad\qquad x^n+y^n-z^n\equiv0\ (\text{mod }n^2)$$

for x, y, z prime to n. We may set $x<n$, $y<n$, $x+y=z$. Multiply (16) by ρ_1^n and ρ_2^n in turn, where $\rho_1x\equiv1$, $\rho_2y\equiv1$ (mod n). Hence the solvability of (16) implies that of

$$(17)\qquad 1+b^n-(b+1)^n\equiv0,\qquad c^n+1-(c+1)^n\equiv0\quad(\text{mod }n^2),$$

where $b\equiv\rho_2x$, $c\equiv\rho_1y$, whence $bc\equiv1$ (mod n). These conditions continue to hold after b is replaced by $b-n$, and c by $c-n$. We get

$$1+(n-t-1)^n-(n-t)^n\equiv0,\qquad t=b\text{ or }c.$$

Since these have the form of (17), it is stated that $(n-b-1)(n-c-1)\equiv1$, whence $b+c+1\equiv0$ (mod n), by a false analogy, as no proof had been given that, for every pair of solutions b, c of (17), we have $bc\equiv1$.

Admitting $b+c+1\equiv0$, $bc\equiv1$, $b\neq c$, we have $n=6m+1$. Solutions b, c then exist and are tabulated for n a prime $\leqq307$. But (p. 274) for n a prime $6m-1$, (16) has no solutions prime to n.

J. Németh[219] noted that $x^k+y^k=z^k$, $x^l+y^l=z^l$ have no common sets of positive solutions if k, l are distinct positive integers.

J. Kleiber[220] stated that if n is an odd prime, x, y, z are relatively prime, and y, z not divisible by n, $x^n+y^n=z^n$ implies that

$$x+\epsilon^iy=(p+\epsilon^iq)^n\qquad(i=0,1,\cdots,n-1;\ \epsilon^n=1),$$

which readily give $y=0$. But he had assumed that the laws of factorization of integers hold for numbers involving ϵ, had not specified the kind of

<hr>

[215] Wiskundige Opgaven, Amsterdam, 10, 1909, 273–4.

[216] Math. Naturw. Blätter, 6, 1909, 61–3.

[217] Giornale di mat., 47, 1909, 219–268. See Cornacchia[188] and the references under Libri.[34]

[218] Zeitschr. Math. Naturw. Unterricht, 40, 1909, 265–275.

[219] Math. és Phys. Lapok, Budapest, 18, 1909, 229–230 (Hungarian).

[220] Zeitsch. Math. Naturw. Unterricht, 40, 1909, 45–47.

quantity whose nth power is $x + \epsilon y$, and in giving the quantity the notation $p + \epsilon q$ had not specified the nature of p and q.

Welsch[221] repeated a proof due to Catalan.[121]

D. Mirimanoff[222] considered the relation of $F = x^l + y^l + z^l = 0$ to cubic congruences. Let x, y, z be the roots of $t^3 - s_1 t^2 + s_2 t - s_3 = 0$. Thus $F = \phi(s_1, s_2, s_3)$, where ϕ is a polynomial of degree l with integral coefficients. We have $s_1 \equiv 0 \pmod{l}$. Let x, y, z be prime to l. By Legendre,[17] $s_1^l - F$ is divisible by $l(x+y)(x+z)(y+z) = l(s_1 s_2 - s_3)$; call the quotient $P(s_1, s_2, s_3)$. Since $s_1 s_2 - s_3$ is prime to l, and since s_1^l is divisible by l^l, $F = 0$ gives $P(0, s_2, s_3) \equiv 0 \pmod{l}$. Hence if $F = 0$ has solutions prime to l,

$$t^3 + s_2 t - s_3 \equiv 0 \pmod{l},$$

subject to $P \equiv 0$, has three roots. For $l = 3$, then $P = 1$ and $F = 0$ is impossible in integers prime to $l = 3$. For $l = 5$, $P = -s_2$; but if $s_2 \equiv 0$, the discriminant of the cubic congruence is $-27 s_3^2$, a quadratic non-residue of l, so that it does not have three roots. The same argument applies to $l = 11$. For $l = 17$, the discriminant is a residue and there are three roots or no root; the first case is excluded by the fourth criterion of Cailler (*ibid.*, 10, 1908, 486; see p. 255 of Vol. I of this History) for cubic congruences. The method fails for $l = 3m + 1$, since we may now have $s_2 \equiv 0$.

Mirimanoff[223] employed Euler's expression for $1 - 2^{p-2} + 3^{p-2} - \cdots \pm y^{p-2}$ as a polynomial in y to obtain a short proof of the final congruence used by Wieferich to prove his criterion that $2^{p-1} \equiv 1 \pmod{p^2}$.

B. Lind[224] proved that $x^2 + y^3 = z^6$ is impossible in integers. If $x^n + y^n = z^n$ is impossible, so are $Z^{2n} - X^2 = 4Y^n$ and $s(2s+1) = t^{2n}$. The last equation implies $s = t_1^{2n}$, $2s + 1 = t_2^{2n}$, $t_1 t_2 = t$, whence $t_2^{2n} - 1 = 2(t_1^2)^n$, a case of Liouville's[32] equation. For a simpler proof, see Kempner.[281]

J. Westlund[225] noted that, if n is an odd prime,

$$x^n + y^n = (x+y-y)^n + y^n = (x+y)^n - n(x+y)^{n-1} y + \cdots$$

is divisible by n^2 if by n. Hence $x^n + y^n = n z^n$ is impossible if z is prime to n.

R. D. Carmichael[226] proved that, if p and q are primes, $p^m - q^n = 1$ only for $m = 1$, $q = 2$, $p = 2^n + 1$; $m = q = 2$, $n = p = 3$; $n = 1$, $p = 2$, $q = 2^m - 1$.

A. Fleck[227] distinguished cases A and B according as none or one (say x) of the integral solutions $\neq 0$ of $x^p + y^p + z^p = 0$ is divisible by the odd prime p. Set $s = x + y + z$. Then

(A) $y + z = a^p$, $z + x = b^p$, $x + y = c^p$, $s = -abcp^3 GM$,

(B) $y + z = p^{2p-1} a^p$, $z + x = b^p$, $x + y = c^p$, $s = -abcp^2 GM$.

He considered the six quantities

$$y^2 + yz + z^2 = GJ, \qquad x^2 - yz = GJ_1,$$
$$z^2 + zx + x^2 = GK, \qquad y^2 - zx = GK_1,$$
$$x^2 + xy + y^2 = GL, \qquad z^2 - xy = GL_1,$$

[221] L'intermédiaire des math., 16, 1909, 14–15.
[222] L'enseignement math., 11, 1909, 49–51.
[223] *Ibid.*, 11, 1909, 455–9. Summary by Dickson,[288] p. 183.
[224] Archiv Math. Phys., (3), 15, 1909, 368–9.
[225] Amer. Math. Monthly, 16, 1909, 3–4.
[226] *Ibid.*, 38–9. Special cases by G. B. M. Zerr, 15, 1908, 237. See Gerono.[92]
[227] Sitzungsber. Berlin Math. Gesell., 8, 1909, 133–148, with Archiv Math. Phys., 15, 1909.

and proved that (i) s has no factor other than a divisor of G in common with one of these six expressions; (ii) any two of the six have no common factor other than a divisor of G, so that $J, \cdots, L_1$ are relatively prime in pairs; (iii) $J, \cdots, L_1$ are products of primes of the form $6\mu p+1$; (iv) $x^{3p} \equiv y^{3p} \equiv z^{3p} \pmod{GJKLJ_1K_1L_1}$.

G. Frobenius[228] gave a simple proof of the criterion of Wieferich,[214] using Mirimanoff's[180] formulation of Kummer's criterion to show that

$$\sum_{r,\,s=0}^{\lambda-1} (-1)^{r-s}(r-s)^{\lambda-2}t^s$$

is congruent modulo λ, for every $t \neq 0, \pm 1$, to both

$$c = \phi_{p-1}(1), \qquad \frac{1+t}{1-t}c,$$

whence $c \equiv 0 \pmod{\lambda}$, so that $2^{\lambda-1} \equiv 1 \pmod{\lambda^2}$.

A. Gérardin[229] gave a brief history and extensive bibliography of the subject. He conjectured that Fermat's last theorem could be proved by showing that the difference or the sum of two nth powers $(n > 2)$ is always comprised between two consecutive nth powers.

P. Bachmann[230] gave an account of results obtained by elementary methods, chiefly those by Abel,[16] Legendre,[17] Wendt,[152] and Dickson.[195-9] The remark (p. 461) that all primes < 100 are regular was corrected on p. 480.

H. Stockhaus[231] gave a lengthy exposition of known methods for exponents 3, 5, 7, with suggestions of doubtful value on the general case.

* K. Rychlik[232] gave a proof for exponents 3, 4, 5.

* Ed. Barbette[233] proved some inequalities.

F. Bernstein[234] proved Fermat's theorem under assumptions milder than those of Kummer.[76] The second case (that in which one of the three numbers is divisible by the prime exponent l) is proved by means of the assumption that the class number of the field $k(Z)$ of the l^2th roots of unity is divisible by l, but not by l^2; and again by means of the assumption that $k(Z)$ contains no class belonging to the exponent l^2, while the class number of $k(\zeta+\zeta^{-1})$ is prime to l, where $\zeta^l=1$. The first case (that in which the three numbers are prime to l) is proved from the assumptions (i) that the second factor h_2 of the class number of $k(\zeta)$ is divisible by l, and (ii) if l^μ is the highest power of l dividing h_2, then in the "Teilklassenkörper" of the l^μth degree every ideal of $k(\zeta)$, whose lth power is a principal ideal in $k(\zeta)$, is itself a principal ideal. [See Vandiver's[296] criticisms.]

---

[228] Sitzungsber. Akad. Wiss. Berlin, 1909, 1222–4. Reprinted in Jour. für Math., 137, 1910, 314–6.

[229] Historique du dernier théorème de Fermat, Toulouse, 1910, 12 pp. Extract in Assoc. franç. av. sc., 39, I, 1910, 55–6. All of his references are found in the present chapter.

[230] Niedere Zahlentheorie, 2, 1910, 458–476.

[231] Beitrag zum Beweis des Fermatschen Satzes, Leipzig, 1910, 90 pp.

[232] Casopis, Prag, 39, 1910, 65–86, 185–195, 305–317 (Bohemian).

[233] Le dernier théorème de Fermat, Paris, 1910, 19 pp.

[234] Göttingen Nachrichten, 1910, 482–488, 507–516.

Ph. Furtwängler[235] proved, in extension of Kummer's[76] work, that if $\alpha^l + \beta^l + \gamma^l = 0$, where α, β, γ are numbers, prime to $L = (\zeta - 1)$, of the field $k(\zeta)$, $\zeta^l = 1$, and if $\alpha \equiv a$, $\beta \equiv b$, $\gamma \equiv c \pmod{L}$, where a, b, c are rational, and if $k(\zeta)$ contains no ideal belonging* to the exponent $2j+1$ modulo L, then, if x, y are any two of a, b, c,

$$\left[\frac{d^{2j+1} \log (x + e^v y)}{dv^{2j+1}} \right]_{v=0} \equiv 0 \pmod{l}.$$

By Mirimanoff,[180] this congruence can not hold when $j = 1, 2, 3$ or 4. Hence if $k(\zeta)$ does not contain ideals belonging to each of the exponents 3, 5, 7, 11, Fermat's equation is impossible in numbers prime to l in $k(\zeta)$. The same conclusion holds if the class number H is at most divisible by l^3.

E. Hecke[236] proved that $x^l + y^l + z^l = 0$ is impossible in integers x, y, z, each not divisible by the odd prime l, if the first factor h_1 of the class number H of the field defined by an lth root of unity is divisible by l, but not by l^2.

D. Mirimanoff,[237] making use of his[180] criterion, proved that if $x^p + y^p + z^p = 0$ has solutions prime to p, each of the six ratios x/y, $\cdots$ is a root t of

$$\prod_{i=1}^{m-1} (t + \alpha_i) \sum_{i=1}^{m-1} \frac{R_i}{t + \alpha_i} \equiv 0 \pmod{p}, \qquad R_i = \frac{\phi_{p-1}(-\alpha_i)}{(1 - \alpha_i)^{p-1}},$$

where $\alpha_1, \cdots, \alpha_{m-1}$ are the roots $\neq 1$ of $z^m = 1$. For $m = 2$ or 3, at least two of the six ratios are incongruent, so that our congruence, being of degree < 2, is an identity; taking $t = -1$ and applying

$$q(m) = \frac{m^{p-1} - 1}{p} \equiv \sum_{i=1}^{m-1} \frac{R_i}{1 - \alpha_i} \pmod{p},$$

we get $q(m) \equiv 0$. Besides Wieferich's $q(2) \equiv 0$, we have $q(3) \equiv 0$. Thus the initial equation is impossible in integers prime to p for all prime exponents p such that either $q(2)$ or $q(3)$ is not divisible by p; in particular, for all prime exponents of the form $2^a 3^b \pm 1$ or $\pm 2^a \pm 3^b$.

G. Frobenius[238] proved the last two criteria and deduced (13) from (8) more simply than had Mirimanoff.[180] Set $b^{2n} = (-1)^{n-1} B_n$, $b^{2n+1} = 0$, $b^1 = -\frac{1}{2}$, so that the Bernoullian numbers are given symbolically by $(b+1)^n - b^n = 0$ $(n > 1)$. Set

$$F(x, y) = \sum_{r=0}^{p-1} \binom{y}{r} (x-1)^r,$$

$$\underline{F}(x, y) x^m = \sum_r \binom{y+m}{r} (x-1)^r + (x-1)^p G(x, y),$$

$$mx G_m(x) = G(x, mb) - G(0, mb) \frac{x^m - 1}{x - 1},$$

$$mF(x) = F(x, mb) - \{F(0, mb) - mpq\}(x-1)^{p-1}.$$

---

* An ideal Q, prime to $L = (\zeta - 1)$, is said to belong to the exponent n modulo L if Q^l is a principal ideal (κ) such that $\kappa \equiv r_1 \pmod{L^n}$, while there exists no unit η in the field $k(\zeta)$ such that $\eta\kappa \equiv r_2 \pmod{L^{n+1}}$, where r_1 and r_2 are rational numbers.

[235] Göttingen Nachrichten, 1910, 554–562.
[236] *Ibid.*, 420–4.
[237] Comptes Rendus Paris, 150, 1910, 204–6.   Reproduced.[246]
[238] Sitzungsber. Akad. Wiss. Berlin, 1910, 200–8.

Then

$$F(x)(x^m-1)+\sum_{n=1}^{p-1}\frac{1-x^n}{n}=(x-1)^p x G_m(x),$$

from which the results of the paper follow. The six ratios of the three solutions prime to p of Fermat's equation satisfy the congruence $G_m(x)\equiv 0$ (mod p) of degree $m-2$. Hence, if $m=2$ or 3, G_m vanishes identically. But $G_m(1)\equiv(1-m^{p-1})/p$.

A. Fleck[239] proved, as an extension of his[227] theorem (iii), that the prime factors of J_1, K_1, L_1 are of the form $6\nu p^2+1$. Hence $J,\cdots,L_1$ are all of the form $6\mu p^2+1$. For any prime factor j of the form $6\mu p+1$ of J, $(ty)^{6\mu}\equiv(tz)^{6\mu}\equiv 1$ (mod j), where $t=1$ in case A, $t=p$ in case B. A like result is said to hold for the prime factors of K or L.

E. Dubouis[240] defined, in honor of Sophie Germain, a " sophien " of a prime n to be a prime θ, necessarily of the form $kn+1$, for which $x^n\equiv y^n+1$ (mod θ) is impossible in integers prime to θ. He stated that Pepin[109] proved that the sophiens of n are finite in number, whereas Pepin proved this only for $n=3$. If the resultant of $a^k=1$, $(a+1)^k=1$ is not divisible by θ, then θ is a sophien of n [Wendt[152]].

B. Lind[241] gave an exposition of various papers dealing with Fermat's last theorem without the use of complex integers or ideals, but unfortunately interpolated careless remarks of his own. Of the results claimed by Lind to be novel, equations (19)–(26) are correct, but long known, while (27) is not proved, viz., that $x+y-z\equiv 0$ (mod 9) if $x^n+y^n=z^n$, it being proved only for modulus 3. This error gave rise to later errors in his inequalities (p. 32) and his equations (95), (106b). His attempt (pp. 61–5) to prove by use of congruences Fermat's last theorem contains several serious errors besides the dependence on (27). The bibliography is quite extensive.

J. Joffroy[242] noted that, if $F=x^{37}+y^{37}-z^{37}=0$ for integers $x<y<z$, then $x>P+1=1919191$. For, $x^{37}-x=Pm$, $P=2\cdot3\cdot5\cdot7\cdot13\cdot19\cdot37$; so that

$$F+Pm_1=x+y-z, \qquad m_1>0.$$

T. Hayashi[243] proved that if, for n an odd prime, $x^n+y^n=nz^n$, or if $x^n+y^n=z^n$ for z divisible by n, then $b_0+b_1+\cdots+b_s\equiv 0$ (mod n^2), where $s=(n-1)/2$, and the b's are the coefficients of the polynomial Y satisfying the identity

$$4\frac{\xi^n-1}{\xi-1}=Y^2-(-1)^s nZ^2,$$

where

$$Y=b_0\xi^s+b_1\xi^{s-1}+\cdots+b_s, \qquad Z=c_0\xi^{s-1}+\cdots+c_{s-1},$$

[239] Sitzungsber. Berlin Math. Gesell., 9, 1910, 50–3 (with Archiv Math. Phys., 16, 1910).

[240] L'intermédiaire des math., 17, 1910, 103–4.

[241] Abh. Geschichte Math. Wiss., 26, II, 1910, 23–65. Reviewed adversely by A. Fleck, Archiv Math. Phys., (3), 16, 1910, 107–9; 18, 1911, 107–8.

[242] Nouv. Ann. Math., (4), 11, 1911, 282–3. Reproduced, Oeuvres de Fermat, IV, 165–6.

[243] Jour. Indian Math. Soc., Madras, 3, 1911, 16–22; 111–4. Same in Science Reports of Tôhoku University, 1, 1913, 43–50, 51–54.

while

$$\eta = b_0 y^s - b_1 y^{s-1} x + \cdots + (-1)^s b_s x^s,$$
$$x\zeta = c_0 x y^{s-1} - c_1 x^2 y^{s-2} + \cdots + (-1)^{s-1} c_{s-1} x^s$$

are such that $\eta^2 - (-1)^s n (x\zeta)^2$ has as divisors only 2 and numbers of the form $r^2 - (-1)^s n t^2$. The initial equations are both impossible if $n = 5$ or 13.

A. E. Pellet[244] considered for a prime $p = hn + 1$, having g as a primitive root, the number hN_3 of times that

$$g^{in} + g^{jn} + g^{kn} \equiv 0 \quad (\bmod\ p) \quad (i,\ j,\ k = 0,\ 1,\ \cdots,\ h-1).$$

By use of the equation for the n periods of the pth roots of unity it is shown that pN_3 has the limits $h^2 \pm \sqrt{(p-h)^3}$, whence [error[245]] the inferior limit is positive if $h > n\sqrt{n}$. Hence in that case, $x^n + y^n + 1 \equiv 0 \pmod{p}$ has solutions prime to p. Cf. Libri.[24]

D. Mirimanoff[246] reproduced his[237] paper and used his first formula to obtain results concerning $q(5)$ and $q(7)$. Also he proved that $\phi_{p-1}(t)$ is divisible by p not only when t is one of the six ratios $\tau = x/y, \cdots$, but also for $t = -\tau$ and $t = -\tau^2$. Finally, he proved Sylvester's formula for $q(m)$ [Vol. I, Ch. IV of this History].

A. Thue[247] proved that, if n is a prime > 3, and ϵ is an imaginary nth root of unity, and each B_i is an integer numerically $\leq K > 0$,

$$\left| B_0 + B_1\epsilon + \cdots + B_{n-2}\epsilon^{n-2} \right| \geq \frac{\tan\ \pi/(2n)}{\{(2n-3)K\}^{(n-3)/2}},$$

if not every $B_i = 0$. Next, for R an integer, let $PQ = R^n$, where

$$P = \sum_{i=0}^{n-2} A_i\epsilon^i, \qquad Q = \Sigma B_i\epsilon^i, \qquad |A_i| \leq S, \qquad |B_i| \leq T.$$

Then for a suitably chosen k and integers f_i, g_i such that

$$|f_i| < 2\{k[(2n-3)T]^{1/n}+1\}, \qquad |g_i| < 2\{k[(2n-3)S]^{1/n}+1\},$$

we have $P/R = -B/A$, where $A = \Sigma f_i\epsilon^i$, $B = \Sigma g_i\epsilon^i$. It is stated that application can be made to Fermat's equation

$$a^n = c^n - b^n = \Pi(c - \epsilon^i b).$$

If $a^n + b^n = c^n$ for relatively prime integers (p. 15), we can find positive integers p, q, r, each $< \sqrt{3c}$, such that $pa + qb = rc$. Hence

$$(ar)^n + (br)^n = (pa+qb)^n,$$

whence $q^n - r^n$ is divisible by a.

Thue[248] proved that if $y^n = x^n + 1$, $n > 3$, the most general solution of

$$A^n + B^n = (c_0 + c_1 y + \cdots + c_{n-1}y^{n-1})^n,$$

where A, B and each c are integral functions of x, is

$$f^n + (fx)^n = (fy)^n$$

where f is an arbitrary integral function of x.

---

[244] L'intermédiaire des math., 18, 1911, 81–2.
[245] This deduction fails if $n = 5$, $h = 20$.
[246] Jour. für Math., 139, 1911, 309–324.
[247] Skrifter Videnskapsselskapet I Kristiania (Math.), 1, 1911, No. 4.
[248] Ibid., 2, 1911, No. 12, 13 pp. For his paper, ibid., No. 20, see[173] Ch. XXIII.

* D. N. Ranucci wrote a pamphlet, Risoluzione dell'equazione

$$x^n - Ay^n = \pm 1,$$

con una nuova dimostrazione dell' ultimo teorema di Fermat, Roma, 1911, 23 pp.

F. Mercier[248a] noted that we may take $x<y<z$ if $n>1$, whence

$$x^n = z^n - y^n = (z-y)(z^{n-1}+yz^{n-2}+\cdots) > (z-y)\ ny^{n-1} > ny^{n-1},$$

$n/x < (x/y)^{n-1} < 1$, $n<x$. This lemma, instead of helping him to prove Fermat's last theorem, led him to commit the error of saying that $3^n + y^n = z^n$ is solvable when n is any integer >1 because it is solvable when $n=2$.

Ph. Furtwängler[249] proved by use of Eisenstein's law of reciprocity for residues of lth powers, where l is an odd prime, that every integral divisor r of x_i satisfies

$$(18) \qquad\qquad r^{l-1} \equiv 1 \quad (\mathrm{mod}\ l^2)$$

if x_1, x_2, x_3 are relatively prime solutions $\neq 0$ of $x_1^l + x_2^l + x_3^l = 0$ and x_i is prime to l. Since one of the x's is divisible by 2, we have the criterion of Wieferich. Next, every factor r of $x_i \pm x_k$ satisfies (18) if $x_i + x_k$ and $x_i - x_k$ are prime to l. Since one of the x's is divisible by 3 unless all three are congruent modulo 3, it follows from the two theorems that, if the x's are all prime to l, (18) holds for $r=3$, which is the criterion of Mirimanoff.

S. Bohniček[250] proved that integral numbers of the domain of the 2^nth roots of unity do not satisfy Fermat's equation with the exponent 2^{n-1}, $n>2$.

H. Berliner[251] considered $x^p = y^p + z^p$ for x, y, z not divisible by the prime $p>2$. In Abel's formulas $2x = a^p + b^p + c^p$, $\cdots$, we may take $a>b>c$. Then $a = b+c \pm 2^k ep$, where 2^k is the highest power of 2 dividing abc, while ep is an odd multiple of 3. For every p, $a<3(b+c)$; for $p \geq 5$, $a<3b$; for $p \geq 31$, $a < 3^{1/5}(b+c)$; for $p \geq 37$, $a < 3^{2/9}b$. If $p \geq 5$, $b>3p$; if $p \geq 37$, $b > 6p+1$.

L. Carlini[252] proved that $x^n + y^n = z^n$ $(n>2)$ is not satisfied by three binary forms in u, v, identically in the variables u, v. Hence a like result holds for polynomials in one or more variables.

J. Plemelj[253] proved $x^5 + y^5 + z^5 = 0$ impossible in $R(\sqrt{5})$ more simply than had Dirichlet.[20]

* B. Bernstein[254] gave some properties of numbers satisfying $x^n + y^n = z^n$. The latter is proved impossible under certain assumptions on x, y, z.

R. D. Carmichael[255] proved that, if $x^p + y^p + z^p = 0$ has integral solutions each not divisible by the odd prime p, there exists a positive integer $s < (p-1)/2$ such that

$$(s+1)^{p^2} \equiv s^{p^2} + 1 \quad (\mathrm{mod}\ p^3).$$

[248a] Mém. Soc. Nat. Sc. Nat. et Math. de Cherbourg, 38, 1911–12, 729–44. Cf. Grunert.[73]
[249] Sitzungs. Akad. Wiss. Wien (Math.), 121, IIa, 1912, 589–592.
[250] *Ibid.*, 727–742.
[251] Archiv Math. Phys., (3), 19, 1912, 60–3.
[252] Periodico di Mat., 27, 1912, 83–8.
[253] Monatshefte Math. Phys., 23, 1912, 305–8.
[254] Math. Unterr., 1912, No. 3, 111–5; No. 4, 150–1 (Russian).
[255] Bull. Amer. Math. Soc., 19, 1912–3, 233–6.

We may (pp. 402–3) replace this condition by the simpler one[264]

$$(s+1)^p \equiv s^p+1 \quad (\mathrm{mod}\ p^3),$$

as noted by G. D. Birkhoff. The test fails for $p=6n+1$ since the congruence has a root. He[255a] stated that $x^6 \pm y^6 \neq \square$.

N. Alliston[256] noted that $x^r \pm y^r = z^m$ has integral solutions if r, m are relatively prime positive integers. R. Norrie (pp. 33–4) treated the same problem.

R. Niewiadomski[257] considered $d_n = z^n - x^n - y^n$. If $d_n = 0$ for n an odd prime, then d_{2n+1} is divisible by $(x+y)(z-x)(z-y)$. He gave linear relations between d_{n+1}, d_n, d_{n-1} and expressions for d_n when $d_1 \equiv 0 \ (\mathrm{mod}\ n^k)$ and when $d_2 = 0$. G. Métrod (pp. 215–6) treated the latter case.

E. Landau[258] noted that the assumptions

$$x^{p-1} \equiv y^{p-1} \equiv 1 \quad (\mathrm{mod}\ p^2), \qquad x+y = mp,$$

where p is an odd number >1 not dividing m, lead to a contradiction. In fact,

$$1 \equiv x^{p-1} \equiv (mp-y)^{p-1} \equiv -(p-1)mpy^{p-2}+1 \quad (\mathrm{mod}\ p^2)$$

requires that p divide $(p-1)my^{p-2}$ and hence also m.

E. Miot[259] gave a false expression for the g.c.d. of 2^x-1, 3^x-1.

H. Kapferer[260] proved Fermat's theorem for the exponents 6 and 10 by showing by descent that $t^2 = (z^2 \pm y^2)^2 - (yz)^2$ is impossible.

H. C. Pocklington[261] noted that $x^{2n}+y^{2n}=z^2$ is impossible for all values of n for which $x^n+y^n=z^n$ is impossible. For, if the former has solutions, it has solutions with x prime to y and with y even. Thus $x^n = u^2-v^2$, $y^n = 2uv$. Hence $u+v = \alpha^n$, $u-v = \beta^n$ and u, v equal $2^{n-1}\gamma^n$, δ^n in some order. Thus $\alpha^n \pm \beta^n = (2\gamma)^n$.

J. E. Rowe[262] proved that if $x^n+y^n=z^n$, where x, y, n are odd, then $x+y$ is divisible by 2^n [evident since the quotient of x^n+y^n by $x+y$ is composed of n terms and hence is odd]. From this main theorem II′ we obtain his theorem I′ by changing the sign of y.

Ph. Maennchen[263] reported on the history of the theorem. Several (p. 294) proved that 2^n+1 is an exact power only for $2^3+1=3^2$.

W. Meissner[264] proved that $x^p+y^p=z^p$ is impossible in integers not divisible by the odd prime p if there exists no integer $v<p$ for which

$$(v+1)^p - v^p \equiv 1 \quad (\mathrm{mod}\ p^3), \qquad v^3 \not\equiv 1 \quad (\mathrm{mod}\ p)$$

[cf. Carmichael[255]]; also if $p=3^k 2^m \pm 1$ or $3^k \pm 2^m$; also if p, but not p^2, is a

---

[255a] Bull. Amer. Math. Soc., 20, 1913, 80.
[256] Math. Quest. Educ. Times, new series, 23, 1913, 17–18.
[257] L'intermédiaire des math., 20, 1913, 76, 98–100.
[258] Ibid., 206.
[259] Ibid., 112. Error noted pp. 183–4, 228.
[260] Archiv Math. Phys., (3), 21, 1913, 143–6.
[261] Proc. Cambridge Phil. Soc., 17, 1913, 119–120.
[262] Johns Hopkins University Circular, July, 1913, No. 7, 35–40; abstract in Bull. Amer. Math. Soc., 20, 1913, 68–69.
[263] Zeitschr. Math. Naturw. Unterricht, 45, 1914, 81–93.
[264] Sitzungsber. Berlin Math. Gesell., 13, 1914, 101–104. See Vol. I, Ch. IV,[39] of this History.

divisor of a number of one of these four forms; and if p^2 divides one of the four forms, provided k and m are divisible by p.

The congruence $5^x+7^y+11^z\equiv0$ (mod 13) was solved by several writers.[265]

T. Suzuki[266] found the 12 sets of solutions of $5^x+8^y+11^z\equiv0$ (mod 13).

L. Aubry[267] noted that, if m is prime to n, $x^m+y^m=z^n$ has the solution $x=A^u a$, $y=A^u b$, $z=A^v$, where $nv-mu=1$, $a^m+b^m=A$. For $m=3$, $n=2$, he gave a solution involving two parameters.

A. Gérardin[267a] gave integral solutions of $x^3-y^2=z^n$ for $2\leqq n\leqq8$.

H. S. Vandiver[268] wrote $q(r)$ for $(r^{p-1}-1)/p$ and proved that if

$$x^p+y^p+z^p=0$$

is satisfied by integers not divisible by the prime p, then

$$q(5)(t-1)(t+2)(t+\tfrac{1}{2})\equiv0 \quad (\text{mod } p)$$

is satisfied by each of the six values $t=x/y$, $\cdots$, z/y, and either $q(2)\equiv0$ (mod p^3), $q(3)\equiv0$ (mod p), or else $q(2)\equiv q(3)\equiv q(5)\equiv0$ (mod p) and, if $p\equiv2$ (mod 3), $q(7)\equiv0$ (mod p).

E. Swift[269] proved that neither of $x^6\pm y^6$ is a square.

H. S. Vandiver[270] proved that if $x^p+y^p+z^p=0$ is satisfied in integers prime to p, then $q(5)\equiv0$ (mod p) and $1+\tfrac{1}{2}+\tfrac{1}{3}+\cdots+1/[p/5]\equiv0$ (mod p).

G. Frobenius[271] proved that, if Fermat's equation has integral solutions each prime to the prime exponent p, then $q(m)$ is divisible by p for $m=11$ and $m=17$, and, in case $p\equiv5$ (mod 6), also for $m=7$, 13, 19. Moreover,

$$\sum_{l=1}^{m-1}\left\{\frac{(l/m+h)^{p-1}-h^{p-1}}{p-1}\right\}x^l$$

vanishes identically modulo p for $m\leqq22$ and $m=24$, 26. Here the symbolic power h^λ is to be replaced by the Bernoullian number b_λ.

J. G. van der Corput[272] proved the impossibility of $x^5+y^5=Az^5$ for $A=1$ and other values.

R. Guimarães[273] gave a bibliography and discussed the history of Fermat's last theorem, including Wronski's[66] pretentions.

N. Alliston[274] proved that Fermat's theorem for odd exponents implies that $b^{4n+2}+c^{4n+2}=\square$ is impossible if $n>0$.

<hr>

[265] Math. Quest. Educ. Times, new series, 26, 1914, 101–3.

[266] Tôhoku Math. Jour., 5, 1914, 48–53. Further report in Ch. XXIII.[105]

[267] L'intermédiaire des math., 21, 1914, 19–20.

[267a] Sphinx-Oedipe, 9, 1914, 136–9. For $7^3-10^2=3^5$, ibid., 6, 1911, 91.

[268] Trans. Amer. Math. Soc., 15, 1914, 202–4.

[269] Amer. Math. Monthly, 21, 1914, 238–9; 23, 1916, 261.

[270] Jour. für Math., 144, 1914, 314–8.

[271] Sitzungsber. Akad. Wiss. Berlin, 1914, 653–81.

[272] Nieuw Archief voor Wiskunde, 11, 1915, 68–75.

[273] Revista de la Sociedad Mat. Española, 5, 1915, No. 42, pp. 33–45. There is a great number of confusing misprints. Both Crelle's Journal and Comptes Rendus Paris are cited as C.r., the second being once cited as Cr., Berlin!

[274] Math. Quest. Educ. Times, new series, 29, 1916, 21.

P. Montel[275] proved that if m, n, p are integers for which $1/m+1/n+1/p$ <1, it is impossible to find three integral functions of a variable such that $x^m+y^n+z^p=0$; in particular, $x^m+y^m+z^m \neq 0$ if $m>3$.

P. Kokott[276] proved that $x^{11}+y^{11}+z^{11}=0$ is impossible in integers prime to 11, using residues modulo 11 of symmetric functions of x, y, z.

W. Mantel[277] proved that if $n>3$ and p are primes, $x^n+y^n+z^n \equiv 0$ (mod p) is impossible in integers prime to p unless $p=(6kn-n-3)/(n-3)$.

E. T. Bell stated and F. Irwin[278] proved that if x^n-y^n is a prime 2^r+1 for r a prime >2 and $n>2$, then $n=3$, $x=2$, $y=1$.

A. Gérardin[279] proved that $10^k+1=z^n$ is impossible in integers if $n>1$.

H. H. Mitchell[280] treated the solution of $cx^\lambda+1=dy^\lambda$ in a Galois field.

A. J. Kempner[281] gave a simple proof that $a^{2n}-1=2b^n$ has only the integral solutions $a=\pm1$, $b=0$ [Liouville,[32] Lind[224]].

A. Korselt[282] proved, without using integrals as had R. Liouville,[105] that $x^m+y^n+z^r=0$ is not solvable in relatively prime integral rational functions of a variable t if each exponent exceeds 2 or if one exponent is 2 and the others exceed 3, the case[282a] $x^3+y^5+z^2=0$ not being decided. In all the remaining cases, the initial equation is solvable　Cf. Velmine,[177] Montel.[275]

* J. Schur[283] gave a simpler proof of Dickson's[199] theorem.

L. Aubry[284] proved that $a \cdot 10^k+1 \neq z^n$ if $0<a<10$, $k>1$, and n is a prime >1.

E. Maillet[285] considered $a^m+b^m=c^m$ for $m=n/p$, where n, p are relatively prime positive integers and $p>1$. It has integral solutions each $\neq 0$ if and only if

$$a_2^m a_1^n+b_2^m b_1^n=c_2^m c_1^n$$

has integral solutions each $\neq 0$ such that a_1, b_1, c_1 are prime to p and relatively prime in pairs, while a_2, b_2, c_2 are relatively prime in pairs and have no prime factors other than those of p. The last equation can be given a similar form in a_1^1, b_1^1, c_1^1, a_2^1, b_2^1, c_2^1, which are relatively prime in pairs, while any prime factor λ of a_2^1, b_2^1 or c_2^1 is a divisor of p such that $m \leqq 1/(\lambda-1)$. In particular, if $m>1/(\mu-1)$, where μ is the least prime factor of p, Fermat's equation with the exponent m is equivalent to one with the exponent n. This is also the case if one of a_2, b_2, c_2, a_2^1, b_2^1, c_2^1 is an exact pth power and hence if p has at most two distinct prime factors. Corresponding results hold for $a^{m_1}+b^{m_2}=c^{m_3}$, with any fractional exponents, and with a, b, c relatively prime in pairs.

[275] Annales sc. l'école norm. sup., (3), 33, 1916, 298–9.
[276] Archiv Math. Phys., (3), 24, 1916, 90–1.
[277] Wiskundige Opgaven, 12, 1916, 213–4.
[278] Amer. Math. Monthly, 23, 1916, 394.
[279] L'intermédiaire des math., 23, 1916, 214–5; Sphinx-Oedipe, 1917.
[280] Trans. Amer. Math. Soc., 17, 1916, 164–177; Annals of Math., 18, 1917, 120–131.
[281] Archiv Math. Phys., (3), 25, 1916–7, 242–3.
[282] *Ibid.*, 89–93.
[282a] This equation is satisfied by the fundamental invariants of the icosaeder group, *ibid.*, 27, 1918, 181–3.
[283] Jahresber. d. Deutschen Math.-Vereinigung, 25, 1916, 114–7.
[284] L'intermédiaire des math., 24, 1917, 16–17.
[285] Bull. Soc. Math. France, 45, 1917, 26–36.

For reports on $q_u = (u^{p-1}-1)/p$, see Ch. IV of Vol. I of his History. There are additional notes by * E. Haentzschel[286] on $2^{p-1} \equiv 1 \pmod{p^2}$, $p = 1093$, and H. E. Hensen[287] on the computation of q_u.

L. E. Dickson[288] gave an account of the history of Fermat's last theorem and the origin and nature of the theory of algebraic numbers.

F. Pollaczek[289] proved that, if $x^p + y^p + z^p = 0$ has integral solutions prime to p, then q_u is divisible by p if $u \leqq 31$ for all primes p except a finite number; also, $x^2 + xy + y^2 \equiv 0 \pmod{p}$ is impossible.

W. Richter[290] proved Korselt's[282] result for the special case $m = n = r$. There exist rational integral functions x, y, z of t satisfying $f \equiv x^n + y^n + z^n = 0$ if and only if the genus $\frac{1}{2}(n-1)(n-2) - d - r$ of the curve is zero, where d is the number of double points and r the number of cusps. But $d = r = 0$ since $\partial f/\partial x = 0$, etc., hold only for $x = y = z = 0$. Hence $n = 1$ or 2.

H. S. Vandiver[291] gave an expression for the residue modulo λ^n of Kummer's[61] first factor h_1 of the number of classes of ideals in the domain defined by a λth root of unity. In terms of Bernoulli numbers we can infer necessary and sufficient conditions that h_1 be divisible by any given power of λ. He[292] stated that if $x^p + y^p + z^p = 0$ holds for integers not divisible by the prime p, then $23^{p-1} \equiv 1 \pmod{p^2}$ for $p \not\equiv 1 \pmod{11}$, and that the Bernoulli number B_s is divisible by p^2 for $s = (tp+1)/2$, $t = p-4$, $p-6$, $p-8$, $p-10$.

A. Arwin[293] gave a method to solve $(x+1)^p - x^p \equiv 1 \pmod{p^2}$, p a prime.

Vandiver[294] derived from one source the theorems of Furtwängler[249] and the criterion of Kummer[76] for solutions prime to p of $x^p + y^p = z^p$.

P. Bachmann[295] gave an almost complete reproduction of the papers by Abel,[16] Legendre,[17] Dirichlet,[20] Kummer,[61] Wendt,[152] Mirimanoff,[180, 246] Dickson,[195-6, 199] Wieferich,[214] Frobenius,[228, 238] and Furtwängler.[249]

Vandiver[296] employed the first factors h_1 and k of the class numbers of the fields of the p^nth and p^{n-1}th roots of unity respectively, and the value of $k_1 = h_1/k$ due to J. Westlund,[297] and proved that k_1 is divisible by p if and only if at least one of the first $(p-3)/2$ Bernoulli numbers is divisible by p. Bernstein's[234] first assumption in his second case therefore implies that $p = l$ is a regular prime (so that his result forms no extension over Kummer[61]), while the assumptions in his first case do not as claimed include those of Kummer.[76] It is shown that 101, 103, 131, 149, 157 are the only irregular primes between 100 and 167.

[286] Jahresber. d. Deutschen Math.-Vereinigung, 25, 1916, 284.

[287] L'enseignement math., 19, 1917, 295–301.

[288] Annals of Math., (2), 18, 1917, 161–87.

[289] Sitzungsber. Akad. Wiss. Wien (Math.) 126, IIa, 1917, 45–59.

[290] Archiv Math. Phys., (3), 26, 1917, 206–7.

[291] Bull. Amer. Math. Soc., 25, 1919, 458–61.

[292] *Ibid.*, 24, 1918, 472.

[293] Acta Math., 42, 1919, 173–190.

[294] Annals of Math., 21, 1919, 73–80.

[295] Das Fermat Problem, Verein Wiss. Verleger, W. de Gruyter & Co., Berlin and Leipzig, 1919, 160 pp.

[296] Proc. National Acad. Sc., May, 1920.

[297] Trans. Amer. Math. Soc., 4, 1903, 201–212.

The Encyclopédie des sc. math., I, 3, p. 473, cited the criteria $q(2) \equiv 0$, $q(3) \equiv 0 \pmod{p}$, without stating that the unknowns are prime to p.

On $u^3 + v^3 = hp^r$, where h is a prime, see Baer[224] of Ch. XXI. Thue[236] of Ch. XXI proved that $x^6 + y^6 \neq z^6$, also that $x^6 + y^3 \neq z^2$ if z is not divisible by 3.

References (all included in the present account) on Fermat's last theorem occur in the following places: Nouv. Corresp. Math., 5, 1879, 90; Zeitschrift Math.-naturw. Unterricht, 23, 1892, 417–8; Ball's Math. Recreations and Essays, 1892, 27–30; ed. 4, 1905, 37–40; l'intermédiaire des math., 2, 1895, 26, 117–8, 359, 427; 12, 1905, 11–12; 13, 1906, 99; 14, 1907, 258; 15, 1908, 217; 17, 1910, 34, 278; 18, 1911, 255.

AUTHOR INDEX.

The numbers refer to pages. Those in parenthesis relate to cross-references. Those in brackets refer to editors or translators (usually listed only in the index of the first chapter in which they occur). The other numbers refer to actual reports.

Ch. I. Polygonal, Pyramidal and Figurate Numbers.

"Alemtejano, U.," 38
Alliston, N., 25, 36
Anonymous, 36
Anton, H., 39
Apianus, P., 4
Arabs, 4
Archimedes, 4 (1)
Arnaudeau, A., 38
Aryabhatta, 4
Ast, F., [1]
Aubry, L., 34, 38, 39

Bachet, C. G., 5, 6, 8 (8, 9, 15, 29)
Bachmann, P., 32, 34
Bahier, E., 36
Baines, J., 20
Ball, W. W. R., 1
Barbette, E., 30, 34, 35 (39)
Barlow, P., 17, 18 (22)
Barniville, J. J., 33
Barruel, E., 18
Bastien, L., 35
Beguelin, N., 13–15 (22)
Benzius, J., 5
Berdellé, C., 37
Berger, A., 31
Bernerie, A., 20
Bible, 5
Biernatzki, K. L., [4]
Bills, S., 24
Blaikie, J., 33
Boethius, A. M. S., 5
Bouniakowsky, V., 23 (19, 39)
Boutin, A., 30, 31, 36, 39
Brady, C., 17
Brassinne, E., [6]
Brocard, H., 25, 27, 34, 38 (26, 39)
Bullialdi, I., [1]
Burali-Forti, C., 34

Cantor, 34
Cantor, M., 4 [1, 3, 4]
Cardan, H. C., 5
Casinelli, A., 20
Catalan, E., 26, 28, 29
Cauchy, A., 18–20 (29, 32)
Cesàro, E., 26–28
Child, J. M., 25, 36
Christie, R. W. D., 32, 33, 37
Chu Shih-chieh, 4
Coste, L. M. P., 19
Crofton, 32

Cunningham, A., 32–34 (39)
Curjel, H. W., 30 (29)

Deidier, Abbé, 10
De Joncourt, Elie, 12
De Montmort, P. R., 9
De Muris, J., 5
De Rocquigny, G., 28, 29, 31, 37, 39 (30)
Descartes, R., 6, 7
De Sluse, René F., 8, 9
Dilling, A., 22
Diophantus, 3, 8 (1, 5–7, 9, 16, 36)
Dostor, G., (27)

Emmerich, A., 32 (35)
Engelhard, N., 12
Engler, E. A., 35
Escott, E. B., 33, 37
Euler, L., 10–17 (7, 22, 25, 27, 35–37, 39)
———, J. A., 12 (14)

Faber Stapulensis, 5
Faulhaber, J., 5
Fauquembergue, E., 25, 28, 36, 37
Fermat, P., 6–8 (10, 11, 13, 14, 17, 18, 29, 30, 32, 33)
Fontana, M., [5]
Frenicle, (8)
Fuss, N. (I), 12 (38)
———, P. H., [10]

Gallimard, M., 12
Gauss, C. F., 17 (23)
Gérardin, A., 9, 34–39 (7, 36–7)
Gerbert (Pope Sylvester II), 4.
Gergonne, J. D., 18
Gerhardt, C. I., [9]
Gill, C., 21, 22 (20)
Glaisher, J. W. L., 26 (39)
Goldbach, Chr., 10, 11
Göring, W., 26
Gough, J., 17, 18
Goulard, A., 32
Grigorief, E., 33

Hayashi, T., 34
Heath, T. L., [3]
Henischiib, G., 5

Henry, C., 26
Heppel, G., 25
Hiller, E., [1]
Hindus, 4
Hoche, R., [2]
Hochheim, A., 24
Hoefer, F., [1]
Hromádko, F., 33
Huntington, J., 20
Hutton, C., 9
Hypsicles, 1 (3)

Iamblichus, 3

Jacobi, C. G. J., 20 (19)
Jolivald, P., 38 (39)

Kästner, A. G., 12 [5]
Klügel, G. S., 17
Kraft, G. W., 12
Kronecker, L., 33

Lagrange, J. L., (15, 19)
Laisant, C. A., 28
Lebesgue, V. A., 22–24
Le Cointe, J. L. A., 39
Legendre, A. M., 17–19 (11, 15, 23, 25, 29)
Leibniz, G. W., 9 (39)
Lemoine, E., 32
Leybourn, T., 17 (25) [18]
Lionnet, E., 24, 27, 34 (36)
Liouville, J., 23, 24
Lipschitz, R., (39)
Loria, G., 8
Lucas, E., 25–27, 29 (17) [4]
Ludwig, H. F. T., 22

Maillet, E., 30 (13)
Marchand, D., 26
Mariares, F., 35
Marpurg, F. W., 16
Martinus, I., 5
Maser, H., [17]
Massoutié, G., [3]
Mathieu, H. B., 38 (39)
Maurolycus, F., 5, 6
Mayer, F. C., 9
Métrod, G., 35, 36
Meyl, A., 25 (39)
Mikami, Y., [4]
Minetola, S., 36
Monteiro, A. S., 38
Moret-Blanc, 25, 27

CH. II. LINEAR DIOPHANTINE EQUATIONS AND CONGRUENCES.

CH. III. PARTITIONS.

Ch. V. Triangles, Quadrilaterals and Tetrahedra with Rational Sides.

Ch. VI. Sum of Two Squares.

CH. VII. SUM OF THREE SQUARES.

CH. VIII. SUM OF FOUR SQUARES.

CH. IX. SUM OF n SQUARES.

CH. X. NUMBER OF SOLUTIONS OF QUADRATIC CONGRUENCES IN n UNKNOWNS.

CH. XI. LIOUVILLE'S SERIES OF EIGHTEEN ARTICLES.

CH. XII. PELL EQUATION; $ax^2 + bx + c$ MADE A SQUARE.

Ch. XIII. Further Single Equations of the Second Degree.

Ch. XIV. Squares in Arithmetical or Geometrical Progression.

CH. XV. TWO OR MORE LINEAR FUNCTIONS MADE SQUARES.

CH. XVI. TWO QUADRATIC FUNCTIONS OF ONE OR TWO UNKNOWNS MADE SQUARES.

CH. XVII.   SYSTEMS OF TWO EQUATIONS OF DEGREE TWO.

CH. XVIII.   THREE OR MORE QUADRATIC FUNCTIONS OF ONE OR TWO UNKNOWNS MADE SQUARES.

CH. XIX. SYSTEMS OF THREE OR MORE EQUATIONS OF DEGREE TWO IN THREE OR MORE UNKNOWNS.

CH. XX. QUADRATIC FORM MADE AN nTH POWER.

CH. XXI. EQUATIONS OF DEGREE THREE.

CH. XXII. EQUATIONS OF DEGREE FOUR.

CH. XXIII. EQUATIONS OF DEGREE n.

CH. XXIV.   SETS OF INTEGERS WITH EQUAL SUMS OF LIKE POWERS.

Ch. XXV.   WARING'S PROBLEM AND RELATED RESULTS.

CH. XXVI.   FERMAT'S LAST THEOREM, $ax^r + by^s = cz^t$, AND THE CONGRUENCE
$$x^n + y^n \equiv z^n \ (\mathrm{MOD}\ p).$$
SUBJECT INDEX

For classification of Diophantine equations, see Table of Contents.

A CATALOG OF SELECTED
DOVER BOOKS
IN SCIENCE AND MATHEMATICS

Astronomy

CHARIOTS FOR APOLLO: The NASA History of Manned Lunar Spacecraft to 1969, Courtney G. Brooks, James M. Grimwood, and Loyd S. Swenson, Jr. This illustrated history by a trio of experts is the definitive reference on the Apollo spacecraft and lunar modules. It traces the vehicles' design, development, and operation in space. More than 100 photographs and illustrations. 576pp. 6 3/4 x 9 1/4. 0-486-46756-2

EXPLORING THE MOON THROUGH BINOCULARS AND SMALL TELESCOPES, Ernest H. Cherrington, Jr. Informative, profusely illustrated guide to locating and identifying craters, rills, seas, mountains, other lunar features. Newly revised and updated with special section of new photos. Over 100 photos and diagrams. 240pp. 8 1/4 x 11. 0-486-24491-1

WHERE NO MAN HAS GONE BEFORE: A History of NASA's Apollo Lunar Expeditions, William David Compton. Introduction by Paul Dickson. This official NASA history traces behind-the-scenes conflicts and cooperation between scientists and engineers. The first half concerns preparations for the Moon landings, and the second half documents the flights that followed Apollo 11. 1989 edition. 432pp. 7 x 10.
0-486-47888-2

APOLLO EXPEDITIONS TO THE MOON: The NASA History, Edited by Edgar M. Cortright. Official NASA publication marks the 40th anniversary of the first lunar landing and features essays by project participants recalling engineering and administrative challenges. Accessible, jargon-free accounts, highlighted by numerous illustrations. 336pp. 8 3/8 x 10 7/8. 0-486-47175-6

ON MARS: Exploration of the Red Planet, 1958-1978--The NASA History, Edward Clinton Ezell and Linda Neuman Ezell. NASA's official history chronicles the start of our explorations of our planetary neighbor. It recounts cooperation among government, industry, and academia, and it features dozens of photos from Viking cameras. 560pp. 6 3/4 x 9 1/4. 0-486-46757-0

ARISTARCHUS OF SAMOS: The Ancient Copernicus, Sir Thomas Heath. Heath's history of astronomy ranges from Homer and Hesiod to Aristarchus and includes quotes from numerous thinkers, compilers, and scholasticists from Thales and Anaximander through Pythagoras, Plato, Aristotle, and Heraclides. 34 figures. 448pp. 5 3/8 x 8 1/2.
0-486-43886-4

AN INTRODUCTION TO CELESTIAL MECHANICS, Forest Ray Moulton. Classic text still unsurpassed in presentation of fundamental principles. Covers rectilinear motion, central forces, problems of two and three bodies, much more. Includes over 200 problems, some with answers. 437pp. 5 3/8 x 8 1/2. 0-486-64687-4

BEYOND THE ATMOSPHERE: Early Years of Space Science, Homer E. Newell. This exciting survey is the work of a top NASA administrator who chronicles technological advances, the relationship of space science to general science, and the space program's social, political, and economic contexts. 528pp. 6 3/4 x 9 1/4.
0-486-47464-X

STAR LORE: Myths, Legends, and Facts, William Tyler Olcott. Captivating retellings of the origins and histories of ancient star groups include Pegasus, Ursa Major, Pleiades, signs of the zodiac, and other constellations. "Classic." – *Sky & Telescope.* 58 illustrations. 544pp. 5 3/8 x 8 1/2. 0-486-43581-4

A COMPLETE MANUAL OF AMATEUR ASTRONOMY: Tools and Techniques for Astronomical Observations, P. Clay Sherrod with Thomas L. Koed. Concise, highly readable book discusses the selection, set-up, and maintenance of a telescope; amateur studies of the sun; lunar topography and occultations; and more. 124 figures. 26 halftones. 37 tables. 335pp. 6 1/2 x 9 1/4. 0-486-42820-6

Browse over 9,000 books at www.doverpublications.com

Chemistry

MOLECULAR COLLISION THEORY, M. S. Child. This high-level monograph offers an analytical treatment of classical scattering by a central force, quantum scattering by a central force, elastic scattering phase shifts, and semi-classical elastic scattering. 1974 edition. 310pp. 5 3/8 x 8 1/2. 0-486-69437-2

HANDBOOK OF COMPUTATIONAL QUANTUM CHEMISTRY, David B. Cook. This comprehensive text provides upper-level undergraduates and graduate students with an accessible introduction to the implementation of quantum ideas in molecular modeling, exploring practical applications alongside theoretical explanations. 1998 edition. 832pp. 5 3/8 x 8 1/2. 0-486-44307-8

RADIOACTIVE SUBSTANCES, Marie Curie. The celebrated scientist's thesis, which directly preceded her 1903 Nobel Prize, discusses establishing atomic character of radioactivity; extraction from pitchblende of polonium and radium; isolation of pure radium chloride; more. 96pp. 5 3/8 x 8 1/2. 0-486-42550-9

CHEMICAL MAGIC, Leonard A. Ford. Classic guide provides intriguing entertainment while elucidating sound scientific principles, with more than 100 unusual stunts: cold fire, dust explosions, a nylon rope trick, a disappearing beaker, much more. 128pp. 5 3/8 x 8 1/2. 0-486-67628-5

ALCHEMY, E. J. Holmyard. Classic study by noted authority covers 2,000 years of alchemical history: religious, mystical overtones; apparatus; signs, symbols, and secret terms; advent of scientific method, much more. Illustrated. 320pp. 5 3/8 x 8 1/2. 0-486-26298-7

CHEMICAL KINETICS AND REACTION DYNAMICS, Paul L. Houston. This text teaches the principles underlying modern chemical kinetics in a clear, direct fashion, using several examples to enhance basic understanding. Solutions to selected problems. 2001 edition. 352pp. 8 3/8 x 11. 0-486-45334-0

PROBLEMS AND SOLUTIONS IN QUANTUM CHEMISTRY AND PHYSICS, Charles S. Johnson and Lee G. Pedersen. Unusually varied problems, with detailed solutions, cover of quantum mechanics, wave mechanics, angular momentum, molecular spectroscopy, scattering theory, more. 280 problems, plus 139 supplementary exercises. 430pp. 6 1/2 x 9 1/4. 0-486-65236-X

ELEMENTS OF CHEMISTRY, Antoine Lavoisier. Monumental classic by the founder of modern chemistry features first explicit statement of law of conservation of matter in chemical change, and more. Facsimile reprint of original (1790) Kerr translation. 539pp. 5 3/8 x 8 1/2. 0-486-64624-6

MAGNETISM AND TRANSITION METAL COMPLEXES, F. E. Mabbs and D. J. Machin. A detailed view of the calculation methods involved in the magnetic properties of transition metal complexes, this volume offers sufficient background for original work in the field. 1973 edition. 240pp. 5 3/8 x 8 1/2. 0-486-46284-6

GENERAL CHEMISTRY, Linus Pauling. Revised third edition of classic first-year text by Nobel laureate. Atomic and molecular structure, quantum mechanics, statistical mechanics, thermodynamics correlated with descriptive chemistry. Problems. 992pp. 5 3/8 x 8 1/2. 0-486-65622-5

ELECTROLYTE SOLUTIONS: Second Revised Edition, R. A. Robinson and R. H. Stokes. Classic text deals primarily with measurement, interpretation of conductance, chemical potential, and diffusion in electrolyte solutions. Detailed theoretical interpretations, plus extensive tables of thermodynamic and transport properties. 1970 edition. 590pp. 5 3/8 x 8 1/2. 0-486-42225-9

Browse over 9,000 books at www.doverpublications.com

Engineering

FUNDAMENTALS OF ASTRODYNAMICS, Roger R. Bate, Donald D. Mueller, and Jerry E. White. Teaching text developed by U.S. Air Force Academy develops the basic two-body and n-body equations of motion; orbit determination; classical orbital elements, coordinate transformations; differential correction; more. 1971 edition. 455pp. 5 3/8 x 8 1/2. 0-486-60061-0

INTRODUCTION TO CONTINUUM MECHANICS FOR ENGINEERS: Revised Edition, Ray M. Bowen. This self-contained text introduces classical continuum models within a modern framework. Its numerous exercises illustrate the governing principles, linearizations, and other approximations that constitute classical continuum models. 2007 edition. 320pp. 6 1/8 x 9 1/4. 0-486-47460-7

ENGINEERING MECHANICS FOR STRUCTURES, Louis L. Bucciarelli. This text explores the mechanics of solids and statics as well as the strength of materials and elasticity theory. Its many design exercises encourage creative initiative and systems thinking. 2009 edition. 320pp. 6 1/8 x 9 1/4. 0-486-46855-0

FEEDBACK CONTROL THEORY, John C. Doyle, Bruce A. Francis and Allen R. Tannenbaum. This excellent introduction to feedback control system design offers a theoretical approach that captures the essential issues and can be applied to a wide range of practical problems. 1992 edition. 224pp. 6 1/2 x 9 1/4. 0-486-46933-6

THE FORCES OF MATTER, Michael Faraday. These lectures by a famous inventor offer an easy-to-understand introduction to the interactions of the universe's physical forces. Six essays explore gravitation, cohesion, chemical affinity, heat, magnetism, and electricity. 1993 edition. 96pp. 5 3/8 x 8 1/2. 0-486-47482-8

DYNAMICS, Lawrence E. Goodman and William H. Warner. Beginning engineering text introduces calculus of vectors, particle motion, dynamics of particle systems and plane rigid bodies, technical applications in plane motions, and more. Exercises and answers in every chapter. 619pp. 5 3/8 x 8 1/2. 0-486-42006-X

ADAPTIVE FILTERING PREDICTION AND CONTROL, Graham C. Goodwin and Kwai Sang Sin. This unified survey focuses on linear discrete-time systems and explores natural extensions to nonlinear systems. It emphasizes discrete-time systems, summarizing theoretical and practical aspects of a large class of adaptive algorithms. 1984 edition. 560pp. 6 1/2 x 9 1/4. 0-486-46932-8

INDUCTANCE CALCULATIONS, Frederick W. Grover. This authoritative reference enables the design of virtually every type of inductor. It features a single simple formula for each type of inductor, together with tables containing essential numerical factors. 1946 edition. 304pp. 5 3/8 x 8 1/2. 0-486-47440-2

THERMODYNAMICS: Foundations and Applications, Elias P. Gyftopoulos and Gian Paolo Beretta. Designed by two MIT professors, this authoritative text discusses basic concepts and applications in detail, emphasizing generality, definitions, and logical consistency. More than 300 solved problems cover realistic energy systems and processes. 800pp. 6 1/8 x 9 1/4. 0-486-43932-1

THE FINITE ELEMENT METHOD: Linear Static and Dynamic Finite Element Analysis, Thomas J. R. Hughes. Text for students without in-depth mathematical training, this text includes a comprehensive presentation and analysis of algorithms of time-dependent phenomena plus beam, plate, and shell theories. Solution guide available upon request. 672pp. 6 1/2 x 9 1/4. 0-486-41181-8

Browse over 9,000 books at www.doverpublications.com

HELICOPTER THEORY, Wayne Johnson. Monumental engineering text covers vertical flight, forward flight, performance, mathematics of rotating systems, rotary wing dynamics and aerodynamics, aeroelasticity, stability and control, stall, noise, and more. 189 illustrations. 1980 edition. 1089pp. 5 5/8 x 8 1/4. 0-486-68230-7

MATHEMATICAL HANDBOOK FOR SCIENTISTS AND ENGINEERS: Definitions, Theorems, and Formulas for Reference and Review, Granino A. Korn and Theresa M. Korn. Convenient access to information from every area of mathematics: Fourier transforms, Z transforms, linear and nonlinear programming, calculus of variations, random-process theory, special functions, combinatorial analysis, game theory, much more. 1152pp. 5 3/8 x 8 1/2. 0-486-41147-8

A HEAT TRANSFER TEXTBOOK: Fourth Edition, John H. Lienhard V and John H. Lienhard IV. This introduction to heat and mass transfer for engineering students features worked examples and end-of-chapter exercises. Worked examples and end-of-chapter exercises appear throughout the book, along with well-drawn, illuminating figures. 768pp. 7 x 9 1/4. 0-486-47931-5

BASIC ELECTRICITY, U.S. Bureau of Naval Personnel. Originally a training course; best nontechnical coverage. Topics include batteries, circuits, conductors, AC and DC, inductance and capacitance, generators, motors, transformers, amplifiers, etc. Many questions with answers. 349 illustrations. 1969 edition. 448pp. 6 1/2 x 9 1/4.
0-486-20973-3

BASIC ELECTRONICS, U.S. Bureau of Naval Personnel. Clear, well-illustrated introduction to electronic equipment covers numerous essential topics: electron tubes, semiconductors, electronic power supplies, tuned circuits, amplifiers, receivers, ranging and navigation systems, computers, antennas, more. 560 illustrations. 567pp. 6 1/2 x 9 1/4. 0-486-21076-6

BASIC WING AND AIRFOIL THEORY, Alan Pope. This self-contained treatment by a pioneer in the study of wind effects covers flow functions, airfoil construction and pressure distribution, finite and monoplane wings, and many other subjects. 1951 edition. 320pp. 5 3/8 x 8 1/2. 0-486-47188-8

SYNTHETIC FUELS, Ronald F. Probstein and R. Edwin Hicks. This unified presentation examines the methods and processes for converting coal, oil, shale, tar sands, and various forms of biomass into liquid, gaseous, and clean solid fuels. 1982 edition. 512pp. 6 1/8 x 9 1/4. 0-486-44977-7

THEORY OF ELASTIC STABILITY, Stephen P. Timoshenko and James M. Gere. Written by world-renowned authorities on mechanics, this classic ranges from theoretical explanations of 2- and 3-D stress and strain to practical applications such as torsion, bending, and thermal stress. 1961 edition. 560pp. 5 3/8 x 8 1/2. 0-486-47207-8

PRINCIPLES OF DIGITAL COMMUNICATION AND CODING, Andrew J. Viterbi and Jim K. Omura. This classic by two digital communications experts is geared toward students of communications theory and to designers of channels, links, terminals, modems, or networks used to transmit and receive digital messages. 1979 edition. 576pp. 6 1/8 x 9 1/4. 0-486-46901-8

LINEAR SYSTEM THEORY: The State Space Approach, Lotfi A. Zadeh and Charles A. Desoer. Written by two pioneers in the field, this exploration of the state space approach focuses on problems of stability and control, plus connections between this approach and classical techniques. 1963 edition. 656pp. 6 1/8 x 9 1/4.
0-486-46663-9

Browse over 9,000 books at www.doverpublications.com

Mathematics–Bestsellers

HANDBOOK OF MATHEMATICAL FUNCTIONS: with Formulas, Graphs, and Mathematical Tables, Edited by Milton Abramowitz and Irene A. Stegun. A classic resource for working with special functions, standard trig, and exponential logarithmic definitions and extensions, it features 29 sets of tables, some to as high as 20 places. 1046pp. 8 x 10 1/2.   0-486-61272-4

ABSTRACT AND CONCRETE CATEGORIES: The Joy of Cats, Jiri Adamek, Horst Herrlich, and George E. Strecker. This up-to-date introductory treatment employs category theory to explore the theory of structures. Its unique approach stresses concrete categories and presents a systematic view of factorization structures. Numerous examples. 1990 edition, updated 2004. 528pp. 6 1/8 x 9 1/4.   0-486-46934-4

MATHEMATICS: Its Content, Methods and Meaning, A. D. Aleksandrov, A. N. Kolmogorov, and M. A. Lavrent'ev. Major survey offers comprehensive, coherent discussions of analytic geometry, algebra, differential equations, calculus of variations, functions of a complex variable, prime numbers, linear and non-Euclidean geometry, topology, functional analysis, more. 1963 edition. 1120pp. 5 3/8 x 8 1/2.   0-486-40916-3

INTRODUCTION TO VECTORS AND TENSORS: Second Edition--Two Volumes Bound as One, Ray M. Bowen and C.-C. Wang. Convenient single-volume compilation of two texts offers both introduction and in-depth survey. Geared toward engineering and science students rather than mathematicians, it focuses on physics and engineering applications. 1976 edition. 560pp. 6 1/2 x 9 1/4.   0-486-46914-X

AN INTRODUCTION TO ORTHOGONAL POLYNOMIALS, Theodore S. Chihara. Concise introduction covers general elementary theory, including the representation theorem and distribution functions, continued fractions and chain sequences, the recurrence formula, special functions, and some specific systems. 1978 edition. 272pp. 5 3/8 x 8 1/2.   0-486-47929-3

ADVANCED MATHEMATICS FOR ENGINEERS AND SCIENTISTS, Paul DuChateau. This primary text and supplemental reference focuses on linear algebra, calculus, and ordinary differential equations. Additional topics include partial differential equations and approximation methods. Includes solved problems. 1992 edition. 400pp. 7 1/2 x 9 1/4.   0-486-47930-7

PARTIAL DIFFERENTIAL EQUATIONS FOR SCIENTISTS AND ENGINEERS, Stanley J. Farlow. Practical text shows how to formulate and solve partial differential equations. Coverage of diffusion-type problems, hyperbolic-type problems, elliptic-type problems, numerical and approximate methods. Solution guide available upon request. 1982 edition. 414pp. 6 1/8 x 9 1/4.   0-486-67620-X

VARIATIONAL PRINCIPLES AND FREE-BOUNDARY PROBLEMS, Avner Friedman. Advanced graduate-level text examines variational methods in partial differential equations and illustrates their applications to free-boundary problems. Features detailed statements of standard theory of elliptic and parabolic operators. 1982 edition. 720pp. 6 1/8 x 9 1/4.   0-486-47853-X

LINEAR ANALYSIS AND REPRESENTATION THEORY, Steven A. Gaal. Unified treatment covers topics from the theory of operators and operator algebras on Hilbert spaces; integration and representation theory for topological groups; and the theory of Lie algebras, Lie groups, and transform groups. 1973 edition. 704pp. 6 1/8 x 9 1/4.   0-486-47851-3

Browse over 9,000 books at www.doverpublications.com

A SURVEY OF INDUSTRIAL MATHEMATICS, Charles R. MacCluer. Students learn how to solve problems they'll encounter in their professional lives with this concise single-volume treatment. It employs MATLAB and other strategies to explore typical industrial problems. 2000 edition. 384pp. 5 3/8 x 8 1/2. 0-486-47702-9

NUMBER SYSTEMS AND THE FOUNDATIONS OF ANALYSIS, Elliott Mendelson. Geared toward undergraduate and beginning graduate students, this study explores natural numbers, integers, rational numbers, real numbers, and complex numbers. Numerous exercises and appendixes supplement the text. 1973 edition. 368pp. 5 3/8 x 8 1/2. 0-486-45792-3

A FIRST LOOK AT NUMERICAL FUNCTIONAL ANALYSIS, W. W. Sawyer. Text by renowned educator shows how problems in numerical analysis lead to concepts of functional analysis. Topics include Banach and Hilbert spaces, contraction mappings, convergence, differentiation and integration, and Euclidean space. 1978 edition. 208pp. 5 3/8 x 8 1/2. 0-486-47882-3

FRACTALS, CHAOS, POWER LAWS: Minutes from an Infinite Paradise, Manfred Schroeder. A fascinating exploration of the connections between chaos theory, physics, biology, and mathematics, this book abounds in award-winning computer graphics, optical illusions, and games that clarify memorable insights into self-similarity. 1992 edition. 448pp. 6 1/8 x 9 1/4. 0-486-47204-3

SET THEORY AND THE CONTINUUM PROBLEM, Raymond M. Smullyan and Melvin Fitting. A lucid, elegant, and complete survey of set theory, this three-part treatment explores axiomatic set theory, the consistency of the continuum hypothesis, and forcing and independence results. 1996 edition. 336pp. 6 x 9. 0-486-47484-4

DYNAMICAL SYSTEMS, Shlomo Sternberg. A pioneer in the field of dynamical systems discusses one-dimensional dynamics, differential equations, random walks, iterated function systems, symbolic dynamics, and Markov chains. Supplementary materials include PowerPoint slides and MATLAB exercises. 2010 edition. 272pp. 6 1/8 x 9 1/4. 0-486-47705-3

ORDINARY DIFFERENTIAL EQUATIONS, Morris Tenenbaum and Harry Pollard. Skillfully organized introductory text examines origin of differential equations, then defines basic terms and outlines general solution of a differential equation. Explores integrating factors; dilution and accretion problems; Laplace Transforms; Newton's Interpolation Formulas, more. 818pp. 5 3/8 x 8 1/2. 0-486-64940-7

MATROID THEORY, D. J. A. Welsh. Text by a noted expert describes standard examples and investigation results, using elementary proofs to develop basic matroid properties before advancing to a more sophisticated treatment. Includes numerous exercises. 1976 edition. 448pp. 5 3/8 x 8 1/2. 0-486-47439-9

THE CONCEPT OF A RIEMANN SURFACE, Hermann Weyl. This classic on the general history of functions combines function theory and geometry, forming the basis of the modern approach to analysis, geometry, and topology. 1955 edition. 208pp. 5 3/8 x 8 1/2. 0-486-47004-0

THE LAPLACE TRANSFORM, David Vernon Widder. This volume focuses on the Laplace and Stieltjes transforms, offering a highly theoretical treatment. Topics include fundamental formulas, the moment problem, monotonic functions, and Tauberian theorems. 1941 edition. 416pp. 5 3/8 x 8 1/2. 0-486-47755-X

Browse over 9,000 books at www.doverpublications.com

Mathematics–Logic and Problem Solving

PERPLEXING PUZZLES AND TANTALIZING TEASERS, Martin Gardner. Ninety-three riddles, mazes, illusions, tricky questions, word and picture puzzles, and other challenges offer hours of entertainment for youngsters. Filled with rib-tickling drawings. Solutions. 224pp. 5 3/8 x 8 1/2. 0-486-25637-5

MY BEST MATHEMATICAL AND LOGIC PUZZLES, Martin Gardner. The noted expert selects 70 of his favorite "short" puzzles. Includes The Returning Explorer, The Mutilated Chessboard, Scrambled Box Tops, and dozens more. Complete solutions included. 96pp. 5 3/8 x 8 1/2. 0-486-28152-3

THE LADY OR THE TIGER?: and Other Logic Puzzles, Raymond M. Smullyan. Created by a renowned puzzle master, these whimsically themed challenges involve paradoxes about probability, time, and change; metapuzzles; and self-referentiality. Nineteen chapters advance in difficulty from relatively simple to highly complex. 1982 edition. 240pp. 5 3/8 x 8 1/2. 0-486-47027-X

SATAN, CANTOR AND INFINITY: Mind-Boggling Puzzles, Raymond M. Smullyan. A renowned mathematician tells stories of knights and knaves in an entertaining look at the logical precepts behind infinity, probability, time, and change. Requires a strong background in mathematics. Complete solutions. 288pp. 5 3/8 x 8 1/2.

0-486-47036-9

THE RED BOOK OF MATHEMATICAL PROBLEMS, Kenneth S. Williams and Kenneth Hardy. Handy compilation of 100 practice problems, hints and solutions indispensable for students preparing for the William Lowell Putnam and other mathematical competitions. Preface to the First Edition. Sources. 1988 edition. 192pp. 5 3/8 x 8 1/2. 0-486-69415-1

KING ARTHUR IN SEARCH OF HIS DOG AND OTHER CURIOUS PUZZLES, Raymond M. Smullyan. This fanciful, original collection for readers of all ages features arithmetic puzzles, logic problems related to crime detection, and logic and arithmetic puzzles involving King Arthur and his Dogs of the Round Table. 160pp. 5 3/8 x 8 1/2. 0-486-47435-6

UNDECIDABLE THEORIES: Studies in Logic and the Foundation of Mathematics, Alfred Tarski in collaboration with Andrzej Mostowski and Raphael M. Robinson. This well-known book by the famed logician consists of three treatises: "A General Method in Proofs of Undecidability," "Undecidability and Essential Undecidability in Mathematics," and "Undecidability of the Elementary Theory of Groups." 1953 edition. 112pp. 5 3/8 x 8 1/2. 0-486-47703-7

LOGIC FOR MATHEMATICIANS, J. Barkley Rosser. Examination of essential topics and theorems assumes no background in logic. "Undoubtedly a major addition to the literature of mathematical logic." – *Bulletin of the American Mathematical Society.* 1978 edition. 592pp. 6 1/8 x 9 1/4. 0-486-46898-4

INTRODUCTION TO PROOF IN ABSTRACT MATHEMATICS, Andrew Wohlgemuth. This undergraduate text teaches students what constitutes an acceptable proof, and it develops their ability to do proofs of routine problems as well as those requiring creative insights. 1990 edition. 384pp. 6 1/2 x 9 1/4. 0-486-47854-8

FIRST COURSE IN MATHEMATICAL LOGIC, Patrick Suppes and Shirley Hill. Rigorous introduction is simple enough in presentation and context for wide range of students. Symbolizing sentences; logical inference; truth and validity; truth tables; terms, predicates, universal quantifiers; universal specification and laws of identity; more. 288pp. 5 3/8 x 8 1/2. 0-486-42259-3

Browse over 9,000 books at www.doverpublications.com

Mathematics–Algebra and Calculus

VECTOR CALCULUS, Peter Baxandall and Hans Liebeck. This introductory text offers a rigorous, comprehensive treatment. Classical theorems of vector calculus are amply illustrated with figures, worked examples, physical applications, and exercises with hints and answers. 1986 edition. 560pp. 5 3/8 x 8 1/2. 0-486-46620-5

ADVANCED CALCULUS: An Introduction to Classical Analysis, Louis Brand. A course in analysis that focuses on the functions of a real variable, this text introduces the basic concepts in their simplest setting and illustrates its teachings with numerous examples, theorems, and proofs. 1955 edition. 592pp. 5 3/8 x 8 1/2. 0-486-44548-8

ADVANCED CALCULUS, Avner Friedman. Intended for students who have already completed a one-year course in elementary calculus, this two-part treatment advances from functions of one variable to those of several variables. Solutions. 1971 edition. 432pp. 5 3/8 x 8 1/2. 0-486-45795-8

METHODS OF MATHEMATICS APPLIED TO CALCULUS, PROBABILITY, AND STATISTICS, Richard W. Hamming. This 4-part treatment begins with algebra and analytic geometry and proceeds to an exploration of the calculus of algebraic functions and transcendental functions and applications. 1985 edition. Includes 310 figures and 18 tables. 880pp. 6 1/2 x 9 1/4. 0-486-43945-3

BASIC ALGEBRA I: Second Edition, Nathan Jacobson. A classic text and standard reference for a generation, this volume covers all undergraduate algebra topics, including groups, rings, modules, Galois theory, polynomials, linear algebra, and associative algebra. 1985 edition. 528pp. 6 1/8 x 9 1/4. 0-486-47189-6

BASIC ALGEBRA II: Second Edition, Nathan Jacobson. This classic text and standard reference comprises all subjects of a first-year graduate-level course, including in-depth coverage of groups and polynomials and extensive use of categories and functors. 1989 edition. 704pp. 6 1/8 x 9 1/4. 0-486-47187-X

CALCULUS: An Intuitive and Physical Approach (Second Edition), Morris Kline. Application-oriented introduction relates the subject as closely as possible to science with explorations of the derivative; differentiation and integration of the powers of x; theorems on differentiation, antidifferentiation; the chain rule; trigonometric functions; more. Examples. 1967 edition. 960pp. 6 1/2 x 9 1/4. 0-486-40453-6

ABSTRACT ALGEBRA AND SOLUTION BY RADICALS, John E. Maxfield and Margaret W. Maxfield. Accessible advanced undergraduate-level text starts with groups, rings, fields, and polynomials and advances to Galois theory, radicals and roots of unity, and solution by radicals. Numerous examples, illustrations, exercises, appendixes. 1971 edition. 224pp. 6 1/8 x 9 1/4. 0-486-47723-1

AN INTRODUCTION TO THE THEORY OF LINEAR SPACES, Georgi E. Shilov. Translated by Richard A. Silverman. Introductory treatment offers a clear exposition of algebra, geometry, and analysis as parts of an integrated whole rather than separate subjects. Numerous examples illustrate many different fields, and problems include hints or answers. 1961 edition. 320pp. 5 3/8 x 8 1/2. 0-486-63070-6

LINEAR ALGEBRA, Georgi E. Shilov. Covers determinants, linear spaces, systems of linear equations, linear functions of a vector argument, coordinate transformations, the canonical form of the matrix of a linear operator, bilinear and quadratic forms, and more. 387pp. 5 3/8 x 8 1/2. 0-486-63518-X

Browse over 9,000 books at www.doverpublications.com

Mathematics–Probability and Statistics

BASIC PROBABILITY THEORY, Robert B. Ash. This text emphasizes the probabilistic way of thinking, rather than measure-theoretic concepts. Geared toward advanced undergraduates and graduate students, it features solutions to some of the problems. 1970 edition. 352pp. 5 3/8 x 8 1/2. 0-486-46628-0

PRINCIPLES OF STATISTICS, M. G. Bulmer. Concise description of classical statistics, from basic dice probabilities to modern regression analysis. Equal stress on theory and applications. Moderate difficulty; only basic calculus required. Includes problems with answers. 252pp. 5 5/8 x 8 1/4. 0-486-63760-3

OUTLINE OF BASIC STATISTICS: Dictionary and Formulas, John E. Freund and Frank J. Williams. Handy guide includes a 70-page outline of essential statistical formulas covering grouped and ungrouped data, finite populations, probability, and more, plus over 1,000 clear, concise definitions of statistical terms. 1966 edition. 208pp. 5 3/8 x 8 1/2. 0-486-47769-X

GOOD THINKING: The Foundations of Probability and Its Applications, Irving J. Good. This in-depth treatment of probability theory by a famous British statistician explores Keynesian principles and surveys such topics as Bayesian rationality, corroboration, hypothesis testing, and mathematical tools for induction and simplicity. 1983 edition. 352pp. 5 3/8 x 8 1/2. 0-486-47438-0

INTRODUCTION TO PROBABILITY THEORY WITH CONTEMPORARY APPLICATIONS, Lester L. Helms. Extensive discussions and clear examples, written in plain language, expose students to the rules and methods of probability. Exercises foster problem-solving skills, and all problems feature step-by-step solutions. 1997 edition. 368pp. 6 1/2 x 9 1/4. 0-486-47418-6

CHANCE, LUCK, AND STATISTICS, Horace C. Levinson. In simple, non-technical language, this volume explores the fundamentals governing chance and applies them to sports, government, and business. "Clear and lively ... remarkably accurate." – *Scientific Monthly*. 384pp. 5 3/8 x 8 1/2. 0-486-41997-5

FIFTY CHALLENGING PROBLEMS IN PROBABILITY WITH SOLUTIONS, Frederick Mosteller. Remarkable puzzlers, graded in difficulty, illustrate elementary and advanced aspects of probability. These problems were selected for originality, general interest, or because they demonstrate valuable techniques. Also includes detailed solutions. 88pp. 5 3/8 x 8 1/2. 0-486-65355-2

EXPERIMENTAL STATISTICS, Mary Gibbons Natrella. A handbook for those seeking engineering information and quantitative data for designing, developing, constructing, and testing equipment. Covers the planning of experiments, the analyzing of extreme-value data; and more. 1966 edition. Index. Includes 52 figures and 76 tables. 560pp. 8 3/8 x 11. 0-486-43937-2

STOCHASTIC MODELING: Analysis and Simulation, Barry L. Nelson. Coherent introduction to techniques also offers a guide to the mathematical, numerical, and simulation tools of systems analysis. Includes formulation of models, analysis, and interpretation of results. 1995 edition. 336pp. 6 1/8 x 9 1/4. 0-486-47770-3

INTRODUCTION TO BIOSTATISTICS: Second Edition, Robert R. Sokal and F. James Rohlf. Suitable for undergraduates with a minimal background in mathematics, this introduction ranges from descriptive statistics to fundamental distributions and the testing of hypotheses. Includes numerous worked-out problems and examples. 1987 edition. 384pp. 6 1/8 x 9 1/4. 0-486-46961-1

Browse over 9,000 books at www.doverpublications.com

Mathematics–Geometry and Topology

PROBLEMS AND SOLUTIONS IN EUCLIDEAN GEOMETRY, M. N. Aref and William Wernick. Based on classical principles, this book is intended for a second course in Euclidean geometry and can be used as a refresher. More than 200 problems include hints and solutions. 1968 edition. 272pp. 5 3/8 x 8 1/2. 0-486-47720-7

TOPOLOGY OF 3-MANIFOLDS AND RELATED TOPICS, Edited by M. K. Fort, Jr. With a New Introduction by Daniel Silver. Summaries and full reports from a 1961 conference discuss decompositions and subsets of 3-space; n-manifolds; knot theory; the Poincaré conjecture; and periodic maps and isotopies. Familiarity with algebraic topology required. 1962 edition. 272pp. 6 1/8 x 9 1/4. 0-486-47753-3

POINT SET TOPOLOGY, Steven A. Gaal. Suitable for a complete course in topology, this text also functions as a self-contained treatment for independent study. Additional enrichment materials make it equally valuable as a reference. 1964 edition. 336pp. 5 3/8 x 8 1/2. 0-486-47222-1

INVITATION TO GEOMETRY, Z. A. Melzak. Intended for students of many different backgrounds with only a modest knowledge of mathematics, this text features self-contained chapters that can be adapted to several types of geometry courses. 1983 edition. 240pp. 5 3/8 x 8 1/2. 0-486-46626-4

TOPOLOGY AND GEOMETRY FOR PHYSICISTS, Charles Nash and Siddhartha Sen. Written by physicists for physics students, this text assumes no detailed background in topology or geometry. Topics include differential forms, homotopy, homology, cohomology, fiber bundles, connection and covariant derivatives, and Morse theory. 1983 edition. 320pp. 5 3/8 x 8 1/2. 0-486-47852-1

BEYOND GEOMETRY: Classic Papers from Riemann to Einstein, Edited with an Introduction and Notes by Peter Pesic. This is the only English-language collection of these 8 accessible essays. They trace seminal ideas about the foundations of geometry that led to Einstein's general theory of relativity. 224pp. 6 1/8 x 9 1/4. 0-486-45350-2

GEOMETRY FROM EUCLID TO KNOTS, Saul Stahl. This text provides a historical perspective on plane geometry and covers non-neutral Euclidean geometry, circles and regular polygons, projective geometry, symmetries, inversions, informal topology, and more. Includes 1,000 practice problems. Solutions available. 2003 edition. 480pp. 6 1/8 x 9 1/4. 0-486-47459-3

TOPOLOGICAL VECTOR SPACES, DISTRIBUTIONS AND KERNELS, François Trèves. Extending beyond the boundaries of Hilbert and Banach space theory, this text focuses on key aspects of functional analysis, particularly in regard to solving partial differential equations. 1967 edition. 592pp. 5 3/8 x 8 1/2. 0-486-45352-9

INTRODUCTION TO PROJECTIVE GEOMETRY, C. R. Wylie, Jr. This introductory volume offers strong reinforcement for its teachings, with detailed examples and numerous theorems, proofs, and exercises, plus complete answers to all odd-numbered end-of-chapter problems. 1970 edition. 576pp. 6 1/8 x 9 1/4. 0-486-46895-X

FOUNDATIONS OF GEOMETRY, C. R. Wylie, Jr. Geared toward students preparing to teach high school mathematics, this text explores the principles of Euclidean and non-Euclidean geometry and covers both generalities and specifics of the axiomatic method. 1964 edition. 352pp. 6 x 9. 0-486-47214-0

Browse over 9,000 books at www.doverpublications.com

Mathematics–History

THE WORKS OF ARCHIMEDES, Archimedes. Translated by Sir Thomas Heath. Complete works of ancient geometer feature such topics as the famous problems of the ratio of the areas of a cylinder and an inscribed sphere; the properties of conoids, spheroids, and spirals; more. 326pp. 5 3/8 x 8 1/2.　　　　　0-486-42084-1

THE HISTORICAL ROOTS OF ELEMENTARY MATHEMATICS, Lucas N. H. Bunt, Phillip S. Jones, and Jack D. Bedient. Exciting, hands-on approach to understanding fundamental underpinnings of modern arithmetic, algebra, geometry and number systems examines their origins in early Egyptian, Babylonian, and Greek sources. 336pp. 5 3/8 x 8 1/2.　　　　　0-486-25563-8

THE THIRTEEN BOOKS OF EUCLID'S ELEMENTS, Euclid. Contains complete English text of all 13 books of the Elements plus critical apparatus analyzing each definition, postulate, and proposition in great detail. Covers textual and linguistic matters; mathematical analyses of Euclid's ideas; classical, medieval, Renaissance and modern commentators; refutations, supports, extrapolations, reinterpretations and historical notes. 995 figures. Total of 1,425pp. All books 5 3/8 x 8 1/2.

Vol. I: 443pp.　0-486-60088-2
Vol. II: 464pp.　0-486-60089-0
Vol. III: 546pp.　0-486-60090-4

A HISTORY OF GREEK MATHEMATICS, Sir Thomas Heath. This authoritative two-volume set that covers the essentials of mathematics and features every landmark innovation and every important figure, including Euclid, Apollonius, and others. 5 3/8 x 8 1/2.

Vol. I: 461pp.　0-486-24073-8
Vol. II: 597pp.　0-486-24074-6

A MANUAL OF GREEK MATHEMATICS, Sir Thomas L. Heath. This concise but thorough history encompasses the enduring contributions of the ancient Greek mathematicians whose works form the basis of most modern mathematics. Discusses Pythagorean arithmetic, Plato, Euclid, more. 1931 edition. 576pp. 5 3/8 x 8 1/2.

0-486-43231-9

CHINESE MATHEMATICS IN THE THIRTEENTH CENTURY, Ulrich Libbrecht. An exploration of the 13th-century mathematician Ch'in, this fascinating book combines what is known of the mathematician's life with a history of his only extant work, the Shu-shu chiu-chang. 1973 edition. 592pp. 5 3/8 x 8 1/2.

0-486-44619-0

PHILOSOPHY OF MATHEMATICS AND DEDUCTIVE STRUCTURE IN EUCLID'S ELEMENTS, Ian Mueller. This text provides an understanding of the classical Greek conception of mathematics as expressed in Euclid's Elements. It focuses on philosophical, foundational, and logical questions and features helpful appendixes. 400pp. 6 1/2 x 9 1/4.　　　　　0-486-45300-6

BEYOND GEOMETRY: Classic Papers from Riemann to Einstein, Edited with an Introduction and Notes by Peter Pesic. This is the only English-language collection of these 8 accessible essays. They trace seminal ideas about the foundations of geometry that led to Einstein's general theory of relativity. 224pp. 6 1/8 x 9 1/4.　0-486-45350-2

HISTORY OF MATHEMATICS, David E. Smith. Two-volume history – from Egyptian papyri and medieval maps to modern graphs and diagrams. Non-technical chronological survey with thousands of biographical notes, critical evaluations, and contemporary opinions on over 1,100 mathematicians. 5 3/8 x 8 1/2.

Vol. I: 618pp.　0-486-20429-4
Vol. II: 736pp.　0-486-20430-8

Physics

THEORETICAL NUCLEAR PHYSICS, John M. Blatt and Victor F. Weisskopf. An uncommonly clear and cogent investigation and correlation of key aspects of theoretical nuclear physics by leading experts: the nucleus, nuclear forces, nuclear spectroscopy, two-, three- and four-body problems, nuclear reactions, beta-decay and nuclear shell structure. 896pp. 5 3/8 x 8 1/2. 0-486-66827-4

QUANTUM THEORY, David Bohm. This advanced undergraduate-level text presents the quantum theory in terms of qualitative and imaginative concepts, followed by specific applications worked out in mathematical detail. 655pp. 5 3/8 x 8 1/2. 0-486-65969-0

ATOMIC PHYSICS AND HUMAN KNOWLEDGE, Niels Bohr. Articles and speeches by the Nobel Prize–winning physicist, dating from 1934 to 1958, offer philosophical explorations of the relevance of atomic physics to many areas of human endeavor. 1961 edition. 112pp. 5 3/8 x 8 1/2. 0-486-47928-5

COSMOLOGY, Hermann Bondi. A co-developer of the steady-state theory explores his conception of the expanding universe. This historic book was among the first to present cosmology as a separate branch of physics. 1961 edition. 192pp. 5 3/8 x 8 1/2. 0-486-47483-6

LECTURES ON QUANTUM MECHANICS, Paul A. M. Dirac. Four concise, brilliant lectures on mathematical methods in quantum mechanics from Nobel Prize-winning quantum pioneer build on idea of visualizing quantum theory through the use of classical mechanics. 96pp. 5 3/8 x 8 1/2. 0-486-41713-1

THE PRINCIPLE OF RELATIVITY, Albert Einstein and Frances A. Davis. Eleven papers that forged the general and special theories of relativity include seven papers by Einstein, two by Lorentz, and one each by Minkowski and Weyl. 1923 edition. 240pp. 5 3/8 x 8 1/2. 0-486-60081-5

PHYSICS OF WAVES, William C. Elmore and Mark A. Heald. Ideal as a classroom text or for individual study, this unique one-volume overview of classical wave theory covers wave phenomena of acoustics, optics, electromagnetic radiations, and more. 477pp. 5 3/8 x 8 1/2. 0-486-64926-1

THERMODYNAMICS, Enrico Fermi. In this classic of modern science, the Nobel Laureate presents a clear treatment of systems, the First and Second Laws of Thermodynamics, entropy, thermodynamic potentials, and much more. Calculus required. 160pp. 5 3/8 x 8 1/2. 0-486-60361-X

QUANTUM THEORY OF MANY-PARTICLE SYSTEMS, Alexander L. Fetter and John Dirk Walecka. Self-contained treatment of nonrelativistic many-particle systems discusses both formalism and applications in terms of ground-state (zero-temperature) formalism, finite-temperature formalism, canonical transformations, and applications to physical systems. 1971 edition. 640pp. 5 3/8 x 8 1/2. 0-486-42827-3

QUANTUM MECHANICS AND PATH INTEGRALS: Emended Edition, Richard P. Feynman and Albert R. Hibbs. Emended by Daniel F. Styer. The Nobel Prize–winning physicist presents unique insights into his theory and its applications. Feynman starts with fundamentals and advances to the perturbation method, quantum electrodynamics, and statistical mechanics. 1965 edition, emended in 2005. 384pp. 6 1/8 x 9 1/4. 0-486-47722-3

Browse over 9,000 books at www.doverpublications.com

Physics

INTRODUCTION TO MODERN OPTICS, Grant R. Fowles. A complete basic undergraduate course in modern optics for students in physics, technology, and engineering. The first half deals with classical physical optics; the second, quantum nature of light. Solutions. 336pp. 5 3/8 x 8 1/2. 0-486-65957-7

THE QUANTUM THEORY OF RADIATION: Third Edition, W. Heitler. The first comprehensive treatment of quantum physics in any language, this classic introduction to basic theory remains highly recommended and widely used, both as a text and as a reference. 1954 edition. 464pp. 5 3/8 x 8 1/2. 0-486-64558-4

QUANTUM FIELD THEORY, Claude Itzykson and Jean-Bernard Zuber. This comprehensive text begins with the standard quantization of electrodynamics and perturbative renormalization, advancing to functional methods, relativistic bound states, broken symmetries, nonabelian gauge fields, and asymptotic behavior. 1980 edition. 752pp. 6 1/2 x 9 1/4. 0-486-44568-2

FOUNDATIONS OF POTENTIAL THERY, Oliver D. Kellogg. Introduction to fundamentals of potential functions covers the force of gravity, fields of force, potentials, harmonic functions, electric images and Green's function, sequences of harmonic functions, fundamental existence theorems, and much more. 400pp. 5 3/8 x 8 1/2.
0-486-60144-7

FUNDAMENTALS OF MATHEMATICAL PHYSICS, Edgar A. Kraut. Indispensable for students of modern physics, this text provides the necessary background in mathematics to study the concepts of electromagnetic theory and quantum mechanics. 1967 edition. 480pp. 6 1/2 x 9 1/4. 0-486-45809-1

GEOMETRY AND LIGHT: The Science of Invisibility, Ulf Leonhardt and Thomas Philbin. Suitable for advanced undergraduate and graduate students of engineering, physics, and mathematics and scientific researchers of all types, this is the first authoritative text on invisibility and the science behind it. More than 100 full-color illustrations, plus exercises with solutions. 2010 edition. 288pp. 7 x 9 1/4. 0-486-47693-6

QUANTUM MECHANICS: New Approaches to Selected Topics, Harry J. Lipkin. Acclaimed as "excellent" (*Nature*) and "very original and refreshing" (*Physics Today*), these studies examine the Mössbauer effect, many-body quantum mechanics, scattering theory, Feynman diagrams, and relativistic quantum mechanics. 1973 edition. 480pp. 5 3/8 x 8 1/2. 0-486-45893-8

THEORY OF HEAT, James Clerk Maxwell. This classic sets forth the fundamentals of thermodynamics and kinetic theory simply enough to be understood by beginners, yet with enough subtlety to appeal to more advanced readers, too. 352pp. 5 3/8 x 8 1/2. 0-486-41735-2

QUANTUM MECHANICS, Albert Messiah. Subjects include formalism and its interpretation, analysis of simple systems, symmetries and invariance, methods of approximation, elements of relativistic quantum mechanics, much more. "Strongly recommended." – *American Journal of Physics.* 1152pp. 5 3/8 x 8 1/2. 0-486-40924-4

RELATIVISTIC QUANTUM FIELDS, Charles Nash. This graduate-level text contains techniques for performing calculations in quantum field theory. It focuses chiefly on the dimensional method and the renormalization group methods. Additional topics include functional integration and differentiation. 1978 edition. 240pp. 5 3/8 x 8 1/2.
0-486-47752-5

Physics

MATHEMATICAL TOOLS FOR PHYSICS, James Nearing. Encouraging students' development of intuition, this original work begins with a review of basic mathematics and advances to infinite series, complex algebra, differential equations, Fourier series, and more. 2010 edition. 496pp. 6 1/8 x 9 1/4. 0-486-48212-X

TREATISE ON THERMODYNAMICS, Max Planck. Great classic, still one of the best introductions to thermodynamics. Fundamentals, first and second principles of thermodynamics, applications to special states of equilibrium, more. Numerous worked examples. 1917 edition. 297pp. 5 3/8 x 8. 0-486-66371-X

AN INTRODUCTION TO RELATIVISTIC QUANTUM FIELD THEORY, Silvan S. Schweber. Complete, systematic, and self-contained, this text introduces modern quantum field theory. "Combines thorough knowledge with a high degree of didactic ability and a delightful style." – *Mathematical Reviews.* 1961 edition. 928pp. 5 3/8 x 8 1/2. 0-486-44228-4

THE ELECTROMAGNETIC FIELD, Albert Shadowitz. Comprehensive undergraduate text covers basics of electric and magnetic fields, building up to electromagnetic theory. Related topics include relativity theory. Over 900 problems, some with solutions. 1975 edition. 768pp. 5 5/8 x 8 1/4. 0-486-65660-8

THE PRINCIPLES OF STATISTICAL MECHANICS, Richard C. Tolman. Definitive treatise offers a concise exposition of classical statistical mechanics and a thorough elucidation of quantum statistical mechanics, plus applications of statistical mechanics to thermodynamic behavior. 1930 edition. 704pp. 5 5/8 x 8 1/4.
0-486-63896-0

INTRODUCTION TO THE PHYSICS OF FLUIDS AND SOLIDS, James S. Trefil. This interesting, informative survey by a well-known science author ranges from classical physics and geophysical topics, from the rings of Saturn and the rotation of the galaxy to underground nuclear tests. 1975 edition. 320pp. 5 3/8 x 8 1/2.
0-486-47437-2

STATISTICAL PHYSICS, Gregory H. Wannier. Classic text combines thermodynamics, statistical mechanics, and kinetic theory in one unified presentation. Topics include equilibrium statistics of special systems, kinetic theory, transport coefficients, and fluctuations. Problems with solutions. 1966 edition. 532pp. 5 3/8 x 8 1/2.
0-486-65401-X

SPACE, TIME, MATTER, Hermann Weyl. Excellent introduction probes deeply into Euclidean space, Riemann's space, Einstein's general relativity, gravitational waves and energy, and laws of conservation. "A classic of physics." – *British Journal for Philosophy and Science.* 330pp. 5 3/8 x 8 1/2. 0-486-60267-2

RANDOM VIBRATIONS: Theory and Practice, Paul H. Wirsching, Thomas L. Paez and Keith Ortiz. Comprehensive text and reference covers topics in probability, statistics, and random processes, plus methods for analyzing and controlling random vibrations. Suitable for graduate students and mechanical, structural, and aerospace engineers. 1995 edition. 464pp. 5 3/8 x 8 1/2. 0-486-45015-5

PHYSICS OF SHOCK WAVES AND HIGH-TEMPERATURE HYDRO DYNAMIC PHENOMENA, Ya B. Zel'dovich and Yu P. Raizer. Physical, chemical processes in gases at high temperatures are focus of outstanding text, which combines material from gas dynamics, shock-wave theory, thermodynamics and statistical physics, other fields. 284 illustrations. 1966–1967 edition. 944pp. 6 1/8 x 9 1/4.
0-486-42002-7